Ship Structural Analysis and Design

SHIP STRUCTURAL ANALYSIS AND DESIGN

by Owen F. Hughes and Jeom Kee Paik
with Dominique Béghin, John B. Caldwell, Hans G. Payer and Thomas E. Schellin

Published by
The Society of Naval Architects
and Marine Engineers
99 Canal Center Plaza, Suite 310
Alexandria, VA 22314

Library of Congress Card Catalog No. 88-62642
ISBN No. 978-0-939773-78-3

PREFACE

For a structure as large and as complex as a ship there are three levels of structural design, the second and most central of which is the subject of this book. Concept design deals with the topology or overall geometry of the structure; preliminary design establishes the scantlings (structural dimensions) of all principal structural members; and detail design is concerned with local aspects such as joints, openings, and reinforcements. Overall structural geometry is generally determined by overall design requirements rather than by structural requirements, while detail design is largely guided and constrained by fabrication methods and requirements. Also, since local structural details are numerous and basically similar among various structures they lend themselves to standardization and to design from handbooks and structural codes. Thus, it is in preliminary design where the structural designer has the largest number of significant decisions and options, and the greatest scope for optimizing the structure so that it best fulfills the objectives and satisfies all of the various constraints and requirements.

Rationally-based design is design from first principles using the tools of modern engineering science: computers and the methods of structural analysis and optimization which computers have made possible. Thus, the rationally-based approach is ideally suited for preliminary structural design, and it is this approach and this level of design that is the subject of this book. As shown by some examples in Section 1.3, this type of design offers substantial benefits to all parties concerned: owner, designer, builder, and operator.

Designing from first principles requires two separate and very extensive analyses: a response analysis to ascertain the true and complete response of the structure to all loads and load combinations, and a limit state analysis to ascertain all of the possible limit or failure values of these responses. Taken together these two analyses are by far the dominant part of rationally-based design, and this is reflected in this text in which 15 of the 17 chapters are devoted to various aspects of analysis. Because of this predominance of analysis, rationally-based design is necessarily computer based and this is the key to many of its benefits: speed, accuracy, thoroughness, economy, easy modification, and so forth. Also, as explained in Section 1.3, the necessary computer programs are already available and the hardware and software costs are quite moderate.

Of the many different topics and aspects in preliminary structural design some are an inherent part of rationally-based design (e.g., the aspects pertaining to response analysis and limit analysis) while others are more distinct and external (e.g., the selection of materials) or are simply constraints in the optimization process (e.g., the avoidance of some natural frequency). One of the advantages of the rationally-based approach is that it unifies and coordinates these many different aspects. Even for the more distinct or external aspects the rationally-based approach provides a framework by which each can be better coordinated with the other aspects.

PREREQUISITES, LEVELS OF STUDY, AND TIME REQUIREMENTS

The material in this book is suitable for either graduate or undergraduate study, or a combination of both. The methods and practices presented in this book will also be useful for practicing engineers and engineers-in-training. The only prerequisites are knowledge of mechanics of solids, strength of materials, and the basic aspects of matrix algebra and of statistics. If necessary, the latter two could be covered in a few introductory classes or in outside reading. The total time required to cover all of the topics in this book is about nine semester hours.

ACKNOWLEDGMENTS

We are very pleased to acknowledge all of those individuals and organizations that helped make this book possible.

The predecessor to this book was published in 1983, and it would never have appeared without the kind help and encouragement that the editor received from many people throughout the many years of its preparation. Special thanks are due to Professor Farrokh Mistree of the University of Oklahoma and Professor Vedran Žanić of the University of Zagreb in Croatia for their suggestions and help, and in particular for the many valuable contributions which they made in the computer implementation of the design method. Thanks also to Mrs. Diane Augee for her flawless typing of the entire manuscript.

For this new book we are deeply grateful to the chapter authors who have generously contributed so much new and valuable material: Dominique Béghin, John B. Caldwell, Hans G. Payer and Thomas E. Schellin. It will be evident to all readers that their expertise has simultaneously broadened the scope of the book and greatly enhanced its quality. Our thanks are also due to the SNAME Publications Director, Ms. Susan Evans Grove, and the book's composition manager, Mrs. Pat Mrozek, of Maryland Composition Co. In many ways, this book would not have been possible without their untiring efforts and assistance.

The second editor, Jeom Kee Paik, is pleased to acknowledge the support of the Lloyd's Register Educational Trust (LRET) which is an independent charity working to achieve advances in transportation, science, engineering and technology education, training and research worldwide for the benefit of all. Special thanks are due to Mr. Michael Franklin, Director of the LRET, for his help and encouragement. He also wishes to thank Dr. Yung Sup Shin of ABS Houston, a coauthor of the most important paper that contributes to the content of Chapter 10. Finally, he takes this opportunity to thank his wife Yun Hee, his son Myung Hoon and his daughter Yun Jung for their unfailing patience and support.

Owen F. Hughes, Virginia Tech
Jeom Kee Paik, Pusan National University

AUTHOR BIOGRAPHIES

Prof. Owen Hughes of Virginia Tech is best known for his development of a pioneering computer-based "first principles" method for the structural design of ships, combining finite element analysis, structural failure analysis and optimization. In 1983 he implemented this method in the MAESTRO computer program, which has been used by 13 navies and hundreds of shipyards and ship design agencies around the world. His earlier book, Ship Structural Design, has been used in many countries and has been translated into Russian and Chinese. He has been a continuously serving committee member or chair of the ISSC since 1970, and has published over 100 papers in technical journals.

Prof. Jeom Kee Paik is Director of the Lloyd's Register Educational Trust Research Centre of Excellence at Pusan National University, Korea. He has been chairman of ISSC committees on ultimate strength, condition assessment of aged ships, and ship collision and grounding. He is convenor of ISO code 18072 development on limit state assessment of ship structures, and Editor-in-Chief of Ships and Offshore Structures. He is author of the books: Ultimate limit state design of steel-plated structures; Ship-shaped offshore installations; and Condition assessment of aged structures.

Dominique Béghin graduated from the French Naval Architecture Institute "Ecole Nationale Supérieure du Génie Maritime", Paris, worked 20 years for "Chantiers de l'Atlantique", Saint-Nazaire, where he designed the structure of various types of merchant ships built by CDA, among them the ULCC "Batillus" launched in 1976 for "Societé Maritime Shell". In 1982 he joined BUREAU VERITAS where he developed rationally-based rules for the Classification of Steel Ships. He completed his career in 1998 as Scientific Director of the Marine Branch. He has been a member and chairman of several IACS Working Groups and chairman from 1993 to 1997 of the ISSC Committee IV-1 "Design Principles and Criteria".

John Caldwell's early research convinced him that the largely empirical ways of designing ships' structures could be replaced by more logical scientific procedures, as used in aircraft design. In his subsequent research, teaching and publications, he progressed from new analytical methods towards design synthesis, with particular emphasis on design for production and for safety. Professor Caldwell was Head of Newcastle University's School of Marine Technology, and participated actively in the first twelve ISS Congresses. He was President of the Royal Institution of Naval Architects from 1984–87.

In 2004 Dr. Hans G. Payer retired from Germanischer Lloyd's Executive Board and CEO of GL Maritime Services and now works as a naval architect and shipping consultant. The Austrian-born Dr. Payer holds an M.S. in Naval Architecture from the University of Vienna, Austria, and a Ph.D. in engineering from the University of California, Berkeley. He has done pioneering work for the use of the finite element method in ship structural analysis and has contributed to the technical development of containerships. He has published more than 100 papers on ship structures, ship hydrodynamics and ship vibrations. He has received the David Taylor Medal of the US Society of Naval Architects and Marine Engineers, SNAME.

Dr. Thomas E. Schellin is consultant naval architect at Germanischer Lloyd after his retirement in 2004. Born in Germany, he obtained his M.S. in Naval Architecture from the Massachusetts Institute of Technology and his Ph.D. from Rice University, Houston. He developed hydrodynamic analysis methods for ships and offshore structures, coordinated nationally and internationally sponsored research projects, and formulated class rules for fast ships, offshore structures, and offshore service vessels. He was GL's representative in the IACS AHG on Mooring and Anchoring, served on the 2001 and 2003 ISSC Committee I.2 Loads, and was the official discusser of the 2009 ISSC I.2 Loads Committee Report. He has published widely and is a Fellow of the American Society of Mechanical Engineers.

CONTENTS

Ship Structural Analysis and Design

RATIONALLY-BASED STRUCTURAL DESIGN

Owen Hughes
Professor, Virginia Tech
Blacksburg, VA, USA

Hans G. Payer
Germanischer Lloyd, Hamburg, Germany (ret)

1.1 INTRODUCTION

Throughout history, shipping has played a central role in transportation and trade. Even today, about 95% of internationally traded goods is carried by ships. The remarkable expansion of world trade and manufacturing over the past 50 years with distributed manufacturing, just-in-time delivery, and other features of our modern world was possible only with a reliable and dependable shipping network distributing all kinds of goods throughout the world, from basic commodities and semiproducts to finished goods.

Simultaneously, with the growth in demand for ships and an increase in their complexity, ship structural design and calculation procedures have advanced considerably. Earlier, ships were designed and dimensioned solely on the basis of prescriptive rules from classification societies, which were themselves largely based on experience and feedback from ships in service; in the final quarter of the last century, rational analysis and design methods were introduced. The development and introduction of the finite element method brought completely new possibilities to deal with complex structural tasks. Just as it would not have been possible to design and construct the drastically new jumbo aeroplane, the Boeing 747, in the 1960s without detailed rational analyses, many of the new ship types introduced during the past 40 or 50 years would not exist without the extensive calculation procedures and analysis possibilities mostly based on the finite element method. This includes liquefied natural gas carriers, modern containerships, large passenger ships, as well as large fast ferries with catamaran or trimaran hull forms. The structural design and analysis of modern naval ships, too, is quite different today.

The history of the containership is a suitable example. Figure 1.1 is an example of a finite ele-

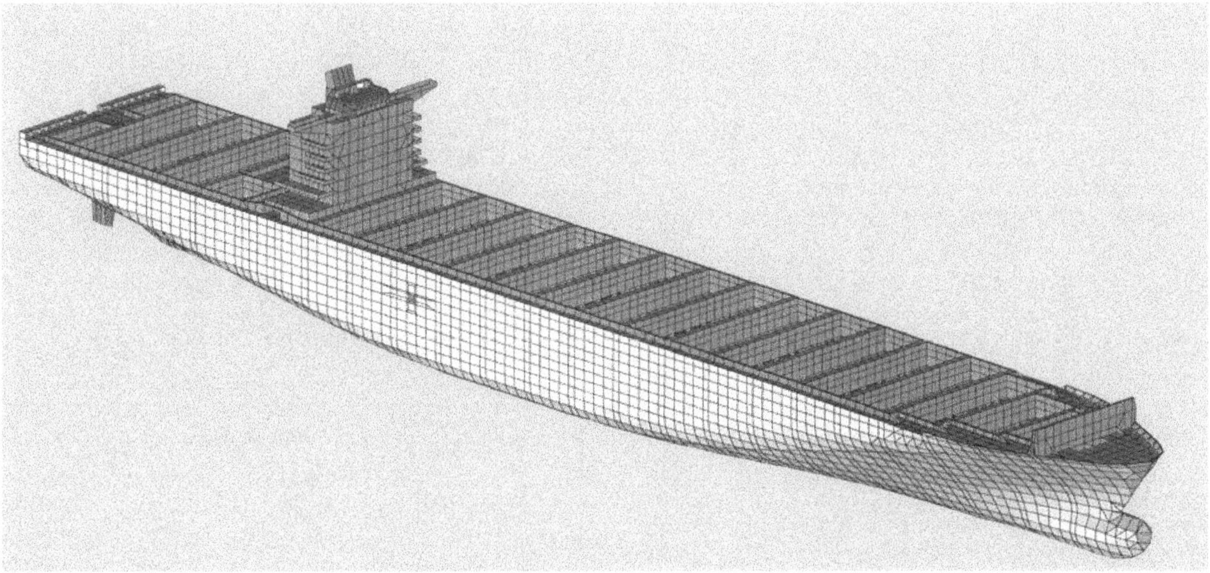

Figure 1.1 Finite element model of a 9200 TEU containership.

ment model of a medium-sized containership. The evolution from the first container carriers with large deck openings of the 1960s, with a carrying capacity of up to 1000 twenty-foot equivalent units (TEU), to the ultralarge container carriers of today, with a carrying capacity of beyond 13,500 TEU, was possible only because of the ever increased analysis possibilities of classification societies and design offices. Improvements of each new class of this ship type were always worked out close to the technically feasible. Ships of that size are characterized by specific aspects that need special technical attention. This involves their static and fatigue strength, their structural flexibility, as well as their behaviour in waves. But it is not the big ships alone that have to be carefully designed and analyzed. Modern container feeder ships, too, are optimized to efficiently carry a maximum number of containers for their individual size, and so careful design and analysis is also needed for these smaller vessels. Similar aspects can be observed for cruise ships, bulk carriers and tankers.

The complexities of modern ships and the demand for greater reliability, efficiency, and economy require a scientific, powerful, and versatile method for their structural design. In the past, ship structural design was largely empirical, based on accumulated experience and ship performance and expressed in the form of structural design codes or "rules" published by various ship classification societies. These rules provide simplified and easy-to-use formulas for structural dimensions, or *scantlings*,* of a ship. This approach saves time in the design process and is still the basis for the preliminary structural design of most ships.

There are, however, several disadvantages and risks to a completely "rulebook" approach to design. First, the modes of structural failure are numerous, complex, and interdependent, and with such simplified formulas the margin against failure remains unknown. Thus, one cannot distinguish between structural adequacy and overcapacity. Therefore, such formulas cannot give a truly efficient design. In some cases, the extra steel may represent a significant cost penalty throughout the life of the ship.

Second, these formulas only aim to avoid structural failure, but there are usually several ways of achieving this, and the particular implied in the formulas may not be the most suitable regarding specific goals of the ship owner over the life of the ship or its particular purpose or economic environment. A true design process must be capable of accepting an objective, of actively moving toward it, and of achieving it to the fullest extent possible.

Third, and most important, these formulas involve a number of simplifying assumptions. They can be used only within certain limits. Outside of this range, they may be inaccurate. The history of structural design abounds with examples of structural failures—in ships, bridges, and aircraft—that occurred when a standard, time-honored method or formula was used, unknowingly, beyond its limits of validity.

For these reasons, there has been a general trend toward "rationally-based" structural design ever since the 1970s or 1980s, which may be defined as a "design directly and entirely based on structural theory and computer-based methods of structural analysis and optimization to achieve an optimum structure based on a designer-selected measure of merit." Thus, a complete rationally-based design involves a thorough and accurate analysis of all factors affecting safety and performance of the structure throughout its life and a synthesis of this information, together with the goal or objective the structure is intended to achieve. The aim is to produce the design that best achieves this objective and that provides adequate safety. This process involves far more calculation than conventional methods and can only be achieved by extensive use of computers. For this reason, rationally-based structural design is necessarily a computer-based and often semiautomated design.

Rationally-based design was first developed and applied for aircraft and aerospace structures. It continues to have its greatest application in these areas because of the predominant economic significance of structural weight, and hence structural efficiency, coupled with the obvious need for high structural reliability. In land-based structures, the move toward this approach was given strong impetus in the 1970s by a series of structural failures of steel box girder bridges. These failures showed that for larger and more slender bridges, the existing empirically-based design codes were inadequate. In the ocean environment, an elementary form of this approach has been used for the design of offshore structures from the beginning, partly because there was little or no previous experience on which to rely and partly because of the high economic stakes and risks in case of failure. In this area, as well as in ship structures, the classification societies encouraged and contributed greatly to the development of rationally-based methods. Since first publication of this book, analysis methods of classification societies have changed and moved considerably towards what may be called rationally-based design.

**Scantlings* is an old but still useful naval architecture term that refers to all local structural sizes, such as thicknesses, web heights, flange breadths, bracket sizes, etc.

Rationally-based ship structural design is definitely not fully automated design, that is, a "black box" process, where the designer's only role is to supply the input data and whereupon the process presents the designer with a finished design. This type of design would require that all design decisions—objectives, criteria, priorities, constraints, and so on—must be made before the design commences. Many of these decisions would have to be built into the program, making it difficult for the designer to even be aware of the influence of the objectives, much less to have control over them. Rather, of its very nature rationally-based design must be interactive. The designer must always remain in charge and be able to make changes and decisions—with regard to objectives, criteria, constraints, priorities, and so on—in light of intermediate results. Therefore, a rationally-based design process should allow the designer to interrupt, go back, make changes, call for more information, skip some steps if they are not relevant at the time, and so forth.

Rationally-based design gives the designer much more scope, capability, and efficiency than ever before. But it does require a basic knowledge of structures and structural analysis (e.g., fundamentals of finite element analysis and basic types of structural failure) together with some experience in structural design. Given these requirements, the deciding factor in choosing the rationally-based approach is whether and to what extent a product and/or a performance (economic, operational, or both) is desired that goes beyond what is obtainable from the rule-based approach. The latter is simpler, but it may not be optimal and is nonadaptable. Thus, the two approaches are complementary, and a good designer will use whichever is more appropriate for a given situation.

1.1.1 Preliminary Design and Detail Design

As in most structures, the principal dimensions of a ship design are usually not determined by structural considerations, but rather by more general requirements, such as beam and draft limitations, required cargo capacity, and so on. For this reason, structural design usually begins with the principal dimensions already established. The designer must determine the complete set of scantlings that provide adequate strength and safety for least cost (or whatever other objective is chosen). Structural design consists of two distinct levels:

1. Preliminary design to determine location, spacing, and scantlings of principal structural members*
2. Detail design to determine geometry and scantlings of local structures (brackets, connections, cutouts, reinforcements, etc.)

Of these two levels, the rationally-based approach has more relevance and more potential benefits regarding preliminary design because of the following.

- This level has the greatest influence on structural design and, hence, offers large potential savings.
- This level provides the input to detail design. Benefits of good detail designs are strongly dependent on the quality of this input.

In fact, rationally-based preliminary design offers several kinds of potential benefits. The economic benefits are illustrated by the tanker example quoted in Section 1.3, in which the rationally-based approach gives a 6% savings in ship structural cost compared with current standard designs (which, for a large tanker, represents a savings of over 1 million dollars) and an even greater amount of extra revenue from increased cargo capacity arising from weight savings. Naval vessels can obtain greater mission capability by a reduction of weight. Ship designers gain a large increase in design capability and efficiency and are able to concentrate more on the conceptual and creative (and more far-reaching and rewarding) aspects of design. And finally, there are also substantial benefits to be gained in ship structural safety and reliability.

This is not meant to imply that detail design is less important than preliminary design; it is equally important for obtaining an efficient, safe, and reliable ship. Also, there are many benefits to be gained by applying modern methods of engineering science, but the applications are different from preliminary design and the benefits are likewise different. Since the items being designed are much smaller, it is possible to do full-scale testing and, since they are more repetitive, it is possible to obtain benefits of mass production, standardization, methods engineering, and so on. In fact, production aspects are of importance primarily in detail design.

Also, most of the structural items that come under detail design are similar from ship to ship, and so in-service experience provides a sound basis for their design. In fact, because of the large number of such items, it is inefficient to attempt to design all of them from first principles. Instead, it is generally more efficient to use design codes and standard designs proven by experience. In other words, detail design is an area where a rule-based approach is appropriate, and rules published by various ship classification societies contain a great deal of useful

*For naval vessels, this is termed "contract design."

information on the design of local structures, structural connections, and other structural details.

1.1.2 Aims and Scope of the Book

Now that we have defined the term "rationally-based" and noted the distinction between preliminary design and detail design, we can give a specific statement of the two aims of this book:

- To present structural analysis theory required for rationally-based preliminary ship structural design in a complete and unified treatment that assumes only basic engineering subjects, such as mechanics and strength of materials
- To present a method for rationally-based design that is practical, efficient, and versatile and that has already been implemented in a computer program and that has been tested and proven

This book is entirely self-sufficient and self-contained; that is, it covers all basic aspects of rationally-based design required by a designer. Even basic aspects such as the finite element method, column buckling, and plate buckling are included. This has been done for two reasons.

First, because this book is intended primarily as a textbook, and in the field of ship structures such books are few and far between. Because of the greater complexity and sophistication of rationally-based design, lack of a unified and comprehensive text would constitute a correspondingly greater difficulty for students and a serious obstacle to further progress in this field.

The second reason is that rationally-based design, both in general and in the particular method presented here, is radically different from the traditional rule-based method and, although many of its features are familiar to experienced designers (such as finite element analysis and elastic buckling), other features are either relatively new (such as nonlinear finite element theory and statistical prediction of wave loads) or totally new (such as new techniques for structural modeling and new methods for ultimate strength analysis of a stiffened panel and of an entire hull girder).

For this reason, the book is also intended for practicing designers who wish to become more familiar with this alternative method of design. In fact, the book's role is of particular importance because rationally-based design of its very nature requires at least a basic knowledge of its underlying theory and methods. Once this is acquired, the method's enormous capability (some of which is demonstrated in Section 1.3 and in the references given there) becomes available to the designer. Moreover, the method's breadth of application and the benefits gained from its use increase in proportion to the knowledge presented here. It is the authors' hope that the presentation of the underlying theory and analysis methods in this text will assist designers to obtain the maximum possible benefits from this new approach.

Also, the authors emphasize that the design method presented herein is not the only possible method, at least not regarding the particular component methods for achieving the basic tasks, such as structural modeling techniques and methods of member limit analysis. The methods presented herein were selected or developed on the basis of their suitability for rationally-based design, but this type of design involves so many different areas that there are bound to be some particular methods and techniques that are as good or better than those given here. Moreover, as further progress is made in such areas as structural theory, numerical methods, and computer hardware and software, still better methods will be developed. But the important point is that now, as the result of many years of effort by many persons and organizations both inside and outside of the field of ship structures, all of the required ingredients for rationally-based design are available.

1.1.3 Applicability to Naval Design

The design method presented herein applies equally well to naval vessels and commercial vessels. Because basic classification rules are intended for commercial vessels and are not suitable for warships, various navies and naval design agencies developed their own methods of structural design. Like classification rules, these methods evolved over a long period and many of them were systematized and codified into some form of *design manual*—a sort of naval counterpart to the rules. Recently, some classification societies in cooperation with a Navy developed rules for naval vessels. Because of the need for greater structural efficiency and other special requirements, naval design methods are generally more thorough and rigorous than rule-based design methods of commercial ships, and they show a stronger trend toward a rationally-based approach. Thus, in addition to design manuals many current methods of naval design already include some of the basic features of rationally-based design.

Section 1.2 gives a brief overview of basic features of rationally-based design, including a discussion of the different aims, measures of merit, and design criteria in commercial ships and naval ships. Section 1.3 considers capabilities, applications, and some sample results. Once these aspects are treated, it becomes apparent that the method presented herein applies equally well to both ship types and that it matches the needs and challenges of naval designs particularly well. Because commercial vessels are more numerous, most of the explanations and figures in this text refer to this type. Therefore, it seems desirable at this point to briefly consider why the rationally-based method presented herein is so well suited for naval design, even though a full appreciation can only be obtained after covering Sections 1.2 and 1.3.

Naval ship structures are subject to many special requirements and constraints. For example, they must be capable of withstanding specified levels of blast, shock, and other special loads. Also, they must be damage tolerant, that is, capable of sustaining some structural damage without loss of main functions. Since rationally-based design considers each limit state explicitly, it can accommodate these special constraints. As discussed in Section 1.2, mission requirements of naval vessels make it extremely important to minimize the weight and vertical center of gravity (VCG) of the structure to the extent allowed by the various constraints (such as cost, adequate strength and safety, and damage tolerance). Hence, there is a paramount need for structural optimization, which is one of the basic features of rationally-based design.

Finally, it is worth noting that the ability of rationally-based design to deal with both commercial and naval ships can also help to unify the field of ship structural design, which until now has been largely split into two separate areas.

1.1.4 Applicability to Other Types of Structure

In this text, the rationally-based approach is described purely in terms of ships. However, because this approach represents the most fundamental and most general type of engineering design, the material presented herein is also applicable to a wide variety of other steel structures, both fixed and floating.* All of the basic principles and most of the analysis methods for other steel structures are the same as for ships, and the scope of this text could have been extended to include these other structures without requiring fundamental change of approach. However, this would have required the extension of the consideration to the specifics of other structures, such as of additional types of loads and failure modes, plus some new examples to illustrate these other applications. This would have increased the book's length unduly.

*For example, in Hughes, Mistree, and Davies (1977), the method presented herein was used for the structural optimization of a large steel box girder bridge.

1.1.5 Practicality of the Method

Rationally-based design is necessarily computer-based. This raises a number of practical questions in regard to computer implementation, accuracy, cost-effectiveness, availability, ease of use, documentation, and so on. These are important questions, and they are dealt with fully in Section 1.3. But, since practicality is so essential in a design method, it is appropriate to mention here that this method, ever since its first version in 1975, has been developed and improved continuously, and that during this same period a computer program based on it has likewise been continuously developed and improved. This program, called MAESTRO†, has now been used for hundreds of ship structural analyses and designs. In addition to its use for optimum design, the analysis portion of MAESTRO can be used to evaluate a given design, to assess proposed changes to a design or to an existing ship, or to evaluate the seriousness of damage incurred by a vessel. The program is also a valuable tool for ship structural research and for the teaching of ship structural design. Further details of all these aspects are given in Section 1.3 and in the references cited there.

1.1.6 International Maritime Organization Goal-Based Standards and IACS Common Structural Rules

As noted earlier, ships have historically been designed and dimensioned on the basis of rules of a ship classification society. These rules were largely based on structural mechanics principles as well as on the extensive experience individual classification societies gathered over the years with ships in service. With their worldwide network of surveyors, classification societies looked after their classified ships not only from the time of initial design to the construction in the shipyards, but also throughout the ship's lifetime up to decommissioning and scrapping. When weaknesses were found in a ship or in a class of ships indicating a lack of strength, the rules were adjusted. This is sound practice followed even today. Competition between classification (or "class") societies was and is a strong driving force to support innovation. The International Association of Classification Societies (IACS) looked after a certain degree of alignment between rules of member societies and a common minimum standard, a situation that was

†**M**odeling, **A**nalysis, **E**valuation and **STR**uctural **O**ptimization.

important particularly when ships changed class during their lifetime.

One of the areas where it was difficult for classification societies to agree on common standards in the past is corrosion. Different societies follow different philosophies on how to treat corrosion during the lifetime of a ship: some have explicit corrosion allowances added to scantlings determined by their rules; others take care of corrosion implicitly within their rules. This works well as long as ships stay within the same class from beginning to end. It does, however, cause confusion and difficulties of interpretation when a ship changes from a class following one philosophy to a class with another procedure. Such problems arose particularly with tankers and bulk carriers, ships that by nature of their trade are especially prone to corrosion. In the 1980s and 1990s, some of the more spectacular accidents with older ships, where heavily-loaded bulkers disappeared during a storm or where tankers floundered and broke apart with severe pollution to the sea and coast, could at least partly be traced back to this state of affairs.

It was agreed in maritime circles that this had to change, and this was supported by strong political pressure. Therefore, the International Maritime Organization (IMO) and IACS set out to improve the situation.

1.1.6.1 Goal-Based Standards

The concept of goal-based ship construction standards (GBS) was introduced at the IMO in 2002, suggesting that IMO plays a larger role in determining overall standards to which new ships are built. The IMO agreed to develop the basis for ship construction standards that permit innovation in design but, at the same time, ensure that ships are constructed in such a manner that, if properly maintained, they remain safe throughout their economic life. These standards should eventually be applied to seagoing ships of all types worldwide.

Consequently, the Maritime Safety Committee (MSC) of IMO developed GBS at first for hull construction of bulk carriers and oil tankers. The procedures are based on vast practical experience gained with these ship types over the years, mostly collected by classification societies. At the same time, GBSs advocate the application of a rational holistic approach, such as presented in this book. This includes, first, defining a procedure for a risk-based evaluation of the current safety level based on existing mandatory regulations related to the safety of these ships and, second, considering ways forward to establish future risk acceptance criteria using Formal Safety Assessment. It is expected that over time, GBS will also be developed for other ship types.

The MSC agreed on the basic principles of IMO GBS in conformity with other GBS to be developed by IMO. A five-tier system was agreed for GBS, comprising goals (Tier I), functional requirements (Tier II), verification of compliance (Tier III), regulations and rules for ships such as classification rules, IMO requirements, and relevant national requirements (Tier IV), and applicable industry standards and practices (Tier V). The five tiers are shown in Figure 1.2.

The first three tiers basically constitute the IMO GBS, whereas Tiers IV and V contain detailed prescriptive provisions developed or to be developed by classification societies (recognized by flag states), the IMO and national administrations, and industry organizations. Thus, IMO's GBS establish *rules for rules*, as opposed to rules for ships.

Verification of compliance of ship construction rules with GBS will be carried out by an international Group of Experts established by IMO's Secretary General in accordance with Guidelines for verification of compliance with GBS, which are currently under consideration by the Committee. These Guidelines foresee that national administrations (i.e., flag states) submit requests for verification of their ship construction rules or, in most cases, those developed by an organization recognized by the administration (in most cases, classification societies) to the Secretary General of IMO, who will forward these requests to the Group of Experts for a verification of information submitted through an independent review. The final report of the group with relevant recommendations will then be forwarded to the MSC for consideration and approval and circulated to the IMO membership by appropriate means, such as MSC circulars.

At the time of finalizing this book, some further developments are necessary before GBS will be implemented. Although the Working Group on GBS recommended that amendments to the International Convention for the Safety of Life at Sea be approved and that the GBS be considered for approval, neither of these actions was agreed in IMO plenary. It appears that there are still several issues that need resolution before that step can be taken. Particularly, the GBS verification process is not yet agreed on, and alternative methods are being considered.

1.1.6.2 Common Structural Rules

In the early years of this century, IACS developed two sets of common structural rules (CSR) which entered into force on April 1, 2006. They apply to all

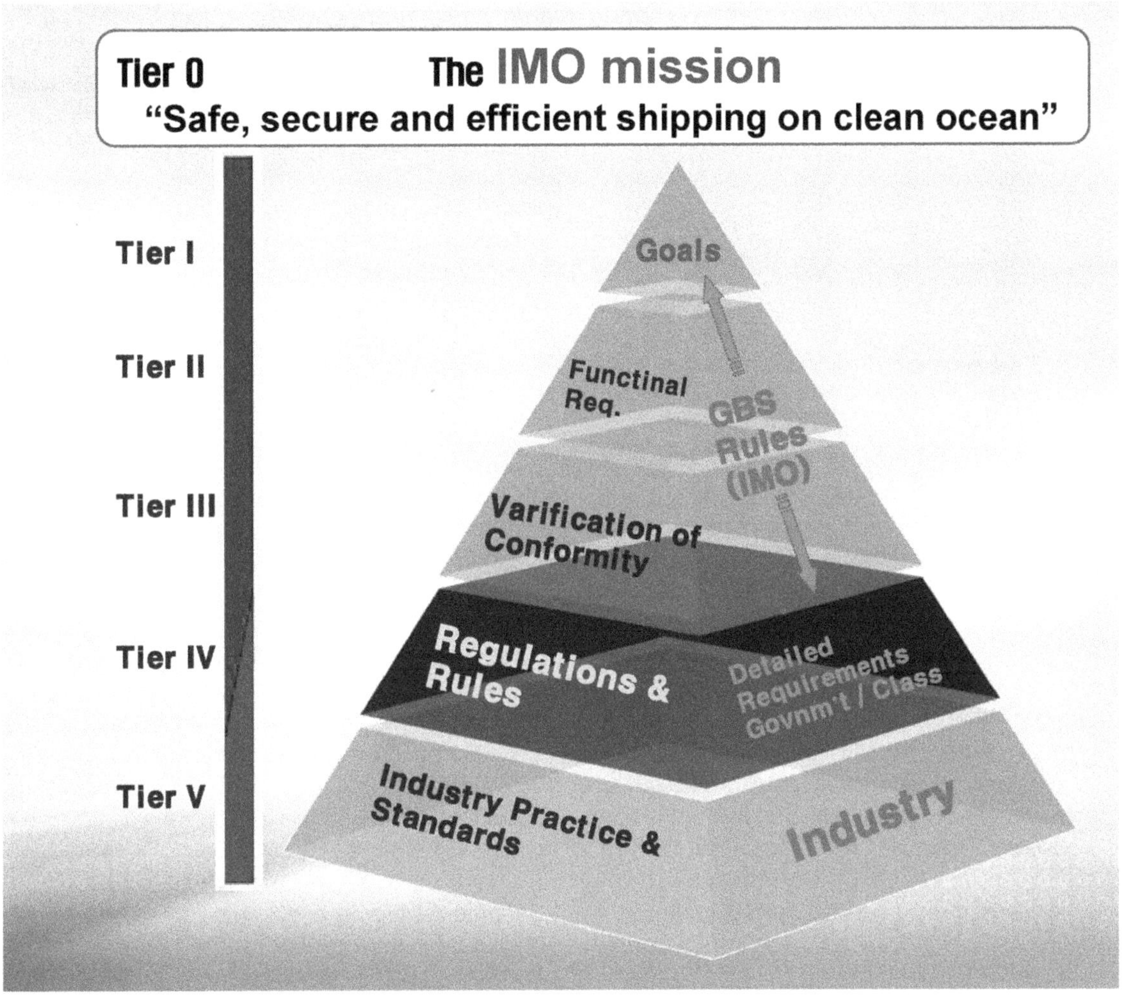

Figure 1.2 The five tiers of the IMO vision for ship construction standards. (Source: Oh, K.-G. [2009]. Recent status of rules and regulations for ships. Keynote lecture, *17th International Ship and Offshore Structures Congress*, Seoul, Korea.)

bulk carriers with length above 90 m and all double hull oil tankers with length above 150 m.

Basic considerations in the CSR include:

- Design life of 25 years
- Net scantlings approach
- Dynamic loading in North Atlantic environmental conditions
- Buckling
- Fatigue life
- Ultimate Limit State of the hull girder

Background and basis for the development of common structural rules is discussed, for instance, by Løvstad and Guttormsen (2007).

Since first entering into force, a few amendments to the CSR were made in an effort to harmonize the CSR for tankers and bulk carriers. Additionally, IACS published common interpretations for the rules to assist its member societies and industry in implementing the CSR in a uniform and consistent manner. There is also a long-term plan in place to further increase harmonization between tanker and bulk carrier common structural rules.

CSR for tankers and bulk carriers initially considered different approaches for corrosion additions, and this was identified as an issue that required harmonization in the short term. The aim was to apply corrosion additions in a way common to both CSR for tankers and bulk carriers. In summary, the corrosion harmonization is as follows.

- A corrosion propagation model based on probabilistic theory for each structural member was developed, and corrosion diminution was estimated at the cumulative probability of 95% for 20 years using the corrosion propagation model.
- Corrosion additions were determined for each structural member and for the corrosion environment.

Figure 1.3 shows how, according to CSR, net scantling thicknesses and corrosion additions are to be

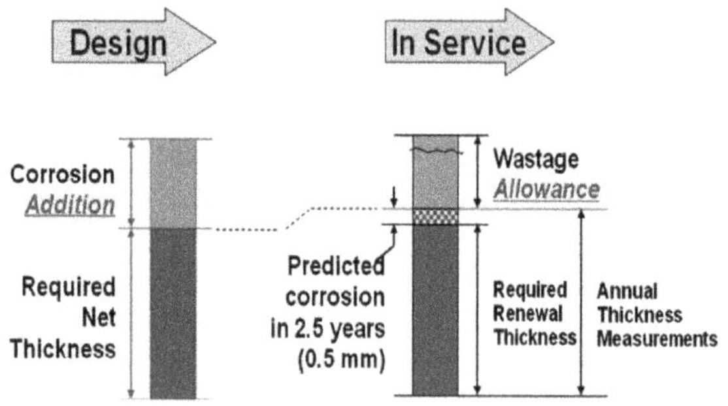

Figure 1.3 IACS CSR net scantling approach to be adopted during design and in service. (Source: Aksu, S., et al. (2009). Technical committee II.1—Quasi-static response. *Proc. of the 17th International Ship and Offshore Structures Congress*, Seoul, Korea, Vol. 1.)

adopted during design and in-service conditions. The corrosion addition approach in CSR is more rational than prescriptive corrosion allowance requirements as practiced in the past. The CSR do not necessarily call for corrosion additions as structural design requirements. If net thicknesses, as determined by the rules, are accepted as representing the minimum acceptable value, it should be the owner's choice to adopt a variety of techniques to determine corrosion additions for the ship. Even zero corrosion addition is a possibility if, for instance, structural scantlings are maintained during the life of the ship by using advanced coatings, by aggressive inspection and repair regimes, and by other ways.

When CSR for tankers and bulk carriers were implemented in 2006, the shipbuilding industry had to cope with two completely new design standards. While rule development in the past was a slow process of evolution, the introduction of CSR represented a step change in assessing the adequacy of a structural design. Tougher strength and fatigue requirements were introduced, and more extensive design calculations were made mandatory to fulfil both prescriptive scantling requirements and direct strength assessments using finite element analysis. All existing designs that had been developed over years and were the basis for new buildings offered by the shipyards had to be reassessed, redesigned, and documented for compliance with the new rules.

A detailed overview of the contents and introduction of GBS and CSR is given in the ISSC report by Aksu et al. (2009).

1.2 BASIC ASPECTS OF STRUCTURAL DESIGN

One of the most fundamental concepts in engineering is that any object of interest is regarded as a system, which may be anything from a simple device to a vast multilevel complex of subsystems.

A ship is an example of a relatively large and complex system, and itself is a part of an even larger system including the ocean environment, port facilities, etc. The ship consists of several subsystems, each essential to the whole system. Examples of subsystems are the propulsion machinery and the cargo handling gear. The structure of the ship can be regarded as a subsystem, providing physical means whereby other subsystems are integrated into the whole and given adequate protection and a suitable foundation for their operation.

In general terms, the design of an engineering system may be defined as, "The formulation of an accurate model of the system to analyze its response—internal and external—to its environment and the use of an optimization method to determine the system characteristics that best achieve a specified objective, while also fulfilling certain prescribed constraints on the system characteristics and the system response." Translating this to the case of preliminary ship structural design, a rationally-based design procedure can be described as follows.

1. External loads are predicted as accurately as possible, taking account of their stochastic nature.
2. Load effects and limit values of load effects are calculated accurately throughout the structure for all load conditions and load cases.
3. Minimum required margins between the load effects and their limit values are selected on the basis of a required degree of safety.
4. The resulting strength requirements are expressed in the form of mathematical constraints on the design variables (in most cases, nonlinear constraints).
5. The designer is left free to specify the measure of merit of the structure, that is, the criteria that are to be used in achieving the *best* structure and the influence of each design variable on the measure of merit. Also, the designer is able to specify any number of other constraints on the design, of any form whatsoever, in addition to the strength-related constraints.
6. An optimization method automatically and efficiently solves for the values of the design variables that yield maximum value of the measure of merit while also satisfying all of the constraints.

From this description of a rationally-based design procedure, it is possible to identify six essential tasks:

1. Calculation of environmental loads
2. Overall response analysis

3. Substructure response analysis
4. Limit state analysis
5. Formulation of reliability-based structural constraints
6. Solution of a large nonlinear optimization problem

Figure 1.4 illustrates the overall design process, consisting of these six tasks. It is also a flowchart of the MAESTRO program. All of these tasks are extensive, especially for structures as large as ships. The principal difficulty or challenge in developing a rationally-based design procedure is to develop methods that can perform these tasks to the required degree of accuracy and thoroughness within acceptable amounts of total man-hours and computational efforts. To define the program more precisely and to explain broadly what it entails, each of these tasks is now considered briefly.

1.2.1 Calculation of Environmental Loads

Environmental loads are loads, both static and dynamic, that come from the ship's environment (mainly because of gravity and fluid pressures) and from its motion. Most of these loads are relatively independent of the structural design, that is, they are not much affected by the structural layout or by the scantlings. Rather, they are more a function of hull shape, the type and distribution of cargo, and other nonstructural factors. Therefore, although calculation of these loads is the first step of structural design

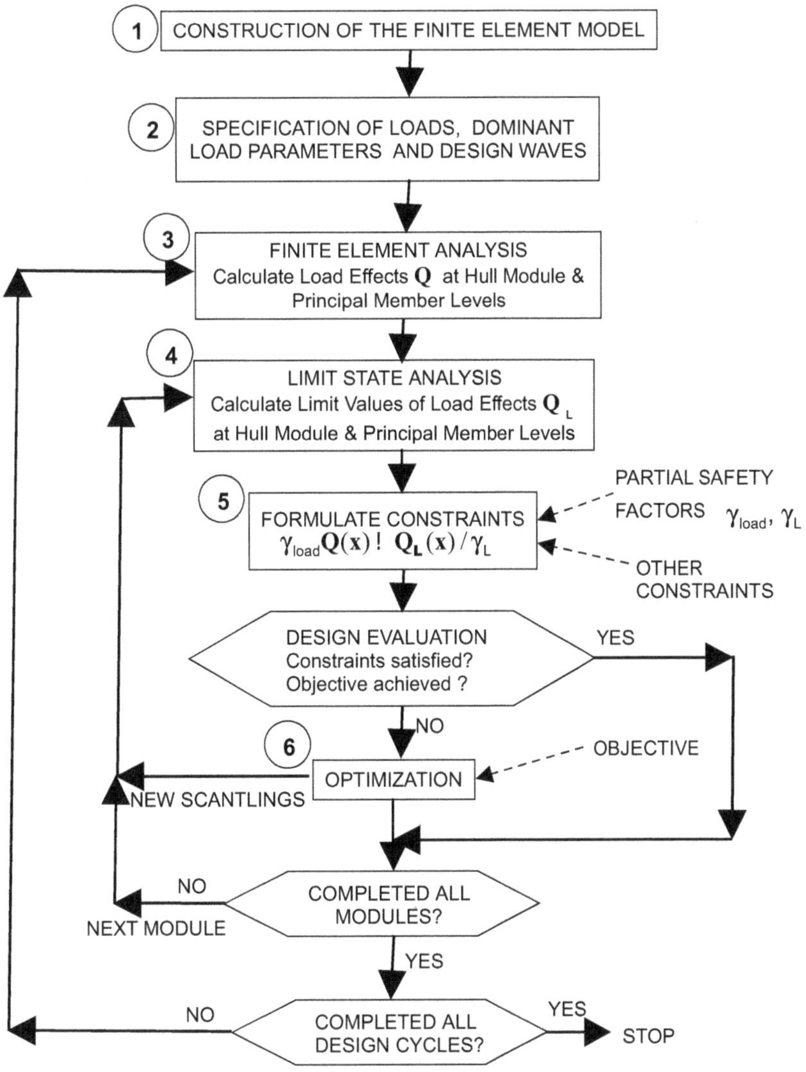

Figure 1.4 Rationally-based structural design.

and one of the most crucial aspects of the entire process, it is essentially a separate initial task. Some of the loads can be readily calculated and controlled by the designer (e.g., those arising from the lightship mass and cargo distribution). Other loads, particularly wave loads including inertia loads and other hydrodynamic loads (slamming, sloshing, etc.) are sufficiently complex that their calculation is not regarded as part of the designer's task, but rather that of hydrodynamicists and other specialists. In contrast, the other two types of calculations—response analysis and limit state analysis—are inherently and totally structural in nature.

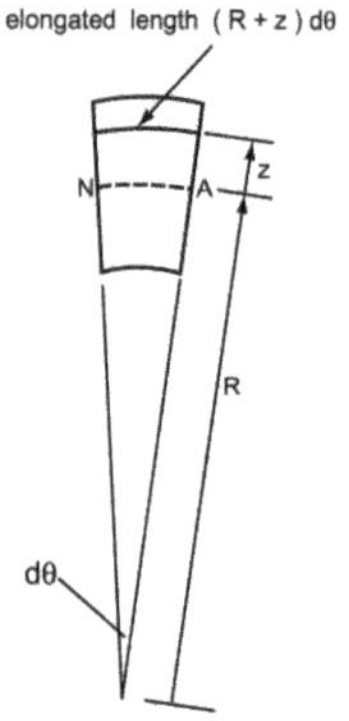

Figure 1.5 Strain distribution in simple beam theory.

1.2.2 Overall Response Analysis

Overall response* analysis entails calculation of the effects of the environmental loads on the overall structure (bending moment, deflection, stress, etc.). For reference, load effects will be represented by the symbol Q (or Q_i if referring to the ith load effect).

For a ship, the overall structure is regarded essentially as a beam—a floating box girder internally stiffened and subdivided. For vertical bending, the decks and bottom structure are flanges and the side shell and any longitudinal bulkheads are the webs. The hull girder analysis deals only with those longitudinally integrated forces and moments that are dealt with in beam theory: vertical shear force, F_z, longitudinal bending moment in the (ship's) vertical and horizontal planes, M_y and M_z, and longitudinal twisting moment, M_x. Of these, the most significant is the vertical bending moment M_y, that is, bending about the Y-axis in Fig. 1.7. This load effect is caused mainly by the unequal distributions of weight and buoyancy along the length of the ship, accentuated by waves. Horizontal bending (i.e., bending about the Z-axis) occurs when the ship is in an inclined condition, as a result of rolling, and this situation also arises from quartering seas where wave crests on one side of the ship are in phase with troughs on the other. In most ships, the maximum value of M_z is smaller than the maximum value of M_y (typically 20% or less), but in large tankers and containerships, for instance, it can rise to as high as 50% of vertical bending. For simplicity, here we only consider vertical bending; horizontal bending and its relation to vertical bending is considered in Section 3.6.6.

The bending moment varies along the length of the ship, being zero at the ends and having a maximum value that usually occurs near the midlength of the ship. The maximum value of hull girder bending moment is the single most important load effect in the analysis and design of ship structures. Hull girder bending is referred to as either "hogging" or "sagging," depending on the sense of curvature which it causes in the hull, as shown in Fig. 1.7.

The hull girder analysis assumes that hull girder bending satisfies simple beam theory (Bernoulli-Euler), which implies the following assumptions.

1. Plane cross-sections remain plane.
2. The beam is essentially prismatic (no openings or discontinuities).
3. Other modes of response to the loads (e.g., transverse and longitudinal deflection and distortion caused by shear and/or torsion) do not affect hull girder bending and may be treated separately.
4. The material is homogeneous and elastic.

The first assumption is illustrated in Fig. 1.5. Under the action of a bending moment, a beam undergoes curvature of radius R locally and, if plane cross-sections remain plane, the longitudinal strain ε_x in a cross-section varies linearly in the vertical direction and is related to R as follows.

$$\varepsilon_x = \frac{(R+z)d\theta - R\,d\theta}{R\,d\theta} = \frac{z}{R}$$

The horizontal surface where z, and hence also the strain, is zero is referred to as either the neutral surface or, regarding the beam problem as one-dimensional, the neutral axis. The material is assumed to be homogeneous and elastic, and so the longitudinal stress is

$$\sigma_x = E\,\varepsilon_x = E\frac{z}{R} \qquad (1.2.1)$$

If there is no external axial force, equilibrium requires

*The "response" of a structure is simply the group of load effects caused by all types of loading; the two terms are essentially the same, and they are used interchangeably throughout this text.

$$\int_A \sigma_x \, dA = 0$$

which reduces to

$$\int_A z \, dA = 0$$

and this indicates that the neutral surface is the horizontal axis passing through the cross-section.

Equilibrium of moments requires that the external moment, M_y, is balanced by the moment of the internal stress forces

$$M_y = \int_A z \, \sigma_x \, dA$$

which from (1.2.1) reduces to

$$M_y = \frac{EI}{R} \qquad (1.2.2)$$

where I is the moment of inertia of the cross-section, defined by

$$I = \int_A z^2 \, dA$$

Equation 1.2.2 relates the curvature to the external bending moment and, if this is used to eliminate R from (1.2.1), the result is the familiar equation for bending stress at a height z from the neutral axis

$$\sigma_x = \frac{M_y z}{I} \qquad (1.2.3)$$

1.2.2.1 Section Modulus

Equation 1.2.3 indicates that the longitudinal bending stress σ_x is greatest when z is greatest, that is, at the extreme upper or lower edge of the section. When z corresponds to one of these extreme values, the quantity I/z is called the *section modulus* and is denoted herein as Z. Since the neutral axis is not generally at half-depth, there will be two extreme values of Z: Z_D for the deck and Z_K for the keel, and there will thus be two values of Z: $Z_D = I/z_D$ and $Z_K = I/z_K$. Because of hydrostatic and hydrodynamic loads, the bottom structure is usually sturdier (heavier scantlings) than the deck and so the neutral axis is usually below the half-depth. A height of 0.4 D above the keel is typical, but the location varies widely between different ship types and designs. Thus, the largest hull girder stress usually occurs in the deck rather than the bottom. More precisely, it occurs in the uppermost member which is longitudinally effective, that is, which is of sufficient length and has a sufficiently rigid attachment to the rest of the hull girder to act as part of the hull girder. In most cases this is a deck, and that deck constitutes the uppermost flange of the hull girder. If the side shell extends up to this deck, then it is referred to as the *strength deck*.

Section modulus is also useful whenever it is desired to assess or control the maximum hull girder stress (wave-induced, stillwater, or total) by itself, separately from the stresses arising from hull module and principal member response. For example, because the wave-induced hull girder stress is cyclic, it is necessary to restrict its amplitude to guard against fatigue failure.

1.2.2.2 Departure from Simple Beam Theory

Equation 1.2.3 states that stress is constant across horizontal decks and varies linearly in the sides. There are several factors that can cause the actual stress distribution to differ from this idealized distribution. Because of transverse shear, there is some longitudinal distortion of the cross-section of the hull girder. Torsional loading will cause further distortions, particularly if there are large openings in the deck; this longitudinal distortion of the cross-section out of its original plane is referred to as "warping" of the cross-section. This means that the first assumption is not fulfilled, at least not perfectly. Likewise, the second and third assumptions are not fulfilled because the hull girder is not prismatic (except in the "parallel midbody," if there is one) and it may have hatches, other openings and discontinuities, and discretely occurring elements such as transverse bulkheads. Also, it is a complex assembly of intersecting members, transverse as well as longitudinal, and there are several other modes of response, in addition to warping, that affect the hull girder bending response. For ships with no major changes in cross-section other than in-line hatches (for which the intermediate portions of deck may be ignored) the longitudinal stress resulting from hull girder bending generally follows the idealized distribution quite closely (ignoring stress concentrations and other local effects) as shown in Fig. 1.6. For such ships, the effects of shear and of other responses, of transverse structure, and even of openings and discontinuities, can be calculated separately (or at least estimated) to assess their importance and to apply corrections where necessary. In some cases, superposition can be used. This structurally prismatic type of hull girder is considered in Chapter 3. For ships with significant changes in cross-section,

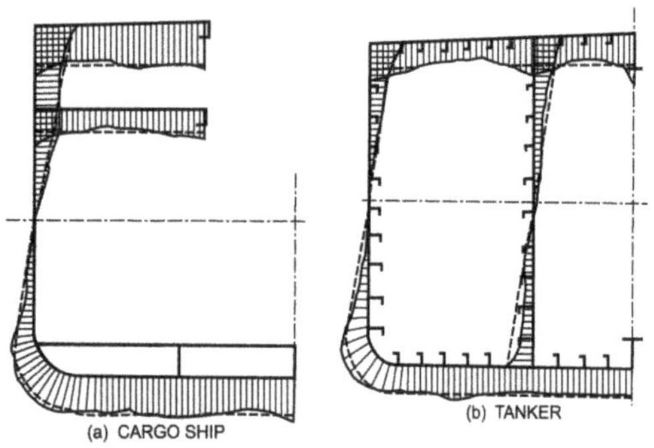

Figure 1.6 Typical hull girder bending stress distribution for structurally prismatic ships.

the load effects are best obtained by a full ship finite element analysis.

Shear force can be significant with some cargo types and distributions (especially bulk cargoes), and both shear force and twisting moment can be significant if the hull girder has low torsional rigidity, as in container ships. Shear force and torsion are treated in Sections 3.7, 3.10, and 3.11.

1.2.3 Levels of Structural Modeling and Analysis

1.2.3.1 Definition and Use of Modules in Analysis and Optimization

In the early 1980s, when MAESTRO first became available, the limited computing capability meant that the finite element model could be only a portion of the ship, and this was called a *hull module*. It could be one cargo hold, as in Fig. 1.7, or several, as in Fig. 1.8. Loads at ends of the model were obtained from the hull girder analysis, as shown in Fig. 1.7. With today's computing power there is no such limitation, and the finite element model usually idealizes the entire ship.

However, the term "hull module" (or just module) is still useful because a ship hull usually does consist of a series of distinct segments: cargo holds in commercial ships and compartments in naval ships and submarines. Other nonhull parts also constitute distinct modules, such as an accommodation block or a funnel.

A finite element model of an entire ship is a large model, and its construction needs to be done in carefully planned levels and sequences. Modules are helpful for this because they are ideal high-level building blocks. Moreover, in the parallel midbody of a tanker, bulker, or submarine, only one module (one cargo hold or compartment) needs to be built and then copied. If there is a need to build the model quickly, several people can create different modules simultaneously.

Also, in the creation of such a large finite element model, it is advisable to test the model as it is being built, so as to catch modeling errors early. A convenient occasion for testing is after the addition of a new module or group of modules. This requires additional temporary data: restraints to prevent rigid body movement and hull girder loads at the ends of the model.

Another consideration is that for such a large structure as a ship, optimization involves so many simultaneous changes that it is difficult to keep track of them and to appreciate which of the many loads, limit states, and designer-specified constraints are driving the design. Therefore, in MAESTRO, although several modules can be optimized in one *design cycle* (the outer loop in the flow chart of Fig. 1.4), each module is optimized in isolation (the inner loop). That is, steps 4, 5, and 6 are performed for just one module at a time. This is permissible because a module is sufficiently large (at least one complete cargo hold) that the limit states (failure modes) do not involve structures longer than one module. Even for the largest and most serious failure mode—hull girder collapse—the failed structure occurs within one cargo hold. The optimization of each module in isolation means that, within each design cycle, the optimization of one module cannot influence or be influenced by other modules. However, in the next design cycle, the finite element analysis of the overall model (step 3) will reflect all of the changes that were made in the previous cycle. For this reason, it is advisable to optimize only a few modules in each run and to choose those modules that are considered to be critical, either because they are most heavily loaded (amidships for bending moment, quarter-length locations for vertical shear force) or have large openings. After these modules have been optimized, then a new run is made in which the optimized modules are "frozen," and a few other modules, in between the frozen ones, are optimized. Thus, the optimization is not an overall, automated, and instantaneous process, and it does not produce a unique "overall optimum" design. Rather, it is a gradual process requiring many runs and the careful involvement of the designer. This is actually an advantage because sometimes the earlier runs may give results or reveal features (influences, sensitivities, tradeoffs, etc.) that were not anticipated and may require new constraints or that give the designer a better understanding of the structure and perhaps some new ideas.

For best results, optimization should be performed using a full ship length model. If it is not the full length, then a module adjacent to a cut end should not be optimized, because at the cut end there

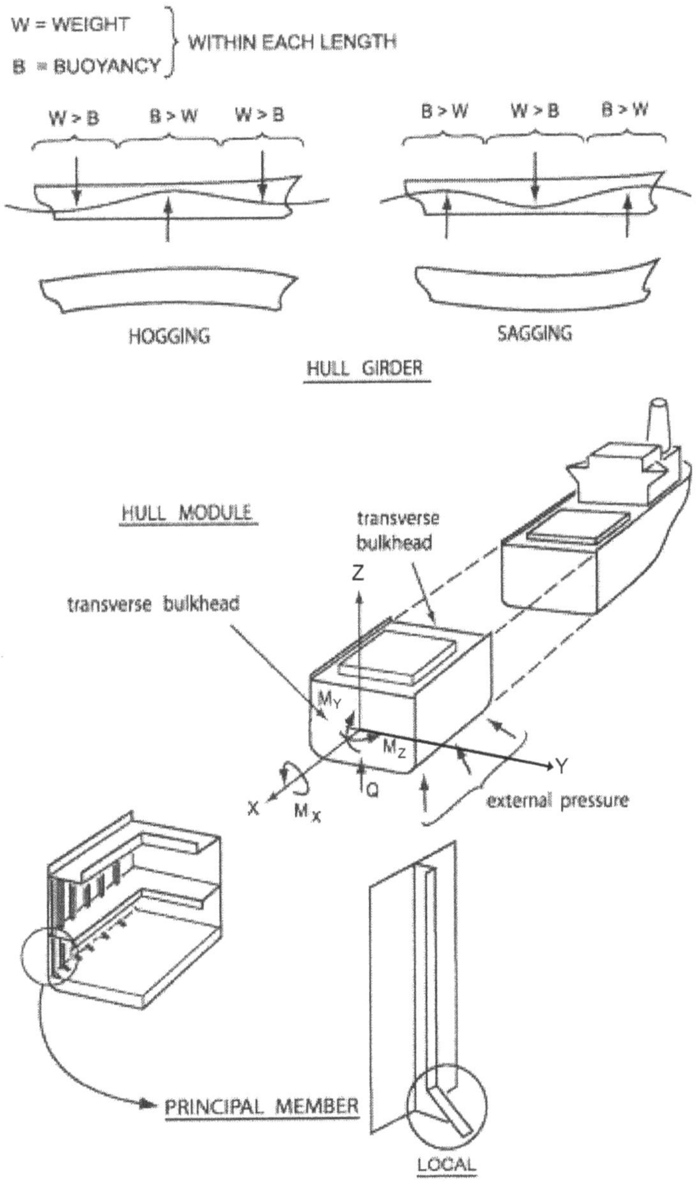

Figure 1.7 Levels of structural analysis.

are hull girder loads and physical restraints, and both of these can cause local distortion and overstressing. A partial length model should never be used for a containership, because its torsional response depends on the longitudinal distribution of both the torsional loading and the hull torsional stiffness over the whole length of the ship.

1.2.3.2 Principal Members

As shown in Fig. 1.7, the next level of structure is that of "principal member." The most common of these is a stiffened panel, which is the basic unit for all decks, sides, double bottoms, and bulkheads of the module. But the panels must be held in place, and this is the purpose of the framing system of a hull, made up of individual beams (transverse frames) as shown in Fig. 1.7. These beams provide bending rigidity in the ship's transverse plane. In this role of supporting the stiffened panels, the plating to which they are welded constitutes one of the two "flanges" of the beam. A transverse bulkhead is likewise made up of stiffened panels, and it too is supported by a framing system. If it carries a large pressure load as in a tanker, this framing system will consist of deep beams, running both vertically and horizontally and forming a "grillage." If a smooth surface is needed as in a dry bulk carrier, corrugated

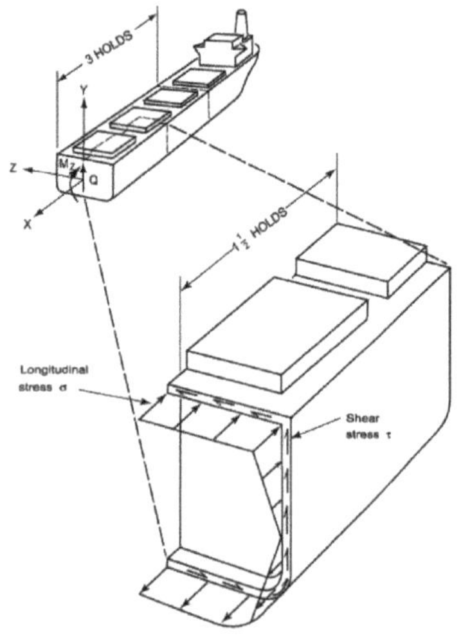

Figure 1.8 Application of hull girder load effects.

plating is used. Pillars are another type of principal member. They are used extensively in ferries and other ships having wide internal spaces because they reduce the span of the beams. They are also used extensively in naval ships because in reducing the span, they provide weight savings.

1.2.3.3 Local Structure

Finally, there is the local structure: brackets, connections, reinforcements, foundations, fittings, and so on. Basically, a structural element can be classified as local if it does not have any appreciable effect on the load distribution within the hull module; in other words, it is local if it does not affect the magnitude and overall distribution of internal forces in the principal members, but has only a local effect on its immediate surroundings.

Because of the irregular geometry of a local structure, its analysis may represent a significant computational task. Analysis and design of a local structure can only be done after the structural dimensions of the principal members have been determined. As illustrated in Fig. 1.4, the design of the principal members is an iterative process, and it would be inefficient to include the design of local structure as part of the preliminary design process. Rather, the design of a local structure—detail design—is a separate step coming after the preliminary design shown in Fig. 1.4. Lamb (2003) contains a great deal of information on detail design, as do the rules and other publications of the classification societies. Moreover, most of the items that come under the heading of local structure are not unique to ships, and there are many design manuals and handbooks for land-based steel structures that contain useful information.

1.2.4 Limit State Analysis

A *limit state* is any condition in which a structure or a structural member has become unfit for one of its intended roles because of one or more loads and/or load effects.* There are two broad categories of limit states: the *ultimate* or *collapse* limit states, in which the structure or member has failed in its primary, load-carrying role; and the *serviceability* limit states, which involve the deterioration or loss of other, mostly less vital functions. The *limit values* are the values of the loads or load effects which produce or correspond to a limit state. A limit value is denoted by the symbol Q_L. The symbol Q_L represents the values of a group of loads and/or load effects which produce a limit state. A limit state analysis consists of the calculation of the limit values, perhaps in various combinations and sequences, which correspond to a specified limit state, either in a member or in the overall structure. An ultimate limit state is often referred to as the *ultimate strength* of the structure or member, and the two terms are used interchangeably throughout this text.

Serviceability limit states arise from the fact that some members are designed on the basis of a form of failure other than structural failure. For example, as shown in Chapter 9, laterally-loaded plating is usually designed on the basis of a maximum allowable "permanent set" (plastic dishing of the plating). The limit value is the load which causes this limit state, whereas the ultimate load is that value beyond which the plate can no longer support the load.

There are three basic types of structural failure: plastic deformation, instability, and fracture. Within these there are several different modes of failure, some of which are more serious than others; these are explained in Section 2.4. Moreover, these various failure modes can combine and can interact, depending on member properties, function, and loading. There are generally several different loading arrangements and load combinations that must be considered (hogging and sagging, deep draft and light draft, various distributions of cargo, etc.). Hence, for each structural member there are usually several limit states, not all of which have the same degree of seriousness. In general, rationally-based design requires that each and every relevant limit

*For the overall structure, it is loads, whereas for a member it is usually load effects. For simplicity, we often use the term "load" even when the term "load effect" might be more accurate. The symbol Q denotes whatever agent is causing a limit state; hence Q can represent either a load or a load effect. Where a distinction is important, the symbol F will be used for a load and the symbol Q for a load effect.

state be examined, and those that interact should be examined together. There should be no *a priori* assumptions as to whether some limit state will or will not govern the design.

Thus, to take a simple example, the laterally-loaded plate referred to above should be examined for both types of limit state. The limit load for ultimate failure is much larger than the limit load for allowable permanent set. But, because of the greater degree of seriousness, there must be a greater margin between the ultimate load and the expected service load, and so either requirement may govern the required thickness of plating. To determine which is the governing requirement, it is necessary to perform both limit state analyses.

The level of seriousness of a limit state usually corresponds approximately to the level or extent of structure which has failed: overall, hull module, principal member, or local. The first two overlap because a hull module is always a complete segment of the hull girder, and so failure of a hull module is failure of the overall ship. Hence, in this text "overall failure" refers to failure at the hull module level, unless noted otherwise. Failure of local structure is not sufficiently serious to be included with the other levels. As noted earlier, this level of structure is usually dealt with in detail design rather than preliminary design. If needed, the local structure can be locally strengthened, usually without effect on other structural components.

Thus, there are two levels of structure that can reach a limit state: the structure as a whole (the hull module) and the principal members. Ideally, the limit analysis of the overall structure should include the limit analysis of the individual principal members. However, the limit analysis of a hull module is an extensive computational task. If necessary, the total amount of computation can be reduced by performing the two separately: a hull module limit analysis using a simplified structural model of the hull module and a separate limit analysis of each different principal member for each different load combination which that member faces. The member limit analyses provide the values of a member's ultimate strength which are used in the hull module limit analysis. It specifies the load combinations which are to be used in each member limit analysis. The determination of these load combinations is crucial for rationally-based ship structural design. At the member level, it is often not possible to adequately account for the interaction between members. Hence, it is not possible to know the true loads that are acting on each member as the structure approaches collapse. Moreover, most large structures have a high degree of static indeterminacy and, therefore, alternative paths through which loads can be transmitted once one member fails. It is unusual—and undesirable—for large structures to have a member that is so vital that collapse of the member would result in collapse of the structure. In most cases, overall collapse requires a large number of individual failures in various members. Some of those failures occur within the same members and cause them to collapse; others are more widely distributed among different members and, therefore, do not cause member collapse. It is even possible for a structure to collapse by a mechanism involving several members, none of which has undergone complete collapse. Hence, it is absolutely necessary to examine the strength of the structure as a whole to identify any and all mechanisms which may cause collapse.

An example of member collapse is the collapse of a stiffened panel in the deck of a ship. In this case, the load is the hull girder bending stress, σ_x. The collapse could be caused by any of the three basic types of failure (for simplicity, we here ignore combinations and interactions). Hence, there are (at least) three separate limit values of σ_x, and the panel collapses when σ_x reaches the lowest of these three values. The magnitude of each of these limit values is determined by the design of the panel: its geometry, scantlings, material, and so on. Expressing this in more general terms, we say that each limit value Q_L is a function of the design variables $\mathbf{X}$ and, when we wish to indicate this dependency, we shall write $Q_L(\mathbf{X})$. Thus, the limit values are under the control of the designer, and the safety of the structure is achieved mainly by choosing $\mathbf{X}$ such that each of the limit values $Q_L(\mathbf{X})$ exceeds the corresponding load Q by a satisfactory margin.

1.2.5 Safety, Uncertainty, and Structural Constraints

1.2.5.1 Strength Constraints

Almost every design involves constraints, that is, conditions or requirements which must be satisfied. In structural design the most important constraints are the strength constraints—those aimed at providing adequate safety and serviceability. Structural safety is inherently probabilistic; it is the probability that a structure will not fail. The risk of failure arises from the various uncertainties which are involved: uncertainties in loads, load effects, and limit values of load effects, which are results of variations in material thickness and quality, workmanship, fabrication, and so on. There are two broad types of uncertainty: statistical and nonstatistical. Statistical uncertainty arises from genuine statistical randomness. Nonstatistical uncertainty arises from subjective elements and from events which are not truly random but are difficult to predict. Wave loads and material properties are examples of statistical uncertainties; these can be dealt with adequately by statistical methods. Examples of nonstatistical uncertainty are those arising from the operation of the ship, such as operating errors (improper loading, mishandling, etc.) or a fundamental change of service.

A rationally-based design procedure must be able to deal with both types of uncertainty in such a way that the required degree of safety (which is ultimately decided by society as a whole through the medium of regulatory authorities and insurance rates) is achieved in a clear and explicit manner. The first type—statistical uncertainty—is dealt with by statistical methods. For all loads, load effects, and limit values which are probabilistic, statistical theory is used to estimate suitable values to be used for design. If the quantity involves a large number of peak values, such as wave-induced bending moment, then the calculation is based on *extreme values* of that quantity, and the particular extreme value which is selected for design is referred to as the *characteristic value*. In this text, an extreme value is denoted by the symbol ^ placed above the quantity (e.g., $\hat{M}_w$ for wave-induced bending moment) and the characteristic value is denoted by a subscript c (e.g., $\hat{M}_{w,c}$).

Besides dealing with statistical uncertainty, a rationally-based design procedure must also provide some means whereby the designer (or the regulatory authority) can explicitly allow for the other uncertainties. This is done by specifying a minimum value of the margin between Q and Q_L. In practice, this margin is usually specified in terms of a *safety factor*, γ_0, which is the minimum factor by which Q_L must exceed $\hat{Q}_c$. In terms of γ_0, the constraint is of the form

$$\gamma_0 \hat{Q}_c(\mathbf{X}) \leq Q_L(\mathbf{X}) \qquad (1.2.4)$$

In addition to accounting for uncertainty, it is also necessary to utilize some further safety factors to allow for the degree of seriousness of each type of failure, both in regard to safety (loss of life) and serviceability (loss of revenue or reduced mission capability). Likewise, it is also necessary to apply some factors to account for particular circumstances, such as the type of ship (passenger, cargo, naval, carrying hazardous cargo, etc.), its costs, and the operational importance of the ship. These various factors are known as *partial safety factors*. The required degree of safety is provided by the total factor of safety, which is the product of the partial safety factors. Thus, in (1.2.4), γ_0 denotes the total factor of safety.

Strength constraints are often nonlinear for a variety of reasons. First, two of the three basic types of failure are generally nonlinear: instability in typical ship structural members is usually followed by inelastic response or collapse, and plastic deformation is inherently nonlinear. Therefore, most of the limit value expressions $Q_L(\mathbf{X})$ are nonlinear and, hence, most of the structural constraints, even those involving only one load, are nonlinear. Modes of failure that involve more than one load and/or more than one structural member are even more nonlinear. Also, in a statically indeterminate structure the load effect in a member, $Q(\mathbf{X})$, can be a nonlinear function of the design variables $\mathbf{X}$.

Failure Involving Multiple Loads. In our discussion thus far, we have mostly considered limit states which involve only one load. For a limit state which involves two or more loads, one of the loads is selected as the principal independent variable, and an expression for its limit value is obtained as a function not only of the design variables, but also of the other loads. For example, in the collapse (ultimate failure) of deck plating resulting from plate buckling, the primary load is the longitudinal compressive stress, σ_x. The limit or ultimate value is the value of σ_x that causes collapse; this is referred to as the "ultimate" stress, $(\sigma_x)_{ult}$. If there is also a transverse stress σ_y acting on the plate, this constitutes a second load which influences the value of $(\sigma_x)_{ult}$. From plate ultimate strength theory (Chapter 13), one can obtain an expression for this influence; in general form it is

$$(\sigma_x)_{ult} = f(\mathbf{X}, \sigma_y)$$

Hence, the constraint equation takes the form

$$\gamma_0(\sigma_x)_{ult} \leq f(\mathbf{X}, \sigma_y)$$

In the design of the plating, the design variables must be such as to satisfy this inequality.

1.2.5.2 Other (Non-Strength-Related) Constraints

As shown in Fig. 1.4, there are other constraints on the structural design besides the strength constraints arising from 1) operational requirements (e.g., minimum size of hatches, limitations on distortion and on vibration, etc.) and 2) fabrication considerations (e.g., maximum plate thickness for cold rolling, minimum spacing between stiffeners for welding, etc.).

These other constraints are relatively straightforward and usually can be expressed directly in terms of the structural design variables. For instance, minimum or maximum values of design variables or ratios of design variables can be specified. An example is the design constraint that, in light of fabrication, the height h_s of a stiffener which passes through a transverse frame of height h_f must not be so large that the cutout interferes with the flange of the frame. Thus, for example, if it were desired to restrict the stiffener height to no more than 80% of the frame height, then the constraint would be $h_s \leq 0.8\, h_f$. Constraints of this type are important, but there is no need to give them further treatment in this book, because they are straightforward and are already contained in structural design manuals and structural codes, such as the rules of the classification societies. Also, since they have a simple mathematical form (linear inequality), it is a simple matter to incorporate them in any mathematical algorithm or computer program for rationally-based structural design.

1.2.6 Definition of the Objective in Structural Optimization

Rationally-based design, of its very nature, must have a goal or objective, and there must be some measure for assessing the merit of a design *vis-à-vis* that objective. Hence, in a rationally-based design process, the designer must be able to define and quantify the objective of the design. The design process must then be capable of actively and automatically achieving this objective to the fullest extent possible, subject to the constraints. This in turn means that the design process must include an optimization method which is capable of solving an optimization problem involving a large number of constraints of various types (linear and nonlinear, equality and inequality) and in which the measure of merit is totally flexible. That is, the optimization method should not restrict the measure of merit to linear expressions or to special cases (such as "least weight") since these may not suit the designer's needs.

Mathematical optimization of any kind requires that the measure of merit be defined as a mathematical quantity which is to be maximized (or minimized) and which is expressed as a function of the design variables. The measure of merit is also referred to as the "objective function." In the overall design of a ship, the structural design interacts with the other aspects of design, such as operational aspects, and something which is beneficial from a structural point of view may be detrimental in some other regard. Therefore, the structural design objective should reflect the overall goal, and the objective function should account for the results of interactions between the structural design and the other aspects of ship design. The goal depends, first of all, on the basic purpose of the ship, and in this regard the two principal categories are commercial vessels and naval vessels. In this section, we consider the measure of merit for a ship structure, first for commercial vessels (while keeping in mind that many of the factors relating to this are also relevant to naval vessels) and then for naval vessels.

1.2.6.1 Commercial Vessels

For commercial vessels, the objective is profitability, either of the ship itself or of some larger system. The principal factors which determine a ship's profitability are shown (in greatly simplified form) in Fig. 1.9, taken from Evans (1975). The quantities that are strongly influenced by the structural design are outlined, and it is clear that the structural design can affect profitability at various levels and in various categories: payload, initial cost, operating cost, and so on.

The choice of the objective function also depends on which person or agency has the authority to decide; that is, it depends on whose behalf the designer is acting. In most cases, it is the ship owner, but it may be the ship operator, the shipyard, or the controller of some larger system in which the ship is to operate. For example, a shipyard which is responsible for the design as well as the construction would probably give greater importance to initial cost than would a ship owner, whereas the latter would have a greater interest in operational aspects and life cycle economics.

Also, the factors and influences shown in Fig. 1.9 have different degrees of importance, and not all of them need to be included in the objective function. In many cases, the only strong influence which the scantlings have on profitability is their effect on initial cost, and in such cases "least initial cost" is a sufficiently accurate objective.

Alternatively, with small weight-critical vessels, such as hydrofoils and surface-effect ships, profitability or performance is determined almost entirely by hull weight because decreased structural weight allows a direct and corresponding increase in payload. In this case, the weight of the structure is, therefore, included in the objective function as a different type of cost. Moreover, even large vessels can

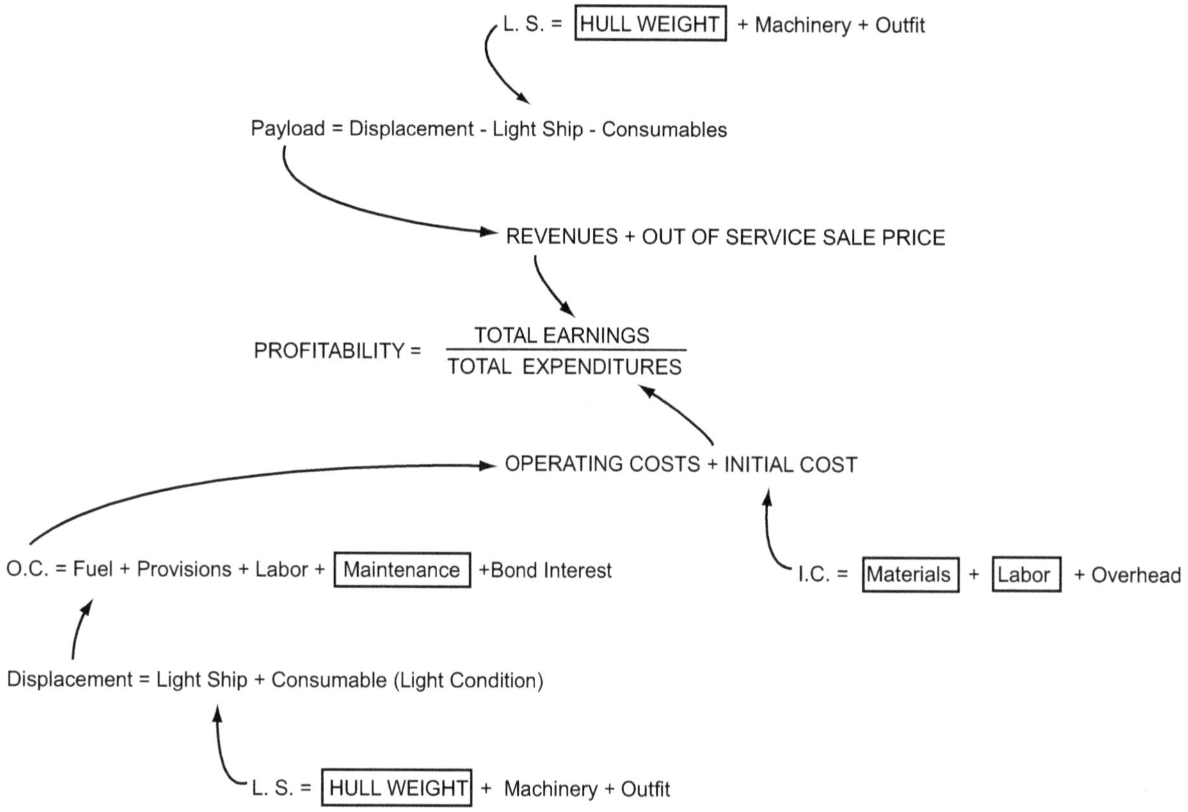

Figure 1.9 Principal factors in ship profitability.

be weight sensitive, such that a saving in hull weight gives an increase in payload as well as reductions in cost. The ways in which this occurs are indicated in Fig. 1.9.

The question then arises as to what is the proper combination of the two goals of weight reduction and initial cost reduction. This question can only be answered by a careful study of the economic life cycle of each ship, to determine the tradeoff between initial cost and increased revenue from weight savings. As an approximate means of allowing for this combination, Caldwell (1971) proposed a useful nondimensional objective function which combines weight and cost in the form

$$U = v\frac{W}{W_0} + (1-v)\frac{C}{C_0} \qquad (1.2.5)$$

where W_0 and C_0 are, respectively, the weight and initial cost of some basis or standard design, W and C are, respectively, the weight and initial cost of a proposed design, and v is a number which varies between zero (where least initial cost is the objective) and unity (where weight-saving is the objective). It can be shown that if the weight saved in structure can be taken up by cargo, then the best value of the weighting factor v in (1.2.5) is

$$v = \frac{1}{1 + R_c/R_w} \qquad (1.2.6)$$

in which

$$R_w = \frac{\text{weight of structure}}{\text{weight of cargo}}$$

in the basis design and

$$R_c = \frac{\text{annual costs arising from initial cost}}{\text{total annual cost}}$$

in the basis design.

1.2.6.2 Naval Vessels

For naval ships, the objective is to obtain the maximum possible mission capability over the life of the ship, subject to budget limitations. Cost is here a constraint instead of an objective. The structural designer's greatest influence on mission capability is the weight of the structure. Weight saving permits either a higher speed, more mission-related equipment (weapons, sensors, etc.), or increased range and endurance, or some combination of these.

The mission capability is also strongly linked to the ship's vertical center of gravity (VCG). A low VCG of the hull structure is of great benefit since most of the important weapons and sensor systems involve large topside weight. In fact, the provision of adequate stability is often the limiting factor on the number or size of such systems, particularly as a vessel gets older and it becomes necessary to fit more modern systems. However, VCG is determined primarily by the basic layout of the ship (e.g., the number of decks) and secondarily by the choice of material (e.g., aluminum versus steel). Both of these are decisions that are made prior to the structural design process of Fig. 1.4. If they are changed, then the structural design must be redone. The structural design variables (scantlings of principal members, denoted by **X** in Fig. 1.4) have only a slight influence on VCG, and that influence could be either an increase or a decrease, depending on the member's vertical location in the ship.

Hence, for a naval vessel, the optimization "objective" in step 6 of Fig. 1.4 is usually "least weight" and does not include VCG. This objective tends to produce a structure that is more intricate and involves less material. Hence, for naval vessels the structural cost (i.e., the cost that is attributable to structure and is a function of the structural design variables) is mainly fabrication cost; the material cost is smaller and, for a given material, it has little influence in determining the final optimum design.* Thus, naval design involves a tradeoff between weight and fabrication cost. The designer seeks to determine the number, arrangement, and size of structural members which will give the lowest possible weight, subject to cost limitations and to a variety of other constraints requiring satisfactory strength, reliability, endurance, and functioning of the vessel. The constraint on cost is somewhat different from the other constraints. Rather than being an absolute limit, it is a somewhat elastic barrier in which the rigidity of the resistance to further increase in cost depends on the cost:benefit ratio, that is, how much benefit the increase in cost will yield. Nevertheless, besides the cost:benefit type of constraint, there can also be an explicit upper limit on total cost.

*The benefits of using a different material (e.g., an aluminum or composite superstructure) would be investigated by making a separate optimum design using that material and then judging whether the weight/VCG savings is worth the extra cost.

1.2.7 Overall Procedure for Ship Structural Optimization

The final basic aspect of the rationally-based design procedure shown in Fig. 1.4 is the structural optimization process. This aspect consists of a mathematical method which utilizes the information provided by the other aspects and generates the design (i.e., the set of scantlings for all principal members) which satisfies all constraints and maximizes the objective. The aim of the present section is twofold: to give an overview of structural optimization and also, now that we have discussed all aspects of the overall design procedure of Fig. 1.4, to explain briefly how that procedure works. For this, it is not necessary to have a detailed knowledge of mathematical optimization theory. It is sufficient to know the basic features. Since the primary aim of this text is to present and explain the method—both theory and practice—of rationally-based structural design, no attempt is made to cover mathematical optimization theory or to give a complete coverage of structural optimization. Only those aspects will be treated here which are needed by a designer. The coverage is in three parts:

1. In this section, a brief summary of the basic features of structural optimization, presented as part of a simple example of the overall design process of Fig. 1.4.
2. In the next section, a brief summary of the broad classes of optimization methods, some comments on these in relation to the requirements of preliminary ship structural design, and references where more detailed information on these methods may be found.
3. In Section 2.7, a summary of a "dual level" optimization strategy, which permits the efficient optimization of large structures in which some constraints apply to the structure as a whole.

1.2.7.1 Sample Application of the Procedure for Rationally-Based Preliminary Structural Design

We now present a simple example of the rationally-based design procedure of Fig. 1.4. The example should illustrate the various steps of the procedure and show how the structural optimization step brings together and utilizes the results of the other steps. The procedure is intended for the structural design of an entire ship. To have a simple example, we will here apply it to just one small part of the structure—a stiffened panel in the strength deck of a vessel, as shown in Fig. 1.10. We will also make simplifying

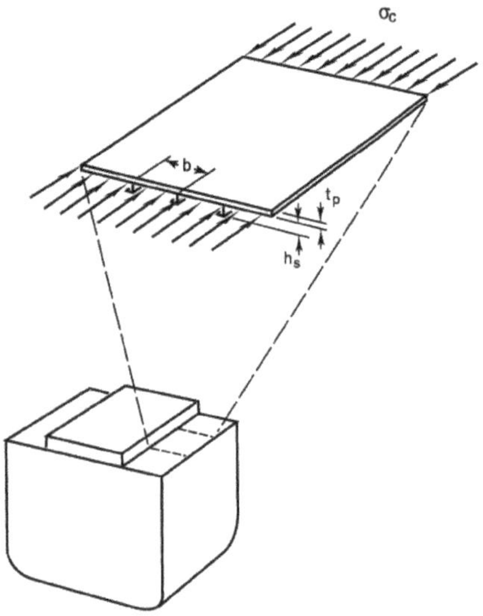

Figure 1.10 Example of a stiffened panel.

assumptions which would not be made in the actual procedure. For example, let us say that the panel has only two design variables: the plate thickness, t_p, and the stiffener height, h_s. In reality, the panel would have several more design variables: the number of stiffeners (or their spacing) and the web thickness and flange area of the stiffeners. But for this example, let us say that the number and type of stiffeners have already been selected, perhaps because of a need to match some existing structure.

1.2.7.2 Specification of Loads, Dominant Load Parameters, and Design Waves

The first step in structural design is to investigate the maximum or extreme value of each load so as to determine a suitable extreme value to be used for design, that is, the *characteristic load* Q_c. For simplicity, let us say that the only significant load is hull girder bending moment, M, and that the only significant load at the panel level is the hull girder bending stress, σ. In reality, there would also be other loads acting on the panel (lateral pressure, shear stress, etc.), and the magnitude of these loads would be obtained by the finite element analysis.

As will be discussed in Chapter 4, the wave-induced portion of the hull girder bending moment, M_w, is probabilistic and, therefore, statistical methods must be used to establish a *characteristic extreme value* for it, denoted as $\hat{M}_{w,c}$. For standard types of ships, a value for $\hat{M}_{w,c}$ is available from classification societies, having been determined by research and by at-sea measurements for that type of vessel (see Section 3.5.1).

1.2.7.3 Finite Element Analysis

The next step is the finite element analysis. Since there are two values of maximum wave bending moment (hogging and sagging) and several values of stillwater bending moment (corresponding to different cargo and/or ballast configurations), it is necessary to perform the hull girder analysis for several load combinations.

We note that, to perform the finite element analysis, it is necessary to have some initial or starting value of the ship's scantlings. For the design variables of the panel, let us denote the initial values as t_{p1} and h_{s1} or, in vector form, as $\mathbf{x}_1$. These initial values are arbitrary; they are required simply because any computer calculation or analysis (such as steps 3, 4, and 6 of Fig. 1.4) requires specific numerical values for all quantities. These values do not require calculation by the designer: he/she can either select standard values, arbitrary values, or values from some other design. This will be discussed further when considering the structural optimization step.

The finite element analysis provides values of individual load effects in each of the principal members, for each load case. We are examining only one principal member in this example—a deck panel—and we are saying, purely for simplicity, that the only load effect is the hull girder stress. Thus we are, in effect, skipping over the finite element analysis.

1.2.7.4 Hull Module Limit State Analysis

We now begin the inner loop to perform the hull module design, starting with the limit analysis, for the initial scantlings, $\mathbf{x}_1$. As shown in Fig. 1.4, the hull module design cycle must be performed repeatedly because at the end of each cycle the values of the design variables (which comprise all of the scantlings of the hull module) are altered by the optimization process. Once the modules have been designed, we return to the outer loop and repeat the finite element analysis, using the new values of $\mathbf{x}$. For reference, we will denote the current values of $\mathbf{x}$ as $\mathbf{x}_i$ (or t_{pi} and h_{si}); that is, $\mathbf{x}_i$ represents the scantlings which are used during the ith design cycle.

Let us assume that there are five limit states for the panel: three types of compressive collapse, with collapse being initiated by 1) plate buckling, 2) torsional buckling of the stiffeners, or 3) flexural buckling of plating and stiffeners acting together, and also 4) large plastic deformation under tensile load, and 5) fracture because of fatigue. Hence, the five limit values are the three buckling stresses denoted by $\sigma_{bj}(\mathbf{x})$ with $j = 1, 2$, and 3, the yield stress σ_Y, and the fatigue-derived limit on wave-induced stress, $(\hat{\sigma}_{w,c})_L$, corresponding to the expected number of hogging and sagging cycles in the life of the ship (which is estimated at about 10^8 for a 25-year life).

The purpose of the limit state analysis is to calculate the limit values for the current values of $\mathbf{x}$. In this example, the fourth and fifth limit values are more or less material properties and do not depend on $\mathbf{x}_i$, so the limit analysis here consists of the calculation of the three buckling stresses, $\sigma_{bj}(\mathbf{x}_i)$. For simplicity, we are assuming that the buckling is elastic. The theory for this is presented in Chapters 12 and 14. In the notation, we are using the following equations.

Local plate buckling $\sigma_{b1} = K_1 E \left(\dfrac{t_p}{b}\right)^2$

Torsional buckling of stiffeners $\sigma_{b2} = K_2 E \left(\dfrac{t_p}{h_s}\right)^2$

Overall panel buckling $\sigma_{b3} = K_3 E \left(\dfrac{t_p}{b}\right)\left(\dfrac{h_s}{b}\right)$

1.2.7.5 Formulation of Constraints

The next step is to formulate the constraints for the optimization problem. The general form of a constraint was given in (1.2.4). In our example, this leads to the following equations for the three buckling constraints.

$$\begin{aligned} \gamma_0 \hat{\sigma}_c &\le K_1 E \left(\frac{t_p}{b}\right)^2 \\ \gamma_0 \hat{\sigma}_c &\le K_2 E \left(\frac{t_p}{h_s}\right)^2 \\ \gamma_0 \hat{\sigma}_c &\le K_3 E \left(\frac{t_p}{b}\right)\left(\frac{h_s}{b}\right) \end{aligned} \qquad (1.2.7)$$

To plot these constraints in Fig. 1.11, we assume the following values: $E = 200{,}000$ MPa, $b = 500$ mm, $\hat{\sigma}_c = 160$ MPa, $\gamma_0 = 1.25$, $K_1 = 4$, $K_2 = 0.1$, and $K_3 = 0.5$.

As mentioned above, an actual design involves several load cases, which here would mean several different values of $\hat{\sigma}_c$. Hence, for each constraint it is necessary to use whichever value of $\hat{\sigma}_c$ is critical for that type of limit state. Moreover, in reality each load case usually involves several loads acting on the member, in various different combinations, and so the search for the decisive combination must be systematic and thorough.

As explained in Section 1.5, the total factor of safety, γ_0 , is made up of several partial safety factors which are chosen in accordance with 1) the degree of (nonstatistical) uncertainty which exists in regard to both the load and the limit value, and 2) the degree of seriousness of the limit state. For this example, the degree of seriousness is about the same for all constraints since they all refer to collapse rather than unserviceability. Although there would be differing degrees of uncertainty in σ_{bj}, σ_Y, and $(\hat{\sigma}_{w,c})_L$ let us say, again purely for simplicity, that γ_0 is the same for all five constraints.

In Fig. 1.11, the axes are the design variables t_p and h_s. The three buckling constraints of (1.2.7) are plotted as curves of t_p versus h_s. In this type of diagram, any specific combination of t_p and h_s—that is, any specific panel design—is a particular point on the diagram. For example, point A represents the initial or starting design, corresponding to t_{p1} and h_{s1}. The plane of the diagram represents all possible designs and is referred to as the *design space* (or hyperspace; the concept may be extended to any number of design variables). The constraint equations are inequalities and, therefore, each curve is the boundary between all designs that satisfy that constraint and all that do not. In Fig. 1.11, the impermissible side of each constraint is indicated by shading the impermissible side.

The two constraints corresponding to tensile yield of the panel and fatigue fracture cannot be drawn in

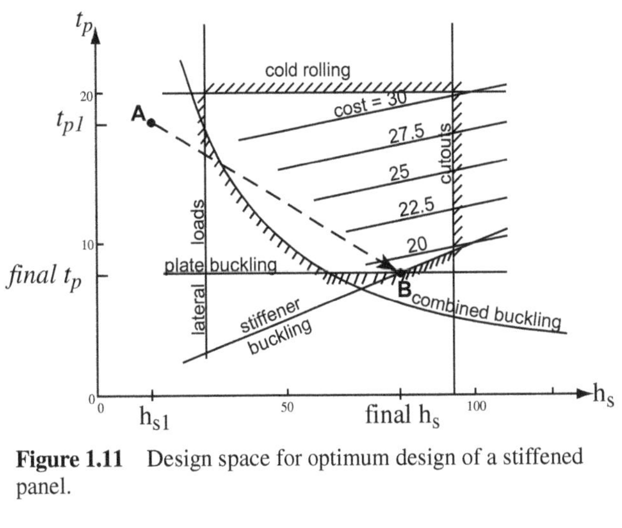

Figure 1.11 Design space for optimum design of a stiffened panel.

the figure because the limit values are material properties and the limit states are, therefore, essentially independent of the design variables. The only way in which the design variables **x** are involved is through σ_c because σ_c is inversely proportional to the hull girder section modulus, Z. But the plate thickness of a single panel has only a small influence on Z, and h_s has even less influence. That is, even a large change in t_p would cause only a small increase in section modulus and, hence, only a small decrease in σ_c. If either of these constraints were not satisfied, such that it would be necessary to reduce σ_c, this would require increasing the plate thickness of *all* deck panels and possibly also of all bottom panels. In other words, these two constraints relate to the overall structure, via section modulus, and not just to this panel. Constraints of this type are discussed in Section 2.5. Also, the determination of the optimum combination of thickness changes for all panels requires that all of the panels be redesigned together in a coordinated manner.

We have already seen that a principal member cannot be analyzed in isolation, but only in conjunction with the other principal members through the medium of the finite element analysis. We now see that in optimization, the situation is similar—a principal member cannot be designed in isolation, but only in conjunction with the other principal members. For the present example, let us assume that the current values of deck and bottom thicknesses are already sufficient to give a reasonable value of σ_c, such that these two constraints are not violated. In that case, they will not become violated because of a reduction in thickness of just this one panel, and so we will not consider them further in this example.

Besides the strength constraints, there are other constraints on the structural design arising from operational requirements and fabrication considerations. For the deck panel, for example, there might be a maximum plate thickness for cold rolling or for weldability, a minimum stiffener height (for a given stiffener spacing b) to support lateral loads, and a maximum stiffener height (for a given height of the transverse deck beams, h_f) so that the cutout for the stiffener leaves sufficient web area in the beam.

We note that in a more realistic example, the stiffener spacing and the frame height would also be design variables and, hence, the latter two constraints would only restrict the ratios h_s/b and h_s/h_f, not the absolute value of h_s.

These additional constraints are also drawn in Fig. 1.11. Taken as a group, the constraints define a region of the design space in which none of the constraints are violated. This region is known as the *feasible* design space, and corresponds to the shaded outline in the figure.

1.2.7.6 Objective Function

Let us assume that in our example least initial cost has been chosen as the objective for this design. The cost of a panel would depend mainly on the amount of steel used and the cost of fabrication. In the present example, t_p and h_s are the only variables, and the amount of steel is linearly proportional to each of them, but in general the amount of steel is the product of two design variables; for example, in a stiffener web it is $h_s t_w$, where t_w is the web thickness. The fabrication cost is related to the design variables in a completely different way from the material cost and, therefore, the sum of the two costs will be even more complex. The cost function $C(\mathbf{X})$ is, like many of the constraints, a nonlinear function of the design variables. In Fig. 1.11, a typical cost function for the panel is indicated by means of contour lines of constant cost.

1.2.7.7 Structural Optimization

The figure also shows two particular combinations of t_p and h_s; that is, two specific panel designs. Design A represents the initial design, that is, the starting point for the optimization process. It may correspond to an actual panel in some existing ship, or it may be a purely arbitrary first assumption. For any good optimization process, the latter is sufficient. Thus, the starting design need not be feasible (and often will not be, even if it does correspond to an actual design, since the loads and/or limit values for the current design are different). Design A is not only infeasible (it violates the minimum h_s and the combined buckling constraint), but also expensive. Design B is the optimum design; it is the point within the feasible region which has the least cost. The task of the mathematical optimization process is to find this optimum point, from any starting point, and to do so with as little computation as possible.

1.2.7.8 Postoptimality Information

In Fig. 1.11, the optimum design is governed by two constraints: combined buckling of stiffener and plating and the minimum plate thickness to prevent plate buckling. Another feature of a good optimization method is that, if requested, it can inform the designer as to all the circumstances relating to the optimum design, such as what are the governing constraints. Other useful information that should be available includes how steep or flat the optimum is; that is, how much extra cost would be incurred in moving away from the optimum design point in various directions; what would be the effect on the cost of relaxing or tightening any of the active constraints; and, if relaxation were possible, which direction would give the greatest further decrease in cost.

1.2.8 Optimization Methods for Large Structures

Mathematical optimization is purely a mathematical procedure and therefore it can and should be a fully automated process. That is, in the overall design process of Fig. 1.4, the optimization step is simply a "black box" which performs a specific task: it accepts as input an objective function and a set of constraints, and it returns as output the specific optimal values of the design variables, that is, the values that maximize the objective while satisfying all of the constraints. Since there is only one optimum point in the design space, the optimization task is straightforward and unambiguous; there is no need or reason for any intervention by the designer. The only requirement is that the method must find the optimum rapidly and efficiently. The particular manner in which it does this is not important, and any optimization method that meets the requirement can be used in the overall design process of Fig. 1.4, without any need for the designer to have a detailed knowledge of it or of mathematical optimization theory. Therefore, in this section we merely describe the broad classes of optimization methods, identify those methods which have been proven to be capable of meeting the requirements of ship structural optimization, and provide references from which detailed information about the methods may be obtained.

1.2.8.1 Types of Nonlinear Optimization Methods

Many optimization methods that have nonlinear capability are available. The majority may be grouped into three categories:

1. Fully nonlinear methods, such as mathematical programming methods and various algorithms for unconstrained minimization
2. Special purpose methods, such as fully-stressed design, the optimality criteria methods, and geometric programming
3. Methods based on sequential application of linear programming

In all, there are many different optimization methods, and nearly all of them can be used for structural optimization in some application or other. However, most of them tend to be suitable only for a particular type of problem. Gallagher and Zienkiewicz (1973) give a basic explanation of all of the principal methods and demonstrate their applications.

In general, the majority of the methods are suitable either for small nonlinear problems (i.e., which have only a few design variables) or for slightly nonlinear large problems. As far as the authors are aware, only the third type of method—sequential linear programming—has been shown to be capable of solving the large nonlinear problem which is involved in ship structural optimization with sufficient speed, computational efficiency, and generality. The fully nonlinear methods perform satisfactorily for small problems (say five or six design variables), but the amount of computation increases sharply with problem size, and for a structure as large and complex as a hull module, these methods are not feasible. The special purpose methods are rapid and efficient, but they are too restricted; they cannot handle constraints that are highly nonlinear and/or involve many design variables, and the first two (fully-stressed design and optimality criteria) cannot handle an arbitrary (user-specified) nonlinear objective.

1.2.8.2 Methods Based on Sequential Linear Programming

In this type of method, all of the nonlinear functions [the objective function and the limit values $Q_L = f(\mathbf{x})$] are replaced by linear approximations, and the linearized problem is solved rapidly by the well-known method of linear programming, using the simplex algorithm. This process is repeated sequentially, with the linear approximations being recalculated at each new design point. The original version of sequential linear programming was developed by Kellog (1960) and Griffith and Stewart (1961). In this first version, the linearized form of each function f was simply the linear terms of its Taylor series expansion, which involves the various first derivatives of f with respect to each design variable: $\partial f/\partial x_i$. It was found that, unless all functions were only slightly nonlinear, the linearized problem was too different from the actual nonlinear problem, and the process would not converge. Hence, for many years, the method was limited to problems which were only moderately nonlinear. But it was subsequently shown that this limitation can be overcome by using some second derivative terms in formulating the linearizations. Various second-order methods have been developed. Hughes and Mistree (1976) presented the SLIP2 method, which uses the second derivatives of the nonlinear functions (but only terms of the form $\frac{\partial^2 f}{\partial x_i^2}$; mixed derivatives $\frac{\partial^2 f}{\partial x_i \partial x_j}$ are not required) to obtain a more accurate linearization of both the objective function and the constraints. This method was developed specifically to meet the requirements of rationally-based ship structural design, and it is this method that is used in the MAESTRO computer program. As shown in the examples and references given in Section 1.3, SLIP2 is able to solve problems involving a large number of constraints of various types (linear and nonlinear, equality and inequality) and in which the objective may be any user-specified nonlinear function of the design variables. Most importantly, the method is rapid and cost-effective. The SLIP2 method is not limited to structures; it is a general purpose method that can be used for commercial, industrial, or other optimization applications. The complete mathematical algorithm and logic structure for SLIP2 are given in Mistree, Hughes, and Phuoc (1981).

Another second-order version of sequential linear programming was developed by Murtagh and Saunders (1980). In Murtagh and Saunders (1983), this method is demonstrated for several large-scale optimization problems in commerce and industry. Although these examples do not include structural optimization, it is clear that the method is also suitable for this application.

Pedersen (1973) presented a systematic method for using "move limits" in sequential linear programming, which overcame most of the convergence problems referred to above.

1.2.9 Coverage and Plan of the Book

This new edition of the book is a major update of the original 1983 publication. The biggest change is to involve multiple authors and editors. It has taken many years and is still not quite finished. But rather than delay publication any further, all of the available new and revised chapters have been inserted in this edition. To show what the final work will include, this section gives a summary of all of the chapters and identifies the two chapters that have yet to be written.

The rest of Chapter 1 is devoted to some further basic aspects of structural safety, probabilistic design, and the use of partial safety factors. Specification of safety factors is primarily the responsibility of classification societies and, therefore, this text does not seek to determine or recommend any specific combination of safety factors or any specific values for them. Instead, these are described in general terms, and the combinations and values given are purely for illustration.

Chapter 2 summarizes the four major analysis tasks in rationally-based ship structural design: calculation of loads, the structure's response to the loads, the various limit values of each response, and optimization. Thus, Chapters 1 and 2 give the overall method of rationally-based structural design.

Chapter 3 presents the traditional hull girder analysis based on beam theory. This is the oldest and best established aspect of preliminary structural design. The chapter covers only those topics which continue to be relevant for rationally-based design. Chapter 4 summarizes the theory and techniques for obtaining a more precise estimate of wave loads on ships, when account is taken of the probabilistic, dynamic, and nonlinear aspects of these loads.

Chapter 5 presents the reliability-based approach to structural design, which is particularly appropriate for ships since their primary loading (forces resulting from waves and ship motions) are best obtained and presented in statistical terms.

Chapters 6 and 7 present the basic features of finite element analysis, starting with frame analysis and introducing some basic two-dimensional elements. Chapter 8 presents the basics of nonlinear finite element analysis.

Chapters 9 through 15 deal with the limit analysis of the principal members: columns, beam-columns, plates, and stiffened panels. In each case, the elastic aspects are covered first and then the inelastic. For computer-based analysis, it is necessary to have either explicit expressions or numerical procedures for calculating limit values, and only these types of methods are presented. Some methods are new, such as that for clamped beam-columns in Section 11.3 and the ultimate strength algorithms for plates and stiffened panels, presented in Chapters 13 and 15.

Chapter 16 deals with the limit state analysis of the hull module. Because of the structural complexity of a hull module and the complex interaction which often occurs between instability and plastic deformation, this analysis requires an *incremental load-deflection* approach, which traces the history of the collapse.

Chapter 17 deals with fatigue of structural details. Two further chapters are intended for future editions. Chapter 18 will deal with a relatively recent type of sandwich panel, consisting of steel faces and an elastomer core. These panels have extraordinary in-plane and bending strength, and they provide excellent protection against projectile impact and fire. They also provide good vibration and acoustic damping. Algorithms will be given for the ultimate strength of these panels under in-plane compression and lateral pressure. Finally, Chapter 19 will give a summary of the available computer programs for ship structural design.

1.3 PRACTICALITY AND PERFORMANCE OF THE METHOD

As explained above, the aims of this text are to present a method for rationally-based design and to explain the basic theory and analysis methods which are required for that method. But the first requirement of a design method is that it be practical. A method which lacks this characteristic is of no real value in practical design, no matter how "rational" it may be. Thus, when a method is proposed which has a much larger theoretical content and is entirely computer-based, the questions which immediately arise are: what stage has it reached regarding implementation, availability, and actual use; what are its benefits and its performance characteristics; what does it require; and in short, how practical is it? Therefore, before proceeding further with the theory and method, it is necessary to demonstrate that the method of rationally-based design presented herein is truly as powerful, practical, and beneficial as implied in the previous section and to provide factual information that will answer the questions just raised.

This can readily be done because the method was developed over a long period of time (since 1972), and throughout all of this period the practical aspects were given just as much attention as the theoretical aspects. As each portion of the method was developed, it was computer-implemented and tested before being accepted. Each portion received many further tests as other portions were added or modified. If at any stage, some portion was found to be impractical, it was promptly discarded, and work was begun on a replacement. In the first decade or so, there were many discards.

The first version of the method was completed in 1975, for which the computer program was called AUSTROSHIP. Under the sponsorship of the American Bureau of Shipping, a second version was completed in 1978, for which the program was called SHIPOPT. This was followed by a series of validation tests; the finite element portion was validated against DAISY, a large general purpose finite element program owned by the American Bureau of Shipping, and the structural optimization portion was validated by a series of formal, full-scale design studies involving four ship types: a 14,000 deadweight (dwt) general purpose cargo vessel (Hughes, Mistree, & Žanić, 1980), a 96,000 dwt segregated ballast tanker, a 140,000 dwt bulk carrier (Liu, Hughes, & Mahowald, 1981), and a destroyer (Hughes, Wood, & Janava, 1982). In 1983, the complete method was implemented in the MAESTRO computer program.

The practicality and performance of the method, and also of MAESTRO, may be judged from the results of the validation tests and design studies. All of them showed a similar performance, and since Liu, Hughes, and Mahowald (1981) is the most comprehensive study, most of the results quoted here are from that reference.

Since this section deals mainly with the features and performance of a particular computer program, it should be noted that the subject matter of the book is not a computer program, but rather the underlying theory of and a general method for rationally-based structural design, which is necessarily computer-based. The theory and method presented in this book can serve as the foundation for various computer programs, some more general and some more specific. It is not limited to one particular program.

1.3.1 Use of MAESTRO for Structural Evaluation

Being a program for rationally-based design, MAESTRO is organized along the lines of the design process of Fig. 1.4. Thus, corresponding to steps 3, 4, and 5, it contains:

1. A special high-speed design-oriented finite element method which calculates the load effects **Q** (deflections and stresses) in all of the principal members for all load cases.
2. A set of coordinated subroutines which perform limit state analysis, examining all relevant types of failure and calculating $\mathbf{Q}_L$, the limit values of the load effects, for each different principal member and for all load cases.
3. Other subroutines which formulate the constraints against each type of limit state for each different principal member. This involves searching the values of **Q** and $\mathbf{Q}_L$ to find and use the currently worst combinations of these two quantities. The pro-

gram then makes a note of the corresponding lowest value of the margin of safety for each limit state and the location and load case where each lowest value occurs.

These three features make the program a powerful tool for a variety of structural analysis and evaluation purposes in addition to optimum design. When used in this mode, the program executes only one cycle and stops just short of the optimization step. Because the finite element portion is extremely rapid and because a given set of input data is easily modified, the designer can quickly determine the effect of a proposed design change. Moreover, the structural evaluation portions of the program have many valuable applications outside of, or immediately following, the design of a ship. Some examples are:

1. To check the structural adequacy of a proposed design
2. To investigate proposed structural modifications
3. To assess the seriousness of structural damage and the degree of urgency of repair
4. To assess the structure after an actual or a projected corrosion wastage

A common example of the first application is the checking of a proposed design prior to or as part of the classification approval process. For large and/or nonstandard ships, classification societies usually require that a finite element analysis be made of the hull structure to check whether the general stress levels are satisfactory and whether there are any particular locations of overstressing. The analysis is usually performed in two stages: a three-dimensional "coarse mesh" analysis of the ship, followed by a series of separate, mostly two-dimensional "fine mesh" analyses of selected areas. MAESTRO allows both of these to be done within a single model. Any portions of the "coarse mesh" model can be converted to fine mesh. As shown in the bulk carrier example, the coarse mesh portion of MAESTRO is approximately 12 to 15 times faster than conventional finite element programs. Also, MAESTRO uses more sophisticated finite elements and, therefore, in spite of the coarse mesh, it yields all important stress values in all principal structural members of the ship. Moreover, besides calculating the stresses, MAESTRO also performs the complete limit state analysis just described and produces a color-coded graphical display of the vessel's structural adequacy for all of the principal members.

These features, together with a graphical user interface, make MAESTRO ideal for obtaining a rapid and yet thorough evaluation of the adequacy of a proposed design. This is useful for two purposes:

1. To assess a design in which there are some nonstandard aspects, but not sufficiently unusual to warrant a detailed (fine mesh) finite element analysis
2. For designs that are nonstandard, to determine whether there are regions which require a fine mesh analysis and, if so, to immediately perform these analyses within the same model (no need to construct separate, stand-alone models, with the associated problem of producing accurate boundary conditions)

1.3.2 Selection of Design Objective

MAESTRO leaves the designer free to specify how the optimization objective is to be measured. For commercial vessels, the usual objective is maximum profitability over the ship's lifetime. Factors which most affect profitability are initial cost and operating revenue. Initial cost is a combination of material cost and fabrication cost, and to a first approximation these two may be expressed in terms of the volume of material and the total length of welding (the combined lengths of all of the girders, frames, and stiffeners). These are the parameters that were used in Liu, Hughes, and Mahowald (1981). The cost algorithm contains four factors:*

1. Cost per unit volume for stiffened panels
2. Cost per unit length of stiffening for stiffened panels
3. Cost per unit volume for web frames and girders
4. Cost per unit length for web frames and girders

The other principal aspect of profitability—operating revenue—is determined mainly by cargo capacity. In bulk carriers, for instance, a saving in hull steel weight gives a corresponding increase in cargo deadweight and, hence, revenue. The additional revenue resulting from weight savings (or, on a cost basis, the extra cost [lost revenue] resulting from weight increase) can be allowed for by increasing the volumetric cost factors.

The four cost factors are part of the data input and, if desired, can be different for different regions of the ship. The lineal (cost per unit length) factors would reflect such items as welding costs and would influence the optimum number of stiffeners in each strake of plating. These factors might vary accord-

*As shown therein, the costs are not, and do not need to be, actual dollar costs, but rather cost indicators, or indices, which portray the correct relative proportions of the various costs.

ing to which shipyard is building the ship. For example, one yard might have better automatic welding machines, so that its cost per unit length of stiffener weld might be less than for another yard. In this case, the optimum design would probably have more stiffeners and less steel than at the yard with higher welding costs. The type of ship could also influence the cost function. In a double bottom bulk carrier, one would want to penalize increased double bottom height since this reduces volumetric cargo capacity. In this way, the program would automatically look first at other structural changes before increasing the double bottom height.

1.3.3 Example 1—96,000 DWT Oil Tanker

The first example from Liu, Hughes, and Mahowald (1981) is a single skin, medium-sized oil tanker. This was before the requirement that tankers must have a double hull. Since one of the main purposes of this study was to assess the economic benefits of rationally-based design, all of the design specifications (principal dimensions, geometry, loads, etc.) were the same as for an actual, rule-based, manually-produced design. As explained therein, steps were taken to avoid any bias in favor of MAESTRO and to remove the rather uncertain question of corrosion allowance from the comparison. In rationally-based design, there is a clear distinction between steel that is required for adequate strength and steel that is provided in order to allow for corrosion. MAESTRO provides only the former; the latter must be added on after the optimization.

The transverse section of the basis ship is shown in Fig. 1.12. For the three cargo tank lengths which comprised the MAESTRO structural model, the cost of the basis design was 9708 cost units, and the structural weight (which is automatically calculated by MAESTRO) was 8050 tons. As mentioned earlier, the initial scantlings for MAESTRO are completely arbitrary and so in this case, in order to provide a direct and graphic comparison between the rule-based design and the MAESTRO design, the scantlings of the former were used as the initial scantlings for MAESTRO. The performance of MAESTRO is shown in Fig. 1.13.*

The solution for the optimum design required 11 design cycles, which today involves only a few seconds of computer time. The resulting optimum design had a total life cycle cost (in which increased revenue from weight savings is converted to and subtracted from initial cost) of 8477 cost units,

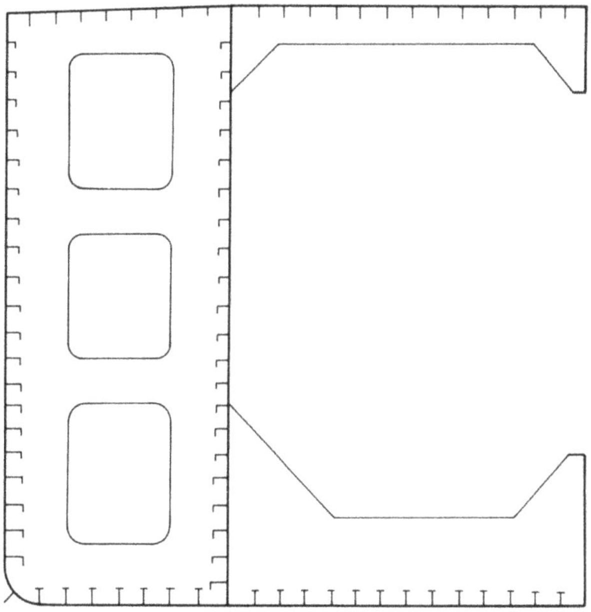

Figure 1.12 Basis ship: 96,000 dwt tanker.

which is a 13% improvement on the basis or current practice design. The savings in initial cost was 6%, which for a tanker of this size represents a savings of the order of 1 million dollars (in 1981 values).

1.3.3.1 Effect of Using Standard Sections

The foregoing savings will be decreased slightly by the need to use standard plate thicknesses and standard rolled sections for the stiffeners. MAESTRO initially treats these design variables, and also the stiffener spacing, as continuous variables in order to avoid the enormous computation and complexity of discrete variable optimization. Then, in the final design cycle, it converts to standard sizes, based on a list of standard sections and available thicknesses that the designer specifies in the input data. The designer can also specify the location and extent of whatever interstrake (see Fig. 2.15) uniformity he or she wishes to impose, in order to limit the total number of different sections and thicknesses. Moreover, the designer can make these choices *after* seeing the idealized optimum, which provides a great deal of insight and guidance. In the final design cycle, the program does not merely round off all scantlings to the next larger standard size, but rather looks for tradeoffs between rounding up and rounding down, subject to the overriding requirement that all constraints (safety, fabrication, etc.) must remain satisfied. As this requirement is essentially "one way," it is unavoidable that the standardized design will be a few percentage points away from the "ideal" (but impractical) optimum of the nonstandard design. In this example, the final design was 8670 cost units, and so the savings over

*Including solutions with two other quite different sets of initial scantlings.

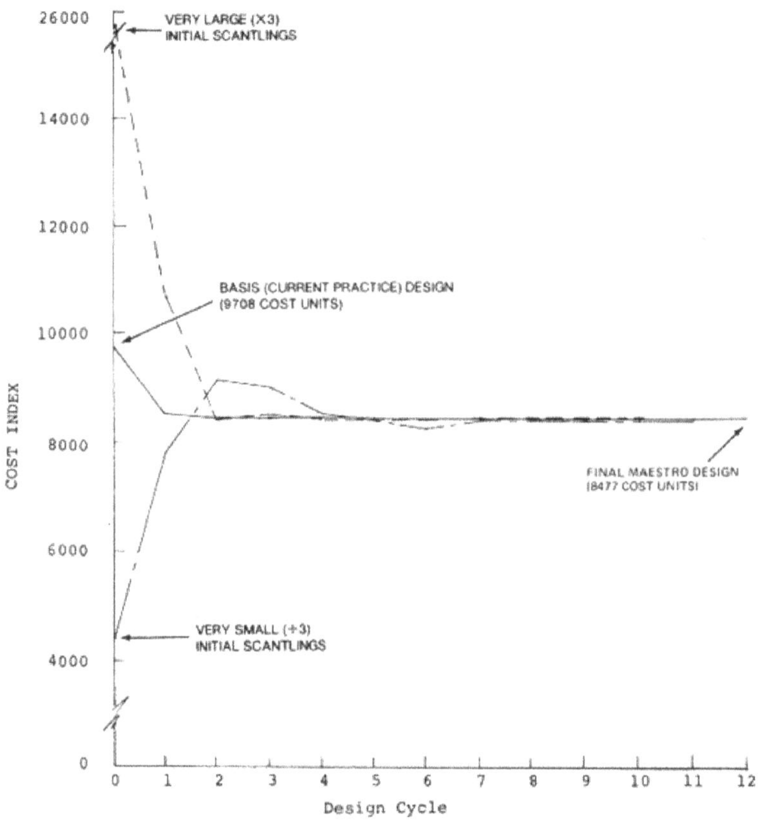

Figure 1.13 Optimization results for 12 design cycles.

the rule-based design decreased from 13% to 11%, and the savings in initial costs became 5%. This is still of the order of 1 million dollars, since the hull construction cost for such a tanker is at least $20 million (in 1981 values). Moreover, this savings can be regarded as realistic because the design itself is both realistic and "production-friendly": standard sizes and a limited number of different sections and thicknesses.

1.3.3.2 Ability to Repeat a Preliminary Design

Since the preliminary structural design, by definition, does not examine local effects, it often happens that in the detail design, it is necessary to increase some of the scantlings because of local loads, cutouts, and so on. This will happen more frequently when the preliminary design is an optimum design, since there is no excess steel in such a design. In most cases, the cost increase is small. However, if there are many such loads or other influences, such that the subsequent cost increase is found to be large, then these are not really local effects; they are general effects and the preliminary design should be redone with these effects included. With manual design, this is a large and time-consuming task; in many cases, the more expensive locally adjusted design would have to be accepted. With MAESTRO, it is a relatively easy matter to add the loads or make an approximate modeling of the local geometry, and rerun the program. In that case, the extra design requirement will be satisfied in the optimal way, and so the cost increase (over the previous preliminary design) will be much less than the cost increase associated with making local adjustments. It will also be less than that obtained by redoing a manual preliminary design.

A similar situation arises when an important design requirement is changed after a preliminary design has been completed. To fulfill the new requirement properly would mean repeating the design. A designer using standard manual methods must choose between making another costly and time-consuming design or only partially fulfilling the requirement by making local changes.

1.3.3.3 Examination of Alternative Structural Configurations

Another principal benefit of the program is that it allows the designer to compare the optimum designs for alternative structural configurations. In order to demonstrate this, a study was made of an alternative tanker design having three longitudinal girders in the center tank; that is, three parallel "ring frames" around the inside of the tank in a vertical longitudinal plane, instead of only one ring frame in the ship's centerplane. The data for this alternative configuration took about one man day to prepare. The results showed that the optimum design based on this configuration was slightly more costly than that of the single-girder model, and hence that there was no point in adding the extra girders. To have performed such a study without a program like MAESTRO would have constituted a major research project, occupying several man months.

1.3.4 Example 2—Bulk Carrier

The other ship type which is investigated in Liu, Hughes, and Mahowald (1981) is a 140,000 dwt bulk carrier. This also was an actual current practice design, but in this case the design chosen was not the final design but an earlier version in which a coarse mesh finite element stress check (by DAISY, a general purpose finite element program) had found stresses exceeding yield in the bottom and inner bottom plating. The yielding was caused by a combination of stresses from the local cargo bending moment and the overall hull girder bending moment (see Figs. 1.14 and 2.3). This early version of the design was chosen to illustrate how MAESTRO can be used to check a tentative design and also, if desired, to produce optimal corrections. For this model, it was necessary to analyze five hold lengths because of the ballast tank arrangement. Since the model was structurally symmetric longitudinally, it was only necessary to model two and a half hold lengths of the structure.

1.3.4.1 Structural Evaluation

At first, MAESTRO was used in the "evaluation mode" to compare its results with those obtained

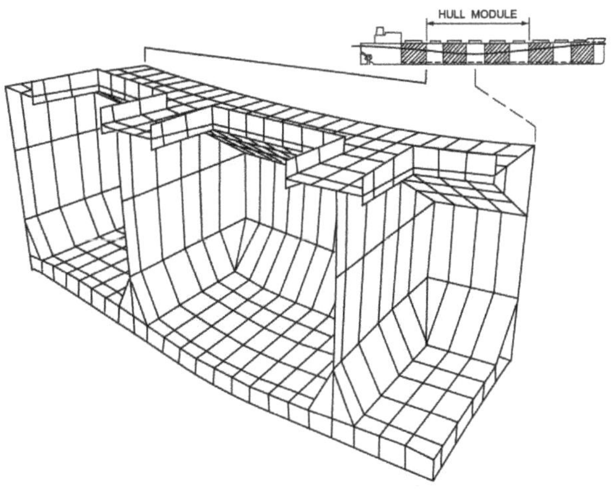

Figure 1.14 MAESTRO deflection plot–full load sagging conditions; heavy ore cargo.

earlier in the DAISY three-dimensional coarse mesh analysis. All of the deflections and all of the stresses given by the MAESTRO analysis were in good agreement with those given by DAISY.

1.3.4.2 Optimization

After the structural evaluation of the existing design, MAESTRO was used to produce an optimum (least cost) design while also correcting the deficiencies. The cost factors were basically the same as those used in Hughes, Mistree, and Žanić (1980).

With the height of the double bottom fixed at its original value of 2.3 m, the optimum design produced by MAESTRO had thicker plating but smaller longitudinals. The resulting cost was 7.5% and the weight 6.4% less than the starting design. Next, the double bottom height was allowed to vary. This produced a design that had a lower double bottom height (1.9 m) and thicker bottom plating (24 mm). Compared to the original design, the cost was reduced 8.8%. These results indicate that the design having a smaller double bottom height is of slightly lower cost. In addition, this design is able to carry more cargo.

The effect of varying the double bottom floor spacing in the bulk carrier was also examined. The original design had a floor spacing of 2.4 m (23 bays in the MAESTRO model). From this original model, two others were developed: one with a spacing of 2.208 m (25 bays) and one with a spacing of 2.76 m (20 bays). The overall lengths of all three designs were identical. The length of the cargo holds, however, did vary because of the differing floor spacing, with the variation not more than 14%. One man day was required to generate these two additional models.

Running MAESTRO for these three floor spacings without restricting the double bottom height produced the results shown in Fig. 1.15. Because there were no restrictions on double bottom height, each floor spacing resulted in a different double bottom height. From the cost index curve, it appears that the original spacing of 2.4 m was a reasonable choice. To determine the minimum value of the curve, it would be necessary to make additional runs with larger frame spacings. Notice that the weight is lowest at the original frame spacing.

To eliminate the variation in double bottom height as a factor in determining the optimum frame spacing, the three models were run with the double bottom height fixed at 1.9 m. The weight and cost index curves for these runs are shown in Fig. 1.16. In this case, it appears that of the three spacings, a frame spacing of 2.76 m is still the optimum.

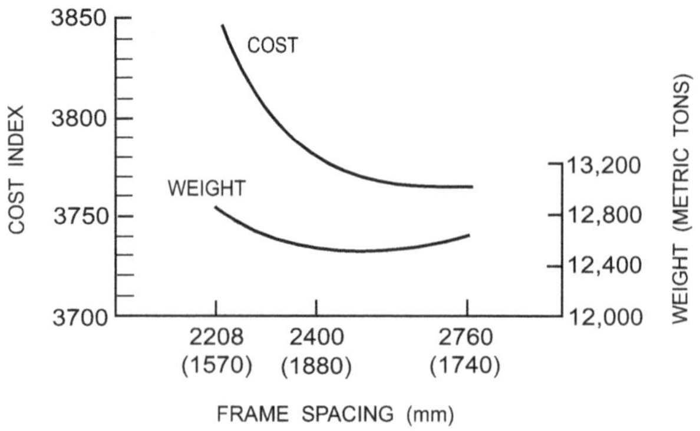

Figure 1.15 Cost and weight versus frame spacing—variable double bottom height. Note: numbers in parentheses are the double bottom height (mm).

1.3.5 Use of MAESTRO in Teaching and Research

Since rationally-based design deals with the actual characteristics of a structure, including the complex and interrelated response of all of its members and the calculation of the various limit values of these responses, and since the program is a mathematical model of these characteristics, it is very helpful in teaching both the theory and the practice of ship structural design. First, because of its structural analysis and structural evaluation features, the program assists in gaining a deeper understanding of the complex and interrelated characteristics of a ship structure, and in learning what types of structural arrangement are most efficient. Second, because it also has an optimization capability, the program is in effect a "ship design simulator," in which a designer can try out various ideas and possibilities, and can learn more about various aspects of structural design, such as the relative cost efficiency of different structural arrangements and the optimum proportions of structural members for different structural arrangements.

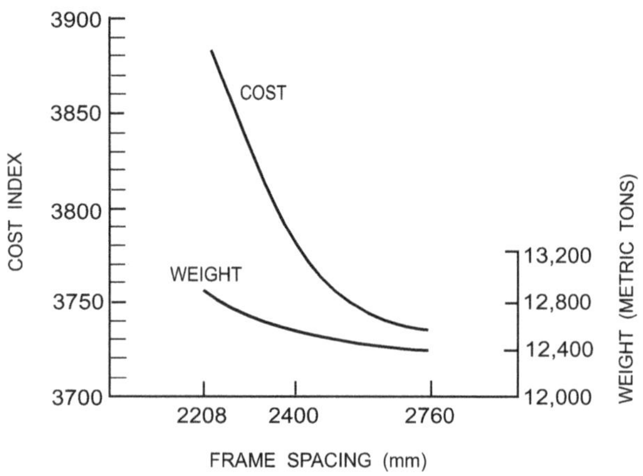

Figure 1.16 Cost and weight versus frame spacing—1880 double bottom height.

The program also has many useful applications in ship structural research. For example, it may be used to examine the effect of corrosion on the strength of the hull girder and to develop improved corrosion criteria (initial allowance, permissible wastage—both thickness and extent, etc.). The program may also be used to perform a series of "hindcasting" analyses of various types of ships in order to determine the approximate magnitude of the safety factors that are inherent in current design practice, and to see how the factors compare for various locations in the ship, types of principal structural members, types and modes of failure, and types of ship.

1.3.6 Other Practical Aspects

MAESTRO includes several features that make it versatile and easy to use, such as a graphical interface and menu system, an interactive modeler for rapid structural modeling, and color graphics that display complete information about the structural model and the various results of a MAESTRO job: stresses, types, and locations of structural failures, degree of structural adequacy (safety margins) of each member for each loadcase, and optimum scantlings. Special methods of color coding allow the designer to quickly review, quantify, and comprehend the wealth of information that is obtained. Thus the designer always remains in control of the overall process.

The use of the program does not require any specialist knowledge about computing or about optimization theory. The program has a comprehensive Help File including a User Guide and tutorials, all of which can be downloaded and printed. In addition, this book itself serves as a very complete type of theoretical manual for the program. MAESTRO can be run on ordinary laptop computers using any recent version of the Windows operating system. As of 2009, it has been used by 13 navies, various structural safety authorities (Coast Guard agencies, classification societies, etc.), and by hundreds of structural designers and shipyards throughout the world. Distribution and technical support is provided by Advanced Marine Technology Center, DRS Defense Solutions, LLC, 160 Sallitt Drive, Suite 200, Stevensville MD, 21666, USA (www.orca3d.com/maestro) and by Design Systems & Technologies, Antibes, France (www.ds-t.com).

1.4 STRUCTURAL SAFETY

1.4.1 Uncertainty, Risk, and Safety

In the design of ocean structures, there are many uncertainties to be dealt with. First, there is the uncertainty of the loads, especially those arising from waves. The ocean environment is severe, complex, and continuously varying. Ocean waves are essentially random and can be adequately defined only by means of probabilistic methods and statistics. Second, there are uncertainties regarding material properties such as yield stress, fatigue strength, notch toughness, and corrosion rate. For example, in ordinary steel which has not had special quality control, the yield stress can vary by as much as 10%; it is also dependent on the rate of loading and on the effects of welding. Third, there is inevitably some degree of uncertainty in the analysis of a structure as complex as a ship. Both the response analysis and the limit state analysis necessarily involve assumptions, approximations, and idealizations in formulating mathematical models of the physical environment and of the structure's response to that environment. Fourth, there can be variations and hence uncertainties in the quality of construction, and this factor may have a particularly strong influence on the strength of a structure. Finally, there are uncertainties of operation, such as operating errors (improper loading, mishandling, etc.) or a change of service.

Wherever there are uncertainties, there is risk of failure. For a structure, the risk of failure is the probability of a load reaching or exceeding its limit value. That is, for each limit state

$$\text{risk} = P_f = \text{prob}\,(Q \geq Q_L) \qquad (1.4.1)$$

in which, for simplicity, we are here considering a limit state which involves only one load. The safety of a structure is the converse—the probability that it will not fail. Hence

$$\text{safety} = \text{prob}\,(Q < Q_L) = 1 - P_f \qquad (1.4.2)$$

Since there are always some uncertainties, and hence some risks of failure, it is impossible to make a structure absolutely safe. Instead it can only be made "sufficiently safe," which means that the probability of failure can be brought down to a level that is considered by society to be acceptable for that type of structure. Therefore, if a structural design process is to be rationally based, the whole question of safety must be dealt with on a probabilistic basis, and the process must provide the means whereby the designer can ensure that the degree of safety meets or exceeds the required level. The calculation of the probability of a particular type of failure involves the probability density functions of the relevant load and of the limit value of that load. If these probability density functions are denoted by $p_Q(\cdot)$ and $p_{Q_L}(\cdot)$ respectively, then the probability of this particular type of failure occurring is

$$P_f = \int_0^{\infty} \left[\int_0^{\eta} p_{Q_L}(\xi)\, d\xi \right] p_Q(\eta)\, d\eta \qquad (1.4.3)^*$$

This is illustrated in Fig. 1.17, which shows that even though $\bar{Q}_L$, the mean of the limit value, is well above the mean load, there is still some overlap of the curves and hence some possibility of failure. (Note: The probability of failure is *not* equal to the area of the overlap, but this area nevertheless provides a useful visual and qualitative indication of P_f.) The figure also shows that the important regions of the distribution curves are the tails, because this is where the overlap occurs. Unfortunately, it is this portion of a distribution curve which is most difficult to obtain with any precision, mainly because one is dealing with rare events. However, it will be shown that there are ways of obtaining satisfactory estimates and upper limits of P_f, even though the tail portions of the distribution curves are not known precisely.

1.4.2 Levels of Safety

The required level of safety varies according to the type of failure and the seriousness of its consequences. Because these levels are ultimately determined by society, there are no precise values or exact rules for determining them, but they can be estimated by surveys and by examining the statistics regarding failures, particularly those types in which the failure rate is considered by the public to be generally satisfactory, in the sense that the costs and resource usages that would be required to further reduce the failure rate are considered to be unwarranted when balanced against other needs. For example, in regard to occupational risk, Flint and

*This equation assumes that Q and Q_L are independent random variables.

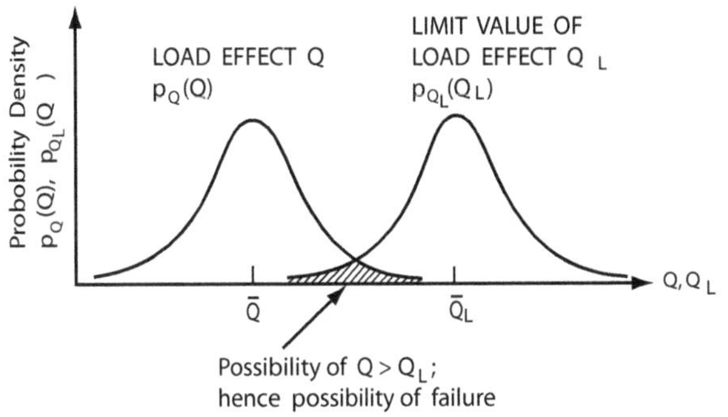

Figure 1.17 Probability distributions of load effect and of limit value of load effect.

Baker (1977) reviewed a range of activities and obtained the results shown in Table 1.1. From the results of a study of merchant ship losses, Lewis et al. (1973) estimated a value of between 0.003 and 0.006 as the lifetime probability of overall structural failure that has been tacitly accepted for large ocean-going ships. From studies of this type, it would appear that the value of P_{fT}, the total annual failure probability per structure (ship, aircraft, drill rig, etc.), ranges from 10^{-3} or less for failures that have moderately serious consequences (substantial economic loss but no fatalities) to 10^{-5} or less for catastrophic failures, such as the crash of a passenger aircraft.

A different approach to the question of required level of safety is the use of economic criteria (Construction Industry Research and Information Association, 1977). This is particularly appropriate for cases in which loss of life is not involved. For a large number of similar structures, the total annual cost C_T of each of them can be formulated as

$$C_T = C_i + P_f[C_f]_{EQ} \qquad (1.4.4)$$

where C_i is the initial cost, converted to an annual depreciation cost; P_f is the annual probability of failure; and $[C_f]_{EQ}$ is the equivalent failure cost in present worth.

The equivalent failure cost involves a discounting of future damage costs to present worth by appropriate interest rates. Failure costs should include all costs involved such as salvage operation, pollution abatement, cleanup, and lost production. It could also include loss of reputation and public confidence. In Fig. 1.18, a typical relationship between annual cost and annual probability of failure is demonstrated. The figure shows that the most important parameter is the ratio ρ between failure cost and initial cost.

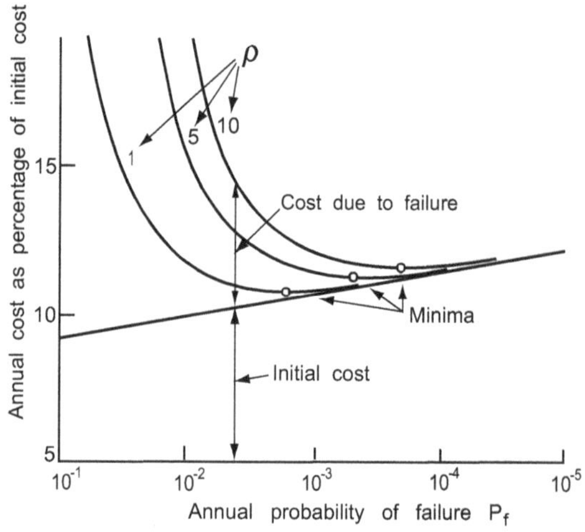

Figure 1.18 Typical relationship between cost and total probability of failure.

Many structures are comprised of, or can be divided into, a set of members, each of which is sufficiently important that failure of any one member is regarded as failure of the structure. If the failure

Table 1.1 Comparative Annual Probability of Death per 10,000 Persons

Activity	Annual Exposure (hr)	Annual Risk per 10^4
Offshore operations (including diving and vessels)		65
Mountaineering (international)	100	27
Distant water trawling (1958–1972)	2900	17
Offshore operations (other than construction, diving, or vessels)		11
Air travel (crew)	1000	12
Coal mining	1600	3.3
Car travel	400	2.2
Construction site	2200	1.7
Air travel (passengers)	100	1.2
Home accidents (all persons)	5500	1.1
Home accidents (able-bodied)	5500	0.4
Manufacturing	2000	0.4
Structural failure (land-based structures)	5500	0.001
All causes (England and Wales, 1960–1963)		
Male, age 30	8700	13
Female, age 30	8700	11
Male, age 50	8700	73
Female, age 50	8700	44
Male, age 53	8700	100

modes of the members are independent, the total failure probability is

$$P_{fT} = \sum P_{fj} \tag{1.4.5}$$

where P_{fj} is the failure probability of the jth member. Since the total probability of failure consists of several components, there is an opportunity to optimize the way in which this total required level is obtained. Under some simplified assumptions the optimal design will come close to satisfying the following equation (Moses, 1978).

$$\frac{C_{ij}}{C_i} = \frac{P_{fj}}{P_{fT}} \tag{1.4.6}$$

where C_{ij} is the cost of member j. This approach can be used to decide on the relative safety margins of the members.

1.5 PROBABILISTIC DESIGN METHODS

This topic is covered at length in Chapter 5. The purpose of this brief section is merely to introduce some of the basic notions and to give some of the early history. Chapter 5 was originally published as a separate document, and so for completeness it includes a few of the figures and tables from this section. Also, there are a few small differences of notation.

The question of a design procedure based on a probabilistic model for loads and strength has received a great deal of attention in the field of structural engineering. Pugsley (1942) and Freudenthal (1947) were the pioneers for aircraft and civil structures in the 1940s. They demonstrated how a relationship can be derived between safety factors and probability of failure, provided that the statistical distributions are known. In subsequent years, these methods were further developed and were increasingly incorporated in structural design codes, both for steel and for concrete. For concrete, this approach has been particularly successful because it accounts for the large variability in the strength of this material. A probabilistic design code is currently being developed for the design of offshore structures, stimulated by the higher risks and the higher economic stakes involved in that field.

In contrast, in the field of ship structures, the probabilistic approach is still at the early stages, in spite of the obvious probabilistic nature of wave loads. But in fairness, it should be pointed out that the load and response analysis is much more complicated for ships than it is for fixed structures or for aircraft, because it must deal with the exceedingly complex interaction between wave excitation and ship motions merely to compute the loads.

1.5.1 Exact and Approximate Probabilistic Methods

The task of achieving a specified level of safety can be pursued at various levels of mathematical rigor. We shall first show that a fully rigorous method requires the gathering of a great deal of information and is simply not justified in the majority of cases. Then we shall present two approximate methods, with an emphasis on the second one—the Partial Safety Factors Method.

1.5.1.1 Fully Probabilistic Design

The most rigorous and most general type of probabilistic design is that which utilizes the complete probability distribution functions of all relevant quantities (loads, load effects, and limit values) to calculate P_f from (1.4.3) for each load and for each type of failure. These values are then combined into an overall probability of failure which is then adjusted, by making modifications to the design, until it falls within the stipulated acceptable overall risk. This approach requires the determination of all of the probability distributions, either by measurement (of the complete phenomena or of their separate constituent aspects, and either full-scale or model) or by theoretical considerations, all of which is a very large task. Since the most highly probabilistic loads are those arising from waves, and since hull girder bending moment is the most important load effect, early research efforts were concentrated on obtaining the probability distribution for the extreme value of wave-induced hull girder bending moment. Sufficient data regarding waves have now been collected and processed statistically to produce some approximate probability distributions for this load effect; these are discussed in Section 4.3. The probability distributions of other loads and load effects are less known, and much work remains to be done. Likewise, the distributions of limit values are not easy to obtain since they arise from so many separate variations (material properties, accuracy of analysis, and quality of construction), each of which requires the collection of a great deal of statistical information. In areas where this information is not yet available, it is necessary to use less rigorous techniques.

Moreover, the availability of information is not the only factor that should be considered; another important question is whether the application really requires the complexity of the fully probabilistic method, because a design method should always be as simple as the circumstances permit. A complex method always introduces more likelihood of errors in its use. Also, greater complexity usually increases both the cost and the time required for the design. Therefore, it is important to consider whether the added accuracy of a rigorous but complex method is really justified, in regard to both safety and economy, for the particular application. For aerospace structures it often is justified, but for ship structures this is less likely.

In this regard, it is also relevant to examine what proportion of ship accidents are due to structural failure. Figure 1.19 from Gran (1978) presents the results of a survey which showed that in a given sample of ship casualties, only about 7% (0.138 × 0.54 × 100%) of severe accidents were caused by structural failure. In view of the many other causes of severe accidents and the relative infrequency of structural failure, it is clear that even a large increase in the rigor and accuracy of structural design would not improve the overall risk of casualty very much. Resources used for this purpose could be used more effectively for improvements in areas of other risks involved. Hence, there is need for moderation in regard to the statistical complexity of the structural design method.

1.5.1.2 Approximate Probabilistic Methods

The desire to reduce the complexity of the fully probabilistic approach has led to the development of simplified methods which retain the basic statistical foundation but which require only the mean and the variance and not the complete probability distribution curves.

Two alternative approximate methods are available and they are basically similar, having the following two fundamental features.

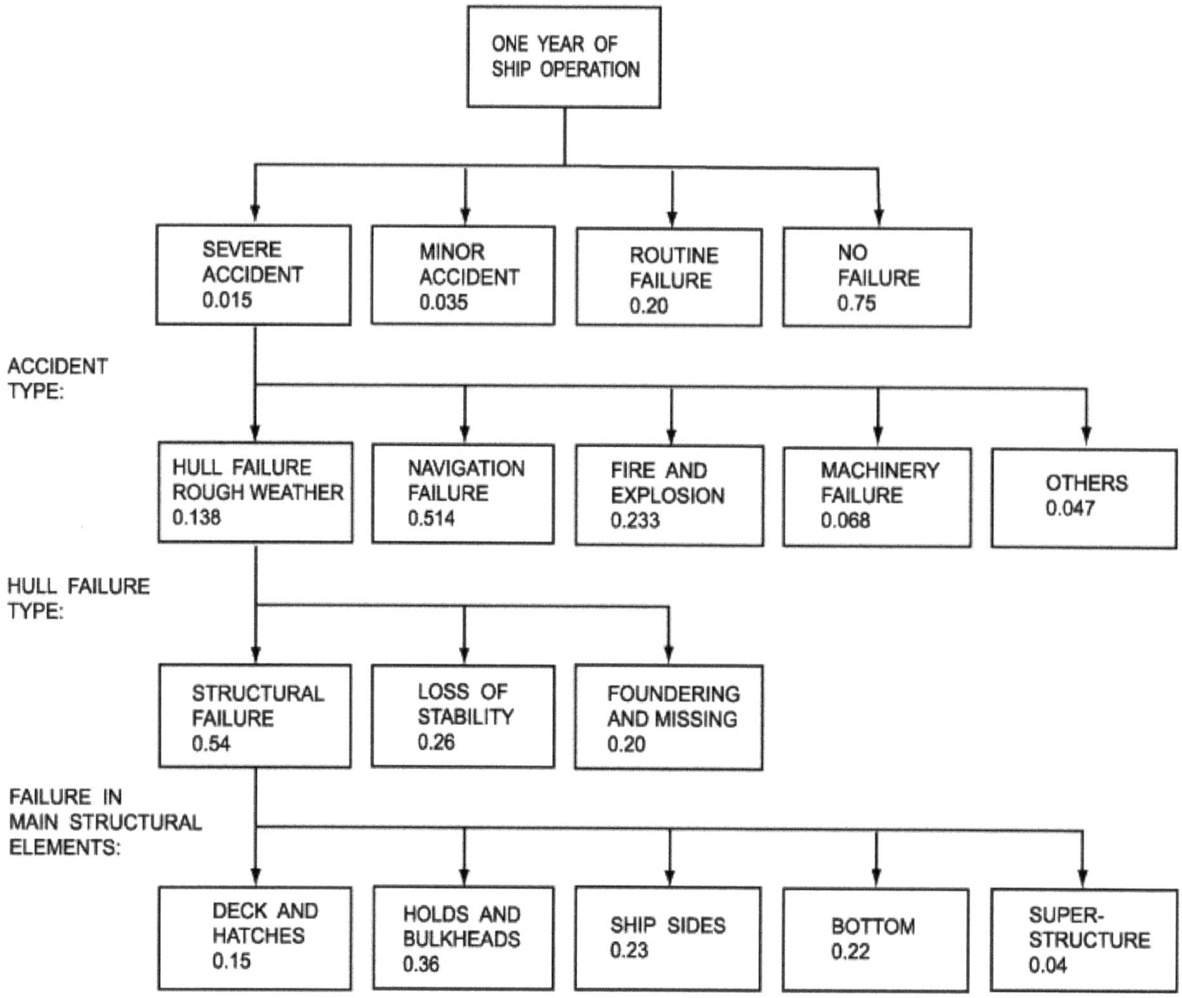

Figure 1.19 Empirical distribution of ship casualties.

1. All failure modes which are independent are treated separately. This greatly simplifies the process, but it requires that a value of acceptable risk must be defined separately for each type of failure (although in practice the same value can be used for all types which have the same degree of seriousness) and it precludes the possibility of combining the separate risks. Therefore, it requires approximations, which must necessarily be on the conservative side, in order to deal with combinations of loads of differing probability and combinations of interactive modes of failure.
2. The basic probability distributions (Gaussian, lognormal, and Rayleigh) are characterized by their first two moments, that is, by the mean and the variance (see Section 4.1 for a brief summary of basic statistical definitions and theorems). For this reason, these methods are sometimes referred to as second moment methods. If these two parameters have been established for the load effects and for the limit values, it is possible for the relevant safety authority to specify the required level of safety in terms of a set of deterministic (i.e., nonstatistical) safety factors from which the designer can immediately calculate the required strength (limit values) which the structure must have.

1.5.2 Safety Index Method

The Safety Index Method is the earlier of the two (Freudenthal, 1956), but it has not been as widely adopted as the second, the Method of Partial Safety Factors. Nevertheless, it will be briefly described here because it introduces basic concepts common to both methods and because the Safety Index itself is a very useful tool in establishing suitable values for the partial safety factors.

The degree of safety is directly related to the margin between the actual value of the load effect and the limit value

$$M = Q_L - Q$$

and failure occurs when the margin becomes negative. Since these are both random variables, M will be likewise, having a probability density function $p_M(M)$, as shown in Fig. 1.20. Therefore, the degree of safety depends not only on the separation of the two curves as measured by the distance between their mean values

$$\overline{M} = \overline{Q}_L - \overline{Q}$$

but also it depends inversely on the "spread" of the two curves, as measured for example, by their coefficients of variation. Therefore, it will also bear some inverse relationship to V_M, the coefficient of variation of M. If V_M is large, the degree of safety will be correspondingly less, and vice versa. The probability of failure is

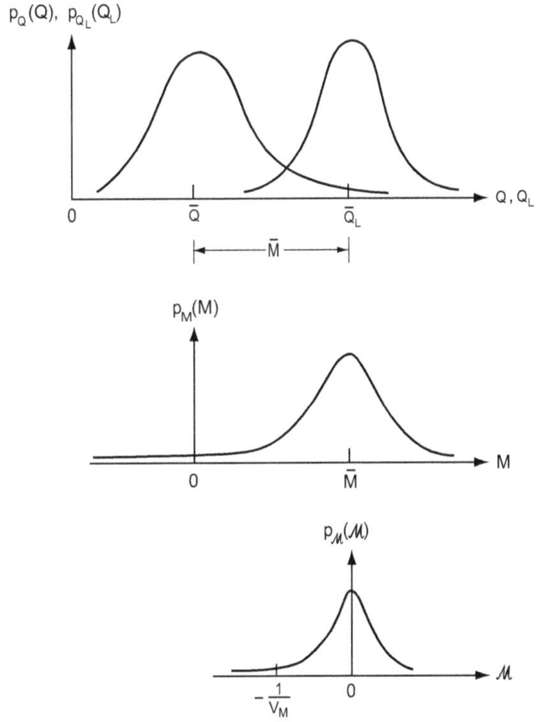

Figure 1.20 Probability distributions of the safety margin.

$$P_f = \text{prob}[M < 0]$$

Subtracting $\overline{M}$ from both sides of the inequality and normalizing by means of the standard deviation σ_M gives

$$P_f = \text{prob}\left[\frac{M - \overline{M}}{\sigma_M} < -\frac{\overline{M}}{\sigma_M}\right] \qquad (1.5.1)$$

By definition, the coefficient of variation is $V_M = \sigma_M / \overline{M}$ and therefore

$$P_f = \text{prob}\left[\frac{M - \overline{M}}{\sigma_M} < -\frac{1}{V_M}\right] \qquad (1.5.2)$$

The left-hand term within the brackets is the normalized margin, for which the distribution has zero mean and unit variance. Let us denote this normalized margin as $\mathcal{M}$ and let $P_{\mathcal{M}}(\cdot)$ be its probability distribution

$P_{\mathcal{M}}(\cdot)$ = cumulative probability
distribution of $\mathcal{M}$ = prob [$\mathcal{M} < \bullet$]

$$P_f = \text{prob}\left[\mathcal{M} < -\frac{1}{V_M}\right] \qquad (1.5.3)$$

The occurrence of $1/V_M$ on the right-hand side within the brackets confirms that the degree of safety depends on the inverse of the coefficient of variation. We therefore give this quantity the name Safety Index and denote it by the symbol β

$$\beta = \frac{1}{V_M}$$

In terms of the Safety Index, (1.5.3) is

$$\begin{aligned} P_f &= \text{prob}[\mathcal{M} < -\beta] \\ &= P_{\mathcal{M}}(-\beta) \end{aligned} \qquad (1.5.4)$$

This last expression shows that there is a direct correspondence between the Safety Index β and the probability of failure. The larger the Safety Index, the smaller the probability of failure, that is, the safer the structure. If the complete distribution $P_{\mathcal{M}}(\cdot)$ is known (i.e., if the distributions of Q and Q_L are known), then the exact value of P_f corresponding to a given value of β can be determined.* But even if the exact $P_{\mathcal{M}}(\cdot)$ is not known, designing on the basis of a specified value of β produces a consistent degree of safety from one design to another, for each type of structure. For ship structures, a suitable value of β can be determined for each type of failure by analyzing the statistics regarding ships which have proven to be reasonably efficient and which also have a satisfactory safety record. This has been done, for example, for extreme hull girder bending moment in Mansour (1974) and Faulkner and Sadden (1979).

One of the principal advantages of the Safety Index and also of the Method of Partial Safety Factors is that the provision of adequate safety, which is a probabilistic quantity, is converted and expressed deterministically in terms of a specific "design" value of load and a specific limit value which the structure must possess. The Safety Index Method makes use of mean values and in particular it involves their ratio which is referred to as the central safety factor, γ_C

$$\gamma_C = \frac{\overline{Q}_L}{\overline{Q}}$$

It can be seen that γ_C corresponds to the familiar single safety factor of deterministic design.

The relationship between β and γ_C can be derived as follows, starting with the definition of β,

$$\beta = \frac{1}{V_M} = \frac{\overline{M}}{\sigma_M} = \frac{\overline{Q}_L - \overline{Q}}{\sqrt{\sigma_L^2 + \sigma_Q^2}} = \frac{\gamma_C - 1}{\sqrt{\gamma_C^2 V_L^2 + V_Q^2}}$$

Upon rearranging for γ_C

$$\gamma_C = \frac{1 + \beta\sqrt{V_L^2 + V_Q^2 - \beta^2 V_L^2 V_Q^2}}{1 - \beta^2 V_L^2} \qquad (1.5.5)$$

In the Safety Index Method of design, the appropriate safety authority specifies the values of β, V_Q, and V_L, depending on the type of structure and the degree of seriousness of the limit state. For each limit state, the designer calculates the central safety factor from (1.5.5) and then calculates the mean value (best estimate) of the relevant load effect $\overline{Q}$ by performing a response analysis. Alternatively, for those structures for which the loads and load effects are well established, the safety authority may provide a formula for a less exact but more universal "design value" of $\overline{Q}$, together with a larger value of β which must be used with it. Knowing $\overline{Q}$, the designer then applies the factor γ_C to obtain $\gamma_C\overline{Q}$, and then designs the structure such that $\overline{Q}_L$ equals or exceeds $\gamma_C\overline{Q}$. This requirement constitutes one of the *strength constraints* which the design must satisfy. Stated mathematically, the constraint is

$$\gamma_C \overline{Q} < \overline{Q}_L \qquad (1.5.6)$$

This procedure is carried out for each limit state, thus producing the complete set of strength constraints which govern the design.

1.5.3 Partial Safety Factor Method

The Partial Safety Factor Method has been adopted in several areas of structural design, from simple building codes for civil structures to designs for aircraft and aerospace structures. It has two advantages over the Safety Index Method. First, it makes a more explicit distinction between *statistical uncertainty*—that which arises purely from genuine statistical randomness and which can therefore be properly and adequately assessed using statistical theory—and *approximational uncertainty*—that which arises from the assumptions, approximations, and judgments that are necessarily involved in any structural design task.* Second, in the Partial Safety Factor Method, each principal circumstance affecting the seriousness of the failure, and each principal source of approximational uncertainty, is accounted for

*For example, for the normal distribution the Safety Index is the same as the standard deviation.

explicitly by means of a separate safety factor, and this clarifies matters and permits greater precision and consistency.

The following paragraphs give a fuller explanation of statistical and approximational uncertainty. It is shown that the first can be accounted for by using characteristic values (instead of mean values), whereas the second can be accounted for by using safety factors, with a separate factor for each source of uncertainty.

1.5.3.1 Statistical Uncertainty

The uncertainty that arises because of the randomness of the variable (Q or Q_L) can and should be assessed by means of basic statistical theory. To do this, it is necessary to establish what type of distribution (normal, Poisson, etc.) is involved. In some cases, this is known from theoretical considerations. In other cases, it is possible to determine by observation which basic type most nearly resembles the actual distribution. Once the type is known, the uncertainty can be calculated by means of the basic laws and relationships of statistics. A very useful way of dealing with statistical uncertainty is in terms of a *characteristic value*, which is the value corresponding to a specified percentage of the area under the probability density curve, that is, to a specified probability of exceedance. For example, Fig. 1.21a illustrates a characteristic value of load Q_c corresponding to a 5% probability of exceedance. Figure 1.21b illustrates a *characteristic limit value* of load, in which case the 5% probability refers to nonexceedance.

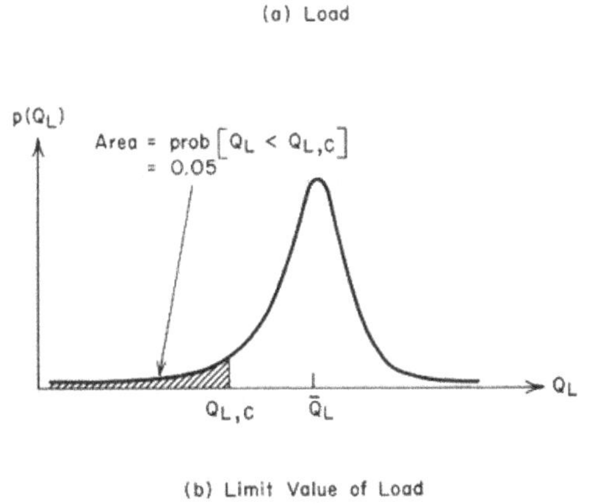

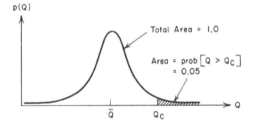

Figure 1.21 Illustration of characteristic values.

In contrast to the Safety Index Method, which uses the mean values $\overline{Q}$ and $\overline{Q}_L$, the Partial Safety Factor Method uses characteristic values Q_c and $Q_{L,c}$, thereby automatically accounting for the statistical probability of failure. Thus, if the only uncertainty was purely statistical (i.e., if Q and Q_L exactly followed their assumed distribution function), there would be no need for any safety factors, and the strength constraint would be simply

$$Q_c \leq Q_{L,c} \qquad (1.5.7)$$

in which the characteristic values would be selected so as to provide whatever degree of safety was required.

To illustrate this, let us take the idealized triangular probability distribution of Fig. 1.22 for both Q and Q_L. Let us define the characteristic values Q_c and $Q_{L,c}$ as those values which correspond to a 2% probability of exceedance and nonexceedance, respectively. We superimpose the two distributions such that their characteristic values coincide, thereby just fulfilling the constraint which is expressed by (1.5.7). For this case, (1.4.3) becomes

$$P_f = \int_0^{0.5}\left[\int_0^{\eta} p_{Q_L}(\xi)\, d\xi\right] p_Q(\eta)\, d\eta \qquad (1.5.8)$$

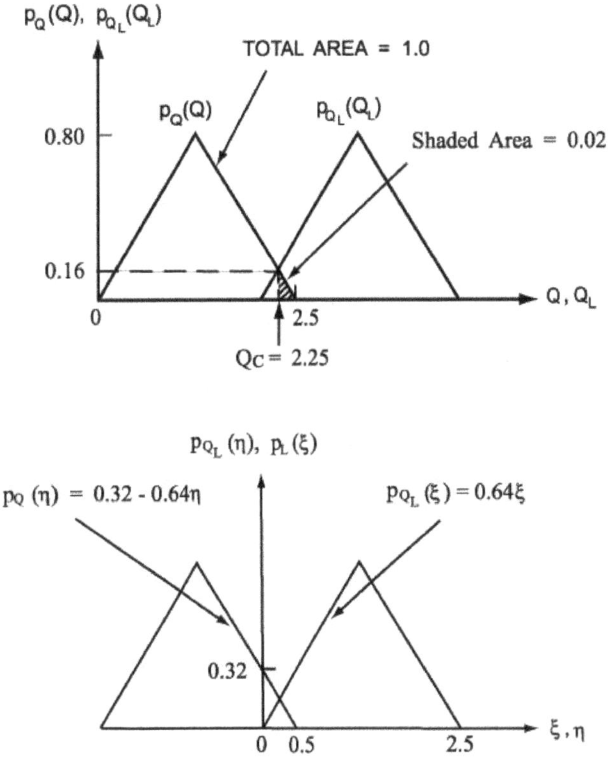

Figure 1.22 Idealized probability distributions.

in which the limits of integration and the origin of the dummy variables ξ and η reflect the fact that the integration need only be performed over the length of the overlap. We now take further advantage of this in writing simple expressions for p_Q and p_L, which are valid within this range

$$p_{Q_L}(\xi) = 0.64\xi$$
$$p_Q(\eta) = 0.32 - 0.64\eta$$

Substituting into (1.5.8) and integrating gives

$$P_f = \int_0^{0.5} 0.64\frac{\eta^2}{2}(0.32 - 0.64\eta)\, d\eta$$

and integrating again gives

$$P_f = \frac{(0.64)^2}{384} = .00107$$

or a failure probability of about 0.1%. Hence, if the required safety corresponded to a failure probability of 0.1%, then the two-percentile characteristic values would be the appropriate values, and the strength constraint of (1.5.7) would give the required safety without needing any factor of safety.

1.5.3.2 Approximational Uncertainty

In reality, of course, there is always some additional uncertainty besides the purely statistical uncertainty, which is either not statistical in nature (e.g., uncertainty arising from value judgements, from approximations, or from legal, political, or other nontechnical influences on design) or which, although statistical, cannot be included in that category because sufficient information is not available. This uncertainty is here called *approximational* because most of it arises from the approximations that are inevitable in structural design. Indeed, even the use of statistical theory to describe ocean waves involves some assumptions and approximations. But in dealing with any statistical aspect in design, the goal should be first to obtain sufficient information so as to be able to use statistical theory and to account for most of the uncertainty by this means, even though some approximations are still required, and then to seek to improve the information so as to further reduce the approximational uncertainty. Thus the amount of approximational uncertainty is reduced as more information becomes available. However, it will never be entirely eliminated; some approximations will always be necessary, and in addition there will always be some sources of uncertainty for which statistics are not entirely adequate; of that, at least, there is no uncertainty.

Because of approximational uncertainty, the characteristic values that account for statistical uncertainty are not sufficient in themselves. It is necessary to further increase the separation of the Q and Q_L curves, by some amount that can only be estimated and that requires judgement, in order to retain the required degree of safety. There are two different ways of doing this, and in describing these we shall use as an example the case of an approximational uncertainty in the load Q. The two approaches are as follows.

1. Artificially increase the variance V_Q. If the margin between Q and Q_L is being specified in terms of mean values and a central safety factor γ_C, as in the Safety Index Method, this produces a larger value for γ_C from (1.5.5), which causes the limit value distribution curve to be displaced to the right, thereby increasing the margin. If the margin is being specified in terms of characteristic values, this produces a larger characteristic load Q_c, and from (1.5.7) this again causes the limit value curve to be displaced to the right.

It is emphasized that as the uncertainty being accounted for is either not statistical or has unknown statistical properties, the amount by which the variance should be increased can only be estimated; it is a matter of judgment and is therefore somewhat subjective and arbitrary. This is emphasized because variance is a statistical quantity and so the stratagem of increasing the variance, as is done in the Safety Index Method, might make the method appear to be entirely statistical and objective, with no subjective or arbitrary element. Hence, the increase in the variance is described as "artificial" at the beginning of this description.

2. Apply a factor of safety γ_0, the value of which is likewise a matter of judgment in exactly the same degree as the increase of variance. However, in this approach, statistical aspects remain unchanged, and the approximational uncertainty is accounted for in a more explicit manner. For example, if statistical uncertainty is being accounted for by means of characteristic values, as in the example involving the triangular distributions, then the strength constraint of (1.5.7) would become

$$\gamma_0 Q_c \leq Q_{L,c} \qquad (1.5.9)$$

The second approach is the basis for the Partial Safety Factor Method, which is presented next. We note here that since the statistical uncertainty is already accounted for by the characteristic values, the magnitude of γ_0 will be much smaller than the value of γ_C in the Safety Index Method.

1.5.3.3 Partial Safety Factors

Regardless of which of the two semiprobabilistic methods is used, and regardless of which technique is used to account for approximational uncertainty, it is absolutely essential to be able to specify different levels of safety for different types of failures, in accordance with the degree of seriousness of the failures. That is, in addition to and quite apart from the need to account for uncertainty, it is also necessary to adjust the separation between the curves of Q and Q_L to account for the degree of seriousness of the particular type of failure which is being considered. Thus, there is a need for some simple and explicit method for adjusting the separation between Q and Q_L; the safety factor γ_0, which relates the two characteristic values Q_c and $Q_{L,c}$, provides just such a method. Therefore, instead of regarding this factor as a single quantity, we will regard it as the product of several *partial safety factors*. These factors can then be used for two main purposes:

1. To account for the degree of seriousness of the particular limit state in regard to safety and serviceability (for a commercial ship, the latter refers mainly to the economic consequences of the failure) taking into account any special circumstances (purpose of the ship, type of cargo, interaction of this limit state with others, etc.). Since safety and serviceability are not the same, it is best to use two independent partial safety factors for this task.
2. To account for the approximational uncertainties such as:
    - Deviation of the probability distribution of the loads from the assumed distribution, due to unforeseen actions or conditions, and consequent deviation of the load effects
    - Assumptions and approximations in the response analysis and in the limit state analysis
    - Deviation of the limit value from its assumed distribution due to unpredictable factors (e.g., poor workmanship)
    - Other matters requiring estimation and judgement

The number of partial safety factors varies from as few as three to eight or more, depending on the type of structure and on what level of detail is preferred for their specification. Factors are sometimes further subdivided if this gives greater precision or consistency. In this text, we will use four factors:

- γ_{S1} and γ_{S2}, which account for the seriousness, in regard to safety and serviceability, of the type of failure under consideration
- γ_Q, which accounts for the approximational uncertainties in the loads and load effects, including the discrepancy between the structure's actual load effect and the value predicted by the (necessarily) idealized response analysis
- γ_L, which accounts for the approximational uncertainties in the estimated limit value

The first three of these factors are applied to the load (or load effect) in the same way as γ_0 is applied in (1.5.9). But since the fourth factor refers specifically to the limit value, it is customary to transfer it to the denominator of the right-hand side and to apply it to $Q_{L,c}$ as a dividing factor. For this reason, γ_{S1}, γ_{S2}, and γ_Q are often referred to as *load factors* while γ_L is called a *limit value reduction factor*, or, in some cases a *usage factor*.

Thus in the Partial Safety Factor Method, the strength constraints are of the form

$$\gamma_{S1}\gamma_{S2}\gamma_Q Q_C \leq Q_{L,c}/\gamma_L \qquad (1.5.10)$$

The situation is illustrated in Fig. 1.23.

The Partial Safety Factor Method has been adopted for civil engineering design codes in nearly all of Europe and in Canada, Australia, and other countries. It has been adopted by the American Institute of Steel Construction (AISC) as an alternative to its existing code. These codes are actually a synthesis of the Safety Index Method and the Partial Safety Factor Method because the former was used to establish suitable values for the partial safety factors. The Safety Index is ideal for measuring and comparing the relative safety of different structures and structural members. Structures or members designed to the same Safety Index will have essentially the same degree of safety. From a survey of past designs, it is possible to calculate values of β for structures that were designed using earlier codes and that have proven satisfactory. For example, the AISC found that $\beta = 3$ gave a reasonable correlation with its previous code. Once a satisfactory value of β is established, it is possible to calculate partial safety factors for the various loads, load effects, and limit values, such that they all have a consistent degree of safety, even though they may have quite different degrees of uncertainty. This process is basically the responsibility of the safety authorities rather than the designers, and so it is beyond the scope of this text. We here merely mention a few basic and simplified aspects.

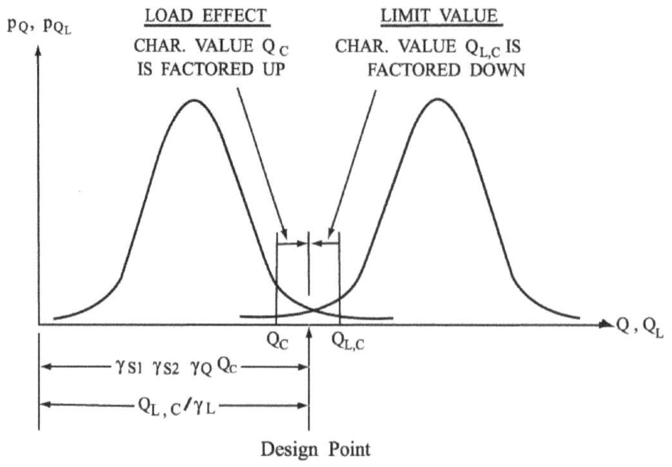

Figure 1.23 Use of partial safety factors.

In general, the distribution that is found to give the best fit to measured values of loads, load effects, and limit values is the lognormal distribution, defined in Section 4.1. From the definition of the Safety Index

$$\beta = \frac{1}{V_M} = \frac{\overline{M}}{\sigma_M} = \frac{\overline{Q}_L - \overline{Q}}{\sqrt{\sigma_L{}^2 + \sigma_Q^2}}$$

it may be shown that for the lognormal distribution

$$\beta = \frac{\dfrac{\overline{Q}_L}{\overline{Q}}\sqrt{\dfrac{1+V_Q^2}{1+V_L^2}}}{\sqrt{ln[(1+V_Q^2)(1+V_L^2)]}} \tag{1.5.11}$$

where V_Q and V_L are the coefficients of variation of the load effect and of the limit value. If V_Q and V_L are 0.3 or less, (1.5.11) can be simplified quite accurately to

$$\beta = \frac{ln\left(\dfrac{\overline{Q}_L}{\overline{Q}}\right)}{\sqrt{V_Q^2 + V_L^2}} \tag{1.5.12}$$

This expression can be further simplified by introducing a "splitting" constant, α, and making the approximation

$$\sqrt{V_Q^2 + V_L^2} \cong \alpha(V_Q + V_L)$$

If, for example, V_Q and V_L have about the same magnitude, then a choice of $\alpha = 0.7$ gives a good approximation for a fairly wide range of magnitudes. With this approximation, (1.5.12) can be rearranged into the form of a strength constraint

$$\gamma_Q \overline{Q} \le \overline{Q}_L / \gamma_L \tag{1.5.13}$$

in which

$$\begin{aligned} \gamma_Q &= \exp\,(\alpha\,\beta V_Q) \\ \gamma_L &= \exp\,(\alpha\,\beta V_L) \end{aligned} \tag{1.6.1a}$$

This equation states that if the mean value (best estimate) of the load effect multiplied by γ_Q is less than the mean (best estimate) of the limit value divided by γ_L, then a Safety Index greater than β will be achieved. Note that in this formulation the mean values $\overline{Q}$ and $\overline{Q}_L$ are used instead of the characteristic values, and so the factors γ_Q and γ_L account for both statistical and approximational uncertainty. This is done if neither Q nor Q_L are amenable to a statistical representation and nearly all of the uncertainty is approximational.

The establishment of a suitable set of partial safety factors for ship structures is a large task, but it is absolutely essential for progress in this field, and it will be of benefit to all in the shipping world: owners, designers, builders, operators, insurers, and others. The new civil engineering codes provide a good start, but there is need for further effort by persons and agencies directly concerned with ship structures. The American Petroleum Institute and other agencies are currently engaged in a similar task for offshore structures.

1.5.4 Distinction Between Stillwater Loads and Wave Loads

The hull girder loading on a ship may be divided into two broad categories: stillwater loads and wave loads. The first relate mainly to cargo loading and other controllable factors, and they are, for the most part, specific, deliberate, and directly calculable; they are basically deterministic, and for the statistical fluctuations which do occur, it is a relatively simple matter to calculate characteristic values. Therefore, the calculation of the corresponding characteristic load effects is relatively straightforward, requiring only a single response analysis. Wave loads are the opposite: they are essentially probabilistic and the calculation of hull girder bending moment in the at-sea condition is a major task, primarily because there is a quite complicated interaction between the load and the response: the seaway loading causes ship motion and the motion influences the seaway loading. Because of this cross-coupling in the response, the characteristic value of load effect (such as hull girder bending moment which is the most important load effect) cannot be obtained by defining a single characteristic value of load and then performing a single, static response analysis. On the contrary, the response analysis for wave-induced loads is quite complex, involving statistical theory, hydrodynamics, and systems analysis, and it can only be done by means of rather sophisticated computer programs that have been developed for the purpose. A brief summary of this analysis is presented in Section 2.2, and it is treated more fully in Chapter 4.

For an individual designer, such an analysis represents a complex and time-consuming task, and yet it is the only way of obtaining accurate values for the wave-induced load effects for a specific ship design. However, for ships of standard geometry and proportions, there is a much easier method, which is made possible by the following observations.

1. The load effects of the stillwater loads can be calculated separately; the stillwater loads have only an indirect influence on the wave loads.
2. The stillwater load effects depend on the particular geometry and the structural weight and cargo distribution, but the wave-induced characteristic load effects, being long-term statistical quantities, are more general and universal, and are applicable to a whole class of ships having the same geometry and proportions.

Ship classification societies and other researchers have utilized computer programs for wave-induced response analysis to calculate the characteristic value of hull girder bending moment for ships of standard geometry and proportions. Then, by subtracting the stillwater portion of the bending moment, they obtained characteristic values for the wave-induced bending moment. Other information has been obtained from long-term measurements of bending strain in ships. As a result, it has been possible to develop expressions for these characteristic values in terms of the principal dimensions of the ship. These expressions are given in the rules of the various classification societies. Some examples are given in Section 3.5. Thus for ships of standard geometry and proportions, the designer can concentrate on the stillwater response analysis and load effects, which depend on internal layout and cargo distribution.

For ships of nonstandard geometry or proportions, the characteristic value of total bending moment should be calculated for the most important cargo configurations using a wave response analysis program. These programs have now been developed to the point where the user need only supply basic information such as the ship's lines (or offsets) and the cargo distribution. They are available at the computer bureaus of various classification societies, and they run easily on personal computers.

1.6 LOAD FACTORS AND DEGREES OF SERIOUSNESS OF FAILURE

As mentioned earlier, three of the four partial safety factors are applied to the load; hence, these three are commonly referred to as load factors.* It will be convenient to also define a "total" load factor, γ_{load}, which is simply the product of these three partial load factors.

$$\gamma_{\text{load}} = \gamma_{S1}\gamma_{S2}\gamma_Q \qquad (1.6.1)$$

The first two refer to the degree of seriousness of the failure, and in this section we present some qualitative definitions of degrees of seriousness and some sample values of load factors. The fourth factor, γ_L, is applied as a divisor to the limit value of the load effect, and its value is determined mainly by material properties and fabrication considerations.

1.6.1 Degrees of Seriousness of Failure

In order to assess the degree of seriousness of a structural failure, we must examine the conse-

*Strictly speaking, they should be called "partial load factors," but the word "partial" is often omitted.

quences: what are the losses and how severe are they? We have seen that the two principal attributes by which the fitness of a ship is measured are *safety* and *serviceability*. Accordingly, we may distinguish two different types of losses.

1. Loss of life and other serious and irreparable noneconomic losses
2. Loss of main function which, for a commercial vessel, means economic loss because of lost revenue, cost of repair or replacement, environmental damage, lawsuits, and so on.

The foregoing categories also apply to noncommercial vessels in which the main function is the performance of some mission or service that has no direct relationship with economic factors. For such vessels, the performance can be quantified by means of a performance index; in fact, as pointed out in Section 1.2, for a rationally-based design the objective has to be specified and its dependency on the design variables must be quantified. The same performance index that serves as the objective function can also be used to assess the degree of seriousness of a failure that adversely affects the performance.

Although safety and serviceability have much in common, they are clearly distinct; there are some failures that can cause fatalities without causing loss of main function and vice versa. Also, they have different relative importance in different situations. For example, in naval vessels the main function is the performance of a mission and therefore serviceability—the accomplishment of the mission—has greater importance relative to safety than it has for commercial vessels.

Since there are two separate attributes for measuring seriousness—safety and serviceability—we use two separate factors, γ_{S1} and γ_{S2}. Even with this distinction between safety and serviceability, the degree of seriousness is still difficult to assess and to quantify. There are any numbers of degrees of seriousness; it is a continuous rather than a discrete quantity. Nevertheless, for our purpose we must nominate and define a few specific degrees of seriousness. As an example, we will herein distinguish four degrees of seriousness which we will call *extreme*, *severe*, *moderate*, and *slight*. These must be defined in terms of their likely consequences in regard to safety and serviceability. For the attribute of safety, the degree of seriousness of a failure corresponds to its consequences in regard to loss of life. Table 1.2 describes in general terms the sorts of consequences that are associated with the four degrees of seriousness. Similarly, for the attribute of serviceability the seriousness is measured by loss of main function. Table 1.3 describes the sorts of consequences that would correspond to these four degrees. It is emphasized that the values of partial load factors given in the tables are merely sample values, given for illustration only. The way in which they

Table 1.2 Degrees of Seriousness of Structural Failure in Regard to Safety

Level of Structure	Degree of Seriousness of Failure	Consequences in Regard to Loss of Life	Sample Range of $\gamma_{S1}\gamma_Q$	γ_{S1} for $\gamma_Q = 1.1$
Hull module collapse	Extreme	Some fatalities likely; may include all personnel if there is another failure or harsh conditions or mismanagement	1.30–1.54	1.18–1.40
Principal member collapse	Severe	Small but definite risk that failure may cause a few fatalities at occurrence; risk of subsequent fatalities very small unless there is another failure, harsh conditions, or mismanagement	1.20–1.40	1.09–1.27
	Moderate	No appreciable risk of fatalities but the structure is weakened and a slight risk would arise if there is another failure, harsh conditions, or mismanagement	1.10–1.30	1.0–1.18
	Slight	No risk of fatalities, but the resulting local damage constitutes a slight risk of injury (e.g., warped deck plating)	1.0–1.20	1.0–1.09

Table 1.3 Degrees of Seriousness of Structural Failure in Regard to Serviceability

Level of Structure	Degree of Seriousness of Failure	Consequences in Regard to Loss of Main Function*	Sample Range of γ_{S1}
Hull module collapse	Extreme	Complete loss (ship out of service) for a long period; may be permanent loss (i.e., total loss of ship) if there is another failure, harsh conditions, or mismanagement	1.2–1.4
Principal member collapse	Severe	Complete loss for short period or partial loss (ship operational but severely handicapped); repair costly and urgent	1.1–1.2
	Moderate	Ship operational but inefficient; loss of some secondary functions; repair as soon as practicable	1.05–1.1
	Slight	Main function impaired; some inconvenience or inefficiency at the secondary level; repair as soon as convenient	1.0–1.05

*Revenue earning or mission performance.

were obtained is described in the next section, which also points out the need to take into account particular factors such as the type of ship (passenger, cargo, naval, hazardous cargo, etc.) and the cost and the operational importance of the ship.

Since the primary aim of structural constraints is to provide adequate safety and serviceability, the most important limit state is that of ultimate failure or collapse of the hull girder (which in practice is performed for hull modules that are each a complete segment of the ship). The other limit states are merely stages toward collapse. The provision of an adequate degree of safety against structure collapse automatically provides a proportional degree of safety against less serious forms of failure, and this is usually sufficient. But the converse is not true; the provision of adequate safety against lesser forms of failure does not necessarily provide sufficient safety at the overall level—which is where it is most required.

In the past, because a true ultimate strength analysis of the overall structure was not possible, the only alternative was to define certain ways in which overall failure could occur, in terms of certain combinations of member failures. This has obvious limitations, the most serious being that structure collapse can in fact be caused by the collapse of one member or of many, depending entirely on the geometry and proportions of the structure and on the loading. The correlation of structural collapse with member collapse requires long-term experience and carefully documented information concerning actual failures of that particular type of structure. If the structure is of a different type, or even of different proportions, it may very well be susceptible to a certain combination of member failures which has not occurred before. This is precisely what has happened in many structural failures.

It is better to determine what the actual forms of failure are for the structure in question, under the various combinations of loads which are expected. Since the ultimate strength testing of a complete hull module is usually out of the question, this can only be done by means of a model—either a physical model or a mathematical model. With modern computing power, mathematical modeling is not only possible but also just as reliable and is far easier and more efficient than the former. As is shown briefly in Section 2.5, and in more detail in Chapter 16, the mathematical calculation of the ultimate strength of a hull module requires an *incremental* or *load-deflection* approach, that is, the formulation of a mathematical model and the determination of the structure's actual load-deflection relationship by an incremental analysis of the individual failures which lead to collapse of the structure. There is no other way of determining the true forms of collapse and the true (i.e., lowest) collapse load for a large structure which is subjected to a variety of load combinations. A load-deflection approach is normally a rather large computational task, but Chapter 16 presents a modeling strategy and a method of analysis that makes it quite economical with present-day computers.

If a true ultimate strength analysis is performed for a specific hull module, there is no need to attempt to anticipate the form of the collapse. Hence, there is no need to define intermediate stages of failure in terms of numbers and types of member failures, or to classify these according to their degree of seriousness and assign partial safety factors for them. Therefore, the ultimate strength analysis of the overall structure logically involves only one level of seriousness—the extreme level. The other levels shown in Tables 1.2 and 1.3 are mainly intended for the ultimate strength analysis of principal members, which we will consider next. But we note in passing that if there is some particular intermediate form of failure for which it is desired to have a specified degree of safety, then these other levels can be used for this purpose.

The other type of limit state analysis is that of the principal members. In most cases, these are sufficiently simple and their characteristics are sufficiently well known that it is possible to define all of the possible limit states for each type of member (the member types being distinguished on the basis of the topology, geometry, and material of the member). The information required is the member's structural dimensions and material properties, and *the loads acting on it*. The last point is emphasized because the true loads acting on a principal member can only be obtained by a full ship finite element analysis. Even so, some approximations and idealizations may be required. For example, a tapered stiffened panel is usually idealized as rectangular.

Since there are several limit states and since some of them may be known to have more serious consequences than others (again because of safety or serviceability or both), it is desirable to have two or three levels of seriousness of member failure, and that is the purpose of the other three categories—severe, moderate, and slight—of Tables 1.2 and 1.3. Each member failure is assigned to one of these categories depending on which of the consequences described in the tables best matches the consequence which that limit state would have on the safety and serviceability of the ship.

1.6.2 Sample Values of Load Factors

At the hull module level, the most significant load effect is hull girder bending moment. Because of the complexity of this load effect, there is as yet no standard or universally accepted value for the percentile level to be used in calculating the characteristic value of wave bending moment, or for the load factors γ_Q and γ_{S1}. The specification of these values is the prerogative of the classification societies and other ship structure authorities. These bodies are currently engaged in the determination of suitable values. In order to reach a worldwide uniform safety climate for shipping, the IMO in cooperation with IACS, within the development of GBS as described in Section 1.1.6 above, is in the process of defining overall safety goals and acceptance criteria for ships. These will be the basis for future rules of the different classification societies with a more uniform safety level. Finally, this will also have an influence on the selection of safety factors and load factors in ship structural design.

In this section, we present some sample values of load factors which are intended merely to illustrate the discussion and to give some indication of the approximate magnitude and range of these factors.

As we have seen, the degree of seriousness of a structural failure, from the point of view of safety, is accounted for by γ_{S1}. But this degree of seriousness depends on many factors and circumstances relating to the nature of the ship and its service. For example, a given type of structural failure is more serious in a vessel carrying passengers, not only because of the added number of people, but also because, unlike a crew, they are uninvolved and untrained. Therefore, although we can define certain degrees of seriousness in general terms as in Table 1.2, it would not be sufficient to assign one specific value of γ_{S1} for each of these degrees. Instead, it is necessary to give a set of values that covers all of the important circumstances relating to safety. In order to give at least a qualitative indication of this, Table 1.2 gives a sample range of values of $\gamma_{S1}\,\gamma_Q$ for each of the defined levels of seriousness. The reason for giving the product instead of the separate load factors is explained later.

The factor γ_{S2} accounts for the degree of seriousness in regard to serviceability. The seriousness of a given failure depends here on the importance of the ship's main function, or on the economic scale of the ship and of the system in which it is operating. For example, a failure that causes 3 weeks of ship immobility for repairs is more serious for a fast or expensive or specialized ship than for a slow-speed, low-cost, easily-substituted ship. Thus, the choice of γ_{S2} is normally made not by regulatory authorities but by whomever is responsible for the overall system in which the ship is to operate, on the basis of economic and operational criteria. Hence the value of γ_{S2} can vary over an appreciable range, and to illustrate this, Table 1.3 gives a range of values. But here again, as in Table 1.2, the values are for illustration only, and the specified upper and lower values are not intended as limits in way.

The sample values of $\gamma_{S1}\,\gamma_Q$ for the "extreme" level of seriousness in Table 1.2 are based on Faulkner and Sadden (1979), who analyzed several actual and "rule-based" naval vessels. Working backward from available data (load estimates, strain records, etc.), the authors estimated that the percentile level for the characteristic hull girder bending moment which is implicit in current design practice is of the order of 2%, that is, a wave bending moment which is sufficiently large that there is only a 2% chance of it being exceeded in the ship's lifetime.* However, this figure of 2% is only approximate because there are still some aspects of ship response analysis theory in which greater accuracy is required, such as some of the nonlinear effects of the ship motion response analysis, the spectral distribution function for ocean waves, and the statistical prediction of long-term values.

These are all examples of approximational uncertainty regarding the load; therefore, they can be accounted for by γ_Q. If the values of the other two load factors were known, it would be possible to deduce the value of γ_Q which is implicit in present-day design practice. But, since there are as yet no officially established or standard values for either of these factors Faulkner and Sadden (1979) could only obtain values of the total load factor, γ_{load}, defined in (1.6.1).†

The results indicated that for hull module collapse (specifically, collapse of the strength deck) which corresponds to the "extreme" level of seriousness, the value of γ_{load} varied from 1.56 to 1.83 for the rule-designed naval vessels, for which the values of γ_{S1} would resemble those which are implicit in merchant ship design. Since all of the ships considered were of the same type, γ_{S2} is constant and so the product $\gamma_{S1}\,\gamma_Q$ is proportional to the load factor. Although the value of γ_{S2} is unknown, it is probably rather low since these were naval vessels. In order to give some indication of the value of the product γ_{S1} γ_Q, let us choose a value of 1.2 for γ_{S2} (this corresponds to the lower end of the sample range of γ_{S2} in Table 1.3). Dividing this into the values of γ_{load} gives a range for $\gamma_{S1}\,\gamma_Q$ from 1.30 to 1.54. This is the range shown in the top row of the table. The other three ranges were chosen such that all four are similar in size and have a small overlap.

From the above discussion, it is clear that the total load factor is influenced by all three partial factors, each of which has a different purpose and is determined according to different criteria. Since the value of γ_{S2} does not relate to safety and is chosen by persons other than the safety authorities, the values of γ_{S1} and γ_Q must be sufficiently large such that the overall level of safety is satisfactory even when γ_{S2} is 1.0, as it might be for a low-cost, easily-replaceable vessel. Obviously, a value of γ_{S2} less than 1.0 cannot be allowed. In principle, this same restriction applies to the other partial safety factors as well. However, circumstances can arise where there might be grounds for allowing a factor to be less than 1.0, such as when a standard characteristic load is being used which is too severe for the vessel in question (e.g., ocean wave bending moment for a vessel intended only for semisheltered waters).

*This is not the probability of failure; it is merely the definition of the characteristic value of the load effect, Q_c in Fig. 1.21a. The probability of failure is much less than 2% because it also involves the characteristic value of the *limit* bending moment $Q_{L,c}$ which is itself defined in terms of a small percentage chance of nonexceedance (typically 5%), and the margin between Q_c and $Q_{L,c}$ which is provided by the partial safety factors.

†Actually, Faulkner and Sadden (1979) presented values of the overall safety factor, $\gamma_0 = \gamma_{S1}\,\gamma_{S2}\,\gamma_Q\,\gamma_L$, but from the information given in the paper it is possible to deduce that γ_L is approximately 1.08.

1.7 REFERENCES

Aksu, S., et al. (2009). Technical committee II.1—Quasi-static response. *Proc. of the 17th International Ship and Offshore Structures Congress*, Seoul, Korea, Vol. 1.

Caldwell, J. B. (1971). Theory and synthesis of thin-shell ship structures. *International Association for Shell Structures Symposium*, Hawaii.

Construction Industry Research and Information Association. (1977). *Rationalization of safety and serviceability factors in structural codes.* CIRIA Report 63. London, UK: Author.

Evans, J. H. (Ed.). (1975). *Ship structural design concepts.* Centerville, MD: Cornell Maritime Press.

Faulkner, D., and Sadden, J. A. (1979). Toward a unified approach to ship structural safety. *Trans. RINA*, Vol. 121, 1–38.

Flint, A. R., and Baker, M. J. (1977). Risk analysis for offshore structures—the aims and methods. *Proc. Conf. on Design and Construction of Offshore Structures*, I.C.E.

Freudenthal, A. M. (1947). The safety of structures. *Trans. ASCE*, Vol. 102.

Freudenthal, A. M. (1956). Safety and probability of structural failure. *Trans. ASCE*, Vol. 121.

Gallagher, R. H., and Zienkiewicz, O. C. (Eds.). (1973). *Optimum structural design.* New York, NY: John Wiley & Sons.

Gran, S. (1978). *Reliability of ship hull structures.* Report No 78-215. Oslo, Norway: Det Norske Veritas.

Griffith, R. E., and Stewart, R. A. (1961). A nonlinear programming technique for the optimization of continuous processing systems. *Manage. Sci.*, Vol. 7, 379–392.

Hano, H., Olamoto, Y., Takeda, Y., and Okada, T. (2009). *Design of ship structures—A practical guide for engineers.* Berlin, Germany: Springer.

Horn, A. M., et al. (2009). Fatigue and fracture. *Report of ISSC-Committee III.2, 17th International Ship and Offshore Structures Congress*, Seoul, Korea.

Hughes, O. F., and Mistree, F. (1976). An automated ship structural optimization method. In Jacobssen, A., et al. (Eds.). *Computer applications in the automation of shipyard operation and ship design* (pp. 203–212). Amsterdam, The Netherlands: North Holland Publishing Co.

Hughes, O. F., Mistree, F., and Davies, J. (1977). Automated limit state design of steel structures. *Proc. of the Sixth*

Australasian Conference on the Mechanics of Structures and Materials, Christchurch, August, 98–106.

Hughes, O. F., Mistree, F., and Žanić, V. (1980). A practical method for the rational design of ship structures. *J. Ship Research*, Vol. 24, Issue 2, 101–113.

Hughes, O. F., Wood, W. A., and Janava, R. (1982). SHIPOPT—A CAD system for rationally-based ship structural design and optimization. *4th International Conference on Computer Applications in Ship Design*, Annapolis, June.

Kellog, H. J. (1960). The cutting plane method for solving complex programs. *SIAM J.*, Vol. 8, 703–712.

Lamb, T. (Ed.). (2003). *Ship design and construction.* Jersey City, NJ: Society of Naval Architects and Marine Engineers.

Lewis, E. V., et al. (1973). *Load criteria for ship's structural design.* SSC Report SSC-240. Glen Cove, NY: Webb Institute of Naval Architecture.

Liu, D., Hughes, O. F., and Mahowald, J. E. (1981). Applications of a computer-aided, optimal preliminary ship structural design method. *Trans. SNAME*, Vol. 89.

Løvstad, M., and Guttormsen, V. (2007). Impact of common structural rules on software for ship design and strength assessment. *Proc. of RINA Int. Conf. on Developments in Classification and International Regulations*, London, 119–132.

Mansour, A. (1974). Approximate probabilistic method of calculating ship longitudinal strength. *J. Ship Research*, Vol. 18, Issue 3, 203–213.

Mistree, F., Hughes, O. F., and Phuoc, H. B. (1981). An optimization method for the design of large, highly constrained complex systems. *Engineer. Optimiz.*, Vol. 5, Issue 3.

Moses, F. (1978). Safety and reliability of offshore structures. In Holand, I., et al. (Eds.). *Safety of structures under dynamic loading*. Trondheim, Germany: Tapir Publ.

Murtagh, B. A., and Saunders, M. A. (1980). *Systems optimization laboratory report 80-1.* Stanford, CA: Stanford University.

Murtagh, B. A., and Saunders, M. A. (1983). A projected lagrangian algorithm and its implementation for sparse nonlinear constraints. In Buckley, A. G, and Goffin, J. L. (Eds.). Mathematical programming study on constrained minimization. New York, NY: North-Holland.

Oh, K.-G. (2009). Recent status of rules and regulations for ships. Keynote lecture, *17th International Ship and Offshore Structures Congress*, Seoul, Korea.

Pedersen, P. (1973). Optimal joint positions for space trusses. *ASCE*, Vol. 99, Issue ST12, 2459–2476.

Petinov, S. (2003). Fatigue analysis of ship structures. Dover, NJ: Backbone Publishing.

Pugsley, G. (1942). *A philosophy of aeroplane strength factors.* Report and Memo No. 1906. London, UK: British Aeronautical Research Committee.

CHAPTER

TWO

LOADS, STRUCTURAL RESPONSE, LIMIT STATES, AND OPTIMIZATION

Owen Hughes
Professor, Virginia Tech
Blacksburg, VA, USA

Hans G. Payer
Germanischer Lloyd, Hamburg, Germany (ret)

This chapter takes a closer look at four of the main aspects of rationally-based design: loads, structural response, limit states, and optimization. The aim is to identify their principal components and to help the reader to gain a clear overall picture by classifying these components according to their types, levels, and the relationships between them. The logical place to start is with the loads, and so we begin with a brief summary of the principal types of loads which act on a ship.

2.1 LOADS ON SHIPS

One way of classifying loads on ships is according to the level of structure at which they act because some loads influence the structure at just one of four levels: hull girder, hull module, principal member, and local (see Fig. 1.5). But other loads have an influence at more than one level, and the most fundamental load—external pressure on the hull—has an influence at all four levels. Nevertheless, the loads can be classified approximately in this way, and it is important to have a clear concept of the levels at which the various loads act or at which they have their principal influence.

Another way of classifying loads is according to how they vary with time: static, slowly varying, or rapidly varying. In calculating load effects, there are three types of structural analysis that more or less correspond to these: static, quasistatic, and dynamic. In a dynamic analysis, effects of time variation of loading are fully accounted for. Almost any irregular dynamic loading can be represented as a combination of regularly varying loads. If the force–displacement relation is linear or only slightly nonlinear, then the problem of calculating load effects can be solved "in the frequency domain," with frequency as the principal independent variable instead of time, which greatly simplifies the calculations. The frequency-based distribution of a load or a load effect (response) is called a *spectrum*, and so we speak of a *wave spectrum* and a *response spectrum*. If the force–displacement relation is nonlinear, then the problem must be solved "in the time domain," with time as the independent variable.

A quasistatic analysis is simply a static analysis in which the motions are estimated and their effect on the structure is accounted for approximately by including some inertia forces. Since there is no essential difference between static and quasistatic analysis, we will in this text usually speak of just "static" and "dynamic," and only use the term "quasistatic" when it is desired to emphasize that some motion effects are being allowed for in the static analysis. In most cases, this will be clear from the context.

Slowly varying loads are those for which even the shortest component period is appreciably longer than the fundamental (longest) natural period of vibration of the structure. In most cases, slowly varying loads can be dealt with by means of static analysis with only a small loss of accuracy, whereas rapidly varying loads usually require a dynamic analysis for sufficient accuracy.

Whenever possible, to minimize computation time, static and dynamic analyses are performed separately, with the latter dealing only with the fluctuation in load, that is, the departure from the static load. Total response of the structure is then obtained by superimposing the two results.

In terms of the three load types just defined, the principal loads on ships are:

Static (or essentially static) loads

1. All "stillwater" loads: external and internal pressures (buoyancy and bulk cargo); all weights
2. Drydocking loads
3. Thermal loads

Slowly varying loads

1. Wave-induced dynamic pressure distribution on the hull resulting from the combination of wave encounter and the resulting ship motion
2. Sloshing of liquid cargoes
3. Shipping of green seas on deck
4. Wave slap on sides and on foredecks
5. Inertia loads, especially on masts and other elongated structures, and also on decks and frames at the attachment points for containers (lashing loads) and other heavy objects
6. Launching and berthing loads
7. Ice-breaking loads (at the hull girder level)

Rapidly varying loads

1. Slamming
2. Forced (mechanical) vibration; pressure pulses from the propeller
3. Other dynamic loads, as discussed later

Static loads are relatively straightforward and do not require special explanation. Calculation of stillwater hull girder load effects is dealt with in Chapter 3. Slowly varying and rapidly varying loads are more complex, and the following sections define and describe the various loads in these categories.

The treatment here is largely qualitative. The aim is to cover the main aspects and to give an overview of the various types of loads and the ways in which they are accounted for.

As computing costs decrease, it is foreseeable that advanced computational methods will be used more frequently in the future to predict dynamic loads. For large containerships in high waves, recently developed techniques based on solving the Reynolds-averaged Navier-Stokes equations in the time-domain have been successfully employed to assess, for example, effects of slamming-induced loads on large-scale elastic hull girder whipping or effects of green water loads on wave breakers. However, it must be stressed that these loads need to be validated against measurements. For ship structural designs, we are often only interested in average loads obtained by integrating local pressures over structural plate fields and not local peak pressures acting over small control volume face fields. Slamming design loads are typically obtained in this way, and predicted slamming loads compare favorably with test measurements. In contrast, it is difficult to validate predicted sloshing loads because, to assess their effects, it may well be necessary to account for concentrated pressure peaks acting on small areas inside partially filled tanks.

2.1.1 Slowly Varying Loads

The most important load in this category is the wave-induced dynamic pressure, and the most important effect is the wave-induced hull girder bending moment, M_w. The term "wave-induced" means the difference or departure from the stillwater value of that load (pressure, bending moment, shear force, etc.), such that the total value at any point is the sum of the two. Ideally, since it is influenced by the ship's motions, the pressure on the hull should be calculated as part of the ship motion analysis, and the wave bending moment obtained from it by integration. Research in ship motion analysis and in the statistical description of ocean waves permitted the development of approximate but sufficiently accurate expressions for the characteristic value of extreme wave-induced (vertical) bending moment, $M_{w,c}$. The accuracy and applicability of these expressions is being steadily improved by theoretical analyses and by model and full-scale testing. In the hull girder analysis, the corresponding wave-induced hull girder (vertical bending) stress, σ_w, is obtained from the section modulus formula $\sigma_w = M_{w,c} / Z$.

In the (static) hull module analysis, these wave-induced hull girder stresses, together with the stillwater hull girder stresses, constitute the loads at the ends of the module. The other hull module loads are an equivalent static pressure distribution on the hull representing the actual dynamic pressure distribution and the various gravity loads (cargo, steel weight, etc.) with an allowance for inertia effects if these are considered important.

Of the other slowly varying loads listed above, loads (2) through (5) should also, ideally, be derived from a ship motion analysis. The first three of these loads are, however, highly nonlinear and hence the computation is difficult. Sloshing requires a dynamic analysis for satisfactory accuracy, and Section 4.6.2 provides some information on this. Chapter 10 provides some closed-form expressions for the deformations resulting from sloshing.

For green seas on deck, the downward pressure loading is approximately equal to the hydrostatic pressure for that height of water (i.e., negligible dynamic effect). The horizontal pressure (on deckhouse fronts, etc.) has both static and dynamic components, and it can be represented approximately by an equivalent static pressure; a value of 50 kPa is typical. Wave slap is similar and is of the same order of magnitude. Inertia loads require knowledge of the ship's peak acceleration at the point in question; a precise value requires a ship motion analysis, but an approximate value can be obtained from the experimental data and ship motion studies that are available in the technical literature. Some information on launching, berthing, and ice-breaking loads can also be found in the literature.

2.1.2 Rapidly Varying Loads

There are several types of load that have short periods and that usually require a dynamic response analysis.

2.1.2.1 Slamming

Slamming can occur in three locations: the forward portion of the bottom, especially if it is flat, the bow if it is flared outward, and the stern if it has a large overhang, as in containerships.*

Bottom slamming occurs when, because of pitching and heaving, possibly combined with the occurrence of a wave trough, the ship's bottom emerges from the water and subsequently, because of its relatively flat area and the combined speed of ship and wave surface, undergoes a severe hydrodynamic impact on reentry. The impact is sufficiently rapid and intense to generate a high-intensity pressure pulse on the bottom plating, of very short duration (typically 0.1 to 1 sec), which is often accompanied by a loud booming or slamming sound.

Bow flare (or stern) slamming is the plunging of the upper flared portion of the bow (or overhanging stern) deeper into the water. This is a somewhat more gradual phenomenon (usually without any sound unless the flare is very concave), but it also imparts a relatively sudden and intensive force to the forward (or aft) part of the ship. Circumstantially, the two types of slamming are quite different; bottom slamming requires emergence of the ship's forefoot, whereas bow or stern slamming occur at other times in the ship's pitching cycle. They are independent phenomena, and either can occur without the other. But in spite of their external differences, these phenomena all have the same fundamental cause—the relatively sudden change in the breadth of the immersed cross-section† of the ship as it moves downward. This change of breadth causes a change in the momentum of the surrounding water, and it does so by changing both velocity and mass (i.e., the amount) of water which is involved. The velocity change—both in direction and in magnitude—is easy to understand because the immersed section of the ship is growing in size and changing in shape. But the motion of a body in a fluid involves simultaneous movement of the fluid, and this makes the body behave as if it had additional or added mass; more force is required to accelerate the body. If the submerged body grows rapidly in size, as happens in slamming, the added mass also grows rapidly and resists the body's motion, which causes a rapid increase in local pressure.

Slamming has important effects at two different levels of structure:

- *Hull Girder Level. Bottom* and *bow* slamming cause a sudden vertical acceleration and deflection of the bow and excite flexural vibration of the hull girder, mainly in the fundamental two-node mode but to a lesser degree also in higher modes. This hull girder flexural vibration is referred to as "whipping." As shown in Fig. 2.1, the vibratory hull girder bending stress, referred to as "whipping stress," is of much higher frequency than the wave-induced stress, and is effectively superimposed on it. The period of the fundamental vibration mode excited by slamming is usually in the range from 0.5 to 2 sec. The hull girder aspects of slamming are discussed further in Section 4.6.1, which deals with the nonlinear ship motion and response analysis.
- *Principal Member Level.* The shell plating and its supporting structure (stiffeners, frames, webs, etc.) are subject to high-impact pressure forces, which accelerate and deflect all of this structure and set up vibrations, particularly in the plating. Damage may occur in the form of permanent deformation of plating and other structure. Analysis of this level of slamming response requires detailed information on the pressure

*Some authors prefer to use the term slamming only for bottom slamming and to refer to the second type as bow flare (or stern) impact.

†Strictly speaking, it is the waterplane *area* of the immersed *volume* which is changing; that is, the phenomenon is actually three-dimensional. But since a ship is a generally prismatic body, the flow is idealized as being two-dimensional.

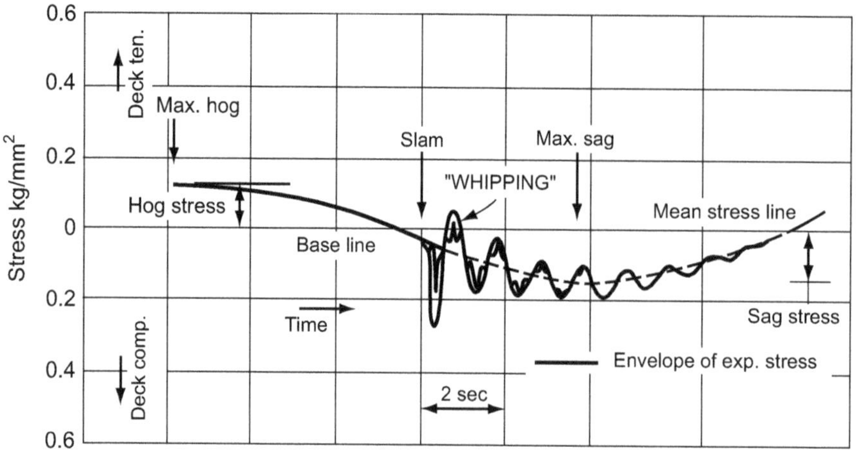

Figure 2.1 Whipping stress resulting from slamming.

distribution, which varies with time and space. The calculation of the permanent deformation is treated in Section 9.5 and Chapter 10.

Bottom slamming is a particularly complex phenomenon. Magnitude and duration of impact pressure depend strongly on, and are sensitive to, the angle and the relative shape of the ship's bottom and the water surface and also on the relative speed of approach. This type of slamming has been the subject of an enormous amount of study and research. Some classical works are Mansour and d'Oliveira (1975), Ochi and Motter (1973), Kawakami et al. (1977), and Evans (1982). Sections 4.6.1 and 4.6.3 summarize more recent methods for predicting the impact pressure and the whipping stresses resulting from bottom slamming.

2.1.2.2 *Forced Vibration*

For the complete hull girder or at the various substructure levels, forced vibrations may be excited by main or auxiliary engines, fluctuating hull pressure loads because of propellers (especially if cavitating), or other sources of excitation. Although these are not severe loads, they can influence the structural design in two ways:

1. By requiring the redesign of a structural member to avoid resonance
2. By requiring thicker plating (to reduce the stress level) to avoid fatigue damage

2.1.2.3 *Other Dynamic Loads*

Ice-impact (at the local or other substructure level)

Underwater explosion (e.g., minehunter)

Certain idealized collision or grounding loads that may be specified by safety authorities for some ship types; the ship structure must be capable of absorbing these without undergoing certain specified type(s) of failure, such as failure of cargo tank bulkheads adjacent to those structures ruptured by collision.

2.1.3 Springing

For most ships, the period of wave encounter is longer than the longest natural period of hull girder vibration. But for ships that are relatively flexible (e.g., some of the very long Great Lakes vessels) the period of two-node vibration is sufficiently long (of the order of 1 sec) that it can be excited by the shorter period components of encountered waves. This phenomenon is known as "springing." Since it depends on the period of encounter, springing is more likely at higher ship speeds. Springing is undesirable for two reasons. First, it produces, at the fore and aft ends of the ship, a noticeable rise and fall of the deck, with a period of the order of 1 sec, which is distracting and even uncomfortable. It seldom lasts for more than a few cycles because it depends on so many coincidental factors, but it is nonetheless an undesirable phenomenon, particularly in ships with accommodation aft. Second, since springing occurs at a higher frequency than ordinary wave-induced bending, it increases the number of stress cycles in the ship's lifetime. Thus, if it were to occur frequently, it would be a contributing factor to hull girder fatigue.

2.2 TYPES OF STRUCTURAL RESPONSE ANALYSIS

2.2.1 Static Only or Static and Dynamic

Figure 2.2 illustrates the principal types of loads and load effects that are involved in the four levels of structural response analysis, from hull girder to local. The figure shows that at each level, the

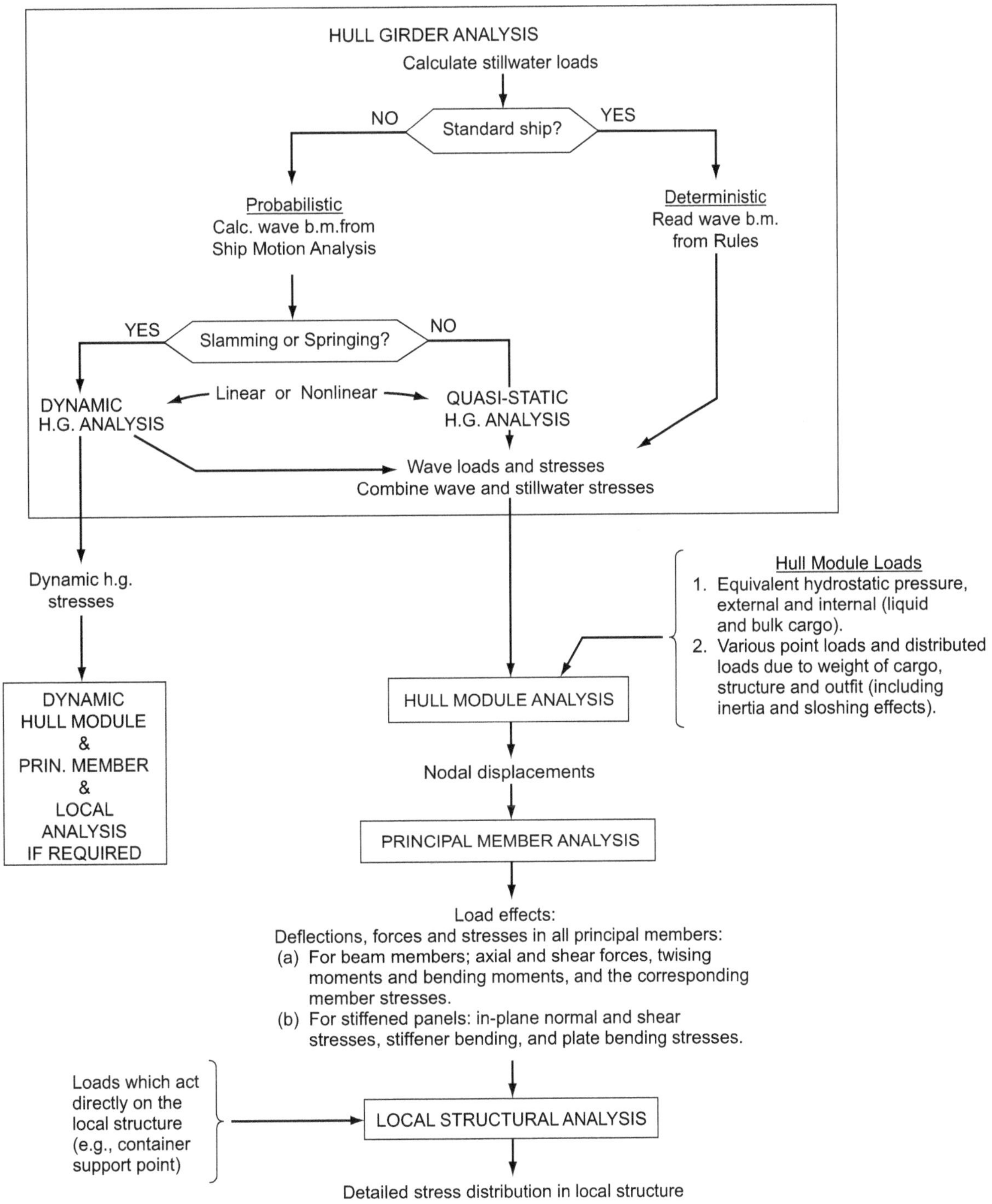

Figure 2.2 Levels of response analysis.

analysis may need to include a dynamic structural analysis. The need depends entirely on whether that level of structure is subject to any significant rapidly varying loads, that is, loads for which the shortest component period is the same order of magnitude or shorter than the longest natural period of that level of structure. Since the latter differs markedly for different levels, with longer periods for larger levels of structure, a load that is slowly varying at a lower level may constitute a rapidly varying load at a higher level (providing that it qualifies as a significant load at that higher level). Springing is an example of this; wave loads have too long a period to cause any excitation at the hull module level, but they can cause vibration of the hull girder if the ship is relatively flexible. An impulsive or high-frequency load, if it is large enough, will cause a vibratory response at several levels. Slamming, in particular, can cause a response at all four levels. But most rapidly varying loads have their principal influence at the local and/or principal member level and are not large enough to induce vibration of an entire hull module or hull girder. Of course, vibration can be transmitted from one member to the next and thus extend over large regions of the ship's structure, but that is still member vibration.

In summary, we may say that at the hull girder and hull module levels a wave-excited dynamic analysis is not required for most ship types. However, a calculation of hull girder natural frequency is nearly always performed. At the principal member and local levels, a vibration analysis may be required if there is some significant and unavoidable source of excitation (propellers, machinery, etc.). In most cases, the only requirement is to calculate the natural frequencies and to design the structure so as to avoid resonance as far as possible.

On the other hand, there are some instances when a dynamic analysis is required at the hull girder and/or hull module levels. For relatively flexible ships, a dynamic hull girder analysis should be performed to check for springing. Also, a dynamic analysis may be required at one or both levels for "high performance" vessels, that is, vessels that have greater structural efficiency and hence are more flexible, particularly at the hull module level, and which at the same time must face significant dynamic loads at these levels because of high-speed operation, unusual hull geometry, or other reasons. Some containerships and some naval vessels fall into this category.

2.2.2 Probabilistic or Deterministic

In addition to the choice between static and dynamic, there are also two different types of response analysis depending on whether an explicit statistical approach is used to define loads and to calculate load effects:

- *Probabilistic**—Characteristic values of load effect are calculated explicitly for the particular structure and load.

- *Deterministic*—Characteristic values are obtained from approximate expressions derived previously by means of a systematic series of probabilistic analyses.

Probabilistic analysis should be used for ships for which the hull girder loads are not already well established, as there are no prederived characteristic values of wave-induced bending moment. If these are well established and characteristic values are available, then deterministic analysis is sufficient. Most types of cargo ships belong to this latter category.

If a probabilistic response analysis is necessary, it is usually required only at the hull girder level because the most uncertain load is usually the wave load. At more detailed levels, there are seldom loads that are so random as to permit probabilistic analysis. Hence the analysis at lower levels is usually deterministic, using the characteristic values of load effects obtained from the hull girder analysis and deterministic estimates for the maximum values of all loads that occur at these lower levels.

For limit state analysis, the situation is reversed. Here, it is the limit value that is uncertain, and the principal source of the (statistical) uncertainty arises at the local level (especially material properties) and the principal member level (especially connections and fittings). Hence, the limit analysis is probabilistic at these levels and produces a characteristic value, $Q_{L,c}$, for the relevant limit value Q_L. At higher levels, the limit analysis is deterministic, combining these characteristic values together to calculate limit values for modes of failure that occur at those higher levels.

2.2.3 Linear or Nonlinear

Finally, wave response analysis may also be classified according to whether it is linear or nonlinear. The former is easier, but for severe sea states there are several sources of nonlinearity: the waves themselves, the governing hydrodynamic equations, and the ship geometry. A linear analysis is based on several simplifying assumptions, of which the principal ones are:

1. The irregular wave surface of the ocean can be represented as the linear sum of a large number of individual regular waves of different heights and frequencies.
2. Hydrodynamic forces on a ship hull can be obtained by considering each transverse section

*Strictly speaking, this should be called semiprobabilistic, but we have seen in Chapter 1 that a fully probabilistic analysis is presently not feasible and is probably not justified. Hence, in this text, the distinction is not required, and from this point on we shall use the terminology defined above.

of the ship separately and combining the results linearly.
3. The wave force acting on each section is linearly proportional to the "emergence" at that section, that is, the difference between the local wave height and the ship's still waterplane is wall-sided.

The accuracy of these assumptions is discussed in Section 4.5. The accuracy of the first two assumptions is generally satisfactory, and the third assumption is valid for ships that are approximately wall-sided in the waterplane region. If this is not so, or if there is any other source of nonlinearity, a nonlinear method of response analysis may be required.

From what has been said previously, it can be seen that wave response analysis can be complex, involving probabilistic, dynamic, and nonlinear aspects. These are discussed in Chapter 4, after first covering the deterministic aspects of hull girder response analysis in Chapter 3. But for purposes of this section, where we are seeking to present an overall view, it may be helpful to list the principal steps involved for the simplest type of wave response analysis, which is based on *linear* ship response theory (see Fig. 4.27).

1. A ship motion response analysis is performed repeatedly for a complete range of deterministic wave loads, each consisting of regular waves of a single specific frequency ω and unit height. This yields the transfer function $|\Phi(\omega)|^2$ for the relevant response (e.g., wave-induced hull girder bending moment, M_w).
2. A typical ocean storm condition is represented by a group of selected ocean wave spectra (i.e., frequency-based distributions of wave height). For each wave spectrum, $S(\omega)$, the corresponding response spectrum [e.g., $M_w(\omega)$] is the product of $|\Phi(\omega)|^2$ and the wave spectrum.
3. The results for the various wave spectra are then combined, according to the proportion in which each spectrum is present in a particular sea state.
4. The analysis is repeated for various sea states (and also for various ship headings and speeds). The resulting short-term response spectra are then combined statistically to obtain long-term characteristic values of load effects, such as the value of bending moment for which there is a sufficiently high probability of nonexceedance in the ship's lifetime.

Price and Bishop (1974) wrote a classical text on the linear probabilistic theory of ship dynamics. Chapter 4 covers this subject thoroughly.

2.3 FURTHER CONSIDERATIONS ABOUT LOADS

Some of the loads on ships have an effect at several levels. For example, in a bulk carrier loaded with a dense cargo, such as iron ore, only about half of the internal volume can be filled because of the high weight of the cargo. Also, in a bulk carrier to be loaded with cargo of normal density, such as grain or coal, each hold should be either full or empty to prevent cargo shifting. To avoid excessive hull girder bending moment, the empty holds must be interspersed between full holds, as shown in Fig. 2.3.

This arrangement causes large shear forces between holds, which cannot be adequately examined by beam theory at the hull girder level, because of the complex interaction between the side shell, double bottom, and transverse bulkhead. Also, as shown in Fig. 2.3, there is a large amount of interbulkhead bending of the double bottom, which cannot be examined in isolation (i.e., by considering only one cargo hold) because the boundary support of the double bottom within each hold is provided largely by the double bottom structure in adjacent holds.

Similarly, in tankers and other liquid bulk carriers, the cargo tanks are either full or empty (with cargo or ballast) to avoid free surfaces and to minimize tank cleaning. Here again, to avoid excessive hull girder bending, full and empty tanks are interspersed, both longitudinally and transversely, in a checkerboard pattern, as shown in Fig. 2.4. This arrangement produces a complex pattern of shear force between all tanks, in both longitudinal and transverse bulkheads, and it also produces alternating hogging and sagging in each combination of three in-line tanks, both longitudinally and transversely. This intertank loading and response is not represented or analyzed adequately by beam theory. In fact, the hull structure of a tanker is outside the scope of beam theory because it has extra "webs"—longitudinal bulkheads and double-hull side structures—of different vertical deflections. Also, these longitudinal bulkheads have approximately the same in-plane rigidity as the transverse bulkheads, and their vertical deflection relative to the side shell depends almost entirely on the vertical shear distortion of the transverse bulkhead. Hence, for all bulk carriers, dry or liquid, the finite element model must include the entire cargo block. Since the bow and the stern blocks have a different internal geometry from the cargo block, their interaction with the cargo block is complex, and the only way to achieve reliable results is to also include them in the finite element model.

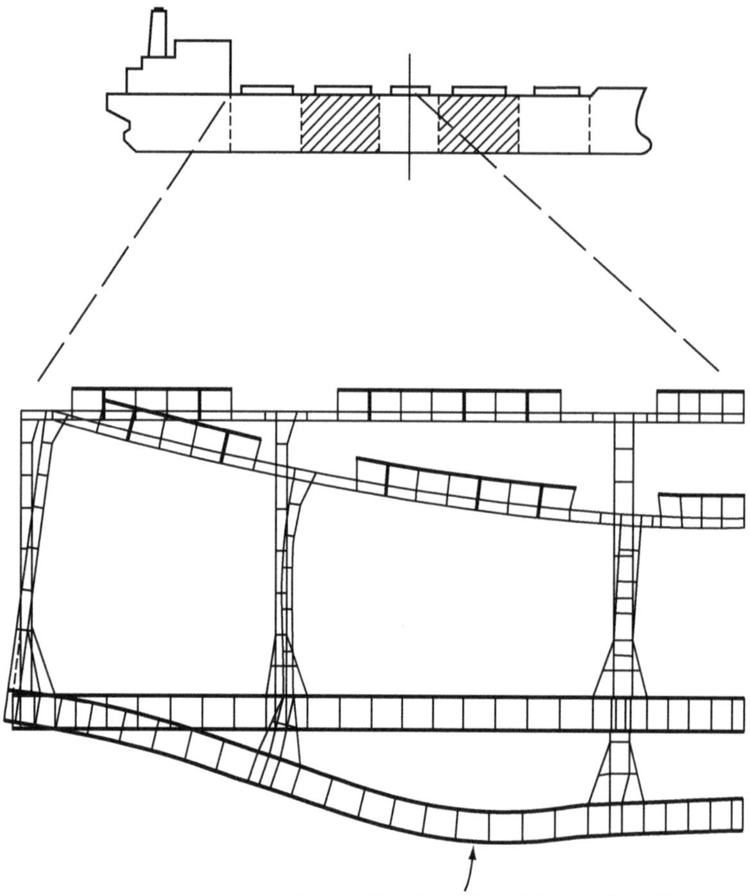

Figure 2.3 Hull module loading and response with a dense cargo.

When the transverse bulkheads are not full depth or are almost entirely absent (e.g., ro-ro ships and train and car ferries), the hull can undergo significant transverse distortion, known as "racking" (Fig. 2.5). For such ships, the structural model should again extend over the full ship length, and at least one loadcase should include a combination of loads representing the "worst case" for racking.

Finally, even if there are relatively rigid transverse bulkheads, the hull girder may be subject to significant rotation if the ship has low torsional rigidity, such as containerships and other ships with large deck openings. Here again, the structural model should extend over the full ship length.

2.4 BASIC TYPES OF STRUCTURAL FAILURE

In contrast to a response analysis dealing with the linear elastic response of a structure to prescribed loads, a limit state analysis seeks to determine those levels and combinations of loads that cause structural failure, both of individual members and of the overall structure. Structural failure is nearly always nonlinear—either a geometric nonlinearity (buckling, or any other large deflection) or a material nonlinearity (yielding and plastic deformation). Also, it is possible for both types of nonlinearity to occur together, and so, in general, limit state analysis is more complex than linear response analysis. In fact, it is probably the most complicated aspect of rationally-based structural design, and for this reason the entire second half of this text—Chapters 9 through 17—is devoted to limit state analysis. Because of its importance, we here give a brief and qualitative review of the basic types of member failure. The next two sections deal with basic aspects of limit state analysis at the overall structure level.

For steel members, the three basic types of failure and their subdivisions are as follows.

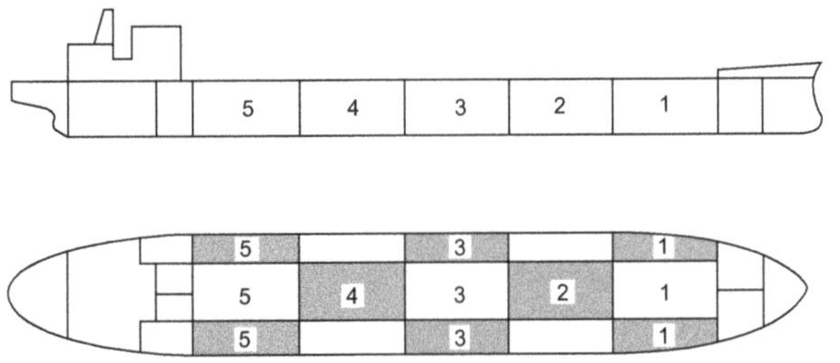

Figure 2.4 Ballasted or partly loaded liquid bulk carrier.

1. Large local plasticity
2. Instability
   - Bifurcation
   - Nonbifurcation
3. Fracture
   - Direct (tensile rupture)
   - Fatigue
   - Brittle

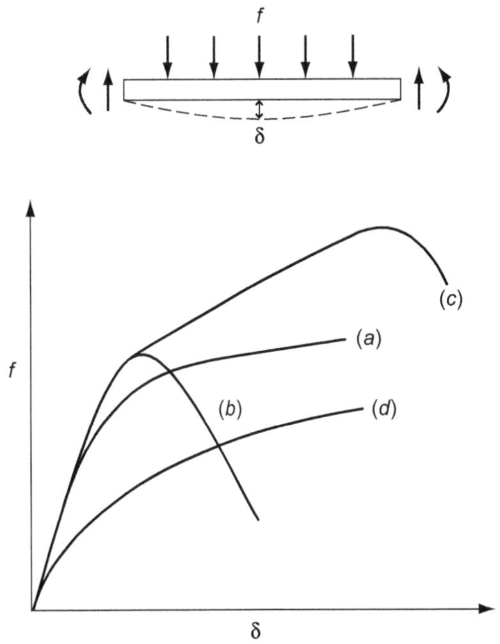

Figure 2.6 Load-deflection curves: (**a**) failure by plastic deformation, (**b**) bifurcation buckling of beams and columns, (**c**) bifurcation buckling of plates, and (**d**) nonbifurcation buckling.

In practice, an individual failure in a structural member often involves a combination of these basic types, particularly the first and second types. For example, for sturdy members, instability is preceded and accompanied by plasticity. Also, the occurrence of local plasticity can seriously diminish the stability of a member and even convert it into a (hinged) mechanism. Nevertheless, each of these three failures is a distinct type of failure, and since we here wish to review only their basic aspects, we shall consider each of them separately. Their interactions are considered in Chapters 9 through 16.

To discuss the basic types of failure and to appreciate the difference between them, it is necessary to examine the relationship between load and deflection for each of them. Figure 2.6 presents a sample of load-deflection curves for individual members, illustrating the variety of shapes the curves have, depending on the type of member and the type of loading and support it receives. The curves are for typical ship structural members, having the following features: 1) they contain local imperfections, including eccentricities and locked-in stresses resulting from forming and welding; 2) they are made of good quality steel, such that premature fracture does not occur and ultimate fracture is preceded by large plastic deformation; and (3) they are of relatively sturdy proportions, such that buckling is not purely elastic but rather involves some yielding. Although the curves differ, they have some basic aspects in common. In most cases, the curve consists of an elastic portion, a region of transition from mainly elastic to mainly plastic behavior, and a plastic region where the slope becomes small, such that the deflection increases greatly for only a small increase in load. The slope of the curve is the instantaneous *stiffness* of the member, indicating its ability to carry additional load. It is also a measure of the *stability* of the member. An unstable member is one that can undergo a large increase in deflection with little or no increase in load.

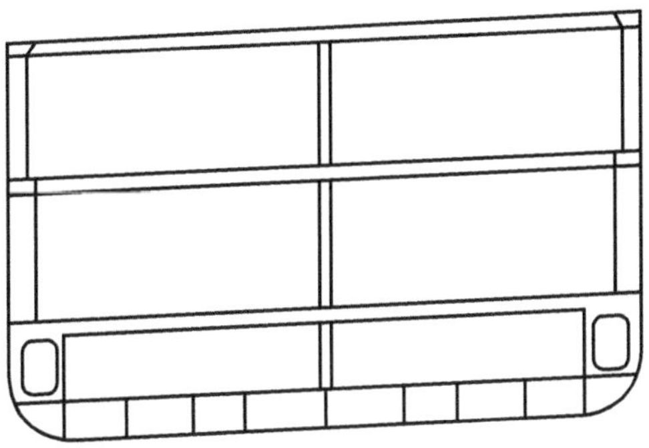

Figure 2.5 Transverse shear or "racking" deformation.

2.4.1 Local Plastic Deformation

We first consider a member that is not susceptible to instability, either because all axial compression loads are small, the member is of sturdy proportions, or it is completely braced and supported. For this type of member, the load-deflection curve resembles curve (a) in Fig. 2.6. In the elastoplastic region of the curve, local regions of plastic deformation occur progressively at the most highly-stressed points, and this gradually decreases member stiffness. In the plastic region of the curve, when this local plastic deformation has grown larger or has occurred at several different points, the stiffness has become quite small and the deflection increases rapidly, eventually becoming so large that the member is considered to have failed; it is no longer fulfilling its main purpose. There is no obvious or precise point on the curve where failure occurs. The failure load (or collapse load or ultimate strength) is usually taken as the load at the beginning of the plastic region, where the deflection first begins to grow rapidly. Although the member's actual loss of function may not occur until the deflection reaches a larger value, the reserve of strength is too small to warrant consideration. This type of structural failure is essentially because of material failure at high stress levels. Hence, this type of failure requires a *stress analysis*.

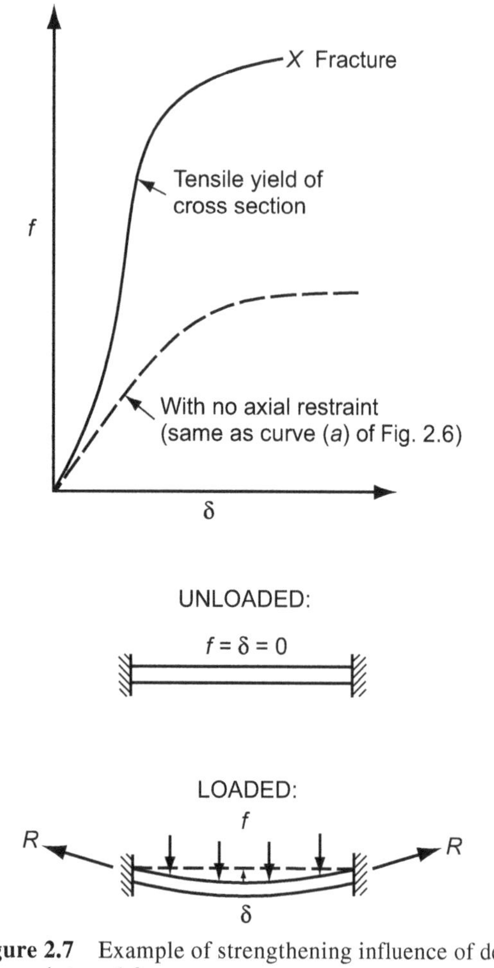

Figure 2.7 Example of strengthening influence of deflection on internal force.

The foregoing discussion assumes implicitly that deflection and deformation do not significantly alter either the geometry of the member or the equilibrium equations. Hence, it is sufficient to perform all calculations using the initial geometry and the initial equilibrium equations. This is a *first-order stress analysis*, and it is by far the most common type of analysis. In some cases, however, the effect of deflection and deformation on the geometry may be important. The effect can cause either a strengthening or a weakening of the member. For example, as shown in Fig. 2.7, a strengthening effect occurs in a laterally loaded beam whose ends are completely restrained against axial deflection ("pull-in") because the relatively small change in geometry resulting from the lateral deflection causes part of the load to be carried by the action of membrane tension instead of by bending. In such a case, stresses in the member have to be calculated with regard to the effect of this deflection on the equilibrium of the member. This is a *second-order stress analysis*.

More often, the change in geometry because of deflection produces weakening effects. For example, in the beam-column of Fig. 2.8, the lateral deflection represents some additional eccentricity and this change of geometry, in conjunction with the axial load f_x, causes additional bending moment in the member. This particular effect is a fundamental characteristic of all geometric *instability* problems; that is, in such problems, the lateral deflection always affects the bending moment in the member and hence it alters the equilibrium conditions. Therefore, geometric instability is essentially a second-order phenomenon and *stability analysis*, which is distinct from stress analysis, is necessarily a second-order analysis.

2.4.2 Bifurcation Buckling

Instability, or buckling, can occur in any member or part of a member that carries an axial or in-plane compressive load. There are two types of buckling: bifurcation and nonbifurcation. The most common example of bifurcation buckling is the buckling of a simple column. The general shape of the load-deflection curve for such a member is shown in Fig. 2.9. For an elastic column, there is some axial load at which an alternative

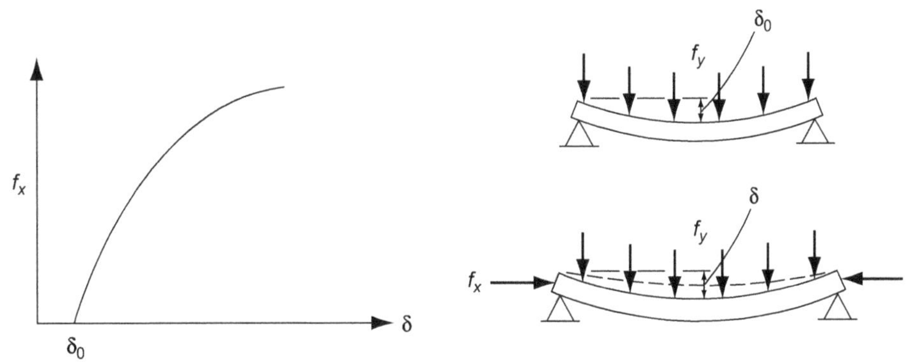

Figure 2.8 Example of weakening influence of deflection on internal force.

equilibrium position exists, corresponding to a bent shape, and this load is the *buckling* or *bifurcation* load of the member. In a typical column that contains some initial eccentricity, the axial load induces bending of the column, and this in turn causes some further lateral deflection. For low levels of load, this effect is negligible, and the initial portion of the curve is approximately linear. But as the axial load increases, the lateral deflection becomes significant and induces additional bending, which in turn further increases the lateral deflection. The result is a relatively rapid loss of stiffness. Also, the large bending stress, in combination with the axial stress, may cause yielding of the compression flange, and this further decreases member stiffness. This loss of stiffness causes the curve to reach a peak and then to fall off. The falling portion of the curve is only obtained if the load on the member is quickly decreased, so that it follows the curve. If the load is not decreased, then, since the member has zero stiffness, the deflection increases suddenly to some large value, that is, the member collapses.

In addition to ordinary flexural buckling, a beam or column of open section, such as an I-beam, can undergo *flexural-torsional buckling* if it is subject to bending in the plane of its web and if it is not constrained to remain in this plane. The bending can be due to either a distributed lateral load, bending moments applied to the ends, or an eccentric axial compressive load. Initially, at low levels of bending load, the member deflects by bending in the plane of the web. However, at a certain value of load, the compression flange may become unstable and buckle laterally, while the tension flange remains stable. Hence, the cross-section undergoes a combination of twist and lateral deflection, as shown in Fig. 2.10. The curve of load versus vertical deflection is illustrated by curve (a) in Fig. 2.11. Alternatively, if sufficient lateral support is provided for the compression flange, such that twist of the cross-section cannot occur, then the load-deflection curve continues upward until failure occurs either by plastic deformation—curve (b)—or, possibly, if there is an axial compressive load and if the member is slender, by flexural buckling in the plane of the web—curve (c).

In many cases, the prebuckling deflection is relatively small, and this permits an idealized approach, known as the *eigenvalue approach*, that simplifies the calculation of the buckling load. In this approach, an *ideal* or *perfect* member and loading conditions are considered. The member is assumed to have no geometric or material

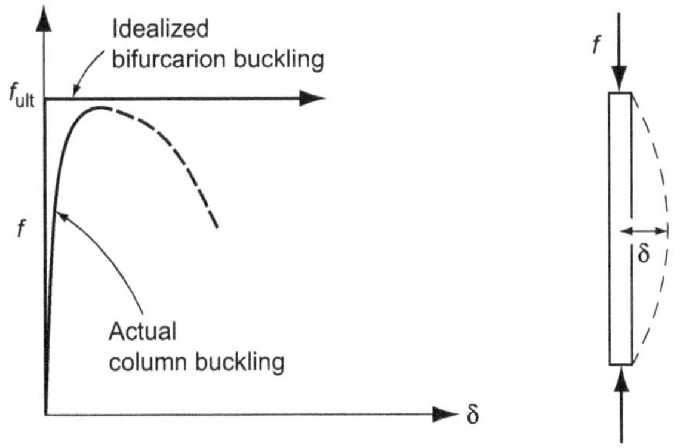

Figure 2.9 Column buckling.

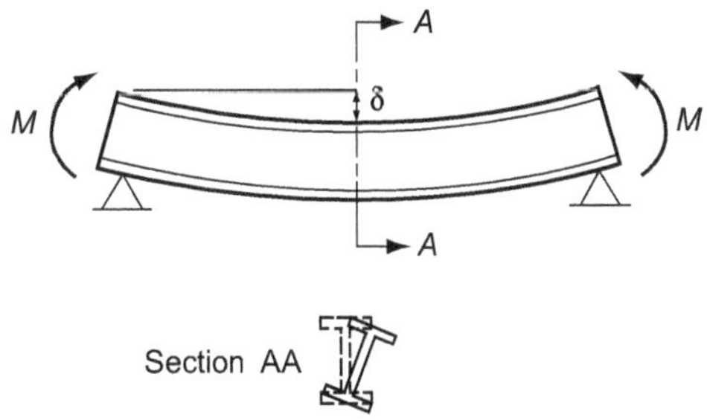

Figure 2.10 Flexural–torsional buckling.

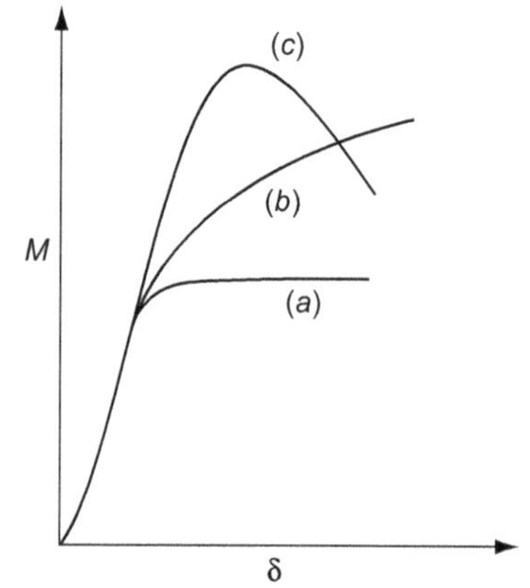

Figure 2.11 Load-deflection curves for a beam-column under various load and support conditions.

imperfections and to be loaded in such an ideal manner that the only deflections that occur prior to buckling are those in the direction of the applied loads. That is, in the case of a concentrically loaded column, the load does not produce transverse deflection until a load of a unique magnitude, the critical load, is reached. At that point, the member has two possible equilibrium states—undeflected and deflected—and it adopts the deflected shape because the strain energy of the undeflected shape just exceeds the strain energy of the deflected shape. The load-deflection behaviour is shown by the heavy line in Fig. 2.9. The essential feature of this idealized form of buckling is that the member can undergo indefinite lateral deflection at no increase in axial load. This feature makes it possible to calculate the buckling load or, more precisely, the *bifurcation load* by means of an eigenvalue analysis.

However, the eigenvalue approach is only appropriate when buckling is of the bifurcation type and when the member fulfills the idealizing assumptions reasonably closely. This is particularly important because in nonideal conditions (eccentric load, eccentric member, local yielding) the buckling load is always *less* than that calculated by an eigenvalue analysis. Hence, if the assumptions are not met to a reasonable degree, the result is a potentially serious overestimate of the buckling load or of the stiffness of the member under a large load.

In plates and stiffened panels, there is some postbuckling reserve strength; that is, the ultimate or collapse load is greater than the buckling load, as illustrated by curve (c) in Fig. 2.6. The buckling analysis and ultimate strength analysis of these members are taken up in various chapters later in the book.

2.4.3 Nonbifurcation Buckling

In nonbifurcation buckling, the load-deflection curve is similar to curve (d) of Fig. 2.6. The lateral deflection commences as soon as the axial load is applied, and it increases progressively, at an increasing rate, causing the member to progressively lose its stiffness from the beginning of the loading until, finally, the stiffness becomes zero. Hence, this type of buckling is a gradual phenomenon. It occurs whenever one of the deflections, which is caused by and which increases with the applied load, is such as to have a weakening effect on the member from the beginning of the loading. This can occur in members (beam-columns, plates, and stiffened panels) that are subject to relatively large lateral loads that increase with axial load. A notable example is a beam-column subject to bending in two planes (biaxial bending).

As can be seen in Fig. 2.6, the load-deflection curve for nonbifurcation buckling resembles that for failure by local plastic deformation. There is no obvious buckling point or any precise peak value of load. As with failure by plastic deformation, the gradual loss of stiffness may produce such large deflections that the member is deemed to have failed on that account. Hence, the failure load is defined as the value corresponding to some limit value of deflection or stiffness.

2.4.4 Static Fracture: Ductile and Brittle

Provision of adequate safety against failure by static fracture is achieved by keeping the stress level throughout each member sufficiently below σ_{UTS}, the ultimate tensile strength of the material. Thus, the constraint against failure by fracture is

$$\gamma\sigma \leq \sigma_{\text{UTS}} \tag{2.4.1}$$

The partial safety factor is chosen according to the degree of uncertainty of σ (if not already factored out) and the importance of the member and the seriousness of the consequence of the fracture. The static fracture constraint is simpler than the constraints against plastic deformation and instability because the limit value, instead of depending strongly and in a complicated manner on the member's dimensions and proportions, is virtually independent of them. Once the material has been selected, the limit value is known.

The term "brittle fracture" refers to the fact that below a certain temperature, the ultimate tensile strength of most steels diminishes sharply. The value of this transition temperature depends almost entirely on the chemical composition of the steel and its metallurgical processes. For ship structures, a good quality steel ductile at prevailing temperatures is absolutely necessary and, in most cases, sufficient to avoid brittle fracture.

2.4.5 Fatigue Fracture

In steel and other metals, a fluctuating stress can initiate microscopic cracks. If the metal is welded or extruded, such cracks are already present. In either case, a fluctuating stress causes these cracks to gradually lengthen until, after a large number of cycles, they become so large that fracture occurs. Fatigue fracture is distinct from static tensile fracture. In fatigue, the most important parameter is the *stress range*, S, which is the total (peak to trough) variation in the cyclic stress, as shown in Fig. 2.12. Since fatigue damage (crack growth) is cumulative, the occurrence of fracture depends on the magnitude and duration (number of cycles) of the individual cyclic loads acting on the structure throughout its life. For an individual cyclic load of constant amplitude, S, the least number of cycles, N, required for fatigue fracture is established experimentally for each type of steel and other materials. This information is usually presented in "S-N diagrams" of the type shown in Fig. 2.13, where the horizontal distance to the sloping line is the *fatigue life* (number of cycles to failure) at that level of S for a certain type of structural specimen (geometry, direction of weld, and quality of weld). Each line is obtained by testing to failure a series of identical structural specimens at different levels of S and then performing linear regression analysis on a log-log plot of the data with, say, a 95% confidence limit. Thus each line (or "S-N curve," as it is commonly called) represents an exponential relationship of the form

$$S_N = \left(\frac{C}{N}\right)^{1/m} \tag{2.4.2}$$

which in logarithmic form becomes

$$\log N = \log C - m \log S_N$$

where N = number of cycles to failure for a constant amplitude stress range S_N
m = the negative slope of the log-log plot of the S-N curve
S_N = the constant amplitude stress range for failure at N cycles
$\log C$ = the life intercept of the S-N curve

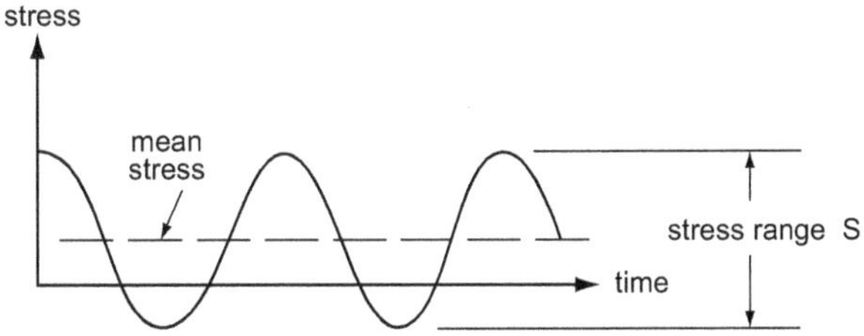

Figure 2.12 Stress range.

For most materials, there is a threshold level of stress range, S_∞, below which fatigue damage does not occur, regardless of the number of cycles. This is commonly referred to as the *fatigue limit.**

In a ship structure, there are three main sources of cyclic stresses: wave-induced loads, especially bending of the hull girder; the alternation between loaded and ballasted conditions, a situation that occurs in tankers and some other ships; and mechanical sources, such as the engine and the propellers. The number of wave bending cycles in a ship's life (say, 20 years) is of the order of 10^8.

In general, fatigue failure is prevented by controlling the cyclic stress amplitude, and in most cases an efficient way to control stress is to either increase the local scantlings and/or modify the local geometry so as to reduce stress concentrations, eccentricities, and discontinuities. Hence, in the overall process of ship structural design, the prevention of fatigue falls mainly within the scope of detail design. However, for all cyclic stresses that are not locally controllable, the matter of fatigue must be dealt with at the preliminary design stage. The principal example is the wave-induced hull girder bending stress, σ_w, which is proportional to section modulus and is relatively unaffected by local changes in scantlings. Therefore, one of the most important tasks in preliminary design is to ensure that σ_w

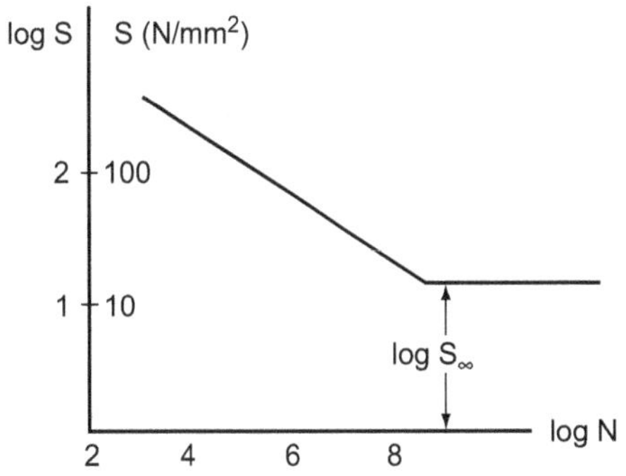

Figure 2.13 Typical S-N diagram.

*A better term would be "fatigue threshold," since fatigue only commences at this level of S.

is sufficiently small to not cause fatigue in any portion of the hull girder, taking into account all unavoidable and non–locally-treatable influences. One way of preventing fatigue fracture is to prevent the occurrence and accumulation of fatigue damage by keeping cyclic stresses below the fatigue limit. This approach is adopted when the number of stress cycles is extremely large, as in mechanically-produced stresses. Such stresses are usually local and, therefore, relatively easy to control. However, this approach is inefficient (if not impossible) for more extensive stresses arising, for instance, from full/ballasted cargo variations and hull girder bending. Instead, it is necessary to design ships in such a way that over their lifetime they can sustain some cumulative fatigue damage without appreciable risk of fracture. To do this, it is necessary to know how individual amounts of fatigue damage caused by various sequences of cyclic stresses of different magnitude, duration, and mean value interact and accumulate to finally cause fracture. This is an extremely complex question.

There are two methods to deal with fatigue: one is based on fracture mechanics and the other based on fatigue tests together with the hypothesis of linear damage accumulation. The fracture mechanics method is more detailed, examining crack growth and calculating the number of load cycles that are needed for small initial defects (which are always present in welds and in the structure adjacent to welds) to grow into cracks large enough to cause fracture. The growth rate is proportional to the stress range, expressed in terms of a *stress intensity factor* K that accounts for stress magnitude, weld and joint geometry, and the current crack size. The equation for crack growth rate is of the form

$$\frac{da}{dN} = C\,(\Delta K)^m \qquad (2.4.3)$$

where a is crack depth, C and m are crack propagation parameters associated with fracture mechanics, and ΔK is the range of K corresponding to the stress range.

The other method is based on fatigue test data (S-N curves) together with the hypothesis, commonly known as Miner's Rule, that fatigue damage accumulates linearly.* According to this hypothesis, total fatigue life under a variety of stress ranges is the weighted sum of individual lives at constant S, as given by the S-N curves, with each being weighted according to the fractional exposure to that level of stress range. To apply this hypothesis, the long-term distribution of stress range is replaced by a stress histogram, consisting of a convenient number of constant amplitude stress range blocks, ΔS_i, and a number of stress cycles, n_i. The constraint against fatigue fracture is then expressed in terms of a nondimensional damage ratio, η:

$$\sum_{i=1}^{B} \frac{n_i}{N_i} \le \eta_L \qquad (2.4.4)$$

where B = number of stress blocks
ni = number of stress cycles in stress block i
N_i = number of cycles to failure at constant stress range S_i
η_L = limit damage ratio

The limit damage ratio, η_L, depends on the maintainability, that is, the possibility for inspection and repair, as well as the importance of the particular structural detail. For important details exposed to seawater, a typical value of η_L is 0.3 if there is good access and maintainability and 0.1 if not.

As stated above, fatigue relates mainly to structural details, and fatigue analysis and prevention is primarily a part of detail design. But, since it is so important, it is covered in Chapter 17.

2.5 OPTIMIZATION OF LARGE STRUCTURES

In nonlinear optimization, the amount of computation increases exponentially with the number of design variables to be optimized simultaneously. A direct solution to a typical ship structure optimization problem requires a prohibitively extensive computational effort. The computation can be substantially reduced if the overall problem is subdivided into a number of subproblems. In fact, if each structural member were to be optimized separately, the total amount of computation would be relatively small. But, in a large structure such as a ship, member-by-member optimization is neither desirable nor possible. It is undesirable because the objective function is a nonlinear function of the design variables, and member-by-member optimization would not allow any tradeoffs between members, which is where a large part of the benefits of optimization come from. In any case, member-by-member

*Originally proposed by Palmgren (1924) and later by Miner (1945); it is usually known under the latter's name because he developed it on a logical basis by considering the work done during each loading cycle.

optimization is not possible, because some limit states relate to the overall structure and cannot be dealt with at the member level.

In a structure as large as a hull module, there are typically 100 to 200 design variables, and a nonlinear optimization problem of this size requires too much computational effort to be solved as a single problem, even with a rapid and efficient optimization algorithm such as one of the sequential linear programming methods discussed in Section 1.2. Therefore, one of the prerequisites for rationally-based ship structural design is a method or strategy to subdivide the overall problem while still retaining true overall optimization and the capability of dealing with overall constraints. In this section, we present a brief outline of a *dual level* strategy that meets these requirements.

2.5.1 Subdivision of the Overall Optimization Problem

Subdivision of the overall optimization problem is based on the concept of a submodule, which is a region of a structure where a sufficient number of the scantlings are linked, either by fixed structural geometry or by explicit constraints linking two or more scantlings, such that the region forms a logical entity from an optimization point of view. The characteristics that most clearly make a region suitable as a submodule are geometric uniformity and identical, repetitive structural members. Such characteristics are frequently imposed on portions of structure to simplify the design and to thereby gain increased economy and efficiency in nearly all aspects of the ship's existence: design, fabrication, outfitting, operation, and maintenance. Modularity of cargo is another common reason to impose structural uniformity. Regardless of the reason, such uniformity provides an opportunity to reduce the amount of computation, and the choice of submodules should reflect and take full advantage of uniformity. In most cases, the choice is obvious and straightforward because it is a natural extension of the designer's decisions regarding geometric uniformity and member repetition. In making such decisions, the designer, in effect, imposes various constraints on the design—constraints requiring that some scantlings or geometric features in one member or region bear a fixed relationship (in most cases identical constraints) to those of another member or region.* For these constraints to be properly accommodated, the optimization problem should extend over and include all of the linked members or regions. Hence, the physical size of the submodule depends mainly on the extent of the linking.

Figure 2.14 shows the most common type of submodule, which, for simplicity, shall be called a "strake." It consists of a longitudinal row of panels, transverse frame segments, and, if applicable, longitudinal girder segments.† In most ships, these members are uniform and repetitive in the longitudinal direction over large distances—nearly always over the full length of a hold and often over several hold lengths (or compartments for a naval vessel). Local changes in geometry or scantlings (cutouts, reinforcements, etc.) are disregarded because they are dealt with in the detail design. The length of the uniformity is taken as the length of the submodule. Because the panels, frame segments, and girder segments in the submodule are identical, the total number of design variables is small; it is the same number as for just one member of each type. In Fig. 2.14, the strake submodule has 14 design variables: 6 for the stiffened panel, 4 for the frame segments, and 4 for the girder segment (if the stiffener is a rolled section, not all of its dimensions are design variables).

Submodules also reduce the number of limit values that have to be calculated because identical members have the same value (if their pattern of internal load effects is the same). Hence, there is just one constraint for each mode of member failure, and to formulate each constraint it is necessary only to scan for and utilize the worst combination of internal load effects.

As mentioned above, the amount of computation increases exponentially with the number of design variables. Experience with the MAESTRO program showed that the total amount of computation (for the complete hull module) remains reasonable as long as the number of design variables in each submodule does not exceed 20. Thus, a simple and general rule to follow is to make submodules as large as possible, subject to the limit of 20 design variables. If a proposed submodule has more than 20 design variables, it should be divided into two submodules or, alternatively, some further uniformity should be specified.

Not all scantlings in a submodule need to be uniform. It frequently happens that in a particular

*For simplicity, the discussion here is in terms of identity constraint, but the principles apply to any simple and direct type of linking, such as a fixed proportion of scantlings or geometry between members.

†In ship construction terminology, a "strake" is a lengthwise strip of plating (e.g., keel strake, sheer strake, etc.). The distinction between this "strake" and a strake submodule will be clear from the context.

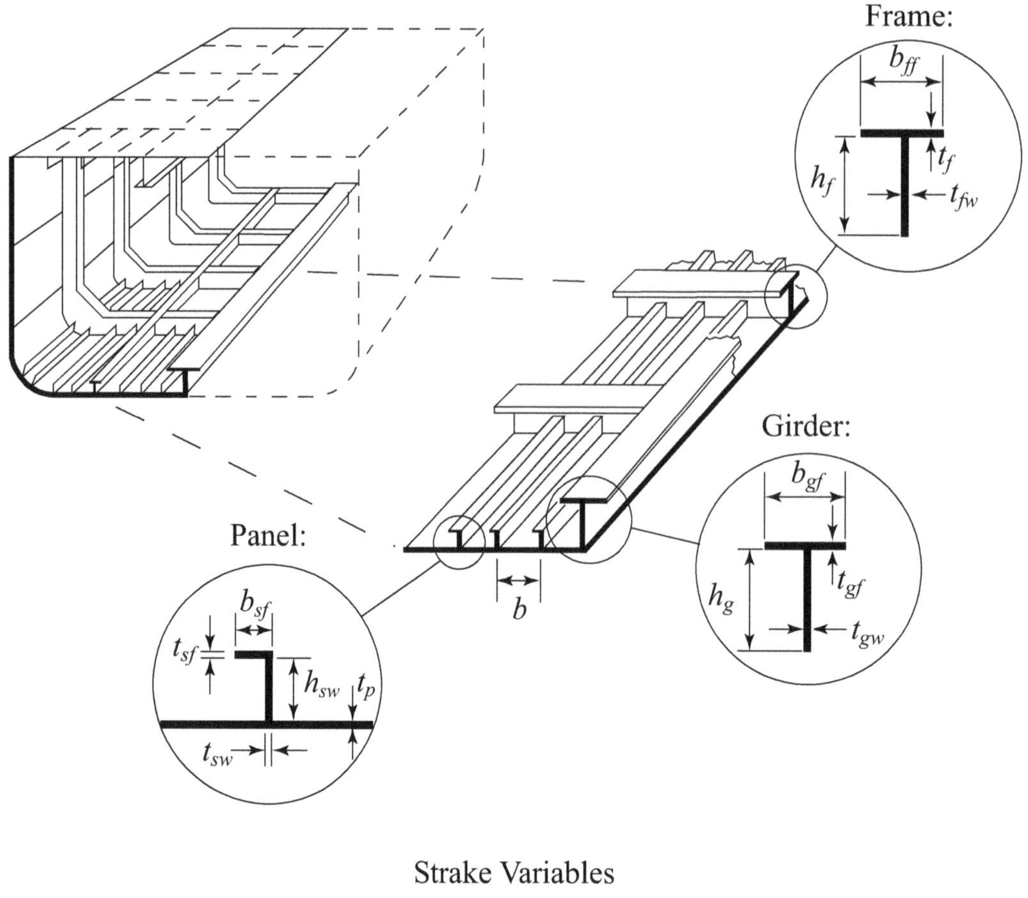

Figure 2.14 Strake submodule.

portion of structure, it is desired that most members of a certain type should be identical, but there are a few members that ought to be allowed to differ, perhaps because they carry special loads. For example, Fig. 2.15 illustrates a hull module that corresponds to one complete cargo hold with, say, eight transverse frames, the second and seventh of which coincide with the forward and aft ends of the hatch. In each of the side and deck strakes, the second and seventh frame segments carry larger loads than the other frame segments. If these two segments were included in the strake, the final optimum scantlings of the other six frame segments would be dictated by the requirements of these two. Therefore, the two hatch end segments should be defined as a separate submodule. The fact that they are not continuous is not a problem, because the grouping of members in a submodule is done for optimization purposes only and not for analysis. Hence, there is no need for a submodule to be a single or continuous portion of structure.

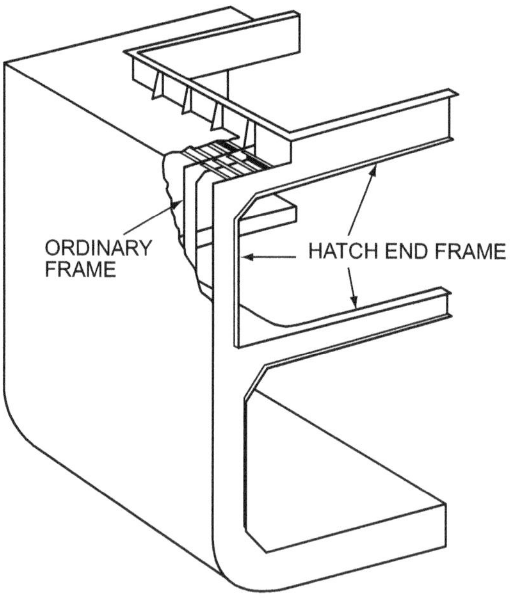

Figure 2.15 Variations in frame size.

2.6 REFERENCES

Evans, J. H. (1982). Preliminary design estimation of hull girder response to slamming. *Trans. SNAME*, Vol. 90, 55–83.

Kawakami, M., Michimoto, J., and Kobashi, K. (1977). Prediction of long-term whipping vibration stress due to slamming of large full ship in rough seas. *International Shipbuilding Progress*, Vol. 24, Issue 272, 83–110.

Mansour, A., and d'Oliveira, J. M. (1975). Hull bending moment due to ship bottom slamming in regular waves. *Journal of Ship Research*, Vol. 19, Issue 2.

Miner, M. A. (1945). Cumulative damage in fatigue. *Journal of Applied Mechanics*, Vol. 12, A159.

Ochi, M. K., and Motter, L. E. (1973). Prediction of slamming characteristics and hull responses for ship design. *Trans. SNAME*, Vol. 81, 144–176.

Palmgren, A. (1924). Die Lebensdauer von Kugellagern (Life expectancy of roller bearings). *ZVDI*, Vol. 68, 339.

Price, W. G., and Bishop, R. E. D. (1974). *Probabilistic theory of ship dynamics*. London, UK: Chapman and Hall.

CHAPTER

THREE

HULL GIRDER RESPONSE ANALYSIS — PRISMATIC BEAM

Owen Hughes
Professor, Virginia Tech
Blacksburg, VA, USA

In this chapter we consider the overall or primary level of ship structural loading and response, in which the ship is idealized as a hollow thin-wall box beam, referred to as the "hull girder." At this level of consideration we can make several simplifying assumptions and approximations, the principal one being that the hull girder acts in accordance with simple beam theory. In Section 3.8, some corrections for these approximations will be given. For the sake of clarity and overall perspective, the principal assumptions are listed hereunder.

1. There is only one independent variable, longitudinal position, and loads and deflections have only a single value at any cross section.

2. The hull girder remains elastic, its deflections are small, and the longitudinal strain due to bending varies linearly over the cross section, about some transverse axis of zero strain (neutral axis).

3. Dynamic effects may be either neglected or accounted for by equivalent static loads. Hence static equilibrium may be invoked.

4. Since the bending strain is linear, the horizontal and vertical bending of the hull girder may be dealt with separately and superimposed. Since they are similar, and since vertical bending (bending in the vertical plane of the hull girder) predominates, we shall deal mainly with it.

3.1 BASIC RELATIONSHIPS: LOAD, SHEAR FORCE, BENDING MOMENT

As shown in Fig. 3.1, overall static equilibrium requires that the total buoyancy force equals the weight of the ship (considered as a force; e.g., megaNewtons (MN), *not* tonnes) and that these two vertical forces coincide; that is, the longitudinal center of buoyancy (l.c.b.) must coincide with the longitudinal center of gravity (l.c.g.). The notation to be used herein is shown in Fig. 3.1. Using this notation the first requirement is

$$\rho g \int_0^L a(x)dx = g\int_0^L m(x)dx = g\Delta \qquad (3.1.1)$$

where $a(x)$ = immersed cross-sectional area
$m(x)$ = mass distribution (mass per unit length)
ρ = mass density of sea water (or fresh water, if appropriate)
g = gravitational acceleration
Δ = displacement.

The factor g is retained on both sides to emphasize that it is forces that are involved.

Similarly, equilibrium of moments requires that

$$\rho g \int_0^L a(x)x\,dx = g\int_0^L m(x)x\,dx = g\Delta l_G \qquad (3.1.2)$$

where l_G = distance from origin to l.c.g.

3.1.1 Application of Beam Theory

In elastic, small-deflection beam theory the governing equation for the bending moment $M(x)$ is

$$\frac{d^2M}{dx^2} = f(x)$$

in which the right hand side, $f(x)$, is the loading on the beam, expressed as a distributed vertical force.

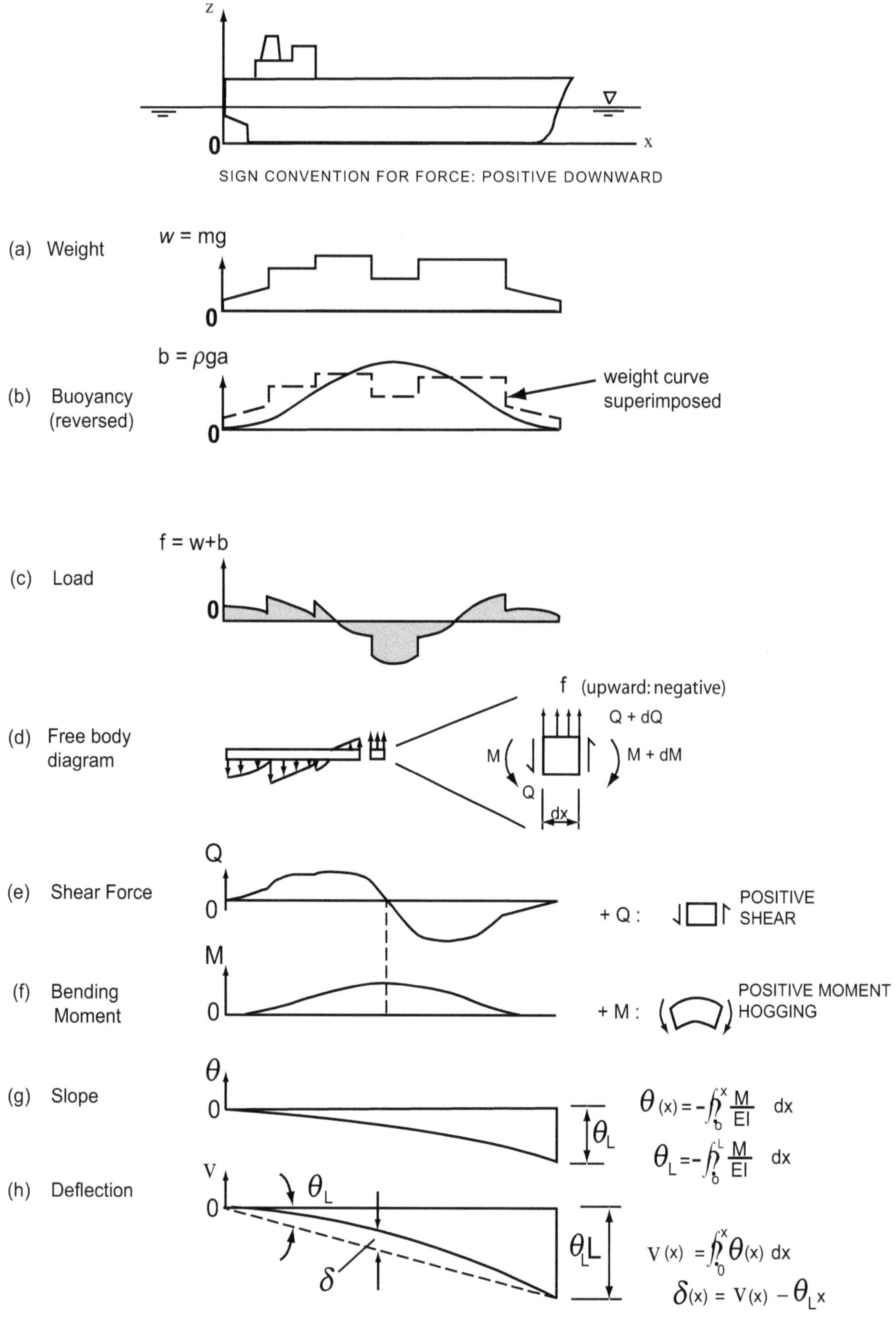

Figure 3.1 Summary of hull girder bending.

Like civil engineering, naval architecture is a very old branch of engineering.*[1] In land-based structures the dominant force is the structure's weight, which always acts downward. Hence in early times forces were taken as positive downward, and ship loading still follows this (nowadays backward) convention. Thus for a ship the loading $f(x)$ is the net resultant of the positive weight force $w(x)$ and the negative buoyancy force $b(x)$, as shown in Fig. 3.1*c*. The solution for $M(x)$ requires two integrations. The first yields the transverse shear force $Q(x)$, and is obtained by imposing vertical force equilibrium of a differential element considered as a free body, as shown in Fig. 3.1*d*.

$$Q + f dx - Q - dQ = 0$$

or

$$f = \frac{dQ}{dx} \tag{3.1.4}$$

from which

$$Q(x) = \int_0^x f(x)\,dx + \not{C}^{\,0} \tag{3.1.5}$$

For ships the constant of integration is always zero because the hull girder is a "free-free" beam, with zero shear force at the ends: $Q(0) = 0$.

As shown in Fig. 3.1*d* equilibrium of moments (say, about the right hand end of the element, and positive clockwise) yields

$$M + Q\,dx + f\,dx\frac{dx}{2} - M - dM = 0$$

The dx^2 term is of second order and therefore

$$Q = \frac{dM}{dx} \tag{3.1.6}$$

from which

$$M(x) = \int_0^x Q(x)\,dx + \not{C}^{\,0} \tag{3.1.7}$$

* The first "engineers" were mainly engaged in military applications – bridges, siege towers, catapults – and those who designed other large land-based structures (churches, castles, etc.) were called "architects" (which means master builder). Thus the designers of ships became known as naval architects. Eventually, as structures became more varied and more complex (not just buildings) the term "civil engineer" emerged.

The sign conventions for shear force and bending moment are shown in Fig. 3.1*e*, *f*. Shear force at any point is positive if the integral, or net accumulation, of load up to that point is positive. If we define a "positive face" as the cross section that one sees when looking in the positive x-direction, then the shear force is positive downward when acting on a positive face, and positive upward when acting on a negative face. As shown in Fig. 3.1*e*, an alternative way of expressing this is to say that positive shear causes counterclockwise rotation of an element.

In a similar manner, the bending moment at any point is positive if the integral, or net accumulation, of shear force up to that point is positive. It can easily be shown that with this definition, positive bending moment corresponds to beam curvature that is convex upward. This condition or state of bending is referred to as "hogging," and the opposite state, concave upward, is referred to as "sagging."

In discussing a sign convention for shear force and bending moment it is important to realize that we are actually considering each of them as a state of loading, rather than a specific force or couple. These two states of loading have their own sign convention, quite apart from the sign convention adopted for specific forces and couples. Thus, for example, the differential element in Fig. 3.1*d* is in a state of positive shear, even though one of the two end forces acting on it is downward.

3.1.2 Characteristics of Shear Force and Bending Moment Curves

As we have seen, both the shear force and the bending moment must be zero at the ends. Since the load is the derivative of the shear force Q, a point of zero load corresponds to a local maximum or minimum value of Q, as shown in Fig. 3.1. In most cases the loading is approximately similar forward and aft of amidships. Under these conditions the shear force is approximately asymmetric, passing through zero somewhere near amidships and having maximum values, positive and negative, near the quarter points.

Similarly, since the shear force and bending moment are related by $Q = dM/dx$, a point of zero Q corresponds to a local maximum or minimum value of bending moment, as shown by the dashed lines joining Fig. 3.1*e* and *f*. In general, therefore, the bending moment, will be a maximum, positive or negative, near amidships, but if the loading is very asymmetric the maximum bending moment may be some distance from amidships. Since shear force is zero at both ends, the bending moment curve will have zero slope at both ends, and its value will usually be small forward and aft of the quarter points.

3.1.3 Distinction between Still Water Loading and Wave Loading

In order to calculate the load on the hull girder, it is necessary to first calculate both the distributed buoyancy force and the distributed weight force. In regard to the former, it is useful to distinguish between the buoyancy force in still water and the additional, and quite different, buoyancy force that occurs as a result of waves. The still water buoyancy is a completely static quantity, and it depends mainly on the shape of the immersed hull. It therefore most logically forms part of the hydrostatic calculations. The additional buoyancy force due to waves is markedly different from the buoyancy force in still water, being essentially both dynamic and probabilistic. Therefore, in order to simplify the analysis, the buoyancy distribution in waves is calculated separately and is superimposed on the static and deterministic still water buoyancy force.

In order to calculate the still water buoyancy distribution, the location of the still waterline of the vessel must be determined from the two overall equilibrium requirements of (3.1.1) and (3.1.2). Therefore, it is also necessary to know the weight (i.e., mass) distribution $m(x)$, or at least the overall Δ weight and the location of the l.c.g. Thus, once the lines of a ship have been specified, the still water buoyancy is fixed and calculable, and hence the still water load, shear force, and bending moment depend entirely on the weight distribution. We therefore begin by examining some steps and techniques for estimating this distribution.

3.2 ESTIMATION OF WEIGHT DISTRIBUTION

The calculation of the longitudinal distribution of weight or mass $m(x)$ is a difficult process, partly because $m(x)$ is made up of discrete items rather than being a continuous and regular curve, and partly because at the design stage many of the individual weights are known only approximately. The calculation cannot be as readily or as thoroughly automated as for the buoyancy distribution, but it can be greatly facilitated and even largely automated by means of a systematic approach and by the use of suitable approximate methods.

The weights in a ship fall into two main categories: those that are relatively unchanging, such as the ship's own structural weight; and those that do change, such as cargo, fuel, stores, and ballast—see Fig. 3.4. The first group constitutes the "lightweight" of a ship, that is, the weight when it is without cargo, fuel, and so on (this condition is referred to as the "lightship" condition). The second group is called the "deadweight" (equivalent to "payload" in modern terminology). The deadweight changes with each different cargo loading, and hence there are usually several loading conditions that need to be investigated. The two most common conditions are "full load" and "ballast."

In general, the information regarding weights is of a discrete nature and must be gathered together and entered into a "Table of Weights" or some other suitable form of information storage. In most cases, the following information is specified for each logically distinct item:

1. Total weight.
2. Vertical* and longitudinal center of gravity (l.c.g.)
3. Longitudinal extent.
4. The type of distribution over this extent.

Once this information is available the rest of the calculations can be done by computer: the lightweight curve and, for each loading condition, the deadweight curve, the total weight curve, the displacement, and the l.c.g. Knowing the latter two items makes it possible to calculate the still water buoyancy curve for each loading condition, usually by means of a comprehensive hydrostatics program. Then, for each loading condition the appropriate buoyancy and weight curves are combined to get the load curve, and this is then integrated twice to get the shear force and bending moment curves.

In specifying the extent and distribution of individual weights, it is helpful and even necessary to use some approximations and idealizations. Nearly all items can be represented in terms of one or more of three basic types of distribution: point, uniform distribution, and trapezoidal distribution. Also, for cargo and ballast, an alternative approach is possible. For these items the weight per unit length is related to the cross-sectional area of the relevant cargo or ballast space, and their weight distribution may be taken as the product of the sectional area curve of the relevant space times the mass density of the cargo or ballast.

*For calculation of stability and ship motions.

The idealization of a weight as a trapezoidal, uniform, or even as a point mass, does not introduce any appreciable error into the calculation of shear force and bending moment as long as it is done reasonably and carefully. After all, both of these quantities are themselves idealizations: they arise from and are a measure of the overall behavior of the hull girder, not the local response; they have only a single value at any cross section. Moreover, they are obtained by integrating the load, and integration is a smoothing process.

Typical examples of point loads are machinery (one point load at each foundation point), masts, winches, and transverse bulkheads. Examples of uniform loads are hull steel within the parallel midbody and cargo, fuel, ballast, and other homogenous weights within prismatic spaces. Outside of the parallel midbody and particularly toward the ends of the ship, a trapezoidal distribution is appropriate, although even here some items can be accurately represented as uniform loads, such as superstructure. For a trapezoid, say of length l, the relevant information may be specified in two different ways: either as total mass, M_0, with a specified position of center of gravity within this length (say a distance $\bar{x}$ from the center; see Fig. 3.2) or in terms of mass per unit length at the forward and after ends: m_f and m_a. The formulas for converting from one form to another are:

$$\left.\begin{aligned} \bar{x} &= \frac{l}{6}\left[\frac{m_f - m_a}{m_f + m_a}\right] \\ M_0 &= \frac{l(m_f + m_a)}{2} \end{aligned}\right\} \quad (3.2.1)$$

and

$$\left.\begin{aligned} m_a &= \frac{M_0}{l} - \frac{6M_0\bar{x}}{l^2} \\ m_f &= \frac{M_0}{l} + \frac{6M_0\bar{x}}{l^2} \end{aligned}\right\} \quad (3.2.2.)$$

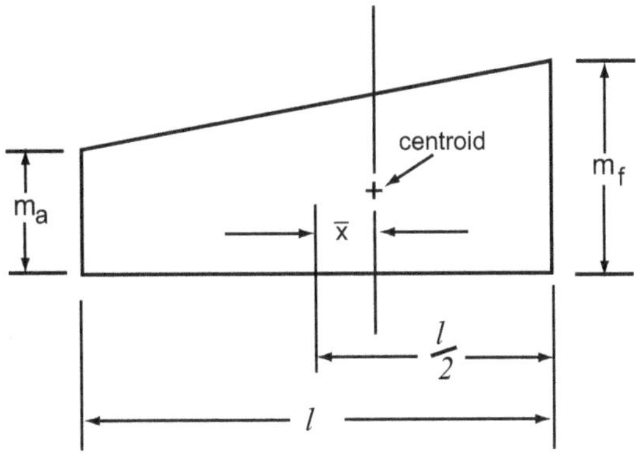

Figure 3.2 Trapezoidal representation of a weight.

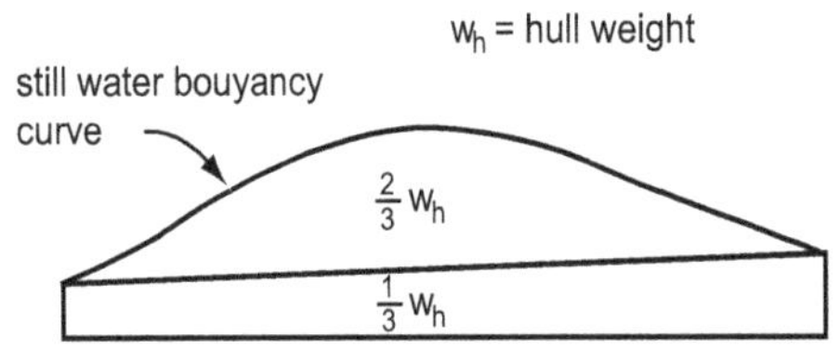

Figure 3.3 Approximation for hull weight distribution.

One of the major items of the weight distribution is the hull itself, and this will sometimes be required before the structural design of the hull has been completed. A useful first approximation to the hull weight distribution is obtained by assuming that two-thirds of its weight follows the still water buoyancy curve and the remaining one-third is distributed in the form of a trapezoid, with end ordinates such that the center of gravity of the entire hull is in the desired position, as shown in Fig. 3.3. An alternative approach is to use a uniform weight distribution over the parallel midbody portion and two trapezoids for the end portions, with end ordinates again chosen such that the l.c.g. of the hull is in the desired position.

A sample weight curve is given in Fig. 3.4. By commencing with "lightship" components (hull steel, machinery, equipment, and outfit), the designer can determine the distribution of variable weight items (cargo, fuel oil, ballast, etc.) for those loading conditions that will be most serious in both hogging and sagging.

The most significant loading condition is the "full and down" condition, with the ship loaded with sufficient cargo to float at its "loadline," the water line corresponding to the minimum permissible freeboard. In some cases, beam and draft restrictions may be the overriding constraints. For simplicity, the cargo is usually assumed to be homogeneous and of a density that just produces the "full and down" condition. Lighter-density cargoes reduce displacement (and hence the loading), whereas heavier densities permit flexibility, and hence optimization of load distribution, as the cargo weight remains essentially limited by the statutory draft limitation. Regardless of whether the maximum bending moment in the

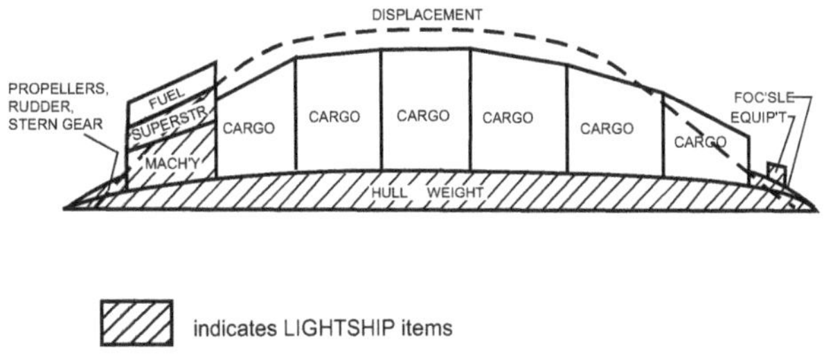

Figure 3.4 Typical weight distribution.

homogeneous cargo case is hogging or sagging, it can be reduced by concentrating the more dense cargo either amidships or in the ends, respectively.

Because the cargo is the largest item of weight and because there are so many possible variations in its distribution, there are often some distributions and combinations that would cause excessive values of bending moment and that therefore must be avoided. This is particularly the case with bulk carriers in which, for various reasons (e.g., tank cleaning, avoidance of free surface or the shifting of dry bulk cargo, very dense cargo such as iron ore) it is preferable to have the cargo holds or tanks either completely full or completely empty. Given such extreme differences it is important that they be spread out and interspersed, rather than grouped together, because the latter would give excessive shear force and/or bending moment, as shown in Fig. 3.5. Therefore, in addition to restrictions arising from stability requirements, the loading manual of a cargo ship may also contain restrictions on the permissible cargo combinations and distributions, arising from longitudinal strength requirements.

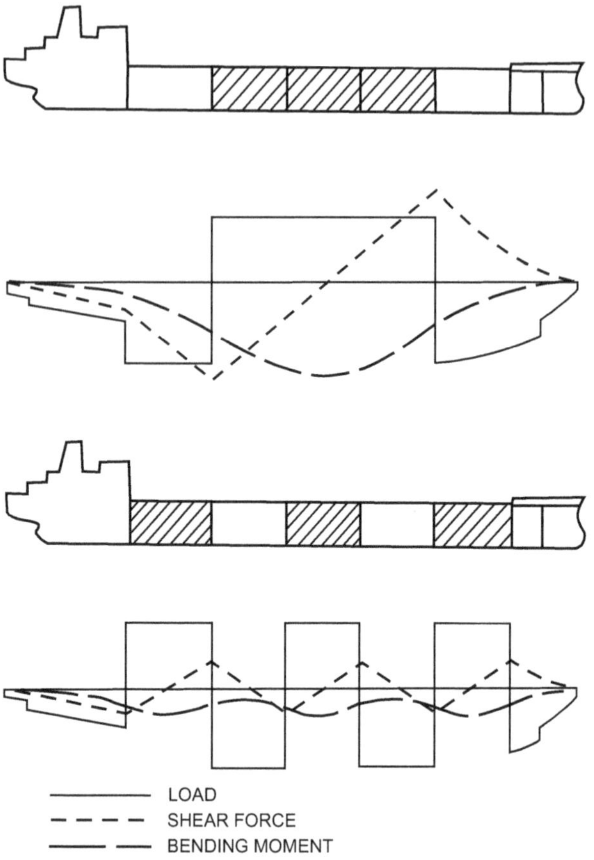

Figure 3.5 Effect of cargo distribution.

3.3 CALCULATION OF STILL WATER BENDING MOMENT

From the foregoing discussion it is clear that the calculation of the still water bending moment is, conceptually at least, a straightforward task; it is simply a double integration of the sum of the buoyancy force and the weight force. The calculations are straightforward but tedious, and so there are many computer routines available for this. The largest part of the calculation is finding the equilibrium values of draft and trim for each loading condition. Since this is a basic part of the hydrostatic calculations, nearly all hydrostatics programs include a routine for calculating the still water load, shear force, and bending moment distributions along the ship length.

At the very least, it is necessary to obtain the maximum bending moment M_{max} and in some cases this may not occur at amidships. Some typical cases are:

1. Vessels with unusual internal arrangements (e.g., combination of oil cargo and ordinary cargo, as in a naval replenishment ship).
2. Tankers and bulk carriers with some empty cargo holds.

For example, in tankers with empty spaces amidships $M_{\text{)(}}$ may be quite small and there may be two peak values of bending moment, at or near the two quarter points. Also, with a heavy cargo such as iron ore, a bulk carrier may be loaded in alternate holds, and in this case there is a peak value of bending moment in the way of each alternate hold. Moreover, with bulk carriers, the shear force Q can be quite significant and it will have several peak values. Therefore, even in preliminary design it is best to calculate Q and M over the full length of the vessel.

3.4 CORRECTION FOR CHANGES IN WEIGHT

Because of the wide variety of possible loading conditions, the ship will rarely be in the same condition as was assumed for the still water bending moment calculation. It is therefore important to be able to calculate, as simply as possible, the effect of the addition or removal of weight on the hull girder bending moment. A useful technique for this is to construct an influence line diagram. An influence line shows the effect on the maximum bending moment of the addition of a unit weight at any position x

along the ship length. The height of the line at x represents the effect on M_{max} of the addition of a unit weight at x. Two influence lines are normally drawn, one for the maximum hogging and one for the maximum sagging condition. Influence lines could, of course, be drawn to show the effect of additions on other bending moment values (i.e., other than M_{max}), but these would be of less interest.

Let us take the case of a weight P which is added at a distance x_p forward of amidships, as shown in Fig. 3.6. Other relevant quantities are defined in the figure. As a result of this addition the ship will undergo a parallel sinkage v and a nondimensional trim t (total trim divided by the total length L). If A_W and I_L are the area and the longitudinal moment of inertia of the waterplane about the Longitudinal Center of Flotation (LCF), then (assuming that the change in the waterplane is small) we have

$$v = \frac{P}{\rho g A_W} \quad \text{and} \quad t = \frac{P(x_P - x_F)}{\rho g I_L}$$

where x is positive forward of amidships. Let R denote the position of maximum bending moment, M_{max}, located at a distance x_R from amidships. The total change in M_{max} can be determined by taking moments about R, either forward or aft. Choosing the forward side, we see that the change in M_{max} is the sum of the negative moment of buoyancy of the parallel sinkage forward of R minus the moment of buoyancy of the wedge forward of R plus the moment of added weight. (This is consistent with the convention that downward force is positive and hogging bending moment is positive). These three quantities are:

1. Moment of buoyancy of parallel sinkage forward of R about R

$$= -\rho g v \mathcal{M}_R = -\frac{P\mathcal{M}_R}{A_W}$$

where $\mathcal{M}_R$ is the moment about R of the waterplane area forward of R (the shaded area A_R in Fig. 3.6).

2. Moment of buoyancy of wedge forward of R

$$= -\rho g \int_0^{FP} 2z\xi\left[\xi + (x_R - x_F)\right] t \, d\xi$$

$$= -\rho g\left[I_R + \mathcal{M}_R(x_R - x_F)\right] t$$

$$= -\frac{P(x_P - x_F)}{I_L}\left[I_R + \mathcal{M}_R(x_R - x_F)\right]$$

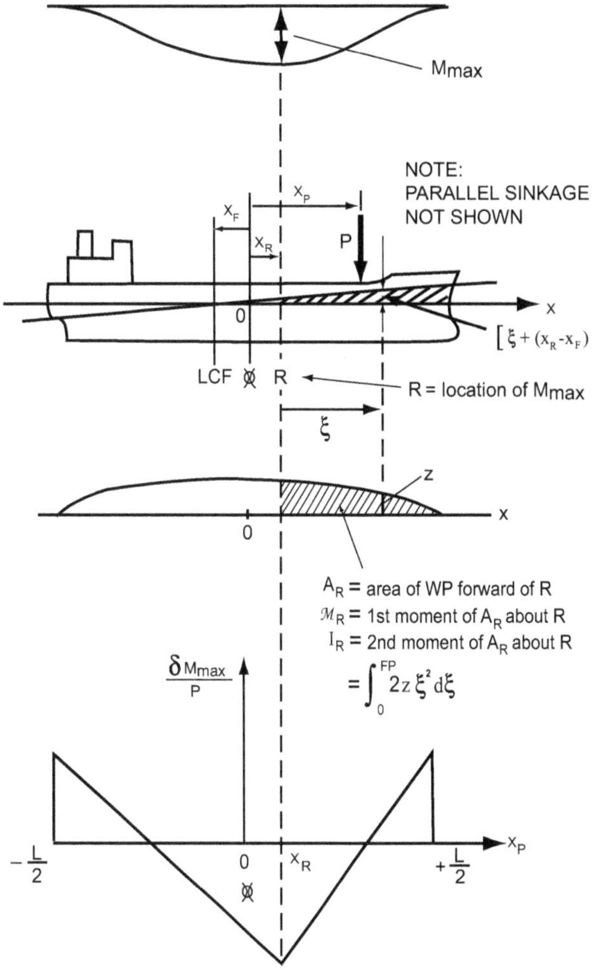

Figure 3.6 Influence line for change in M_{max} due to added weight.

3. Moment of added weight = $P(x_P - x_R)$.

Note: since we are calculating moments forward of R, this third term is only included if the added weight is forward of R; if not, it is omitted. To assist in remembering this, we shall write this third term as $P\langle x_P - x_R\rangle$, in which the pointed brackets indicate that whenever the quantity within the brackets becomes negative its value is taken as zero.

Therefore, the net effect on the bending moment at R (i.e., on M_{max}) due to the addition of a weight P is

$$\delta M_{max} = P\left\{-\frac{\mathcal{M}_R}{A_W} - \frac{(x_P - x_F)}{I_L}\left[I_R + \mathcal{M}_R(x_R - x_F)\right] + \langle x_P - x_R \rangle\right\} \quad (3.4.1)$$

This equation is also valid for negative values of x_P (i.e., for P aft of amidships), provided that, once again, if the expression in pointed brackets is negative it is taken as zero. A discontinuity occurs at R, the position of maximum bending moment. As shown in Fig. 3.6, the influence lines are straight lines that cross the axis at approximately the quarter points of the vessel. Therefore, a weight added within this length causes an increased sagging moment and an added weight outside this length causes an increased hogging moment. To construct an influence line diagram, (3.4.1) should be evaluated for three values of x_P: $-L/2$, x_R, and $+L/2$.

By making suitable approximations it is possible to simplify (3.4.1). It can be shown that if both LCF and R are taken as being at amidships, then $x_R = x_F = 0$, $I_R = I_L/2$, and $\mathcal{M}_R = ½A_W\, x_{(½)WP}$, where $x_{(½)WP}$ is the distance from amidships to the centroid of the forward half waterplane. With these simplifications (3.4.1) becomes

$$\delta M_{\text{⊠}} \cong P\left\{-\frac{x_{(1/2)WP}}{2} - \frac{x_P}{2} + \langle x_p \rangle\right\}$$

Thus, if the weight is added forward of amidships the result is

$$\delta M_{\text{⊠}} \cong P\left\{\frac{x_P - x_{(1/2)WP}}{2}\right\}$$

and this has a direct and relatively simple physical interpretation: it is half of the moment of P about the centroid of the half waterplane area (and the slope of the influence line is ½.) Moreover, for this simplified case, it is possible to define the terms in such a way that a single expression applies, regardless of whether P is forward or aft of amidships. To do this, we define l as the distance between the added weight and the *nearer* centroid of the half-waterplane area, as shown in Fig. 3.7, and we adopt the sign convention that if P lies between this centroid and amidships, the moment change is negative (sagging) or is otherwise positive (hogging).

The approximate expression for the moment change is then simply

$$\delta M_{\text{⊠}} \cong \frac{Pl}{2} \tag{3.4.2}$$

Note that this approximation can only be used for the admidships value of a symmetric bending moment distribution. Of course, if a weight is removed, then $-P$ replaces P in all of the above formulas.

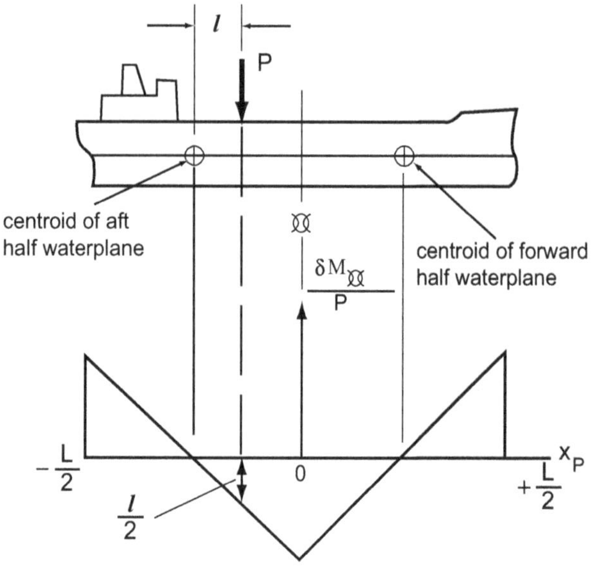

Figure 3.7 Simplified influence line: change in $M_{\text{⊠}}$ due to added weight.

In addition to their use in design, influence lines are a helpful tool for the ship operator and are sometimes provided as a part of the loading manual. However, it should be noted that they are intended for small weight changes only, certainly not more than 5% of the displacement. If the change exceeds this amount, a new bending moment calculation should be done. Obviously, the approximate formula of (3.4.2) is even less accurate and it should not be used at all if either the LCF or the location of M_{max} is not close to amidships.

3.5 APPROXIMATE DESIGN VALUES OF WAVE LOADS

Because of the complexity of ocean waves and of the dynamic interaction between ship and waves, the direct calculation of an appropriate design value of wave loading for a given ship is a very complex task, and one that has occupied the attention of several generations of naval architects. After many years of investigation significant progress has been made, due largely to simultaneous and complementary advances in several fields: ocean wave data collection, statistical theory, system response analysis, free surface hydrodynamics, and above all, computing. Thanks to these advances, methods are now available that are more rationally-based, more accurate, and more comprehensive than the methods of only a few years ago. Being computer-based, these new methods also eliminate much of the tedium of the older methods.

But as mentioned in Chapter 1, the design values for wave-induced loads are long-term statistical quantities and are therefore applicable to all ships of a given size, speed, hull geometry, and distribution of mass. Therefore, over the past several years the classification societies have gathered much information concerning wave loads from the computer-based methods, from model studies and from direct full-scale measurements. From this they have developed explicit formulas for the characteristic extreme values for standard ship types, expressed in terms of the principal dimensions of the ship. Because of the complexity of wave loading and because it is probabilistic there is a good deal of scatter in the measured values, and the formulas developed from them are only approximate and somewhat conservative. Nevertheless, the explicit specification of wave loads is of great convenience and usefulness to the designer, and this helps to offset the approximations and possible extra safety margin that it contains.

For special ship types the characteristic extreme values of wave loads should be calculated explicitly, using the computer-based methods. This does not require specialist knowledge or involve an undue amount of calculation, because these computer-based methods are currently being incorporated into user oriented packages, which require for their use only a broad familiarity with the basic concepts of statistics, ship motions, and system response analysis, all of which are presented in Chapter 4.

Beginning in the 1970s, the International Association of Classification Societies (IACS) has compiled, updated and published a set of Unified Requirements governing the design of ships and their equipment. Chapter S deals with Ship Structure. In 1989, IACS published Section S11 Longitudinal Strength Standard. In the remainder of this section we present a set of characteristic wave loads from IACS (1989, revised through 2006) and Bureau Veritas Rules (2000), which may be used for the design of standard ships. These characteristic values correspond to a probability of exceedencc of the order of 10^{-8}.

3.5.1 Wave Vertical Bending Moment

The vertical wave bending moments at any hull transverse section are, in units of Nm:

$$(M_{wv})_{sag} = -110\, F_M C L^2 B (C_B + 0.7) \qquad (3.5.1a)$$

$$(M_{wv})_{hog} = +190 F_M C L^2 B C_B \qquad (3.5.1b)$$

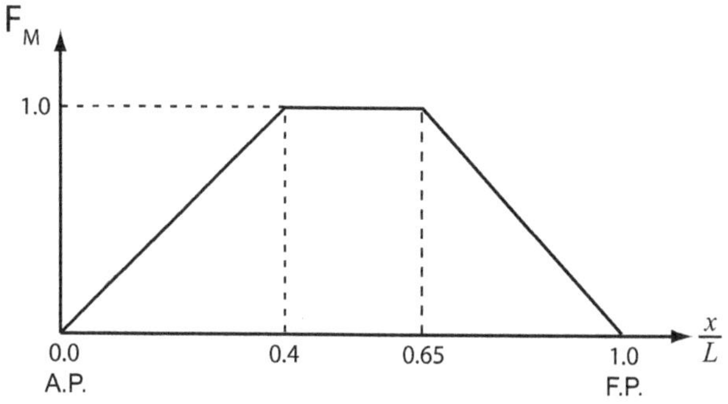

Figure 3.8 Distribution factor F_M.

where:

F_M = Distribution factor given in Fig. 3.8

L = Length of the ship in meters

B = Greatest molded breadth in meters

C_B = Block coefficient, not to be taken less than 0.6

C = Wave parameter

$$C = \begin{cases} 10.75 - \left(\dfrac{300 - L}{100}\right)^{1.5} & \text{for } 90 \le L < 300 \\ \text{or } 10.75 & \text{for } 300 \le L < 350 \\ \text{or } 10.75 - \left(\dfrac{L - 350}{150}\right)^{1.5} & \text{for } 350 \le L \le 500 \end{cases} \qquad (3.5.2)$$

The effects of bow flare impact are to be considered where all the following conditions occur:

(1) $120 \text{ m} \le L \le 200\text{m}$
(2) $V \ge 17.5$ knots
(3) $100 F_D A_S > LB$

where:
A_S = twice the shaded area shown in Fig. 3.9, which is to be obtained, in m^2, from the following formula:

$$A_S = b\, a_0 + 0.1L\,(a_0 + 2\, a_1 + a_2)$$

where:
a_0, a_1, a_2, and b are the distance, in m, shown in Fig. 3.9. F_D is given in Table 3.1.

For multideck ships, the upper deck shown in Fig. 3.9 is to be taken as the deck (including superstructures), which extends up to the extreme forward end of the ship and has the largest breadth forward of 0.2L from the fore perpendicular.

To account for the dynamic effects of bow flare impact, the sagging bending moment at any hull transverse section, defined in Equation (3.5.1a), is to be multiplied by the coefficient F_D from Table 3.1.

Table 3.1 Coefficient F_D

Hull transverse section location	Coefficient F_D
$0 < x < 0.4L$	1
$0.4L < x < 0.5L$	$1+10(C_D-1)(x/L - 0.4)$
$0.5L < x < L$	C_D
$C_D = 262,5 \dfrac{A_s}{CLB\,(C_B + 0,7)} - 0,6$ with $1.0 \leq C_D \leq 1.2$	

When at least one of the three conditions does not occur, F_D may be taken as 1.

3.5.2 Wave Horizontal Bending Moment

NOTE: The material in this section and section 3.5.3 is from Bureau Veritas Rules (2000).

The horizontal bending moment at any hull transverse section is obtained, in Nm, from the following formula:

$$M_{wh} = 420 F_M H L^2 T C_B \quad \text{(Nm)} \qquad (3.5.3)$$

where:

F_M = Distribution factor given in Fig. 3.8
H = Wave parameter
T = Summer draft

$$H = 8.13 - \left(\frac{250 - 0.7L}{125}\right)^3$$

without being taken greater than 8.13

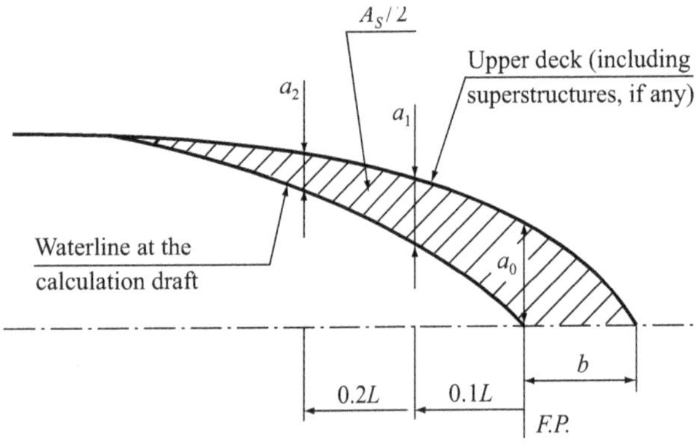

Figure 3.9 Area A_S.

3.5.3 Wave Torsional Moment

The wave torque at any hull transverse section is to be calculated considering the ship in two different conditions:

Condition 1: ship direction forming an angle of 60° with the prevailing sea direction

Condition 2: ship direction forming an angle of 120° with the prevailing sea direction.

The values of the wave torques in these conditions, calculated with respect to the section center of torsion, are obtained, in Nm, from the following formula:

$$M_{w,torsion} = 250 HL\,(F_{TM} C_M + F_{TQ} C_Q d) \qquad (3.5.4)$$

where:

F_{TM}, F_{TQ} = Distribution factors, defined in Figs. 3.10 and 3.11 for ship conditions 1 and 2.
H = Wave parameter defined in Section 3.5.2.
C_M = Wave torque coefficient
= $0.45\ B^2\ C_w^{\ 2}$
C_Q = Horizontal wave shear coefficient
= $5T\ C_B$
C_w = Waterplane coefficient, to be taken not greater than the value obtained from the following formula:
C_w = $0.165 + 0.95\ C_B$
where C_B is to be assumed not less than 0.6. In the absence of more precise determination, C_w may be taken equal to the value provided by the above formula.

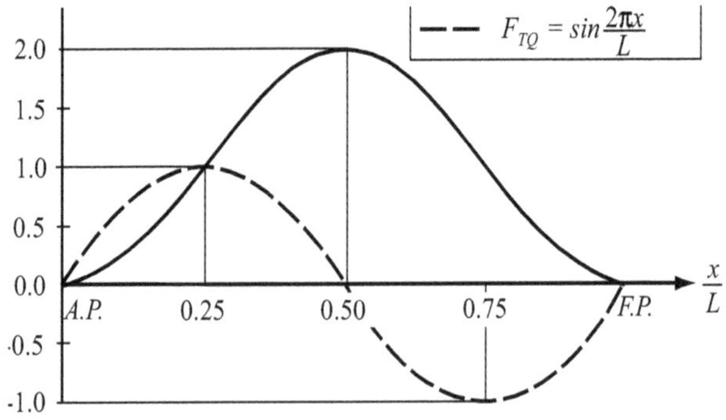

Figure 3.10 Distribution factors F_{TM} and F_{TQ} for condition 1.

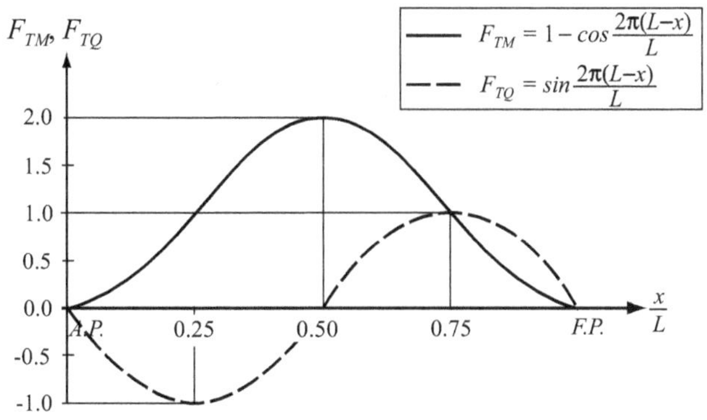

Figure 3.11 Distribution factors F_{TM} and F_{TQ} for condition 2.

d = Vertical distance, in m, from the center of torsion to a point located 0.6T above the baseline.

3.5.4 Wave Vertical Shear Force

The vertical wave shear force at any hull transverse section is obtained, in N, from the following formula:

$$Q_{wv} = 300 F_Q C L B (C_B + 0.7) \quad \text{(N)} \tag{3.5.5}$$

where:
F_Q = Distribution factor given in Fig. 3.12 for positive and negative shear forces.
C = Wave parameter defined in (3.5.2).

Since shear force is the first derivative of bending moment, the magnitude of wave shear force is directly proportional to the magnitude of the wave bending moment. In (3.5.5) the quantity $CLB(C_B + 0.7)$ is the same as in the sagging wave bending moment in (3.5.1b) except that, being a derivative, it has L instead of L^2. Examining the factor F_Q in Fig. 3.12 we see that it is a pure number for the forward positive peak (1.0) and the aft negative peak (0.92). That means that these peak values of wave shear force come directly from the sagging wave. Conversely, the hogging wave bending moment in (3.5.1b) has C_B in place of $(C_B + 0.7)$. In Fig. 3.12 the value of F_Q for the other two peaks has the ratio $C_B / (C_B + 0.7)$, so that here the wave shear force is proportional to the hogging wave bending moment, and these peak values are produced by the hogging wave. Although each shear force distribution—sagging and hogging—goes through zero somewhere near amidships, the location varies widely among various waves, and it would be unwise to try to take advantage of the decrease. Hence in the amidships region $(0.4 < x/L < 0.6)$ the envelope of all the possible distributions is taken as a horizontal line at $F_Q = \pm 0.7$.

3.6 HULL GIRDER BENDING STRESS

As explained earlier, section modulus $Z = I/z$ is a coefficient that converts bending moment into maximum bending stress, either in the strength deck $(z = z_D)$ or in the bottom $(z = z_B)$. It is a convenient grouping of the two factors in the bending stress equation that are determined by the geometry and scantlings of the hull girder cross section. In other words, section modulus is the quantity through which the designer can control the maximum hull girder stress (stillwater, wave-induced or total) by itself, separately from the stresses arising from hull module and principal member response.

3.6.1 Reduction Factor k for Higher Yield Strength Steels

As we have seen, modern ship structural design is based on explicit consideration of the failure

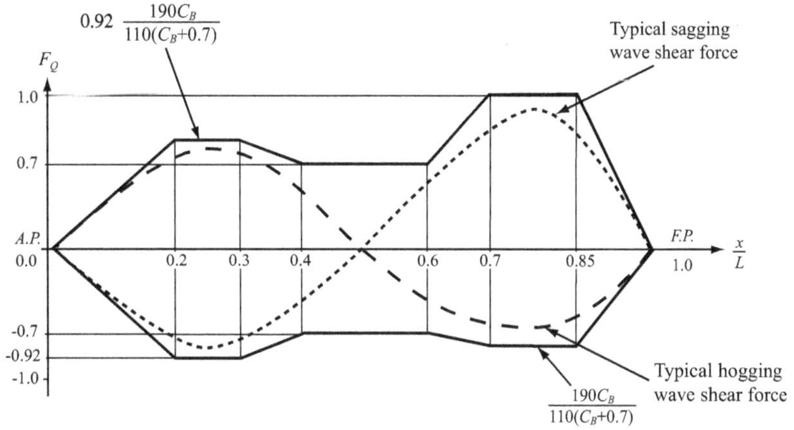

Figure 3.12 Distribution factors F_Q.

modes of yielding, buckling, and fatigue, but in earlier times only yield was explicitly considered. Before the 1960s there was no need or incentive to use a higher yield steel for commercial ships. It was expensive, difficult to weld, and ship sizes were such that it was not difficult to keep the maximum hull girder stress sufficiently below the yield stress of ordinary steel, which is about 235 MPa. Therefore hull structural design was based on keeping the maximum hull girder stress below an "allowable stress" set at 190 MPa, or 81% of yield. In practice this limit was imposed indirectly, by requiring a specified minimum value of section modulus; this will be discussed in the next two sections. Placing a limit on the maximum stress, based on yield, had two further benefits: it kept the compressive stress low enough to avoid buckling, and it kept the cyclic (wave-induced) stress low enough to avoid fatigue.

In the 1960s oil tanker size began to grow, causing an increase in maximum hull girder bending moment. Oil tankers are "weight critical ships"—heavier scantlings (more steelweight) requires a corresponding reduction in oil cargo weight, and therefore a reduction in revenue. To avoid heavier scantlings, ship designers began using higher yield steels, which meant that the allowable stress could be increased and still maintain the same safety factor against yield. But the classification societies realized that increasing the stresses would lower the safety margins against buckling and fatigue, and these margins were not known with any precision. Therefore the allowable stress was not kept at the same fraction of the yield stress (0.81) but rather at a slowly decreasing fraction specified by a Higher Strength Steel Factor, denoted as *k*. This factor is defined in the third column of Table 3.2, adapted from Table 6.1.1 of Section 6, Paragraph 1.1.4 of ABS (2008).

In practice this strategy was not sufficient. The increased stresses caused severe problems of buckling and fatigue, and the classification societies had to introduce additional and more explicit requirements: a control over the slenderness of plating and stiffeners to prevent buckling, and a limit on cyclic stress to prevent fatigue. This was the beginning of the gradual change away from the "allowable stress" type of design and toward the modern "limit state" design, in which buckling and fatigue are dealt with explicitly. For the yield limit state, the notion of allowable stress remains valid, and we will see in the next two sections that there is still a specified minimum value of section modulus, and an allowable stress specified by the factor *k*.

3.6.2 Limit on Combined Bending Stress

Section 8 Paragraph 1.2.3 of ABS (2008), specifies the following requirement for section modulus (for bending moment in Nm)

$$Z \geq \left| M_s + M_w \right| \frac{k}{190 \times 10^6} \qquad (m^3) \qquad (3.6.1)$$

NOTE: The value of 190 in (3.6.1) is for Z calculated using net thicknesses after subtracting the corrosion allowance; in IACS (2006) the value is 175 because there Z is based on gross thicknesses.

Since Z is a coefficient that relates bending moment and maximum bending stress, this requirement can be interpreted as an explicit limit on the total bending stress, using the 10^6 factor to get N/mm^2 or MPa

$$\left| \sigma_s + \sigma_w \right| \leq \frac{190}{k} \qquad \text{(MPa)} \qquad (3.6.2)$$

In other words, there can be a linear tradeoff between the still water and wave stresses, as long as the total does not exceed 190 MPa (for

Table 3.2 Higher Strength Steel Factor, *k*

Steel Type	Yield Stress	k	Allowable Stress 190/k	Allowable / Yield
AH24	235	1.00	190	0.81
AH27	265	0.93	204	0.77
AH32	315	0.78	244	0.77
AH35	340	0.74	257	0.76
AH36	355	0.72	264	0.74
AH40	390	0.68	279	0.72

ordinary steel, for which $k = 1$). This combined limit is the 45 degree line in Fig. 3.13, to be discussed in the next section.

3.6.3 Minimum Value of Section Modulus and Its Relation to Sagging Wave Bending

Besides the required value given in (3.6.1) Section 8 Paragraph 1.2.2.2 of ABS (2008) specifies another formula for the minimum value of the section modulus.

$$Z_{\min} = 0.9CL^2B(C_B + 0.7)k \times 10^{-6} \quad (\mathrm{m}^3) \qquad (3.6.3)$$

Although not at first evident, all the terms on the right hand side (except 0.9 and $k \times 10^{-6}$) are the same as in (3.5.1.a), which is the IACS formula for the design value of the amidships sagging wave bending moment, $M_{w,sag}$. The only term that does not appear in (3.6.3) is the 110 factor. That is, the above formula is equivalent to

$$Z_{\min} = 0.9\left(M_w\right)_{sag}\left(\frac{k \times 10^{-6}}{110}\right)$$

The ratio $M_{w,sag}/Z_{\min}$ is of course the corresponding maximum sagging wave bending stress. Therefore the above formula for minimum value of section modulus can be converted into an upper limit on $\sigma_{w,sag}$ (using the 10^{-6} factor to get N/mm² or MPa).

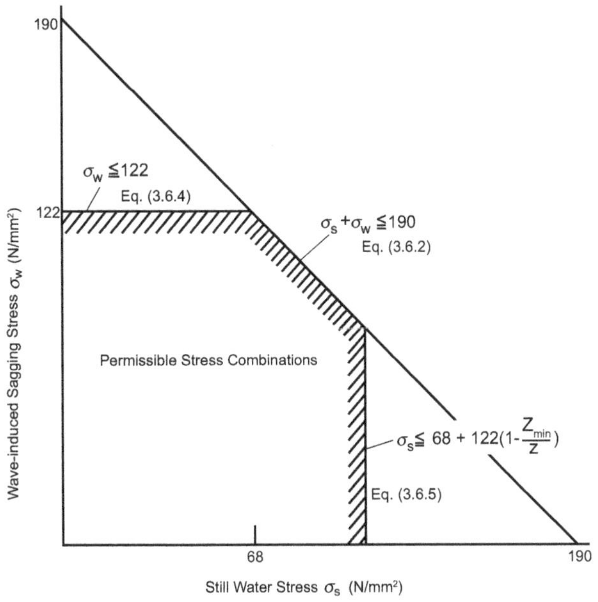

Figure 3.13 Permissible combination of stresses due to still water bending moment and wave in-duced bending moment in sagging of hull girder at midships.

$$\sigma_{w,sag} \le \frac{110}{0.9k}$$

$$\text{or} \quad \sigma_{w,sag} \le \frac{122}{k} \quad (\mathrm{MPa}) \qquad (3.6.4)$$

For ordinary steel, the yield stress is about 235 MPa and $k = 1$, and so for that case this equation is saying that $\sigma_{w,sag}$ must not exceed 122 MPa, or about 52% of the yield stress. For higher strength steels $1/k$ increases at a slower rate than the yield stress, and so the limit value of $122/k$ MPa is an even smaller percentage of yield stress (e.g. 48% for AH36 steel). Thus the underlying reason for the $Z_{\min}$ imposed by (3.6.3) is to limit the sagging wave bending stress to about half of yield at all times, even for a ship that does not have any appreciable sagging still water bending moment. For ordinary steel, the limit is 122 MPa.

There is also a reason for choosing sagging bending moment. An important feature of wave-induced bending is that it is not symmetric—the sagging moment is larger than the hogging moment, because of bow flare and other factors that are explained in Section 4.4.3.

The formula applies to both the deck and the bottom. Since the hull girder neutral axis is usually lower than 0.5D (D = depth of hull girder), the bending stress in the strength deck is larger than in the bottom, and so the formula is more likely to be an active constraint in the design of the strength deck.

Thus for the sagging case we have three limitations on the combined still water and wave bending stress. To simplify the logic, we will here treat sagging stress as positive and we will assume ordinary steel, for which $k = 1$.

(1) The limit on wave-induced stress in (3.6.4)

$$\sigma_{w,sag} \le 122$$

(2) The limit on the combined bending stress in (3.6.2)

$$\sigma_s + \sigma_{w,sag} \le 190$$

(3) Since (3.6.2) limits the total stress to 190, the maximum still water stress depends on the actual wave-induced stress. For example, if σ_w is at its maximum value of 122, the limit on σ_s is $\sigma_s < 68$. For other values of $\sigma_{w,sag}$ we have

$$\sigma_s \le 68 + 122\left(1 - \frac{Z_{\min}}{Z}\right) \qquad (3.6.5)$$

These three limits are plotted in Fig. 3.13.

SECTION MODULUS REDUCTION AWAY FROM AMIDSHIPS

Because the ship will meet a wide variety of stillwater and wave loads during its lifetime, the envelope curve of all of the various bending moment distributions is relatively flat throughout the amidships region. Therefore, the required value of section modulus needs to be maintained at almost its full magnitude for some appreciable length forward and aft of amidships. For simplicity, and in order to avoid having to calculate the envelope curve, it is common procedure to maintain the full value of section modulus throughout the middle 40% or so of the ship length. Outside of this length the scantlings may be reduced but the reduction must be gradual, for example, to one-half of the full value at $0.15L$ from the ends.

If the bending moment envelope curve is used, items included in the hull girder section amidships may be reduced in size or eliminated as the bending moment decreases, but always in a gradual manner and always such that the maximum permissible stress is not exceeded.

3.6.4 Allowable Area for Section Modulus

In general, the following items are included in the calculation of the section modulus, provided they are longitudinally continuous:

Deck plating (strength deck and other effective decks).
Shell and inner-bottom plating.
Longitudinal bulkheads and girders.
Longitudinal stiffeners.

Some items may be partly effective, depending on their length, manner of attachment, and longitudinal stiffness:

1. Length: To be effective, a member must extend over a sufficient length such that some portion of the longitudinal stress field enters the member. An approximate rule-of-thumb is that the longitudinal stress can diffuse at an angle of 15° on each side. Thus, as shown in Fig. 3.14*a*, a longitudinal bulkhead would become fully effective in its midlength region only if it extended over a length of at least $D/(2 \tan 15°) = 1.866D$, where D is the depth of the bulkhead; at locations closer to its ends, it would be only partly effective. Similarly, if at a certain point along the ship length the height of the side shell increases by H, as in a full width superstructure, the additional side plating only becomes fully effective after a length of $H/\tan 15° = 3.73H$. An even longer length is required for the entire superstructure to become fully effective (see Fig. 3.14*c*). This same rationale can also be used to estimate and allow for the ineffective plating immediately forward or aft of an opening. As shown in Fig. 3.14*d*, the material that lies within the "shadow area" subtended by the two 15° rays is ineffective. This shadow area rule is useful for dealing with openings that occur close to, but not quite within, the same transverse plane. It also indicates that the material between hatches is seldom very effective, as shown in Fig. 3.14*d*. Thus if there were other openings, such as sideports, located at other points in the ship's transverse plane between the hatches, it could happen that this plane would be the critical one.

2. Manner of attachment: The attachment must be capable of transmitting shear force without undergoing longitudinal slip. In welded structures this is generally not a problem.

3. Physical continuity: If the member consists of lengthwise segments that are joined by welding, the joints must be butt welds and not fillet welds. For example, in order to be effective, a longitudinal stiffener must pass through the web of a transverse frame or beam, by means of a cutout in the latter, as shown in Fig. 3.14*e*. If the longitudinal stiffener is "intercostal"; i.e., if it does not pass through but rather stops at the web and is simply fillet welded to it, it cannot be counted.

4. Longitudinal stiffness: To be fully effective, a member must have the same (or greater) Young's modulus as the rest of the hull girder and must have parity of strain; that is, it must undergo the same elongation or compression as is occurring in the rest of the hull girder at that same elevation. Thus there are two further ways in which a member might be only partly effective: by having a smaller material stiffness (e.g., aluminum) or by deflecting vertically so as to adopt a larger radius of curvature, thereby reducing its longitudinal strain and thus "shirking" part of the stress load it would otherwise carry. This phenomenon can occur, for example, in superstructures that are not full width, as illustrated in Fig. 3.14*f*. The question of superstructure is discussed in Section 3.9.

In most cases the critical hull girder cross section will be the section that contains the least amount of effective material—that is, the section containing the largest hatches or other

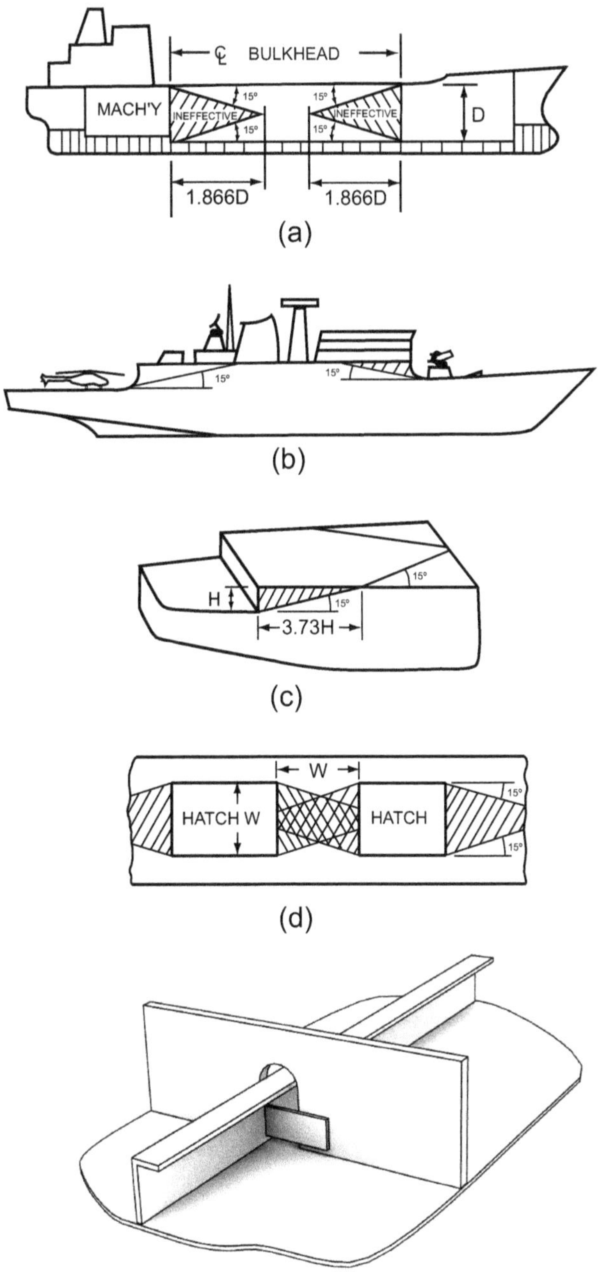

Figure 3.14 Effective longitudinal material.

openings—but it also depends on the distance of these from the neutral axis. If there is any doubt, the section modulus should be calculated for all of the potentially critical sections.

In general, the net sectional areas of longitudinal members are to be used in the section modulus calculation, except that small isolated openings need not be deducted, provided the openings and the shadow area breadths of other openings in any one transverse section do not reduce the section modulus by more than a few percent.

3.6.5 Calculation of Hull Girder Moment of Inertia

Figure 3.15 is an example of a longitudinal structural section showing (only) the longitudinally effective material. The two quantities to be calculated are the position of the neutral axis of the section and the moment of inertia I about the neutral axis. If done by hand, the calculation is best carried out in tabular form, as shown in Table 3.3. The dimensions and the vertical height of the centroid of each item are entered, and then the area and the first and second moments of area of each item are calculated about some axis *yy*. The choice of the axis *yy* is arbitrary but the keel position is probably the most appropriate because the location of the neutral axis is usually quoted as height above the keel. To obtain the moment of inertia of an item about the *yy* axis, it is necessary to add to ah^2 the moment of inertia of the item about its own neutral axis, which is denoted as i. This is entered in the last column. Most of the items can be treated as rectangles so that the moment of inertia is simply $1/12ad^2$, where a is the area and d the depth. Note that, as shown in Fig. 3.15, this expression is also valid for inclined sections of plating.

It will be noted that there are only a few entries in the last column; this is because the moment of inertia of what may be called "horizontal" material about its own neutral axis is sufficiently small to be negligible. If, for example, a panel of plating is of thickness t and width $100t$, the value of i is 10^6t^4 if the panel is vertical but is only 10^2t^4 if it is horizontal; since the values are to be summed, the latter quantity is clearly negligible.

The distance of the neutral axis above the keel is

$$h_{NA} = \frac{\Sigma\, a_i h_i}{\Sigma\, a_i}$$

Finally, from the parallel axis theorem, the moment of inertia about the neutral axis is

$$I = I_{yy} - Ah_{NA}^2 \qquad (3.6.6)$$

where I = moment of inertia about the neutral axis

I_{yy} = moment of inertia about the baseline

A = total area

and h_{NA} = distance from baseline to neutral axis.

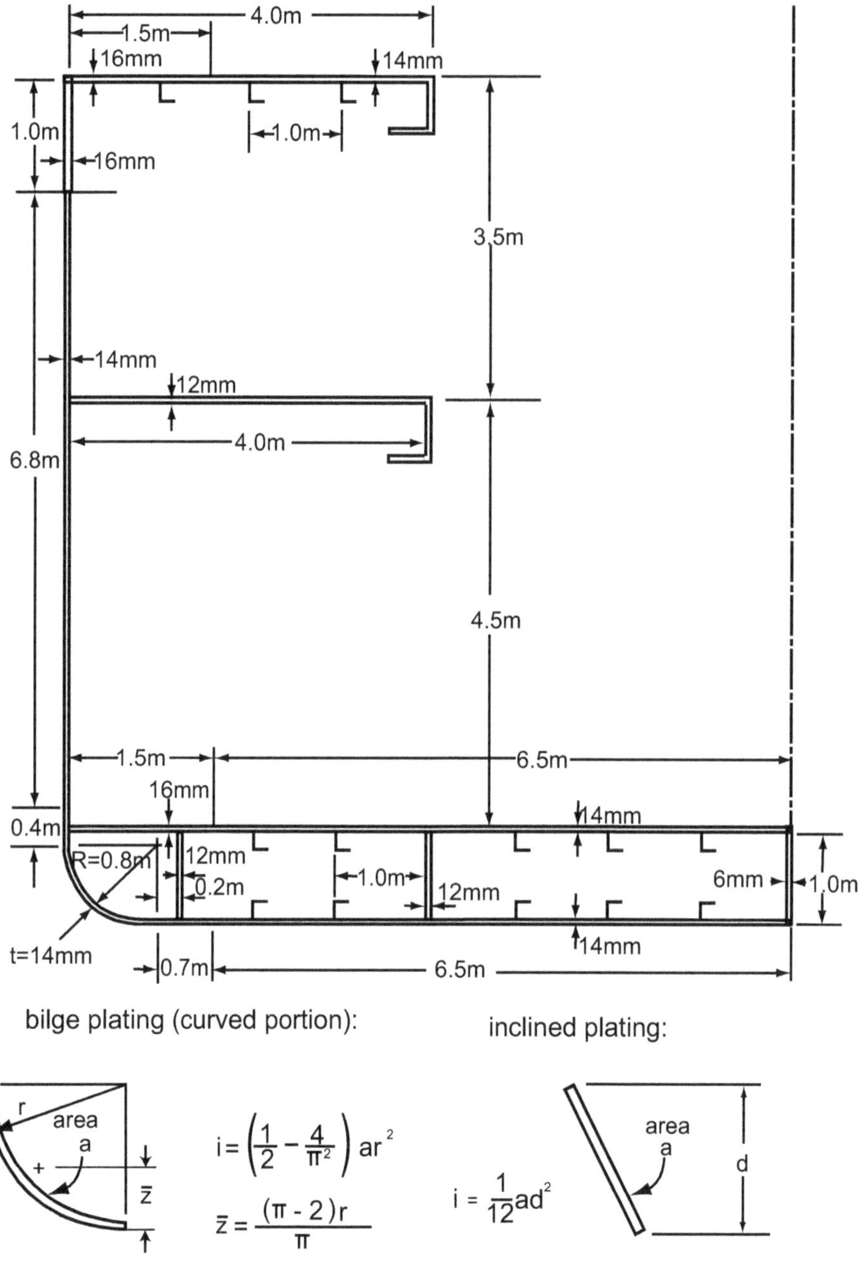

Figure 3.15 Example of longitudinally effective material.

As shown in the example of Table 3.3, the calculation is usually carried out for one side of the ship only, and therefore the results have to be multiplied by two.

For open flange-web beams (tees, angles, etc.) an explicit formula for moment of inertia is given in Section 3.12.

For calculating horizontal moment of inertia (about the centerline) add three more columns to the table

Distance from centerline, g
The second moment ag^2
The horizontal centroidal moment of inertia of the member i_{CL} (if relevant).

Then use the sums of the last two columns to calculate $\frac{1}{2}I_{CL}$, and double it to get I_{CL}. Finally, calculate Z_{CL}, the section modulus for a location at the ship's side.

3.6.6 Combined Vertical and Horizontal Bending

The calculation just described assumes that the ship is upright and that the bending moment is in the ship's vertical plane. This is referred to as vertical bending. If the ship is at an angle of heel due to rolling, it will also be subjected to horizontal bending, that is, a bending moment

Table 3.3 Section Modulus Calculation

Item	Scantlings m	mm		Area a (m²)	Height a (m²)	Moment ah	2nd Moment ah^2	Local 2nd Moment i (m⁴)
Strength Deck Plating		2.5	14	0.0350	9.0	0.3150	2.835	
S.D. stinger plate		1.5	16	0.0240	9.0	0.2160	1.944	
S.D. longitudinals (mm)	W160	14; F40	14	0.0084 for 3	8.9	0.0748	0.666	
Sheer stake		1.0	16	0.0160	8.5	0.1360	1.156	0.001
Side plating		7.2	14	0.1008	4.4	0.4435	1.951	0.435
2nd deck plating		4.0	12	0.0480	5.5	0.2640	1.452	
Bilge (curved portion)		R = 0.8; t =	14	0.0176	0.29	0.0051	0.001	0.001
Inner bottom plating		6.5	14	0.0910	1.0	0.0910	0.091	
I.B. margin plate		1.5	16	0.0240	1.0	0.0240	0.024	
I.B. longitudinals (mm)	W200	10; F66	15	0.0150	0.86	0.0129	0.011	
Side girders		1.0	12	0.0240	0.5	0.0030	0.001	
Center girder (1/2)		1.0	6	0.0060	0.5	0.0030	0.001	0.001
Bottom plating		7.2	14	0.1008	0	0	0	
Bottom longitudinals (mm)	W200	10; F66	15	0.0150	0.14	0.0021	0.000	
Upper hatch side girder	W0.5	25; F0.4	25	0.0225	8.64	0.1944	1.680	
Lower hatch side girder	W0.5	25; F0.4	25	0.0225	5.14	0.1157	0.595	
Totals (for half section)				0.5706		1.9095	12.413	0.440

Height of neutral axis: $h_{NA} = \dfrac{\Sigma a_i h_i}{\Sigma a_e} = \dfrac{1.9095}{0.5706} = 3.346$ m

$$\frac{1}{2} I_{yy} = 12.413 + 0.440 \quad = \quad 12.853 \text{ m}^4$$

$$\text{Parallel axis term} = -(\Sigma a_i) h_{NA}^2 \quad = \quad -6.390 \text{ m}^4$$

$$\frac{1}{2} I = \quad 6.463 \text{ m}^4$$

Full values: $A = 1.142$ m²

$I = 12.93$ m⁴

$$Z_D = \frac{I}{h_D - h_{NA}} = \frac{12.93}{9.000 - 3.346} = 2.287 \text{ m}^3$$

$$Z_K = \frac{I}{h_{NA}} = \frac{12.93}{3.346} = 3.864 \text{ m}^3$$

M_z acting in the ship's horizontal plane (see Fig. 3.16). For this bending moment the neutral axis is the ship's vertical centerline.

Let us first take the case in which M_z is entirely due to inclination of the vessel, say to an angle θ. In this case M_y and M_z are directly related, being components of the total bending moment M (which acts in the true vertical plane):

$$M_y = M\cos\theta$$
$$M_z = M\sin\theta \tag{3.6.7}$$

If y and z are the coordinates of any point in the cross section and I_{NA} and I_{CL} are the moments of inertia about the horizontal axis in the upright condition and about the centerline respectively, then the stress at (y,z) is

$$\sigma = \sigma_V + \sigma_H = \frac{M\cos\theta}{I_{NA}} z + \frac{M\sin\theta}{I_{CL}} y \tag{3.6.8}$$

When $\sigma = 0$ then y and z are on the neutral axis (y_N, z_N) and it follows that

$$\frac{z_N \cos\theta}{I_{NA}} + \frac{y_N \sin\theta}{I_{CL}} = 0 \qquad (3.6.9)$$

or

$$z_N = -\frac{I_{NA}}{I_{CL}} \tan\theta \; y_N \qquad (3.6.10)$$

This gives the equation of the neutral axis in the inclined condition. Note that in Fig. 3.16 y_N is negative, and so the z_N given by (3.6.10) is positive, as in the figure. If we had chosen a neutral axis point on the other side, z_N would have been negative. This gives the equation of the neutral axis in the inclined condition. The angle ψ of the new neutral axis relative to its original position is given by

$$\tan\psi = \frac{z_N}{y_N} = -\frac{I_{NA}}{I_{CL}} \tan\theta \qquad (3.6.11)$$

In the figure, the ship is rotating through a positive angle θ relative to global coordinates Y, Z, whereas the neutral axis is rotating through a negative angle ψ relative to ship coordinates y, z. If the vessel were such that $I_{NA} = I_{CL}$, then $\tan\psi = -\tan\theta$, and the neutral axis would remain horizontal. In general this is not so, I_{CL} being larger than I_{NA} , and so the neutral axis rotates less than the ship and is inclined to the horizontal. This is a well-known feature of unsymmetric bending. In a ship there is one axis of symmetry, the centerline, and therefore I_{CL} and I_{NA} are the maximum and minimum moments of inertia respectively.

Referring to (3.6.8), the angles of heel at which the greatest and least stresses occur may be obtained by putting $d\sigma/d\theta = 0$.

Hence,

$$\frac{d\sigma}{d\theta} = -\frac{M\sin\theta}{I_{NA}} z_i + \frac{M\cos\theta}{I_{CL}} y_i = 0 \qquad (3.6.12)$$

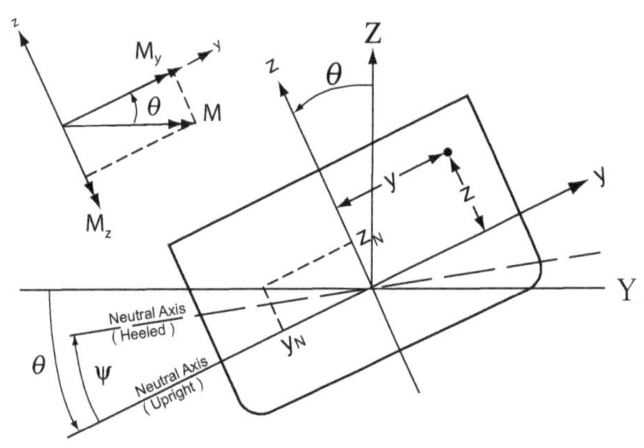

Figure 3.16 Neutral axis with simultaneous horizontal and vertical bending.

or

$$\tan\theta = \frac{y_i}{z_i}\frac{I_{NA}}{I_{CL}} = \frac{Z_{NA}}{Z_{CL}} \qquad (3.6.13)$$

where $i = D$ or B (deck or bottom)

The greatest and least stresses will also be associated with the maximum values of y and z and this means that these stresses will occur at the corners of the section. Figure 3.17 is a qualitative illustration of the stresses at the four corners of a ship section, for which the maximum and minimum stresses occur at a heel angle of about 30°.

In practice the horizontal and vertical bending moments are not so directly coupled and do not necessarily occur simultaneously. Their relationship varies with different sea conditions and depends mainly on ship heading. The situation is summarized in Fig. 3.18, based on International Ship Structures Congress (1976), which shows the results of theoretical calculations of the characteristic value of wave bending stress in a 300-m tanker (We recall from Section 1.2 that the characteristic value is the extreme value corresponding to some specified probability of exceedence in the ship's lifetime.) The figure shows the stresses at the deck edge due to vertical and horizontal bending, $\hat{\sigma}_V$ and $\hat{\sigma}_H$, and also the total stress, $\hat{\sigma}$, over a range of ship headings. This type of calculation, and the underlying statistical principles, is discussed in Chapter 4. But to interpret Fig. 3.18, it should be mentioned here that the degree of simultaneity or correlation between two random variables may be measured by a correlation coefficient ε. If the variables are independent, then $\varepsilon = 0$; if they are completely linked, such that they always occur simultaneously, then $\varepsilon = 1$. Horizontal and vertical bending stresses are random variables and for this reason their characteristic values cannot be just added together, as is done for deterministic quantities. Instead, the combined characteristic stress depends on the extent to which the variables are correlated. If $\hat{\sigma}_V$ and $\hat{\sigma}_H$ denote the individual

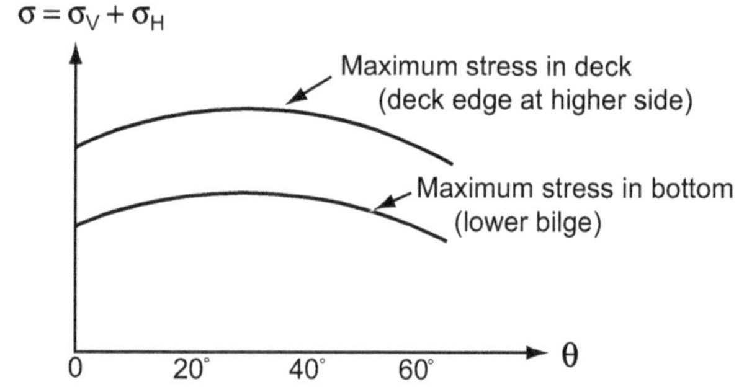

Figure 3.17 Variation of maximum total stress with angle of heel.

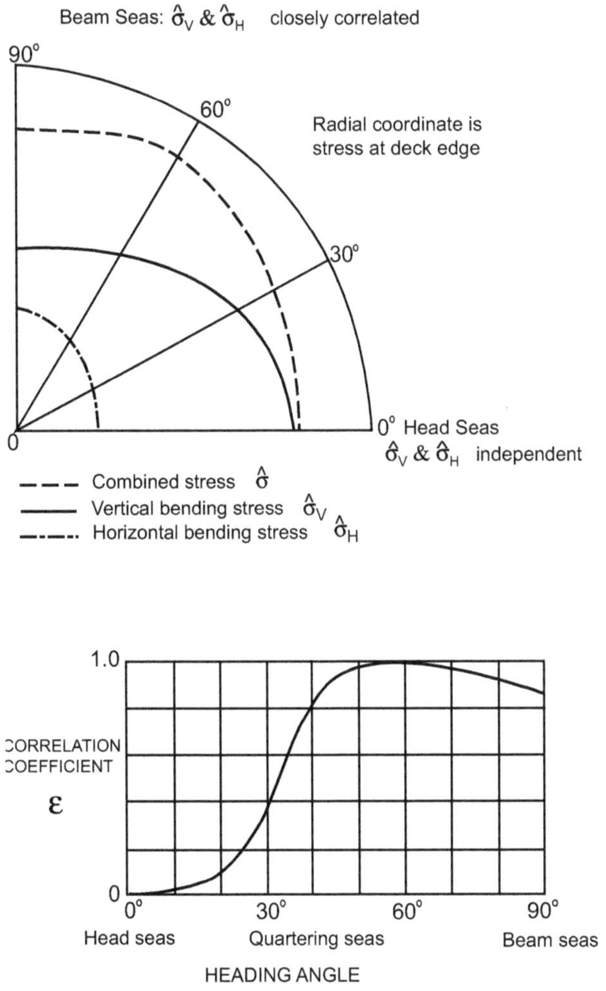

Figure 3.18 Combination of long-term vertical and horizontal bending stress.

values, it can be shown that the magnitude of the combined stress $\hat{\sigma}$ is given by

$$\hat{\sigma} = \sqrt{\hat{\sigma}_H^2 + 2\varepsilon\hat{\sigma}_H\hat{\sigma}_V + \hat{\sigma}_V^2} \qquad (3.6.14)$$

Thus, if the bending moments are independent, their stresses will be uncorrelated ($\varepsilon = 0$) and the magnitude of the long-term total stress is

$$\hat{\sigma} = \sqrt{\hat{\sigma}_H^2 + \hat{\sigma}_V^2} \qquad (3.6.15)$$

If on the other hand the bending moments are perfectly correlated, such that they always coincide, then $\varepsilon = 1$ and the total long-term stress is simply the sum of the two separate values:

$$\hat{\sigma} = \hat{\sigma}_H + \hat{\sigma}_V \qquad (3.6.16)$$

Figure 3.18 presents the values of $\hat{\sigma}_V$ and $\hat{\sigma}_H$, their correlation coefficient ε, and the total stress $\hat{\sigma}$, as given by (3.6.14), plotted for a range of ship headings. For beam seas and quartering seas ε is nearly unity, indicating that $\hat{\sigma}_V$ and $\hat{\sigma}_H$ are closely correlated. This is because for beam seas and quartering seas the horizontal bending is due mainly to rolling of the ship. Likewise the largest value of $\hat{\sigma}$ occurs with beam seas because rolling is usually a maximum for this heading. Thus the horizontal and vertical bending moments are, for the most part, components of the same overall bending moment and so their respective stresses are closely correlated. Therefore, the total stress calculated from (3.6.14) is nearly the same as the direct sum of $\hat{\sigma}_V$ and $\hat{\sigma}_H$.

For head seas the situation is the reverse: here the horizontal stress has its smallest value and it is nearly independent of the vertical stress. This is because there is less rolling and much of the horizontal bending is caused by irregular seas that give different load distributions along the two sides of the ship. The total stress as given by (3.6.14) is only slightly greater than the vertical stress, and for this heading $\hat{\sigma}_H$ could be ignored.

Figure 3.18 shows that as the heading angle increases, $\hat{\sigma}_V$ diminishes and $\hat{\sigma}_H$ increases. Also, because of the change in their correlation, the total stress is approximately constant, with a peak value that is about 12% larger than the maximum (head seas) value of $\hat{\sigma}_V$. This indicates that for design purposes it is sufficient (and conservative) to calculate the two stress values separately, using characteristic values of wave bending moment such as those given by (3.5.1) and (3.5.3), and then to simply add them together. Note that this gives the stress at the deck edge; the stress in the deck at the centerline would be $\hat{\sigma}_V$ because $\hat{\sigma}_H$ is always zero at this location.

3.6.7 Changes in Hull Girder Cross Section

The provision of the required section modulus is necessarily an iterative process, particularly if it is to be done efficiently, avoiding an overly large value. As the design progresses, it will be necessary to add or remove material in the hull girder cross section. A typical situation is shown in Fig. 3.19, in which an area δA is added at a height z above the neutral axis of a vessel having moment of inertia I, total area A, and distances z_D and z_K to the deck and keel. The effect of the addition is to raise the neutral axis a distance δh and to increase the moment of inertia to a value $I + \delta I$ (about the new neutral axis). The net effect on the deck and bottom can vary, depending on the location of δA. For example,

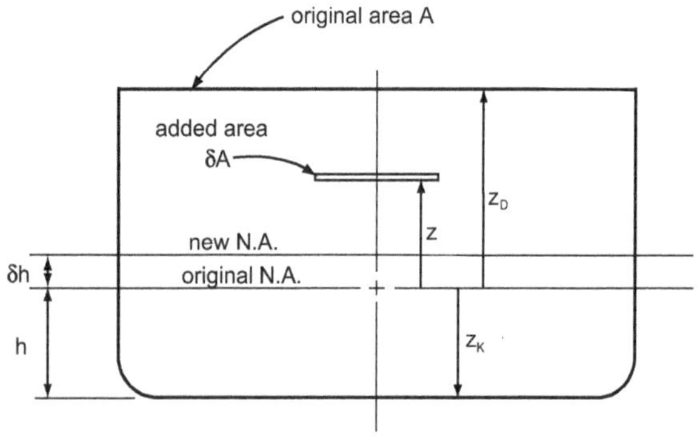

Figure 3.19 Effect of adding area. *Notes:* 1. z values are positive above the original N.A., and negative if below. 2. If material is *removed,* δA is negative.

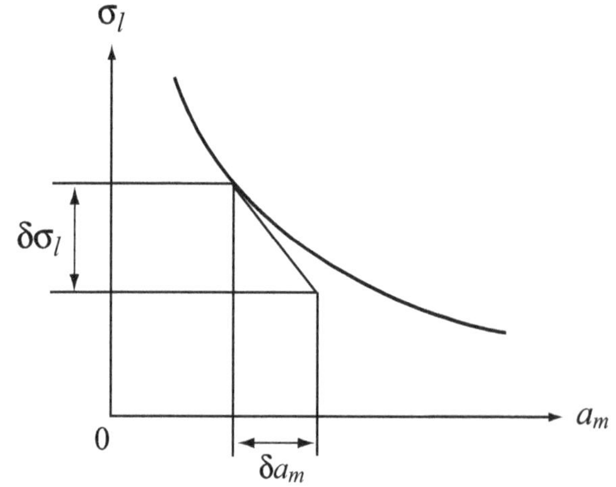

Figure 3.20 Variation of hull girder stress at location l with sectional area of member m.

although the addition shown would reduce the deck stress (because it increases I and decreases z_D) it might increase the keel stress because z_K is increased, and this might outweigh the increase in I. In addition to this complication, there is also the fact that the calculation of I is a lengthy computation, and it would not be desirable to have to repeat it for every change of area, even within a computer-aided design program. Therefore it would be helpful to have some means for quickly estimating the effect of adding material. To this end we first derive expressions for δh and δI.

With the addition of the area δA, the rise of the neutral axis is

$$\delta h = \frac{\delta A\, z}{A + \delta A} \tag{3.6.17}$$

and the additional moment of inertia is

$$\delta I = \delta A z^2 + i - (A + \delta A)(\delta h)^2$$

If the material is added below the original neutral axis, the value of z is negative. If the material is removed, then the value of δA is negative, and also i is negative in the foregoing equation.

Substitution for δh from (3.6.17) gives

$$\delta I = \delta A z^2 - \frac{\delta A^2 z^2}{A + \delta A} + i = \frac{A\,\delta A z^2}{A + \delta A} + i \tag{3.6.18}$$

3.6.8 Derivative of Hull Girder Stress

For design purposes a method is needed for quickly estimating the incremental change in hull girder stress due to a change in effective material. Figure 3.20 shows how this can be done by extrapolating along the slope of the stress-area curve, which is the derivative $\partial\sigma_l / \partial a_m$. That is, if the area a_m of any member m is changed by a small amount δa_m, the change in hull girder stress σ_l at any arbitrary location l is given by

$$\delta\sigma_l = \frac{\partial\sigma_l}{\partial a_m}\delta a_m \tag{3.6.19}$$

and so if a general expression were available for the stress derivative, $\partial\sigma_l / \partial a_m$, the change in stress could be readily calculated.

Figure 3.21 illustrates some of the basic parameters relating to the member being changed: area a_m, height of centroid above the keel h_m, distance from neutral axis z_m (positive upward) and vertical depth d_m. A subscript l denotes the location where the stress change is to be calculated. For generality the expression to be derived requires a rigorous sign convention for z, the distance from the neutral axis, and so we adopt the usual choice of positive upward. We thus have $z_l = h_l - h_{NA}$, and $z_m = h_m - h_{NA}$.

In order to obtain an expression for $\partial\sigma_l / \partial a_m$ it will be necessary to have the partial derivatives of h_{NA} and I with respect to a_m. These can be obtained from the expressions developed earlier for δh and δI—the changes in h_{NA} and in I due to adding some material. In the present case, the added material is δa_m. The centroidal moment of inertia of the added material, i_m, is usually negligible except when the member m is oriented vertically. In this case, assuming that it is a straight, thin-walled member of vertical depth d_m, i_m is equal to $\delta a_m d^2_m/12$ (see the bottom of Fig. 3.15). With these changes (3.6.17) and (3.6.18) become

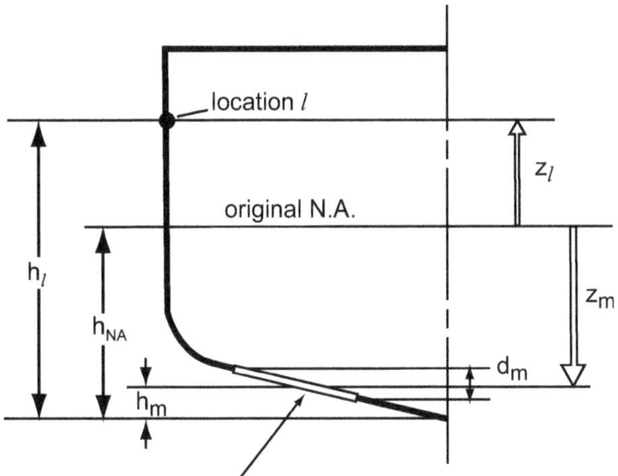

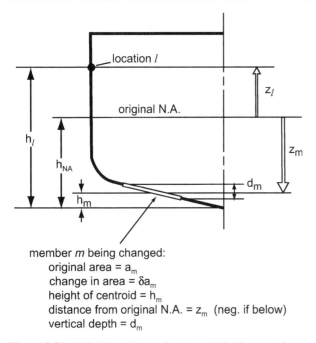

Figure 3.21 Definition of terms for stress derivative equation.

$$\delta h_{NA} = \frac{z_m \delta a_m}{A + \delta a_m} \tag{3.6.20}$$

$$\delta I = \frac{A z_m^2 \delta a_m}{A + \delta a_m} + \frac{\delta a_m d_m^2}{12} \tag{3.6.21}$$

We may obtain the desired derivatives by dividing these changes by the increment of area δa_m and taking the limit.

$$\frac{\partial h_{NA}}{\partial a_m} = \lim_{\delta a_m \to 0} \left[\frac{\delta h_{NA}}{\delta a_m} \right] = \lim_{\delta a_m \to 0} \left[\frac{z_m}{A + \delta a_m} \right]$$

$$= \frac{z_m}{A} \tag{3.6.22}$$

$$\frac{\partial I}{\partial a_m} = \lim_{\delta a_m \to 0} \left[\frac{\delta I}{\delta a_m} \right]$$

$$= \lim_{\delta a_m \to 0} \left[\frac{A z_m^2}{A + \delta a_m} + \frac{d_m^2}{12} \right] \tag{3.6.23}$$

$$= z_m^2 + \frac{d_m^2}{12}$$

We may now proceed to derive the stress derivative that is required in (3.6.19). The hull girder bending stress at z_l is

$$\sigma_l = \frac{M z_l}{I} \tag{3.6.24}$$

and its derivative is

$$\frac{\partial \sigma_l}{\partial a_m} = M \frac{I \left(\frac{\partial z_l}{\partial a_m} \right) - z_l \left(\frac{\partial I}{\partial a_m} \right)}{I^2}$$

Since $z_l = h_l - h_{NA}$, the first derivative in the numerator is $-\partial h_{NA} / \partial a_m$, which is given by (3.6.22), and the other derivative is given by (3.6.23). Substitution yields

$$\frac{\partial \sigma_l}{\partial a_m} = -\frac{M z_l}{I} \left[\frac{z_m}{A z_l} + \frac{1}{I} \left(z_m^2 + \frac{d_m^2}{12} \right) \right]$$

and from (3.6.24) this is

$$\frac{\partial \sigma_l}{\partial a_m} = -\sigma_l \left[\frac{z_m}{A z_l} + \frac{1}{I} \left(z_m^2 + \frac{d_m^2}{12} \right) \right] \tag{3.6.25}$$

In the simplest case when $d_m << z_m$, this equation becomes

$$\frac{\partial \sigma_l}{\partial a_m} = -\sigma_l \left[\frac{z_m}{A z_l} + \frac{z_m^2}{I} \right] \tag{3.6.26}$$

This expression is to be used in conjunction with (3.6.19). Of course, if the change in area is large—say, more than 5% of the total area—then the moment of inertia should be recalculated as in Table 3.3.

3.7 CALCULATION OF HULL GIRDER SHEAR STRESS

3.7.1 Shear Stress in Open Sections

In the hull girder, as in any beam loaded by transverse vertical forces, there is a vertical shear force Q acting on the cross section. In a thin-walled section such as an I-beam or a box girder, it is important to know how the total shear force Q is distributed across the section so that the wall thicknesses can be adequately sized. That is, it is necessary to determine the distribution of the shear stress τ around the entire cross section.

Figure 3.22 shows a thin-walled symmetric box girder subjected to a vertical shear force

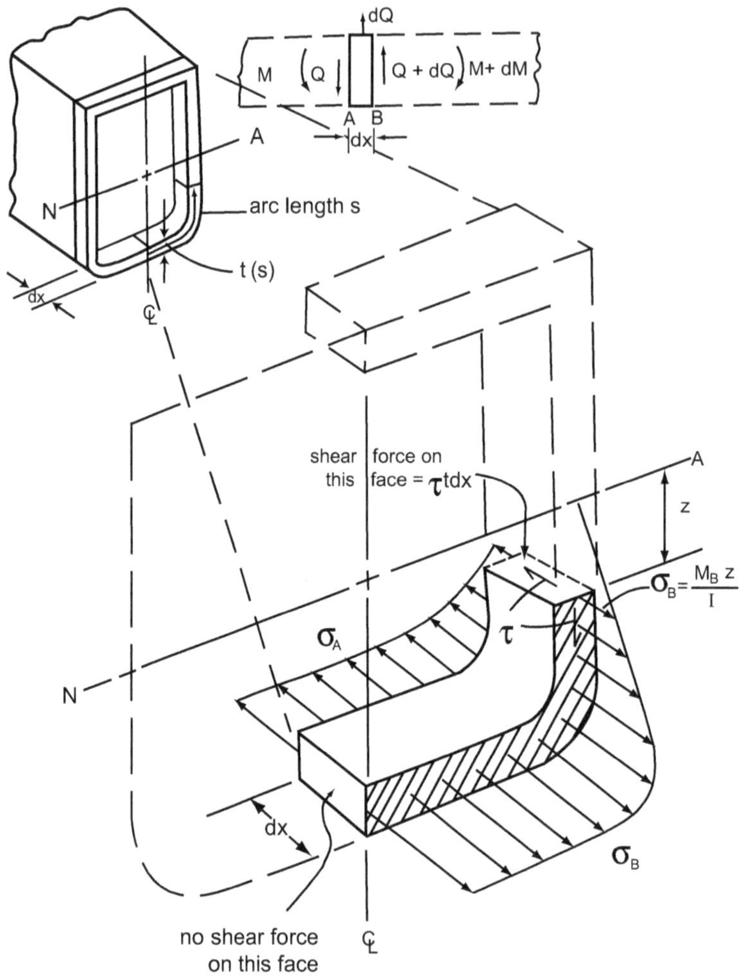

Figure 3.22 Free body diagram for transverse shear.

Q. From elementary beam theory it is known that over a differential segment of length dx, Q causes a change of bending moment given by

$$dM = Qdx$$

Due to this change in the bending moment, the bending stresses σ_A and σ_B on the two faces of the differential segment are not equal. Therefore if we isolate a portion of the differential segment by making two cuts, one at the centerline and the other at an arc length s from the centerline, the imbalance in the longitudinal normal stress forces must be counterbalanced by longitudinal shear stress forces across the cut sections. However, because of symmetry, there can be no shear stress in the centerplane cut and hence the balancing force must come entirely from the shear stress τ at the other cut. Longitudinal equilibrium therefore states that

$$\tau t\,dx = \int_0^s \sigma_B t\,ds - \int_0^s \sigma_A t\,ds \qquad (3.7.1)$$

Substituting $\sigma = Mz / I$ on both faces

$$\tau t\,dx = \frac{M_B - M_A}{I}\int_0^s zt\,ds$$

$$= \frac{dM}{I}\int_0^s zt\,ds$$

and substituting $dM = Qdx$ gives

$$\tau t = \frac{Q}{I}\int_0^s zt\,ds \qquad (3.7.2)$$

The integral on the right hand side is a function of the geometry of the section and of position s around the section. For convenience we assign the symbol m to this quantity:

$$m(s) = \int_0^s zt\,ds \qquad (3.7.3)$$

and we note that m is the first moment about the neutral axis of the cumulative section area starting from the "open" end (shear-stress-free end) of the section.

Substituting for m in (3.7.2) and solving for τ.

$$\tau = \frac{Qm}{tI} \tag{3.7.4}$$

The product τt has special significance in the torsion of thin-walled sections, and has some analogies to the flow of an ideal fluid within a closed pipe. This product is therefore referred to as the "shear flow" and is assigned the symbol q:

$$q = \tau t \tag{3.7.5}$$

The shear flow is also a useful quantity in the present case, in which the shear stress is due to a transverse load. From (3.7.4) we observe that the shear flow is given by

$$q = \frac{Qm}{I} \tag{3.7.6}$$

Since Q and I are constants for the entire section, the shear flow is directly proportional to m. In fact, the ratio Q/I may be regarded as simply a scaling factor, and once the distribution of m has been calculated, the shear flow distribution is identical to it but with different units. Still another advantage of q is that its value does not vary abruptly with local thickness changes, as does τ (see Fig. 3.26).

It may be seen from the derivation of (3.7.4) that the values of Q and I are normally those for the entire hull girder cross section, whereas the calculation of m is performed using a half-section. This is the convention that is usually followed. Obviously, the Q and I values for a half-section could also be used since their ratio is then unchanged. However, *m should always be calculated for a half-section*; if both port and starboard halves had been used in deriving (3.7.4), then there would have been two symmetric and identical cut faces, each carrying an axial shear force of $\tau\ t\ dx$. In this case the denominator of (3.7.4) would have to be $2tI$. The convention of using a half-section for the calculation of m is also more appropriate because in practice half-sections are used for structural drawings.

The calculation of m is illustrated in Fig. 3.23 for an idealized hull girder. For horizontal portions, the moment arm z is constant, and m therefore increases linearly with arc length. This occurs in the deck and bottom if there is no camber or deadrise. For instance, in the deck

$$m(s_1) = gt_D s_1$$

and

$$m_A = m(b) = gt_D b$$

In the side shell m is parabolic

$$m(s_2) = m_A + \int_0^{s_2} zt_s ds_2 = m_A + \left(gs_2 - \frac{1}{2}s_2^2\right)t_s$$

In ships of normal proportions, the parabola is very flat and hence the shear flow q is almost constant vertically.

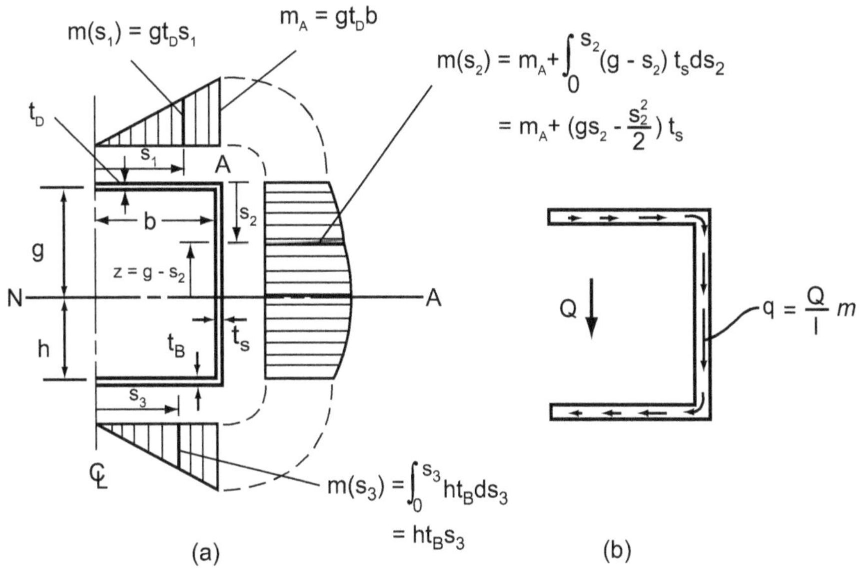

Figure 3.23 Calculation of moment term m by integrating along branches.

Because of changes in the orientation and thickness of the plates that make up the hull girder cross section, the integration for m is usually performed in segments. The integration is always commenced at the "open" end of any branch. As shown in Fig. 3.24, this need not be on the centerline; it may be at the edge of a hatch or other opening.

Figure 3.24 also shows the effect of multiple branches—for example, additional decks. If an imaginary cut were made at point C, the shear force at that point would have to balance the net imbalance in bending stress forces in the second deck and all plating above it. Therefore, all of this area must be included in the calculation of m at point C. The new area that is incurred in passing from B to C is the area of the second deck, and therefore the increment in m is equal to the total value of m for the second deck. That is, $m_C = m_A + m_B$ and, since q is directly proportional to m, $q_C = q_A + q_B$. This illustrates one of the reasons for the use of the term "shear flow": at any junction or branchpoint, the increment in the shear flow is equal to the flow contributed or taken away by the branch, as shown in Fig. 3.25. It should be noted that since deck and side plating may be of different thickness, this rule of continuity of shear flow does not hold for τ. Figure 3.26 illustrates how τ changes with changes in thickness.

It is well known (e.g., from Mohr's circle) that because of equilibrium the shear stress in a differential element at any point occurs in the form of two equal and opposite stress couples, one positive and the other negative. Since they are equal it makes little difference which is which, and so there is no need for a rigorous sign convention for either m or τ or q. The direction of the shear flow may be determined by inspection,

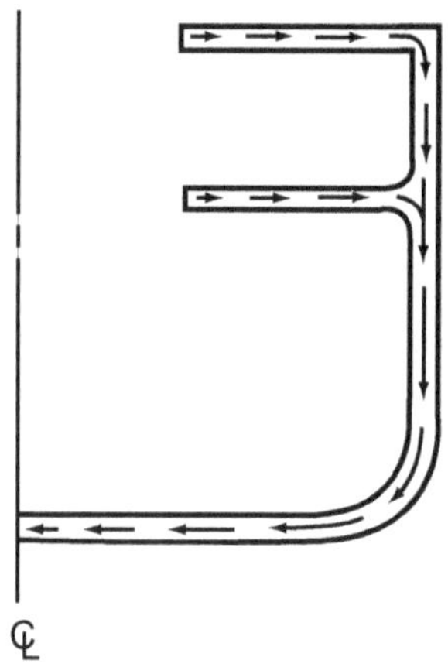

Figure 3.25 Sample diagram indicating direction of shear flow.

because in the webs of the hull girder it has the same direction—upward or downward—as the overall shear force, Q. In "open" sections such as that of Fig. 3.25 the flow is a "straight-through" flow; there is rarely a reversal in the flow direction and so usually it is only the magnitude of m, or of q, which is of interest, and its sign will usually be positive. Also, the integration for m is always commenced at the open end of each branch because that simplifies the computation. Therefore, the moment arm z is always taken as positive at the beginning of each branch, regardless of which side of the neutral axis the branch commences, and it only becomes negative if and when that particular branch crosses the neutral axis. For this reason, it is best to stop at the neutral axis and to finish that branch by starting from the other end. If this is not possible then the integration can proceed across the neutral axis, provided that a negative moment arm is used for all points on the other side.

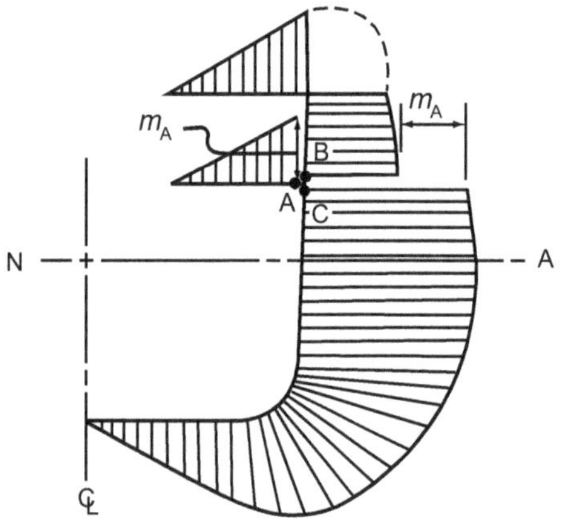

Figure 3.24 Conservation of shear flow at corners and branch points.

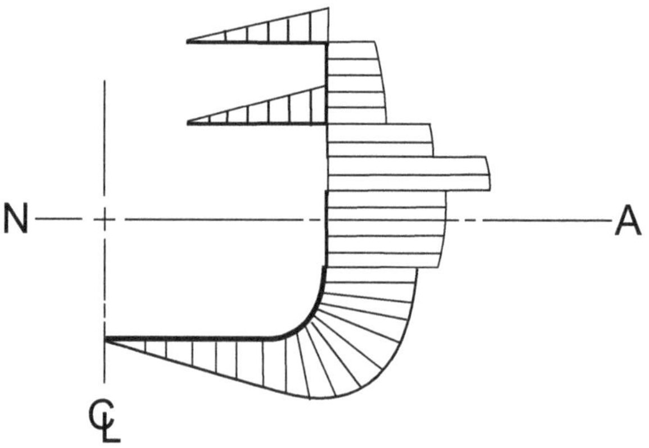

Figure 3.26 Change of τ due to change of thickness.

3.7.2 Shear Stress in Multicell Sections

In the definition of m in (3.7.3), it was assumed that the line integral was always commenced at a point of zero shear flow. Thus, as noted earlier, if there are any branch points (such as a second deck and side shell connection) then each branch must have zero shear flow at its end. This is equivalent to requiring that there should be no closed loops or "cells" within the overall half-section of the hull girder. Consequently, for the tanker of Fig. 3.27 the value of m can only be calculated along AB and FE; it cannot be calculated anywhere around the perimeter of the wing tank BCDEB. The difficulty arises from the fact that the shear flow divides at point B (and reunites at point E) and the separate components cannot be determined from simple statics. The problem is *statically indeterminate* and, as usual in such problems, additional information must be obtained from a consideration of geometric compatibility.

Simple beam theory assumes that plane cross sections remain plane. The *warping* of a cross section is the net axial deformation pattern that occurs when a cross section deforms out of plane. It does not include the axial displacement caused by a pure rotation of the section (about the neutral axis, due to bending). Rather it is the *net* axial displacement pattern over the cross section after the rotation due to bending is subtracted out. Warping is caused by shear stress, arising either from transverse shear force or from torsion. In the next section (3.8), we will examine how transverse shear force causes warping, and what errors warping causes in the predictions of simple beam theory. Then in Section 3.10 we will examine warping caused by torsion. It is clear that if we move around the perimeter of a closed cross section and record the warping at each point, when we come back to the starting point the warping must return to its original value. In Section 3.10 we will see that the net warping over the full perimeter must be zero, and this is the geometric compatibility condition that we need in order to solve for the shear stress in a cross section that includes closed cells. Therefore we will defer the treatment of such cross sections until Sections 3.10 and 3.11.

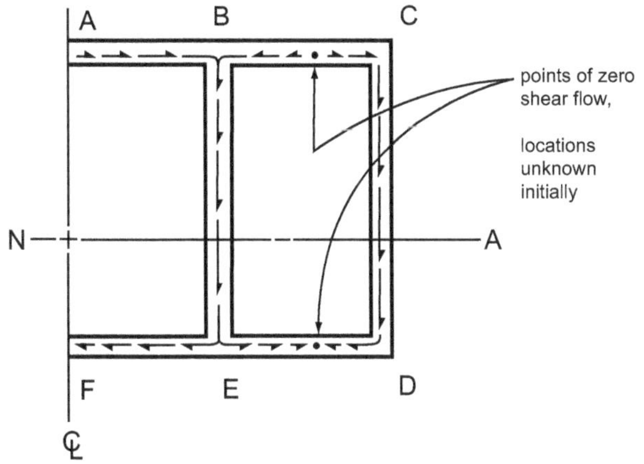

Figure 3.27 Shear flow in a multicell section.

3.8 SHEAR EFFECTS AND OTHER DEPARTURES FROM SIMPLE BEAM THEORY

3.8.1 Shear Lag

Simple beam theory assumes that plane cross sections remain plane, and that therefore the bending stress is directly proportional to the distance from the neutral axis. Thus in any flange-and-web type of beam, the stress should be constant across the flanges. However, in most cases the bending is not caused by the application of a pure couple to the ends of the beam; rather, it is caused by vertical loads, and these loads are absorbed by the webs of the beam and not by the flanges. That is, even for a hull girder, in which the vertical loads may initially act on the flanges (e.g., pressure on the bottom), they are immediately transferred to the webs by transverse beams and frames; the plating of the flanges can only take longitudinal in-plane loads (we are discussing principal loads, not small local loads). Therefore, the vertical loads act on the webs and cause them to deflect to some radius of curvature, thus inducing maximum strain in the flanges. Since they carry maximum strain, and hence maximum stress, the flanges make the largest contribution to the bending stiffness. But, it is important to note that this maximum strain comes initially from the webs and only reaches the flanges by shear. This is illustrated in Fig. 3.28, which shows a portion of a box girder cantilever loaded by a vertical force F. The force is reacted by, or carried by, the webs, which deflect to some radius of curvature such that the upper and lower edges of the web are elongated and shortened. For simplicity, the curvature is not shown; only the change in length. At the upper edge the elongated web pulls the flange plating with it, through shear forces, and this sets up shear stresses in the flange; these were discussed fully in the previous section. The bending and shear stresses cause stretching and in-plane distortion of the flange. On the left and right sides of the figure, an element is shown before and after this stretching and distortion.

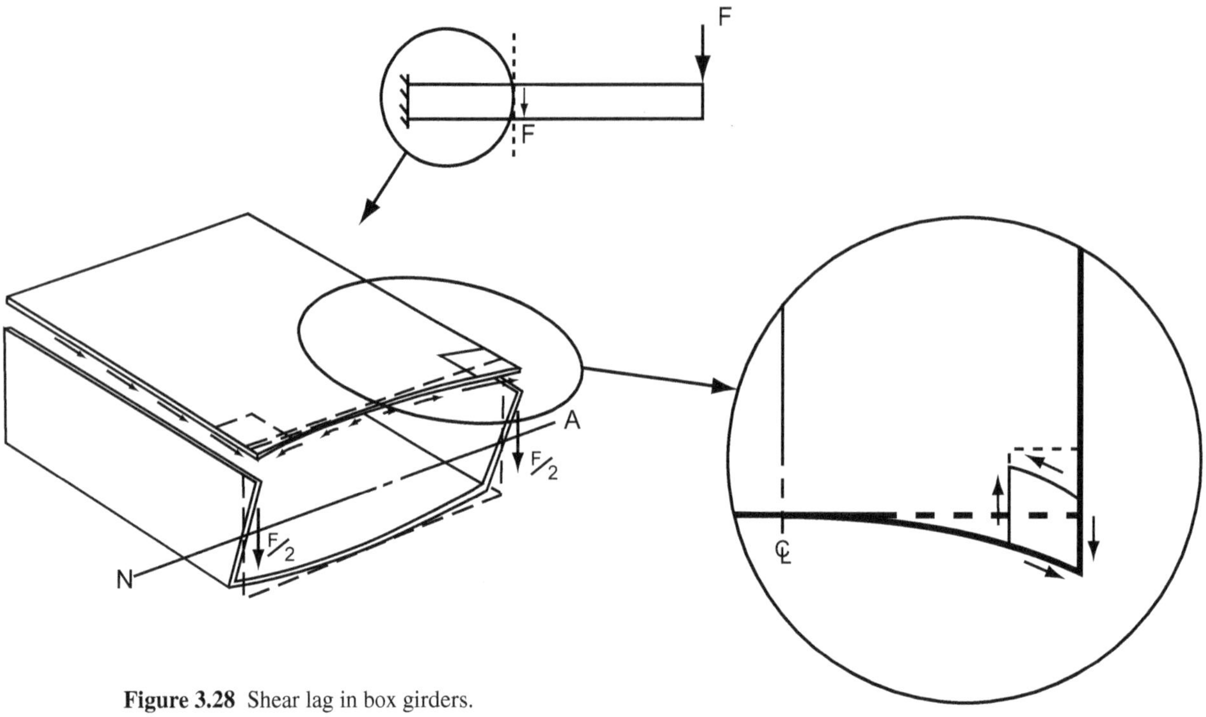

Figure 3.28 Shear lag in box girders.

The shear distortion is such that the inner edge of the element does not have to stretch as much as the outer edge; that is, the longitudinal strain is less at the inner edge and therefore so is the longitudinal stress. This same phenomenon will occur at each element, from the edges to the centerline, although it will diminish steadily and disappear at the centerline because the shear stress diminishes to zero. The overall result is that the flange undergoes in-plane longitudinal distortion and therefore *plane cross sections do not remain plane* when shear stress is present. This distortion is commonly referred to as "warping" of the cross section. The most significant aspect of the shear distortion is that the inner portion of the flange carries less bending stress, and is therefore less effective than the outer portion. That is, due to shear effects, the bending stress remote from a web "lags behind" the stress near the web. The phenomenon is therefore termed the "shear lag" effect. The same effect occurs in the compression flange since here the shear distortion allows fibers that are remote from a web to avoid some of the shortening and hence to carry less compressive stress. Shear lag occurs in any wide-flanged section that carries a lateral load in the manner of a beam. In an open section, such as the standard single web beam of Fig. 3.29*a*, it is the outer edges of the flange which are less effective.

The exact distribution of stress in a wide-flanged section can be found using the mathematical theory of elasticity, but this analysis, involving the use of stress functions, is too complex for design calculations. The exact analysis shows that the magnitude of the shear lag effect (i.e., the extent to which the distribution of bending stress differs from simple beam theory) is dependent on:

1. The ratio of width to length of the flange.
2. The distribution of lateral loading along the beam.
3. The relative proportions of web and flanges.
4. The type of section (single or multiple web, symmetric or unsymmetric, etc.).
5. The position along the beam. The shear lag effect in general varies from point to point along the length, and is maximum at maximum shear force gradient (concentrated loads).

Shear lag is of importance in beams having very wide flanges and shallow webs, such as aircraft wings. In steel box girders the effect is much smaller, even in box girder bridges (Dowling et al, 1977), which have large concentrated loads due to the point supports. In the hull girder bending of ships, the shear lag effect is usually only a few percent. It is more important in the consideration of the effects of superstructures and of the effective breadth of plating in local strength problems, which will be discussed later.

3.8.2 Effective Breadth Due To Shear Lag

Rather than using a mean value of flange stress as a way of allowing for shear lag, it is preferable to retain

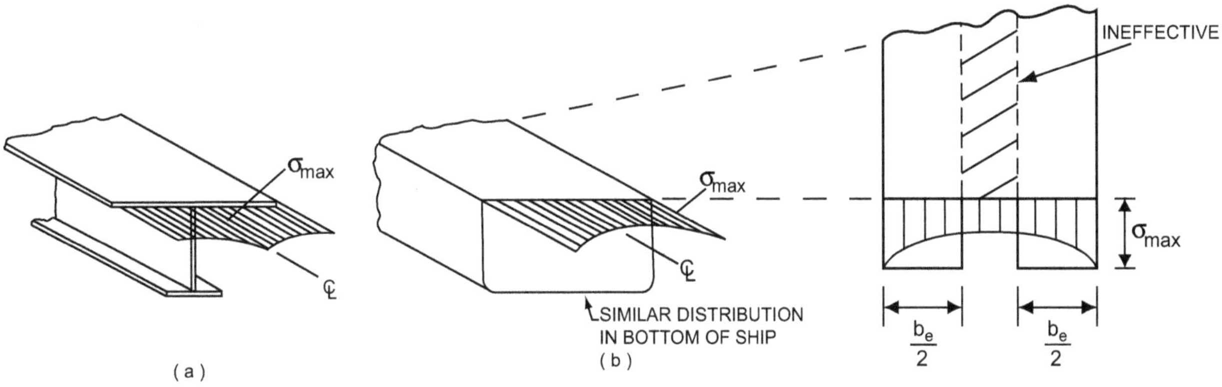

Figure 3.29 Shear lag in beam flanges.

the value of the maximum stress σ_{max} at the junction of the flange and web, so that it can be allowed for in designing the flange. Therefore, the usual way of allowing for shear lag is to use an "effective breadth" of flange, b_e, which is defined as:

> The breadth of plating which, when used in calculating the moment of inertia of the section, will give the correct maximum stress σ_{max} across the effective width of the flange, using simple beam theory.

Also, the effective breadth must be such that the total longitudinal force in the flange is equal in the actual and simplified cases. Equating forces

$$b_e t \sigma_{max} = b_e t \int_0^b \sigma_x \, dy$$

or

$$b_e = \frac{1}{\sigma_{max}} \int_0^b \sigma_x dy \tag{3.8.1}$$

Effective breadth is illustrated in Fig. 3.30. In Schade (1951, 1953) effective breadths are calculated for a wide variety of structures and loading conditions. These references should be consulted if it is required to investigate shear lag effects in detail.

The main conclusions from such analyses are:

1. The most important parameter that determines effective breadth of plating is the ratio of flange width b to the length L_o between points of zero bending moment. For simply supported beams, L_o = beam span. A low L_o/b ratio results in a small ratio of b_e/b.
2. Effective breadth varies from point to point along the span of a beam, being smallest at points of concentrated loading on the beam, where there is a discontinuity in the shear force curve. Conversely, there is no shear lag effect in pure bending (no shear).
3. Shear lag occurs in both tension and compression flanges equally, provided that in the latter case buckling does not occur.

For design purposes the effective breadth at the section of maximum bending moment is of most importance. Figure 3.31, derived from Schade (1951, 1953) enables effective breadths to be found at points of maximum bending moment. These breadths should then be used in calculating the effective moment of inertia I_e of the section, and hence the maximum bending stress in the beam.

$$\sigma_{max} = \frac{M_y z_{max}}{I_e}$$

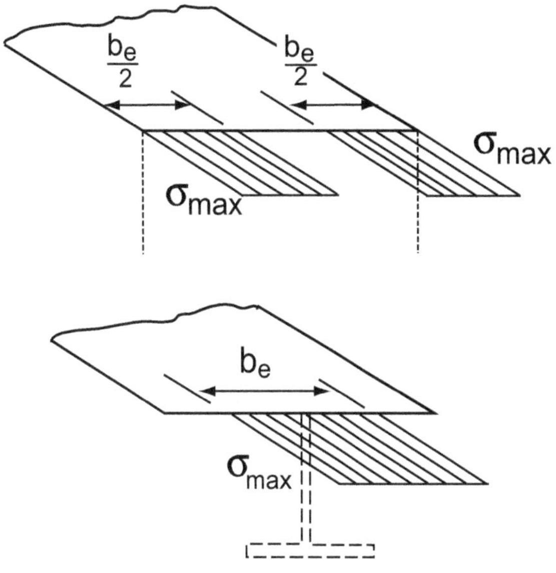

Figure 3.30 Effective breadth of flanges.

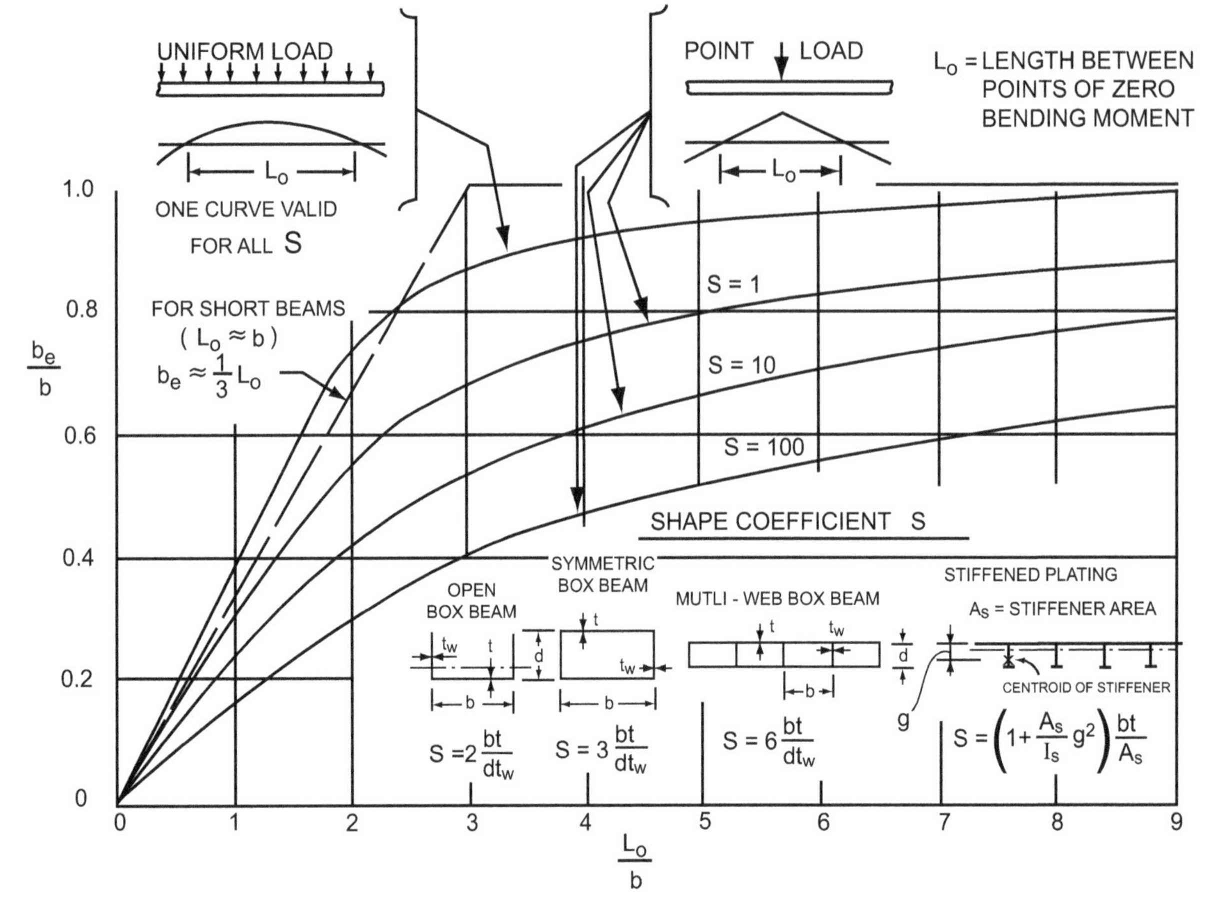

Figure 3.31 Effective breadth at maximum bending moment.

3.9 HULL-SUPERSTRUCTURE INTERACTION

The interaction between hull and superstructure (or deckhouse) is highly three dimensional and quite complex, and therefore it cannot be dealt with adequately by beam theory, but rather only by a 3D finite element model of the full ship. This is especially true for large passenger ships, where the superstructure is almost full length and many decks high. For example, Fig. 3.32 is shows a MAESTRO finite element model of a cruise ship. In Fig. 3.32*a* the colors indicate material type. In Fig. 3.32*b* the colors indicate panel type; i.e they reflect such properties as plate thickness and stiffener spacing. The increased variety of colors gives a better appreciation of the 3D complexity of the structure, showing that it could never be analyzed by beam theory.

Despite the complexity, there are some basic topological features that influence the degree to which the superstructure participates in the hull girder bending, and we therefore consider these briefly and qualitatively. This topic has long been of concern to naval architects but it was only when the finite element method was developed in the early 1960s that an adequate analysis became possible. Paulling and Payer (1968) was a landmark paper, and since then many other contributions have been made, particularly by naval designers such as Mitchell (1978) and McVee (1980), because naval vessels have relatively long superstructures, while at the same time, it is important to keep the weight of the superstructure as small as possible because of the criticality of vertical center of gravity in naval vessels.

3.9.1 House Side Continuous with Ship Side

In essence, a superstructure participates in hull girder bending to the extent that its webs (sides) are forced to act as a vertical continuation of the ship's sides and to undergo in-plane bending to the same radius as the hull. This will be maximal if the house sides are in the same plane as the ship's sides. Figure 3.33*a* illustrates the deflected shape and the distribution of longitudinal strain ε_x for such a case. The bottom fibers of the house side undergo the same extension as the top of the hull. At the midlength the upper deckhouse

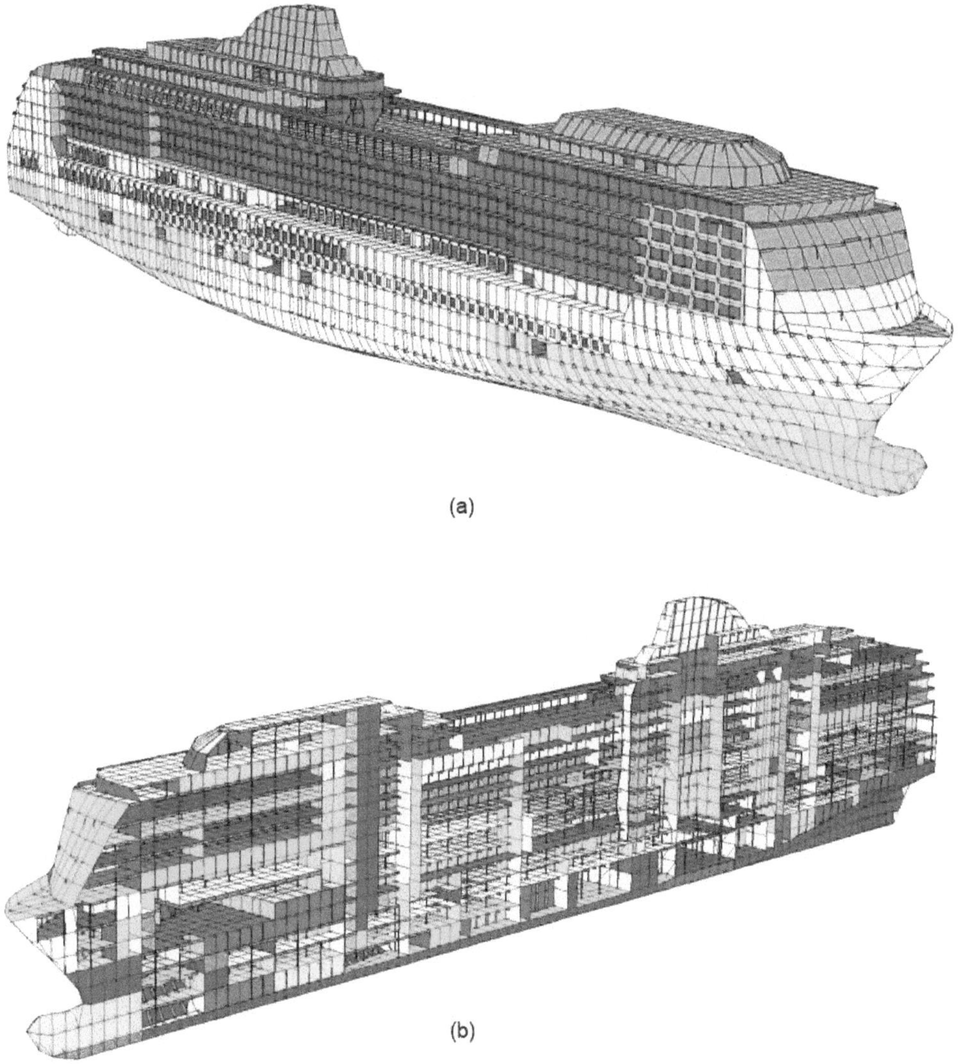

Figure 3.32 MAESTRO model of a cruise ship. (Picture courtesy of Kvaemer Masa Marine, Vancouver, BC, Canada, V6J 1T5).

fibers also contribute, and so the axis of zero strain is higher and the curvature $1/R$ is smaller than if the deckhouse was not present. Hence the vertical rate of increase in strain is smaller. But away from the midlength the upper deckhouse fibers undergo progressively less extension, and at the ends of the house ε_x will be zero.

Thus even though the bottom of the house side has the same curvature as the hull, the vertical distribution of strain is not linear because of the longitudinal shear deformation (warping) of the house side; that is, a plane cross section of house and hull does not remain plane. The figure shows typical strain distributions at the midlength and quarter points, and also the lengthwise variation of ε_x in the upper fibers of the house side. The sharp reentrant corners would be avoided in practice since they cause very high stress concentration.

3.9.2 House Side Offset from Ship Side

Figure 3.33*b* illustrates the effect if the house sides are not in line with the ship's sides. Because of the relative flexibility of the deck beams, the house sides are able to adopt a much larger radius of curvature and thereby escape from a good deal of the bending. (No matter how sturdy a deck beam may be, its flexural stiffness is only a small fraction of the in-plane stiffness of plating.) In this situation the strain is small even at the midlength, and the deckhouse is largely independent of the hull in regard to primary bending.

As the ship undergoes hogging and sagging, the bottom comers of the deckhouse will alternately pull upward and push downward on the deck, and sufficient area of attachment must be provided to keep the cyclic stresses sufficiently small in order to avoid fatigue. Also it is usually advisable to

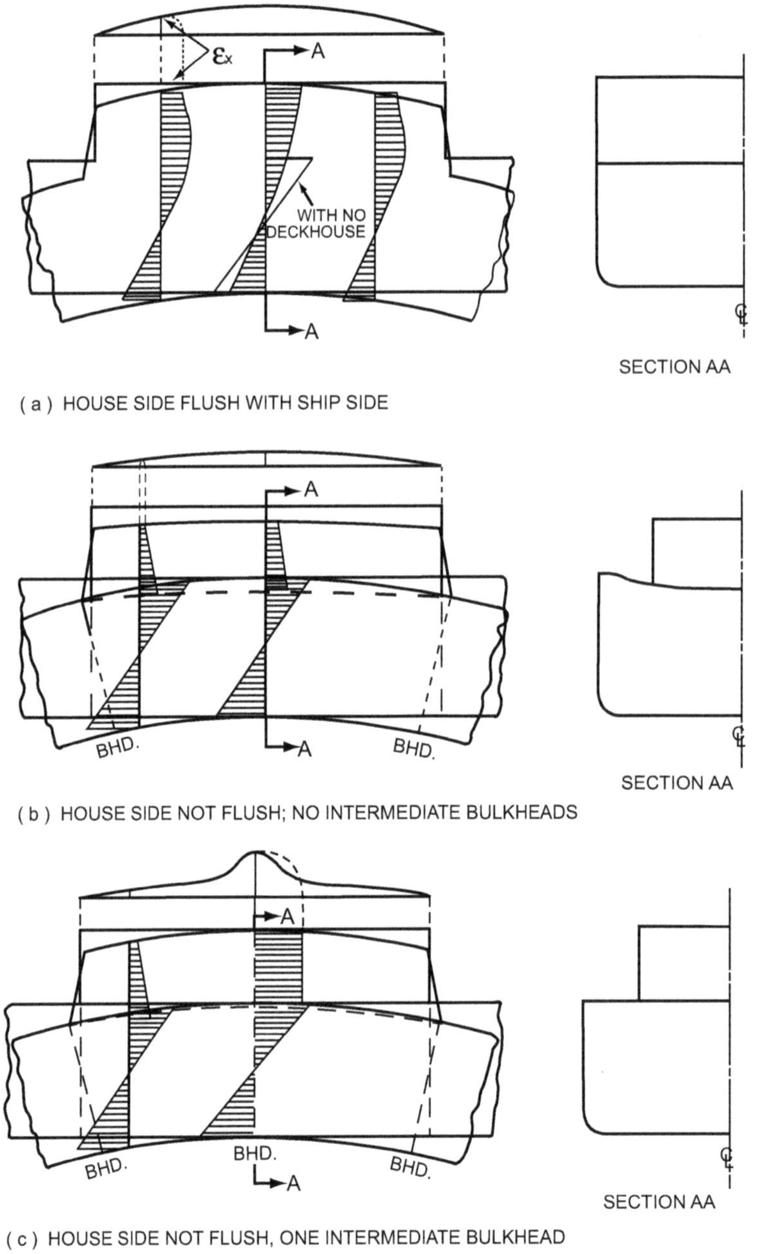

Figure 3.33 Interaction between hull and deckhouse.

terminate the deckhouse at transverse bulkheads in order to avoid excessive cyclic deflections and cyclic stresses in the deck structure.

3.9.3 Amidships Superstructure in Naval Ships

A particular challenge in the design of the superstructure of a naval ship is the need for Replenishment At Sea (RAS) operations. Because of pitch motions the location must be close to amidships, which is precisely where the hull girder bending moment is largest. RAS operations require a large open area on the main deck, on both sides of the ship. Consequently the superstructure in this region must be narrower than elsewhere. The difficulty is further increased if the superstructure is welded aluminum, which is more prone to fatigue than steel.

Figure 3.34 provides examples of the superstructure design for three classes of US Navy ships. Figure 3.34*a* is a MAESTRO finite element model of a Perry class frigate. Color indicates material type, the ship having an aluminum superstructure (the hull has more than one type of steel). The superstructure is relatively long, continuous and of essentially constant height. It would therefore be strongly affected by hull girder bending, especially in

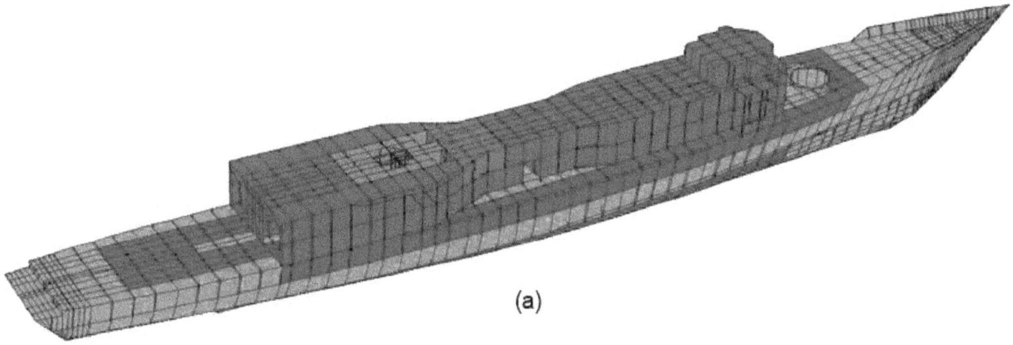

Figure 3.34 (a) MAESTRO finite element model of a Perry class frigate. *Continued on next page.*

the amidships region, where because of RAS operations it is narrower.

Figures 3.34*b,c* are a photo and a MAESTRO finite element model of a Ticonderoga class cruiser, which also has an aluminum superstructure. In this ship each end of the superstructure consists of a massive full-width block, which would follow the hull girder deflection and slope quite closely, and could cause large stresses in the amidships region. However, in this region the superstructure is lower and very narrow. It is possible that this portion of the superstructure might act as a relatively "soft" portion, thereby reducing the potentially large hull girder bending stress. If so, this region would have been a very challenging design.

Figure 3.34*d* is a photo of an Arleigh Burke class destroyer. Here the superstructure has been split into two entirely separate portions. Also, the portions forward and aft of the amidships gap are mostly the nearly void spaces of intakes and uptakes. The only significant superstructure is the full-width eight-sided "pyramid" at the forward end, and this structure is relatively short. Thus in this ship the superstructure (which is steel) would participate only slightly in hull girder bending.

3.10 TORSION OF PRISMATIC THIN-WALLED BEAMS

Torsion of noncircular sections causes longitudinal deformation of the cross section, called *warping*. In solid sections this has a negligible effect, and the assumption may be made that a plane cross section remains plane during torsion. But in thin-walled beams the warping deformation can induce considerable longitudinal normal stresses, called warping stresses. Although these are self-balancing, they cannot be considered as local stresses. The theory of torsion of thin-walled beams was developed by numerous authors, mainly in Germany. Classical and comprehensive texts include Vlasov (1961), Kolbrunner and Basler (1969), Kolbrunner and Hajdin (1972 & 1975). This theory is based on three assumptions:

1. Since the beam is everywhere thin-walled, the warping and the shear flow are assumed to be constant through the thickness.

2. During torsion the beam cross section undergoes rotation and longitudinal warping, but its shape is assumed to remain unchanged; that is, there is no distortion in the tranverse or cross-sectional plane.

3. The longitudinal warping causes additional (or "secondary") stresses of both types, normal stress and shear stress, but the theory ignores the effect of the latter; that is, it ignores the additional shear deformation caused by the secondary shear stress.

In their overall topology, most ship hulls have a closed cross section, and the hatches are simply openings. But a containership hull is essentially a beam of open cross section, and the provision of adequate torsional strength is vital and difficult. For thin-walled beams the torsional characteristics differ markedly, depending on whether the section is open or closed. Open sections have much less torsional stiffness than closed sections; that is, they undergo more rotation for a given twisting moment. Also, they exhibit much more warping, that is, non-uniform axial deformation u such that an initially plane cross section is no longer plane (see Fig. 3.35). In fact, the low rotational stiffness of open sections is directly linked to their ability to warp, and if warping of any section is prevented or partly restrained, then that section becomes effectively stiffer. Therefore, in our discussion of torsion we will deal separately with open sections and with closed sections, and within each of these

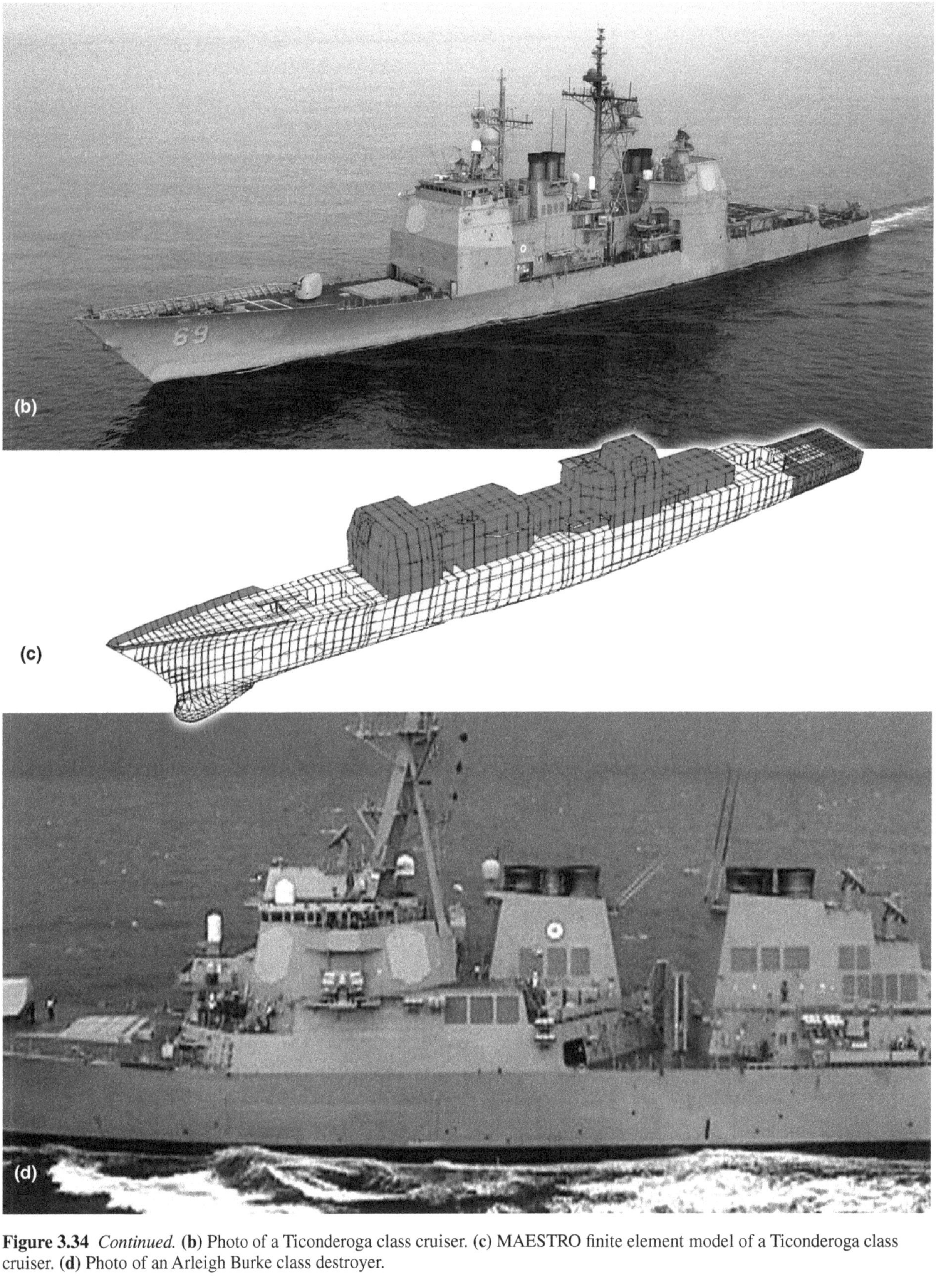

Figure 3.34 *Continued.* (**b**) Photo of a Ticonderoga class cruiser. (**c**) MAESTRO finite element model of a Ticonderoga class cruiser. (**d**) Photo of an Arleigh Burke class destroyer.

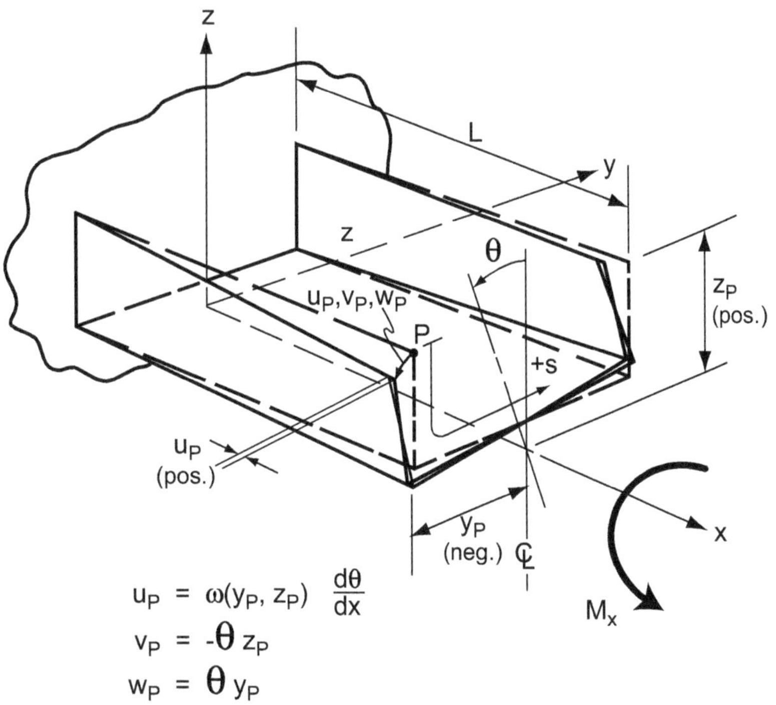

Figure 3.35 Torsional deflection of a prismatic thin-wall beam of open section.

we will first consider the case of free warping and then the effect of warping restraint. As noted previously, the theory deals only with prismatic members and it assumes that during the twisting the cross section retains its shape; that is, that it rotates in rigid body fashion about a point, which is referred to as the center of twist, and does not deform in its own plane. Although such deformation does occur due to transverse shear, its magnitude relative to the size of the cross section is very small (in prismatic members). Under these conditions the displacements of any point (y,z) in the cross section are (see Fig. 3.35)

$$\begin{aligned} u &= \omega_n(y,z)\theta' = \omega_n(s)\theta' \\ v &= -\theta z \\ w &= \theta y \end{aligned} \tag{3.10.1}$$

where $\theta' = d\theta\,/\,dx$ = rate of twist.

That is, the distribution of the warping around the cross section is given by a (normalized) *warping function* $\omega_n(y,z)$, which does not change in the axial direction x. Also, as will be shown, the magnitude of the warping is directly proportional to the rate of twist. In thin-wall sections the warping is constant through the thickness, and hence it can be expressed as a function of arc length s instead of y and z. If the cross section has an axis of symmetry then the center of twist* lies on this axis, and the warping is antisymmetric about this axis.

If there is no warping restraint, then torsion of any section—open or closed—is governed by the first-order differential equation

$$M_x = GJ\frac{d\theta}{dx}$$

where M_x = the twisting moment
G = the shear modulus
J = St. Venant's torsional constant

Figure 3.35 shows the sign convention that is used herein. The twisting moment M_x, the twist angle θ, and the arc length s are all positive in the clockwise direction when looking along the $+x$-axis. As noted earlier, open and closed sections differ markedly in regard to both their stress distribution and the degree of warping. We shall now describe each of these briefly, before going into detail.

Figure 3.36 shows the shear stress distribution in a member having an open section. The shear stress

* The center of twist of any section—open or closed—is identical to the shear center of that section.

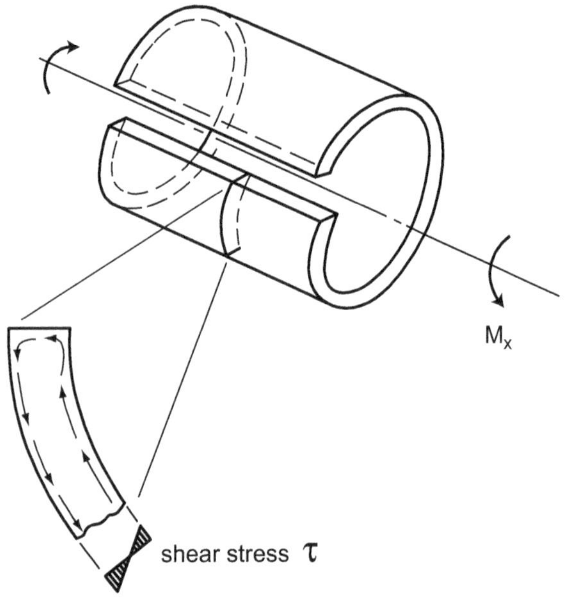

Figure 3.36 Chear stress in open sections.

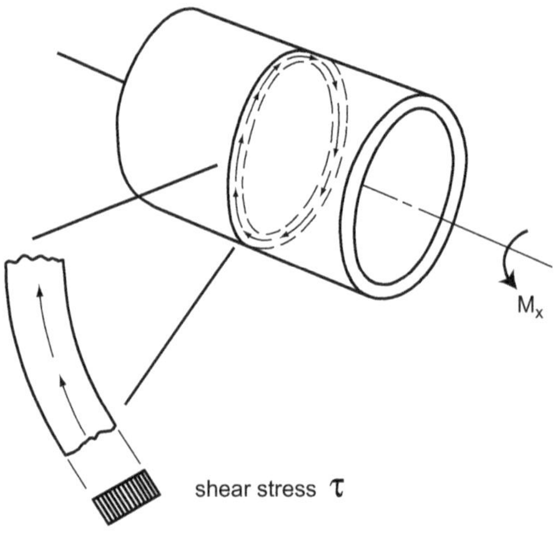

Figure 3.37 Shear stress in closed sections.

varies linearly through the thickness and is zero at the mid-thickness. Such a member, regardless of its shape, has essentially the same stress distribution as a flat slab of width b and thickness t, where b is the total arc length of the open cross section of the member. Thus in reality the member has a very slender "solid" cross section, and the shear stress can "circulate" (thus balancing the applied twisting moment) only within this confined area. Also, the shear stress is maximum along the outside edges and much lower everywhere inside, and this is a quite inefficient distribution. The main problem is that because of the slenderness of the member, the moment of these internal stresses is small, and so equilibrium between this internal moment and the applied twisting moment necessarily involves a large twisting angle and large peak values of shear stress.

In a closed cross section (see Fig. 3.37) the "circulation" of shear stress occurs around the closed path formed by the section, and this has two advantages: the stress is constant through the thickness, thus giving a better utilization of material, and it has a much larger moment arm about the center of twist. The result is that a closed section involves much less rotation and shear stress for a given twisting moment. Also, as will be shown subsequently, in a closed section the *shear flow* q ($q = \tau t$) is constant around the section.

We turn now to the question of warping of the two types of sections. In an open section, warping is not caused by shear. In fact, as we have seen, the mean value of shear stress (averaged through the thickness) is everywhere zero. As a result, the warping (as long as it is not restrained) is caused entirely by rigid body rotation of the member wall within its own plane. This rotation is the direct kinematic complement to the axial rotation of the member. Figure 3.38 shows that for a differential member of length dx, the tangential

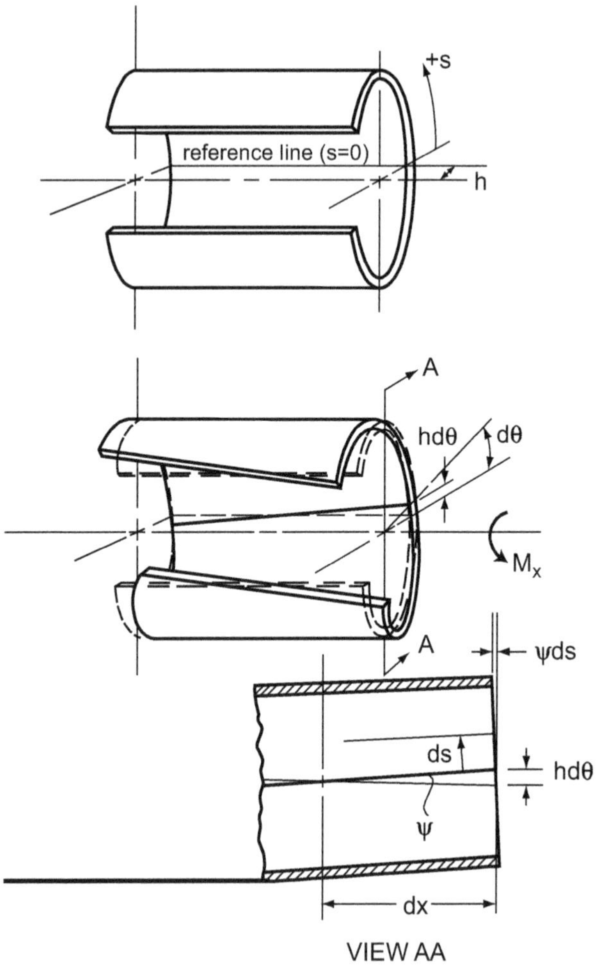

Figure 3.38 Warping in an open section.

displacement of any point around the section is $hd\theta$, where h is the distance from the center of twist to the tangent to the point in question. This distance shall be considered positive when, in conjunction with the positive direction of the arc length coordinate s, it would correspond to a positive twist. The tangential displacement causes a "spiral" angle $\psi = hd\theta / dx = h\theta'$. Since the net shear stress is zero through the wall thickness, there is no in-plane shear deformation of the wall. Instead, the wall rotates in rigid body fashion and so the cross section rotates out of its plane by this same angle, ψ, and this causes a local warping displacement $du = -\psi\, ds = -h\theta'\, ds$. Therefore, the warping displacement of any point in the cross section, as a function of arc length s around the section, is proportional to $\omega(s)$ which is defined as

$$\omega(s) = \int_0^s h(s)\,ds \qquad (3.10.2)$$

This function is related to the normalized warping function by

$$\omega_n(s) = -\omega(s) + \omega_0 \qquad (3.10.3)$$

The quantity ω_0 corresponds to a constant of integration. As will be shown subsequently, it is determined by axial equilibrium requirements.

In a closed section, warping is necessarily much smaller because regardless of how it may vary as a function of arc length around the (closed) cross section, it must return to the same value after a complete cycle. Hence for a circular section there is no warping whatever. The underlying reason for the reduction in warping is the presence of shear stress. As we have seen, in a closed section the shear stress τ is nonzero and is essentially constant through the thickness. The shear strain $\gamma = \tau / G$ causes shear deformation of the member walls which, as shown in Fig. 3.39, corresponds to a local rotation of the cross section that is opposite to the rigid body rotation ψ. Hence the net warping of a differential element of arc length ds is

$$du = (\gamma - \psi)ds = \left(\frac{\tau}{G} - h\theta'\right)ds \qquad (3.10.4)$$

There is always some cancellation between these two components and hence there is less warping with closed sections.

If warping is restrained then there is some additional resistance to the twisting moment, and it will be shown that for all sections—open

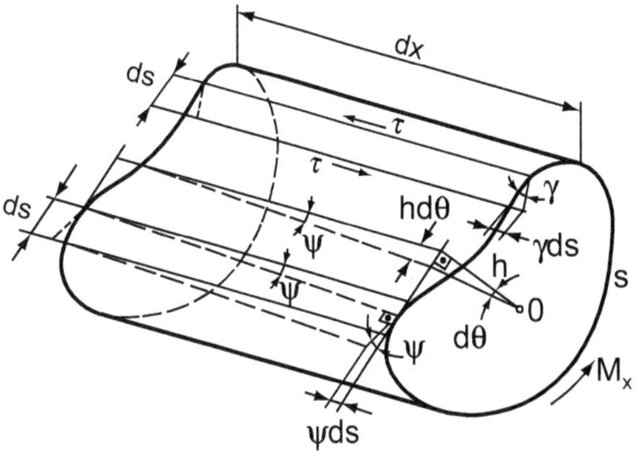

Figure 3.39 Warping in a closed section.

or closed—this gives rise to an additional term in the differential equation such that it becomes

$$M_x = GJ\frac{d\theta}{dx} - EI_\omega \frac{d^3\theta}{dx^3}$$

in which I_ω is the warping torsional stiffness of the section, to be defined subsequently.

3.10.1 Open Sections: Free Warping

SHEAR STRESS

As noted earlier, for free warping the relationship between twisting moment and rate of twist is

$$M_x = GJ\theta'$$

and it may be shown (see, for example, Kolbrunner & Basler [1969]) that for open sections St. Venant's torsional constant is given by

$$J = \frac{1}{3}\int_0^b t^3 ds \qquad (3.10.5a)$$

where b is the total arc length of the cross section. For a section composed of straight portions, a more convenient form of equation is

$$J = \frac{1}{3}\sum_{i=1}^{n} t_i^3 b_i \qquad (3.10.5b)$$

where b_i and t_i are the breadth and thickness of the individual portions of the section.

The shear stress at any point is maximum at the outer surfaces and zero at the mid-thickness. The maximum value is given by

$$\tau_{\max} = \frac{M_x t}{J}$$

WARPING

As explained earlier, warping in open sections is due entirely to the transverse rotation of the cross section which necessarily accompanies the axial rotation so as to give rigid body rotation because the net shear stress is zero everywhere around the section. The distribution of the warping around the cross section is given by the normalized warping function defined in (3.10.2) and (3.10.3). The value of ω_0 and the location of the shear center are determined from the conditions that the net warping must be zero (since otherwise there would be extension of the member) and that the net first moments of the warping, both horizontal and vertical, about the center of twist must be zero (because to be otherwise would require a horizontal or vertical bending moment acting on the member; see Kolbrunner & Basler (1969) for details. Stated mathematically these three conditions are

$$\int_0^b \omega_n(s)t\,ds = 0 \qquad (3.10.6a)$$

$$\int_0^b y\omega_n(s)t\,ds = 0 \qquad (3.10.6b)$$

$$\int_0^b y\omega_n(s)t\,ds = 0 \qquad (3.10.6c)$$

In the absence of symmetry the procedure to evaluate ω_0 and to determine the location of the center of twist is to define a set of reference axes η,ζ with the origin at the centroid, as shown in Fig. 3.40*a*, and then calculate the following quantities

$$\left.\begin{aligned} \omega_c(s) &= \int_0^s h_c\,ds \\ I_\zeta &= \int_0^b \eta^2 t\,ds \\ I_\eta &= \int_0^b \zeta^2 t\,ds \end{aligned}\right\} \qquad (3.10.7)$$

Then calculate

$$(\omega_c)_0 = \frac{1}{A_s}\int_0^b \omega_c(s)t\,ds \qquad (3.10.8)$$

in which A_s is the total area of the cross section:

$$A_s = \int_0^b t\,ds$$

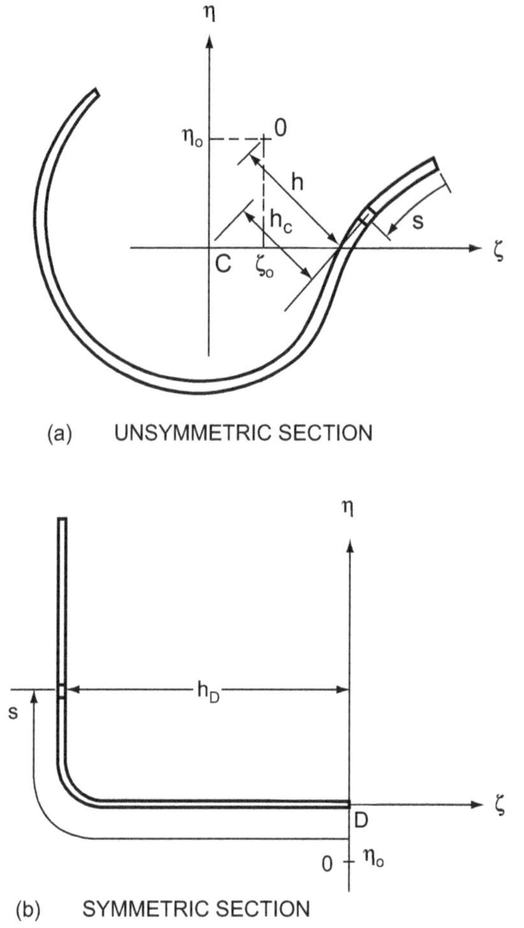

Figure 3.40 Calculation of center of twist.

Thus, $(\omega_c)_0$ is the mean value of the "centroidal" warping function $\omega_c(s)$.

The location of the center of twist is

$$\left.\begin{aligned} \zeta_0 &= \frac{1}{I_\zeta}\int_0^b \eta\left[\omega_c(s) - (\omega_c)_0\right] t\,ds \\ \eta_0 &= \frac{+1}{I_\eta}\int_0^b \zeta\left[\omega_c(s) - (\omega_c)_0\right] t\,ds \end{aligned}\right\} \qquad (3.10.9)$$

Then calculate $\omega(s)$ from (3.10.2), calculate its mean value ω_0 as in (3.10.8) but now with ω in place of ω_c, and then obtain $\omega_n(s)$ from (3.10.3).

If the section has an axis of symmetry, then the center of twist lies on that axis and the foregoing calculations may be simplified. Since ships are usually symmetric about their vertical centerplane, we shall describe the calculations for the case when the *z*-axis is an axis of symmetry. First, when there is symmetry it is not necessary to place the origin of the η,ζ axes at the centroid. Instead it may be placed at any convenient point on the axis of symmetry, say *D*. If the ship has a flat bottom it is advantageous

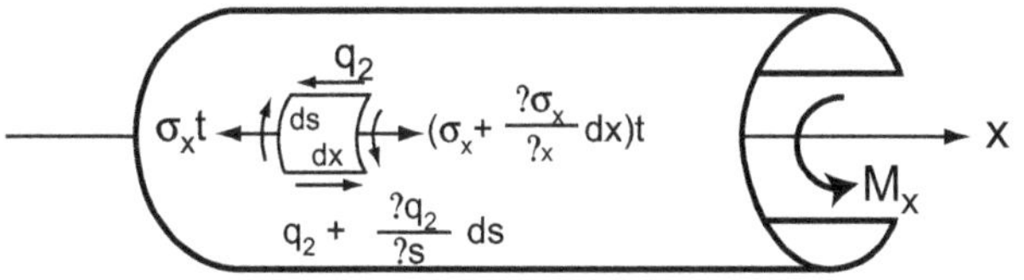

Figure 3.41 Stresses due to warping restraint.

to place the origin there because then h_D, the distance to the tangent line of any point, is zero for all points in the bottom. Also, s should be measured from the centerline because this greatly simplifies matters. Note that s always follows a right hand rule. In Fig 3.40 +x is into the page, and so +s is clockwise. With these conventions the resulting equations are

$$\omega_D(s) = \int_0^s h_D ds \tag{3.10.10}$$

$$\eta_0 = \frac{1}{I_\eta} \int_0^b \zeta \omega_D(s) t\, ds \tag{3.10.11}$$

$$\omega_n(s) = -\omega_D(s) + \eta_0 \zeta \tag{3.10.12}$$

3.10.2 Open Sections: Warping Restrained

If a beam is composed of a series of prismatic but differing segments, each segment will have a different warping response and hence will interfere to some extent with the warping of adjacent segments. When warping in a prismatic member is restrained, axial stresses σ_x are set up in the member and, as we shall see, these stresses are not constant in the axial direction. Moreover, since they are not constant they are accompanied by a secondary distribution of shear stress τ_2 over and above the primary distribution due to the (free warping) torsion. Since the warping is constant through the thickness, so also are σ_x and τ_2, and therefore it will be more convenient to deal with shear flow, $q_2 = \tau_2 t$. Note that in this regard the secondary shear stress differs from the primary shear stress τ_1 that (for open sections) varies linearly through the thickness. The total shear stress in the member is the sum of τ_1 and τ_2.

We may obtain expressions for q_2 and σ_x by considering axial equilibrium of a differential element, as shown in Fig. 3.41.

$$\left(\sigma_x t + \frac{\partial \sigma_x t}{\partial x} dx\right) ds - \sigma_x t\, ds + \left(q_2 + \frac{\partial q_2}{\partial s} ds\right) dx - q_2 dx = 0$$

or

$$\frac{q_2}{s} + t\frac{\sigma_x}{x} = 0$$

Therefore

$$q_2(s,x) = -\int_0^s \frac{\partial \sigma_x}{\partial x} t\, ds + (q_2)_0 \tag{3.10.13}$$

The constant of integration $(q_2)_0$ will be zero, provided that $q_2 = 0$ at $s = 0$; this will be satisfied as long as the integration is commenced at one of the free edges of the member. Moreover, it is automatically satisfied if the integration is performed over the entire section, as it will be in some of the derivations that follow.

The value of σ_x is obtained as follows:

$$\sigma_x = E\varepsilon_x = E\frac{\partial u}{\partial x} = E\frac{\partial}{\partial x}\left[\omega_n(s)\theta'\right]$$

or

$$\sigma_x = E\omega_n \frac{d^2\theta}{dx^2} \tag{3.10.14}$$

Substituting this into (3.10.13) gives

$$q_2(s,x) = -\int_0^s E\frac{d^3\theta}{dx^3}\omega_n t\, ds = -E\frac{d^3\theta}{dx^3}\int_0^s \omega_n t\, ds \tag{3.10.15}$$

with s measured from one of the free edges of the member. The secondary shear flow q_2 sets up a secondary twisting moment M_{x2} that may be evaluated as follows:

$$M_{x2} = \int_0^b q_2 h\, ds$$

and from (3.10.15) this is

$$M_{x2} = -EI_\omega \frac{d^3\theta}{dx^3} \tag{3.10.16}$$

where

$$I_\omega = \int_0^b \left(\int_0^s \omega_n t\, ds\right) h\, ds$$

The latter may be simplified by rearranging as follows:

$$I_\omega = \int_0^b \omega_n t\, ds \int_0^b h\, ds - \int_0^b \int_0^s (h\, ds)\, \omega_n t\, ds$$

and observing that the first integral is zero from (3.10.6a). Also, from (3.10.2) and (3.10.3) we see that the inner integral of the second term is

$$\int_0^s h\, ds = \omega_0 - \omega_n$$

Hence the expression becomes

$$\begin{aligned} I_\omega &= \int_0^b (\omega_n - \omega_0)\, \omega_n t\, ds \\ &= \int_0^b \omega_n^2 t\, ds - \omega_0 \int_0^b \omega_n t\, ds \end{aligned}$$

and again (3.10.6a) means that the second integral is zero. Therefore we finally obtain

$$I_\omega = \int_0^b \left[\omega_n(s)\right]^2 t\, ds \qquad (3.10.17)$$

This quantity is referred to as the *warping torsional stiffness*. The product EI_ω that occurs in (3.10.16) is the *warping torsional rigidity* and, as the equation indicates, the additional resistance to twisting that arises when warping is restrained is directly proportional to EI_ω and to the third derivative of the twisting angle. The general equation governing the torsion of open sections is

$$M_x = GJ\frac{d\theta}{dx} - EI_\omega \frac{d^3\theta}{dx^3} \qquad (3.10.18)$$

As we shall now see, this equation is also valid for closed sections, but the expression for J is different.

3.10.3 Closed Sections: Free Warping

SHEAR STRESS

In a closed section the shear stress τ is constant through the thickness. We seek now to determine how τ varies around the section. Figure 3.42*b* shows an element of varying thickness, and from the requirement that the net axial force is zero we have

$$(\tau_b t_b - \tau_a t_a)\,dx = (q_b - q_a)\,dx = 0$$

and therefore the primary or St. Venant shear flow q is constant around the section.

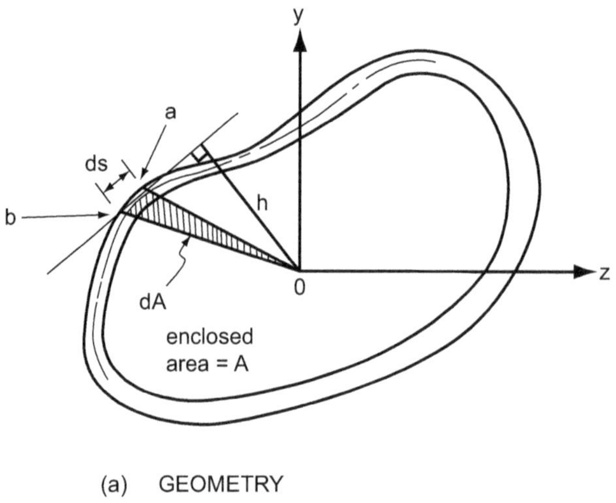

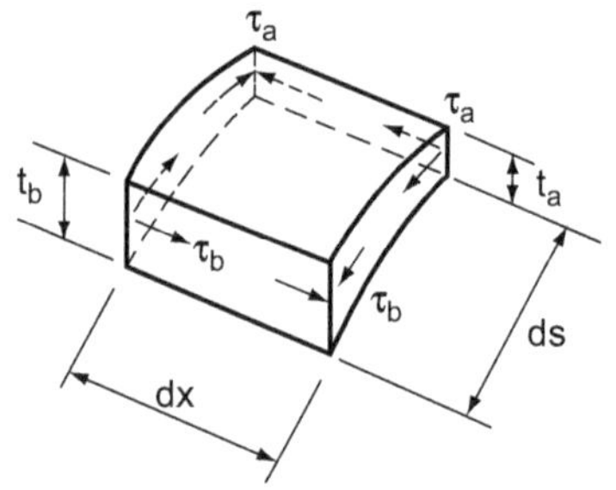

Figure 3.42 Geometry and equilibrium of stresses in a closed section.

Let us now consider overall equilibrium between the applied twisting moment M_x and the moment of the internal shear stress. Figure 3.42*a* shows that the force due to an element *ds* is *q ds* and the moment it exerts about the center of twist is

$$dM_x = hq\, ds$$

From the geometric properties of a triangle the sector area subtended by *ds* (shown shaded in Fig. 3.42a) is

$$dA = \frac{1}{2} h\, ds$$

and so dM_x can be expressed in terms of this sector area

$$dM_x = 2q\, dA$$

The total moment is obtained by integrating around the full section which gives

$$M_x = 2qA \qquad (3.10.19)$$

where A is the total area enclosed by the perimeter of the closed section. The shear flow around the section is then

$$q = \frac{M_x}{2A} \qquad (3.10.20)$$

For a typical member segment of length dx, we now invoke the principle of strain energy, which states that the work done by M_x equals the internal strain energy:

$$\tfrac{1}{2} M_x d\theta = dx \oint \tfrac{1}{2} \frac{\tau^2}{G} t\, ds$$

and therefore, from (3.10.20) and $q = \tau t$

$$M_x = GJ \frac{d\theta}{dx} = GJ\theta' \qquad (3.10.21)$$

where

$$J = \frac{4A^2}{\oint \frac{ds}{t}} \qquad (3.10.22)$$

This value of J for closed sections differs markedly from the value for open sections given by (3.10.5). For most section shapes, it greatly exceeds the latter. For example, in a square box section of side $600t$ (typical of ships) the ratio of the two values is 270,000! Hence in a closed section that has attached open section members, such as longitudinal stiffeners on a hull, the latter's St. Venant torsional stiffness J may be ignored.

WARPING

As discussed earlier, and as shown in Fig. 3.39, the warping of a closed section involves the combination and partial cancellation of the shear deformation and the "rotational" or rigid body warping. The net warping distribution is

$$u(s) = \int_0^s \left(\frac{\tau}{G} - h\theta' \right) ds + \omega_0 \theta' \qquad (3.10.23a)$$

By means of $\tau = q/t$ and equations (3.10.19), (3.10.20), and (3.10.21), this may be converted to

$$u(s) = \left[\omega_n(s)\right]\theta' = \left(\frac{J}{2A} \int_0^s \frac{ds}{t} - \int_0^s h\, ds + \omega_0 \right)\theta' \qquad (3.10.23b)$$

3.10.4 Closed Sections: Warping Restrained

Here again, as for open sections, any restraint on the warping will introduce a nonuniform axial stress distribution $\sigma_x(s, x)$ and an accompanying secondary shear flow $q_2(s, x)$. The derivation of q_2 exactly parallels the derivation given for open sections earlier in this Section, with the warping function now given by (3.10.23). The only difference is that the constant of integration $(q_2)_0$ in (3.10.13) can no longer be automatically set to zero because there is now no free edge at which q_2 is known to be zero, and which would provide a convenient starting point for the integration. Thus the equation corresponding to (3.10.15) is

$$q_2(s,x) = -E \frac{d^3\theta}{dx^3} \int_0^s \omega_n t\, ds + (q_2)_0 \qquad (3.10.24)$$

In order to evaluate $(q_2)_0$ we make use of the fact that since the section is closed, $q_2(s)$ must be such that the net axial displacement strain that it causes over one full cycle of the section is zero. Therefore we make a cyclic evaluation of the warping caused by q_2, and to do so we divide q_2 into two parts: one part, say q_2*, which corresponds to the first term of (3.10.24) and for which the value at the starting point of the integration is taken to be zero; and a second part that is the true, but unknown, starting value $(q_2)_0$. The latter is constant and hence the foregoing condition is

$$\frac{1}{G} \oint \frac{q_2^*}{t} ds + \frac{(q_2)_0}{G} \oint \frac{ds}{t} = 0$$

from which

$$(q_2)_0 = -\frac{\oint \frac{q_2^*}{t} ds}{\oint \frac{ds}{t}} \qquad (3.10.25)$$

After solving for $(q_2)_0$, the additional resisting moment that results from q_2 is found in the same way as for an open section

$$M_{x2} = \oint q_2 h\, ds = -EI_\omega \frac{d^3\theta}{dx^3}$$

where

$$I_\omega = \int_0^b \omega_n^2 t\, ds$$

and ω_n is given by (3.10.23*b*). The foregoing equations are the same as (3.10.16) and (3.10.17), and so these equations apply to both open and closed sections, with the warping function being given by (3.10.3) or (3.10.23b), respectively. The integration for I_ω may be done piecewise, in any order and in any direction, such that each portion of the section is traversed once only. Therefore (3.10.18) is the governing differential equation for both types of sections, with J being calculated from either (3.10.5) or (3.10.22) as appropriate.

3.10.5 Multiple Cell and Mixed Sections

MULTIPLE CELL: FREE WARPING

In a section containing n cells, each cell comprises a closed section and so the overall shear flow consists of the superposition of n separate circulating shear flows, one in each cell, each of which is constant around the perimeter of its cell. Along the interface between any two cells the shear flow is the algebraic sum of the two circulating shear flows. From (3.10.19) the torque transmitted by each cell is

$$M_{xi} = 2A_i q_i$$

and the total torque transmitted by the section is the sum of these values

$$M_x = \sum_{i=1}^{n} 2A_i q_i \qquad (3.10.26)$$

Since there are n unknown values of q, the problem is statically indeterminate and we must invoke geometric compatibility. In a closed section the geometric condition is that there must be zero net warping in one complete circuit of each cell. From (3.10.23*a*) this condition is

$$\frac{1}{G}\oint \tau\, ds - \theta' \oint h\, ds = 0$$

in which ω_0 does not appear because the integral is exactly one cycle. Rearrangement gives

$$\frac{1}{\theta'}\oint \frac{q}{t} ds = 2AG$$

As the integration proceeds around the ith cell, the shear flow along each wall of the cell will be the value associated with that cell, q_i, plus the flow from the adjacent cell, if any. For example, in Fig. 3.43 there are two walls that adjoin other cells, designated $i - 1$ and $i + 1$. For convenience we regard all of the circulating shear flows as circulating in the positive (i.e., counterclockwise) direction. If the solution gives a negative value for any particular shear flow, this indicates that the shear flow around that cell is clockwise. With this convention the contribution of adjacent cells will always be opposite to q_i and must be subtracted. Thus for the ith cell we have the equation

$$\frac{q_i}{\theta'}\oint \frac{ds}{t} - \frac{q_{i-1}}{\theta'}\int_{i-1,i} \frac{ds}{t} - \frac{q_{i+1}}{\theta'}\int_{i+1,i} \frac{ds}{t} = 2A_i G$$

Since there is no transverse deformation all cells undergo the same rate of twist, θ'. Therefore although θ' is an unknown in the foregoing equation, it will have the same value in all such equations and so it can be regarded as a normalizing factor that is applied to all of the unknown values of q. Therefore we define a normalized shear flow

$$\bar{q}_i = \frac{q_i}{\theta'} \qquad (3.10.27)$$

Also, we note that the integrals in the equation may all be evaluated from the given geometry, and so they constitute known coefficients, which will be represented by the symbol C. The equation then becomes

$$C_i \bar{q}_i - C_{i-1}\bar{q}_{i-1} - C_{i+1}\bar{q}_{i+1} = 2A_i G$$

This equation is written for all cells, resulting in a system of n equations

$$[C]\{\bar{q}\} = 2G\{A\} \qquad (3.10.28)$$

which can be solved for $\{\bar{q}\}$. The total torque is, from (3.10.26),

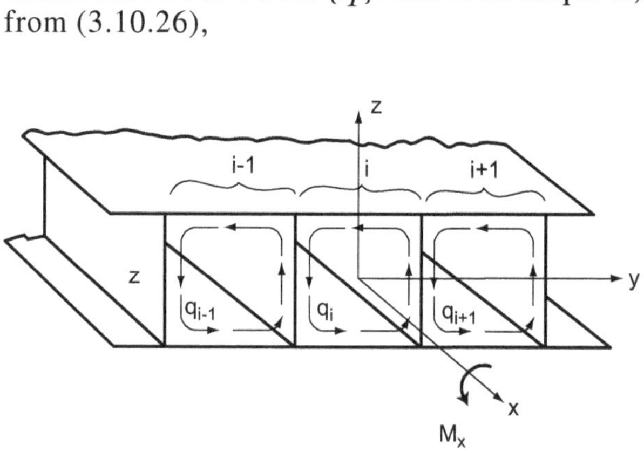

Figure 3.43 Shear flow in a multi-cell section.

$$M_x = 2\{A\}^T\{q\} = 2\theta'\{A\}^T\{\bar{q}\}$$

and therefore

$$\theta' = \frac{M_x}{2\{A\}^T\{\bar{q}\}}$$

From the latter equation we see that the denominator is the torsional rigidity of the overall section, corresponding to GJ for a single closed section. That is, for a multicell section

$$GJ = 2\{A\}^T\{\bar{q}\} \qquad (3.10.29)$$

After solving for θ', the values of q may be calculated from the normalized values.

MULTIPLE CELL: WARPING RESTRAINT

The effect of warping restraint is accounted for by the warping torsional stiffness, $I_\omega = \omega_n{}^2 t\, ds$, and for multiple cells the integration must be done in such a way that no wall is traversed more than once. Some typical examples are illustrated in Fig. 3.44. The figure also shows that if the cells are slender relative to the overall dimensions, such as a double-wall side shell, a double bottom, or an isolated "torsion box," then for the calculation of I_ω it is sufficiently accurate to combine the two walls and to lump other areas such as cross webs. Stiffeners and other small members may be ignored.

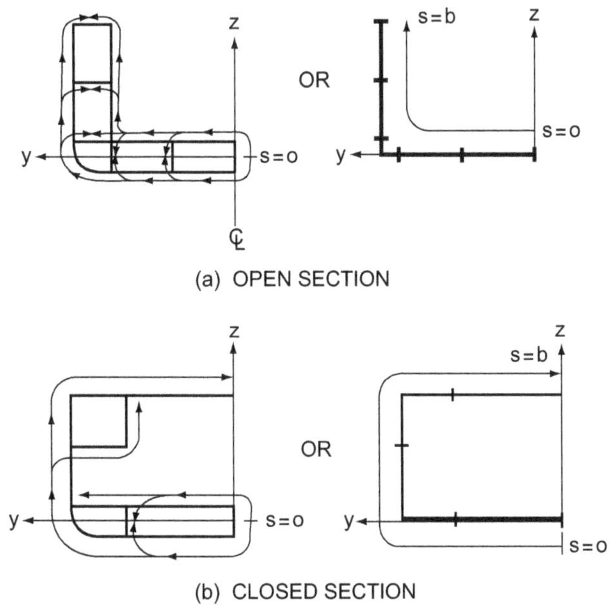

Figure 3.44 Integration paths for calculating I_ω.

MIXED SECTIONS

If the overall cross section contains both closed portions and open portions, the total St. Venant stiffness is the sum of J for these two portions, and the total warping stiffness is the sum of I_ω for the two portions.

As mentioned earlier, the value of J for open sections is negligible compared to that for closed sections, unless the open section is very long or very thick-walled. In fact, in a mixed section the only significant contribution of open section portions is their contribution to the warping torsional stiffness I_ω. This can be estimated with sufficient accuracy by assuming that the warping is constant within that portion and is equal to the value in the closed section at the point of attachment to the latter. With this assumption the area of the open portion may be lumped at the point of attachment, as shown in Fig. 3.45*a*.

Figures 3.44 and 3.45 illustrate some modeling assumptions and idealizations which may be used, depending on whether the overall section is open or closed. Because of symmetry the calculations are usually performed for a half section, as shown in the figures.

3.10.6 Numerical Calculation of the Warping Function

For hand calculation $\omega_n(s)$ is calculated from the equations just derived but for automated or computer-based applications a numerical approach is more suitable. As shown by Herrmann (1965) the distribution of $\omega_n(s)$ can be represented in a discretized manner by defining "nodes" at which ω_n has a specific value and by assuming a linear variation between nodes. Since the value of ω_0 is not known beforehand, the problem is formulated in terms of the general warping function ω, rather than ω_n. The method produces a system of linear equations for the nodal values of ω. After solving for these the value of ω_0 is calculated by numerical integration (within the same computer program) and added to each value, thus giving ω_n. The warping torsional stiffness I_ω can then be computed, again using numerical integration. In this approach the overall cross section is modeled as a series of straight-line "elements," each of constant thickness and of unit depth in the axial (x) direction. Along the element, the warping function varies linearly as shown in Fig. 3.46. The elements can be any length and so it is possible to model quite complicated cross sections with good accuracy. In terms of the nodal values ω_1 and ω_2, the warping within an element is

Model for calculating J

$$J = \frac{4A^2}{\oint \frac{ds}{t}} = \frac{16\,(A/2)^2}{2\int_o^b \frac{ds}{t}} = \frac{2A^2}{\int_o^b \frac{ds}{t}}$$

open sections (e.g. mach'y flat)
and small cells are ignored
when calculating J.

Model for calculating I_ω

$$I_\omega = 2 \int_o^b \omega_n^2 \, t \, ds$$

where $\omega_n = \frac{J}{2A} \int_o^s \frac{ds}{t} - \int_o^s h \, ds + \omega_o$

(a) CLOSED SECTIONS

Model for calculating J

Model for calculating I_ω

$$GJ = 4 \sum A_i \bar{q}_i$$

values of $\bar{q}_i$ are obtained from

$$[\,c\,]\{\bar{q}\} = 2G\,\{A\}$$

$$\sigma_x = E\omega_n\theta''$$

$$I_\omega = 2 \int_o^b \omega_n^2 \, t \, ds$$

where $\omega_n = -\int_o^s h \, ds + \omega_o$

$$q_2 = -E \left[\int_o^s \omega_n \, t \, ds \right] \theta'''$$

(b) OPEN SECTIONS

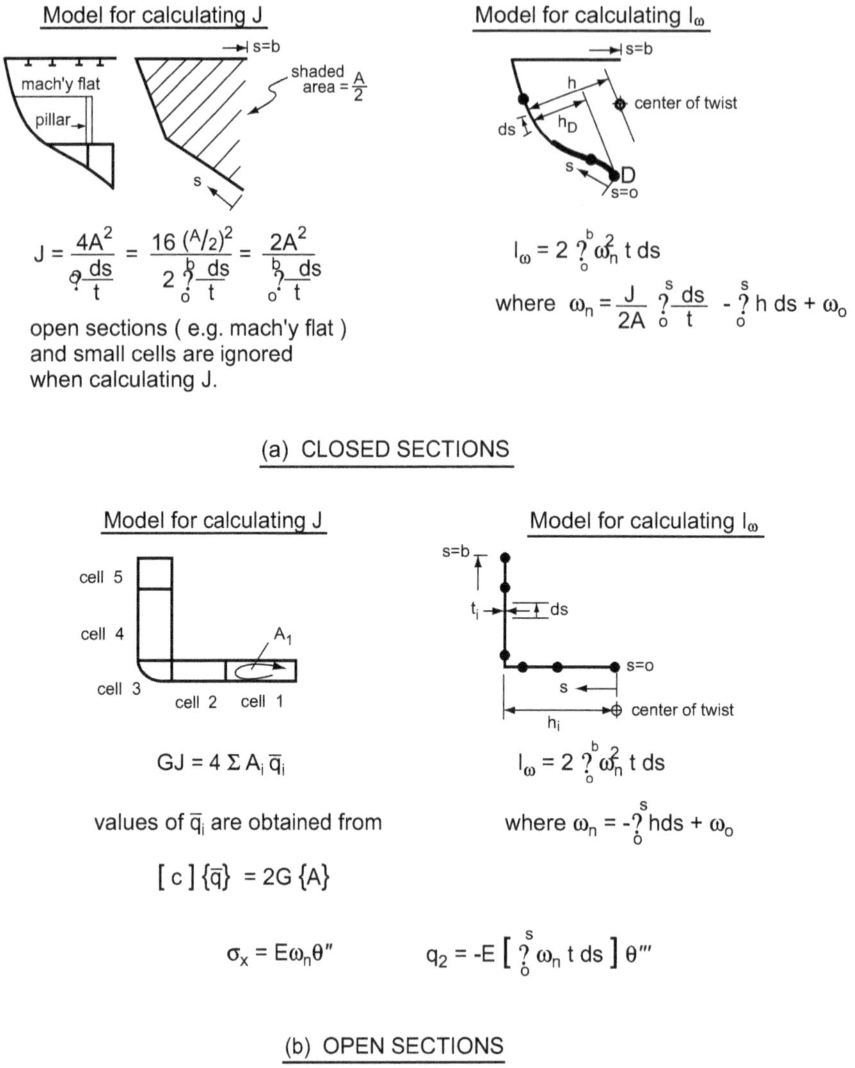

Figure 3.45 Structural models for calculation of torsional stiffnesses.

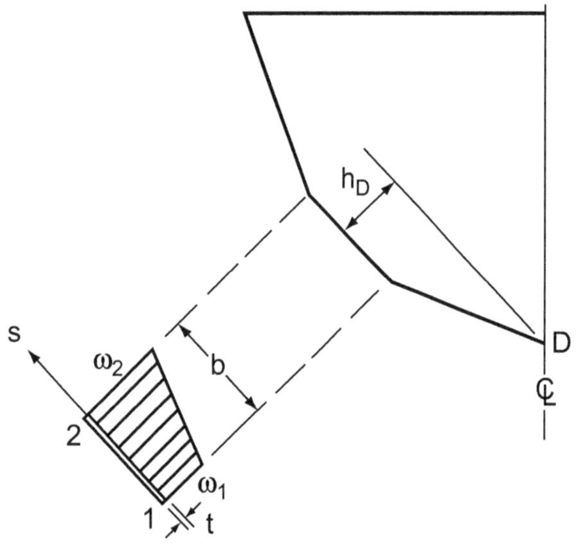

Figure 3.46 Segment or calculation of warping function distribution.

$$\omega(s) = \omega_1 + (\omega_2 - \omega_1)\frac{s}{b}$$

in which the $+s$-direction for the element is the same as for the overall section. From (3.10.4) the shear strain in the element is

$$\gamma = \frac{\partial u}{\partial s} + h\theta'$$

and from the definition of the warping function we have $u = (-\omega + \omega_0)\theta'$ and therefore

$$\gamma = \left(-\frac{\partial \omega}{\partial s} + h\right)\theta'$$

In terms of strain the potential energy of element e is

$$U_e = \frac{Gt_e L}{2} \int_0^{b_e} \left[\gamma(s)\right]^2 ds$$

where L is the length of the element in the x-direction; that is, the length of the overall prismatic member. From the above equation for γ the expression for U_e becomes

$$\sum_{e=1}^{N_e} \pm t_e \left(\frac{\omega_1 - \omega_2}{b_e} + h_e \right) = 0$$

The total potential energy of the structure is the sum of that for each of the elements, taken over all of the elements, say N_e in number.

$$U_{total} = \frac{GL}{2}(\theta')^2 \sum_{e=1}^{N_e} t_e \int_0^{b_e} \left[\frac{\omega_1 - \omega_2}{b_e} + h_e \right]^2 ds$$

The overall distribution of ω, as defined by the set of nodal values $(\omega_1, \ldots, \omega_i, \ldots, \omega_N)$ must be such as to minimize the total potential energy.* That is

$$\frac{\partial U_{total}}{\partial \omega_i} = 0 \qquad i = 1, \ldots, N$$

which yields a system of N equations, one for each node. The differentiation can be done inside the integral and the common terms $(GL / 2)(\theta')^2$ can be dropped. The result is

$$\sum_{e=1}^{N_e} t_e \int_0^{b_e} 2 \left[\frac{\omega_1 - \omega_2}{b_e} + h_e \right] \left(\frac{\pm 1}{b_e} \right) ds = 0$$

in which the $\pm$ arises from the differentiation. The plus sign applies if ω_i corresponds to ω_1 of element e, and the minus sign applies if it corresponds to ω_2. Integrating along the element length gives

$$\sum_{e=1}^{N_e} \pm t_e \left(\frac{\omega_1 - \omega_2}{b_e} + h_e \right) = 0 \qquad (3.10.30)$$

There is one such equation for each nodal value ω_i. In each equation the summation need not be performed over all N_e elements but only on the M elements that are attached to node i. Let us use the subscript m to denote each of these attached elements, such that $m = 1, 2, \ldots, M$. Also, for each element, let us denote the node remote from i by the subscript r. Then as the summation process at node i proceeds, taking each of the M elements in turn, the values of ω_1 and ω_2 of each element will either be ω_i and $\omega_{r,m}$ or the reverse, depending on the orientation of element m. If i corresponds to node 1, the term in brackets in (3.10.30) would be $(\omega_i - \omega_{r,m} + h_m b_m)/b_m$. If i corresponds to node 2 the term would be

$$-(\omega_{r,m} - \omega_i + h_m b_m) / b_m$$
$$= (\omega_i - \omega_{r,m} - h_m b_m) / b_m$$

Hence the system of equations may be written as

$$\sum_{m=1}^{M} t_m \left(\frac{\omega_i - \omega_{r,m}}{b_m} \pm h_m \right) = 0 \qquad i = 1, \ldots, N \qquad (3.10.31)$$

in which the minus sign applies whenever i corresponds to node 2 of the mth element.

In most cases the location of the center of twist is not known beforehand and so in the foregoing solution the values of h are measured relative to an assumed location D (which we are here presuming to be on the vertical centerline) and the result is ω_D, as in (3.10.10). In this case the normalized warping function ω_n can be calculated directly from (3.10.11) and (3.10.12) using numerical integration; there is no need to calculate ω_0. The warping torsional stiffness I_ω can then be calculated from (3.10.17), again using numerical integration.

3.11 PRACTICAL CALCULATION OF HULL GIRDER SHEAR EFFECTS

In Section 3.7 the shear stress due to transverse shear loads was calculated by considering equilibrium. For a multicell section it was necessary to invoke geometric compatibility, which requires that the net warping of each closed cell must be zero. This approach requires careful consideration of the connectivity of the section and of the pattern of the shear flow—its direction in the various cell walls and the way in which it divides at branch points. Such a method is not well suited for computer implementation and so in this section we present an alternative method that is formulated not in terms of shear flow but rather in terms of the warping due to transverse shear. This approach is simpler because warping is everywhere single-valued, even at branch points.

We have now seen that when transverse shear is present, plane cross sections do not remain

* The work done by the load is independent of ω, and therefore does not need to be included in the total energy.

plane, as assumed in simple beam theory, but undergo deformation out of this plane. This type of deformation is referred to as *warping*, and occurs both in torsion and in transverse shear. These two types of warping are distinct, having separate causes. If there is both twisting and transverse shear, both types of warping will exist, and if the warping is restrained the two types will interact. For simplicity this section deals only with warping due to transverse shear.

In a prismatic beam that is not subject to twist the shear strain in the cross section is simply the derivative of the warping:

$$\gamma_{xy} = \frac{\partial u}{\partial y}; \qquad \gamma_{xz} = \frac{\partial u}{\partial z}$$

In a thin-walled beam it is more convenient to use the tangential or arc length coordinates within the cross section, for which the relationship is;

$$\gamma = \frac{\partial u}{\partial s} \tag{3.11.1}$$

and there is no need for subscripts. This relationship may also be obtained by setting the spiral angle ψ to zero in (3.10.4). If end effects are ignored the warping in a prismatic beam maintains the same cross-sectional pattern along the length of the beam, and the magnitude at any section is proportional to the transverse shear force $Q(x)$ at that section. It is therefore convenient to introduce a warping function $\omega_Q(s)$, which describes the transverse distribution of warping, and to express the warping as a product

$$u(s,x) = \omega_Q(s)Q(x) \tag{3.11.2}$$

Since the shear strain γ (and hence also the shear stress, $\tau = G\gamma$) is simply the derivative of the warping, the distribution of transverse shear stress can be determined by first solving for the warping distribution and then differentiating. This approach was developed by Mason and Herrmann (1968) and is similar to the latter's method for calculating the torsional warping function that was presented in Section 3.10. It was first applied to ship hulls by Kawai (1973). The method is essentially an application of the classical Rayleigh-Ritz technique. First the distribution of ω_Q is represented in a discretized manner by defining "nodes" at which ω_Q has a specific value and by assuming a linear variation between nodes. Then the total potential energy of the system Π is expressed in terms of the nodal values ω_Q, and the derivative of Π with respect to each value of ω_Q is set equal to zero, thus obtaining a system of equations for the ω_Q values. After solving for these, the complete distribution of shear stress can be readily calculated.

The derivation is presented here for a prismatic segment of the hull of length L. As shown in Fig. 3.47, the segment may consist of any number of flat rectangular elements of constant thickness t_e, breadth b_e, and length L. The transverse shear force Q can vary linearly along the length of the element, as occurs if the overall load on the hull segment is a uniform distributed load. If the magnitude of this load is f_z (see Fig. 3.47) the shear force at any section is:

$$Q(x) = Q_A\left(1 + \phi\frac{x}{L}\right) \tag{3.11.3}$$

where

$$\phi = \frac{f_z L}{Q_A}$$

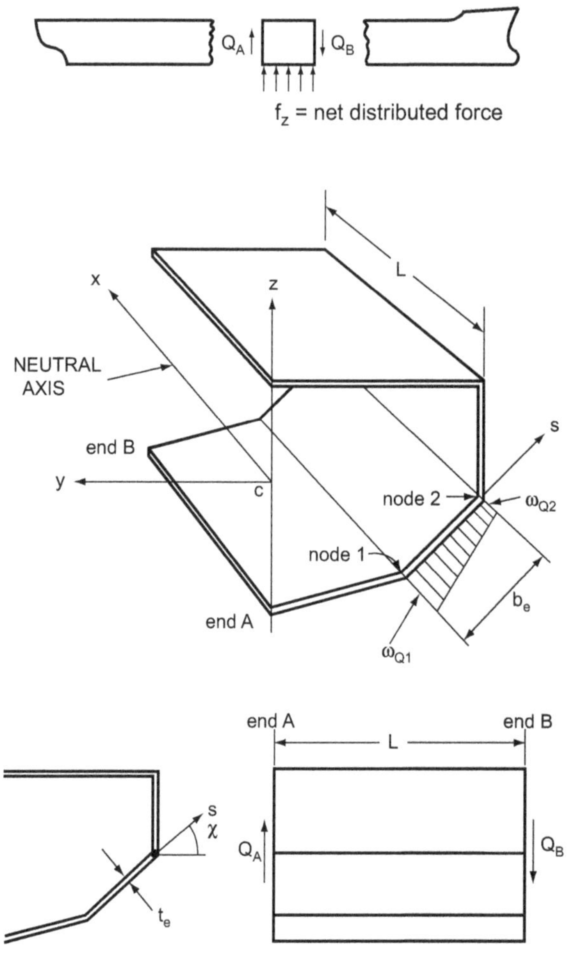

Figure 3.47 Warping due to transverse shear.

and Q_A is the shear force at end A of the segment. The warping is then:

$$u(s,x) = \omega_Q(s) Q_A \left(1 + \phi \frac{x}{L}\right) \qquad (3.11.4)$$

Since the longitudinal variation of ω_Q is independent of the transverse variation, the problem can be formulated entirely in terms of the warping at end A, and so the problem is essentially a two-dimensional shear flow problem, in which the shear flow $q(s)$ (at end A) is to be determined.

The assumption of a linear variation of $\omega_Q(s)$ within each element implies constant shear flow in the element, whereas it is known from Section 3.7 that the shear flow varies linearly in horizontal portions of the ship's cross section and parabolically in all other portions. It will be shown that the constant value of shear flow which is obtained for each element is the mean value for that element, and that the other part of the total shear flow (a linear or parabolic distribution, having a zero mean value) can be obtained separately for each element and superimposed on the mean value.* Thus the assumption of a linear distribution of warping within the element does not constitute an approximation, and so a single element can be used for any straight, constant thickness portion of the cross section, even if its breadth b_e is quite large.

In terms of nodal values ω_{Q1} and ω_{Q2} the warping within an element is:

$$\omega_Q(s) = \omega_{Q1} + \left(\omega_{Q2} - \omega_{Q1}\right)\frac{s}{b_e} \qquad (3.11.5)$$

in which the $+s$-direction is from element node 1 to element node 2. The orientation of the element is defined by the angle χ which the $+s$-axis makes with the horizontal axis of the hull girder, as shown in Fig. 3.47.

In terms of strain the potential energy of an element is:

$$U_e \frac{Gt_e}{2} \int_0^L \int_0^{b_e} \left[\gamma(s,x)\right]^2 ds\,dx$$

Substituting from (3.11.1) and (3.11.4) and integrating along the length gives:

$$U_e = \frac{1}{2} C_Q Q_A^2 G t_e L \int_0^{b_e} \left(\frac{\partial \omega_Q}{\partial s}\right)^2 ds \qquad (3.11.6)$$

* This approach was suggested to the author by Dr. Vedran Žanić of the University of Zagreb.

where

$$C_Q = 1 + \phi + \frac{1}{3}\phi^2 \qquad (3.11.7)$$

Substituting the linear expression for $\omega_Q(s)$ and summing over all the elements (say N_e in number):

$$U = \frac{C_Q Q_A^2 G L}{2} \sum_{e=1}^{N_e} \int_0^{b_e} \left(\frac{\omega^e_{Q^2} - \omega^e_{Q^1}}{b_e}\right)^2 t_e ds \qquad (3.11.8)$$

The other component of the total potential energy is the negative of the work done by the loads as a result of the warping. For the differential element in Fig. 3.48 the work done is the product of the net longitudinal force in the element

$$dF = \frac{\partial \sigma_x}{\partial x} dx\, t_e\, ds$$

times the distance through which the element moves, which is the local warping displacement $u(s, x)$. That is, the work done throughout the element is:

$$W = \int_{vol} u(s,x)\, dF \qquad (3.11.9)$$

(There is no ½ factor because the stress is caused by the external loads and not by the warping.) The longitudinal stress can come either from hull girder bending or from a restraint on the *torsional* warping if the segment is undergoing twist. For bending stress dF may be related to the change in bending moment and hence to the transverse shear force Q, as was done in deriving (3.7.1) and (3.7.2) (and as illustrated in Fig. 3.22). The steps are:

$$\begin{aligned} dF &= \frac{\partial}{\partial x}\left(\frac{Mz}{I}\right) dx\, t_e\, ds \\ &= \frac{\partial M}{\partial x}\frac{z}{I} dx\, t_e\, ds \\ &= \frac{Q(x)}{I} z\, dx\, t_e\, ds \end{aligned}$$

After substituting this expression and (3.11.4) and (3.11.5) into (3.11.9), integrating along the length of the element, and finally summing over all of the elements, the result is:

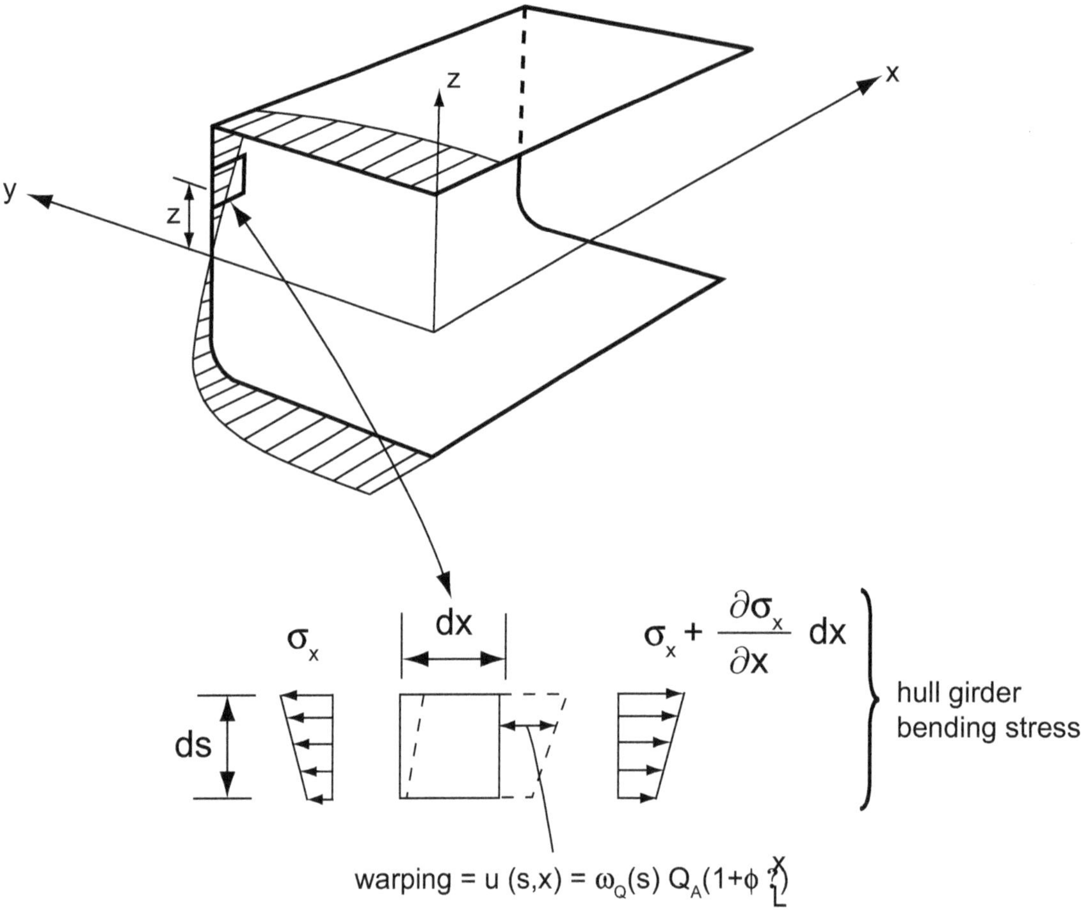

Figure 3.48 Distribution of warping.

$$W_Q = \frac{C_Q Q_A^2 L}{I} \sum_{e=1}^{N_e} t_e \int_0^{b_e} \left[\omega_{Q1}^e + \frac{\left(\omega_{Q2}^e - \omega_{Q1}^e\right)}{b_e} \right] z(s)\, ds \qquad (3.11.10)$$

where C_Q is defined in (3.11.7). The value of $z(s)$ can be expressed in terms of the value at element node 1 and the orientation angle χ (see Fig. 3.47)

where
$$\left. \begin{aligned} z(s) &= z_1 + s\mu \\ \mu &= \sin \chi \end{aligned} \right\} \qquad (3.11.11)$$

We have seen in Section 3.10.2 that if a section that is undergoing twist is even partially restrained from warping, this causes a longitudinal stress σ_x which varies in the axial direction. From (3.10.14)

$$\sigma_x = E \omega_n \frac{d^2\theta}{dx^2}$$

and so the net axial force in an element of width ds and length dx is:

$$\begin{aligned} dF &= \frac{\partial \sigma_x}{x} dx\, t_e ds \\ &= E \omega_n \frac{d^3\theta}{dx^3} dx\, t_e ds \end{aligned}$$

Again substituting into (3.11.9) and integrating lengthwise, the following expression is obtained for the work done by the *torsional* warping stress:

$$W_\theta = E Q_A t_e \int_0^L \frac{d^3\theta}{dx^3} \int_0^{b_e} \left(\omega_n \omega_Q ds \right) \left(1 + \phi \frac{x}{L} \right) dx \qquad (3.11.12)$$

Thus if there is torsion and if the torsional warping is restrained so as to give a nonzero $d^3\theta/dx^3$, then the two types of warping interact. For most ships transverse shear Q is a much more severe load than torsion, and the shear flow q is due almost entirely to Q. Therefore the remainder of this section deals with transverse (vertical) shear only.

The total potential energy of the system is $\Pi = U - W$, and the system of equations for the nodal values of ω is obtained by requiring that Π must be a minimum with respect to each ω_i, as was done in Section 3.10.6 for torsional warping. That is

$$\frac{\partial \Pi}{\partial \omega_i} = 0 \qquad i = 1, \ldots, N_n$$

where N_n is the number of nodes. The steps are straightforward and the resulting system of equations is

$$G \sum_{m=1}^{M} t_m \left(\frac{\omega_{Q,i} - \omega_{Q,r}^{m}}{b_m} \right) =$$

$$\frac{1}{2I} \sum_{m=1}^{M} t_m b_m \left(z_{1m} + \frac{n_m}{3} \mu_m b_m \right) \qquad i = 1, \ldots, N_n \tag{3.11.13}$$

In these equations the subscript e has been replaced by m to indicate that since each equation refers to one node (the ith node) the summation over the elements need only include the M elements that touch that node. For each of these M elements the subscript r denotes the node which is *remote* from node i, and the symbol n_m is the *element* node number (either 1 or 2) which corresponds to node i. The symbols z_1 and μ were defined in (3.11.11).

It can be shown that axial equilibrium requires that $\omega_Q = 0$ at all points which lie on the neutral axis. This requirement can be imposed by placing nodes at all such points and setting these values of ω_Q equal to zero.* This effectively divides the system of equations into two independent subsystems.

In order to demonstrate the simplicity of this method and to show how the final distribution of shear flow is obtained, it will be applied to the small idealized tanker shown in Fig. 3.49. For brevity the calculation will be performed only for the upper portion. As shown in Fig. 3.49 there are only three nodal values of ω_Q to be determined, and only four elements are required. For elements a and b the parameter values are:

$$t = 0.032, \quad b = 10, \quad z_1 = 12, \quad \chi = 0^\circ, \quad \mu = 0$$

while for elements c and d they are:

$$t = 0.032, \quad b = 12, \quad z_1 = 0, \quad \chi = 90^\circ, \quad \mu = 1$$

*If this is not done the system of equations is singular.

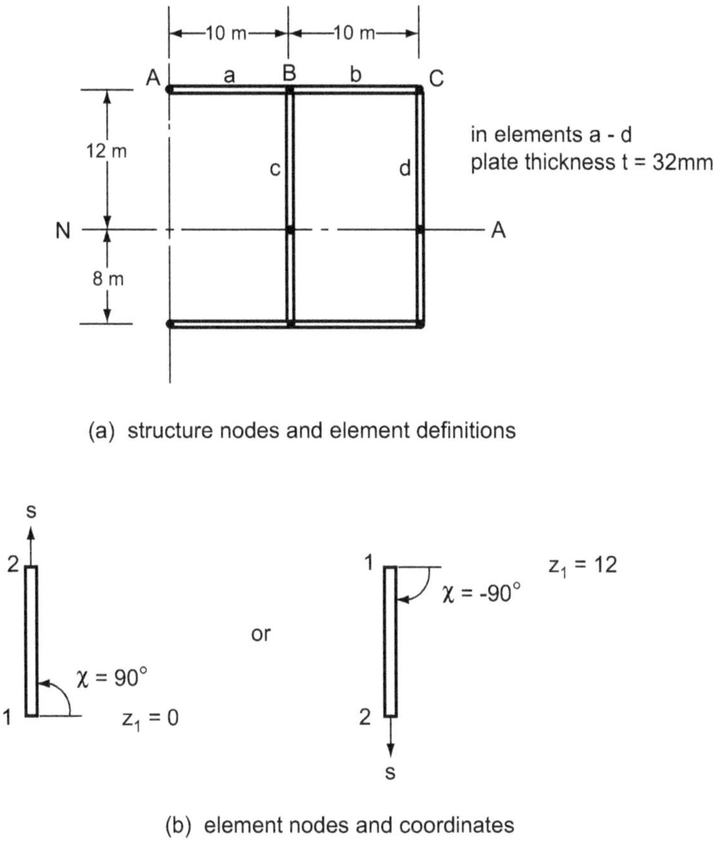

Figure 3.49 Sample calculation of warping due to transverse shear.

These values correspond to a choice of the lower end as element node 1. It would be equally valid to choose the upper end, as shown in Fig. 3.49*c*.

At node A only element a is involved and so (3.11.13) becomes:

$$t_a \frac{\omega_{Q,A} - \omega_{Q,B}}{b_a} = \frac{1}{2GI} t_a b_a \left(z_{1a} + \frac{1}{3} \mu_a b_a \right)$$

Substituting the foregoing values gives:

$$0.0032\omega_{Q,A} - 0.0032\omega_{Q,B} = \frac{0.32}{2GI}(12)$$

At node B elements a, b, and c are involved and the equation (with zero values already inserted) is

$$\left(\frac{t_a}{b_a} + \frac{t_b}{b_b} + \frac{t_c}{b_c} \right) \omega_{Q,B} - \frac{t_a}{b_a}\omega_{Q,A} - \frac{t_b}{b_b}\omega_{Q,C} - \frac{t_c}{b_c}0$$

$$= \frac{1}{2GI}\Bigg[t_a b_a (z_{1a} + 0)$$

$$+ t_b b_b (z_{1b} + 0) + t_c b_c \left(0 + \frac{2}{3} b_c \right) \Bigg]$$

which becomes:

$$0.009067\omega_{Q,B} - 0.0032\omega_{Q,A} - 0.0032\omega_{Q,C} = \frac{10.752}{2GI}$$

At node C elements b and d are involved and the equation is:

$$\left(\frac{t_b}{b_b} + \frac{t_c}{b_c} \right) \omega_{Q,C} - \frac{t_b}{b_b}\omega_{Q,B} - \frac{t_d}{b_d}0$$

$$= \frac{1}{2GI}\left[t_b b_b (z_b + 0) + t_d b_d \left(0 + \frac{2}{3} b_d \right) \right]$$

which becomes:

$$0.005867\omega_{Q,C} - 0.0032\omega_{Q,B} = \frac{6.912}{2GI}$$

The solution to these three equations is:

$$GI\omega_{Q,A} = 2827.8$$

$$GI\omega_{Q,B} = 2227.8$$

$$GI\omega_{O.C} = 1804.2$$

The mean value of shear stress in each element is given by:

$$\bar{\tau}_e = G\frac{\partial u}{\partial s} = GQ\frac{\partial \omega_Q^e}{\partial s} = GQ\frac{\omega_2^e - \omega_1^e}{b_e} \qquad (3.11.14)$$

Since Q and I are constant for a given section it is convenient to deal with scaled values of τ which correspond to a unit value of Q/I, and then apply this factor at the end.

Thus for element a the mean value of the (scaled) shear stress is:

$$\bar{\tau}_a = \frac{\omega_{Q,A} - \omega_{Q,B}}{b_a} = \frac{2827.8 - 2227.8}{10} = 60$$

The other values obtained from (3.11.14) are:

$$\bar{\tau}_b = 42.35, \quad \bar{\tau}_c = 185.65, \quad \bar{\tau}_d = 150.35$$

Once the mean value of τ is known in any element the local variation can be found from (3.7.2), which gives the local or incremental distribution of τ in any element, over and above the value at the end where the integration is started (node 1). We require a local distribution which has a zero *mean* value, instead of a zero value at node 1, and this can be obtained by subtracting the mean value. Thus for each element the local distribution $\tau^1_e(s)$ which is to be superimposed on the mean value obtained from a warping solution is given by:

$$\tau^l_e(s) = \frac{Q}{t_e I}\left[m_e(s) - \bar{m}_e \right]$$

where $m(s)$ is the first moment of area, defined in (3.7.3). Expressing $z(s)$ in terms of z_1 and μ as in (3.11.11), this equation becomes, after integration:

$$\tau^l_e(s) = \frac{Q}{I}\left[z_{1e}\left(s - \frac{b_e}{2} \right) + \frac{\mu}{2}\left(s^2 - \frac{b_e^2}{3} \right) \right] \qquad (3.11.15)$$

Since this expression comes from the shear flow equations of Section 3.7, it follows the same rules in regard to sign convention. That is, it assumes that the s-axis is in the direction of increasing shear flow (or shear stress). If this is not so, then the sign of $\tau^1_e(s)$ must be reversed.

In the tanker example the resulting local distributions for elements a and c are:

$$\tau^l_a(s) = \frac{Q}{I}\left[12(s-5) \right]$$

$$\tau^l_c(s) = -\frac{Q}{I}\left[0 + \tfrac{1}{2}\left(s^2 - \frac{144}{3} \right) \right] = \frac{Q}{I}\left[24 - \tfrac{1}{2}s^2 \right]$$

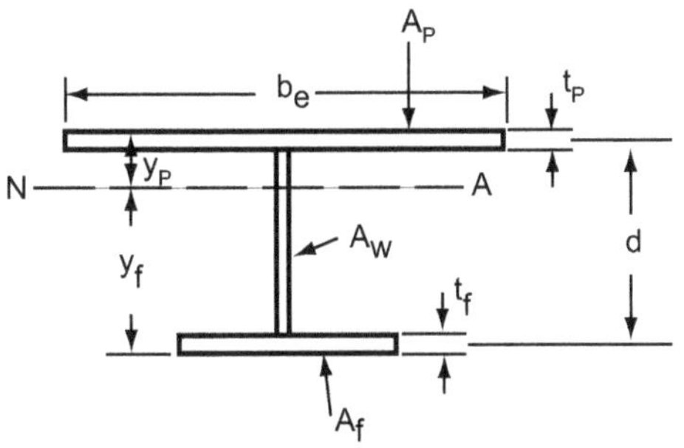

Figure 3.50 Geometric properties of beams attached to plating.

Note that for $\tau^1_e(s)$ the sign has been reversed because element node 1 is at the neutral axis ($z_{1c} = 0$) and the s-axis points upward ($\mu = 1$), which is the direction of decreasing τ. The maximum value of τ occurs at the neutral axis and for element c this value is

$$\begin{aligned}\tau(z=0) &= \bar{\tau}_c + \tau^l_c(0) \\ &= (185.6 + 24)Q/I \\ &= 209.6Q/I\end{aligned}$$

3.12 GEOMETRIC PROPERTIES OF BEAMS ATTACHED TO PLATING

In ship structures it is quite common for a beam to be welded to plating, such that the plating acts as a second flange (the "plate flange") to the beam. The breadth b of the plate flange is usually much wider than the beam's own flange, and so the neutral axis of the combined section is usually close to the plating. Because of "shear lag" (Section 3.8.1) the bending stress in the plate flange may vary across the width, having a maximum value at the web. To allow for this the plate flange may be assigned an "effective breadth" b_e. In this section we present some simple and useful formulas for the neutral axis location and the moment of inertia of the combined cross section. These formulas are sufficiently accurate for design use provided that the section is a thin-wall section.

We define the following quantities (see Figure 3.50)

Then it can be shown that:

A_p = effective plate area = $b_e t_p$

A_w = web area

A_f = flange area

A_T = total area = $A_p + A_w + A_f$

t_p = plate thickness

t_f = flange thickness

d = distance from midthickness of plate to midthickness of flange

$$\left.\begin{aligned} y_f &= \frac{1}{2}t_f + dC_2 \\ y_p &= \frac{1}{2}t_p + d(1 - C_2) \\ I_e &= A_T d^2 C_1 \end{aligned}\right\} \qquad (3.12.1)$$

in which

$$C_1 = \frac{A_w\left(\dfrac{A_T}{3} - \dfrac{A_w}{4}\right) + A_f A_p}{A_T^2}$$

$$C_2 = \frac{\dfrac{A_w}{2} + A_p}{A_T}$$

REFERENCES

Bureau Veritas. (2001). *Rules for the classification of steel ships*. Part B, Chapter 5, Section 2, Hull Girder Loads, April.

Det Norske Veritas. (1977). *Rules for design, construction and installation of offshore structures.* Oslo, Norway, Author.

Dowling, P. J., Harding, J. E., and Frieze, P.A. (1977). *Steel plated structures.* London: Crosby Lockwood Staples.

Fain, R. A., and Booth, E. T. (1979). Results of the first five "data years" of extreme stress scratch gauge data collected aboard Sea Land's SL 7's. *SSC*, Vol. 286.

Herrmann, L. R. (1965). Elastic torsional analysis of irregular shapes. *J. of Eng. Mechanics Division*, Vol. EM6, 11–19.

Hoffman, D., and Lewis, E. V. (1969). Analysis and interpretation of full-scale data on midship bending stresses of dry cargo ships. *SSC Report SSC-196*, June.

International Ship Structures Congress. (1976). *Report of Committee I-3*. 6th ISSC, Boston, August.

Kamerrer, J. T. (1966). A design procedure for determining the contribution of deckhouses to the longitudinal strength of ships. Assoc. of Senior Engineers, Bureau of Ships, 3rd Annual Technical Symposium.

Kolbrunner, C. F., and Basler, K. (1969). *Torsion in structures*. Berlin: Springer-Verlag.

Kolbrunner, C. F., and Hajdin, N. (1972). *Dünnvandige stäbe*, band 1. Berlin: Springer-Verlag.

Kolbrunner, C. F., and Hajdin, N. (1975). *Dünnvandige stäbe*, band 2. Berlin: Springer-Verlag.

Lewis, E. V., and Zubaly, R. B. (1975). Dynamic loadings due to wave and ship motions. *STAR Symposium, SNAME.*

Little, R. S., Lewis, E. V., and Bailey, F. C. (1971). A statistical study of wave induced bending moments on large oceangoing tankers and bulk carriers. *Trans. SNAME*, 117–168.

Mason, W. E., Jr., and Herrmann, L. E. (1968). Elastic shear analysis of general prismatic beams. *J. of Eng. Mechanics Division*, Vol. EM4, 965–983.

McVee, J. D. (1980). A finite element study of hull-deckhouse interaction. *Computers and Structures*, Vol. 12, 371–393.

Mitchell, G. C. (1978). Analysis of structural interaction between a ship's hull and deckhouse. *Trans. RINA*, Vol. 120, 121–134.

Munse, W. (1981). Fatigue criteria for ship structural details. *Extreme Loads Response Symposium, SNAME*, 231–247.

Paulling, J. R., and Payer, H. G. (1968). Hull-deckhouse interaction by finite element calculations. *Trans. SNAME*, Vol. 76, 281–296.

Schade, H. A. (1951). The effective breadth of stiffened plating under bending loads. *Trans. SNAME*, Vol. 59, 403–420.

Schade, H. A. (1953). The effective breadth concept in ship structure design. *Trans. SNAME*, Vol. 61, 410–430.

Schade, H. A. (1969). Hull strength. In D'Arcangelo, A. M. (Ed.). *Ship Design and Construction*. New York: SNAME.

Welding Institute Research Bulletin. (1976). Vol. 17, Issue 5.

Vlasov, V. Z. (1961). *Thin-walled elastic beams*. Washington, DC: Israel Program for Scientific Translations.

CHAPTER

FOUR

WAVE LOADS—STATISTICAL, DYNAMIC, AND NONLINEAR ASPECTS

Owen Hughes
Professor, Virginia Tech
Blacksburg, VA, USA

Thomas Schellin
Germanischer Lloyd, Hamburg, Germany (ret)

In the context of ship structural analysis, the primary purpose of computing wave-induced loads is to be able to furnish an essential part of the input for a finite element model in terms of forces acting on selected nodal points of the ship's structure. This involves various specialized branches of knowledge, including probability theory and extreme value statistical theory, statistical data—both short-term and long-term—regarding ocean storm waves, hydrodynamics of the flow around a ship in the presence of a free surface, and numerical methods employed in ship seakeeping analyses. Obviously, it is a complex field, and a complete coverage is beyond the scope of this book and also beyond what is normally required for rational analysis methods used for the structural design of ships. As mentioned in Chapter 3, a deterministic approach is often sufficient, whereby the wave loads, such as the wave bending moment, are obtained from approximate formulas given by classification societies. These formulas are the result of a variety of statistical analyses of theoretical and experimental studies and full-scale measurements, and they are adequate for most standard kinds of ships. Nevertheless, to estimate loads is one of the most crucial aspects of structural design and, hence, it is important for ship structural designers to know at least the basics of the theory and technique for a reliable prediction of wave loads.

Moreover, there are some kinds of ships for which it is advisable to perform an explicit estimate of wave loads. This is especially so for large, modern containerships. The unique hull form of these vessels with pronounced bow flare and large, flat overhanging stern coupled with high service speed introduces nonlinear ship motions and wave loads. These nonlinear sea loads can result in significantly higher wave-induced bending moments, shear forces, and torsional loads than have been considered in formulation of traditional prescriptive rule values. To better predict motions and structural behavior of these ships, a hydrodynamic sea load approach that accounts for the significant nonlinear effects must be used and integrated with a full ship finite element structural analysis to augment standard classification review. Numerical methods, based on a first-principles approach for ship structures, are increasingly becoming indispensable to determine design loads, the structure's response (stress, deflection, etc.) to those loads, and assessment of the response compared to acceptance criteria. These computations are performed by means of comprehensive computer codes developed expressly for that purpose, and it is important for the designer to be familiar with at least the basic aspects, in order to use these codes correctly and to maximum advantage.

Probability theory is an indispensable standard tool to assess design values of wave-induced loads, that is, values that have a specified probability of non-exceedence in the ship's lifetime. For linear wave-induced response, theoretical methods to estimate short- and long-term probability distributions are well established. For nonlinear response, it is difficult to obtain sufficiently accurate extreme value estimates. However, the extreme nonlinear response is usually related in time to the extreme linear response. Thus, to illustrate the overall method of obtaining design values based on the use of probability theory, it is essential to be familiar with the particular topics and formulas for linear systems. Section 4.1 of this chapter contains a brief summary of these aspects. Section 4.2 presents information regarding the extreme values of random processes, both for responses that can be obtained from a linear as well as a nonlinear treatment of the wave-ship system. Section 4.3 deals with the mathematical and statistical representation of ocean waves, showing how a typical sea state may be presented in terms of a family of spectral density functions or, more commonly, wave spectra, and it also presents information on short-term and long-term statistics of these sea states.

Parts of these sections are either left unchanged or amended from the 1988 SNAME edition of Ship Structural Design. Of course, the numbering of the equations is adapted to fit the reworked text. Specifically, Section 4.1, dealing with the basics of probability theory and random processes, is essentially unaltered, and Section 4.2, concerned with the prediction of extreme values, is supplemented by part 4.2.2 for nonlinear systems, whereas part 4.2.1 for linear systems is not changed. In Section 4.3, treating the statistical representation of the sea surface, only parts 4.3.1 and 4.3.4, describing the mathematical treatment of ocean waves and the duration of sea states, respectively, are left the same, whereas parts 4.3.2 and 4.3.3, documenting ocean wave spectra and their families, are modified. Finally, in Section 4.4, describing the computation of wave-induced loads, only part 4.4.2 dealing with linear computations is left as is.

Section 4.4 presents a brief perspective of ship motion theory as it relates to seakeeping computations for structural wave-induced load predictions. Section 4.5 deals with the determination of equivalent regular design waves that are based on linear frequency-domain computations corrected for the major nonlinear effects. The resulting nonlinearly corrected (pseudo) transfer functions of the critical loads are formally treated like transfer functions of linear systems, and an almost standard stochastic analysis procedure based on these pseudo transfer functions yields long-term wave-induced loads. Section 4.6 discusses special load effects of slamming, sloshing, and hull girder whipping as these loads are important for the design of local ship structures.

Finally, Section 4.7 is devoted to the treatment of nonlinear response simulation for the prediction of wave loads, as numerical simulation is the sole computational method that can handle strong nonlinearities and is, therefore, the appropriate tool to investigate strongly nonlinear ship response in severe seas.

4.1 BASICS OF PROBABILITY AND RANDOM PROCESSES

4.1.1 Probability Density Functions

In the general theory of statistics a random variable X is an event, or an outcome, among all possible outcomes. If the possible outcomes form a continuous "space," $-\infty < x < \infty$, and each event is some portion of this space, then the probability of an event occurring is simply the probability that X will lie within that portion of x. This probability is specified by *a probability density function* $p_X(x)$ as shown in Fig. 4.1.1.

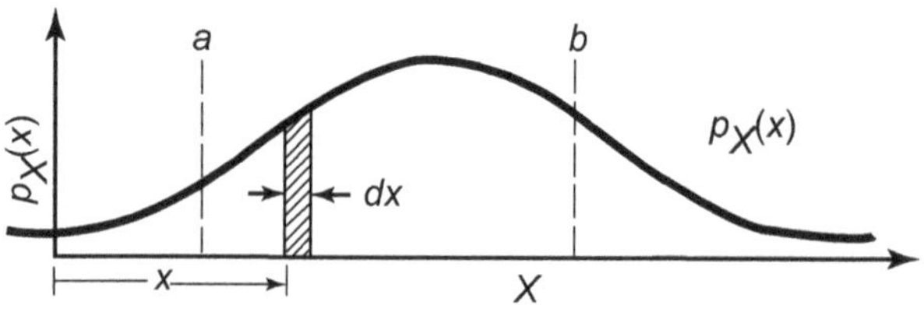

Figure 4.1.1 Probability density function.

In terms of this function, the probability that X lies between x and $x + dx$ is $p_X(x)dx$. That is

$$\text{Prob}[x \leqslant X \leqslant x + dx] = p_X(x)dx \quad (4.1.1)$$

In general, the probability of X lying within any interval is equal to the area under the $p_X(x)$ curve within that interval.

$$\text{Prob}[a \leqslant X \leqslant b] = \int_a^b p_X(x)dx \quad (4.1.2)$$

and hence the total area under the curve is equal to unity:

$$\text{Prob}[-\infty < X < \infty] = \int_{-\infty}^{\infty} p_X(x)dx = 1 \quad (4.1.3)$$

The *cumulative probability distribution function, $P_X(x)$*, often referred to as simply the "probability distribution," is the indefinite integral of the probability density function

$$P_X(x) = \int_{-\infty}^{x} p_X(x)dx \quad (4.1.4)$$

We next define $E[X]$, the *average* or the *expected value,* either of a random variable X or, more generally, of any function of X, $f(x)$, as follows:

$$E[x] = \int_{-\infty}^{\infty} x\, p_X(x)dx \quad (4.1.5)$$

$$E[f(X)] = \int_{-\infty}^{\infty} f(x)p_X(x)dx \quad (4.1.6)$$

Equation 4.1.3 states that, for any function $f(X)$ of a random variable having a probability density function $p_X(x)$, the *average of* $f(X)$ is equal to the *moment of* $p_X(x)$, with $f(X)$ taken as the moment arm. For the special case when $f(X)$ is X itself, the average is simply the *mean,* or *mean value,* of X, that is, the most direct and most familiar type of average. We shall use the symbol μ to indicate this mean value, thus:

$$\mu = E[X] = \int_{-\infty}^{\infty} x\, p_X(x)dx \quad (4.1.7)$$

We next examine the average or expected value of some simple functions of X, namely, powers of

X. In this case, the moments are second moments, third moments, and so on, of $p_X(x)$. For example, the average of X^2, or the *mean square* of X, is the expected value of X^2. From (4.1.6) this is

$$E[X^2] = \int_{-\infty}^{\infty} x^2 p_X(x)dx \qquad (4.1.8)$$

which is the second moment of $p_X(x)$, taken about $x = 0$.

In defining the averages for higher powers of X, it is convenient to introduce the *deviation* from the mean, $X - \mu$, and to take powers of the deviation instead of X because this corresponds to taking moments about the mean μ.

With this definition the second moment is a measure of the spread or dispersion of $p_X(x)$ and is known as the *variance*, σ^2:

$$\begin{aligned}\sigma^2 &= E[(X - \mu)^2] \\ &= \int_{-\infty}^{\infty} (x - \mu)^2 \, p_X(x)dx \\ &= m_2 \end{aligned} \qquad (4.1.9)$$

The symbol m_2 is introduced for this second moment to be consistent with the more general relationship to be defined in (4.1.11).

The only difference between the variance and the earlier form of second moment, the mean square, is the use of a different moment arm, and hence they are closely related:

$$\begin{aligned}\sigma^2 &= \int_{-\infty}^{\infty} x^2 p_X(x)dx - 2\mu \int_{-\infty}^{\infty} x p_X(x)dx \\ &\qquad + \mu^2 \int_{-\infty}^{\infty} p_X(x)dx \\ &= E[X^2] - \mu^2 \end{aligned} \qquad (4.1.10)$$

Since $p_X(x)dx$ is a pure number, the units of variance are those of X^2. This is not convenient for some purposes and, therefore, the measure of dispersion is usually taken as the positive square root of σ^2. This quantity σ is referred to as the *standard deviation*. It is also the root mean square of the deviation.

It may be helpful to note that if $p_X(x)$ were visualized as the mass distribution of a rod, then μ would equal the distance of the center of mass from $x = 0$, and σ^2 would equal the moment of inertia about the center of mass. Also, (4.1.10) is simply a statement of the parallel axis theorem.

We next proceed to define moments of higher order, denoting a moment of order k as m_k:

$$m_k = E[(X - \mu)^k] = \int_{-\infty}^{\infty} (x - \mu)^k p_X(x)dx \quad (4.1.11)$$

Each *of* these moments is a kind of weighted average of $p_X(x)$; that is, each moment is a parameter that characterizes $p_X(x)$, and it would be possible to define or describe a probability density function in terms of its various moments. However, a complete and unique definition would, in general, require moments of all orders. This is impractical, and we usually deal only with the first two orders: the mean and the variance. As we shall see, for some particular probability distributions, this information is sufficient to completely define the distribution.

GAUSSIAN DISTRIBUTION

The Gaussian or normal distribution

$$p_X(x) = \frac{1}{\sqrt{2\pi}\,\sigma} e^{-(x-\mu)^2/2\sigma^2} \; (-\infty < x < +\infty) \qquad (4.1.12)$$

is by far the most common distribution. Many random processes, be they physical or mathematical, exhibit this relationship. One of the features of this distribution is that it is defined entirely and explicitly in terms of its mean value μ and its variance σ^2. Some typical Gaussian distributions are shown in Fig. 4.1.2 for $\mu = 0$ and $\sigma = 1$, 2, and 3. (Where a slash (/) is used in an exponent of this and other probability density functions below, all of the terms to the right of the slash are in the denominator, unless indicated otherwise by parentheses.)

LOG-NORMAL DISTRIBUTION

A positive random variable X is said to have a log-normal distribution if its natural logarithm, $Z = \ln X$, has a normal (Gaussian) distribution. That is,

$$p_Z(z) = \frac{1}{\sqrt{2\pi}\,\sigma_Z} e^{-(z-\mu_Z)^2/2\sigma_Z^2} \qquad (4.1.13)$$

where μ_Z and σ^2_Z are the mean and the variance of $\ln X$, respectively. The probability density function of X is then

$$\begin{aligned} p_X(x) &= p_Z(z)\frac{dZ}{dX} = \frac{p_Z(x)}{x} \\ &= \frac{1}{\sqrt{2\pi}\,\sigma_Z x} e^{-(\ln x-\mu_Z)^2/2\sigma_Z^2} \quad (x \geqslant 0) \end{aligned} \qquad (4.1.14)$$

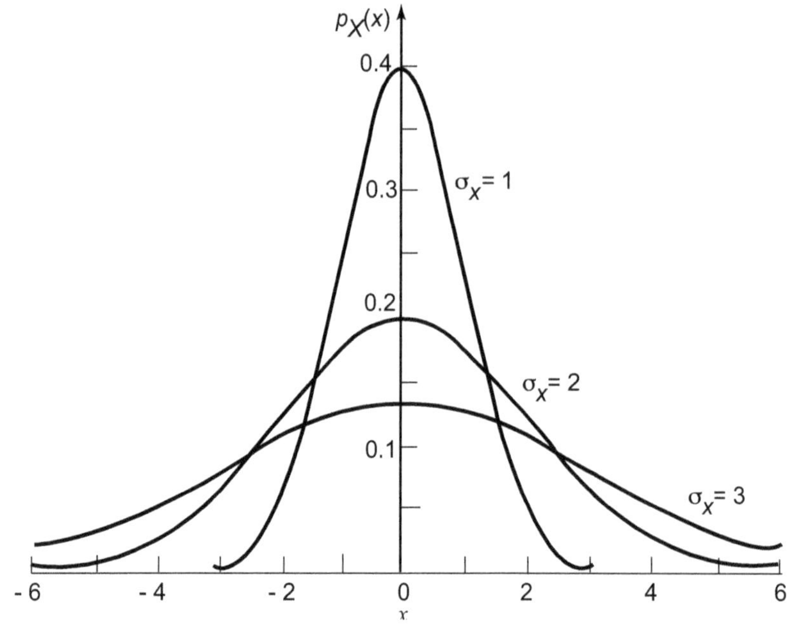

Figure 4.1.2 Gaussian pobability density function, with $\sigma_X = 1, 2,$ and 3, and with $\mu = 0$.

It has been found from ocean wave data that the two parameters that characterize a sea state—significant height and modal period—follow a lognormal distribution. This is discussed in Section 4.3.

RAYLEIGH DISTRIBUTION

The surface elevation X at any point in the ocean is a random variable with a Gaussian distribution and a zero mean. In Section 4.1.2 it will be shown that the *peak values* of X, denoted here as $\tilde{X}$, have a Rayleigh probability density function which is

$$p_{\tilde{X}}(\tilde{x}) = \frac{\tilde{x}}{\sigma_x^2} e^{-\tilde{x}^2/2\sigma_x^2} \quad (\tilde{x} > 0) \qquad (4.1.15)$$

or, in nondimensional form, with $Y = \tilde{x}/\sigma_X$

$$p_Y(y) = p_{\tilde{X}}(y) \frac{dX}{dY} = \frac{y}{\sigma_X} e^{-y^2/2} \sigma_X$$
$$= y\, e^{-y^2/2} \qquad (4.1.16)$$

where σ_x^2 is the variance of X. This is illustrated in Fig. 4.1.3.

All three of the foregoing distributions are defined in terms of no more than two parameters: μ and σ^2. In such cases the first two moments, $m_1 = \mu = E[X]$ and $m_2 = \sigma^2 = E[X^2]$, are sufficient to completely define the distribution; if they are known, then the full distribution can be created. For other distributions, further information would be required.

CENTRAL LIMIT THEOREM

The central limit theorem states that, if a random variable Z is the sum of n independent random variables

$$Z = \sum_{i=1}^{n} X_i \qquad (4.1.17)$$

and if all of the X_i have the same distribution, then the probability density function $p_Z(z)$ for

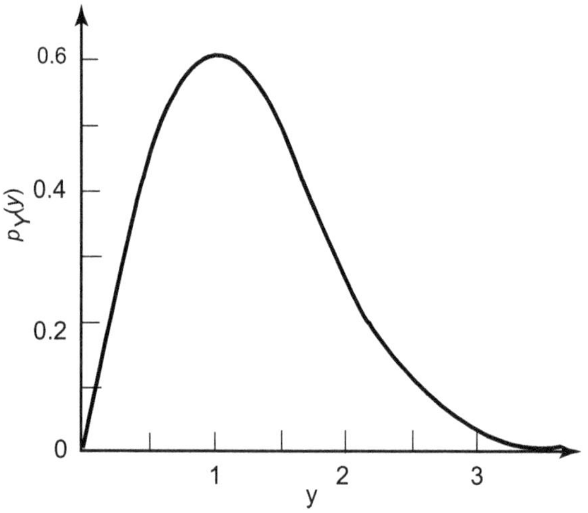

Figure 4.1.3 Rayleigh probability density function of the nondimensional random variable $y = \tilde{x}/\sigma_X$.

Z approaches the Gaussian distribution as n approaches infinity, regardless of the type of distribution of the X_i variables. That is, a random variable that is the sum of similar random effects also has a Gaussian distribution.

PROBABILITY DENSITY FUNCTION FOR TWO VARIABLES

If there are two events or random variables X and Y, the corresponding probability density function expresses the probability of both events occurring together:

$$\text{Prob}[(x \le X \le x + dx) \cap (y \le Y \le y + dy)] = p_{XY}(x, y)dx\,dy \quad (4.1.18)$$

A typical probability density function for two variables is shown in Fig. 4.1.4, illustrating that the joint probability is the volume under that part of the $p_{XY}(x, y)$ surface bounded by the two intervals.

$$\text{Prob}[(a \le X \le b) \cap (c \le Y \le d)] = \int_a^b \int_c^d p_{XY}(x, y)dx\,dy \quad (4.1.19)$$

Also, the total volume under the complete surface is unity.

In the special case when the two random variables are independent, then the joint probability density function is simply the product of the two separate distributions

$$p_{XY}(x, y) = p_X(x)\, p_Y(y) \quad (4.1.20)$$

For a joint distribution we can again take moments as before. The expected value of the product XY corresponds to a mixed second moment:

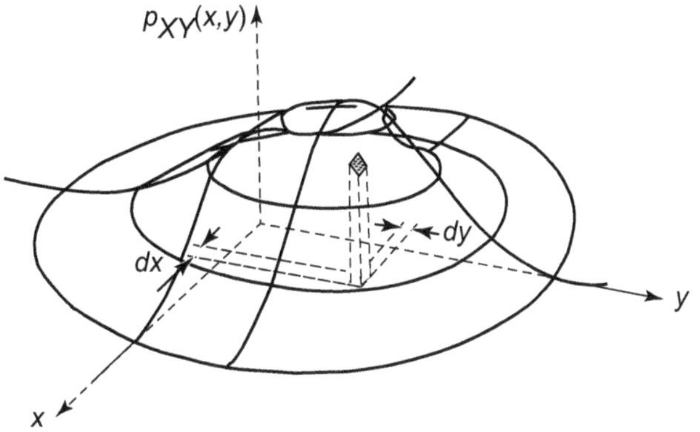

Figure 4.1.4 Second-order probability density function.

$$E[XY] = \int_{-\infty}^{\infty} \int_{-\infty}^{\infty} xy\, p_{XY}(x, y)dx\,dy \quad (4.1.21)$$

If we take moments about the respective mean values μ_X and μ_Y (computed independently from the one-dimensional distributions) we have what is known as the *covariance* C_{XY} of the random variables X and Y:

$$\begin{aligned} C_{XY} &= E[(X - \mu_X)(Y - \mu_Y)] \\ &= \int_{-\infty}^{\infty} \int_{-\infty}^{\infty} (x - \mu_X)(y - \mu_Y)p_{XY}(x, y)dx\,dy \\ &= E[XY] - E[X]E[Y] \end{aligned} \quad (4.1.22)$$

When $X = Y$, the covariance C_{xx} reduces to the variance $\sigma^2 x$ of the random variable X.

The *correlation coefficient* ρ_{XY} of the random variable is defined as

$$\rho_{XY} = \frac{C_{XY}}{\sigma_X \sigma_Y} \quad (4.1.23)$$

It can be shown that the correlation coefficient lies in the range $-1 \le \rho_{XY} \le 1$. When $\rho_{XY} = 0$, the random variables X and Y are said to be "uncorrelated." If two random variables X and Y are independent, then they are also uncorrelated because

$$\begin{aligned} E[XY] &= \int_{-\infty}^{\infty} \int_{-\infty}^{\infty} xy\, p_{XY}(x, y)dx\,dy \\ &= \int_{-\infty}^{\infty} xp_X(x)dx \int_{-\infty}^{\infty} yp_Y(y)dy \\ &= E[X]E[Y] \end{aligned} \quad (4.1.24)$$

and $\quad C_{XY} = E[XY] - E[X]E[Y] = 0 = \rho_{XY}$

However, uncorrelated random variables are not necessarily independent random variables.

The expected value of a linear combination of random variables $Z = aX + bY$ is

$$\begin{aligned} E[Z] &= E[(aX + bY)] \\ &= \int_{-\infty}^{\infty} \int_{-\infty}^{\infty} (ax + by)\, p_{XY}(x, y)dx\,dy \\ &= a \int_{-\infty}^{\infty} x \int_{-\infty}^{\infty} p_{XY}(x, y)dy\,dx \\ &\quad + b \int_{-\infty}^{\infty} y \int_{-\infty}^{\infty} p_{XY}(x, y)dx\,dy \\ &= a \int_{-\infty}^{\infty} xp_X(x)dx + b \int_{-\infty}^{\infty} yp_Y(y)dy \\ &= a\mu_X + b\mu_Y \end{aligned} \quad (4.1.25)$$

and the variance is

$$\begin{aligned}\sigma_Z^2 &= E[\{(aX + bY) - (a\mu_X + b\mu_Y)\}^2] \\ &= E[\{a(X - \mu_X) + b(Y - \mu_Y)\}^2] \\ &= a^2E[(X - \mu_X)^2] + 2abE[(X - \mu_X)(Y - \mu_Y)] \\ &\quad + b^2E[(Y - \mu_Y)^2] \\ &= a^2\sigma_X^2 + 2abC_{XY} + b^2\sigma_Y^2 \end{aligned} \tag{4.1.26}$$

If X and Y are independent random variables, then the covariance $C_{XY} = 0$ and

$$\sigma_Z^2 = a^2\sigma_X^2 + b^2\sigma_Y^2 \tag{4.1.27}$$

We may extend these results to the set of n independent random variables $X_1, X_2, \ldots, X_n$ with individual expected values μ_i and variances $\sigma^2 i$, $(i = 1, 2, \ldots n)$. It may be shown that the summed random variable

$$z = \sum_{i=1}^{n} X_i \tag{4.1.28}$$

has the expected value

$$\mu_Z = \sum_{i=1}^{n} \mu_i \tag{4.1.29}$$

and the variance

$$\sigma_Z^2 = \sum_{i=1}^{n} \sigma_i^2 \tag{4.1.30}$$

4.1.2 Random Processes

A *random process* is a random function of a *time* parameter. Figure 4.1.5 shows several

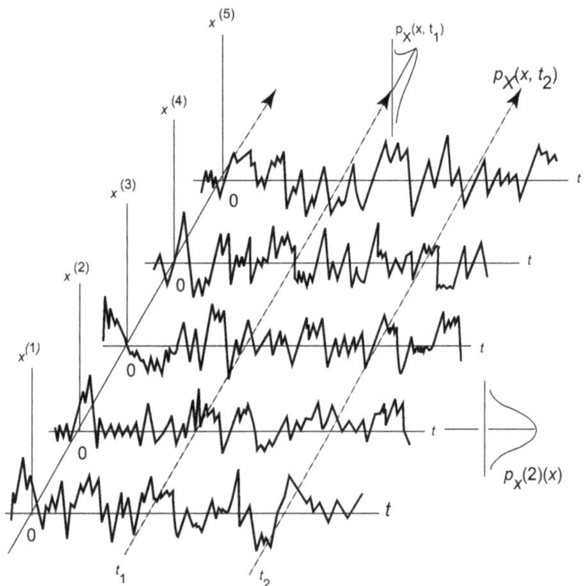

Figure 4.1.5 Schematic representation of a random process $X(t)$ (each $X^{(j)}(t)$ is a sample function of the ensemble).

samples or observations of a random, time-dependent quantity $X(t)$. In each case the time t is the time from the commencement of sampling. Since X is a random variable, each curve $X^{(k)}(t)$ represents just one sample out of an infinite number of possible samples or measurements. The collection of such samples, theoretically infinite in number, is known as an "ensemble."

Of course, there may be more than one independent variable; for example, the process may depend on location as well as time. Whatever the dependency, it is necessary that each sample be taken under identical conditions, at least in regard to the factors that affect $X(t)$, so that $X(t)$ *is* a truly random process. If variations in some such factor cannot be avoided, then that factor must be included as a second independent variable, in addition to t, and the process would then be "two-dimensional." In the case of ocean waves, where $X(t)$ is the sea surface elevation, it is possible to identify and isolate these factors sufficiently well so that, for short-term observations (up to a few hours in duration), $X(t)$ is essentially one-dimensional.

STATIONARY AND ERGODIC RANDOM PROCESSES

If the statistical characteristics of a random process do not change with time t, we say that the process is *stationary*. Thus, for example, if the process in Fig. 4.1.5 were a stationary random process, then the probability density functions at t_1, t_2, and so on would all be the same. Now, the statistical quantities used in the analysis of random processes are averages of $X(t)$ and of various functions of $X(t)$, such as its square, as is done for a simple, non–time-dependent variable X. If the process is stationary, then the averages and moments are invariant over time, and they therefore have the same definition and meaning as given earlier for simple, non–time-dependent random variables. That is, the mean, the variance, and the higher moments of the process continue to be given by (4.1.7), (4.1.9), and (4.1.11).

It has been found experimentally that, over the short term (as defined before), the sea surface elevation is a stationary random process. Hence, all of the earlier definitions of averages and moments are still applicable, even though the random variable X is now a function of time, i.e., $X(t)$.

Because a random process is a function of time, there are in fact two different ways of calculating averages. The averages may be taken over all of the samples of the ensemble at the same instant of time, say t_1, or they may be taken for a particular sample, say $X^{(1)}(t)$ over all time from $-\infty$ to ∞. These two alternatives are referred to as *ensemble averages* and *temporal* (or *time*) *averages*, respectively. All of the averages defined thus far have been ensemble averages. For instance, the mean μ, the variance σ^2, and the higher moments m_k are the ensemble averages of X, $(X - \mu)$, and $(X - \mu)^k$, respectively. Likewise, the ensemble average of any function $f(X)$ of X is given by (4.1.6).

The temporal averages, on the other hand, are computed for a particular sample $X(t) = X^{(1)}(t)$ [we will omit the (1) superscript from here on] over some length of time—a *long* time for sufficient accuracy. Thus, for example, the *temporal mean* is

$$\mu = \langle X(t) \rangle = \lim_{T\to\infty} \frac{1}{T} \int_{-T/2}^{T/2} X(t)dt \qquad (4.1.31)$$

where brackets $\langle\ \rangle$ indicate a temporal averaging. Likewise, the *temporal mean square* is

$$\langle X^2(t) \rangle = \lim_{T\to\infty} \frac{1}{T} \int_{-T/2}^{T/2} X^2(t)dt \qquad (4.1.32)$$

It is often useful to represent a time-dependent variable as the sum of a temporal mean value plus a fluctuating component. The fluctuating component is described by the *variance* σ^2, which is from (4.1.9)

$$\begin{aligned}\sigma^2 &= \langle [X(t) - \mu]^2 \rangle \\ &= \lim_{T\to\infty} \frac{1}{T} \int_{-T/2}^{T/2} [X(t) - \mu]^2 dt \qquad (4.1.33)\end{aligned}$$

In practice, of course, some finite length of time must be used in place of the limit $T \to \infty$, and so the above equalities become approximations, or estimates, of the mean, the mean square, and the variance.

In general, these time averages are different from those obtained by averaging "across the ensemble." However, many random processes, including ocean waves, are such that time averages formed from a single sample over a time interval are, in the limit, equal to the ensemble averages. Such processes are known as *ergodic* processes. In qualitative terms, an ergodic process is one where a single sample $X(t)$ is sufficiently typical to represent the entire process. Obviously, an ergodic process must be stationary, but a stationary random process is not necessarily ergodic. Ergodicity implies that all of the various expectations are equal to, and may be replaced by, the corresponding temporal averages. This is important because, although most of the theoretical relationships are defined or derived in terms of ensemble averages, in practice these are difficult, if not impossible, to obtain, whereas calculating time averages from a single observation is relatively easy. If the process is ergodic, these can be used in place of the ensemble averages.

As noted earlier, there are two distinct methods of specifying a random process $X(t)$:

1. Through its probability density function, or
2. Through various averages that correspond to various moments of the probability density function.

The first of these methods is impractical, because it involves an enormous amount of information. The second is practical, providing that the averages can be measured. Hence, ergodicity is important. With this property, the required averages may be computed from measurements of a single observation. Ergodicity is assumed in virtually all practical applications, particularly in the estimation of parameters of empirical models. Fortunately, oceanographers and statisticians established that the surface elevation at a given location in the ocean can be considered, at least for engineering purposes, as an ergodic process with a Gaussian distribution. This will be taken up in Section 4.3.

AUTO-CORRELATION FUNCTION FOR AN ERGODIC RANDOM PROCESS

The averages used to analyze linear systems by statistical methods are those that represent or measure the degree of association between values of the random variable $X(t)$ at times differing by a specified time interval τ. These averages are called *correlation functions*. The most basic of them is the *auto correlation function*, τ, which is the average, or expected value, of the product of any two values of X: $X_1 = X(t_1)$ and $X_2 = X(t_2) = X(t_1 + \tau)$:

$$\begin{aligned}R(\tau) &= E[X(t)X(t+\tau)] \\ &= E[X_1X_2] \\ &= \int_{-\infty}^{\infty}\int_{\infty}^{\infty} x_1x_2p_{XX}(x_1, x_2)dx_1dx_2 \qquad (4.1.34)\end{aligned}$$

For a stationary process, $R(\tau)$ is independent of t because $E[X(t_1)X(t_1 + \tau)] = E[X(t_2)X(t_2 + \tau)]$. Also, since the product X_1X_2 is commutative, we have

$$\begin{aligned}R(\tau) &= E[X_1X_2] = E[X_2X_1] = E[X(t_2)X(t_2 - \tau)] \\ &= R(-\tau) \qquad (4.1.35)\end{aligned}$$

indicating that $R(\tau)$ is an even function of τ. The value at the origin is simply the mean square

$$R(0) = E[X^2] \tag{4.1.36}$$

To get some idea of the shape of $R(\tau)$, let us consider an ensemble of functions $X(t)$ with zero mean. (This zero mean restriction is not necessary for the theory developed in this section. However, as it is characteristic of ocean waves and as it also simplifies the development, it is introduced at this stage.) If t and τ are such that there is little or no association between $X(t)$ and $X(t + \tau)$, then a particular value of $X(t)$ is just as likely to be associated with a positive value of $X(t + \tau)$ as with negative value. For this reason we can expect the average value of the product $X(t)\, X(t + \tau)$ to be near zero. On the other hand, the association between the two values may be such that both tend to have the same sign, either positive or negative. In this case the product will have an average that is positive. Again, if $X(t)$ and $X(t + \tau)$ tend to have opposite signs, the average of the product will be negative.

To clarify matters, let us consider two extreme cases. On the one hand, values of a function $X(t)$ of the ensemble at different times may be completely unrelated, however close the times may be. This kind of an ensemble, called *white noise*, generally has no continuity; it represents the characteristic of an erratic variation in the extreme. In this case, $R(\tau) = 0$ for all values of τ except zero, where it equals the mean square.

On the other hand, if each sample $X(t)$ were identical, thus exhibiting perfect correlation, then $R(\tau)$ would be a constant equal to the mean square.

Usually the situation is intermediate between these extremes. Thus, if τ is small, $X(t + \tau)$ can be expected to lie in a range of values that do not differ greatly from $X(t)$. On the other hand, if τ is large, there will be very little association between $X(t)$ and $X(t + \tau)$. Consequently, provided that there are no periodic components present in the random process, $R(\tau)$ tends to zero since products will be equally positive and negative. That is, $R(\tau) \to 0$ as $\tau \to \infty$. With a nonzero mean, $R(\infty) = \mu^2$. Two typical auto correlation functions are shown in Fig. 4.1.6.

Instead of the ensemble average of X_1X_2, we can form the time average; the *temporal auto-correlation* function is

$$R(\tau) = \langle X(t)X(t + \tau)\rangle$$
$$= \lim_{T\to\infty} \frac{1}{T} \int_{-T/2}^{T/2} X(t)X(t + \tau)dt \tag{4.1.37}$$

For an ergodic process, this will be identical to the ensemble average of (4.1.34), and in the case of a physical process, the time average is much easier to obtain. Hence, (4.1.37) is taken as the definition of $R(\tau)$.

Some further insight into the concept of auto-correlation may be gained by examining a typical method for measuring the auto-correlation of a random signal (for example, the heave motion of a buoy for measuring wave heights). Figure 4.1.7 shows a block diagram for the procedure. The sample function is recorded on a tape, and a tape recorder with *two* heads is used. The spacing between the heads is adjustable. If the speed of the tape is V, the time interval between readings is $\tau = a/V$. The values $f(t)$, $f(t + \tau)$ are then multiplied and averaged over a long time period T. For a finite averaging time the measured quantity $y(t)$ fluctuates slowly. In the limit $T \to \infty$ the reading y approaches the temporal auto-correlation function $R(\tau)$ and, if the process is ergodic, this is also equal to the ensemble auto-correlation function. To obtain the auto-correlation function as a function of τ, it is necessary to perform the average just described for a whole

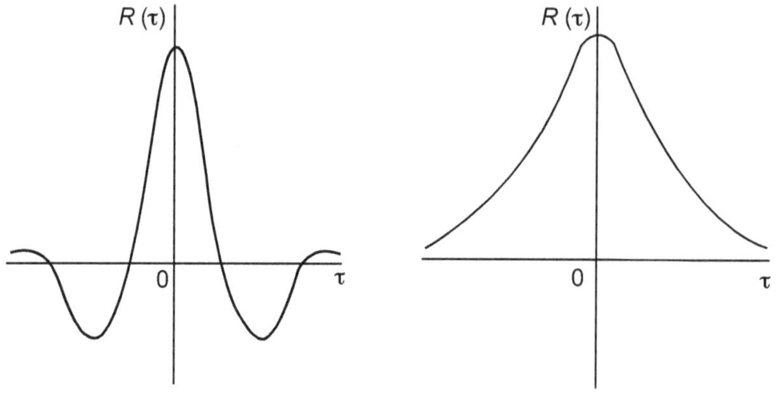

Figure 4.1.6 Possible forms of the auto-correlation function.

range of values of τ; that is, for a whole range of values of a.

FREQUENCY ANALYSIS OF AN ERGODIC PROCESS

In dealing with fluctuating time variations, it is usually more convenient to work in terms of frequency. This is obviously true for *periodic* variations, but it is also true for random variations, particularly if the process is ergodic.

Before considering a random function $X(t)$, it is useful to recall the Fourier expansion of a *deterministic* time dependent function $g(t)$. When $g(t)$ *is periodic with period T, or frequency* $\omega_0 = 2\pi/T$, *we have*

$$g(t) = \sum_{n=-\infty}^{\infty} g(n)e^{in\omega_0 t} \tag{4.1.38}$$

where $$g(n) = \frac{1}{T}\int_{-T/2}^{T/2} g(t)e^{-i\omega t}\, dt$$

The quantity $g(n)$ is referred to as the "Fourier transform" of the periodic function $g(t)$. The function $g(t)$ and the quantity $g(n)$ are said to constitute a "Fourier transform pair."

When g(t) *is a nonperiodic function, it may, under fairly general conditions, still be represented in a transformed manner as a "Fourier integral:"*

$$g(t) = \int_{-\infty}^{\infty} G(\omega)e^{i\omega t}\, d\omega \tag{4.1.39}$$

in which $$G(\omega) = \frac{1}{2\pi}\int_{-\infty}^{\infty} g(t)e^{-in\omega_0 t}\, dt$$

is the Fourier transform of the function $g(t)$. Here again, the functions $g(t)$ and $G(\omega)$ are a Fourier transform pair, constituting inverses of each other. For deterministic time-dependent functions, the transformation to the "frequency domain" usually simplifies and facilitates the analysis.

For a random process $X(t)$ a similar frequency-based or spectral representation can be achieved. Although it is not possible to transform $X(t)$ directly, it can be shown that the auto correlation function $R(\tau)$ fulfills all of the requirements to define a Fourier transform. The Fourier transform of $R(\tau)$ and its inverse for a stationary random process $X(t)$ are, respectively,

$$S(\omega) = \frac{1}{2\pi}\int_{-\infty}^{\infty} R(\tau)\cos\omega\tau\, d\tau \tag{4.1.40}$$

$$R(\tau) = \frac{1}{2\pi}\int_{-\infty}^{\infty} S(\omega)\cos\omega\tau\, d\omega \tag{4.1.41}$$

These two equations are called the "Wiener Khintchine Relations." The quantity $S(\omega)$ is called the "spectral density function" because its domain is the spectrum $-\infty < \omega < \infty$. Like the auto-correlation function, the spectral density function is a measure, or at least an indicator, of the repetitiveness or harmonic content of the random process $X(t)$ and of its distribution over the frequency range. If $X(t)$ exhibits some approximate repetition having an approximate frequency ω_0, then $S(\omega)$ will have a local peak in the vicinity of ω_0.

The full name of this function is the "*mean square* spectral density function," or the variance spectrum for $E[x(t)] = 0$. It is called this because the area under the $S(\omega)$ curve is the mean square

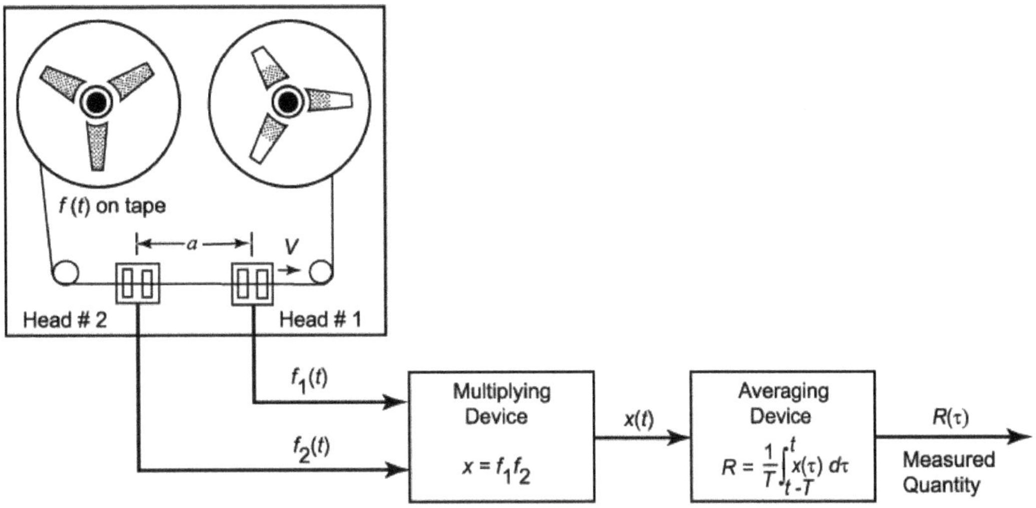

Figure 4.1.7 Schematic diagram for measurement of autocorrelation function $R(\tau)$ for sample function $f(t)$. The adjustable distance a between tape heads is proportional to lag τ.

value of $X(t)$. This may be shown from (4.2.8) in which, if we set $\tau = 0$, the right-hand side becomes the area under the $S(\omega)$ curve and the left-hand side becomes $R(0)$, which is simply the mean square of the process, as was shown earlier in (4.1.36). That is,

$$\text{Area under } S(\omega) = \int_{-\infty}^{\infty} S(\omega)d\omega = R(0) = E[X^2(t)] = \langle X^2(t)\rangle \quad (4.1.42)$$

Hence, in an analogous manner to the probability density function, the area $S(\omega)d\omega$ that lies within the interval or "band" from ω to $\omega + d\omega$ is the contribution which is made to the mean square by components whose frequencies lie within that band.

Other properties of $S(\omega)$ may be deduced from its definition. Since $R(\tau)$ is real and even, it follows that $S(\omega)$ will be a real and even function of ω. Also, since $S(\omega)$ is the spectral density of the mean square, it is nonnegative. An example of a typical spectral density function is shown in Fig. 4.1.8. In defining $S(\omega)$ we adopted, simply for mathematical convenience, the range or spectrum $-\infty < \omega < \infty$. Since we wish to apply this theory to physical processes, it is necessary to use a modified, one-sided spectrum: $\omega \geq 0$. For this purpose we define the one-sided mean square spectral density function

$$S^+(\omega) = \begin{cases} 2S(\omega) & \text{for } \omega \geq 0 \\ 0 & \text{otherwise} \end{cases} \quad (4.1.43)$$

as shown in Fig. 4.1.8. Since both the auto-correlation function and the spectral density function are real and even, we have

$$R(\tau) = \int_0^{\infty} 2S(\omega) \cos \omega\tau \, d\omega \quad (4.1.44)$$

$$R(\tau) = \int_0^{\infty} 2^+(\omega) \cos \omega\tau \, \mathrm{d}\omega \quad (4.1.45)$$

and so, for a one-sided spectrum, $S^+(\omega)$ bears the same relationship to $R(\tau)$ as did $S(\omega)$ for the two-sided spectrum. Since S and S^+ are so closely related, and since the latter is used in all practical applications, we will not bother using the + sign to distinguish them. We will use the two-sided spectrum only when it is more convenient mathematically, and its use in such instances will be clear from the context.

Since the spectral density function is the mathematical transform of the auto-correlation function

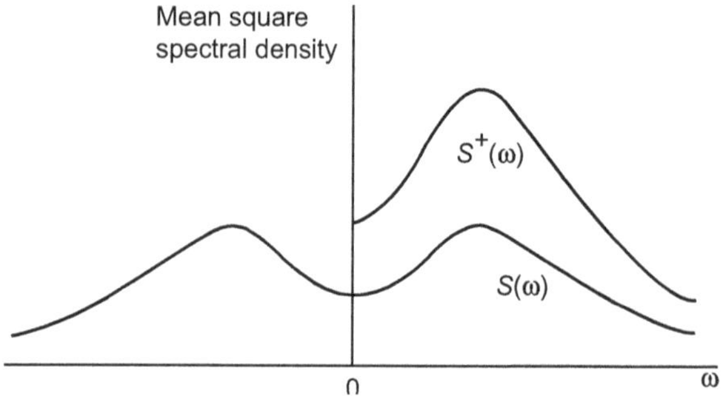

Figure 4.1.8 One- and two-sided mean square spectral density functions of a random process.

of $X(t)$, the units of $S(\omega)$ depend on the units of $X(t)$. If $X(t)$ is wave height in meters, then $S(\omega)$ has units of m^{2}s. It may be shown that for any kind of wave—gravitational, electromagnetic, and so on—the area $S(\omega_0)\delta\omega$ within a bandwidth $\delta\omega$ is directly proportional to the total energy of all of the components that lie within the band (ω_0 – ½ $\delta\omega$, ω_0 + ½ $\delta\omega i$). Because of this direct relationship between energy and the mean square spectral density function and because of the attraction of a shorter name, the terms "energy spectrum" or "wave spectrum" are commonly used in place of the full name. Sometimes this is even further shortened to simply "spectrum" although, strictly speaking, this word means a range of frequencies rather than a function defined within that range.

One of the principal advantages of the spectral density function is that all basic characteristics of a random process can be expressed in terms of moments of this function:

$$m_n = \int_0^{\infty} \omega^n S(\omega)d\omega \quad (4.1.46)$$

where n can be any integer. We have already seen in (4.1.42) that the mean square value is the "zero order moment." That is

$$\langle X^2(t)\rangle = m_0 = \int_0^{\infty} S(\omega)d\omega \quad (4.1.47)$$

Some other basic characteristics that may be expressed in terms of spectral moments are the following:

Average mean period

$$T_p = 2\pi \frac{m_0}{m_1} \quad (4.1.48)$$

Average zero crossing period

$$T_{p0} = 2\pi \sqrt{\frac{m_0}{m_1}} \quad (4.1.49)$$

Period between maxima, regardless of magnitude,

$$T_{pc} = 2\pi\sqrt{\frac{m_0}{m_2}} \qquad (4.1.50)$$

Average crest-to-crest period for a general random process

$$T_{pc} = 4\pi\left(\frac{\sqrt{1-\varepsilon^2}}{1+\sqrt{1-\varepsilon^2}}\right)\sqrt{\frac{m_0}{m_2}} \qquad (4.1.51)$$

and for a narrow band random process

$$T_{pc} = 2\pi\sqrt{\frac{m_2}{m_4}} \qquad (4.1.52)$$

Broadness (or "bandwidth")

$$\varepsilon = \sqrt{1-\frac{m_2^2}{m_0 m_4}} \qquad (4.1.53)$$

PROPERTIES OF A NARROW-BAND RANDOM PROCESS

Figure 4.1.9 shows two extreme kinds of random process and their corresponding energy spectra. The two kinds are:

1. *Narrow-Band Process*—A process that is made up of components whose frequencies lie within a narrow band or range, whose width is small compared with the magnitude of the center frequency of the band, ω_0. This periodicity produces regularly spaced peaks in the auto correlation function and a single narrow peak at ω_0 in the energy spectrum.

2. *Wide-Band Process*—In this case the process contains components of many different frequencies, so that there is little or no periodicity and the auto-correlation function is almost zero. The energy spectrum is therefore quite wide.

Analysis of ocean wave data showed that for a fully developed, wind-generated, mid-ocean sea state (i.e., no growth or decay, no coastal effects, and no swell) the wave spectrum is relatively narrow-banded. Of course, high-frequency wave components do occur, but they correspond to waves that are small—both in height and in length—and waves that have little effect on the ship. In fact, the ship acts as a filter, such that the spectra of ship motions and of the hull girder load effects are even more narrow-banded than the wave spectrum. Also, like the waves, these various responses have distributions that are Gaussian and stationary (in the short term, i.e., for a given sea state). This is an important point because a process of this kind is much easier to analyze than a general, wide-band process. In this section we present some of the principal characteristics of a narrow-band process, $X(t)$; these will have application not only to waves, but also to the various wave-induced load effects in ships, especially hull girder bending moment.

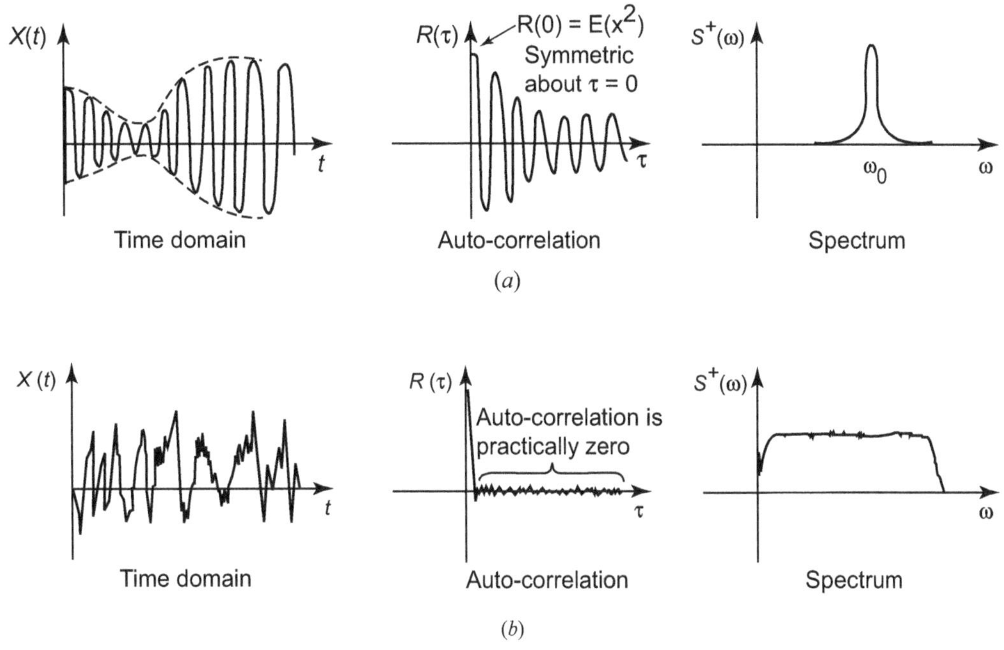

Figure 4.1.9 Narrow-band (*a*) and wide-band (*b*) spectrum.

For simplicity we shall assume that $X(t)$ has a zero mean. This is, of course, true for the ocean surface elevation, and it is also true for wave-induced motions and load effects if they are calculated from the linearized theory presented in Section 4.4. The more accurate nonlinear motions and load effects have nonzero mean values, but this in itself does not present any difficulty. Rather, the difficulty with nonlinear processes is that a frequency-based response analysis is not possible; the analysis must be performed in the time domain. This is discussed in Section 4.4.

For design purposes we are interested in peak values, $\tilde{X}$, rather than the full range of values of $X(t)$. Peak values of a random process are a special subgroup and, therefore, they have a probability density function of their own, different to that of $X(t)$. If we denote this function as $p_{\tilde{X}}(\tilde{x})$, the probability that a peak value, chosen at random, will exceed the value $\tilde{X}$ is

$$\text{Prob}\{\text{peak exceeds } \tilde{X}\} = \int_{\tilde{X}}^{\infty} p_{\tilde{X}}(x)dx \quad (4.1.54)$$

This is illustrated in Fig. 4.1.10.

As shown originally by Rice (1945) and in standard texts such as Crandall and Mark (1963), $p_{\tilde{x}}(\tilde{x})$ may be derived by examining the probability of "positive crossings" of $X = \tilde{x}$; that is, the average frequency $\omega_{\tilde{x}}$ with which X exceeds a specified magnitude $\tilde{x}$. If $X(t)$ is Gaussian, the frequency is

$$\omega_{\tilde{x}} = \frac{1}{2\pi}\frac{\sigma_{\dot{X}}}{\sigma_X} e^{-\tilde{x}^2/2\sigma_X^2} \quad (4.1.55)$$

where σ_X and $\sigma_{\dot{X}}$ are the standard deviations of $X(t)$ and $\dot{X}(t)$ (= dX/dt), respectively. Setting $\tilde{x} = 0$ gives the average frequency for the process, ω_0. In a narrow band process, $X(t)$ crosses the axis before and after each peak, with only rare exceptions. Therefore, the expected number of positive peak values is approximately equal to the number of cycles. In time T there will be, on average, $\omega_0 T$ cycles. Of these, the number of cycles with peak values exceeding $\tilde{x}$ will be $\omega_{\tilde{x}}T$. Hence, the probability that any peak value, chosen at random, exceeds $\tilde{x}$ is equal to $\omega_{\tilde{x}}/\omega_0$; that is,

$$\frac{\omega_{\tilde{x}}}{\omega_0} = \int_{\tilde{x}} p_{\tilde{x}}(x)dx \quad (4.1.56)$$

Substitution of (4.1.55) and differentiation with respect to $\tilde{x}$ gives

$$p_{\tilde{X}}(\tilde{x}) = \frac{\tilde{x}}{\sigma_X^2} e^{-\tilde{x}^2/2\sigma_X^2} \quad (4.1.57)$$

which is the Rayleigh distribution of (4.1.15). This shows that for a random process which is Gaussian, ergodic, and narrow-banded, peak values follow a Rayleigh distribution. For a process with zero mean, the variance σ_x^2 is equal to the mean square, and from (4.1.47) this is equal to m_0, the area under the spectral density function of the process. In terms of m_0, (4.1.57) is

$$p_{\tilde{X}}(\tilde{x}) = \frac{\tilde{x}}{m_0} e^{-\tilde{x}^2/2m_0} \quad (4.1.58)$$

The cumulative probability distribution function is

$$P_{\tilde{X}}(\tilde{x}) = \int_0^{\tilde{x}} p_{\tilde{X}}(x)dx = 1 - e^{-\tilde{x}^2/2m_0} \quad (4.1.59)$$

Thus, the probability of $\tilde{X}$ exceeding a specified value $\tilde{X}_p$ is

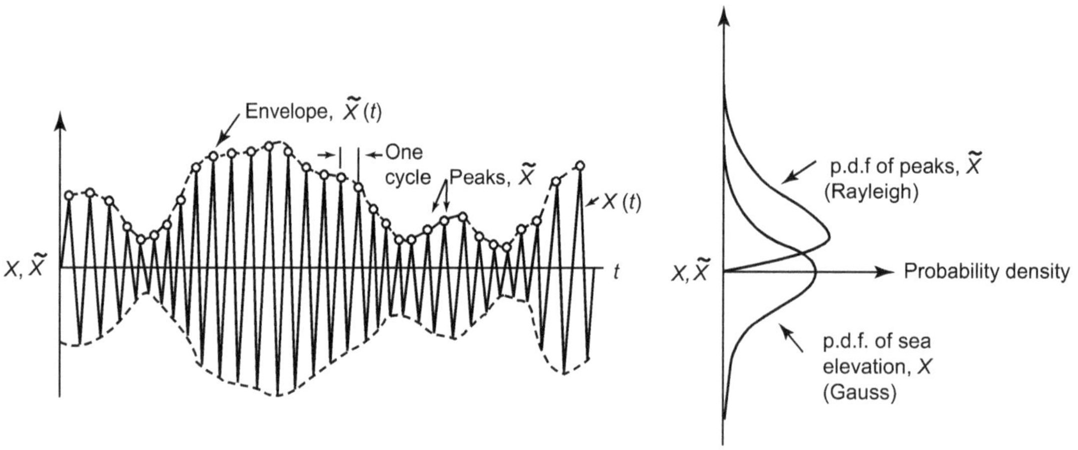

Figure 4.1.10 Distribution of peak values in a narrow-band process.

$$\text{Prob}\{\tilde{X} > \tilde{X}_p\} = 1 - P_{\tilde{X}}(\tilde{X}_p) = e^{-\tilde{x}_p^2/2m_0} \quad (4.1.60)$$

By inverting this we can define any value $\tilde{X}_p$ in terms of its probability of being exceeded. For example, as illustrated in Fig. 4.1.11a, we can define $\tilde{X}_{1/3}$ as the value of $\tilde{X}$ for which there is a probability of exceedance of 1/3. This implies

$$\text{Prob}\{\tilde{X} \geq \tilde{X}_{1/3}\} = e^{-\tilde{X}_{1/3}^2/2m_0} = \frac{1}{3} \quad (4.1.61)$$

from which we obtain

$$\tilde{X}_{1/3} = \sqrt{2m_0 \ln 3} = 1.482\sqrt{m_0} \quad (4.1.62)$$

The average of all of the values above the one-third value is called the *significant value*, and it is denoted $\tilde{X}_S$. It is the horizontal distance to the centroid of the area under $p_{\tilde{x}}(\tilde{x})$ beyond $\tilde{x} = \tilde{X}_{1/3}$, which is given by

$$\tilde{X}_S = 3\int_{\tilde{X}_{1/3}}^{\infty} \tilde{x} p_{\tilde{x}}(x)dx \quad (4.1.63)$$

The result is

$$\tilde{X}_S = \{\sqrt{\ln 3} + 3\sqrt{\pi}\,[1 - \phi(\sqrt{2\ln 3})]\}\,\sqrt{2m_0} \quad (4.1.64)$$

which, to three significant figures, is

$$\tilde{X}_S = 2.00\sqrt{m_0} \quad (4.1.65)$$

where

$$\phi(\xi) = \frac{1}{2}\left[1 + \text{erf}\left(\frac{\xi}{\sqrt{2}}\right)\right]$$

and $erf(\xi/\sqrt{2})$ is the error function or the probability integral

$$\text{erf}\left(\frac{\xi}{\sqrt{2}}\right) = \frac{2}{\sqrt{\pi}}\int_0^{\xi/\sqrt{2}} e^{-\xi^2} d\xi$$

Similarly, the average of the highest "one-nth" values of a random process that has a Rayleigh distribution is

$$\bar{X}_{1/n} = \left\{\sqrt{\ln(n)} + n\sqrt{\pi}\left[1 - \phi\left(\sqrt{\ln(n)}\right)\right]\right\}\sqrt{2m_0} \quad (4.1.66)$$

When n becomes large, this reduces to

$$\bar{X}_{1/n} = \sqrt{2m_0 \ln(n)} \quad (4.1.67)$$

The average 1/nth value provides information about the magnitude of the larger peaks of a random process and, hence it characterizes the "severity" of the process. In particular, the *significant value*, $\tilde{X}_S$, is used to measure and specify the severity of sea states. In this case the random variable is the trough-to-crest wave height, $\tilde{H}$, which is twice the amplitude of $\tilde{X}$. Therefore, the preceding expression must be multiplied by 2, giving

$$\tilde{H}_S = 4.01\sqrt{m_0} \quad (4.1.68)$$

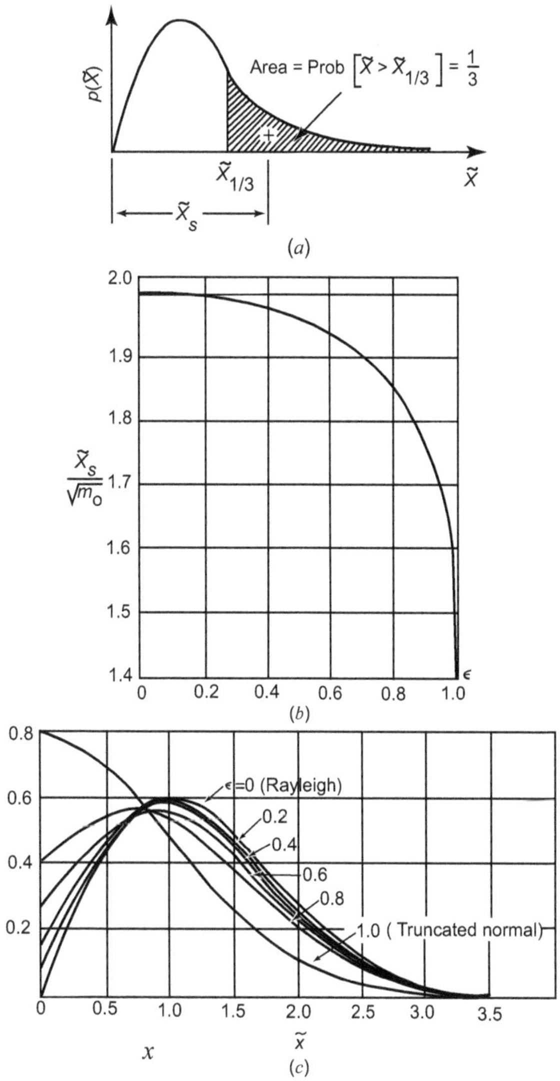

Figure 4.1.11 (a) Definition of significant value, $\tilde{X}_S$; (b) effect of bandwidth on $\tilde{X}_S$; and (c) effect of bandwidth on p.d.f. of peak values.

The statistical properties of a general nonnarrow band process were investigated by Cartwright and Longuet Higgins (1956) and by Huston and Skopinski (1956). Figure 4.1.11*b* from Ochi (1982) shows the results of computations of significant wave heights (in dimensionless form) for various values of the bandwidth parameter ε, defined in (4.1.53). As can be seen in the figure, there is no appreciable difference in the significant wave heights for ε up to 0.5. The difference only becomes significant for ε greater than 0.8. Since the ε-value for ocean waves is generally within the range from 0.4 to 0.8, the significant wave height evaluated on the assumption of a narrow-band process is only slightly overestimated, of the order of 5% or so.

Fig. 4.1.11*c* shows the probability density function of the peak values of various nonnarrow-band random processes as a function of ε. For a totally wide-band process the function is simply the Gaussian curve.

4.2 PREDICTION OF EXTREME VALUES

4.2.1 Extreme Values for Linear Responses

As explained in Chapter 1, the rationally-based design of ships requires the consideration of the largest value (extreme value) of wave loading, especially the wave-induced bending moment, which is expected to occur within the ship's lifetime and, in particular, the prediction of a *characteristic value* associated with a certain probability of nonexceedance in that time. To achieve this, we must first examine the concept of an *extreme value* of a random process. This topic was first treated by Fisher and Tippett (1928) and was later systematized by Gumbel (1966).

In defining a design value, it might seem sufficient to use the $1/n$th value, for which the probability of being exceeded is $1/n$, and to simply choose an appropriate value of n. It is true that the probability of exceeding the value $\tilde{Q}_{1/n}$ is, on average, once in n observations (peaks). (The symbol Q is chosen because in the majority of applications of the results obtained here, the random process $\tilde{Q}_{1/n}$ is a response or a load effect, which in this text is denoted by the symbol Q.) However, there is no assurance that $\tilde{Q}_{1/n}$ will occur once in n observations. For design purposes we need information concerning the *extreme value,* that is, the largest peak value that will occur in the life of the ship, and also the probability (or risk) that this largest value will exceed a specified magnitude. The *characteristic value* is then that magnitude of extreme value that has an appropriate probability of exceedance. Since the statistical properties of negative peak values are essentially the same as those of positive peak values, only the latter are considered in the following discussion (see Fig. 4.2.1).

For linear systems, there are three different methods to calculate extreme values, depending on whether or not the probability density function of the peak values is known analytically. In discussing extreme value statistics (for linear systems) this function is referred to as the *initial* probability density function, and its integral is referred to as the *initial* probability distribution. If the former is known analytically, then an exact analytical expression can be derived for the probable extreme value by applying *order statistics.* This situation exists, for example, in regard to the short-term peak values of wave height because the initial probability density function for this process is the Rayleigh distribution.

If the initial probability density function is not known analytically and the only information available is some measured (or observed) data, then there are two other possibilities, depending on the nature of the data:

1. If enough data are available to allow an approximate initial probability distribution $P_{\tilde{Q}}(\tilde{Q})$ to be constructed, then the extreme value for a longer period of time can be established by plotting the distribution in the form $1/[1 - P_{\tilde{Q}}(\tilde{Q})]$ and extrapolating the curve. Obviously, this method possesses the usual uncertainties associated with any process of extrapolation. It is used, for example, to estimate the

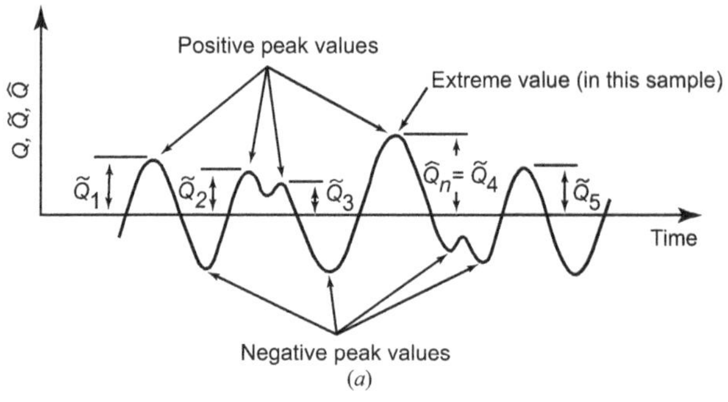

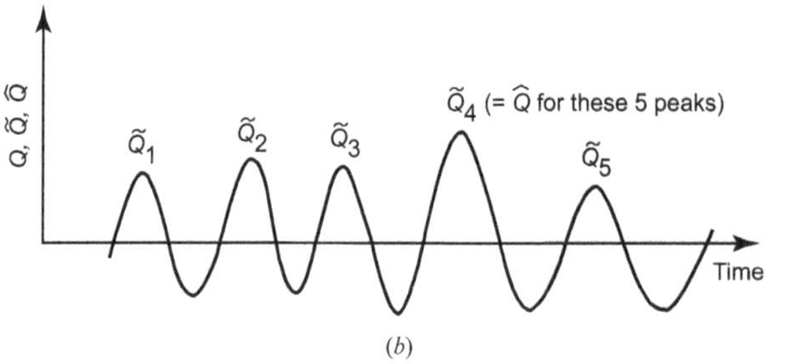

Figure 4.2.1 Random processes: (*a*) wide-band, (*b*) narrow-band

most severe sea state likely to be encountered over a long period of time.

2. If the available data consists of a large number of short-term extreme values (e.g., the largest wave measured or observed each day over a period of many years), then the extreme value likely to be encountered over a longer period can be estimated by means of the asymptotic distributions that were originally developed by Fisher and Tippet and later systematized by Gumbel.

ANALYTICAL EVALUATION OF EXTREME VALUES

Let $p_{\tilde{Q}}(\tilde{Q})$ be the probability density function that governs the positive peak values $\tilde{Q}$ of a random process $Q(t)$ and let $P_{\tilde{Q}}(\tilde{Q})$ be the corresponding cumulative probability function, as shown in Fig. 4.2.2. The function $1 - P_{\tilde{Q}}(\tilde{Q}_0)$ gives the probability that any one peak value, selected at random, exceeds a specified magnitude, $\tilde{Q}_0$. However, for design purposes we wish to know the probability that, out of a total of n peak values that are encountered, at least one peak will exceed a specified magnitude. We will derive this in two stages. First, we will obtain the probability density function for the *extreme value,* that is, the largest peak value among a sample of *n,* peak values, and then we will use this to relate the magnitude of the extreme value to its probability of occurrence.

Let $(\tilde{Q}_1, \tilde{Q}_2, ..., \tilde{Q}_n)$ be a typical sample of size n where each $\tilde{Q}_i$ is an observed peak value of $Q(t)$, as in Fig. 4.2.1. Since none of the members of the sample are specially defined or distinguished in any way, all of them have the same probability density function, $p_{\tilde{Q}}\tilde{Q}$. Let us now arrange this random sample into an ordered sample $(\hat{Q}_1, \hat{Q}_2, ..., \hat{Q}_n)$ in ascending order of magnitude, such that $\hat{Q}_1 < \hat{Q}_2 < ... < \hat{Q}_n$. We choose a new symbol, $\hat{Q}$, to emphasize that each of these random variables now has its own separate definition (largest peak, second largest peak, etc.) and, therefore, each has its own probability density function $p_{\hat{Q}}(\hat{Q}_i)$ different from $p_{\tilde{Q}}(\tilde{Q}_i)$. The variable of interest is the extreme value, that is, the largest peak value in the ordered sample. Since the other values are not of interest, we will dispense with the subscript and denote the extreme value as $\hat{Q}$. When we wish to associate $\hat{Q}$ with the sample size, we will write $\hat{Q}_n$.

We wish to obtain the probability density function of $\hat{Q}$. Now, for any one member to be the *largest* requires that all of the other $n - 1$ peak values must be less than $\hat{Q}$, and the probability of this is $[P_{\tilde{Q}}(\hat{Q})]^{n-1}$ (assuming the peaks are statistically independent). But since there are n members, any one of which could be the largest, we must multiply this probability by n. Hence, the probability density function of $\hat{Q}$ is

$$p_{\hat{Q}}(\hat{Q}) = p_{\tilde{Q}}(\hat{Q})n[P_{\tilde{Q}}(\hat{Q})]^{n-1} \qquad (4.2.1)$$

A typical extreme value probability density function is illustrated in Fig. 4.2.3. The value of $\hat{Q}$ for which the probability density function $p_{\hat{Q}}(\hat{Q})$ is maximum is the extreme value that is most likely to occur in n observations, and this is called the "probable extreme value" and is denoted by $\hat{Q}_p$. The probable extreme value is useful because the extreme value that actually occurs in n peak values of wave loading is close to this value. It is obtained as the solution of the following equation:

$$\frac{d}{d\hat{Q}}[p_{\hat{Q}}(\hat{Q})] = 0 \qquad (4.2.2)$$

which can be expanded to

$$p'_{\tilde{Q}}(\hat{Q})P_{\tilde{Q}}(\hat{Q}) + (n-1)[p_{\tilde{Q}}(\hat{Q})]^2 = 0 \qquad (4.2.3)$$

For a narrow-band process the initial distributions are the Rayleigh distributions

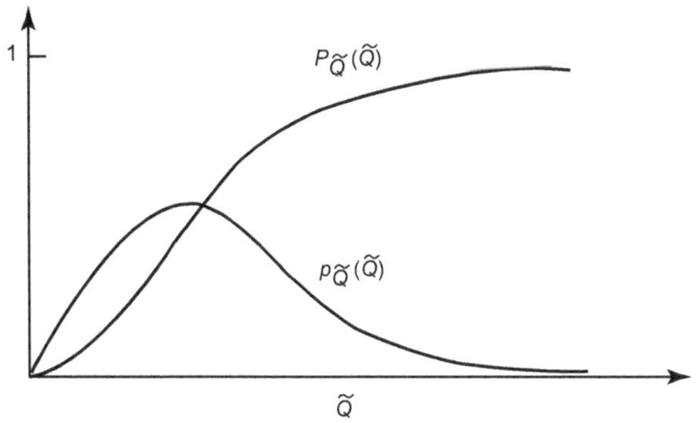

Figure 4.2.2 Distributions of peak values.

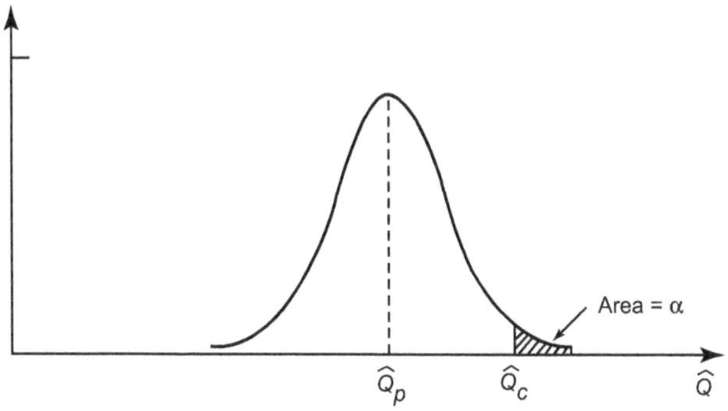

Figure 4.2.3 Typical probability density function of extreme values.

$$p_{\tilde{Q}}\left(\tilde{Q}\right)=\frac{2\tilde{Q}}{R}e^{-\tilde{Q}^2/R} \tag{4.2.4}$$

and

$$P_{\tilde{Q}}=1-e^{-\tilde{Q}^2/R} \tag{4.2.5}$$

where

$$R=\begin{cases}2\sigma_Q^2=2m_0 \text{ for amplitude}\\ 8\sigma_Q^2=8m_0 \text{ for height (crest-to-trough)}\end{cases} \tag{4.2.6}$$

Substituting these into (4.2.3) gives

$$2\frac{\hat{Q}^2}{R}\left(ne^{-\hat{Q}^2/R}-1\right)-\left(e^{-\hat{Q}^2/R}-1\right)=0 \tag{4.2.7}$$

The second term may be neglected for large n. Since $\hat{Q}$ is an amplitude, we can write

$$R=2\,m_0 \tag{4.2.8}$$

The result is

$$\hat{Q}_p=\sqrt{R\ln(n)} \tag{4.2.9}$$

Ochi (1982) showed that, for a random process having a bandwidth parameter $\varepsilon \leq 0.9$ (which easily applies to wave-induced bending moment and most other wave-induced loads), the solution to (4.2.3) is

$$\hat{Q}_p=\sqrt{2m_0\ln\left\{\frac{2\sqrt{1-\varepsilon^2}}{1+\sqrt{1-\varepsilon^2}}n\right\}} \quad \text{for } \varepsilon\leq 0.9 \tag{4.2.10}$$

However, the probability that the largest peak value may exceed the probable extreme value is large and, hence, it is not appropriate to use this value for engineering design. For example, Ochi showed that, for a perfectly narrow-banded process (ε = 0) the probability that the extreme value exceeds $\hat{Q}_p$ is $1 - e^{-1} = 0.632$. For purposes of structural design, we must obtain an extreme value for which the probability of being exceeded is some acceptably small value, α, chosen by the designer—a typical value is $\alpha = 0.01$. As explained in Chapter 1, α is not the probability of failure, since that also involves the probability distribution function of the limit value of the wave-induced load or load effect. Rather, α is a *risk parameter* by means of which the designer explicitly controls the probability, or risk, of the design load being exceeded in a total of n peak values. Hence, we wish to obtain the extreme value $\hat{Q}_c$, here referred to as the *characteristic value*, for which there is a probability α of being exceeded. That is, $\hat{Q}_c$ must be such that

$$\text{Prob}\{\text{extreme value among } n \text{ peak values} > \hat{Q}_c\} = \alpha$$

An equivalent requirement is that there must be a probability of $1 - \alpha$ that all of the n peak values will be less than or equal to $\hat{Q}_c$. For each peak value this probability is $P_{\tilde{Q}}(\hat{Q}_c)$, and the joint probability of this being satisfied for all peak values simultaneously is the product of the n individual probabilities, that is, $[P_{\tilde{Q}}(\hat{Q}_c)]^n$. Thus we have

$$[P_{\tilde{Q}}(\hat{Q}_c)]^n = 1-\alpha \tag{4.2.11}$$

Considering that α is small and n is large, we have

$$\begin{aligned}P_{\tilde{Q}}(\hat{Q}_c) &= (1-\alpha)^{1/n}\\ &\approx 1-\frac{\alpha}{n}+O(\alpha^2)\end{aligned} \tag{4.2.12}$$

where $O(\alpha^2)$ indicates terms of the order of α^2. By developing further the results of Cartright and Longuet-Higgins (1956), Ochi derived an expression for $P_{\tilde{Q}}(\tilde{Q})$ which, in combination with (4.2.12), gives the following expression for $\hat{Q}_c$:

$$\hat{Q}_c=\sqrt{2m_0\ln\left(\frac{\sqrt{1-\varepsilon^2}}{1+\sqrt{1-\varepsilon^2}}\frac{2n}{\alpha}\right)} \quad \text{for } \varepsilon\leq 0.9 \tag{4.2.13}$$

This equation expresses the design value as a function of the number of peak values, n, of the random process under consideration, such as the wave-induced bending moment. For practical purposes it is preferable to express the design value in terms of time rather than as a function of the number of peak values. Ochi further showed that the number of peak values can be expressed as a function of time by

$$n=(60)^2\left(\frac{T}{4\pi}\right)\left(\frac{1+\sqrt{1-\varepsilon^2}}{\sqrt{1-\varepsilon^2}}\right)\sqrt{\frac{m_2}{m_0}} \tag{4.2.14}$$

where T is time in hours. Substituting in (4.2.13) gives

$$\hat{Q}_c = \sqrt{2m_0 \ln\left\{\frac{1800T}{\pi\alpha}\sqrt{\frac{m_2}{m_0}}\right\}} \tag{4.2.15}$$

It is seen that the characteristic value $\hat{Q}_c$ expressed in terms of time T is no longer a function of the bandwidth parameter ε, but it is a function only of the area m_0 and the second moment m_2 of the spectral density function of the random process. Therefore, (4.2.15) applies for any Gaussian process, regardless of its bandwidth.

From (4.2.10) and (4.2.14), the most probable extreme value, $\hat{Q}_p$, can be also expressed as a function of time by

$$\hat{Q}_p = \sqrt{2m_0 \ln\left\{\frac{1800T}{\pi}\sqrt{\frac{m_2}{m_0}}\right\}} \tag{4.2.16}$$

ESTIMATION OF EXTREME VALUES BY EXTRAPOLATION OF MEASURED DATA

The previous section presented an analytical expression for extreme values when the initial probability distribution of peak values is known analytically. In practice, the initial distribution is often not known with such precision. For example, the probability distribution that is applicable for the long-term sea severity (as measured by significant wave height H_s) is not known analytically. The cumulative probability distribution $P(H_s)$ is derived empirically from analysis of observed (or measured) data accumulated over a certain period of time, and there seems to be no single mathematically defined probability distribution function that applies to this quantity. In fact, there is probably no way of theoretically deriving such a distribution, as contrasted with the Rayleigh probability function applicable for the height of individual waves in a given sea. For design purposes we need to estimate extreme values of H_s, particularly the probable extreme value $H_{s,p}$ corresponding to a certain period, and so we must use an approximate method for this purpose.

Strictly speaking, significant wave height is a peak value quantity, and in the notation of the previous section it would have been written as $\tilde{H}_s$. The symbols ˜ and ^ for peak value and extreme value, respectively, are necessary when discussing the general case, but in most applications the fact that a quantity is a peak value or an extreme value is obvious from the context. For example, a *significant value* is always a peak value by definition. Therefore, significant wave height will generally be written as H_s; the symbol ˜ is not required. Likewise, since the subscripts p and c indicate the probable extreme value and the characteristic extreme value, respectively, there is no need to also use the symbol ^ for these quantities. Therefore, the probable extreme significant wave height will be denoted as $H_{s,p}$.

For long-term statistics it is possible to estimate the general asymptotic form of the cumulative probability distribution and to use the theory of asymptotic distribution of extreme values, as this theory is applicable to any probability function if certain conditions are met (David, 1970). In this method the initial cumulative probability distribution is assumed to be of the form

$$P_{\tilde{Q}}(\tilde{Q}) = 1 - e^{-f(\tilde{Q})} \tag{4.2.17}$$

The function $f(\tilde{Q})$ specifies the precise manner in which $P_{\tilde{Q}}(\tilde{Q})$ approaches its asymptote. The probability density function is the derivative of $P_{\tilde{Q}}(\tilde{Q})$:

$$\begin{aligned} p_{\tilde{Q}}(\tilde{Q}) &= \frac{d}{d\tilde{Q}}[1 - e^{-f(\tilde{Q})}] \\ &= e^{-f(\tilde{Q})} f'(\tilde{Q}) \end{aligned} \tag{4.2.18}$$

In our intended application, $\tilde{Q}$ is significant wave height H_s; however, for generality we retain the symbol $\tilde{Q}$. Likewise, the symbol for the extreme value of $\tilde{Q}$ in a sample of size n is $\hat{Q}$. As shown in the previous section, the probable extreme value $\hat{Q}_p$ is the solution of (4.2.3). In the present case, this equation becomes

$$\frac{f''(\hat{Q})}{[f'(\hat{Q})]^2}[1 - e^{-f(\hat{Q})}] + ne^{-f(\hat{Q})} - 1 = 0 \tag{4.2.19}$$

For large n the first term is small in comparison with the other terms, and the equation becomes

$$e^{-f(\hat{Q})} \cong n \tag{4.2.20}$$

and $\hat{Q}_p$ is the particular value of $\hat{Q}$ that satisfies this equation. Equation 4.2.17 holds for any value of $\hat{Q}$, including the probable extreme value, $\hat{Q}_p$. When (4.2.29) is substituted back into (4.2.17) the assumed functional expression for the asymptote drops out, leaving simply

$$\frac{1}{1 - P_{\tilde{Q}}(\hat{Q}_p)} = n \tag{4.2.21}$$

This result shows that the probable extreme value expected to occur in n observations can be evaluated from the initial cumulative distribution function and the number of observations. Although the initial cumulative distribution function $P_{\tilde{Q}}(\tilde{Q})$ is not known analytically, it can be constructed from the observed data. Then the left-hand side of (4.2.21) can be plotted versus $\hat{Q}_p$, and the extreme value for a number of occurrences larger than n can be estimated by extrapolating the curve. The left-hand side of (4.2.21) is known as the *return period* because it corresponds to the number of occurrences in between extreme values of a certain magnitude. The use of this technique to estimate the probable extreme sea state is shown in Section 4.3.

ESTIMATION OF EXTREME VALUE FROM OBSERVED SHORT-TERM EXTREME VALUES

The observed short-term extreme value considered here is the largest value that is either measured or observed during a certain period of time. As an example, let us assume that the data of the largest wave height measured every day are available. If the number of measured data is large, then it is possible to estimate the wave height expected to occur in a long time period of, say, 20 or 40 years. The estimate is based on the asymptotic distribution of extreme values developed by Gumbel (1966). This theory shows that, if the initial distributions are exponential, meaning that they satisfy the following relationship

$$\frac{p_{\tilde{Q}}(\tilde{Q})}{1 - P_{\tilde{Q}}(\tilde{Q})} = -\frac{p'_{\tilde{Q}}(\tilde{Q})}{p_{\tilde{Q}}(\tilde{Q})}, \quad Q \to \infty \qquad (4.2.22)$$

[which is satisfied, for example, by the Rayleigh distribution and by (4.2.17)] then a suitable asymptotic approximation to the exact cumulative probability function of the extreme value is given by

$$P_{\hat{Q}}(\hat{Q}) = e^{-e^{-\kappa(\hat{Q}-\hat{Q}_p)}} \qquad (4.2.23)$$

where

$$\kappa = \frac{\pi}{\sqrt{6\ \mathrm{Var}[\hat{Q}]}}$$

$$\hat{Q}_p = E[\hat{Q}] - \frac{\gamma}{\kappa}$$

Here $\mathrm{Var}[\hat{Q}]$ and $E[\hat{Q}]$ are, respectively, the variance and the expected value of the available data (i.e., the observed short-term extreme values) and γ is Euler's constant, 0.577.

Following Gumbel (1966) Loukakis and Grivas (1980) showed that, if the initial distribution is the Rayleigh distribution [see (4.2.4)] then the parameter κ in (4.2.23) is given by

$$\kappa = 2\sqrt{\frac{\ln(n)}{R}} \qquad (4.2.24)$$

4.2.2 Extreme Values for Nonlinear Responses

Usually, it is the statistical properties of the simulated load response that are of practical interest and not the results from the deterministic response history for a limited duration as such. Sufficiently accurate extreme value estimates are difficult to obtain from deterministic simulations. In principle, extreme value predictions can be obtained by carrying out simulations for a sufficiently long duration and to estimate the extreme values directly from the simulated record. Several possibilities exist. Most commonly, a global statistical model is applied to the entire response history by either fitting a Weibull model to the local maxima (Winterstein and Torhaug 1996) or a Hermite model to the first four statistical moments of the response (Winterstein 1988). Alternatives used are tail-fit models that either fit a Gumbel model to the largest peaks within a number of subintervals of the response or that only use peaks over a certain threshold in the fitting of a Weibull model. However, from an engineering standpoint, the extensive computer resources needed make it impractical to carry out sufficiently long simulations to generate long time histories to directly obtain converging extreme value estimates.

It is especially difficult to obtain long-term extreme value statistics in this way. For the strength assessment of ship structures, it is necessary to estimate the long-term maximum value of wave-induced loads during the ship's operational life. Based on this value, large deflection and limit state strength analyses of ship structures are performed. For a fatigue strength analysis, the time history of wave-induced loads is also required. For linear wave-induced response, theoretical methods to estimate short- and long-term probability distributions are well established. The Rayleigh distribution, for example, describes the short-term distribution of wave-induced loads, assuming that linear superposition holds and that the response spectrum is narrow-banded. A weighted sum of the short-term descriptions that consider various wave headings, ship speeds, and routes then yields the long-term distribution. However, no theoretical distribution of peak values exists for the nonlinear wave-induced response. Methods to obtain short- and long-term probability distributions for nonlinear wave-induced loads, therefore, are continuously being developed.

SHORT-TERM DISTRIBUTION

Many theoretical and experimental methods were proposed to estimate the probability distribution of nonlinear response in an irregular short-term wave condition characterized by a large significant wave height. These proposals showed that the statistical distributions depend not only on the ship type, but also on the kind of loading. Table 4.2.1 summarizes probability distribution functions frequently used to predict the nonlinear short-term extreme of the load responses (ISSC 2006) for specific ship types. Table 4.2.1 also lists the procedure used to validate these functions.

According to Wang and Moan (2004), the generalized Weibull distribution appears to be suitable to represent the wave load peak value statistics for all types of ships, sea states, and ship speeds. They found this to be the case after systematically analyzing the statistics of nonlinearly simulated wave loads on seven ship models under various short-term sea state and ship speed conditions. In the upper tail area the fit was favorable and the statistics uncertainty of the estimated extremes was small.

Kapsenberg et al. (2003) performed model tank tests of a cruise ship and found that a three-parameter Wiebull distribution favorably represented the extreme values of the wave frequency component of the vertical wave bending moment, the whipping moment, and the total vertical bending moment in head seas. Using a nonlinear, hydroelastic strip theory, Baarholm and Jensen (2004) studied the effect of slam-induced whipping on the extreme (design) value of midship vertical bending moment for the S-175 containership. For a good fit, they chose the three-parameter Weibull distribution for the wave-induced loads and the exponential distribution for the whipping maxima.

Based on a strip theory for high speed ships, Wu and Moan (2004) performed nonlinear simulations of a high-speed pentamaran in regular and short-crested irregular waves. They estimated the short-term exceedance probabilities of wave bending moments by fitting the generalized gamma distribution to the histograms of extreme values extracted from the simulations. To validate their numerical predictions, they compared computed results with model test data in regular and irregular waves and observed favorable agreement.

To develop a practical prediction method for green water loading on a ship's bow and deck, Ogawa (2003) conducted a series of model tests for a standard Japanese tanker and a cargo ship in regular and long-crested irregular head seas. He expressed the probability function for the maximum value of the green water load in terms of the probability function of the relative water height at the ship's stem. By assuming that this water height follows a Rayleigh distribution, he expressed the probability function of the green water load by the truncated Rayleigh distribution.

CRITICAL WAVE EPISODE

Torhaug et al. (1998) defined so-called critical wave episodes that are used as input to a selected number of nonlinear simulations. This method is based on the assumption that short wave episodes chosen as the waves that produce the largest linear responses efficiently produce the largest nonlinear responses. Prior knowledge of the ship's behavior in waves at different speeds and wave headings, such as precomputed transfer functions, helps to identify the critical wave episodes. If the primary properties of the response process are known, critical wave episodes can also be identified directly from the simulated surface elevation of the seaway. If, for instance, the relevant response is the midship bending moment, wave episodes with wave lengths close to the ship length and wave heights above some specified level may well be suitable as critical wave episodes. For the torsional strenghth assessment of containership structures, Iijima et al. (2004) introduced critical design sea states and specified a dominant regular

Table 4.2.1 Probability distribution functions for nonlinear short-term load response

Distribution function	Load response	Ship type	Validation procedure
Weibull	Vertical bending moment	Containership, tanker, frigate, destroyer	Numerical simulation
Weibull	Bending moment, whipping moment	Cruise ship	Tank test
Weibull for bending moment, exponential for whipping moment	Bending moment, whipping moment	Containership	Numerical simulation
Generalized gamma	Bending moment	High speed ship	Numerical simulation
Truncated Reyleigh	Green water load	Cargo ship, tanker	Tank test
Weibull	Green water load	Containership	Tank test

wave condition for which the torsional response of a containership is largest. The length of this regular wave is 35 percent of the ship's length and its wave heading is 120° (180° denotes head waves).

MOST LIKELY EXTREME RESPONSE

By combining the ideas of Fries-Hansen and Nielsen (1995), Taylor et al. (1995), Adegeest et al. (1998), and Dietz et al. (2004) proposed the most likely response wave (MLRW) to estimate the entire nonlinear extreme value distribution for a selected operational profile, given the amplitude and phase information from linear transfer functions (ISSC, 2006). The most likely extreme response (MLER) method of Adegeest et al. accounts for the correct response memory and reduces the number of uncertain variables needed as input for an extreme response analysis. They systematically investigated extreme values of the midship vertical wave bending moment for a Panamax containership in an irregular seaway of 1500 s duration. The ship advanced at a median speed corresponding to a Froude number of 0.145. Table 4.2.2 lists the ship's principal particulars.

Comparable model test measurements for the 1500 s full-scale duration were available for comparison. A significant wave height of 4.8 m and a zero-crossing period of 8.0 s characterized the irregular seaway. The spectral density of the wave spectrum as generated in the model basin is shown in Fig. 4.2.4. This wave spectrum was used to specify the irregular seaway for the numerical simulations. The segmented backbone model allowed measurement of global sectional loads. The maximum measured sagging and hogging midship bending moments were $4.46 \cdot 10^5$ and $-2.31 \cdot 10^5$ kNm, respectively.

Adegeest et al. (1998) also estimated the extreme response in the 1500 s sea state according to linear theory, using narrow-band short-tem statistics. The midship bending moment was calculated on the basis of the encounter frequency transfer functions in head waves at a probability level corresponding to 234 response cycles. This resulted in a mean response period of 6.4 s and an expected linear extreme bending moment of $3.17 \cdot 10^5$ kNm. Using the panel

Table 4.2.2 Principal particulars of sample containership

Length bet. perpendiculars	160.00 m
Breadth	24.65 m
Draft	8.93 m
Displacement	20491 t
Block coefficient	0.57
Waterplane area coefficient	0.71
Pitch radius of gyration	39.39 m

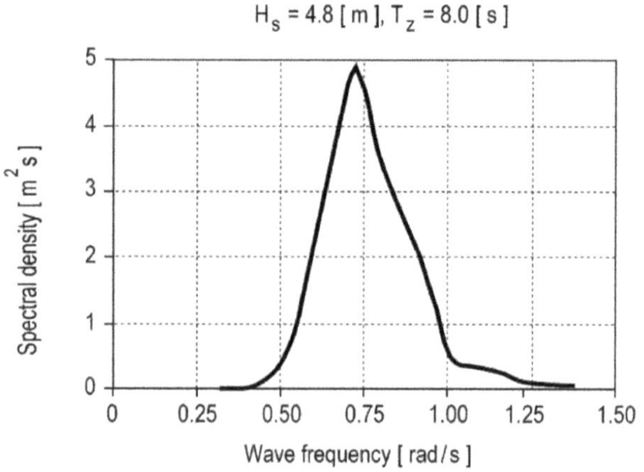

Figure 4.2.4 Measured wave spectrum of model tests (*Source:* Adegeest et al. 1998, National Academies Press; used with permission).

code SWAN (Kring et al., 1997) in its nonlinear mode, they also performed one long simulation of 1500 s full-scale duration in random irregular waves, using a deterministic wave amplitude distribution with phases distributed randomly between 0 and 2π. The simulation yielded the extreme midship sagging and hogging bending moments of $4.81 \cdot 10^5$ and $-2.17 \cdot 10^5$ kNm, respectively. They also applied the regular design wave approach, but this did not yield accurate predictions of extremes, mainly because the results were sensitive to the selected wave period, which is not well defined in the regular design wave procedure for short-term analyses.

Using statistical moments calculated from the SWAN simulation in irregular waves, they also applied a Hermite moment-based model documented by Winterstein (1988) to obtain the distribution of extreme midship vertical bending moments in sagging and hogging. The length of the time series comprised 30,000 samples. Therefore, statistical measures were calculated for three separate data blocks of 10,000 samples each, and Hermite distributions were obtained for each data block as well as for the complete time series. Figure 4.2.5 shows the calculated Hermite distributions for the three different data blocks and for the complete time series as well as the distribution from the measurements. The agreement between the Hermite distribution and the distribution from model test measurements is favorable only for the distribution based on the statistical moments valid for the complete time series. The variations in the calculated tails of the distributions were large for the different sets of data blocks, implying that long time histories are required for accurate extreme predictions by the Hermite model because only then will the required statistical measures become more stable. The Hermite model for the complete time series yielded

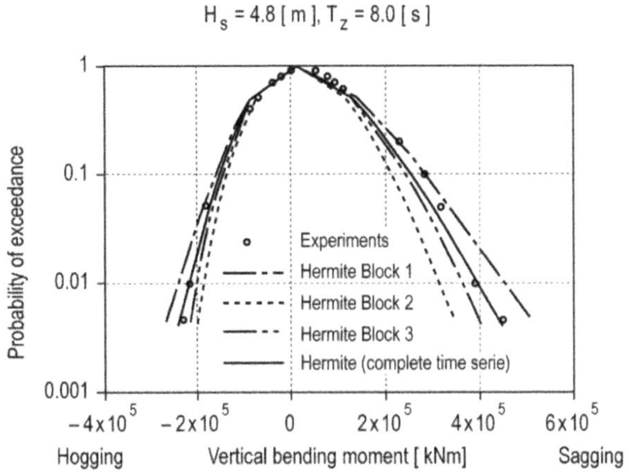

Figure 4.2.5 Hermite distributions of extreme midship vertical bending moment based on different data blocks and comparative distribution from model test measurements (*Source:* Adegeest et al. 1998, National Academies Press; used with permission).

extreme sagging and hogging moments of $4.46 \cdot 10^5$ and $-2.41 \cdot 10^5$ kNm, respectively.

The advantage of the most likely extreme response (MLER) method is that it accounts for memory effects while reducing the number of uncertain variables necessary to specify the input procedure for a nonlinear time domain simulation. The method as documented by Adegeest et al. used the theory that generates the so-called most likely wave (MLW) profiles with conditioned amplitudes (Tromans et al., 1991) and frequencies (Friis-Hansen and Nielsen, 1995) and applied this theory to response spectra. The amplitude and phase information of the frequency response functions serves to derive the irregular wave train that causes the MLER, and successive nonlinear simulations can be performed with this irregular wave train as input. Adegeest et al. (1998) validated this procedure for the estimation of extreme vertical bending moments and maximum wave heights of green water on deck. Of course, the MLER approach can produce only one estimate at a particular probability level per simulation. Separate simulations have to be performed in different wave conditions, derived from different conditioned responses, to obtain additional data points needed to specify the probability distribution function. For a mean response frequency and a conditioned linear extreme bending moment in sagging of $3.17 \cdot 10^5$ kNm and in hogging of $-3.17 \cdot 10^5$ kNm, Figs. 4.2.6 and 4.2.7 show time histories of the nonlinearly computed most likely extreme midship vertical bending moment for the sample containership in sagging and hogging, respectively. The resulting maximum values were $4.57 \cdot 10^5$ in sagging and $-2.36 \cdot 10^5$ kNm in hogging. The figures also depict time histories of the underlying wave profile as well as the linearly computed MLER.

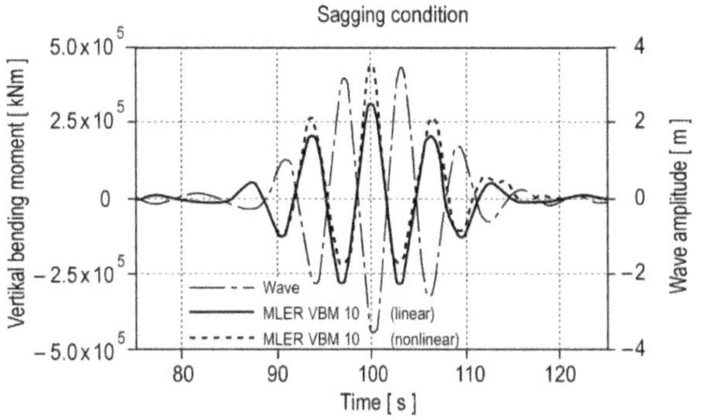

Figure 4.2.6 Time histories of linear and nonlinear most likely extreme midship vertical bending moment for the sagging condition (*Source:* Adegeest et al. 1998, National Academies Press; used with permission).

Results of the estimated extreme vertical bending moments are summarized in Table 4.2.3. They show that the different methods may produce estimates of the extreme response that deviate significantly from each other and from extremes based on measurements from towing tank model tests. These results show that the MLER predictions agreed most favorably with the experiments. Sagging and hogging extreme bending moments predicted with this method deviated less than 3 percent form comparable measurements, and only 200 s of real time had to be nonlinearly simulated. The MLER method is based on the application of the MLW theory to response spectra. Therefore, the resulting extremes are a statistically correct product of well-defined quantities, such as a linear transfer function, a wave spectrum, the expected mean response period, the expected duration of exposure, and the expected extreme according to linear short-term statistics. However, the method is based on the assumption that the extreme response, including all nonlinear effects, is related in time to the extreme linear response.

Pastoor (2002) also presented a summary of computational procedures to determine extreme

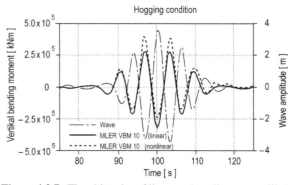

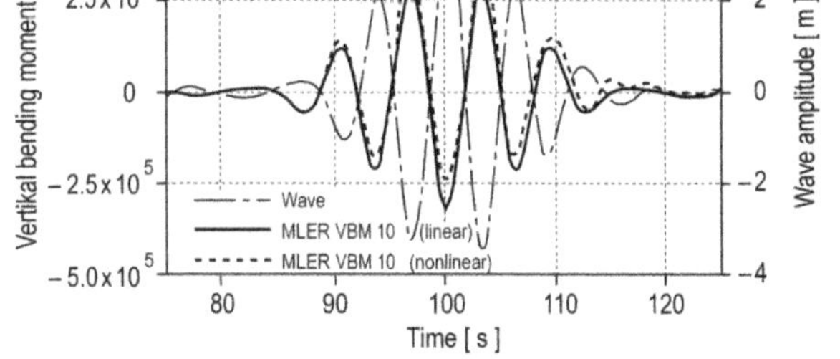

Figure 4.2.7 Time histories of linear and nonlinear most likely extreme midship vertical bending moment for the hogging condition (*Source:* Adegeest et al. 1998, National Academies Press; used with permission).

Table 4.2.3 Comparative extremes of midship vertical bending moments of the sample containership at $F_n = 0.145$ in a 1500 s seaway characterized by $H_s = 4.8$ m and $T_z = 8.0$ s (Adegeest et al. 1998)

Method	Midship vertical bending moment in sagging [kNm]	Midship vertical bending moment in hogging [kNm]
Model test measurements	$4.46 \cdot 10^5$	$-2.31 \cdot 10^5$
Linear (Rayleigh)	$3.17 \cdot 10^5$	$-3.17 \cdot 10^5$
Nonlinear simulation	$4.81 \cdot 10^5$	$-2.17 \cdot 10^5$
Hermite model (complete time series)	$4.46 \cdot 10^5$	$-2.41 \cdot 10^5$
Nonlinear MLER	$4.57 \cdot 10^5$	$-2.36 \cdot 10^5$

responses using nonlinear simulations. When the expected extreme is of interest, the MLER method is applicable; however, when large nonlinear response amplitudes are calculated by correcting linear responses, a so-called extended version of the MLER procedure yields more accurate predictions. He documented results from model tests carried out with a segmented naval frigate and investigated the response-conditioning technique more extensively. He found that it was possible to generate the conditioned incident waves and, by tuning a control mechanism of the carriage, to synchronize the transient wave profile with the moving ship model. He conducted a series of conditioned model tests in severe wave conditions with large amounts of green water on deck and accurately predicted the bending moment amplitude probability function. He compared his results with results from other techniques and demonstrated that the extrapolation based on fitting a mathematical function to the tail of the response is critical when it comes to safety and reliability. The advantage of the response-conditioning technique is that it accurately predicts the behavior in severe conditions, and it is this behavior that defines the tail of the probability distribution.

SIMPLIFIED DESIGN WAVE

From an engineering standpoint, the extensive computer resources needed make it impractical to carry out sufficiently long simulations to generate long time histories to directly obtain converging extreme value estimates. Simplified design wave conditions, where the equivalent long-term extreme loads can be generated, can improve the efficiency of obtaining long-term extreme values from numerical simulations.

The simplest method, the regular design wave, is still widely used and can be considered to be a standard engineering tool. To determine the design wave, the first step consists of using standard methods to determine the long-term extreme value based on linear theory. Next, the period of the design wave (for a given load response, wave heading, vessel speed, etc.) is chosen as the period corresponding to the peak of the transfer function of the load response under consideration. Finally, the amplitude of the design wave is obtained by dividing the extreme value by the value of the transfer function at the period just found. Corrections may have to be made if the amplitude conflicts with wave steepness limits. This design wave can be used as input to a nonlinear simulation, and the resulting response is then taken to be the extreme value estimate.

The concept of equivalent waves, modeled with a simplified geometry (e.g., Folso and Rizzuto, 2003), approximates complex wave patterns that lead to a linear extreme response in the long-term distribution. This approximation can differ substantially from reality, particularly if the simplified wave is sinusoidal with characteristics that differ significantly from those of extreme waves. Calculations carried out for a 128 m Ro-Ro fast ferry, for instance, indicate that a set of three or four equivalent head, beam, and bow waves with different wave lengths is sufficient to cover the chosen responses.

LONG-TERM DISTRIBUTION

The complete long-term distribution of the nonlinear response is still outstanding (ISSC, 2006). However, Minoura and Naito (2004) proposed a stochastic process model for the long-term statistics of ship responses. They investigated the correlation between significant wave height and standard deviation of ship responses at sea by analyzing the monitored ship response data on board a containership and a bulk carrier for three years. They observed that these standard deviations follow a Markov process, regress to an equilibrium, and fluctuate linearly. Stochastic differential equations describe the stochastic process model based on these properties, and the Fokker-Plank equation can be applied to yield the probability density function. Their long-term predictions based on this model agreed favorably with the monitored data. Shin et al. (2004) presented a method for computing correlation factors to combine the long-term dynamic stress components of ship structures from various loads in irregular seas. This method, based on the stationary ergodic narrow-band Gaussian processes, expresses the total combined stress in short-term sea states as a linear summation of the component stresses with the corresponding combination factors. The long-term total stress is then similarly expressed by linear summation of component stresses with appropriate combination factors. They found that the combination factors strongly depend on wave period and wave heading in short-term sea states and that these factors are not sensitive to the selected probability level of the long-term stress.

A long-term analysis may need to account for nonlinear as well as transient elastic effects on the midship vertical bending moment. The so-called contour line approach can be applied to obtain the long-term extreme value in such cases. Baarholm and Jensen (2004), for example, obtained long-term extreme values using this method. They compared their results with various simplified methods as well as with classification society rules and found that the contour line approach yields satisfactory results.

At times, the limiting wave condition that a ship can navigate is considered to estimate the probability distribution of the nonlinear extreme wave bending moment. Kawabe et al. (2005), for instance, used this approach by selecting pitch angle, bottom slamming, and deck wetness as the relevant parameters for the limiting wave condition. Their nonlinear analysis under the most severe short-term wave condition yielded a distribution of extreme bending moment that is about 10 percent lower in sagging and about 40 percent lower in hogging than a distribution obtained by linear analysis.

4.3 STATISTICAL REPRESENTATION OF THE SEA SURFACE

The surface of the ocean—that is, the pattern of surface elevation—is highly irregular and totally random (nonrepeating) even under relatively calm conditions. Fortunately for naval architects, oceanographers found that the irregularity of the ocean surface can be represented as the superposition of a large number of regular waves having different heights, lengths, directions, and random phase. This finding is important because it allows the ocean surface to be described mathematically, and it also permits the use of statistical methods to predict the maximum wave loads in a ship's lifetime.

4.3.1 Mathematical Representation of Ocean Waves

The first major contributions were made by Pierson (1952) and Pierson et al. (1955), who proposed that the completely irregular and nonrepeating pattern of the ocean surface can be represented as the sum of an infinite number of regular sinusoidal waves, of all frequencies, each of which satisfies the governing hydrodynamic equation for gravity waves. For simplicity, we begin with a "long-crested" sea, that is, a sea in which the waves are all parallel. The surface profile of a typical component wave is

$$\zeta(x, t) = a \cos(-kx - \omega t + \theta) \tag{4.3.1}$$

This and other quantities and terms are defined as follows (see Fig. 4.3.1):

a = *wave amplitude,* measured from the mean water surface, which is also the location of the x-axis.
λ = *wave length,* the horizontal distance between successive crests or troughs.
$k = 2\pi/\lambda$ = *wave number*
T = *wave period,* the time between two successive crests to pass a fixed point on the x-axis or the time between a crest to travel a distance equal to one wavelength.
$\omega = 2\pi/T$ = *wave frequency*
θ = *phase angle*

For deep water, the wave number and the wave frequency are related by

$$k = \frac{\omega^2}{g} \tag{4.3.2}$$

The total energy (kinetic and potential) per unit area of water surface is

$$\Omega = \frac{1}{2}\rho g a^2 \tag{4.3.3}$$

Pierson et al. (1955) were the first to propose that the surface elevation $h(x, t)$ of an irregular sea could be represented as

$$h(x, t) = \lim_{n \to \infty} \sum_{i=1}^{n} a_i \cos \psi_i \tag{4.3.4}$$

where

$$\psi_i = -k_i x - \omega_i t + \theta_i \tag{4.3.5}$$

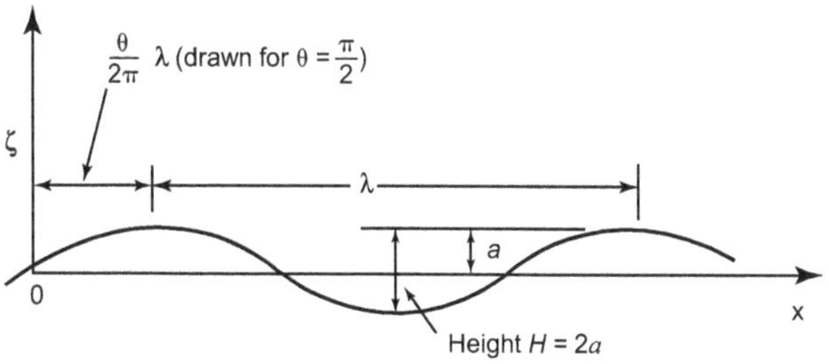

Fig. 4.3.1 Geometry of a regular wave.

and where the phase angle θ_i is a random variable, equiprobably distributed over the range $(0, 2\pi)$.

Each component wave must satisfy the governing differential equation for gravitational waves. This equation is most conveniently expressed in terms of a velocity potential $\phi(x, y, t)$, defined such that its derivatives correspond to velocities:

$$u = -\frac{\partial\phi}{\partial x}, \qquad v = -\frac{\partial\phi}{\partial y} \tag{4.3.6}$$

In terms of the potential function, the governing equation is the Laplace equation

$$\nabla^2\phi = 0 \tag{4.3.7}$$

and the pressure at any point is

$$p(x, y, t) = \rho\left[\frac{\partial\phi}{\partial t} + gy + \frac{1}{2}(\nabla\phi)^2\right] \tag{4.3.8}$$

The problem is nonlinear because the free surface boundary condition is $p = 0$ and the pressure $p(x, y, t)$ is a nonlinear function of ϕ. The problem is further complicated because the boundary—the free surface—is an undulating time-dependent surface instead of a fixed boundary. If the wave height is small in comparison to the wave length, then the problem may be linearized, and it may be shown that the solution for the linearized problem is the velocity potential

$$\phi(x, y, t) = \frac{a\omega}{k} e^{-ky} \sin\psi \tag{4.3.9}$$

The surface elevation at a particular location is found by setting $x = 0$

$$h(t) = \lim_{n\to\infty}\sum_{i=1}^{n}\zeta_i = \lim_{n\to\infty}\sum_{i=1}^{n} a_i\cos(-\omega_i t + \theta_i) \tag{4.3.10}$$

It is easily verified that each of the component random processes $\zeta_i(t)$ is stationary and, hence, $h(t)$ is stationary. Also, since these processes are independent with zero mean values, terms like $E[\zeta_i(t_1)\zeta_j(t_2)]$ are all zero for $i \neq j$ and are only non-zero when $i = j$, so that the mean square wave height is given by

$$E[h^2(t)] = E\left[\sum_{i=1}^{n}\zeta_i^2(t)\right] = \sum_{i=1}^{n} E[\zeta_i^2(t)] = \frac{1}{2}\sum_{i=1}^{n} a_i^2 \tag{4.3.11}$$

and the auto-correlation function of the summed process is

$$\begin{aligned} R(\tau) &= E[h(t_1)h(t_2)] \\ &= \sum_{i=1}^{n}\sum_{j=1}^{n} E[\zeta_i(t)\zeta_j(t+\tau)] \\ &= \sum_{i=1}^{n} R_i(\tau) \end{aligned} \tag{4.3.12}$$

Because the surface elevation $h(t)$ is the sum of a large number of independent variables, it will, by the central limit theorem (Section 4.1), have a Gaussian, or normal, probability density function. For the case of a zero mean, this is

$$p(h) = \frac{1}{\sigma\sqrt{2\pi}} e^{-(h/\sigma)^2/2} \tag{4.3.13}$$

and, since the mean is zero, the variance equals the mean square:

$$\sigma^2 = E[h^2] \tag{4.3.14}$$

Also, because $h(t)$ is Gaussian, the variance is sufficient to uniquely define the entire process.

As indicated earlier, it is advantageous to describe a random process in terms of its mean square spectral density function $S(\omega)$, which is the Fourier transform of $R(\tau)$ given by (4.1.44). For all practical applications we use the one-sided form of this function. As we have seen, the area under this curve is the mean square or, in this case, the variance:

$$\sigma^2 = \int_0^\infty S(\omega)d\omega \tag{4.3.15}$$

From (4.3.3) the energy of each component wave is $\Omega_i = ½\rho g a_i^2$ and, hence, the total energy is

$$\Omega = \sum\Omega_i = \tfrac{1}{2}\rho g\sum a_i^2 \tag{4.3.16}$$

Using (4.3.11) and noting that the variance equals the mean square, the foregoing can be rewritten as

$$\Omega = \rho g\sigma^2 \tag{4.3.17}$$

and hence, as noted earlier, the spectral density function is often referred to as the *energy spectrum* because the area under this curve is directly proportional to the total energy of the waves per unit area of surface:

$$\Omega = \rho g\int_0^\infty S(\omega)d\omega \tag{4.3.18}$$

Again, for each component wave the relationship is

$$\tfrac{1}{2}\rho g a_i^2 = \rho g S(\omega_i)\delta\omega \tag{4.3.19}$$

and so the relationship between the wave spectrum and the amplitude of each component wave is

$$a_i = \sqrt{2S(\omega_i)\delta\omega} \qquad (4.3.20)$$

In terms of $S(\omega)$, the surface elevation is

$$h(t) = \lim_{\substack{n\to\infty \\ \delta\omega\to\infty}} \sum_{i=1}^{n} \sqrt{2S(\omega_i)\delta\omega}\cos(\omega_i t + \theta_i) \qquad (4.3.21)$$

This relationship is illustrated in Fig. 4.3.2 for $n = 14$.

4.3.2 Ocean Wave Spectra

Over many years oceanographers and other researchers gathered and tabulated both visual and measured data concerning ocean waves in various parts of the world. One of the most comprehensive sets of data is the atlas of wave statistics presented by Hogben, Dacunha and Olliver (1986), where data for 104 ocean areas around the world are tabulated for different seasons for wave periods varying roughly from 0.6 to 22.5 s over ten increments. These data are based on 55 million visual observations from ships on passage between 1854 and 1984. Wave statistics were correlated with measurements, and unrealistic data were eliminated. The data are presented as scatter diagrams, subdivided into different wave directions, and comprise joint frequency of occurrence of combinations of significant wave height and zero-crossing period occurring simultaneously. The frequency of occurrence of waves from each of the specified directions is given as "percentage of observations" at the top of each scatter diagram.

Further data on the North Atlantic were presented by Roll (1953) and Walden (1964). Data for the North Pacific were provided by Yamanouchi et al. (1965). All of these data are visual observations and, hence, they are not entirely consistent, being based on judgements made by many different observers. Measured wave data are more precise, but these kinds of data are limited in quantity and geographical location when compared with the vast accumulation of visual observations. Most of the measured data was obtained from British weather ships in the North Atlantic. Additional wave data were generated by a hindcast technique (i.e., Chen et al., 1979). Bales et al. (1981) documented a comprehensive summary of standardized wave and wind conditions, and Michel (1999) comprised a compendium of updated spectrum formulations, Rayleigh factors, and associated wave height and period relationships prepared for easy understanding and application.

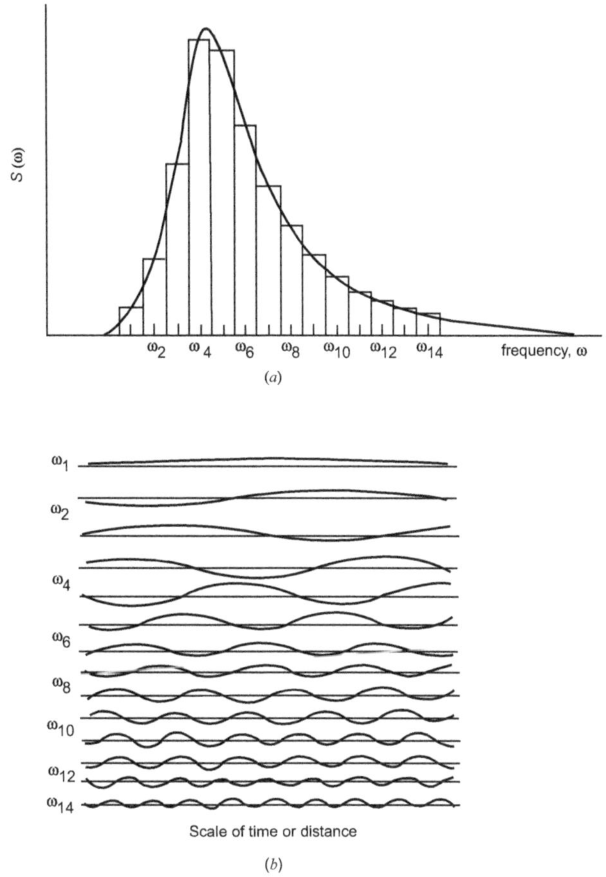

Figure 4.3.2 Typical energy spectrum showing approximation by a finite sum of components: (*a*) spectrum, (*b*) component waves.

THE MODIFIED PIERSON-MOSKOWITZ SPECTRUM

The most widely recognized theoretical function representing the state of the sea in the form of an energy spectrum is the two-parameter spectrum developed by Bretschneider (1959). Bretschneider was the first to propose that the wave spectrum for a given sea state can be represented in terms of two parameters that are characteristic of that sea state, such as average wave height, $\bar{H}$, and average wave period, $\bar{T}$. Various other formulas in addition to Bretschneider's were proposed, such as those of Pierson-Moskowitz, the International Towing Tank Committee (ITTC), and the International Ship Structure Congress (ISSC). In some cases the average frequency $\bar{\omega}(= 2\pi/\bar{T}$ if expressed in radians/s) is used instead of $\bar{T}$. In terms of $\bar{H}$ and $\bar{\omega}$, the general form of these two-parameter formulas is

$$S(\omega) = A\left(\frac{\bar{\omega}}{\omega}\right)^k \frac{\bar{H}}{\omega} e^{-B(\bar{\omega}/\omega)^l} \qquad (4.3.22)$$

where coefficients A and B and exponents k and l are selected to fit the data and the system of units. Some formulas make use of an alternative pair of parameters: significant wave height, H_s, (shown as $\tilde{X}_s$ in Fig. 4.1.11*a*) and modal wave frequency, ω_m, the frequency where the wave spectrum has its maximum height. It may be shown that, for a narrow-band spectrum, these two parameters are related to the first two as follows:

$$H_s = \sqrt{\frac{8}{\pi}}\bar{H} = 1.60\,\bar{H}$$

$$\omega_m = \frac{(0.8)^{1/4}}{\Gamma\left(\frac{3}{4}\right)}\bar{\omega} = 0.77\bar{\omega} \qquad (4.3.23)$$

where $\Gamma(\ )$ denotes the Gamma function. In the work to follow we use a two-parameter formula, now adopted by the ITTC. This modified version of Bretschneider's formula, expressed in terms of H_s and ω_m, is also labelled the modified Pierson-Moskowitz spectrum and has gained acceptance through usage:

$$S(\omega) = 173\,\frac{H_S^2}{\bar{T}^4\omega^5}\exp\left[-\frac{691}{\bar{T}^4\omega^4}\right]$$

$$= 0.313\,\frac{H_S^2\omega_m^4}{\omega^5}\exp\left[-\frac{5}{4}\left(\frac{\omega_m}{\omega}\right)^4\right] \qquad (4.3.24)$$

The Pierson-Moskowitz spectrum was first introduced in 1964 as a one-parameter spectrum, that is, in terms of the wind speed measured at a height of 19.5 m above sea level (the height of the anemometers used on ships that provided the data). Some time later it was realized that its proper use is restricted to fully developed seas as generated by relatively moderate seas over large fetches. Pierson-Moskowitz spectra for wind speeds of 20 to 50 knots, measured at a height of 19.5 m above sea level, are shown in Fig. 4.3.3. The peak or modal frequency decreases with increasing wind speed, and the magnitude of the spectral density function $S(\omega)$, or the energy of the sea state (area under the

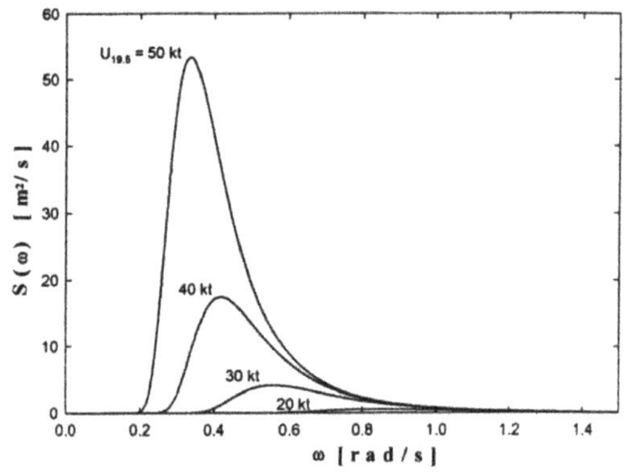

Figure 4.3.3 Pierson-Moskowitz spectra for various wind speeds (*Source:* Lewandowski, 2004, World Scientific Publishing Co. Pte. Ltd., Singapore; used with permission).

spectrum), increases substantially with wind speed.

The general two-parameter formulation was needed to account for the more prevalent conditions of high winds over relatively short fetches, which conditions produce spectra covering lower periods for a given wave height. In such areas as the North Sea, many spectral shapes are so highly peaked as to require a multi-parameter treatment.

THE JONSWAP SPECTRUM

In the early 1970s the Joint North Sea Wave Project (JONSWAP) was organized to systematically record North Sea wave patterns. Spectral methods analyzed the data, and the resulting spectra were parameterized in an equation that accommodated spectral shapes ranging from those sharply peaked to those representing the fully developed Pierson-Moskowitz limit. A number of wave-recording stations were positioned along a course from the German coast extending west for about 100 miles. Then, with a wind coming directly from shore, the various fetch distances and wind speeds were established at each station. Knowing that the generated waves had no prior history to corrupt the resulting data, it was anticipated that the wave spectra would start out sharply peaked and gradually ease towards the fully developed spectrum according to Pierson-Moskowitz. Reportedly, some 2000 wave records were analyzed in the course of the project. On this basis, the spectrum parameters were evaluated and the JONSWAP spectrum was developed.

The results indicated that the sharply peaked spectra are common when waves are generated in the North Sea, and data from other North Sea locations under more severe conditions substantiate this. Surprisingly, however, the spectra did not "settle down" to the fully developed Pierson-Moskowitz formulation as fetch increased. Several opinions expressed that perhaps such fully-developed conditions may never occur.

In view of this latter observation, a simpler formulation for easier analysis and evaluation could have been introduced for the JONSWAP spectral formulation to eliminate the Pierson-Moskowitz dependency. However, with all the data having been processed in the existing JONSWAP form, that spectrum function was retained, and it continues to be specified for many North Sea operations. It is formulated as follows:

$$S(\omega) = 155\,\frac{H_S^2}{\bar{T}^4\omega^5}\exp\left[-\frac{944}{\bar{T}^4\omega^4}\right]\nu^{\exp\left[-\left(\frac{0.191\omega\bar{T}-1}{\sqrt{2}\,\sigma}\right)^2\right]}$$

$$= 0.205\,\frac{H_S^2\omega_m^4}{\omega^5}\exp\left[-\frac{5}{4}\left(\frac{\omega_m}{\omega}\right)^4\right]\nu^{\exp\left[-\frac{(\omega-\omega_m)^2}{2\sigma^2\omega_m^2}\right]}$$

(4.3.25)

This is a peak-enhanced Pierson-Moskowitz spectrum. The factors ν and σ control the height and width of the peak, respectively. The parameters were determined by analysis of North Sea wave data and depend on the fetch (in this case the distance from the lee shore) and the mean wind velocity measured at a height of 10.0 m above the sea surface. Figure 4.3.4 shows the evolution with fetch of a JONSWAP spectrum for a wind speed of 10.0 m/s and the corresponding Pierson-Moskowitz spectrum. The JONSWAP modal frequency decreases with increasing fetch, and the peak of the spectrum increases noticeably. Eventually, when no further energy storage is possible, the sea is "saturated," and the area under the spectrum stops growing.

In general, where site-specific data are unavailable, the following average values obtained in the JONSWAP experiment are considered:

ν = peakedness parameter
= 3.3 for a mean spectrum
σ = spectral shape parameter
= 0.07 if $\omega \leq 5.24/\bar{T}$
= 0.09 if $\omega > 5.24/\bar{T}$

Where spectrum data were recorded at a specific site, the values of several JONSWAP parameters probably need to be changed to obtain a reasonable fit with the data as the formulation has no analytic basis that allows ready correction.

PERIOD RELATIONSHIPS OF SPECTRA

At times the period in these spectrum equations are replaced with representative periods that may be more visually apparent in the seaway. Before defining these periods, it is convenient to recall the moments of the spectrum:

$$m_n = \int_0^\infty \omega^n S(\omega) d\omega \qquad (4.3.26)$$

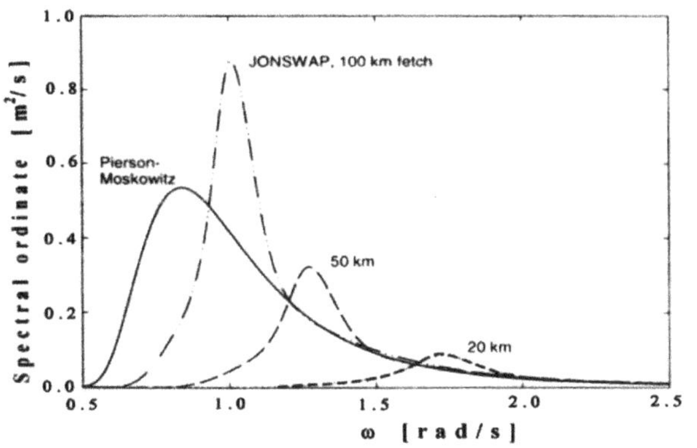

Figure 4.3.4 Evolution of JONSWAP spectrum with fetch and comparative modified Pierson-Moskowitz spectrum (*Source:* Lewandowski, 2004, World Scientific Publishing Co. Pte. Ltd., Singapore; used with permission).

The zeroth moment m_0 is equal to the area under the spectrum, i.e., the mean square wave elevation. Moments m_2 and m_4 correspond to the mean square values of velocity and acceleration of the wave surface.

Several periods are in use; the most common ones are defined as follows:

T_m The modal period is used with the frequency spectrum to denote the reciprocal of the peak frequency:

$$T_m = \frac{2\pi}{\omega_m} \qquad (4.3.27)$$

$\bar{T}$ The true average period is the period of the elemental waves in the spectrum:

$$\bar{T} = 2\pi \frac{m_0}{m_1} = 0.857 T_m \qquad (4.3.28)$$

T_z The mean zero-crossing period is the average period between successive crests in the wave record. This is the period most readily determined from the wave record or from model test data:

$$T_z = 2\pi \sqrt{\frac{m_0}{m_2}} = 0.710 T_m \qquad (4.3.29)$$

T_S The significant period is defined as the average period of the one-third highest waves in the record. This period is determined from measured or estimated times between significant crests, involving personal judgment in deciding what significant crests are. The value of the significant period cannot be mathematically determined from the spectrum function. Nevertheless, most oceanographic data were obtained by these methods. Bretschneider proposed the following value:

$$T_S = 0.946 T_m \qquad (4.3.30)$$

T_1 This period was adopted by ITTC in 1969 as their standard reference, labeled characteristic period.

$$T_1 = 0.772 T_m \qquad (4.3.31)$$

T_v The visually estimated period from shipboard observations. Earlier, ISSC suggested that the characteristic period T_1 adopted by ITTC might be equivalent to this visually estimated period:

$$T_v \approx T_1 \qquad (4.3.32)$$

The relative positions of the various periods for the frequency spectrum are shown in Fig. 4.3.5. The modal period T_m (at the peak frequency) is dominant and is thus often proposed as the standard period parameter for the frequency spectrum. The mean

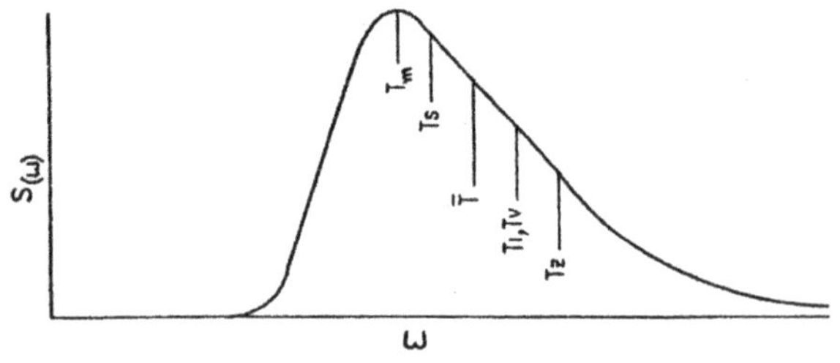

Figure 4.3.5 Relative positions of periods for the frequency spectrum.

zero-crossing period T_z is important because it can be directly evaluated from the wave record as well as from spectral moments.

RELATIONSHIP BETWEEN WAVE FREQUENCY AND SIGNIFICANT WAVE HEIGHT

Both of the modal frequency, ω_m, as well as the significant wave height, H_S, are random processes and, to use any of the two-parameter formulas, it is necessary to know the probability density function and the interdependence, if any, of the two parameters. Customarily, H_s is chosen as the primary parameter, and it then becomes necessary to account for the dependency of ω_m (or $\bar{\omega}$) on it. Until recently, this was done by attempting to fit a deterministic expression for ω_m as a function of H_s, but as pointed out by Ochi (1982), this is not appropriate, because both quantities are random processes, and ω_m is neither totally dependent nor totally independent of H_s. The dependency can only be accounted for by establishing the statistical relationship between these two parameters. Specifically, what is needed is the probability density function of the wave frequency for a given value of H_s, that is, the *conditional* probability density function of ω. From an analysis of North Atlantic data, Ochi showed that H_s follows the log-normal probability density function given in (4.1.14) (and not the Weibull distribution as others have proposed). Only large values of H_s (the uppermost 1 percent, corresponding roughly to values exceeding 10.0 m) depart from this law, and for such extreme values the probability density function is obtained in a different manner, that is, by means of the asymptotic approximation of the cumulative probability distribution function as discussed in Section 4.2. Ochi also showed that, if the data are expressed in terms of wave period instead of wave frequency, the modal period, T_m, also follows a log-normal probability density function. Therefore, since both H_s and T_m are log-normally distributed, the wave period for a given wave height follows the conditional log-normal probability density function given by

$$p_T\left(T_m \middle| H_s\right) = \frac{1}{\sigma_T^* \sqrt{2\pi} T_m} e^{-\frac{1}{2}[(\ln T_m - \mu_T^*)/\sigma_T^*]^2} \tag{4.3.33}$$

where

$$\mu_T^* = \mu_T + \rho \frac{\sigma_T}{\sigma_H}\left(\ln H_s - \mu_H\right)$$

$$\sigma_T^* = \sqrt{1-\rho^2}\ \sigma_T$$

In this expression the two pairs of parameters — μ_H, σ_H and μ_T, σ_T— are the mean and the standard deviation of the probability density function for $\ln(H_s)$ and $\ln(T_m)$, respectively, and ρ is the correlation coefficient between the two random variables H_s and T_m. Table 4.3.1 gives values of these five parameters that Ochi obtained by averaging values calculated from data obtained at seven weather stations in the North Atlantic.

It was mentioned previously that most of the available data on wave height and wave period are visually observed values, H_v and T_v. For example, the data of Walden (1964) are values observed at

Table 4.3.1 Parameters associated with bivariate log-normal probability distribution for various locations in the North Atlantic

Weather Station	A	B	C	D	I	J	K
Significant height							
μ_H	0.946	0.910	1.024	0.968	1.112	1.053	0.748
σ_H	0.619	0.588	0.571	0.588	0.562	0.565	0.680
Modal period							
μ_T	2.505	2.462	2.494	2.483	2.588	2.594	2.600
σ_T	0.218	0.218	0.216	0.209	0.142	0.147	0.174
Correlation coefficient							
ρ	0.498	0.594	0.578	0.586	0.358	0.339	0.331

nine locations in the North Atlantic over a period of 10 years. The observations were made regularly by trained oceanographers at 3½-hour intervals on average. Wave heights were visually estimated, and wave periods were counted by a stopwatch from visually observed wave crests. Therefore, the observed wave height H_v represents neither the significant nor the average wave height. Also, the observed wave period T_v represents neither the zero-crossing nor the average wave period. Hence, to use these data, it is necessary to convert them to the statistical quantities $\bar{H}$ and T_0, or H_s and T_m. Having shown that both the visually observed data and the measured data follow a log-normal distribution, Ochi obtained the following relationships between them

$$H_s = H_v^{1.08} \tag{4.3.34}$$

$$T_m = 2.99T_v^{0.73} \tag{4.3.35}$$

SHORT-CRESTED WAVES

The mathematical spectral formulas and wave spectra previously discussed are spectra of the sea at a fixed point, that is, they are one-dimensional spectra. This can be thought of as describing a long-crested irregular sea. A more complete representation of the sea is given by a two-dimensional *directional spectrum*, $S(\omega, \theta)$, as shown in Fig. 4.3.6, that indicates the direction θ as well as the frequencies of the wave components and thus accounts for the typical short-crestedness of ocean storm waves. The most common method for approximating $S(\omega, \theta)$ is to use the form

$$S(\omega, \theta) = S(\omega)f(\theta) \tag{4.3.36}$$

The function $f(\theta)$ is referred to as the *spreading function.* A function commonly used is

$$f(\theta) = \begin{cases} \dfrac{2}{\pi}\cos^2\theta & \left(-\dfrac{\pi}{2} \le \theta \le \dfrac{\pi}{2}\right) \\ 0 & \text{otherwise} \end{cases} \tag{4.3.37}$$

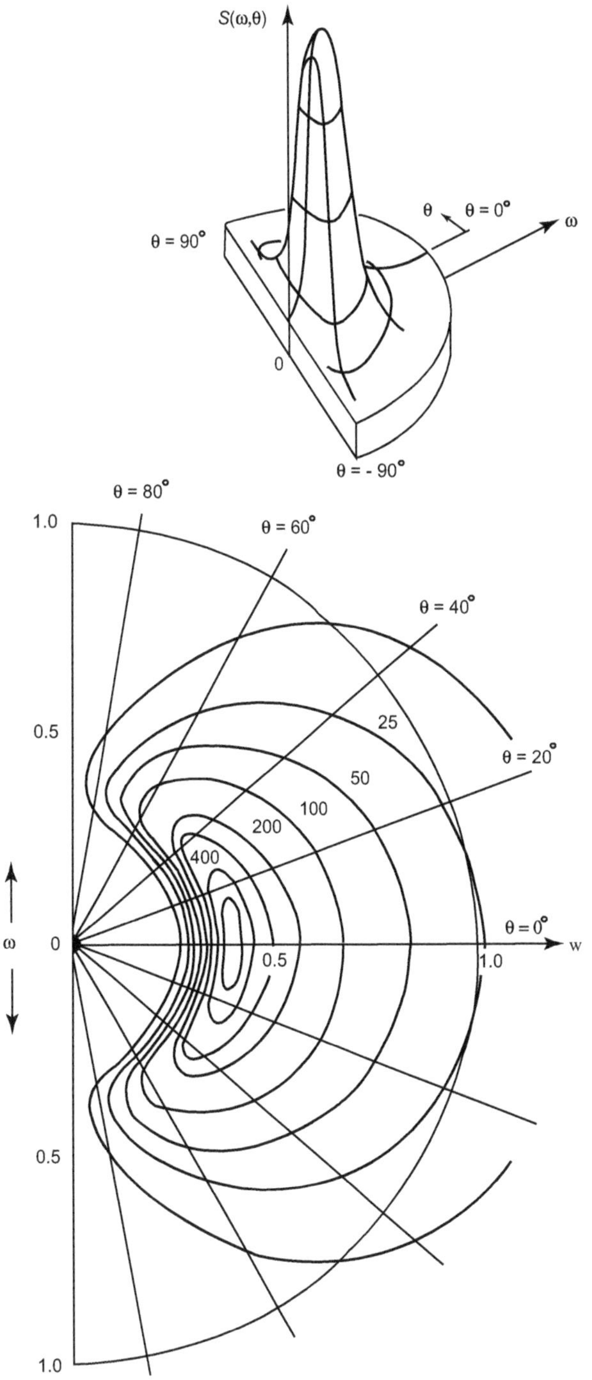

Figure 4.3.6 Idealized directional sea spectrum.

Figure 4.3.6 shows a contour plot of $S(\omega, \theta)$ after applying the spreading function.

4.3.3 Families of Wave Spectra

The spectra given by a two-parameter formula such as (4.3.24) are idealized and simplified. In reality, the shape of wave spectra observed in the ocean varies considerably (even for the same significant wave height) depending on the geographical location, duration and fetch of wind, stage of growth or decay of a storm, and existence of swell. Since a ship encounters an infinite variety of wave conditions and since the magnitude of the response is significantly influenced by the shape of the wave spectrum, it is necessary to have a method that accounts for this variety of wave spectra. For this purpose researchers developed families of wave spectra that consist of groups of some ten or twelve spectra for each of several levels of sea state severity, that is, for each of several significant wave heights. Examples are the H-family of Lewis (1967), the extension of it by Hoffman (1975), and the two-parameter and six-parameter families of Ochi and Hubble (1976). The approach of Ochi and Hubble is chosen for illustration because it is based on the probabilistic relationship between ω_m and H_s expressed in (4.3.33). The six-parameter family is more comprehensive and versatile, but it involves more computation. Both families are briefly summarized here.

TWO-PARAMETER WAVE SPECTRA

There are nine members of this family. Each family is generated for a different value of ω_m (or, equivalently, T_m) from the two-parameter formula adopted by the ITTC, which is a modified version of Bretschneider's formula, expressed in terms of H_s and ω_m:

$$S(\omega) = \frac{5}{16}\left(\frac{\omega_m}{\omega}\right)^4 \frac{H_s^2}{\omega} e^{-1.25(\omega_m/\omega)^4} \tag{4.3.38}$$

Since the probability density function of T_m for any given H_s is known, the nine values of T_m can be chosen so as to give a complete and balanced representation of the variation in wave period occurring in each level of sea state. This is achieved by choosing the most probable value of T_m and four pairs of values on either side of it, corresponding to confidence coefficients 0.95, 0.85, 0.75, and 0.50. The resulting expressions for ω_m as a function of H_s are given in Table 4.3.2, which also gives the weighting function by which each of the nine wave spectra is multiplied to reflect the differing probabilities of occurrence. Figure 4.3.7 shows the family of two-parameter wave spectra for a significant wave height of 3.0 m.

Table 4.3.2 Modal frequencies for the (mean) North Atlantic wave spectra as a function of specific wave height (ω_m in rps, H_s in meters)

Confidence Coefficients	Value of ω_m	Weighting Factor
Lower ω_m		
0.95	$0.048(8.75 - \ln H_s)$	0.0500
0.85	$0.054(8.44 - \ln H_s)$	0.0500
0.75	$0.061(8.07 - \ln H_s)$	0.0875
0.50	$0.069(7.77 - \ln H_s)$	0.1875
Most probable	$0.079(7.63 - \ln H_s)$	0.25
Upper ω_m		
0.50	$0.099(6.87 - \ln H_s)$	0.1875
0.75	$0.111(6.67 - \ln H_s)$	0.0875
0.85	$0.119(6.65 - \ln H_s)$	0.0500
0.95	$0.134(6.41 - \ln H_s)$	0.0500

THREE- AND FOUR-PARAMETER WAVE SPECTRA

The spectrum given by the two-parameter Pierson-Moskowitz formula (4.3.24) is idealized and simplified. In reality, the shape of wave spectra observed in the ocean varies considerably (even for the same significant wave height) depending on the geographical location, duration and fetch of wind, stage of growth or decay of a storm, and existence of swell. Since a ship encounters an infinite variety of wave conditions and since the magnitude of the response is significantly influenced by the shape of the wave spectrum, it is necessary to have a

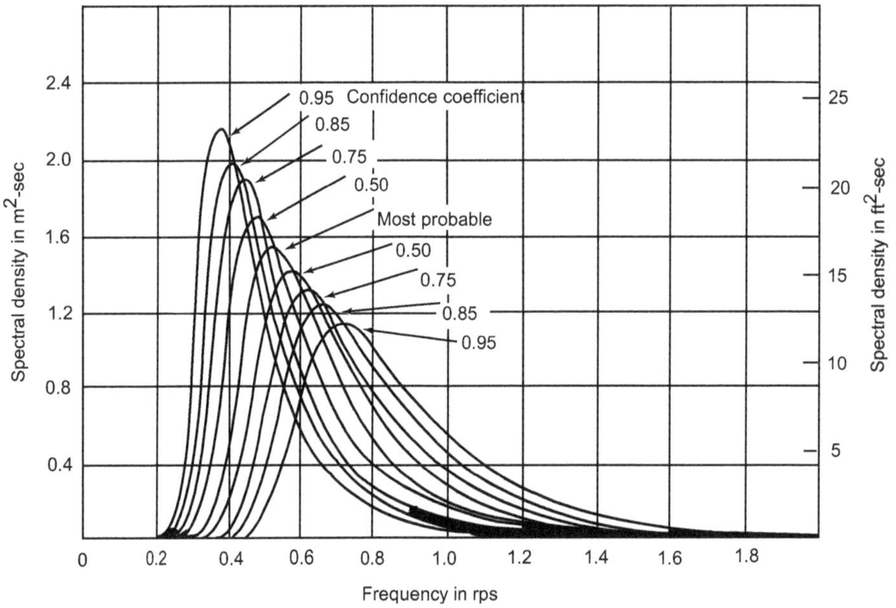

Fig. 4.3.7 Family of two-parameter wave spectra for a significant wave height of 3.0 m.

method that accounts for this variety of wave spectra. For this purpose researchers developed multi-parameter spectra, based on the general form (4.3.22). For convenience, we shall rewrite this general formulation, resulting in the standard form of Michel (1999):

$$S(\omega) = \alpha\omega^{-l}\, e^{-\beta\omega^{-n}} \qquad (4.3.39)$$

For sharply peaked spectra, such as the spectrum (4.3.25) addressed by JONSWAP, more tractable equations can be obtained using this standard from. The complete integral of $S(\omega)$ includes a gamma function. At the frequency ω_0 the slope of the spectral function is zero, which leads to the spectrum's peak value:

$$S(\omega_0) = \frac{nH_s^2}{16}\,\frac{\left(\frac{l}{n}\right)^{\frac{l-1}{n}}}{\Gamma\left(\frac{l-1}{n}\right)}\,\omega_0^{-1}\, e^{-\frac{l}{n}} \qquad (4.3.40)$$

The complete equation for the spectrum may then be written as follows:

$$S(\omega) = \frac{nH_s^2}{16}\,\frac{\left(\frac{l}{n}\right)^{\frac{l-1}{n}}}{\Gamma\left(\frac{l-1}{n}\right)}\,\omega_0^{l-1}\omega^{-l} e^{-\frac{l}{n}\left(\frac{l-1}{\omega}\right)^n} \qquad (4.3.41)$$

When $l = 5$ and $n = 4$, these two relationships reduce to those of the standard two parameter spectral form.

Ochi (1978, 1993) proposed an effective three-parameter spectrum by letting the factor n = 4 as a constant value:

$$S(\omega) = \frac{H_s^2}{4}\left(\frac{l}{4}\right)^{\frac{l-1}{4}}\,\frac{\omega_0^{l-1}\omega^{-l}}{\Gamma\left(\frac{l-1}{4}\right)}\, e^{-\frac{l}{4}\left(\frac{\omega_0}{\omega}\right)^4} \qquad (4.3.42)$$

with a peak value of

$$S(\omega_0) = \frac{H_s^2}{4}\left(\frac{l}{4}\right)^{\frac{l-1}{4}}\,\frac{\omega_0^{-1}}{\Gamma\left(\frac{l-1}{4}\right)}\, e^{-\frac{l}{4}} \qquad (4.3.43)$$

Michel (1999) documented an alternative three-parameter spectrum that provides a simpler yet effective formulation by keeping the relationship $l = n + 1$:

$$S(\omega_0) = \frac{H_s^2}{16}(n+1)\omega_0^n\omega^{-(n+1)} e^{-\frac{n+1}{n}\left(\frac{\omega_0}{\omega}\right)^n} \qquad (4.3.44)$$

with a peak value of

$$S(\omega_0) = \frac{H_s^2}{16}(n+1)\omega_0^{-1} e^{-\frac{n+1}{n}} \qquad (4.3.45)$$

Either of these three-parameter spectra can be readily approximated from the characteristics of a measured sea spectrum (ω_0, $S(\omega_0)$, H_s) by evaluating the product

$$\frac{\omega_0 S(\omega_0)}{H_s^2}$$

The appropriate value of l can be found directly from Fig. 4.3.8.

Taking the factors l and n as independent variables in (4.3.41) produces a four-parameter spectrum, wherein the characteristic mean zero-crossing period, T_z, is represented as the forth index of a given spectrum, along with ω_0, $S(\omega_0)$, and H_s. In 1976 a recorded North Sea spectrum was compared with theoretical spectra (Det Norske Veritas 1976). The three-parameter spectra conformed most closely to the record, whereas the more complete four-parameter spectrum showed least agreement. It became apparent that the three-parameter spectra represent the wave spectrum satisfactorily without concern about the relative value of the zero-crossing period. This example may help support the general conclusion that the three-parameter formulations are more representative of sharply peaked spectra than the two- or four-parameter spectra and that the zero-crossing period is not significant in this application.

SIX-PARAMETER WAVE SPECTRA

Ochi's six-parameter family accounts for two additional sources of variation in ocean wave spectra. The first is the shape, or degree of sharpness, of the spectrum peak. To account for this feature, Ochi added a shape parameter λ ($0 \leq \lambda \leq \infty$) to the basic

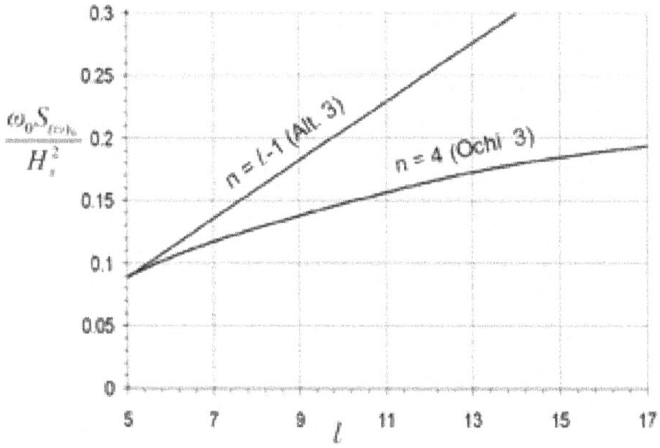

Figure 4.3.8 Evaluation of factor l for the three-parameter spectra (*Source:* Michel, 1999, SNAME; used with permission).

two-parameter formula of (4.3.38); the generalized formula is

$$S(\omega) = \frac{1}{4\Gamma(\omega)}\left[\frac{4\lambda + 1}{4}\left(\frac{\omega_m}{\omega}\right)^4\right]^\lambda \frac{H_s^2}{\omega} e^{-[(4\lambda+1)/4](\omega_m/\omega)^4} \quad (4.3.46)$$

where $\Gamma(\ )$ is the Gamma function. The shape of the wave spectrum becomes sharper with increasing λ. For $\lambda = 1$, (4.3.46) reduces to the two-parameter wave spectrum formula of (4.3.38). The derivation of (4.3.46) is documented in Ochi and Hubble (1976).

Another characteristic of actual wave spectra is that there often exists a plateau, or even a second peak, at a higher frequency as shown in Fig. 4.3.9. This may arise because of swell coexisting with wind-generated waves or because of the growth or decay of a storm. Although the wave energy at the higher frequencies is usually less than that at the lower frequencies, its contribution to ship response may be significant, depending on the ship size and speed.

Thus, it is highly desirable to represent the shape of the entire spectrum as closely as possible, and this may be achieved by separating the spectra into two parts, one that includes primarily the lower-frequency components of the energy and another that covers primarily the higher-frequency components of the energy as shown in Fig. 4.3.9. This gives a six-parameter spectral formula:

$$S(\omega) = \frac{1}{4}\sum_{j=1,2}\frac{1}{\Gamma(\lambda_j)}\left[\frac{4\lambda_j + 1}{4}\left(\frac{\omega_{mj}}{\omega}\right)^4\right]^{\lambda_j} \frac{H_{sj}^2}{\omega} e^{-[(4\lambda_j+1)/4](\omega_{mj}/\omega)^4} \quad (4.3.47)$$

where $j = 1,2$ stands for the lower and higher frequency components, respectively.

To obtain expressions for the six parameters, Ochi first grouped the 800 observed North Atlantic wave spectra documented by Moskowitz et al. (1963), Bretschneider et al. (1962), and Miles (1972) into

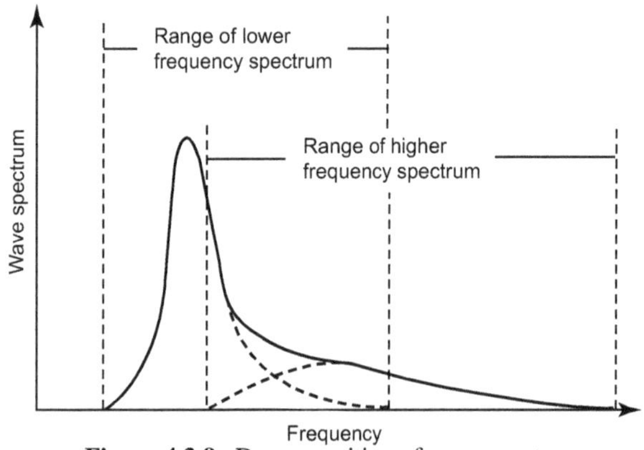

Figure 4.3.9 Decomposition of wave spectra.

ten groups, according to severity, and then for each group he performed a statistical analysis, of which the principal steps are given next. In this analysis it was possible to deal with only five parameters by working in terms of the ratio of the two significant wave heights: $r_H = H_{s1}/H_{s2}$. The steps are:

1. Probability density functions were established for each parameter. For example, it was found that the parameters λ_1 and λ_2 both follow the gamma probability law for all ten groups.
2. Three values were determined from the probability density function for each parameter, namely, the modal value and upper and lower values corresponding to a confidence coefficient of 0.95.
3. For each value of a parameter, values of each of the other parameters were determined from the original data by taking their respective averages in the region of ±5 percent. Thus, a total of 15 spectra were established for a given sea severity.
4. Of these 15 spectra, five are associated with the modal value of the five parameters. It was found, however, that these five spectra had nearly the same shape; therefore, the spectrum associated with the modal value of the parameter r_H was chosen as representative, and this spectrum is called the most probable spectrum for a given sea severity. Thus, a total of 11 spectra ware derived as a family of wave spectra for a specified sea severity.

In Table 4.3.3 the values of the six parameters for this family are expressed as functions of significant wave height H_s, and from these expressions the complete family of spectra for the desired sea can be generated from (4.3.47).

The weighting factor for each member of the family is

Most probable spectrum	0.50
All other spectra	(each) 0.05

The weight given to the most probable wave spectrum is higher than that for the other spectra. This is so, as stated above, because the most probable spectrum also represents four other spectra associated with the modal value of the parameters.

Figure 4.3.10 shows the family of six-parameter spectra for a significant wave height of 3.0 m, and this family may be compared with the family of two-parameter spectra shown in Fig. 4.3.7. It can be seen that the members of the six-parameter family have a wider variety of shapes than the members of the two-parameter family. Some members of the six-parameter family have double peaks, and the majority of the spectra have sharper peaks than those of the two-parameter family.

Table 4.3.3 Values of six parameters as a function of significant wave height (H_s in meters)

	H_{s1}	H_{s2}	ω_{m1}	ω_{m2}	λ_1	λ_2
Most probable spectrum	$0.84\ H_s$	$0.54\ H_s$	$0.70\ e^{-0.046\ H_s}$	$1.15\ e^{-0.039\ H_s}$	3.00	$1.54\ e^{-0.062\ H_s}$
	□$0.95\ H_s$	$0.31\ H_s$	$0.70\ e^{-0.046\ H_s}$	$1.50\ e^{-0.046\ H_s}$	1.35	$2.48\ e^{-0.102\ H_s}$
	□$0.65\ H_s$	$0.76\ H_s$	$0.61\ e^{-0.039\ H_s}$	$0.94\ e^{-0.036\ H_s}$	4.95	$2.48\ e^{-0.102\ H_s}$
	□$0.84\ H_s$	$0.54\ H_s$	$0.93\ e^{-0.056\ H_s}$	$1.50\ e^{-0.046\ H_s}$	3.00	$2.77\ e^{-0.112\ H_s}$
	□$0.84\ H_s$	$0.54\ H_s$	$0.41\ e^{-0.016\ H_s}$	$0.88\ e^{-0.026\ H_s}$	2.55	$1.82\ e^{-0.089\ H_s}$
0.95 confidence spectra	□$0.90\ H_s$	$0.44\ H_s$	$0.81\ e^{-0.052\ H_s}$	$1.60\ e^{-0.033\ H_s}$	1..80	$2.95\ e^{-0.102\ H_s}$
	□$0.77\ H_s$	$0.64\ H_s$	$0.54\ e^{-0.039\ H_s}$	0.61	4.50	$1.95\ e^{-0.082\ H_s}$
	□$0.73\ H_s$	$0.68\ H_s$	$0.70\ e^{-0.046\ H_s}$	$0.99\ e^{-0.039\ H_s}$	6.40	$1.78\ e^{-0.069\ H_s}$
	□$0.92\ H_s$	$0.39\ H_s$	$0.70\ e^{-0.046\ H_s}$	$1.37\ e^{-0.039\ H_s}$	0.70	$1.78\ e^{-0.069\ H_s}$
	□$0.84\ H_s$	$0.54\ H_s$	$0.74\ e^{-0.052\ H_s}$	$1.30\ e^{-0.039\ H_s}$	2.65	$3.90\ e^{-0.085\ H_s}$
	□$0.84\ H_s$	$0.54\ H_s$	$0.62\ e^{-0.039\ H_s}$	$1.03\ e^{-0.039\ H_s}$	2.60	$0.53\ e^{-0.069\ H_s}$

4.3.4 Duration of Sea States

The extreme value of a response is a function of the number of peak values encountered and, hence, it is necessary to know the duration of each sea state. Ideally, we wish to know the length of time a ship is subjected to a given sea condition. In many instances a ship encounters a particular sea state some time after that sea state arose, or it departs from the relevant area of the ocean before the sea state has subsided. However, it is always in the midst of some sea state, and the transfers from one to another largely cancel out. Hence, we can assume that the exposure time to a given sea state is equal to the duration of that state. This is conservative because of modern forecasting and long-range weather radar a ship can often avoid the worst regions of a storm area. Figure 4.3.11, taken from Ochi and Motter (1974), is an envelope curve of the longest recorded durations of every 1.52 m interval of significant wave height, estimated from analysis of the data documented by Moskowitz et al. (1962-65). For example, significant wave heights between 6.0 and 7.5 m can be expected to persist for a maximum of 40 hours. At the upper and lower extremities of sea severity, the maximum duration tends toward a constant value of approximately 3 hours and 45 hours, respectively.

ESTIMATION OF MOST PROBABLE EXTREME SEA STATE

The second method of Section 4.2.1 can be used to estimate the probable extreme value of significant wave height, $H_{s,p}$, from available wave data. Ochi (1978) did this, using the visually observed data from ten weather stations collected over a period of 10 years. As an example, Fig. 4.3.12 shows the results for Station J, which generally encountered the most severe conditions

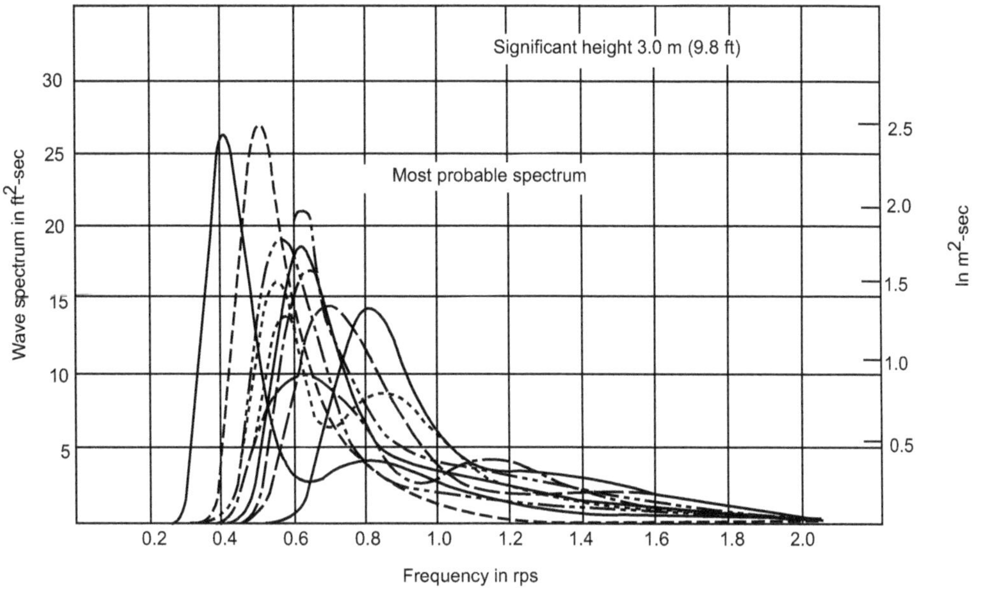

Figure 4.3.10 Family of six-parameter wave spectra for a significant wave height of 3.0 m.

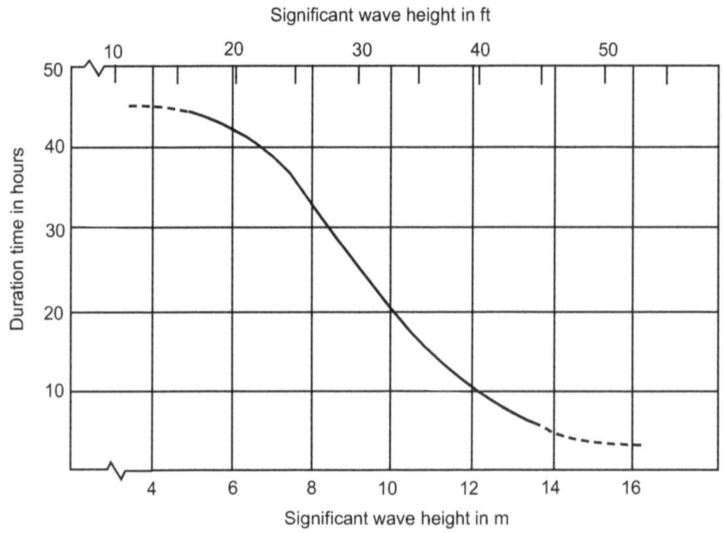

Figure 4.3.11 Significant wave height and its persistence in the North Atlantic every 1.52 m interval.

of the ten stations. The cumulative distribution of H_s, $P(H_s)$, was evaluated from the data, and then the left-hand side of (4.2.21) was plotted in logarithmic form. The resulting points lie approximately along a straight line, showing that the long-term distribution of H_s is approximately exponential; that is, that $f(H_s)$ in (4.2.17) is approximately linear in H_s.

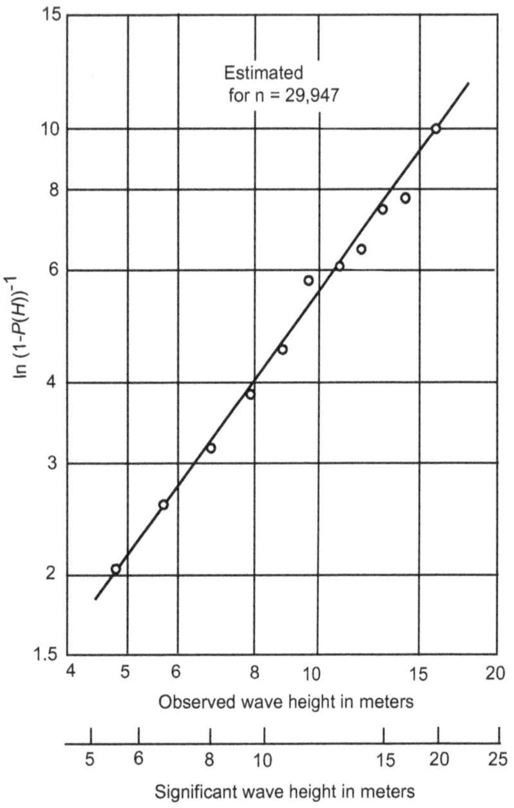

Figure 4.3.12 Probable extreme significant wave height in 10 years at Station J (Walden, 1964).

In estimating the probable extreme values from the line drawn in the figure, the number of observations involved in the data has to be considered since the magnitude of the extreme value depends on this number. For Station J the number of observations in 10 years was n = 24,947. Hence, the right-hand side of (4.2.21) is ln(24,947) = 10.12, and this point on the line corresponds to a visually observed wave height of 15.4 m. Hence, this is the probable extreme value for a 10-year period. Using the conversion given in (4.3.34), the probable extreme significant wave height expected to occur in 10 years is 19.2 m. Ochi also showed that these estimates obtained from visually observed wave data agree well with the estimates calculated from measured values of H_s, provided the comparison is made for the same sample size, that is, for the same value of n.

Figure 4.3.13 gives the combined results for all 10 stations and covers the full range of H_s, including the smaller values [below $P(H_s) = 0.99$] which follow the log-normal probability density function as discussed earlier.

The data used for Fig. 4.3.12 cover a 10-year period, whereas we wish to know the probable extreme significant wave height expected in the lifetime of the ship of, say, 20 years. If we assume that the statistical characteristics of the extreme values expected to occur in 20 years are the same as those observed in the data accumulated in 10 years, then the value for 20 years may be obtained by extending the line to a point corresponding to a sample size that is twice as large. The ordinate of Fig. 4.3.12 is $\ln[\{1 - P(H_{s,p})\}^{-1}]$, and from (4.2.21) this is equal to the natural logarithm of n. Hence, the new ordinate is ln(2n) or 10.82. By extrapolating the straight line up to this point, we obtain a value

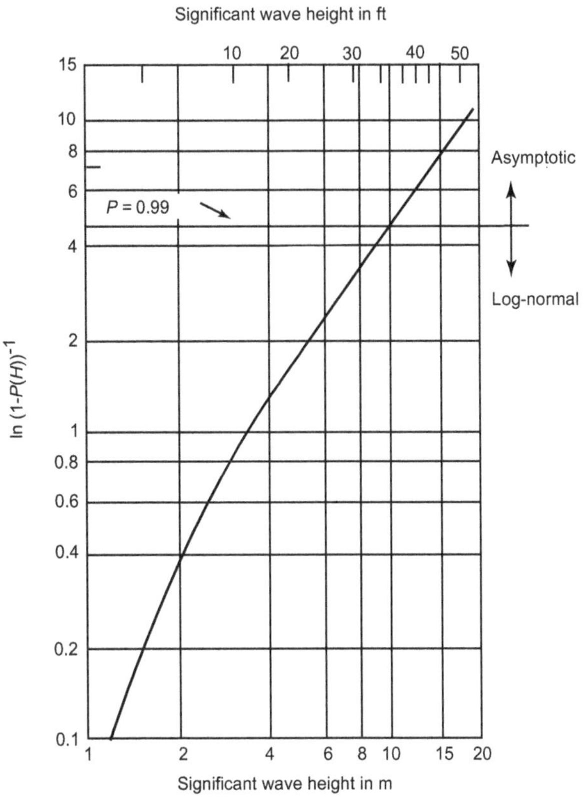

Figure 4.3.13 Cumulative distribution of significant wave height for the (mean) North Atlantic.

of 20.2 m for $H_{s,p}$. Thus, the 20-year value is about 5 percent greater than the 10-year value.

FREQUENCY OF OCCURRENCE OF VARIOUS SEA STATES

It is necessary, or at least advisable, to perform a response analysis not only for the probable extreme significant wave height $H_{s,p}$, but also for one or two smaller values of H_s. To maintain a uniform risk level, it is necessary to decrease the value of the risk parameter for these values in proportion to their more frequent occurrence. Hence, we need to know the frequency of occurrence of the various sea states. This is even more necessary if a response analysis is being performed for the full range of H_s and over the complete life of the ship, as might be done if a rigorous hull girder fatigue calculation is required.

The frequency of occurrence of various sea severities can be obtained from the available wave data. For small and moderate sea states the governing probability density function is the log-normal function. This applies up to a cumulative probability distribution of 0.99 (i.e., values of H_s that have at least a 1 percent likelihood of occurring). For severe sea states the calculation must be based on the asymptotic extreme distribution. Ochi (1978) performed the calculations for the average North Atlantic data and obtained the results shown in Table 4.3.4, listing the frequencies for each 1-meter interval of significant wave height.

Table 4.3.4 Long-term frequency of occurrence of various sea states in the (mean) North Atlantic

Significant Wave Height (Meters)	Frequency of Occurrence	Significant Wave Height (Meters)	Frequency of Occurrence
<1	0.0503	9–10	0.0079
1–2	0.2665	10–11	0.0054
2–3	0.2603	11–12	0.0029
3–4	0.1757	12–13	0.0016
4–5	0.1014	13–14	0.00074
5–6	0.0589	14–15	0.00045
6–7	0.0346	15–16	0.00020
7–8	0.0209	16–17	0.00012
8–9	0.0120	17<	0.00009

4.4 COMPUTATION OF WAVE-INDUCED LOADS

4.4.1 Computational Methods

Beck et al. (1996) and Beck and Reed (2001) comprehensively reviewed computational methods to solve marine hydrodynamic problems. Here we shall limit our discussion of computational seakeeping methods that are used in the context of load generation for finite element structural analysis. Therefore, only the case of ships in waves with forward speed in infinitely deep water is of interest.

The nonlinearities are the major difficulties in seakeeping computations. The free surface causes the ship to behave nonlinearly because of the nature of the free surface boundary conditions and the nonlinear characteristics of the incident waves. The time-dependent change of position and wetted surface of the ship in waves often cause nonlinear hydrostatic restoring forces and nonlinear force contributions at the free-surface intersection. There are nonlinearities associated with viscous force contributions that depend quadratically on the water velocity and introduce velocity squared terms in the pressure equation. However, because of forward speed, ships are generally long and slender with smooth shape variations along their length. This geometric feature of typical ships allowed a significant amount of progress to date and is the basis of many approximations.

To obtain wave loads and ship motions, compressibility effects and cavitation can be ignored. Neglecting these two effects reduces the mathematical problem to the extent that only the three fluid velocity components and the fluid pressure have to be determined. These four unknowns are determined by solving the four governing equations, namely, the continuity equation and the three-component Navier-Stokes equations.

Conditions that reflect the physical situation, so-called boundary conditions, need to be satisfied to obtain a unique solution to a particular problem. Applying the kinematic boundary condition ensures that no fluid passes through the hull surface. On the free surface a kinematic condition of no flow through the surface as well as a dynamic condition of constant pressure must be applied, leading to highly nonlinear free-surface boundary conditions. Satisfying radiation boundary conditions prevents wave reflection in the far field. Additional conditions of no-slip boundary condition on rigid surfaces and no tangential shear stress on the free surface are needed for viscous fluids.

The general solution requires solving nonlinear, partial differential equations with nonlinear boundary conditions and temporal instabilities. At the present time, the computational capacity to solve these equations is not available. For a particular problem, therefore, some approximations are necessary, such as reducing the governing equations to make them easier to solve and/or simplifying the boundary conditions. The Navier-Stokes equations can be simplified by assuming the viscosity is zero, yielding the so-called Euler equations. The flow field in such an ideal flow is then irrotational, meaning that the three velocity components can be determined from the gradient of a scalar potential and that the continuity equation reduces to the Laplace equation. This so-called potential flow is probably the most widely used assumption in ship hydrodynamics. The Navier-Stokes equations can be integrated to yield the Bernoulli equation for the pressures, and mathematical tools are available to solve the Laplace equation, which is a linear partial differential equation. Assuming the flow to be ideal and irrotational thus reduces the original problem of four coupled nonlinear partial differential equations for the four unknowns (three velocity components and pressure) to one linear partial differential equation for the unknown scalar velocity potential. The gradient of the known velocity potential yields the velocity components; the Bernoulli equation, the pressure.

For ship-related problems, the free-surface boundary conditions can be linearized about the plane of the undisturbed surface. The transverse dimensions of most ships are significantly less than the longitudinal dimensions, and the cross-sectional shape varies smoothly along the length. This so-called slender-body assumption allows the complete three-dimensional potential flow problem to be subdivided into a series of two-dimensional potential problems that are solved in the transverse plane. The two-dimensional solutions are then combined to approximate the three-dimensional problem. The slender-body theory most widely used in ship hydrodynamics is probably strip theory. Strip theory gives useful results for normal ships up to moderate forward speeds in head seas. At higher speeds, predictions can be poor for hull forms with large shape changes because forward speed effects in the free-surface boundary conditions and three-dimensional effects are not properly accounted for.

Theories were developed to overcome the deficiencies of slender-body theory. They retain the linearized free-surface boundary condition, but satisfy the body boundary condition on what would be the wetted surface of the hull at rest in calm water. For small motions, this mean position is close to the exact wetted hull surface that changes with time. However, for large amplitude motions, the linear assumption breaks down because the actual hull surface can be significantly displaced from the mean position.

So-called panel codes are typical linear codes that account for three-dimensional effects. They divide the hull into a large number of small surface elements (panels). Over these panels, singularities are distributed that satisfy the Laplace equation in the fluid domain, the linearized free-surface boundary condition, and the boundary conditions at infinity. Wave Green functions are singularities that satisfy these conditions (Wehausen and Laitone, 1960). On each panel, the strength of the singularities has to satisfy the boundary conditions of each panel, which leads to a set of M simultaneous linear equations, where M is the number of panels or, equivalently, the number of unknown singularity strengths.

These linear numerical methods can be extended by introducing some nonlinearity into the linear problem. The idea is to compute the hydrostatic restoring forces and the Froude-Krylov part of the exciting forces using the instantaneous submerged part of the hull as theses forces are relatively easy to compute. Hydrodynamic terms that are difficult to compute are retained as linear, and the equations of motion are then solved using both the linear as well as the nonlinear terms, yielding large ship motions and wave loads in finite amplitude waves. This approach gives useful design loads as the hydrostatic and Froude-Krylov terms are the largest nonlinearities. Beck and Magee (1991) and Lin and Yue (1990) extended this method by computing the hydrodynamic terms using the complete body boundary condition applied on the exact instantaneous wetted surface, but retaining the linearized free-surface boundary condition. However, this method is computationally intensive because the body surface constantly changes, so that a fully nonlinear computation can be performed with approximately the same computational effort.

For fully nonlinear computations, the viscosity in the governing Navier-Stokes equations must be included. Of course, the flow is then no longer irrotational, and the velocity potential alone cannot be used. This means solving four nonlinear partial differential equations with four unknowns. Direct solutions to this problem are too computer intensive and not practical for ship design purposes. Thus, further approximations must be made. This is done by computing the average flow in a viscous fluid flow. For load predictions, the approach based on the Reynolds-averaged Navier-Stokes (RANS) equations has typically been used. As the flow about a ship with forward speed in a seaway is invariably turbulent over essentially the entire hull surface, the RANS are derived by assuming that all velocity components can be approximated by a mean component plus a highly oscillatory, small amplitude, zero-mean component that represents turbulence. After substituting these velocities into the Navier-Stokes equations, they are time-averaged over a suitable time scale. The resulting equations describe the mean flow. They are identical to the original Navier-Stokes equations except for the addition of so-called Reynolds stress terms that represent the influence of turbulence on the mean flow. Recently developed RANS codes incorporate the nonlinear free surface, and work is in progress on unsteady RANS codes that include incident waves and ship motions.

FREQUENCY DOMAIN CODES

Frequency-domain codes obtain solutions for periodic ship motions using wave Green functions for both zero speed and constant forward speed. Both of these Green functions satisfy their respective linearized free-surface boundary conditions and the appropriate radiation conditions. The zero-speed frequency-domain Green function poses relatively few numerical difficulties. However, the constant forward speed frequency-domain Green function is computationally complex because it must also capture the Kelvin wave field created by the steady flow with a uniform stream as the basis flow. This problem has a computationally difficult Green function, known as the Havelock singularity that satisfies both the linearized steady flow free-surface boundary condition and the radiation condition; see, for example, Ba and Guilbaud (1995) and Iwashita and Okhusu (1992). Consequently, the zero-speed frequency-domain Green function is used regularly while engineering applications of the constant forward-speed frequency-domain Green function are rare. For a literature review of these methods, see ISSC (1994).

An internationally well-known seakeeping code capable of solving frequency domain problems is SWAN 1 (Sclavounos et al., 1993). It is based on linearizing the flow about a double body to numerically model wave propagation and ship dynamics, using a three-dimensional Rankine panel method for potential flows based on a linear, frequency-domain formulation for steady and unsteady ship motions. Some Green function methods consider the forward speed under the so-called encounter frequency approach, where the boundary conditions on the ship are evaluated with the Green function evaluated only at zero speed. This saves a huge amount of computational effort, which is the reason why such methods are widely used for many routine design applications; see, for example, Rathje et al. (2000). Especially for ships with large bow flare and stern overhang, where three-dimensional effects become significant, such methods are replacing strip theory-based methods. Even for fast ships up to speeds corresponding to a Froude number of 0.4, these methods yield practical useful results, albeit only for relatively small amplitude waves (Schellin et al., 2003). However, this approach must be used with caution and needs to be validated for critical wave situations and specific ship types.

For many engineering applications, useful design loads can be obtained by correcting linear predictions for nonlinear effects. For large motions that do not involve bow emergence or water on deck, the nonlinearity of a ship's response is mainly caused by the nonvertical sides. To account for this nonlinearity, Hachmann (1986) formulated an approximation for the hydrodynamic pressure between still water level and wave contour. With this method, linear theory load predications can be corrected for this nonlinear effect, yielding realistic results for many standard applications (Hachmann, 1991). As this approach is computationally efficient, it has been used to obtain large amplitude wave-induced design loads that are then part of the input for finite element structural analyses of many modern ships (e.g., Payer and Fricke, 1994; Rathje and Schellin, 1997).

TIME DOMAIN CODES

If the motion response of a ship in waves and the associated wave-induced loads are highly nonlinear with respect to the wave amplitude, the ship should not be investigated in elementary regular waves, because these waves do not appear in nature, and the nonlinear response of the ship in a natural seaway cannot be deduced from the response in elementary waves. For these nonlinear cases, simulation in the time-domain is the appropriate tool for numerical predictions. Simulations performed in the time-domain facilitate the inclusion of important nonlinear effects, such as hydrostatics (wave profile) and roll damping.

Time-domain codes use their own wave Green functions. As in boundary element solutions in the frequency-domain, these singularities are distributed over the wetted hull surface, and the solutions are integrated over time as well as over the surface of the body. Alternatively, simple singularities can be distributed over both the hull and the undisturbed free surface. However, simple singularities must be employed to obtain solutions of nonlinear free-surface problems.

An internationally well known time-domain code with deep water potential flow assumptions is the Large Amplitude Motion Program (LAMP), developed to compute ship motions and wave loads under the assumption of weak scattering (Lin and Yue, 1990; Lin et al., 1994; Lin et al., 1996). A unique feature of the LAMP code is its multi-level degree of sophistication, allowing analyses with increasing complexity. The newest approach of LAMP is that, instead of satisfying the boundary condition on that portion of the hull that is below the mean free surface, the body boundary condition is satisfied on the actual instantaneous wetted hull under the incident wave profile. At each time step, local free-surface elevations are used to transform the body geometry into a computational domain with a deformed hull and a flat free surface. By linearizing the free-surface boundary conditions about this incident wave surface, the problem is solved using a linearized free-surface transient Green function. In this way, the correct hydrostatic and Froude-Krylov wave forces are automatically included.

The time-domain code SWAN2, the second code in the SWAN family, was extended to apply to nonlinear wave ship interactions (Nakos et al., 1993; Sclavounos, 1996). As with SWAN1, it employs a three-dimensional Rankine panel method for potential flows based on Green's third identity, and the radiation condition is enforced by introducing a dissipative beach.

The methods described above are generally adequate to predict ship motions and the associated global loads for a large number of situations. However, these methods have their limitations. They cannot handle breaking waves or green water effects, and the flow around sharp edges (such as the hull-deck intersection) is usually not well modeled. Furthermore, effects of viscous damping have to be implemented artificially, relying, for instance, on damping coefficients that may have to be linearized. Other methods are needed, therefore, to predict not only global loads in severe seas, but also local water-impact related loads.

It has long been recognized that such loads can be accurately predicted only if the free-surface flow is correctly simulated. Interface-capturing techniques of the VOF type proved to be most suitable for handling strong nonlinearities and are today the obvious choice for computing complex free-surface shapes with breaking waves, sprays and air trapping. These techniques are suitable to also analyze related problems, such as sloshing loads in partially filled tanks.

The computer code COMET (CD-adapco, 2002), for instance, a code that implements interface-capturing techniques of the volume-of-fluid (VOF) type, proved to be suitable for handling strong nonlinearities. Today, this kind of code is the obvious choice for computing complex free-suface shapes with breaking waves, sprays, and air trapping, hydrodynamic phenomena that should be considered to predict impact-related slamming and sloshing pressures. The conservation equations for mass and momentum in their integral form serve as the starting point. The solution domain is subdivided into a finite number of control volumes that may be of arbitrary shape. The integrals are numerically approximated using the midpoint rule. The mass flux through the cell face is taken from the previous iteration, following a simple Picard iteration approach. The unknown variables at the center of the cell face are determined by combining a central differencing scheme (CDS) with an upwind differencing scheme (UDS). The spatial distribution of each of the two fluids (air and water) is obtained by solving an additional transport equation for the volume fraction of the water. To accurately simulate the convective transport of the two immiscible fluids, the discretization must be nearly free of numerical diffusion and must not violate the boundedness criteria (Ferziger and Peric, 1996). For this purpose, the high resolution interface capturing (HRIC) scheme is used (Muzaferija and Peric, 1998). The scheme is a nonlinear blend of upwind and downwind discretization, and the blending is a function of the distribution of the volume fraction and the local Courant number. The free surface is smeared over two to three control volumes. Fluid–structure interaction effects are presently not accounted for, i.e., the body is assumed to be rigid. The fluid is assumed to be viscous and incompressible.

For special cases, it may be opportune to use an extended RANS solver; for example, when effects of slamming pressures that cause significant hull girder whipping are to be analyzed. The nonlinear equations of ship motions are solved and coupled with the RANS solver (Brunswig and El Moctar, 2004). The computational procedure consists of four main steps. First, flow around the ship is computed, taking into account viscosity, flow turbulence, and deformation of the free surface. Second, the hydro- and aerodynamic forces and moments acting on the ship are calculated by integrating the pressure and friction stresses over the ship's surface. Third, the nonlinear rigid body motion equations are solved for the six degrees of motion, and subsequent time integration yields accelerations, velocities, and displacements. Fourth, by updating the position of the ship and again computing the fluid flow for the new position and integrating this procedure over time, the trajectory of the ship is obtained.

POTENTIAL FLOW FORMULATION

For potential flow computations it is assumed that the water is invisced, homogeneous, incompressible, and of constant density. The surface tension on the free surface is neglected. Considered is a ship advancing at a steady mean forward speed U. A right-handed orthogonal coordinate system, $Oxyz$, is fixed with respect to the mean position of the ship and translates in the positive x-direction relative to an earth-fixed frame. The $z = 0$ plane of this $Oxyz$ system corresponds to the calm water level with z directed positive upwards as shown in Fig. 4.4.1. Translatory ship motions surge, sway and heave are denoted by η_1, η_2 and η_3; angular ship motions roll, pitch and yaw, by η_4, η_5 and η_6, respectively. The time-dependent ship speed is denoted by $U(t)$; at steady forward speed, $U(t) = -U$. A ship-bound coordinate system, $Ox'y'z'$, defines the hull shape of the ship itself. Ship motions are measured in terms of the translation and rotation of the ship-fixed axes relative to $Oxyz$.

The governing equations and boundary conditions are presented in the time domain. For frequency domain computations, the dependence on the time t is to be replaced by $\exp(i\omega t)$, and it is understood that only the real part is to be used. The total velocity potential of the flow, Φ, is separated into a time-independent steady contribution caused by the ship's forward speed U and a time-dependent part associated with the incident wave system and the oscillating ship motions (Beck and Reed, 2001):

$$\Phi(x, y, z; t) = -U(t)x + \phi(x, y, z; t) \qquad (4.4.1)$$

where $\phi(x, y, z; t)$ is the perturbation potential. Both potentials Φ and ϕ must satisfy the Laplace equation in the fluid domain:

$$\nabla^2\Phi = 0 \qquad (4.4.2)$$

On all surfaces surrounding the fluid, boundary conditions must be satisfied. On the instantaneously wetted hull surface (S_H), the kinematic body boundary condition is applied as follows:

$$\frac{\partial \phi}{\partial n} = -U(t)n_1 + \vec{V}_H \cdot \vec{n} \quad \text{on } S_H \qquad (4.4.3)$$

where $\vec{V}_H$ is the vector of velocity relative to the moving coordinate system (including rotational effects) of a point on the hull surface, and $\vec{n} = (n_1, n_2, n_3)$ is the unit normal vector out of the hull surface (into the fluid). The kinematic boundary condition must also be satisfied on the bottom. For infinitely deep water, this condition becomes

$$\nabla\phi \rightarrow 0 \quad \text{as} \quad z \rightarrow -\infty \qquad (4.4.4)$$

On the instantaneous free surface (S_F), the kinematic as well as the dynamic boundary condition must be satisfied. The kinematic condition is

$$\frac{\partial \eta}{\partial n} = \frac{\partial \phi}{\partial z} - \nabla\phi \cdot \nabla\eta - U(t)\frac{\partial \eta}{\partial t} \quad \text{on } S_F \qquad (4.4.5)$$

where $z = \eta(x, y; t)$ is the free-surface elevation. The dynamic boundary condition requires that the pressure everywhere on the free surface equals the ambient pressure, p, normally set equal to zero. Use of the Bernoulli equation for unsteady flow leads to the dynamic boundary condition:

$$\frac{\partial \phi}{\partial t} = -g\eta - \frac{1}{2}\nabla\phi \cdot \nabla\phi - U(t)\frac{\partial \phi}{\partial x} - \frac{p}{\rho} \quad \text{on } S_F \qquad (4.4.6)$$

where ρ is the fluid density and g the acceleration of gravity.

In the time domain, the initial values of the potential and the free-surface elevation must be specified. Normally, the computations start from rest, such that $\phi_s = \phi_T = \eta = 0$ for time $t < 0$. In the frequency domain, no initial conditions are necessary.

Boundary conditions must also be satisfied at infinity. In the time domain, for an initial value problem with no incident waves, it is also necessary for the fluid disturbance to vanish at infinity:

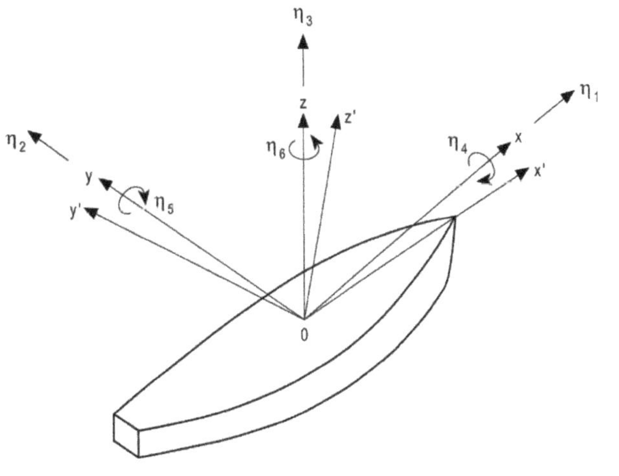

Figure 4.4.1 Coordinate system.

$$\nabla\phi \rightarrow 0 \quad \text{as} \quad R = \sqrt{x^2 + y^2} \rightarrow \infty \tag{4.4.7}$$

In the frequency domain, waves caused by the hull disturbance, including diffracted waves, must be outgoing towards infinity.

Hydrodynamic forces acting on the hull are found by integrating the pressure over the instantaneous wetted surface. The generalized force acting on the hull in the jth direction of the hull-bound coordinate system is thus given by the following expression:

$$F_j = \iint_{S_H} n_j \, p ds \tag{4.4.8}$$

Here n_j is a component of the generalized outward unit normal $\vec{n}$ at the hull surface, defined as

$$(n_1, n_2, n_3) = \vec{n} \text{ and } (n_4, n_5, n_6) = \vec{r} \times \vec{n} \tag{4.4.9}$$

and $\vec{r} = (x', y', z')$ is the position vector of a point on the hull surface (S_H) referred to the ship-fixed coordinate system. Subscripts j = 1, 2, 3 correspond to the force directions and subscripts j = 3, 4, 5 to the moments about the ship-fixed coordinate system, respectively.

Applying the Bernoulli equation for unsteady flow yields the pressure in the moving coordinate system:

$$p = -\rho \left[\frac{\partial\phi}{\partial t} + U(t)\frac{\partial\phi}{\partial x} + \frac{1}{2}\nabla\phi \cdot \nabla\phi + gz \right] \tag{4.4.10}$$

VISCOUS FLOW FORMULATION

As mentioned above, when computing the viscous flow about a ship, the velocity potential alone cannot be used because the flow is no longer irrotational. The viscous flow is then governed by the Navier-Stokes equations and the continuity equation, and loads are predicted using RANS solvers. The resulting RANS and the continuity equation are of the following form (Beck and Reed 2001):

$$\frac{\partial \bar{u}_i}{\partial t} + \bar{u}_j \frac{\partial \bar{u}_j}{\partial x} = -\frac{\partial \bar{p}}{\partial x_i} + \rho g_i + \nu\nabla^2\bar{u}_i - \frac{\partial \tau_{ij}}{\partial x_j} \tag{4.4.11}$$

$$\frac{\partial \bar{u}_i}{\partial x_i} = 0 \tag{4.4.12}$$

Here the velocities and pressure are expressed as $u_i = \bar{u}_i + u_i'$ and $p = \bar{p} + p'$, respectively, where the overbar represents a Reynolds average taken over a time/special scale large relative to the scale of turbulence, and the primed quantities account for the velocities and pressure at turbulent scale. The u_i for i = 1, 2, 3 are the x-, y-, and z-components of the velocity, p is the pressure, g_i is the ith component of the gravitational acceleration g in the x_i-direction. Double subscripts within a term imply summation over that index, and $\tau_{ij} = \bar{u}_i\bar{u}_j$ is the Reynolds stress tensor.

The RANS equations must be solved by satisfying boundary conditions on the hull surface, the free surface, the fluid boundary far away from the ship, and on the ocean bottom. On the hull surface, a kinematic boundary condition and a no-slip condition hold. On the free surface, a kinematic boundary condition assures that no fluid passes through the surface, and a dynamic boundary condition requires that the fluid pressure equals the atmospheric pressure and that no shear is acting. Conditions on the far fluid surface boundary must assure that there is either no disturbance or no wave reflection. On the (infinite) bottom, the disturbance must vanish.

Equations 4.4.11 and 4.4.12 represent four equations with 13 unknowns, namely, the three velocity components, the pressure, and the nine components of the Reynolds stress tensor. To find a solution, the turbulent kinetic energy and an equation relating the turbulent kinetic energy to the mean velocities and the eddy viscosity is introduced. This eddy viscosity relates the Reynolds stress tensor to the mean velocities.

At present, the use of RANS solvers for seakeeping problems is in its beginning stage. For vertical plane motions and wave loads, RANS and potential flow predictions generally compare favorably with experimental measurements.

STRIP THEORY

Strip theory methods are the standard tool for seakeeping computations. The development of such methods started about 50 years ago. The steady potential ϕ_s is omitted completely and the unsteady potential ϕ_T is approximated for each strip independently of the other strips. The essence of strip theory thus is to reduce the three-dimensional hydrodynamic problem to a series of two-dimensional boundary value problems that are easier to solve. The actual free-surface condition has to be simplified as well. The principle is to divide the underwater part of the ship into a number of strips (usually about 20) as shown in Fig. 4.4.2. The two-dimensional flow about an infinite cylinder of the same cross section as the ship at the strip's position determines hydrodynamic forces. The two-dimensional forces for each strip are combined to obtain the forces for the entire ship. Strip theory implies that the variation of flow in the cross-sectional plane is much larger than the variation of flow in the longitudinal direction. This is not the case at the ends of the hull.

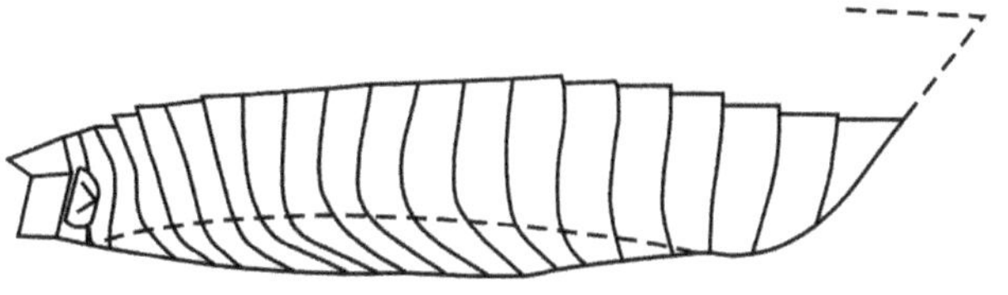

Figure 4.4.2 Strip theory idealization of a ship's hull (*Source:* Faltinsen, 1990, Cambridge University Press, New York; used with permission).

The original work on strip theory was documented by Korvin-Kroukovsky and Jacobs (1957). Most of today's strip methods are variations of the strip method of Salvesen, Tuck, and Faltinsen (1970), and these methods are generally known as STF strip methods. Analytical or panel methods are used to solve the two-dimensional problem for each strip. Analytical approaches rely on conformal mapping techniques to transform semi-circular cross sections to cross sections resembling ship sections (Lewis forms). Although this transformation cannot deal with submerged sections, such as a bulbous bow, it still yields results of similar quality as panel based (close-fit approach) strip methods.

Strip methods are fast and cheap and give reasonably accurate results over a wide range of parameters, and recent developments have shown improved comparison with experiments. However, they are still not entirely satisfactory. Although strip methods are today considered to be the most practical design tool to assess global wave-induced loads, it is important to be aware of their limitations. Strip theory is basically a high-frequency theory. Strip methods fail for waves shorter than about one-third of the ship length. Thus, it is more applicable in head and bow waves than in following and quartering seas for a ship with forward speed. Furthermore, strip theory is a low Froude number theory. It does not properly account for the interaction between the steady wave system and the oscillatory effects of ship motions. Another limitation is the assumption of linearity between response and incident wave amplitude. Therefore, it is questionable to apply strip theory for severe sea states.

UNIFIED THEORY

A theory that is theoretically applicable at all frequencies is the unified theory (Newman 1978). It uses the slenderness of the ship hull to justify coupling a two-dimensional flow in the near field to a three-dimensional flow in the far field. Distributing singularities along the ship's centerline generates the far field flow. Although its theoretical treatment is more consistent, for real ships results based on unified theory are not significantly more accurate than results from strip theory. Therefore, the method is not generally accepted in practice.

HIGH-SPEED STRIP THEORY

For fast ships with speeds at Froude numbers greater than about 0.4, the high-speed strip theory was developed, initially by Chapman (1975). For lower speeds, it is inappropriate. The theory is often referred to as being two-and-a-half-dimensional because, at a particular location along the ship's length, it only considers the effect of upstream sections on the flow and not the effect of downstream sections. Boundary conditions at the free surface and at the hull are satisfied to obtain the velocity potential and the wave elevation, and numerical differences between strips determine derivatives in the longitudinal direction. By marching downstream from strip to strip, the computation ends at the stern just before the transom.

GREEN FUNCTION METHOD

Green function methods discretize the average wetted hull surface into a large number of small surface elements (panels). Some methods use a slightly submerged surface inside the hull. Usually, the calm-water floating position defines the wetted surface, neglecting dynamic trim and sinkage as well as the steady wave profile. For each panel, a Green function defines the velocity potential. Usually, these potentials are sources that model the displacement effect of the ship. If lift plays a significant role, such as for yawing or maneuvering ships, additional vortices and dipoles are employed to model lift effects. All these potentials automatically fulfill the Laplace equation, the radiation condition, and the linearized free-surface condition, leading to an integral equation for the potentials (source strengths). To determine the unknown potentials, the integral equation is replaced by a set of linear equations, such that the no-penetration condition is satisfied at the collocation points of each panel.

When the ship is excited by elementary waves, it is customary for panel methods to evaluate ship responses in the frequency domain. When the ship is excited impulsively, an alternative to the solution in the frequency domain is the formulation in the time domain. Evaluation of highly oscillating integrands is then avoided; however, other difficulties related to the proper treatment of the time history of the flow (memory effect) by means of so-called convolution integrals are introduced. As the problem is linear, the superposition of both frequency as well as time domain solutions is possible to obtain the response under an arbitrary excitation, such as a natural seaway.

All Green function methods are fundamentally simplified in that they completely neglect the potential ϕ_s for steady flow. This omission can introduce significant errors in the prediction of local pressures, especially in the bow region.

RANKINE SINGULARITY METHOD

The Rankine singularity method includes the potential ϕ_s for steady flow. In addition, more complicated boundary conditions on the free surface and the hull are considered. However, the free surface surrounding the hull as well as the hull itself must be discretized by panels. In this way, all waves are accounted for. The main difficulty of this method is to avoid physically unrealistic reflections of waves at the outer (artificial) boundaries of the computational domain. A comprehensive overview of various Rankin singularity methods for seakeeping is documented by Bertram and Yasukawa (1996).

4.4.2 Linear Computations

Computations of linear seakeeping properties of a ship in elementary waves are of immense practical value because these results, in combination with statistical methods, can describe the ship's response more broadly. In practice, potential flow solvers are used almost exclusively to compute linear seakeeping properties of a ship in elementary waves. In addition to the neglect of viscosity (Euler solvers) potential flow assumes that the flow is irrotational. Assuming irrotational flow does not introduce a major loss in the physical model, because rotation is created by the water adhering to the hull, and this information is already lost in the Euler flow model. Of relevance for practical application is that potential flow solvers are so much faster than Euler and RANS solvers because only one linear differential equation needs to be solved when dealing with potential flows instead of four nonlinear coupled differential equations. Potential flow solvers are usually based on boundary element methods and therefore need only to discretize the boundaries of the fluid domain, not the entire fluid space. The effort to generate grids is considerably reduced. However, potential flow solvers require a simple continuous free surface. Flows involving breaking waves and splashes cannot be analyzed using potential flow solvers.

In certain classes of seakeeping problems, viscosity becomes significant and cannot be neglected, especially if the boundary layer periodically separates from the hull, which is the case for roll and yaw motions. This results unavoidably in nonlinear differential equations. In practice, empirical corrections are introduced. The problem also arises when the flow separates at sharp edges in the aftbody, as it does at transom sterns or rudders. Usually, a Kutta condition can enforce a smooth detachment of the flow from edges.

For many practical problems, linear theory adequately describes the wave-induced motions of and sea loads on ships. However, in severe sea states nonlinear effects become important and need to be considered to obtain reliable predictions. Linear theory considers a ship advancing at constant speed in regular waves of small amplitude and small wave steepness. Linear theory implies that wave-induced motions and loads are proportional to the wave amplitude.

Section 4.3 showed that linear theory is used to simulate an irregular (natural) sea and to obtain statistical estimates. As shown therein, the method of St. Denis and Pierson (1953) relies on two critical assumptions: first, an ergodic, Gaussian random process with zero mean describes the sea surface elevation and, second, a linear system represents the ship. The first assumption ensures that the area under the spectral density of the ship responses, i.e., the variance, completely characterizes the probability density function of the ship responses. From the probability density function for a given response, all the desired response statistics can be determined and, by multiplying the incident wave spectrum by the square of the transfer function of the response, the spectral density of any given response can be found. A transfer function specifies the amplitude and phase of the desired response of the ship subject to regular incident waves at a given frequency.

The use of the St. Denis and Pierson approach requires that the wave spectrum and the transfer functions of the ship are known. Reliable wave spectral information is critical. Usually, oceanographers are called upon to supply this information. Here, it will be assumed that necessary wave spectra are available. The development of analytical methods to determine the transfer functions started in the 1950's, neglecting viscosity and using potential flow.

RESPONSE IN A NATURAL SEAWAY

To examine the interaction between the ship and the waves so as to be able to calculate the wave-induced structural loading on the ship, it is helpful to consider the interactive process between the wave and the ship as a system, as illustrated in Fig. 4.4.3. The input to the system is the irregular and randomly varying elevation of the sea surface. We have seen

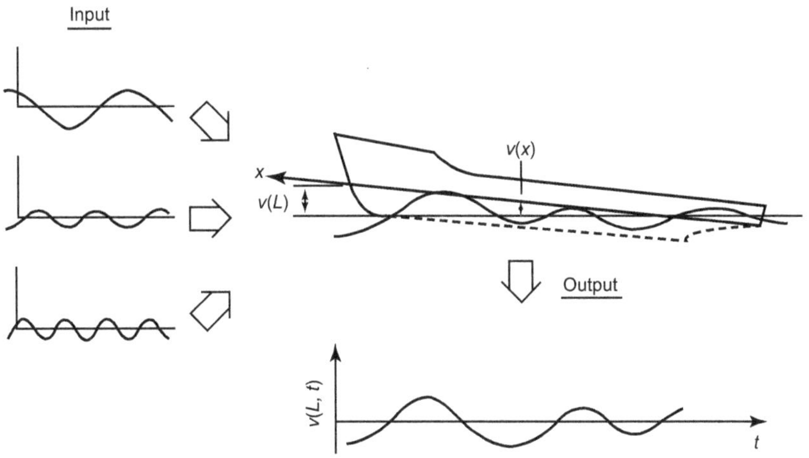

Figure 4.4.3 Wave-ship system. (Here $v(L)$ represents vertical displacement of the ship's bow.)

that the usual practice is to represent this irregular sea surface as a linear superposition of a large number of regular waves of various amplitudes and frequencies, and that the resulting combination is conveniently described in terms of a wave spectrum. The output of the system is whichever form of ship response we are interested in, such as ship motion, hull pressure distribution, and bending moment. In the preceding sections we obtained all of the necessary information about the input that, being a random process, requires the use of statistical methods for its determination.

A useful consequence of linearly predicted response is that results from regular waves of different amplitudes, wave lengths, and propagation directions can be superimposed to obtain the response in natural seaways made up of a large number of regular waves of different length and height. Furthermore, the time-varying processes can be represented in terms of a spectral density function. There are many areas of engineering, such as vibration theory and communications, that deal with linear systems where the input and the output are time-varying functions, say $X(t)$ and $Y(t)$. We now show that in all such cases the response analysis is simplified if it is performed using the spectral representation of $X(t)$ and $Y(t)$ because, if the system is linear, the two spectra are directly rclated by a *frequency response function* $H(\omega)$, commonly referred to as a *transfer function.* The transfer function depends on the characteristics of the system, and it can be determined either by mathematical analysis (providing the governing equations for the system are known) or by experiment, or by a combination of the two. To establish the basic properties of this function, let us consider the response to a sinusoidal wave input. If the input is a constant amplitude cosine wave of fixed frequency

$$x(t) = x_0 \cos \omega t \tag{4.4.13}$$

then it can easily be shown that the output also is a steady state cosine wave of constant amplitude, having the same frequency ω and a phase difference θ. That is

$$y(t) = y_0 \cos(\omega t - \theta) \tag{4.4.14}$$

Information about the amplitude ratio y_0/x_0 and the phase angle θ defines the transmission characteristics or transfer function of the system at the fixed frequency ω. Instead of thinking of amplitude ratio and phase angle as two separate quantities, it is customary to use a single complex number to represent both quantities. This is called the *(complex) frequency response function* $H(\omega)$ and is defined such that its magnitude is equal to the amplitude ratio and the ratio of its imaginary part to its real part is equal to the tangent of the phase angle. If

$$H(\omega) = A(\omega) - iB(\omega) \tag{4.4.15}$$

where $A(\omega)$ and $B(\omega)$ are real functions of ω, then

$$|H(\omega)| = \sqrt{A^2 + B^2} = \frac{y_0}{x_o} \tag{4.4.16}$$

and

$$\frac{\text{Imaginary part}}{\text{Real part}} = \frac{B}{A} = \tan\theta \tag{4.4.17}$$

Using complex exponential notation, we can now say that if the input is a harmonic wave of amplitude x_0,

$$\begin{aligned} x(t) &= x_0 \cos \omega t = x_0[\text{the real part of } e^{i\omega t}] \\ &= x_0 \operatorname{Re}\{e^{i\omega t}\}, \end{aligned} \tag{4.4.18}$$

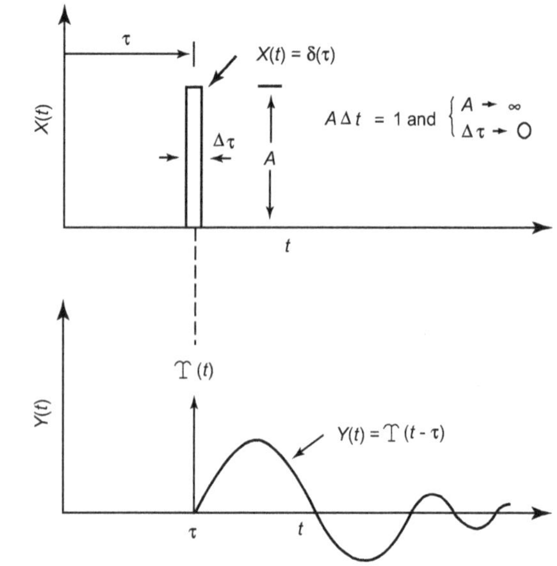

Figure 4.4.4 Response to a unit impulse.

then the corresponding harmonic output will be

$$y(t) = x_0 \operatorname{Re}\{H(\omega)e^{i\omega t}\} \tag{4.4.19}$$

We now determine the transfer function for a general input, $X(t)$. Let $\Upsilon(t-\tau)$ be the response $Y(t)$ of the system for the special case when the input is a unit impulse at time $t = \tau$; that is, $\Upsilon(t-\tau)$ is the response to $X(t) = \delta(\tau)$, as shown in Fig. 4.4.4. To obtain the response for a general input function $X(t)$, the latter is divided into a series of impulses, as shown in Fig. 4.4.5, and the response from each impulse is superimposed, yielding

$$Y(t) = \sum_i [X(\tau_i)\Delta\tau]\Upsilon(t-\tau_i) \tag{4.4.20}$$

In the limit, as $\Delta\tau \to 0$,

$$Y(t) = \int_{-\infty}^{\infty} X(\tau)\Upsilon(t-\tau)d\tau \tag{4.4.21}$$

This integral is known as the Duhamel or convolution integral. By an appropriate change of variables,

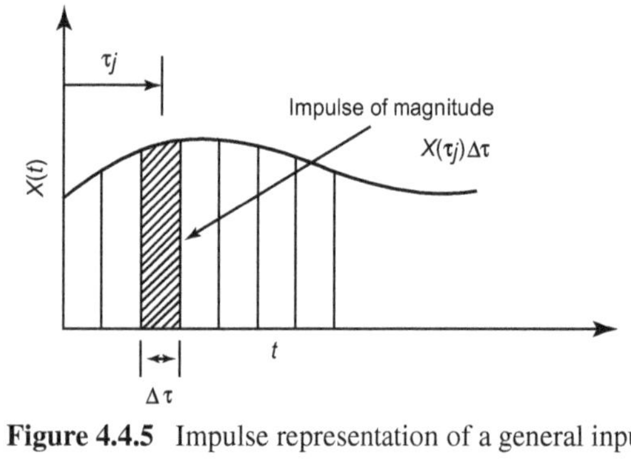

Figure 4.4.5 Impulse representation of a general input.

it may be written in an alternative form more convenient for mathematical purposes:

$$Y(t) = \int_{-\infty}^{\infty} X(t-\tau)\Upsilon(\tau)d\tau \tag{4.4.22}$$

It may be shown that the frequency response function $H(\omega)$ and the impulse response function $\Upsilon(t)$ constitute a "Fourier transform pair" and are related by

$$H(\omega) = \frac{1}{2\pi}\int_{-\infty}^{\infty} \Upsilon(t)e^{-i\omega t}\,dt \tag{4.4.23}$$

and

$$\Upsilon(t) = \frac{1}{2\pi}\int_{-\infty}^{\infty} H(\omega)e^{i\omega t}\,d\omega \tag{4.4.24}$$

We have seen that the analysis of a fluctuating process such as ocean surface elevation is simplified by using a spectral representation, that is, by describing the process in terms of its frequencies, and we are therefore led to rewrite the relationship between $X(t)$ and $Y(t)$ of (4.4.22), expressing it in terms of their spectral density functions $S_X(\omega)$ and $S_Y(\omega)$. The details of this step are available in standard texts, i.e., Crandall and Marc (1963). The result is the strikingly simple relationship

$$S_Y(\omega) = |H(\omega)|^2 S_X(\omega) \tag{4.4.25}$$

Thus, for a linear system the spectral density of the response is simply the spectral density of the input multiplied by a single scalar function of ω: the square of the amplitude of the transfer function. Once $S_Y(\omega)$ is known, the averages and expected maxima can be computed from its moments, using the formulas presented in Section 4.1.1. The quantity $|H(\omega)|^2$ is often called the *response amplitude operator* (RAO) of the system. We showed that $|H(\omega)|$ is the amplitude of the steady state response to a unit sinusoidal input of frequency ω. Thus, one way of determining the transfer function for ship motions and loads is to subject a ship model to a series of sinusoidal waves of unit amplitude and various frequencies and to measure the response amplitude for each frequency. However, this is only valid for small waves and small response amplitudes, so that the wave-ship system is linear.

Thus, if the wave-ship system is linear, the calculation of wave bending moment M_w or other wave-induced response is both simple and rapid. We here summarize the procedure to calculate the response for a given wave spectrum $S_w(\omega)$ and a given ship heading and ship speed (and, hence, a given wave encounter frequency ω_e):

1. Calculate the transfer function $H(\omega)$ by performing a ship motion analysis for the response to regular waves of unit height over a complete range of wave encounter frequency.
2. Express the wave spectrum in terms of encounter frequency and apply the transfer function to obtain the response spectrum $S_M(\omega_e)$, as in (4.4.25).
3. By calculating the appropriate moments m_0, m_2, and so on of $S_M(\omega_e)$, determine whatever response values are required, such as

bandwidth: $$\varepsilon = \sqrt{1 - \frac{m_2^2}{m_0 m_4}}$$

significant value: $$M_{w,s} = 2\sqrt{m_0} \qquad (4.4.26)$$

probable extreme value: $$\hat{M}_{w,p} = \sqrt{2\ \ln\left(\frac{1800T}{\pi}\right)\sqrt{\frac{m_2}{m_0}}}$$

Recall that T stands for the time in hours.

Use of this procedure to obtain the characteristic extreme value of bending moment is illustrated schematically in Fig. 4.4.6.

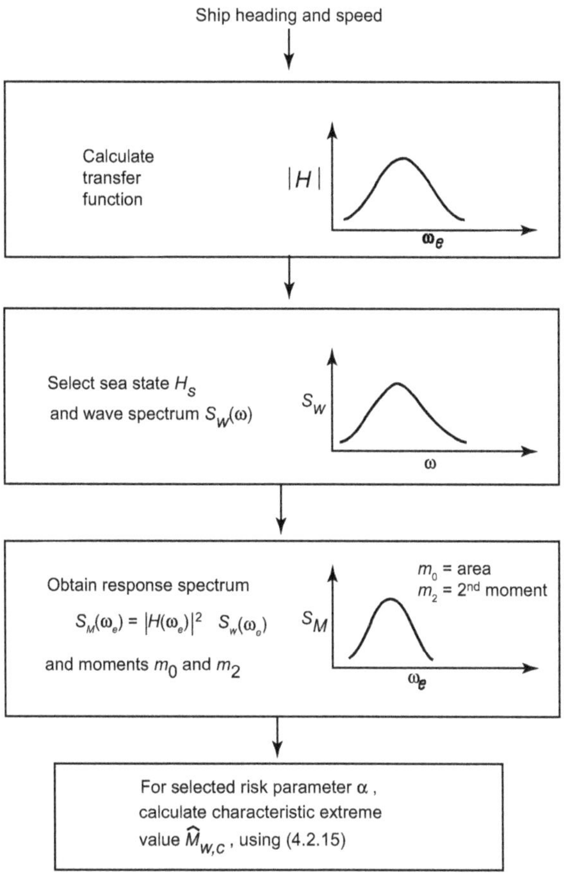

Figure 4.4.6 Determination of the characteristic extreme value of bending moment, $\hat{M}_{w,c}$, for a linear wave-ship system.

4.4.3 Linear Response in Regular Waves

Consider a ship advancing at a steady mean forward speed U in a train of regular waves of small amplitude moving in six degrees of freedom. The angle β, measured between the direction of U and the direction of wave propagation, defines the ship's heading (β = 180° for head waves). It is assumed that both the wave excitation forces and the resultant oscillatory motions are linear and harmonic, acting at the frequency of encounter, ω_e, expressed as follows:

$$\omega_e = \left|\omega_0 - \frac{\omega_0^2}{g}\, U \cos\beta\right| \qquad (4.4.27)$$

where ω_0 is the circular frequency of the incident waves, and g is the acceleration of gravity.

The coordinate system shown in Fig. 4.4.1 defines the six degree of freedom ship motions. A set of six coupled linear differential equations, with six unknowns, must be solved simultaneously to describe the ship motions. The fundamental motion equation is written as follows:

$$(M + A)\ddot{s} + B\dot{s} + Cs = F_a e^{-j\omega_e t} \qquad (4.4.28)$$

where M represents the mass matrix, A the hydrodynamic added mass matrix, B the matrix of potential damping, and C the stiffness matrix of restoring forces, with F_a defined as the complex amplitude of the exciting force or moment. In linear theory the harmonic response of the ship is proportional to the amplitude of the exciting force or moment. It oscillates at the same frequency, but with a phase shift. Solutions of (4.4.28) are the harmonic ship motions, $s = s_a(\omega_e)e^{-j\omega t}$, where $s_a(\omega_e)$ is a complex vector containing the six motion amplitudes.

The ship's mass (displacement) defines components of the mass matrix M, and hydrostatic buoyancy effects define components of the restoring force matrix C. Well-known classical ship theory can be used to define components of the added mass matrix A, the damping matrix B, and the exciting force and moment vector F. However, methods based on the use of advanced hydrodynamic theory yield more accurate results.

The velocity potential Φ is separated into a time-independent steady contribution caused by the ship's forward speed U and a time-dependent part associated with the incident wave system and the oscillating ship motions:

$$\Phi = (-Ux + \phi_s) + \phi_T e^{-j\omega_e t} \qquad (4.4.29)$$

Here $[-Ux + \phi_s]$ is the steady contribution and ϕ_T the complex amplitude of the unsteady potential. The potentials can be superimposed because the fundamental field equation (Laplace) is linear with respect to Φ. Various approximations are used for Φ and ϕ_s, and they affect computational effort and accuracy of results. The most important linear methods are strip theory methods, Green function methods, and Rankine singularity methods. These methods were discussed near the end of section 4.4.1.

For a ship advancing at constant mean forward speed in a seaway, the incident waves undergo a certain amount of scattering (diffraction), leading to a diffraction wave potential with complex amplitude ϕ_7. Induced by the incident wave, this diffraction wave potential oscillates harmonically and creates a wave field of the same frequency radiating away from the ship. Furthermore, the ship moving in the wave field itself generates body motion potentials corresponding to translational and angular motions. These body potentials and the diffraction potential are superimposed on the incident wave potential with complex amplitude ϕ_0. Accordingly, the complex amplitude of the unsteady potential is expressed as follows:

$$\phi_T = \phi_0 + \phi_7 + \sum_{j=1}^{6} \phi_j s_j \qquad (4.4.30)$$

The local potential function ϕ_j with $j = 1, 2, \ldots, 6$ depends only on the hull geometry and is, therefore, independent of the as yet unknown hull velocity. Index j stands for the six degrees of freedom ship motions: surge s_1, sway s_2, heave s_3, roll s_4, pitch s_5, and yaw s_6. Each of the above potentials must satisfy the Laplace equation in the fluid domain as well as appropriate boundary conditions.

Components of the wave excitation loads F_j are separated into the incident wave load part, F_j^I, and the diffracted wave load part, F_j^D:

$$F_j = F_j^I + F_j^D \qquad (4.4.31)$$

where $j = 1, 2, \ldots 6$ refer to the directions of the six degrees of freedom motions. To determine the incident wave part, the complex amplitude of the incident wave velocity potential ϕ_0 is substituted into the expression of the incident wave excitation load. After incorporating the frequency of encounter, an expression is found for the incident wave load, also known as the Froude-Krylov forces. To obtain the diffraction part of the wave exciting load, the complex amplitude of the diffracted wave velocity potential is substituted into the expression of the diffracted wave load. The resulting expression can be split up into a speed-independent part and a speed-corrective part. After applying a variant of Stokes theorem (Salvesen et al., 1970) to the speed-corrective part, the relationship is found for the diffraction wave loads.

Added mass and damping loads are steady-state hydrodynamic forces and moments caused by forced harmonic rigid hull motions. These motions generate outgoing waves, resulting in oscillating fluid pressures on the hull surface. Integrating these pressures over the hull surface yields forces and moments acting on the ship. Because the ship oscillates with circular frequency of encounter, added mass and damping have to be evaluated for the frequency of encounter.

THE ENCOUNTER FREQUENCY BASED PANEL METHOD

The numerical solution to the formulated boundary value problems is approached by means of the Green function G, representing a known velocity potential at field point (x, y, z) of a source at (ξ, υ, ζ) of unit strength. It is possible to show that all solutions of ϕ_j are of the form

$$\phi_j(x, y, z) = \iint_S Q_j(\xi, \upsilon, \zeta) \cdot G(x, y, z, \xi, \upsilon, \zeta)\, dS \qquad (4.4.32)$$

These are integrals over the hull surface S, with Q_j being the unknown source strengths and G the Green function. The source strengths are found by satisfying the body boundary conditions, leading to linear integral equations of the second Fredholm type for each of the source strength contributions. These equations are solved numerically by replacing them with a system of linear equations. By a discretization procedure, the wetted hull is divided into a finite number of small triangular or rectangular surface patches (panels), capable of representing a curved surface and avoiding "leakage" gaps. The source strength distribution is taken to be uniform over each panel, with the boundary condition being satisfied at the center of the panel. Subdividing into M panels replaces the integral equations by seven systems of M linear equations, corresponding to the diffraction potential and the six local potentials of the six-degree-of-freedom body motions. From these systems the desired source strengths are determined:

$$\phi_j = \sum_{\mu=1}^{M} Q_{j\mu} G_{j\mu}\, \Delta S_\mu \quad \text{for } j = 1, 2, \ldots 7 \text{ and } \mu = 1, 2, \ldots M \qquad (4.4.33)$$

After the velocity potentials are derived from the Laplace equation and the appropriate boundary conditions, the dynamic pressure follows from the linearized Bernoulli equation. Proper integration over the hull surface then yields the desired wave excitation forces and moments F_j from potentials ϕ_0 and ϕ_7 and the added mass and damping coefficients A_{jk} and B_{jk} from potentials ϕ_j. The equations of motion can be solved, resulting in the six degree of freedom ship motions. Now the total velocity potential is obtained, enabling the computation of fluid velocities, accelerations, and pressures at any desired point in the fluid domain. Integration of pressures over appropriate parts of the hull yields the wave-induced global sectional loads.

NONLINEAR CORRECTION FOR ROLL

Although motions are generally small if the ship is stable and if the incident wave amplitude is relatively small, roll resonance in beam seas is an exception and must, therefore, be treated specially. Experimental and theoretical investigations showed that roll can be handled satisfactorily by equivalent linear approximations (Himeno, 1981). An additional viscous roll damping term, B_v, is included to account for the viscous resistance to rolling. This term is assumed to comprise four effects, namely, hull friction B_F, eddy making resistance B_E, normal force damping of bilge keels B_{BKN}, and hull pressure damping of bilge keels B_{BKH}:

$$B_v = B_F + B_E + B_{BKN} + B_{BKH} \qquad (4.4.34)$$

As these effects are nonlinear with respect to the roll velocity, they cannot be introduced directly into the motion equations. Based on the principle of harmonic balance, a quasi-linear viscous roll damping coefficient, $B_{44}^v = B_v \dot{s}_4$, can be obtained which, when added to the linear (potential) roll damping coefficient B_{44}, yields the total damping coefficient.

To obtain accurate predictions for twin-hull ships, it is necessary to consider viscous effects of lift and damping on both hulls and, if the ship is equipped with stabilizing fins, lift and drag effects on these fins (Rathje and Schellin, 1997). This is because twin-hull ships generally have a relatively small waterplane area and do not generate large waves when oscillating in the vertical plane. Consequently, potential (wave making) damping is small compared to that of monohull ships, and viscous lift and drag therefore contribute significantly to the overall damping of twin-hull ships.

NONLINEAR CORRECTION FOR HULL SHAPE

One consequence of using linear methods is that predicted wave-induced vertical loads have the same magnitude in sagging as well as in hogging. However, past full-scale measurement programs (e.g., Smith, 1966) demonstrated that the resulting magnitudes of vertical bending moment in sagging and hogging are not equal, which should be the case for linear signals. The sag/hog ratios of vertical bending moment magnitudes tend to be larger for slender ships, such as modern container ships, than for fuller ships, such as tankers and bulkers.

Jensen and Pedersen (1981) performed one of the first truly nonlinear analyses of a VLCC tanker and a containership, both operating in the fully loaded condition in a moderate sea state. The hull shape of the containership was typical in that it featured a large bow flare, a large stern overhang, a relatively small block coefficient, and a relatively high operating speed. Applying their perturbational strip method (Jensen and Pedersen, 1979) for the containership, they theoretically obtained the probability distribution of the extreme values of sagging and hogging wave bending moment over a short-term (1 hour) period. Their results showed that the sagging bending moment is about 20 percent larger than the value predicted by linear theory and that the hogging bending moment is about 10 percent smaller. They also showed that complementary measurements of short-term extreme values agree favorably with theory. This clearly indicates that for a containership a nonlinear method should be used. In contrast, the VLCC results showed that for this kind of ship, at deep draft, the nonlinear effects are small, and linear theory is adequate.

For flared or high-performance vessels, nonlinear effects should always be accounted for, especially in an investigation of springing, where the item of principal interest is the ship's flexible response to the waves. Linear theory appears to be adequate for a tanker or bulker in the deep draft condition. For a tanker in the ballast condition, it is preferable to rely on nonlinear theory because the shallow draft forward introduces nonlinear effects (Guedes Soares and Schellin, 1998).

For large amplitude wave-induced ship motions that do not involve bow emergence or water on deck, Hachmann (1986) developed a practical procedure to correct linear results for the nonlinear effect caused by the nonvertical sides. Today his method is still used to obtain realistic results for many standard applications (Hachmann, 1991), including modern containerships as well as naval vessels. The linear loads can be based either on a strip theory or a panel method. The corrected loads are processed to yield global wave-induced design loads that then become part of the input for a finite element structural analysis (e.g., Payer and Fricke, 1994; Rathje and Schellin, 1997).

Table 4.4.1 Principal particulars of the tested containership

Length between perpendiculars	294.4 m
Breadth	42.8 m
Draft	12.5 m
Block coefficient	0.64

To validate Hachmann's method, the Hamburg Ship Model Basin (HSVA) carried out systematic model tests (Blume, 1999) of a containership advancing in large amplitude regular head waves. Table 4.4.1 lists principal particulars (full-scale values) of the tested ship. The ship is a modern 6700 TEU containership designed by Kvaerner Warnow Werft characterized by large bow flare and large stern overhang. The self-propelled free running model of this ship was segmented at the foreship by a sectional cut located 0.8 *L* from its aft perpendicular, where *L* is the ship's length between perpendiculars (Fig. 4.4.7). At this cut, the ship experienced a relatively large increase in vertical bending moment over the linear results.

Model tests corresponding to a ship advancing at a speed of 20 knots (Froude number 0.20) in regular head waves were conducted for two wave heights of 6.9 and 13.4 m. The wave length in both cases equaled 1.1 times the ship's length between perpendiculars. All measurements were affected by elastic, high frequency oscillations of the model. After filtering the model's elastic response, the first harmonics of measured global dynamic hull girder loads compared favorably with nonlinearly corrected GLPANEL (Papanikolaou and Schellin, 1992) predictions for all cases investigated. Figure 4.4.8 shows two representative samples of the measured midship vertical bending moment, VBM, together with the corresponding computed bending moment (Beiersdorf and Rathje, 2000). Results are presented as time histories over one wave encounter period, here denoted by *T*. At time $t = 0$, the location of the wave crest was close to the ship's center of buoyancy.

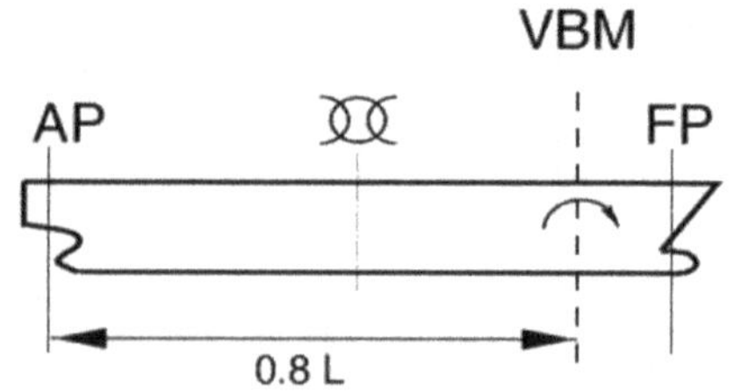

Figure 4.4.7 Schematic of model tested containership.

4.5 LOAD GENERATION FOR FINITE ELEMENT ANALYSIS

The analysis of complex ship structures often requires the inclusion of the dynamic response of the hull girder, and it may even be necessary to perform such an analysis as part of the ship motion appraisal. A dynamic analysis is generally not required for standard ships although the hull girder vibration frequencies should be determined to check against resonance with main engines, propellers, etc. If a dynamic analysis is desired (for example, to investigate slamming), it can be performed separately after computing wave-induced loads. Structural response, be it of low frequency such as wave-induced bending or of high frequency such as whipping, has no significant effect on the ship's overall motion. Therefore, from the standpoint of load specification, the ship can be considered rigid.

Today, rational dimensioning of complex ship structures is frequently based on refined finite element (FE) analyses of the entire ship, e.g., Payer and Fricke (1994), and Shi et al. (2005). This allows a realistic application of loads and an accurate analysis of stresses, even for complex ship structures. Unlike the traditional rule formula-based design, this method realistically accounts for loads experienced by the ship.

The equivalent regular wave approach represents a consistent rational procedure that employs a direct analysis for the particular hull structure

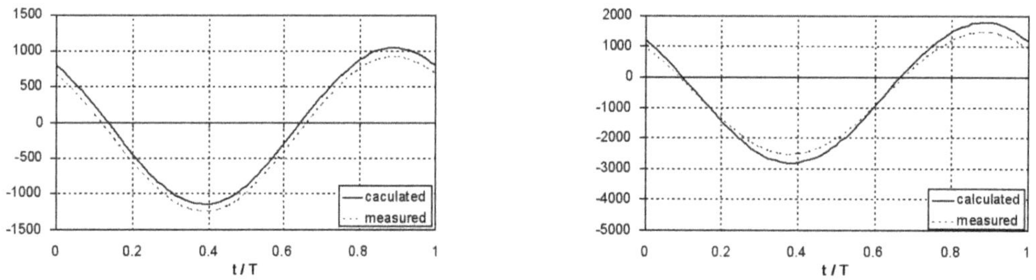

Figure 4.4.8 First harmonic of measured and computed, nonlinearly corrected dynamic midship vertical bending moment of the tested containership model in regular head waves of 6.9 m height *(left graph)* and 13.4 m height *(right graph).*

being considered. It is a practical compromise between the design loading condition approach and the physical load approach. By defining loads for the FE analysis, it mitigates modeling uncertainties that are introduced when using rule scantling formulas. Development of rule formulas necessarily relied on simplifications to readily account for applied loads, structural response, and material strength. Thus, the equivalent regular wave approach provides more reliable structural analysis results as well as improved insight of structural system behavior.

The underlying assumption is that if the ship is investigated to resist loadings caused by selected (equivalent) regular waves, it will resist all loads expected during its lifetime. Loads generated for this kind of analysis constitute extreme loads and must be based on a return period of at least 20 years. In assessing dynamic loads, it is necessary to consider a range of sea conditions and headings that cause a critical response of the structure. The resulting loads are then incorporated within an FE analysis to determine the resulting stresses experienced by the hull structure. Applied loads include static as well as dynamic loads.

When assessing hull girder strength against extreme loads that are exceeded only once during the ship's lifetime, the equivalent wave approach relies on long-term load predictions. However, the equivalent wave approach is also useful to define rule-based loads suitable for dimensioning structural components, e.g., Rörup et al. (2008). The basic difference is that rule-based loads instead of long-term load predictions are taken into account to determine amplitudes of equivalent regular waves. Major classification societies provide guidelines and/or software tools suitable for the generation of rule-based global loads. These rule-based loads are nominal (design) loads used to determine minimum scantlings of structural components. The resulting stresses must always be less than the material's permissible stress.

If all possible wave situations are to be systematically analyzed, extensive computations are required. The selection to find the most relevant load cases from the large number of possible wave situations can be reduced by choosing so-called dominant load parameters (DLPs). Specified by the classification society to expedite the analysis, DLPs represent critical wave conditions. They are based on previous experience with similar ships and include hull girder loads that cause maximum stresses in structural components and/or large deformations of the hull structure. In any case, for containerships the following two loading conditions are always investigated:

1. The ship at its maximum displacement, with a distribution of containers that subject the ship to the maximum allowable vertical still-water bending moment in hogging.
2. The ship at its maximum displacement, with a distribution of containers that subject the ship to the maximum allowable vertical still-water bending moment in sagging or to the minimum allowable still-water bending moment in hogging.

For tankers, the following five loading conditions, typically found in the loading manual, generally need to be considered (Liu at al., 1992):

1. Homogeneous full load condition at design draft
2. Normal ballast load condition at light draft
3. Partial load conditions, 33 percent full load
4. Partial load conditions, 50 percent full load
5. Partial load conditions, 67 percent full load

However, the Common Structural Rules for Bulk Carriers (IACS, 2006a) and the Common Structural Rules for Double Hull Oil Tankers (IACS, 2006b) specify only the first two of these loading conditions.

For fast ships, special effects of slamming need to be considered to generate global loads because motions of fast ships in high waves may be so large that the ship's forefoot and propeller are exposed. This occurs most frequently at high speed in head waves although it also happens in other conditions. Reentry of the keel after emergence may result in slamming as the ship's bottom strikes the water surface. Therefore, for fast ships class rules generally dictate that special effects of slamming need to considered for global loads. Specifically, this means that the procedure to determine amplitudes of equivalent regular waves has to account for an additional slamming-induced bending moment in sagging, e.g., Schellin and Perez de Lucas (2004).

4.5.1 Prediction of Wave-induced Loads

The physically most realistic numerical methods to predict wave-induced loads directly solve the Reynolds-averaged Navier-Stokes (RANS) equations in the time domain. By relying on the interface-capturing technique of the volume-of-fluid (VOF) type, for example, this technique accounts for highly nonlinear wave effects in that it computes the two-phase flow of water and air to describe the physics associated with complex free-surface shapes with breaking waves and air trapping, e.g., Schellin and el Moctar (2007). Such simulations need to be carried out for all wave situations that might occur during the operating life of the ship. In addition, results obtained from the FE analysis have to be post-processed. In practice, these are prohibitively time consuming and expensive tasks. Furthermore, such physically realistic loads do not automatically represent reliable loads for design, because they model only physical effects of simulated wave conditions and not loads that subject the ship to experience-based design bending moments, shear forces, or torsional moments.

In contrast to a physically rigorous load approach, magnitudes of wave-induced loads based on the equivalent regular wave approach mainly depend on experience-based design loading conditions and not on capturing all physical effects by computational fluid dynamics (CFD) simulations. Moreover, increased accuracy of these results is relatively unimportant, because loads finally applied to the FE model are calibrated in accordance with extreme or rule-based loads.

Extensive design experience with frequency domain codes exists. Therefore, thoroughly validated linear codes suffice. Major nonlinear effects can be accounted for. One nonlinear effect is roll resonance in beam seas. As already discussed in Section 4.4.2, realistic results at resonance conditions in roll can be obtained by equivalent linear approximations. Specifically, this consists of including an additional viscous roll damping term in the motion equations to account for viscous resistance to rolling.

An important consequence of using linear seakeeping methods is that predicted wave-induced vertical shear forces and bending moments have the same magnitude in hogging as well as sagging. Especially for modern containerships that are often characterized by extreme bow flare and strong stern overhang, buoyancy forces on the ship in a position different from still-water equilibrium vary nonlinearly with displacement from the still-water position. Hence, amplitudes of hogging and sagging global midship bending moments differ, a result not predicted by linear theory. There is evidence that for moderately large motions that do not involve water on deck, the nonlinearity of the ship's response due to variation of the ship's cross section caused by nonvertical sides explains most of the differences between hogging and sagging stresses.

To compensate for errors of linear theory caused by truncating the pressure distribution at the still-water level, linear seakeeping models can be made to incorporate a nonlinear correction to account for the hydrodynamic pressure between still-water level and wave contour. Hachmann (1991) developed such a correction method that, when implemented within a linear model, allows not only the assessment of extreme wave-induced load effects, but also the routine generation of rule-based global loads for strength analyses of ships using finite element techniques.

The method starts with linear computations in the frequency domain to obtain ship motions and dynamic pressures acting on the ship's hull. Hachmann developed a revised pressure formula based on hull bound steady perturbation flow to calculate the hydrodynamic pressure. Hachmann (1986) also introduced a procedure to define the wetted profile along the ship's sides and to extrapolate the pressure distribution up to this profile or to limit the resulting pressures to zero. The imbalance of forces that generally results when integrating pressures is corrected for by small changes of the ship's accelerations. In contrast to linear theory techniques, this pressure integration yields realistically reduced hogging and increased sagging moments.

Hachmann's revised pressure formula accounts for the steady perturbation of the forward speed flow, which is generally neglected by common linear methods. For small-amplitude waves, his hull-bound perturbation flow is a concept identical to common strip theory models, such as the one of Salvesen et al. (1970), with the pressure p given by

$$p = \rho\left[\left(-\frac{\partial}{\partial t} + U\frac{\partial}{\partial x}\right)\phi - g\zeta\right] \tag{4.5.1}$$

where U is the ship's constant forward speed, ϕ is the velocity potential, ζ is the vertical ship displacement at the location in question, t is time, x is the shipbound length coordinate, ρ is density of water, and g is acceleration of gravity. Hachmann extended the unsteady part of the velocity potential to approximately account for the violation of the hull surface condition associated with the forward speed flow variation potential. His revised formula includes an additional term to read

$$p = \rho\left[\left(-\frac{\partial}{\partial t} + U\frac{\partial}{\partial l}\right)\phi - g\zeta + W_s\zeta_r^*\right] \tag{4.5.2}$$

where l stands for the hull-bound stream function coordinate, ζ_r^* is the ship's unsteady velocity relative to the fluid at the location in question, and W_s is the velocity field of the steady perturbation flow which, for slender bodies, can be approximated by its transverse components. For slender bodies, $\partial l \approx \partial x$. Detailed analysis of the flow field's gradient is avoided, and relevant pressure terms are exposed to enable efficient numerical evaluation. Figure 4.5.1 shows a schematic diagram of the pressure distribution on a cross section of a ship's hull positioned in a wave crest. Coordinate r proceeds along a tangent to the ship's still waterline (into the paper in Fig. 4.5.1), and coordinate s, orthogonal to r, extends from the still-water level positive upward along the side of the ship. The origin of the r,s coordinate system is located at the intersection of the still-water level and the considered ship section. Angle α, measured between the vertical and the ship's side, designates the flare of the ship section.

To obtain the wetted height along the ship's side and the corresponding pressure extrapolation above and pressure reduction below the still waterline, Hachmann's method defines a time harmonic velocity potential of the flow field above the still waterline. In regular waves, the wetted height ζ at every station along the ship's length is assumed to vary harmonically with the encounter frequency, ω. Applying Bernoulli's equation together with the kinematic boundary condition at the ship's hull yields hydrodynamic pressure p_s as a function of position s, with s measured tangent to the ship's side:

$$p_s = p_{WL} - \rho(g \cos \alpha - \omega^2\zeta)s \qquad (4.5.3)$$

where p_{WL} denotes hydrodynamic pressure at still-water level obtained from Hachmann's revised pressure formula. This pressure formula does not account for hydrodynamic effects caused by the displacement of the ship's hull relative to the still-water level. Therefore, these effects need to be added to obtain the total pressure.

The gradient of the hydrodynamic pressure at still-water level

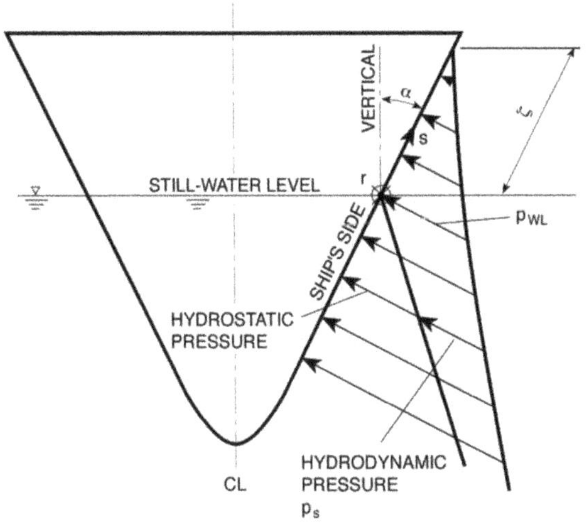

Figure 4.5.1 Pressure distribution on ship cross section.

$$\frac{dp_s}{ds} = -\rho(g \cos \alpha - \omega^2\zeta) \qquad (4.5.4)$$

shows the influence of the Smith effect. At the wave crest ($\zeta > 0$), the pressure gradient is reduced; at the wave trough ($\zeta < 0$), it is increased.

Above still-water level ($s > 0$), the pressure is always specified according to (4.5.3); below still-water level ($s < 0$), however, the calculated hydrodynamic pressure $p_s(s)$ becomes unrealistic when its absolute value exceeds the hydrostatic pressure. If this is the case, the hydrodynamic pressure is set equal to the negative value of the hydrostatic pressure, causing the superposed total pressure to vanish.

Along the wave contour (at $s = \zeta$) the pressure $p_s(s)$ is zero. Applying this condition to (4.5.3) yields the following expression for the wetted height along the side of the ship:

$$\zeta = \frac{g \cos \alpha}{2\omega^2}\left[1 - \sqrt{1 - \frac{p_s(0)}{\rho}\left(\frac{2\omega}{g \cos \alpha}\right)^2}\right] \qquad (4.5.5)$$

This wetted height attains its maximum when the hydrodynamic pressure equals $\rho(g \cos \alpha/2\omega)^2$, resulting in a crest 'amplitude' of $\zeta_{max} = g \cos \alpha/2\omega^2$. Half a wave period later, the dynamic pressure becomes $-\rho(g \cos \alpha/2\omega)^2$, resulting in a trough 'amplitude' of $\zeta_{min} = 0.414\ \zeta_{max}$. Accordingly, the maximum wetted height that can be obtained with this procedure equals $\zeta_{max} - \zeta_{min} = 1.414\ \zeta_{max}$. For a deepwater wave not influenced by the presence of a ship, assuming that the wave breaks against a vertical wall ($\cos \alpha = 1$), the wetted height equals the wave height, and the ratio of maximum wave height to wave length turns out to be about 1/9. The theoretical limit of this ratio is around 1/7, indicating that wave elevations predicted with this method are within realistic bounds.

If the position s along the ship's side exceeds the wetted height ζ, the hydrodynamic pressure is zero. Furthermore, if the hydrodynamic pressure at the still-water level exceeds the theoretical limit for a breaking wave, i.e., if $p_s(0) = \rho(g \cos \alpha/2\omega)^2$, the hydrodynamic pressure is obtained by substituting ζ_{max} for ζ into (4.5.3). The resulting pressure corrections are schematically illustrated in Fig. 4.5.2, depicting the ship cross section in the wave crest and in the wave trough. For the wave crest case, the nonlinear correction is seen adjusting the total pressure at mean water level to zero at the wave contour. For the wave trough case, it is seen that total (adjusted) pressure is zero at the wave contour when the hydrodynamic pressure is added to the hydrostatic pressure.

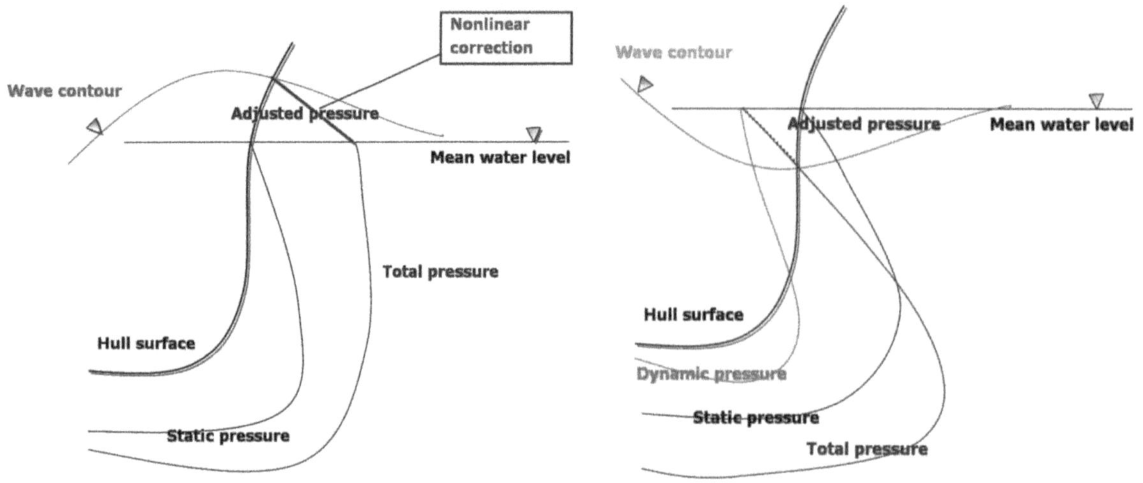

Figure 4.5.2 Schematic presentation of pressure correction for a ship cross section in a wave crest *(left)* and in a wave trough *(right).*

4.5.2 Extreme Global Loads

Extreme loads are used to investigate ultimate strength, at both hull girder level and individual structural member level. Accurate assessment of ultimate strength is important not only for the initial design, but also for the operation, maintenance, repair, and modification of the structure. Of course, rule-based design comes first because calculating ultimate strength requires known scantlings. Ultimate strength assessment is basically a post-design safety check, and values must exceed the extreme load effects by some specified margin.

As computational power increases, it is becoming feasible to perform direct analyses of the hull structure by combining hydrodynamic computations with structural FE modeling. This modeling is well suited for an elastic longitudinal global strength assessment of ship structures. In general, however, this kind of analysis is suitable only for an elastic stress analysis and not for a collapse analysis.

Long-term predictions are called for to assess hull girder strength against extreme loads that are exceeded only once during the ship's lifetime. These extreme loads, generally based on a probability of exceedance of about 10^{-8}, correspond to the Recommendation No. 34 of IACS (2001) for a return period of at least 20 years. This recommendation generally serves as an accepted standard of wave statistics to predict long-term (extreme) loads for operation in unrestricted waters over the service life of the ship. It is based on wave statistics for the North Atlantic and documents wave data as a so-called scatter diagram shown in Table 4.5.1. The wave climate is modeled as an ergodic succession of short-term stationary sea states,

Table 4.5.1 IACS recommended wave climate for the North Atlantic

Hs/Tz	1.5	2.5	3.5	4.5	5.5	6.5	7.5	8.5	9.5	10.5	11.5	12.5	13.5	14.5	15.5	16.5	17.5	18.5	SUM
0.5	0.0	0.0	1.3	133.7	865.6	1186.0	634.2	186.3	36.9	5.6	0.7	0.1	0.0	0.0	0.0	0.0	0.0	0.0	3050
1.5	0.0	0.0	0.0	29.3	986.0	4976.0	7738.0	5569.7	2375.7	703.5	160.7	30.5	5.1	0.8	0.1	0.0	0.0	0.0	22575
2.5	0.0	0.0	0.0	2.2	197.5	2158.8	6230.0	7449.5	4860.4	2066.0	644.5	160.2	33.7	6.3	1.1	0.2	0.0	0.0	23810
3.5	0.0	0.0	0.0	0.2	34.9	695.5	3226.5	5675.0	5099.1	2838.0	1114.1	337.7	84.3	18.2	3.5	0.6	0.1	0.0	19128
4.5	0.0	0.0	0.0	0.0	6.0	196.1	1354.3	3288.5	3857.5	2685.5	1275.2	455.1	130.9	31.9	6.9	1.3	0.2	0.0	13289
5.5	0.0	0.0	0.0	0.0	1.0	51.0	498.4	1602.9	2372.7	2008.3	1126.0	463.6	150.9	41.0	9.7	2.1	0.4	0.1	8328
6.5	0.0	0.0	0.0	0.0	0.2	12.6	167.0	690.3	1257.9	1268.6	825.9	386.8	140.8	42.2	10.9	2.5	0.5	0.1	4806
7.5	0.0	0.0	0.0	0.0	0.0	3.0	52.1	270.1	594.4	703.2	524.9	276.7	111.7	36.7	10.2	2.5	0.6	0.1	2586
8.5	0.0	0.0	0.0	0.0	0.0	0.7	15.4	97.9	255.9	350.6	296.9	174.6	77.6	27.7	8.4	2.2	0.5	0.1	1309
9.5	0.0	0.0	0.0	0.0	0.0	0.2	4.3	33.2	101.9	159.9	152.2	99.2	48.3	18.7	6.1	1.7	0.4	0.1	626
10.5	0.0	0.0	0.0	0.0	0.0	0.0	1.2	10.7	37.9	67.5	71.7	51.5	27.3	11.4	4.0	1.2	0.3	0.1	285
11.5	0.0	0.0	0.0	0.0	0.0	0.0	0.3	3.3	13.3	26.6	31.4	24.7	14.2	6.4	2.4	0.7	0.2	0.1	124
12.5	0.0	0.0	0.0	0.0	0.0	0.0	0.1	1.0	4.4	10.0	12.8	11.0	6.8	3.3	1.3	0.4	0.1	0.0	51
13.5	0.0	0.0	0.0	0.0	0.0	0.0	0.0	0.3	1.4	3.5	5.0	4.6	3.1	1.6	0.7	0.2	0.1	0.0	21
14.5	0.0	0.0	0.0	0.0	0.0	0.0	0.0	0.1	0.4	1.2	1.8	1.8	1.3	0.7	0.3	0.1	0.0	0.0	8
15.5	0.0	0.0	0.0	0.0	0.0	0.0	0.0	0.0	0.1	0.4	0.6	0.7	0.5	0.3	0.1	0.1	0.0	0.0	3
16.5	0.0	0.0	0.0	0.0	0.0	0.0	0.0	0.0	0.0	0.1	0.2	0.2	0.2	0.1	0.1	0.0	0.0	0.0	1
SUM	0	0	1	165	2091	9280	19922	24879	20870	12898	6245	2479	837	247	66	16	3	1	100000

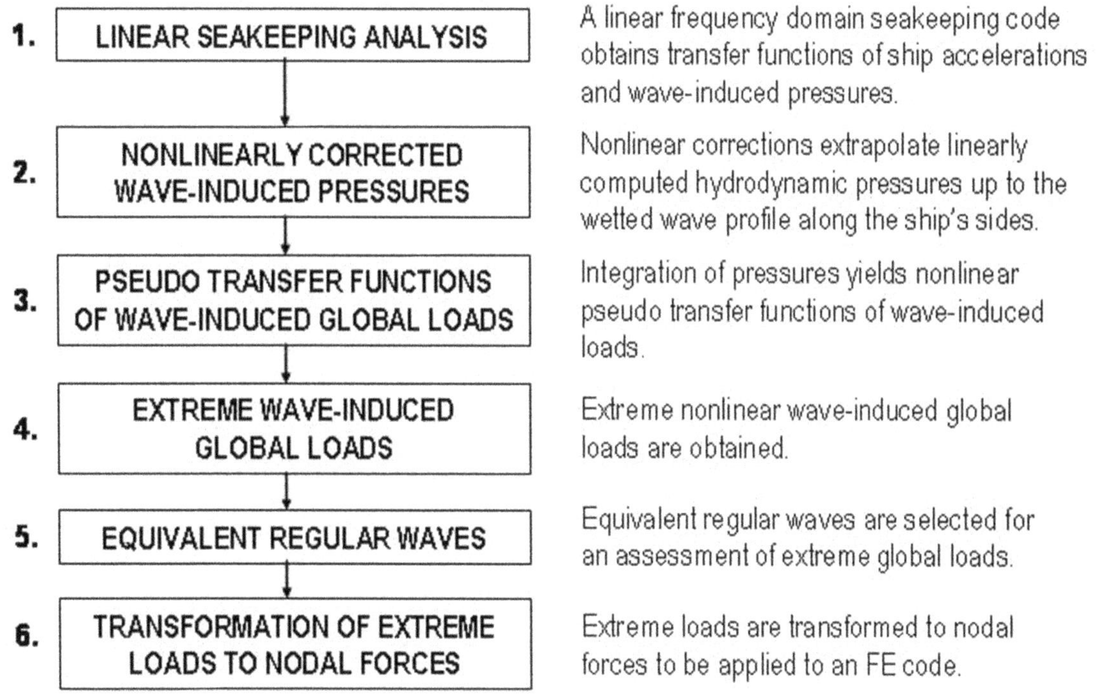
1. LINEAR SEAKEEPING ANALYSIS
A linear frequency domain seakeeping code obtains transfer functions of ship accelerations and wave-induced pressures.
2. NONLINEARLY CORRECTED WAVE-INDUCED PRESSURES
Nonlinear corrections extrapolate linearly computed hydrodynamic pressures up to the wetted wave profile along the ship's sides.
3. PSEUDO TRANSFER FUNCTIONS OF WAVE-INDUCED GLOBAL LOADS
Integration of pressures yields nonlinear pseudo transfer functions of wave-induced loads.
4. EXTREME WAVE-INDUCED GLOBAL LOADS
Extreme nonlinear wave-induced global loads are obtained.
5. EQUIVALENT REGULAR WAVES
Equivalent regular waves are selected for an assessment of extreme global loads.
6. TRANSFORMATION OF EXTREME LOADS TO NODAL FORCES
Extreme loads are transformed to nodal forces to be applied to an FE code.

Figure 4.5.3 Flow chart to determine extreme global loads.

where each short-term sea state is characterized by the two-parameter Pierson-Moskowitz seaway spectrum with a significant wave height (H_S) and an average zero up-crossing period (T_Z).

Computations start with the use of a linear frequency domain code to obtain wave-induced pressure distributions below the calm waterline. These codes are either based on strip theory, or they rely on the so-called panel method. Strip theory codes are usually sufficiently accurate; however, for ships with large bow flare and stern overhang, where three-dimensional effects become significant, panel codes often replace strip theory codes. When panel codes are used, they almost always account for forward speed effects under the so-called encounter frequency approach, where boundary conditions used to solve for pressures acting on the ship's hull are obtained with the Green function evaluated for the ship advancing at zero forward speed. This saves a huge amount of computational effort, which is the reason why this approach is widely used in panel method codes, e.g., Rathje et al. (2000). The next steps consist of nonlinear correction of hydrodynamic pressures, integration of pressures to obtain global loads, specifying extreme loads, selecting the corresponding equivalent regular waves, and transforming the extreme loads to nodal forces for an FE code. The flow chart in Fig. 4.5.3 presents the major steps:

1. A linear frequency domain seakeeping computer code analyzes ship motions in regular unit amplitude waves of different lengths and directions to obtain transfer functions of rigid body ship accelerations and wave-induced pressures acting on the ship advancing at constant forward speed.
2. Based on these transfer functions, nonlinear corrections are performed. Hachmann's method is employed to modify and extrapolate the linearly computed hydrodynamic pressures to the wetted wave profile along the ship's sides. This is done for a number of different wave amplitudes extending up to the wave amplitude that will not be exceeded.
3. Integration of these wave-induced pressures yields wave-induced loads. The sum of wave loads and inertial forces is generally not in balance. Therefore, the linear equations are resolved to retain equilibrium. Repeating these load computations for different amplitudes and frequencies and dividing the loads by the corresponding wave amplitude results in nonlinearly corrected pseudo transfer functions of wave-induced global loads, corresponding to each of the different wave amplitudes. This correction artificially transforms a linear load response into two responses, namely, a response of decreased amplitude for hogging and increased amplitude for sagging. Linear theory is valid only for infinitesimally small amplitude waves. Consequently, linear theory yields small amplitude motions about the ship's floating position in still water, resulting in wave loads of equal amplitude in hogging and sagging, an effect that is realistic only if the ship is wall-sided.
4. Long-term nonlinear wave-induced global loads are obtained. Their formulation is based

on the formulation of linear wave-induced global loads, recognizing that the wave surface elevation is a nonstationary Gaussian stochastic process that can be discretized in a sequence of periods of time over which the process is stationary. This stochastic analysis yields cumulative distributions of long-term wave-induced extreme loads.

5. Equivalent regular waves are selected for an assessment of extreme global loads. Where appropriate, still water loads are first added to long-term wave-induced loads. Amplitudes, wave lengths, and wave crest positions of the selected equivalent regular waves are then determined from their corresponding pseudo transfer functions.

6. Extreme loads are transformed into nodal forces that can be applied to any standard FE code.

4.5.2.1 Extreme Wave-Induced Load Formulation

For the linear case, extreme wave-induced loads can be efficiently evaluated applying linear potential flow hydrodynamic procedures in the frequency domain together with spectral analysis and a weighted summation of short-term Rayleigh distributions of maxima. However, at least for ships with small block coefficient, such as containerships, wave-induced loads are highly nonlinear. The asymmetry of global wave-induced loads, especially the vertical bending moment, is the visible nonlinear characteristic of this response. In these cases, the linear procedure cannot be applied, and the extreme wave loads must be based on load predictions that account for nonlinearites together with appropriate extreme vale distributions.

Several approaches have been proposed, and presently it is not clear which is the best. Several procedures, also discussed in Section 4.2.2, are based on the assumption that the linear model is a good identifier of conditions under which extreme wave loads occur. Recall that Aadegeest et al. (1998), for example, presented their extreme regular wave method, whereby a regular wave is first determined from linear long-term predictions and then a nonlinear simulation is performed for that particular wave to determine structural loads. Baarholm and Moan (2002) employed the so-called coefficients of contribution method to investigate the nonlinear vertical bending moment on a containership in the Northern North Sea by using linear long-term results to identify those sea states of the scatter diagram that contribute most to the probability of exceedance of the structural loads occurring during the ship's lifetime. They then applied a nonlinear simulation program to a selected small number of sea states only.

Also recall the critical wave episodes method of Torhaug et el. (1998) mentioned in Section 4.2.2, which is adequate for the most advanced and time consuming hydrodynamic codes. By performing a linear time domain analysis for each relevant stationary sea state, this method identifies the random incident wave sequences that result in the extreme ship response. The wave sequences resulting in the largest linear responses are then applied to the nonlinear analysis. In the most likely response method of Pastoor et al. (2003), a linear frequency analyses first determines the most likely extreme response. Then, using the theory of Gaussian processes near a maximum, the corresponding deterministic wave elevation is produced and applied to the nonlinear simulation. Another technique is the so-called contour line approach. Based on a long-term analysis of the wave climate, this method defines a set of sea states that include the most severe environmental conditions corresponding to a specific return period. These sea states lie on an enclosed contour of the scatter diagram, and the extreme response is the most probable extreme value determined within all short-term sea states on the contour line.

Schellin et al. (1996) and Guedes Soares and Schellin (1998) generalized the linear long-term prediction procedure to account for the nonlinear asymmetry of the vertical bending moment by using form functions that transform the amplitude of linear transfer functions to nonlinear transfer functions associated with sea states of different intensity. Based on this approach, Guedes Soares et al. (2004) and Schellin and Perez de Lucas (2004) took advantage of a strip theory based time domain seakeeping code to obtain the extreme wave-induced vertical bending moment for a fast monohull operating in the North Sea. First, they obtained transfer functions for a range of wave amplitudes, and then they directly computed response spectra with these transfer functions for all sea states in the scatter diagram.

The procedure applied here to compute extreme wave-induced loads is similar as it is also based on the use of pseudo transfer functions. For the ship in a regular wave with a wave height equal to the sea state's significant wave height, linearly computed transfer functions are replaced by pseudo transfer functions, nonlinearly corrected according to Hachmann's method (Section 4.5.1) to account for the hydrodynamic pressure between still-water level and wave contour. Amplitudes of pseudo transfer functions for hogging and sagging, which are now different, are then determined separately by dividing the corresponding peak response values by wave amplitude. This procedure is somewhat inconsistent from a strictly mathematical standpoint of nonlinear systems. This is because the response, although periodic, is no longer harmonic. However, for practical purposes it has been shown to be more than adequate, e.g., Rathje et al. (2000).

For strength assessment of the hull structure, it is necessary to estimate the extreme values of wave-induced loads during a ship's average working lifetime. In computing the long-term response it is necessary to decide on a probability level for the extreme response. For example, if a lifetime of 20 years is assumed for a ship and an average wave period is given as 6 s, the long-term probability of exceedance (once in 20 years) is found to be $6/(20 \times 365 \times 24 \times 60 \times 60) = 10^{-8}$.

Then, all short-term response spectra for the given wave spectra and the pseudo transfer functions must be computed. Each wave spectrum, $S_w(\omega)$, is characterized in terms of the significant wave height, H_s, and the average period between successive crests, T_z, as listed, for example, in Table 4.5.1. Each pseudo transfer function, $H(\omega, H_s)$, is valid only for the particular sea state's significant wave height. Then, for each of these wave spectra a short-term response spectrum, $S_R(\omega, H_s)$, is obtained from the input wave spectrum, $S_w(\omega)$, and the pseudo transfer function, $H(\omega, H_s)$:

$$S_R(\omega, H_s) = S_w(\omega, H_s, T_z) \cdot |H(\omega, H_s)|^2 \quad (4.5.6)$$

The variance of the response, $\sigma_{\tilde{Q}}^2$, defines its statistical properties and is obtained by integrating its spectrum as follows:

$$\sigma_{\tilde{Q}}^2 = \int_0^\infty S_R(\omega, H_s)d\omega \quad (4.5.7)$$

Since the wave height is assumed to be Rayleigh distributed, so is the response amplitude, $\tilde{Q}$. The probability density function of $\tilde{Q}$ is

$$p_{\tilde{Q}}(\tilde{Q}, \sigma_{\tilde{Q}}) = \frac{\tilde{Q}}{\sigma_{\tilde{Q}}^2}\exp\left(-\frac{\tilde{Q}^2}{2\sigma_{\tilde{Q}}^2}\right) \quad (4.5.8)$$

and the short-term probability of exceeding the value $\tilde{Q}$ is

$$Q_S(Q > \tilde{Q}|\sigma_{\tilde{Q}}) = \exp\left(-\frac{\tilde{Q}^2}{2\sigma_{\tilde{Q}}^2}\right) \quad (4.5.9)$$

Thus, the probability of a value Q exceeding a certain short-term response amplitude $\tilde{Q}$ depends only on the variance of the response, $\sigma_{\tilde{Q}}^2$.

At a random time during its lifetime, the ship experiences a sea state that causes a response of variance $\sigma_{\tilde{Q}}^2$, which itself is a random variable. Thus, the probability of exceeding amplitude $\tilde{Q}$ in the long-term is obtained in multiplying the long-term probability of exceeding the value $\tilde{Q}$ by the probability of occurrence of each sea state variance and integrating over all possible variances:

$$Q_L(Q > \tilde{Q}|\sigma_{\tilde{Q}}) = \int_0^\infty Q_S(Q > \tilde{Q}|\sigma_{\tilde{Q}}) \cdot [p_{\tilde{Q}}(\tilde{Q}, \sigma_{\tilde{Q}})]_L d\sigma_{\tilde{Q}} \quad (4.5.10)$$

where $[p_{\tilde{Q}}(\tilde{Q}, \sigma_{\tilde{Q}})]_L$ is the general form of the complete long-term probability density function of peak values of response for the ship's lifetime. It is a weighted sum of the various short-term probability density functions, each of which is for a particular combination of specified conditions and carries weighting factors to account for the relative frequency of the particular combination. The expression is as follows:

$$[p_{\tilde{Q}}(\tilde{Q})]_L = \frac{\sum_i\sum_j\sum_k\sum_l\sum_m \bar{n}f_if_jf_kf_lf_m[p_{\tilde{Q}}(\tilde{Q}, \sigma_{\tilde{Q}})]_{ijklm}}{\sum_i\sum_j\sum_k\sum_l\sum_m \bar{n}f_if_jf_kf_lf_m} \quad (4.5.11)$$

where $[p_{\tilde{Q}}(\tilde{Q}, \sigma_{\tilde{Q}})]_{ijklm}$ is the probability density function for the peak values of the short-term response, conditionally dependent on certain relevant physical parameters; $\bar{n} = (1/2\pi)\sqrt{m_2/m_0}$ is the average number of responses per second for the short-term response; m_0 and m_2 are moments of the short-term response spectrum; and f_i, f_j, f_k, f_l, and f_m are the weighting factors for these physical parameters, corresponding to severity of the sea state (from Table 4.5.1), wave spectrum shape, wave heading (usually taken to be uniform), ship speed, and the ship's loading condition, respectively. Other variations of this comprehensive formulation can also be found in the literature. The total number of responses expected over the lifetime of the ship is then

$$n_L = \left(\sum_i\sum_j\sum_k\sum_l\sum_m \bar{n}f_if_jf_kf_lf_m\right) \times T \times (60)^2 \quad (4.5.12)$$

where T is the total exposure time to the sea, in hours.

This evaluation of extreme cumulative distributions of wave-induced loads over a long period of time is an application of the so-called Lifetime Weighted Sea method (Hughes 1988). In this method the total lifetime response history of the ship may be thought of as a series of short-term response episodes, whereby all of the short-term responses are combined by a procedure that takes into account the relative amount of exposure to the various levels of sea severity.

4.5.2.2 Demonstrative Example

Let us demonstrate the generation of global extreme loads with an example. The ship we consider is a typical modern containership with principal particulars listed in Table 4.5.2. We chose this particular ship because of its large flare at the bow and strong overhang at the stern. We wish to generate these

Table 4.5.2 Principal particulars

Length at design waterline	321.0 m
Rule length	316.9 m
Molded breadth	45.6 m
Draft	15.0 m
Depth	27.2 m
Displacement	147088 t
Block coefficient	0.64
Metacentric height	2.93 m
Service speed	26.9 kn

loads as part of the input for an FE strength analysis of the hull structure. We assume the ship's lifetime to be 20 years, which in waves with an average period of 6 s corresponds to a long-term probability of exceedance of 10^{-8}. The finite element model of the ship's structure, shown in Fig. 4.5.4, idealizes all major structural components.

STEP 1—SHIP RESPONSE IN REGULAR WAVES

A panel seakeeping code computes the ship's motions, accelerations, and associated wave-induced pressures to unit amplitude regular waves of different frequencies and headings. Wave frequencies correspond to ratios of wave length to ship length ranging from 0.2 to about 5.6. Wave headings range from head seas (β = 180 deg) to following seas (β = 0 deg) at 30 deg increments. Results, when plotted against ratios of wave length to ship length, yield transfer functions. Seakeeping computations are based on discretizing the hull surface of the ship by a total of 4460 small quadrilateral and triangular surface panels; of these, 2922 panels idealize the wetted hull up to the design waterline.

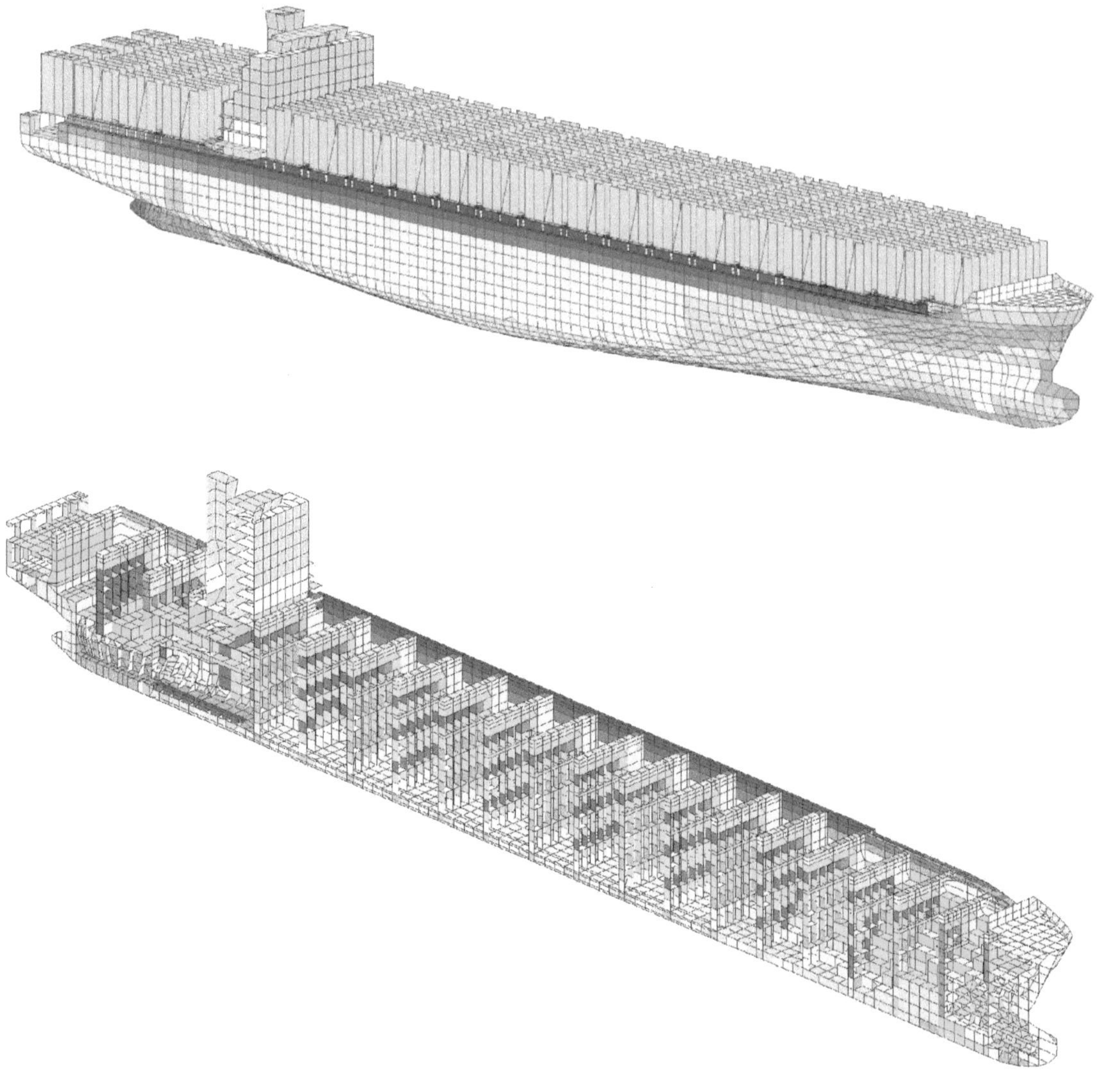

Figure 4.5.4 Global FE model of the ship.

STEP 2—NONLINEAR CORRECTIONS TO WAVE-INDUCED PRESSURES

The ship's wave-induced pressures and inertial loads caused by resonant rolling are realistically accounted for by increasing the roll damping in the calculation of roll response, based on the equivalent linear approximations documented by Himeno (1981). The resulting maximum roll amplitudes are found to be between 9 and 15 deg to both port and starboard. Hachmann's method is applied to obtain nonlinearly corrected wave-induced pressures up to the wave contour, accounting for the altered wetted surface caused by the ship's changed position in finite amplitude waves. Amplitudes of the regular waves for which these nonlinear corrections are performed extend up to 8.5 m in 0.5 m increments. For each wave, 50 different equidistant wave phases are considered to assess critical global loads for maximum values.

A sample of the computed pressure distributions on the hull in a 7.0 m amplitude regular head wave with a wave length of 264 m (0.83 of the ship's rule length) is shown in Fig. 4.5.5. Darker (blue) panels designate high pressure zones; lighter (green and yellow) panels, low pressure zones, viewed from below. Wave pressures as well as wave contours along the ship's side are seen to correspond to wave phases that represent hogging and sagging conditions. In the sagging condition, breaking of wave causes unsteady pressure transitions under the flared bow.

STEP 3—PSEUDO TRANSFER FUNCTIONS

Integration of external pressures yields wave-induced loads. The imbalance of forces that generally results when summing the integrated pressures and inertial forces is corrected for by small changes of the ship's accelerations, obtained by resolving the linear motion equations. With all forces in balance, this results in nonlinearly corrected load responses relevant for the wave amplitude and frequency under consideration. Repeating these load computations for all wave amplitudes, frequencies, and headings and dividing these loads by the corresponding wave amplitude yields nonlinearly corrected pseudo transfer functions of global hull girder loads. As these pseudo transfer functions depend on wave amplitude, they no longer satisfy an important criterion of linear dynamics, which stipulates that the output signal amplitude (ship response) at any given frequency is linearly proportional to the input signal amplitude (wave amplitude). Of course, these pseudo transfer functions are valid only for one or, from a practical point of view, a certain limited range of wave amplitudes.

This correction artificially transforms a sinusoidal load response into two responses, namely, a response of decreased amplitude for hogging and increased amplitude for sagging. Of course, only half of the cycles are of interest, positive ones in hogging and negative ones in sagging since, ultimately, the interest is stresses in deck and bottom deck structures, respectively. This procedure is acceptable for long-term predictions because the probability distribution functions consider only peaks of the amplitudes. Thus, linear wave-induced global loads are transformed into two signals of differing amplitude, and these signals are used as a basis to construct two long-term distribution functions of peak values.

Samples of (linear) transfer functions together with pseudo transfer functions are shown in Fig. 4.5.6 as absolute values of vertical bending moment (VBM) at the ship's midship section for four different wave headings, here for 7.0 m amplitude waves. As expected, absolute values of nonlinearly corrected bending moments in hogging are less than and, in sagging, greater than linear bending moments. These results are typical for other wave headings as well.

STEP 4—CUMULATIVE GLOBAL WAVE-INDUCED LOADS

Long-term computations are performed for cumulative distributions of global wave-induced loads

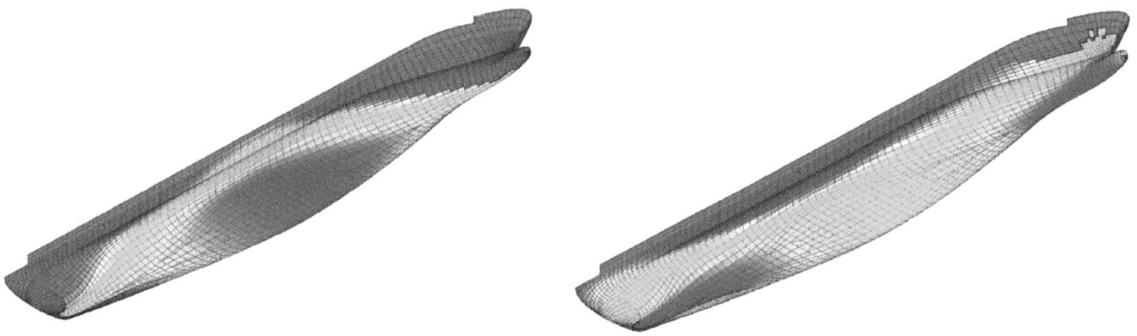

Figure 4.5.5 Wave pressure distribution on hull in regular head waves for hogging *(left)* and sagging *(right)*.

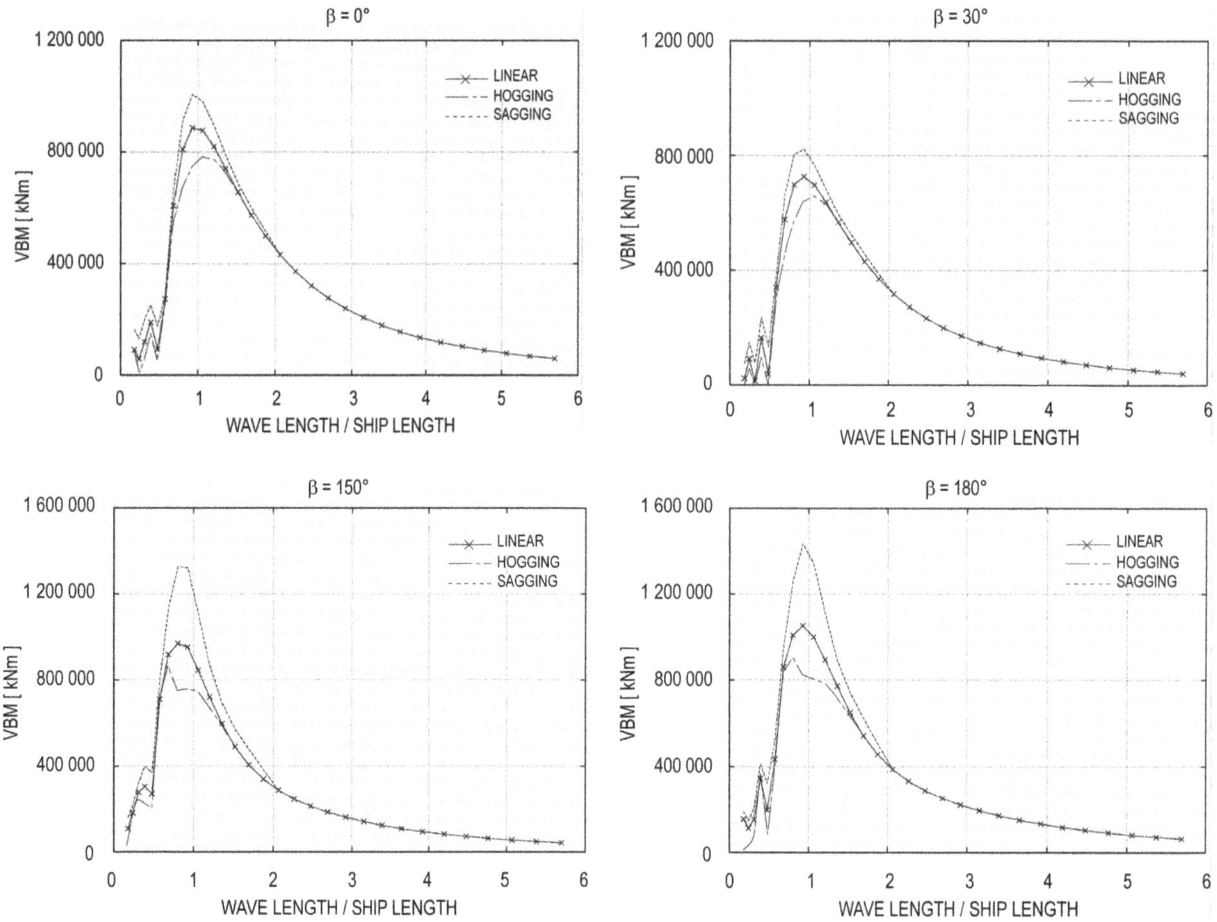

Figure 4.5.6 Transfer functions (linear) and pseudo transfer functions of wave-induced midship vertical bending moment in regular 7.0 m amplitude stern waves ($\beta = 0°$), quartering waves ($\beta = 30°$), bow waves ($\beta = 150°$), and head waves ($\beta = 180°$).

according to the method described above in Section 4.5.2.1. The severity of the sea states considered and their probability of occurrence are taken from Table 4.5.1, where each short-term sea state is characterized by the two-parameter Pierson-Moskowitz seaway spectrum with a significant wave height, H_S, and an average zero up-crossing period, T_Z. To account for the short-crestedness of each seaways, the short-term response spectra are multiplied by the cosine squared spreading function (4.3.37). The range of wave headings is divided into equal 30 deg intervals. To account for speed reduction in waves, the ship is analyzed while advancing at constant two-thirds service speed. The ship's loading condition corresponds to its maximum displacement, with a distribution of containers that subject the hull girder to the maximum allowable vertical still-water bending moment in hogging. Samples of the resulting cumulative distributions of extreme loads acting at the ship's midship section, plotted as functions of the probability level, are shown in Fig. 4.5.7 for wave-induced shear forces and in Fig. 4.5.8 for wave-induced bending and torsional moments, respectively. Distributions based on nonlinearly corrected values are significantly different than distributions based on linear values.

Of course, cumulative distributions are obtained for the other extreme load components as well, and this is done also at ship stations other than amidships. These distributions, when used to determine extreme loads acting at different stations, generally result in smooth envelope curves of shear forces, bending moments, and torsional moments over the ship's length, approximating the loading resulting from systematically varying the ship's position in waves. For the probability level of 10^{-8}, Figs. 4.5.9 and 4.5.10 depict envelopes of the resulting wave-induced shear forces and wave-induced bending moments, respectively.

Modeling extreme loads is useful to assess the elastic longitudinal ultimate global strength of the hull girder. Therefore, extreme values must exceed rule-based values. Recall that rule-based values are nominal loads used to specify ship scantlings. Thus, a direct comparison with rule-based values is inappropriate.

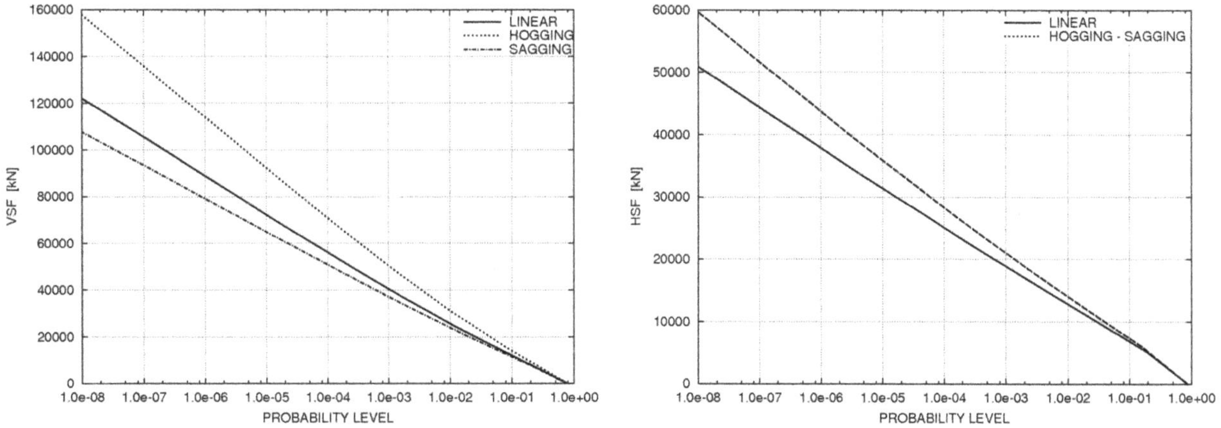

Figure 4.5.7 Absolute values of cumulative distributions of three-quarter-length values of vertical shear force *(left)* and horizontal shear force *(right)*.

STEP 5—EQUIVALENT REGULAR EXTREME WAVES

Bending moments acting on the hull reflect gross effects of waves, and the midship bending moment is one of the most important global parameters. Therefore, let us illustrate the identification of extreme regular waves by focusing on maximum vertical hull girder bending moments in hogging and sagging. Large amplitude waves with lengths close to ship length will typically cause extreme response. This is because in hogging, the fore and aft ends of the ship are located simultaneously in wave troughs with the midship located in the wave crest and, in sagging, the fore and aft end of the ship are located simultaneously in wave crests with midship located in the wave trough. Theses waves are characterized by wave amplitude, wave length, wave heading, and wave phase (wave crest position referenced to amidships). The wave amplitude of such an equivalent extreme wave is determined by dividing the long-term value of the load under consideration by the value of its pseudo transfer function occurring at the wave frequency and wave heading corresponding to the maximum amplitude of the pseudo transfer function. The wave length is obtained from the functional relationship to the wave frequency; the wave crest position, from the wave phase.

Let us demonstrate how to determine the amplitude of the equivalent regular extreme waves. We

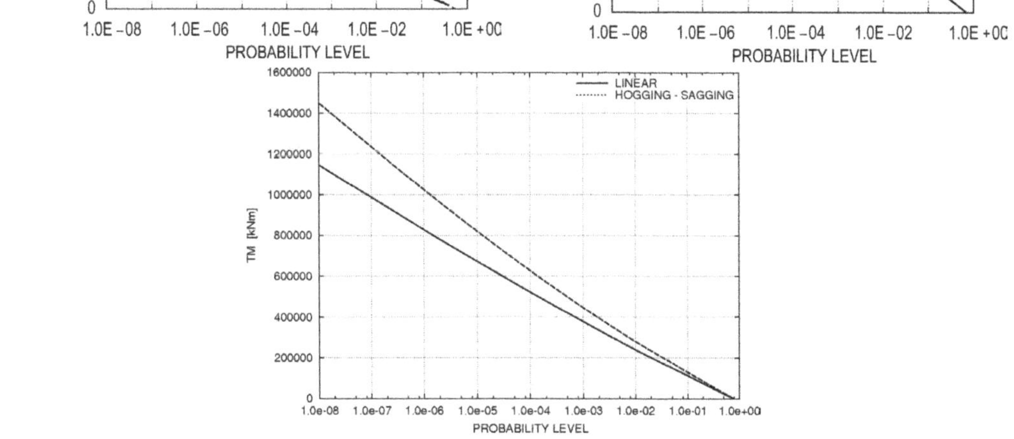

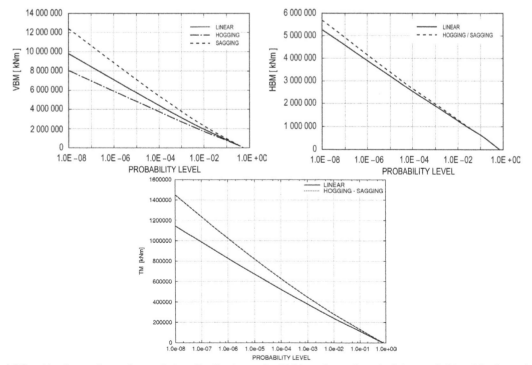

Figure 4.5.8 Absolute values of cumulative distributions of midship values of vertical *(upper left)* and horizontal bending moment *(upper right)* and three-quarter-length values of torsional moment *(lower)*.

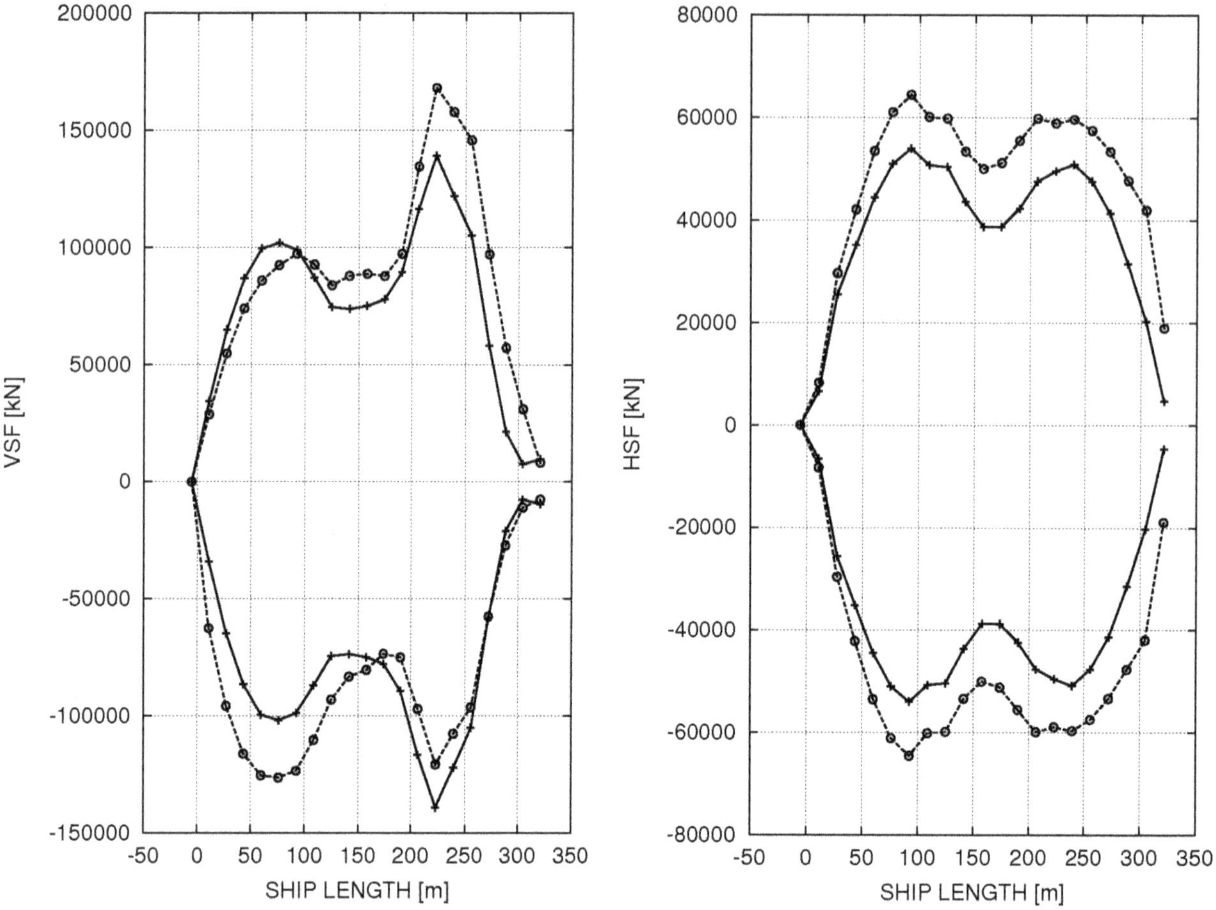

Figure 4.5.9 Nonlinearly (-----) and linearly (——) computed wave-induced extreme values of vertical *(left)* and horizontal *(right)* shear force.

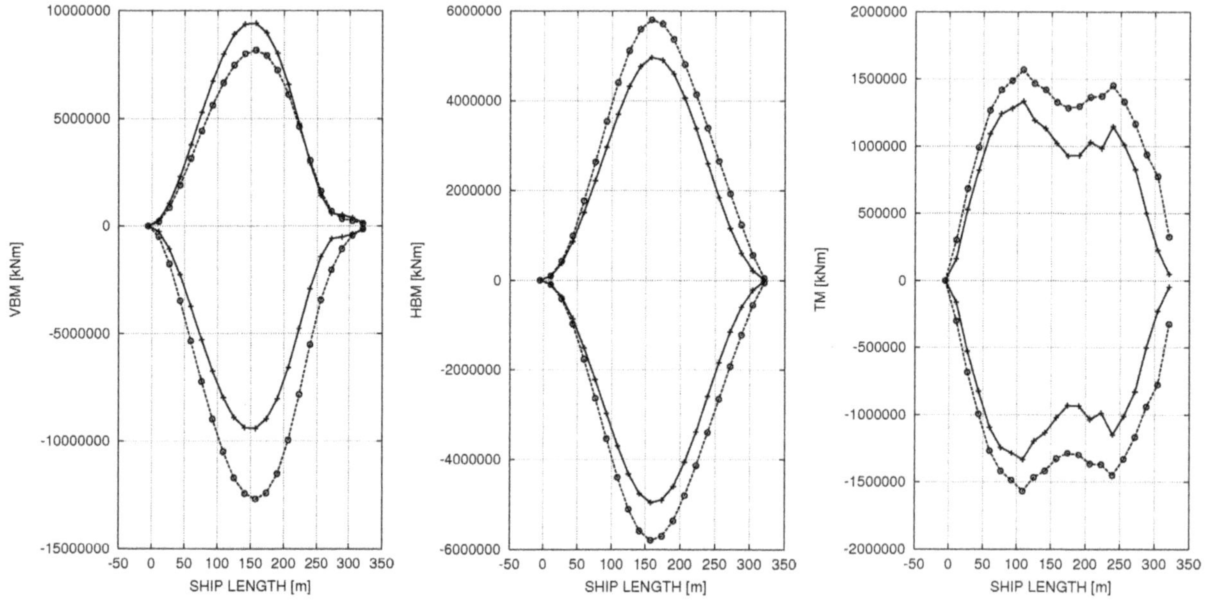

Figure 4.5.10 Nonlinearly (-----) and linearly (——) computed wave-induced extreme values of vertical bending moment *(left)*, horizontal bending moment *(center)*, and torsional moment *(right)*.

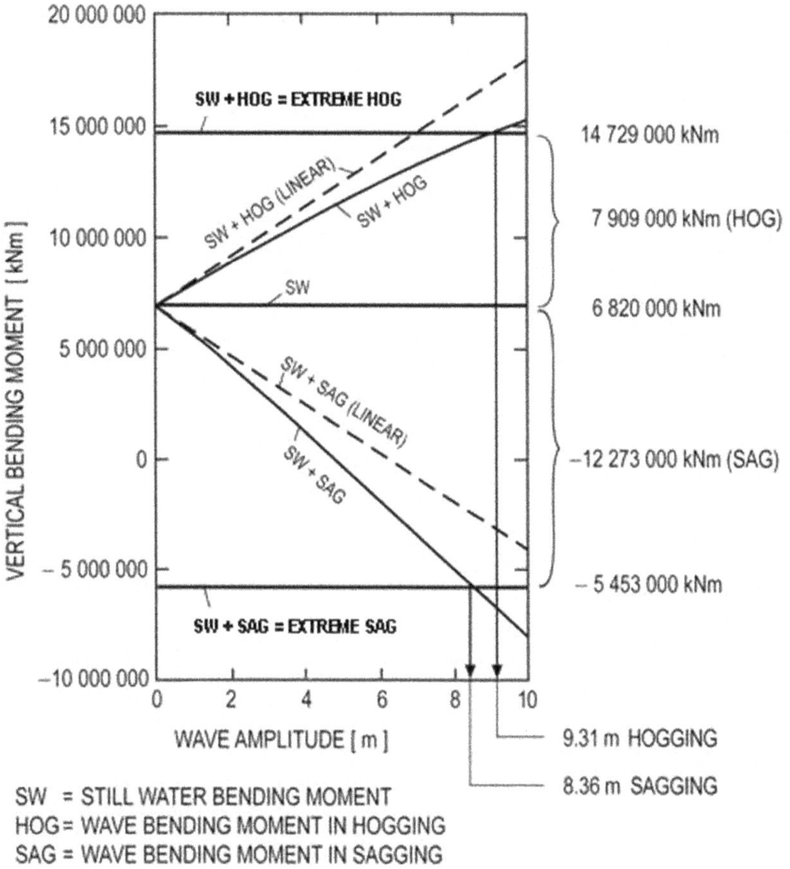

Figure 4.5.11 Determination of equivalent regular extreme wave amplitudes.

want to obtain amplitudes for the hogging condition as well as for the sagging condition. Figure 4.5.11 shows the sum of still-water and wave-induced midship vertical bending moments as functions of wave amplitude for the hogging and sagging conditions. Positive values stand for bending moments in hogging; negative values, for bending moments in sagging. Linearly computed bending moments are seen to be straight (dashed) lines. Nonlinearly corrected values deviate from linear values. In accordance with their pseudo transfer functions (Fig. 4.5.6), absolute values are smaller in hogging and larger in sagging. For the loading condition under consideration, the midship still-water bending moment has a value of 6,820,000 kNm. The horizontal line in the middle of Fig. 4.5.11 depicts this still-water bending moment. From the cumulative distributions in Fig. 4.5.8 (upper left graph), here for the probability level of 10^{-8}, the extreme wave-induced vertical midship bending moments are seen to be 7,909,000 kNm for the hogging condition and –12,273,000 kNm for the sagging condition. By adding the still-water bending moment to the long-term wave bending moments, we obtain the extreme values of total midship vertical bending moments in hogging of 14,729,000 kNm and in sagging of –5,453,000 kNm. As seen in Fig. 4.5.11, the corresponding amplitudes of the two equivalent regular extreme waves turn out to be 9.31m for hogging and 8.36m for sagging condition.

STEP 6—TRANSFORMATION OF EXTREME LOADS TO NODAL FORCES

The external pressures caused by the extreme wave-induced loads are added to inertial loads and transformed to the FE model as nodal forces. Nodal loads are given preference over surface loads because, for a global strength analysis, nodal loads yield sufficiently accurate results and their application is straightforward.

4.5.3 Global Wave-Induced Loads According to Class Rules

Reliable computation of loads is crucial for an accurate global FE analysis of a ship. Classification society rules (e.g., Germanischer Lloyd, 2006) require the ship to withstand global loads that subject the hull girder to given (rule-based) shear forces, bending moments, and torsional loads. Accordingly, major classification societies publish guidelines that are specially suited for a structural analysis of ships. Generally, these guidelines are based on the equivalent regular wave approach to obtain load combinations relevant for dimensioning the ship structure. The guideline Global Strength Analysis for Container Vessels of Germanischer Lloyd (2007), for example, uses this approach. In contrast to the loading approaches in the Common Structural Rules for Bulk Carriers (IACS, 2006a) and Common Structural Rules for Double Hull Oil Tankers (IACS, 2006b), ship accelerations as well as wave-induced pressures are obtained from first principle hydrodynamic computations of the ship's behavior in regular waves.

The interactive software package GL ShipLoad (Rörup et al., 2008) was developed as an aid to assess the global structural integrity of containerships. Using the equivalent regular wave approach, it constitutes the standard tool to generate rule-based loads for a global FE analysis of containerships. The global FE model of the ship's structure serves as input, and nodal forces that can be applied to that same FE model are its output. Performing a structural analysis on this basis comprises several tasks (Eisen and Cabos, 2007), schematically presented by the flow chart in Fig. 4.5.12.

1. An FE mesh is generated to idealize all major structural components, such as decks, transverse and longitudinal bulkheads, walls, floors, web frames, and shell plating.
2. Global loading conditions are selected from load cases specified in the ship's stability booklet. One selected load case subjects the ship to the maximum allowable vertical still-water bending moment in hogging. Another load case subjects the ship to the maximum allowable vertical still-water bending moment in sagging or to the minimum allowable still-water bending moment in hogging. The distribution of containers for these two load cases causes the ship to float at its maximum displacement.
3. To facilitate convenient access and reuse for different loading conditions, components of the ship's basic masses are typically grouped into assembled mass items made up of reusable mass components. These grouped masses are added to the FE model.
4. A large number of sea states, characterized by different wave heights, wave lengths, and wave headings are investigated systematically for a realistic representation of wave-induced loads. Containerships, having a high deck opening ratio, may need special consideration because load conditions in oblique seas often are decisive from a structural strength point of view. In such cases, it is not enough to separately analyze vertical, horizontal, and torsional hull girder loads; such effects have to be combined in a phase correct manner to achieve realistic design loads. A strip theory-based code solves the linear problem of a ship advancing at constant speed in waves, considering a sufficiently wide range of wave frequencies and wave headings. Viscous roll damping is added and hydrodynamic pressures

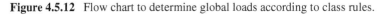

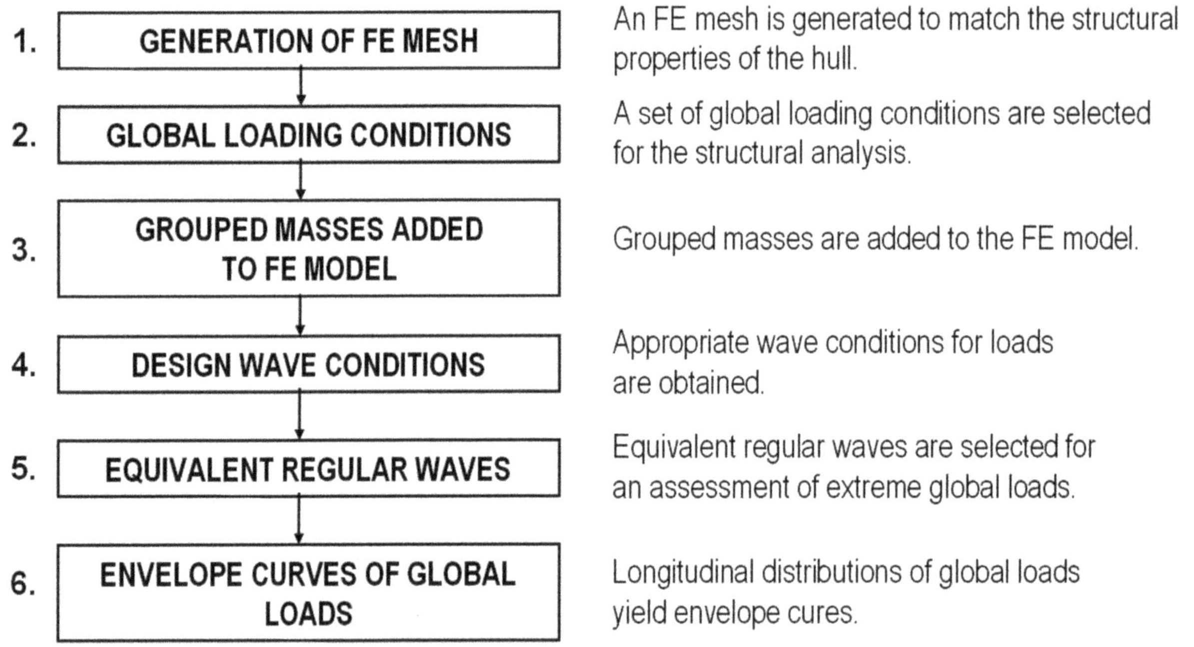
1.
GENERATION OF FE MESH
An FE mesh is generated to match the structural properties of the hull.
2.
GLOBAL LOADING CONDITIONS
A set of global loading conditions are selected for the structural analysis.
3.
GROUPED MASSES ADDED TO FE MODEL
Grouped masses are added to the FE model.
4.
DESIGN WAVE CONDITIONS
Appropriate wave conditions for loads are obtained.
5.
EQUIVALENT REGULAR WAVES
Equivalent regular waves are selected for an assessment of extreme global loads.
6.
ENVELOPE CURVES OF GLOBAL LOADS
Longitudinal distributions of global loads yield envelope cures.

Figure 4.5.12 Flow chart to determine global loads according to class rules.

in finite amplitude waves are corrected according to Hachmann (1986). The resulting pressures for rule-based equivalent regular (design) waves are generally specified for a lower probability level of about 10^{-6}, which is less than the probability level of 10^{-8} for extreme loads. These resulting pressures are integrated to obtain nonlinear pseudo transfer functions of wave-induced global loads. Generally, these computations are extensive. Therefore, to expedite computations, one or more dominant load parameters (DLPs) are usually specified by the classification society to represent design wave loading conditions.

5. From a large number of sea states, a smaller number of regular equivalent design waves are selected which subject the hull girder to the required design loads. For head and stern wave cases, the selected wave heights subject the hull girder to rule-based bending moments. For other wave headings, wave length, wave phase, and roll angle are systematically varied, such that the resulting equivalent regular design waves subject the hull girder to the other global design loads, such as shear forces torsional moments. For each wave, some 50 different equidistant wave phases are considered to assess critical loads for maximum values.
6. The longitudinal distributions of global loads result in envelope curves.

The resulting pressures and inertial forces corresponding to the design load cases are transferred as nodal forces to the FE model, ready as input for the structural analysis of hull components.

Performing most of these tasks is state of the art. Codes based on various numerical methods exist that can obtain wave-induced loads and perform a global FE strength analysis, and preprocessors are commercially available to assist in setting up the required meshes. However, modeling cargo loads efficiently is not addressed by standard tools, because it is specific to the subject ship design. Even if software components were available for all the above tasks, performing a structural analysis based on computer generated loads remains complex and time consuming. Codes from various vendors need to be interfaced and executed in a coordinated manner, and experts from different departments need to cooperate, which can be an organizational challenge.

Software tool GL ShipLoad was developed to address these problems. It integrates all algorithms necessary to assess and combine the wave-induced pressures and the ship's structure to generate appropriate external loads for the FE model. In this way it facilitates the application of ship and cargo masses as well as external, wave-induced loads to the FE model. Structural loads come from the acceleration of masses (inertial loads) and from external (wave-induced) pressures. The program provides support in modeling the mass distribution of the ship and its cargo, in computing hydrostatic and hydrodynamic pressures from waves, and in combining both kinds of loads to obtain balanced quasi-static load cases.

Read-in files are processed for the FE model and the hull description, write-out files are prepared for the results (loads), storage files are created for user input data, and files are set up for communicating with external programs (e.g., NASTRAN).

Data of nodes, elements, and materials used in any standard FE program are loaded from the FE model into a binary file format. The resulting nodal loads consist of inertial loads that result from the acceleration of the mass distribution and from static and dynamic wave-induced pressures. These nodal loads are either appended to the FE model file or directly identified as output files of nodal loads.

In addition to detailed geometric information contained in the FE model, principal ship particulars have to be entered, mainly to evaluate the prescribed global loads, such as rule-based vertical bending moments. It is also possible to specify a frame table, allowing the longitudinal positions to be addressed by frame number rather than by a length coordinate.

Normally, hatch covers are not explicitly modeled by finite elements, because hatch covers must not contribute to the overall stiffness of the model. For containerships, however, appropriate features of hatch covers need to be entered as input, inasmuch as deck container loads are applied at hatch covers.

Elements representing the hull need to be specified to compute the vessel's trim and to transform computed hydrostatic pressures into nodal loads. This can be done automatically by specifying the height up to which the FE model represents the watertight hull.

To generate the appropriate external loads for the FE model, all algorithms necessary to assess and combine wave-induced pressures and the ship's structure are implemented. A graphical user interface is provided to control the load generation process in a time- and cost-saving manner. The screen shot in Fig. 4.5.13 shows a sample view of this interface (Cabos et al., 2006), consisting of user input windows that can be opened by clicking on separated tree items in the work space. The interface is subdivided into areas "tree" (on the left), "output" (at the bottom), and "workspace." User input windows are opened in the workspace by clicking on the tree items. Additional symbols for actions that are specific to the active window appear on the tool bar. Some windows have a "preview" for graphical feedback of the user input in the white input fields. The section of the FE model in Fig. 4.5.13 shows

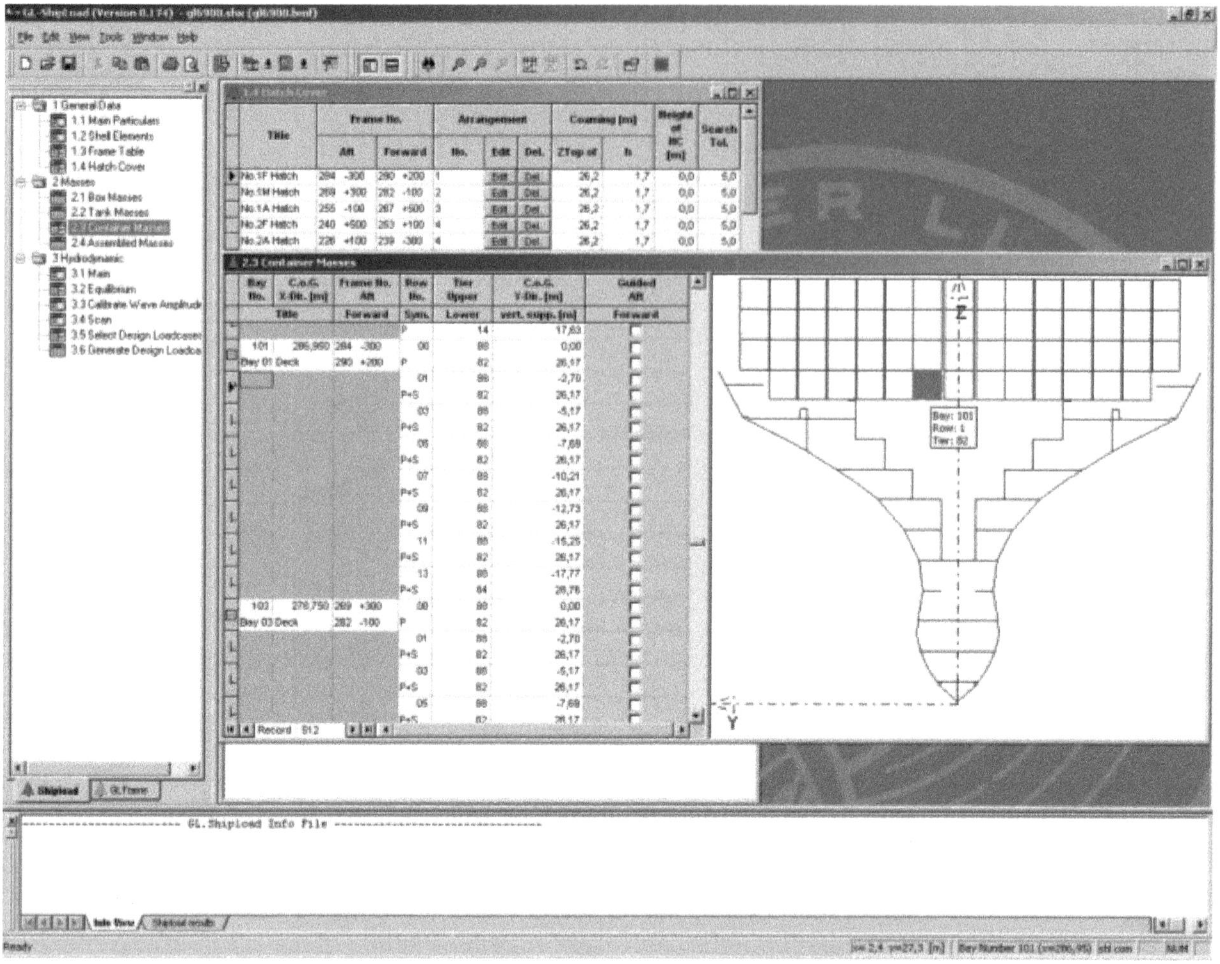

Figure 4.5.13 Graphical user interface showing user input windows.

the currently examined container bay. At the mouse position the tool tip displays bay, row, and tier number of the container.

Tree items are arranged such that, by proceeding from top to bottom, the user is guided through all required steps, beginning with input of the principal dimensions and ending with generation of FE loads. A progress bar monitors the workflow of lengthier computations, and a persistent log traces the program run, displaying information, warning, and error messages.

4.5.3.1 Load Groups

In GL ShipLoad all loads for each load case comprise a linear combination of load groups. This leads to an efficient storage of loads for many different wave conditions. Loads on the hull structure result from acceleration of masses (inertial loads) and from external (hydrostatic and hydrodynamic) pressures of the surrounding water. Loads applied to the FE model for each load case are sorted into the following load groups:

1. Hydrostatic buoyancy loads
2. Static weight loads
3. Static tank loads
4. Six inertial unit load groups, resulting from three translational and three rotational rigid body accelerations of all masses except tanks
5. Six inertial unit load groups, resulting from three translational and three rotational rigid body accelerations of tanks (Tanks are grouped separately because the fluid distribution inside the tanks depends on the ship's floating position.)
6. One hydrodynamic load group for each selected wave pressure distribution

Any load case applied to the FE model is a combination of these load groups. Combining the first three load groups yields balanced hydrostatic load cases. To obtain balanced hydrodynamic load cases, factors for unit load groups are computed, based on the condition that no residual forces and moments remain when combining hydrostatic and hydrodynamic load groups. These factors then represent rigid body accelerations. For any chosen load case, longitudinal distributions of sectional forces and moments are immediately displayed.

4.5.3.2 Mass Distribution

Components of the ship's basic masses are grouped into assembled mass items made up of reusable mass components, such as steel weight of the hull, equipment, and accommodations (lightship weight), fuel oil, fresh water, and other consumables (bunkering), water ballast, and cargo. While some basic mass components differ for each loading condition, such as bunkering masses at departure and arrival, other basic mass components remain the same for each loading condition, such as lightship weight. For a containership, Fig. 4.5.14 schematically shows typical masses grouped into assembled mass items.

So-called mass matrices represent basic as well as assembled mass items. A mass matrix assigns nodal loads to nodal accelerations, which are derived from computed rigid body accelerations. Translational accelerations are directly applied to all nodes; rotational accelerations are converted to translational accelerations.

Lightship weight of the hull is obtained by applying a material density to the finite elements. It is common practice to scale element masses to account for structural components not included in the model, such as brackets. To match a specified center of gravity position for the hull structure weight, different material densities can be used for individual element groups. The remainder of lightship weight (machinery, hatch covers, and outfitting) and consumables are represented by a distribution of nodal masses in relevant regions according to their locations and centers of gravity. The mass of each weight group is adjusted to achieve the correct mass distribution and the position of the center of gravity. Use of negative nodal masses is not acceptable. The entire mass model must comply with the considered lightship weight distribution. Water ballast and tanks are represented by distributing nodal masses to the surrounding structure.

ELEMENT GROUP MASSES

Elements of the FE model are divided into element groups. Each element of the FE model belongs to one element group, and all elements of a specific element group are of the same element type, namely, truss, beam, plain stress, shell, or boundary element. Normally, elements in an element group share some common properties if, for instance, they represent the same structural member, such as the main deck. Structural masses, therefore, are represented by element group mass items. They are computed from the element geometry and the associated material density.

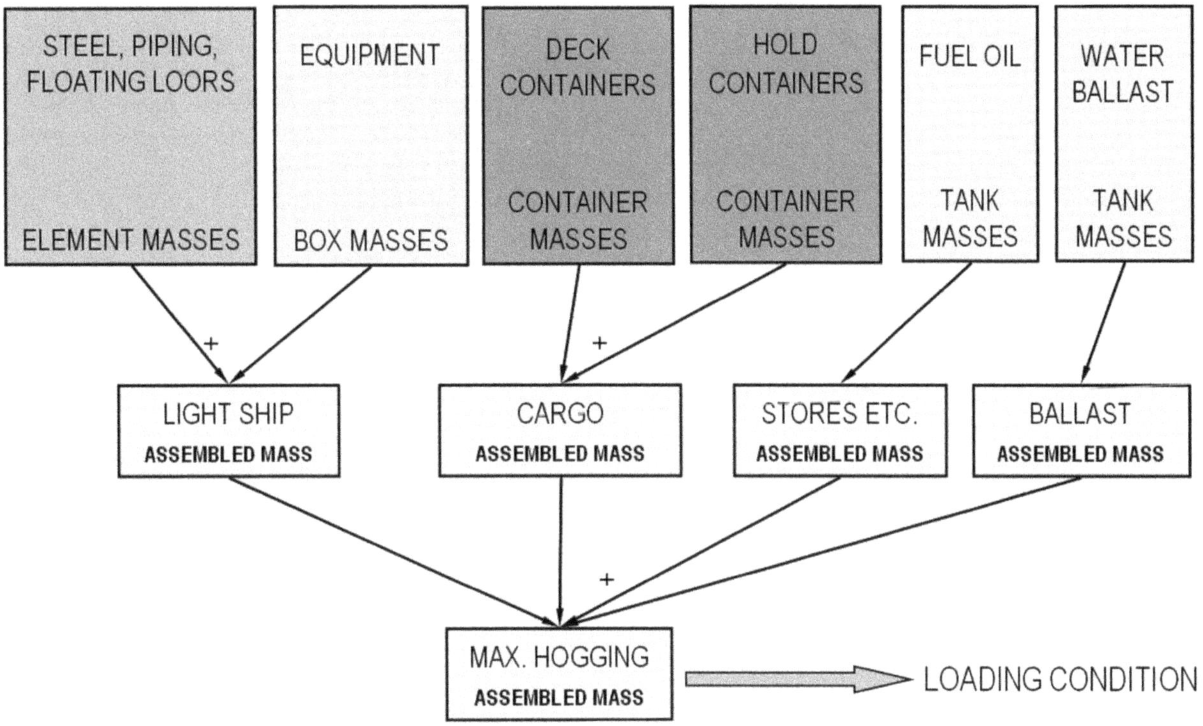

Figure 4.5.14 Typical containership basic masses grouped in assembled mass items.

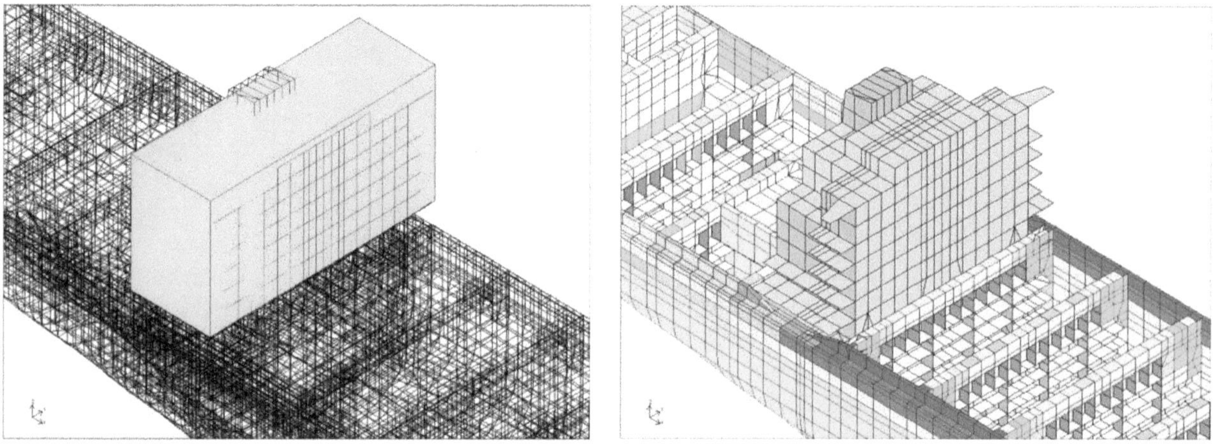

Figure 4.5.15 Mass box *(left)* enclosing the superstructure of a containership *(right)*.

BOX MASSES

Spatial regions enclosing a mass item, so-called mass boxes, are defined. Figure 4.5.15 depicts such a mass box enclosing the superstructure of a containership. Two diagonally opposing points specify a mass box. The mass enclosed by such a mass box is concentrated at its center of gravity, which must be located inside the mass box. The mass inside a mass box is to be distributed as homogeneously as possible. Nodes in regions of higher nodal density are allocated less mass per node than nodes in regions of lower nodal density. It is possible to distribute the box mass onto so-called hard nodes. Such nodes are characterized by a higher stiffness in all three spatial directions. In this way, unrealistically large local deflections of inner nodes are avoided. Typically, box mass items are used to model the weight of the ship's outfitting.

TANKS

Tank loads are applied automatically by identifying (topologically) closed cells of the FE model. Tanks are defined by composite boxes, analogous to mass boxes, either as surrounding boxes or as enclosed boxes. In case of a surrounding box, the combination of all closed cells that lie completely inside the box constitutes the tank; in case of an enclosed box, the smallest closed region that completely encloses the box constitutes the tank.

Often, tank volumes found in this way differ somewhat from tank volumes designated, for example, by the loading manual. The reason is that normally the FE model does not perfectly represent the actual ship geometry. Therefore, if the user designates the tank volume, the computed tank volume is scaled accordingly.

The mass distribution within a tank is computed by finding the position of its free surface with the ship floating in calm water. Nodal forces acting perpendicular to tank walls transfer the resulting hydrostatic pressures to the FE model. Dynamic tank pressures, resulting from the ship's acceleration, are transferred to the model as nodal masses. Mass boxes as described above are used to define spatial regions enclosing the tanks, see Fig. 4.5.16.

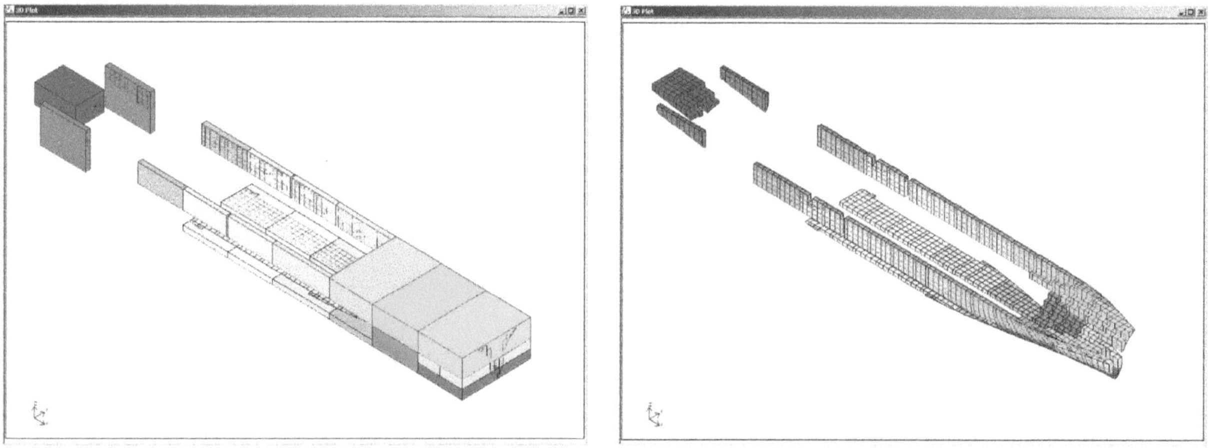

Figure 4.5.16 Mass boxes *(left)* enclosing tanks of a containership *(right)*.

CONTAINERS

GL ShipLoad can deal with standard size containers of 20 and 40 foot length. Container bay numbers denote their location. For each container bay, mass items represent container masses in holds or on deck. Container bays are defined by their longitudinal center of gravity and fore and aft positions at which horizontal loads are applied to the ship's structure. Container stacks are defined by their lateral center of gravity, their vertical supports, and their vertical locations of upper and lower tiers. Later, as part of the input of assembled mass items, the individual masses of each container bay must be defined by a weighting factor.

The general procedure of transferring container masses to the ship's structure is the same for hold and deck containers; however, nodal degrees of freedom differ. For hold containers, all lateral loads are applied to the fore and aft transverse bulkheads. For deck containers, hatch cover specifications determine which loads are applied to structural nodes. Vertical loads are applied at fore and aft edges of hatches; specifically, at lateral positions of corners of the container stack. Horizontal loads are applied at stopper positions. Loads resulting from containers that completely or partially overlap hatch covers are applied to the nearest node on deck, as their support is generally not explicitly modeled in the FE model.

ASSEMBLED MASSES

The combination of the ship's structural mass and a so-called assembled mass item defines the mass distribution for the particular loading condition at hand. As discussed above, an assembled mass item consists of an ensemble of element group masses, box masses, tank masses, and container masses. The center of gravity of partially filled tanks, which depends on the ship's static trim, is updated for each trim computation. Control factors for tank mass items are part of the input to account for the changing center of gravity of the tank mass, for different tank fill levels, and for densities of the fluid carried inside a tank. For container mass items, such control factors account for adjustments of the vertical center of gravity of a container.

Computation of total mass, center of gravity, and inertia tensor for one or all assembled mass items considers all dependencies on other assembled mass items. If any cyclic dependencies are detected, an error message is issued. Element group and box mass factors can be adjusted to correspond to the ship's total mass and its center of gravity. A typical sample view of the graphical user interface shown in Fig. 4.5.17 lists basic mass items combined into a mass distribution. Here, items comprise predefined assembled masses Light Ship, Bunkering, Cargo, and various tank masses with water ballast tanks of specified fill levels and densities.

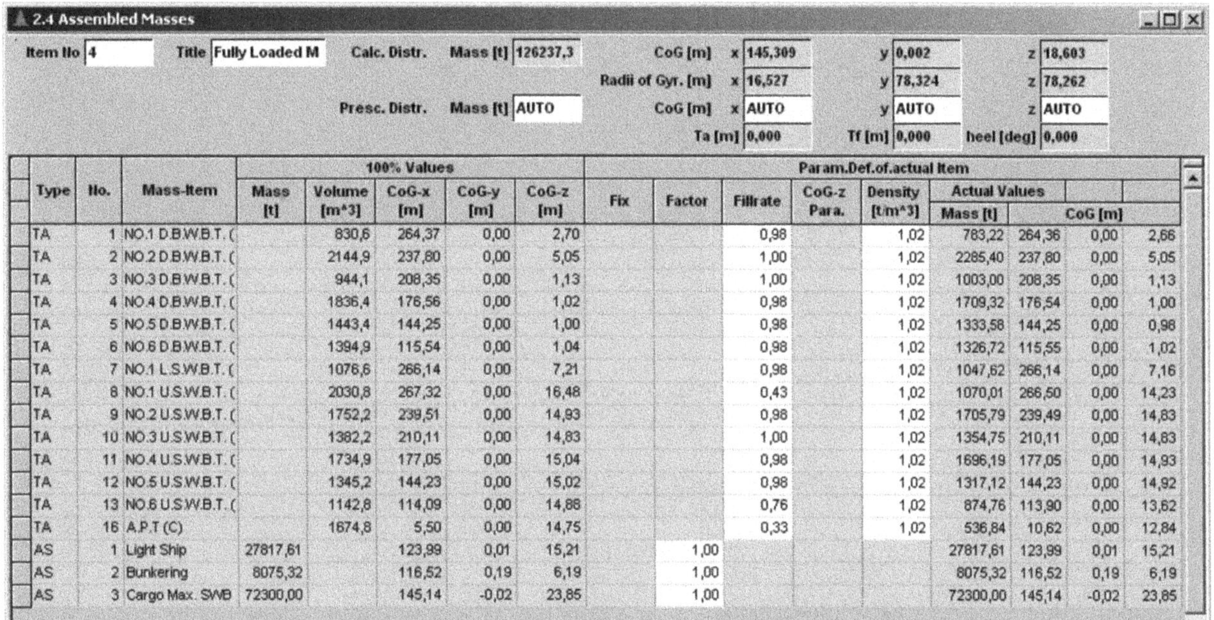

Type	No.	Mass-Item	100% Values Mass [t]	Volume [m^3]	CoG-x [m]	CoG-y [m]	CoG-z [m]	Param.Def.of.actual Item Fix	Factor	Fillrate	CoG-z Para.	Density [t/m^3]	Actual Values Mass [t]	CoG [m]		
TA	1	NO.1 D.B.W.B.T. (		830,6	264,37	0,00	2,70			0,98		1,02	783,22	264,36	0,00	2,66
TA	2	NO.2 D.B.W.B.T. (		2144,9	237,80	0,00	5,05			1,00		1,02	2285,40	237,80	0,00	5,05
TA	3	NO.3 D.B.W.B.T. (		944,1	208,35	0,00	1,13			1,00		1,02	1003,00	208,35	0,00	1,13
TA	4	NO.4 D.B.W.B.T. (		1836,4	176,56	0,00	1,02			0,98		1,02	1709,32	176,54	0,00	1,00
TA	5	NO.5 D.B.W.B.T. (		1443,4	144,25	0,00	1,00			0,98		1,02	1333,58	144,25	0,00	0,98
TA	6	NO.6 D.B.W.B.T. (		1394,9	115,54	0,00	1,04			0,98		1,02	1326,72	115,55	0,00	1,02
TA	7	NO.1 L.S.W.B.T. (		1076,6	266,14	0,00	7,21			0,98		1,02	1047,62	266,14	0,00	7,16
TA	8	NO.1 U.S.W.B.T. (		2030,8	267,32	0,00	16,48			0,43		1,02	1070,01	266,50	0,00	14,23
TA	9	NO.2 U.S.W.B.T. (		1752,2	239,51	0,00	14,93			0,98		1,02	1705,79	239,49	0,00	14,83
TA	10	NO.3 U.S.W.B.T. (		1382,2	210,11	0,00	14,83			1,00		1,02	1354,75	210,11	0,00	14,83
TA	11	NO.4 U.S.W.B.T. (		1734,9	177,05	0,00	15,04			0,98		1,02	1696,19	177,05	0,00	14,93
TA	12	NO.5 U.S.W.B.T. (		1345,2	144,23	0,00	15,02			0,98		1,02	1317,12	144,23	0,00	14,92
TA	13	NO.6 U.S.W.B.T. (		1142,8	114,09	0,00	14,88			0,76		1,02	874,76	113,90	0,00	13,62
TA	16	A.P.T (C)		1674,8	5,50	0,00	14,75			0,33		1,02	536,84	10,62	0,00	12,84
AS	1	Light Ship	27817,61		123,99	0,01	15,21		1,00				27817,61	123,99	0,01	15,21
AS	2	Bunkering	8075,32		116,52	0,19	6,19		1,00				8075,32	116,52	0,19	6,19
AS	3	Cargo Max. SWB	72300,00		145,14	-0,02	23,85		1,00				72300,00	145,14	-0,02	23,85

Figure 4.5.17 Typical sample of graphical user interface showing user assembled masses.

4.5.3.3 External Pressures

The ship's hydrostatic equilibrium in calm water determines its trim and heel. To achieve hydrostatic equilibrium, GL ShipLoad relies on a Newton iteration of draft, trim, and heel until buoyancy forces and moments are in balance with the mass distribution, whereby a finite difference scheme computes the Jacobian matrix required for the Newton iteration. Integration of hydrostatic pressures over shell elements idealizing the hull yields buoyancy forces, and multiplying distributed masses with the gravity vector (in ship coordinates) determines gravity forces.

Idealization of the ship's hull accounts for the computed trim. For linear hydrodynamic computations, shell elements at the waterline are clipped at the calm water plane. For nonlinear pressure extrapolations, redefined frame distances that extend above the calm water plane are part of the hull idealization. A thoroughly validated strip theory code is implemented in GL ShipLoad. Major nonlinear effects are accounted for by an additional viscous damping term that is added when solving for resonant roll motions in beam seas, see Section 4.4.3, and by nonlinearly correcting hydrodynamic pressure distributions for wave-induced loads in large amplitude waves, see Section 4.5.1.

Integration of the resulting pressures yields instantaneously acting wave induced loads. These computations are repeated for different wave periods to obtain pseudo transfer functions corrected for nonlinear effects.

4.5.3.4 Load Cases

The determination of load cases used for the finite element analysis is based on loading conditions, critical load parameters, and wave conditions. Both static and dynamic loads are included. The former are modeled quasi-statically as the product of the mass involved and the local acceleration. Using these loads, the finite element analysis determines the resulting stresses in the hull structure. In assessing dynamic loads, it is necessary to consider a range of sea conditions and headings that produce a critical response of the structure. Typical critical wave load cases for a strength analysis are pictured in Fig. 4.5.18.

For containerships, the following five load parameters are typically identified as dominant load parameters:

1. Vertical bending moment (VBM)
2. Horizontal bending moment (HBM)
3. Torsional moment (TM)
4. Vertical shear force (VSF)
5. Horizontal shear force (HSF)

Of these, load cases associated with bending and torsion are more important than load cases associated with shear forces.

It seems appropriate to again list two loading conditions, together with additional comments, that always have to be investigated for containerships as specified in guidelines published by classification societies:

1. The maximum displacement (at scantling draft) with maximum permissible vertical hogging stillwater bending moment (Max SWBM) must be considered. Provided the ship's stability is not endangered, in all bays a homogeneous weight distribution shall be used for this loading condition with a high stack load for the containers on deck. This loading condition leads to a relatively low metacentric height.
2. The maximum displacement (at scantling draft) with minimum possible vertical hogging or maximum possible sagging still water bending moment (Min SWBM) must be considered. Hold containers are to be loaded in all bays. Deck containers are to be arranged in the midship area as needed to achieve the Min SWBM. For hold containers as well as deck containers a relatively high uniform weight is to be used. This loading condition leads to a relatively high metacentric height.

If considered necessary, an optional third loading condition is specified. It is similar to the first

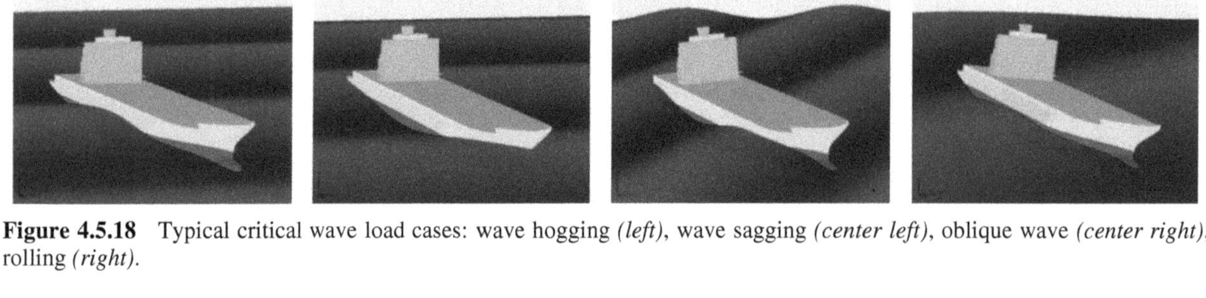

Figure 4.5.18 Typical critical wave load cases: wave hogging *(left)*, wave sagging *(center left)*, oblique wave *(center right)*, rolling *(right)*.

Table 4.5.3 Principal particulars

Length between perpendiculars	366.0 m
Molded breadth	54.2 m
Molded depth	27.7 m
Molded scantling (design) draft	15.0 m
Design speed	26.0 kn

condition (Max SWBM), but one 40 ft. bay in a hold of the midship area is empty. Additional load parameters may be specified. For example, if slamming loads affect the design, vertical accelerations at the ship's ends must be considered dominant load parameters. Local loads, such as wave impact due to slamming, are not included. Such loads must be treated in accordance with current rule requirements or by a separate analysis.

4.5.3.5 Demonstrative Example

As a demonstration, let us generate sectional loads according to GL ShipLoad for another generic containership design of principal particulars listed in Table 4.5.3. Let us take the midship vertical wave bending moment for the ship in head seas as the dominant load parameter (DLP).

STEP 1—FE MESH

An FE mesh is generated to idealize the structural components of the ship. The location of primary structural members governs the mesh fineness. Plane stress membrane elements idealize all decks, transverse and longitudinal bulkheads, walls, floors, web frames, and shell plating. Plane stress elements model beam-like primary members that contain additional degrees of freedom to improve an in-plane bending behavior. Beam or truss elements model secondary members, such as deck beams, stanchions, longitudinals, and stiffeners. Figure 4.5.19 shows the global FE model for this ship.

STEP 2—GLOBAL LOADING CONDITIONS

The ship is assumed floating at scantling draft and advancing at two-third design speed. The strip theory-based analysis is performed for the ship advancing at constant two-third design speed in waves. Viscous roll damping is added according to Himeno (1981) and hydrodynamic pressures in finite amplitude waves are nonlinearly corrected according to Hachmann (1986).

STEP 3—GROUPED MASSES

Loads applied to the FE model are broken down into appropriate load groups according to the concept of GL ShipLoad. The ship's cargo load group comprises deck containers, cargo hold containers, ballast and fuel oil, and other smaller items that affect the FE analysis. Forces and moments resulting from motions of deck containers are applied at rolling and pitching stopper positions located on deck; forces and moments caused by motions of cargo hold containers, at the container cell guide system inside the holds. Forces caused by inertial loads of liquid cargoes are applied to nodes at tank wall elements directly below and/or above the tank's free surface.

STEP 4—DESIGN WAVE CONDITIONS

As shown in Fig. 4.5.20, vertical bending moments are sensitive to the wave length chosen. (In Fig. 4.5.20, wave length is designated by Lw; ship length between perpendiculars, by Lpp.) This figure shows that bending moments are largest in waves of length equal to 0.94 times the ship's length.

According to global strength analysis guidelines (Germanischer Lloyd 2007), the smallest wave heights that results in maximum wave bending moments (VBM_{WH}) have to be selected as equivalent regular design wave heights. For this ship,

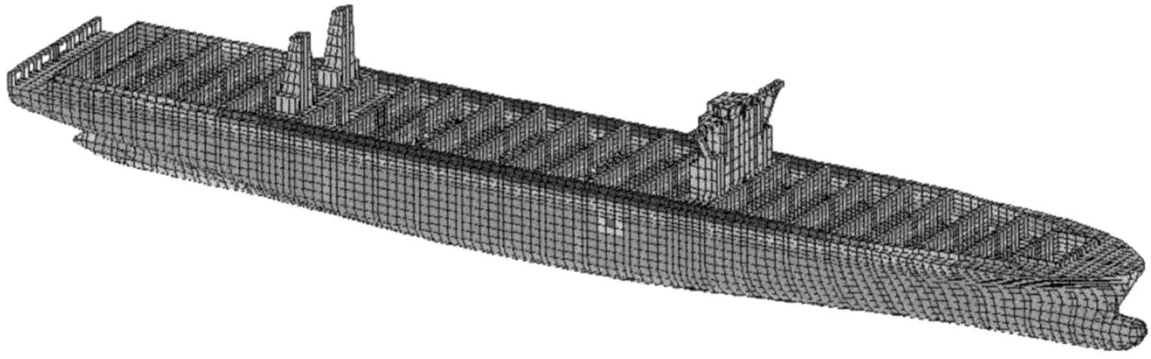

Figure 4.5.19 Global FE model.

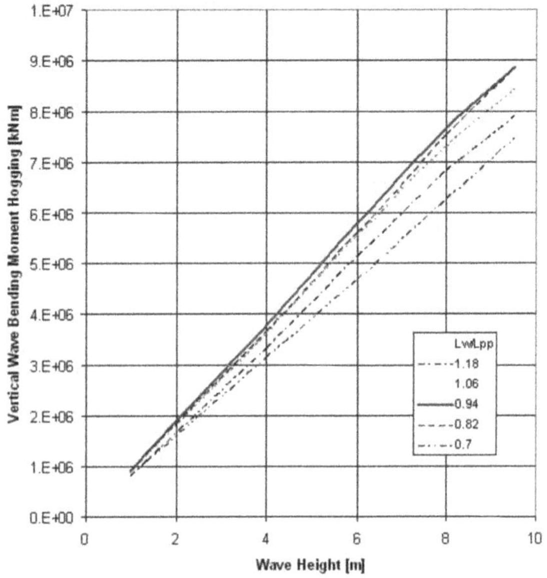

Figure 4.5.20 Midship vertical wave bending moment in hogging vs. wave height.

rule values of vertical midship bending moments are $VBM_{WH\ HOG} = 7.38 \cdot 10^6$ kNm in hogging and $VBM_{WH\ SAG} = 8.93 \cdot 10^6$ kNm in sagging (Germanischer Lloyd, 2007). Design wave heights depend on the functional relationships of bending moment vs. wave height, in this case for the largest bending moments, that is, for waves with lengths equal to 0.94 times the ship's length, here plotted in Fig. 4.5.20 for the hogging condition. The two curves in Fig. 4.5.21 plot this bending moment not only for the hogging condition, but also for the sagging condition. Design wave heights are obtained by selecting (reading off) wave heights that correspond to the rule-based bending moments. They turn out

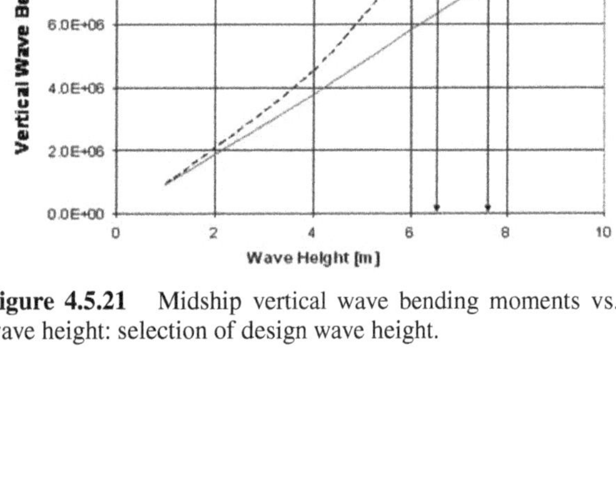

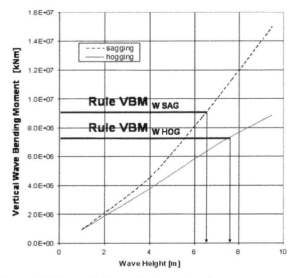

Figure 4.5.21 Midship vertical wave bending moments vs. wave height: selection of design wave height.

to be 7.82 and 6.34 m for the hogging and sagging conditions, respectively.

STEP 5—EQUIVALENT REGULAR DESIGN WAVES

Wave lengths that are analyzed to obtain global loads in regular waves range from 0.35 to 1.2 times ship length. Wave headings range from 0 to 180 deg at 30 deg intervals. For each combination of wave length and wave heading, 50 equidistant wave crest positions over the ship's length are considered. From a total of 9500 situations of the ship in regular waves, 40 design load cases are selected for two static loading conditions, namely, the maximum hogging condition and the minimum hogging condition. They are selected by matching global wave-induced loads to design values of classification society rules. Typical results of load computations are shown in Figs. 4.5.22 and 4.5.23 for the containership subject to the DLP that defines wave loads causing maximum vertical bending and to the DLP that defines maximum torsion of the hull girder, respectively. Both figures present hydrodynamic pressures acting at 44 cross sections of the ship, longitudinal distributions of total sectional loads (shear forces, bending moments, and torsional moments) that also include still-water loads, and the corresponding hull deformation.

Large open hatch areas, characteristic of modern containerships, give rise to weak torsional strength over the length of the cargo holds. Therefore, an enforced roll angle is specified as an input parameter for GL ShipLoad to account for effects of an additional torsional moment due to roll. Extreme roll angles and maximum vertical bending moments do not occur simultaneously. Therefore, the ship is analyzed in reduced amplitude waves. Here, in compliance with global strength analysis guidelines (Germanischer Lloyd, 2007), the analysis was performed for the two enforced roll angles of 9.2 and 16.0 deg to port as well as to starboard. For these roll angles, the wave amplitude is reduced to 86 and 50 percent of the design wave amplitude, respectively.

If we compare Figs. 4.5.22 and 4.5.23, we see that the ship's hull structure experiences significantly different global loads. For the ship subject to maximum vertical bending (Fig. 4.5.22), a symmetric distribution of hydrodynamic pressures about the ship's centerline is consistent with global hull girder loading and hull girder deformation. For the ship subject to maximum torsion (Fig. 4.5.23), although horizontal shear and horizontal bending are no longer zero, maximum vertical

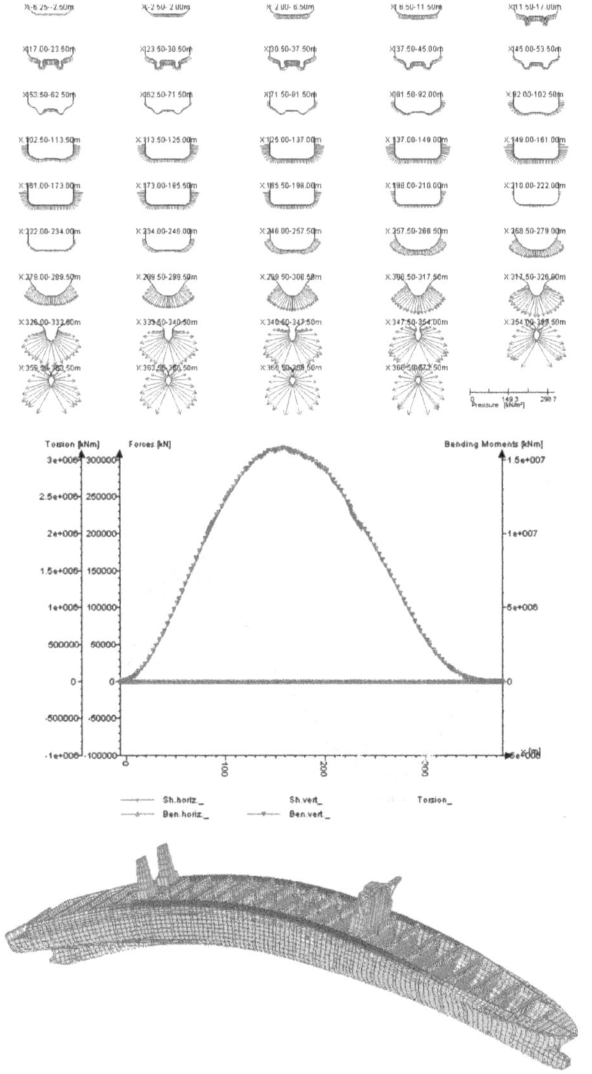

Figure 4.5.22 Hydrodynamic pressures, sectional loads, and hull deformation for maximum vertical bending moment in hogging.

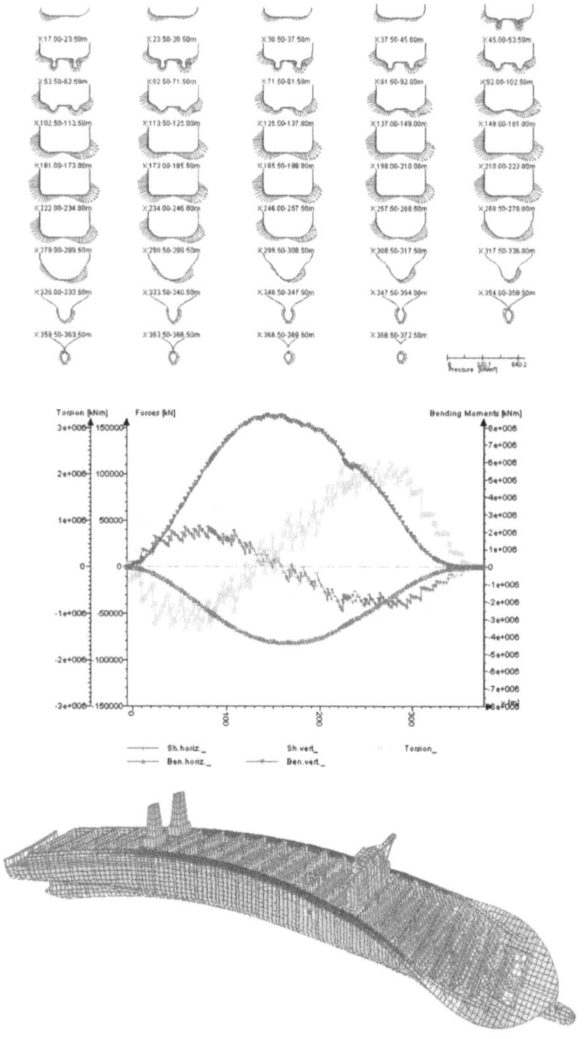

Figure 4.5.23 Hydrodynamic pressures, sectional loads, and hull deformation for the ship under maximum torsion and 16 deg enforced roll to starboard.

bending moment is reduced by over 60 percent, and this maximum occurs not near amidships, but at a location about one-third of the ship's length from the aft perpendicular.

STEP 6—ENVELOPE CURVES OF GLOBAL LOADS

Sectional loads are defined as forces and moments that have to be applied at a cut through a formerly balanced model to maintain equilibrium. They are computed by summing all forces and moments up to the position of the cut, comprising nodal forces caused by the accelerating mass distribution and all wave-induced pressure forces.

Envelope curves of sectional loads in Fig. 4.5.24 depict longitudinal distributions of vertical shear force (VSF), horizontal shear force (HSF), vertical bending moment (VBM), horizontal bending moment (HBM), and torsional moment (TM). Solid lines represent GL ShipLoad results; dashed lines, rule-based design values. Here the abscissa is marked by the ratio x/L, which stands for the fraction of ship length measured from the ship's aft perpendicular. As anticipated, these envelope curves show that global loads generated using GL ShipLoad closely approximate global rule-based design loads.

4.6 SPECIAL LOADS

4.6.1 Slamming

In rough seas wave impact related slamming may occur. The associated water impact loads are larger than other wave loads. They may cause local

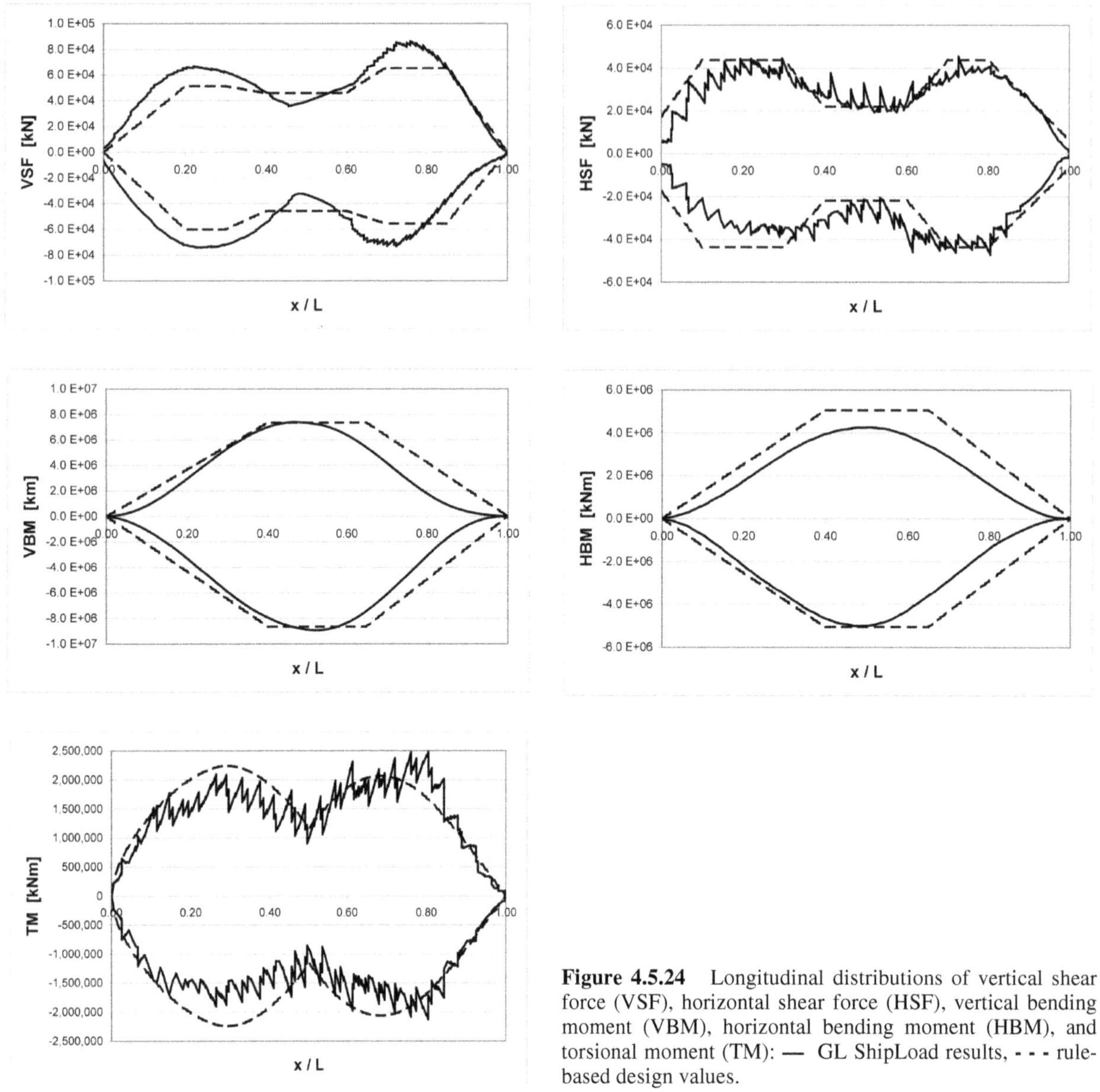

Figure 4.5.24 Longitudinal distributions of vertical shear force (VSF), horizontal shear force (HSF), vertical bending moment (VBM), horizontal bending moment (HBM), and torsional moment (TM): — GL ShipLoad results, - - - rule-based design values.

damage or large-scale buckling of deck structures. Even if each impact load is relatively small, frequent impact loads accelerate fatigue failures of hulls. This phenomenon is important especially for high-speed ships, where slamming loads may threaten the safety of ships. Therefore, rational and practical methods to estimate wave impact loads are important for the safe design of ships.

Wave impact is a strongly nonlinear phenomenon and a random process that is sensitive to relative motion and contact angle between body and free surface. As the duration of wave impact is short, dynmaic effects are large. In addition, because of air trapping, the wave impact phenomenon is difficult to describe theoretically. Slamming is a three-dimensional process that is characterized by highly peaked local pressures of short duration. Hence, applying slamming peak pressures over ship structural plate fields is not appropriate for design purposes.

Wave impact-related slamming on a ship can be roughly classified into the four kinds as shown in Fig. 4.6.1 (Bertram, 2000):

1. Bottom slamming occurs when, because of pitching, the ship's bottom emerges from the water and subsequently, because of its relatively flat area, undergoes a severe hydrodynamic impact on reentry.
2. Bow-flare slamming occurs when the upper flared portion of the bow plunges deeper into the water at a high relative velocity between the flared bow and the water surface.
3. Breaking wave impacts occur because of the superposition of an incident wave and the bow wave hitting the bow of a blunt ship, even for small ship motions.

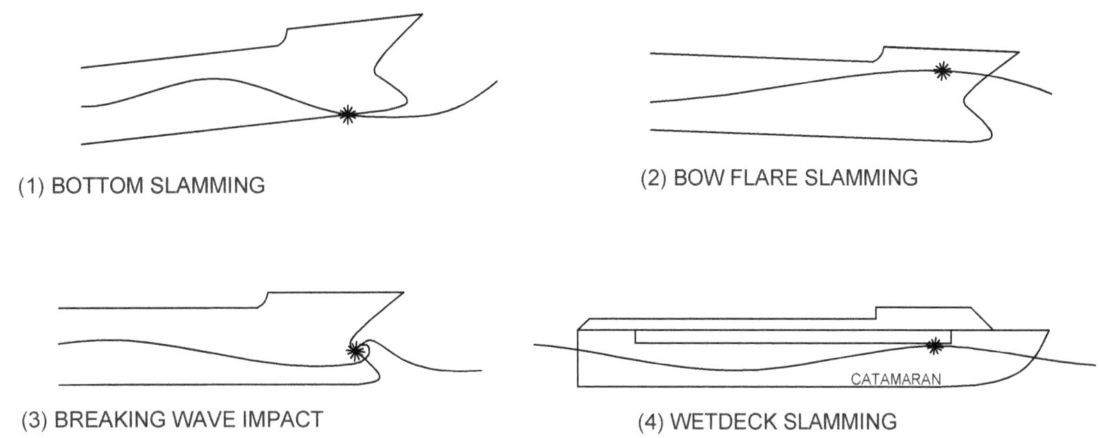

Figure 4.6.1 The four kinds of slamming impact on a ship.

4. Wetdeck slamming, a phenomenon known to be experienced by high-speed catamarans, occurs when the relative heaving amplitude exceeds the height of the catamaran's wetdeck.

Classical theories approximate the fluid as inviscid, irrotational, incompressible, and free of surface tension. Assuming that effects of gravity are negligibly small, the problem can be treated analytically in the framework of potential theory. For bodies of small deadrise angle, the problem can be linearized. Von Karman (1929) was the first to theoretically study water impact, idealizing the two-dimensional entry of a wedge into calm water to estimate impact loading on a seaplane during landing. He obtained the impact pressure by equating the momentum before impact with the sum of the wedge momentum and the added mass momentum. Being based on momentum conservation, von Karman's model is usually referred to as momentum impact. As it neglects the water surface elevation, the added mass and impact loads are underestimated, particularly for small deadrise angles.

Wagner (1932) derived a more realistic water-impact theory. As long as the deadrise angle is large enough not to trap air, Wagner's theory can be applied to arbitrarily shaped bodies. Wagner's theory is simple and useful, but it introduces a singularity at the edge of the wedge, resulting in negative infinite pressure there. With a correction for peak pressure based on measurements, Wagner's theory gives accurate peak impact pressures for practical use.

The singularity of Wagner's theory can be removed by taking spray into account. An inner solution for the wedge is asymptotically matched to an outer solution of the spray region, as proposed by, for example, Watanabe et al. (1986). Results for constant entry velocity are free from singularities. Despite this theoretical improvement, Watanabe's and Wagner's theories predict basically the same peak impact pressure.

Recently, real progress was achieved with numerical procedures based on computational fluid dynamics (CFD), using surface capturing methods of the volume-of-fluid (VOF) type (Schellin and el Moctar, 2007). The computer code COMET (CD-adapco, 2002) is based on this technique. For special cases, a so-called multi-stage approach can yield spatial mean slamming pressures that can be applied as equivalent static design loads to determine scantlings of hull structural elements. This is accomplished by using a chain of seakeeping codes (el Moctar et al., 2004). The procedure consists of the following steps:

1. A linear, frequency-domain Green function panel code computes the ship response in unit amplitude regular waves. Wave frequency and wave heading are systematically varied to cover all possible combinations that are likely to cause slamming. Results are then linearly extrapolated to obtain responses in wave heights that represent severe conditions, here characterized by steep waves close to breaking. Under such conditions, a conservative one-third (voluntary or involuntary) speed reduction usually is assumed.
2. Regular design waves are selected on the basis of maximum magnitudes of relative normal velocity between ship critical areas and wave, averaged over the critical area.
3. The nonlinear strip method simulates large-amplitude rigid body motions of the ship under design conditions by time-domain integration of the motion equations, thereby accounting for the ship's forward speed, the swell-up of water in finite amplitude waves, as well as the ship's wake that influences the wave elevation at the ship's sides, particularly at moderate to high forward speed.

4. The nonlinearly computed, time-domain ship motions constitute part of the input for the RANS code COMET. Numerical volume grids surround the ship. A large enough domain of the mesh has to be chosen to allow for the ship motions. When examining head wave conditions, grid density near the ship hull and ahead of the ship must be high to resolve the wave, whereas aft of the ship the grid becomes course to dampen the waves. Front, side, bottom and top flow boundaries are specified as inlets of known velocities and known void fraction distributions defining water and air regions. On the hull surface a no-slip condition is enforced on fluid velocities and on the turbulent kinetic energy. At the outlet boundary a zero gradient pressure boundary condition has to be satisfied. The wake flow boundary is specified as a zero-gradient pressure boundary.

The procedure is here applied to predict slamming loads for a reference ship that features a flared bow with a pronounced bulb and a flat overhanging stern, a hull shape typical of modern offshore supply vessels. Table 4.6.1 lists principal particulars. Slamming loads are examined at the ship's forefoot. Specifically, areas under the flared bow are considered. The favorable agreement of computed results with model test measurements validated this method.

Numerical volume grids surrounding the ship comprising about one and a half million hexahedral control volumes are generated. To avoid flow disturbances at outer grid boundaries, these boundaries are located at a distance of about one ship length ahead of the bow, two ship lengths aft of the stern, one ship length beneath the keel, and one ship length above the deck. The large domain of the mesh, especially below the keel and above the deck, is chosen to allow large pitch motions in head waves. Figure 4.6.2 shows part of the numerical grid domains surrounding the reference ship.

Front, side, bottom, and top flow boundaries are specified as inlets of known velocities and known void fraction distributions defining water and air regions. On the hull surface a no-slip condition is enforced on fluid velocities and on the turbulent kinetic energy. The wake flow boundary is specified as a zero-gradient pressure boundary (hydrostatic pressure). All computations are performed using the RNG-k-ε turbulence model with wall functions (Speziale and Thangam, 1992). The time step size is chosen such that the Courant number is smaller than unity on average. The momentum equations are discretized using 85 percent central differences and 15 percent upwind differences. Ship motions are accounted for by moving the entire grid at each time step. Thus, all boundary conditions are newly computed at each time step.

Volume fractions and velocities that initialize the flow field arise from superposition of ship speed and orbital particle velocities of the design waves. The influence of numerical damping on the wave height has to be taken into account. Numerical diffusion caused by the coarse grid aft of the ships dampens the incident wave to such an extent that no significant wave reflection occurs at the outlet boundary. For runs in regular waves, simulation of the flow field must continue until a periodic solution is reached. After a simulation time of two to five encounter periods, depending on ship motions, ship speed, wave height, and position of the investigated plate fields, periodically converging solutions are obtained. For each time step up to fifteen outer iterations are needed.

The Hamburg Ship Model Basin (HSVA) conducted systematic seakeeping model tests of this hull to experimentally verify the computations. The model was constructed with a separated bow segment, located above the still waterline, enabling measurements of forces and moments acting on this segment. A pressure sensor was installed on each side of this bow segment to record local pressures. A gyro unit recorded pitch motions, three accelerometers recorded vertical accelerations, and three wave probes recorded wave elevations. Measurements of loads on the bow section were corrected for inertial effects. The free-running, self-propelled model, constructed at a scale of 1:10, was hand operated by the helmsman accommodated on the carriage of HSVA's large towing tank. Tests were run in regular waves of heights ranging from 2.0

Table 4.6.1 Principal particulars of the reference ship

Length between perpendiculars	70 m
Molded breadth	15 m
Draft	5 m
Displacement	1550 t
Service speed	16 kn

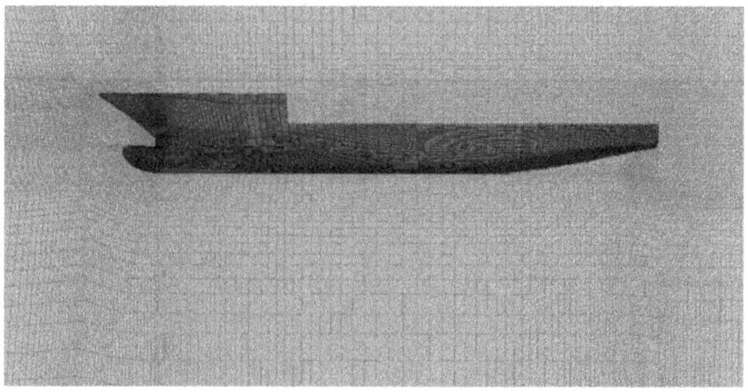

Figure 4.6.2 Part of numerical grid domains surrounding the hull.

Table 4.6.2 Amplitudes of motions and accelerations

	Computations	Model tests
Heave	1.05 m	1.10 m
Pitch	6.15°	6.30°
Acc. at aft perp.	4.40 m/s²	4.25 m/s²
Acc. at fwd. perp.	6.50 m/s²	6.70 m/s²

to 3.5 m and wave length to ship length ratios ranging from 0.8 to 1.4 at propeller turning rates corresponding to calm water speeds of 14 and 16 kn. As head waves were investigated, only pitch and heave motions were considered. Computed results generally compared favorably with model test measurements. As an example, Table 4.6.2 lists average amplitudes of heave and pitch motions and accelerations obtained from HSVA test run no. 5, where full-scale test conditions corresponded to 3.5 m high head waves at a wave length to ship length ratio of $\lambda/L = 1.0$ and a ship speed of v = 14 kn.

The location of the separated bow section and the critical plate fields 1 and 2 are shown in Fig. 4.6.3. Pressure sensors were located at a height of 5.0 m (full-scale) above the baseline, with the port one positioned 2.0 m (full-scale) aft and the starboard one 0.5 m (full-scale) aft of the forward perpendicular. Under the HSVA test conditions, the RANS solver performed computations of wave-induced slamming loads and local pressures acting on the separated bow section of the ship. Simulations of the flow field were performed over two consecutive encounter periods, while measurements lasted over several periods. Over the first two periods, the computed vertical force on the separated bow compared favorably with measurements (Fig. 4.6.4). For the corresponding slamming pressures, averaged over the plate field areas, the functional relationship of computed values compared favorably with experimental data, but peak values differed. This deviation was

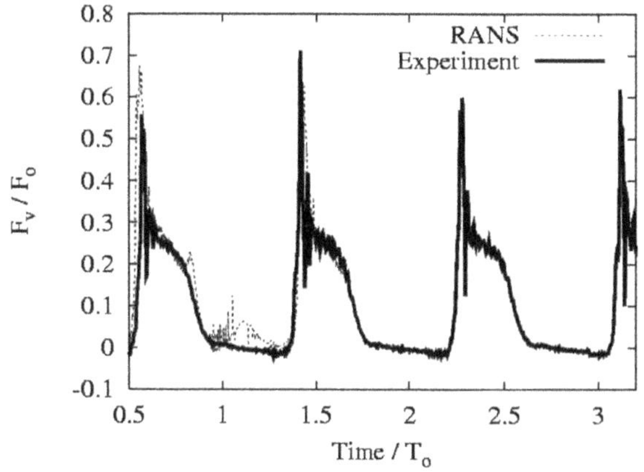

Figure 4.6.4 Time histories of measured and computed vertical force on bow section.

largely attributed to the relatively strong variation of the measured peaks, most likely caused by the inability of the model to attain steady state conditions during tests in regular waves. At both plate fields, the time histories of measured and computed pressures were similar. In Fig. 4.6.5 these histories are shown for plate field 1. The force and pressure histories in Figs. 4.6.4 and 4.6.5 are presented as non-dimensional values. Force (F_V) was normalized by a maximum value of F_0 = 2350 kN; pressure (p_{sl}), by a maximum value of p_0 = 265 kPa; and time (Time), by the wave encounter period of T_0 = 6.0 sec.

Only head wave conditions were investigated because these conditions were rated most critical from the standpoint of slamming loads on the forebody. First, systematic GLPANEL seakeeping computations were performed, yielding transfer functions of relative normal velocity at selected critical locations under the flared bow for severe waves. In addition, appropriate weather data, past experience with similar ships, and results of the HSVA model tests were used to define design

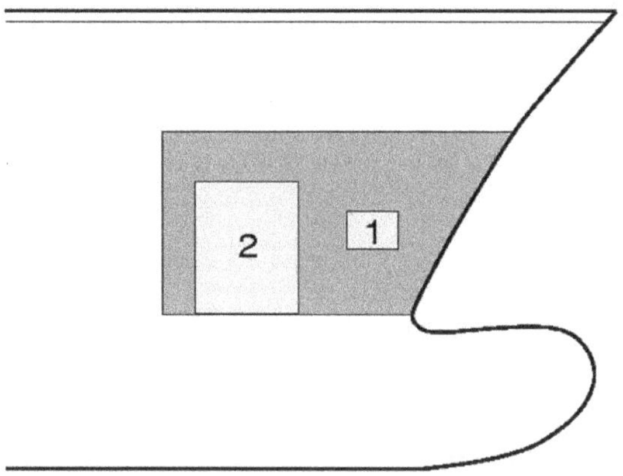

Figure 4.6.3 Locations of separated bow section *(dark shading)* and selected plate fields *(light shading)*.

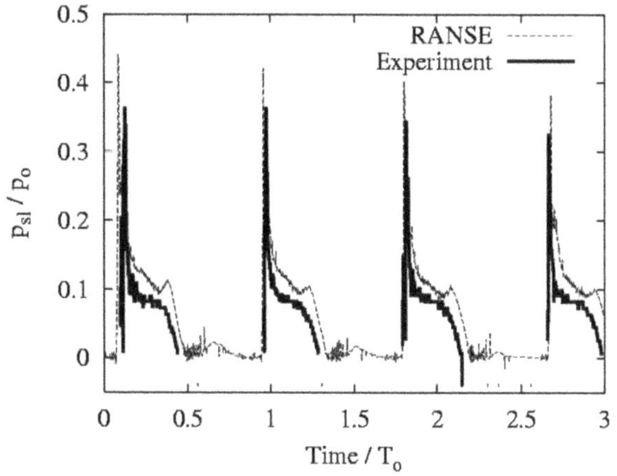

Figure 4.6.5 Time histories of measured and computed pressures on plate field 1.

Table 4.6.3 Design wave conditions

Wave height	7.0 m
Wave length	112.8 m
Wave length / ship length	1.6
Wave length / wave height	16.1
Wave encounter period	5.8 sec
Ship speed	12.1 kn

wave conditions as summarized in Table 4.6.3. It was assumed that the ship advances at three-quarters service speed.

GLSIMBEL computations provided ship motions at design wave conditions. For this ship, it was possible to use time harmonic functions to accurately describe all motion characteristics. Heave and pitch amplitudes were 4.6 m and 11.1 deg, respectively. Four areas under the flared bow were considered critical for structural design, and the computed slamming pressures were averaged over these areas. Their locations are shown in Fig. 4.6.6, where they are identified as plate fields numbered 1 to 4 having areas of 0.2, 4.5, 5.1, and 4.5 m^2, respectively.

The three-dimensional flow field surrounding the hull under the influence of design wave conditions was computed as a transient process. Computed pressures hardly changed after the first period. Over the third period, computed time histories of these averaged slamming pressures are shown in Fig. 4.6.7. Short duration peak pressures are seen to be highest at plate fields 1 and 2, reaching peak values of about 180 kPa. A typical pressure distribution over the ship's hull, corresponding to the time of maximum pressure peak occurrence, is shown in Fig. 4.6.8.

Computed slamming pressures averaged over the area of these plate fields, albeit for a different but similar ship (el Moctar et al., 2004), were performed to demonstrate that slamming pressures depend on ship speed, wave height, and wave length. As seen by the time histories in Fig. 4.6.9, highest pressures occurred at the largest ship speed (v = 18 kn), in the highest waves (H_W = 11 m), and in waves of length equal to ship length (λ/L = 1.0). Pressures here are normalized against the maximum value of p_0 = 270 kPa and times against the wave encounter period of T_0 = 6.0 sec.

Accurate prediction of slamming loads continues to be a difficult undertaking, mainly because of the influence of a large number of parameters involved. The multi-stage procedure presented here is capable of predicting slamming loads suitable for design of a ship's structure. It attempts to combine the physics of a ship in a seaway together with the use of advanced numerical techniques. From a practical standpoint, the procedure represents a compromise between attainable accuracy and computational effort. However, before using the RANS solver, it is necessary to identify wave conditions likely to cause severe slamming and to obtain reliable predictions of ship motions under such conditions. Ship motions in such a seaway should not be predicted

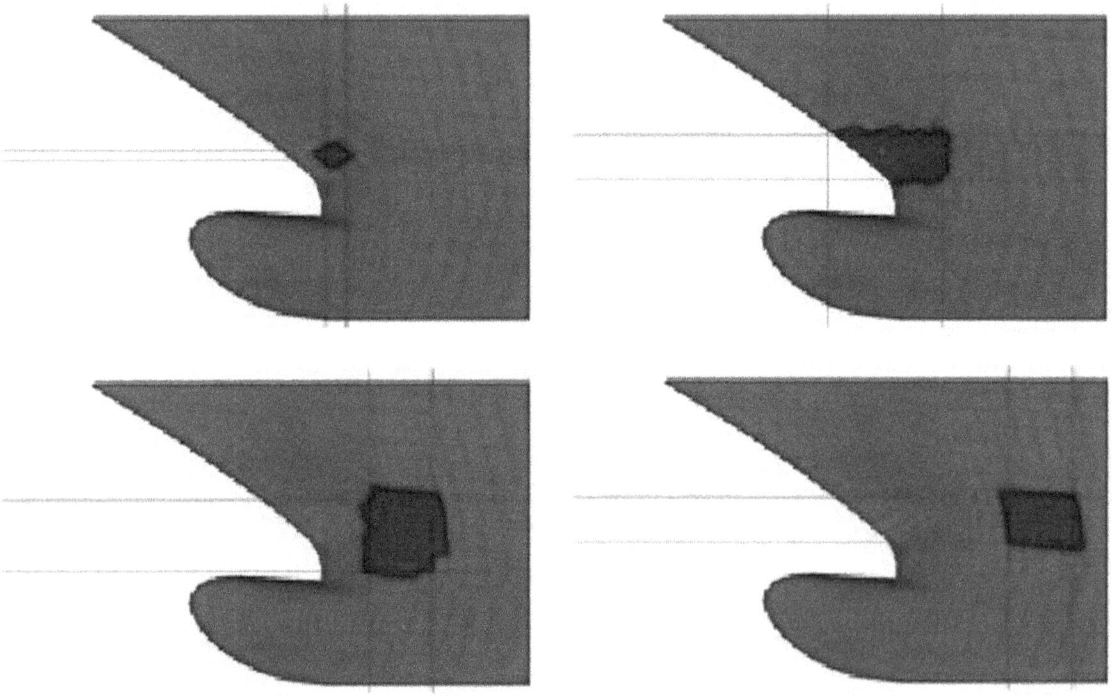

Figure 4.6.6 Locations of critical plate field 1 *(upper left)*, field 2 *(upper right)*, field 3 *(lower left)*, and field 4 *(lower right)*.

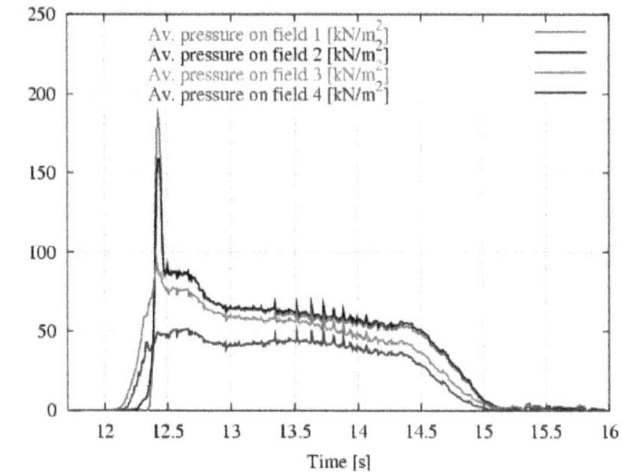

Figure 4.6.7 Time histories of slamming pressures at plate fields 1 to 4.

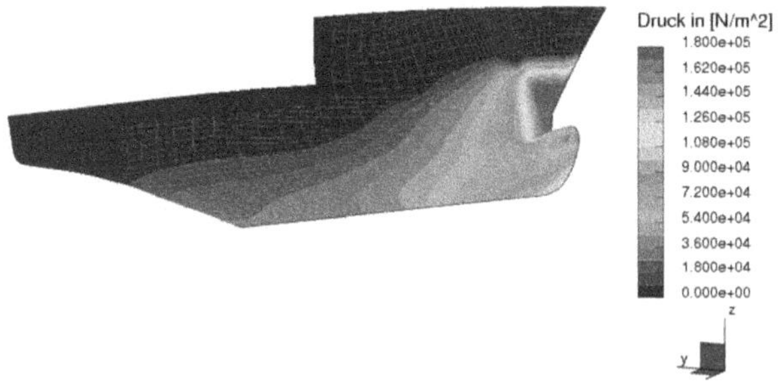

Figure 4.6.8 Typical predicted pressure distribution during slamming.

from linear methods alone. A subsequent nonlinear analysis should be performed to obtain sufficiently accurate motion predictions that are then part of the input for the RANS solver (see the four steps listed earlier, just before the example).

To model the seaway, the procedure has to rely on defining an equivalent regular design wave. This approach has been successfully applied to many kinds of sea load predictions for ships. Breaking wave criteria together with wave scatter diagrams of operating sea areas together with past experience with similar ships serves as a basis to identify design wave conditions. Although a design wave only roughly approximates the real wave conditions when slamming, there is at present no alternative available for practical application. Modeling the natural seaway by superimposing a large number of regular waves would have made it necessary to deal with a prohibitively large number of variables.

4.6.2 Sloshing

Large fluid motions inside partially filled tanks are excited whenever the period of the moving ship is close to the natural period of the fluid inside the tanks. This phenomenon is called sloshing. Sloshing can cause large structural loads that affect the structural integrity of tanks. Bulkheads and tank tops as well as baffles and other internal structures are subject to time-dependent pressure loads that can lead to collapse or fatigue failure. To ensure the safety of such structures, sloshing phenomena must be analyzed accurately by means of up-to-date rules and procedures. To assess the dangers of sloshing, it is first necessary to get an estimate of the natural period of the flow inside a tank. The next step consists of calculating the exciting period that causes the largest ship motions. Finally, the natural period of the fluid and the exciting period of the ship are compared to check for resonance conditions. Classification societies publish rules from which both of these periods may be obtained. The rule formulas for sloshing induced pressures, however, may be used only if the natural period of fluid motion is well outside the resonance range. For critical fill levels with a natural period of fluid motion close to the period of the moving ship, direct computations are recommended to simulate the arbitrarily deformed free surface flow inside the tank.

The research community has long been aware that loads due to sloshing can only be accurately predicted if the multi-phase free-surface flow

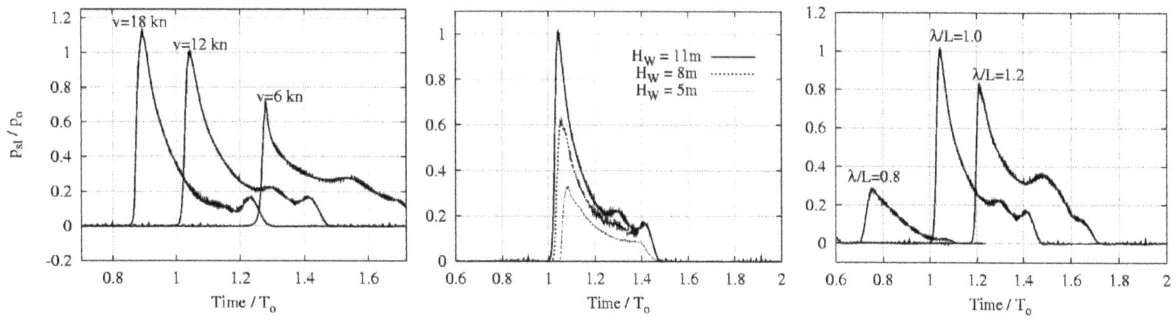

Figure 4.6.9 Influence of ship speed *(left)*, wave height *(center)*, and wave length *(right)* on computed slamming pressure acting at a critical plate field under the flared bow.

inside the tank is correctly simulated. Field methods, such as computational fluid dynamics (CFD) are able to simulate such flows correctly. Of these, the interface-capturing techniques proved to be most suitable for handling strong nonlinearities and are today the obvious choice for computing complex free-surface shapes with breaking waves, sprays, and air trapping, phenomena that need to be dealt with in sloshing computations. A comparative study on sloshing loads conducted by the International Ship Structure Committee (ISSC, 2003) showed that CFD methods are effective to simulate viscous multi-phase flows inside tanks.

Modern computational methods suitable for the prediction of hydrodynamic impact-related loads have been developed (e.g., Sames and Schellin, 2001). They generally consist of the following steps:

- Relevant tanks are identified, based on their geometry and position in the ship.
- An appropriate seakeeping code is employed to compute the exciting periods of the ship under different loading conditions, wave directions, and ship speeds. Based on these results and the wave climate appropriate for the ship's operating environment, the maximum ship motions for sloshing in critical tanks are determined.
- The fluid natural period for different tank fill ratios between 5 and 95 percent are estimated. Usually about 20 fill ratios are investigated. Fill ratios that produce matching fluid natural periods within the relevant range of ship motions are identified as critical fill levels, and these cases are analyzed by numerically simulating fluid motions inside the tanks.
- The analysis requires use of an appropriate RANS solver; for example, one that is based on the finite volume method. For each critical fill level, time traces of computed pressures acting on tank walls are obtained, and the resulting maximum loads are compared with each other for all critical fill levels to determine overall maximum loads.

Table 4.6.4 Principal particulars of sample product carrier

Length overall	184.9 m
Length between perpendiculars	175.2 m
Molded breadth	28.0 m
Molded depth	16.8 m
Draft in ballast	6.0 m
Draft fully laden	11.0 m
Deadweight	34,578 t
Service speed	15.9 kn

The characteristic of this computational procedure is that a RANS solver is employed in combination with classical potential theory methods to analyze worst-case sloshing scenarios. A significant advantage of this kind of direct computations is that, compared to model tests, scale effects can be adequately accounted for because uncertain methods to extrapolate model test results to full-scale need not be applied. Furthermore, such computations constitute a cost-efficient tool to optimize tank geometries in the design stage and, if necessary, to modify the tank design during the operational phase.

As an example, a typical product carrier is investigated by following the above procedure. Table 4.6.4 lists principal particulars and Fig. 4.6.10 shows the general layout and tank arrangement of the product carrier. Ten cargo tanks are symmetrically arranged on both sides of a central longitudinal bulkhead. With this arrangement, all tanks are relatively long and slender and, with the exception of the two forward tanks, nearly of rectangular shape. Sloshing due to roll motions, therefore, is not as relevant as sloshing due to pitch motions. For this sloshing investigation, pitch is the critical mode of motion. For the pitching ship in waves, the forward tanks experience the severest motions because these tanks are located furthest

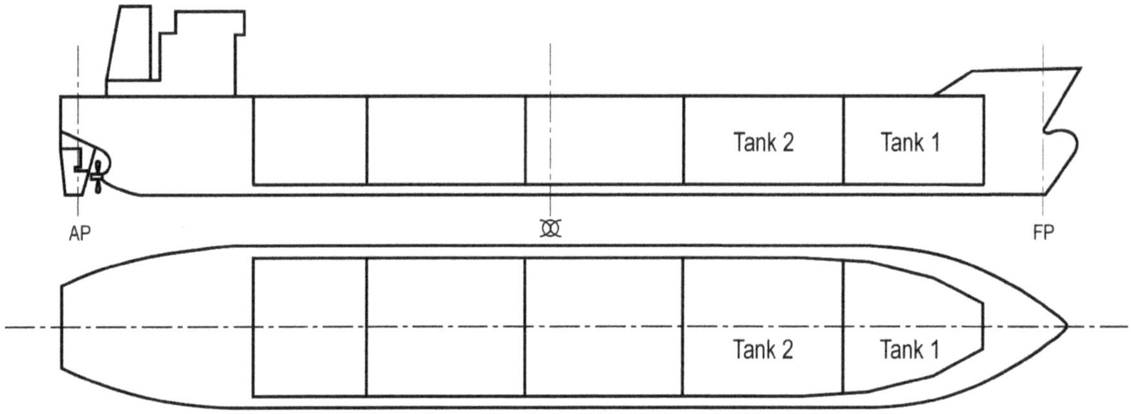

Figure 4.6.10 Side view and top view of product carrier.

from their centers of rotations. Thus, the two most forward tanks on one side of the ship are deemed critical and, therefore, are investigated for sloshing. Table 4.6.5 summarizes the main dimensions of these tanks. The density of the fluid carried in these tanks is 1.3 t/m^3. Considered are two loading conditions, namely, ship in ballast and ship fully laden, and two ship speeds, namely, one-third service speed and two-third service speed.

A linear frequency domain seakeeping code is employed first to yield pitch motions of the tanker in regular unit amplitude waves of different headings. The resulting response amplitude operators (RAOs) are obtained for two ship speeds and two drafts. Typical RAOs of pitch motion are plotted in Fig. 4.6.11 as a function of wave encounter period, here for the fully laden ship sailing at two-third design speed. The curves depict results for seven different wave headings, ranging from 0 deg (stern waves) to 180 deg (head waves) at 30 deg intervals. They show that maximum pitch motions occur at a wave encounter period of around T_S = 10.0 sec and that, at periods lower than 6.0 sec, pitch motions are relatively small. These results are typical also for the other cases considered, i.e., for cases of the ship in ballast sailing at one-third design speed. Thus, lower bounds for the RANS sloshing computations, marked in Fig. 4.6.11, are the wave encounter period of 6.0 sec, which also agrees with the rules, and 0.38 deg/m for the pitch amplitude, which corresponds to relatively small pitch amplitudes where the motion of the fluid inside the tanks is unlikely to significantly affect sloshing pressures. The operational profile of the tanker leads to a maximum wave amplitude of 8.0 m for the worst case sloshing scenario. The corresponding pitch amplitudes are obtained from the pitch RAOs.

The next step consists of estimating the natural sloshing period, T_L, for the different fill levels of

Table 4.6.5 Main dimensions of cargo critical tanks

	Tank 1	Tank 2
Maximum length	25.35 m	28.90 m
Maximum breadth	11.45 m	11.45 m
Maximum height	15.42 m	15.42 m

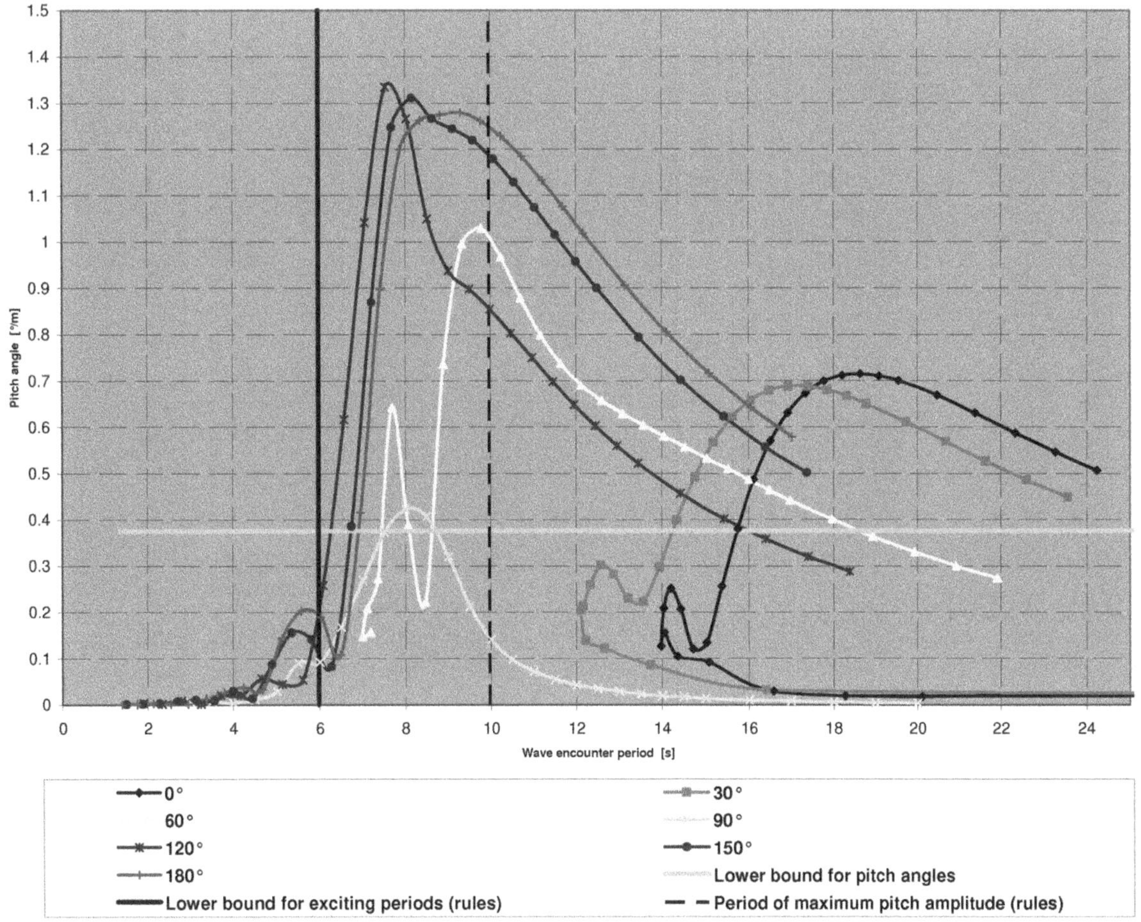

Figure 4.6.11 Typical pitch RAOs as a function of wave encounter period for the fully laden product carrier at two-third design speed.

the fluid inside the tanks. The following formula, valid for a tank of rectangular shape, is used:

$$T_L = 1.132\sqrt{\frac{e_t}{\tanh\left(\frac{\pi h}{e_t}\right)}} \qquad (4.6.1)$$

where e_t is the characteristic tank dimension, in this case the tank length, and h the fill height. (Dimensions of tank length and fill height are in meters.) The estimated natural sloshing periods are plotted in Fig. 4.6.12 against fill levels for tanks 1 and 2. Fill levels of less than ten percent do not have enough energy to induce significant sloshing pressures and, therefore, need not be considered for the RANS sloshing analysis. A vertical red line marks the lower bound for fill levels; a horizontal blue line, the lower bound for sloshing periods.

Now a RANS solver is employed to compute the flow fields in the tanks as a transient process. Here the code COMET is used. The fluid domain is idealized by a volume grid comprising about 734,000 cells for the three-dimensional discretizations: 356,000 cells for the foremost tank 1 and 378,000 cells for tank 2 located directly behind this tank (Fig. 4.6.13). The time step size is chosen such that the Courant number is unity on average. The impulse equations are discretized using 90 percent central differences and 10 percent upwind differences. As usual, the transport equations for the turbulent kinetic energy and its dissipation rate are discretized using upwind differences. The entire flow field is initialized by the hydrostatic pressure and zero velocity. On the tank walls, no-slip conditions are enforced on fluid velocities and on the turbulent kinetic energy.

The tank motions at the estimated relevant natural sloshing periods are used as input for the RANS solver. The tank motions are directly obtained from the ship motion RAOs. Screen shots of computed instantaneous two-phase flow and the corresponding pressure distribution inside this tank are shown in Fig. 4.6.14. Time histories of maximum pressure acting at selected critical locations inside the tank are presented in Fig. 4.6.15. The locations where these pressures act are shown as red dots in Fig. 4.6.16. A summary of the maximum values of all computed pressures for the different fill levels considered are collectively depicted in Fig. 4.6.17. A maximum computed pressure of 250 kPa results for a fill level of 25 percent. The corresponding rule values are shown as well.

The above investigation typifies the general procedure of a sloshing investigation. Such a study

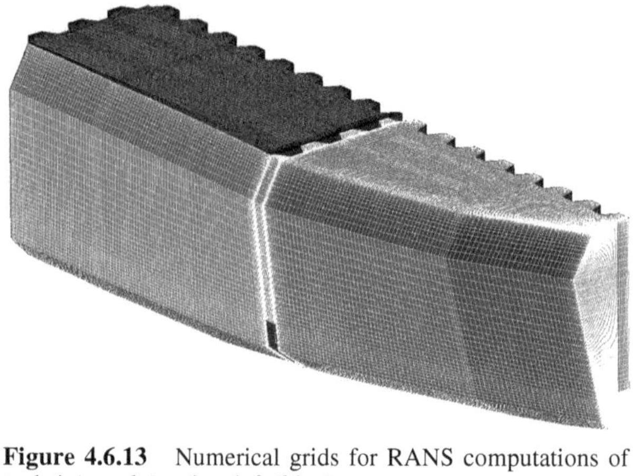

Figure 4.6.13 Numerical grids for RANS computations of tank 1 *(purple)* and tank 2 *(blue)*.

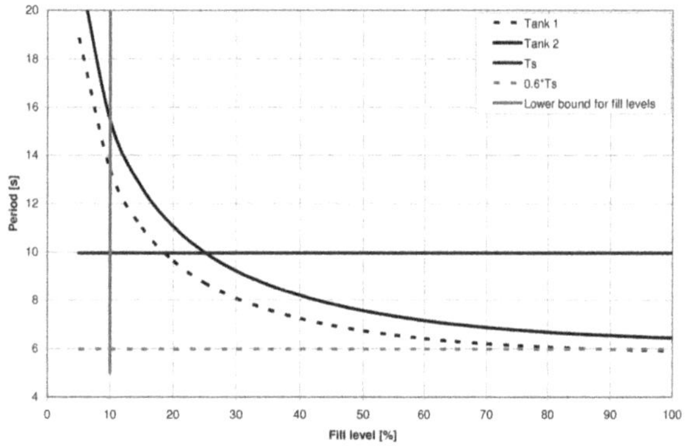

Figure 4.6.12 Natural sloshing period as a function of critical tank fill level.

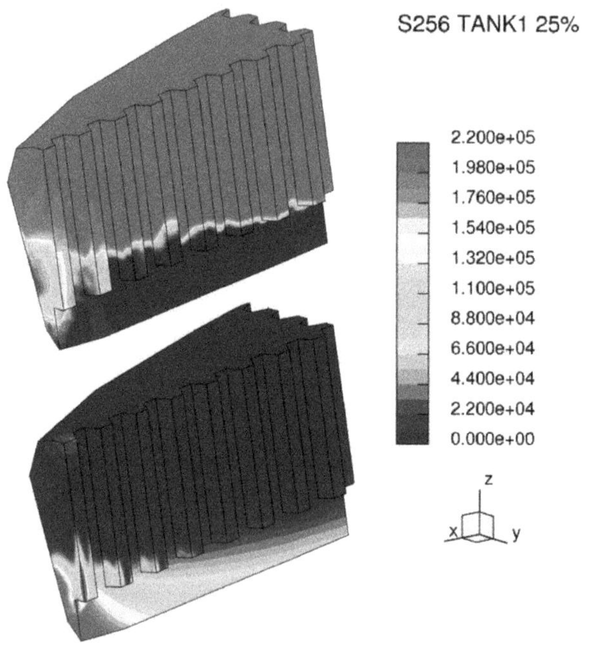

Figure 4.6.14 Screen shots of instantaneous two-phase flow *(left)* and the corresponding pressure distribution for 25 percent fill level *(right)* of tank 1.

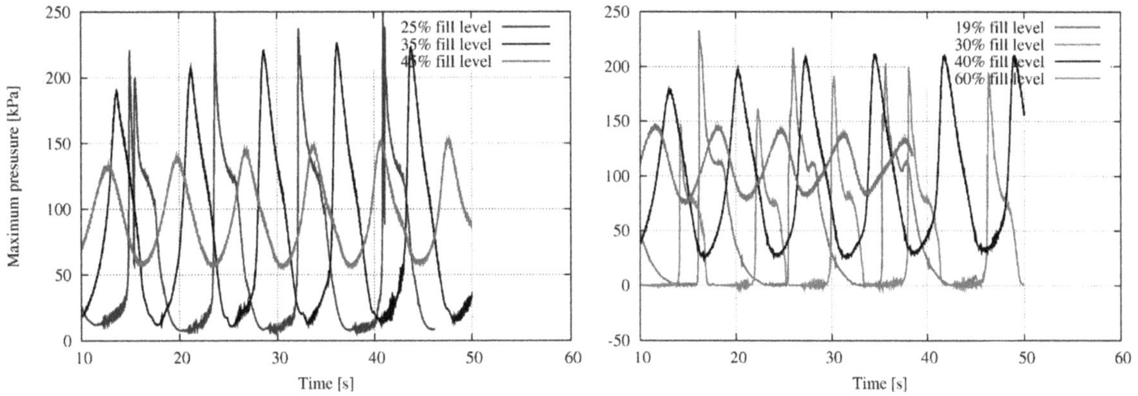

Figure 4.6.15 Time histories of maximum pressure in tank 1 for different fill levels.

generally starts with the identification of the critical cargo tanks of the ship. Next, ship motions as well as natural periods of the fluid corresponding to different fill levels in these critical tanks are assessed to enable the identification of relevant fill levels. Finally, a RANS solver computes the corresponding sloshing-related pressure loading caused by fluid motion in these partially filled tanks, and these pressures can then be compared to rule-based values. For the above product carrier, results show that the highest computed pressures occur at fill levels between 20 and 45 percent. At these fill levels, sloshing-induced pressures are critical as they exceed class rule values, particularly at the foremost bulkhead of tank 1.

4.6.3 Whipping

Impact-related wave loads on the forefoot, bow flare, and other parts of the hull structure may induce transient vibration of the entire hull in which the principal contribution comes from the fundamental two-noded vertical mode. This slamming-induced vibration is known as whipping, and it may result in vibratory stress intensities that are equal in magnitude to the wave-induced low-frequency bending stresses. There are two factors that simplify the treatment of whipping. First, the structural response of a slamming load usually occurs at a much higher frequency than wave-induced responses. Second, for large and medium-sized ships the overall mass and the overall damping are sufficiently large to not appreciably affect the ship's heaving and pitching motions. Therefore, the ship's motion and all of its wave-induced responses can be calculated without considering the extra complexities of impact-related slamming and green water effects. As demonstrated in the previous section 4.6.1, slamming can be investigated based on a separate systematic seakeeping analysis of ship motions and corresponding relative velocities of the ship and the water surface. The whipping response is then superimposed on the wave-induced response calculated for the rigid ship.

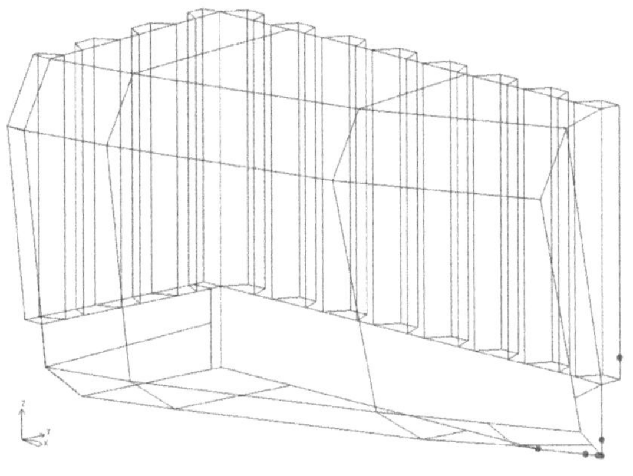

Figure 4.6.16 Locations of computed maximum pressures of tank 1.

Various theoretical and experimental studies (e.g., Yamamoto et al., 1980) confirmed that the maximum bending moment due to slamming can be obtained accurately enough by using rigid ship theory to calculate ship motions and other parameters affecting slamming, then performing a dynamic analysis of the elastic hull girder to yield the whipping response to slamming, and finally superimposing this whipping response on the wave-induced response. However, slamming strongly influences the motions of smaller ships, such as a naval patrol boat. Consequently, an accurate investigation of slamming can be performed only as part of a ship motion analysis that includes the highly nonlinear slamming terms.

Recent investigations of wave-induced whipping are based on coupling CFD and FE methods to predict motions and wave-induced global loads as well as the structural response for ships in a

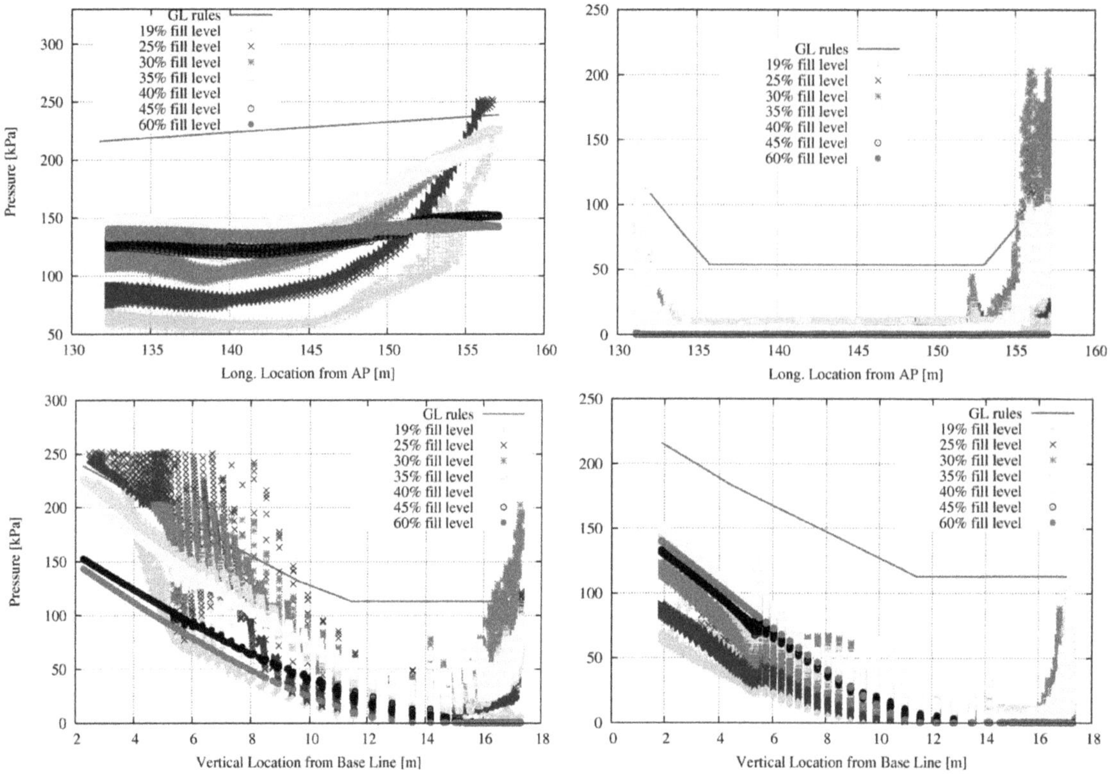

Figure 4.6.17 Maximum computed pressures and comparative rule based pressures at bottom *(upper left)*, top *(upper right)*, forward bulkhead *(lower left)*, and aft bulkhead *(lower right)* of tank 1.

seaway. Usually, it suffices to account for the influence of hydrodynamic pressures on hull deformation and to neglect the effect of hull deformation on hydrodynamic pressures as this latter effect usually is relatively small. El Moctar et al. (2006), for example, did just that by one-way coupling an extended version of the RANS solver COMET to the FE code ANSYS (Müller and Groth, 2002). COMET performed the CFD computations and ANSYS the structural analysis. For a large generic 13,000 TEU containership with principal particulars listed in Table 4.6.6, they investigated the global structural behavior under slamming critical head wave conditions.

Extending the RANS solver COMET meant solving the nonlinear ship motion equations. The computational procedure comprises the following steps:

- The flow around a ship's hull is computed, taking into account viscosity, flow turbulence, and deformation of the free surface.
- Hydro- and aerodynamic forces and moments acting on the ship are calculated by integrating the pressure and friction stresses over the ship's surface.
- The nonlinear equations of the rigid body six-degrees-of-freedom motions are solved in the time domain, taking account of all forces acting on the body. Integration yields motions, accelerations, velocities, and translational and rotational displacements.
- After updating the position of the ship and again computing the fluid flow for the new position and integrating over time, the trajectory of the ship is obtained.

The equations of motion are solved in the Newtonian reference frame (G, x, y, z) fixed with respect to the mean position of the ship. Its origin is located at the center of gravity of the ship, G. The coordinate transformation between the Newtonian and the ship-fixed local reference frame (G, ξ, η, ζ) is defined as follows:

$$\vec{x} = \vec{x}_G + S\vec{\xi} \quad (4.6.2)$$

where S denotes the transformation matrix and $\vec{\xi}$ the rotational position vector. The transformation

Table 4.6.6 Principal Particulars

Length btw. perp.	382.0 m
Molded breadth	54.2 m
Depth	27.7 m
Scantling draft	15.0 m
Service speed	26.0 knots

matrix results from consecutive rotations about the vertical axis (yaw), the new transverse axis (pitch), and the new longitudinal axis (roll). The governing equations of motions are expressed as follows:

$$\vec{F} = m \cdot \ddot{\vec{x}}_G \quad (4.6.3)$$

$$\vec{M} = SI_L S^T \dot{\vec{\omega}} + \vec{\omega} \times SI_L S^T \vec{\omega} \quad (4.6.4)$$

where $\vec{F}$ denotes the resulting force vector, m the mass of the ship, $\ddot{\vec{x}}_G$ the translational acceleration vector of G, $\vec{M}$ the resulting moment vector with respect to G, I_L the inertia tensor of the ship about the axes of the ship-fixed (local) reference frame, and $\vec{\omega}$ the angular velocity vector. Integration in the time domain relies on the use of the so-called explicit trapezoidal method (Brunswig and el Moctar, 2004).

At each time step, fluid-structure interaction between wave-induced loads and the elastic FE model of the ship's hull structure is accounted for. The computational approach, implemented within ANSYS, yields the dynamic response of the elastic hull structure by incorporating a technique called transient dynamic analysis, also known as the time-history analysis. This technique determines the dynamic response of the hull structure under the action of time dependent slamming loads.

For the containership investigated, the time scale of the loading was such that inertia and damping effects had to be considered. Two percent of critical damping is appropriate for this kind of ship. Lewis-forms at 100 equidistant ship stations yielded added mass coefficients for two-dimensional cross sections and, based on these results, were used to obtain added masses for the whole ship. Appropriate reduction factors that depend on the kind of mode shape as well as on the ratio of ship length to ship breadth accounted for the three-dimensional flow. For the two-node vertical bending mode investigated, added masses of 144 and 27 percent of the ship's displacement were obtained for the vertical motion and for the transverse motion, respectively. These added masses, when distributed as additional nodal masses on the FE model, resulted in a natural frequency of 0.37 Hz of the two-node vertical hull girder bending mode shown in Fig. 4.6.18.

At successive time points t, the transient analysis solves the following basic equation of motion:

$$(M)\{\ddot{U}\} + (B)\{\dot{U}\} + (C)\{U\} = \{F(t)\} \quad (4.6.5)$$

where (M) is the mass matrix, (B) is the damping matrix, (C) is the stiffness matrix, $\{\ddot{U}\}$ is the nodal acceleration vector, $\{\dot{U}\}$ is the nodal velocity vector, $\{U\}$ is the nodal displacement vector, and $\{F(t)\}$ is the load vector. The ANSYS code employs the Newmark time integration method to solve this motion equation at discrete time points in the ship-bound coordinate system. The integration time step size is 0.02 sec.

El Moctar et al. (2006) investigated the ship advancing at two-third service speed and under slamming-critical wave conditions, that is, in 9.5 m amplitude head waves of 400 m wave length. The assumed loading condition of the ship corresponded to the design draft of 15.0 m. Figure 4.6.19 shows the resulting time histories of vertical acceleration at the forward perpendicular (FP) and at the center of gravity (G). These wave conditions caused time-varying dynamic amplifications of strains and stresses in the hull structure. Figure 4.6.20 shows the resulting maximum deformation of the hull girder at a time step where the deformation of the forebody hull structure is greatest. Here the relative deflection between amidships and the bow is about a 1.0 m. The shadings in Fig. 4.6.20 represent the resulting von Mises stress levels.

Comparable (normalized) stresses were also computed for the inelastic hull girder. Figure 4.6.21 presents the corresponding time histories of longitudinal hatch coaming stresses at two different locations, namely, at amidships (A) and at the forebody (B). These stresses are mainly bending stresses caused by the vertical bending of the hull girder. Positive values represent tensile stresses; negative values, compressive stresses.

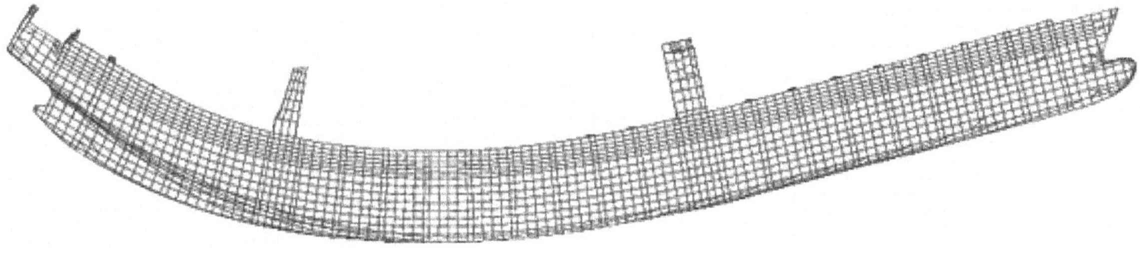

Figure 4.6.18 Two-node vertical bending mode of the containership.

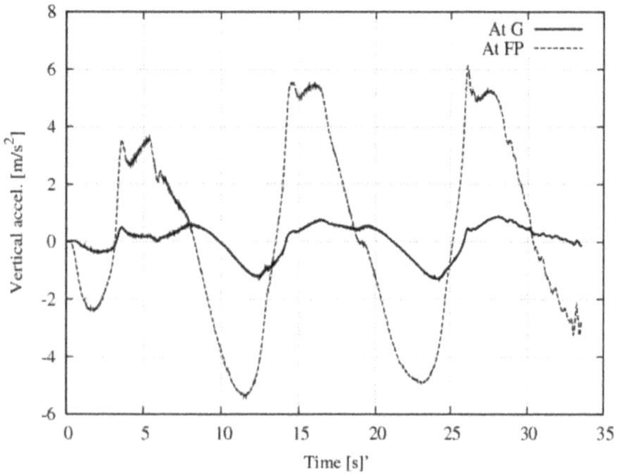

Figure 4.6.19 Time history of vertical acceleration at *G* and at FP in 9.0 m amplitude head waves.

The huge mass and inertia of the investigated ship resulted in wave-induced ship motions that are relatively insensitive to slamming effects. This is seen, for example, by the time histories of vertical acceleration shown in Fig. 4.6.19. The acceleration peaks at time points 14.5 and 26.0 sec were caused by slamming. At these time points, the acceleration peaks at the ship's forward perpendicular (FP) are mainly due to pitch, whereas the hardly noticeable acceleration peaks at the ship's center of gravity (G) are mainly due to heave. Thus, in the ship's bow region, pitch had an expectedly greater influence on vertical acceleration than in the midship region.

In this example the dynamic stress amplification due to hull girder whipping was caused by bow slamming. The main concern was longitudinal stresses caused by the vertical bending of the hull girder. As expected, the highest stresses occurred in the foreship, see Fig. 4.6.20. The two positions A and B in Fig. 4.6.21 represent hatch coaming locations amidships and at the foreship, respectively, where stress levels generally are largest. The two time histories of stresses shown in Fig. 4.6.21 illustrate the comparable stress history at these two locations for the fully dynamically and the quasi-statically analyzed hull girder. At both locations, it is seen that by treating the hull girder dynamically, the longitudinal stress increased, albeit significantly more so for the midship location A, where the increase is about 20 percent. Although, for this worst-load case, maximum values of all the resulting stresses did not exceed the yield stress, the analysis demonstrates that it may be important to account for the dynamic stress amplification.

The stresses in Fig. 4.6.20 are mainly longitudinal stresses caused by the vertical bending of the hull girder in head waves. The dynamic slamming impact excited the two-node vertical bending mode of the hull girder. The dynamic stresses oscillate about the static stresses at the ship's whipping frequency of 0.37 Hz. This effect is more pronounced at the midship section. Time histories shown in Fig. 4.6.21 start at time $t = 0$ sec with the ship in the hogging position. At this instant the hatch coaming is subject to tensile stresses. At time $t = 5.5$ sec, when the wave progresses to where the crest reaches the forefoot, that is, after about one half of an encounter period, slamming occurred, subjecting the hull girder to the sagging condition. Now the hatch coaming experiences high compression (negative) stresses.

The resulting stresses and structural hull deformations are somewhat conservative because the RANS-computed pressures were only one-way coupled to the FE structural model. To fully account for the fluid-structure interaction, the iteration at each time step would have had to consider also

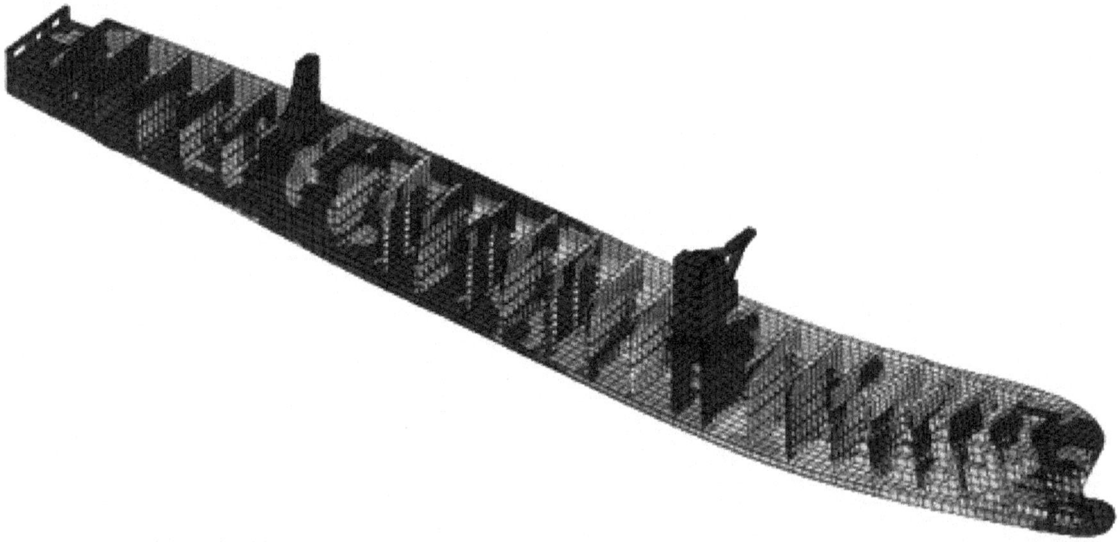

Figure 4.6.20 Deformed FE model and von Mises stresses of hull in sagging condition.

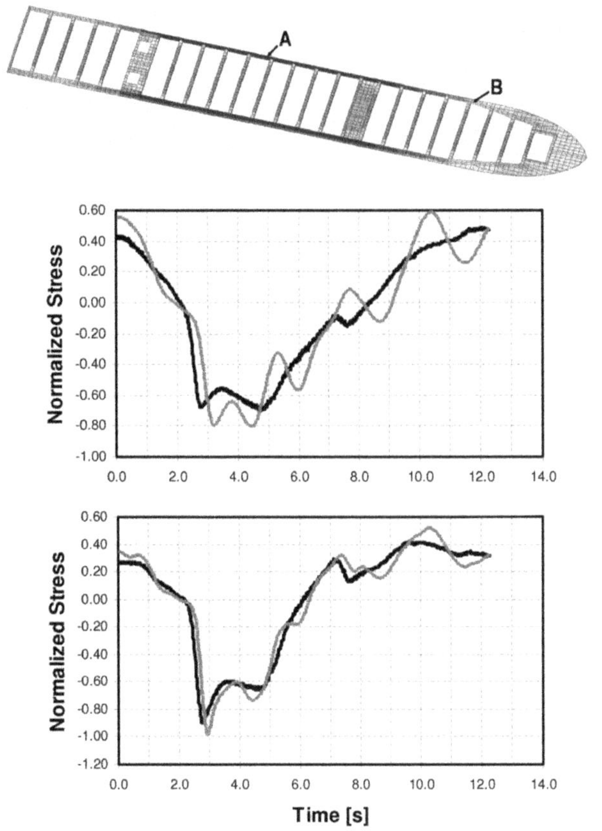

Figure 4.6.21 Time histories of longitudinal hatch coaming stresses at location A *(upper graph)* and location B *(lower graph)* for the quasi-static analysis *(dark curves)* and the fully dynamic analysis of the hull girder *(light curves)*.

the influence of the structural deformation on the RANS-computed hydrodynamic pressures. However, such effects were considered to be relatively insignificant. From a practical standpoint, the applied procedure represents a compromise between attainable accuracy and computational effort.

4.7 NONLINEAR RESPONSE SIMULATION

Frequency domain computations based on linear (small amplitude) theory have limits for predicting motions and loads of a ship in waves. Time domain simulation is the sole computational method that can handle strong nonlinearities and is, therefore, the appropriate tool to investigate strongly nonlinear ship response in severe seas. However, no fully nonlinear code is available to the industry at present. There are a number of codes that implement various levels of sophistication.

Simulation methods primarily developed for load predictions generally simulate all six degrees of freedom rigid-body ship motions by solving the nonlinear motion equations at small successive time steps, starting from a realistically chosen initial position and velocity of the ship. The following external forces (and moments, which are always included) are considered:

- Ship weight
- Froude-Krylov forces (i.e., forces caused by the pressure field of the waves unaffected by the ship)
- Radiation and diffraction forces (i.e., forces caused by the ship's motion on the pressure field)
- Maneuvering forces
- Viscous roll damping
- Rudder and appendage forces
- Propeller thrust
- Wind forces

The derivation of the motion equations is based on the conservation of linear and angular momentum. The motion equations describe the momentum for translations and rotations and comprise translatory and rotational accelerations, both of which are three-component vectors suitably expressed in the ship-fixed coordinate system. Although initially expressed in the inertial (earth-fixed) reference system, these equations can be transformed into a rotating, ship-fixed system by applying Euler's method. These equations form a system of six scalar, coupled ordinary second-order differential equations that can be transformed into a system of 12 first-order differential equations, and these equations can be integrated numerically using a fourth-order Runge-Kutta method.

FROUDE-KRYLOV FORCES

The Froude-Krylov forces result from the integration of the total water pressures induced by the incident waves as if the ship were not present. These pressures cause changes in the righting arm curve due to the changing position of the wave crests relative to the ship. Some simulation methods determine the righting moment due to this pressure by integrating the pressure over the wetted surface, while other methods interpolate between stored righting arms hydrostatically precomputed for the ship situated in a wave-shaped water surface. The latter procedure is faster, but less accurate.

The waves are treated linearly according to Airy theory. This means that nonlinear effects in steep seaways, such as higher and steeper wave crests than wave troughs, are neglected. The natural seaway is linearized by superposing small amplitude elementary Airy waves. To obtain reliable probability distributions of wave and response data, the frequencies of individual component waves must not be integer multiples of a minimum wave frequency $\omega_{\min}$. If this were the case, the seaway would repeat itself after a time of $2\pi/\omega_{min}$. To simulate a natural seaway, the phase angle associated with each component wave should be generated randomly between 0 and 2π. Wave frequencies, wave encounter angles, and phase angles selected before the simulation begins should not be changed during the simulation.

RADIATION AND DIFFRACTION FORCES

Hydrodynamic radiation forces (added mass and damping) are generated by ship motions, whereas diffraction forces arise from the difference between the forces acting on the stationary ship in waves and the Froude-Krylov forces. This difference is caused by the change of the wave pressure field due to the presence of the stationary ship. For instance, if the ship partly emerges from the water, the diffraction component of the emerged part must be zero and thus cannot be computed for the non-moving ship. An effective way to reliably determine the sum of radiation and diffraction forces is to rely on the relative motion hypothesis, that is, to assume that the force depends on the motion of the ship minus the motion of the water. The nonlinear strip theory based simulation code GLSIMBEL (Pereira, 1995), for example, makes use of this hypothesis by assuming that, at the considered ship cross section, the orbital velocity of the wave motion averaged over a ship cross section determines the fluid flow and the force at that cross section. Summing the contributions from all ship sections (strips) yields the total force on the ship. This code treats longitudinal interactions caused by the ship's forward speed in a simplified manner similar to the linear strip method. Since its development, initiated by Söding (1982), Böttcher (1986), and Pereira (1988), GLSIMBEL was extended to also account for the effects of water flowing through a damaged hull, sloshing of liquids in partially filled tanks, and green water on deck.

A special difficulty in simulating the ship's behavior in a natural seaway is the frequency dependence of the radiation forces (added mass and damping) because, in a natural seaway, a multitude of frequencies act at the same time and also, because of the nonlinearity, it is invalid to superimpose the contributions from individual frequencies. Time domain simulations must account for this frequency dependence of the added mass and damping matrices because the pressure distribution depends not only on the ship's momentary position, velocity, and acceleration, but also on the past history of the ship's motion, which is reflected in the wave pattern. This effect is especially strong for heave and pitch motions. In frequency domain computations, this so-called memory effect is expressed in the frequency dependence of added mass and damping. In time domain simulations, one solution to this problem is to use impulse-response functions by formulating the hydrodynamic memory forces and moments, $\vec{F}$, as convolution integrals that account for the dependency of these forces on the accelerations at different time steps (e.g., Bertram, 2000):

$$\vec{F}(t) = \int_{-\infty}^{u} K(\tau)\vec{u}(\tau)d\tau \qquad (4.7.1)$$

where the matrix $K(\tau)$ is determined from potential flow computations of the individual strips at different immersion drafts and inclination angles of the waterline. The velocity $\vec{u}$ contains not only ship motions, but also the incident wave particle velocities. The waves are generated by the ship at time $(t - \tau)$. As they are still present at time t, they will continue to exert forces and moments on the ship.

Although time domain simulations of ship response can be carried out with memory effects being updated at each time step by recalculating all convolution integrals, it is numerically far more efficient to use a finite state space approximation as advocated by Schmiechen (1973). The GLSIMBEL code makes use of this alternative. Instead of relying on the proportionality between force and acceleration, a relation is established between acceleration and a few of its time derivatives on the one hand and of the force and a few of its time derivatives on the other hand:

$$A_0\ddot{u} + A_1\frac{\partial}{\partial t}\ddot{u} + A_2\frac{\partial^2}{\partial t^2}\ddot{u} + \cdots = B_0 + B_1\dot{f} + B_2\ddot{f} + \cdots \qquad (4.7.2)$$

where f is the force per unit length, and $\ddot{u}$ is the acceleration. Both f and $\ddot{u}$ are three-component column vectors. Matrices A_0, A_1, A_2, ... and B_0, B_1, B_2, ... are 3 × 3 complex matrices that do not depend on frequency, but they still depend on section shape, submergence depth, and heel, and thus implicitly on time. They are computed from the frequency dependent added mass and damping coefficients by regression analysis. Three degrees of freedom have to be considered for each strip, namely, the transverse and vertical translations and the roll motion. Thus, to enhance the numerical efficiency of simulations, these matrices are determined for a sufficient number of (encounter) frequencies, immersion drafts, and inclination angles before starting the simulations. During a simulation, values for the actual immersion and waterline inclination are obtained by interpolation.

Some nonlinear strip theory based simulation methods treat the radiation and diffraction forces linearly. Comparison between linear and nonlinear predictions indicated that this computationally efficient approach is capable of accounting for the predominant nonlinearities in wave-induced motions and loads. McTaggart and de Kat (2000), for example, applied the FREDYN code for a capsize risk analysis to investigate the safe design and operation of intact naval frigates in severe seas. This code expresses the radiation and diffraction forces linearly as follows:

$$f_{jk}(t) = -a_{jk}^{\infty}\ddot{x}_k(t) - c_{jk}x_k(t) + \int_0^{\infty} K_{jk}(t-\tau)\dot{x}_k(\tau)d\tau \tag{4.7.3}$$

Here $f_{jk}(t)$ is the force (or moment) contribution in direction j due to a force (or moment) from direction k, a_{jk}^{∞} is the infinite frequency added mass, c_{jk} is the frequency independent restoring force (or moment) coefficient, $x_k(t)$ is the time dependent ship displacement for direction k, $\dot{x}_k(t)$ is the time dependent ship velocity for direction k, and $\ddot{x}_k(t)$ is the time dependent ship acceleration for direction k. Indices j and k range from 1 to 6, corresponding to the surge, sway, heave, roll, pitch, and yaw directions.

The memory function K_{jk} accounts for the past history of the ship's motion. It may be computed using the inverse Fourier cosine function of the frequency dependent damping coefficients $b_{jk}(\omega_e)$ as follows:

$$K_{jk}(t) = \frac{2}{\pi}\int_0^{\infty} b_{jk}(\omega_e)\cos(\omega t)d\omega_e \tag{4.7.4}$$

where ω_e is wave encounter frequency. This formulation is not unique. Alternative expressions involving convolutions with the displacement or acceleration of the ship can also be developed (e.g., Bingham et al., 2009). In practice a_{jk}^{∞} is usually not known; however, damping coefficients $b_{jk}(\omega_e)$ can be used to compute K_{jk}, and a_{jk}^{∞} can then be obtained as follows:

$$a_{jk}^{\infty} = a_{jk}(\omega_e) - \frac{1}{\omega}\int_0^{\infty} K_{jk}(t)\sin(\omega t)dt \tag{4.7.5}$$

To carry out the time domain simulations, the memory effects must be updated at each time step by recalculating all convolution integrals in (4.7.3).

The hull resistance can be regarded as part of the radiation and diffraction forces. Numerous methods are employed to determine this force. They are generally based on the calm water characteristics and depend on the instantaneous ship speed and sinkage. For higher speeds in shallow waters, the squat of the ship has to be taken into account.

MANEUVERING FORCES

Because the ship may perform rudder-induced and wave-induced yaw and sway motions, the simulation methods must include hull forces due to drift and yaw motions. Here quasi-steady formulations used for maneuvering simulations in still water define hull forces in the horizontal plane. Included are empirical relations that consist of linear as well as nonlinear terms to account for viscous forces caused by flow separation.

VISCOUS ROLL DAMPING

Simulation methods almost always include additional roll damping terms to account for viscous effects caused by the hull and the bilge keels. Thus, the damping moment d_D comprises a linear term, proportional to the angular roll velocity, as well a quadratic term, proportional to the square of the angular roll velocity:

$$d_D = -b_L\dot{\varphi} - b_Q\dot{\varphi}|\dot{\varphi}| \tag{4.7.6}$$

The coefficient b_L contains a small speed-independent part caused by wave generation of the rolling ship and another (in most cases larger) part that is proportional to the forward speed of the ship and is caused by lift forces generated by hull, propeller, and rudder. Either model test data or slender body theory computations serve to approximate this coefficient. The coefficient b_Q accounts for bilge keel effects. If the ship is equipped with active (steered) roll stabilizing fins, an additional term must be included. Fixed fins are treated as an appendage.

PROPELLER THRUST

The propeller thrust is estimated from a functional relationship of the propeller rate of rotation, the propeller diameter, the thrust-deduction coefficient at the propeller, the propeller thrust coefficient, and the instantaneous advance coefficient. A motion equation for the propulsion system determines the propeller rate of rotation. A polynomial regression can effectively determine the longitudinal propeller thrust coefficient for series propellers. The determination of the advance coefficient accounts for the longitudinal component of the orbital velocity and the time-dependent ship speed. The propeller thrust is reduced in case of propeller emergence.

In oblique flow caused by yaw and drift motions of the ship and by orbital motion in waves, a propeller exerts a substantial transverse force. This transverse force can be estimated if reliable data are available from, for example, measurements in cavitation tunnels, CFD computations, or still water maneuvering simulations.

RUDDER AND APPENDAGE FORCES

As in maneuvering simulations, forces on the rudder are determined by considering the orbital fluid motion of the undisturbed wave and the time-varying emergence of the upper part of the rudder in the seaway. Usually, either a lifting line method or a panel method is employed. The propeller slipstream and the maximum lift (stall angle) need to be known. The rudder angle from an autopilot can be specified. Appendages are treated in a similar way as the rudder.

WIND FORCES

Wind forces are important for ships with large lateral areas above the still water line. Standard semiempirical methods are generally used to model wind forces. Wind tunnel tests are the preferred choice for more accurate estimates. Such tests can be performed quickly at relatively low cost. Several prototype applications demonstrated the capability of CFD computations to predict air flow about complex ship geometries. However, CFD computations are not yet competitive, because grid generation is too time consuming and expensive for most applications.

THREE-DIMENSIONAL NONLINEAR POTENTIAL FLOW METHODS

Potential flow-based time domain methods that use Green functions provide computationally efficient tools with various levels of approximation in analyzing nonlinear phenomena. Generally, these methods only evaluate the Froude-Krylov and restoring forces exactly, while maintaining the linear aspect of radiation and diffraction forces and moments. Originally, Lin and Yue (1991) included the nonlinear effects in the radiation and diffraction components, using the transient Green function. This is incorporated in the LAMP software, Version 4. In this "body-nonlinear" approximation the free surface boundary condition is linear, but the body boundary condition is satisfied on the instantaneous wetted surface (ISSC, 2006). Kataoka et al. (2002) used this approximation. Although they focused on improvements of numerical efficiency, their studies also showed that the influence of nonlinearities on the pressure were largely due to the hydrostatic restoring forces, thus providing confirmation as to why this simple approximation works.

Solution of the fully nonlinear potential flow problem is still in the development stage. Methods based on solving the radiation and diffraction problem in a linear way provide a computationally efficient method. Sen (2002) applied this approach to Wigley and Series 60 hull forms for a range of forward speeds and wave characteristics. Comparison between linear and nonlinear predictions indicated that this approach is capable of accounting for the predominant nonlinearities in wave-induced motions and loads (ISSC, 2006). The WASIM code is an example of a code widely used by the industry that solves the radiation and diffraction problem in a linear way while evaluating the Froude-Krylov and restoring forces exactly. It was developed in cooperation between MIT and DNV and was previously known as DNV-SWAN (Kring et al., 1997). This code exists at different levels of sophistication and is, at its most advanced level, based on a large-amplitude three-dimensional time domain solution with forward ship speed.

The WASIM code is based on a three-dimensional Rankine panel method (Pastoor and Tveitnes, 2003). Panels are situated on the hull and the free surface, and a numerical beach is situated at the far ends of the free surface. Viscous damping is included, usually tuned by empirical methods that rely on data from model tests. The motion equations are solved in an Eulerian frame, thus allowing for large amplitude motions. The incident waves are modeled according to linear wave theory. Long- as well as short-crested irregular seaways can be simulated.

A special feature of WASIM is the inclusion of slamming effects. Preprocessed calculations for a set of two-dimensional strips of the hull are performed, based on the two-dimensional boundary element method developed under the MARIN-CSR research program and documented by Zhao et al. (1996). Typically, the

Table 4.7.1 Main Particulars of Sample Containership

Capacity	8400 TEU
Length overall	332.0 m
Length between perpendiculars	317.3 m
Beam	43.2 m
Draft at aft perpendicular	14.9 m
Draft at forward perpendicular	14.1 m
Ship mass	141023 t
Longitudinal distance of center of gravity from aft perpendicular	150.3 m
Vertical distance of center of gravity from base line	18.4 m
Metacentric height	2.0 m

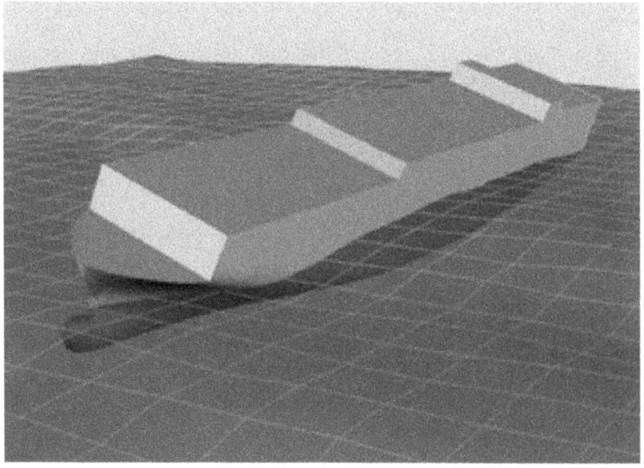

Figure 4.7.1 Screen shot of sample containership in the simulated seaway.

slamming sensitive part of the hull is subdivided into strips of width similar to the panel dimensions. The code uses a simplified free surface boundary condition that makes it possible to describe the impact pressure as the sum of a time-invariant potential solution and a pressure coefficient term, where both terms are functions of only the position on the hull and the depth of immersion. In this way, slamming computations need to be performed only for a large number of immersion depths, but not for different entrance velocities. As WASIM treats the ship as a rigid body, a hull girder flexible analysis is required to obtain realistic load contributions from slamming.

SAMPLE RESULTS

To illustrate the kind of results obtained from nonlinear simulations, let us present computed wave-induced motions and loads of a typical fully loaded modern containership in an irregular short-crested seaway, characterized by a Pierson-Moskowitz spectrum having a significant wave height of 10.0 m and a mean period of 13.9 sec. The predominant wave direction relative to the ship's heading is 170 deg (180 deg. refer to head seas). A cosine squared spreading function, which is independent of the frequency, describes the seaway's short-crestedness. The ship advances at constant mean forward speed of 15.6 knots, corresponding to a Froude number of 0.144. The ship s rudder is held constant at its zero position. Table 4.7.1 lists the main particulars of the ship, and Fig. 4.7.1 shows a screen shot of the ship in the simulated seaway. Figure 4.7.2 shows the time history of the simulated wave elevation at the ship's midship section, and Fig. 4.7.3 shows time histories of the resulting motion response. Figures 4.7.4 to 4.7.6 show the corresponding sectional loads acting at the three ship sections located, respectively, at about 25 percent of the ship's length from the aft perpendicular, at amidships, and at about 75 percent of the ship's length from the aft perpendicular. These results were obtained using the nonlinear strip theory code GLSIMBEL.

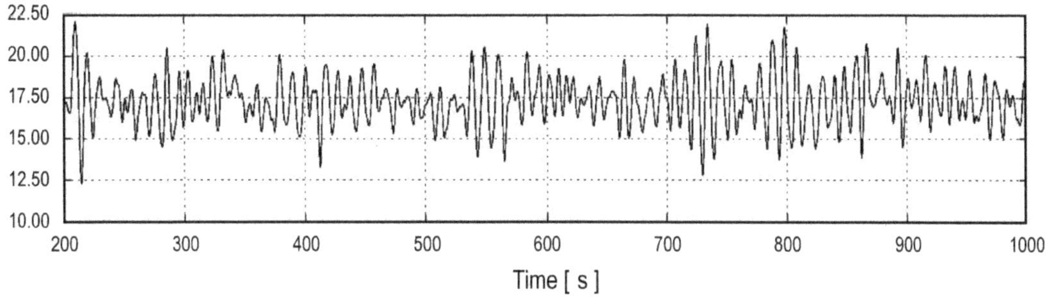

Figure 4.7.2 Time series of simulated wave elevation (m) at the ship's midship section.

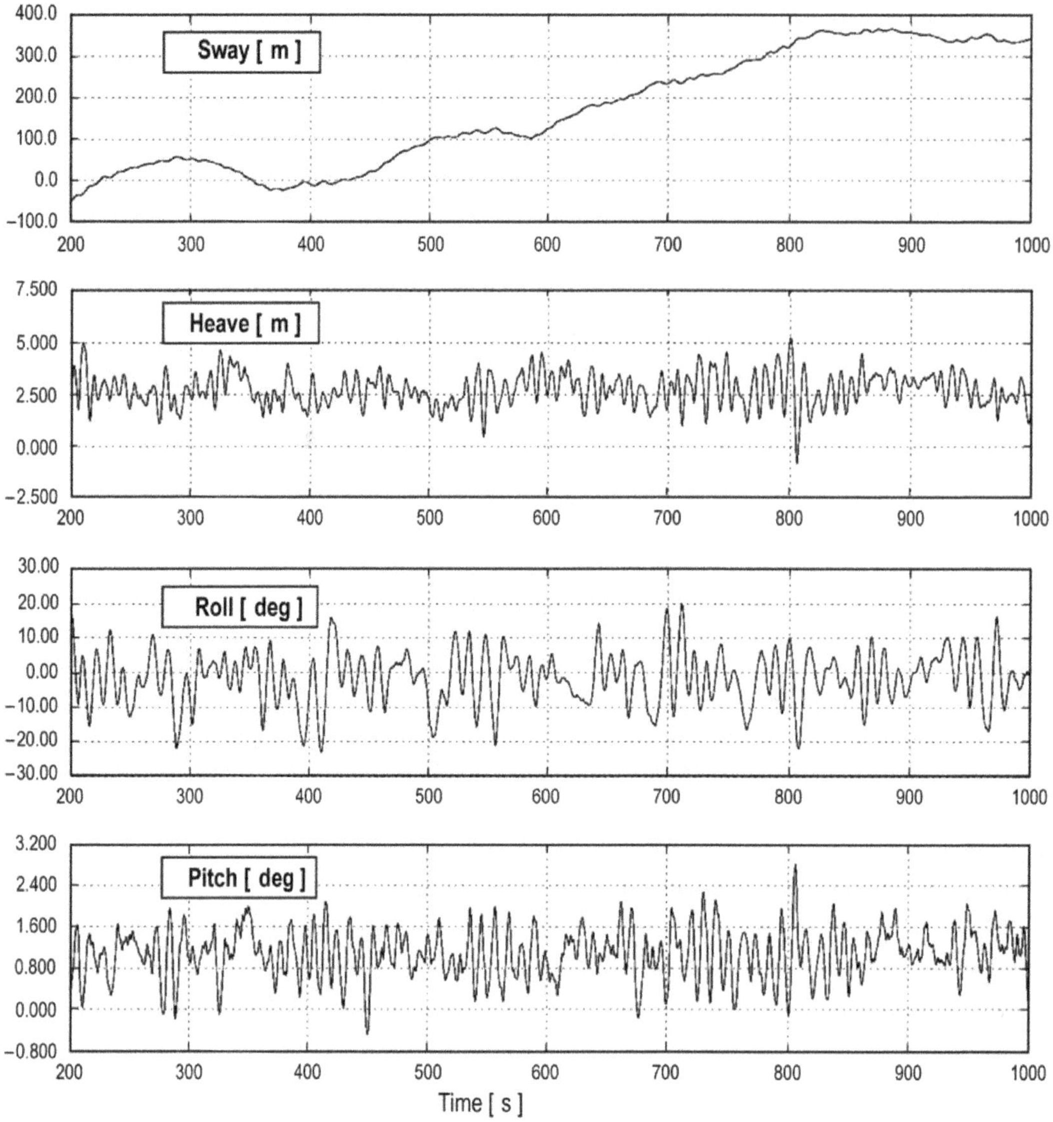

Figure 4.7.3 Time series of simulated ship motions.

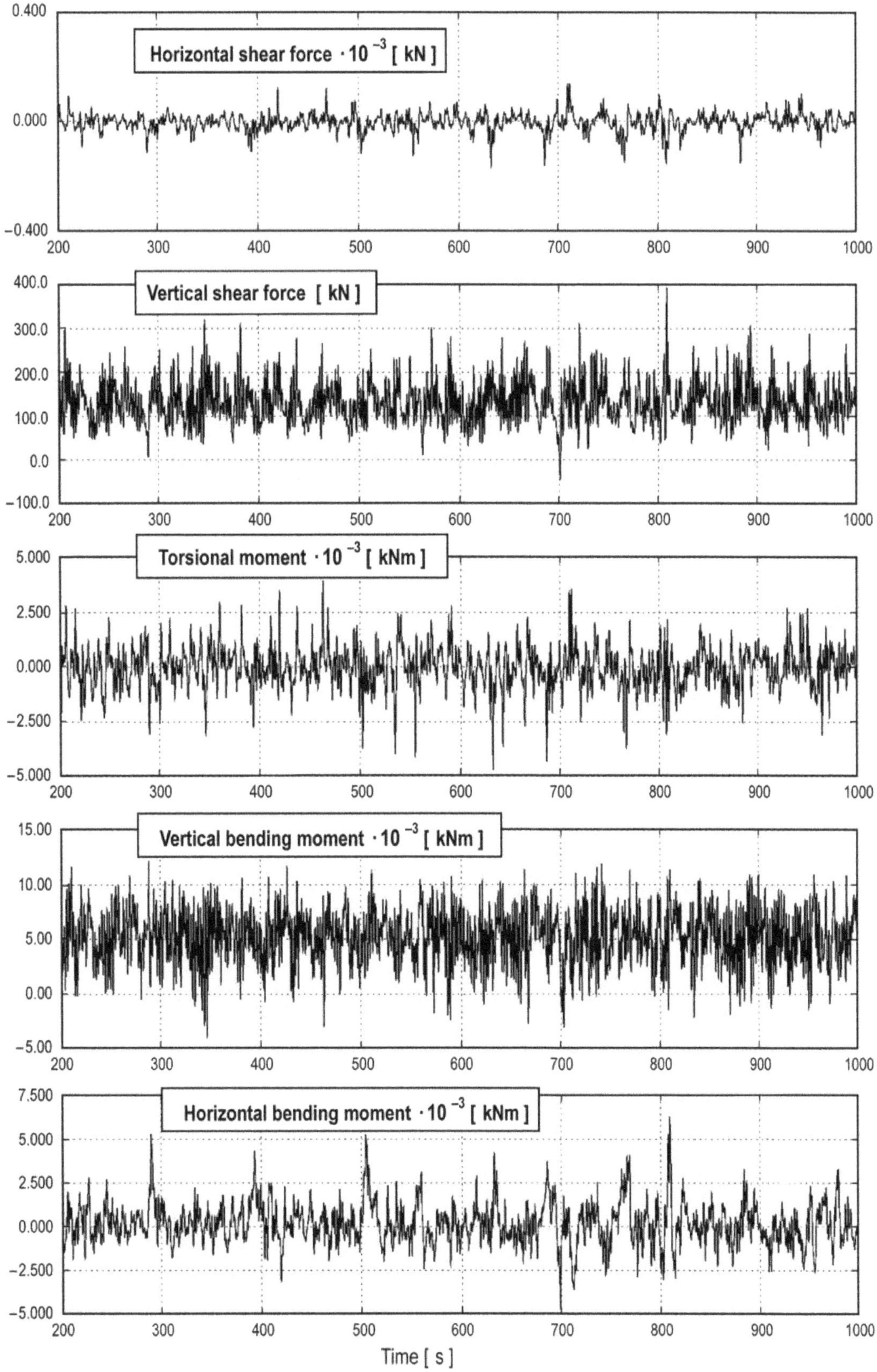

Figure 4.7.4 Time series of sectional loads at the distance of 78.0 m from aft perpendicular.

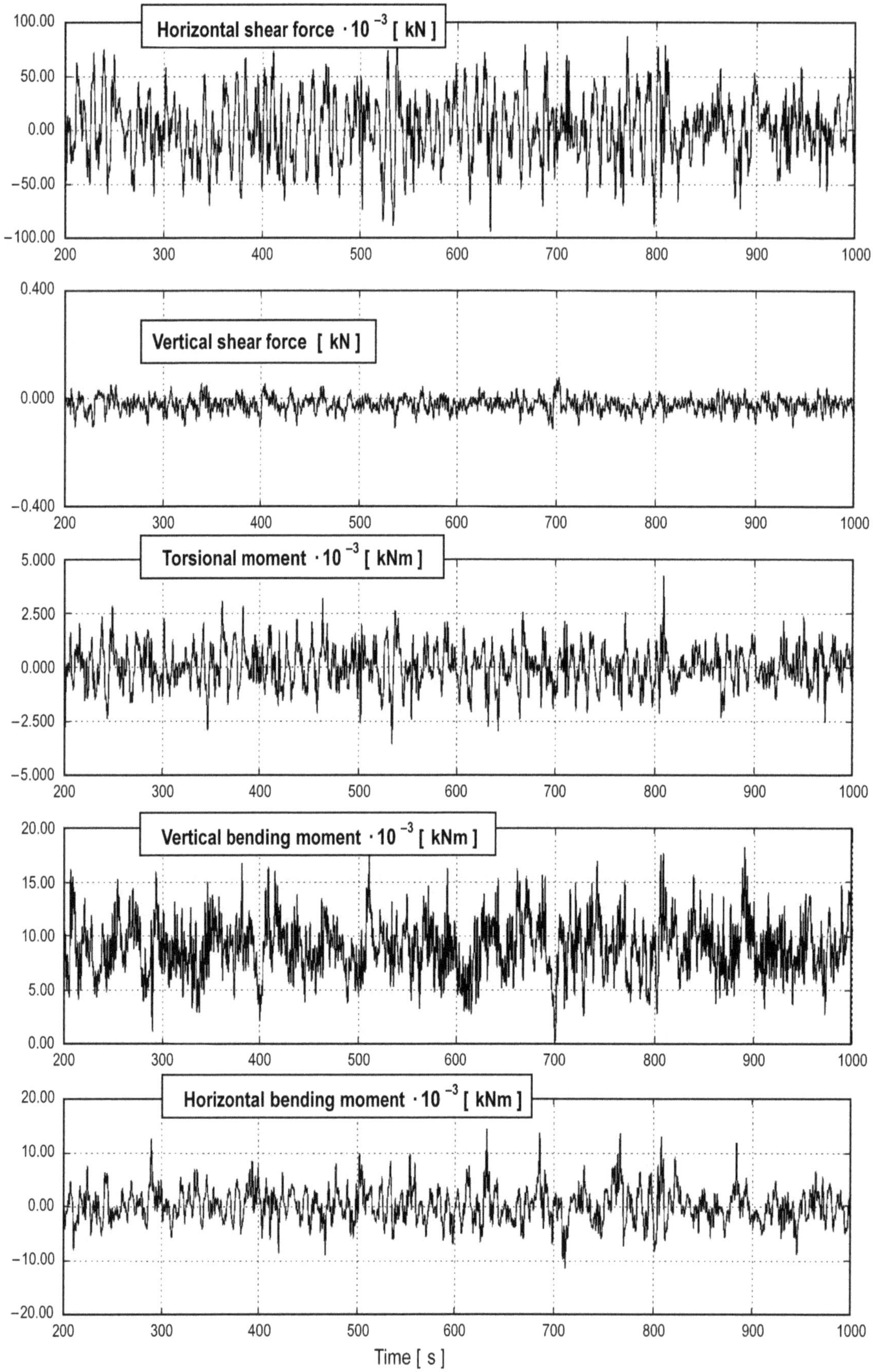

Figure 4.7.5 Time series of sectional loads at amidships.

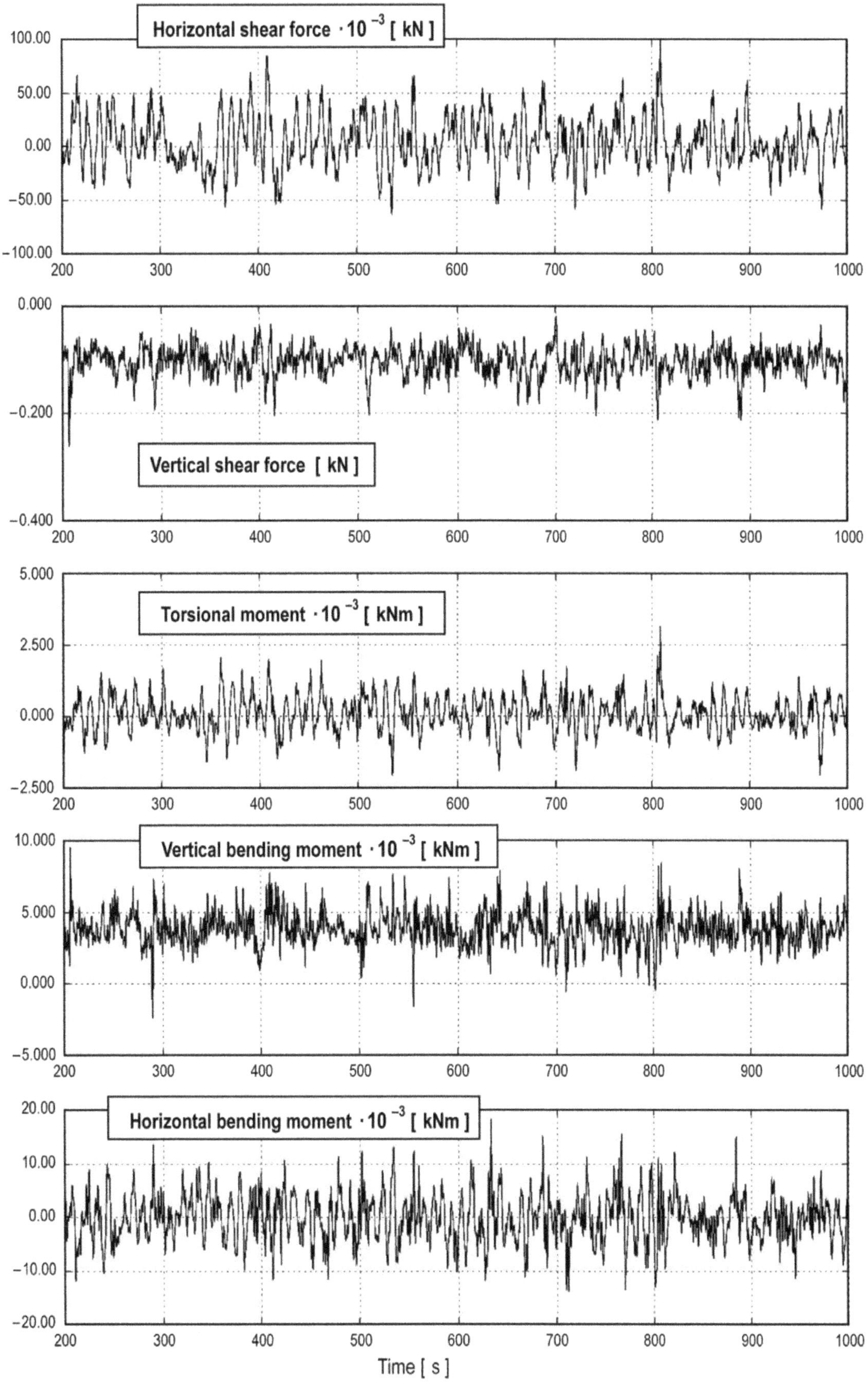

Figure 4.7.6 Time series of sectional loads at the distance of 238.0 m from aft perpendicular.

REFERENCES

Adegeest, L. J. M., Braathen, A., and Vada, T. (1998). Evaluation of methods for estimation of extreme nonlinear ship response based on numerical simulations and model tests. *Proc. 22nd Symp. on Naval Hydrodynamics*. Washington, DC: National Academy Press, 70–84.

Ba, M., and Guilbaud, M. (1995). A fast method of evaluation for the translating and pulsating green function. *J. Ship Technology Research*, Vol. 42, 68–80.

Baarholm, G. S., and Jensen, J. J. (2004). Influence of whipping on long-term vertical bending moment. *Journal of Ship Research*, Vol. 48, 261–272.

Baarholm, G. S., and Moan, T. (2002). Efficient estimation of extreme long-term stresses by considering a combination of longitudinal bending stresses. *J. Marine Science and Technology*, Vol. 6, Issue 3, 122–134.

Bales, S. L., Lee, W. T., and Voelker, J. M. (1981). *Standardized wave and wind environments for NATO operational areas*. DTNSRDC Report SPD-0919-01.

Beck, R. F., and Reed, A. M. (2001). Modern computational methods for ships in a seaway. *SNAME Trans.*, Vol. 109, 1–51.

Beck, R. F., Reed, A. M., and Rood, E. P. (1996). Application of modern numerical methods in marine hydrodynamics. *SNAME Trans.*, Vol. 104, 519–537.

Beck, R. F., and Magee, A. R. (1991). Time-domain analysis for predicting ship motions. In Price, W. G., Temarel, P., and Keane, A. J. (Eds.). *Dynamics of marine vehicles and structures in waves*. Maryland Heights, MO: Elsevier Science Publishers, 49–64.

Beiersdorf, C., and Rathje, H. (2000). *Loads for the bow region of a 6700 TEU containership design*. Report FG 2000.060A. Hamburg: Germanischer Lloyd (in German).

Bertram, V. (2000). *Practical ship hydrodynamics*. Oxford, England: Butterworth-Heinemann.

Bertram, V., and Yasukawa, H. (1996). Rankine source methods for seakeeping problems. *Jahrbuch Schiffbautechnische Gesellschaft*, Vol. 90, 411–425.

Bingham, H. B., Madsen, P. A., and Fuhrman, D. R. (2009). Velocity potential of highly accurate bouissinesq-type models. *Coastal Engineering*, Vol. 56, Issue 4, 467–478.

Blume, P. (1979). Experimentally determined coefficients for effective roll damping and application to estimate extreme roll angles. *J. Ship Technology Research*, Vol. 26, 3–23 (in German).

Blume, P. (1999). *Measurement of hydrodynamic forebody forces of a containership model in waves*. Report S 384/99. Hamburg: HSVA.

Böttcher, H. (1986). Ship motion simulation in a seaway using detailed hydrodynamic force coefficients. *Proc. 3rd Conf. on Stability of Ships and Ocean Vehicles* (STAB '86), Gdansk.

Bretschneider, C. L., Darbyshire, J., Neumann, G., Pierson, W. J., Walden, H., and Wilson, B. W. (1962). Data for high wave conditions observed by the OWS "Weather Reporter" in December 1959. *Deutsche Hydrographische Zeitschrift*, Vol. 15, Issue 6, 243–255.

Bretschneider, C. L. (1959). *Wave variability and wave spectra for wind-generated gravity waves*. Technical Memorandum No. 118. Washington, DC: Beach Erosion Board, U.S. Army Corps of Engineers.

Brunswig, J., and el Moctar, O. (2004). Prediction of ship motions in waves using RANSE. *Proc. 7th Numerical Towing Tank Symposium*, Hamburg.

Cabos, C., Eisen, H., and Krömer, M. (2006) GL ShipLoad: An integrated load generation tool for FE analysis. *Proc. of 5th Int. Conf. on Computer Application and Information Technology in the Maritime Industries (COMPIT)*, Delft University of Technology, Oegstgeest, The Netherlands, 199–210.

Cartwright, D. E., and Longuet-Higgins, M. S. (1956). The statistical distribution of the maxima of a random function. *Proc. Royal Society*, Series A, Vol. 237, 212–232.

CD-adapco. (2002). User Manual COMET, Version 2.0, Nürnberg.

Chapman, R. B. (1975). Free-surface effects for yawed surface piercing plates. *J. Ship Research*, Vol. 11, 190–198.

Chen, H., Chen, H. H., and Hoffman, D. (1979). The implementation of the 20-year hindcast wave data in the design and operation of marine structures. *Proc. Offshore Technology Conf. (OTC)*, Houston.

Crandall, S., and Marc, W. (1963). *Random vibration in mechanical systems*. New York: Academic Press.

David, H. A. (1970). *Order statistics*. New York: John Wiley & Sons.

Det Norske Veritas. (1976). *Environmental conditions of the Norwegian continental shelf*. Oslo, Norway: Petroleum Directorate.

Dietz, J. S., Friis-Hansen, P., and Jensen, J. J. (2004). Most likely response waves for estimation of extreme value ship response statistics. *Proc. 9th Int. Symp. on Practical Design of Ships and Other Floating Structures, PRADS 2004*, Lübeck-Travemünde, Germany, 286–293.

Eisen, H., and Cabos, C. (2007). Efficient generation of CFD-based loads for the FEM-analysis of ship structures. *Proc. Int. Conf. on Applications in Shipbuilding (ICCAS)*, Vol. II, Portsmouth, U.K., 91–98.

el Moctar, O., Schellin, T. E., and Priebe, T. (2006). CFD and FE methods to predict wave loads and ship structural response. *Proc. 26th Symp. on Naval Hydrodynamics*, Rome, Italy. Washington, DC: National Academies Press.

el Moctar, O., Brehm, A., and Schellin, T. E. (2004). Prediction of slamming loads for ship structural design using potential flow and RANSE codes. *Proc. 25th Symp. on Naval Hydrodynamics*, St. John's. Washington, DC: National Academies Press, Vol. 4, 116–129.

Faltinsen, O. M. (1990). *Sea loads on ships and offshore structures*. Cambridge: Cambridge University Press.

Ferziger, J., and Peric, M. (1996). *Computational methods for fluid dynamics*. Berlin: Springer-Verlag.

Fisher, R. A., and Tippett, L. H. C. (1928). Limiting forms of the frequency distribution of the largest or smallest member of a sample. *Proc. Cambridge Philosophical Society*, Vol. 24, 180–190.

Folso, L., and Rizzuto, E. (2003). Equivalent waves for sea loads on ship structures. *Proc. 23rd Int. Conf. on Offshore Mechanics and Arctic Eng. (OMAE)*, Paper No. 37439.

Fries-Hansen, P., and Nielsen, L. P. (1995). On the new wave model for the kinematics of large ocean waves. *Proc. 14th Int. Conf. on Offshore Mechanics and Arctic Eng. (OMAE)*, Vol. 1A, 17–24.

Germanischer Lloyd. (2006). Rules for classification and construction, I—ship technology, part 1—sea-going ships, chapter 1—hull structures. Hamburg: Author.

Germanischer Lloyd. (2007). Rules for classification and construction, V—analysis techniques, part 1—strength and stability, chapter 1—guidelines for strength analyses for ship structures with the finite element method, section 2—global strength analysis for containership structures. Hamburg: Author.

Guedes Soares, C., Fonseca, N., and Pascoal, R. (2004). Long-term prediction of nonlinear vertical bending moments on a fast monohull. *J. Applied Ocean Research*, Vol. 26, Issue 6, 288–297.

Guedes Soares, C., and Schellin, T. E. (1996). Long-term distribution of nonlinear wave-induced vertical bending moments on a containership. *J. Marine Structures*, Vol. 9, Issues 3–4, 333–352.

Guedes Soares, C., and Schellin, T. E. (1998). Nonlinear effects on long-term distributions of wave-induced loads for tankers. *J. Offshore Mechanics and Arctic Engineering*, Vol. 120, 65–70.

Gumbel, E. (1966). *Statistics of extremes*. New York: Columbia University Press.

Hachmann, D. (1991). The calculation of pressures on a ship's hull in waves. *J. Ship Technology Research*, Vol. 38, 111–133.

Hachmann, D. (1986). Determination of the wave elevation at ship sections based on pressure variations at the design waterline under the influence of the Smith Effect. Report MTK 325 II, Hamburg: Germanischer Lloyd (in German).

Himeno, Y. (1981). *Prediction of ship roll damping—state of the art*. Report No. 239, Dept. of Naval Architecture and Marine Engineering, University of Michigan, Ann Arbor.

Hogben, N., Dacunha, N. M., and Olliver, G. F. (1986). *Global wave statistics*. Teddington, England: British Maritime Technology Ltd..

Hoffman, D. (1975). *Wave data application for ship response predictions*. Glen Cove, NY: Webb Institute of Naval Architecture.

Hughes, O. F. (1988). *Ship structural design*. Jersey City, New Jersey: SNAME.

Huston, W. B., and Skopinski, T. H. (1956). *Probability and frequency characteristics of some flight buffet loads*. NACA Tech. Note 3733.

International Association of Classification Societies Ltd. (2001). *Recommendation No. 34, standard wave data*. London, U.K.: International Association of Classification Societies Ltd.

International Association of Classification Societies Ltd. (2006a). *Common structural rules for bulk carriers*. London, U.K.: International Association of Classification Societies Ltd.

International Association of Classification Societies Ltd. (2006b). *Common structural rules for double hull oil tankers*. London, U.K.: International Association of Classification Societies Ltd.

Iijima, K., Shigemi, T., Miyake, R., and Kumano, A. (2004). A practical method for torsional strength assessment of containership structures. *J. Marine Structures*, Vol. 17, 355–384.

International Ship and Offshore Structures Congress (ISSC). (1994). Technical Committee I.2: Loads. *Proc. 12th Int. Ship and Offshore Structures Congress,* University of Newfoundland, St. John's.

International Ship and Offshore Structures Congress (ISSC). (2003). Technical Committee I.2: Loads. *Proc. 15th Int. Ship and Offshore Structures Congress.* Oxford: Elsevier Science Ltd.

International Ship and Offshore Structures Congress (ISSC). (2006). Technical Committee I.2: Loads. *Proc. 16th Int. Ship and Offshore Structures Congress,* University of Southampton, Southampton.

Iwashita, H., and Ohkusu, M. (1992). The Green Function method for ship motions at forward speed. *J. Ship Technology Research*, Vol. 39, 3–21.

Jensen, J. J., and Pedersen, P. T. (1979). Wave-induced bending moments in ships—a quadratic theory. *RINA Trans.*, Vol. 121, 141–157.

Jensen, J. J., and Pedersen, P. T. (1981). Bending moments and shear forces in ships sailing in irregular waves. *J. Ship Research*, Vol. 4, 243–251.

Kapsenberg, G. K., van't Veer, A. P., Hackett, J. P., and Levadou, M. M. D. (2003). Aftbody slamming and whipping loads. *SNAME Trans.*, Vol. 111, 213–231.

Kataoka, S., Sueyoshi, A., Arihara, K., Iwashita, H., and Takaki, M. (2002). A study on body nonlinear effects on ship motion in waves. *Trans. of the West-Japan Society of Naval Architects*, Vol. 104, 111–120.

Kawabe, H., Ohtani, H., Maeno, Y., Fujii, Y., Iijima, K., and Yao, T. (2005). Probabilistic assessment of ultimate hull girder strength in longitudinal bending. *Proc. 15th Int. Offshore and Polar Eng. Conf. (ISOPE)*, Vol. 4, 773–744.

Korvin-Kroukovsky, B. V., and Jacobs, W. R. (1957). Pitching and heaving motions of a ship in regular waves. *SNAME Trans.*, Vol. 65, 590–632.

Kring, D., Huang, Y.-F., Sclavounos, P., Vada, T., and Braathen, A. (1997). Nonlinear ship motions and wave-induced loads by a rankine method. *Proc. 21st Symp. on Naval Hydrodynamics*. Washington, DC: National Academies Press, 45–63.

Lewandowski, E. M. (2004). The dynamics of marine craft, maneuvering and seakeeping. Singapore: World Scientific Publ. Co.

Lewis, E. V. (1967). Predicting long-term distribution of wave-induced bending moments on ship hulls. *Proc. SNAME Spring Meeting*, Montreal.

Lin, W.-M., Zhang, S., and Yue, D. K. P. (1996). Linear and nonlinear analysis of motions and loads of a ship with forward speed in large-amplitude waves. *Proc. 11th Int. Workshop on Water Waves and Floating Bodies*, Hamburg.

Lin, W.-M., Meinhold, M. J., Salvesen, N., and Yue, D. K. P. (1994). Large-amplitude motions and wave loads for ship design. *Proc. 20th Symp. on Naval Hydrodynamics*, Santa Barbara. Washington, DC: National Academies Press, 205–226.

Lin, W.-M., and Yue, D. K. P. (1991). Numerical solutions for large-amplitude ship motions in the time domain. *Proc. 18th Symp. on Naval Hydrodynamics*, Ann Arbor. Washington, DC: National Academies Press, 41–46.

Liu, D., Spencer, J., Itoh, T., Kawachi, S., and Shigematsu, K. (1992). Dynamic load approach in tanker design. *SNAME Trans.*, Vol. 100, 143–172.

Lukakis, T. A., and Grivas, S. B. (1980). A method for establishing ship design wave bending moment and its comparison with classification societies' rules. *Ocean Engineering*, Vol. 7, 357–371.

McTaggart, K., and de Kat, J. O. (2000). Capsize risk of intact frigates in irregular seas. *SNAME Trans.*, Vol. 108, 147–177.

Michel, W. H. (1999). Sea spectra revisited. *Marine Technology*, Vol. 36, Issue 4, 211–227.

Miles, M. (1972). *Wave spectra estimates from a stratified sample of 323 North Atlantic wave records*. Report 118A, Division of Mechanical Engineering, National Research Council, Ottawa, Canada.

Minoura, M., and Naito, S. (2004). A stochastic model for evaluation of seakeeping performance. *Proc. 14th Int. Offshore and Polar Eng. Conf. (ISOPE)*, Vol. 4, 331–338.

Moskowitz, L., Pierson, W. J., and Mehr, E. (1962). Wave spectra estimated from wave record obtained by OWS "Weather Explorer" and OWS "Weather Reporter." Report I, New York University, College of Engineering, Dept. of Meteorology and Oceanography, New York.

Moskowitz, L., Pierson, W. J., and Mehr, E. (1963). Wave spectra estimated from wave record obtained by OWS "Weather Explorer" and OWS "Weather Reporter." Report II, New York University, College of Engineering, Dept. of Meteorology and Oceanography, New York.

Moskowitz, L., Pierson, W. J., and Mehr, E. (1965). Wave spectra estimated from wave record obtained by OWS "Weather Explorer" and OWS "Weather Reporter." Report III, New York University, College of Engineering, Dept. of Meteorology and Oceanography, New York.

Muzaferija, S., and Peric, M. (1998). Computation of free-surface flows using interface-tracking and interface-capturing methods. In Mahrenholtz, O., and Markiewicz, M. (Eds.). *Nonlinear Water Wave Interaction*. Southampton, U.K.: Computational Mechanics Publ., 59–100.

Müller, G., and Groth, C. (2002). *Practical application of FEM Code ANSYS, Vol. 1: Basics*, (7th ed.). Berlin: Expert Verlag (in German).

Nakos, D. E., Kring, D. C., and Sclavounos, P. D. (1993). Rankine panel methods for transient free-surface flows. *Proc. 6th Int. Conf. on Numerical Ship Hydrodynamics*, Iowa City, 613–634.

Newman, J. N. (1978). The theory of ship motions. *J. Applied Mechanics*, Vol. 18, 222–283.

Ochi, M. K. (1978). Wave statistics for the design of ships and ocean structures. *SNAME Trans.*, Vol. 86, 47–76.

Ochi, M. K. (1982). Stochastic analysis and probabilistic prediction of random seas. *Advances in Hydroscience*, Vol. 13. Washington, DC: National Academic Press, 217–375.

Ochi, M. K., and Hubble, E. N. (1976). On six-parameter wave spectra. *Proc. 15th Coastal Engineers Conf.*, ASCE.

Ochi, M. K., and Motter, L. E. (1974). Prediction of extreme ship responses in rough seas of the North Atlantic. *Proc. Symp. on the Dynamics of Marine Vehicles and Structures in Waves.*

Ogawa, Y. (2003). Long-term prediction method for the green water load and volume for an assessment of the load line. *J. Marine and Science Technology*, Vol. 7, 137–144.

Papanikolaou, A. D., and Schellin, T. E. (1992). A three-dimensional panel method for motions and loads of ships with forward speed. *J. Ship Technology Research*, Vol. 39, 147–156.

Pastoor, L. W., Bloch, H. J., and Bitner-Gregersen, E. (2003). Time simulation of ocean-going structures in extreme waves. *Proc. 22nd Int. Offshore and Arctic Eng. Conf.*, ASME paper OMAE 03-37490, Cancun, Mexico.

Pastoor, L. W., and Tveitnes, T. (2003). Rational determination of nonlinear design loads for advanced vessels. *Proc. 7th Int. Conf. on Fast Sea Transportation (FAST 2003)*, Vol. 2, Univ. of Naples, Ischia, 41–48.

Pastoor, L. W. (2002). On the assessment of nonlinear ship motions and loads. Ph.D. Thesis, Delft University of Technology, Delft.

Payer, H. G., and Fricke, W. (1994). Rational dimensioning and analysis of complex ship structures. *SNAME Trans.*, Vol. 102, 395–417.

Pierson, W. J., Neumann, G., and James, R. W. (1955). Practical methods for obtaining and forecasting ocean waves by means of wave spectra and statistics. Washington, DC: Hydrographic Office Publication No. 603.

Pierson, W. J. (1952). *A unified mathematical theory for the analysis of propagation and refraction of storm-generated ocean surface waves, Parts I and II.* New York: New York University, College of Engineering, Dept. of Meteorology and Oceanography,

Pereira, R. (1995). Program SIMBEL V.6.1—user manual. Report 651/7/1041-001, MTG Marinetechnik GmbH, Hamburg (in German).

Pereira, R. (1988). Simulation of nonlinear sea loads. *J. Ship Technology Research*, Vol. 35, 173–193 (in German).

Rathje, H., Schellin, T. E., Otto, S., and Östergaard, C. (2000). Predicting nonlinear wave-induced design loads for ships. *Proc. 19th Int. Offshore and Arctic Eng. Conf.,* ASME paper OMAE 00-6122, New Orleans.

Rathje, H., and Schellin, T. E. (1997). Viscous effects in seakeeping prediction of twin-hull ships. *J. Ship Technology Research*, Vol. 44, 44–52.

Rörup, J., Schellin, T. E., and Rathje, H. (2008). Load generation for structural strength analysis of large containerships. *Proc. 27th Int. Offshore and Arctic*

Eng. Conf., ASME paper OMAE 2008-57121, Estoril, Portugal.

Roll, H. U. (1953). Height, length, and steepness of sea waves in the North Atlantic and dimensions of sea waves as functions of wind force. T & R Bulletin No. 1-19, SNAME.

Salvesen, N., Tuck, E. O., and Faltinsen, O. (1970). Ship motions and sea loads. *SNAME Trans.*, Vol. 78, 250–287.

Sames, P. C., and Schellin, T. E. (2001). Assessment of sloshing loads for tnkers. *Proc. 8th Int. Symp. on Practical Design of Ships and Other Floating Structures (PRADS 2001)*, Shanghai, 637–643.

Schellin, T. E., and el Moctar, O. (2007). Numerical prediction of impact-related wave loads on ships. *J. Offshore Mechanics and Arctic Engineering*, Vol. 129, 39–47.

Schellin, T. E., and Perez de Lucas, A. (2004). Longitudinal strength of a high-speed ferry. *J. Applied Ocean Research*, Vol. 26, 298–308.

Schellin, T. E., Beiersdorf, C., Chen, X.-B., Fonseca, N., Guedes Soares, C., Maron Loureiro, A., et al. (2003). Numerical and experimental investigation to evaluate wave-induced global design loads for fast ships. *SNAME Trans.*, Vol. 111, 437–461.

Schellin, T. E., Östergaard, C., and Guedes Soares, C. (1996). Uncertainty assessment of low frequency load effects for containerships. *J. Marine Structures*, Vol. 9, Issues 3–4, 313–352.

Schmiechen, M. (1973). On state space models and their application to hydrodynamic systems. Univ. of Tokyo, Dept. of Naval Arch., NAUT Report 5002.

Sclavounos, P. D. (1996). Computations of wave ship interactions. In Ohkusu, M. (Ed.). *Advances in Marine Hydrodynamics*. Southhampton: Computational Mechanics Publ., 233–278.

Sclavounos, P. D., Nakos, D. E., and Huang, Y. (1993). Seakeeping and wave induced loads on ships with flare by a Rankine panel method. *Proc. 6th Int. Conf. on Numerical Ship Hydrodynamics*, Iowa City, 57–78.

Sen, D. (2002). Time-domain computation of large amplitude 3D ship motions with forward speed. *J. Ocean Engineering*, Vol. 29, 973–1002.

Shi, B., Liu, D., and Wiernicki, C. (2005). Dynamic loading approach for structural evaluation of ultra large container carriers. *SNAME Trans.*, Vol. 113, 402–417.

Shin, C. H., Ha, T. B., and Choi, S. J. (2004). A comparison study of measured hull girder stress using a hull monitoring system on board and theoretically calculated ones. *Proc. 9th Int. Symp. on Practical Design of Ships and Other Floating Structures (PRADS)*, Vol. 1, 93–97.

Smith, C. S. (1966). Measurement of service stresses in warships. *Proc. Int. Conf. Stresses in Service,* Inst. of Civil Engineers, London, 1–8.

Söding, H. (1982). *Damaged stability in waves.* Report 429. Hamburg: Institut für Schiffbau, Univ. of Hamburg (in German).

Speziale, C. G., and Thangam, S. (1992). Analysis of an RNG based turbulence model for separated flows. *J. Engineering Science*, Vol. 30, 1379–1388.

St. Denis, M., and Pierson, W. J. (1953). On the motions of ships in confused seas. *SNAME Trans.*, Vol. 61, 280–357.

Taylor, P. H., Jonathan, P., and Harland, L. A. (1995). A time domain simulation of jack-up dynamics with extremes of a gaussian process. *Proc. 14th Int. Conf. on Offshore Mechanics and Arctic Eng. (OMAE)*, ASME, Vol. 1A, 313–319.

Torhaug, R., Winterstein, S. R., and Braathen, A. (1998). Nonlinear ship loads: Stochastic models or extreme response. *J. Ship Research*, Vol. 42, 46–55.

Tromans, P. S., Anaturk, A. H. R., and Hagemeijer, P. (1991). A new model for the kinematics of large amplitude ocean waves—application as a design wave. *Proc. 1st Int. Offshore and Polar Eng. Conf. (ISOPE)*, Vol. 3, 64–71.

von Karman, T. (1929). *The impact on seaplane floats during landing. Washington, DC: NACA TN 321.*

Wagner, H. (1932). *Landing of seaplanes.* Washington, DC: NACA TN 622.

Walden, H. (1964). Die eigenschaften der meereswellen im Nordatlantischen Ozean. Deutscher Wetterdienst, Einzelveröffentlichung No. 41 (in German).

Wang, L., and Moan, T. (2004). Probalistic analysis of nonlinear wave loads on ships using Weibull, generalized gamma, and Pareto distribution. *J. Ship Research*, Vol. 48, 202–217.

Watanabe, I., Tanizawa, K., and Savada, H. (1986). An observation of bottom impact phenomena by means of high speed video and transparent model. *J. Society of Naval Architecture of Japan*, Vol. 164, 120–126.

Wehausen, J. V., and Laitone, E. V. (1960). *Surface waves.* Berlin: Springer, 46–778.

Winterstein, S. R., and Torhaug, R. (1996). Extreme jackup response: Simulation and nonlinear analysis methods. *J. Offshore Mechanics and Arctic Engineering*, Vol. 118, 103–108.

Winterstein, S. R. (1988). Nonlinear vibration models for extremes and fatigue. *J. Engineering Mechanics*, Vol. 114, 1772–1790.

Wu, M. K., and Moan, T. (2004). Direct calculation of design wave loads in a high-speed pentamaran. *Proc. 9th Int. Symp. on Practical Design of Ships and Other Floating Structures (PRADS)*, Vol. 2, 679–688.

Yamamoto, Y., Fujino, M., and Fukasasawa, T. (1980). Motion and longitudinal strength of a ship in head sea and the effects of nonlinearities. *Naval Arch. and Ocean Engineering, Society of Naval Architecture of Japan Trans.*, Vol. 18, 91–100.

Yamanouchi, Y., Unoki, S., and Kanda, T. (1965). *On the winds and waves on the Northern North Pacific Ocean and South adjacent seas of Japan as the environmental conditions of the ship.* Ship Research Institute, Tokyo.

Zhao, R., Faltinsen, O. M., and Aarsnes, J. (1996). Water entry of arbitrary two-dimensional sections with and without flow separation. *Proc. 21st Symp. on Naval Hydrodynamics*, Trondheim. Washington, DC: National Academies Press, 408–423.

RELIABILITY-BASED STRUCTURAL DESIGN

Dominique Béghin
Bureau Veritas, Paris, France (ret)

5.1 PROBABILISTIC APPROACH AND SHIP DESIGN

5.1.1 General

The U. K. House of Lords Select Committee (1992) highlighted that "modern science and technology were not being adequately applied in many of the fields which affect the safety of ships" and that "newer industries were approaching safety regulation in new and better ways." In that respect, nuclear, aeronautics, and space industries compelled to introduce increasingly sophisticated technologies have encouraged the emergence of systems of growing complexity. This made it necessary to not only control the performance but also to provide for malfunctions of any system from design to operation. This necessity gave rise to the emergence of a new science, risk analysis, based on the assessment of

1. *Reliability*, i.e., the ability of a system to ensure its primary role.
2. *Availability*, i.e., the capability of a system to fulfill its function on demand.
3. *Maintainability*, i.e., the ability of a system to be inspected and repaired.
4. *Safety*, i.e., to make evidence that a system complies with the level of safety as defined by the regulator.

The same report proposed to base ship safety on this new RAMS concept and to set up a safety regime based on

1. "Primary safety goals for all aspects of operation," including standards for
   a. Structural strength.
   b. Stability.
   c. Maneuverability and performance in a seaway.
   d. Operational competence and safety management.

Using the same approach as for aeronautics or space industries, these standards should be based on the RAMS concept, including
   a. The determination of acceptable risks. Risk is the danger undesired events represent for humans, the environment, and economic values and may be defined as the product of the probability of occurrence of an adverse event, and the consequences that this undesired event produces.
   b. Quantified risk assessment.
   c. Analysis of costs and benefits.
2. "A safety case for every ship trading commercially produced by the operator and approved and audited by the flag state." In particular, the safety case has to demonstrate that the ship is operated in accordance with the primary goals. In other words, application of the RAMS approach to a complex system enables
   a. Development of the design according to quantified objectives expressed by the permissible probabilities of failure.
   b. Verification of the safety of the system and its environment over its service life.
   c. Cost optimization.

The capacity of a system to operate without major failures, as characterized by its reliability, is therefore ensured by defining performance standards rather than design criteria. This last remark has been pointed out by the U.S. National Academy of Science (1990), which emphasized that "performance standards rather than design criteria should be developed for spill free tanker." This means that we have to move from design for compliance to design for performance.

5.1.2 Review of Risk Analysis Techniques

Quantified risk analyses aim at identifying the hazards and assessing the risks associated with them and include the following six steps:

1. Definition of reliability objectives.
2. Preliminary risk analysis for identification of
   a. Hazards, for example, fire, explosions, collisions, groundings, typhoons, heavy weather conditions, unsafe human behavior, poor communication, defective equipment, etc. Various qualitative techniques and procedures exist for hazard identification, such as failure mode and effect analysis (FMEA), hazard and operability analysis (HAZOP).
   b. Risks associated with those hazards.
   c. Accidental scenarios considering events, breakdowns, and errors leading to hazardous situations or accidents. Various methods exist for scenario identification, such as event tree analysis (ETA), fault tree analysis (FTA), cause-consequence analysis (CCA), escape evacuation and rescue analysis (EER).
   d. Necessary preventive measures for controlling the identified risks.
3. Risk assessment for comparison with acceptable levels of risk:
   a. *Intolerable*, requiring improvement of the design,
   b. *Tolerable but not negligible*, subject to reduction or mitigation on the basis of cost-benefit analysis. This requires further action to make the risk "as low as reasonably practicable," refer to the ALARP diagram of Figure 5.1 as provided by the U.K. Health and Safety Executive (1992).

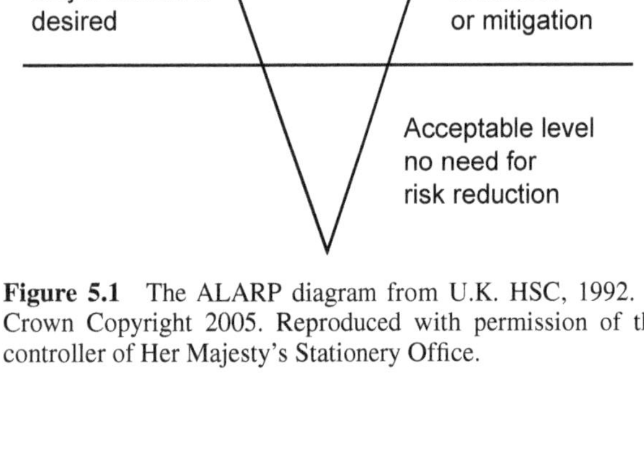

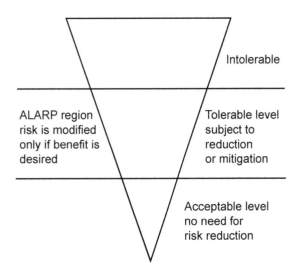

Figure 5.1 The ALARP diagram from U.K. HSC, 1992. © Crown Copyright 2005. Reproduced with permission of the controller of Her Majesty's Stationery Office.

   c. *Negligible or broadly acceptable*, with no modifications.
4. Risk reduction and mitigation measures generally requiring changes in the design.
5. Cost-benefit analysis for implementing remedial measures and comparison with the benefits of risk mitigation.
6. Selection of the optimum design and mitigation measures.

5.1.3 Reliability-Based Structural Design

UNCERTAINTY, RISK, AND SAFETY

An ocean structure is a complex, thin, stiffened shell with randomly disposed fabrication imperfections due to material and workmanship quality and subjected to random loads resulting from the action of winds, waves, currents, ice, temperature, etc. Consequently, many uncertainties are to be dealt with in the design of ocean structures. First, there is the uncertainty of the loads, especially those arising from waves. The ocean environment is severe, complex, and continuously varying. Ocean waves are essentially probabilistic and can be adequately defined only by means of statistics. Second, there are uncertainties regarding material properties, such as yield stress, fatigue strength, notch toughness, and corrosion rate. For example, in mild steel that has not had special quality control, the yield stress can vary by as much as 10% and is also dependent on the rate of loading and the effects of welding. Third, there is inevitably some degree of uncertainty in the analysis of a structure as complex as a ship. Both the response analysis and the limit state analysis necessarily involve assumptions, approximations, and idealizations in formulating mathematical models of the physical environment and the structure's response to that environment. Fourth, there can be variations and hence uncertainties in the quality of construction, and this factor also has a significant influence on the strength of a structure. Finally, there are uncertainties of operations, such as operating errors resulting from human action (improper loading, mishandling, etc.) or a change in service, and wherever there are uncertainties, there is a risk of failure.

Since there are always uncertainties, and hence some risk of failure, it is impossible to make a structure absolutely safe. Instead it can be made only "sufficiently safe," which means that the risk can be brought down to a level that is considered by society to be acceptable for that type of structure. It is clear that an objective evaluation of the strength of a given structure is an impossible task. As more and more attention is paid to safety at sea as well as to economic considerations, what every designer aims at is an optimized combination of safety and cost. Therefore, a rational design should rely on a statistical basis and the structural design process must provide the means whereby the designer can ensure that the degree of safety meets or exceeds the required level. Application of reliability analysis techniques to the design of ship structures should improve the ship safety by providing

1. Clear distinction between resistance and loads.
2. Better knowledge of the safety through appraisal of uncertainties.
3. Better coherency of calculations.
4. Quantification of risks.

However, there are yet many difficulties in applying reliability concepts to ships, among which are

1. Lack of statistical data on modes of failure, damage, workmanship, maintenance, extreme load value.
2. Influence of corrosion and minor damage that contribute to the reduction of capability with time.
3. Does the probability of failure increase with time and how should this affect the initial choice of probability of failure?

LEVELS OF SAFETY

The required level of safety varies according to the type of failure and the seriousness of its consequences. Because these levels are ultimately determined by society, there are no precise values or exact methods for determining them, but they can be estimated by surveys and examining the statistics regarding failures, particularly those types in which the failure rate is considered by the public to be generally satisfactory, in the sense that the costs and resource usages that would be required to further reduce the failure rate are considered to be unwarranted when balanced against other needs. From the results of a study of merchant ship losses Lewis et al. (1973) estimated a value of between 0.003 and 0.006 as the lifetime probability of overall structural failure, which has been tacitly accepted in the past for large oceangoing ships. In this regard, it is relevant to examine what proportion of ship accidents are due to structural failure. Figure 5.2 from Gran (1978) presents the results of a survey which showed that in a given sample of ship casualties only about 7% (0.138 × 0.54 × 100%) of severe accidents were caused by structural failure.

Based on more recent statistics of losses of propelled seagoing merchant ships of more than 100 GT, as given in Table 5.1 from Lloyd's Register of Shipping (1999), and assuming that only about 10% of severe accidents are caused by structural failure, it may be concluded that the lifetime probability of overall structural failure would range from 0.004 to 0.005. From similar studies for other types of structures it would appear that the total annual failure probability per structure (aircraft, drill rig, etc.) ranges from 10^{-3} or less for failures that have moderately serious consequences (substantial economic loss but no fatalities) to 10^{-5} or less for catastrophic failures, such as the crash of a passenger aircraft.

A different approach to the question of required level of safety is the use of economic criteria, as proposed by the Construction Industry Research and Information Association (CIRIA 1977) and the International Ship and Offshore Structures Congress (ISSC 1994a and 1997). This is particularly appropriate for cases in which loss of life is not involved. For a large number of similar structures the total cost C_T of each of them is of the form

$$C_T = C_I + P_F \cdot C_F = C_I + R \qquad (5.1.1)$$

where C_I = initial cost.
P_F = probability of failure over the expected lifetime of the structure.
C_F = present value of expected cost of failure, including repair cost and economic consequences of nonoperability, salvage operation, pollution abatement, and cleanup as well as loss of reputation and public confidence.
R = cost of failure defined as the product of the probability of failure and consequences that this failure produces (R quantifies the risk).

From the owner's point of view, the required probability of failure is that which minimizes R, while the regulatory bodies have in addition to take into account the consequences of catastrophic structural damage in regard to human life and protection of the environment. Taking into account that there are various causes of failure the cost of failure R may be written as

$$R = P_F \cdot C_F = \sum_j P_{Fj} \cdot C_{Fj} = \sum_j R_j \qquad (5.1.2)$$

in which P_{Fj} is the probability of occurrence of a particular mode of failure and C_{Fj} the expected cost associated with that mode of failure. The designer has first to determine the various modes of failure of the structure and their economic consequences, including accidents such as collision or grounding that may have dramatic consequences on human life

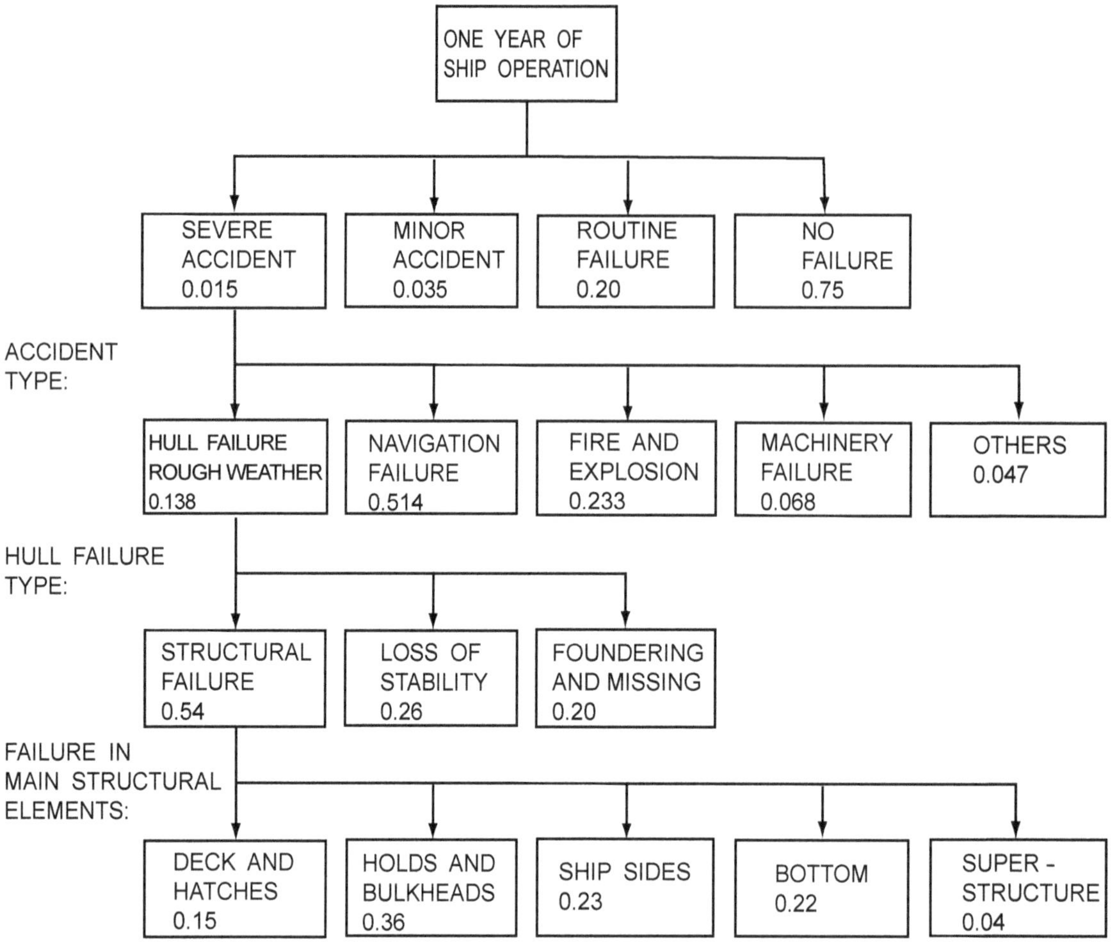

Figure 5.2 Empirical distribution of ship casualties (from Gran, 1978).

Table 5.1 Total 1998 Loss Rates (Per 1000 Ships at Risk) by Ship Type, Category, and Incident Type (from Table 4b of Lrs, 1999)

Ship Type	Foundered/Missing	Fire/Explosion	Contact Types[a]	Overall
LPG tanker	1.0		1.0	2.1
Chemical carrier	0		0.4	0.8
Oil product tanker	0.6		0.4	1.0
Bulk dry	1.4	0.6	1.2	3.4
General cargo	1.9	0.2	1.7	4.3
Container	0.4		0.4	1.3
Refrigerated cargo		1.4		1.4
Ro-ro cargo	0.6	0.6	0.6	2.3
Passenger ro-ro cargo	0.8	0.8		1.6
Passenger (cruise)	3.0			3.0
Passenger ship	0.4			0.4
Fishing vessel	1.3	0.2	0.5	2.2
Overall	1.0	0.2	0.7	2.0

[a]Contact types include collision, wrecked/stranded, and contact.

and environment, and second to minimize the cost of failure R. Optimizing the risk R resulting from the individual risks R_j requires quantifying the safety levels for each hazard, taking into account that the target probability of failure P_{Fj} associated with the risk j depends not only on the consequences of that particular risk but also on that of the other risks.

These overall considerations show how an economic approach can be a useful complementary tool to decide on the relative safety margins of the various structural members.

5.2 RELIABILITY-BASED DESIGN PROCEDURES

5.2.1 General

Probabilistic methods have been used in civil engineering for many years now. Pugsley (1942) and Freudenthal (1947) were the pioneers for aircraft and civil engineering in the 1940s. They demonstrated how a relationship can be derived between safety factors and probability of failure, provided that the statistical distributions of the random variables are known. In subsequent years, these methods were further developed and were increasingly incorporated in structural design codes, both for steel and concrete. In the latter case, this approach has been particularly successful because it accounts for the large variability in the strength of this material. Probabilistic design codes have been developed for the design of offshore structures stimulated by the higher risks and the higher economic stakes involved in that field.

In contrast, it is only recently that the probabilistic approach has been introduced in ship structural design procedures, in spite of the obvious probabilistic nature of wave loads. The load and response analysis is much more complicated for ships than it is for fixed structures or for aircraft, because it must deal with the exceedingly complex interaction between wave excitation and ship motions merely to compute the loads. This analysis capability has only recently become available and still has limited usefulness.

5.2.2 Reliability Procedures

GENERAL

Reliability procedures are generally classified according to the share given to probabilistic calculations:

1. *Level I procedures* (first-moment methods). In level I procedures, such as the partial safety factor method, "characteristic" or nominal values of the various random variables are used and safety factors covering the uncertainties in the variables are introduced in the limit state equations, whose values are based on the results of level II reliability analyses.
2. *Level II procedures* (second-moment methods). In level II procedures, the various random variables are represented by their mean value and standard deviation; where the random variables are correlated, there is also a need to measure their degree of correlation. Among these procedures, the reliability index method is the most frequently used.
3. *Level III procedures* (full probabilistic). These procedures utilize the complete probability distribution functions of all relevant quantities (loads, load effects and limit values) for calculation of the probability of failure P_f associated to each load and each mode of failure. These probabilities are then combined into an overall probability of failure.
4. *Level IV procedures*. These methods combine both event probabilities of failure and the associated benefits and costs. These procedures are used for particular structures whose failure may have dramatic consequences.

Whatever procedure is used, the designer must define the capacity C of the structure or of a structural element, that is to say, its ability to withstand the load effects (or demand D) that it may be subjected to. Also, a criterion must be chosen representing the limit above which the structure is considered as to have failed. This criterion should be independent of loads and should be a specific characteristic of the material or the geometry of the structure; for example,

1. Von Mises criterion.
2. Limit value of deformations.
3. Critical buckling stress.
4. Limit value of crack length.
5. Limit value of acceleration.

Each of the two parameters C and D depends on design variables that are randomly distributed. Their probability density functions and at least their mean and standard deviation are to be defined. This is one of the most difficult problems to solve prior to promoting reliability-based ship structural design, since reliability analyses carried out in the past on various types of ship structures have shown how the calculated level of safety is very sensitive to the choice of probability distributions adopted for the various stochastic variables.

LIMIT STATES

First, it is necessary to define the various modes of failure or limit states that may deteriorate the structure. A *limit state* is defined as a condition for which a particular structural member or a complete structure is unable to perform the function for which it has been designed. According to the ISO (1994), there are four types of limit states:

1. Serviceability limit states involving deterioration of less vital functions and including
   - *Local damage that may reduce the durability of the structure or affect the efficiency of structural or nonstructural elements.*
   - *Unacceptable deformations that affect the efficient use of structural or nonstructural elements or the functioning of equipment.*
   - *Excessive vibrations that cause discomfort to people or affect nonstructural elements or the functioning of equipment.*
2. Ultimate limit states leading to the collapse of the structure and including
   - *Loss of equilibrium of the structure or part of the structure, considered as a rigid body (e.g., overturning or capsizing).*
   - *Attainment of the maximum resistance capacity of sections, members, or connections by gross yielding, rupture, or fracture.*
   - *Instability of the structures or part of it, such as buckling of columns, plates, shells, and stiffened panels.*
3. Fatigue limit state resulting from damage accumulation under the action of cyclic loads.
4. Accidental limit states, such as collision or grounding.

LIMIT STATE FUNCTION

Any failure criterion as obtained from application of first engineering principles may be expressed by a limit state function $g(x) = g(x_1, x_2, ..., x_n)$ in which the x_i's represent the design parameters:

$g(x) = g(x_1, x_2, ..., x_n)$
$g(x) =$ Capacity of the structure – Load effects

This limit state function characterizes the condition of the structure and defines in the x-space two domains of safety separated by the failure surface:

$g(x) < 0$ in the unsafe domain
$g(x) > 0$ in the safe domain
$g(x) = 0$ on the failure surface or limit state surface

If we replace in the function $g(x)$ the parameters x_i by the corresponding random variables X_i, we obtain a random variable, which is called the *safety margin* $M = g(X)$. Where the resistance function C and the load effect function D are independent random variables, the safety margin M may be expressed as the difference between C and D:

$$M = C2 - D \tag{5.2.1}$$

MODELING OF UNCERTAINTIES

According to the ISO (1994) and Nikolaidis and Kaplan (1991), the uncertainties of any basic variable may be classified as follows:

1. Statistical or random uncertainties that arise purely from genuine statistical randomness and can therefore be properly and adequately assessed using statistical theory.
2. Approximational or modeling uncertainties that arise from the assumptions, approximations and judgments necessarily involved in any design task. These approximational uncertainties can be reduced by improving the knowledge.

(A) STATISTICAL UNCERTAINTIES

The uncertainties that arise because of the randomness of the variables can and should be assessed by means of basic statistical theory. To do this, it is necessary to establish what type of distribution (normal, Poisson, etc.) is involved. In some cases, this is known from theoretical considerations. In other cases, it is possible to determine by observation which basic type most nearly resembles the actual distribution. Once the type of distribution is known, the uncertainty can be calculated by means of the basic laws and relationships of statistics. When applying reliability methods that require only the mean value $E(X)$ and the standard deviation σ_X of the random variable X, these values may be determined from the results of measurements without any assumption on the probability distribution:

$$E(X) = \frac{1}{n}\sum_{i=1}^{i=n} x_i \tag{5.2.2}$$

$$\sigma_X^2 = \frac{1}{n-1}\sum_{i=1}^{i=n}\left[x_i - E(X)\right]^2 \tag{5.2.3}$$

where the x_i's are the measured values of the random variable X.

A very useful way of dealing with statistical uncertainty is in term of a *characteristic value*, which is the value corresponding to a specified percentage of

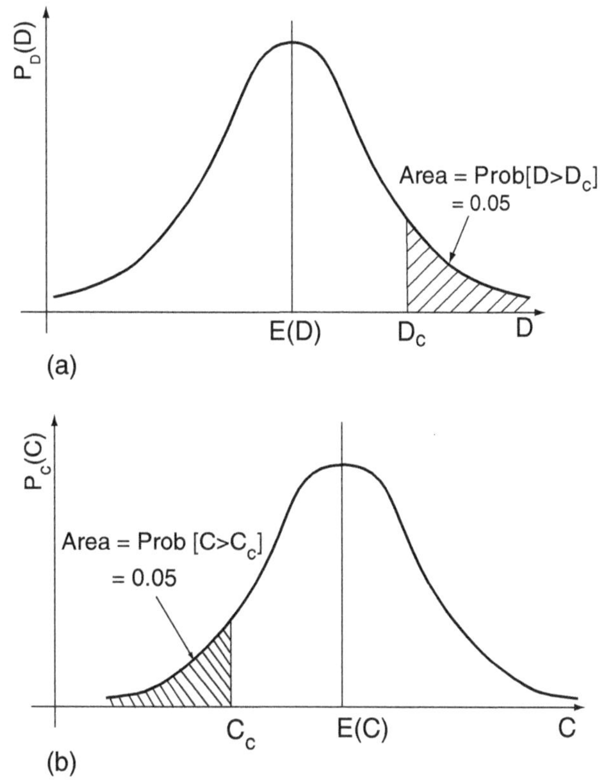

Figure 5.3 Illustration of characteristic values.

the area under the probability density curve, that is, a specified probability of exceedance. For example, Figure 5.3(a) illustrates a characteristic value of the load effects D_c corresponding to a 5% probability of exceedance. Figure 5.3(b) illustrates a characteristic limit value of the capacity C_c, in which case the 5% probability refers to nonexceedance.

In contrast to the safety index method, which uses the mean values $E(C)$ and $E(D)$ of resistance and load effects, the partial safety factor method uses "characteristic" values, thereby automatically accounting for the statistical probability of failure. Thus, if the only uncertainty was purely statistical (e.g., if C and D followed exactly their assumed distribution function), there would be no need for any safety factor, and the strength constraint would be simply

$$C_c \geq D_c \tag{5.2.4}$$

where the characteristic values C_c and D_c would be selected so as to provide whatever degree of safety is required.

(B) APPROXIMATIONAL UNCERTAINTIES

In reality, of course, besides the purely statistical uncertainty, there is always some additional uncertainty, which is either not statistical in nature (e.g., uncertainty arising from value judgments, from approximations, or from legal, political, or other nontechnical influences on design) or which, although statistical, cannot be included in that category because sufficient information is not available. This uncertainty is here called *approximational* because most of it arises from the approximations that are inevitable in structural design. Indeed, even the use of statistical theory to describe ocean waves involves some assumptions and approximations. But, in dealing with any statistical aspect in design, the goal should be first to obtain sufficient information so as to be able to use statistical theory and to account for most of the uncertainty by this means, even though some approximations are still required, and then to seek to improve the information so as to further reduce the approximational uncertainty. Thus, as more information becomes available, the amount of approximational uncertainty is reduced. However, it will be never entirely eliminated; some approximations will always be necessary, and in addition there will be always some sources of uncertainty for which statistics are not entirely adequate; of that, at least, there is no uncertainty. This is particularly the case for blunders that include neglect and human errors, considered responsible for about 70% of structural failures. In conclusion, at each step of the structural design:

1. Description of the environment.
2. Calculation of the design loads.
3. Combination of loads and load effects.
4. Structural response analysis.
5. Selection of the failure criteria.

there are approximational uncertainties that have to be identified and properly modeled.

(C) MODELING OF UNCERTAINTIES

According to the method proposed by Ang and Cornell (1974) for modeling statistical and approximational uncertainties, any random variable of actual value X can be expressed as follows:

$$X = B_I B_{II} X_0 \tag{5.2.5}$$

where X_0 = value of the random variable X as given by a design code

$$B_I = \frac{X_P}{X_0}.$$

X_P = theoretically predicted value of the variable X

$$B_{II} = \frac{X}{X_P}.$$

B_I measures the statistical uncertainty while B_{II} measures the approximational uncertainty. Assuming that B_I and B_{II} are independent random variables, the mean value $E(B)$ and the coefficient of variation V_B of the random variable $B = B_I B_{II}$ are given by (refer to Appendix 5–A)

$$E(B) = E(B_I)E(B_{II}) \tag{5.2.6}$$

$$V_B = \frac{\sigma_B}{E(B)} = \sqrt{V_{B_I}^2 V_{B_{II}}^2 + V_{B_I}^2 + V_{B_{II}}^2}$$
$$\cong \sqrt{V_{B_I}^2 + V_{B_{II}}^2} \tag{5.2.7}$$

where σ_B is the standard deviation of B.

If the various approximational uncertainties B_i are identified, B_{II} may be expressed as $B_{II} = \prod_{i=1}^{i=n} B_i$ and the total uncertainty B is

$$B = B_I \prod_{i=1}^{i=n} B_i \tag{5.2.8}$$

where the B_i's represent all the approximational uncertainties occurring in determination of the random variables. If the various uncertainties are assumed uncorrelated, the mean value $E(B)$ and the coefficient of variation V_B of the random variable B are given by

$$E(B) = E(B_I)\prod_{i=1}^{i=n} E(B_i) \tag{5.2.9}$$

$$V_B \cong \sqrt{V_{B_I}^2 + V_{B_{II}}^2} = \sqrt{V_{B_I}^2 + \left[\prod_{i=1}^{i=n}\left(1 + V_{B_i}^2\right) - 1\right]}$$
$$\cong \sqrt{V_{B_I}^2 + \sum_{i=1}^{i=n} V_{B_i}^2} \tag{5.2.10}$$

STRUCTURAL RELIABILITY

As already mentioned, the reliability of a system defines its ability to perform its primary role over a specified period of time. This general definition may be expressed in a probabilistic manner as "*the probability of a device performing its function over a specified period of time and under specified operating conditions.*"

The reliability of any structural component with respect to a given mode of failure is therefore defined as the probability that it will not fail:

$$R = P\left[g(x) > 0\right]$$
$$= \int_{g(x)>0} p_{X_1,X_2,\ldots,X_n}(x_1, x_2, \ldots, x_n)\ dx_1, dx_2 \ldots dx_n \tag{5.2.11}$$

where $p_X(x)$ is the "joint" probability function of the random variables X and the domain of integration includes all values of the X_i's where the safety margin is positive. The probability of failure P_f is consequently given by

$$P_f = 1 - R = P\left[g(x) \le 0\right]$$
$$= \int_{g(x)\le 0} p_{X_1,X_2,\ldots,X_n}(x_1, x_2, \ldots, x_n)dx_1 dx_2 \ldots dx_n \tag{5.2.12}$$

The calculation of the probability of a particular type of failure involves the probability density functions of the relevant random variables. If it is assumed that the capacity C of the structure and the demand D or load effects are independent random variables represented by two probability density functions, respectively, $p_C(\cdot)$ and $p_D(\cdot)$, as shown in Figure 5.4, then the probability of this particular type of failure occurring is

$$P_f = \int_0^\infty \left[\int_0^\eta p_C(\xi)d\xi\right] p_D(\eta)d\eta$$
$$= \int_0^\infty F_C(\eta)p_D(\eta)d\eta \tag{5.2.13}$$

where $F_C(\cdot)$is the cumulative distribution function of C. This is illustrated in Figure 5.4, which shows that even though the mean value $E(C)$ of the capacity is well above the mean value $E(D)$ of the load effects there is still overlap of the curves and hence some possibility of failure. (Note: The probability of failure is not equal to the area of overlap, but this area nevertheless provides a useful visual and qualitative indication of P_f). The figure also shows that the important regions of the distribution are the tails, because this is where the overlap occurs. Unfortunately,

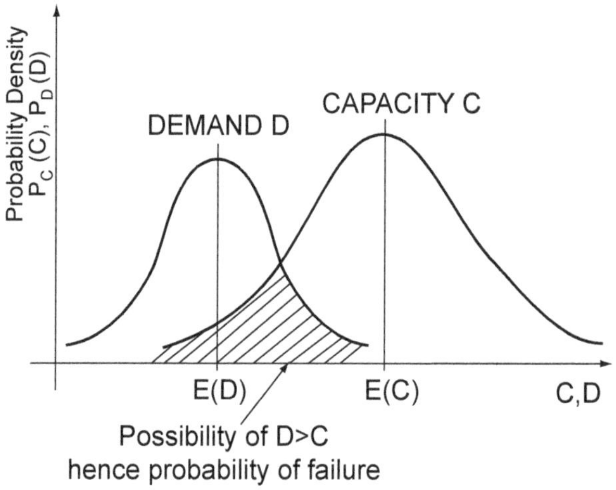

Figure 5.4 Probability distributions of the capacity and load effects

it is this portion of a distribution curve that is most difficult to obtain with any precision, mainly because one is dealing with rare events.

TIME-DEPENDENT RELIABILITY

Both capacity C of the structure and load effects D are generally time-dependent random variables. For example, the capacity C may continuously deteriorate over the ship's life under the effect of corrosion or damage accumulation, such as fatigue and crack propagation. Time-dependent reliability problems consist of determining the time $t = T$ when the limit state function $g[x(t)]$ becomes negative; this time T, called the *time to failure*, is a random variable. The probability that $g[x(t)] \leq 0$ is called the *first passage* or *outcrossing probability*. Additional information is given in Section 5.5.1.

If the effects of deterioration with time of the capacity C are neglected, taking into account that the structural reliability is maintained over the ship's life thanks to a system of periodical surveys, the reliability of ship structures can be determined by applying the theory of stochastic process for calculation of the extreme value distribution of the load effects. Stochastic variables are then modeled by the distribution of their extreme values over the lifetime period and introduced as time-invariant random variables in the limit state functions. Moreover, where the structure is subjected to two or more time-dependent loads the theory of stochastic load combination has to be used, bearing in mind that, for independent stochastic variables, it is unlikely that their maximum respective values occur at the same time. This is particularly the case for the various wave-induced loads acting on the ship structure. Examples of load combination methods are given in Section 5.4.3.

5.3 PROBABILISTIC DESIGN

The task of achieving a specified level of safety can be pursued at various levels of rigor. We shall first present various approximate methods based on level I and II procedures. Then we shall show that a fully rigorous method requires the gathering of a great deal of information and is simply not justified in the majority of cases, and finally we shall present examples of simplified level III procedures.

5.3.1 Partial Safety Factor Method (Level I Procedure)

The basic idea of the partial safety factor concept can be understood on the example of the load and resistance factor design (LRFD) format, introduced by the American Institute of Steel Construction (AISC 1994) or classification societies for the design of steel constructions or ship structures. The old format, called *working stress design* (WSD), was expressed as

$$\frac{R}{\gamma} \geq F_D + F_L + F_W \qquad (5.3.1)$$

where R = nominal resistance.
γ = safety factor.
F_D = nominal gravity load.
F_L = nominal live load.
F_W = nominal environmental load.

Designs based on the WSD format have proven to provide reliable structures without considering explicitly uncertainties and probability distributions of the random variables. However, this format was not able to design structures with uniform safety levels because one safety factor cannot cover all the numerous uncertainties in the design variables. In the LRFD format, e.g., refer to Ayyub et al. (2002), or Assakaf et al. (2002), the design equation is expressed as

$$\gamma_R C \geq \sum_{i=1}^{i=n} \gamma_i F_i \qquad (5.3.2)$$

where γ_R = resistance reduction factor.
F_i = nominal load effect due to load component i.
γ_i = amplification factor for load component i.

The coefficients γ_R and γ_i depend on the accuracy of the method considered for calculation of loads and resistance and are based on the results of probabilistic models, measured data, and past satisfactory experience.

In level I procedures based on the partial safety factor (PSF) concept, the design parameters are considered random variables, represented by their characteristic or nominal values. For each limit state, a limit state function is determined from application of first principles, and the safety of the structure is expressed in terms of partial safety factors (PSF), which take into account all the uncertainties that affect the determination of the design variables and whose values are based on the results of reliability analyses (refer to Section 5.3.4). If the capacity C and demand D are independent random variables, the design equation expressed in the PSF format is

$$\frac{C_k}{\gamma_m \gamma_c} \geq \gamma_0 D\left(\gamma_f F_k\right) \qquad (5.3.3)$$

where F_k = characteristic values of load effects, corresponding to a specified percentage of the area under the probability curve.

C_k = characteristic value of the capacity of the structure, corresponding to a specified percentage of the area under the probability curve for the limit state being investigated.

$D(\gamma_f F_k)$ = value of the demand calculated from the characteristic values of loads and weighed by the partial safety factor γ_f covering approximational uncertainties in loads resulting from approximations and assumptions in the description of the environment.

γ_m = partial safety factor covering uncertainties in the material characteristics.

γ_c = partial safety factor covering approximational uncertainties in the actual capacity of the structure, such as assumptions and approximations in the response analysis and the limit state analysis.

γ_0 = additional partial safety factor taking into account the degree of seriousness of the particular limit state in regard to safety and serviceability.

Equation (5.3.3) may be generalized to multiple types of loads:

$$\frac{C_k}{\gamma_m \gamma_c} = \gamma_0 \sum_i D_i \left(\gamma_{f_i} k_{li} F_{ki}\right) \tag{5.3.4}$$

All F_{ki} are not simultaneous and, depending on the structural component, a load combination factor k_{li} has to be considered to take into account that extreme values of wave-induced loads do not occur at the same time (refer to Section 5.4.3).

Because of approximational uncertainties the characteristic values that account for statistical uncertainty are not sufficient in themselves. It is necessary to further increase the separation of the capacity and demand curves, by some amount that can only be estimated and requires judgment, in order to retain the required degree of safety. Thus, there is a need for some simple and explicit method for adjusting the separation between these two curves, and the coefficient γ_0 provides just a method. Therefore, instead of regarding this factor as a single quantity, we regard it as the product of several partial safety factors, depending on the type of structure and the level of detail preferred for their specification. These safety factors are used for two main purposes:

1. To account for the degree of seriousness of the particular limit state in regard to safety and serviceability (for a commercial ship, the latter refers mainly to the economic consequences of the failure), taking into account any special circumstances (purpose of the ship, type of cargo, interaction of this limit state with others, etc.). Since safety and serviceability are not the same, it is best to use two independent partial safety factors for this task.
2. To account for the approximational uncertainties that are not purely statistical and therefore cannot be properly modeled (refer to Section 5.2.2):
   a. Deviation of the probability distribution of the loads, due to unforeseen actions or conditions, and consequent deviation of the load effects.
   b. Deviation of the limit value from its assumed distribution due to unpredictable factors, e.g., poor workmanship.
   c. Others matters requiring estimation and judgment.

5.3.2 Second-Moment Methods (Level II Procedures)

In second-moment methods, all failure modes are independent and treated separately. This greatly simplifies the process but it requires that a value of the acceptable risk must be defined separately for each mode of failure (although in practice the same value can be used for all types that have the same degree of seriousness) and it precludes the possibility of combining the separate risks. Therefore, it requires approximations, which must necessarily be on the conservative side, in order to deal with combinations of loads of differing probability and combinations of interactive modes of failure.

CORNELL SAFETY INDEX

This concept is the earlier method and was initially introduced by Freudenthal (1956). Subsequently Cornell (1969) defined a reliability safety index as

$$\beta = \frac{E(M)}{\sigma(M)} \tag{5.3.5}$$

where $E(M)$ = mean value of the safety margin.
$\sigma(M)$ = standard deviation of the safety margin.

If the capacity C of the structure and the load effects D are assumed to be independent random variables, the safety margin M may be defined as the difference between C and D

$$M = C - D \tag{5.3.6}$$

and failure occurs when the margin becomes negative. According to equation (5.3.5) the safety index β is given by

$$\beta = \frac{E(C)-E(D)}{\sqrt{\sigma_C^2+\sigma_D^2}} \tag{5.3.7}$$

Since C and D are random variables, M will be likewise, having a probability density function $p_M(M)$ as shown in Figure 5.5(a).

Therefore, the degree of safety depends not only on the separation of the two curves, as measured for example, by the distance between their mean values $E(M) = E(C) - E(D)$, but also inversely on the spread of the two curves, as measured for example, by their coefficients of variation. Therefore, it will also bear some inverse relationship to V_M, the coefficient of variation of M. If V_M is large, the degree of safety will be correspondingly less and vice versa. The probability of failure is

$$P_f = \text{Prob } [M < 0] \tag{5.3.8}$$

Subtracting $E(M)$ from both sides, and normalizing by means of the standard deviations $\sigma(M)$ gives

$$P_f = \text{Prob}\left[\frac{M-E(M)}{\sigma(M)} < -\frac{E(M)}{\sigma(M)}\right] \tag{5.3.9}$$

Figure 5.5(a) shows that the safety index measures, in standard deviation units, the distance from the mean safety margin to the failure region.

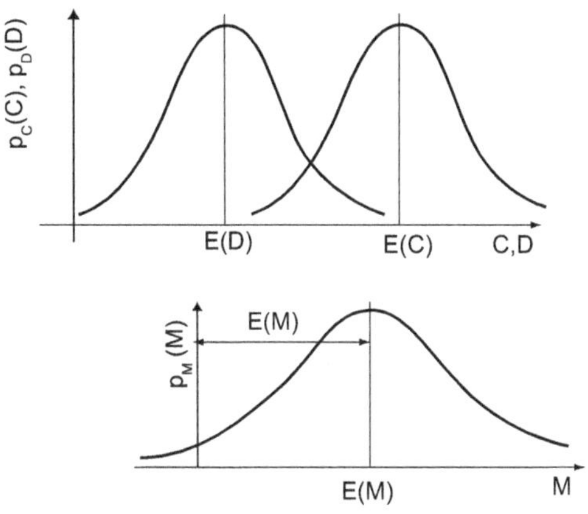

Figure 5.5(a) Probability density function of the safety margin.

By definition, the coefficient of variation (COV) is

$V_M = \dfrac{\sigma(M)}{E(M)}$ and therefore

$$P_f = \text{Prob}\left[\frac{M-E(M)}{\sigma(M)} < -\frac{1}{V_M}\right] \tag{5.3.10}$$

The left-hand term is the normalized margin, for which the distribution has zero mean and unit variance. Let us denote this normalized margin as M_0 and let $P_{M_0}(\cdot)$ be its cumulative probability distribution:

$$M_0 = \frac{M-E(M)}{\sigma(M)}$$

$P_{M0}(.)$ = cumulative probability distribution of M_0 as shown in Figure 5.5(b)
= Prob ($M_0 \leq .$)

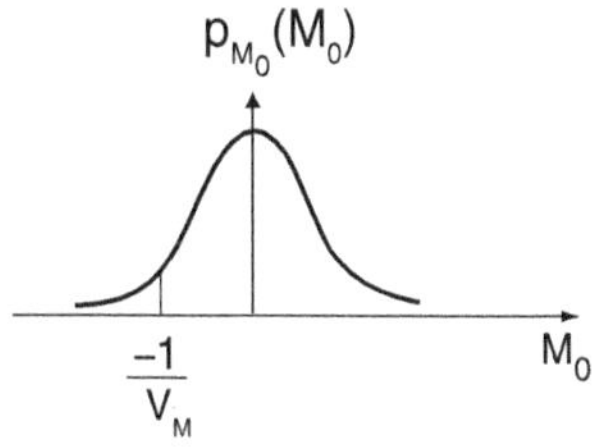

Figure 5.5(b) Probability density function of the normalized safety margin.

Equation (5.3.10) becomes

$$P_f = \text{Prob}\left[M_0 < -\frac{1}{V_M}\right] \tag{5.3.11}$$

The occurrence of $1/V_M$ on the right-hand side within the brackets confirms that the degree of safety depends on the inverse of the coefficient of variation of the safety margin. We therefore give this quantity the name *safety index* and denote it β

$$\beta = \frac{1}{V_M} \tag{5.3.12}$$

In terms of the safety index, equation (5.3.11) becomes

$$P_f = \text{Prob}\left(M_0 \leq -\beta\right) = P_{M_0}\left(-\beta\right) \tag{5.3.13}$$

This last expression shows that there is a direct correspondence between the safety index β and the probability of failure. The larger the safety index β, the smaller the probability of failure, that is, the safer the structure. If the complete distribution of M is known (i.e., if the distributions of C and D are known), then the exact value of P_f corresponding to a given value of β can be determined. For example, if C and D are assumed normal random variables, the relationship between the probability of failure and the safety index is given by

$$\beta = -\,\Phi^{-1}\left(P_f\right) \tag{5.3.14}$$

where F is the standard normal cumulative distribution function. Table 5.2 defines, in that case, the relationship between β and P_f.

Table 5.2 Relationship Between β and p_f

P_f	10^{-1}	10^{-2}	10^{-3}	10^{-4}	10^{-5}	$10^{-5.}10^{-7}$		10^{-8}
β	1.3	2.3	3.1	3.75	4.2	4.7	5.2	5.6

From the definition of β, it appears that the safety index depends on how the failure surface is defined. Assuming that the random variables C and D are lognormally distributed, instead of $M = C - D$, the safety margin could be defined as

$$M = \ln \frac{C}{D} = \ln C - \ln D \qquad (5.3.15)$$

Taking into account that

$$E(\ln X) = \ln E(X) - \ln\sqrt{1+V_X^2}$$

$$\sigma_{\ln X} = \sqrt{\ln\left(1+V_X^2\right)}$$

the safety index β would become

$$\beta = \frac{E(\ln C) - E(\ln D)}{\sqrt{\sigma_{\ln C}^2 + \sigma_{\ln D}^2}} \cong \frac{\ln\left(\frac{E(C)}{E(D)}\sqrt{\frac{1+V_D^2}{1+V_C^2}}\right)}{\sqrt{\ln\left(1+V_C^2\right)\left(1+V_D^2\right)}} \qquad (5.3.16)$$

For V_D and V_C less than 0.3,

$$\beta = \frac{\ln\frac{E(C)}{E(D)}}{\sqrt{V_C^2 + V_D^2}} \qquad (5.3.17)$$

Obviously equations (5.3.7) and (5.3.16) give different values for the safety index β.

The preceding approach may be generalized to cases where the limit state function $g(x)$ is a linear function of the design parameters x_i. In that case, the safety margin M is expressed as

$$M = b + \sum_i a_i X_i \qquad (5.3.18)$$

or in matrix notation $M = a^T X + b$, where a^T is a row matrix and X the column matrix of the random variables. Based on the definition of the safety index β, we may write (refer to Ditlevsen and Madsen (1996) and Appendix 5–B)

$$\beta = \frac{E(M)}{\sigma(M)} = \frac{E(M)}{\sqrt{\sum_{i=1}^{i=n}\sum_{j=1}^{j=n} a_i\, a_j \operatorname{Cov}\left(X_i, X_j\right)}} \qquad (5.3.19)$$

where $\operatorname{Cov}(X_i, X_j)$ is the covariance of X_i and X_j defined as

$$\operatorname{Cov}(X_i, X_j) = E\left[\left(X_i - \bar{X}_i\right)\left(X_j - \bar{X}_j\right)\right],$$

where $\bar{X} = E(X)$ (5.3.20)

In matrix notation, the safety index is

$$\beta = \frac{E(M)}{\sigma(M)} = \frac{a^T E(X) + b}{\sqrt{a^T C_X a}} \qquad (5.3.21)$$

where C_X is the matrix of covariances defined as $C_X = E\left[(X-\bar{X})(X-\bar{X})^T\right]$.

If the random variables are independent, the safety index becomes

$$\beta = \frac{E(M)}{\sigma(M)} = \frac{E(M)}{\sqrt{\sum_i a_i^2\, \sigma_{X_i}^2}} \qquad (5.3.22)$$

where the safety margin is nonlinear, it may be linearized using a Taylor series around a point x:

$$M(X) = g(x) + \sum_{i=1}^{i=n} \frac{\delta g(x)}{\delta x_i}\left(X_i - x_i\right) \qquad (5.3.23)$$

and the corresponding safety index, defined as a "first-order second-moment reliability index," is given by

$$\beta = \frac{g(x) + \sum_{i=1}^{i=n} \frac{\delta g(x)}{\delta x_i}\left[E(X_i) - x_i\right]}{\sqrt{\sum_{i=1}^{i=n}\sum_{j=1}^{j=n} \frac{\delta g(x)}{\delta x_i}\frac{\delta g(x)}{\delta x_j} \operatorname{Cov}\left(X_i X_j\right)}} \qquad (5.3.24)$$

If the linearization point is the mean-value point, the safety index is

$$\beta = \frac{g\left[E(X)\right]}{\sqrt{\sum_{i=1}^{i=n}\sum_{j=1}^{j=n} \frac{\delta g\left[E(X)\right]}{\delta x_i}\frac{\delta g\left[E(X)\right]}{\delta x_j} \operatorname{Cov}\left(X_i X_j\right)}}$$

(5.3.25)

The safety index as obtained from equation (5.3.24) depends not only on how the failure surface is defined but also on the choice of the linearization point.

Designing on the basis of a specified value of β produces a consistent degree of safety from one design to another, for each type of structure. For ship structures, a suitable value of β can be determined for each type of failure by analyzing the statistics regarding ships that have proven to be reasonably efficient and also have a satisfactory record (refer to Sections 5.1.3 and 5.4.6). One of the principal advantages of the safety index method is that the provision of adequate safety, which is a probabilistic quantity, is converted and expressed deterministically in terms of a specific "design" value of load and a specific limit value that the structure must possess. The safety index method makes use of mean values and in particular involves their ratio, which is referred to as the *central safety factor* γ_C

$$\gamma_C = \frac{E(C)}{E(D)} \tag{5.3.26}$$

where γ_C is the familiar single safety factor of the deterministic design. The relationship between β and γ_C can be derived as follows, starting from the definition of β and assuming that the safety margin is $M = C - D$:

$$\beta = \frac{1}{V_M} = \frac{E(C) - E(D)}{\sqrt{\sigma_C^2 + \sigma_D^2}} = \frac{\gamma_C - 1}{\sqrt{\gamma_C^2 V_C^2 + V_D^2}}$$

On rearranging for γ_C,

$$\gamma_C = \frac{1 + \beta\sqrt{V_C^2 + V_D^2 - \beta^2 V_C^2 V_D^2}}{1 - \beta^2 V_C^2} \tag{5.3.27}$$

In the Cornell safety index method of design, the appropriate safety authority specifies the expression of safety margin and the values of target safety index and variances of random variables depending on the type of structure and the degree of seriousness of the limit state. For each limit state, the designer calculates the central safety factor from (5.3.27). He then calculates the mean value (best estimate) of the relevant load effect $E(D)$ by performing a response analysis. Alternatively, for those structures for which the loads and load effects are well established, the safety authority may provide a formula for a less exact and more universal design value of $E(D)$ (e.g., vertical wave bending moment as given by the rules of classification societies) together with a larger value of β, which must be used with it. Knowing $E(D)$, the designer then applies the factor γ_C to obtain $\gamma_C E(D)$, and he must then design the structure such that $E(C)$ equals or exceeds $\gamma_C E(D)$. This requirement constitutes one of the *strength constraints* that the design must satisfy. Stated mathematically, the constraint is $\gamma_C E(D) < E(C)$, and this procedure is carried out for each limit state, thus producing the complete set of constraints that govern the design. The safety index is ideal for measuring and comparing the relative safety of different structures and structural members.

HASOFER AND LIND SAFETY INDEX

As shown in the preceding paragraph, the Cornell safety index is not an invariant measure of the safety but depends on how the limit state function is defined. To solve this problem and avoid this lack of consistency, Hasofer and Lind (1974) introduced a geometrical concept of the safety index, acknowledging that, for a one-random-variable criterion, as shown in Figure 5.6, or for the case where the capacity and demand are independent random variables (refer to Figure 5.5(a)), the Cornell safety index measures the number of standard deviations from the mean value of the safety margin to the failure surface. They generalized these particular cases to n random variables as follows:

1. All the random variables X_i are transformed into a set of independent and reduced normal variables U_i defined in matrix notation as

$$E(U) = 0 \tag{5.3.28}$$

$$C_U = E(U\,U^T) = I \tag{5.3.29}$$

The transformed variables U_i may be written as $U = T\ (X - \bar{X})$, where T is the transformation matrix. From the definition of U, we may write $UU^T = T(X - \bar{X})(X - \bar{X})^T T^T$, which gives

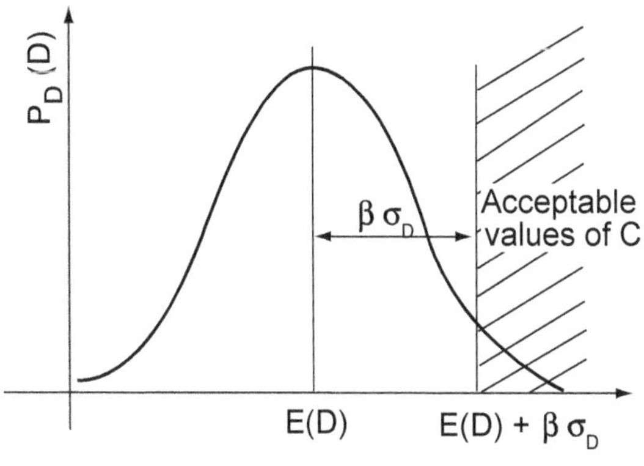

Figure 5.6 One-variable safety index.

$$E(UU^T) = TE\left[(X-\bar{X})(X-\bar{X})^T\right]T^T = TC_XT^T = I \quad (5.3.30)$$

where C_X is the covariance matrix of the random variables X. In the transformation, the mean-value point is transformed into the origin of coordinates in the reduced space.

2. The failure surface separating the safe and unsafe domains is then expressed in the reduced space as $g'(u) = 0$.

3. It can be shown that the distance $\beta(u)$ in the u-space from the origin to any point on the failure surface $g'(u) = 0$. given in matrix notation by, $\beta(u) = \sqrt{U^T U}$, is equal to the number of standard deviations from the mean-value point to the corresponding point on $g(x) = 0$. The safety index is defined in the reduced space as the minimum distance from the origin to a point u^* on the failure surface, as shown in Figure 5.7.

$$\beta_{HL} = \min\left(\sqrt{U^TU}\right) \text{ with } g'(u) = 0 \quad (5.3.31)$$

or in the x-space $\beta_{HL} = \min\left[\sqrt{U^TU}\right]$ $= \min\sqrt{(X-\bar{X})^T T^T T(X-\bar{X})}$. Taking into account that $TC_x T^T = I$, it can be shown that $T^T T = C_x^{-1}$, which gives

$$\beta_{HL} = \min_{g(x)=0}\sqrt{(X-\bar{X})^T C_X^{-1}(X-\bar{X})} \quad (5.3.32)$$

For example, if we assume that $X_i's$ are independent random variables the transformation matrix T is a diagonal matrix whose elements are equal to $1/\sigma_i$ and the distance $\beta(u)$ is

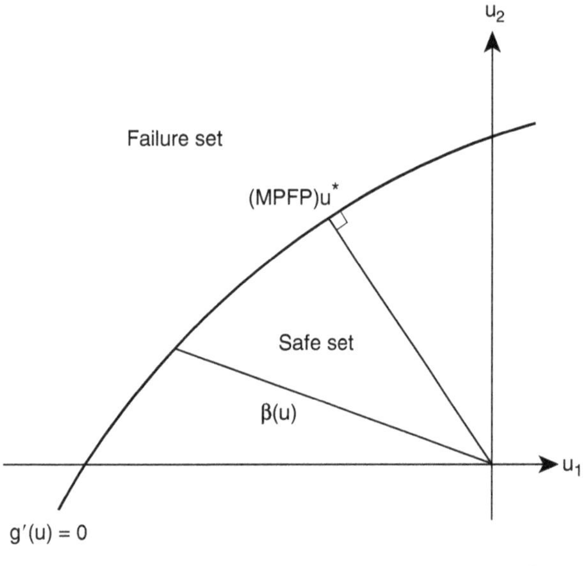

Figure 5.7 Geometrical definition of the safety index β_{HL}.

$$\beta(u) = \sqrt{\sum_{i=1}^{i=n}\left[\frac{x_i - E(X_i)}{\sigma_i}\right]^2}$$

where $x_i - E(X_i)/\sigma_i$ measures, on the x_i-axis, the number of standard deviations from the mean value $E(X_i)$ to the point of coordinate x_i.

The point u^* on the failure surface $g'(u) = 0$ corresponding to that minimum distance represents the point where the probability of failure is maximum. This point, called the *most probable failure point* (MPFP), is the solution of an optimization problem. In particular, the MPFP is used to define the partial safety factors, as indicated in Section 5.3.4. As pointed out by Hasofer and Lind (1974), the proposed criterion does not depend on the exact analytical form of the limit state function but on the boundary of the failure region in the vicinity of the MPFP. For example, in the case of two independent variables, the failure function may be taken as $M = C - D$ or $M = (C/D) - 1 = 0$ or $\ln(C/D) = 0$, but for each of these three failure functions, the failure surface corresponds to $C - D = 0$.

If $(u_1^*, u_2^*, \ldots u_n^*)$ are the coordinates of the MPFP in the reduced space, the safety index β_{HL} is

$$\beta_{HL} = \sqrt{\sum_{i=1}^{i=n} u_i^{*2}} \quad (5.3.33)$$

and the vector u^* may be written as

$$u^* = \beta_{HL}\,\alpha \quad (5.3.34)$$

where $\alpha = (\alpha_1, \alpha_2, \ldots \alpha_n)$ is the unit vector normal to the failure surface at the design failure point and the $\alpha_i's$ are the direction cosines of the normal vector directed toward the failure surface and given by

$$\alpha_i = \frac{\dfrac{\delta g'}{\delta u_i}}{\sqrt{\sum\left(\dfrac{\delta g'}{\delta u_i}\right)^2}} = -\frac{\dfrac{\delta g'}{\delta u_i}}{\left|\nabla g'(u)\right|} \quad (5.3.35)$$

where $\nabla g'(u)$ is the gradient vector of $g'(u)$

Figure 5.7 shows that the safety index β_{HL} has the same value for the failure surface as for the tangent hyperplane at the design point, and if the random variables are normally distributed, the probability of failure P_f may be approximated by $P_f = \Phi(-\beta)$, noting that when the failure surface is a hyperplane, the Cornell and Hasofer-Lind safety indices are identical (refer to Appendix 5–C). This approximation is valid provided there are no other extrema of the safety index in the vicinity of the design point and the principal curvatures of the failure surface at the design point are not too large, which means, in other words, that the failure surface is well approximated by its tangent hyperplane at the design point.

In conclusion, let us describe briefly the iterative procedure that can be used for determination of the Hasofer and Lind safety index, as proposed by Hasofer and Lind (1974) and Madsen, Krenk, and Lind (1986), when the safety margin is not a linear function of the random variables. The procedure is based on the linearization of the failure surface at each step of the iterative process. If $u^{(m)}$ is the solution of step m, the failure surface $g'(u)=0$ at step $m+1$ is linearized around that point $u^{(m)}$ according to equation (5.3.23):

$$g'\left(u^{(m)}\right)+\sum_{i=1}^{i=n}\frac{\delta g'\left(u^{(m)}\right)}{\delta u_i}\left(u_i-u_i^{(m)}\right)=0 \qquad (5.3.36)$$

and the point $u^{(m+1)}$ is defined as the intersection of the hyperplane, defined by equation (5.3.36) with the normal to this hyperplane drawn from the origin. The coordinates of $u^{(m+1)}$ are given by (refer to Appendix 5–C)

$$u_i^{(m+1)}=\frac{g'\left(u^{(m)}\right)}{\left|\nabla g'\left(u^{(m)}\right)\right|}\alpha_i^{(m)}+\left(\sum_{i=1}^{i=n}\alpha_i^{(m)}u_i^{(m)}\right)\alpha_i^{(m)} \qquad (5.3.37)$$

where $\alpha_i^{(m)}=-\dfrac{\dfrac{\delta g'\left(u^{(m)}\right)}{\delta u_i}}{\left|\nabla g'\left(u^{(m)}\right)\right|}$

and the safety index $\beta^{(m+1)}$ is given by

$$\beta^{(m+1)}=\frac{g'\left(u^{(m)}\right)}{\left|\nabla g'\left(u^{(m)}\right)\right|}+\sum_{i=1}^{i=n}\alpha_i^{(m)}u_i^{(m)} \qquad (5.3.38)$$

The starting point may be the origin of coordinates in the reduced space and the procedure is continued until convergence of the safety index. At the design point u^*, the failure surface is approximated by its tangent hyperplane:

$$\sum_{i=1}^{i=n}\frac{\delta g'(u^*)}{\delta u_i}\left(u_i-u_i^*\right)=0 \qquad (5.3.39)$$

The Hasofer and Lind safety index generally gives a satisfactory ordering of structures, provided the failure surface at the design point may be well approximated by a hyperplane i.e., if the radius of curvature is large compared to the safety index. On the contrary, for cases where the radius of curvature is on the same order as the safety index or when the safety index function has several minima, Ditlevsen (1979) introduced the concept of a generalized safety index, which gives a more accurate measure of the structural reliability for nonlinear failure surfaces than the Hasofer and Lind safety index.

A last point deserves to be mentioned. From equation (5.3.33), we may write

$$\frac{\delta\beta(u)}{\delta u_i}=\frac{\delta}{\delta u_i}\left(\sqrt{\sum_{i=1}^{i=n}u_i^2}\right)=\frac{u_i}{\beta}=\alpha_i \qquad (5.3.40)$$

When carrying out a reliability analysis for any mode of failure, it is necessary to select the random variables and the variables that can be considered as deterministic. In that respect, equation (5.3.40) shows that the direction cosines α_i of the normal vector at the most probable failure point, expressed in the reduced space, measure the sensitivity of the safety index to each random variable. The α_is are frequently called the *sensitivity factors*. Sensitivity analysis that aims at determining the influence on the safety index of a variation of each variable enables the selection of the variables that have to be considered as random in a reliability analysis. Variables that have a small value of α may be considered as deterministic variables. Moreover, where α_i is negative, the safety index is a decreasing function of u_i, and on the contrary, where α_i is positive, the safety index is an increasing function of the random variable u_i. There are other probabilistic sensitivity factors, as mentioned by Mansour and Wirsching (1995), that quantify the influence of the various statistical parameters on the safety index or probability of failure, among which are

- Sensitivity factor to the mean value of the random variable X_i: $\delta_i=\delta\beta/\delta\mu_i$
- Sensitivity to the uncertainty V_{X_i} of the random variable X_i : $\eta_i=\delta\beta/\delta\sigma_i$

(where σ_i is the standard deviation of the variable X_i).

In particular, the sensitivity factor η_i quantifies, for any random variable X_i, the influence of a variation of the uncertainty V_{X_i} on the value of the safety index. More generally, sensitivity analysis is a useful tool for selection of the random variables that have the largest influence on the probability of failure and the actions that have to be taken by the designer to improve safety.

5.3.3 Full Probabilistic Methods (Level III Procedures)

GENERAL

The most rigorous and general type of probabilistic design is that which utilizes the complete probability distribution functions of all the random variables (loads, load effects, and limit values) to calculate the probability of failure P_f associated with each load and each mode of failure. These probabilities are then combined into an overall probability of failure that is then adjusted, by making modifications to the design, until it falls within the stipulated acceptable overall risk. This approach requires the determination of all the probability distributions, either by measurements (of the complete phenomena or their separate constituent aspects, and either full-scale or model) or by theoretical considerations, all of which is a very large task. Since the most highly probabilistic loads are those arising from waves and since hull girder bending moment is the most important load effect, research efforts during the past 20 years have concentrated on obtaining the probability distribution for the extreme value of wave-induced hull girder bending moment. Sufficient data regarding waves have now been collected and processed statistically to produce some approximate probability distributions for this load effect. The probability distributions of other loads and load effects are less known and much work remains to be done. Likewise, the distributions of limit values are not easy to obtain, since they arise from so many separate variations (material properties, accuracy of analysis, and quality of construction), each of which requires the collection of a great deal of statistical information. In areas where this information is not yet available it is necessary to use less rigorous techniques.

Moreover, the availability of information is not the only factor that should be considered; another important question is whether the application really requires the complexity of the fully probabilistic method, because a design method should always be as simple as the circumstances permit. A complex method always introduces more likelihood of errors in its use. Also, greater complexity usually increases both the cost and the time required for the design. Therefore, it is important to consider whether the added accuracy of a rigorous but complex method is really justified, in regard to both safety and economy, for the particular application. For aerospace structures, it often is justified, but for ship structures, this is less likely, all the more as each ship is generally a prototype. In view of the many causes of severe accidents and the relative infrequency of structural failure, as shown in Figure 5.2, it is clear that even a large increase in the rigor and accuracy of structural design would not improve the overall risk of casualty very much. Resources used for this purpose could be used more efficiently to improve the other risks involved. Hence, there is a need for moderation in regard to the statistical complexity of the structural design method.

In conclusion, the complexity of the fully probabilistic approach led to the development of simplified methods that retain the basic statistical foundation but do not require the analytical computation of integral (5.2.12). Among these methods, we present the first-order reliability methods (FORM) and the second-order reliability methods (SORM).

FIRST-ORDER RELIABILITY METHODS

As we can see, the Hasofer and Lind method does not take into account the distributions of the random variables but only the mean value $E(X_i)$ and the standard deviation σ_{X_i} of each random variable X_i as well as the covariances $\mathrm{Cov}(X_i, X_j)$. Obviously, as shown in Figure 5.4, the probability of failure depends on the tails of the capacity and load effect probability distributions. This effect is not taken into account in the Hasofer-Lind method, which shows that structures may have the same safety index while they have significant differences on their actual probabilities of failure. To avoid these inconsistencies, improved analytical methods that take into account actual probability distributions of the random variables have been proposed for calculation of the safety index. First-order reliability methods (FORM) are an extension of the Hasofer-Lind method. All the random variables whose distribution functions are assumed to be defined are transformed into a set of independent and standard normal variables (for more information refer to Madsen et al. 1986 and Melchers 1987). For example, when the random variables are independent, the following procedure is used:

1. All basic independent random variables $X_i's$ are transformed into a set of reduced normal variables according to

$$F_{X_i}=\Phi(u_i) \quad i=1, 2, ..., n \tag{5.3.41}$$

where $F_{X_i}(x_i)$ is the cumulative distribution function of X_i. The relationship between the basic variables and the transformed variables is

$$x_i = F_{X_i}^{-1}\left[\Phi(u_i)\right] \tag{5.3.42}$$

$$u_i = \Phi^{-1}\left[F_{X_i}(x_i)\right] \tag{5.3.43}$$

Note: For correlated variables, the Rosenblatt (1952) transformation may be used, as suggested by Hohenbichler and Rackwitz (1981).

2. The failure surface $g'(u)=0$ is then expressed in terms of the reduced variables:

$$g(x) = g\left\{F_X^{-1}\left[\Phi(u)\right]\right\} = g'(u) = 0 \quad (5.3.44)$$

3. The most probable failure point (MPFP), which is the point on the failure surface with the highest probability of failure, i.e., closest to the origin of reduced coordinates, is the solution of an optimization problem. For example, the iterative procedure described in Section 5.3.2 for determination of the Hasofer-Lind safety index and based on linearization of the failure surface may be used. At each step of the process, we have to calculate $g'(u^{(m)})$ from equation (5.3.44) and the gradient vector $\nabla g'(u^{(m)})$ of $g'(u^{(m)})$. As the random variables are assumed to be uncorrelated, we can write

$$\frac{\delta g'\left(u^{(m)}\right)}{\delta u_i} = \frac{\delta g\left(x^{(m)}\right)}{\delta x_i}\frac{\delta x_i}{\delta u_i} = \frac{\delta g\left(x^{(m)}\right)}{\delta x_i}\frac{\phi(u_i)}{f_{X_i}(x_i)} = \frac{\delta g\left(x^{(m)}\right)}{\delta x_i}\frac{\phi\left[\Phi^{-1}\left(F_{X_i}(x_i)\right)\right]}{f_{X_i}(x_i)} \quad (5.3.45)$$

$\left(\text{according to equation (5.3.41)}\ \frac{\delta x_i}{\delta u_i} = \frac{\phi(u_i)}{f_{X_i}(x_i)}\right)$.

4. At the design point u^* (MPFP) the failure surface $g'(u)=0$ is approximated by its tangent hyperplane:

$$\sum_{i=1}^{i=n}\frac{\delta g'(u^*)}{\delta u_i}(u_i - u_i^*) = \sum_{i=1}^{i=n}\frac{\delta g(x^*)}{\delta x_i}\frac{\phi\left[\Phi^{-1}\left(F_{X_i}(x_i)\right)\right]}{f_{X_i}(x_i)}\left(u_i - u_i^*\right) = 0 \quad (5.3.46)$$

The coordinates of the MPFP are given by $u^* = \beta\alpha^*$ where β = safety index defined as the distance from the origin to the failure point,
α^* = unit vector normal to the failure surface at $u = u^*$,
and the probability of failure is approximated by, $P_f \cong \Phi(-\beta)$ which corresponds to the linearization of the failure surface at the design point.

The following presents the so-called iterative "*normal tail approximation method*" for the case of independent variables, but the method is also applicable to correlated variables (refer to Ditlevsen 1981). The method requires changing the distribution function of each random variable X_i in a normal distribution function $\Phi((x_i - \mu_i)/\sigma_i)$:

1. The parameters of the normal distribution (μ_i, σ_i) are determined so that the cumulative distributions and probability density functions of both actual and normal distributions are identical at a given point x_0 of the x space:

$$F_{X_i}(x_{0i}) = \Phi\left(\frac{x_{0i} - \mu_i}{\sigma_i}\right) \quad (5.3.47)$$

$$f_{X_{0i}}(x_{0i}) = \frac{1}{\sqrt{2\pi}\sigma_i}\exp\left\{-\frac{1}{2}\left(\frac{x_{0i} - \mu_i}{\sigma_i}\right)^2\right\} = \frac{1}{\sigma_i}\phi\left(\frac{x_{0i} - \mu_i}{\sigma_i}\right) \quad (5.3.48)$$

where ϕ is the standard normal probability density function with zero mean and unit variance. The solution of equations (5.3.47) and (5.3.48) is

$$\mu_i = x_{0i} - \sigma_i\ \Phi^{-1}\left(F_{X_i}(x_{0i})\right) \quad (5.3.49)$$

$$\sigma_i = \frac{\phi\left[\Phi^{-1}\left\{F_{X_i}(x_{0i})\right\}\right]}{f_{X_i}(x_{0i})} \quad (5.3.50)$$

2. An iterative procedure, similar to that defined in Section 5.3.2, can be used for determination of the coordinates of the most probable failure point. At step 1 of the procedure, a point $x^{(0)}$ has to be selected, usually the mean value point, and the normal tail parameters $\mu_i^{(1)}$ and $\sigma_i^{(1)}$ of the distributions are calculated according to equations (5.3.49) and (5.3.50). Reduced normal variables $u_i=(x_i-\mu_i^{(1)})/\sigma_i^{(1)}$ are introduced in the limit state function $g(x) = g(\mu_1^{(1)} + \sigma_1^{(1)}u_1, \mu_2^{(1)} + \sigma_2^{(1)}u_2, ..., \mu_n^{(1)} + \sigma_n^{(1)}u_n) = g-(u) = 0$ and the coordinates of the point $u^{*(1)}$ closest to the origin are calculated. The corresponding point in the x-space whose coordinates are $x_i^{(1)} = u_i^{*(1)}\sigma_i^{(1)} + \mu_i^{(1)}$ is generally different from $x^{(0)}$. A second iteration is performed with new normal tail parameters $\mu_i^{(2)}$ and $\sigma_i^{(2)}$ calculated for the point $u^{(1)}$, and the coordinates of the point $u^{*(2)}$ closest to the origin are calculated. Assuming that, at step m, the coordinates in the u-space of the closest point to the origin are $u^{*(m)}$, at step $m + 1$, the parameters of the normal distributions $\mu_i^{(m+1)}$ and $\sigma_i^{(m+1)}$ are given by

$$\sigma_i^{(m+1)} = \frac{\phi\left[\Phi^{-1}\left\{F_{X_i}\left(x_i^{(m)}\right)\right\}\right]}{f_{X_i}\left(x_i^{(m)}\right)}$$

$$\mu_i^{(m+1)} = x_i^{(m)} - \sigma_i^{(m)}\Phi^{-1}\left(F_{X_i}\left(x_i^{(m)}\right)\right)$$

with $x_i^{(m)} = \mu_i^{(m)} + \sigma_i^{(m)} u_i^{*(m)}$

and the corresponding point in the x-space is $x^{(m+1)}$. The procedure is continued until convergence of the point $x^{(m+1)}$ in the x-space. At each step of the process the coordinates of the point $u^{*(m+1)}$ closest to the origin can be calculated according to the iterative procedure described in Section 5.3.2. For example, let us assume that the failure surface is a linear function of the random variables X_i given by at step $g(x) = b + \Sigma_{i=1}^{i=n} a_i x_i = 0$, at step $m + 1$ of the iterative process, the coordinates of $u^{(m+1)}$ are

$$u_i^{(m+1)} = -\frac{a_i\sigma_i^{(m+1)}}{\sqrt{\sum_{i=1}^{i=n}\left(a_i\sigma_i^{(m+1)}\right)^2}} \frac{\sum_{i=1}^{i=n} a_i\mu_i^{(m+1)}}{\sqrt{\sum_{i=1}^{i=n}\left(a_i\sigma_i^{(m+1)}\right)^2}} \tag{5.3.51}$$

and the safety index $\beta^{(m+1)}$ is

$$\beta^{(m+1)} = \frac{\sum_{i=1}^{i=n} a_i\mu_i^{(m+1)}}{\sqrt{\sum_{i=1}^{i=n}\left(a_i\sigma_i^{(m+1)}\right)^2}} \tag{5.3.52}$$

SECOND-ORDER RELIABILITY METHODS

Second-order reliability methods are a refinement of the description of the failure surface aiming at improving the accuracy of the location of the failure design point, especially where there are several minimum distances to the origin. At the design point, the failure surface is approximated by a curvature-fitted hyperparaboloid at the design point:

$$\sum_{i=1}^{i=n} \frac{\delta g'(u^*)}{\delta u_i}\left(u_i - u_i^*\right) + \frac{1}{2}\sum_{i=1}^{i=n}\sum_{j=1}^{j=n}\frac{\delta^2 g'(u^*)}{\delta u_i \delta u_j}\left(u_i - u_i^*\right)\left(u_j - u_j^*\right) = 0 \tag{5.3.53}$$

An iterative procedure similar to the one used in FORM and based on the approximation of the failure surface by a curvature-fitted hyperparaboloid at each step of the process can be used, and the SORM reliability safety index is given by

$$\beta_{SORM} = -\Phi^{-1}\left(P_f\right) \tag{5.3.54}$$

For large values of β, Breitung (1984) has shown that the probability of failure is asymptotically given by

$$P_f = \Phi(-\beta)\prod_{j=1}^{i=n-1}\frac{1}{\sqrt{\left(1-\beta\kappa_j\right)}} \tag{5.3.55}$$

where β = Hasofer and Lind safety index.
$\kappa_i = n - 1$ principal curvatures of the limit state surface at the design point.

SIMULATION METHODS

When the failure probability is large or the limit state function is highly nonlinear, analytical methods may lead to inaccurate results. In such a case, it is generally advisable to use Monte Carlo simulation methods for calculation of the failure probability. A random sampling is considered to simulate experiments or calculations. For each simulation cycle, the limit state function is calculated and compared to failure, i.e., $g(x) \leq 0$, and the calculations are repeated until convergence of the probability of failure given by

$$P_f = \frac{N_f}{N} \tag{5.3.56}$$

in which N_f is the number of cases for which $g(x) \leq 0$ and N the number of simulation cycles.

A large number of calculations is necessary to obtain the probability of failure. More advanced methods have therefore been developed, aiming at reducing the variance of the estimate of P_f, e.g., the "importance sampling," where the sample points are distributed in the vicinity of the most likely failure points as obtained from a FORM or SORM analysis.

5.3.4 Reliability-Based Partial Safety Factors

THEORETICAL EXPRESSIONS

Let us define the partial safety factor γ_i^* as

$$\gamma_i^* = \frac{X_i^*}{X_{in}} \tag{5.3.57}$$

where X_i^*s = coordinates of the most probable failure point (refer to Section 5.3.2), given by $X^* - E(X) = T^{-1}U^*$ in matrix notation.

$X'_{in}s$ = nominal or characteristic values of the radom variables.

By introducing the X_i^* in the limit state function $g(X) = 0$, we obtain the following design equation expressed in terms of the PSF:

$$g\,(\gamma_i^* X_{in}) = 0 \qquad (5.3.58)$$

Assuming that the required target safety index β is known for any particular mode of failure (refer to Section 5.4.6), it is then possible to calculate the partial safety factors for the various loads, load effects, and limit values from the results of reliability analyses (e.g., FORM or SORM), so that all the structures and structural members have a consistent degree of safety:

$$\gamma_i^* = \frac{X_i^*}{X_{in}} = \frac{T^{-1} u_i^* + E(X_i)}{X_{in}} \qquad (5.3.59)$$

with u_i^* = coordinates of the MPFP in the reduced space.

$$u_i^* = -\beta_{HL} \frac{\dfrac{\delta g'}{\delta u_i}}{\sqrt{\sum \left(\dfrac{\delta g'}{\delta u_i}\right)^2}}.$$

CALIBRATION PROCEDURE

Determination of the PSF requires performing reliability analyses for the various types of ships, structural elements and modes of failure. The code calibration procedure, as shown in Figure 5.8 from the ISSC (1997), is divided into three main steps:

1. *Definition of the main assumptions:*
   a. Design code definition (applicability domain, limit state design equation expressed in terms of

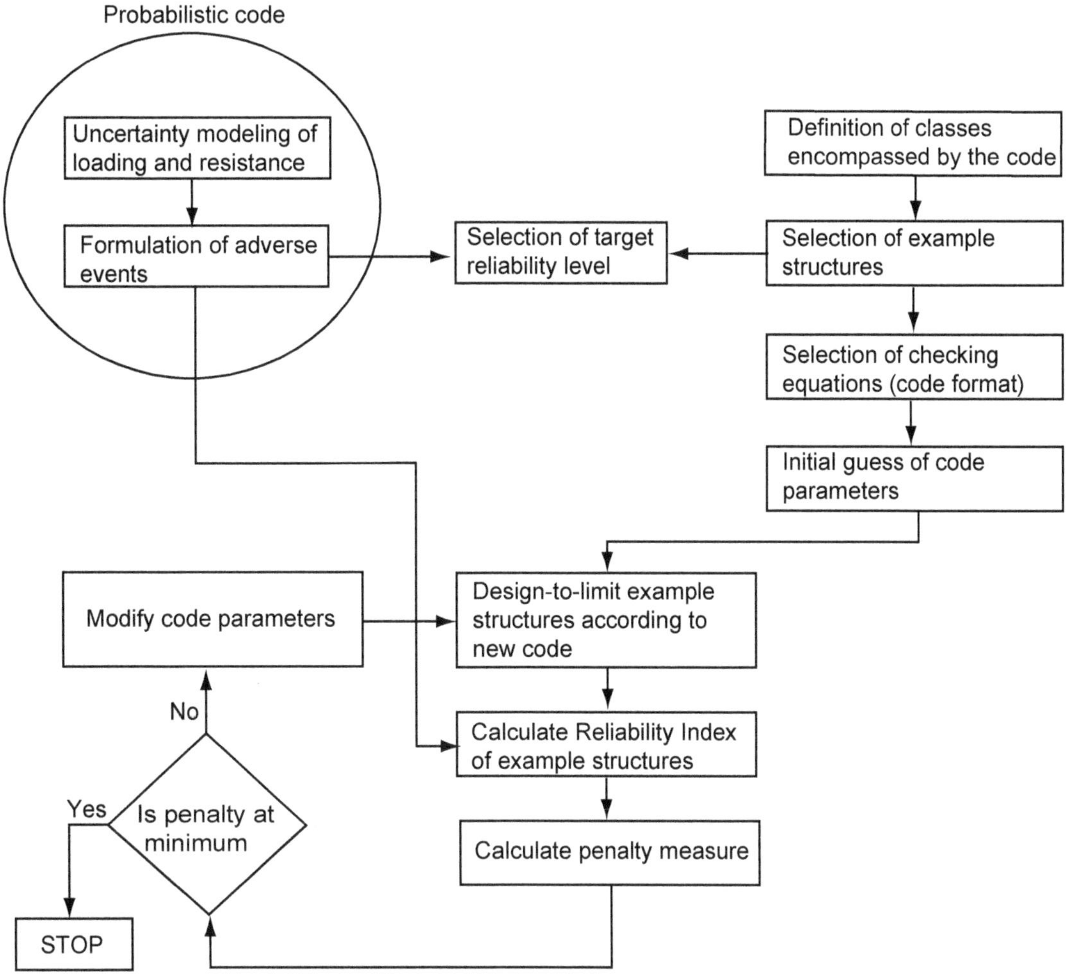

Figure 5.8 PSF calibration procedure. Reprinted from ISSC (1997), Report of Committee IV-1 Design Principles and Criteria, Vol. 1, page 382, © 2005. with permission from Elsevier.

PSF, safety objectives, i.e., target safety index or allowable probabilities of failure and objective function).
b. Select a set of PSFs and design of a representative set of example structures according to the code.
2. *Reliability analysis:*
a. Select the probability distributions for the various random variables.
b. Calculate the safety index for each structure within the class.
3. *Optimization of the objective function and selection of the PSF.*

Since a large number of combinations of PSFs satisfies the condition for the target reliability index β_T, another criterion for the choice of PSF is to be fixed. This criterion is related to the uniformity of the safety level, i.e., the new code should produce structures with the most consistent and uniform safety index. Mathematically, this problem is expressed as an optimization problem. For example, the objective function to be minimized can be taken as

$$OF(\beta,\beta_T)=\sum_{i=1}^{i=n}(\beta-\beta_T)^2+\sum_{i=1}^{i=n}(\gamma_i-\gamma_i^*)^2 \quad (5.3.60)$$

where n is the number of example structures.

The first term of the function requires that the safety indices are located close to the target safety index β_T, while the second term prefers solutions closer to the partial safety factors γ_i^*corresponding to the most probable failure point. The following objective function, as proposed by Friis Hansen and Terndrup Pedersen (1994), can also be used:

$$OF(\beta,\beta_T)=c(\beta-\beta_T)+e^{-c(\beta-\beta_T)}-1 \qquad (5.3.61)$$

where c is a properly chosen constant.

As a concluding remark, let us say that the PSF method has two advantages over the safety index method. First, it makes a more explicit distinction between statistical and approximational uncertainties. Second, in the PSF method, each principal circumstance affecting the seriousness of the failure and each principal source of uncertainty is accounted for explicitly by means of a separate factor, and this clarifies matters and permits greater precision and consistency. The partial safety factor method has been adopted for civil engineering codes in nearly all of Europe and in Canada, Australia, and other countries (refer to the American Institute of Steel Construction) and for aircraft and aerospace structure design codes. The PSF method has also been used by API and most of the classification societies in their rules for the design and construction of offshore structures and is progressively implemented in their rules applicable to ships.

5.4 SHIP STRUCTURAL RELIABILITY ANALYSIS

5.4.1 General

Ship structures are composed of many structural elements that may be classified as follows:

1. Hull girder.
2. Primary structure (e.g., transverse web frames, longitudinal girders, horizontal girders and vertical webs on transverse bulkheads, grillages, orthotropic plate panels).
3. Stiffened panels.
4. Unstiffened plates.
5. Structural details.

Assessment of the ship structural reliability requires classifying the various structural members either as single components for which the governing modes of failure are known or as a combination of m single components or systems, which requires carrying out system reliability analyses. There are two different types of systems:

1. Series system of m single components for which failure occurs when one or more of the m components collapses (e.g., in the transverse web frames of oil tankers considered as an assemblage of beams, collapse occurs by formation of a sufficient number of plastic hinges).
2. Parallel system of m single components for which the failure occurs when all the elements collapse (e.g., collapse of a stiffened panel).

Moreover, a series system may be made of several parallel subsystems and a parallel system may be made of several series subsystems.

The general deterministic procedure considered for determination of the hull scantlings is based on the hierarchy of structural elements: structural details, unstiffened plates, stiffened panels, primary structure, and finally, the hull girder. Unstiffened plates and stiffened panels may be considered as single components and their reliability assessed accordingly. On the contrary, reliability analysis of the primary structure requires performing linear or nonlinear FEM structural analyses, which is the current practice nowadays, and the results of these analyses are to be coupled with a reliability code. To date, this coupling requires the development of appropriate interfaces between FEM programs and reliability codes prior to introducing reliability analyses of primary structures in the standard design process. According to the present practice, the hull girder response is assessed by considering that the ship may be idealized as a hollow thin-walled box beam acting in accordance with the simple beam theory. This approach allows us to define global strength parameters, such as section modulus or shear area, and express the load effects by bending moments, torsional moments, and shear forces. If we assume that this approach that has proven to be successful in the past and is applicable to the hull girder reliability analysis, the limit states can be expressed by considering that the hull girder acts as a component. In conclusion, with the exception of the primary structure for which assessment of the reliability requires additional research, the other structural members, including the hull girder, can be considered acting as single components for assessment of their structural reliability.

Based on the preceding considerations, any structural reliability analysis includes the following steps:

1. Identification of the possible modes of failure and selection of a failure criterion for each limit state identified.
2. Definition of the limit state functions associated with the identified failure modes.
3. Calculation of the various loads applied on the structure (e.g., static loads, transient loads, low- and high-frequency steady-state wave-induced loads, vibrational loads, impact loads, and residual stresses).
4. Load and load effect combinations (e.g., still-water and wave-induced bending moments, global and local loads).
5. Structural response analysis for determination of load effects (e.g., stresses and deformations).
6. Identification and statistical modeling of the random variables. The latter is the more difficult and crucial problem as reliability analyses performed for various types of ships show how the safety index depends on that choice. In that respect, for the advancement of probabilistically based ship design, there is an urgent need for standardization of the probability distributions of the various random variables.
7. Modeling of the uncertainties associated with the capacity C of the structure and load effects D.
8. Structural reliability analysis (e.g., calculation of the safety index β or probability of failure P_f) and comparison with the target safety index.
9. Sensitivity analysis.

5.4.2 Failure Modes and Limit States

GENERAL

As already stated, a limit state is defined as a condition for which a particular structural member or structure is unable to ensure its function for which it has been designed. There are two types of limit states:

- *Serviceability limit states*, involving deterioration of less vital functions under normal service loads.
- *Ultimate limit states*, leading under extreme loads to the collapse of the structure.

Structural failure is generally a nonlinear phenomenon due to either a geometrical nonlinearity (buckling, or any other large deflection), material nonlinearity (yielding and plastic deformation), or a combination of these two types. For steel members, there are three basic types of failure and their subdivisions are as follows:

1. Large local plasticity.
2. Instability
   - Bifurcation.
   - Nonbifurcation.
3. Fracture
   - Direct (tensile rupture),
   - Fatigue,
   - Brittle.

In practice an individual failure in a structural member often involves a combination of these basic types, particularly the first and second types. To appreciate the differences among the basic types of failure, it is necessary to examine the relationship between load and deflection. Figure 5.9 presents a

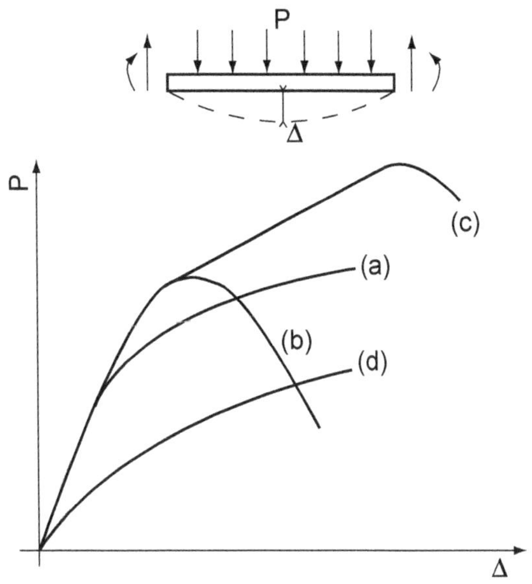

Figure 5.9 Load-deflection curves: (a) failure by plastic deformation, (b) bifurcation buckling of beams and columns, (c) bifurcation buckling of plates, and (d) nonbifurcation buckling.

sample of load-deflection curves for individual members, illustrating the variety of shapes the curves have, depending on the type of member and the type of loading and support it receives. Analysis of the load-deflection curve enables generally the determination of the collapse load (i.e., the load for which the stiffness of the member becomes zero or for which the deflection increases greatly for a small increase in load).

The following presents a brief review of some of the most significant limit states for the various ship structural elements.

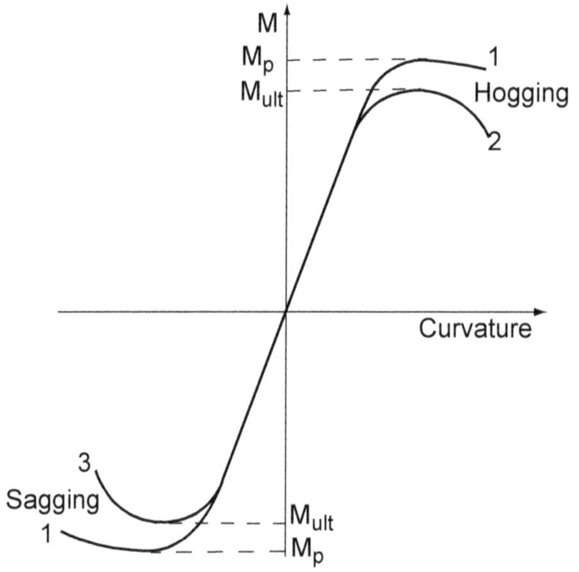

Figure 5.10 Elasto-plastic hull girder response.

HULL GIRDER

Assuming that, for analysis of the ship response under global loads, the ship structure may be idealized as a hollow, thin-walled box beam (the decks and bottom structure are flanges and the side shell and any longitudinal bulkhead are the webs) acting in accordance with the simple beam theory, the following limit states can be identified for the hull girder:

1. *First yielding*. Although the yielding criterion is not satisfactory, it is given because it represents the current design practice. This limit state occurs as soon as the hull girder stresses under normal service loads exceed the yield stress σ_Y. Depending on the ship's type, the following load effects,
   a. Still water bending moment,
   b. Vertical wave bending moment,
   c. Horizontal wave bending moment,
   d. Torsional moment (open-deck ships),
   e. Shearing forces, especially for ships in alternate loading conditions,

are to be taken into account and combined.

2. *Ultimate strength*. Beyond occurrence of the first yielding, there is a reserve of strength characterized by the maximum hull girder bending moment for which the flexural stiffness of the hull girder becomes zero. As shown in Figure 5.10, the collapse occurs either by full yielding of the section (curve 1) or by buckling (curves 2 or 3). The same load effects as for first yielding may have to be taken into account and combined.
3. *Brittle fracture*. Below a given temperature. known as the *transition temperature*, steels lose their ductility and become "brittle." Under even low stresses, cracks may appear suddenly and propagate rapidly. The value of the transition temperature depends on the chemical composition and metallurgic process. Thanks to the use of good-quality steels with a controlled toughness, in particular for sheer strake and bilge, this type of failure may be generally disregarded.

PRIMARY STRUCTURE

Collapse of the primary structure may be due to

1. Loss of overall stiffness and load-carrying ability.
2. Extensive yielding, buckling, or combination of the two.
3. Fracture.

In this type of collapse, involving combined types of failure and nonlinear interaction among various members, a rigorous and accurate value of the limit loads can be obtained only by calculating the complete load-deflection relationship using an incremental or stepwise approach. The load-deflection curve depends on the type of structure; it gives generally precise information on the behavior of the structure and enables identification of the various limit states. Particular attention has to be paid to the limit states of girders, grillages, orthotropic plates; these are

1. Serviceability limit states.
   a. *First yielding*.
   b. *Elastic buckling* under various loading combinations (longitudinal or transverse compression, edge shear, and combination of these elementary modes of buckling).
2. Ultimate limit states. Depending on the type of structure, they combine axial or biaxial loads, edge shear, and lateral pressure.

a. *Overall collapse.*
b. *Biaxial compressive collapse.*
c. *Beam-column type collapse.*

STIFFENED PANELS

The limit states of stiffened panels subjected to lateral pressure or in-plane loads refer to the interframe failure of secondary stiffeners under lateral loads, uniform compression, or a combination of the two types of loading, assuming that the strength of the primary supporting structure is sufficient to prevent its collapse prior to that of the secondary stiffeners; these are

1. Serviceability limit states.
   a. *First yielding.*
   b. *Elastic buckling* (column buckling, flexural-torsional buckling, local buckling).
2. Ultimate limit states of axially or laterally loaded stiffeners, including effects of end conditions and initial distorsions).
   a. *Inelastic buckling.*
   b. *Flexural collapse.*
   c. *Combination of the two.*

UNSTIFFENED PLATES

The limit states of unstiffened plates subjected to lateral pressure or in-plane loads refer to the failure of the plate panels between secondary stiffeners under lateral loads, uniform compression, or a combination of the two types of loading; these are

1. Serviceability limit states.
   a. *First yielding.*
   b. *Elastic and inelastic buckling* (uniaxial compression, biaxial compression, shear, biaxial compression and shear) including effect of restraints at sides, lateral pressure, residual stresses, and openings.
   c. *Formation of plastic hinges* (when lateral pressure increases beyond p_Y corresponding to the first yielding, plastic hinges form at edges and then at mid-span).
2. Ultimate limit states. Laterally loaded plates have a large reserve of strength after first yielding, as shown in Figure 5.11. For large pressures, membrane action occurs thanks to lateral restraint given by the surrounding plating. Specific ultimate limit state functions have to be developed to represent the behavior of axially and laterally loaded plates after formation of plastic hinges and taking into account, in particular, the influence of residual stresses,

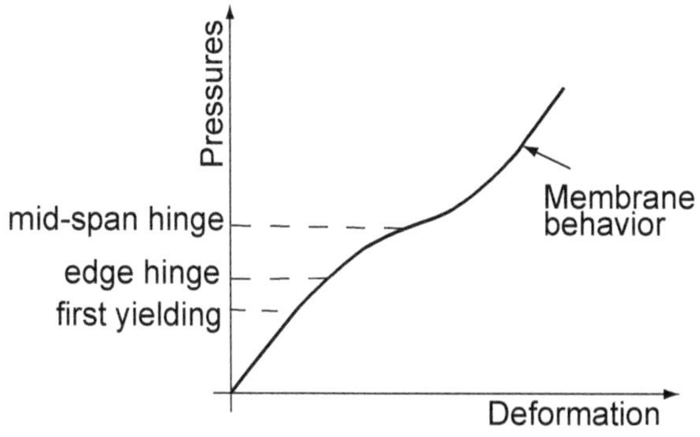

Figure 5.11 Elasto-plastic behavior of plates.

restraints at sides, aspect ratio, initial deformations, etc. Collapse may be due to
   a. *Gross yielding.*
   b. *Large deformations.*
   c. *Combination of the two.*

STRUCTURAL DETAILS

Most of the ship structural damage occurs on structural details and is due to fatigue or corrosion. It may be said that fatigue cracking occurs generally on welded structural details subjected to fluctuating stresses, due to either incorrect prediction of cyclic loads, improper design, or bad workmanship. Moreover, depending on the type of structural detail, fatigue cracking may have dramatic consequences on the ship safety or environment (e.g., knuckles of double hull oil tankers or LNG carriers). These general considerations highlight the need for assessment of the fatigue strength of structural details and reliability analyses are particularly suitable in that case, taking into account the large number of uncertainties involved in this particular limit state.

5.4.3 Loads and Load Effect Combinations

GENERAL

Loads applied on ships may be categorized as follows:

1. Static loads.
2. Transient loads such as thermal stresses.
3. Low- and high-frequency (e.g., springing) steady-state wave-induced loads.
4. Vibratory loads resulting from main engine or propeller vibratory forces.
5. Impact loads (e.g., bottom slamming, bow flare impact [whipping], sloshing and shipping of green seas.

6. Residual stresses resulting from the process of fabrication.

With the exception of transient and vibratory loads, which are specific to particular types of ships (e.g., asphalt carriers and passenger vessels) as well as springing loads (e.g., Great Lakes Bulk Carriers), the static, wave-induced, and impact loads and, in a lesser degree, residual stresses are the main loads or load effects that govern the ship design. Whatever concept is used for determination of the scantlings (i.e., deterministic or probabilistic), the designer is facing the difficult problem of the combination of the various loads or load effects acting on the structure, taking into account that they are generally time dependent and their extreme values do not occur at the same time. The loads or load effects that have to be combined depend on the limit state and structural element considered and can be decomposed into

1. Global loads acting on the hull girder (static loads, wave-induced loads, and impact loads) and their load effects (still water bending moment, vertical and horizontal wave-induced bending moments, shear forces, torsional moment, impact bending moment).
2. Local loads acting on single components (static pressures, external sea pressures, inertial cargo loads, and impact pressures) and their load effects (stresses and deformations).

From the review of the various failure modes of ship structures (refer to Section 5.4.2), the following load effects have to be combined:

1. Hull girder load effects
   - Vertical (VWBM) and horizontal (HWBM) wave-induced bending moments.
   - VWBM, HWBM, and torsional wave-induced moment (applicable to open-deck ships).
   - VWBM and springing bending moment.
   - VWBM and slamming or whipping bending moment.
   - SWBM and wave-induced bending moments including impact bending moment, where applicable.
   - Still-water and wave-induced bending stresses combined with still-water and wave-induced shear stresses.
2. Local load effects for transverse primary and secondary structures, such as static and wave induced local pressure effects. Note: Impact loads can be considered separately and it does not seem necessary to take into account this type of loads in reliability analyses.
3. Hull girder and local load effects for longitudinal primary and secondary structures. Still-water and wave-induced hull girder stresses combined with static and wave-induced local pressure effects. The influence of impact load effects may also have to be taken into account.

COMBINATION OF WAVE-INDUCED LOAD EFFECTS

Mansour and Thayamballi (1994) developed a method for combination of two or three wave-induced load effects. The method assumes that the seaway and loads are Gaussian processes and the ship is considered a set of multiple linear time-invariant systems, each of them representing a particular load. The stresses for each load are then added with the correct phase at any location of the ship structure. Main recommendations of this research work follow for the case of two and three correlated wave-induced load effects. If the load effects are expressed in terms of stresses, the combined stress σ_c is

Two correlated stresses: $\sigma_c = \sigma_1 + K\sigma_2$ (5.4.1)

Three correlated stresses:

$$\sigma_c = \sigma_1 + K_{12}\sigma_2 + K_{13}\sigma_3 \quad (5.4.2)$$

where σ_c = combined extreme stress.
σ_i = time-dependent extreme stresses.
K = load combination factor given by

$$K = \frac{1}{r}\left[\sqrt{1 + r^2 + 2\rho_{12} r} - 1\right]. \quad (5.4.3)$$

$$r = \frac{\sigma_2}{\sigma_1} < 1.$$

ρ_{12} = correlation coefficient between the stress components 1 and 2 as obtained from the results of a ship motion and load analysis.

Coefficients K_{12} and K_{13} depend on the stress ratios $r_2 = \frac{\sigma_2}{\sigma_1} < 1$, $r_3 = \frac{\sigma_3}{\sigma_1} < 1$ and on the correlation coefficients ρ_{ij} between the stresses.

Where a direct analysis is not carried out, Mansour and Thayamballi (1994) give, for some significant cases, approximate values for the load combination factors that can be used for the design of large ocean-going ships (refer to Table 5.3).

COMBINATION OF VWBM AND SLAMMING BENDING MOMENT

Combining slamming and vertical wave bending moments is not an easy task and has been studied by various authors, among them Kaplan (1972), Kaplan

Table 5.3 Load Combination Factors

Two-load effects	**K**	
Vertical and horizontal bending stresses	0.5.	
Vertical bending and local plate or beam stresses	0.70	
Three-load effects	**K_{12}**	**K_{13}**
Vertical and horizontal bending and local plate or beam stresses	0.40	0.55

and Raff (1986), and Ferro and Mansour (1985). Based on the method described by Kaplan for calculating and combining the vertical wave-induced and slamming bending moments, Nikolaidis and Kaplan (1991) calculated the maximum combined slamming and vertical wave bending moments for a large number of wave elevation time histories and compared the results with those obtained with standard methods, that is, Turkstra's rule, the peak coincidence method, and the square root of sum of squares rule (SRSS). Results of these calculations, based on the assumption that the calculated theoretical values represent the actual ones, are summarized in Table 5.4 and show that the SRSS rule gives the best approximation by comparison with the predicted values.

Table 5.4 Bias and Cov of Combined Slamming and Bending Moments

Method	**Bias**	**COV**
Turkstra's rule	1.17	0.11
Peak coincidence method	0.72	0.11
SRSS rule	1.01	0.12

Mansour and Thayamballi (1994) proposed, on their side, a simplified method for calculation of the combined load effect. The combined extreme stress is given by

$$\sigma_c = \sigma_1 + K\sigma_2 \tag{5.4.4}$$

where σ_c = combined extreme stress.
σ_1 = extreme vertical wave-induced bending stress.
σ_2 = extreme slamming stress.

$$r = \frac{\sigma_2}{\sigma_1} < 1.$$

Assuming that the stresses σ_1 and σ_2 are uncorrelated (in terms of frequency and not intensity), which seems confirmed by Friis Hansen (1994), the load combination factor K is

$$K = \frac{1}{r}[\sqrt{1+r^2} - 1] \tag{5.4.5}$$

which is equivalent to the SRSS method

$$\begin{aligned}\sigma_c &= \sigma_1 + K\sigma_2 \\ &= \sigma_1 + \frac{\sigma_2}{r}\left[\sqrt{1+r^2} - 1\right] \\ &= \sigma_1\sqrt{1+r^2}.\end{aligned}$$

COMBINATION OF SWBM AND VWBM

Particular attention has been paid to the combination of still-water bending moment (SWBM) and vertical wave-induced bending moment, since they govern the overall structural ship design and, contrary to the cases considered by Mansour and Thayamballi, may be considered as uncorrelated. Various procedures may be used for combination of the still-water and wave-induced bending moments acting on the ship structure:

1. Stochastic methods that combine the stochastic processes directly (e.g., Ferry-Borges and Castenheta 1971 and Moan and Jiao 1988 methods), thus enabling one to determine the combined bending moment corresponding to a given probability of exceedance and, consequently, the load combination factors. Guedes Soares (1984) demonstrated that stochastic methods provide exact solutions for combining still-water and wave-induced bending moments.
2. Deterministic methods that combine the characteristic values of the stochastic processes (e.g., peak coincidence method, Turkstra's rule, square root of the sum of squares, Söding 1979 method). Wang and Moan (1996) showed that the simplified Söding formula gives a good approximation of the combined bending moment for production ships.

The following gives an overview of these deterministic methods:

1. *The peak coincidence method* assumes that the maximum value over the lifetime T of a linear combination of independent modal responses $X(t) = \Sigma_i a_i x_i(t)$ occurs when each of the individual random process is maximum:

$$\begin{aligned}X_{\max,T} &= \max_T X(t) \\ &= \max_T\left[\sum_i a_i X_i(t)\right] \\ &= \sum_i a_i \max_T X_i(t)\end{aligned} \tag{5.4.6}$$

Current rules of classification societies are based on this peak coincidence method, which is generally conservative.

2. *Turkstra's rule* assumes that the value of the sum of two independent random processes $X_1(t)$ and $X_2(t)$ is maximum when one of the two variables is maximum:

$$\max\left[X_1(t)+X_2(t)\right]= \max\left[(X_1)_{\max}+E(X_1),E(X_1)+(X_2)_{\max}\right] \quad (5.4.7)$$

3. *The SRSS method.* The root mean square of a linear combination of independent modal responses $X(t)=\sum_i a_i X_i(t)$ in a time period T is approximately given by

$$\sqrt{E\left[X(T)^2\right]}=\sqrt{\sum_i a_i^2 E\left[X_i(T)^2\right]} \quad (5.4.8)$$

Assuming that the ratio between the root mean square and maximum values is the same for all the responses:

$$E(X_{\max})=p\sqrt{E\left[X(T)^2\right]}$$

$$E(X_{i,\max})=p\sqrt{E\left[X_i(T)^2\right]}$$

the expected value of the maximum response $E(X_{\max})$ is given by

$$E(X_{\max})=\sqrt{\sum_i a_i^2 E\left(X_{i,\max}\right)^2} \quad (5.4.9)$$

The SRSS method, which assumes that the load effects to be combined are 90° apart in phase, seems to be quite appropriate for combining either the wave-induced vertical and horizontal bending moments or the wave-induced vertical bending moment and torsional moment.

4. *Söding rule.* Assuming that the SWBM follows a normal distribution and the VWBM an exponential distribution, Söding (1979) obtained the following relationship:

$$M_t = M_{vw,1}+E\left(M_{sw}\right)+\frac{\sigma_{M_{sw}}^2 \ln N}{2\, M_{vw,1}} \quad (5.4.10)$$

where $M_{vw,1}$ = extreme wave bending moment as given by equation (5.4.21).

N = number of cycles over the period of time considered.

$\sigma_{M_{sw}}$ = standard deviation of the still-water bending moment.

Statistical Modeling of Random Variables

Although reliability methods have been developed for more than 30 years, they are not yet used as a standard design tool for ship structures. The main reason is the difficulty that we face for establishing rational and reliable statistical models for most of the random variables: wave loads, static loads, and resistance. Preliminary reliability analyses show that structures belonging to the same class of ships have not necessarily the same level of safety although their scantlings are based on the same requirements. This is due mainly to the great sensitivity of the safety index to the choice of the probability distribution functions and, in particular, to the tails of the distributions.

STATIC LOADS AND LOAD EFFECTS

Still-water bending moment is a static effect whose magnitude depends on the loading condition and cargo distribution. If the cargo distribution is known, the still-water bending moment can be calculated accurately. Methods of calculation of the SWBM are well established and, provided the actual cargo loads are known, it may be considered that approximational uncertainties are negligible and only statistical uncertainties are to be taken into account, since during the design, it is nearly impossible to predict all distributions of cargo that would be realized during the ship's life.

Statistical analysis of still-water data has shown that, in most of the cases, the SWBM is well below the design moment, but in some cases, the design value is exceeded. Guedes Soares and Moan (1988) reviewed the statistical distribution of the SWBM and shear forces for about 2000 voyages of about 100 ships. The following ships were analyzed: 3 dry cargo ships, 15 container ships, 14 bulk carriers, 7 ore/bulk/oil carriers, 6 chemical tankers, 4 ore/oil carriers, and 39 oil tankers. Data used in that study were the result of analysis of the information given by the loading instruments installed onboard. Results of this analysis were used to calculate the lifetime extreme still-water bending moment. Table 5.5 gives the mean and coefficient of variation of the most probable extreme SWBM as obtained by Guedes Soares and Moan (1988) for the different classes of ships considered.

Considering the example of tankers with a mean of –0.7 and a COV of 0.4 and assuming that the still-water bending moment is represented by a normal distribution, we may write

Table 5.5 Most Probable Extreme Still-Water Bending Moment

Type of Ship	Most Probable Extreme SWBM[a]	COV
Cargo	1.27	0.16
Container ship	1.16	0.14
Bulk carrier	–0.84[b]	0.27
OBO	1.13	0.31
Chemical carrier	–0.5.	0.31
Ore/oil carrier	–1.04	0.32
Tanker	–0.70	0.40

[a]The mean value is normalized by the design SWBM as given by the classification societies.
[b]The negative sign means sagging.

$$\frac{(M_{sw})_{design} - E(M_{sw})}{\sigma_{M_{sw}}} = 1.07$$

$$= \Phi^{-1}\left[P\left(M_{sw} < (M_{sw})_{design}\right)\right] \qquad (5.4.11)$$

which gives for the probability of non exceedance of the design still-water bending moment $P = 0.86$.

Note that the most probable extreme SWBM as given by Table 5.5 for container ships and cargo ships should be reviewed with more recent data, bearing in mind that a loading instrument is now required for all container ships and, depending on their design, cargo ships.

In conclusion, from the results of other studies carried out by Guedes Soares (1990) and Guedes Soares and Dias (1996) on the suitable probabilistic models for the SWBM, it seems appropriate to characterize the still-water bending moment by a normal distribution.

Static sea pressures are well monitored, since the actual draught cannot exceed the freeboard draught and therefore can be considered as deterministic variables.

For the same reasons as for the still-water bending moment, approximational uncertainties in static cargo loads are negligible and only statistical uncertainties are to be taken into account. In service, it is frequently not possible to know precisely the content of cargo in each hold. For instance, high loading rates make difficult the monitoring of actual weight of cargo inside holds of bulk carriers, leading to errors on cargo pressures applied on the structure. On the contrary, errors in the level of filling of cargo tanks are generally small.

As for the SWBM, internal cargo loads may be represented by a normal distribution with mean value taken as the design load and COV varying from 0.05 for liquid cargoes to 0.15 for bulk cargoes.

WAVE-INDUCED LOADS AND LOAD EFFECTS

The general procedure for calculation of wave-induced loads and load effects may be summarized as follows:

1. Calculation of the transfer functions of loads and load effects for regular waves of unit amplitude and for a range of wave periods, heading angles, and ship speeds.
2. Determination of the response spectra of loads and load effects for various wave spectra and heading angles (each sea state is represented by a two-dimensional directional wave spectrum defined in terms of two parameters, significant wave height, and modal wave frequency).
3. Determination of the short-term ship response for various sea states and heading angles.
4. Construction of the long-term distribution of loads and load effects giving the probability $P(X_0)$ of the load effect exceeding X_0 by combining
   a. The short-term probability of X exceeding a specified value X_0.
   b. The probability of encountering each sea state. The wave data considered correspond generally to a worldwide service.
   c. The probability of encountering the heading angle ϕ.
   d. The probability of encountering the maximum speed or a reduced speed.

The long-term distribution of the wave-induced bending moment is well approximated by the two-parameter Weibull distribution, as concluded from at-sea measurements carried out by Little, Lewis, and Bailey (1971), Lewis and Zubaly (1975), and Fain and Booth (1979). For this distribution, the probability density function is

$$p_X(X) = \frac{\xi}{\sigma_{Weib}}\left(\frac{X}{\sigma_{Weib}}\right)^{\xi-1} e^{-(X/\sigma_{Weib})^{x}} \qquad (5.4.12)$$

where ξ = Weibull shape parameter.
σ_{Weib} = characteristic value of the load effect X given by $\sigma_{Weib} = X_p/(\ln N)^{1/\xi}$.
N = number of cycles corresponding to the probability of exceedance of $1/N$.
X_p = wave-induced load effect at the probability of exceedance of $1/N$.

The cumulative distribution function is the integral of $p_X(X)$, which is

$$F(X) = 1 - e^{-(X/\sigma_{Weib})\xi} \qquad (5.4.13)$$

where $F(X)$ = probability that the wave induced bending moment of amplitude X will not be exceeded.

More generally, wave-induced load effects can be assumed to follow the Weibull distribution, and equation (5.4.13) represents the probability that the amplitude of a wave-induced load effect is less than a given value X at any one of the N cycles encountered. More important is to use the extreme value distribution, as proposed by Faulkner (1981), giving the probability that the load effect amplitude is less than a given value X_e over the N cycles:

$$F(X_e) = \left(1 - e^{-(X_e/\sigma_{Weib})^{\xi}}\right)^N = \text{Prob}(X \le X_e) \qquad (5.4.14)$$

The probability density function of the extreme value, which is the derivative of the cumulative distribution, is maximum for $X_e = X_e^* = X_p$, where X_e^* represents the *most probable value* of the extreme value distribution. If the N cycles are assumed to be independent and sufficiently large, it can be shown that the extreme value distribution as given by equation (5.4.14) converges to the Gumbel distribution:

$$F_{Gumb}(X_e) = e^{-\exp\left[-(X_e - X_e^*)/\alpha\right]} \qquad (5.4.15)$$

where

$$\alpha = \frac{\sigma_{Weib}}{\xi}\left[\ln N\right]^{(1-\xi)/\xi} = \frac{X_e^*}{\xi \ln N} \qquad (5.4.16)$$

The mean value and standard deviation of any random variable X_e distributed according to the Gumbel distribution are

$$E(X_e) = X_e^* + 0.577\alpha = X_e^*\left[1 + \frac{0.577}{\xi \ln N}\right] \qquad (5.4.17)$$

$$\sigma_{X_e} = \frac{\pi}{\sqrt{6}}\alpha = \frac{\pi}{\sqrt{6}}\frac{X_e^*}{\xi \ln N} \qquad (5.4.18)$$

$$V_{X_e} = \frac{\sigma_{X_e}}{E(X_e)} = \frac{\pi}{\sqrt{6}(0.577 + \xi \ln N)} \qquad (5.4.19)$$

The extreme bending moment ($X_e = M_{vw,1}$) over the N cycles encountered at the probability of exeedance of 5% is given by

$$e^{-\exp\left[-(M_{vw,1} - X_p)/\alpha\right]} = 0.95$$

which gives

$$\left(M_{vw,1} - X_p\right) = 2.97\,\alpha$$

or

$$M_{vw,1} = 2.97\alpha + M_{vw,0} = \left[1 + \frac{2.97}{\text{x} \ln N}\right] M_{vw,0} \qquad (5.4.20)$$

$$M_{vw,1} = E(M_{vw}) + 1.865\,\sigma_{M_{vw}} \qquad (5.4.21)$$

where $M_{vw,0}$ = design vertical wave-induced bending moment at the probability of exceedance of $1/N$.

Figure 5.12 shows the probability density function of the extreme value superimposed to the Weibull long-term distribution.

MATERIAL PROPERTIES

The normal or lognormal distribution is generally adopted for representing the material properties (yield stress, ultimate strength, and Young's modulus). The mean value of the yield stress of hull steels is about two standard deviations greater than characteristic material strength, due to acceptance criteria consisting of rejecting samples with strength less than the minimum specified value, and the COV is between 0.06 and 0.1. Assuming that the yield stress is represented by a lognormal distribution and that the minimum yield stress $(\sigma_Y)_{min}$ is guaranteed with a probability of 99%, we can write

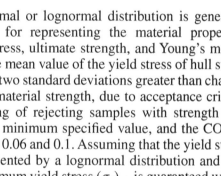

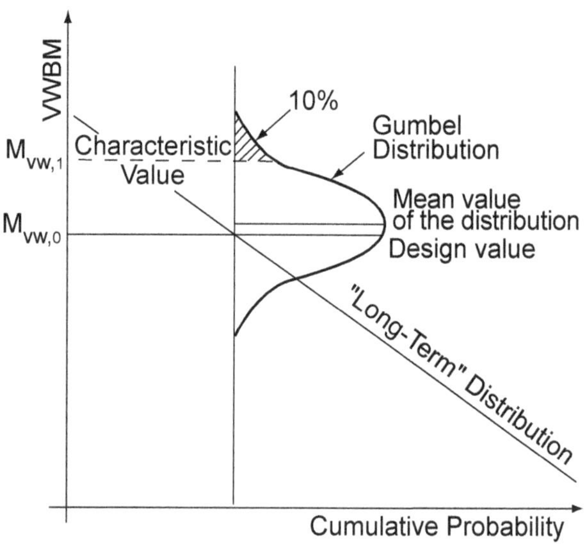

Figure 5.12 Wave-induced bending moment distribution.

$$F\left[(\sigma_Y)_{\min}\right] = \text{Prob}\left[\sigma_Y \le (\sigma_Y)_{\min}\right] = \Phi\left(\frac{\ln(\sigma_Y)_{\min} - E(\ln\sigma_Y)}{\sigma_{\ln\sigma_Y}}\right) = 0.01 \quad (5.4.22)$$

$$\frac{\ln(\sigma_Y)_{\min} - \ln E(\sigma_Y) + \ln\sqrt{\left(1+V_{\sigma_Y}^2\right)}}{\sqrt{\ln\left(1+V_{\sigma_Y}^2\right)}} = \Phi^{-1}(0.01) = -2.326$$

$$\ln E(\sigma_Y) = \ln(\sigma_Y)_{\min} + 0.189 \quad \text{with } V_{\sigma_Y} = 0.08$$

$$E(\sigma_Y) = 1.21(\sigma_Y)_{\min} \quad (5.4.23)$$

For a normal distribution we would obtain $E(\sigma_Y) = 1.23(\sigma_Y)_{\min}$.

CONSTRUCTIONAL PARAMETERS

Constructional parameters (e.g., main dimensions of the hull and thicknesses) are also random variables. Therefore, the strength parameters (e.g., plating thicknesses and section modulus of the hull girder or of any beam) are random variables. However, taking into account the continuous improvement of the methods of construction and implementation of quality control procedures in shipyards and steel works, uncertainties in the main dimensions of the hull and thicknesses become more and more negligible. Therefore, constructional parameters can be considered as deterministic variables.

On the contrary, for ships prone to corrosion, thicknesses vary with time and strength parameters become time-dependent random variables (refer to Section 5.5.1). Corrosion introduces a new random variable, that is, the corrosion rate. The Tanker Structure Co-operative Forum (1997) provides useful information on the corrosion rates observed in cargo and ballast tanks of single-hull tankers. Classification societies give also in their rules information on the corrosion rates applicable to various types of ships.

The reduction in the plate thickness calculated according to Paik et al (1998) is

$$\Delta t_i(t) = c_i\left(y_a - y_0\right)^{c_2} = c_i\, f(t) \quad (5.4.24)$$

where y_a = age of the ship in years.
y_0 = life of coating in years.
c_i = random variable characterizing the corrosion rate.
c_2 = exponent ranging between 0.3 and 1.

Based on analysis of experimental data, Yamamoto, Kumano, and Matoba (1994) concluded that the corrosion rate can be represented by the Weibull distribution.

CONCLUSION

The following probability distributions, as given in Table 5.6, may be recommended for the main random variables:

1. Still-water bending moment: normal distribution.
2. Static pressures: normal distribution.
3. Extreme wave-induced bending moment: Gumbel distribution. Note: The Gumbel distribution may also be used for impact bending moments.
4. Extreme wave-induced pressures: Gumbel distribution.
5. Material properties (yield stress, ultimate strength, and Young's modulus): normal or lognormal distribution.
6. Load combination factors: normal distribution.
7. Corrosion rate: Weibull distribution.

5.4.5 Modeling Errors in Loads and Load Effects

Irrespective of the method considered for determination of the ship scantlings, that is, deterministic or probabilistic, the designer is facing the following problems:

1. Describing the wave environment.
2. Determining the loads and load combinations resulting from action of the environment.
3. Calculating the ship structural response, that is, stresses and deformations.
4. Selecting the failure criteria.

Each of these steps involves many assumptions and subjective decisions and introduces modeling errors resulting from the lack of knowledge and accuracy of the calculation procedures. In the past two decades, much effort has been devoted to identifying and quantifying the various modeling errors with emphasis given to the hull girder wave-induced loads. The following presents a brief summary of relevant results of the recent research works.

WAVE-INDUCED LOADS AND LOAD EFFECTS

Besides uncertainties in modeling the wave environment as examined by Guedes Soares (1984) and

Table 5.6 Characteristics of the Probability Distributions

Probability Law	Probability Density Function	Probability Distribution	Mean Value	Standard Deviation
Normal	$f(x) = \frac{1}{\sigma\sqrt{2\pi}} e^{-\frac{(x-m)^2}{2\sigma^2}}$	$\Phi(x) = \frac{1}{\sigma\sqrt{2\pi}} \int_{-\infty}^{x} e^{-\frac{(x-m)^2}{2\sigma^2}} ds$	m	ϕ
Lognormal	$f(z) = \frac{1}{\sigma_z\sqrt{2\pi}} e^{-\frac{(z-m_z)^2}{2\sigma_z^2}}$ $z = \ln(x)$	$\Phi(z) = \frac{1}{\sigma_z\sqrt{2\pi}} \int_{-\infty}^{z} e^{-\frac{(z-m_z)^2}{2\sigma_z^2}} ds$	$m_z = \ln \frac{E(x)}{\sqrt{(1+V_x^2)}}$	$\sigma_z = \sqrt{\ln(1+V_x^2)}$
Weibull	$f(x) = \frac{\xi}{\sigma_w}(x/\sigma_w)^{\xi-1} e^{-x/\sigma_w}$	$F(x) = 1 - e^{-(x/\sigma_w)^{\xi}}$	$\bar{x} = \sigma_w \Gamma(1+1/\xi)$	$\sigma_w \begin{bmatrix} \Gamma(1+2/\xi) \\ -\Gamma^2(1+1/\xi) \end{bmatrix}^{1/2}$
Gumbel		$F(x) = e^{-\exp-(x-x^*)/\alpha}$	$\bar{x} = x^* + 0.577\alpha$ $\alpha = \frac{\sigma_w}{\xi}(\ln N)^{(1-\xi)\xi}$ $\sigma_w = \frac{x^*}{(\ln N_R)^{1/\xi}}$	$\sigma_x = \frac{\pi}{\sqrt{6}}\alpha$

Nikolaidis and Kaplan (1991), there are several other sources of modeling errors in the calculation of the long-term vertical and horizontal wave-induced bending moments:

1. Uncertainties in response amplitude operators (RAO). As pointed out by Kaplan et al. (1984) and Guedes Soares (1984 and 1996), uncertainties come from three different sources:
   a. Model uncertainty, that is, differences between actual and calculated transfer functions for moderate wave heights.
   b. Nonlinear effects, that is, differences between transfer functions in hogging and sagging for larger wave heights.
   c. Differences resulting from the use of different versions of programs based on the linear strip theory, leading to different predictions of transfer functions.
2. Uncertainties in the wave scatter diagram. Guedes Soares (1996) examined the influence of wave data on the long-term vertical wave bending moment for one container ship and two tankers and found large differences depending on the wave scatter diagram selected, which is confirmed by the results of similar calculations carried out by classification societies. Table 5.7 summarizes the results of calculations carried out by Guedes Soares using IST (Instituto Superior Tecnico) transfer functions and various sources of North Atlantic wave data, either visually observed or obtained from fixed measurements or hindcasts. The values of the vertical wave bending moment (VWBM) at a probability of exceedance of 10^{-8} are normalized by the smaller one computed from Hogben and Lumb wave data.
3. Long-term approximational uncertainties resulting from the various assumptions introduced in the calculation of the long-term wave-induced bending moment. Table 5.8 gives the bias and coefficients of variation for approximational uncertainties in the long-term vertical and horizontal wave-induced bending moments as obtained by Faulkner (1981), Guedes Soares (1984 and 1996), and ISSC (1985 and 1991) for various types of ships and block coefficients.

From Table 5.8 an average bias equal to unity (rules of classification societies make the distinction between sagging and hogging bending moments) and a COV of 0.10 to 0.15 may be considered to cover

Table 5.7 Influence of the Wave Scatter Diagram

VWBM ($P = 10^{-8}$)	Container Ship (L = 270 m)		Tanker (L = 270 m)		Tanker (L = 15. m)	
	Absolute	Relative[a]	Absolute	Relative[b]	Absolute	Relative[b]
Walden	0.246	1.23	0.252	1.23	0.217	1.13
David Taylor	0.25.	1.34	0.287	1.40	0.214	1.12
Hogben-Lumb	0.200	1.00	0.206	1.00	0.192	1.00
Global wave	0.215	1.07	0.221	1.07	0.195	1.02
IACS	0.277	1.39	0.294	1.43	0.227	1.19
IACS Rules	0.151	0.75	0.177	0.86	0.155	0.81

[a]WBM is normalized by the smaller one.
[b]Global Wave Statistics are assumed to give the best estimate for VWBM.

approximational uncertainties in the vertical wave-induced bending moment, including errors due to simplifications, idealizations and nonlinearities.

In conclusion, it is worth mentioning that ships are generally designed according to the rules of classification societies that give necessary information for calculation of loads and load effects, thus avoiding the need for direct calculations. For example, the hull girder strength is verified according to the IACS Unified Requirement S11, which gives the values of the vertical wave-induced bending moment to be considered for calculation of the minimum section modulus of the transverse sections in the midbody area. Values of the extreme bending moment calculated according to UR-S11 are significantly less than those obtained from direct calculations using various wave scatter diagrams, as shown in Table 5.7 (for ships considered by Guedes Soares 1996, the IACS bending moment is 70% of the mean of calculated values). Although it is not clearly stated, the IACS UR-S11 implicitly takes into account that ships designed for worldwide service do not encounter the most extreme sea states of the North Atlantic and this should be considered in direct calculations. Moreover, in heavy weather conditions ship masters can take appropriate countermeasures, such as reduction of speed and modification of the ship's route, to reduce the load effects.

Based on the satisfactory experience of ships in service, it seems appropriate to calculate the mean

Table 5.8 Long-Term Wave-Induced Bending Moment–Modeling Uncertainties

Long-Term WBM[b]	Bias[a]	COV	References
VWBM	—	0.10	Faulkner (1981)
VWBM[c]:			Guedes Soares (1984)
Tankers[d] in hogging or sagging	1.13	0.04	
Container ships in hogging	0.88	0.05	
In sagging	1.28	0.04	
Any ship in hogging	1.00	0.15	
In sagging	1.20	0.08	
VWBM:			ISSC (1991)
In hogging condition	0.75.		
In sagging condition	1.035		
VWBM	0.85[e]	0.15[e]	ISSC (1985)
HWBM	0.95[e]	0.10[e]	

[a]Bias is the actual/predicted value.
[b]This includes uncertainties in nonlinear effects and various assumptions introduced in calculation of the long-term bending moment.
[c]Bias and COV are obtained by comparing predictions based on linear strip theory and model test data.
[d]Influence of nonlinearities should also be taken into account for tankers.
[e]Bias and COV correspond to the total uncertainties.

value of the extreme VWBM over the ship's life ($N = 10^8$ cycles) from the IACS design value $M_{vw,0}$ according to equation (5.4.17).

Little data are available on modeling uncertainties in wave-induced local loads, that is, sea pressures and inertial cargo loads. As for hull girder loads, an average bias equal to unity, and a covariance of 0.10 to 0.15 could be considered to cover approximational uncertainties in wave-induced local loads, covering errors due to simplifications, idealizations, and nonlinearities. It is well known that calculations based on the linear strip theory do not represent properly the distribution of external sea pressures, especially in the vicinity of the waterline, for the following reasons:

1. Influence of nonlinearities, especially near the waterline.
2. Three-dimensional effects, especially at the ship's ends.
3. Differences resulting from the use of different versions of programs.

The use of 3D hydrodynamic programs should improve the accuracy of these calculations and reduce the level of uncertainties.

LOAD COMBINATION FACTORS

The combination of loads or load effects introduces new modeling errors. The load combination factors are themselves random variables assumed to be normally distributed. Table 5.9 gives the bias (actual/predicted value) and COV for the associated modeling errors, as proposed by Mansour (1995) for the case of load combination factors as obtained from direct analysis.

Table 5.9 Bias and Cov of Load Combination Factors

Combined Loads	Bias	COV
Wave-induced load effects	0.9	0.15
VWBM and slamming bending moment	1	0.15
SWBM and VWBM	1	0.15

STRENGTH CAPABILITY

As pointed out by various authors, e.g., Ang and Ellingwood (1971) and Hess et al. (2002), the source of uncertainties in capability can be categorized as either "objective" or "subjective." Objective uncertainties are more concerned with mechanical characteristics of the materials or constructional parameters (e.g., yield stress, fracture toughness, main dimensions of the hull, thicknesses, residual stresses), which can be measured, thus enabling us to define more and more precisely, as input data are collected, statistical distributions of the various random variables. Hughes et al. (1994) highlighted that approximational or modeling uncertainties are more concerned with subjective uncertainties that result mainly from lack of knowledge or information. For example, regarding the physical phenomena, many assumptions are made resulting in imperfect analytical models and limit state functions.

The following "subjective" uncertainties in strength models can be identified:

1. Simple beam theory in ship primary bending.
2. Modeling of the failure mechanisms.
3. Numerical errors in strength analysis.
4. Finite-element analysis (FEA):
   a. Structural idealization (extent of the 3D finite-element model, type of elements, boundary conditions, etc.) requiring engineering judgment due to the complexity of ship structures. The comparative study carried out by ISSC (1994a) on a side structure of a middle-size tanker shows clearly how the results depend on the engineering judgment.
   b. Numerical solution given by the various FEM (finite element method) codes. Error indicators have been developed to assess the error introduced by the FEM solution, which gives useful information to select adequate FEM codes.
   c. Human error. In that respect, guidelines for finite-element analysis of ship structures have been recently developed by classification societies and national regulatory agencies, aimed at keeping this type of uncertainty within insignificant limits.

Bias and coefficients of variation representing the various uncertainties in strength models are to be defined for each limit state, depending on the nature of assumptions adopted for building the analytical model and for definition of the limit state function. From comparison and analysis of FEM calculations carried out for other engineering structures, Nikolaidis and Kaplan (1991) concluded that the average bias should be taken equal to unity and the COV between 0.1 and 0.15.

5.4.6 Target Reliability Levels

Regardless of which of the methods is used and which technique is used to account for approximational uncertainties, it is absolutely essential to be able to specify different levels of safety for different types of failures, depending on their degree of seriousness. In order to assess the degree of seriousness of a structural failure, we must examine the consequences: What are the losses and how severe are they? We have seen that the two principal attributes by which the fitness of a ship is measured are *safety* and *serviceability*. Accordingly, we may distinguish two different types of losses:

1. Loss of life and other serious and irreparable noneconomic losses, such as the destruction of the environment.
2. Loss of main functions, which for a commercial ship, means economic loss due to loss of revenue, cost of repair or replacement, lawsuits, and so on.

The foregoing categories also apply to noncommercial vessels, in which the main function is the performance of some mission or service that has no direct relationship with economic factors. For such vessels, the performance can be quantified by means of a performance index; in fact, a design cannot be said to be rationally based unless the objective is specified and its dependency on the design variables is quantified. The same performance index that serves as the objective function can also be used to assess the degree of seriousness of a failure that adversely affects the performance.

Although safety and serviceability have much in common, they are distinct; some failures can cause fatalities without causing loss of main functions and vice versa. Also they have different relative importance in different situations. For example, in naval vessels, the main function is the performance of a mission, and therefore serviceability (i.e., the accomplishment of the mission) has greater importance relative to safety than it has for commercial vessels.

There is any number of degrees of seriousness; it is a continuous rather than a discrete quantity. Nevertheless, for the purpose of defining target reliability safety indices, it is necessary to define few specific degrees of seriousness. As an example, we herein distinguish three degrees of seriousness, which we call *extreme*, *severe*, and *moderate*. These must be defined in terms of their likely consequences in regard to safety and serviceability. For the attribute of safety, the degree of seriousness of a failure corresponds to its consequences in regard to loss of life and protection of the environment. Similarly, for the attribute of serviceability, the seriousness is measured by loss of main function and economic consequences. Table 5.10 describes in general terms the sort of consequences that would correspond to these three degrees.

Since the primary aim of structural constraints is to provide adequate safety and serviceability, the most important limit state is that of ultimate failure of the hull girder. The other limit states are merely stages toward structure collapse. The provision of adequate safety against structure collapse automatically provides a proportional degree of safety against less serious forms of failure, and this is usually sufficient. But the converse is not true; the provision of adequate safety against lesser forms of failure does not necessarily provide sufficient safety at the overall level, which is where it is required most. Therefore, first, the possible modes of failure, under the various combinations of loads that are expected, are to be defined for each type of structure (hull girder, primary structure, stiffened panels, and unstiffened plates); second, each member failure is to be assigned to one of the three levels of seriousness, depending on which of the consequences described in Table 5.10 best matches the consequence that limit state would have on the safety and serviceability of the ship. These considerations are summarized in Table 5.11.

Once the criticality of the various possible modes of failure is defined, the next task—and more difficult—is to select the target probabilities of failure or the target safety indices. This has to take into account the past experience of ships in service and can be based on

1. Recommended values given by regulatory bodies (e.g., American National Standard, AISC, API, Canadian Standard Association, A. S. Veritas).
2. Design code calibration by comparison with existing codes that have proven satisfactory, see Melchers (1987) for more information.
3. Economic value analysis. The safety indices are selected to minimize the present value of construction plus maintenance costs during the expected ship's life.

Based on the review of proposals made by various regulatory bodies and analysis of the results of reliability analyses performed for the last 30 years, Mansour et al. (1996 and 1997), see Table 5.12, recommend target safety indices for hull girder (primary), stiffened panels (secondary), and unstiffened plates (tertiary) modes of failure as well as for fatigue failure.

The initial yield criterion for the hull girder is included only because, for many years, it has been the criterion used in the classification society rules and still is one of the criteria used to verify the

Table 5.10 Degrees of Seriousness of Structural Failures in Regard to Safety and Serviceability

Degree of Seriousness of Failure	Safety (consequences in regard to loss of life or main functions)	Serviceability (consequences in regard to loss of less vital functions)
Extreme	Some fatalities or significant pollution likely, may include all personnel if there is another failure or harsh conditions or mismanagement. Ship out of service for a long period. May be permanent loss (e.g., due to hull girder collapse) if there is another failure or harsh conditions or mismanagement.	Ship efficiency seriously impaired with economic consequences (e.g., permanent deformations of hull girder). Repair urgent.
Severe	Small but definite risk that the failure may cause a few fatalities or pollution at occurrence; risk of subsequent fatalities very small unless there is another failure or harsh conditions or mismanagement. Ship out of service for short period or ship operational but seriously handicapped (e.g., fracture of primary structure). Repair urgent.	Ship operational but reduced efficiency (e.g., unacceptable deformations or vibrations). Loss of some secondary functions. Repair as soon as practicable.
Moderate	No appreciable risk of fatalities but the structure is weakened (e.g., buckling of unstiffened plates) and a slight risk would arise if there is another failure or harsh conditions or mismanagement.	Main function unimpaired, some inconvenience or inefficiency at the secondary level (e.g., excessive vibrations affecting comfort). Repair as soon as convenient.

Table 5.11 Degree of Seriousness of Structural Failures

Structural Member	Yielding	Instability	Fracture (tensile rupture)	Fatigue
Hull girder	Serviceability: Extreme	Safety: Extreme	Safety: Extreme	—
Primary structure	Serviceability: Severe	Safety: Severe	Safety: Severe	— —
Stiffened panels	Serviceability: Severe	Safety: Severe	Serviceability: Severe	—
Unstiffened plates	Serviceability: Moderate	Safety: Moderate	Serviceability: Severe	
Structural details	—	—	—	Safety: Severe to Moderate depending on the criticality of the detail.

Table 5.12 Target Safety Indices—Mansour's Proposal

Failure Mode	**Commercial Ships**		**Naval Ships**	
	P_f	β_0	P_f	β_0
Primary (initial yield)	2.9×10^{-7}	5.0	1.0×10^{-9}	5.0
Primary (ultimate strength)	2.3×10^{-4}	3.5	3.2×10^{-5}	4.0
Secondary (ultimate strength)	5.2×10^{-3}	2.5	1.4×10^{-3}	3.0
Tertiary (ultimate strength)	2.3×10^{-2}	2.0	5.2×10^{-3}	2.5
Fatigue				
Very serious	1.4×10^{-3}	3.1	2.3×10^{-4}	3.5
Serious	5.2×10^{-3}	2.5	1.4×10^{-3}	3.0
Not serious	$1.5.\times 10^{-1}$	1.0	5.7×10^{-2}	1.5

strength of the hull girder. It was introduced over a half century ago to deal with hull girder bending of steel ships, and Vedeler (1965) was one of the pioneers. At that time, classification societies (CS) were aware that other, more serious types of failure could occur, notably buckling. Consequently, in an effort to avoid the other failures, they deliberately required a minimum section modulus of the hull girder $Z_{\min}$ sufficiently high that the probability of buckling, even though it is much greater than that of initial yield, would nevertheless be sufficiently small. Since then, CS and ship designers have gradually improved the efficiency of ship hulls and additional requirements have been introduced in CS rules to prevent the possibility of buckling of structural members. However, classification societies continued to base the minimum section modulus $Z_{\min}$ on initial yield but with some relaxation, resulting from satisfactory experience and better assessment of design loads. For more than 30 years, two separate requirements have coexisted:

1. Initial yield criterion, $\sigma_b = \dfrac{M_{sw} + M_{vw}}{Z_{\min}} \leq \dfrac{175}{k}$,

where k is the material factor. Note: In addition, for higher-strength steels, this material factor does not take full benefit of the increase in yield stress to maintain a satisfactory level of safety with respect to fatigue. For $\sigma_Y = 355$ MPa, $k = 0.72$ instead of $k' = 235/355 = 0.66$).

2. Buckling criterion for individual members, $\sigma_{comp} \leq \sigma_{crit}$.

More recently, additional requirements on the ultimate strength of the hull girder have been introduced.

Assuming that the yield stress follows a normal distribution, as shown in Figure 5.13, the hull girder bending design stress of 175 MPa, which includes a safety factor to account for stresses resulting from the bending of primary structure and secondary stiffeners, is far below the "mean" yield stress and "minimum" guaranteed value (235 MPa), which corresponds to a probability of nonexceedance extremely small ($P(\sigma_Y \leq 175) \cong 10^{-9}$). This explains why, as shown in Table 5.12, Mansour found that the implied probability of initial yield is $P_f = 2.9 \times 10^{-7}$. This is a paradoxical value because it is far less than the value for ultimate strength of the hull girder ($P_f = 2.3 \times 10^{-4}$), which is a much more serious failure. However, this situation is currently changing. The theory and software tools to perform an accurate and yet practical hull girder ultimate strength analysis have become available. Therefore, it is likely that the hull girder ultimate strength criterion that explicitly considers member buckling will become the prevailing criterion, while the minimum section modulus requirement and individual member buckling criterion will be used for determination of the initial scantlings of members contributing to the longitudinal strength.

If the acceptable lifetime probability of overall structural failure is about 10^{-3}, see Section 5.1.3, the target safety index $\beta = 3.5$ as proposed by Mansour et al. (1997) should be reduced to 3.1. Moreover, based on the results of previous reliability analyses, a safety index of 4.5 for initial yield of the hull girder would be more reasonable (refer also to Section 5.5.2). This is, moreover, the value adopted by Mansour et al. (2000) for calculation of the partial safety factors for the yield strength formulation. In

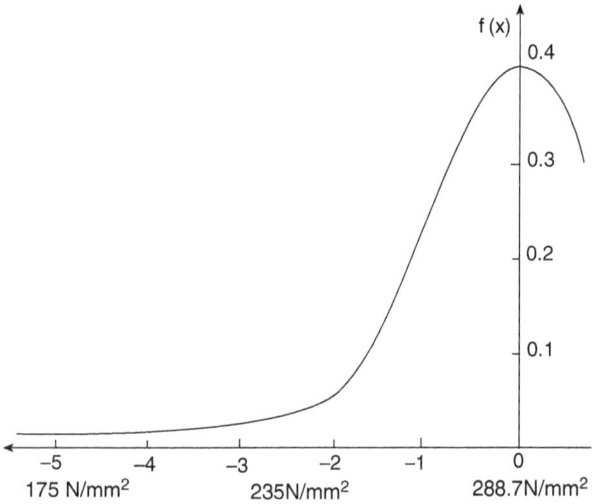

Figure 5.13 Probability distribution of the yield stress (mild steel).

addition, the degree of seriousness of fatigue failures has to be assessed on a case-by-case basis, depending on their consequences with regard to safety. For example, fatigue failures of knuckle joints of double-hull oil tankers combined with accelerated corrosion due to breakdown of the coating may lead to oil leakage in void spaces and increase the risk of explosion. The seriousness of such a fatigue failure is extreme.

According to the previous comments, Table 5.13 summarizes the target safety indices that could be considered for a reliability-based ship design.

Table 5.13 Target Safety Indices

	Safety		Serviceability	
	P_f	β_0	P_f	β_0
Extreme	1.0×10^{-3}	3.1	5.1×10^{-6}	4.5
Severe	5.2×10^{-3}	2.5	3.1×10^{-5}	4.1
Moderate	2.3×10^{-2}	2.0	1.2×10^{-4}	3.7

In conclusion, let us note that, prior to introducing reliability-based design codes as the standard practice, a large effort has yet to be devoted

1. To carry out systematic reliability analyses on a large sample of existing ships designed according to present or past rules with a view to calibrating the target safety indices.
2. To agree on the statistical properties of the various random variables (distributions, mean values and variances).
3. To agree on the reliability procedure to carry out these analyses.

5.5 LIMIT STATE FUNCTIONS OF SHIP COMPONENTS

Once the failure modes and limit states are identified for any individual structural member, the limit state functions $g(x)$ can be defined from application of first engineering principles. In the following, some limit state functions are given for typical structural members and failure modes. Note: These functions are based on simplified equations, bearing in mind that this section aims mainly at giving the methodology that can be used for developing the limit state function once the mode of failure is identified.

In most of these examples, the safety margin is a nonlinear function of the random variables X_i. The iterative procedure, as described in Section 5.3.2 or 5.3.3, based on linearization of the failure surface at each step of the process, can be used for calculation of the Hasofer and Lind safety index and comparison with the target safety index β_0. For determination of the reliability-based partial safety factors, the calibration procedure as described in Section 5.3.4 can be used.

5.5.1 Hull Girder

INITIAL YIELDING

The yielding criterion of the hull girder can be expressed as

$$\frac{M_{sw} + K_w M_{vw}}{Z} \leq \sigma_Y \qquad (5.5.1)$$

leading to the following definition of the limit state function:

$$g(x) = Z\sigma_Y - M_{sw} - K_w M_{vw} \qquad (5.5.2)$$

where σ_Y = yield stress of the material.
Z = section modulus of the transverse section at strength deck or bottom.
M_{sw} = still-water bending moment.
M_{vw} = vertical wave-induced bending moment.
K_w = load combination factor between the still-water bending and the vertical wave-induced bending moment.

Note: Equation (5.5.1) corresponds to the basic equation used by design codes, $\frac{Z\sigma_Y}{\gamma} \geq M_{sw} + M_{vw}$ with $K_w = 1$.

The safety margin with respect to initial yielding is obtained by replacing the design parameters in equation (5.5.2) with the corresponding random variables:

$$M = Z\sigma_Y - B_{M_{sw}} M_{sw,0} - K_w B_{M_{vw}} M_{vw,0} \qquad (5.5.3)$$

where $M_{sw,0}$ = design still-water bending moment.
$M_{vw,0}$ = design vertical wave-induced bending moment at the probability of exceedance of $1/N$.
N = number of cycles corresponding to the probability of exceedance of $1/N$.
$B_{M_{sw}}$ = uncertainties in the still-water bending moment. According to Section 5.4.4, $B_{M_{SW}}$ includes only statistical uncertainties ($B_{M_{sw}} = B_{I_{sw}}$).
$B_{M_{vw}}$ = uncertainties in the vertical wave-induced bending moment, as defined by equation (5.2.5); that is, $B_{M_{vw}} = B_{I_{vw}} B_{II_{vw}}$.

The safety margin as given by equation (5.5.3) may be approximated by the following linear function:

$$M = g(X) = Z X_1 - M_{sw,0} X_2 - M_{vw,0} X_3 \qquad (5.5.4)$$

where $X_i's$ = random variables assumed to be independent.
$X_1 = \sigma_Y$.
$X_2 = B_{M_{sw}}$.
$X_3 = K_w B_{M_{vw}} = K_w B_{I_{vw}} B_{II_{vw}}$.

Note: For ships prone to corrosion, the section modulus Z also should be considered a random variable (refer to Section 5.5.1).

Since the safety margin is assumed to be a linear function of the independent random variables $X_i's$, the Cornell and Hasofer-Lind safety indices are equal and given by

$$\beta = \frac{Z E(X_1) - M_{sw,0} E(X_2) - M_{vw,0} E(X_3)}{\sqrt{\sum_{i=1}^{i=n} a_i^2 \sigma_{X_i}^2}} \qquad (5.5.5)$$

For determination of the PSF the random variables $X_i's$ are transformed into a set of reduced normal variables $U_i's$. Since the $X_i's$ are independent random variables, the transformation matrix T is a diagonal matrix whose elements are equal to $1/\sigma_i$. Therefore, the limit state function expressed in terms of the reduced variables is

$$g'(u) = Z\left[E(X_1) + u_1\sigma_{X_1}\right] - M_{sw,0}\left[E(X_2) + u_2\sigma_{X_2}\right] - M_{vw,0}\left[E(X_3) + u_3\sigma_{X_3}\right]$$

and the coordinates u_i^* of the MPFP are

$$u_i^* = -\begin{pmatrix} -\beta \dfrac{Z\sigma_{X_1}}{\sqrt{\sum a_i^2 \sigma_{X_i}^2}}, \beta \dfrac{M_{sw,0}\sigma_{X_2}}{\sqrt{\sum a_i^2 \sigma_{X_i}^2}}, \\ \beta \dfrac{M_{vw,0}\sigma_{X_3}}{\sqrt{\sum a_i^2 \sigma_{X_i}^2}} \end{pmatrix} \qquad (5.5.6)$$

According to equation (5.3.60), the partial safety factors are given by

$$\frac{1}{\gamma_1^*} = \frac{X_1^*}{(X_1)_{nom}} = \frac{E(X_1) + u_1^*\sigma_{X_1}}{(X_1)_{nom}} = \frac{1}{(\sigma_Y)_{\min}}\left(E(X_1) - \beta \frac{Z\sigma_{X_1}^2}{\sqrt{\sum a_i^2 \sigma_{X_i}^2}}\right) \qquad (5.5.7)$$

$$\gamma_2^* = \frac{X_2^*}{(X_2)_{nom}} = \frac{E(X_2) + u_2^*\sigma_{X_2}}{(X_2)_{nom}} = E(X_2) + \beta \frac{M_{sw,0}\sigma_{X_2}^2}{\sqrt{\sum a_i^2 \sigma_{X_i}^2}}, \quad \text{with } (X_2)_{nom} = 1 \qquad (5.5.8)$$

$$\gamma_3^* = \frac{X_3^*}{(X_3)_{nom}} = \frac{E(X_3) + u_3^*\sigma_{X_3}}{(X_3)_{nom}} = E(X_3) + \beta \frac{M_{vw,0}\sigma_{X_3}^2}{\sqrt{\sum a_i^2 \sigma_{X_i}^2}}, \quad \text{with } (X_3)_{nom} = 1 \qquad (5.5.9)$$

The design equation expressed in terms of the partial safety factors is given by

$$g = Z X_1^* - M_{sw,0} X_2^* - M_{vw,0} X_3^* = Z\frac{(\sigma_Y)_{\min}}{\gamma_1^*} - \gamma_2^* M_{sw,0} - \gamma_3^* M_{vw,0} \geq 0$$

or, in a conventional form,

$$Z \geq \gamma_R \frac{\gamma_{M_{sw}} M_{sw,0} + \gamma_{M_{vw}} M_{vw,0}}{(\sigma_Y)_{\min}} \quad (5.5.10)$$

Where the influence of the horizontal wave-induced bending moment cannot be neglected and if it is assumed that the combined wave-inducing bending moment can be calculated according to the SRSS method, the limit state function may be defined as

$$\begin{aligned} g(x) &= \sigma_Y - \sigma_{sw} - K_w \sigma_{cw} \\ &= \sigma_Y - \sigma_{sw} - K_w \sqrt{\sigma_{vw}^2 + \sigma_{hw}^2} \\ &= \sigma_Y - \frac{M_{sw}}{Z} - K_w \sqrt{(M_{vw}/Z)^2 + (M_{hw}/Z_H)^2} \end{aligned}$$

or

$$\begin{aligned} g(x) &= Z\sigma_Y - M_{sw} \\ &\quad - K_w \sqrt{M_{vw}^2 + (Z/Z_H)^2 M_{hw}^2} \end{aligned} \quad (5.5.11)$$

where σ_{cw} = combined wave-induced hull girder bending stress calculated according to the SRSS method $\left(\sigma_{cw} = \sqrt{\sigma_{vw}^2 + s_{hw}^2}\right)$.

σ_{vw} = hull girder bending stress due to the vertical wave-induced bending moment.

σ_{hw} = hull girder bending stress due to the horizontal wave-induced bending moment.

M_{hw} = horizontal wave-induced bending moment.

K_w = load combination factor between the still-water bending moment and the combined wave-induced bending moment calculated according to the SRSS method $\left(M_{cw} = \sqrt{M_{vw}^2 + M_{hw}^2}\right)$.

Z_H = horizontal section modulus.

The safety margin with respect to initial yielding is obtained by replacing the design parameters in equation (5.5.11) with the corresponding random variables:

$$\begin{aligned} M &= Z\sigma_Y - M_{sw,0} B_{M_{sw}} \\ &\quad - K_w \sqrt{M_{vw,0}^2 B_{M_{vw}}^2 + (Z/Z_H)^2 M_{hw,0}^2 B_{M_{hw}}^2} \end{aligned}$$

or

$$\begin{aligned} M &= ZX_1 - M_{sw,0} X_2 \\ &\quad - X_5 \sqrt{M_{vw,0}^2 X_3^2 + (Z/Z_H)^2 M_{hw,0}^2 X_4^2} \end{aligned} \quad (5.5.12)$$

where $M_{hw,0}$ = design horizontal wave-induced bending moment corresponding to the probability of exceedance of $1/N$.

$B_{M_{hw}}$ = uncertainties in the horizontal wave-induced bending moment ($B_{M_{hw}} = B_{I_{hw}} B_{II_{hw}} = X_4$)

X_5 = K_w.

The design equation expressed in terms of the partial safety factors is given by

$$\begin{aligned} g &= Z X_1^* - M_{sw,0} X_2^* \\ &\quad - X_5^* \sqrt{M_{vw,0}^2 (X_3^*)^2 + (Z/Z_H)^2 M_{hw,0}^2 (X_4^*)^2} \geq 0 \end{aligned} \quad (5.5.13)$$

or, in a conventional form,

$$Z \geq \gamma_R \times \frac{\gamma_{sw} M_{sw,0} + \gamma_{K_w} \sqrt{\gamma_{M_{vw}} M_{vw,0}^2 + \gamma_{M_{hw}} (Z/Z_H)^2 M_{hw,0}^2}}{(\sigma_Y)_{\min}} \quad (5.5.14)$$

More sophisticated expressions have to be developed, where the influence of shear stresses has to be taken into account and combined with the bending stresses. For example, the following simplified equation gives the safety margin with respect to initial shear yielding of the hull girder:

$$\begin{aligned} M &= g(X) \\ &= \frac{I \sum_i t_i}{S} \frac{\sigma_Y}{\sqrt{3}} - B_{Q_{sw}} Q_{sw,0} - K_w B_{Q_{vw}} Q_{vw,0} \end{aligned}$$

or

$$\begin{aligned} M &= g(X) \\ &= \frac{I \sum_i t_i}{S} \frac{X_1}{\sqrt{3}} - Q_{sw,0} X_2 - Q_{vw,0} X_3 \end{aligned} \quad (5.5.15)$$

where S = first moment of the transverse section about the neutral axis.

I = moment of inertia of the transverse section about the neutral axis.

$\Sigma_i t_i$ = minimum thickness of side shell and lonitudinal bulkhead plating.

$Q_{sw,0}$ = design still-water shear force.
$Q_{vw,0}$ = design wave-induced shear force corresponding to the probability of exceedance of 1/N.

The design equation expressed in terms of the Partial Safety Factors is

$$\frac{I\sum_i t_i}{S\sqrt{3}} \geq \frac{Q_{sw,0}\, X_2^* + Q_{vw,0}\, X_3^*}{X_1^*} \quad (5.5.16)$$

or

$$\frac{I\sum_i t_i}{S\sqrt{3}} \geq \gamma_R \frac{\gamma_{Q_{sw}}\, Q_{sw,0} + \gamma_{Q_{vw}}\, Q_{vw,0}}{(\sigma_Y)_{\min}} \quad (5.5.17)$$

ULTIMATE STRENGTH

A similar approach may be considered for assessment of the ultimate strength of the hull girder. In that case, the safety margin may be given by

$$M = g(X) = B_{M_{vu}}\, M_{vu,0} - B_{M_{sw}}\, M_{sw,0} - K_w \sqrt{B_{M_{vw}}^2\, M_{vw,0}^2 + B_{M_{dw}}^2\, M_{dw,0}^2}$$

or

$$M = g(X) = M_{vu,0}\, X_1 - M_{sw,0}\, X_2 - X_5 \sqrt{M_{vw,0}^2 X_3^2 + M_{dw,0}^2 X_4^2} \quad (5.5.18)$$

where $M_{vu,0}$ = design ultimate vertical bending moment.
$M_{dw,0}$ = design dynamic bending moment.
K_w = load combination factor between the still-water bending moment and the combined wave-induced bending moment calculated according to the SRSS method ($M_{cw} = \sqrt{M_{vw}^2 + M_{dw}^2}$)
$X_i's$ = random variables.
$X_1 = B_{M_{dw}}$.
$B_{M_{vu}}$ = uncertainties in the ultimate vertical bending moment.
$X_4 = B_{M_{dw}}$.
$B_{M_{dw}}$ = uncertainties in the dynamic bending moment.
$X_5 = K_w$.

Assuming that the combined wave-induced bending moment is obtained according to the SRSS method, the design equation expressed in terms of the partial safety factors is

$$M_{u,0} \geq \frac{\left(X_2^*\, M_{sw,0} + X_5^* \sqrt{(X_3^*)^2\, M_{vw,0}^2 + (X_4^*)^2\, M_{dw,0}^2}\right)}{X_1^*} \quad (5.5.19)$$

or

$$M_{u,0} \geq \gamma_R \left(\gamma_{M_{sw}}\, M_{sw,0} + \gamma_{K_w} \sqrt{\gamma_{M_{vw}}\, M_{vw,0}^2 + \gamma_{M_{dw}}\, M_{dw,0}^2}\right) \quad (5.5.20)$$

As for first yielding, equation (5.5.20) may be extended to cases where the horizontal wave-induced bending moment cannot be neglected. In such a case, the following interaction formula proposed by Paik and Thayamballi (2000) can be used:

$$g(x) = 1 - \left(\frac{M_{sw} + M_{vw}}{M_{vu}}\right)^{1.85} - \left(\frac{M_{hw}}{M_{hu}}\right) \geq 0 \quad (5.5.21)$$

and the safety margin becomes

$$M = g(X) = 1 - \left(\frac{B_{M_{sw}}\, M_{sw,0} + K_{vw}\, B_{M_{vw}}\, M_{vw,0}}{B_{M_{vu}}\, M_{vu,0}}\right)^{1.85} - \frac{B_{M_{hw}}\, M_{hw,0}}{B_{M_{hu}}\, M_{hu,0}} \quad (5.5.22)$$

or

$$M = g(X) = 1 - \left(\frac{M_{sw,0}\, X_2 + M_{vw,0}\, X_3}{M_{vu,0}\, X_1}\right)^{1.85} - \frac{M_{hw,0}\, X_4}{M_{hu,0}\, X_5} \quad (5.5.23)$$

where $M_{hu,0}$ = design ultimate horizontal bending moment.

K_{vw} = load combination factor between the still-water bending moment and the vertical wave-induced bending moment.

$B_{M_{hu}}$ = uncertainties in the ultimate horizontal bending moment.

$X_3 = K_{vw}\, B_{M_{vw}}$.

$X_4 = B_{M_{hw}}$.

X_5 = random variable covering uncertainties in horizontal ultimate bending moment ($X_5 = B_{M_{hu}}$).

The design equation expressed in terms of the partial safety factors is given by

$$1-\left(\frac{M_{sw,0}\,X_2^* + M_{vw,0}\,X_3^*}{M_{vu,0}\,X_1^*}\right)^{1.85} - \gamma_{Rh}\,\frac{M_{hw,0}\,X_4^*}{M_{hu,0}\,X_5^*} \geq 0 \qquad (5.5.24)$$

or

$$1-\gamma_{Rv}\left(\frac{\gamma_{M_{sw}}\,M_{sw} + \gamma_{M_{vw}}\,M_{vw}}{M_{vu,0}}\right)^{1.85} - \gamma_{Rh}\,\frac{\gamma_{M_{hw}}\,M_{hw,0}}{M_{hu,0}} \geq 0 \qquad (5.5.25)$$

5.5.2 Primary Structure

Determination of the various limit states of primary members requires nonlinear FEM structural analyses for determination of the load-deflection curve. As mentioned previously, the load-deflection relationship can be obtained only by using an incremental approach.

For example, Bureau Veritas (2000) defines four basic load cases for determination of the scantlings of primary members, and for each of these load cases, the design pressure p_{des} is

$$p_{des} = p_{st,0} + p_{w,0} \qquad (5.5.26)$$

where $p_{st,0}$ = design static pressure.

$p_{w,0}$ = design external or inertial wave-induced pressure corresponding to the probability of exceedance of $1/N$.

Since only the wave-induced component of the total pressure varies, increments are applied to the wave-induced pressure and at step n of the process the pressure is

$$p = p_{st,0} + p_{w,0} + n\,\Delta p_{w,0} = p_{st,0} + \left(1 + n\,\Delta p_{w,0}/p_{w,o}\right)p_{w,0}$$

Finally, the ultimate pressure $p_{\lim}$ corresponding to the limit state considered (serviceability or ultimate) may be expressed as

$$p_{\lim} = p_{st,0} + \lambda_{\lim}\,p_{w,0} \qquad (5.5.27)$$

where $\lambda_{\lim}$ = dynamic load factor given by $(1 + n_{\max}\,\Delta p_{w,0}/p_{w,0})$.

For the limit state considered, the limit state function is

$$g(x) = p_{\lim} - p_{des} \qquad (5.5.28)$$

The safety margin is obtained by replacing the design parameters in equation (5.5.28) with the corresponding random variables:

$$M = g(X) = \left(p_{st,0} + B_\lambda\,\lambda_0\,p_{w,0}\right) - \left(B_{p_{st}}\,p_{st,0} + B_{p_w}\,p_{w,0}\right)$$

or

$$M = g(X) = p_{st,0}\left(1 - X_1\right) + \left(\lambda_0\,X_3 - X_2\right)p_{w,0} \qquad (5.5.29)$$

where λ_0 = calculated dynamic load factor.

B_λ = random variable covering uncertainties in the dynamic load factor.

$B_{p_{st}}$ and B_{p_w} = random variables covering uncertainties in static and dynamic pressures.

Since the safety margin is a linear function of the random variables, the safety index is given by

$$\beta = \frac{E(M)}{\sigma_M} = \frac{\left[1 - E(X_1)\right]p_{st,0} + \left[\lambda_0\,E(X_3) - E(X_2)\right]p_{w,0}}{\sqrt{p_{st,0}^2\,\sigma_{X_1}^2 + \sigma_{X_2^2}\,p_{w,0}^2 + \lambda_0^2\,p_{w,0}^2\sigma_{X_3^2}}}$$

or

$$\beta = \frac{E(M)}{\sigma_M} = \frac{\left[1 - E(X_1)\right] + \left[\lambda_0\,E(X_3) - E(X_2)\right]\left(p_{w,0}/p_{st,0}\right)}{\sqrt{\sigma_{X_1}^2 + \left(\sigma_{X_2^2} + \lambda_0^2\,\sigma_{X_3^2}\right)\left(p_{w,0}/p_{st,0}\right)^2}} \qquad (5.5.30)$$

and is to be compared to the target safety index β_0 for the limit state considered.

For primary members contributing to the longitudinal strength, not only the wave-induced pressure but also the hull girder wave-induced stress has to be incremented and a relationship has to be established between increments in the wave-induced pressure and hull girder stress for performing the nonlinear analyses. Although the peak coincidence method is on the conservative side, increments may be calculated so that the resulting hull girder bending moment and wave-induced local pressure correspond to the same probability of exceedance.

5.5.3 Stiffened Panels

INITIAL YIELDING OF AXIALLY AND LATERALLY LOADED STIFFENERS

The elastic behavior of uniformly laterally loaded stiffeners subjected to axial compression is governed by the following differential equation:

$$EI\frac{d^4w}{dx^4}+N_x\frac{d^2w}{dx^2}=ps \qquad (5.5.31)$$

Yielding of the flange of laterally loaded stiffeners subjected to normal stresses occurs as soon as

$$\sigma_f=\sigma_n+\frac{M_{max}}{Z_S}\geq\sigma'_Y \qquad (5.5.32)$$

leading to the following definition of the limit state function:

$$g(x)=\sigma'_Y-\sigma_n-\frac{M_{max}}{Z_S} \qquad (5.5.33)$$

where σ'_Y = equivalent yield stress, as given in Table 5.14.
σ_n = normal stress = $N_x/\left(A_S+s\,t_p\right)$.
Z_S = section modulus of the stiffener with attached plate.
$M_{max}=\phi\frac{ps\ell^2}{m}$ in which the various coefficients are given in Table 5.14.

$$\sigma_E=\frac{\pi^2EI}{(A_S+s\,t_p)\,\ell^2}\,. \qquad (5.5.34)$$

N_x = axial compressive load.
A_S = cross-sectional area of the stiffener without attached plate.

Table 5.14 Axially and Laterally Loaded Stiffeners

	Axial Compressive Stress	
	Simply Supported Stiffeners	**Fixed Stiffeners**
Pressure	**Acting on the stiffener side**	**Acting on the plating side**
σ'_Y	σ'_Y	$\sigma_Y\sqrt{1-3\left(\tau/\sigma\right)_Y^2}$
τ	0	$\tau=\frac{ps\ell}{2A_w}$
m	8	12
ϕ	$\frac{1}{1-\sigma_n/\sigma_E}$	$\frac{1}{1-0.18\sigma_n/\sigma_E}$

p = lateral pressure applied on the stiffener, $p=p_{st}+p_w$.
ℓ = stiffener span.
s = stiffener spacing.
t_p = thickness plating.
E = Young's modulus.
I = moment of inertia of the stiffener with attached plating.

At the intersection of the web and the faceplate the shear stress is

$$\tau=\frac{Q\,A_f}{Z_S\,t_w} \qquad (5.5.35)$$

where Q = shear force.
A_f = faceplate cross-sectional area.
t_w = web thickness.

The safety margin with respect to the flange yielding of axially and laterally loaded stiffeners is obtained by replacing the design parameters in equation (5.5.33) with the corresponding random variables. For stiffeners subjected to hull girder bending stresses $\left(\sigma_n=\frac{M_{sw}+M_{vw}}{Z}\right)$, we obtain

$$M=g(X)=\sigma'_Y-\frac{B_{M_{sw}}M_{sw,0}+K_wB_{M_{vw}}M_{vw,0}}{Z}-\frac{\phi\frac{\left(B_{p_{st}}p_{st,0}+B_{p_w}p_{w,0}\right)s\ell^2}{m}}{Z_S}$$

$$M = g(X) = Z_S X_1' - \frac{Z_S}{Z}\left(M_{sw,0} X_2 + M_{vw,0} X_3\right) - X_2' \frac{(p_{st,0} X_4 + p_{w,0} X_5)\ s\ \ell^2}{m} \quad (5.5.36)$$

where Z = section modulus of the transverse section at the longitudinal stiffener considered.

$X_4 = B_{p_{st}}$
$X_5 = B_{p_w}$

$$X_1' = X_1\sqrt{1-3\left(\tau/X_1\right)^2} = X_1\sqrt{1-3\left(\frac{A_f}{Z_S\, t_w}\frac{\left(p_{st,0}X_4 + p_{w,0}X_5\right)s\,\ell}{2X_1}\right)^2}.$$

$$X_2' = \frac{1}{1-0.18\dfrac{M_{sw,0} X_2 + M_{vw,0} X_3}{Z\sigma_E}}.$$

$$X_2' = \frac{1}{1-\dfrac{M_{sw,0} X_2 + M_{vw,0} X_3}{Z\sigma_E}}.$$

The design equation expressed in terms of the partial safety factors is given by

$$g = Z_S\left(X_1'\right)^* - \frac{Z_S}{Z}\left(M_{sw,0} X_2^* + M_{vw,0} X_3^*\right) - \left(X_2'\right)^* \frac{\left(p_{st,0} X_4^* + p_{w,0} X_5^*\right) s\,\ell^2}{m} \geq 0 \quad (5.5.37)$$

Similar equations may be developed for laterally loaded stiffeners subjected to in-plane tension with $\phi = 1$.

BUCKLING

The buckling limit state function of stiffened panels subjected to compressive loads is given by

$$g(x) = \sigma_{cr} - \sigma_n \quad (5.5.38)$$

where σ_{cr} = critical buckling stress.
σ_n = applied compressive stress resulting from the bending of the primary structure or hull girder.

$$\sigma_n = \frac{M_{sw} + K_w\, M_{vw}}{Z} + \sigma_{pm}.$$

σ_{pm} = normal stress due to the bending of primary members.

Beyond the proportional limit stress σ_{ps}, the Johnson-Ostenfeld correction gives the critical buckling stress as

$$\sigma_{cr} = \sigma_E, \qquad \text{for} \quad \sigma_E \leq \sigma_{ps} \quad (5.5.39)$$

$$\sigma_{cr} = \sigma_Y\,(1 - \sigma_Y/4\sigma_E), \qquad \text{for} \quad \sigma_E = \sigma_{ps} \quad (5.5.40)$$

where $\sigma_E = \min(\sigma_{E1}, \sigma_{E2}, \sigma_{E3})$.
σ_{E1} = Euler column buckling stress.
σ_{E2} = Euler torsional buckling stress.
σ_{E3} = Euler plate buckling stress.

The safety margin with respect to buckling of axially loaded stiffened panels is obtained by replacing the design parameters in equation (5.5.38) with the corresponding random variables ($\sigma_{pm} = 0$):

$$M = g(X) = \sigma_Y\left(1 - \sigma_Y/4\,\sigma_E\right) - \frac{B_{M_{sw}} M_{sw,0} + K_w\, B_{M_{vw}}\, M_{vw,0}}{Z}$$

or

$$M = X_1\left(1 - X_1/4\,\sigma_E\right) - \frac{M_{sw,0} X_2 + M_{vw,0} X_3}{Z} \quad (5.5.41)$$

Note: σ_E is assumed to be a deterministic variable.

The design equation expressed in terms of the partial safety factors is given by

$$g = X_1^*\left(1 - X_1^*/4X_4^*\right) - \frac{M_{sw,0} X_2^* + M_{vw,0} X_3^*}{Z} \geq 0 \quad (5.5.42)$$

or

$$\frac{\left(\sigma_Y\right)_{\min}}{\gamma_R}\left(1 - \frac{1}{\gamma_R}\frac{\left(\sigma_Y\right)_{\min}}{\sigma_E}\right) \geq \frac{\gamma_{M_{sw}} M_{sw,0} + \gamma_{M_{vw,0}} M_{vw,0}}{Z} \quad (5.5.43)$$

ULTIMATE STRENGTH OF LATERALLY LOADED STIFFENERS

The distribution of the shear force and bending moment over the span of uniformly laterally loaded stiffeners fixed at both ends, as shown in Figure 5.14, is given by

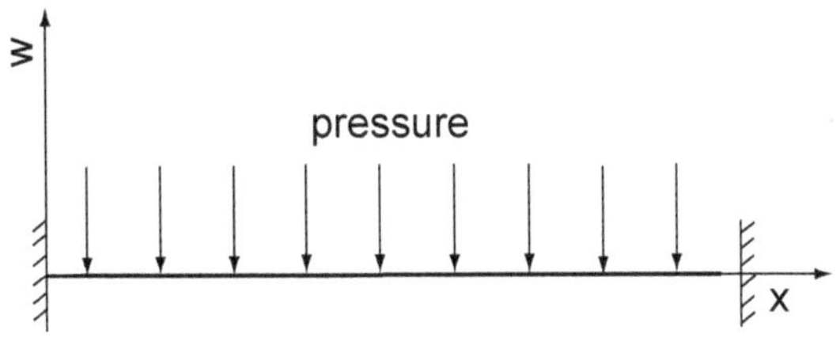

Figure 5.14 Laterally loaded stiffeners

$$T(x) = \frac{(M_B - M_A)}{\ell} + \frac{d\mu(x)}{dx} \quad (5.5.44)$$

$$M(x) = M_A + (M_B - M_A)\frac{x}{\ell} + \mu(x)$$

$$= M_A + T_A\,x - \frac{p\,s\,x^2}{2} \quad (5.5.45)$$

where $\mu(x)$ = bending moment, assuming the stiffener is simply supported at both ends.
T_A = end reaction = $T_B = ps\ell/2$.
M_A and M_B end bending moments.

Collapse of the stiffener occurs when the pressure leads to the formation of three plastic hinges at both ends and mid-span. The bending moment at mid-span is given by

$$M(\ell/2) = M_{pm} = M_A + T_A\frac{\ell}{2} - \frac{p\,s\,\ell^2}{8}$$

$$= \frac{ps\ell^2}{8} + M_A = Z_{pm}\,\sigma_Y \quad (5.5.46)$$

Taking into account that at, collapse, $M_A = M_B = -Z_{pe}\,\sigma_Y$, the ultimate limit state function of laterally uniformly loaded stiffeners fixed at both ends is given by

$$g(x) = (Z_{pm} + Z_{pe})\sigma_Y - \frac{p\,s\,\ell^2}{8} \quad (5.5.47)$$

where Z_{pm} = plastic section modulus at mid-span.
Z_{pe} = end plastic section modulus calculated with a reduced web area, $A_{wr} = A_w\sqrt{1 - 3(\tau/\sigma_Y)^2}$ to account for shear stress.
A_w = web area.

The plastic neutral axis generally falls inside the plate, and the plastic section modulus may be calculated assuming that the neutral axis is at the web-plate intersection:

$$Z_{pm} = A_f\,(h_w + 0.5\,t_f) + 0.5\,h_w\,A_w = (A + B)$$

$$Z_{pe} = A_f\left(h_w + 0.5\,t_f\right) + 0.5\,h_w A_w \sqrt{1 - 3\left(\tau/\sigma_y\right)^2}$$

$$Z_{pe} = A + B\sqrt{1 - 3\left(\tau/\sigma_Y\right)^2}$$

The safety margin of laterally loaded stiffeners fixed at both ends is obtained by replacing the design parameters in equation (5.5.47) with the corresponding random variables:

$$M = g(X) = \left(Z_{pm} + Z_{pe}\right)\sigma_Y - \frac{\left(B_{p_{st}}\,p_{st,0} + B_{p_w}\,p_{w,0}\right)s\ell^2}{8} \quad (5.5.48)$$

$$M = (2A + \lambda B)\,\sigma_Y - \frac{\left(B_{p_{st}}\,p_{st,0} + B_{p_w}\,p_{w,0}\right)s\,\ell^2}{8} \quad (5.5.49)$$

where $\lambda = 1 + \sqrt{1 - 3\left[\frac{\left(B_{p_{st}}\,p_{st,0} + B_{p_w}\,p_{w,0}\right)s\ell}{2A_w\sigma_Y}\right]^2}$.

Assuming that the section modulus Z_{pe} is a deterministic variable given by

$$Z_{pe} = A + B\sqrt{1 - 3\left[\frac{\left[E(p_{st}) + E(p_w)\right]s\,\ell}{2\,A_w\left(\sigma_Y\right)_{\min}}\right]^2}$$

the safety margin is a linear function of the random variables, and the design equation expressed in terms of the partial safety factors is

$$\left(Z_{pm} + Z_{pe}\right)X_1^* - \left(\frac{(p_{st,0}\,X_2^* + p_{w,0}\,X_3^*)s\,\ell^2}{8}\right) \ge 0$$

or, in a conventional form,

$$\left(Z_{pm} + Z_{pe}\right) \ge \frac{\left(\gamma_{p_{st}}\,p_{st,0} + \gamma_{p_w}\,p_{w,0}\right)s\,\ell^2}{8} \quad (5.5.50)$$

ULTIMATE STRENGTH OF AXIALLY AND LATERALLY LOADED STIFFENERS

The collapse of uniformly laterally loaded stiffeners fixed at both ends and subjected to axial compressive stresses, as shown in Figure 5.15, occurs by formation of three plastic hinges (two at ends and one at mid-span). and their elasto-plastic behavior is governed by the following approximate differential equation:

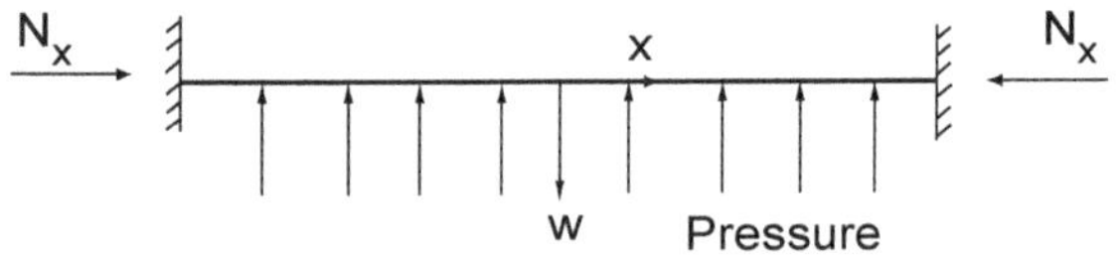

Figure 5.15 In-plane compression and local bending

$$E_t I'_e \frac{d^4 w}{dx^4} + N_x \frac{d^2 w}{dx^2} = ps \qquad (5.5.51)$$

where E_t = structural tangent modulus taken as

$$\frac{E_t}{E} = \frac{(\sigma_e)_x \left[\sigma_Y - (\sigma_e)_x\right]}{\sigma_{ps}(\sigma_Y - \sigma_{ps})}, \text{ for } (\sigma_e)_x \geq \sigma_{ps}.$$

$E_t = E$, for $(\sigma_e)_x < \sigma_{ps}$.
σ_{ps} = structural proportional limit. A typical value of σ_{ps} is $0.6\sigma_y$ for plates and $0.5\sigma_y$ for rolled, wide flange sections.
$(\sigma_e)_x$ = compressive stress in the stiffener.
N_x = compressive axial load.
p = lateral pressure.
s = stiffener spacing.

The solution of equation (5.5.51) is based on the assumption that the effective width of the attached plating may be calculated for any strain level by considering the generalized slenderness of plating β_e as defined by Gordo and Guedes Soares (1993):

$$\beta_e = \frac{s}{t_p}\sqrt{\frac{(\sigma_e)_x}{E}}.$$

where b_e = effective width of attached plating, taken as $b_e = \left(\frac{\lambda}{\beta_e} - \frac{\lambda - 1}{\beta_e^2}\right)s$ with $1.8 \leq \lambda \leq 2.25$. For plates with simply supported edges and average initial distortions Faulkner (1975) proposes $\lambda = 2$. Refer also to Guedes Soares (1988).

The compressive stress in the stiffener with attached plating of width b_e is

$$(\sigma_e)_x = \frac{A_S + s\,t_p}{A_S + b_e t_p}\sigma_n \qquad (5.5.52)$$

where A_S = cross-sectional area of the stiffener without attached plate.
t_p = thickness of attached plating.

Introducing b_e in equation (5.5.52) gives

$$\sqrt{(\sigma_e)_x} = \frac{-\lambda t_p^2 \sqrt{E} + \sqrt{\left(\lambda t_p^2 \sqrt{E}\right)^2 + 4A_S\left[\left(A_S + s\,t_p\right)\sigma_n + (\lambda - 1)\frac{t_p}{s}t_p^2 E\right]}}{2A_S} \qquad (5.5.53)$$

In equation (5.5.51), the moment of inertia of the stiffener I'_e is calculated with an attached plating of width b'_e equal to the tangent effective width $b'_e = s/\beta_e$, refer to Faulkner (1975).

The collapse mechanism, that is, formation of three plastic hinges at both ends and at mid-span, is described by the following equation:

$$M(0) = \left(\frac{M_e}{\cos u} + \frac{ps\ell^2}{4u^2\cos u}\right) - \frac{ps\ell^2}{4u^2} = \alpha Z_{pm}\sigma_Y \qquad (5.5.54)$$

where $M_e = -\alpha Z_{pe}\sigma_Y$.
$\alpha = \text{sgn}(p)$ ($\alpha = -1$ for pressure acting on the plating side).
Z_{pe} = end plastic section modulus calculated with a reduced web area A_{wr}.

$$u = \frac{\pi}{2}\sqrt{\frac{(\sigma_e)_x}{\sigma'_E}} < \frac{\pi}{2}.$$

$$\sigma_E = \frac{\pi^2 E_t I'_e}{(A_S + b_e t_p)\ell^2}$$

Note: The widths of attached plating considered for calculation of the moment of inertia I'_e and plastic section moduli Z_{pm} and Z_{pe} are taken as

$\alpha = -1$ b'_e and b_e (the attached plating is assumed to be buckled).
$\alpha = 1$ $b = s$ (the attached plating is assumed to be not buckled).

From equation (5.5.51), we obtain the limit state function with respect to ultimate strength of laterally loaded stiffeners fixed at both ends and subjected to in-plane compressive stress σ_n:

$$g(x) = Z_{pe}\sigma_Y\left(1 + \frac{Z_{pm}}{Z_{pe}}\cos u\right) - \frac{4(1 - \cos u)}{u^2}\frac{ps\ell^2}{16} \qquad (5.5.55)$$

The limit state function may also be expressed as

$$g(x) = p_{coll} - p_{des} \qquad (5.5.56)$$

where p_{coll} = collapse pressure as obtained from equation (5.5.55) by writing g(x)=0.

$$p_{coll} = \frac{16\, Z_{pe}\, \sigma_Y}{s\, \ell^2} \frac{u^2 \left(1 + \frac{Z_{pm}}{Z_{pe}} \cos u\right)}{4(1 - \cos u)} \tag{5.5.57}$$

The safety margin with respect to ultimate strength of uniformly laterally loaded stiffeners subjected to in-plane compression is obtained by replacing the design parameters in equation (5.5.56) with the corresponding random variables assumed to be positive quantities, which gives

$$M = g(X) = Z_{pe}\left(1 + \frac{Z_{pm}}{Z_{pe}} \cos u\right)\sigma_Y - \frac{4(1-\cos u)}{u^2} \frac{\left[B_{p_{st}} p_{st,0} + B_{p_w} p_{w,0}\right] s\,\ell^2}{16} \tag{5.5.58}$$

where σ_Y, $(\sigma_e)_x$, u, Z_{pm}, Z_{pe}, B_{M0w}, $B_{p_{st}}$ and B_{p_w} are the random variables.

Equation (5.5.58) can be approximated as follows

$$M = g(X) = \left[Z_{pe} + Z_{pm}\left(1 - u^2/2\right)\right]\sigma_Y - \frac{\left[B_{p_{st}} p_{st,0} + B_{p_w} p_{w,0}\right] s\,\ell^2}{8} \tag{5.5.59}$$

or

$$M = g(X) = \left[Z_{pe} + Z_{pm}\left(1 - \frac{\pi^2}{8}\frac{(\sigma_e)_x}{\sigma'_E}\right)\right]\sigma_Y - \frac{\left[B_{p_{st}} p_{st,0} + B_{p_w} p_{w,0}\right] s\,\ell^2}{8} \tag{5.5.60}$$

At collapse, normal stresses in the stiffener are distributed as shown in Figures 5.16 and 5.17 for pressure acting on the plating side. The plastic section modulus Z_{pm} is given by

$$Z_{pm} = \left\{ b_e(t_p - x)\left(\frac{t_p + x}{2} + x_1\right)10^3 + b_f t_f (h_w + 0{,}5\, t_f - x_1) + t_w (h_w - z)\left(\frac{h_w + z}{2} - x_1\right)\right\}10^{-3} \tag{5.5.61}$$

where x, x_1, and z are computed from the following system of linear equations:

$$b_e\left(t_p - x\right) = b_f t_f + t_w\left(h_w - z\right) \tag{1}$$

$$b_e x\left(\frac{x}{2} + x_1\right) + t_w \frac{x_1^2}{2} = t_w \frac{\left(z - x_1\right)^2}{2} \tag{2}$$

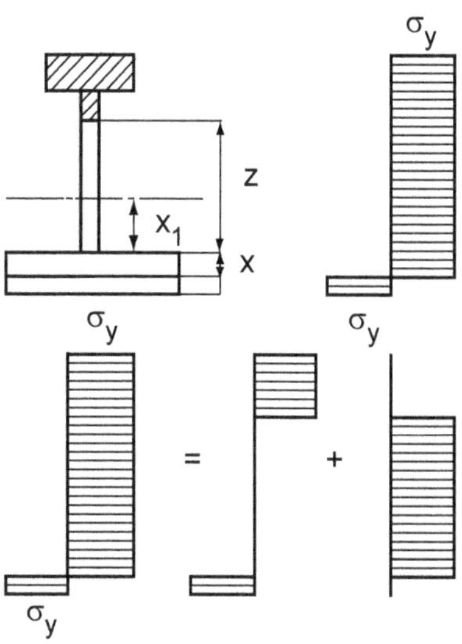

Figure 5.16 Distribution of stresses at end.

$$\sigma_Y\left(b_e\, x + t_w\, z\right) = \left(A_S + b_e\, t_p\right)\left(\sigma_e\right)_x \tag{3}$$

Note: If $z > h_w$, the distribution of stresses is modified and another set of equations has to be developed for calculation of the plastic section moduli Z_{pm} and Z_{pe}.

The plastic section modulus Z_{pe} is to be calculated according to equation (5.5.61) with a reduced web thickness t_{wr} given by

$$t_{wr} = t_w \sqrt{1 - 3\left[\frac{\left[E(p_{st}) + E(p_w)\right] s\,\ell}{2\, A_w\, (\sigma_Y)_{min}}\right]^2} \tag{5.5.62}$$

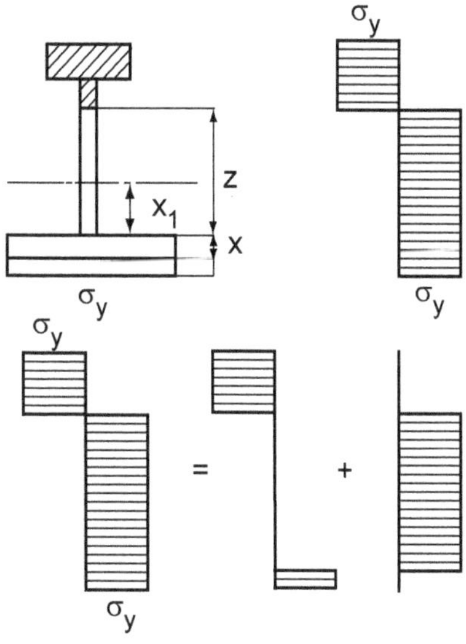

Figure 5.17 Distribution of stresses at mid-span.

The plastic section moduli Z_{pm} and Z_{pe} depend on the ratio $(\sigma_e)_x/\sigma_Y$ and are expressed as

$$Z = A\left(\frac{(\sigma_e)_x}{\sigma_Y}\right)^2 + B\frac{(\sigma_e)_x}{\sigma_Y} + C + D\frac{\sigma_Y}{(\sigma_e)_x} \quad (5.5.63)$$

where A, B, C, and D are constants depending on the geometrical characteristics of the stiffener.

The safety margin as given by equation (5.5.59) is a nonlinear function of the random variables. The iterative procedure, as described in Section 5.3.2 or 5.3.3, based on linearization of the failure surface at each step of the process, can be used for calculation of the Hasofer and Lind safety index and comparison with the target safety index β_0.

5.5.4 Unstiffened Plate Panels

INITIAL YIELDING OF LATERALLY LOADED PLATE PANELS SUBJECTED TO TRANSVERSE IN-PLANE COMPRESSIVE STRESSES

Let us consider infinitely long plates with clamped edges, laterally loaded and subjected to in-plane stresses (uniform compressive stress σ_y acting on the longer sides and shear stress τ_{xy}), such as shell plating. The behavior of the plating may be approximated by the following differential equation:

$$D\frac{d^4w}{dx^4} + N_x\frac{d^2w}{dx^2} = p \quad (5.5.64)$$

and the maximum bending moment, per unit length, occurring on the longer sides is given by

$$M_{\max} = \frac{ps^2}{4u^2}\left(1-\frac{u}{\tan u}\right) = \phi\frac{ps^2}{12} = \sigma_b\frac{t^2}{6} \quad (5.5.65)$$

(Note: In the following units are m and MPa)

where $D = Et_p^3/12(1-\nu^2)$.

$$u = \sqrt{3(1-\nu^2)}\,\frac{s}{t_p}\sqrt{\frac{\sigma_y}{E}} = 1,65\,\frac{s}{t_p}\sqrt{\frac{\sigma_y}{E}}$$

$$= 3.635\times10^{-3}\frac{s}{t_p}\sqrt{\sigma_y}. \quad (5.5.66)$$

ν = Poisson's ratio (ν= 0.3).
E = Young's modulus (E = 2.05.× 10^5 MPa).

$$\phi = \frac{3}{u^2}\left(1-\frac{u}{\tan u}\right).$$

σ_b = plate bending stress

For $u \le 2$, ϕ may be approximated as follows:

$$\phi \cong \frac{1}{1-7.25\times10^{-2}u^2} = \frac{1}{1-9.6\times10^{-7}\sigma_y\left(s/t_p\right)^2} \quad (5.5.67)$$

At yielding, the von Mises equivalent stress is equal to the yield stress:

$$\left(\sigma_y + \sigma_{perm}\right)^2 + 3\tau_{xy}^2 = \sigma_Y^2$$

and the permissible plate bending stress is

$$\sigma_{perm} = \sigma_Y\sqrt{1-3\left(\tau_{xy}/\sigma_Y\right)^2} - \sigma_y$$

Therefore, the yielding limit state function with respect to initial yielding is

$$g(x) = \sigma_{perm} - \sigma_b$$
$$= \sigma_{perm} - \phi\frac{p}{2}\left(s/t_p\right)^2$$
$$= \sigma_Y\sqrt{1-3\left(\tau_{xy}/\sigma_Y\right)^2} - \sigma_y - \phi\frac{p}{2}\left(s/t_p\right)^2 \quad (5.5.68)$$

The limit state function may be also expressed as

$$g(x) = p_{\lim} - p_{des} \quad (5.5.69)$$

where $p_{\lim}$ is obtained from equation (5.5.5.) by writing $g(x) = 0$.

$$P_{\lim} = 2\left[\sigma_r\sqrt{(1-3)(\tau_{xy}/\sigma_Y)^2} - \sigma_Y\right] \times\left[(t_p/s)^2 - 9.6\times10^{-7}\sigma_Y\right] \quad (5.5.70)$$

The safety margin with respect to initial yielding of laterally loaded plate panels subjected to in-plane compressive stresses is obtained by replacing the design parameters in equation (5.5.69) with the corresponding random variables (σ_Y, σ_y, τ_{xy}, $B_{p_{st}}$, B_{p_w}):

$$M = 2\left[\sigma_Y\sqrt{1-3\left(\tau_{xy}/\sigma_Y\right)^2} - \sigma_y\right] \times\left[\left(t_p/s\right)^2 - 9.6\times10^{-7}\sigma_y\right] - \left(B_{p_{st}}p_{st,0} + B_{p_w}p_{w,0}\right)$$

or

$$M = 2\left[X_1\sqrt{1-3\tau_{xy,0}^2\left(X_2/X_1\right)^2} - \sigma_{y,0}X_3\right]$$
$$\times\left[\left(t_p/s\right)^2 - 9.6\times10^{-7}\,\sigma_{y,0}X_3\right]$$
$$-\left(p_{st,0}X_4 + p_{w,0}X_5\right) \qquad (5.5.71)$$

Assuming that the shear stress can be neglected, the design equation expressed in terms of the partial safety factors is

$$g(x) = 2\left[(\sigma_Y)_{\min}/\gamma_R - \gamma_{\sigma_y}\sigma_{y,0}\right]$$
$$\times\left[(t_p/s)^2 - 9.6\times10^{-7}\gamma_{\sigma_y}\sigma_{y,0}\right]$$
$$-\left(\gamma_{p_{st}}p_{st,0} + \gamma_{p_w}p_{w,0}\right) \geq 0$$

or, in a conventional form,

$$\left(t_p/s\right)^2 \geq 9.6\times10^{-7}\gamma_{\sigma_y}\sigma_{y,0} + \frac{\gamma_{p_{st}}p_{st,0} + \gamma_{p_w}p_{w,0}}{2\left((\sigma_Y)_{\min}/\gamma_R - \gamma_{\sigma_y}\sigma_{y,0}\right)} \qquad (5.5.72)$$

If $\sigma_y = 0$, we obtain the well-known formula for laterally loaded plates:

$$\frac{t_p}{s} \geq \gamma_R\sqrt{\frac{\gamma_{p_{st}}p_{st,0} + \gamma_{p_w}p_{w,0}}{2\,(\sigma_Y)_{\min}}} \qquad (5.5.73)$$

Note: A correction factor has to be applied for plates with an aspect ratio less than 3.

BUCKLING

Plates may be subjected to various compressive loadings:

1. Uniaxial compression.
2. Biaxial compression.
3. Shear.
4. Biaxial compression and shear.

and a limit state function has to be established for each of these loadings.

For example, the buckling limit state function for plate panels subjected to in-plane compressive stresses and shear stress, as shown in Figure 5.18, may be given by the following interaction formula as proposed by Paik, Ham, and Ko (1992):

$$g(x) = 1 - \left(\left\{\frac{\sigma_x}{R_{sx}\sigma_{xcr}}\right\}^n + \left\{\frac{\sigma_y}{R_{sy}\sigma_{ycr}}\right\}^n\right) \qquad (5.5.74)$$

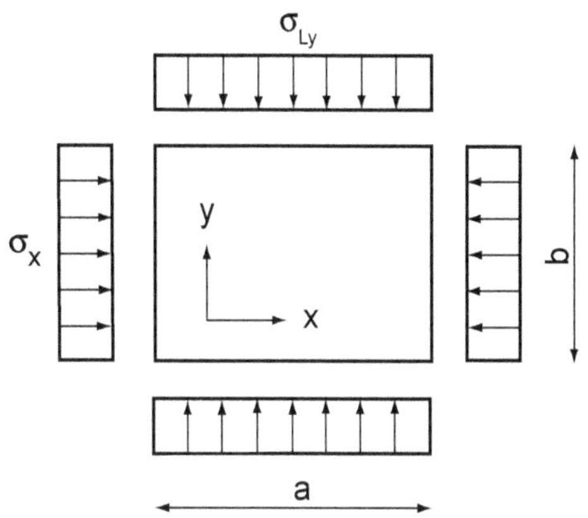

Figure 5.18 Biaxial compression of plate panels.

where $n = 1$, for $\alpha \leq \sqrt{2}$.

$n = 2$, for $\alpha > \sqrt{2}$ and $\sqrt{\sigma_{xcr}^2 + \sigma_{ycr}^2 - \sigma_{xcr}\sigma_{ycr} + 3\tau_{xy}^2} > 0.6\sigma_Y$.

$\alpha = a/b > 1$, with the longer edge a taken in the x direction, as shown in Figure 5.18.

σ_x = in-plane compressive stress in x direction.

σ_y = in-plane compressive stress in y direction.

σ_{xcr} = critical buckling stress in x direction.

σ_{ycr} = critical buckling stress in y direction.

$R_{sx} = 1 - \left(\tau_{xy}/\tau_{cr}\right)^{n_1}$

$R_{sy} = 1 - \left(\tau_{xy}/\tau_{cr}\right)^{n_2}$

τ_{xy} = shear stress acting on the plate panel.

τ_{cr} = critical shear buckling.

$n_1 = 1.08(1+\alpha) - 0.16\alpha^2$, for $\alpha \leq 3.2$.

$n_1 = 2.9$, for $\alpha > 3.2$.

$n_1 = 1.9 + 0.1\,\alpha$, for $\alpha \leq 2$.

$n_1 = 0.7(1+\alpha)$, for $\alpha > 3.2$.

Let us consider the case of biaxial compression of deck and bottom shell. The safety margin is obtained by replacing the design parameters in equation (5.5.71), that is, applied and critical buckling stresses, with the corresponding random variables ($R_{sx} = R_{sy} = 1$). Assuming that $\sigma_x = \dfrac{M_{sw} + K_w M_{vw}}{Z}$, the safety margin is

$$M = g(X) = 1 - \left(\frac{1}{Z}\frac{B_{M_{sw}}M_{sw,0} + K_w B_{M_{vw}}M_{vw,0}}{B_{\sigma_{xcr}}\sigma_{xcr,0}}\right)^n - \left(\frac{B_{\sigma_y}\sigma_{y,0}}{B_{\sigma_{ycr}}\sigma_{ycr,0}}\right)^n$$

or

$$M = g(X) = 1 - \left(\frac{1}{Z} \frac{M_{sw,0} X_2 + M_{vw,0} X_3}{\sigma_{xcr,0} X_5} \right)^n - \left(\frac{\sigma_{y,0} X_4}{\sigma_{ycr,0} X_6} \right)^n \quad (5.5.75)$$

where $Z = Z_d$ or Z_b.
$Z_b = I_v / z_{na}$.
$Z_d = I_v / (D - z_{na})$.
I_v = moment of inertia of the cross section about the neutral axis.
D = depth of the ship.
z_{na} = distance of the neutral axis to the baseline.
B_{σ_y} = uncertainties in the transverse stress $\sigma_y (B_{\sigma_y} = X_4)$.
$B_{\sigma_{xcr}}$ and $B_{\sigma_{ycr}}$ = uncertainties in the critical stresses $(B_{\sigma_{xcr}} = X_5)$ and $(B_{\sigma_{ycr}} = X_6)$.

The design equation expressed in terms of the partial safety factors is

$$g = 1 - \left(\frac{1}{Z} \frac{M_{sw,0} X_2^* + M_{vw,0} X_3^*}{\sigma_{xcr,0} X_5^*} \right)^n - \left(\frac{\sigma_{y,0} X_4^*}{\sigma_{ycr,0} X_6^*} \right)^n \geq 0 \quad (5.5.76)$$

FORMATION OF THREE PLASTIC HINGES IN LATERALLY LOADED PLATE PANELS SUBJECTED TO TRANSVERSE IN-PLANE COMPRESSIVE STRESSES

The elasto-plastic behavior of infinitely long plates with clamped edges, transversely and laterally loaded, is governed by equation (5.5.64). The plastic bending moment, per unit length, corresponding to the formation of three plastic hinges (the plate panel is assumed to be subjected to compressive stresses σ_n acting on the longer sides and shear stress), is given by

$$M_p = \frac{Ps^2}{4u^2} \left(\frac{1 - \cos u}{1 + \cos u} \right) = \phi \frac{Ps^2}{16} \quad (5.5.77)$$

where $u = 1{,}65 \frac{s}{t_p} \sqrt{\frac{\sigma_n}{E}}$.

$$\phi = \frac{4}{u^2} \frac{1 - \cos u}{1 + \cos u}.$$

For $u \leq 2\phi$ may be approximated by

$$\phi \cong \frac{1}{1 - 0.155 u^2} = \frac{1}{1 - 0.42 \left(s/t_p \right)^2 \sigma_n / E}$$

The limit state function corresponding to the formation of three plastic hinges may be written as

$$g(x) = \sigma_{perm} Z_p - \phi \frac{p s^2}{16} = \sigma_{perm} \frac{t_p^2}{4} \left[1 - \left(\sigma_n / \sigma_{perm} \right)^2 \right] - \phi \frac{p s^2}{16} \quad (5.5.78)$$

where Z_p= plastic section modulus of the plate given by $Z_p = \frac{t_p^2}{4} \left[1 - \left(\sigma_n / \sigma_{perm} \right)^2 \right]$ in which $\left[1 - \left(\sigma_n / \sigma_{perm} \right)^2 \right]$ is a reduction factor reflecting the compressive load.

σ_{perm} = permissible plate bending stress, solution of the following equation stating that the von Mises equivalent stress is equal to the yield stress $(\sigma_x^2 - \sigma_x \sigma_y + \sigma_y^2 + 3\tau_{xy}^2 = \sigma_Y^2)$.

Noting that $\sigma_x = \nu \sigma_{perm}$ and $\sigma_y = \sigma_{perm}$ gives

$$(1 - \nu + \nu^2) \sigma^2_{perm} + 3\tau^2_{xy} = \sigma^2_Y$$

or

$$\sigma_{perm} = \frac{\sigma_Y \sqrt{1 - 3 \left(\tau_{xy} / \sigma_Y \right)^2}}{\sqrt{1 - \nu + \nu^2}}$$

The safety margin with respect to the formation of three plastic hinges is obtained by replacing the design parameters in equation (5.5.78) with the correspond ing random variables$(\sigma_Y, \sigma_n, \sigma_{xy}, B_{p_{st}}, B_{p_w})$:

$$M = g(X) = \sigma_{perm} \frac{t_p^2}{4} \left\{ 1 - \left(\sigma_n / \sigma_{perm} \right)^2 \right\} - \phi \frac{\left(B_{p_{st}} p_{st,0} + B_{p_w} p_{w,0} \right) s^2}{16} \quad (5.5.79)$$

The limit state function may also be expressed as

$$g(x) = p_{coll} - p_{des} \quad (5.5.80)$$

where p_{coll} = collapse pressure as obtained from equation (5.5.78) by writing $g(x)=0$:

$$p_{coll} = 4\sigma_{perm} \left[1 - (\sigma_n / \sigma_{perm})^2 \right] \times \left[(t_p / s)^2 - 0.42 \sigma_n / E \right] \quad (5.5.81)$$

The safety margin is obtained by replacing the design parameters in equation (5.5.79) with the corresponding random variables:

$$M = 4\sigma_{perm}\left[1-(\sigma_n/\sigma_{perm})^2\right] \times\left[(t_p/s)^2 - 0.42\sigma_n/E\right] - \left(B_{p_{st}}\, p_{st,0} + B_{p_w}\, p_{w,0}\right) \quad (5.5.82)$$

Where the shear stress can be neglected, such as at deck and bottom $\sigma_{perm} = \sigma_Y/\sqrt{1-\nu+\nu^2} = 1.125\sigma_Y$ and the safety margin is

$$M = 4.5\sigma_Y\left[1-0.79(\sigma_n/\sigma_Y)^2\right] \times\left[(t_p/s)^2 - 0.42\sigma_n/E\right] - \left(B_{p_{st}}\, p_{st,0} + B_{p_w}\, p_{w,0}\right)$$

or

$$M = 4.5\, X_1\left[1-0.79(X_2/X_1)^2\right] \times\left[(t_p/s)^2 - 0.42\, X_2/E\right] - \left(p_{st,0}\, X_3 + p_{w,0}\, X_4\right) \quad (5.5.83)$$

where σ_n, σ_Y, $B_{p_{st}}$, and B_{p_w} are the random variables.

The design equation expressed in terms of the partial safety factors is

$$(t_p/s)^2 \ge 0.42\frac{\gamma_{\sigma_n}\sigma_{n,0}}{E} + \frac{\gamma_R}{4.5(\sigma_Y)_{\min}} \times \frac{\gamma_{p_{st}}\, p_{st,0} + \gamma_{p_w}\, p_{w,0}}{1-0.79\left(\dfrac{\gamma_{\sigma_n}\gamma_R\sigma_{n,0}}{(\sigma_Y)_{\min}}\right)^2} \quad (5.5.84)$$

Similar equations may be obtained for other types of loading conditions:

1. Laterally loaded plates subjected to in-plane tensile stress (acting on the longer sides) and shear stress.
2. Laterally loaded plates subjected to in-plane axial stress (acting on the shorter sides) and shear stress.
3. Laterally loaded plates subjected to in-plane biaxial stresses and shear stress.

5.6 NUMERICAL APPLICATIONS

The following calculations give two practical examples of application of reliability methods to well-known ship structural limit states and do not pretend to give precise results, which would need more refined analyses, taking into account the actual distributions of the random variables. They give a taste of how the design codes should be presented in the future with partial safety factors based on the results of reliability analyses, to permit design of all structures with the same level of safety.

5.6.1 Hull Girder Reliability

GENERAL

For this application the safety index calculations are performed for two types of ships (seven tankers and five bulk carriers) of various dimensions and for the following two limit states:

1. Initial yielding.
2. Ultimate strength.

All ships are assumed designed according to the IACS Unified Requirement UR-S11. In no case is the design still-water bending moment $M_{sw,0}$ less than

$$\text{Sagging condition: } M_{sw,0} \ge 0.59\left(M_{vw,0}\right)_S \quad (5.6.1)$$

Hogging condition:

$$M_{sw,0} \ge 1.59\left(M_{vw,0}\right)_S - \left(M_{vw,0}\right)_H = \frac{122.5 - 15\,C_B}{110(C_B + 0.7)}\left(M_{vw,0}\right)_S \quad (5.6.2)$$

CHARACTERISTICS OF THE RANDOM VARIABLES

1. *Still-water bending moment.* According to Section 5.4.4, the still-water bending moment M_{sw} is assumed normally distributed and its mean value and the coefficient of variation are taken as

Tankers: $E(B_{sw}) = 0.67$ and $V_{Bsw} = 0.25$, which corresponds to a probability of exceedance of the design SWBM of 2.5%, $\Phi^{-1}(0.975) = 1.97 = M_{sw,0} - E(M_{sw})/\sigma M_{sw}$. The actual SWBM of tankers can be easily monitored thanks to the l oading instrument on board.

Bulk carriers: $E(B_{sw}) = 0.75$ and $V_{Bsw} = 0.25$, which corresponds to a probability of exceedance of the design SWBM of 5%, $\Phi^{-1}(0.95)=1.67=M_{sw,0}-E(M_{sw})/\sigma M_{sw}$. As already mentioned, the actual SWBM of bulk carriers may exceed the design value more often than for tankers due to the difficulty of monitoring the loading operations.

2. *Wave bending moment.* The random variable $B_{I_{vw}}$ follows a Gumbel distribution:

$$E\left(B_{M_{vw}}\right)=E\left(B_{I_{vw}}\right)E\left(B_{II_{vw}}\right)=\left(1+\frac{0.577}{\xi \ln N}\right)E\left(B_{II_{vw}}\right)$$

$$V_{B_{vw}}=\sqrt{V_{B_{I_{vw}}}^2+V_{B_{II_{vw}}}^2}$$

where $E(B_{II_{vw}})$ = approximational uncertainties, taken as unity in this numerical application.

N = number of cycles over the period of time considered, taken as 10^8 cycles.

$$\xi = 1.1-0.35\,\frac{L-100}{300}\geq 0.85.$$

V_{B_I} = coefficient of variation of the statistical uncertainties.

$V_{B_{II}}$ = coefficient of variation of the approximational uncertainties.

$$K_w=\frac{M_{\max}-M_{sw,0}}{M_{vw,1}}.$$

$M_{\max}$ = maximum bending moment calculated according to Söding rule.

$M_{vw,1}$ = wave bending moment calculated according to equation (5.4.21).

The mean value and coefficient of variation of the random variable X_3 are

$$E\left(X_3\right)=\left(1+\frac{0.577}{\xi \ln N}\right)E\left(K_w\right) \qquad (5.6.3)$$

$$V_{X_3}=\sqrt{V_{K_w}^2+V_{B_{I_{vw}}}^2+V_{B_{II_{vw}}}^2}$$

$$=\sqrt{0.1^2+V_{B_{I_{vw}}}^2+0.125^2} \qquad (5.6.4)$$

where $V_{B_{I_{vw}}}$ is given by equation (5.4.19)

3. *Slamming bending moment.* Taking into account the type of ships considered (tankers and bulk carriers), the influence of the slamming bending moment can be disregarded for the following two reasons:
   a. In the sagging condition , that is, for laden conditions, there is no risk of slamming.
   b. In the hogging condition, that is, for ballast conditions, the forward draught is generally increased to avoid occurrence of slams; moreover, the slamming bending moment, which is a sagging moment,reduces the total bending moment.

4. *Yield stress.* The mean value and coefficient of variation of the yield stress (refer to Section 5.4.4) are given by

$E(\sigma_Y)=1.209(\sigma_Y)_{\min}$ (the yield stress is assumed to follow a lognormal distribution)

$V_{\sigma_Y}=0.08$

5. *Ultimate bending moment.* Mean values of the ultimate bending moments (refer to Appendix 5-E) are calculated for the mean value $E(\sigma_Y)$ of the yield stress and taken from Beghin, Jastrzebski, and Taczala (1998). The ultimate bending moment M_u is assumed to follow a normal distribution and its coefficient of variation taken as 0.125.

INITIAL YIELDING

As the safety margin is a linear function of the random variables (refer to Section 5.5.1), the Cornell and Hasofer-Lind safety indices are equal and calculated according to equation (5.5.5). The partial safety factors are calculated according to equations (5.5.7) to (5.5.9). The design equation expressed in terms of the partial safety factors is given by equation (5.5.10)

$$Z\geq\gamma_R\frac{\gamma_{sw}M_{sw,0}+\gamma_{vw}M_{vw,0}}{(\sigma_Y)_{\min}} \qquad (5.6.5)$$

Table 5.15 summarizes the results of calculations carried out accordingly for seven tankers and five bulk carriers, whose main particulars are given in Appendix 5-E. Ships considered for this analysis comply strictly with the IACS requirements and the design SWBM is equal to the permissible bending moments as given by equations (5.5.1) and (5.5.2). From these partial results the design section modulus of oil tankers expressed in terms of PSF would be

Hogging:

$$Z\geq 1.13\frac{0.97M_{sw,0}+1.34M_{vw,0}}{(\sigma_Y)_{\min}}$$

$$=\frac{1.1\;M_{sw,0}+1.515M_{vw,0}}{(\sigma_Y)_{\min}} \qquad (5.6.6)$$

Sagging:

$$Z\geq 1.12\frac{0.93M_{sw,0}+1.37M_{vw,0}}{(\sigma_Y)_{\min}}$$

or

$$Z\geq\frac{1.04M_{sw,0}+1.53M_{vw,0}}{(\sigma_Y)_{\min}} \qquad (5.6.7)$$

Note: These PSF are quite different from those obtained by Mansour et al (2001) for $r = (M_{vw,0}/M_{sw,0})=1.67$:

$$Z\geq 0.988\frac{0.764M_{sw,0}+1.708M_{vw,0}}{(\sigma_Y)_{\min}}$$

Table 5.15 Initial Yielding of Bulk Carriers and Oil Tankers

	Seagoing Conditions			
	Bulk Carriers		Tankers	
	Hogging	Sagging	Hogging	Sagging
β	4.55	4.45	4.55	4.45
k_w	0.925	0.925	0.90	0.90
$1/\gamma_R$	0.876	0.888	0.894	0.884
γ_{sw}	0.989	0.958	0.93	0.97
γ_{vw}	1.37	1.38	1.38	1.34

or $$Z \geq \frac{0.755 M_{sw,0} + 1.687 M_{vw,0}}{(\sigma_Y)_{\min}}$$

Contrary to Mansour (2001), these approximate calculations are carried out for only 12 ships, the actual probability distributions of random variables are not taken into account but only the mean and standard deviation, and the safety margin is approximated by a linear expression, which can explain the differences observed on the partial safety factors. Table 5.16 compares the minimum section modulus (in m^3) for the seven tankers, calculated according to Mansour (2001), equations (5.5.6) and (5.5.7), and IACS Unified Requirement S11.

In addition, for ships subjected to high risk of corrosion, it may be necessary to take into account the degradation with time of the cross-sectional properties. For example, the safety margin with respect to initial yielding of the hull girder, as given by equation (5.5.3), becomes

$$g(t) = Z(t)\,\sigma_Y - B_{M_{sw}} M_{sw,0} - K_w B_{M_{vw}} M_{vw,0} \qquad (5.6.8)$$

where $Z(t)$ is a time-dependent random variable.

For calculation of the section modulus at deck and bottom of any cross-transverse section, the reduction in the plate thickness Δt_i of the ith member due to corrosion may be calculated as indicated in Section 5.4.4. The mean and standard deviation of the section modulus are calculated according to the method given in Appendix 5-D.

As proposed by Wirsching, Ferensic, and Thayamballi (1997), the safety index β is calculated from equation (5.5.3) for various values of time $t = T$, assuming that the wave-induced bending moment follows an extreme value distribution, whose mean value and standard deviation are calculated according to equations (5.4.17) and (5.4.18) with $N = 1/T$. The probability of failure at time T may be approximated by $P(f\,|\,T) = \Phi(-\beta)$ and the probability of failure for the ship's lifetime is

$$P = \frac{1}{T_s}\int_0^{T_s} P(f\,|\,t)\,dt \qquad (5.6.9)$$

where $P(f\,|\,t)$ = conditional probability of failure at a random time T, calculated by considering that the extreme wave bending moment occurs at time T.
T_s = ship's lifetime.

ULTIMATE STRENGTH

Calculations of the safety index are performed for the same ships as for initial yielding and according to the same procedure. If we assume that oil tankers and bulk carriers spend half of their lifetime in a sagging condition, when fully laden, and half in hogging condition, when in ballast, the resulting probability of failure is

$$(P_f)_{mean} = 0.5(P_f)_{sag} + 0.5(P_f)_{hog} \qquad (5.6.10)$$

and the corresponding safety index is $\beta = -\Phi^{-1}[(P)_{mean}]$. Table 5.17 summarizes the results of calculations.

Table 5.16 Minimum Section Modulus for Tankers

Tankers	Mansour (2001)		Equations (5.5.5. and (5.5.7)		IACS S11
	Hogging	Sagging	Hogging	Sagging	
1	5.98	7.23	7.20	7.24	7.22
2	9.91	10.29	10.24	10.31	10.27
3	5..50	5..24	5..22	5..38	5..14
4	70.51	72.33	72.36	72.46	72.19
5	70.00	71.96	71.93	72.11	71.83
6	5..04	5..85	5..81	5..97	5..72
7	117.24	119.73	120.01	119.98	119.52

Table 5.17 Ultimate Strength of Bulk Carriers and Oil Tankers

	Seagoing Conditions			
	Bulk Carriers		Tankers	
	Hogging	Sagging	Hogging	Sagging
β_{min}	3.94	2.78	3.44	2.92
β_{max}	4.47	3.03	4.25	3.42
β_{mean}	4.20	2.90	3.85	3.17
P_f	1.335 10^{-5}	$1.885.10^{-3}$	1.597 10^{-5}	7.5.7 10^{-4}
$(P_f)_{mean}$	$9.395.10^{-4}$		3.903 10^{-4}	
β	3.10		3.35	

5.6.2 Reliability of Horizontal Stiffeners of Cargo Tank Transverse Bulkheads

INITIAL YIELDING

Keeping the notations of Section 5.5.3, the safety margin with respect to initial yielding of laterally loaded horizontal stiffeners of cargo tank transverse bulkheads is

$$M = g(X) = Z_S\,\sigma_Y\sqrt{1-3(\tau/\sigma_Y)^2} - \frac{B_{p_{st}}\,p_{st,0}\,s\,\ell^2}{m} - \frac{B_{p_w}\;p_{w,0}\,s\,\ell^2}{m} \tag{5.6.11}$$

where σ_Y, $B_{p_{st}}$ and B_{p_w} are random variables assumed to be independent. $B_{p_{st}}$ and B_{p_w} measure the uncertainties in the static and wave-induced pressures.

Assuming that the reduction factor $\lambda = \sqrt{1-3(\tau/\sigma_Y)^2}$ may be considered as a deterministic variable given by

$$\lambda = \sqrt{1-3\left[\frac{A_f}{Z_S\,t_w}\,\frac{\left[E(p_{st})+E(p_w)\right]s\,\ell}{2(\sigma_Y)_{min}}\right]^2} \tag{5.6.12}$$

the safety margin is a linear function of the random variables, expressed as

$$M = \lambda\,Z_S X_1 - \left(p_{st,0}\,X_2 + p_{w,0}X_3\right)\frac{s\,\ell^2}{12}$$
$$= a_1 X_1 - \frac{p_{st,0}\;s\,\ell^2}{12}X_2 - \frac{p_{w,0}\;s\,\ell^2}{12}X_3 \tag{5.6.13}$$

where $m = 12$ for stiffeners fixed at both ends.
$X_i's$ = random variables.
$X_1 = \sigma_Y$.
$X_2 = B_{p_{st}}$.
$X_3 = B_{p_w}$.

Since the limit state function expressed by equation (5.5.14) is linear, the safety index is given by

$$\beta = \frac{\lambda Z_S\,E(X_1) - a_2\,E(X_2) - a_3\,E(X_3)}{\sqrt{\sum_{i=1}^{i=n} a_i^2\,\sigma^2(X_i)}} \tag{5.6.14}$$

Based on the definitions of Table 5.12, the yielding limit state of transverse bulkhead stiffeners may be considered as a severe serviceability limit state, which according to Table 5.15, gives a target reliability safety index β_0 of 4.1. The partial safety factors are given by equations (5.5.7) to (5.5.8), and the design equation expressed in terms of the PSF is

$$g = a_1\,X_1^* - a_2\,X_2^* - a_3\,X_3^*$$
$$= \frac{\lambda Z_S\,(\sigma_Y)_{min}}{\gamma_R} - \gamma_{p_{st}}\frac{p_{st,0}\;s\,\ell^2}{12} - \gamma_{p_w}\frac{p_{w,0}\;s\,\ell^2}{12}$$
$$\geq 0 \tag{5.6.15}$$

or, in a more conventional form,

$$Z_S \geq \frac{\gamma_R}{\lambda(\sigma_Y)_{min}}\,\frac{\left(\gamma_{p_{st}}\,p_{st,0} + \gamma_{p_w}\,p_{w,0}\right)s\,\ell^2}{12} \geq 0 \tag{5.6.16}$$

Another simplified approach consists in determining the "first-order second-moment reliability index" as given by equation (5.3.25) for uncorrelated random variables $X_i's$. Introducing equation (5.5.13) in (5.5.12) gives

$$M = g(X)$$
$$= Z_S\,\sigma_Y\sqrt{1-3\left[\frac{A_f}{Z_S\,t_w}\,\frac{\left(B_{p_{st}}\,p_{st,0} + B_{p_w}\;p_{w,0}\right)s\,\ell}{2\,\sigma_Y}\right]^2}$$
$$-\frac{B_{p_{st}}\,p_{st,0}\,s\,\ell^2}{12} - \frac{B_{p_w}\,p_{w,0}\;s\,\ell^2}{12}$$

or

$$M = Z_S X_1 \sqrt{1 - \mu \left[\frac{\left(p_{st,0} X_2 + p_{w,0} X_3\right)}{X_1} \right]^2} - \left(p_{st,0} X_2 + p_{w,0} X_3\right) \frac{s \ell^2}{12} \quad (5.6.17)$$

where $\mu = 3\left[\frac{A_f}{Z_S t_w} \frac{s \ell}{2}\right]^2$.

The first-order second-moment reliability index is

$$\beta = \frac{g[E(X)]}{\sqrt{\sum_{i=1}^{i=3} \left(\frac{\delta g[E(X)]}{\delta x_i} \right)^2 \sigma_{X_i}^2}} \quad (5.6.18)$$

where

$$\frac{\delta g[E(X)]}{\delta x_1} = Z_S \sqrt{1 - \mu \left[\frac{\left[p_{st,0} E\left(B_{p_{st}}\right) + p_{w,0} E\left(B_{p_w}\right) \right]}{E(X_1)} \right]^2} + \frac{Z_S}{\left[E(X_1)\right]^2} \frac{\mu \left[p_{st,0} E\left(B_{p_{st}}\right) + p_{w,0} E\left(B_{p_w}\right) \right]^2}{\sqrt{1 - \mu \left[\frac{\left[p_{st,0} E\left(B_{p_{st}}\right) + p_{w,0} E\left(B_{p_w}\right) \right]}{X_1} \right]^2}}$$

$$\frac{\delta g[E(X)]}{\delta x_2} = - \frac{\mu Z_S p_{st,0} \left[p_{st,0} E\left(B_{p_{st}}\right) + p_{w,0} E\left(B_{p_w}\right) \right]}{E(X_1) \sqrt{1 - \mu \left[\frac{\left[p_{st,0} E\left(B_{p_{st}}\right) + p_{w,0} E\left(B_{p_w}\right) \right]}{E(X_1)} \right]^2}} - \frac{p_{st,0} s \ell^2}{12}$$

$$\frac{\delta g[E(X)]}{\delta x_3} = - \frac{\mu Z_S p_{w,0} \left[p_{st,0} E\left(B_{p_{st}}\right) + p_{w,0} E\left(B_{p_w}\right) \right]}{E(X_1) \sqrt{1 - \mu \left[\frac{\left[p_{st,0} E\left(B_{p_{st}}\right) + p_{w,0} E\left(p_w\right) \right]}{E(X_1)} \right]^2}} - \frac{p_{w,0} s \ell^2}{12}$$

NUMERICAL APPLICATION

Calculations are performed for the upper, mid-height, and lower stiffeners of a cargo/ballast tank transverse bulkhead of a VLCC assumed to be fixed at their both ends. The safety margin is given by equation (5.5.16).

1. *Static pressures* are assumed to follow a normal distribution. Calculations are carried out for full tanks. Since the filling ratio of cargo or ballast tanks is easily monitored the mean value and coefficient of variation of static pressures are taken as

Mean value = design value.

Covariance = 0.05.

2. *Wave-induced pressures* are assumed to follow a Gumbel distribution. Their mean value and covariance are given by

$$E(X_3) = E\left(B_{I_w}\right) E\left(B_{II_w}\right) = \left(1 + \frac{0.577}{\xi \ln N}\right) E\left(B_{II_w}\right)$$

$$V_{X_3} = \sqrt{V_{B_{I_w}}^2 + V_{B_{II_w}}^2}$$

where $E(B_{II_w})$ = approximational uncertainties, taken as unity in this numerical application.

N = number of cycles over the period of time considered, taken as 10^8 cycles.

$\xi = 1.4 - 0.044\ \alpha^{0.8} \sqrt{L}$ (refer to ABS 2002 5-1-1/5-5, α=0.8 for transverse bulkheads).

V_{B_I} = coefficient of variation of the statistical uncertainties taken as

$$V_{B_I} = \frac{\pi}{\sqrt{6}\left[0.577 + \xi \ln N\right]}$$

$V_{B_{II}}$ = coefficient of variation of the approximational uncertainties, taken as 0.10.

The mean value and coefficient of variation of the random variable X_3 are

$$E(X_3)=\left(1+\frac{0.577}{\xi\ \ln N}\right)=1.04 \qquad (5.6.19)$$

$$V_{X_3}=\sqrt{V_{B_I}^2+V_{B_{II}}^2}=\sqrt{0.09^2+0.10^2}=0.135 \quad (5.6.20)$$

$$\sigma_{X_3}=0.14$$

Note: Since calculations are carried out for full tanks, sloshing loads are not considered.

3. *Yield stress*. The yield stress is assumed to follow a lognormal distribution and its mean value and coefficient of variation are taken as

$$E(\sigma_Y)=1.209(\sigma_Y)_{\min}$$

$$V_{\sigma_Y}=0.08$$

Table 5.18 summarizes the results of these calculations. The following conclusions can be drawn from this analysis:

1. Upper stiffeners have a level of safety less than that of lower stiffeners, although their scantlings are based on the same requirements. This is due, obviously, to the uncertainties in the wave-induced pressure that have a larger influence on the probability of failure for the upper stiffeners.
2. This calculation shows how a reliability analysis may be used to "put the material at the right place."

For a target safety index of 4.1, the minimum section modulus should be approximately given by

$$Z \geq 1.18\frac{1.05\ p_{st}+1.25\ p_w}{\mu(\sigma_Y)_{\min}}\frac{s\,\ell^2}{12} \qquad (5.6.21)$$

where $\lambda=\sqrt{1-3\left[\frac{A_f}{Z_S\ t_w}\frac{\left[E(p_{st})+E(p_w)\right]s\,\ell}{2(\sigma_Y)_{\min}}\right]^2}$

ULTIMATE STRENGTH

Keeping notations of Section 5.5.3, the safety margin with respect to ultimate strength of laterally loaded horizontal stiffeners fixed at both ends of cargo tank transverse bulkheads is

$$M=g(X)=\left(Z_{pm}+Z_{pe}\right)\sigma_Y-\frac{\left(B_{p_{st}}\ p_{st,0}+B_{p_w}\ p_{w,0}\right)s\,\ell^2}{8} \qquad (5.6.22)$$

where σ_Y, Z_{pe}, $B_{p_{st}}$, and B_{p_w} are random variables assumed to be independent.

Assuming that the plastic section modulus Z_{pe} is a deterministic variable given by

$$Z_{pe}=A+B\sqrt{1-3\left[\frac{\left(E(p_{st})+E(p_w)\right)s\,\ell}{2\,A_w\,(\sigma_Y)_{\min}}\right]^2} \qquad (5.6.23)$$

the safety margin is a linear function of the random variables expressed as

$$M=\left(Z_{pm}+Z_{pe}\right)\sigma_Y-\frac{\left(B_{p_{st}}\ p_{st,0}+B_{p_w}\ p_{w,0}\right)s\,\ell^2}{8}$$
$$=a_1\ X_1-a_2\ X_2-a_3\ X_3 \qquad (5.6.24)$$

Table 5.19 summarizes the results of the calculations carried out for stiffeners whose scantlings are defined in Table 5.18.

Table 5.18 Initial Yielding of Transverse Bulkhead Stiffeners

Stiffener	Z_{rule} (cm³)	$p_{st,0}$ (kN/m²)	$p_{vw,0}$ (kN/m²)	β	"FORI" β	PSF for $\beta = 4.1$		
						$1/\gamma_1^*$	γ_2^*	γ_3^*
Upper	830 300×11.5–100×18	23.55	5..15	2.41	2.25	0.884	1.02	1.38
Mid-height	2775 550×12–145×22	157.3	73.9	4.10	3.825	0.845.	1.05	1.225
Lower	4715 700×13–150×28	281.45	105.95	3.91	3.56	0.841	1.055	1.205

Table 5.19 Ultimate Strength of Transverse Bulkhead Stiffeners

Stiffener	Scantlings	β
Upper	300×11.5–100×18	5.25
Mid-height	550×12–100×18	7.75
Lower	700×13–150×28	7.73

APPENDIX 5A. MEAN AND VARIANCE OF THE QUADRATIC FUNCTION

$$F = b + \sum_{i=1}^{i=n} a_i X_i + XY$$

In the following, the random variables are assumed to be independent random variables.

Linear Function

$$F = b + \sum_{i=1}^{i=n} a_i X_i$$

The expected mean value of the linear function F is

$$E(F) = b + \sum_{i=1}^{i=n} a_i E(X_i) \qquad (5.A.1)$$

and the variance of the linear function F is given by

$$\sigma^2(F) = \iint \left[F - E(F)\right]^2 p(x_i, x_j)\, dx_i\, dx_j$$

or

$$\sigma^2(F) = \iint \sum_{i=1}^{i=n} \sum_{j=1}^{j=n} a_i a_j \left[x_i - E(X_i)\right]\left[x_j - (X_j)\right] p(x_i, x_j)\, dx_i\, dx_j \qquad (5.A.2)$$

Equation (5.A.2) may be written as

$$\sigma^2(F) = \sum_{i=1}^{i=n} \sum_{j=1}^{j=n} a_i a_j \mathrm{Cov}(X_i, X_j) \qquad (5.A.3)$$

where

$$\mathrm{Cov}(X_i, X_j) = \iint \left[x_i - E(X_i)\right] \times \left[x_j - E(X_j)\right] p(x_i, x_j)\, dx_i\, dx_j \qquad (5.A.4)$$

Since the random variables are independent $\mathrm{Cov}(X_i, X_j) = 0$, for $i \neq j$, and the variance of the linear function is

$$\sigma^2(F) = \sum_{i=1}^{i=n} a_i^2 \sigma_{X_i}^2 \qquad (5.A.5)$$

Quadratic Function

$$F = XY$$

The expected mean value and variance of the quadratic function $F = XY$ are given by

$$E(F) = \iint XYp(x,y)\, dxdy = \int X\, p(x) \int Yp(y)\, dy = E(X)E(Y) \qquad (5.A.6)$$

$$\sigma_F^2 = \iint \left[F - E(F)\right]^2 p(x,y)\, dx\, dy = \iint \left[X^2Y^2 - 2E(X)E(Y)XY + \left[E(X)\right]^2\left[E(Y)\right]^2\right] p(x,y)\, dx\, dy$$

$$\sigma_F^2 = \int X^2 p(x)\, dx \int Y^2 p(y)\, dy - 2\, E(X)\, E(Y) \int X\, p(x) dx \int Y\, p(y)\, dy + \left[E(X)\right]^2 \left[E(Y)\right]^2$$

$$\sigma_F^2 = \left(\sigma_X^2 + \left[E(X)\right]^2\right)\left(\sigma_Y^2 + \left[E(Y)\right]^2\right) - \left[E(X)\right]^2\left[E(Y)\right]^2 = \sigma_X^2 \sigma_Y^2 + \left[E(X)\right]^2 \sigma_Y^2 + \left[E(Y)\right]^2 \sigma_X^2$$

$$\sigma_F^2 = \left(V_X^2 V_Y^2 + V_X^2 + V_Y^2\right)\left[E(X)\right]^2\left[E(Y)\right]^2 \cong \sigma_X^2\left[E(Y)\right]^2 + \sigma_Y^2\left[E(X)\right]^2$$

$$V_F^2 = V_X^2 + V_Y^2 \qquad (5.A.7)$$

Combined Quadratic and Linear Function

$$F = b + \sum_{i=1}^{i=n} a_i X_i + XY$$

$$E(F) = b + \sum_{i=1}^{i=n} a_i E(X_i) + E(X)E(Y) \qquad (5.A.8)$$

$$\sigma_F^2 = \sigma_X^2\, E(Y)^2 + \sigma_Y^2\, E(X)^2 + \sum_{i=1}^{i=n} a_i^2\, \sigma_{X_i}^2 \qquad (5.A.9)$$

APPENDIX 5B. LINEAR SAFETY MARGIN

$$M = b + \sum_{i=1}^{i=n} a_i X_i$$

The Cornell safety index is given by

$$\beta = \frac{E(M)}{D(M)}$$

$$= \frac{E(M)}{\sqrt{\sum_{i=1}^{i=n}\sum_{j=1}^{j=n} a_i a_j \operatorname{Cov}(X_i, X_j)}} \quad (5.B.1)$$

If the random variables are independent $\operatorname{Cov}(X_i, X_j) = 0$ for $i \neq j$, and the safety index is given by

$$\beta = \frac{E(M)}{D(M)} = \frac{E(M)}{\sqrt{\sum_{i=1}^{i=n} a_i^2 \sigma_{X_i}^2}} \quad (5.B.2)$$

When the failure surface is a hyperplane, the Cornell and Hasofer-Lind safety indices are identical. The safety margin expressed in terms of the reduced variables is

$$M = b + \sum_{i=1}^{i=n}\left[a_i\left(\sigma_i u_i + E(X_i)\right)\right] \quad (5.B.3)$$

The MPFP is defined as the intersection between the hyperplane and the normal to this hyperplane drawn by the origin. The equation of the normal is

$$\frac{u_1}{-\frac{\delta M}{\delta u_1}} = \frac{u_2}{-\frac{\delta M}{\delta u_2}} = \ldots = \frac{u_n}{-\frac{\delta M}{\delta u_n}} = \lambda$$

and the coefficient λ is given by

$$\lambda = \frac{b + \sum_{i=1}^{i=n} a_i E(X_i)}{\sum_{i=1}^{i=n} a_i^2 \sigma_i^2} = \frac{E(M)}{\sum_{i=1}^{i=n} a_i^2 \sigma_i^2} \quad (5.B.4)$$

The coordinates of the MPFP are given by

$$u_i^* = -\frac{a_i \sigma_i E(M)}{\sum a_i^2 \sigma_i^2} = -a_i \sigma_i \beta$$

and the Hasofer-Lind safety index is

$$\beta_{HL} = \sqrt{\sum_{i=1}^{i=n}\left(u_i^*\right)^2} = \frac{E(M)}{\sqrt{\sum_{i=1}^{i=n} a_i^2 \sigma_i^2}}$$

$$= \frac{E(M)}{D(M)} = \beta_C \quad (5.B.5)$$

This conclusion may be extended to the case where the random variables are correlated. The limit state function expressed in matrix notation is given by

$$M = a^T X + b = a^T T^{-1} u + E(X) + b$$

$$= a^T T^{-1} u + E(M) = 0 \quad (5.B.6)$$

where a^T = row matrix.
X = column matrix of the random variables.
T = transformation matrix defined as $u = T\{x - E(X)\}$.

The co-ordinates of the MPFP are given by

$u_i^* = -\lambda \dfrac{\delta M}{\delta u_i}$ or in matrix notation

$$u^* = -\lambda\left(\frac{\delta M}{\delta u}\right) = -\lambda\left(T^{-1}\right)^T a$$

and the safety index is given by

$$\beta_{HL} = \sqrt{\sum_{i=1}^{i=n}\left(u_i^*\right)^2} = \lambda\sqrt{\sum_{i=1}^{i=n}\left(\delta M/\delta u_i\right)^2}$$

$$= \lambda\sqrt{a^T T^{-1}\left(T^{-1}\right)^T a} \quad (5.B.7)$$

The coefficient λ is obtained from Equation (5. A.13) and given by

$$\lambda = \frac{E(M)}{a^T T^{-1}(\delta M/\delta u)} = \frac{E(M)}{a^T T^{-1}\left(T^{-1}\right)^T a}$$

$$= \frac{E(M)}{a^T C_X a}$$

noting that $T^{-1}(T^{-1})^T = C_X$.

Finally, the Hasofer-Lind safety index is

$$\beta_{HL} = \lambda\sqrt{\sum_{i=1}^{i=n}\left(\delta M/\delta u_i\right)^2}$$

$$= \frac{E(M)}{\sqrt{a^T C_X a}} = \beta_C \quad (5.B.8)$$

APPENDIX 5C. ITERATIVE PROCEDURE FOR DETERMINATION OF THE MPFP

The most probable failure point (MPFP) is obtained as the limit of an iterative procedure based on linearization of the failure surface at each step of the sequence. To start this procedure, an initial approximation point is to be defined (e.g., origin of coordinates in the reduced space) and the process is continued until convergence of the safety index. If we assume that $u^{(m)}$ is the solution of the step m, the failure surface of the step $m + 1$ is replaced by the tangent hyperplan at $u = u^{(m)}$:

$$g'\left(u^{(m)}\right)+\sum_{i=1}^{i=n}\frac{\delta g'\left(u^{(m)}\right)}{\delta u_i}\left(u_i-u_i^{(m)}\right)=0 \qquad (5.C.1)$$

Then, the point $u^{(m+1)}$ of the step $m + 1$ is defined as the intersection between the hyperplane and the normal to this plan drawn by the origin. The equation of the normal is

$$\frac{u_1^{m+1}}{-\frac{\delta g'\left(u^{(m)}\right)}{\delta u_i}}=\frac{u_2^{m+1}}{-\frac{\delta g'\left(u^{(m)}\right)}{\delta u_2}}=\frac{u_n^{m+1}}{-\frac{\delta g'\left(u^{(m)}\right)}{\delta u_n}}=\lambda \qquad (5.C.2)$$

and the coefficient λ is given by

$$\lambda=\frac{g'\left(u^{(m)}\right)-\sum_{i=1}^{i=n}\frac{\delta g'\left(u^{(m)}\right)}{\delta u_i}u_i^{(m)}}{\sum_{i=1}^{i=n}\left(\delta g'\left(u^{(m)}\right)/\delta u_i\right)^2} \qquad (5.C.3)$$

The coordinates of $u^{(m+1)}$ are given by

$$u_i^{(m+1)}=-\frac{\delta g'\left(u^{(m)}\right)}{\delta u_i}\frac{g'\left(u^{(m)}\right)-\sum_{i=1}^{i=n}\frac{\delta g'\left(u^{(m)}\right)}{\delta u_i}u_i^{(m)}}{\sum_{i=1}^{i=n}\left(\delta g'\left(u^{(m)}\right)/\delta u_i\right)^2}$$

$$=\frac{\alpha_i^{(m)}g'\left(u^{(m)}\right)}{\sqrt{\sum_{i=1}^{i=n}\left(\delta g'\left(u^{(m)}\right)/\delta u_i\right)^2}}+\left(\sum_{i=1}^{i=n}\alpha_i^{(m)}u_i^{(m)}\right)\cdot\alpha_i^{(m)}$$

$$u_i^{(m+1)}=\frac{g'\left(u^{(m)}\right)}{\left|\nabla g'\left(u^{(m)}\right)\right|}\alpha_i^{(m)}+\left(\sum_{i=1}^{i=n}\alpha_i^{(m)}u_i^{(m)}\right)\alpha_i^{(m)} \qquad (5.C.4)$$

where $\nabla g'(u^{(m)})$ = gradient vector of $g'(u^{(m)})$ at $u = u^{(m)}$,

$$\alpha_i^{(m)}=-\frac{\frac{\delta g'\left(u^{(m)}\right)}{\delta u_i}}{\sqrt{\sum_{i=1}^{i=n}\left(\delta g'\left(u^{(m)}\right)/\delta u_i\right)^2}}=-\frac{\nabla g'\left(u^{(m)}\right)}{\left|\nabla g'\left(u^{(m)}\right)\right|}$$

At step $m + 1$ the safety index $\beta^{(m+1)}$ is given by

$$\beta^{(m+1)}=\sqrt{\sum_{i=1}^{i=n}\left(u_i^{(m+1)}\right)^2}=\lambda\sqrt{\sum_{i=1}^{i=n}\left(\delta g'\left(u^{(m)}\right)/\delta u_i\right)^2}$$

$$=\frac{g'\left(u^{(m)}\right)}{\left|\nabla g'\left(u^{(m)}\right)\right|}+\sum_{i=1}^{i=n}\alpha_i^{(m)}u_i^{(m)} \qquad (5.C.5)$$

APPENDIX 5D. MEAN VALUE AND STANDARD DEVIATION OF THE HULL GIRDER SECTION MODULUS

The section modulus of the hull girder or any beam is a random variable the mean value and standard deviation, which may be calculated as follows, refer to Wirsching et al. (1997):

1. Area of the section: $A=\sum_{i=1}^{i=n}A_i$

   $$=\sum_{i=1}^{i=n}b_i\left(t_{i,nom}-c_i f(t)\right)$$

   Mean value: $E(A)=\sum_{i=1}^{i=n}b_i\left(t_{i,nom}-E(c_i)f(t)\right)$

   Variance: $\sigma_A^2=\sum_{i=1}^{i=n}\sigma_{A_i}^2$, with $\sigma_{A_i}=b_i f(t)\,\sigma_{c_i}$

2. Moment of area of the section: $M=\sum_{i=1}^{i=n}z_i A_i$

   Mean value: $E(M)=\sum_{i=1}^{i=n}z_i E(A_i)$

   Variance: $\sigma_M^2=\sum_{i=1}^{i=n}\sum_{j=1}^{j=n}z_i^2\sigma_{A_i}^2$

where z_i is the distance from the center of gravity of the ith element to the baseline.

3. Position of the neutral axis: $z_{na}=\frac{M}{A}=\frac{\sum_{i=1}^{i=n}A_i z_i}{\sum_{i=1}^{i=n}A_i}$

4. Inertia of the section: $I = I_b - Az_{na}^2$

$$= \sum_{i=1}^{i=n}\left[A_i z_i^2 + I_{0i}\right] - Az_{na}^2,$$

with $I_{0i} = \dfrac{t_i h_i^3}{12}$

Mean value: $E(I) = E(I_b) - E(A)\left[E(z_{na})\right]^2$

$$E(I) = \sum_{i=1}^{i=n}\left[E(A_i)z_i^2 + E(I_{0i})\right] - E(A)\left[E(z_{na})\right]^2$$

$$\text{Variance}: \sigma_I^2 = \sum_{i=1}^{i=n}\left(z_i^4\sigma_{A_i}^2 + \left(h_i^3/12\right)^2\sigma_{t_i}^2\right) + \left[E(z_{na})\right]^4\sum_{i=1}^{i=n}\sigma_{A_i}^2$$

$$\sigma_{t_i} = f(t)\,\sigma_{c_i}$$

5. Section modulus :

$$E(Z_{bot}) = \frac{E(I)}{E(z_{na})} \quad (5.D.1)$$

$$E(Z_{deck}) = \frac{E(I)}{D - E(z_{na})} \quad (5.D.2)$$

$$\sigma_{Z_{bot}}^2 = \frac{\sigma_I^2}{\left[E(z_{na})\right]^2} \quad (5.D.3)$$

$$\sigma_{Z_{deck}}^2 = \frac{\sigma_I^2}{\left[D - E(z_{na})\right]^2} \quad (5.D.4)$$

More refined equations taking into account that the position of the neutral axis is also a random variable have been proposed by Guedes Soares and Garbatov (1996 and 1977):

1. Neutral axis.

$$\text{Mean value}: E(z_{na}) = \frac{E(M)}{E(A)} = \frac{\sum_{i=1}^{i=n} E(A_i)z_i}{E(A)}$$

$$\text{Variance}: \sigma_{z_{na}}^2 = \frac{\sigma_M^2}{\left[E(A)\right]^2} + \frac{\left[E(M)\right]^2}{\left[E(A)\right]^4}\sigma_A^2$$

Inertia of the section.

$$\text{Mean value}: E(I) = \left\{\sum_{i=1}^{i=n} E(A_i)z_i^2 + E(I_{0i})\right\} - E(A)\left[E(z_{na})\right]^2$$

$$\text{Variance}: \sigma_I^2 = \sigma_{I_b}^2 + \left[E(z_{na})\right]^4\sigma_A^2 + 4\left[E(A)\right]^2\left[E(z_{na})\right]^2\sigma_{z_{na}}^2$$

$$\sigma_{I_b}^2 = \sum_{i=1}^{i=n}\left(z_i^4\,\sigma_{A_i}^2 + \left(h_i^3/12\right)^2\sigma_{t_i}^2\right)$$

3. Section modulus.

$$\text{Variance } \sigma_{Z_{bot}}^2 = \frac{\sigma_I^2}{\left[E(z_{na})\right]^2} + \frac{\left[E(I)\right]^2}{\left[E(z_{na})\right]^4}\sigma_{z_{na}}^2 \quad (5.D.5)$$

$$\sigma_{Z_{deck}}^2 = \frac{\sigma_I^2}{\left[D - E(z_{na})\right]^2} + \frac{\left[E(I)\right]^2}{\left[D - E(z_{na})\right]^4}\sigma_{z_{na}}^2 \quad (5.D.6)$$

APPENDIX 5E. CHARACTERISTICS OF TEST SHIPS

Table 5.20 Main particulars of Bulk Carriers

Ship	L[m]	B[m]	D[m]	C_B	SWBM $M_{sw,0}$ (kN·m)		VWBM $M_{vw,0}$ (kN·m)		M_{ult} (kN·m)	
					Hogging	Sagging	Hogging	Sagging	Hogging	Sagging
1	135	21,7	12,2	0.775	0.378 E06	0.327 E06	0.503 E06	0.554 E06	1.5.1 E06	1.199 E06
2	152	24	13,10	0.844	0.545 E06	0.498 E06	0.794 E06	0.843 E06	2.347 E06	1.818 E06
3	210.49	32,2	18,3	0.812	1.558 E06	1.388 E06	2.179 E06	2.349 E06	—	—
4	211.36	32,2	17,6	0.811	1.573 E06	1.400 E06	2.198 E06	2.370 E06	5.112 E06	5.075.E06
5	255.57	43	23,9	0.857	3.247 E06	2.997 E06	4.822 E06	5.072 E06	1.405 E07	1.05. E07

Table 5.21 Main particular of Oil Tankers

Ship	L[m]	B[m]	D[m]	C_B	SWBM $M_{sw,0}$ (kN·m)		VWBM $M_{vw,0}$ (kN·m)		M_{ul}(kN·m)	
					Hogging	Sagging	Hogging	Sagging	Hogging	Sagging
1	151,32	23,5	12,75	0.801	0.531 E06	0.45. E06	0.732 E06	0.794 E06	2.082 E06	1.5.2 E06
2	15.,92	28,4	13,70	0.790	0.75. E06	0.5.8 E06	1.034 E06	1.130 E06	2.5.3 E06	2.342 E06
3	310,89	56	29,4	0.831	5.402 E06	5.790 E06	9.187 E06	9.799 E06	2.308 E07	2.047 E07
4	323	53,6	25.4	0.840	5.5.5.E06	5.017 E06	9.594 E06	1.018 E07	2.304 E07	2.199 E07
5	324,95	53	28,3	0.831	5.5.0 E06	5.987 E06	9.499 E06	1.013 E07	2.498 E07	2.330 E07
6	327,3	51,82	27,35	0.830	5.55. E06	5.935 E06	9.411 E06	1.004 E07	2.547 E07	2.331 E07
7	400	5.	37,13	0.85.	1.15. E07	1.079 E07	1.739 E07	1.825.E07	4.55. E07	4.234 E07

Table 5.22 Main Characteristics of MidSHIP SECTION

Ship	Bulk Carriers				Oil Tankers			
	$(\sigma_Y)_{deck}$	$(\sigma_Y)_{bot}$	Z_{deck}	Z_{bot}	$(\sigma_Y)_{deck}$	$(\sigma_Y)_{bot}$	Z_{deck}	Z_{bot}
1	235	235	5.5.7	8.179	235	235	8.057	9.478
2	235	235	9.103	11.45.	235	235	10.938	11.443
3	235	235	22.558	27.885	355	355	5..711	72.403
4	235	235	21.729	25.008	315	315	75.991	75.310
5	390	355	31.35.	44.721	315	315	81.779	85.158
6					355	355	74.285	75.383
7					355	355	121.943	139.293

REFERENCES

American Bureau of Shipping. (2002). *Rules for building and classing vessels.* Houston, TX: Author.

American Institute of Steel Construction. (1994). *Load and resistance factor design, manual of steel construction.* Chicago: Author.

Ang, A. H.-S., and Cornell, A. C. (1974). Reliability bases of structural safety and design. *J. of the Structural Division*, Vol. 100, Issue ST9, 1755–1769.

Ang, A. H.-S., and Ellingwood, B. R. (1971). Analysis of reliability principles relative to design. *Conf. on Application of Statistics and Probability to Soils and Structural Engineering*, Hong-Kong.

Assakaf, I. A., Ayyub, B. M., Hess, P. E., and Atua, K. (2002). Reliability-based load and resistance factor design (LFRD)—guidelines for stiffened panels and grillages of ship structures. *Naval Engineers Journal*, Vol. 114, Issue 2, 89–111.

Ayyub, B. M., Assakaf, I. A., Beach, J. E., Melton, W. M., Nappi, N, Jr., and Conley, J. A. (2002). Methodology for developing reliability-based load and resistance factor design (LRFD)—guidelines for ship structures. *Naval Engineers Journal*, Vol. 114, Issue 2, 23–41.

Beghin, D., Jastrzebski, T., and Taczala, M. (1998). Hull girder reliability of bulk carriers. *PRAD's 98*, Delft, Netherlands.

Breitung, K. (1984). Asymptotic approximations for multinormal integrals. *J. of the Engineering Mechanics Division*, Vol. 110, 357–366.

Bureau Veritas. (2000). *Rules for the classification of steel ships.* Part B—Hull and Stability, Chapter 5. Paris: Author.

Construction Industry Research and Information Association (CIRIA). (1977). Rationalization of safety and serviceability factors in structural codes. CIRIA Report 5. London: Author.

Cornell, C. A. (1969). A probability-based structural code. *J. of the American Concrete Institute*, Vol. 16, Issue 12, 974–985.

Ditlevsen, O. (1979). Generalized second-moment reliability index. *J. of Structural Mechanics*, Vol. 7, 435–451.

Ditlevsen, O. (1981). Principle of normal tail approximation. *J. of the Engineering Mechanics Division*, Vol. 107, 1191–1208.

Ditlevsen, O., and Madsen, H. O. (1995). *Structural reliability methods.* New York: John Wiley and Sons.

Fain, R. A., and Booth, E. T. (1979). Results of the first five "data years" of extreme stress scratch gauge data collected aboard Sea Land's SL 7's. *SSC*, Vol. 286.

Faulkner, D. (1975). A review of effective plating for use in the analysis of stiffened plating in bending and compression. *J. of Ship Research*, Vol. 19, 1–17.

Faulkner, D. (1981). Semi-probabilistic approach to the design of marine structures. International Symposium on the Extreme Load Response, *Trans. SNAME*, 213–230.

Ferro, G., and Mansour, A. E. (1985). Probabilistic analysis of the combined slamming and wave-induced responses. *J. of Ship Research*, Vol. 29, Issue 3, 170–188.

Ferry-Borges, J., and Castenheta, M. (1971). *Structural safety.* Lisbon: Laboratoria Nacional de Engenhera Civil.

Freudenthal, A. M. (1947). The safety of structure. *Trans. ASCE*, Vol. 102, 269–324.

Freudenthal, A. M. (1955). Safety and probability of structural failures. *Trans. ASCE*, Vol. 121, 1337–1375

Friis Hansen, P. (1994). On combination of slamming and wave-induced bending responses. *J. of Ship Research*, Vol. 38, Issue 2, 104–114.

Friis Hansen, P., and Terndrup Pedersen, P. (1994). *On the development and calibration of partial safety factors.* SHIPREL-Report No 3.4R-01 (A).

Gordo, J. M., and Guedes Soares, C. (1993). *Approximate load shortening curves for stiffened plates under uniaxial compression. Conference on integrity of offshore structures.* Glasgow: Elsevier.

Gran, S. (1978). *Reliability of ship hull structures.* Report No 78-215. Oslo: Det Norske Veritas.

Guedes Soares, C. (1984). *Probabilistic models for load effects in ship structures.* Report UR-84-38. Trondheim: Dept. of Marine Technology, Norwegian Institute of Technology.

Guedes Soares, C. (1988). Design equation for the compressive strength of unstiffened plate elements with initial imperfections. *J. of Construction Steel Research*, Vol. 9, 287–310.

Guedes Soares, C. (1990). Stochastic modeling of maximum still water load effects in ship structures. *J. of Ship Research*, Vol. 34, Issue 3, 199–205.

Guedes Soares, C. (1995). On the definition of rule requirements for wave induced vertical bending moments. *Marine Structures*, Vol. 9, 409–425.

Guedes Soares, C., and Dias, S. (1995). Probabilistic models of still water load effects in container ships. *Marine Structures*, Vol. 9, 287–312.

Guedes, C., and Garbatov, Y. (1995). Reliability of maintained ship hulls subjected to corrosion. *J. of Ship Research*, Vol. 3, 235–243.

Guedes Soares, C., and Garbatov, Y. (1997). Reliability assessment of maintained ship hulls with correlated corroded elements. *Marine Structures*, Vol. 10, 629–653.

Guedes Soares, C., and Moan, T. (1988). Statistical analysis of still water load effects in ship structures. *Trans. SNAME*, Vol. 95, 129–155.

Hasofer, A. M., and Lind, N. C. (1974). Exact and invariant second moment code format. *J. of the Engineering Mechanics Division*, Vol. 100, Issue EM1, 111–121.

Hess, P. E., Bruchman, D., Assakkaf, I. A., and Ayyub, B. M. (2002). Uncertainties in material strength, geometric and load variables. *Naval Engineers Journal*, Vol. 114, Issue 2, 139–165.

Hohenbichler, M., and Rackwitz, R. (1981). Non normal dependent vectors in structural safety. *J. of the Engineering Mechanics Division*, Vol. 107, 1227–1238.

Hughes, O., Nikolaidis, E., Ayyub, B., White, G., and Hess, P. (1994). *Uncertainty in strength models for marine structures*. Report SSC-375. Washington, DC: Ship Structure Committee.

International Ship and Offshore Structure Congress (ISSC). (1985). *Report of Committee V-I on applied design*. Geneva: Author.

International Ship and Offshore Structure Congress (ISSC). (1991). *Report of Committee V-I on applied design*. China: Author.

International Ship and Offshore Structures Congress (ISSC). (1994a). *Report of Committee IV-1—design principles and criteria*. St. John's, Canada: Author.

International Ship and Offshore Structures Congress (ISSC). (1994b). *Report of Committee II-1—quasistatic load effects*. St. John's, Canada: Author.

International Ship and Offshore Structure Congress (ISSC). (1997). *Report of Committee IV-1—design principles and criteria*. Trondheim, Norway: Author.

International Standardization Organization (ISO). (1994). *General principles on reliability for structures*. Revision of IS 2394. Geneva: Author.

Kaplan, P. (1985). Analysis and prediction of flat bottom slamming impact of advanced marine vehicles in waves. *AIAA, Eighth Advanced Marine Systems Conference*, San Diego, CA.

Kaplan, P., and Raff, A. L. (1972). *Evaluation and verification of computer calculation of water-induced ship structural loads*. Report SSC-229. Washington, DC: Ship Structure Committee.

Kaplan, P., Benatar, M., Bentson, J., and Achtarides, T. A. (1984). *Analysis and assessment of major uncertainties associated with ship hull ultimate failure*. Report SSC-322. Washington, DC: Ship Structure Committee.

Lewis, E. V., and Zubaly, R. B. (1975). Dynamic loadings due to wave and ship motions. *STAR Symposium, SNAME.*

Lewis, E. V., Hoffman, D., Maclean, W. M., van Hooff, R., and Zubaly, R. B. (1973). *Load criteria for ship structural design*. SSC Report-240. Washington, DC: Ship Structure Committee.

Little, R. S., Lewis, E. V., and Bailey, F. C. (1971). A statistical study of wave induced bending moments on large oceangoing tankers and bulk carriers. *Trans. SNAME*, Vol. 79, 117–168.

Lloyd's Register of Shipping. (1999). *World casualty statistics 1998*. London: Author.

Madsen, H. O., Krenk, S., and Lind, N. C. (1985). *Methods of structural safety*. Englewood Cliffs, NJ: Prentice Hall.

Mansour, A. E. (1995). Extreme loads and load combinations. *J. of Ship Research*, Vol. 39, Issue 1, 53–61.

Mansour, A. E., and Thayamballi, A. K. (1994). *Probability based ship design; loads and load combinations*. SSC Report SSC-373. Washington, DC: Ship Structure Committee.

Mansour, A. E., and Wirsching, P. H. (1995). Sensitivity factors and their application to marine structures. *Marine Structures*, Vol. 8, 229–255.

Mansour, A. E., Ayyub, B. M., White, G. J., and Wirsching, P. H. (1995). *Probability based ship design: Implementation of design guidelines*. Report SSC-392. Washington, DC: Ship Structure Committee.

Mansour, A. E., Wirsching, P., Luckett, M., and Plumpton, A. (1997). *Assessment of reliability of ship structures*. Report SSC-398. Washington, DC: Ship Structure Committee.

Mansour, A. E., Spencer, J., Wirsching, P., McGovney, J., and Tarman, D. (2001). Consistent code formulation for ship structural design. *PRAD'S 2001*. China: Elsevier Science.

Melchers, R. E. (1987). *Structural reliability analysis and prediction*. West Sussex, U.K.: Ellis Horwood Limited.

Moan, T., and Jiao, G. (1988). *Characteristic still water load effect for production ships*. Report MK/R 104/88. Trondheim: The Norwegian Institute of Technology.

Nikolaidis, E., and Kaplan, P. (1991). Uncertainties in stress analysis of marine structures. SNAME Structural Inspection, Maintenance and Monitoring Symposium, Arlington, VA, March. SSC Report-35. Washington, DC: Ship Structure Committee.

Paik, J. K., Ham, J. H., and Ko, J. H. (1992). A new plate buckling design formula. *J. of the Society of Naval Architects of Japan*, Vol. 172, 417–425.

Paik, J. K., Thayamballi, A. K., Kim, S. K., and Yang, S. H. (1998). Ship hull ultimate strength reliability considering corrosion. *J. of Ship Research,* Vol. 42, 154–165.

Paik, J. K., and Thayamballi, A. K. (2000). *Ultimate limit state design of steel plated structures*. London: John Wiley and Sons.

Pugsley, A. G. (1942). *A philosophy of aeroplane strength factors*. Report and Memo no. 1905. London: British Aeronautical Research Committee.

Rosenblatt, M. (1952). Remarks on a multivariate transformation. *Annals of Mathematical Statistics*, Vol. 23, Issue 3, 470–472.

Söding, H. (1979). The prediction of still water wave bending moments in containerships. *Schiffstechnik*, Vol. 25, 24–41.

Tanker Structure Co-operative Forum. (1997). *Guidance manual for tanker structures*. London: Witherby and Co.

U.K. Health and Safety Commission. (1992). *The offshore installation regulations (safety case)*. Technical Report. London: Health and Safety Executive.

U.K. House of Lords, Select Committee on Science and Technology. (1992). *Safety aspects of ship design and technology*. HL Paper 30-1. London: Author.

U.S. National Academy of Science. (1990). *Design for spill free oil tanker*. Washington, DC: National Academies of Science.

Vedeler, G. (195.). *Recent Development in Ship Structural Design*. Publication no. 48. Oslo: Det Norske Veritas.

Wang, X., and Moan, T. (1995). Stochastic and deterministic combinations of still water bending moments in ships. *Marine Structures*, Vol. 9, 787–810.

Wirsching, P. H., Ferensic, J., and Thayamballi, A. K. (1997). Reliability with respect to ultimate strength of a corroding ship hull. *Marine Structures*, Vol. 10, 501–518.

Yamamoto, N., Kumano, A., and Matoba, M. (1994). Effect of corrosion and its protection on hull strength. *J. of Society of Naval Architects of Japan*, Vol. 175, 281–289.

FRAME ANALYSIS

Owen Hughes
Professor, Virginia Tech
Blacksburg, VA, USA

In this chapter, we consider structures which are made up of one-dimensional beam or bar elements connected together at their ends. The connection may be either pinned or rigid. Such structures are referred to as frames and the connection points are referred to as *nodes*. Frames are probably the most common class of structure. In a ship, in spite of the plating, the three-dimensional assemblage of deck beams, side frames, and longitudinal girders constitutes a framework, especially in regard to the transverse loads because these act normal to the plating and are carried mainly by the framing system.

6.1 BASIC CONCEPTS

6.1.1 Frame Analysis: Nodal Displacements

Frame analysis is a well-established technique in the field of structural analysis. In this technique, everything is expressed in terms of what happens at the nodes. External forces are applied at the nodes (and only there), and the displacements of the structure are expressed entirely in terms of nodal displacements. The starting point is to determine the response characteristics of each individual member, that is, the relationship between nodal forces and nodal displacements. This relationship is taken to be linear, which means that each nodal force f_i is linearly related to each nodal displacement u_i: $f_i = \sum_j k_{ij} u_j$. The coefficient k_{ij} is referred to as a *stiffness coefficient*. The complete relationship between all nodal forces and all nodal displacements in a member is a system of linear equations, and such systems are best expressed in matrix notation, $\mathbf{f} = \mathbf{ku}$, in which $\mathbf{f}$ and $\mathbf{u}$ are vectors containing the nodal forces and displacements, and $\mathbf{k}$ is a square matrix containing all of the stiffness coefficients; this is the *stiffness matrix* of the member.

6.1.2 Fundamental Laws

In structural analysis, there are three fundamental "laws" or relationships which must be satisfied:

1. Equilibrium of forces (within each member and between members).
2. Compatibility of displacements (within each member and between members).
3. Law of material behavior (stress-strain law) of each member.

As noted, the first two of these must be satisfied at two levels: within each member and also for the structure as a whole (i.e., between members). For one-dimensional members such as bars and beams, which we deal with in this chapter, all three laws are exactly satisfied within each member because they are implicit in the member's nodal force-nodal displacement relationship, that is, the member stiffness. Therefore, this chapter deals mainly with the first two of the laws applied to the structure as a whole. In other words, there must be equilibrium of forces between the external loads and the various member forces, and there must be compatibility in the deformations of the members, such that they continue to fit together.

For a statically determinate structure, the equilibrium requirement is sufficient because the member forces can be calculated directly, whereupon the member displacements and internal forces can be calculated directly. But a frame structure is statically indeterminate, and so in frame analysis, the equilibrium and compatibility requirements are imposed at each node of the structure. These requirements, together with the force-displacement relationships of each member (as embodied in the member's stiffness matrix $\mathbf{k}$) produce a system of equations for the nodal displacements. After solving for these displacements, some previously established relationships between them and the member's internal forces and deformations (that is, stresses and strains) are used to solve for the internal forces.

In a typical three-dimensional frame structure, there may be a large number of nodal displacements, several thousand perhaps, but the solution of a large system of linear equations is a routine task for a computer. Because there are a large number of equations, any discussion of the underlying theory and of the setting up of the equations is greatly facilitated by using matrix notation; hence, this field is often referred to as matrix frame analysis or matrix stiffness analysis.

For one-dimensional members, imposing the requirements of equilibrium and compatibility at the nodes is sufficient to ensure that they are satisfied everywhere, both within members and between members, because the members are connected only at the nodes. For members that are two- or three-dimensional in extent, it is necessary to define a relationship between nodal displacements and internal deformation in such a way that these two laws are satisfied to sufficient accuracy both within the member and between members. This is the key step in finite element analysis, and it is the only major difference between the basic finite element method and frame analysis. Hence, the first three sections of this chapter serve as a foundation for the presentation of the finite element method in the remaining sections.

6.1.3 Stiffness Matrix of a Structure

The device most directly associated with stiffness is a simple elastic spring, such as that shown in Fig. 6.1. It can also be regarded as an example of a structural member because it has the same basic structural characteristics as a pin-ended bar: it undergoes axial displacements, transmits axial forces, and exhibits a linear internal (or material) behavior, that is, there is a linear relationship between internal force $\mathcal{F}$ (axial force) and internal deformation d (elongation or shortening) which is of the form

$$\mathcal{F} = sd$$

and the deformation is

$$d = (u_2 - u_1)$$

The quantity s is termed the *stiffness* of the spring and corresponds to the slope of the internal force-deformation diagram (Fig. 6.1).*

Besides being a structural member, the same spring could be considered as a structure—a one-member structure which is supported at one node and loaded at the other by an applied load F (Fig. 6.2). The load causes a structure displacement U which is linearly proportional to F,

$$F = KU \tag{6.1.1}$$

and the constant of proportionality K is termed the stiffness of the structure. In this case, since the structure is identical to the member, the two stiffnesses are the same: $K = s$. Knowing the value of this stiffness and of the applied load, (6.1.1) may be inverted to give the displacement

$$U = \frac{1}{K} F \tag{6.1.2}$$

For this structure, the response consists of just one displacement. For more realistic structures such as the pin-jointed frame shown in Fig. 6.3, it is necessary to determine the displacements of nodes C, D, E, and F in order to be able to evaluate the member deformations (axial strains) and hence the internal forces (stresses) in the members.

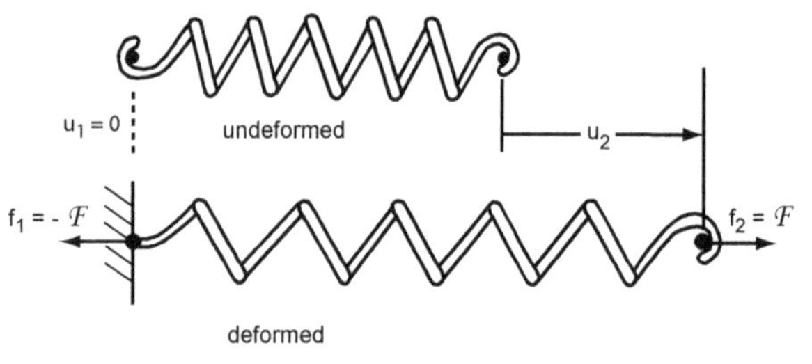

(a)

(*a*) Nodal forces f_1 (negative; $f_1 = -\mathcal{F}$) and f_2 (positive: $f_2 = \mathcal{F}$)
(*b*) Internal force $\mathcal{F}$. Magnitude of $\mathcal{F}$ = magnitude of a pair of equal and opposite nodal forces required to cause a specified deformation, d. Sign of $\mathcal{F}$: positive for elongation.

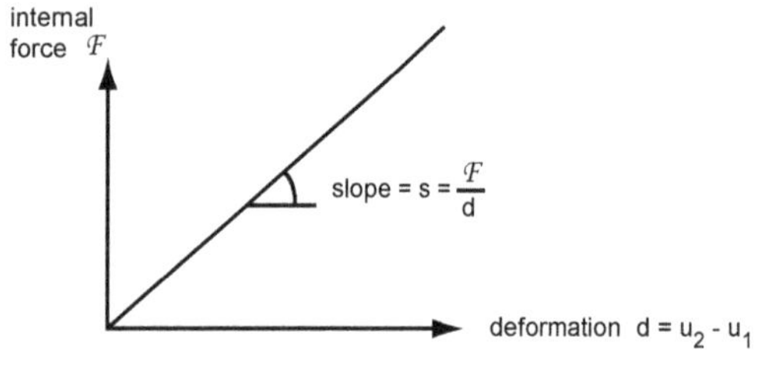

(b)

Figure 6.1 Internal force–deformation relationship for an elastic spring element.

*Note that there is a subtle but important distinction between the internal force in a member and the applied external forces f_1 and f_2. Although their definitions are such that they always have the same magnitude, they are conceptually quite different. The internal force $\mathcal{F}$ is a state or situation inside of the member, and it is defined in terms of the internal deformation. In the present case, it is the magnitude which a *pair* of equal and opposite forces must have in order to cause a specified elongation. It is a scalar quantity (no direction), it is diffused throughout the member, and its sign depends on the type of deformation: positive for elongation. In contrast, the external forces are directional and can be regarded as vectors (although here they have only one component each), they act at specific locations, and their sign depends on their direction.

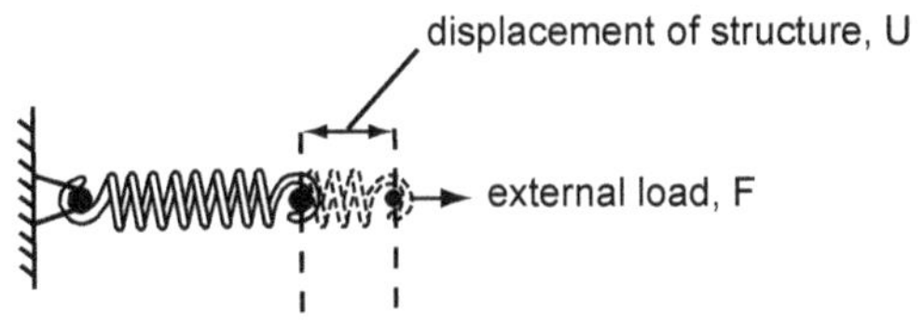

(a)

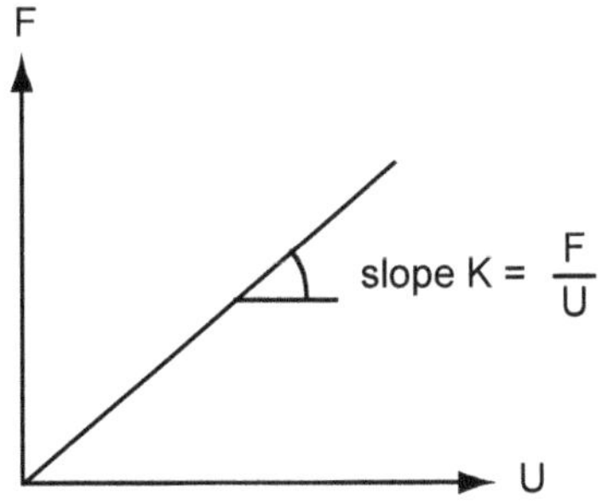

(b)

Figure 6.2 Load–displacement relationship for an elastic spring structure.

Thus, the most basic task of structural analysis is to determine the relationship between the nodal loads and the nodal displacements of the structure. This relationship will be analogous to (6.1.1) for the simple spring structure, but there will be many interconnecting members and therefore many interrelated nodal displacements. Likewise, there can be many external loads because these may occur at any node. Therefore, a general linear relationship between all of the external loads and all of the nodal displacements will consist of a system of simultaneous equations, and such systems are best expressed in matrix notation. The load-displacement relation is therefore of the form

$$\mathbf{F} = \mathbf{KU} \tag{6.1.3}$$

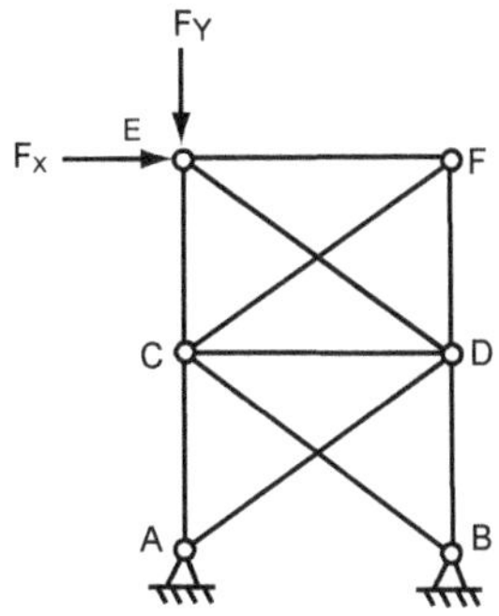

Figure 6.3 Statically indeterminate pin-jointed frame.

in which **F** and **U** are vectors of nodal loads and nodal displacements respectively, and the matrix **K** is termed the "stiffness matrix" of the structure, since it consists of the coefficients relating **F** to **U**. For the very simple structure of Fig. 6.2, which has only one possible displacement, this matrix is of order 1 × 1 and the vectors **F** and **U** contain only one term each.

It can now be seen that the basic task of determining the load-displacement relationship of a structure consists essentially in obtaining the stiffness matrix of the structure because once **K** has been determined, the solution for **U** follows immediately. As will now be shown, the load-displacement relationship for each member or *element* (we will use the latter term from now on) can also be expressed in terms of an element stiffness matrix $\mathbf{k}^e$. Next, it will be shown that the structure stiffness matrix **K** is in fact obtained ("assembled") by a systematic superposition of the element stiffness matrices, taking account of the particular way in which the elements are arranged and connected in the structure.

6.1.4 Stiffness Matrix of a Spring (or Bar) Element

For simplicity, we again consider the simple spring of Fig. 6.1 and we now regard it as a basic element, forming part of a larger structure. In this case, each of its two nodes would be connected to other elements in the structure and, in general, each node would be subjected to a nodal force and would undergo a nodal displacement, as shown in Fig. 6.4. For a pin-jointed bar element, the stiffness is *AE*/*L*. Positive forces and displacements are as shown in the figure. The vector of nodal forces is

$$\mathbf{f} = \begin{Bmatrix} f_1 \\ f_2 \end{Bmatrix}$$

and the vector of nodal displacements is

$$\mathbf{u} = \begin{Bmatrix} u_1 \\ u_2 \end{Bmatrix}$$

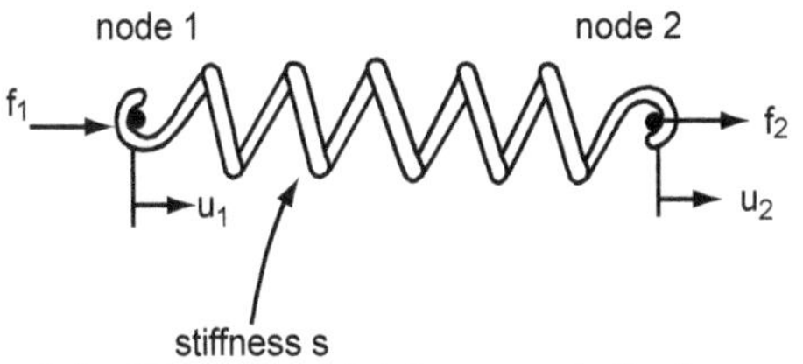

Figure 6.4 Element nodal forces and displacements for the spring element.

Therefore, the stiffness matrix for the spring element is of order 2×2 and the relationship between nodal forces and nodal displacements is of the form

$$\begin{Bmatrix} f_1 \\ f_2 \end{Bmatrix} = \begin{bmatrix} k_{11} & k_{12} \\ k_{21} & k_{22} \end{bmatrix} \begin{Bmatrix} u_1 \\ u_2 \end{Bmatrix} \quad \text{or} \quad \mathbf{f} = \mathbf{k}^e\mathbf{u} \qquad (6.1.4)$$

in which the superscript e indicates that this stiffness matrix is for a single element. We now seek to determine the individual terms of $\mathbf{k}^e$. These can be obtained by considering each nodal displacement in isolation (i.e., keeping the other nodal displacement 0) and using the law of material behavior to evaluate the two nodal forces. For example, let us consider nodal displacement u_1 and keep $u_2 = 0$, as shown in Fig. 6.5a. From equilibrium, the two nodal forces must be equal and opposite, and from the definition of $\mathcal{F}$ their magnitude must be the same as that of $\mathcal{F}$. Since $u_2 = 0$, the deformation is $d = u_2 - u_1 = -u_1$, and hence the material law of the spring requires that $\mathcal{F} = s\,(-u_1)$. Therefore, $f_1 = -\mathcal{F} = su_1$ and $f_2 = -f_1 = -su_1$. If these values are substituted into (6.1.4), the result is

$$\begin{Bmatrix} su_1 \\ -su_1 \end{Bmatrix} = \begin{bmatrix} k_{11} & k_{12} \\ k_{21} & k_{22} \end{bmatrix} \begin{Bmatrix} u_1 \\ u_2 = 0 \end{Bmatrix}$$

from which

$$su_1 = k_{11}u_1 \qquad \text{or} \qquad k_{11} = s$$

and

$$-su_1 = k_{21}u_1 \qquad \text{or} \qquad k_{21} = -s$$

We have thus obtained the terms in the first column of $\mathbf{k}^e$. The terms in the second column are obtained in a similar manner by setting $u_1 = 0$ (Fig. 6.5b) and, by again observing that from equilibrium and the definition of the spring constant, s, the nodal forces are $f_1 = -su_2$ and $f_2 = su_2$. Substituting these values into (6.1.4) and again noting that $u_1 = 0$ gives

$$-su_2 = k_{12}u_2 \qquad \text{or} \qquad k_{12} = -s$$

together with

$$su_2 = k_{22}u_2 \qquad \text{or} \qquad k_{22} = s$$

Therefore, the complete element stiffness matrix for the spring element is

$$\mathbf{k}^e = \begin{bmatrix} s & -s \\ -s & s \end{bmatrix} \qquad (6.1.5a)$$

Hence, the relationship between element nodal forces and element nodal displacements is

$$\begin{Bmatrix} f_1 \\ f_2 \end{Bmatrix} = \begin{bmatrix} s & -s \\ -s & s \end{bmatrix} \begin{Bmatrix} u_1 \\ u_2 \end{Bmatrix} \quad \text{or} \quad \mathbf{f} = \mathbf{k}^e\mathbf{u} \qquad (6.1.5b)$$

It may be seen that the stiffness matrix is symmetric. This is true of all stiffness matrices, whether for a single element or for an entire structure. The symmetry comes from (or is an alternative way of stating) Maxwell's reciprocal theorem. Further examination of $\mathbf{k}^e$ would also reveal that it is singular; the reason for this will be explained subsequently.

It may also be noted that in obtaining $\mathbf{k}^e$, we made use of the laws of material behavior and of equilibrium within the element. For such a simple element, the law of compatibility hardly arises but it is satisfied; the spring's internal deformation (stretching or shortening) is linear and continuous.

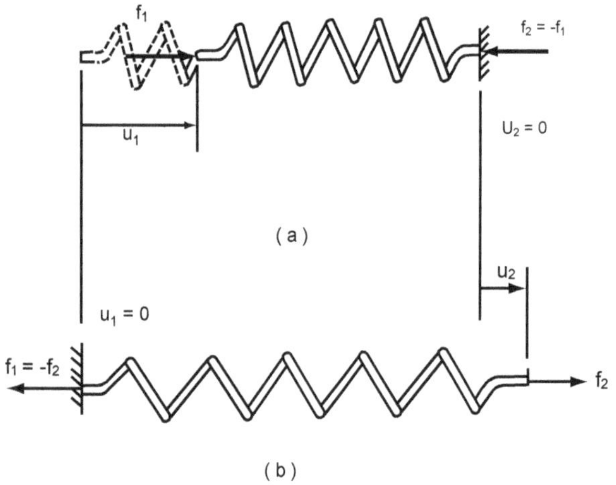

Figure 6.5 Calculation of $\mathbf{k}^e$ by imposing individual nodal displacements.

6.1.5 Assembling the Structure Stiffness Matrix

The next point to consider is how the stiffness matrix for a structure can be obtained from the element stiffness matrices of its constituent elements. To achieve this, we make use of the element force-displacement relationship, (6.1.5), of which the element stiffness matrix is the crucial part, together with the two laws that must be met at the structure level—those of equilibrium and of compatibility. Both laws are to be applied at the nodes of the structure (because everything is being described in terms of nodal values) and so we need a structure node numbering system, quite apart from the element node numbering system of Fig. 6.4. We will also need a sign convention for the applied loads **F** and for the structure nodal displacements **U**.

Consider, for example, the two-element structure of Fig. 6.6. The structure nodes are labeled A, B, and C (letters are chosen instead of numbers to emphasize the distinction between the element system and the structure system) and both the loads **F** and the displacements **U** are positive to the right. Since there are three nodes, there may be up to three applied (external) loads, as shown in the figure, but there could just as well be only one or two.*

Also, each load will act in a particular direction. In the example, loads F_B and F_C act towards the right and load F_A to the left. Hence the numerical value of the latter would be negative.

The law of equilibrium requires that at each node, there must be equilibrium between the external applied load at that node (if any) and the sum of the element forces at that node. Therefore (see Fig. 6.6)

$$
\begin{aligned}
F_A &= f_{a1} \\
F_B &= f_{a2} + f_{b1} \\
F_C &= f_{b2}
\end{aligned}
$$

Next, we use the element force-displacement relationship, that is, the element stiffness expressions of (6.1.5), to express the element nodal forces **f** in terms of the element nodal displacements **u**. The foregoing equations then become

*For the structural analysis, the applied loads are known quantities, although it will often be the task of the structural analyst to calculate them before commencing the analysis.

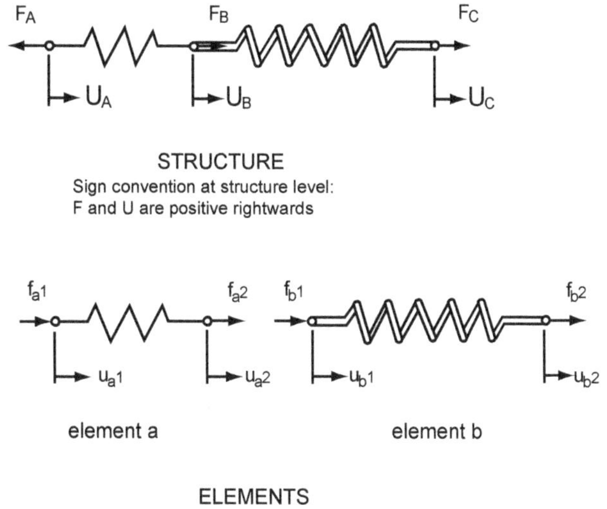

Figure 6.6 Two-element structure.

$$
\begin{aligned}
F_A &= s_a u_{a1} - s_a u_{a2} \\
F_B &= -s_a u_{a1} + s_a u_{a2} + s_b u_{b1} - s_b u_{b2} \\
F_C &= -s_b u_{b1} + s_b u_{b2}
\end{aligned}
$$

The next step is to impose interelement compatibility of the *structure* nodal displacements. This consists essentially in relating the element displacements **u** to the structure displacements **U**. In this example, the relationships are

$$
\begin{aligned}
u_{a1} &= U_A \\
u_{a2} &= u_{b1} = U_B \\
u_{b2} &= U_C
\end{aligned}
$$

Substituting these gives

$$
\begin{aligned}
F_A &= s_a U_A \qquad - s_a U_B \\
F_B &= -s_a U_A + (s_a + s_b) U_B - s_b U_c \\
F_C &= \qquad - s_b U_B + s_b U_C
\end{aligned}
$$

which are the desired equations relating the applied loads to the structure displacements. Therefore, the coefficients of these equations constitute the individual terms of the structure stiffness matrix **K**, introduced in (6.1.3). In matrix form, the above equations are

$$
\begin{Bmatrix} F_A \\ F_B \\ F_C \end{Bmatrix} = \begin{bmatrix} s_a & -s_a & 0 \\ -s_a & s_a + s_b & -s_b \\ 0 & -s_b & s_b \end{bmatrix} \begin{Bmatrix} U_A \\ U_B \\ U_C \end{Bmatrix}
$$

or $$\mathbf{F} = \mathbf{KU} \qquad (6.1.6)$$

It may be seen that **K** is an assemblage which includes each and every one of the individual terms of $\mathbf{k}^a$ and $\mathbf{k}^b$, assembled together in a systematic manner. Hence, the laws of equilibrium and compatibility are equivalent to a rule or procedure by which the terms of the $\mathbf{k}^e$ matrices are assembled to form **K**. If we can specify this procedure in a suitably general and practical form, we can then obtain **K** by a direct assembly process, without having to explicitly impose these laws. To achieve this, we first show that the assembly process is essentially a superposition of the element stiffness matrices, and then we present a procedure for assembling **K** which is computationally practical and efficient.

To demonstrate the superposition of the element stiffness matrices, we first write down the full set of equilibrium equations for each element.

Element a Element b

$$\begin{Bmatrix} f_A \\ f_B \end{Bmatrix} = \begin{bmatrix} s_a & -s_a \\ -s_a & s_a \end{bmatrix} \begin{Bmatrix} U_A \\ U_B \end{Bmatrix} \qquad \begin{Bmatrix} f_B \\ f_C \end{Bmatrix} = \begin{bmatrix} s_b & -s_b \\ -s_b & s_b \end{bmatrix} \begin{Bmatrix} U_B \\ U_C \end{Bmatrix}$$

Although the two $\mathbf{k}^e$ matrices are of the same order, they may not be added directly since they relate to different sets of displacements. However, by inserting rows and columns of zeros, both may be expanded in such a way that they each relate to the three displacements U_A, U_B, and U_C.

$$\begin{Bmatrix} f_A \\ f_B \\ f_C \end{Bmatrix} = \begin{bmatrix} s_a & -s_a & 0 \\ -s_a & s_a & 0 \\ 0 & 0 & 0 \end{bmatrix} \begin{Bmatrix} U_A \\ U_B \\ U_C \end{Bmatrix}$$

$$\begin{Bmatrix} f_A \\ f_B \\ f_C \end{Bmatrix} = \begin{bmatrix} 0 & 0 & 0 \\ 0 & s_b & -s_b \\ 0 & -s_b & s_b \end{bmatrix} \begin{Bmatrix} U_A \\ U_B \\ U_C \end{Bmatrix}$$

We now impose equilibrium of forces which says that the sum of the element forces at each node is equal to the external load at that node (or equals zero if there is no external load). Performing the summation gives

$$\begin{Bmatrix} F_A \\ F_B \\ F_C \end{Bmatrix} = \begin{bmatrix} s_a & -s_a & 0 \\ -s_a & s_a + s_b & -s_b \\ 0 & -s_b & s_b \end{bmatrix} \begin{Bmatrix} U_A \\ U_B \\ U_C \end{Bmatrix}$$

From this, we can see that the way in which the individual terms of $\mathbf{k}^a$ and $\mathbf{k}^b$ are combined together (assembled) to form $\mathbf{K}$ depends entirely on how the physical elements are arranged (or assembled) in the actual structure. In explaining the assembly procedure in more detail, we shall identify each of the individual terms in a stiffness matrix by its row and column, and we note here that in $\mathbf{K}$, these rows and columns correspond to structure nodes, whereas in $\mathbf{k}^a$ and $\mathbf{k}^b$ they correspond to element nodes. Also, the arrangement of the elements in the structure may be described by specifying the structure node where each element node is located. That is, for each node of each element, the corresponding structure node is specified, and this summarizes the basic layout of the structure. For this example, the structure node corresponding to each element node is shown in Fig. 6.7 immediately below the element stiffness matrices.

The foregoing summation of matrices shows that each term in a particular row and column position of $\mathbf{K}$ (corresponding to structure nodes I and J, say) is the sum of all of the $\mathbf{k}^e$ terms whose row and column numbers (i.e., whose element node numbers) correspond to (are located at) structure nodes I and J. Thus, a particular $\mathbf{k}^e$ term, say k_{lm}, is added into the K_{IJ} position of $\mathbf{K}$, where I and J are the structure nodes at which the element nodes l and m are located in the structure.

With this rule, it is no longer necessary to expand each of the element matrices up to the full structure size and then perform matrix addition. Instead, we may think of each of the element matrices as being "broken up" and its individual terms being inserted directly into their proper position in $\mathbf{K}$, according to an "assembly plan" which is simply a list of the structure nodes corresponding to the element nodes of that element. With this procedure, the only information which is required is the list of structure node numbers for each element. This list is usually referred to as the *location vector*. There is one location vector for each element, and it is usually a very short vector since its length equals the number of nodes in the element.

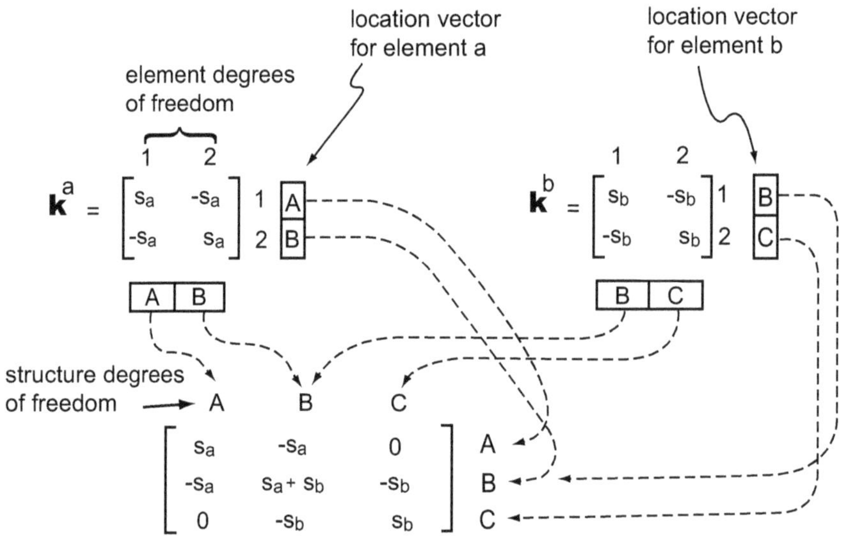

Figure 6.7 Location vectors and assembly of the structure stiffness matrix.

The location vectors for the above example are shown in Fig. 6.7.

Because each element stiffness matrix is symmetric, the resulting structure stiffness matrix is likewise always symmetric. Therefore, only half of it needs to be assembled.

In the simple example presented here, the elements are all in-line and there is only one component of displacement at each node. With more practical elements, there are several components of displacement at each node and so in later sections of this chapter, and also in Chapter 7, the foregoing definitions will need to be generalized. Nevertheless, these simple in-line spring elements illustrate all of the basic principles of matrix stiffness analysis and provide a useful introduction.

6.1.6 Solution Procedure

After the structure stiffness matrix has been fully assembled, an apparent paradox arises—in every case the matrix is singular! This would seem to indicate that the equations for the structure's nodal displacements cannot be solved and that the displacements are indeterminate. This apparent paradox has a simple explanation—the structure has not yet been tied down or given any points of support, and in this situation the applied loads would cause the entire structure to move as a rigid body, thus indeed giving indeterminate displacements. It is then necessary to provide sufficient restraints on the structure to prevent this rigid body motion. These restraints are applied by specifying zero values for some of the structure's displacements. In practice, structures often have more restraints than the minimum number. If so, they must all be specified because the structure would not otherwise be accurately represented. The full set of restraints is referred to as the support conditions. In the simple two-element structure of Fig. 6.6, there is only one possible type of rigid body motion—horizontal movement—and hence only one restraint is required for this structure. For example, let us say that the structure is supported at node A, so that $U_A = 0$. Then (6.1.6) can be rewritten in partitioned form as follows:

$$\begin{Bmatrix} F_A \\ \hline F_B \\ F_C \end{Bmatrix} = \left[\begin{array}{c|cc} s_a & -s_a & 0 \\ \hline -s_a & s_a + s_b & -s_b \\ 0 & -s_b & s_b \end{array}\right] \begin{Bmatrix} U_A = 0 \\ \hline U_B \\ U_C \end{Bmatrix} \qquad (6.1.7)$$

The system of equations contains two unknown displacements, U_B and U_C, and an unknown reaction, F_A. F_B and F_C are known applied loads. Using standard matrix manipulation, we have

$$\{F_A\} = s_a\{U_A\} + [-s_a \quad 0]\begin{Bmatrix} U_B \\ U_C \end{Bmatrix}$$

$$\begin{Bmatrix} F_B \\ F_C \end{Bmatrix} = \begin{Bmatrix} -s_a \\ 0 \end{Bmatrix}\{U_A\} + \begin{bmatrix} s_a + s_b & -s_b \\ -s_b & s_b \end{bmatrix}\begin{Bmatrix} U_B \\ U_C \end{Bmatrix}$$

and since U_A is zero we have

$$\{F_A\} = [-s_a \quad 0]\begin{Bmatrix} U_B \\ U_C \end{Bmatrix} \qquad (6.1.8)$$

together with

$$\begin{Bmatrix} F_B \\ F_C \end{Bmatrix} = \begin{bmatrix} s_a + s_b & -s_b \\ -s_b & s_b \end{bmatrix}\begin{Bmatrix} U_B \\ U_C \end{Bmatrix} \qquad (6.1.9)$$

Equation (6.1.9) consists of two equations in the two unknowns U_B and U_C and may be solved for these values which, when substituted in (6.1.8), give the value of the unknown reaction F_A.

It should be noted that (6.1.9) may be obtained directly from (6.1.7) simply by deleting the rows and columns of **K** corresponding to zero displacements.

Once the displacements have been obtained, the internal forces in the elements may be determined from the law of material (or internal) behavior of the element, that is, the relationship between internal force and internal deformation. For a spring, this relationship is simply $\mathcal{F} = sd$ where $\mathcal{F}$ is the internal force in the spring and d is the change in length of the spring. Hence, for the two elements we have

$$\begin{aligned} \mathcal{F}_a &= s_a(u_{a2} - u_{a1}) \\ \mathcal{F}_b &= s_b(u_{b2} - u_{b1}) \end{aligned} \qquad (6.1.10)$$

Since the elements are all in-line, the required element displacements may be obtained by simply substituting the appropriate structure displacements.

$$\begin{aligned} \mathcal{F}_a &= s_a(U_B - U_A) \\ \mathcal{F}_b &= s_b(U_C - U_B) \end{aligned}$$

This completes the solution process. It will be helpful at this point to summarize the principal steps in matrix stiffness analysis because these remain basically the same regardless of the type of structure and the types of element which are used. The seven principal steps are given in Table 6.1. The transformation referred to in Step 2 was not necessary in the foregoing example. It is introduced and explained in the next section. Step 7 will be explained in Section 6.3.5.

Table 6.1 Principal Steps in Matrix Stiffness Analysis

1. Define the structural model (structure coordinate system, nodes, node numbers, element types and loads).
2. Determine each element stiffness matrix $\mathbf{k}^e$ and transform it to structure coordinates (whereupon it is denoted $\mathbf{K}^e$).
3. Assemble the structure stiffness matrix $\mathbf{K}$ from the individual element stiffness matrices. $\mathbf{K}$ is the coefficient matrix of the system of equilibrium equations.
4. Construct the load vector (right hand side of the system) by applying the external loads to the structure nodes. Then apply the support conditions to the system.
5. Solve for the structure nodal displacements $\mathbf{\Delta}$ and then, if desired, the reaction forces at the supports.
6. For each element: (a) transform the nodal displacements $\mathbf{\Delta}$ from structure coordinates to element coordinates, denoted as $\boldsymbol{\delta}$; and (b) use the internal force (or stress) matrix of that element to calculate the internal forces (or stresses).
7. For each element for which equivalent nodal loads were used to represent distributed loads, superimpose the "fixed-end" or local internal force.

6.1.7 Numerical Example

A simple numerical example is given to illustrate the procedure. The structure shown in Fig. 6.8 consists of three springs and is supported at nodes A and D. If axial loads of 4 kN and 18 kN are applied at nodes B and C, respectively, as shown, determine the displacements at nodes B and C and the reaction forces at A and D.

The element stiffness matrices and location vectors are:

Element a: $\mathbf{K}^a = \begin{array}{c} \begin{array}{cc} A & B \end{array} \\ \begin{bmatrix} 2400 & -2400 \\ -2400 & 2400 \end{bmatrix} \end{array}$

Element b: $\mathbf{K}^b = \begin{array}{c} \begin{array}{cc} B & C \end{array} \\ \begin{bmatrix} 1600 & -1600 \\ -1600 & 1600 \end{bmatrix} \end{array}$

Element c: $\mathbf{K}^c = \begin{array}{c} \begin{array}{cc} C & D \end{array} \\ \begin{bmatrix} 1000 & -1000 \\ -1000 & 1000 \end{bmatrix} \end{array}$

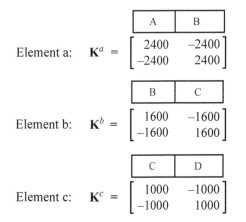

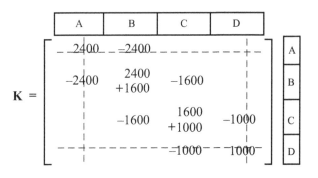

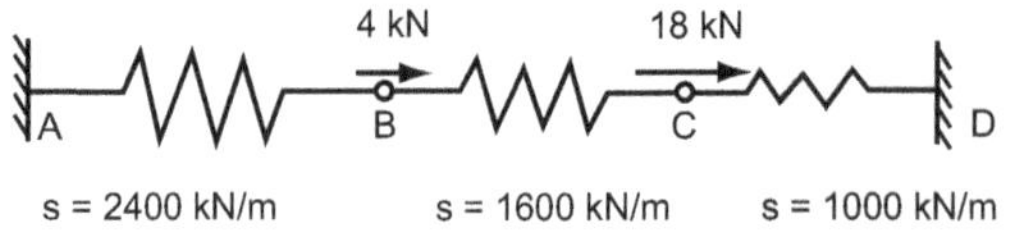

Figure 6.8 Numerical example, three-element structure.

Assembling these matrices to form the structure stiffness matrix:

$$\mathbf{K} = \begin{array}{c} \begin{array}{cccc} A & B & C & D \end{array} \\ \begin{bmatrix} 2400 & -2400 & & \\ -2400 & 2400+1600 & -1600 & \\ & -1600 & 1600+1000 & -1000 \\ & & -1000 & 1000 \end{bmatrix} \end{array} \begin{array}{c} A \\ B \\ C \\ D \end{array}$$

The boundary conditions are $U_A = U_D = 0$. Hence, these two rows and columns of $\mathbf{K}$ can be deleted as shown by the dashed lines in the above matrix. Also, the corresponding forces and displacements are excluded from the equilibrium equations, which are then

$$\begin{Bmatrix} 4 \\ 18 \end{Bmatrix} = \begin{bmatrix} 4000 & -1600 \\ -1600 & 2600 \end{bmatrix} \begin{Bmatrix} U_B \\ U_C \end{Bmatrix}$$

The solution to these equations is $U_B = 0.005$ m and $U_C = 0.010$ m. Knowing the displacements, we can now solve for each of the unknown reaction forces by using the equilibrium equation corresponding to that force. These are the same equations which were deleted from the original system of equations. They are

$$F_A = 2400\,U_A - 2400\,U_B + 0\,U_C + 0\,U_D$$

and

$$F_D = 0\,U_A + 0\,U_B - 1000\,U_C + 1000\,U_D$$

Substituting the displacements gives $F_A = -12$ kN and $F_D = -10$ kN. Knowing all of the forces permits an equilibrium check to be made, so as to verify the solution (not an essential step, but highly recommended):

$$F_A + F_B + F_C + F_D = -12 + 4 + 18 - 10 = 0$$

6.2 PIN-JOINTED FRAMES

6.2.1 Transformation of Coordinates

At the beginning of this chapter, it was shown that a pin-ended bar is equivalent to a spring. Its law of internal force-deformation is

$$\mathscr{F} = \frac{AE}{L} d \qquad (36)$$

where $\mathscr{F}$ is the tension or compression in the bar and d is the change in length of the bar. The physical stiffness is AE/L and so from (6.1.5b), the relationship between nodal forces acting on the element and nodal displacements of the element is

$$\begin{Bmatrix} f_1 \\ f_2 \end{Bmatrix} = \frac{AE}{L} \begin{bmatrix} 1 & -1 \\ -1 & 1 \end{bmatrix} \begin{Bmatrix} u_1 \\ u_2 \end{Bmatrix} \quad \text{or} \quad \mathbf{f} = \mathbf{k}^e \mathbf{u} \qquad (6.2.1)$$

which establishes $\mathbf{k}^e$, the element stiffness matrix for this element. In this equation, the forces and displacements are always axial regardless of how the bar may be orientated in the structure. That is, (6.2.1) is expressed in "local" or "element" coordinates (x, y) in which the x-axis is always aligned with the element, as shown in Fig. 6.9. In this text, all quantities expressed in element coordinates are denoted by a lower case letter (**f**, **k**, **u**, etc.).

In structural frameworks, the members occur at various angles to one another, and it is necessary to allow for this. In particular, in order to assemble the structure stiffness matrix, it is first necessary to express the stiffness matrix of each element of the structure, not in terms of its own element coordinates, but in terms of a "global" or "structure" coordinate system, that is, a coordinate system that is used as the basic reference system for the structure as a whole.

In Fig. 6.9, a pin-ended bar element is inclined at an angle χ to the global system, where χ is positive when measured counterclockwise from the global X-axis to the element x-axis. Axes x and y are the local or element coordinates and X and Y are the structure coordinates. The respective displacements are u and v and U and V, and the respective forces are f_x, f_y, and F_x, and F_y. Since an axial displacement u of an element generally possesses both a U and V component in the structure coordinate system, it is necessary to expand (6.2.1):

$$\begin{Bmatrix} f_{x1} \\ f_{y1} \\ f_{x2} \\ f_{y2} \end{Bmatrix} = \frac{AE}{L} \begin{bmatrix} 1 & 0 & -1 & 0 \\ 0 & 0 & 0 & 0 \\ -1 & 0 & 1 & 0 \\ 0 & 0 & 0 & 0 \end{bmatrix} \begin{Bmatrix} u_1 \\ v_1 \\ u_2 \\ v_2 \end{Bmatrix} \qquad (6.2.2)$$

or

$$\mathbf{f} = \mathbf{k}^e \boldsymbol{\delta}$$

in which the symbols **f** and $\boldsymbol{\delta}$ represent the vectors of nodal forces and nodal displacements. That is

$$\mathbf{f} = \begin{Bmatrix} f_{x1} \\ f_{y1} \\ f_{x2} \\ f_{y2} \end{Bmatrix} \quad \text{and} \quad \boldsymbol{\delta} = \begin{Bmatrix} u_1 \\ v_1 \\ u_2 \\ v_2 \end{Bmatrix} \qquad (6.2.3)$$

Since a pin-ended bar can only carry an axial load, $f_{y1} = f_{y2} = 0$. It may be seen from Fig. 6.10 that the local and global system of forces at node 1 are related by the expressions

$$\begin{aligned} f_{x1} &= F_{X1} \cos\chi + F_{Y1} \sin\chi \\ f_{y1} &= -F_{X1} \sin\chi + F_{Y1} \cos\chi \end{aligned}$$

and similar expressions apply for node 2. Hence if we define

$$\mu = \sin\chi; \qquad \lambda = \cos\chi \qquad (6.2.4)$$

then the full relationship between the two systems of forces is

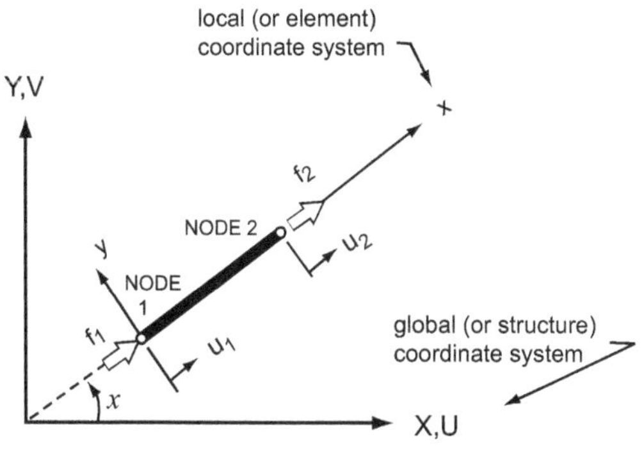

Figure 6.9 Coordinate systems.

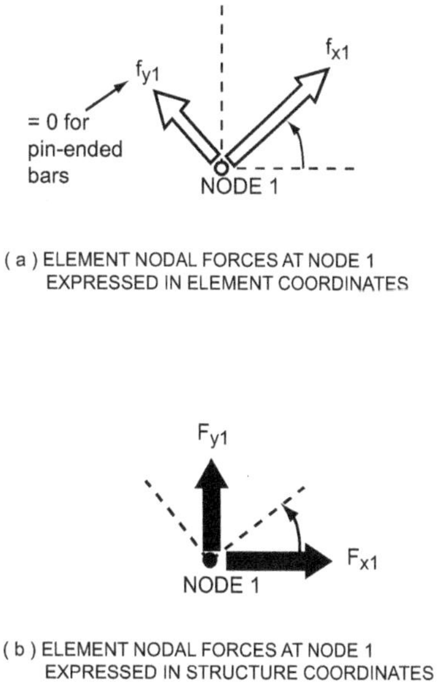

Figure 6.10 Transformation of nodal forces.

$$\begin{Bmatrix} f_{x1} \\ f_{y1} \\ f_{x2} \\ f_{y2} \end{Bmatrix} = \begin{bmatrix} \lambda & \mu & 0 & 0 \\ -\mu & \lambda & 0 & 0 \\ 0 & 0 & \lambda & \mu \\ 0 & 0 & -\mu & \lambda \end{bmatrix} \begin{Bmatrix} F_{X1} \\ F_{Y1} \\ F_{X2} \\ F_{Y2} \end{Bmatrix}$$

or

$$\mathbf{f} = \mathbf{TF} \tag{6.2.5}$$

where **T** is called the transformation matrix. It may be shown that this transformation matrix has the very useful property that its inverse is equal to its transpose, that is,

$$\mathbf{T}^{-1} = \mathbf{T}^T$$

For orthogonal systems, the transformation between local and global displacements is the same as that between the two sets of forces, namely

$$\boldsymbol{\delta} = \mathbf{T}\boldsymbol{\Delta} \tag{6.2.6}$$

where

$$\boldsymbol{\Delta} = \begin{Bmatrix} U_1 \\ V_1 \\ U_2 \\ V_2 \end{Bmatrix}$$

A general transformation involves rotations of three axes x, y, z relative to the structure axes X, Y, Z. The x-axis is at angles θ_{xX}, θ_{xY}, and θ_{xZ} measured from the axes X, Y, and Z as shown in Fig. 6.11, and the cosines of these three angles are the direction cosines of χ with respect to X, Y, and Z. They are denoted by λ_{11}, λ_{12}, and λ_{13}. Similarly, the direction cosines of y and z are λ_{21}, λ_{22}, λ_{23} and λ_{31}, λ_{32}, λ_{33} respectively. Thus, the nodal transformation matrix λ for transforming forces or deflections at each node from structure coordinates to element coordinates is

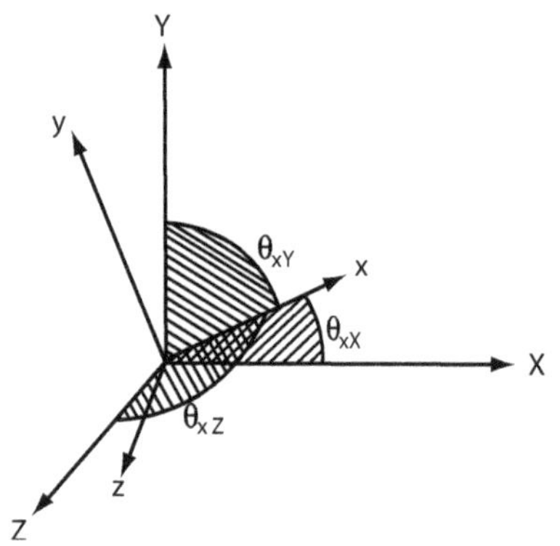

Figure 6.11 General transformation of axes.

$$\boldsymbol{\lambda} = \begin{bmatrix} \cos\theta_{xX} & \cos\theta_{xY} & \cos\theta_{xZ} \\ \cos\theta_{yX} & \cos\theta_{yY} & \cos\theta_{yZ} \\ \cos\theta_{zX} & \cos\theta_{zY} & \cos\theta_{zZ} \end{bmatrix} \tag{6.2.7}$$

which would be used as follows

$$\mathbf{f}_i = \boldsymbol{\lambda}\mathbf{F}_i \quad \text{and} \quad \boldsymbol{\delta}_i = \boldsymbol{\lambda}\boldsymbol{\Delta}_i$$

where i denotes the node number.

As with **T**, we have $\lambda^{-1} = \lambda^T$, and therefore the reverse transformations are

$$\mathbf{F}_i = \boldsymbol{\lambda}^T\mathbf{f}_i \quad \text{and} \quad \boldsymbol{\Delta}_i = \boldsymbol{\lambda}^T\boldsymbol{\delta}_i$$

The transformation matrix θ applies at each node, and hence for a pin-jointed bar element the full transformation matrix is

$$\mathbf{T} = \begin{bmatrix} \boldsymbol{\lambda} & \mathbf{0} \\ \mathbf{0} & \boldsymbol{\lambda} \end{bmatrix}$$

If the bar element lies in the x,y plane in element coordinates and in the X,Y plane in structure coordinates, then θ reduces to the simple 2×2 matrix given in (6.2.5)

$$\boldsymbol{\lambda} = \begin{bmatrix} \cos\theta_{xX} & \cos\theta_{xY} \\ \cos\theta_{yX} & \cos\theta_{yY} \end{bmatrix}$$

$$= \begin{bmatrix} \cos\theta_{xX} & \cos\left(\frac{\pi}{2} - \theta_{xX}\right) \\ \cos\left(\frac{\pi}{2} + \theta_{xX}\right) & \cos\theta_{xX} \end{bmatrix}$$

$$= \begin{bmatrix} \cos\chi & \sin\chi \\ -\sin\chi & \cos\chi \end{bmatrix}$$

$$= \begin{bmatrix} \lambda & \mu \\ -\mu & \lambda \end{bmatrix}$$

Using the transformation matrix **T**, we can derive an expression for the element stiffness matrix in global coordinates, $\mathbf{K}^e$. The basic force-displacement relationship for the element is given by (6.2.2). If we substitute (6.2.5) on the left-hand side and (6.2.6) for $\boldsymbol{\delta}$ on the right-hand side, we obtain

$$\mathbf{TF} = \mathbf{k}^e\mathbf{T}\boldsymbol{\Delta}$$

Premultiplying both sides by $\mathbf{T}^{-1}$ and using the fact that $\mathbf{T}^{-1} = \mathbf{T}^T$ gives

$$\mathbf{F} = \mathbf{T}^T\mathbf{k}^e\mathbf{T}\boldsymbol{\Delta} = \mathbf{K}^e\boldsymbol{\Delta}$$

and therefore the element stiffness matrix in local coordinates $\mathbf{k}^e$ is transformed to the element stiffness matrix in global coordinates $\mathbf{K}^e$ by the compound transformation

$$\mathbf{K}^e = \mathbf{T}^T\mathbf{k}^e\mathbf{T} \qquad (6.2.8)$$

The result of this double multiplication is

$$\mathbf{K}^e = \frac{AE}{L}\begin{bmatrix} \lambda^2 & \lambda\mu & -\lambda^2 & -\lambda\mu \\ \lambda\mu & \mu^2 & -\lambda\mu & -\mu^2 \\ -\lambda^2 & -\lambda\mu & \lambda^2 & \lambda\mu \\ -\lambda\mu & -\mu^2 & \lambda\mu & \mu^2 \end{bmatrix} \qquad (6.2.9)$$

Unless the elements are all in-line, the transformation of the element stiffness matrix from $\mathbf{k}^e$ to $\mathbf{K}^e$ of (6.2.9) must always be performed before the structure stiffness matrix $\mathbf{K}$ can be assembled. Since the arrangement of elements differs for each structure, an element stiffness matrix, if it is to be of general use, must first be expressed in terms of its own local coordinates. Also, since the various elements in a structure generally have different sizes and orientations, the transformation to structure coordinates must be performed element by element. Hence, in the principal steps of matrix stiffness analysis given in Table 6.1, Step 2 consists of the transformation of each element stiffness matrix from $\mathbf{k}^e$ to $\mathbf{K}^e$, that is, from local to global coordinates. For all standard elements, a general expression for $\mathbf{k}^e$ will be available and so the first part of Step 2—determining $\mathbf{k}^e$—is merely a matter of substituting that element's dimensions and physical properties.

After assembling and solving the complete system of equations (Steps 3 to 5 in Table 6.1), the next step in a structural analysis is the calculation of the internal forces in each element, $\mathcal{F} = sd$, where s is the element's material stiffness and d is the deformation. For this it is necessary to convert back to element coordinates in order to determine the element deformation d. For a pin-ended bar, the material stiffness is the axial stiffness AE/L. Hence, (6.1.10) becomes

$$\mathcal{F}_e = \left(\frac{AE}{L}\right)_e (u_2 - u_1)_e \qquad (6.2.10)$$

where u_1 and u_2 are the nodal displacements in element coordinates for element e. Thus, before transforming the nodal displacements of the structure from structure coordinates to element coordinates, it is first necessary to allocate these displacements to the appropriate ends (nodes) of each element. Then, for each element we perform Step 6 of Table 6.1. Step 6(a) is to transform from $\boldsymbol{\Delta}$ to $\boldsymbol{\delta}$ using (6.2.6), and Step 6(b) is to calculate the internal forces (or stresses). Step 6(a) uses (6.2.6), which in expanded form is

$$\begin{Bmatrix} u_1 \\ v_1 \\ u_2 \\ v_2 \end{Bmatrix}_e = \begin{bmatrix} \lambda & \mu & 0 & 0 \\ -\mu & \lambda & 0 & 0 \\ 0 & 0 & \lambda & \mu \\ 0 & 0 & -\mu & \lambda \end{bmatrix} \begin{Bmatrix} U_1 \\ V_1 \\ U_2 \\ V_2 \end{Bmatrix}$$

Then substitute u_1 and u_2 into (6.2.10) to get the axial force, as follows.

$$\mathcal{F}_e = \left(\frac{AE}{L}\right)_e (u_2 - u_1) \qquad (6.2.11)$$

6.2.2 Degrees of Freedom: Demonstration of Complete Method

The complete process will now be demonstrated for the simple three-bar pin-jointed frame illustrated in Fig. 6.12. The frame is supported at A and C and all members have the same cross-sectional area A and Young's modulus E. The frame carries vertical and horizontal loads as shown in the figure. Since there are three nodes and two independent displacements U and V at each node, the vectors of

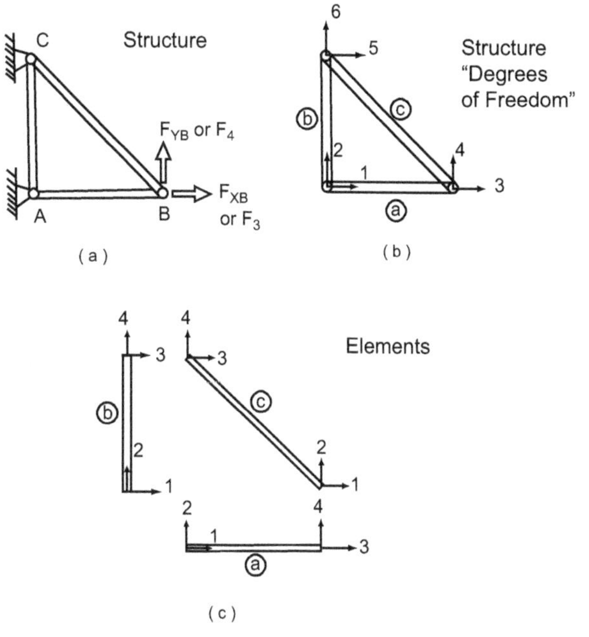

Figure 6.12 Three-member pin-jointed frame. (*a*) Structure; (*b*) structure degrees of freedom; and (*c*) element degrees of freedom.

applied loads and displacements each contain six terms as follows:

$$\mathbf{F} = \begin{Bmatrix} F_{XA} \\ F_{YA} \\ F_{XB} \\ F_{YB} \\ F_{XC} \\ F_{YC} \end{Bmatrix} \quad \text{and} \quad \mathbf{\Delta} = \begin{Bmatrix} U_A = 0 \\ V_A = 0 \\ U_B \\ V_B \\ U_C = 0 \\ V_C = 0 \end{Bmatrix}$$

At this point, it is appropriate to introduce the notion of "degrees of freedom." The degrees of freedom of a structure, or of an element, are basically the same as the nodal displacements, but they are slightly more general in concept. Specifically, the degrees of freedom of a structure or element are all of the independent nodal parameters that are used to describe all of the particular ways or modes in which the structure or element can deflect. There are two reasons for introducing this new term:

1. To allow for additional and more general types of "displacement." The most common example is a degree of freedom which is a derivative of an ordinary displacement. For example, in the case of a beam, it could be the slope dv/dx or the curvature d^2v/dx^2 at the end (or node) of the beam; these additional "displacements" are derivatives of the beam's ordinary lateral displacement v. The use of this form of displacement will be discussed in the next section.
2. To provide a suitable terminology for the continuous sequential numbering of all of the displacements in a structure (or element), as an alternative to the "two-level" system used thus far which specifies first the node number and then the particular nodal force or displacement at that node. The sequential numbering system based on degrees of freedom is more in harmony with standard matrix notation and is also more suitable for the practical computer implementation of frame analysis and finite element analysis. Nevertheless, the two-level node numbering is often more suitable for explanatory purposes and we shall continue to make frequent use of it.

The degree of freedom approach is illustrated in Fig. 6.12b, which shows that the structure degrees of freedom are numbered sequentially within each node, moving from one node to another. Within each node, the numbering should follow some systematic and unchanging sequence (e.g., the X-direction first, then the Y-direction), whereas the nodes may be taken in any sequence. However, we shall see in the next section that the amount of computation can be reduced by a careful choice of node sequence.

Using the numbering system based on structure degrees of freedom, the vectors of applied loads and structure displacements become

$$\mathbf{F} = \begin{Bmatrix} F_1 \\ F_2 \\ F_3 \\ F_4 \\ F_5 \\ F_6 \end{Bmatrix} \quad \text{and} \quad \mathbf{\Delta} = \begin{Bmatrix} \Delta_1 \\ \Delta_2 \\ \Delta_3 \\ \Delta_4 \\ \Delta_5 \\ \Delta_6 \end{Bmatrix}$$

The expression for $\mathbf{K}^e$ for a typical element is given by (6.2.9), and to evaluate this, the direction cosines λ and μ for each element must be determined. Recalling that χ is measured counter-clockwise from the structure X-axis to the element x-axis, these values are given in Table 6.2.

Table 6.2 Terms Needed in (6.2.9)

Member	χ	λ	μ	λ^2	μ^2	$\lambda\mu$
a	0°	1	0	1	0	0
b	90°	0	1	0	1	0
c	135°	–0.707	0.707	½	½	– ½

There must be consistency between the choice of χ and the orientation of the element (i.e., the positioning of element nodes 1 and 2). The element χ axis runs from node 1 to node 2, and therefore χ must be defined at node 1. Thus for member c, if element node 1 is placed at structure node B, then $\chi = 135°$, measured counter-clockwise at node B. If element node 1 is placed at structure node C, then $\chi = 315°$ (or –45°).

Once the structure degrees of freedom have been defined, the location vectors can be determined by inspection. Sometimes, a diagram such as Fig. 6.12c is of assistance. For each element, one proceeds sequentially through the element degrees of freedom, expressed in structure coordinates, and records the corresponding structure degree of freedom. In a hand solution, it is helpful to write the location vector above and to the right of the transformed stiffness matrix for that element. For the three elements of the sample problem, the $\mathbf{K}^e$ matrices and the location vectors are as follows:

Element a:

$$\begin{array}{c} \boxed{1\;|\;2\;|\;3\;|\;4} \\ \mathbf{K}^a = \dfrac{AE}{L}\begin{bmatrix} 1 & 0 & -1 & 0 \\ 0 & 0 & 0 & 0 \\ -1 & 0 & 1 & 0 \\ 0 & 0 & 0 & 0 \end{bmatrix} \begin{array}{|c|} \hline 1 \\ 2 \\ 3 \\ 4 \\ \hline \end{array} \end{array}$$

Element b:

$$\begin{array}{c} \boxed{1\;|\;2\;|\;5\;|\;6} \\ \mathbf{K}^b = \dfrac{AE}{L}\begin{bmatrix} 0 & 0 & 0 & 0 \\ 0 & 1 & 0 & -1 \\ 0 & 0 & 0 & 0 \\ 0 & -1 & 0 & 1 \end{bmatrix} \begin{array}{|c|} \hline 1 \\ 2 \\ 5 \\ 6 \\ \hline \end{array} \end{array}$$

Element c:

$$\begin{array}{c} \boxed{3\;|\;4\;|\;5\;|\;6} \\ \mathbf{K}^c = \dfrac{AE}{2\sqrt{2}L}\begin{bmatrix} 1 & -1 & -1 & 1 \\ -1 & 1 & 1 & -1 \\ -1 & 1 & 1 & -1 \\ 1 & -1 & -1 & 1 \end{bmatrix} \begin{array}{|c|} \hline 3 \\ 4 \\ 5 \\ 6 \\ \hline \end{array} \end{array}$$

These matrices are each "broken up" and their terms are assembled together to form the stiffness matrix for the structure. However, before we can do so it is necessary to redefine the assembly process in more general terms. In the initial presentation, in Section 6.1, the example contained only one displacement at each node and so it was possible to use node numbers to identify the rows and columns of the stiffness matrices—both element and structure. It has now been shown that practical problems have multiple displacements (degrees of freedom) at each node, and therefore it is the degree of freedom number which defines the rows and columns of the stiffness matrices—both element and structure. Hence, the rule given earlier for performing the assembly process is easily generalized by substituting the term "degree of freedom" in place of "node number." Also, in the earlier example, the elements were all in-line and so there was no need to transform $\mathbf{k}^e$ to $\mathbf{K}^e$. Hence, to obtain the general form of the rule it is also necessary to replace $\mathbf{k}^e$ with $\mathbf{K}^e$. The result is the following general rule for assembling the structure stiffness matrix **K**:

> For each transformed element stiffness matrix in turn, say $\mathbf{K}^e$, each term of $\mathbf{K}^e$, say $\mathbf{K}^e_{lm}$, is added into the K_{IJ} position of **K**, where I and J are the structure degrees of freedom corresponding to the element degrees of freedom l and m. The correspondence between I and l and between J and m is indicated by the (previously constructed) location vector for that element.

Applying this rule for the present example gives:

$$\mathbf{K} = \begin{bmatrix} a_{11}+b_{11} & a_{12}+b_{12} & a_{13} & a_{14} & b_{13} & b_{14} \\ a_{21}+b_{12} & a_{22}+b_{22} & a_{23} & a_{24} & b_{23} & b_{24} \\ a_{31} & a_{32} & a_{33}+c_{11} & a_{34}+c_{12} & c_{13} & c_{14} \\ a_{41} & a_{42} & a_{43}+c_{21} & a_{44}+c_{22} & c_{23} & c_{24} \\ b_{31} & b_{32} & c_{31} & c_{32} & b_{33}+c_{33} & b_{34}+c_{34} \\ b_{41} & b_{42} & c_{41} & c_{42} & b_{43}+c_{43} & b_{44}+c_{44} \end{bmatrix}$$

where a_{ij} represents the term in the ith row and the jth column of $\mathbf{K}^a$, and similarly for b_{ij} and c_{ij}. For each term in an element stiffness matrix, the numbers in the location vector above and to the right of that term indicate the column and row in **K** where it is to be inserted. For example, the term b_{34} of $\mathbf{K}^b$ is inserted into the $\mathbf{K}_{56}$ location of **K**. Later, when $\mathbf{K}^c$ is being processed, the term c_{34} is also inserted into the $\mathbf{K}_{56}$ location, being added to b_{34}. In terms of the actual numerical values, the fully assembled structure stiffness matrix is

$$\mathbf{K} = \frac{AE}{L}\begin{bmatrix} 1 & 0 & -1 & 0 & 0 & 0 \\ 0 & 1 & 0 & 0 & 0 & -1 \\ -1 & 0 & 1.35 & -0.35 & -0.35 & 0.35 \\ 0 & 0 & -0.35 & 0.35 & 0.35 & -0.35 \\ 0 & 0 & -0.35 & 0.35 & 0.35 & -0.35 \\ 0 & -1 & 0.35 & -0.35 & -0.35 & 1.35 \end{bmatrix}$$

where for brevity, 0.35 is written in place of $1/[2(2)^{1/2}] = 0.35355$.

The next step is to impose the boundary conditions and to reduce the system of equations accordingly. For this structure, both displacements are zero at node A and at node C. These correspond to degrees of freedom 1, 2, 5, and 6, and therefore these four rows and columns of **K**, and the corresponding terms in **F** and Δ, should be deleted. The reduced system is

$$\begin{Bmatrix} F_3 \\ F_4 \end{Bmatrix} = \frac{AE}{L}\begin{bmatrix} 1.35 & -0.35 \\ -0.35 & 0.35 \end{bmatrix}\begin{Bmatrix} \Delta_3 \\ \Delta_4 \end{Bmatrix}$$

At this point, we must insert the actual values of F_3 and F_4, and for AE/L. Let us say that F_3 and F_4 are both 20 kN, A = 100 mm², E = 200 kN/mm², and for elements a and b, L = 2000 mm, from which AE/L = 10 kN/mm². For element c, AE/L = 7.071 kN/mm². We now have a complete 2 × 2 system of equations which can be solved for the structure displacements by a numerical solution method (such as Gaussian elimination) taking full advantage of the symmetry of **K**. The result is

$$\Delta_3 = 4 \text{ mm and } \Delta_4 = 9.657 \text{ mm}$$

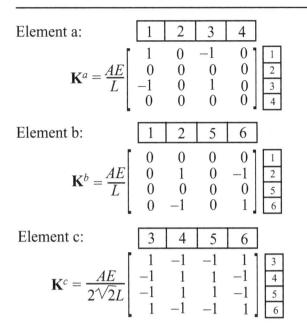

If desired, the reaction forces at nodes A and C can now be obtained by substituting Δ_3 and Δ_4 into the equations corresponding to the deleted rows of **K**. Since Δ_1, Δ_2, Δ_3, and Δ_4 are all zero, all terms in **K** that multiply them may be omitted. The result is

$$\begin{Bmatrix} F_1 \\ F_2 \\ F_5 \\ F_6 \end{Bmatrix} = \begin{bmatrix} -1 & 0 \\ 0 & 0 \\ -0.35 & 0.35 \\ 0.35 & -0.35 \end{bmatrix} \begin{Bmatrix} \Delta_3 = 4 \\ \Delta_4 = 9.657 \end{Bmatrix} = \begin{Bmatrix} -4 \\ 0 \\ 2 \\ -2 \end{Bmatrix} \qquad (65)$$

Now that we have solved for the global displacements **Δ**, we go back "down" to the element level. For each element, we perform Step 6 in Table 6.1, which always has two parts: (a) transforming **Δ** into **δ** using (6.2.6), and (b) calculating the element internal force using (6.2.10). For example for element c:

Step 6(a)

$\lambda = -0.7071$ and $\mu = 0.7071$

Substituting for **T** and using (6.2.6)

$$\begin{Bmatrix} u_1 \\ v_1 \\ u_2 \\ v_2 \end{Bmatrix} = \mathbf{T}_c \Delta_c = 0.7071 \begin{bmatrix} -1 & 1 & 0 & 0 \\ -1 & -1 & 0 & 0 \\ 0 & 0 & -1 & -1 \\ 0 & 0 & -1 & -1 \end{bmatrix} \begin{Bmatrix} \Delta_3 = 4.000 \\ \Delta_4 = 9.657 \\ \Delta_5 = 0.000 \\ \Delta_6 = 0.000 \end{Bmatrix}$$

from which

$u_1 = 0.7071(-1 \times 4.000 + 1 \times 9.657) = 4.000$ and $u_2 = 0$.

Step 6(b)

From (6.2.10)

$$\mathscr{F}_c = \left(\frac{AE}{L}\right)_c (u_2 - u_1)$$
$$= 7.0711(0 - 4) = -28.28 \text{ kN (compression).}$$

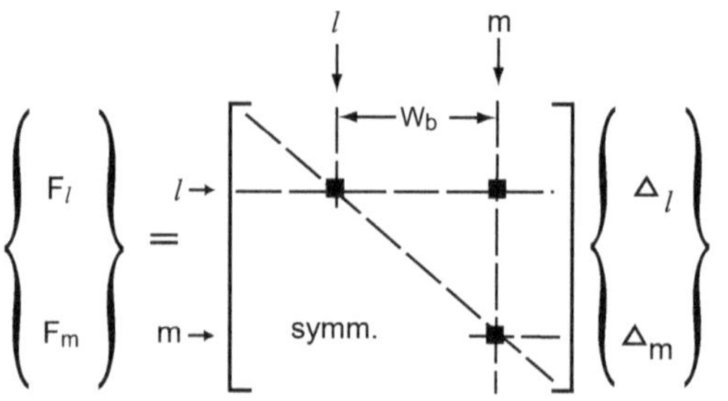

Figure 6.13 Bandwidth of the structural stiffness matrix.

6.2.3 Numbering Sequence of Nodes

If two displacements (or degrees of freedom) in the structure, Δ_l and Δ_m, are linked together by a bar element, this will give rise to terms in the structure stiffness matrix in locations K_{ll}, K_{mm}, K_{lm}, and K_{ml}, as shown in Fig. 6.13. If these two degrees of freedom have widely differing numbers, the equilibrium equations corresponding to F_l will have a wide "bandwidth" w_b, as shown in the figure. In the numerical solution of a system of equations, the amount of computation depends on the product Nw_b^2, where N is the total number of degrees of freedom and w_b is the average bandwidth. For a given structure, N is a constant and so for the present discussion, we may ignore the effect of N. Therefore, in the degree of freedom, numbering the order in which the nodes are considered should be such that the average difference in the degree of freedom numbers is as small as possible. The rule for achieving this is simple: the numbering should always proceed across the shortest width of the structure, thus only gradually moving along the length of the structure. This is illustrated in the 13-bar pin-jointed structure of Fig. 6.14. To simplify the discussion, we can use node numbers instead of degrees of freedom, because this will not alter the conclusion in any way. It simply means that since there are two degrees of freedom at each node, each square in the stiffness matrix diagram represents four terms. In the first case (Fig. 6.14a), this rule is not followed and w_b is 3.9, whereas if the rule is followed (Fig. 6.14b) the value is 2.7. Thus, the first node sequence requires $(3.9/2.7)^2 = 2.09$ times as much computation as the second sequence.

6.3 BEAM ELEMENT: RIGID-JOINTED FRAME ANALYSIS

6.3.1 Flexure-Only Beam Element

For several reasons, the two- and three-dimensional frameworks found in ships are rigid-jointed rather

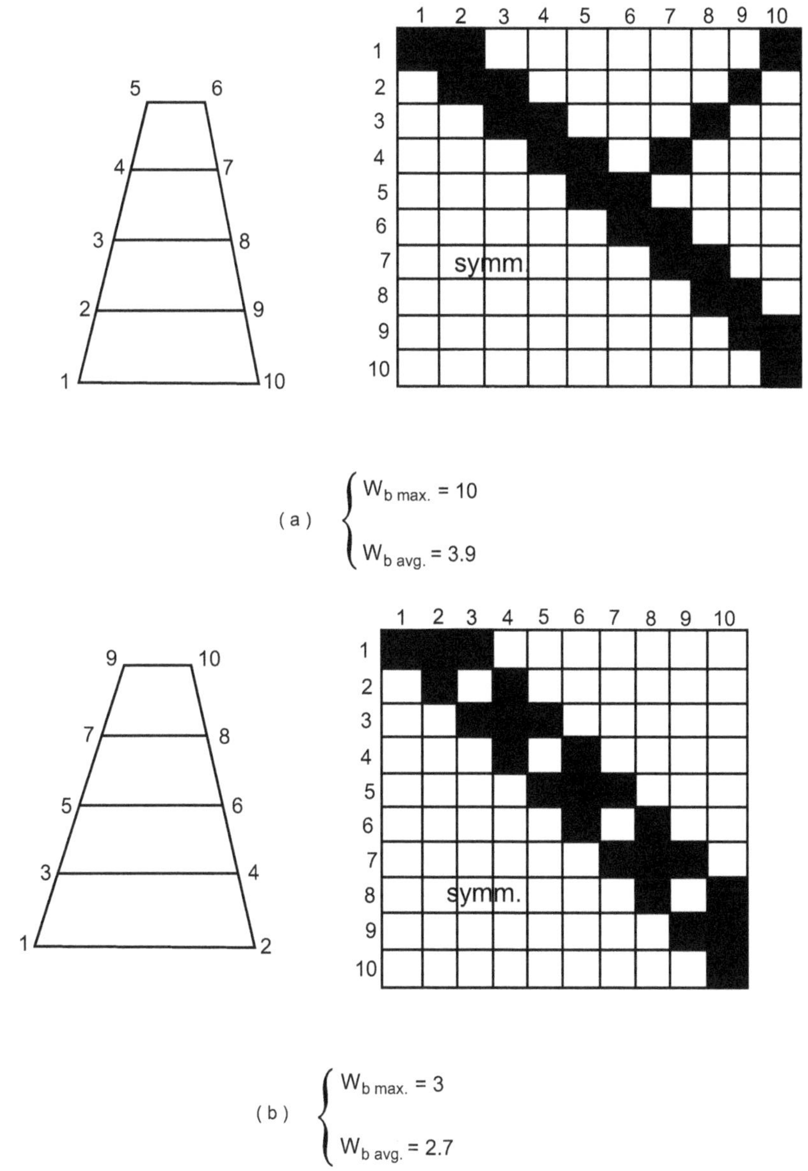

Figure 6.14 Effect of structure node numbering on bandwidth.

than pin-jointed: the need for water-tightness, the use of welding, and the large lateral loads which must be carried. The basic element for a rigid-jointed frame is the beam element shown in Fig. 6.15. The beam is loaded by forces and moments at each node and is assumed to be of uniform flexural rigidity *EI*.

By definition, the nodes of a beam element are the points where the external forces and moments act, and these are assumed to be in equilibrium with the internal forces which act along the beam's neutral axis. Therefore, unless special provision is made for nodal eccentricity, the nodes of a beam element are located at the centroid of the beam's cross section.

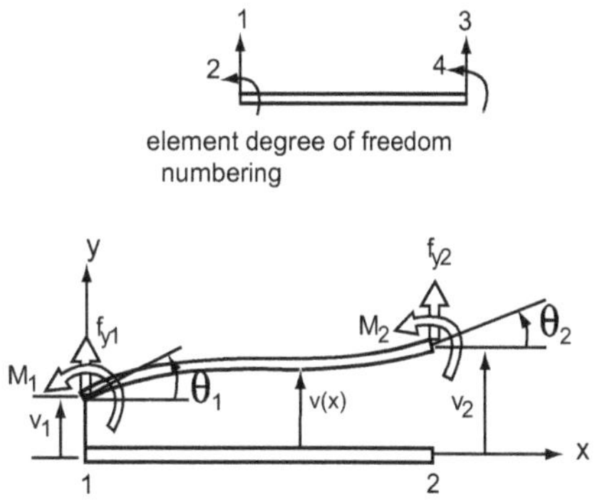

Figure 6.15 Flexure-only beam element.

The element stiffness matrix $\mathbf{k}^e$ may again be obtained by following the same procedure as for the bar element, that is, by imposing each of the nodal displacements in turn, determining the nodal forces set up, and then superimposing the results obtained from each individual case. Alternatively, since the element is a prismatic beam, we may make use of beam theory to obtain the required relationships between the nodal forces and displacements. Specifically, these relationships may be obtained from the well-known slope-deflection equations of beam theory. For the element shown in Fig. 6.15, the slope-deflection equations are

$$M_1 = \frac{6EI}{L^2}v_1 + \frac{4EI}{L}\theta_1 - \frac{6EI}{L^2}v_2 + \frac{2EI}{L}\theta_2$$

$$M_2 = \frac{6EI}{L^2}v_1 + \frac{2EI}{L}\theta_1 - \frac{6EI}{L^2}v_2 + \frac{4EI}{L}\theta_2$$

From vertical force equilibrium

$$f_{y1} = -f_{y2} = \frac{M_1 + M_2}{L}$$

$$= \frac{12EI}{L^3}v_1 + \frac{6EI}{L^2}\theta_1 - \frac{12EI}{L^3}v_2 + \frac{6EI}{L^2}\theta_2$$

In matrix notation, these equations are

$$\begin{Bmatrix} f_{y1} \\ M_1 \\ f_{y2} \\ M_2 \end{Bmatrix} = \frac{EI}{L^3}\begin{bmatrix} 12 & 6L & -12 & 6L \\ 6L & 4L^2 & -6L & 2L^2 \\ -12 & -6L & 12 & -6L \\ 6L & 2L^2 & -6L & 4L^2 \end{bmatrix}\begin{Bmatrix} v_1 \\ \theta_1 \\ v_2 \\ \theta_2 \end{Bmatrix} \quad (6.3.1)$$

or

$$\mathbf{f} = \mathbf{k}^e\,\boldsymbol{\delta}$$

which defines the nodal force vector $\mathbf{f}$, the nodal displacement vector $\boldsymbol{\delta}$, and the stiffness matrix $\mathbf{k}^e$ for a beam element in which only the flexural response is considered. It is important to note that in this element, axial forces and axial deformation are deliberately ignored. Hence, this element could only be used for structures in which the loads are carried purely by beam bending, such as multispan beams and laterally loaded flat grillages. In actual practice, an ordinary beam element would always be used, and this element is presented in the next section. The only purpose of the "flexure-only" beam element is to facilitate the explanation of nodal rotation and to simplify the worked examples.

The nodal displacements given above include the rotation θ at the end of the beam. It should be noted that in small-deflection beam theory, the quantity that is actually involved is the *slope* of the beam, that is, the derivative $v' = dv/dx$; for small deflections, this is equivalent to the angle of rotation, and the latter term is usually chosen because a rotation is easier to visualize as a form of displacement. However, for basic derivations, the derivative term is preferable because it shows that within the element, the deflected state of the beam is described completely by the general displacement function $v(x)$; the slope at any point—including the ends—is simply the derivative of $v(x)$ at that point. Thus the four nodal "displacements" are in reality the nodal values of $v(x)$ and of $v'(x)$, and this illustrates why the name "degree of freedom" is often used instead of "displacement."

For an arbitrarily orientated beam element, it is again necessary to expand $\mathbf{k}^e$ to allow for the element displacements v (which are always normal to the beam) to be transformed into U and V displacements in the global coordinate system. The nodal bending moments are unaffected. Consequently $\mathbf{T}$ takes the form

$$\mathbf{T} = \begin{bmatrix} \lambda & \mu & 0 & 0 & 0 & 0 \\ -\mu & \lambda & 0 & 0 & 0 & 0 \\ 0 & 0 & 1 & 0 & 0 & 0 \\ 0 & 0 & 0 & \lambda & \mu & 0 \\ 0 & 0 & 0 & -\mu & \lambda & 0 \\ 0 & 0 & 0 & 0 & 0 & 1 \end{bmatrix} \quad (6.3.2)$$

As shown in (6.2.8), the transformation of the element stiffness matrix from local to global coordinates is $\mathbf{T}^T\mathbf{k}^e\mathbf{T}$. For the beam element the result is

$$K^e = \frac{EI}{L^3}\begin{bmatrix} 12\mu^2 & & & & & \\ -12\lambda\mu & 12\lambda^2 & & & \text{symm.} & \\ -6\mu L & 6\lambda L & 4L^2 & & & \\ -12\mu^2 & 12\lambda\mu & 6\mu L & 12\mu^2 & & \\ 12\lambda\mu & -12\lambda^2 & -6\lambda L & -12\lambda\mu & 12\lambda^2 & \\ -6\mu L & 6\lambda L & 2L^2 & 6\mu L & -6\lambda L & 4L^2 \end{bmatrix} \quad (6.3.3)$$

From this point on, the use of this element for the structural analysis of a rigid-jointed frame would proceed along the remaining steps of Table 6.1: the overall stiffness matrix $\mathbf{K}$ is assembled, the support conditions are applied, and the resulting set of equations solved. However, before proceeding further in regard to the application of the beam element, we shall use it as an example to demonstrate the general method for defining an element and deriving its stiffness matrix.

6.3.2 General Method for Deriving An Element Stiffness Matrix

In presenting the general method, we choose the flexure-only beam element as the example because this element is quite simple and yet is sufficiently general to illustrate all of the basic steps in the definition of an element and the derivation of its stiffness matrix. In Chapter 7, it will be shown that in finite element theory, this same method is used to derive the element stiffness matrix for a wide variety of two- and three-dimensional finite elements. Hence, this method constitutes one of the basic foundations of finite element theory. In terms of the seven steps of matrix structural analysis given in Table 6.1, this method provides the material which is essential for Step 2, that is, the complete definition of the element properties, and especially the element stiffness matrix $\mathbf{k}^e$. The method consists of five steps, and in order to distinguish them from the seven basic steps of matrix structural analysis, we shall identify them by means of Roman numerals.

Step I: Select Suitable Displacement Function

The first step is to specify a displacement function which uniquely defines the state of displacement at all points within the element in terms of the nodal degrees of freedom. For a beam, the displacement within the element is $v(x)$. This displacement can conveniently be represented by a polynomial expression and, since the aim is to express $v(x)$ in terms of the nodal degrees of freedom $\boldsymbol{\delta}$, the assumed polynomial must contain one unknown coefficient for each degree of freedom possessed by the element. As defined previously, the beam element possesses four degrees of freedom

$$\boldsymbol{\delta} = \begin{Bmatrix} v_1 \\ v_1' \\ v_2 \\ v_2' \end{Bmatrix} \tag{6.3.4}$$

and therefore we choose a cubic polynomial to represent $v(x)$, since this has four coefficients.

$$v(x) = C_1 + C_2 x + C_3 x^2 + C_4 x^3 \tag{6.3.5}$$

In matrix form, this is

$$v(x) = \mathbf{H}(x)\mathbf{C} \tag{6.3.6}$$

where $\mathbf{H}(x) = [1 \quad x \quad x^2 \quad x^3]$

and $\mathbf{C}^T = [C_1 \quad C_2 \quad C_3 \quad C_4]$

Equation (6.3.4) defines the vector of nodal displacements $\boldsymbol{\delta}$. We now construct the *generalized displacement vector* $\boldsymbol{\delta}(x)$ which expresses the values of the displacements $v(x)$ and and slopes $v'(x)$ at any point within the element, in terms of the four (as yet undetermined) coefficients. Differentiating (6.3.5) gives

$$v'(x) = C_2 + 2C_3 x + 3C_4 x^2$$

and by writing this in matrix form and combining it with (6.3.6), we obtain

$$\boldsymbol{\delta}(x) = \begin{Bmatrix} v(x) \\ v'(x) \end{Bmatrix} = \begin{bmatrix} 1 & x & x^2 & x^3 \\ 0 & 1 & 2x & 3x^2 \end{bmatrix} \begin{Bmatrix} C_1 \\ C_2 \\ C_3 \\ C_4 \end{Bmatrix} \tag{6.3.7}$$

Step II: Relate General Displacements Within The Element To Nodal Displacements

The coefficients of the displacement function $\mathbf{C}$ are now expressed in terms of the nodal displacements $\boldsymbol{\delta}$ and hence by substituting into (6.3.7), the displacements at any point within the element are obtained in terms of the nodal displacements $\boldsymbol{\delta}$.

Since $\boldsymbol{\delta}(x)$ represents the displacement at any point x, the nodal displacements can be obtained from it by simply substituting the appropriate nodal coordinates into (6.3.7).

At node 1, $x = 0$ and therefore

$$v_1 = C_1 \quad \text{and} \quad v_1' = C_2$$

At node 2, $x = L$ and therefore

$$v_2 = C_1 + C_2 L + C_3 L^2 + C_4 L^3$$
$$v_2' = C_2 + 2C_3 L + 3C_4 L^2$$

Writing these in matrix form

$$\begin{Bmatrix} v_1 \\ v_1' \\ v_2 \\ v_2' \end{Bmatrix} = \begin{bmatrix} 1 & 0 & 0 & 0 \\ 0 & 1 & 0 & 0 \\ 1 & L & L^2 & L^3 \\ 0 & 1 & 2L & 3L^2 \end{bmatrix} \begin{Bmatrix} C_1 \\ C_2 \\ C_3 \\ C_4 \end{Bmatrix}$$

or $\quad \boldsymbol{\delta} = \mathbf{AC}$

Since the matrix $\mathbf{A}$ is known, this equation may be solved for the vector of unknown coefficients $\mathbf{C}$

$$\mathbf{C} = \mathbf{A}^{-1}\boldsymbol{\delta} \tag{6.3.8a}$$

and it may be shown that

$$\mathbf{A}^{-1} = \begin{bmatrix} 1 & 0 & 0 & 0 \\ 0 & 1 & 0 & 0 \\ \frac{-3}{L^2} & \frac{-2}{L} & \frac{3}{L^2} & \frac{-1}{L} \\ \frac{2}{L^3} & \frac{1}{L^2} & \frac{-2}{L^3} & \frac{1}{L^2} \end{bmatrix} \qquad (6.3.8b)$$

For future use, we note here that (6.3.6) and (6.3.8) can be combined to obtain an expression for the displacement within the element, $v(x)$, in terms of the nodal displacements $\boldsymbol{\delta}$. Combining these two equations gives

$$v(x) = \mathbf{H}(x)\mathbf{A}^{-1}\boldsymbol{\delta}$$

and the result of the matrix multiplication is

$$v(x) = \mathbf{N}(x)\boldsymbol{\delta} \qquad (6.3.9a)$$

in which $\mathbf{N}(x)$ is known as the *shape function* and is given by

$$\mathbf{N}(x) = [(1 - 3\xi^2 + 2\xi^3) \quad x(1 - 2\xi + \xi^2) \quad (3\xi^2 - 2\xi^3) \quad x(-\xi + \xi^2)]$$

$$\text{where } \xi = x/L. \qquad (6.3.9b)$$

Step III: Express the Internal Deformation in Terms of the Nodal Displacements

The definition of "deformation" of an element depends on which particular field of solid mechanics is involved. In plane elasticity, the deformation is strain $\varepsilon(x, y)$, whereas for beam bending, it is the curvature of the beam. In each case, the deformation is some form of derivative of the displacement; in plane elasticity, it is a first derivative ($\varepsilon_x = \partial u/\partial x$, etc.) whereas the curvature of a beam is (for small deflections) the second derivative of the beam displacement, $v''(x) = d^2v/dx^2$. Therefore, from equation (74)

$$v''(x) = 2C_3 + 6C_4x$$

or in matrix form

$$v''(x) = [0 \quad 0 \quad 2 \quad 6x]\begin{Bmatrix} C_1 \\ C_2 \\ C_3 \\ C_4 \end{Bmatrix}$$

From (6.3.8a)

$$v''(x) = [0 \quad 0 \quad 2 \quad 6x]\,\mathbf{A}^{-1}\boldsymbol{\delta}$$

and substitution for $\mathbf{A}^{-1}$ followed by matrix multiplication gives

$$v''(x) = \mathbf{B} \qquad (6.3.10)$$

where

$$\mathbf{B} = \left[\frac{-6}{L^2} + \frac{12x}{L^3} \quad \frac{-4}{L} + \frac{6x}{L^2} \quad \frac{6}{L^2} - \frac{12x}{L^3} \quad \frac{-2}{L} + \frac{6x}{L^2}\right]$$

Step IV: Express the Internal Force in Terms of the Nodal Displacements Using the Element's Law of Elastic Behavior

As with internal deformation, the definition of internal force depends on the class of problem being considered. For problems in two- and three-dimensional elasticity, it consists of the components of stress, whereas for beam bending, it is the internal bending moment in the beam, $\mathcal{M}_z(x)$.*

In this step, the goal is to express the internal force in terms of the nodal displacements. Since a relationship between the internal deformation and the nodal displacements $\boldsymbol{\delta}$ is already known, the internal force can be related to the nodal

*The use of script symbols for internal forces is especially helpful in the case of bending moment, $\mathcal{M}$, because it is necessary to distinguish it from an applied (nodal) moment M; these two quantities are not the same and they have a different sign convention. The term "bending moment" is used to describe the internal state of the beam, that is, its response to the external load. The deformation response is a curvature and so the sign of the internal force response (i.e., bending moment $\mathcal{M}$) is determined by the curvature of the beam: $\mathcal{M}_z(x)$ is positive if $v''(x)$ is positive. A nodal moment M is an applied load, a certain type of force (a couple) that acts at a particular location. It may be either an external load acting on the structure or an element nodal moment acting on a particular element. Either way, its sign is determined by the coordinate system for forces and deflections (using the right-hand rule), either at the structure level or at the element level. Thus, M_z is a moment acting about the z-axis, and it is positive if its rotation would give an advance in the $+z$-direction. Of course, at a node (say element node 1) the internal bending moment $\mathcal{M}_z(0)$, or $\mathcal{M}_{z1}$, would be equal in magnitude to the element nodal moment acting at that node, M_{z1}, but the sign could be different, and this could be a source of confusion and error if these two are not clearly distinguished. In small problems which are solved by hand, the possibility of confusion and error is perhaps not so serious but with large problems and for computer implementation, it is important to distinguish between these two quantities. In addition to using different symbols, in this text the phrase "bending moment" shall be used only for $\mathcal{M}_z(x)$, and never for M. Other terms, such as (element) nodal moment, generalized nodal force (or load), applied moment, and so on, which do not contain the phrase "bending moment," will be used for M.

displacements providing that the element's internal force-deformation relationship is known. For elasticity problems, this relationship is Hooke's Law. For beam bending, it is the moment-curvature relationship

$$\mathcal{M}_z(x) = EIv''(x) \tag{6.3.11}$$

Therefore, the expression for the internal force in terms of nodal displacements is

$$\mathcal{M}_z(x) = EI\,\mathbf{B}\boldsymbol{\delta}$$

From the expression for **B** in (6.3.10), it can be seen that in the beam element the bending moment varies linearly along the beam.

For any element, the expression for internal force is used not only in deriving the element stiffness matrix but also in any structural analysis which makes use of that element, because Step 6 of matrix structural analysis is the calculation of the internal forces in the various elements by transforming the nodal displacements back into element coordinates and substituting them into the expression for internal force. From equation (89), the expression is

$$\mathcal{M}_z(x) = \mathfrak{b}[-6L + 12x \quad -4L^2 + 6Lx \quad 6L - 12x \quad -2L^2 + 6Lx]\begin{Bmatrix} v_1 \\ v_1' \\ v_2 \\ v_2' \end{Bmatrix} \tag{6.3.12a}$$

in which $\mathfrak{b} = \dfrac{EI}{L^3}$

In many cases it is desired, or is sufficient, to obtain the internal forces only at the nodal points. In the beam element, for example, the linear distribution of $\mathcal{M}_z(x)$ means that it is sufficient to calculate $\mathcal{M}_z(x)$ only at the nodes. An expression for these nodal values is easily obtained by simply substituting $x = 0$ and $x = L$ into the foregoing expression. The result is

$$\begin{Bmatrix} \mathcal{M}_z(0) \\ \mathcal{M}_z(L) \end{Bmatrix} = \mathbf{S}\boldsymbol{\delta} \tag{6.3.12b}$$

where

$$\mathbf{S} = \frac{EI}{L^3}\begin{bmatrix} -6L & -4L^2 & 6L & -2L^2 \\ 6L & 2L^2 & -6L & 4L^2 \end{bmatrix}$$

Step V: Obtain the Element Stiffness Matrix $\mathbf{k}^e$ by Relating Nodal Forces to Nodal Displacements

In this step, the internal forces are replaced by statically equivalent nodal forces **f** and hence the nodal forces are related to the nodal displacements $\boldsymbol{\delta}$, thereby defining the required element stiffness matrix $\mathbf{k}^e$.

The principle of virtual work is used to determine the set of nodal forces that is statically equivalent to the internal forces. The condition of equivalence may be expressed as follows: during any virtual displacement imposed on the element, the total external work done by the nodal forces must equal the total internal work done by the internal forces. While the virtual displacement is being imposed, all actual forces, both nodal and internal, are constant. The arbitrary virtual nodal displacements are represented by the symbol $\boldsymbol{\delta}^*$; that is

$$\boldsymbol{\delta}^* = \begin{Bmatrix} v_1^* \\ v_1'^* \\ v_2^* \\ v_2'^* \end{Bmatrix}$$

We begin with the external virtual work. If the above virtual nodal displacements are imposed on an element in which the actual nodal forces are

$$\mathbf{f} = \begin{Bmatrix} f_1 \\ M_1 \\ f_2 \\ M_2 \end{Bmatrix}$$

then the virtual work done externally is

$$W_{\text{ext}} = v_1^* f_1 + v_1'^* M_1 + v_2^* f_2 + v_2'^* M_2 = \boldsymbol{\delta}^{*T}\mathbf{f}$$

Next we evaluate the internal virtual work. If an arbitrary virtual curvature $v''(x)^*$ is imposed on a beam of length L in which the actual internal bending moment is $\mathcal{M}_z(x)$, the internal virtual work is

$$W_{\text{int}} = \int_0^L \left(v''(x)^*\right)^T \mathcal{M}_z(x)\,dx$$

If the virtual curvature is imposed indirectly, in terms of virtual nodal displacements $\boldsymbol{\delta}^*$, the corresponding value of v''^* is

$$(v''(x)^*)^T = (\mathbf{B}\,\boldsymbol{\delta}^*)^T = \boldsymbol{\delta}^{*T}\mathbf{B}^T$$

In matrix algebra, expanding the transpose of a product causes the terms to be reversed. The actual internal bending moment can be expressed in terms of the actual curvature $v''(x)$ by means of (6.3.11) and so the internal work resulting from the virtual nodal displacement is

$$W_{\text{int}} = \int_0^L \boldsymbol{\delta}^{*T}\mathbf{B}^T EI v''(x)\, dx$$

Finally, from (6.3.10) the actual curvature $v''(x)$ can be expressed in terms of the actual nodal displacement $\boldsymbol{\delta}$. The result is

$$W_{\text{int}} = \int_0^L \boldsymbol{\delta}^{*T}\mathbf{B}^T EI\mathbf{B}\ \boldsymbol{\delta}\, dx$$

Equating the internal and external work and setting the arbitrary virtual displacement $\boldsymbol{\delta}^*$ to unity (we can do this because the value is arbitrary) gives

$$\mathbf{f} = \left[\int_0^L \mathbf{B}^T EI\mathbf{B}\, dx\right]\boldsymbol{\delta}$$

$$\mathfrak{a} = \frac{AE}{L} \quad \text{and} \quad \mathfrak{b} = \frac{EI}{L^3} \tag{6.3.14}$$

In terms of $\mathfrak{a}$ and $\mathfrak{b}$, the stiffness matrix of the ordinary beam element is

$$\mathbf{k}^e = \begin{bmatrix} \mathfrak{a} & 0 & 0 & -\mathfrak{a} & 0 & 0 \\ 0 & 12\mathfrak{b} & 6\mathfrak{b}L & 0 & -12\mathfrak{b} & 6\mathfrak{b}L \\ 0 & 6\mathfrak{b}L & 4\mathfrak{b}L^2 & 0 & -6\mathfrak{b}L & 2\mathfrak{b}L^2 \\ -\mathfrak{a} & 0 & 0 & \mathfrak{a} & 0 & 0 \\ 0 & -12\mathfrak{b} & -6\mathfrak{b}L & 0 & 12\mathfrak{b} & -6\mathfrak{b}L \\ 0 & 6\mathfrak{b}L & 2\mathfrak{b}L^2 & 0 & -6\mathfrak{b}L & 4\mathfrak{b}L^2 \end{bmatrix} \tag{6.3.15}$$

which can be recognized as the "flexure-only" stiffness matrix plus the axial stiffness terms.

As before, the element stiffness matrix in structure coordinates is $\mathbf{T}^T\mathbf{k}^e\mathbf{T}$ with $\mathbf{T}$ defined in (6.3.2). The result is

$$\mathbf{K}^e = \begin{bmatrix} \mathfrak{a}\lambda^2 + 12\mathfrak{b}\mu^2 & & & & \text{symmetric} & \\ (\mathfrak{a} - 12\mathfrak{b})\lambda\mu & \mathfrak{a}\mu^2 + 12\mathfrak{b}\lambda^2 & & & & \\ -6\mathfrak{b}L\mu & 6\mathfrak{b}L\lambda & 4\mathfrak{b}L^2 & & & \\ -\mathfrak{a}\lambda^2 - 12\mathfrak{b}\mu^2 & -(\mathfrak{a} - 12\mathfrak{b})\lambda\mu & 6\mathfrak{b}L\mu & \mathfrak{a}\lambda^2 + 12\mathfrak{b}\mu^2 & & \\ -(\mathfrak{a} - 12\mathfrak{b})\lambda\mu & -\mathfrak{a}\mu^2 - 12\mathfrak{b}\lambda^2 & -6\mathfrak{b}L\lambda & (\mathfrak{a} - 12\mathfrak{b})\lambda\mu & \mathfrak{a}\mu^2 + 12\mathfrak{b}\lambda^2 & \\ -6\mathfrak{b}L\mu & 6\mathfrak{b}L\lambda & 2\mathfrak{b}L^2 & 6\mathfrak{b}L\mu & -6\mathfrak{b}L\mu & 4\mathfrak{b}L^2 \end{bmatrix} \tag{6.3.16}$$

Therefore, since this equation relates nodal force $\mathbf{f}$ to nodal displacement $\boldsymbol{\delta}$, the quantity in brackets must be the element stiffness matrix $\mathbf{k}^e$. Substituting for $\mathbf{B}$ and performing the integration produces the same stiffness matrix which was derived earlier using the slope-deflection equations and was presented in (6.3.1).

6.3.3 Ordinary Beam Element

For general frame structures in which the members are not orientated orthogonally, the axial stiffness and the flexural stiffness are not independent and the former must be included in the element stiffness matrix. Hence, the usual beam element has six degrees of freedom, as shown in Fig. 6.16. These are usually taken in the following order.

$$\boldsymbol{\delta} = \begin{Bmatrix} u_1 \\ v_1 \\ \theta_1 \\ u_2 \\ v_2 \\ \theta_2 \end{Bmatrix} \tag{6.3.13}$$

Accordingly, in the 6×6 element stiffness matrix the axial stiffness terms occur in the first and fourth rows and columns, and they are the same as for the pin-ended bar element. For convenience, we define the following axial and flexural parameters

The expression for the internal forces is simply a combination of (6.2.10) for the axial force and (6.3.12) for the bending moment. In the former, we shall divide by the beam cross-sectional area A to obtain axial stress σ_x. If desired, an expression for the shear force can also be included since the shear force is simply the derivative of the internal bending moment: $\mathcal{Q}(x) = (d/dx)\mathcal{M}_z(x)$. From (6.3.10) and (6.3.11)

$$\mathcal{Q}(x) = \frac{EI}{L^3}\,[12 \quad 6L \quad -12 \quad 6L\,]\boldsymbol{\delta} \tag{6.3.17}$$

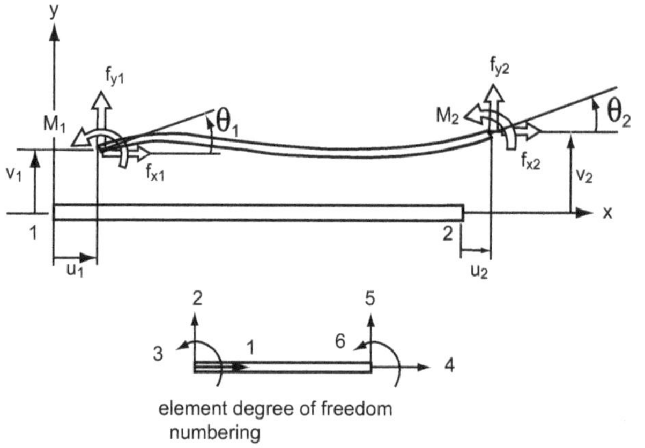

Figure 6.16 Ordinary beam element.

Thus, the shear force is constant along the beam element, as is the axial force. Therefore, it is sufficient to have an expression for the forces only at the nodes. This is

$$\begin{Bmatrix} \sigma_{x1} \\ \mathcal{Q}_1 \\ \mathcal{M}_{z1} \\ \sigma_{x2} \\ \mathcal{Q}_2 \\ \mathcal{M}_{z2} \end{Bmatrix} = \begin{bmatrix} \frac{-E}{L} & 0 & 0 & \frac{E}{L} & 0 & 0 \\ 0 & 12\mathfrak{b} & 6\mathfrak{b}L & 0 & -12\mathfrak{b} & 6\mathfrak{b}L \\ 0 & -6\mathfrak{b}L & -4\mathfrak{b}L^2 & 0 & 6\mathfrak{b}L & -2\mathfrak{b}L^2 \\ \frac{-E}{L} & 0 & 0 & \frac{E}{L} & 0 & 0 \\ 0 & 12\mathfrak{b} & 6\mathfrak{b}L & 0 & -12\mathfrak{b} & 6\mathfrak{b}L \\ 0 & 6\mathfrak{b}L & 2\mathfrak{b}L^2 & 0 & -6\mathfrak{b}L & 4\mathfrak{b}L^2 \end{bmatrix} \begin{Bmatrix} u_1 \\ v_1 \\ \theta_1 \\ u_2 \\ v_2 \\ \theta_2 \end{Bmatrix} \tag{6.3.18}$$

Note that although these are nodal values, the script form is used for $\mathcal{M}$ and $\mathcal{Q}$ in order to indicate that they are internal forces.

6.3.4 Restraints and Specified Displacements

Earlier, when discussing boundary conditions and restraints, it was shown that the imposition of zero displacement at a node can be achieved by simply deleting that row and column from the structure equilibrium equations. This is quite suiTable 6. if the equations are being solved by hand (any reduction in the number of equations is then most valuable) but it is not so suiTable 6. for a computer solution. Deletion of rows and columns from **K** would involve repacking it within the computer, and this is time-consuming. A more efficient technique is to replace the diagonal terms of **K** corresponding to the zero displacement by unity and replace the rest of the terms in the corresponding row and column with zeros.

A second method is available which requires only one multiplication for each degree of freedom that is to be eliminated. In this method, each diagonal term of **K** that corresponds to a zero displacement is multiplied by a large number, such as 10^{25}. For example, suppose the displacement which is degree of freedom number I is to be zero. After the multiplication, the equilibrium equation corresponding to that degree of freedom is

$$F_1 = K_{I1}\Delta_1 + K_{I2}\Delta_2 + \cdots + 10^{25}K_{II}\Delta_I + \cdots + K_{IN}\Delta_N$$

and thus

$$\Delta_I = \frac{F_1 - (K_{I1}\Delta_1 + K_{I2}\Delta_2 + \cdots + K_{IN}\Delta_N)}{10^{25}K_{II}}$$

Since the diagonal term is now so much greater than the off-diagonal terms, the value of the displacement Δ_I is very close to zero.

If a displacement Δ_I is required to have some specified nonzero value d, then this effect too may be allowed for using either of these methods, together with an adjustment of the load vector. If the first method is used, then after placing the 1 and the zeros in **K**, the F_I term in the force vector is replaced by the value of the prescribed displacement d. Alternatively, using the second method, the F_I term is replaced by the diagonal term multiplied by the prescribed displacement. Thus writing out the equation for row I

$$10^{25}K_{II}d = K_{I1}\Delta_1 + K_{I2}\Delta_2 + \cdots + 10^{25}K_{II}\Delta_I + \cdots + K_{IN}\Delta_N$$

Consequently

$$\Delta_1 = \frac{10^{25}K_{II}d - (K_{I1}\Delta_I + K_{I2}\Delta_2 + \cdots + K_{IN}\Delta_N)}{10^{25}K_{II}}$$

Since the diagonal term $10^{25}K_{II}$ is so much larger than the off-diagonal terms, then to a good approximation $\Delta_I = d$.

6.3.5 Distributed Loads: Equivalent Nodal Loads

In frame analysis, and also in finite element analysis, all quantities are specified or represented in terms of their values at the nodes. Until now we have assumed that the loads occur only at nodes, whereas in most frame structures, and certainly in ships, the loads are distributed loads. The technique for dealing with such loads is to divide the analysis into two parts: a local analysis which deals only with the internodal loading and response of the loaded beam, and an overall analysis which deals with the response of the overall structure. A separate solution is obtained for each, and then the solutions are superimposed. For this purpose, the distributed load is divided into two parts: an internal load which is applied within the beam and which accounts for the internodal bending, and an external load which is applied at the boundaries of the loaded beam (i.e., at the nodes) and which acts on the entire structure. Since it is a nodal load, the usual frame analysis method may be used for the solution.* Each load must give an accurate solution within its own field, and the two loads taken together must match the original load. In the case of beam bending, this is achieved by defining the internal load as the original distributed load plus equilibrium end forces and moments that correspond to complete fixity at the boundary, that is, zero deflection and zero rotation at the ends (or nodes) of the loaded beam. In this way, the internal problem satisfies equilibrium and may be dealt with in isolation from the overall solution. Thus, the internal problem is simply the given distributed load acting on a clamped beam, and for standard types of distributed loading, the solution is already available. In particular, the bending moments at the ends of the beam are simply the "fixed-end moments" for that loading. Figure 6.17 gives the fixed-end reaction forces and moments for

*This technique is used in several branches of continuum mechanics. For example, in fluid mechanics, a flow is often represented by the combination of a "near field" flow which accounts for the local aspects (e.g., flow around a body) and a "far field" flow which accounts for the effects in the rest of the field.

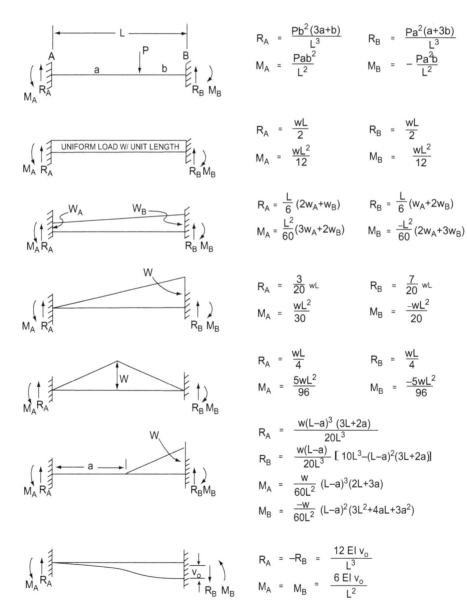

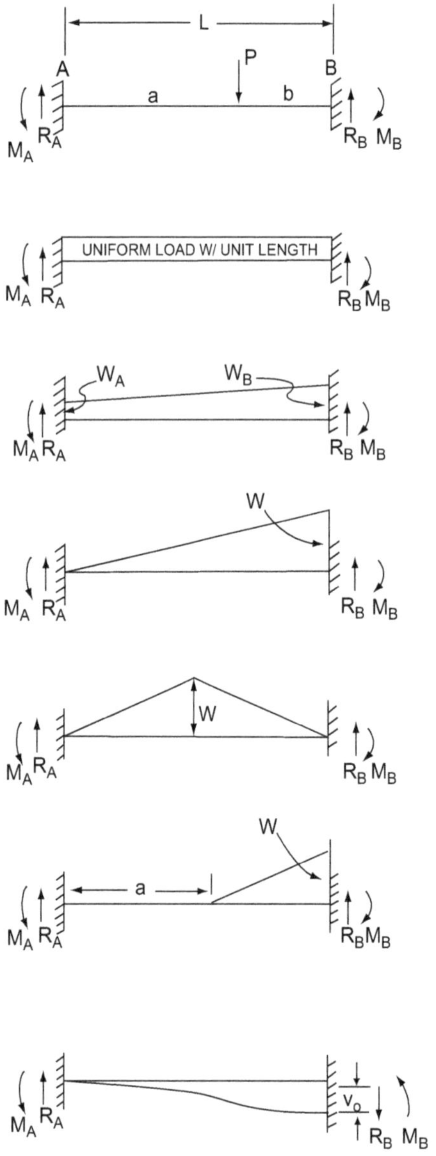

$R_A = \frac{Pb^2(3a+b)}{L^3}$ $\quad R_B = \frac{Pa^2(a+3b)}{L^3}$

$M_A = \frac{Pab^2}{L^2}$ $\quad M_B = -\frac{Pa^2b}{L^2}$

$R_A = \frac{wL}{2}$ $\quad R_B = \frac{wL}{2}$

$M_A = \frac{wL^2}{12}$ $\quad M_B = \frac{wL^2}{12}$

$R_A = \frac{L}{6}(2w_A+w_B)$ $\quad R_B = \frac{L}{6}(w_A+2w_B)$

$M_A = \frac{L^2}{60}(3w_A+2w_B)$ $\quad M_B = \frac{-L^2}{60}(2w_A+3w_B)$

$R_A = \frac{3}{20}wL$ $\quad R_B = \frac{7}{20}wL$

$M_A = \frac{wL^2}{30}$ $\quad M_B = \frac{-wL^2}{20}$

$R_A = \frac{wL}{4}$ $\quad R_B = \frac{wL}{4}$

$M_A = \frac{5wL^2}{96}$ $\quad M_B = \frac{-5wL^2}{96}$

$R_A = \frac{w(L-a)^3(3L+2a)}{20L^3}$

$R_B = \frac{w(L-a)}{20L^3}[10L^3-(L-a)^2(3L+2a)]$

$M_A = \frac{w}{60L^2}(L-a)^3(2L+3a)$

$M_B = \frac{-w}{60L^2}(L-a)^2(3L^2+4aL+3a^2)$

$R_A = -R_B = \frac{12\,EI\,v_o}{L^3}$

$M_A = M_B = \frac{6\,EI\,v_o}{L^2}$

Figure 6.17 Fixed-end reaction forces and moments.

a variety of distributed loads commonly occuring in ship structures.

The external load must satisfy three requirements:

1. It must consist of nodal forces and moments.
2. In regard to its effect on the overall solution, it must be statically equivalent to the actual distributed load.
3. The combination of the internal and external loads must be the actual load.

From requirement 3, the external load must be equal and opposite to the clamping forces and moments which were added to the distributed load in the internal problem. Since these clamping forces and moments equilibrated the distributed load, they were statically equivalent to it but opposite in sign. Hence, the forces and moments in the external problem, being of opposite sign to these clamping forces and moments, will be exactly statically equivalent to the distributed load. Thus in any frame analysis, the statically equivalent nodal loads which are used in place of a distributed load are exactly the same, but opposite in sign, as the fixed-end reaction forces and moments corresponding to that load. However, although these equivalent nodal loads can be used to solve for the structural (nodal) displacements, that solution does not provide the full and final distribution of element internal forces (shear force and bending moment) in those elements which have an internal load. For such elements, the solution to the *internal load model* provides the missing information.

The procedure for a beam with a uniform distributed load is illustrated in Fig. 6.18. In the *internal load model*, the beam is clamped, and therefore has the clamped end forces and moments given in Fig. 6.18. In the *external load model*, the distributed load is replaced by the equivalent nodal forces, which are equal and opposite to the clamped end forces and moments. Thus the sum of the internal and external load models is the same as the original structure. Since the equivalent nodal forces occur only at nodes, they can be part of the load vector and so the structure system of equations can be solved for the nodal displacements. Then the bending moment is obtained at both nodes of all beam elements using (6.3.12b). For all beam elements which do not have an internal load, the bending moment varies linearly from one node to the other. But, for beam elements which have an internal load, this missing internal bending moment must be superimposed on the linearly varying bending moment. This correction constitutes Step 7 in the overall procedure of frame analysis (see Table 6.1).

A more general method for deriving equivalent nodal forces for any finite element is given in Section 7.6.

6.3.6 Example

The portal frame of Fig. 6.19 is used as an example to demonstrate the complete process of frame analysis in terms of the seven steps of Table 6.1. For completeness, the ordinary beam element is used. The numerical values of all dimensions and physical properties are given in the figure. The axial and flexural stiffness parameters are $\mathfrak{a} = AE/L = 100$ MN/m and $\mathfrak{b} = EI/L^3 = 1$ MN/m.

STEP 1. DEFINE THE STRUCTURAL MODEL

The structural modeling is illustrated in Fig. 6.19b. Because of symmetry, only half of the structure is modeled. To permit this, a node is introduced at point C, even though there is no joint there. The geometric conditions arising from symmetry are zero horizontal displacement and zero slope at the point of

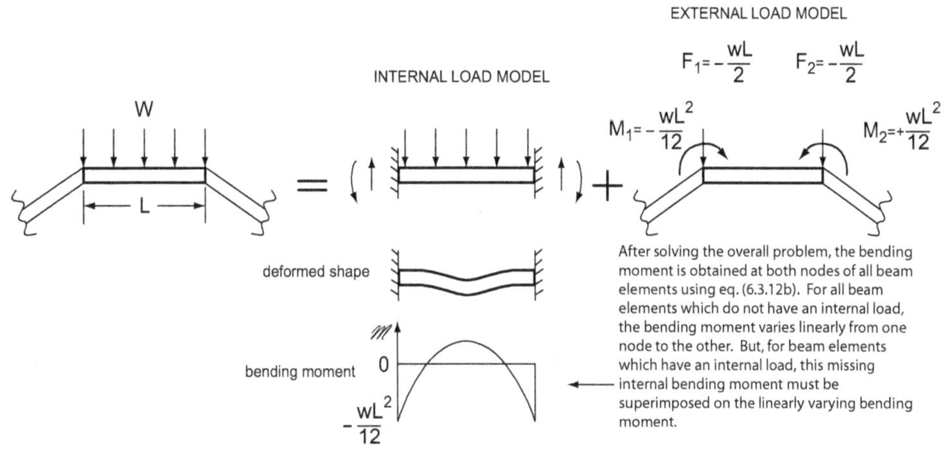

Figure 6.18 Equivalent nodal forces and internal bending moment for a uniform distributed load.

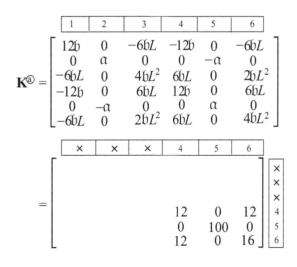

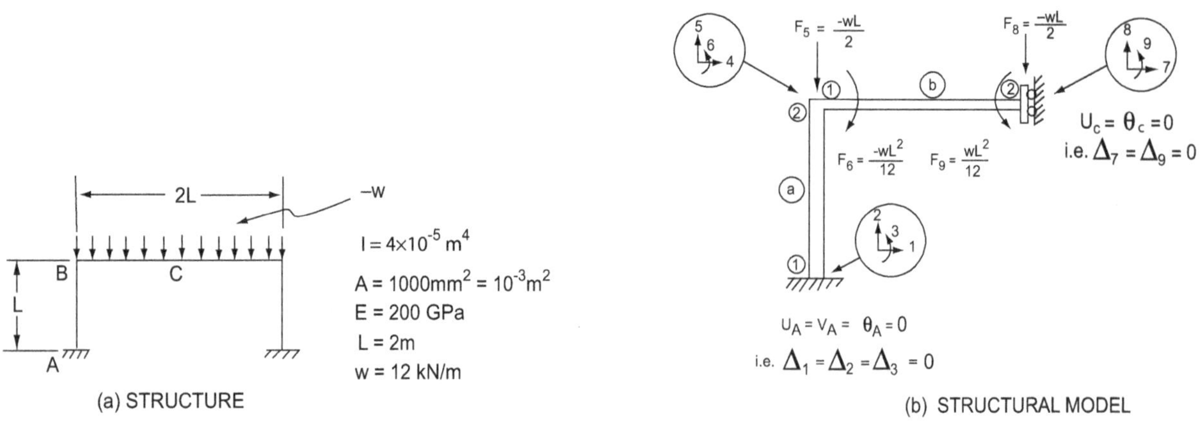

Figure 6.19 Sample problem with distributed load.

symmetry. The clamped support at A means that the three displacements at that point are also zero. The structure degrees of freedom are numbered from 1 to 9 as shown. The uniform distributed load of intensity w is represented by means of the statically equivalent loads $wL/2$ and $wL^2/12$. It may be seen that F_9 will not actually be applied to the structure because degree of freedom 9 is eliminated due to symmetry.

STEP 2. CALCULATE EACH ELEMENT STIFFNESS MATRIX IN STRUCTURE COORDINATES

The general expression for the beam element stiffness matrix at an arbitrary orientation χ is given in equation (6.3.16). The values of λ and μ for the two elements are as follows.

Member	χ	λ	μ
ⓐ	90°	0	1
ⓑ	0°	1	0

The element stiffness matrix for element ⓐ is given. It is suggested that the reader should evaluate at least a few of the terms and then compare them with the following.

$$\begin{array}{cccccc} 1 & 2 & 3 & 4 & 5 & 6 \end{array}$$

$$\mathbf{K}^{ⓐ} = \begin{bmatrix} 12\mathfrak{b} & 0 & -6\mathfrak{b}L & -12\mathfrak{b} & 0 & -6\mathfrak{b}L \\ 0 & \mathfrak{a} & 0 & 0 & -\mathfrak{a} & 0 \\ -6\mathfrak{b}L & 0 & 4\mathfrak{b}L^2 & 6\mathfrak{b}L & 0 & 2\mathfrak{b}L^2 \\ -12\mathfrak{b} & 0 & 6\mathfrak{b}L & 12\mathfrak{b} & 0 & 6\mathfrak{b}L \\ 0 & -\mathfrak{a} & 0 & 0 & \mathfrak{a} & 0 \\ -6\mathfrak{b}L & 0 & 2\mathfrak{b}L^2 & 6\mathfrak{b}L & 0 & 4\mathfrak{b}L^2 \end{bmatrix}$$

$$\begin{array}{cccccc} \times & \times & \times & 4 & 5 & 6 \end{array}$$

$$= \begin{bmatrix} & & & & & \\ & & & & & \\ & & & & & \\ & & & 12 & 0 & 12 \\ & & & 0 & 100 & 0 \\ & & & 12 & 0 & 16 \end{bmatrix} \begin{array}{c} \times \\ \times \\ \times \\ 4 \\ 5 \\ 6 \end{array}$$

The location vector is written above the matrix. In a hand solution such as this, it is often worthwhile to anticipate which degrees of freedom are to be eliminated and to not bother calculating these terms in $\mathbf{K}^e$. For greater clarity, the foregoing matrix has been written out in full initially but in substituting the numerical values, this shortcut has been used. The eliminated degrees of freedom are indicated by ×. The stiffness matrix for element ⓑ is

$$\begin{array}{cccccc} 4 & 5 & 6 & 7 & 8 & 9 \end{array}$$

$$\mathbf{K}^{ⓑ} = \begin{bmatrix} \mathfrak{a} & 0 & 0 & -\mathfrak{a} & 0 & 0 \\ 0 & 12\mathfrak{b} & 6\mathfrak{b}L & 0 & -12\mathfrak{b} & 6\mathfrak{b}L \\ 0 & 6\mathfrak{b}L & 4\mathfrak{b}L^2 & 0 & -6\mathfrak{b}L & 2\mathfrak{b}L^2 \\ -\mathfrak{a} & 0 & 0 & \mathfrak{a} & 0 & 0 \\ 0 & -12\mathfrak{b} & -6\mathfrak{b}L & 0 & 12\mathfrak{b} & -6\mathfrak{b}L \\ 0 & 6\mathfrak{b}L & 2\mathfrak{b}L^2 & 0 & -6\mathfrak{b}L & 4\mathfrak{b}L^2 \end{bmatrix}$$

$$\begin{array}{cccccc} 4 & 5 & 6 & \times & 8 & \times \end{array}$$

$$= \begin{bmatrix} 100 & 0 & 0 & -100 & 0 & 0 \\ 0 & 12 & 12 & 0 & -12 & 12 \\ 0 & 12 & 16 & 0 & -12 & 8 \\ -100 & 0 & 0 & 100 & 0 & 0 \\ 0 & -12 & -12 & 0 & 12 & -12 \\ 0 & 12 & 8 & 0 & -12 & 16 \end{bmatrix} \begin{array}{c} 4 \\ 5 \\ 6 \\ \times \\ 8 \\ \times \end{array}$$

STEPS 3 AND 4. ASSEMBLE THE STRUCTURE STIFFNESS MATRIX AND APPLY THE BOUNDARY CONDITIONS

In this problem, these two steps are combined because the elimination of the degrees of freedom is being achieved by simply not assembling these terms in the structure stiffness matrix.

The result of the assembly process is given next. Here again, it is suggested that the reader should perform the assembly process and then compare the result with that given.

$$\mathbf{K} = \begin{array}{c} \begin{array}{ccccccccc} 1 & 2 & 3 & 4 & 5 & 6 & 7 & 8 & 9 \end{array} \\ \left[\begin{array}{cccc} 112 & 0 & 12 & 0 \\ 0 & 112 & 12 & -12 \\ 12 & 12 & 32 & -12 \\ 0 & -12 & -12 & 12 \end{array}\right] \end{array} \begin{array}{c} 1\\2\\3\\4\\5\\6\\7\\8\\9 \end{array}$$

Thus the equilibrium equations are a 4×4 system of equations. The load vector must be expressed in units which are consistent with **K**. In this problem, the units of $\mathfrak{a}$ and $\mathfrak{b}$, and hence of **K**, are N and m. The distributed load is $w = 0.012$ MN/m. Therefore, the magnitudes of the equivalent nodal loads are

$$\frac{wL}{2} = \frac{12 \times 10^{-3}(2)}{2} = 12 \times 10^{-3}\ \text{MN}$$

and

$$\frac{wL^2}{12} = \frac{12 \times 10^{-3}(4)}{12} = 4 \times 10^{-3}\ \text{MNm}$$

Hence, the system of equations is

$$\begin{Bmatrix} 0 \\ -12 \\ -4 \\ -12 \end{Bmatrix} \times 10^{-3} = \begin{bmatrix} 112 & 0 & 12 & 0 \\ 0 & 112 & 12 & -12 \\ 12 & 12 & 32 & -12 \\ 0 & -12 & -12 & 12 \end{bmatrix} \begin{Bmatrix} \Delta_4 \\ \Delta_5 \\ \Delta_6 \\ \Delta_8 \end{Bmatrix}$$

STEP 5. SOLVE FOR DISPLACEMENTS AND REACTIONS

The solution to this system is

$$\begin{aligned} \Delta_4 &= 0.0917 \times 10^{-3}\ \text{m} \\ \Delta_5 &= -0.2400 \times 10^{-3}\ \text{m} \\ \Delta_6 &= -0.8550 \times 10^{-3}\ \text{rad} \\ \Delta_8 &= -2.0950 \times 10^{-3}\ \text{m} \end{aligned}$$

Reaction Forces. If desired, the support reactions could now be calculated from the appropriate (excluded) equilibrium equations. For example F_3, the reaction moment at A, is given by the third excluded equation

$$\begin{aligned} F_3 &= \mathfrak{b}[-6L \;\; 0 \;\; 4L^2 \;\; 6L \;\; 0 \;\; 2L^2 \;\; 0 \;\; 0 \;\; 0]\,\boldsymbol{\Delta} \\ &= -5.74 \times 10^{-3}\ \text{MNm} \end{aligned}$$

STEPS 6 AND 7. CALCULATE THE ELEMENT INTERNAL FORCES

A. *Element* ⓐ

1. *Element Nodal Displacements*

For a simple structure such as this, the element nodal displacements may be obtained from the structure displacements simply by inspection, but for a more complex structure or in a computer program, it would be necessary to use the location vector for the allocation of the $\boldsymbol{\Delta}$ values and then to use the transformation matrix **T** to obtain the element nodal displacements in element coordinates. For example, for element ⓐ

$$\boldsymbol{\delta}_{ⓐ} = \mathbf{T}_{ⓐ}\boldsymbol{\Delta}_{ⓐ} = \begin{bmatrix} \lambda & \mu & 0 & 0 & 0 & 0 \\ -\mu & \lambda & 0 & 0 & 0 & 0 \\ 0 & 0 & 1 & 0 & 0 & 0 \\ 0 & 0 & 0 & \lambda & \mu & 0 \\ 0 & 0 & 0 & -\mu & \lambda & 0 \\ 0 & 0 & 0 & 0 & 0 & 1 \end{bmatrix} \begin{Bmatrix} \Delta_1 \\ \Delta_2 \\ \Delta_3 \\ \Delta_4 \\ \Delta_5 \\ \Delta_6 \end{Bmatrix}$$

and since $\lambda = 0$ and $\mu = 1$ the result is

$$\boldsymbol{\delta}_{ⓐ} = \begin{Bmatrix} u_1 \\ v_1 \\ \theta_1 \\ u_2 \\ v_2 \\ \theta_2 \end{Bmatrix}_{ⓐ} = \begin{Bmatrix} \Delta_2 \\ -\Delta_1 \\ \Delta_3 \\ \Delta_5 \\ -\Delta_4 \\ \Delta_6 \end{Bmatrix} = \begin{Bmatrix} 0 \\ 0 \\ 0 \\ -0.24 \\ -0.0916 \\ -0.8550 \end{Bmatrix} \times 10^{-3}$$

2. *Element Internal Forces*

For the beam element, the internal forces are axial stress, shear force, and bending moment. Since the first two are constant and the third is linear within the element, it is sufficient to calculate these forces only at the nodes. The expression for this, in terms of nodal displacements, is given by (6.3.18). The results are

Axial stress

$$\begin{aligned} \sigma_{x1} = \sigma_{x2} &= \frac{E}{L}(-u_1 + u_2) \\ &= \frac{200 \times 10^3}{2}(0 - .24 \times 10^{-3}) = -24\ \text{MPa} \end{aligned}$$

Shear force

$$\begin{aligned} \mathcal{Q}_1 = \mathcal{Q}_2 &= \mathfrak{b}[0 \;\; 12 \;\; 6L \;\; 0 \;\; -12 \;\; 6L]\,\boldsymbol{\delta}_{ⓐ} \\ &= -9.16 \times 10^{-3}\ \text{MN} = -9.16\ \text{kN} \end{aligned}$$

Bending moment

$$\begin{Bmatrix} \mathcal{M}_z(0) \\ \mathcal{M}_z(L) \end{Bmatrix} = \mathfrak{b}\begin{bmatrix} 0 & -6L & -4L^2 & 0 & 6L & -2L^2 \\ 0 & 6L & 2L^2 & 0 & -6L & 4L^2 \end{bmatrix} \boldsymbol{\delta}_{ⓐ}$$

which gives

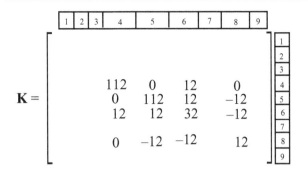

$$\mathcal{M}_z(0) = \quad 5.74 \text{ kNm}$$
$$\mathcal{M}_z(L) = \; -12.58 \text{ kNm}$$

The distribution of $\mathcal{M}_z(x)$ is plotted in Fig. 6.20a to the left of the element.

3. *Element Nodal Forces*

Usually there is no need to calculate the element nodal forces **f** but we will do so in this example to illustrate the distinction between nodal forces F_{y1}, F_{y2} and shear force $\mathcal{Q}$, and between nodal moments M_{z1}, M_{z2} and bending moment $\mathcal{M}_z(x)$. It will also be demonstrated that each element is in equilibrium. The quantity that relates element nodal forces and element nodal displacements is the element stiffness matrix, since $\mathbf{f} = \mathbf{k}^e\,\boldsymbol{\delta}$. Hence, from (6.3.15)

$$\begin{aligned} F_{x1} &= \mathfrak{a}(u_1 - u_2) = 100(0 + 0.24\times10^{-3}) \\ &= 24 \text{ kN} \end{aligned}$$

$$\begin{aligned} F_{y1} &= \mathfrak{b}(12v_1 + 6L\theta_1 - 12v_2 + 6L\theta_2) \\ &= -9.16 \text{ kN} \end{aligned}$$

$$\begin{aligned} M_{z1} &= \mathfrak{b}(6Lv_1 + 4L^2\theta_1 - 6Lv_2 + 2L^2\theta_2) \\ &= -5.74 \text{ kNm} \end{aligned}$$

$$F_{x2} = \mathfrak{a}(-u_1 + u_2) = -24 \text{ kN}$$

$$\begin{aligned} F_{y2} &= \mathfrak{b}(-12v_1 - 6L\theta_1 + 12v_2 - 6L\theta_2) \\ &= +9.16 \text{ kN} \end{aligned}$$

$$\begin{aligned} M_{z2} &= \mathfrak{b}(6Lv_1 + 2L^2\theta_1 - 6Lv_2 + 4L^2\theta_2) \\ &= -12.58 \text{ kNm} \end{aligned}$$

These forces and moments are illustrated in Fig. 6.20b. Equilibrium of forces is obvious, and equilibrium of moments may be verified by taking moments about point A:

$$\sum M_A = -5.74 - 12.58 + 9.16L = 0$$

B. *Element* ⓑ

1. *Element Nodal Displacements*

In a similar fashion, the nodal displacements of element ⓑ are

$$\boldsymbol{\delta}_{ⓑ} = \begin{Bmatrix} u_1 \\ v_1 \\ \theta_1 \\ u_2 \\ v_2 \\ \theta_2 \end{Bmatrix}_{ⓑ} = \begin{Bmatrix} \Delta_4 \\ \Delta_5 \\ \Delta_6 \\ \Delta_7 \\ \Delta_8 \\ \Delta_9 \end{Bmatrix} = \begin{Bmatrix} 0.0916 \\ -0.2400 \\ -0.8550 \\ 0 \\ -2.0950 \\ 0 \end{Bmatrix} \times 10^{-3}$$

2. *Element Internal Forces*

From (6.3.18)

Axial Stress

$$\begin{aligned} \sigma_{x1} = \sigma_{x2} &= \frac{E}{L}(-u_1 + u_2) \\ &= \frac{200\times10^3}{2}(-0.0916\times10^{-3} + 0) \\ &= -9.16 \text{ MPa} \end{aligned}$$

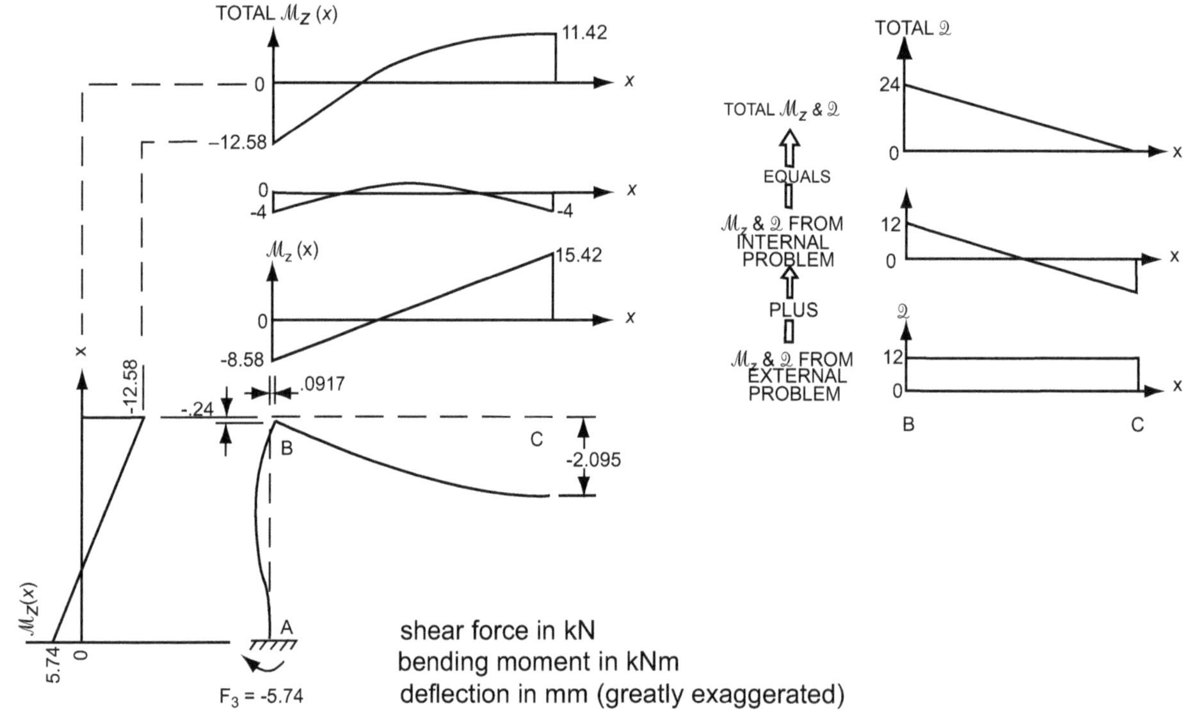

Figure 6.20a Deflection, shear force, and bending moment.

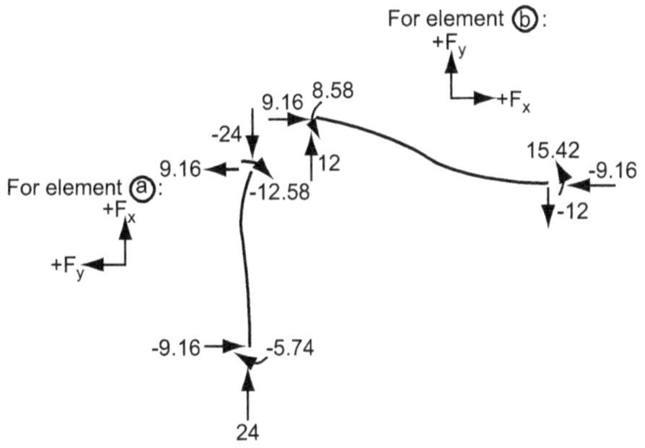

Figure 6.20b Element nodal forces.

Shear force

$$\mathcal{Q}_1 = \mathcal{Q}_2 = \mathfrak{b}[0 \quad 12 \quad 6L \quad 0 \quad -12 \quad 6L]\,\boldsymbol{\delta}_{ⓑ}$$
$$= 12 \times 10^{-3}\,\text{MN} = 12\text{ kN}$$

Bending Moment

$$\begin{Bmatrix} \mathcal{M}_z(0) \\ \mathcal{M}_z(L) \end{Bmatrix} = \mathfrak{b}\begin{bmatrix} 0 & -6L & -4L^2 & 0 & 6L & -2L^2 \\ 0 & 6L & 2L^2 & 0 & -6L & 4L^2 \end{bmatrix} \boldsymbol{\delta}_{ⓑ}$$
$$= \begin{Bmatrix} -8.58 \\ 15.42 \end{Bmatrix} \text{kNm}$$

The distributions of $\mathcal{Q}$ and of $\mathcal{M}_z(x)$ are plotted in Fig. 6.20a immediately above the element. Since this element has a distributed load, these two internal forces must be corrected by superimposing the corresponding forces from the solution of the internal problem. In the present example, the internal problem is simply a uniformly loaded, clamped beam. The corresponding shear force and bending moment distributions are shown just above those obtained from the frame analysis. The final and complete distributions, which are the sum of these two, are drawn at the top of the figure. Note that the total value of $\mathcal{M}_z$ at point B, −12.58, agrees with the value that was calculated in element ⓐ. This is another useful check on the solution.

3. *Element Nodal Forces*

Here again, although not necessary for the solution, we shall calculate the element nodal forces in order to point out some of the relationships and differences between these and other quantities. As before, $[\mathbf{f}_{ⓑ} = \mathbf{k}^e\,\boldsymbol{\delta}_{ⓑ}]$ with $\mathbf{k}^e$ given by (6.3.15). Substitution of $[\boldsymbol{\delta}_{ⓑ}]$ gives

$$F_{x1} = -F_{x2} = 9.16$$
$$F_{y1} = -F_{y2} = 12$$
$$M_{z1} = 8.58; \qquad M_{z2} = 15.42$$

These forces and moments are illustrated in Fig. 6.20b. Once again, equilibrium of forces is immediately evident and equilibrium of moments can be verified by taking moments about point B:

$$\sum M_B = 8.58 + 15.42 - 12L = 0.$$

This summation of moments illustrates the fact that frame analysis takes no account of any eccentricity of axial loads arising from the beam's deflection. This procedure is valid as long as the deflections are small, which they are in this example because the largest deflection is 2.095 mm, which is only 0.1% of the beam length L. The small-deflection assumption is usually valid for the elastic analysis of ships and other plated steel frame structures because the overall stiffness of such structures is usually sufficiently large such that any load which does not cause yielding produces only small deflections. However, the stiffness of a structure depends on the material stiffness, on the member proportions and scantlings (thickness, etc.), and on the arrangement of the members. Hence, when using frame analysis (or any other elastic, small-deflection technique), a check should always be made on the size of the deflections relative to the dimensions of the structure.

Another observation which may be made in the foregoing example is that at point B, the net vertical force of −12 kN is equal to the applied load at that point, which is the equivalent load, $F_4 = -wL/2$. Similarly, the net moment of −4 kNm is equal to the applied moment, $F_6 = -wL^2/12$. However, the net moment at point C (15.42 kNm) is not equal to the equivalent moment, $F_9 = wL^2/12$, because the latter was never applied to the structure; it was excluded by the zero-slope boundary condition at point C. The value of 15.42 is an uncorrected value; when the solution to the internal problem is superimposed, as in Fig. 6.20a, the value becomes 11.42, which is the correct value. If we were to obtain an exact solution of this problem (treating the structure as a continuum rather than a discrete structure with only two elements and three nodes) we would obtain 11.42 as the reaction moment at point C.

BASIC ASPECTS OF THE FINITE ELEMENT METHOD

Owen Hughes
Professor, Virginia Tech
Blacksburg, VA, USA

7.1 AIM, APPROACH, AND SCOPE

The aim of this chapter is to explain as simply and as directly as possible the basic ideas and features of the finite element method. In its totality, the finite element method is very broad and powerful; it includes numerous special features and techniques, and it has a great variety of applications—both structural and nonstructural. Structural applications include not only the small-deflection analysis of two- and three-dimensional elastic continua, but also structural stability (buckling) and structural dynamics (e.g., vibration analysis). Also, with the aid of special iterative techniques, the method can be used in nonlinear solid mechanics, dynamics, and stability problems. Nonstructural applications are also numerous because, although the method was originally derived for structures, it has subsequently been given a broader and more general foundation by means of variational theory. As a result, it can be used for a wide variety of problems for which the solution corresponds to the minimization of a functional. In structural applications, the functional is strain energy, but in the broader fields of continuum mechanics (fluid and solid), the functional can be related to pressure, temperature, fluid velocity, and so on. Likewise, it can be the potential in the various aspects of potential field theory (electrostatics, magnetism, gravity, etc.).

However, such generality would not be helpful or appropriate in a book dealing exclusively with ship structures. Several books (such as Cook, Malkus, & Plesha, 1989) are available which present the finite element method in varying degrees of generality, breadth of application, and level of detail, and these are recommended for readers wishing to explore the method more thoroughly.

The scope and level of treatment of this chapter are intended to explain merely the basic aspects of the method as it applies to ship structures. This is a middle course between two extremes: that of attempting to give a general and mathematically rigorous treatment and to cover a broad spectrum of features and applications; and that of omitting the method entirely on the grounds that the material can be found in other texts. Some specific advantages of this course are:

1. It makes the method easier to learn and does not require knowledge of variational calculus.
2. It avoids requiring the reader to have a previous knowledge of the method.
3. It leads to a more unified and cohesive coverage of ship structural analysis and design, especially in regard to overall philosophy, underlying assumptions, and methodology.
4. It avoids the need for an additional and separate text, with the consequent problems of differing notation, approach, and emphasis.

Furthermore, although the finite element method has now been given a broader and more fundamental theoretical basis, it was originally developed mainly from matrix frame analysis. Hence, it is possible to present the method as a generalization and extension of matrix frame analysis to continuum structures such as plates and shells. Although not all of the historical development actually came from frame analysis, most of the concepts and terminology are similar. Therefore, with the aim of providing greater unity and simplicity of presentation, the method is presented here as an extension of matrix frame analysis presented in the previous three sections.

The key feature of the finite element method and the point of departure from frame analysis is the use of two- and three-dimensional elements for the discrete representation of a continuum. The first paper to clearly and fully implement this idea was Turner, Clough, Martin, & Topp (1956). It should, however, be mentioned that some of the basic features of the method had appeared before this in Hrenikoff (1941), Courant (1943), McHenry (1943), and others. There followed many other contributions by many authors, including a series of papers on matrix structural analysis by Argyris and his collaborators, beginning in 1954 and culminating in a book (Argyris and Kelsey, 1960). The first elements were for plane stress applications. After this, finite elements were developed for three-dimensional solids, plates in bending, thin shells, thick shells, and other structural forms. Once these had been established for linear elastic analysis, work was directed toward more complex areas such as dynamic response, buckling, and material and geometric nonlinearities. This required the introduction of iterative finite element analysis.

In addition to these developments in the structural field, in the early 1960s the method began to be recognized as a form of the Ritz method, and Zienkiewicz and Cheung (1964) showed that it is applicable to all field problems that can be cast in variational form.

For structural applications, a number of general purpose finite element computer programs are now available and this, together with the inherent versatility of the finite element method, has meant that the number and variety of practical applications of the method have grown enormously over time.

7.2 FUNDAMENTALS OF THE METHOD

As mentioned, the basic concept of the finite element method is the same as in matrix frame analysis: namely, that the structure can be represented as an assemblage of individual structural elements interconnected at a discrete number of nodes. This manner of representation is a natural one for frame structures since they consist of discrete beam members connected together at a number of joints. But in a continuum structure such as a panel of plating, a corresponding natural subdivision does not exist; therefore, it is necessary to divide the continuum artificially into a number of elements, connected at their nodes. The artificial elements, or "finite elements," are usually either triangles or quadrilaterals, and the nodes are usually but not necessarily at the corners. For example, Fig. 7.1 shows the web of a deep beam subdivided into triangular elements.

To use matrix methods, the essential requirement is that the structural continuum must be represented in terms of a finite number of discrete variables. These variables are the nodal displacements and, in some cases, their derivatives. If the latter are included, we speak of "degrees of freedom" instead of nodal displacements, but this does not arise for the simple elements which are presented in this chapter. In terms of nodal displacements, the essential requirements are that the internal displacements of the elements must be related to the nodal displacements and all of the interactions between elements must be expressed in the terms of the nodal displacements. In this way, the only unknowns in the problem are the nodal displacements and so the problem becomes discrete instead of continuous. Although there may be a large number of nodal displacements, there are nevertheless a finite number of discrete variables which are interrelated by linear equations which therefore can be conveniently handled by matrix methods.

From the discussion at the beginning of Chapter 6, it is clear that to achieve an exact solution, the finite element representation would have to satisfy the requirements of equilibrium and geometric compatibility and would have to satisfy them everywhere, both within each of the elements and also between elements. This amounts to four separate requirements. By way of illustration, let us consider the requirement of interelement compatibility. In a continuous structure such as the plate web of Fig. 7.1, there is complete continuity of displacement along the common boundaries of the elements. Hence, in the finite element model, it would not be sufficient to satisfy continuity of displacement only at the nodes or to place no conditions on the displacements along the common boundaries. This would make the model much more flexible than the structure since it would allow gaps and overlaps to occur, as shown in Fig. 7.2. One way of reducing the error is to use smaller and more numerous elements because then there are more nodes and hence more points at which compatibility is satisfied.

However, of its very nature, a discrete model can never give an exact representation of a continuum, regardless of the number of discrete variables which it employs. There is always some error, although it may be made negligibly small and localized. Hence, in the finite element model, it is not possible to fulfill all four of the foregoing requirements exactly, even though the elements may be made small and numerous.

However, besides using smaller and more numerous elements, it is also possible to reduce the

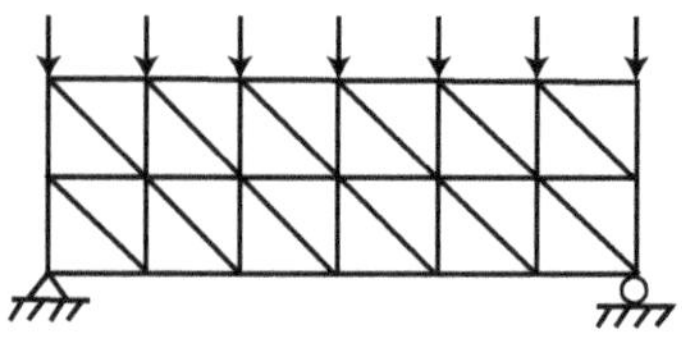

Figure 7.1 Representation of a beam web using triangle elements.

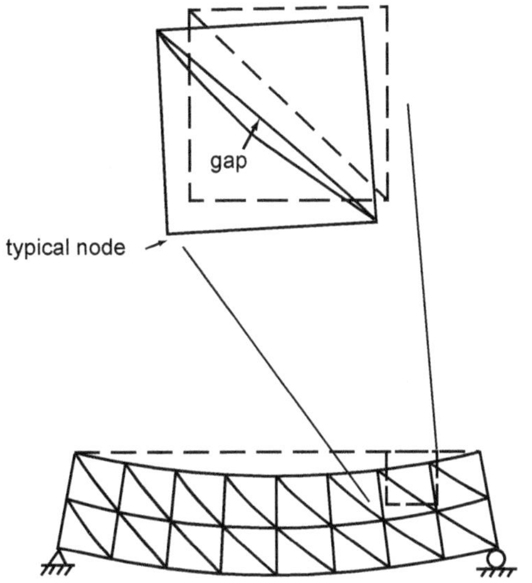

Figure 7.2 Deformation of elements.

error by carefully choosing and defining suitable element properties. This specification of element properties is the most fundamental feature of the finite element method and is the means whereby it can satisfy the four requirements sufficiently well without having to use extremely small elements. The element behavior is described by a specially chosen set of functions which represent either the stresses or the displacements in that element. In other words, each element is constrained to deform in a specific pattern. The result is that even though equilibrium and compatibility are explicitly enforced only at the nodes, the specified internal behavior pattern of the elements ensures that intra- and interelement equilibrium and compatibility are all satisfied to a sufficient degree. For example, if the displacement within each element is specified as being linear, this linearity would also apply along the element edges. Therefore, since two adjacent elements share a common displacement at each of their two common nodes, they must also have equal displacements all along their common boundary. Hence, the specification of linear displacement would cause the requirement of interelement compatibility to be satisfied.

Thus there are two principal features of the finite element method which require discussion: the subdivision of the structure into elements and, more importantly, the choice of the internal behavior function. We first consider some aspects and criteria regarding subdivision. In order to explain and illustrate the choice of the internal behavior function, we then consider some specific elements which are simple, commonly used, and suitable for ship structure applications.

7.2.1 Subdivision of Structure

As mentioned previously, the accuracy of the finite element method increases with the number of elements used. However, increasing the number of elements also increases the computer time required for a solution and therefore increases the cost.

In many cases a "graded mesh," that is, a gradation of element sizes, is adopted to provide a more detailed modeling of regions of the structure where there are stress concentrations such as in the vicinity of openings and cut-outs or near the point of application of concentrated loads, as shown in Fig. 7.3. Mesh grading is a very efficient technique and can give a significant savings in solution time without any loss of accuracy. However, because of the enormous variety of structures and of loads, it is not possible to give a general rule as to the number or size of elements or the type of mesh grading which is required to provide sufficient accuracy. In each case, the subdivision should be based on experience with similar structures. If this is not possible, then a series of appropriate test problems should be solved using different mesh sizes in order to observe the rate of convergence and thereby determine a suitable mesh size for the particular problem. Some examples are given later in this chapter.

The external loads acting on the structure must be represented by an equivalent system of point loads applied at the element nodes. In the case of concentrated loads, the obvious course is to choose the mesh such that there is a node at the point of application of the load. For distributed loading, statically equivalent nodal point loads must be used. Most plane stress elements do not have a rotational degree of freedom and hence there is no possibility of using equivalent nodal moments. This is of little consequence because with the close proximity of nodes, the equivalent nodal forces are in themselves a good approximation to the distributed loading. For example, Fig. 7.4 shows a deep web of the type considered earlier, with a length of 6 m and a uniform distributed load of 10 kN/m. The choice of equivalent nodal loads is obvious

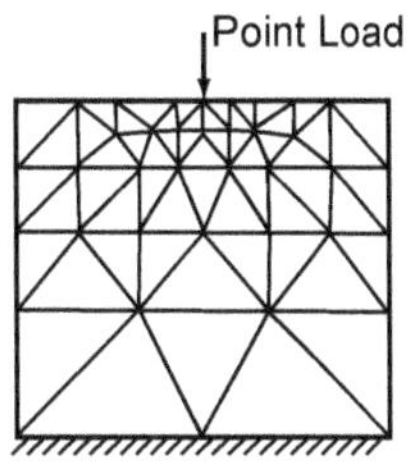

Figure 7.3 Mesh grading.

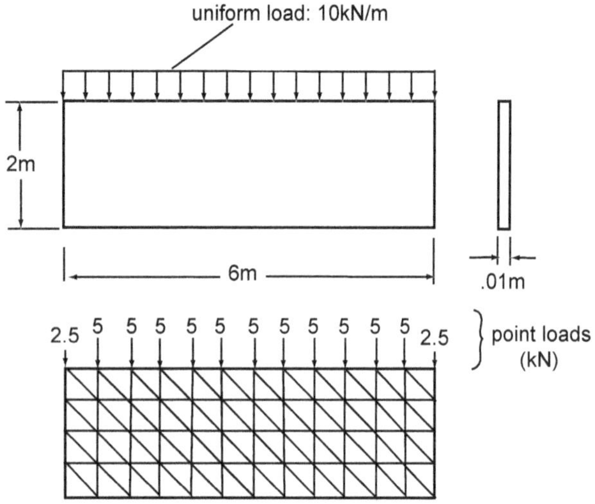

Figure 7.4 Equivalent nodal point loads.

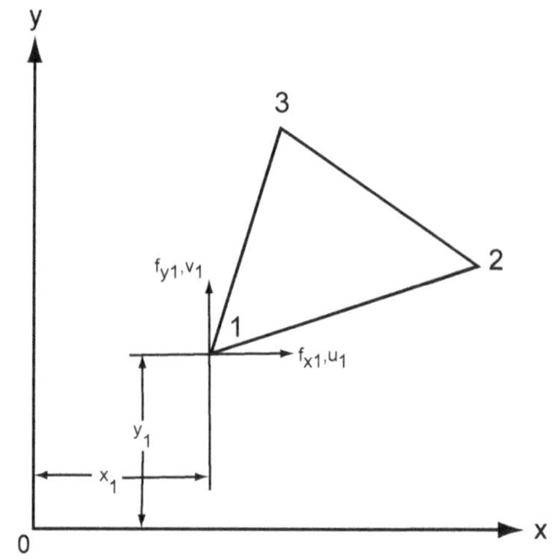

Figure 7.5 Coordinates and node numbering for triangle element.

and the loads can be obtained by inspection, as shown in Fig. 7.4.

7.2.2 Elements for Ship Structural Analysis and Design

Since this is merely an introductory treatment, we consider only two-dimensional plane stress elements: a triangle element and two rectangle elements. Shell elements and three-dimensional elements are required for realistic applications, and these are explained in many available textbooks.

7.3 CONSTANT STRESS TRIANGLE (CST) ELEMENT

As the first example of a finite element, we take the case of a flat triangular-shaped element of constant thickness t and isotropic material properties. This element is of primary importance because of its versatility of geometry; almost any two-dimensional shape can be represented by an assemblage of triangles. Moreover, three-dimensional curved surfaces can also be modeled with this element, providing that the curvature is everywhere gradual and that the height of the surface (or the ratio of true surface area to projected area) is small.

In Section 6.3, the five basic steps in the derivation of the stiffness matrix of an element were presented for the beam element. We now show how the same five steps can be used to derive the stiffness matrix for finite elements, beginning with the triangle element.

The coordinate system and the node numbering are shown in Fig. 7.5. For hand calculations, the nodes must be numbered counterclockwise (in a computer program, this requirement is easily avoided). Since we are dealing with displacements in a plane, the element has two degrees of freedom at each node, as shown in Fig. 7.6, giving a total of six degrees of freedom.

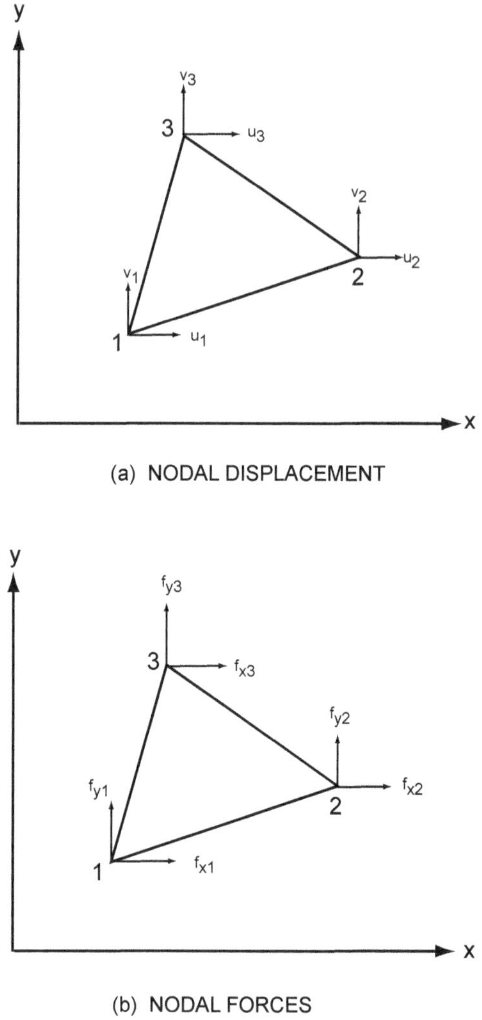

Figure 7.6 Nodal displacements and forces.

$$\boldsymbol{\delta} = \begin{Bmatrix} \boldsymbol{\delta}_1 \\ \boldsymbol{\delta}_2 \\ \boldsymbol{\delta}_3 \end{Bmatrix} = \begin{Bmatrix} u_1 \\ v_1 \\ u_2 \\ v_2 \\ u_3 \\ v_3 \end{Bmatrix} \tag{7.3.1}$$

The corresponding forces are

$$\mathbf{f} = \begin{Bmatrix} \mathbf{f}_1 \\ \mathbf{f}_2 \\ \mathbf{f}_3 \end{Bmatrix} = \begin{Bmatrix} f_{x1} \\ f_{y1} \\ f_{x2} \\ f_{y2} \\ f_{x3} \\ f_{y3} \end{Bmatrix} \tag{7.3.2}$$

Both $\boldsymbol{\delta}$ and $\mathbf{f}$ contain six terms, and therefore the element stiffness matrix $\mathbf{k}_e$ is a 6 × 6 matrix since by definition it is the coefficient matrix in the element nodal force–displacement relationship

$$\mathbf{f} = \mathbf{k}^e\,\boldsymbol{\delta}$$

7.3.1 Step I. Select Suitable Displacement Function

As mentioned at the beginning of this chapter, the key aspect of the finite element method is the specification of the displacement pattern of an element in terms of some mathematical function. A polynomial is a function that is both simple and versatile, since it can be multidimensional (for plane stress only two dimensions, x and y, are required) and it can contain any number of independent parameters (the coefficients) depending on the order of the polynomial. Hence, we choose a polynomial to describe the internal displacements in the triangle. Since there are six degrees of freedom, we must choose a polynomial having six independent parameters or coefficients. Also, since no bias should be shown in either the x- or the y-direction, we should use three coefficients for the u displacement and three for the v displacement. Therefore, the logical choice is to represent each of these displacement components by a linear polynomial in x and y:

$$\begin{aligned} u &= C_1 + C_2 x + C_3 y \\ v &= C_4 + C_5 x + C_6 y \end{aligned} \tag{7.3.3}$$

With this choice, the displacement varies linearly in x and y. Hence, it varies linearly along each boundary of each element, and since the displacement of adjoining elements must agree at their common nodes, the displacements must also agree along the entire boundary. As mentioned earlier, the choice of a linear displacement function, in conjunction with corner nodes, ensures continuity of displacement along all interelement boundaries. As shown subsequently, it does not provide perfect equilibrium along the boundaries but rather only at the nodes. Thus, there will inevitably be some error, but it can be reduced to an acceptable level by making the elements sufficiently small.

Equation 7.3.3 can be written in matrix form as

$$\boldsymbol{\delta}(x,y) = \begin{Bmatrix} u \\ v \end{Bmatrix} = \begin{bmatrix} 1 & x & y & 0 & 0 & 0 \\ 0 & 0 & 0 & 1 & x & y \end{bmatrix} \begin{Bmatrix} C_1 \\ C_2 \\ C_3 \\ C_4 \\ C_5 \\ C_6 \end{Bmatrix} \tag{7.3.4}$$

or

$$\boldsymbol{\delta}(x, y) = \mathbf{H}(x, y)\mathbf{C}$$

7.3.2 Step II. Relate the General Displacement Within the Element to the Nodal Displacements

This step is achieved by substituting the values of the nodal coordinates into (7.3.4), thereby obtaining expressions that can be solved for the unknown coefficients. For example, at node 1

$$\begin{aligned} \boldsymbol{\delta}_1 = \boldsymbol{\delta}(x_1, y_1) &= \mathbf{H}(x_1, y_1)\mathbf{C} \\ &= \begin{bmatrix} 1 & x_1 & y_1 & 0 & 0 & 0 \\ 0 & 0 & 0 & 1 & x_1 & y_1 \end{bmatrix} \mathbf{C} \end{aligned}$$

and for nodes 2 and 3

$$\boldsymbol{\delta}_2 = \begin{bmatrix} 1 & x_2 & y_2 & 0 & 0 & 0 \\ 0 & 0 & 0 & 1 & x_2 & y_2 \end{bmatrix} \mathbf{C}$$

$$\boldsymbol{\delta}_3 = \begin{bmatrix} 1 & x_3 & y_3 & 0 & 0 & 0 \\ 0 & 0 & 0 & 1 & x_3 & y_3 \end{bmatrix} \mathbf{C}$$

Substituting these three matrices into (7.3.1) yields

$$\boldsymbol{\delta} = \begin{Bmatrix} \boldsymbol{\delta}_1 \\ \boldsymbol{\delta}_2 \\ \boldsymbol{\delta}_3 \end{Bmatrix} = \begin{bmatrix} 1 & x_1 & y_1 & 0 & 0 & 0 \\ 0 & 0 & 0 & 1 & x_1 & y_1 \\ 1 & x_2 & y_2 & 0 & 0 & 0 \\ 0 & 0 & 0 & 1 & x_2 & y_2 \\ 1 & x_3 & y_3 & 0 & 0 & 0 \\ 0 & 0 & 0 & 1 & x_3 & y_3 \end{bmatrix} \begin{Bmatrix} C_1 \\ C_2 \\ C_3 \\ C_4 \\ C_5 \\ C_6 \end{Bmatrix} \quad (7.3.5)$$

The matrix of nodal coordinates is known as the *connectivity matrix*. For future reference, we denote it as **A**, and this equation then becomes

$$\boldsymbol{\delta} = \mathbf{AC}$$

The unknown polynomial coefficients **C** are now determined from (7.3.5) by inverting the connectivity matrix.

$$\mathbf{C} = \mathbf{A}^{-1}\boldsymbol{\delta} \quad (7.3.6)$$

The inverse matrix $\mathbf{A}^{-1}$ for the triangle element is given in (7.3.7).

$$\mathrm{A}^{-1} = \frac{1}{2\mathrm{A}_{123}} \begin{bmatrix} x_2y_3 - x_3y_2 & 0 & -x_1y_3 + x_3y_1 & 0 & x_1y_2 - x_2y_1 & 0 \\ y_2 - y_3 & 0 & y_3 - y_1 & 0 & y_1 - y_2 & 0 \\ x_3 - x_2 & 0 & x_1 - x_3 & 0 & x_2 - x_1 & 0 \\ 0 & x_2y_3 - x_3y_2 & 0 & -x_1y_3 + x_3y_1 & 0 & x_1y_2 - x_2y_1 \\ 0 & y_2 - y_3 & 0 & y_3 - y_1 & 0 & y_1 - y_2 \\ 0 & x_3 - x_2 & 0 & x_1 - x_3 & 0 & x_2 - x_1 \end{bmatrix} \quad (7.3.7)$$

$$\text{where} \quad 2A_{123} = \det \begin{vmatrix} 1 & x_1 & y_1 \\ 1 & x_2 & y_2 \\ 1 & x_3 & y_3 \end{vmatrix}$$

$$= (x_2y_3 - x_3y_2) - (x_1y_3 - x_3y_1) + (x_1y_2 - x_2y_1)$$

$$= 2 \times \text{area of triangle } 123$$

Equation 7.3.6 expresses the six unknown coefficients **C** in terms of the nodal displacements **δ**. From (7.3.4), the general displacement **δ**(x, y) at any point (x,y) within the element can now be expressed in terms of the nodal displacements **δ** by substituting for **C**. The result is

$$\boldsymbol{\delta}(x, y) = \mathbf{HA}^{-1}\boldsymbol{\delta} \quad (7.3.8)$$

7.3.3 Step III. Express the Internal Deformation (Strain) in Terms of the Nodal Displacements

In general, the internal deformation of a member is defined as the derivative of its displacement. The most common example is strain, which is the first derivative (or local rate) of displacement. For rigid body deflections, the displacement (translation or rotation) is constant and hence the strain is zero. In Section 6.1, the internal deformation for a one-dimensional bar element was the elongation (or shortening), but this could have been converted to axial strain by dividing by the length. For the beam element, the internal deformation was curvature rather than strain because in that case, the internal "force" (bending moment) is proportional to the second derivative of lateral displacement. Also, in both cases the member was one-dimensional, and therefore the internal deformation was a single quantity, that is, a scalar.

To deal with elastic continua, we must revert to the more general measure of internal deformation, that is, strain. For a two-dimensional continuum, there are three components of strain and so we can define a *strain vector*

$$\varepsilon(x, y) = \begin{Bmatrix} \varepsilon_x \\ \varepsilon_y \\ \gamma \end{Bmatrix} \quad (7.3.9)$$

in which ε_x and ε_y are the direct strains and γ is the shear strain. From the theory of elasticity, the relationship between the strain ε and the displacements u and v is

$$\varepsilon_x = \frac{\partial u}{\partial x}$$
$$\varepsilon_y = \frac{\partial v}{\partial y} \quad (7.3.10)$$
$$\gamma = \frac{\partial u}{\partial y} + \frac{\partial v}{\partial x}$$

Substituting for u and v from (7.3.3)

$$\begin{aligned}\varepsilon_x &= \frac{\partial}{\partial x}(C_1 + C_2 x + C_3 y)\\ &= C_2\\ \varepsilon_y &= \frac{\partial}{\partial y}(C_4 + C_5 x + C_6 y)\\ &= C_6\\ \gamma &= \frac{\partial}{\partial y}(C_1 + C_2 x + C_3 y) + \frac{\partial}{\partial x}(C_4 + C_5 x + C_6 y)\\ &= C_3 + C_5\end{aligned}$$

Thus

$$\boldsymbol{\varepsilon}(x,y) = \begin{Bmatrix}\varepsilon_x\\ \varepsilon_y\\ \lambda\end{Bmatrix} = \begin{Bmatrix}C_2\\ C_6\\ C_3 + C_5\end{Bmatrix}$$

We note that because of the linearity of the chosen displacement function—(7.3.3) or (7.3.4)—all three components of strain are constant within the element. This will be discussed further at the end of the derivation. Although the strain happens to be constant, we shall continue to denote the strain matrix as $\varepsilon(x,y)$ in order to indicate that it expresses the strain throughout the element, rather than only at the nodes.

In terms of the full coefficient matrix, the above equation is

$$\boldsymbol{\varepsilon}(x,y) = \begin{Bmatrix}\varepsilon_x\\ \varepsilon_y\\ \lambda\end{Bmatrix} = \begin{bmatrix}0 & 1 & 0 & 0 & 0 & 0\\ 0 & 0 & 0 & 0 & 0 & 1\\ 0 & 0 & 1 & 0 & 1 & 0\end{bmatrix}\begin{Bmatrix}C_1\\ C_2\\ C_3\\ C_4\\ C_5\\ C_6\end{Bmatrix} \tag{7.3.11}$$

or

$$\boldsymbol{\varepsilon}(x, y) = \mathbf{GC}$$

By substituting for **C** from (7.3.6), we obtain the desired expression for strain in terms of nodal displacements

$$\boldsymbol{\varepsilon}(x, y) = \mathbf{GA}^{-1}\boldsymbol{\delta}$$

and if we define a *strain coefficient matrix*

$$\mathbf{B} = \mathbf{GA}^{-1} \tag{7.3.12}$$

the foregoing expression becomes

$$\boldsymbol{\varepsilon}(x, y) = \mathbf{B}\boldsymbol{\delta} \tag{7.3.13}$$

Because $\mathbf{A}^{-1}$ has been obtained explicitly, the strain coefficient matrix **B** can be obtained by performing the matrix multiplication of (7.3.12). The result is

$$\mathbf{B} = \frac{1}{2A_{123}}\begin{bmatrix}y_2 - y_3 & 0 & y_3 - y_1 & 0 & y_1 - y_2 & 0\\ 0 & x_3 - x_2 & 0 & x_1 - x_3 & 0 & x_2 - x_1\\ x_3 - x_2 & y_2 - y_3 & x_1 - x_3 & y_3 - y_1 & x_2 - x_1 & y_1 - y_2\end{bmatrix} \tag{7.3.14}$$

where, as defined earlier, in (7.3.7), A_{123} is the area of the element.

7.3.4 Step IV. Express the Internal Force (Stress) in Terms of the Nodal Displacements, Using the Element's Law of Elastic Behavior

In an elastic continuum, the internal force is stress, and for two-dimensional members, there are three components of stress: σ_x, σ_y, and τ. Accordingly, the stress vector is

$$\boldsymbol{\sigma} = \begin{Bmatrix}\sigma_x\\ \sigma_y\\ \tau\end{Bmatrix} \tag{7.3.15}$$

For plane stress, the relationship between stress and strain is

$$\begin{aligned}\varepsilon_x &= \frac{\sigma_x}{E} - \frac{\nu\sigma_y}{E}\\ \varepsilon_y &= \frac{-\nu\sigma_x}{E} + \frac{\sigma_y}{E}\\ \gamma &= \frac{\tau}{G} = \frac{2(1+\nu)}{E}\tau\end{aligned}$$

where E is Young's modulus, G is the shear modulus, and ν is Poisson's ratio. In matrix notation, the relationship is

$$\boldsymbol{\varepsilon}(x,y) = \frac{1}{E}\begin{bmatrix}1 & -\nu & 0\\ -\nu & 1 & 0\\ 0 & 0 & 2(1+\nu)\end{bmatrix}\begin{Bmatrix}\sigma_x\\ \sigma_y\\ \tau\end{Bmatrix} \tag{7.3.16}$$

For our purpose, we require the inverse of this relationship which is

$$\boldsymbol{\sigma}(x,y)=\begin{Bmatrix}\sigma_x\\ \sigma_y\\ \tau\end{Bmatrix}=\frac{E}{1-\nu^2}\begin{bmatrix}1 & \nu & 0\\ \nu & 1 & 0\\ 0 & 0 & \dfrac{1-\nu}{2}\end{bmatrix}\begin{Bmatrix}\varepsilon_x\\ \varepsilon_y\\ \gamma\end{Bmatrix} \tag{7.3.17}$$

or

$$\boldsymbol{\sigma}(x, y) = \mathbf{D}\boldsymbol{\varepsilon}(x, y)$$

Finally, upon substituting for $\boldsymbol{\varepsilon}(x, y)$ from (7.3.13), the desired expression for the element stresses in terms of the nodal displacements is obtained.

$$\boldsymbol{\sigma}(x, y) = \mathbf{DB}\boldsymbol{\delta} \tag{7.3.18}$$

The principal steps in matrix structural analysis (i.e., both frame analysis and the finite element method) were outlined in Section 6.1 and are summarized in Table 6.1. As shown there, the element stiffness matrix is used in setting up (assembling) the system of equations for the structure displacements. After solving this system, the next step (Step 6) is to calculate the stresses within each element; to do this, we need an expression for the stresses in terms of the element nodal displacements. This is given by (7.3.18), and since **B** and **D** are both known explicitly for the triangle element, we can perform the matrix multiplication and thus obtain an explicit *stress matrix* **S** = **DB** which provides the stresses directly from the element nodal displacements. The stress matrix for the triangle element is given in (7.3.19).

$$\boldsymbol{\sigma}(x, y) = \mathbf{S}\boldsymbol{\delta}$$

where

7.3.5 Step V. Obtain the Element Stiffness Matrix by Relating Nodal Forces to Nodal Displacements

In this step, the element is given virtual nodal displacements $\boldsymbol{\delta}^*$ and the external work, involving actual nodal forces **f**, is equated to the internal work, involving the virtual strain $\boldsymbol{\varepsilon}^*$ (expressed in terms of $\boldsymbol{\delta}^*$) and the actual stress $\boldsymbol{\sigma}$ (expressed in terms of the actual displacements $\boldsymbol{\delta}$, corresponding to **f**). The virtual nodal displacements are then set to unity, and the resulting matrix relating **f** to $\boldsymbol{\delta}$ is the required element stiffness matrix.

The external virtual work is

$$W_{\text{ext}} = \boldsymbol{\delta}^{*\mathrm{T}}\mathbf{f}$$

and the internal virtual work is

$$W_{\text{int}} = \int_{\text{vol}}\left[\boldsymbol{\varepsilon}^*(x,y)\right]^{\mathrm{T}}\boldsymbol{\sigma}(x,y)\,d\text{vol}$$

where vol is the volume of the element.

Substituting for $\boldsymbol{\varepsilon}^*(x,y)$ from (7.3.13) and for $\boldsymbol{\sigma}(x,y)$ from (7.3.17) gives

$$\begin{aligned}W_{\text{int}} &= \int_{\text{vol}}\left[\mathbf{B}\boldsymbol{\delta}^*\right]^{\mathrm{T}}\mathbf{D}\boldsymbol{\varepsilon}(x,y)\,d\text{vol}\\ &= \int_{\text{vol}}\boldsymbol{\delta}^{*\mathrm{T}}\,\mathbf{B}^{\mathrm{T}}\,\mathbf{D}\mathbf{B}\boldsymbol{\delta}(x,y)\,d\text{vol}\end{aligned}$$

Setting $\boldsymbol{\delta}^*$ equal to unity and equating the internal and external work gives

$$\mathbf{f}=\left[\int_{\text{vol}}\mathbf{B}^{\mathrm{T}}\mathbf{D}\mathbf{B}\,d\text{vol}\right]\boldsymbol{\delta}$$

and therefore the element stiffness matrix is the quantity in the square brackets. The matrices **B** and **D** contain only constant terms, and therefore, they can be taken outside the integral, leaving only $\int d\text{vol}$. For an element of constant thickness this equals the area of the triangle multiplied by its thickness, t. Therefore

$$\mathbf{S}=\frac{E}{2A_{123}(1-\nu^2)}\begin{bmatrix} y_{23} & -\nu x_{23} & -y_{13} & \nu x_{13} & y_{12} & -\nu x_{12}\\ \nu y_{23} & -x_{23} & -\nu y_{13} & x_{13} & \nu y_{12} & -x_{12}\\ \dfrac{-(1-\nu)x_{23}}{2} & \dfrac{(1-\nu)y_{23}}{2} & \dfrac{(1-\nu)x_{13}}{2} & \dfrac{-(1-\nu)y_{13}}{2} & \dfrac{-(1-\nu)x_{12}}{2} & \dfrac{(1-\nu)y_{12}}{2}\end{bmatrix} \tag{7.3.19}$$

and $x_{ij} = x_i - x_j$

and $y_{ij} = y_i - y_j$

$$\mathbf{k}^e = \mathbf{B}^T\mathbf{D}\mathbf{B}A_{123}t$$

where, as before

$$2A_{123} = \det\begin{vmatrix} 1 & x_1 & y_1 \\ 1 & x_2 & y_2 \\ 1 & x_3 & y_3 \end{vmatrix}$$

For the triangle element, **B** has been obtained explicitly and so an explicit expression for $\mathbf{k}^e$ can be obtained by matrix multiplication. For convenience of presentation, the result is given in two components.

$$\mathbf{k}^e = \mathbf{k}_\varepsilon + \mathbf{k}_\gamma \tag{7.3.20}$$

which corresponds to the direct strain terms and the shear strain term of the **D** matrix of (7.3.17). The two components are

$$\mathbf{k}_\varepsilon = \frac{Et}{4A_{123}(1-\nu^2)} \times \begin{bmatrix} y_{32}^2 & & & & & \text{symmetric} \\ -\nu y_{32}x_{32} & x_{32}^2 & & & & \\ -y_{32}y_{31} & \nu x_{32}y_{31} & y_{31}^2 & & & \\ \nu y_{32}x_{31} & -x_{32}x_{31} & -\nu y_{31}x_{31} & x_{31}^2 & & \\ y_{32}y_{21} & -\nu x_{32}y_{21} & -y_{31}y_{21} & \nu x_{31}y_{21} & y_{21}^2 & \\ -\nu y_{32}x_{21} & x_{32}x_{21} & \nu y_{31}x_{21} & -x_{31}x_{21} & -\nu y_{21}x_{21} & x_{21}^2 \end{bmatrix} \tag{7.3.21a}$$

and

$$\mathbf{k}_\gamma = \frac{Et}{8A_{123}(1+\nu)} \times \begin{bmatrix} x_{32}^2 & & & & & \text{symmetric} \\ -x_{32}y_{32} & y_{32}^2 & & & & \\ -x_{32}x_{31} & y_{32}x_{31} & x_{31}^2 & & & \\ x_{32}y_{31} & -y_{32}y_{31} & -x_{31}y_{31} & y_{31}^2 & & \\ x_{32}x_{21} & -y_{32}x_{21} & -x_{31}x_{21} & y_{31}x_{21} & x_{21}^2 & \\ -x_{32}y_{21} & y_{32}y_{21} & x_{31}y_{21} & -y_{31}y_{21} & -x_{21}y_{21} & y_{21}^2 \end{bmatrix} \tag{7.3.21b}$$

where, as before, $x_{ij} = x_i - x_j$ and $y_{ij} = y_i - y_j$.

For the analysis of an individual panel or a group of coplanar panels, the x,y axes are the structure axes and so there is no need for a transformation between element and structure coordinate systems.

We now examine the question as to whether the requirements of equilibrium and compatibility are satisfied within the element and between elements. First, since the assumed displacement field is a continuous function, compatibility is clearly satisfied within the element. Second, in regard to internal equilibrium, the equilibrium equations for a two-dimensional stress field are

$$\begin{aligned} \frac{\partial \sigma_x}{\partial x} + \frac{\partial \tau}{\partial y} &= 0 \\ \frac{\partial \sigma_y}{\partial y} + \frac{\partial \tau}{\partial x} &= 0 \end{aligned} \tag{7.3.22}$$

The element strains are constant because they are obtained by differentiation of the linear displacement field. The stresses, being related to the strains by elastic constants, are also constant, and therefore these equations are satisfied everywhere within the element.

We have already seen that the assumption of a linear displacement field ensures compatibility between elements. Thus, there only remains the question of equilibrium of stresses across the boundary of adjacent elements. Figure 7.7 illustrates such a boundary between elements A and B. The stress matrix for the triangle element was presented in (7.3.19). From this equation, it may be seen that each stress component is a function of the displacements at all three nodes of an individual element. Consequently, although both $\boldsymbol{\sigma}_A$ and $\boldsymbol{\sigma}_B$ are functions of u_1, u_2, v_1, and v_2, $\boldsymbol{\sigma}_A$ is a function of u_3 and v_3, whereas $\boldsymbol{\sigma}_B$ is a function of u_4 and v_4, and therefore the boundary stresses will not in general be equal.

Thus, for the triangular plane stress element, both the equilibrium and compatibility requirements are satisfied within the element, and the requirement of interelement compatibility is also satisfied along the element boundaries. However,

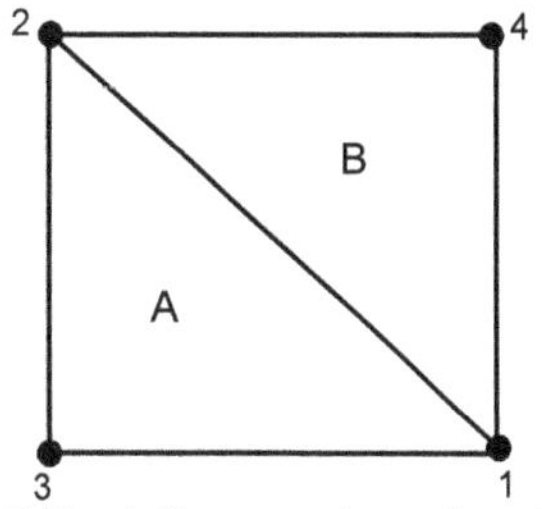

Figure 7.7 Adjacent triangular elements.

interelement equilibrium is not satisfied exactly along the boundaries. Of course, equilibrium between elements is satisfied in a discrete and less detailed manner at the nodes by means of the structure (or global) equilibrium equations which enforce equilibrium of element nodal forces throughout the structure. Therefore, as the elements are made smaller and more numerous, the solution error due to this localized lack of equilibrium will become progressively smaller and more negligible. This is shown in the following example.

7.3.6 Sample Application of the Triangle Element

The deep web beam of Fig. 7.4 will be used as an example of using the triangle element. For this simple example, there is an exact solution from theory of elasticity, and this will allow us to assess the accuracy of this element and the reduction of error with decreasing mesh size. Figure 7.8 shows two different mesh patterns which could be used. Numerical studies (Walz et al. 1968) have shown that the first pattern gives better accuracy. It also has the advantage that automatic generation of the mesh can be accomplished more easily. Another point about mesh creation is that the accuracy is reduced if any of the vertex angles are small. Hence, in choosing the mesh pattern and layout, one should always avoid elongated triangles.

Figure 7.9 shows three finite element models for which solutions have been obtained (Rockey, Evans, Griffiths, & Nethercot, 1975). The element size decreases in the proportion 1.0 to 0.8 to 0.4, and the corresponding number of elements increases from 96 to 150 to 600. The simply supported end condition is modeled by specifying zero vertical displacement of all the nodes at the ends of the beam. Figure 7.10, adapted from Rockey et al. (1975), shows the vertical deflection of the beam for the three cases, together with the solution from simple beam theory and the exact solution from the theory of elasticity. It may be seen that in each case, the finite element values are less than the exact values, but as the mesh size is reduced, the results approach the exact solution. The fact that the deflections are always smaller than the exact values is because in matrix stiffness analysis, the use of an assumed displacement function to satisfy compatibility always causes the finite element model to be stiffer than the actual structure. Since the error is one-sided, the possibility exists of reducing the error by relaxing the compatibility requirement slightly. This technique is illustrated in Sections 7.5 and 7.7.

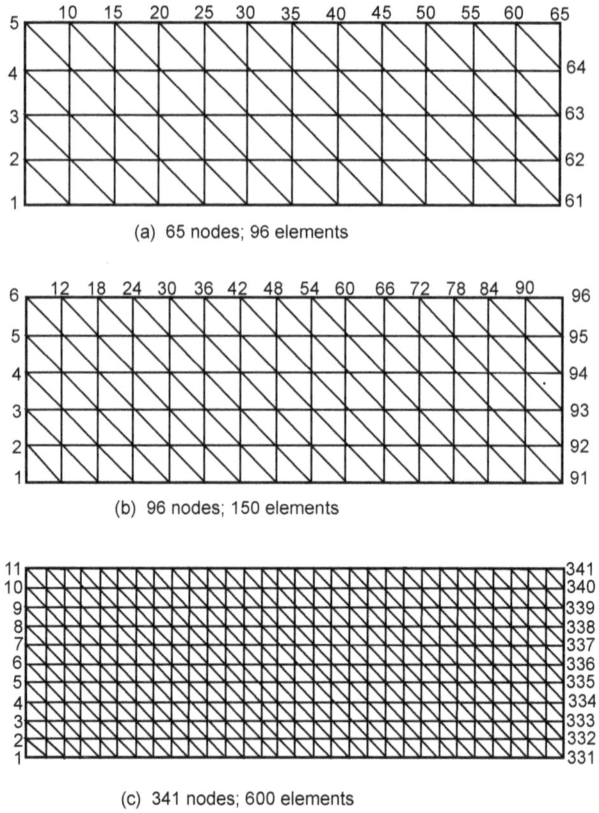

Figure 7.9 Typical finite element models.

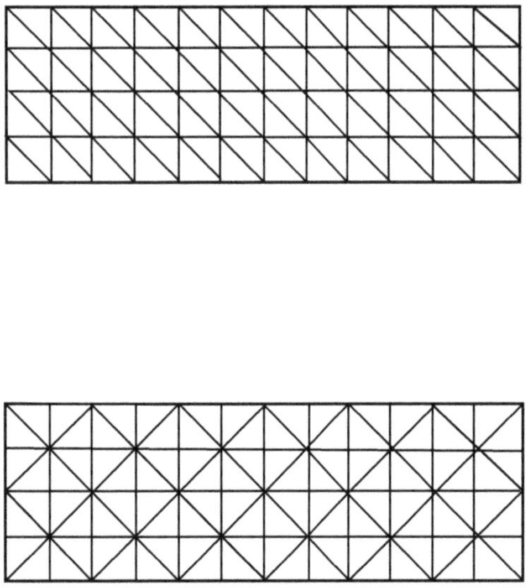
Figure 7.8 Alternative mesh patterns.

Figure 7.11 shows the vertical distribution of longitudinal stress at the midlength of the beam for the same five cases. Since the stresses obtained by the finite element method are constant within each triangle element, each stress value is plotted at the height corresponding to the centroid of the element. Other techniques are possible, such as calculating nodal stress values by averaging the stresses in elements meeting at the node.

Finally, in concluding this presentation of the triangle (or CST) element, it bears mentioning once

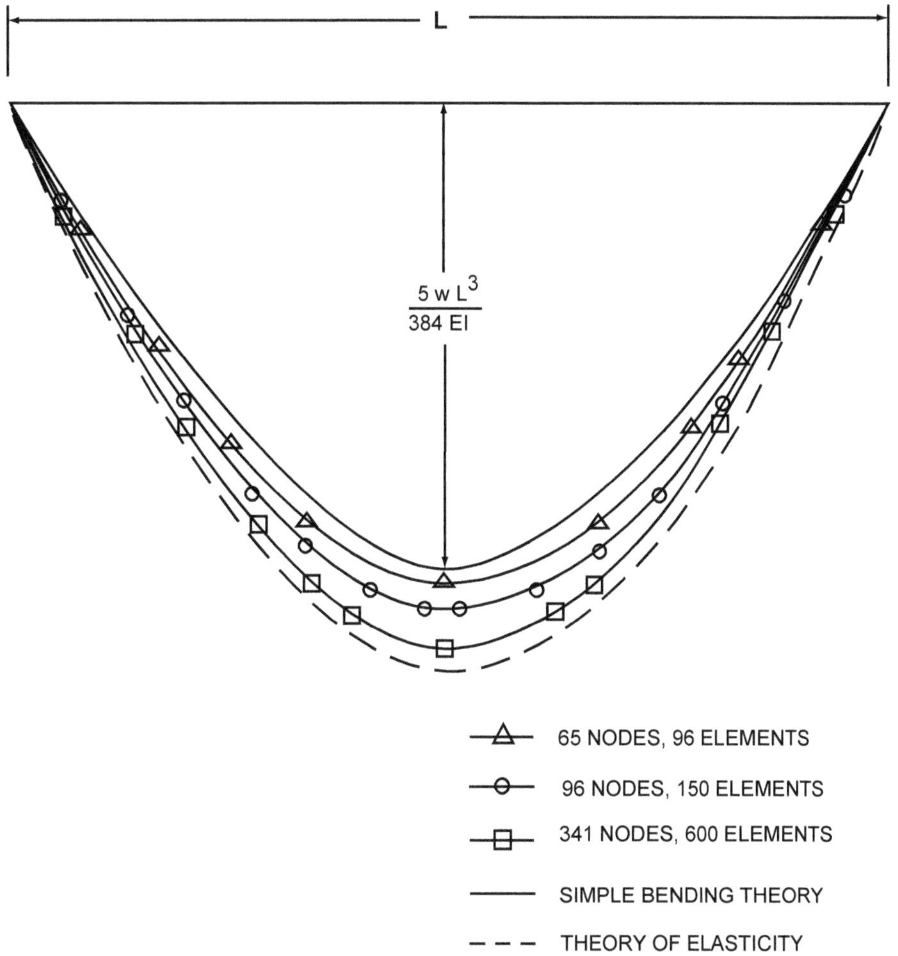

Figure 7.10 Deflection of simply supported deep beam with uniform loading.

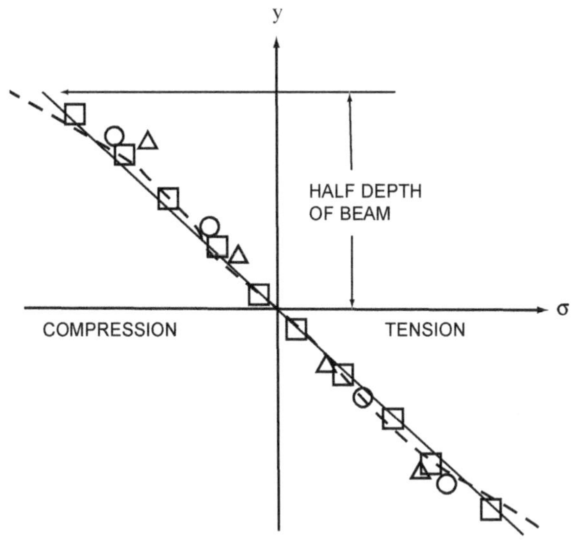

Figure 7.11 Vertical distribution of longitudinal stress, σ.

again that a group of triangles of various shapes and sizes can represent almost any irregularly shaped region. This very useful feature, together with the simplicity of the element, has made it one of the most widely used of elements; this in turn has produced a well-documented relationship between error and mesh size.

7.4 LINEAR STRAIN RECTANGLE (LSR) ELEMENT

In this section, we again use the five basic steps to derive the stiffness matrix of a rectangular plane stress element. The rectangle has sides a and b and thickness t as shown in Fig. 7.12a. The node numbering system is also shown in the figure. Since there are two degrees of freedom at each node, the element has eight degrees of freedom. Fig. 7.12b

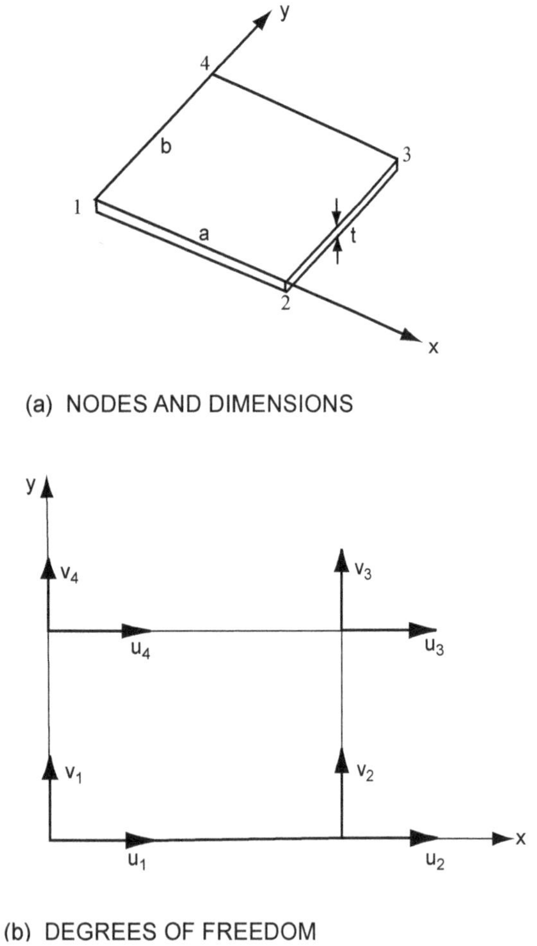

Figure 7.12 Linear strain rectangle.

shows the eight nodal displacements, which in matrix form are

$$\boldsymbol{\delta} = \begin{Bmatrix} u_1 \\ v_1 \\ u_2 \\ v_2 \\ u_3 \\ v_3 \\ u_4 \\ v_4 \end{Bmatrix}$$

7.4.1 Step I. Select Suitable Displacement Function

Since the element has eight degrees of freedom, eight unknown coefficients must be involved in the polynomial representing the displacement pattern. To avoid bias in the x- or y-direction, we use four coefficients for the u displacement and four for the v displacement, and we take an equal number of x and y terms in each. The result is

$$\begin{aligned} u(x,y) &= C_1 + C_2x + C_3y + C_4xy \\ v(x,y) &= C_5 + C_6x + C_7y + C_8xy \end{aligned} \qquad (7.4.1)$$

It may be seen that when x is a constant, both u and v vary linearly with y, and similarly when y is a constant, both displacements vary linearly with x. The displacements thus vary linearly along each side of the element. Since the displacements of two adjacent elements must be equal at their common nodes, the displacements will also agree along the entire common boundary; that is, interelement compatibility is satisfied.

Writing (7.4.1) in matrix form

$$\begin{Bmatrix} u \\ v \end{Bmatrix} = \begin{bmatrix} 1 & x & y & xy & 0 & 0 & 0 & 0 \\ 0 & 0 & 0 & 0 & 1 & x & y & xy \end{bmatrix} \begin{Bmatrix} C_1 \\ C_2 \\ C_3 \\ C_4 \\ C_5 \\ C_6 \\ C_7 \\ C_8 \end{Bmatrix} \qquad (7.4.2)$$

or

$$\boldsymbol{\delta}(x, y) = \mathbf{H}(x, y)\mathbf{C}$$

7.4.2 Step II. Relate the General Displacement Within the Element to the Nodal Displacements

As before, this step is achieved by substituting the values of the nodal coordinates into (7.4.2) four times, once for each node, and then solving for $\mathbf{C}$. The substitution gives

$$\boldsymbol{\delta} = \begin{bmatrix} 1 & 0 & 0 & 0 & 0 & 0 & 0 & 0 \\ 0 & 0 & 0 & 0 & 1 & 0 & 0 & 0 \\ 1 & a & 0 & 0 & 0 & 0 & 0 & 0 \\ 0 & 0 & 0 & 0 & 1 & a & 0 & 0 \\ 1 & a & b & ab & 0 & 0 & 0 & 0 \\ 0 & 0 & 0 & 0 & 1 & a & b & ab \\ 1 & 0 & b & 0 & 0 & 0 & 0 & 0 \\ 0 & 0 & 0 & 0 & 1 & 0 & b & 0 \end{bmatrix} \mathbf{C} \qquad (7.4.3)$$

or

$$\boldsymbol{\delta} = \mathbf{AC}$$

It may be shown that the solution for **C** is

$$\mathbf{C} = \frac{1}{ab}\begin{bmatrix} ab & 0 & 0 & 0 & 0 & 0 & 0 & 0 \\ -b & 0 & b & 0 & 0 & 0 & 0 & 0 \\ -a & 0 & 0 & 0 & 0 & 0 & a & 0 \\ 1 & 0 & -1 & 0 & 1 & 0 & -1 & 0 \\ 0 & ab & 0 & 0 & 0 & 0 & 0 & 0 \\ 0 & -b & 0 & b & 0 & 0 & 0 & 0 \\ 0 & -a & 0 & 0 & 0 & 0 & 0 & a \\ 0 & 1 & 0 & -1 & 0 & 1 & 0 & -1 \end{bmatrix}\boldsymbol{\delta} \qquad (7.4.4)$$

or

$$\mathbf{C} = \mathbf{A}^{-1}\boldsymbol{\delta}$$

From (7.4.2), the expression for the general displacement in terms of the nodal displacements is

$$\boldsymbol{\delta}(x, y) = \mathbf{HA}^{-1}\boldsymbol{\delta} \qquad (7.4.5a)$$

Although it is not required for the derivation of the element stiffness matrix, we next evaluate the product $\mathbf{HA}^{-1}$ to obtain an explicit expression for the element internal displacement $\boldsymbol{\delta}(x,y)$ in terms of the nodal displacements, in order to examine the displacement field of the element more closely. Substituting for **H** and $\mathbf{A}^{-1}$ and multiplying them gives

$$\begin{aligned} u &= (1-\xi)(1-\eta)u_1 + \xi(1-\eta)u_2 + \xi\eta u_3 \\ &\quad + (1-\xi)\eta u_4 \\ v &= (1-\xi)(1-\eta)v_1 + \xi(1-\eta)v_2 + \xi\eta v_3 \\ &\quad + (1-\xi)\eta v_4 \end{aligned} \qquad (7.4.5b)$$

where $\xi = x/a$ and $\eta = y/b$.

Figure 7.13 shows the displacement field which results when a unit horizontal displacement is imposed at node 3 ($u_3 = 1$; $v_3 = 0$) and all other nodes are pinned: $u_i = v_i = 0$, i = 1, 2, 4. A similar pattern would occur if unit displacements were imposed at other nodes. The displacement pattern is extremely simple, which is the reason why interelement compatibility is satisfied, but this simplicity brings with it a large degree of artificiality in the way the element deforms. For example, the pattern implies very large local shear deformation in the vicinity of nodes 2 and 3. The actual deformation of an element of plating subjected to the prescribed nodal displacements of Fig. 7.13 would depend on the total boundary conditions on all of the edges, but the deformation would certainly include some in-plane bending and curved edges, as shown later in Fig. 7.15. Since the linear strain rectangle (LSR) element lacks this degree of versatility, it will inevitably be too stiff and this brings with it some error, in spite of the fact that interelement compatibility is satisfied exactly. As we shall see in the next section, elements can be derived which have this versatility and have greater accuracy, even when compatibility is not satisfied exactly.

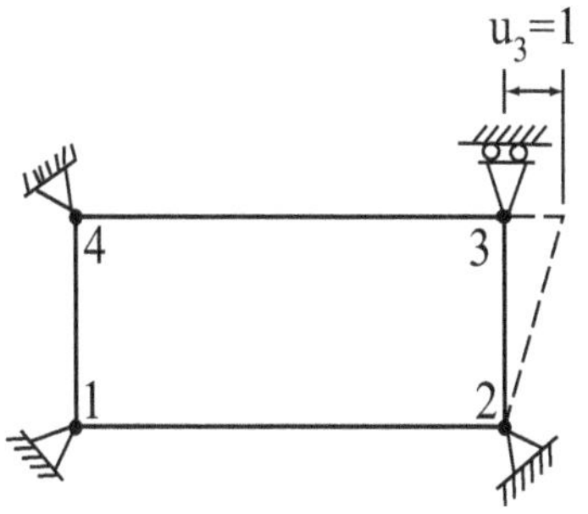

Figure 7.13 Typical unit nodal displacement.

7.4.3 Step III. Express the Internal Deformation (Strain) in Terms of the Nodal Displacements

The definition of strain in a two-dimensional continuum was given in (7.3.10). Substituting for u and v from (7.4.2) and performing the required differentiation gives

$$\boldsymbol{\varepsilon}(x,y) = \begin{bmatrix} 0 & 1 & 0 & y & 0 & 0 & 0 & 0 \\ 0 & 0 & 0 & 0 & 0 & 0 & 1 & x \\ 0 & 0 & 1 & x & 0 & 1 & 0 & y \end{bmatrix}\mathbf{C} \qquad (7.4.6)$$

or

$$\boldsymbol{\varepsilon}(x, y) = \mathbf{GC}$$

As before, we substitute for **C** from (7.4.4) and introduce the strain matrix $\mathbf{B} = \mathbf{GA}^{-1}$, in terms of which (7.4.6) becomes

$$\boldsymbol{\varepsilon}(x, y) = \mathbf{B}\boldsymbol{\delta} \qquad (7.4.7)$$

On performing the required matrix multiplication, the strain matrix is found to be

$$\mathbf{B} = \frac{1}{ab}\begin{bmatrix} -b+y & 0 & b-y & 0 & y & 0 & -y & 0 \\ 0 & -a+x & 0 & -x & 0 & x & 0 & a-x \\ -a+x & -b+y & -x & b-y & x & y & a-x & -y \end{bmatrix} \qquad (7.4.8)$$

From this expression, it may be seen that for any given set of nodal displacements, $\boldsymbol{\delta}$, the ε_x strain is constant in the x-direction and varies linearly with y. Similarly, the ε_y strain is constant in the y-direction, and varies linearly with x. Finally, the shear strain γ varies linearly with both x and y.

7.4.4 Step IV. Express the Internal Force (Stress) in Terms of the Nodal Displacements, Using the Element's Law of Elastic Behavior

Stress is obtained from strain by or $\boldsymbol{\sigma} = \mathbf{D}\boldsymbol{\varepsilon}$, with the $\mathbf{D}$ matrix for plane stress given by (7.3.17). Therefore, this step consists in simply substituting for $\boldsymbol{\varepsilon}(x,y)$ from (7.4.7) to obtain

$$\boldsymbol{\sigma}(x, y) = \mathbf{DB}\boldsymbol{\delta}$$

For the derivation of the element stiffness matrix, this expression is all that is required. If desired, the explicit stress matrix $\mathbf{S}(x,y) = \mathbf{DB}$ could be calculated at this point, so that it would be available when performing a structural analysis with the element. The expression for $\mathbf{S}(x,y)$ would show that all stress components in the rectangle vary linearly in the x- and y-directions. Substitution of these stresses into the equilibrium relations of equation (7.3.22) would reveal that, for an arbitrary set of nodal displacements, equilibrium is not satisfied exactly. It would, however, be satisfied for uniform extension ($u_1 = u_4$, $u_2 = u_3$, $v_1 = v_2$, $v_3 = v_4$) which corresponds to uniform direct strain: $\varepsilon_x =$ constant and $\varepsilon_y =$ constant. The departure from internal equilibrium is therefore proportional to the degree of shear that is present, and so the element can be expected to have less accuracy when large shear stresses are present.

Since the stress distribution within the element depends on all four nodal displacements, two adjacent elements will not have identical stress distributions along their common boundary, and so once again, interelement equilibrium is not satisfied to an exact and detailed degree, but only in an overall node-by-node manner.

Since the stresses vary linearly within the element, the stress matrix $\mathbf{S}(x,y)$ does not give specific numerical values of stress as it did for the triangle element. Therefore, in order to calculate actual stress values, it is necessary to choose some specific location. The usual practice is to choose the nodes, much the same as in the calculation of nodal values of bending moment in a beam element. The nodal stress matrix $\mathbf{S}$ is obtained by making four successive substitutions of the four nodal coordinates into the $\mathbf{B}$ matrix, giving an 8×12 matrix.

$$\mathbf{S} = \begin{bmatrix} \mathbf{DB}(0,0) \\ \mathbf{DB}(a,0) \\ \mathbf{DB}(a,b) \\ \mathbf{DB}(0,b) \end{bmatrix}$$

The detailed results are not given here because in the next section we present another rectangle element which has better accuracy than the simple LSR element presented here.

Since the stress distribution in each element is constrained to be of the form specified by $\mathbf{S}(x,y)$, it gives only an approximate representation of the stress distribution in the actual structure. Hence, at a common node shared by two or more elements, the respective nodal stress values will not be exactly equal. However, the discrepancy is usually small, especially if a small mesh size is used, and a good approximation can be obtained by averaging all of the stress values at a particular node.

7.4.5 Step V. Obtain the Element Stiffness Matrix by Relating Nodal Forces to Nodal Displacements

This step is identical to that of Section 7.3.5; the element stiffness matrix is again given by

$$\mathbf{k}^e = \left[\int_{\text{vol}} \mathbf{B}^{\mathrm{T}}\mathbf{D}\mathbf{B}\, d\text{vol}\right]$$

or, for an element of constant thickness

$$\mathbf{k}^e = t\left[\int_0^b \int_0^a \mathbf{B}^{\mathrm{T}}\mathbf{D}\mathbf{B}\, dx\, dy\right]$$

with $\mathbf{B}$ defined by (7.4.8).

Unlike the triangle element in which $\mathbf{B}$ contained only constant terms, $\mathbf{B}$ here contains both x and y terms and cannot be taken outside the integral. Hence, the product $\mathbf{B}^{\mathrm{T}}\mathbf{DB}$ must be evaluated first, and then the terms of the resulting matrix must be integrated over the area of the element. As with the triangle element, it is convenient to present the result as the sum of two parts:

$$\mathbf{k}^e = \mathbf{k}_\varepsilon + \mathbf{k}_\gamma \tag{7.4.9a}$$

where

$$\mathbf{k}_\varepsilon = \frac{Et}{12(1-\nu^2)}\begin{bmatrix} 4/\alpha &&&&&&& \\ 3\nu & 4\alpha &&&&&& \\ -4/\alpha & -3\nu & 4/\alpha &&&&& \\ 3\nu & 2\alpha & -3\nu & 4\alpha &&&& \\ -2/\alpha & -3\nu & 2/\alpha & -3\nu & 4/\alpha &&& \\ -3\nu & -2\alpha & 3\nu & -4\alpha & 3\nu & 4\alpha && \\ 2/\alpha & 3\nu & -2/\alpha & 3\nu & -4/\alpha & -3\nu & 4/\alpha & \\ -3\nu & -4\alpha & 3\nu & -2\alpha & 3\nu & 2\alpha & -3\nu & 4\alpha \end{bmatrix} \tag{7.4.9b}$$

in which $\alpha = a/b$ and

$$\mathbf{k}_\gamma = \frac{Et}{24(1+\nu)}\begin{bmatrix} 4\alpha &&&&&&& \\ 3 & 4/\alpha &&&&&& \\ 2\alpha & 3 & 4\alpha &&&&& \\ -3 & -4/\alpha & -3 & 4/\alpha &&&& \\ -2\alpha & -3 & -4\alpha & 3 & 4\alpha &&& \\ -3 & -2/\alpha & -3 & 2/\alpha & 3 & 4/\alpha && \\ -4\alpha & -3 & -2\alpha & 3 & 2\alpha & 3 & 4\alpha & \\ 3 & 2/\alpha & 3 & -2/\alpha & -3 & -4/\alpha & -3 & 4/\alpha \end{bmatrix} \tag{7.4.9c}$$

7.5 CONSTANT SHEAR STRESS RECTANGLE (CSSR) ELEMENT

We now present another rectangle element which is generally superior to the LSR element. It is actually one of the oldest elements, having been introduced in the landmark paper by Turner et al. (1956). Instead of using an assumed displacement function to define the internal behavior of the element, we do this by specifying a stress distribution within the element.

7.5.1 Step I. Obtain a Distribution Function From An Assumed Stress Distribution

We choose a simple distribution in which the direct stresses vary linearly and the shear stress is constant (the latter characteristic furnishing the name of the element)

$$\begin{aligned} \sigma_x &= C_1 + C_2 y \\ \sigma_y &= C_3 + C_4 x \\ \tau &= C_5 \end{aligned} \tag{7.5.1}$$

where $C_1, \ldots, C_5$ are constant coefficients, as yet unknown. Figure 7.14 shows the local coordinate system and other details.

This simple stress distribution, unlike that of the LSR element, satisfies the stress equilibrium relationships of (7.3.22) within the rectangle. On the other hand, as we shall see, the resulting displacement distribution does not satisfy interelement compatibility of boundary displacements.

From the stress-strain relationship for plane stress given in (7.3.16), we have

$$\varepsilon_x = \frac{\partial u}{\partial x} = \frac{1}{E}\left(C_1 + C_2 y - \nu C_3 - \nu C_4 x\right) \tag{7.5.2}$$

which, when integrated, becomes

$$u = \frac{1}{E}\left(C_1 x + C_2 xy - \nu C_3 x - \frac{\nu C_4 x}{2}\right) + \frac{f(y)}{E} \tag{7.5.3}$$

where $f(y)$ is an arbitrary function of y. Similarly, starting with the strain ε_y, we can show that

$$v = \frac{1}{E}\left(C_3 y + C_4 xy - \nu C_1 y - \frac{\nu C_2 y^2}{2}\right) + \frac{g(x)}{E} \tag{7.5.4}$$

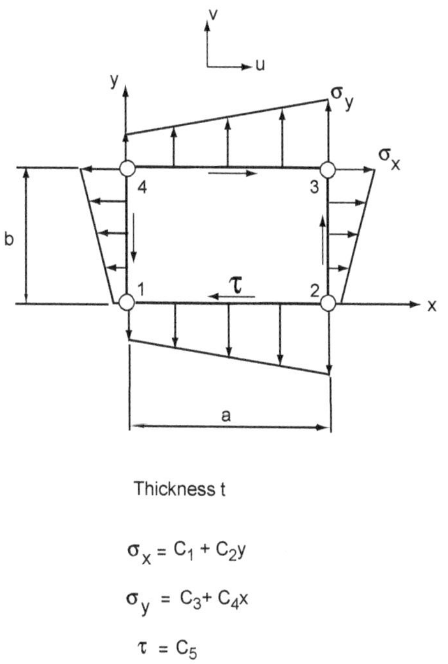

Figure 7.14 Constant shear stress rectangle (CSSR) element.

where $g(x)$ is a function of x only. Also, from the equation for the shear strain, we have

$$\gamma = \frac{\partial u}{\partial y} + \frac{\partial v}{\partial x} = \frac{\tau}{G} = 2(1+\nu)\frac{C_5}{E} \qquad (7.5.5)$$

Substituting (7.5.3) and (7.5.4) into (7.5.5) and rearranging, we have

$$\frac{\partial f}{\partial y} + C_4 y = 2(1+\nu)C_5 - \left[\frac{\partial g}{\partial x} + C_2 x\right] = C_6 \qquad (7.5.6)$$

where C_6 represents a constant, which is the only possible condition that will satisfy (7.5.6). Solving for $f(y)$ and $g(x)$, we obtain

$$f(y) = -\frac{C_4 y^2}{2} + [(1+\nu)C_5 + C_6]y + C_7 \qquad (7.5.7)$$

$$g(x) = -\frac{C_2 x^2}{2} + [(1+\nu)C_5 - C_6]x + C_8 \qquad (7.5.8)$$

The constants of integration C_7 and C_8 represent rigid-body translations, while the previously introduced constants C_5 and C_6 define rigid-body rotation.

If we substitute (7.5.7) and (7.5.8) into (7.5.3) and (7.5.4), we obtain the general displacement in terms of the unknown coefficients

$$\boldsymbol{\delta}(x,y) = \begin{Bmatrix} u \\ v \end{Bmatrix} = \mathbf{H}(x,y)\mathbf{C} \qquad (7.5.9)$$

where

$$\mathbf{H} = \frac{1}{E}\begin{bmatrix} x & xy & -\nu x & -\frac{1}{2}(\nu x^2 + y^2) & (1+\nu)y & y & 1 & 0 \\ -\nu y & -\frac{1}{2}(\nu y^2 + x^2) & y & xy & (1+\nu)x & -x & 0 & 1 \end{bmatrix}$$

Equation 7.5.9 is the displacement function which results from the chosen stress distribution and which fulfills the requirement of geometric compatibility within the element. Hence, the foregoing analysis is an alternative method of accomplishing Step I in the standard procedure for deriving the stiffness matrix of an element. From here on, the remaining four steps are the same as for the previous two elements.

7.5.2 Step II. Relate the General Displacement Within the Element to the Nodal Displacements

The unknown coefficients $\mathbf{C}$ can now be expressed in terms of the element nodal displacements $\boldsymbol{\delta}$ by the usual technique of substituting each pair of nodal coordinates into $\mathbf{H}(x,y)$ in (7.5.9) to give

$$\boldsymbol{\delta} = \mathbf{AC}$$

where

$$\mathbf{A} = \begin{bmatrix} \mathbf{H}(0,0) \\ \mathbf{H}(a,0) \\ \mathbf{H}(a,b) \\ \mathbf{H}(0,b) \end{bmatrix}$$

and then solving for $\mathbf{C}$ by inverting $\mathbf{A}$

$$\mathbf{C} = \mathbf{A}^{-1}\boldsymbol{\delta} \qquad (7.5.10)$$

In the present case the inverse matrix is

$$\mathbf{A}^{-1} = \frac{E}{a}\begin{bmatrix} -\mathscr{C} & -\mathscr{A}\alpha & \mathscr{C} & -\mathscr{A}\alpha & \nu\mathscr{A} & \mathscr{A}\alpha & -\nu\mathscr{A} & \mathscr{A}\alpha \\ \frac{1}{b} & 0 & \frac{-1}{b} & 0 & \frac{1}{b} & 0 & \frac{-1}{b} & 0 \\ -\mathscr{A} & -\mathscr{C}\alpha & \mathscr{A} & -\nu\mathscr{A}\alpha & \mathscr{A} & \nu\mathscr{A}\alpha & -\mathscr{A} & \mathscr{C}\alpha \\ 0 & \frac{1}{b} & 0 & \frac{-1}{b} & 0 & \frac{1}{b} & 0 & \frac{-1}{b} \\ -\mathscr{B}\alpha & -\mathscr{B} & -\mathscr{B}\alpha & \mathscr{B} & \mathscr{B}\alpha & \mathscr{B} & \mathscr{B}\alpha & -\mathscr{B} \\ \frac{-3\alpha}{4} & \frac{3}{4} & \frac{\alpha}{4} & \frac{-3}{4} & \frac{-\alpha}{4} & \frac{1}{4} & \frac{3\alpha}{4} & \frac{-1}{4} \\ a & 0 & 0 & 0 & 0 & 0 & 0 & 0 \\ 0 & a & 0 & 0 & 0 & 0 & 0 & 0 \end{bmatrix}$$

where

$$\mathscr{A} = \frac{\nu}{2(1-\nu^2)} \qquad \mathscr{B} = \frac{1}{4(1+\nu)}$$

$$\mathscr{C} = \frac{2-\nu^2}{2(1-\nu^2)} \qquad \alpha = \frac{a}{b}$$

At this point, we digress momentarily from the derivation of $\mathbf{k}^e$ in order to examine the displacement field of this element more closely. For this purpose, we calculate the explicit expression for the general displacement $\boldsymbol{\delta}(x,y)$ in terms of the nodal displacements. From (7.5.9) and (7.5.10), the matrix expression is

$$\boldsymbol{\delta}(x,y) = \begin{Bmatrix} u \\ v \end{Bmatrix} = \mathbf{H}(x,y)\mathbf{A}^{-1}\boldsymbol{\delta}$$

and, after substituting for $\mathbf{A}^{-1}$ and performing the required matrix multiplication, the final result is

$$\begin{aligned} u &= (1-\xi)(1-\eta)u_1 + \xi(1-\eta)u_2 + \xi\eta u_3 \\ &\quad + (1-\xi)\eta u_4 \\ &\quad + \frac{1}{2}\left[\nu\alpha(\xi-\xi^2) + \frac{1}{\alpha}(\eta-\eta^2)\right] \times \\ &\quad (v_1 - v_2 + v_3 - v_4) \end{aligned} \tag{7.5.11}$$

$$\begin{aligned} v &= (1-\xi)(1-\eta)v_1 + \xi(1-\eta)v_2 + \xi\eta v_3 \\ &\quad + (1-\xi)\eta v_4 \\ &\quad + \frac{1}{2}\left[\alpha(\xi-\xi^2) + \frac{\nu}{\alpha}(\eta-\eta^2)\right] \times \\ &\quad (u_1 - u_2 + u_3 - u_4) \end{aligned}$$

where $\xi = x/a$, $\eta = y/b$, and $\alpha = a/b$.

Figure 7.15 illustrates the deformation of the element for a unit value of u_3 with all other nodal displacements held at zero, as was done for the LSR element in Fig. 7.13. Along the top edge ($0 < \xi < 1$; $\eta = 1$) the displacements are

$$u = \xi; \quad v = \frac{\alpha}{2}(\xi - \xi^2)$$

and along the bottom edge ($0 < \xi < 1$; $\eta = 0$), they are

$$u = 0; \quad v = \frac{\alpha}{2}(\xi - \xi^2)$$

In addition to the elongation of the top edge, there is some sympathetic curvature of the top and bottom edges, which makes the deformed shape more realistic because it avoids the large local shear deformation which is implied in the LSR element. In fact, by definition, the shear stress in the constant shear stress rectangle (CSSR) element is uniform throughout the element.

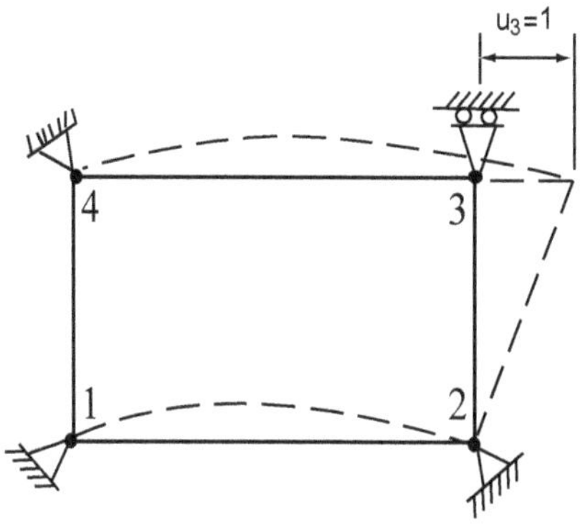

Figure 7.15 Deformation of CSSR element due to displacement u_3.

The v displacement of the upper and lower edges is a quadratic expression, which requires three coefficients to define uniquely, whereas there are only two nodal values of v along each edge. Hence, interelement compatibility is not entirely satisfied by this element.

It would be possible to have quadratically varying edge displacements and to also satisfy interelement compatibility by adding midside nodes along each edge because with three nodes the quadratic would be uniquely defined. Another possibility is that elements may have more than two degrees of freedom per node (e.g., they could involve the derivatives of u and v thereby involving more coefficients and hence higher order displacement functions). However, midside nodes cause complications in mesh generation, and both of these approaches increase the total number of degrees of freedom and, therefore, the solution time and cost. In many cases, these higher order elements give a lower ratio of accuracy to cost than is achieved by simply using more elements. For example, a rectangle with midside nodes has a total of 16 degrees of freedom, whereas using four quarter-size ordinary rectangle elements would involve 18 degrees of freedom, which is only a marginal increase and gives a more detailed solution.

The remaining three steps in the derivation of $\mathbf{k}^e$ for the CSSR element are the same as for the LSR element; therefore, only the results are presented. The stiffness matrix and the equations for nodal values of stress are given in (7.5.12) and (7.5.13).

$$\mathbf{k}^e = \frac{Et}{4(1-\nu^2)} \times$$

$$\begin{bmatrix}
\mathcal{A}\alpha + \mathcal{D}/\alpha & & & & & & & \\
\mathcal{B} & \mathcal{A}/\alpha + \mathcal{D}\alpha & & & & & & \\
\mathcal{A}\alpha - \mathcal{D}/\alpha & \mathcal{C} & \mathcal{A}\alpha - \mathcal{D}/\alpha & & & & & \\
\mathcal{C} & -\mathcal{A}/\alpha + \mathcal{E}\alpha & -\mathcal{B} & \mathcal{A}\alpha + \mathcal{D}/\alpha & & & & \\
-\mathcal{A}\alpha - \mathcal{E}/\alpha & -\mathcal{B} & -\mathcal{A}/\alpha + \mathcal{E}\alpha & \mathcal{C} & \mathcal{A}\alpha + \mathcal{D}/\alpha & & & \\
-\mathcal{B} & -\mathcal{A}/\alpha - \mathcal{E}\alpha & -\mathcal{C} & \mathcal{A}\alpha - \mathcal{D}\alpha & \mathcal{B} & \mathcal{A}\alpha + \mathcal{D}/\alpha & & \\
-\mathcal{A}\alpha - \mathcal{E}/\alpha & -\mathcal{C} & -\mathcal{A}/\alpha - \mathcal{E}\alpha & \mathcal{B} & \mathcal{A}\alpha - \mathcal{D}/\alpha & \mathcal{C} & \mathcal{A}\alpha + \mathcal{D}/\alpha & \\
\mathcal{C} & \mathcal{A}/\alpha - \mathcal{D}\alpha & \mathcal{B} & -\mathcal{A}/\alpha - \mathcal{E}\alpha & -\mathcal{C} & -\mathcal{A}\alpha + \mathcal{E}/\alpha & -\mathcal{B} & \mathcal{A}\alpha + \mathcal{D}/\alpha
\end{bmatrix} \tag{7.5.12}$$

where

$$\mathcal{A} = \frac{1-\nu}{2}; \quad \mathcal{B} = \frac{1+\nu}{2}; \quad \mathcal{C} = \frac{1-3\nu}{2}$$

$$\mathcal{D} = \frac{4-\nu^2}{3}; \quad \mathcal{E} = \frac{2+\nu^2}{3}; \quad \alpha = \frac{a}{b}$$

$$(\sigma_x)_1 = (\sigma_x)_2 = \frac{E}{1-\nu^2}\left[\frac{-u_1+u_2}{a} + \frac{\nu^2(u_1-u_2+u_3-u_4)}{2a} + \frac{\nu(-v_1-v_2+v_3+v_4)}{2b}\right]$$

$$(\sigma_x)_3 = (\sigma_x)_4 = \frac{E}{1-\nu^2}\left[\frac{-u_3+u_4}{a} + \frac{\nu^2(-u_1+u_2-u_3+u_4)}{2a} + \frac{\nu(-v_1-v_2+v_3+v_4)}{2b}\right]$$

$$(\sigma_y)_1 = (\sigma_y)_4 = \frac{E}{1-\nu^2}\left[\frac{-v_1+v_4}{b} + \frac{\nu^2(v_1-v_2+v_3-v_4)}{2b} + \frac{\nu(-u_1+u_2+u_3-u_4)}{2a}\right] \tag{7.5.13}$$

$$(\sigma_y)_2 = (\sigma_y)_3 = \frac{E}{1-\nu^2}\left[\frac{-v_2+v_3}{b} + \frac{\nu^2(-v_1+v_2-v_3+v_4)}{2b} + \frac{\nu(-u_1+u_2+u_3-u_4)}{2a}\right]$$

$$\tau = \frac{E}{4(1+\nu)}\left[\frac{-u_1-u_2+u_3+u_4}{b} + \frac{-v_1+v_2+v_3-v_4}{a}\right]$$

7.6 SHAPE FUNCTIONS AND EQUIVALENT NODAL FORCES

Step II in the derivation of an element stiffness matrix leads to an expression for the displacements $u(x,y)$ and $v(x,y)$ anywhere within the element in terms of the nodal displacements. For example, for the LSR element the expression is given in (7.4.5b), which we now rewrite in matrix notation

$$\boldsymbol{\delta}(x, y) = \mathbf{N}^e(\xi, \eta)\boldsymbol{\delta}$$

or in expanded form

$$\begin{Bmatrix} u(x,y) \\ v(x,y) \end{Bmatrix} = \begin{bmatrix} \mathbf{N}(\xi,\eta) & \\ & \mathbf{N}(\xi,\eta) \end{bmatrix} \begin{Bmatrix} u \\ v \end{Bmatrix} \tag{7.6.1}$$

where

$$\mathbf{N}(\xi, \eta) = [(1-\xi)(1-\eta) \;\; \xi(1-\eta) \;\; \xi\eta \;\; (1-\xi)\eta]$$

and in which $\xi = x/a$ and $\eta = y/b$.

The vector $\mathbf{N}(\xi,\eta)$ is referred to as the *shape function* because it completely specifies the pattern or shape of the displacement everywhere within the element, given the nodal displacements. We will now show how the shape function can be used to obtain a general expression for the equivalent nodal forces for the two types of distributed external loads: "body forces," **b** (forces on a body that are proportional to the body's volume, such as weight and inertia), and "surface forces," **p** (forces which act on the surface, either normally—pressure—or tangentially—shear or "traction"). To do so, we will again utilize the Principle of Virtual Work, but we now apply it to the *overall structure*, rather than to just one element. The expression is

$$\boldsymbol{\Delta}^{*T}\mathbf{P} + \int \boldsymbol{\Delta}^{*}(\mathbf{X})^T\mathbf{b}(\mathbf{X})dV + \int \boldsymbol{\Delta}^{*}(\mathbf{S})^T\mathbf{p}(\mathbf{S})dA = \int \boldsymbol{\varepsilon}^{*}(\mathbf{X})^T\boldsymbol{\sigma}(\mathbf{X})dV \qquad (7.6.2)$$

The left-hand side is the virtual work done by point loads **P**, body forces **b**, and surface forces **p**. V and A refer to the volume and external surface area of the structure, and **X** and **S** are coordinates indicating location within this volume and on this surface, respectively. The right-hand side is the internal virtual work.

We now make the two substitutions that lie at the heart of the finite element method:

1. We divide the structure volume V and the structure boundary S into elements and use integration by parts; that is, the integrals over the structure are each replaced by a sum of integrals over the elements.
2. We represent the displacement within each element by means of shape functions **N(X)** which specify the displacement in terms of the nodal values, $\boldsymbol{\Delta}$:

$$\boldsymbol{\Delta}(\mathbf{X}) = \mathbf{N}(\mathbf{X})\boldsymbol{\Delta} \qquad (7.6.3)$$

With these substutions, (7.6.2) becomes*

$$\boldsymbol{\Delta}^{*T}\mathbf{P} + \sum_e\left[\int_{V^e}(\mathbf{N}\boldsymbol{\Delta}^*)^T\mathbf{b}(\mathbf{X})\,dV^e + \int_{A^e}(\mathbf{N}\boldsymbol{\Delta}^*)^T\mathbf{p}(\mathbf{S})dA^e\right] = \sum_e\int_{V^e}(\mathbf{B}\boldsymbol{\Delta}^*)^T\mathbf{D}\mathbf{B}\boldsymbol{\Delta}dV^e$$

or

$$\boldsymbol{\Delta}^{*T}\left[\mathbf{P} + \sum_e\int_{V^e}\mathbf{N}^T\mathbf{b}(\mathbf{X})dV^e + \sum_e\int_{A^e}\mathbf{N}^T\mathbf{p}(\mathbf{S})dA^e\right] = \boldsymbol{\Delta}^{*T}\left[\sum_e\int_{V^e}\mathbf{B}^T\mathbf{D}\mathbf{B}dV^e\right]\boldsymbol{\Delta} \qquad (7.6.4)$$

in which the virtual displacements $\boldsymbol{\Delta}^*$ cancel. The right-hand side is seen to be $\mathbf{K}\boldsymbol{\Delta}$; the left-hand side is the complete load vector in which the second and third terms are the equivalent discrete nodal forces, in structure coordinates, that are used to represent the distributed body forces and surface forces that act on the structure. These equivalent nodal forces are calculated element by element and summed over all of the structure degrees of freedom, in much the same manner as the structure stiffness matrix is assembled element by element.

7.7 QUADRILATERAL AND ISOPARAMETRIC ELEMENTS

In this text, the process of defining and deriving each new finite element has usually commenced with the selection of a displacement function **H**(x) or $\mathbf{H}(x,y)$, whereupon the shape function **N** is obtained from $\mathbf{N} = \mathbf{H}\mathbf{A}^{-1}$. But an alternative (and more general) approach is to begin by specifying the shape function. Once this is done, all of the element's properties (stiffness matrix, stress matrix, equivalent nodal forces, etc.) can be established uniquely. As shown in such standard texts as Cook et al. (1989), many elements are now available for various specialized applications, and many of these have been developed through the use of shape functions. This is a very large topic and all of it is beyond the scope of this text, but there are four particular topics that are fundamental in developing more general elements, and they are now briefly described.

7.7.1 Natural Coordinates

For a four-sided element, shape functions are conveniently defined in terms of "natural" coordinates s, t which have a unit value at the nodes, as shown in Fig. 7.16a. A set of four shape functions is defined as follows:

$$N_i\,(s,t) = \tfrac{1}{4}(1 + s_is)(1 + t_it) \quad (i = 1,\ldots,4) \qquad (7.7.1)$$

in which s_i and t_i are the values of s and t at node i (either +1 or −1). Thus, each $N_i\,(s,t)$ is a smoothly varying interpolating function which has a unit value at node i and is zero at the other three nodes. The internal displacement can therefore be expressed as the sum of these four functions,

*In reality, the shape function is defined in element coordinates, and (7.6.3) implies a prior transformation to structure coordinates.

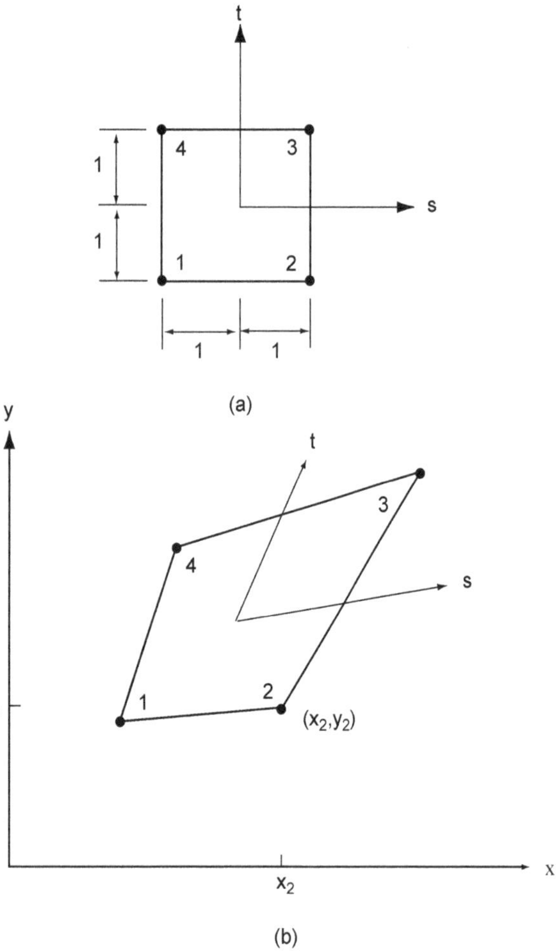

Figure 7.16 (a) Natural coordinates; (b) Isoparametric quadrilateral element.

with each being multiplied by its associated nodal displacement:

$$\delta(s,t) = \sum_{i=1}^{4} \delta_i N_i(s,t) \tag{7.7.2}$$

7.7.2 Nonconforming Displacement Modes

The element displacement defined by (7.7.1) and (7.7.2) is in fact the same shape function as that of the LSR element given in (7.6.1); the only difference is the use of natural coordinates. Section 7.5 showed that the displacement function of the CSSR element is preferable to that of the LSR element because it contains quadratic terms that allow the element to accommodate in-plane bending. Shape functions provide a means for adding such higher order terms explicitly and selectively in order to allow an element to accommodate a particular mode of deformation. For example, this technique may be used to explicitly add some quadratic terms to the LSR displacement, and the resulting element is found to be identical to the CSSR element. This method of derivation was used by Wilson, Taylor, Doherty, Ghabussi (1973), but it was not recognized until some time later that the element was actually the same as the CSSR element. When the quadratic terms are added to the basic shape function of (7.7.2), the resulting displacement function is

$$\boldsymbol{\delta}(s,t) = \sum_{i=1}^{4} \boldsymbol{\delta}_i N_i(s,t) + \sum_{i=5}^{6} \boldsymbol{\delta}_i \overline{N}_i(s,t) \tag{7.7.3}$$

where the additional shape functions are

$$\overline{N}_5(s,t) = 1 - s^2$$

and

$$\overline{N}_6(s,t) = 1 - t^2$$

As noted in the discussion following the derivation of the CSSR element, the presence of the quadratic terms means that interelement compatibility is not perfectly satisfied. Such elements are referred to as "nonconforming" and, if this feature is used properly, it gives greater accuracy because a conforming element is always "too stiff."

7.7.3 Isoparametric Elements: Quadrilateral

Besides their use in describing displacements within elements, shape functions can also be used to define the basic geometry of the elements themselves. This technique allows the development of a wide variety of two- and three-dimensional elements known collectively as *isoparametric elements*. We here consider only one of these: the straight-sided quadrilateral shown in Figure 7.16b. For conceptual purposes, this quadrilateral may be regarded as the result of a smoothly varying, in-plane "distortion" of the basic square. The four nodes are moved from their unit coordinates in the square to the points defined by the quadrilateral's nodal coordinates, $\mathbf{x}_i = (x_i, y_i)$. The latter may be regarded as scaling factors which are applied to the original unit coordinates. To obtain the complete quadrilateral, these scale factors are applied to the unit shape functions, as follows:

$$\mathbf{x} = \sum_{i=1}^{4} \mathbf{x}_i N_i(s,t)$$

Wilson et al. (1973) and Taylor, Beresford, and Wilson (1976) use this shape function approach to derive a quadrilateral element, called the QM6 element, that is similar to the CSSR element and that reduces to it when the quadrilateral is a rectangle.

7.7.4 Numerical Integration

Because of the greater complexity of the QM6 element in both external shape and internal displacement, the integration involved in obtaining the element stiffness matrix must be performed numerically rather than analytically. Therefore, neither the stiffness matrix nor the stress matrix can be obtained in explicit algebraic form, but this is not a problem as the matrices are generated by the computer as required. Although this involves more computation, the increase is quite acceptable when balanced against the great versatility of the element. Also, the use of numerical integration brings with it the possibility of "optimal sampling," a technique which gives improved accuracy in the final stress values.

7.8 DYNAMIC ANALYSIS BY THE FINITE ELEMENT METHOD

The finite element method can also be used for dynamic structural analysis, such as calculating a structure's response to a given dynamic load or determining its natural frequencies of vibration. For a dynamic system such as the spring and point mass of Fig. 7.17, Newton's law

$$m\ddot{u} = P - ku$$

can be written

$$ku = P - m\ddot{u}$$

The term $-m\ddot{u}$ can be considered an "inertial force," acting opposite to the direction of acceleration; this is d'Alembert's Principle. This is an example of a body force, as discussed in Section 7.6, and so in a finite element analysis, it can be accounted for in the same manner. Thus, if an element is undergoing accelerations $\ddot{u}(x, y)$ and $\ddot{v}(x, y)$, the distributed body force $\mathbf{b}$ has components $b_x = -\rho\ddot{u}(x, y)$ and $b_y = -\rho\ddot{v}(x, y)$, where ρ is the mass density of the

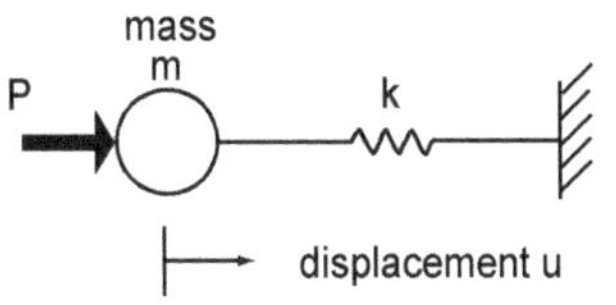

Figure 7.17 Spring-mass system.

element. The equivalent nodal forces are given by the second term of equation (7.6.4).

$$\mathbf{F}_{eq} = \sum_e \int_{V^e} \mathbf{N}^{\mathrm{T}}\, \mathbf{b}(\mathbf{X})\, dV^e \qquad (7.8.1)$$

For a linear elastic body, the natural frequency of vibration may be calculated from Newton's Law with the applied loads set to zero. For the undamped point mass of Fig. 7.17, the equation is

$$m\ddot{u} + ku = 0 \qquad (7.8.2)$$

The solution to this differential equation is the simple harmonic vibration

$$u = a \sin \omega t$$

In all such cases, the acceleration is a scalar multiple of u

$$\ddot{u} = -\omega^2 a \sin \omega t$$

and ω is the natural frequency, given by

$$\omega = \left(\frac{k}{m}\right)^{1/4}$$

In an extended body such as a two-dimensional finite element, the accelerations $\ddot{u}$ and $\ddot{v}$ continue to have the same general form as the displacements u and v. Therefore, the same shape function that describes the element's internal displacement in terms of nodal displacements also gives its internal acceleration in terms of the nodal accelerations, $\ddot{u}$ and $\ddot{v}$; that is

$$\ddot{u}(x, y) = N(x, y)\ddot{u}$$
$$\ddot{v}(x, y) = N(x, y)\ddot{v}$$

Transforming to structure coordinates and substituting into (7.8.1) gives

$$\mathbf{F}_{eq} = -\sum_e \left[\int_{V^e} \mathbf{N}^{\mathrm{T}}\, \boldsymbol{\rho}\, \mathbf{N} dV^e\right] \ddot{\mathbf{\Delta}} \qquad (7.8.3)$$

in which for two-dimensional motion

$$\boldsymbol{\rho} = \begin{bmatrix} \rho & 0 \\ 0 & \rho \end{bmatrix}$$

In (7.8.3) the expression in square brackets is a square matrix which accounts for the inertial effects (i.e., the mass) of each element. It is therefore known as the element mass matrix, $\mathbf{M}^e$, and the assemblage of these is the structure mass matrix: $\mathbf{M} = \sum_e \mathbf{M}^e$. Thus, for a complete structure the equation corresponding to (7.8.2) is

$$\mathbf{M}\ddot{\mathbf{U}} + \mathbf{K}\mathbf{U} = 0$$

The general solution is, again, of the form

$$\mathbf{U} = \bar{\mathbf{U}} \sin \omega t$$

and, after taking second derivative and substituting for $\ddot{\mathbf{U}}$, the equation becomes the *eigenvalue* or *characteristic value* problem.

$$[-\omega^2\mathbf{M} + \mathbf{K}]\bar{\mathbf{U}} = 0$$

for which a nonzero solution requires that the determinant of the coefficients be zero.

$$\det[-\omega^2\mathbf{M} + \mathbf{K}] = 0$$

The values of ω that are obtained from this are the natural frequencies of the structure.

This extremely brief discussion of dynamic structural analysis is intended simply to show that the finite element method is a powerful tool for such analysis, and that it greatly unifies static analysis and dynamic analysis, allowing dynamic aspects (such as the natural frequencies of a hull girder) to be calculated as part of the response analysis. This in turn allows dynamic constraints (such as the avoidance of certain natural frequencies) to become an integral part of the design process, just as much as the static design constraints. The only significant additional steps are the assembly of the structure mass matrix and the calculation of the eigenvalues, for which various computer routines are available.

REFERENCES

Argyris, J., and Kelsey, S. (1960). *Energy theorems and structural analysis*. London, England: Butterworth Scientific Publications.

Cook, R. D., Malkus, D. S., and Plesha, M. E. (1989). *Concepts and applications of finite element analysis*. Hoboken, NJ: John Wiley & Sons.

Courant, R. (1943). Variational methods for the solution of problems of equilibrium and vibration. *Bull. Am. Math. Soc.*, Vol. 49, 1–43.

Davies, J. D. (1976). A finite element to model the in-plane response of ribbed rectangular panels. *M. Eng. Sc. Thesis*, University of New South Wales, Sydney, Australia.

Gallagher, R. H. (1964). *A correlation study of methods of matrix structural analysis*. Report to the 14th Meeting, Structures and Materials Panel, Advisory Group for Aeronautical Research and Development, N.A.T.O., Paris, France, July 6, 1962. New York: Pergamon Press.

Hrenikoff, A. (1941). Solution of problems in elasticity by the framework method. *J. Appl. Mech.*, Vol. 8, 169–175.

McHenry, D. (1943) A lattice analogy for the solution of plane stress problems. *J. Inst. Civil Eng.*, Vol. 21, 59–82.

Rockey, K. C., Evans, H. R., Griffiths, D. W., and Nethercot, D. A. (1975). *The finite element method*. London, England: Crosby Lockwood Staples.

Taylor, R. L., Beresford, P. J., and Wilson, E. L. (1976). A non-conforming element for stress analysis. *Int. J. Num. Meth. Eng.*, Vol. 10, 1211–1220.

Turner, M., Clough, R., Martin, H., and Topp, L. (1956). Stiffness and deflection analysis of complex structures. *J. Aero. Sci.*, Vol. 23, Issue 9, 805–823.

Walz, J. E., Fulton, R. E., and Cyrus, N. J. (1968). Accuracy and convergence of finite element approximations. *Proceedings of 2nd Conference on Matrix Methods in Structural Mechanics* (AFDDL-TR-68-150), Wright-Patterson AFB, Ohio.

Wilson, E. L., Taylor, R. L., Doherty, W. P., and Ghabussi, T. (1973). Incompatible displacement models. In Fenves, S. T. (Ed.). *Numerical and computer methods in structural mechanics*. Amsterdam: Academic Press Inc., 43–57.

Zienkiewicz, O. C., and Cheung, Y. K. (1965). Finite elements in the solution of field problems. *The Engineer*, Vol. 220, 507–510.

NONLINEAR FINITE ELEMENT ANALYSIS

Jeom Kee Paik
Professor, Pusan National University
Busan, Korea

8.1 INTRODUCTION

For the design and safety assessment of structures, quantification of both loads and load effects is equally important. Depending on the characteristics of the loads, the resulting load effects or structural consequences will be either linear or nonlinear. As the environmental and operational conditions for ships and offshore structures become harsher, there is an increasing tendency for nonlinear structural consequences to be involved.

Nonlinear finite element methods are powerful tools in analyzing nonlinear structural consequences that involve geometric and/or material nonlinearities. Today, these methods are considered mature enough to be adopted in the daily practice of structural design and strength assessment.

However, it is very important to realize that nonlinear finite element method solutions may be totally wrong if the structural modeling techniques employed are inadequate in terms of idealizing the real situation that surrounds the problem. A number of textbooks have dealt with nonlinear finite element method theories, but there remains a lack of publications that deal with tips and techniques for the modeling of this method.

In this regard, the present chapter focuses on how to develop successful models of nonlinear finite element methods for the analysis of nonlinear structural consequences. Some illustrative examples of nonlinear finite element method modeling for the analysis of ultimate strength and structural crashworthiness are presented, the former being associated with extreme loads and the latter being associated with accidental actions, such as collisions, grounding, fire, and explosion.

8.2 SOLUTION PROCEDURES FOR NONLINEAR STRUCTURAL MECHANICS

Prior to the application of nonlinear finite element methods, a better understanding of their formulations and solution procedures is required. A comprehensive discussion of the formulations for these methods, however, would require many volumes. A number of such textbooks are available (e.g., Bathe, 1982; Cook et al., 1989; Owen and Hinton, 1980; White, 1985; Zienkiewicz, 1977), so the present chapter provides a brief description and focuses on the procedures used to solve nonlinear time-independent stiffness equations as illustrative examples (Paik and Thayamballi, 2003).

A time-independent problem in finite element analysis is typically expressed by the following stiffness equation.

$$\{R\} = [K]\{U\} \tag{8.2.1}$$

where $\{R\}$ = load vector, $\{U\}$ = displacement vector, and $[K]$ = (secant) stiffness matrix.

The stiffness matrix is a function of the structure's geometric and material properties. If these properties are constant, the problem is linear. If these properties are dependent on either $\{R\}$ or $\{U\}$, the problem is nonlinear.

This section presents some of the fundamental procedures used to solve nonlinear stiffness equations. To begin, we consider the simple problem shown in Fig. 8.1, in which a nonlinear spring is subjected to load P. The relationship between load P and displacement u is given by

$$P = (k_o + k_N)u \tag{8.2.2}$$

where $k = k_o + k_N$ = stiffness of the nonlinear spring, k_o = constant term, and $k_N = f(u)$ = nonlinear term that is a function of displacement.

Displacement u must be computed for a given load P by solving (8.2.2). Because (8.2.2) is a

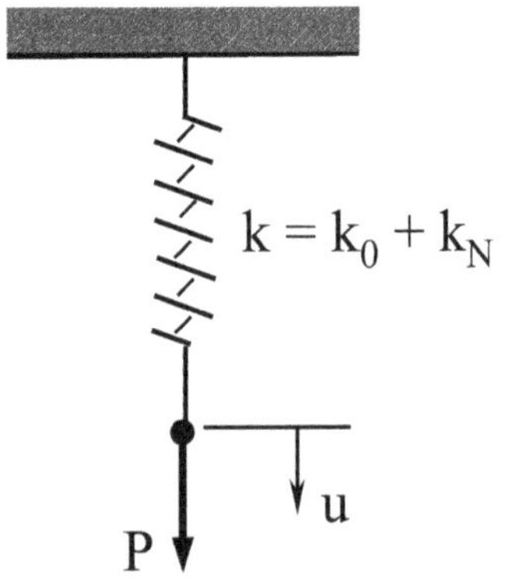

Figure 8.1 A nonlinear spring under load P.

nonlinear function of displacement u, solving this equation is not at all a straightforward matter. Here, we introduce a number of procedures that can be used to solve (8.2.2).

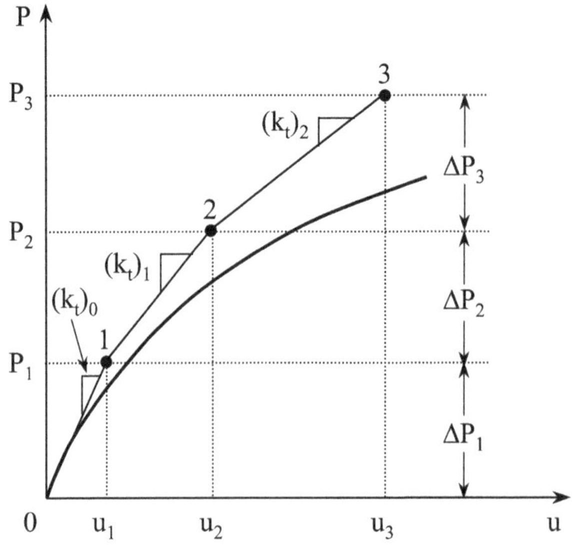

Figure 8.2 Schematic of the incremental method (Cook et al., 1989).

8.2.1 Iterative Approximation

We first determine displacement $u = u_A$ for a given load $P = P_A$ via iterative approximation. The nonlinear term of the stiffness equation in the first iteration is assumed to be zero: i.e., $k_N = 0$. The displacement is obtained from (8.2.2) as $u = P_A/k_o \equiv u_1$. The spring stiffness is then given by $k = k_o + f(u_1)$, where $f(u_1)$ is a function of u_1, at the end of the first iteration.

We can then compute the displacement as $u = [k_o + (k_N)_1]^{-1}P_A \equiv u_2$, where $(k_N)_1 = f(u_1)$, in the second iterative approximation.

After the $(i + 1)$th iteration, via a similar computation process, the displacement can be given by

$$u = [k_o + (k_N)_i]^{-1}P_A \equiv u_{i+1} \tag{8.2.3}$$

where $(k_N)_i = f(u_i)$, and u_i = displacement after the (i)th iteration. The number of iterations depends on the required accuracy.

8.2.2 The Incremental Method

Another approach is the incremental method, a schematic of which is shown in Fig. 8.2. In this method, the applied load is increased incrementally to search for the displacement solutions. The incremental form of (8.2.2) can be given by

$$\frac{dP}{du} = \frac{d}{du}(k_o + k_N)u = k_o + \frac{d}{du}(k_N u) \equiv k_t \tag{8.2.4}$$

where k_t is often termed the tangent stiffness.

Alternatively, (8.2.4) can be rewritten in terms of load increments ΔP versus displacement increments Δu as follows.

$$\Delta u = (k_t)^{-1}\,\Delta P \tag{8.2.5}$$

This incremental process is shown in Fig. 8.2. The tangent stiffness can be readily obtained as $k_t = k_o = (k_t)_o$ from (8.2.4) when $P = 0$ because $u = 0$. Displacement u_1 is then given by $u_1 = (k_t)_0^{-1}\Delta P_1$ in the first step of load increments ΔP_1.

Following the first incremental load step, the tangent stiffness is $k_t = (k_t)_1$, with $u = u_1$. In the second step of load increments ΔP_2, displacement u_2 becomes $u_2 = u_1 + (k_t)_1^{-1}\,\Delta P_2$.

A similar computation process results in displacement u_i at the (i)th step of the load increments as follows.

$$u_i = u_{i-1} + (k_t)_{i-1}^{-1}\,\Delta P_i \tag{8.2.6}$$

where $(k_t)_{i-1}$ = tangent stiffness at the end of the $(i - 1)$th step of the load increments, which is determined from (8.2.4) but as a function of $u = u_{i-1}$.

In the incremental method there are always unbalanced forces between the external forces (applied loads) $\sum \Delta P_i$ and the internal forces $\sum (k_t)_{i-1}(u_i - u_{i-1})$, and the larger the load increments, the more significant they become. Either the Newton-Raphson method or the modified Newton-Raphson method can be used to progressively reduce the unbalanced forces and ensure acceptable tolerance. The use of these two methods is discussed in the following.

8.2.3 The Newton-Raphson Iteration

The Newton-Raphson method employs a number of iterative techniques with updated tangent stiffness to eliminate the unbalanced forces between the external and internal forces after each incremental load step, as illustrated in Fig. 8.3. The stiffness equation with displacement u_A for corresponding load P_A at the end of any such step can be obtained from (8.2.2) as follows.

$$P_A = \{k_o + (k_N)_A\}u_A \qquad (8.2.7)$$

where $(k_N)_A = k_N$ at $u = u_A$.

Because the applied load is a function of the displacement, i.e., $P = f(u)$, its truncated Taylor series expansion is given by

$$f(u_A + \Delta u_1) = f(u_A) + \left(\frac{dP}{du}\right)_A \Delta u_1 \qquad (8.2.8)$$

where $(dP/du)_A \equiv k_t$ = tangent stiffness, as defined by (8.2.4) at $u = u_A$.

Displacement increment Δu_1 at the first iteration can then be computed from (8.2.8).

$$\Delta u_1 = (k_t)_0^{-1}(P_B - P_A) \qquad (8.2.9a)$$

where $P_B = f(u_A + \Delta u_1)$, $P_A = f(u_A)$, $(k_t)_0 = k_t$ in (8.2.4) with $u = u_A \equiv u_0$.

The displacement can be determined after the first iteration as follows.

$$u = u_A + \Delta u_1 \equiv u_1 \qquad (8.2.9b)$$

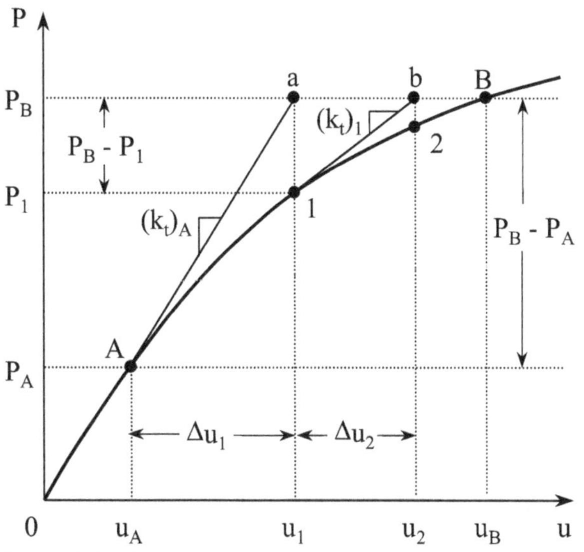

Figure 8.3 Schematic of the Newton-Raphson iteration (Cook et al., 1989).

As can be seen in Fig. 8.3, the unbalanced forces are now $P_B - P_1$, and further iteration is required to eliminate them. A new tangent stiffness $(k_t)_1$ is obtained from (8.2.4) with $u = u_1$ at the next iteration, meaning that displacement increment Δu_2 can be determined from (8.2.8), as follows.

$$\Delta u_2 = (k_t)_1^{-1}(P_B - P_1) \qquad (8.2.10a)$$

Following the second iteration, the displacement is approximated by

$$u = u_1 + \Delta u_2 \equiv u_2 \qquad (8.2.10b)$$

The iterations are continued in a similar fashion until the unbalanced forces are eliminated and acceptable tolerance has been achieved. Following the (i)th iteration, the displacement increment is given by

$$\Delta u_i = (k_t)_{i-1}^{-1}(P_B - P_{i-1}) \qquad (8.2.11a)$$

and the resulting displacement is computed as follows.

$$u = u_B = u_{i-1} + \Delta u_i \qquad (8.2.11b)$$

The Newton-Raphson method employs the updated tangent stiffness for each iteration, as can be seen from Fig. 8.3. The tangent stiffness must thus be recalculated at every iteration process to eliminate the unbalanced forces, which requires considerable computational effort.

8.2.4 The Modified Newton-Raphson Iteration

The modified Newton-Raphson method, in contrast, does not require the tangent stiffness matrix to be updated during the iteration process to eliminate the unbalanced forces, as can be seen in Fig. 8.4, thus avoiding the aforementioned extensive repetitions. As a result, great computational effort will be saved in analysis of nonlinear problems with a large number of degrees of freedom or unknowns. However, the total number of iterative cycles must usually be increased to achieve acceptable tolerance relative to the original Newton-Raphson method.

8.2.5 The Arc Length Method

Some structures exhibit an unstable behavior in their loading or unloading path, particularly after they have reached their ultimate limit states. In this case the foregoing methods may be

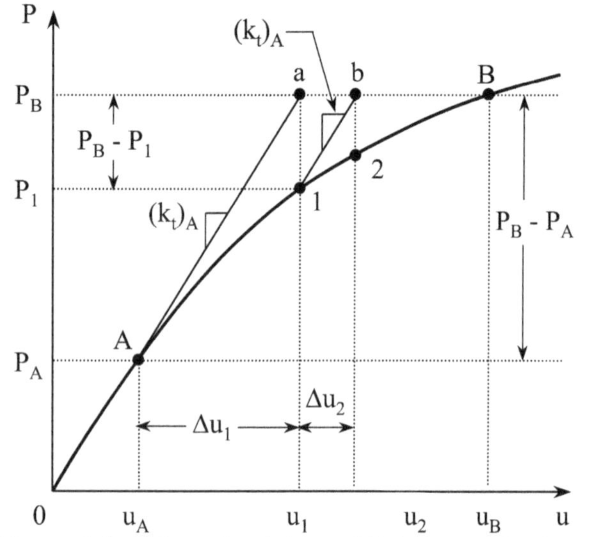

Figure 8.4 Schematic of the modified Newton-Raphson iteration (Cook et al., 1989).

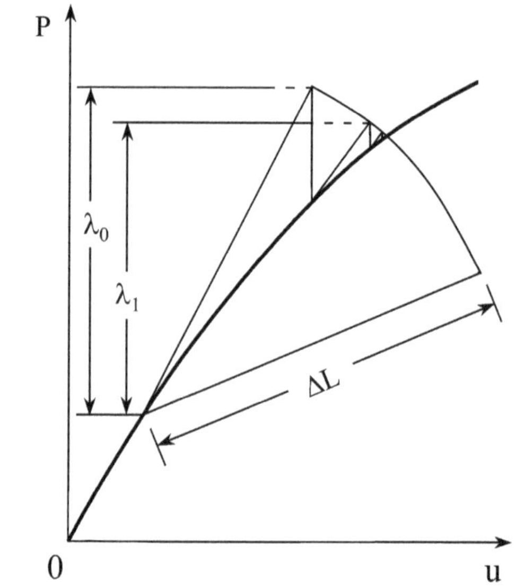

Figure 8.5a Schematic of the arc length method.

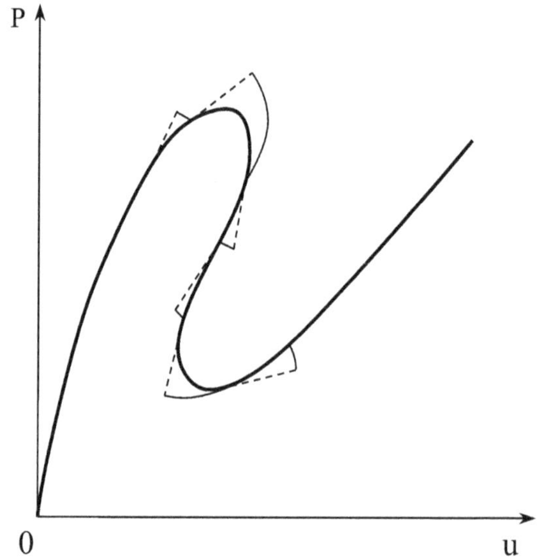

Figure 8.5b Illustrative example of the arc length method applied to the "snap-through" problem.

unsuitable, and the arc length method (Crisfield, 1981) is often used instead. In this method, load increment ΔP is considered to be a variable, even during the iteration process adopted to eliminate the unbalanced forces.

$$\Delta P = \lambda_i \, \Delta P_o \qquad (8.2.12)$$

where ΔP_o = initial load increment, λ_i = load magnification factor at the (i)th step of the iteration, as defined in Fig. 8.5a.

Arc length ΔL, which is associated with load magnification factor λ_i at the (i)th iteration process, can then be defined as follows.

$$\Delta L = \sqrt{\{\Delta u\}_i^T \{\Delta u\}_i} \qquad (8.2.13)$$

where $\{\Delta u\}_i$ = incremental nodal displacement vector at the (i)th iteration process.

The iteration is repeated until acceptable tolerance is achieved through elimination of the unbalanced forces. The arc length method is capable of solving highly nonlinear problems, including the very unstable "snap-through" response shown in Fig. 8.5b.

8.3 TIPS AND TECHNIQUES FOR NONLINEAR ANALYSIS

This section presents a number of useful tips and techniques for nonlinear finite element method modeling in association with an assessment of ultimate limit states and structural crashworthiness. The former are due to buckling, yielding, and collapse, and the latter are related to crushing, yielding, and rupture. This guidance is based on current practice using typical commercial nonlinear finite element method programs.

The accuracy of nonlinear finite element method solutions is governed by the ability of the structural modeling techniques to idealize various factors of influence, including geometric and material properties, load application, boundary conditions, and initial imperfections. This section thus focuses on tips and techniques for the modeling of such factors (Paik, 2008a, 2008b, 2008c; Paik et al., 2009).

8.3.1 Extent of the Analysis

It is desirable to take the entire structure under consideration when performing an analysis. However, if the time or resources available for structural modeling and computation are limited, finite element method modeling may be used to consider only a part of the target structure. In such cases, it is important to realize that an artificial boundary is formed for the target structure and the solution will be satisfactory only if the boundary conditions (loads, supports, etc.) are idealized in an appropriate manner. The extent of the analysis is typically cut out of the target structure with respect to the symmetric envelope in terms of structural deformations and failure modes. A number of illustrative examples for plates and stiffened plate structures under uniaxial compression are shown in Fig. 8.6.

8.3.2 Types of Finite Elements

A variety of finite element types is available, but it is often difficult to establish specific guidelines for which types are best for a given application. For the nonlinear analysis of thin-walled or plated structures, rectangular plate-shell elements are more appropriate than triangular elements because the former make it easier to define the membrane stress components inside each element when the Cartesian coordinate system is applied. This practice is also true for linear structural mechanics and analysis (Paik and Hughes, 2007).

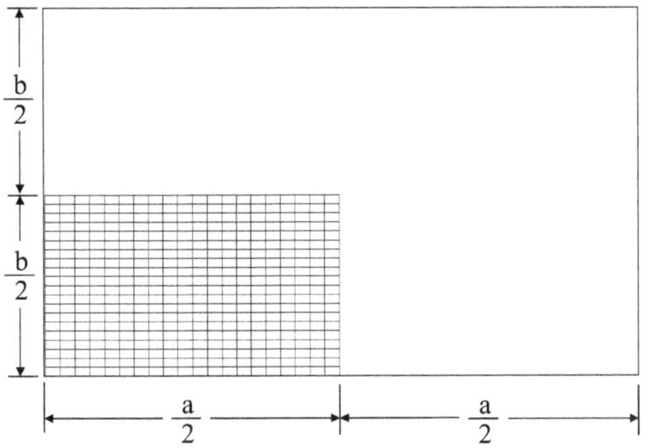

Figure 8.6a Quarter model for a rectangular plate under uniaxial compression.

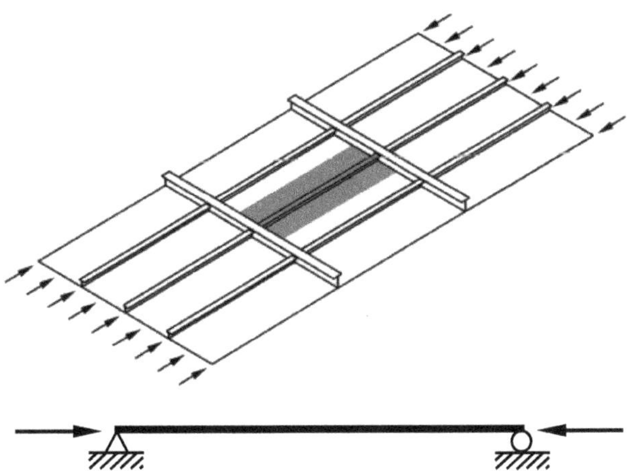

Figure 8.6b One-bay plate-stiffener combination model for a stiffened plate structure under uniaxial compression.

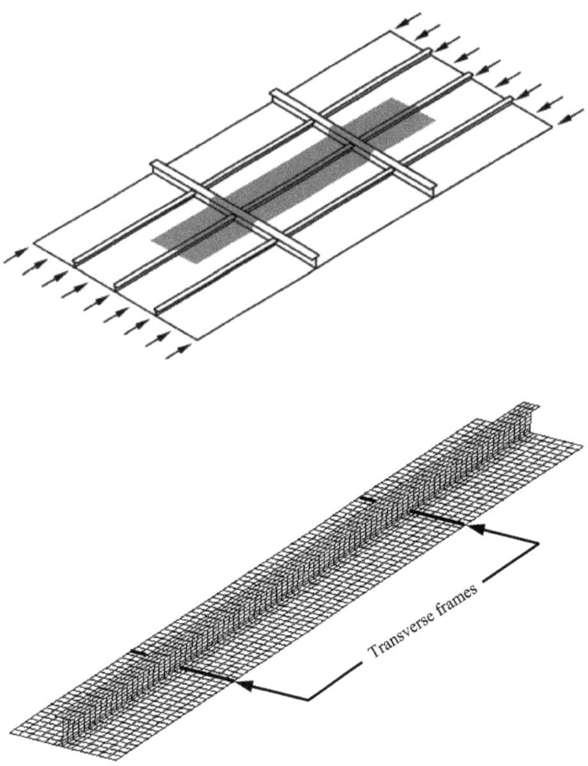

Figure 8.6c Two-bay plate-stiffener combination model for a stiffened plate structure under uniaxial compression.

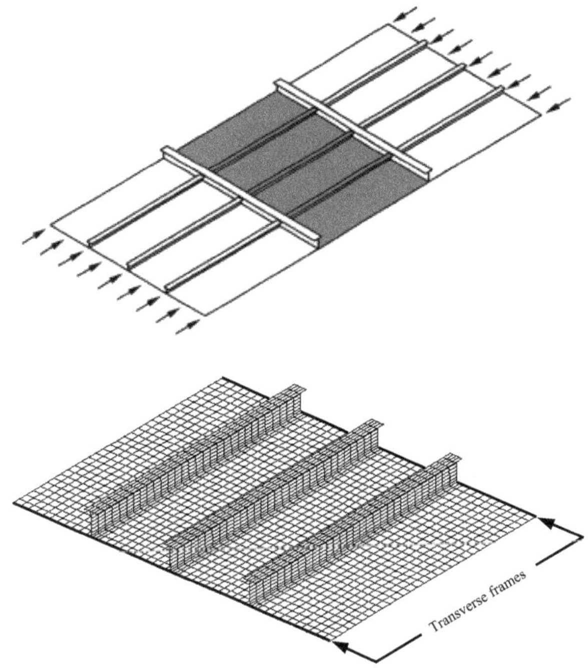

Figure 8.6d One-bay stiffened panel model for a stiffened plate structure under uniaxial compression.

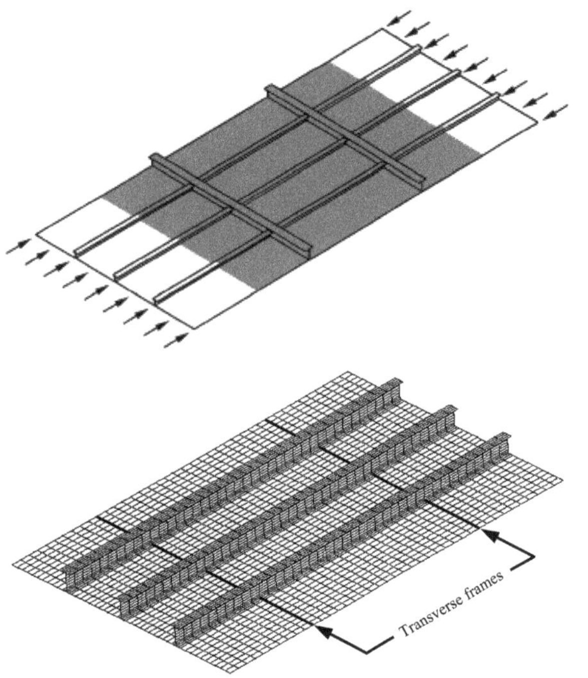

Figure 8.6e Two-bay stiffened panel model for a stiffened plate structure under uniaxial compression.

Thus, for the nonlinear analysis of ships and offshore structures, in association with their ultimate limit states and structural crashworthiness, four-noded plate-shell elements are most often employed. The nodal points in the plate thickness direction are located in the mid-thickness of each element, which indicates that no element mesh is assigned to the thickness layers. To reflect nonlinear behavior more accurately, plate-shell elements should be used for webs and flanges, as well as for the plating. However, beam elements are sometimes more efficient when modeling these supporting members or at least the flanges.

Two types of algorithms (namely, implicit and explicit) are relevant to simulations of dynamic structural crashworthiness using nonlinear finite element methods, depending on the time-integration techniques that are applied. The explicit algorithm computes both the internal and external forces at each nodal point of the finite elements, and the resulting acceleration at each nodal point can be obtained by dividing total nodal force by nodal mass. The stiffness equation is solved at time t by direct time integration, where the maximum time step size must be controlled carefully.

In contrast, in the implicit algorithm a traditional finite element solver is employed to calculate the nodal displacement increments for specified nodal force increments. The stiffness equation is solved at time $t + \Delta t$, where Δt is a time increment (step). The explicit algorithm-based methods are more useful for simulating the dynamic structural consequences that are associated with impact crashworthiness involving dynamic crushing, among other factors.

8.3.3 Finite Element Mesh Size

Although finer mesh modeling certainly results in more accurate solutions, it is not necessarily the best practice. A similar degree of accuracy can be attained with coarser mesh modeling, which requires considerably less computational cost. A convergence study is usually carried out to determine the best size of finite element mesh based on a compromise between computational cost and accuracy. Sample applications of the corresponding nonlinear analysis are undertaken with a variety of element mesh sizes to search for the largest size that provides a sufficient level of accuracy.

Such a convergence study can provide best practice nonlinear finite element method modeling in terms of a determination of the relevant mesh size. However, a convergence study itself requires considerable computational effort. Therefore, guidance is required to define the finite element mesh size without the need for such a study.

For the ultimate strength analysis of stiffened plate structures that involve an elastic-plastic large deflection response, current practice indicates that at least eight four-noded plate-shell elements are required to model the plating in between the small support members (e.g., the longitudinal stiffeners). The size of these plate-shell elements is assigned in the plate length direction to ensure that the aspect ratio of each finite element is near unity, which is desirable. There will probably be at least six elements in the web height direction and at least two elements across the (full) flange breadth when using four-noded plate-shell elements.

In analysis of structural crashworthiness that involves the crushing or folding of thin walls, at least eight four-noded plate-shell elements are required to reflect the folding behavior of the single crushing length of a plate, as shown in Fig. 8.7. Theoretical formulations of the plate crushing length for thin-walled structures under crushing loads are available. For example, the following plate crushing length formula has been derived by Wierzbicki and Abramowicz (1983).

$$H = 0.983b^{2/3}t^{1/3} \tag{8.3.1}$$

where b = plate breadth, t = plate thickness, and H = half-fold length.

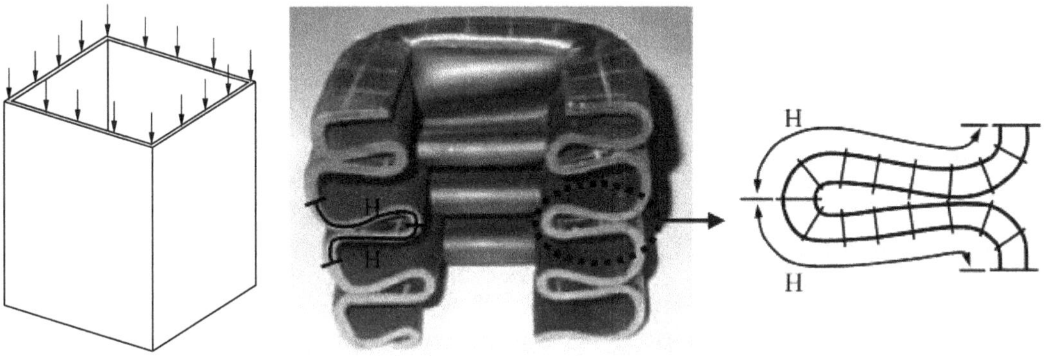

Figure 8.7 Crushing behavior of a thin-walled structure and the necessary size of the finite elements (Paik and Thayamballi, 2003).

Therefore, the mesh size of a single finite element for plate crashworthiness analysis can be determined as the crushing length predicted by (8.3.1) divided by 8. The element size in the plate length direction should be determined to ensure that the element aspect ratio is near unity.

8.3.4 Material Modeling

Nonlinear structural consequences almost always involve material nonlinearity in association with plasticity or yielding, among other factors. For nonlinear finite element analysis, therefore, the characteristics of material behavior should be defined precisely in terms of the stress versus strain relationship.

Characterization of the Engineering Stress-Engineering Strain Relationship

It is of course desirable to determine the realistic relationship between these stresses and strains through tensile coupon testing, which covers pre-yielding behavior; yielding; post-yielding behavior, including the strain-hardening effect; ultimate strength; and post-ultimate strength behavior, including the necking effect. It is interesting to note that the current practice for ultimate limit state assessment in the maritime industry employs a simpler material model, although the realistic characteristics of the aforementioned material have been applied for accidental limit state assessment.

For example, the effects of strain-hardening and necking (strain-softening) are often unaccounted for in ultimate strength analysis. This simplified type of material model is termed the "elastic-perfectly plastic material model" and represents the material's elastic behavior until the yield strength has been reached. Neither strain-hardening nor necking is allowed for in the post-yielding regime. This approximation may be useful for steel when the primary concern is buckling and there is only a moderate amount of strain, in contrast to structural crashworthiness, which involves crushing and rupture with large strains. However, the elastic-perfectly plastic model does not give sufficiently accurate solutions for aluminum alloy materials.

When details of the stress versus strain relationship are unavailable, but such fundamental parameters as elastic modulus E and yield strength σ_Y are known, the relationship between engineering stress and engineering strain can often be approximated using the Ramberg-Osgood equation, which was originally proposed for aluminum alloys (Ramberg and Osgood, 1943).

$$\varepsilon = \frac{\sigma}{E} + \left(\frac{\sigma}{B}\right)^n \tag{8.3.2a}$$

where E = elastic modulus at the origin of the stress versus strain curve, ε = engineering strain, σ = engineering stress, and B and n are constants to be determined through experiments. Equation (8.3.2a) is often simplified as follows (Mazzolani, 1985).

$$\varepsilon = \frac{\sigma}{E} + 0.002\left(\frac{\sigma}{\sigma_{0.2}}\right)^n \tag{8.3.2b}$$

where $\sigma_{0.2}$ = proof stress at 0.2% strain, i.e., with ε_o = 0.002, as shown in Fig. 8.8a, which is usually taken as material yield stress σ_Y, i.e., $\sigma_{0.2} = \sigma_Y$. Exponent n is given as a function of $\sigma_{0.2}$ and $\sigma_{0.1}$, as follows.

$$n = \frac{\ln 2}{\ln\left(\dfrac{\sigma_{0.2}}{\sigma_{0.1}}\right)} \tag{8.3.2c}$$

where $\sigma_{0.1}$ = proof stress at 0.1% strain, with ε_o = 0.001, as shown in Fig. 8.8a.

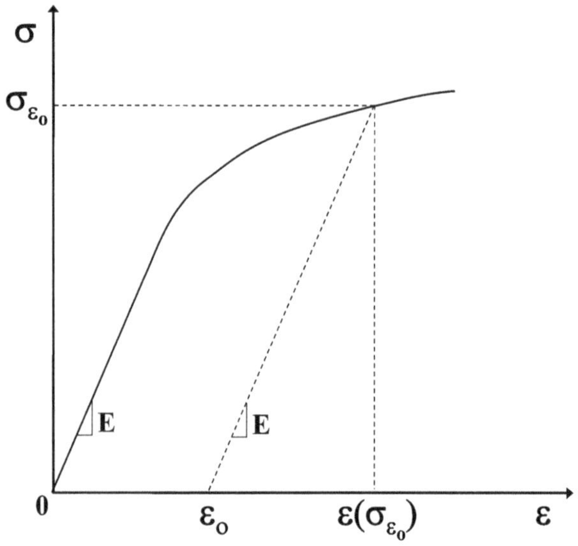

Figure 8.8a Characterization of the engineering stress-engineering strain relationship.

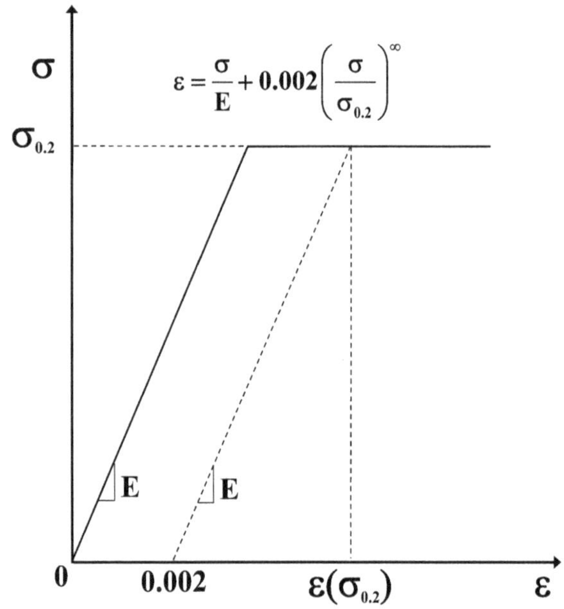

Figure 8.8b Elastic-perfectly plastic model of material.

When the Ramberg-Osgood law is employed, one practical difficulty is the determination of $\sigma_{0.1}$, in addition to E and $\sigma_{0.2}$ ($\approx \sigma_Y$). Without considering the strain-hardening effect, if ratio $\sigma_{0.2}/\sigma_{0.1}$ approaches 1 (or $\sigma_{0.1} = \sigma_{0.2}$), the exponent becomes infinity, i.e., $n = \infty$. This behavior indicates the elastic-perfectly plastic model, as shown in Fig. 8.8b, which is usually adopted for mild steel, so (8.3.2b) is rewritten for mild steel as follows.

$$\varepsilon = \frac{\sigma}{E} + 0.002\left(\frac{\sigma}{\sigma_{0.2}}\right)^{\infty} \quad (8.3.2d)$$

For aluminum alloys, Steinhardt (1971) proposes an approximate method for determining exponent n without the value of $\sigma_{0.1}$ being known, as follows.

$$0.1n = \sigma_{0.2}\ (\text{N/mm}^2) \text{ or } n = 10\sigma_{0.2} \quad (8.3.2e)$$

Characterization of the True Stress-True Strain Relationship

Nonlinear finite element methods actually use the relationship of the material characteristics between the true stresses and strains, which can be approximately estimated from the corresponding relationship between the engineering stresses and strains, namely,

$$\sigma_{true} = \sigma(1 + \varepsilon),\ \varepsilon_{true} = \ln(1 + \varepsilon) \quad (8.3.3)$$

where σ_{true} = true stress, ε_{true}= true strain, σ = engineering stress, and ε = engineering strain.

Figures 8.9a and 8.9b show the engineering stress-engineering strain curve versus the true stress-true strain curve for mild steel and aluminum alloy 5383-H116, respectively.

It is seen from Fig. 8.9 that (8.3.3) overestimates the strain-hardening and necking (strain-softening) effects. To resolve this issue, Paik (2007a, 2007b) has suggested that (8.3.3) be modified by introducing a knock-down factor that is a function of engineering strain, as follows.

$$\sigma_{true} = f(\varepsilon)\sigma(1 + \varepsilon),\ \varepsilon_{true} = \ln(1 + \varepsilon) \quad (8.3.4a)$$

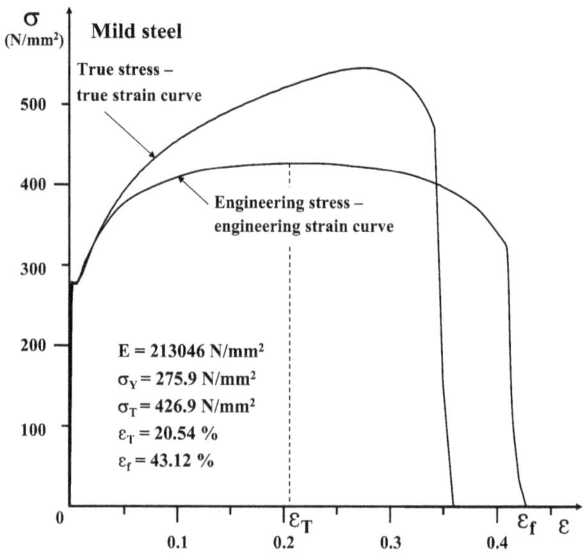

Figure 8.9a Engineering stress-engineering strain curve versus true stress-true strain curve for mild steel at room temperature.

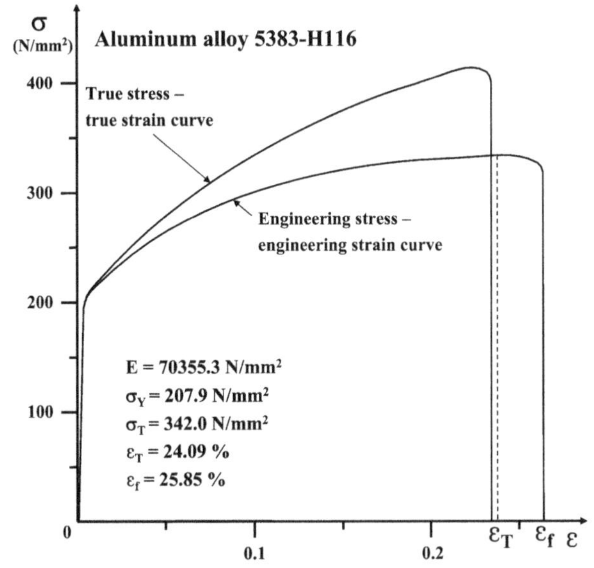

Figure 8.9b Engineering stress-engineering strain curve versus true stress-true strain curve for aluminum alloy 5383-H116 at room temperature.

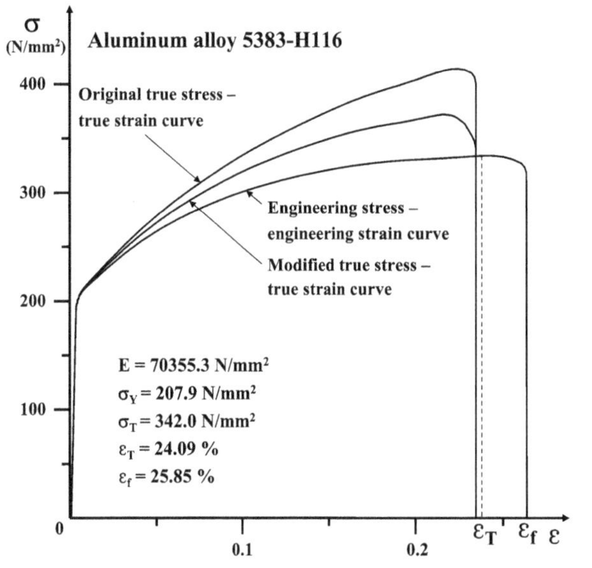

Figure 8.10b The original true stress-true strain curve versus the modified true stress-true strain curve for aluminum alloy 5083-H116 at room temperature.

$$f(\varepsilon) = \begin{cases} \dfrac{C_1 - 1}{\ln(1 + \varepsilon_T)} \ln(1 + \varepsilon) + 1 \text{ for } 0 < \varepsilon \le \varepsilon_T \\ \dfrac{C_2 - C_1}{\ln(1 + \varepsilon_f) - \ln(1 + \varepsilon_T)} \ln(1 + \varepsilon) + C_1 \\ - \dfrac{(C_2 - C_1)\ln(1 + \varepsilon_T)}{\ln(1 + \varepsilon_f) - \ln(1 + \varepsilon_T)} \text{ for } \varepsilon_T < \varepsilon \le \varepsilon_f \end{cases} \tag{8.3.4b}$$

where $f(\varepsilon)$ = knock-down factor as a function of engineering strain, ε_f = fracture strain of the material, ε_T = strain at the ultimate tensile stress, and C_1, C_2 = test constants affected by material type and plate thickness, among other factors.

Although the knock-down factor is governed by the characteristics of the material type and plate thickness, the test constants may be given as C_1 = 0.9 and C_2 = 0.85 for mild and high-tensile steel, respectively (Paik, 2007a, 2007b). Figures 8.10a and 8.10b compare the original true stress-true strain curve (8.3.3) versus the modified (knock-downed) true stress-strain curve (8.3.4a) of mild steel and aluminum alloy, respectively where the constants C_1 = 0.9 and C_2 = 0.85 were applied for both mild steel and aluminum alloy.

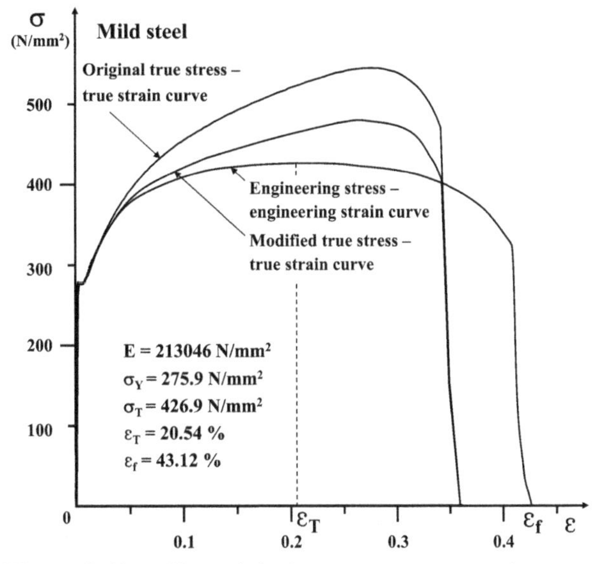

Figure 8.10a The original true stress-true strain curve versus the modified true stress-true strain curve for mild steel at room temperature.

Fracture Strain in Nonlinear Finite Element Analysis

In accidental situations which involve structural crashworthiness with large strains, structures may be exposed to fracture. In this case, fracture behavior must be taken into account. The critical fracture strain of plate-shell-type finite elements is affected by element size and plate thickness, among other factors. Therefore, it is very important to define the critical fracture strain used for the nonlinear finite element analysis. The following formula can be used to predict the critical fracture strain of the material as a function of the finite element size and the plate thickness.

$$\varepsilon_{fc} = \gamma d_1 \left(\frac{t}{s}\right)^{d_2} \varepsilon_f \tag{8.3.5}$$

where ε_f = fracture strain determined on the basis of the tensile coupon test data, ε_{fc} = critical fracture strain taking into account the effect of the element size and the plate thickness in nonlinear finite element analysis, t = plate thickness, s = finite element mesh size (length), γ = correction (knock-down) factor associated with localized bending due to folding, and d_1, d_2 = coefficients.

The coefficients d_1 and d_2 in (8.3.5) can be determined based on a series of finite element simulations with varying the finite element size and the plate thickness with respect to tensile coupon test database in which the finite element simulations must correspond to the test results in terms of the fracture strain. For mild steel at room temperature, $d_1 = 4.1$ and $d_2 = 0.58$ may be used for t = 2mm (Paik et al., 2003). With increase in the plate thickness, the localized bending effect becomes more significant and thus the correction factor γ will take much smaller value than unity such as 0.3-0.4.

Once ε_{fc} is determined from (8.3.5), the true stress-true strain relation used for FEA shall be defined by adjusting (extending or shrinking) the "modified" true stress-true strain relation in terms of the critical fracture strain ε_{fc}, where the ultimate tensile stress will be kept at the constant level beyond the strain corresponding to the ultimate tensile stress until the fracture strain.

Effect of Strain-rate Sensitivity

In structural crashworthiness and/or impact response analysis, strain-rate sensitivity plays an important role. Therefore, material modeling in terms of dynamic yield strength and dynamic fracture strain needs to be considered. The following Cowper-Symonds equation (Cowper and Symonds, 1957) is usually applied for this purpose.

$$\sigma_{Yd} = \left\{1 + \left(\frac{\dot{\varepsilon}}{C}\right)^{1/q}\right\}\sigma_Y \tag{8.3.6a}$$

$$\varepsilon_{fd} = \left\{1 + \left(\frac{\dot{\varepsilon}}{C}\right)^{1/q}\right\}^{-1}\varepsilon_{fc} \tag{8.3.6b}$$

where σ_Y = static yield stress, σ_{Yd} = dynamic yield stress, ε_{fc} = static fracture strain used for nonlinear finite element analysis as defined in (8.3.5), ε_{fd} = dynamic fracture strain, $\dot{\varepsilon}$ = strain rate (1/sec), and C and q = test constants.

In (8.3.6), the test constants are often taken as C = 40.4/sec, q = 5 for mild steel; C = 3200/sec, q = 5 for high-tensile steel; and C = 6500/sec, q = 4 for aluminum alloys (Paik and Thayamballi, 2003, 2007). Strain rate $\dot{\varepsilon}$ can be calculated approximately by assuming that the initial speed V_o of the dynamic loads is linearly reduced to zero until the loading is finished, with average displacement δ, namely,

$$\dot{\varepsilon} = \frac{V_o}{2\delta} \tag{8.3.7}$$

8.3.5 Boundary Condition Modeling

When the target structure has boundaries that are linked to adjacent structures, the condition of these boundaries must be idealized realistically. This problem most often occurs when the extent of the analysis is partial and carried out by cutting a section out of the target structure, thus producing artificial boundaries. A similar situation may occur inside the target structure when certain structural modeling simplifications are attempted. For example, a strong support member that is regarded as undeforming and preventing diplacements and/or rotations can be replaced by rigid restraints, and a weak support member may be ignored (zero restraint). However, when the degree of restraint at the boundaries is neither zero nor infinite, a more detailed set of boundary conditions is required.

It is very important that the reality of these boundaries is clearly understood before idealizations are made. If there is uncertainty about the correct boundary conditions to replace a portion of structure, it is probably better to include that portion in the structural model, even though doing so more computations.

When comparing Fig. 8.6b with Fig. 8.6c or Fig. 8.6d with Fig. 8.6e, which represent the nonlinear finite element method models for stiffened plate structures under axial compressive actions that are applied in the direction parallel to the longitudinal stiffeners, the former (which make use of the one-bay model approach) are relevant only if the restraint at the transverse frame location is either zero or infinity (simply supported or fixed). In reality, the rigidity of these frames is neither zero nor infinite, and the decision depends entirely on the required level of accuracy.

In recent years the latter models (e.g., Fig. 8.6c or Fig. 8.6e), which employ the two-bay model approach, have become more popular because they allow the transverse frames to be included as part of the finite element model. To avoid an even larger model (the structure that supports the frames), they are usually regarded as simply supported. Because they are included in the model, their rotational restraint on the rest of the model is automatically accounted for.

Figure 8.11a provides an illustrative example of the extent of nonlinear finite element analysis for a stiffened plate structure under axial compression. Figure 8.11b presents a nonlinear finite element method model using the two-bay model approach, in which finite element meshes are not assigned for the transverse frames, along which the lateral deformations are restrained.

In the following definitions, T[x, y, z] indicates the translational constraints and R[x, y, z] the rotational constraints around the x-, y-, and z-coordinates. "0" indicates a constraint, and "1" indicates no constraint.

- At boundaries A-C and A′-C′: the edges along the longitudinal girders are modeled as simply supported, i.e., T[1, 1, 0], R[1, 0, 0], with each edge having an equal y-displacement.
- At the transverse frame (floor) intersections: T[1, 1, 0] at the plate nodes and T[1, 0, 1] at the stiffener web nodes.
- At boundaries A-A′ and C-C′: the symmetric conditions with R[1, 0, 0] at all of the plate nodes and stiffener nodes having an equal x-displacement for the present illustrative panel with an odd number of buckling half-waves (e.g., m = 5) in the panel length (x) direction. However, for a panel with an even number of buckling half-waves, only the straight condition or equal x-displacement at the plate nodes may be applied.

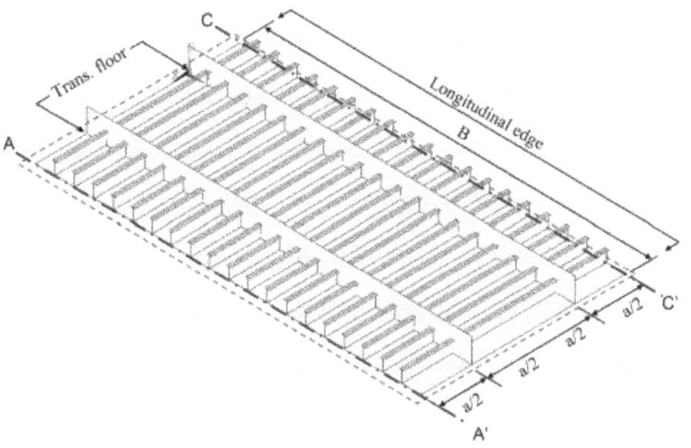

Figure 8.11a Extent of the nonlinear finite element analysis for a stiffened plate structure.

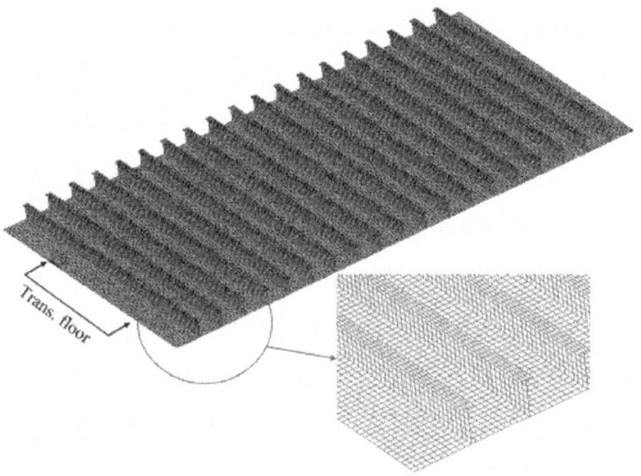

Figure 8.11b Nonlinear finite element method model for a stiffened plate structure.

8.3.6 Modeling of Initial Imperfections

Welded metal structures always have initial imperfections in the form of initial distortions and residual stresses, caused by the successive expansion and shrinkage during the heating and cooling, as shown in Figs. 8.12a and 8.12b (Paik et al., 2006). The width $2b_t$ is the "heat-affected zone" (HAZ), in which the stress is approximately equal to the tensile yield stress (because the molten

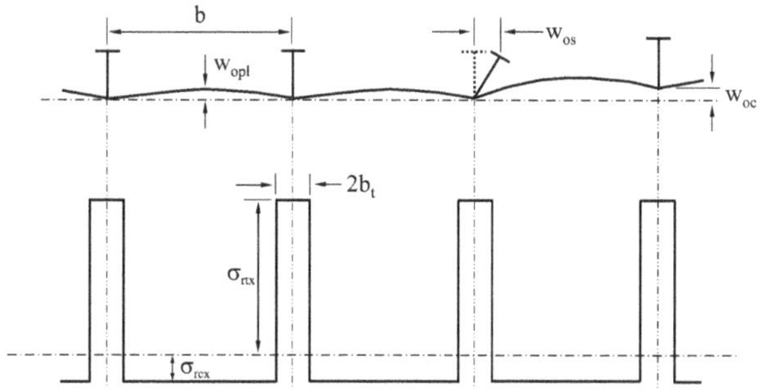

Figure 8.12a Weld-induced initial distortions and residual stresses in a stiffened plate structure.

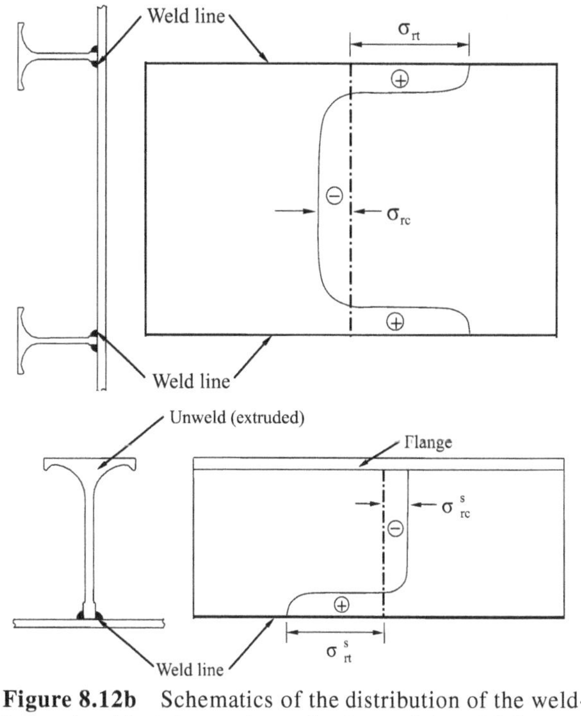

Figure 8.12b Schematics of the distribution of the weld-induced residual stresses in an aluminum plate welded at two edges and a stiffener web welded at one edge (upper: plating; lower: extruded stiffener web; +: tension; -: compression).

metal can expand freely, as a liquid, whereas after welding it quickly reverts to a solid and the shrinkage that occurs during cooling involves "plastic flow"). The compressive residual stress in the rest of the plating is discussed in a later section. Figure 8.12c (Paik et al., 2006) shows that in the HAZ of welded aluminum structures (b_p' and b_s'), in contrast to steel structures, a softening phenomenon occurs in which the yield strength within the HAZ is reduced relative to that of the base metal. This is discussed further in a later section.

Because such fabrication-related initial imperfections may have an effect on the structural properties and load-carrying capacities of structures, they must be dealt with as parameters of influence in structural design and strength assessment. Initial imperfections can significantly reduce ultimate strength and thus must be taken into account in ultimate limit state assessment. However, they do not play a significant role in structural crashworthiness that is related to such accidental phenomena as collisions, grounding, fire, and gas explosions in which the structural consequences are more likely to be governed by large strains. Therefore, the effects of initial imperfections are often neglected in current accidental limit state assessment practice.

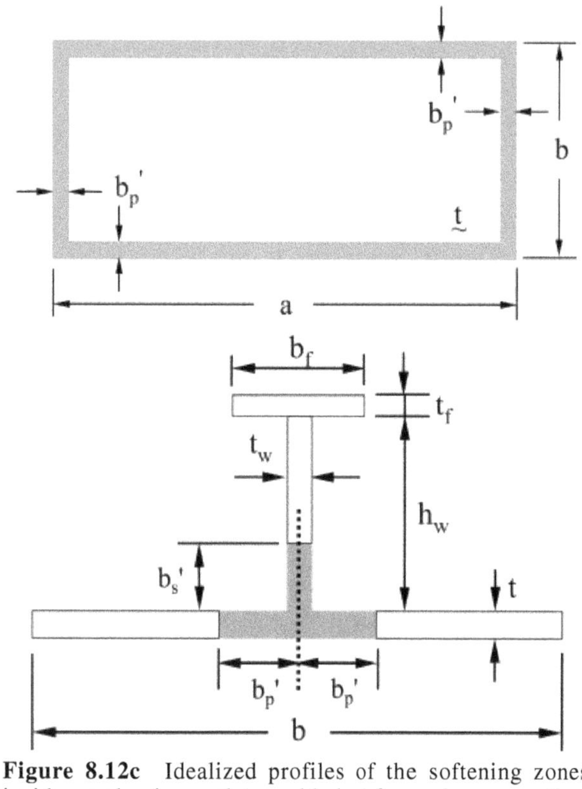

Figure 8.12c Idealized profiles of the softening zones inside an aluminum plate welded at four edges as well as its counterpart in the extruded stiffener attachment to the plating.

A number of textbooks (e.g., Masubuchi, 1980; Paik and Thayamballi, 2003) provide detailed descriptions of the mechanisms and realistic configurations of fabrication-related initial imperfections together with their mathematical idealizations. Thus, this subsection presents a number of guidelines on implementing these imperfections in nonlinear finite element method modeling.

Initial Distortion Modeling

Three types of initial distortions are relevant to welded-metal stiffened-plate structures as follows.

- Initial deflection of the plating between the support members
- Column-type initial distortion of the support members
- Sideways initial distortion of the support members

It is important to remember that both the magnitude and shape of each type of initial distortions play important roles in buckling collapse behavior, so a better understanding of the actual imperfection configurations in the target structures is necessary. In fact, it is desirable to have precise information about the initial distortions of the target structure before structural modeling begins. Considering the significant amount of uncertainty involved in fabrication-related initial imperfections, existing measurements of the initial distortions in welded metal structures (e.g., Paik et al., 2006; Paik et al., 2008) are often useful for developing representative models.

In current maritime industry practice with regard to practical structural design and strength assessment, an average magnitude is often assumed for these initial distortions, with their shape to be the buckling mode because this shape usually has the most unfavorable consequences for the structure until and after the ultimate limit state is reached.

The amplitude or maximum magnitude w_{opl} of plate initial deflection w_o^p is often assumed to be the following.

$$w_o^p = w_{opl} \sin \frac{m\pi x}{a} \sin \frac{\pi y}{b} \qquad (8.3.8a)$$

$$w_{opl} = C_1 b \qquad (8.3.8b)$$

$$w_{opl} = C_2 \beta^2 t \qquad (8.3.8c)$$

where w_o^p = initial deflection of the plate, w_{opl} = maximum magnitude of plate initial deflection, b = plate breadth along the short edge or spacing between the longitudinal stiffeners, t = plate thickness, $\beta = (b/t)\sqrt{\sigma_Y/E}$ = plate slenderness coefficient, E = elastic modulus of the material, σ_Y = yield strength, C_1 and C_2 = constants, and m = buckling half-wave number of the plate.

It is interesting to note that the two alternative formulae, i.e., (8.3.8b) and (8.3.8c), have different usage backgrounds. The former, supported by some classification societies, states that w_{opl} is a function only of plate breadth, whereas Smith et al. (1988) suggest that the latter gives a more precise representation of the plate characteristics. In addition, the use of the former may result in too small initial deflection for very thin plates and too large initial deflection for very thick plates. The latter formula, in contrast, is suitable for both very thin and very thick plates. Nevertheless, the use of the first formula mentioned, i.e., (8.3.8b), remains more popular today in ship and offshore structure construction, as long as a moderate plate thickness is considered.

The constants in (8.3.8b) and (8.3.8c) may be determined based on statistical analyses of the initial deflection measurements of the welded metal plates. The following provides some additional guidance.

$C_1 = 0.005$ for an average level in steel plates (the practice suggested by the Classification Societies).

$$C_1 = \begin{Bmatrix} 0.0032 \text{ for a slight level} \\ 0.0127 \text{ for an average level} \\ 0.0290 \text{ for a severe level} \end{Bmatrix}$$

in aluminum plates (Paik, 2007c).

$$C_2 = \begin{Bmatrix} 0.025 \text{ for a slight level} \\ 0.1 \text{ for an average level} \\ 0.3 \text{ for a severe level} \end{Bmatrix}$$

in steel plates (Smith et al., 1988).

$$C_2 = \begin{Bmatrix} 0.018 \text{ for a slight level} \\ 0.096 \text{ for an average level} \\ 0.252 \text{ for a severe level} \end{Bmatrix}$$

in aluminum plates (Paik et al., 2006).

To determine the shape of the buckling mode initial distortions, eigenvalue computations employing the nonlinear finite element method model are required. Based on these eigenvalue computations, the buckling modes of the stiffened plate structures can then be decomposed into the three aforementioned types of initial distortions. Each type of initial distortion should be amplified to the maximum target value, and the three resulting patterns should then be superimposed to provide a complete picture of the initial distortions. It is here worth discussing the classical theory of structural mechanics, which gives the buckling half-wave number of a simply supported plate element under longitudinal compression alone. This number is predicted as the minimum integer that satisfies the following condition, as discussed in Chapter 12, namely,

$$\frac{a}{b} \le \sqrt{m(m+1)} \tag{8.3.9}$$

where m = number of buckling half-waves of the plate in the longitudinal (long) direction, whereas the number in the transverse (short) direction is assumed to be unity.

The plate buckling half-wave number can then be determined under any combination of longitudinal compression σ_x and transverse compression σ_y, again as a minimum integer, but satisfying the following condition (Paik and Thayamballi, 2003).

$$\frac{(m^2/a^2 + 1/b^2)^2}{m^2/a^2 + c/b^2} \le \frac{[(m+1)^2/a^2 + 1/b^2]^2}{(m+1)^2/a^2 + c/b^2} \tag{8.3.10}$$

where $c = \sigma_y/\sigma_x$ = loading ratio. When $c = 0$, i.e., under longitudinal compression alone, (8.3.10) simplifies to (8.3.9).

For support members, the column-type initial distortion and sideways initial distortion of the stiffeners are often presumed to be as follows.

$$w_o^c = w_{oc} \sin \frac{\pi x}{a} \tag{8.3.11a}$$

$$w_o^s = w_{os} \frac{z}{h_w} \sin \frac{\pi x}{a} \tag{8.3.11b}$$

$$w_{oc} = C_3 a \tag{8.3.11c}$$

$$w_{os} = C_4 a \tag{8.3.11d}$$

where w_o^c = column type initial distortion of the support members, w_o^s = sideways initial distortion of the support members, z = coordinate in the direction of stiffener web height h_w, a = length of the small stiffeners between two adjacent strong support members, and C_3 and C_4 = constants. The constants in (8.3.11c) and (8.3.11d) are often taken to be as follows.

$C_3 = C_4 = 0.0015$ for an average level in steel plates (the practice suggested by the Classification Societies).

$$C_3 = \begin{Bmatrix} 0.00016 \text{ for a slight level} \\ 0.0018 \text{ for an average level} \\ 0.0056 \text{ for a severe level} \end{Bmatrix}$$

in aluminum plates (Paik, 2006).

$$C_4 = \begin{Bmatrix} 0.00019 \text{ for a slight level} \\ 0.001 \text{ for an average level} \\ 0.0024 \text{ for a severe level} \end{Bmatrix}$$

in aluminum plates (Paik et al., 2006).

Welding Residual Stress Modeling

Figure 8.13 shows a typical idealization of the welding-induced residual stress distribution inside the metal plates. The welding residual stress comprises the tensile residual stress block and the compressive residual stress block. In addition, welding residual stresses may also develop in both the longitudinal and transverse directions because the support members are usually attached by welding in these two directions.

The tensile residual stress blocks are equivalent to the HAZ, and their breadth can be estimated from the equilibrium between the tensile and compressive residual stresses, as follows.

$$2b_t = \frac{\sigma_{rcx}}{\sigma_{rcx} - \sigma_{rtx}} b, \; 2a_t = \frac{\sigma_{rcy}}{\sigma_{rcy} - \sigma_{rty}} a \qquad (8.3.12)$$

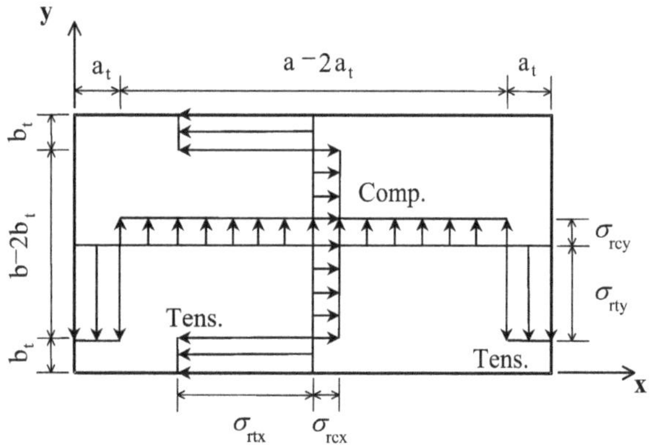

Figure 8.13 Typical idealization of the welding-induced residual stress distribution inside the metal plate element in the x and y directions.

where b_t, a_t= breadths of the tensile residual stress block, σ_{rcx}, σ_{rcy} = compressive residual stresses in the x and y directions, respectively, and σ_{rtx}, σ_{rty} = tensile residual stresses in the x and y directions, respectively.

As noted earlier, for mild steel the tensile residual stress in the HAZ reaches the material yield stress, but for high-tensile steel it is around 80% of the material yield stress. In addition, the compressive residual stress in the x direction of a steel plate is often assumed to be as follows (Smith et al., 1988).

$$\sigma_{rcx} = \begin{cases} -0.05\sigma_Y \text{ for a slight level} \\ -0.15\sigma_Y \text{ for an average level} \\ -0.3\sigma_Y \text{ for a severe level} \end{cases} \qquad (8.3.13a)$$

The counterpart of the compressive residual stress in the y direction may be assumed to be as follows (Paik and Thayamballi, 2003).

$$\sigma_{rcy} = k \frac{b}{a} \sigma_{rcx} \qquad (8.3.13b)$$

where k = correction factor, which may take a value smaller than 1.0. When the residual stress is considered in the x direction alone, $k = 0$.

Once both the tensile and compressive residual stresses have been defined, the breadths of the HAZ can be determined from (8.3.12). In the nonlinear finite element method modeling of plates, the size (breadth) of the finite element meshes located in the HAZ must be adjusted to equal the breadth of the tensile residual stress block. Only one mesh is enough to model the tensile residual stress block in the breadth direction (Paik and Sohn, 2009). Although this may break the aforementioned rule of the unity element-aspect ratio, the related effects may be negligible.

Similar modeling is considered for the stiffener webs, as necessary. If the flange-web junction is welded, the residual stress distribution pattern will be similar to that of the plating. However, rolled or extruded types of support members have an HAZ on one side alone, i.e., along the intersection between plating and support member, as shown in Fig. 8.12b. Built-up T-types of support members may have a distribution of welding residual stresses similar to that of plating surrounded by support members. In nonlinear finite element method simulations, the residual stresses should be dealt with as the initial stresses. Most commercial computer codes provide facilities for allocating the initial stresses in specific finite elements.

Figure 8.14 provides examples of the nonlinear structural behavior of a welded-steel plate under axial compression in the *x* direction, with the welding residual stress features varied, as those obtained by the ANSYS (2009) nonlinear finite element method. It is evident that this stress significantly affects the plate's ultimate strength behavior. It is particularly interesting to note that the residual stress that has developed in the *y* direction does not affect the plate behavior until the ultimate strength has been reached under pure-longitudinal compression, but it significantly affects the post-ultimate strength behavior. It can be surmised that the residual stress in both the *x* and *y* directions affects the plate's ultimate strength behavior before and after the plate reaches its ultimate strength under biaxial compressive actions. It is current maritime industry practice to disregard the residual stress in the *y* direction, i.e., that in the direction of the plate width, but this practice must be reconsidered.

Softening Phenomenon Modeling in the Heat-Affected Zone

In contrast to welded steel structures, the HAZ in welded aluminum structures is "softer," which means it has a reduced material yield stress. The nonlinear finite element method modeling technique used to deal with this softening phenomenon is similar to that used for residual stresses, except that the yield stress in the HAZ must be reduced

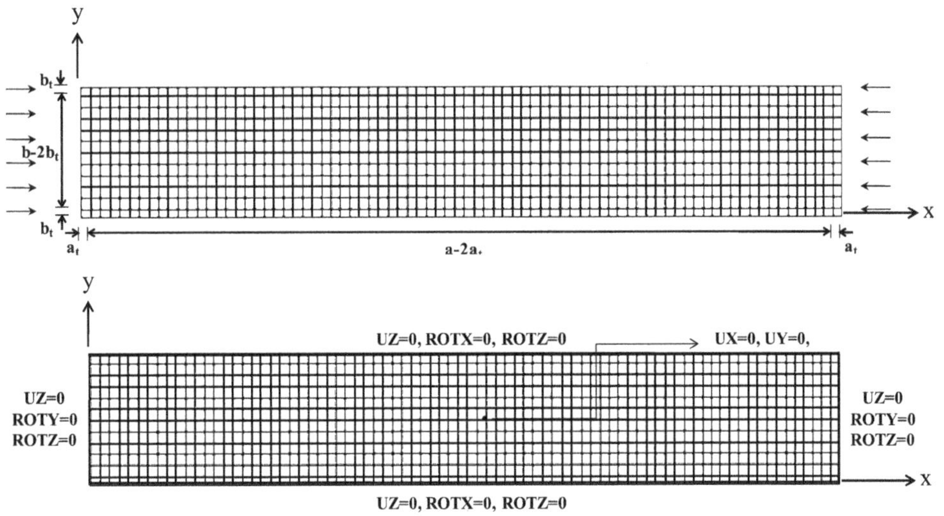

Figure 8.14a Sample of the nonlinear finite element method model with the boundary condition applied.

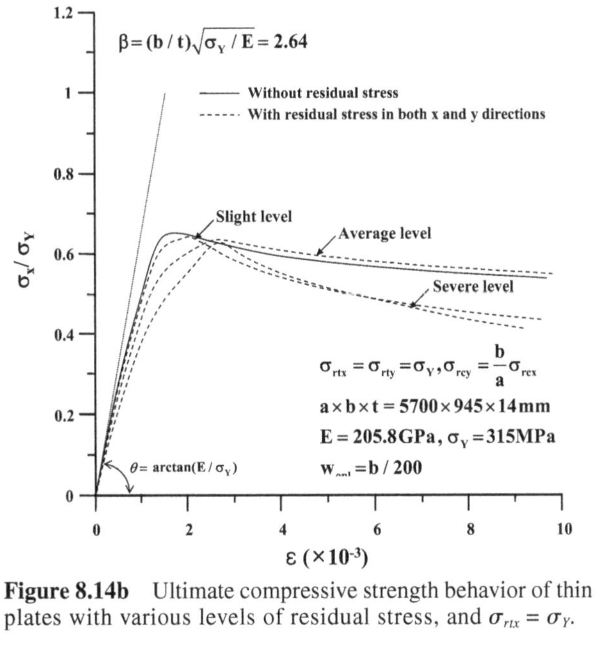

Figure 8.14b Ultimate compressive strength behavior of thin plates with various levels of residual stress, and $\sigma_{rtx} = \sigma_Y$.

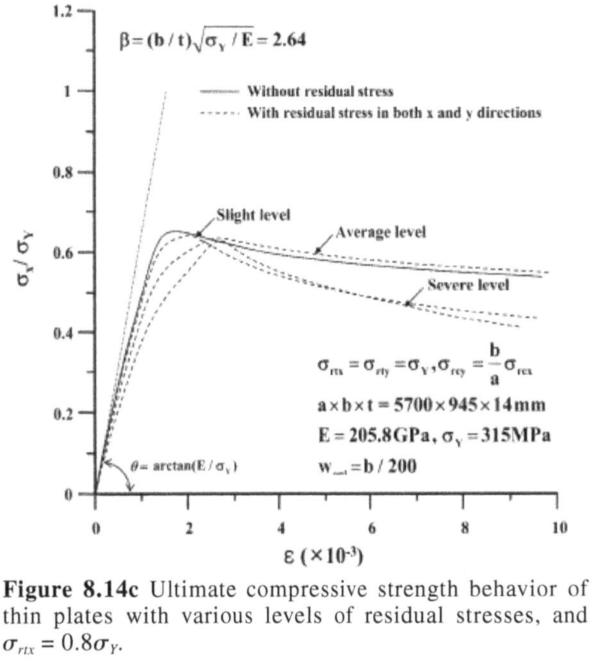

Figure 8.14c Ultimate compressive strength behavior of thin plates with various levels of residual stresses, and $\sigma_{rtx} = 0.8\sigma_Y$.

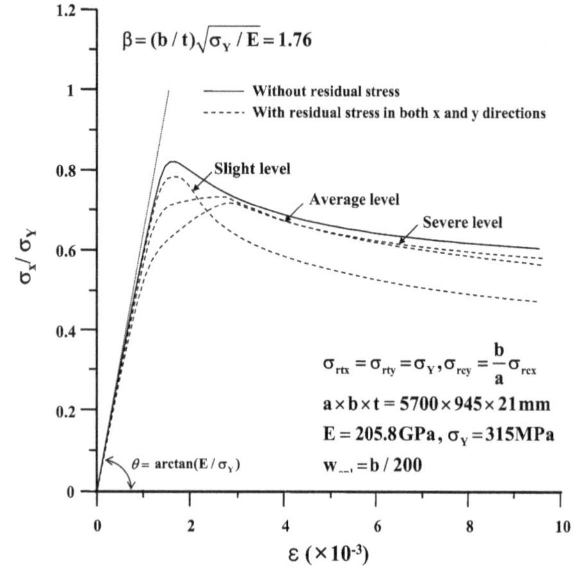

Figure 8.14d Ultimate compressive strength behavior of thick plates with various levels of residual stresses, and $\sigma_{rtx} = \sigma_Y$.

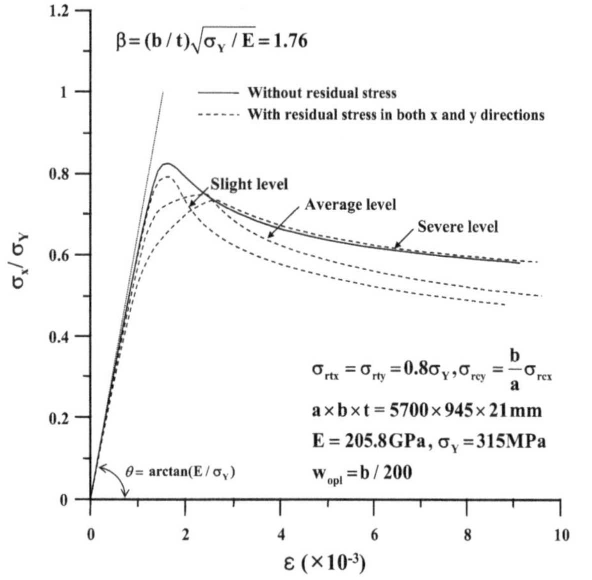

Figure 8.14e Ultimate compressive strength behavior of thick plates with various levels of residual stresses, and $\sigma_{rtx} = 0.8\sigma_Y$.

below that in the base metal because of softening. However, the "locked-in" tensile residual stress rule in the HAZ is still approximately equal to the material yield stress. In welded aluminum plate structures, the breadths of the HAZ with the nomenclature defined in Fig. 8.12c may be determined as follows (Paik et al., 2006).

$$b_p' = b_s' = \begin{cases} 11.3 \text{ mm for a slight level} \\ 23.1 \text{ mm for an average level} \\ 29.9 \text{ mm for a severe level} \end{cases} \quad (8.3.14)$$

The yield strength in the HAZ may be obtained as follows, depending on the type of aluminum alloy (Paik et al., 2006).

(a) Yield stress of the HAZ material for aluminum alloy 5083-H116

$$\frac{\sigma_{YHAZ}}{\sigma_Y} = \begin{Bmatrix} 0.906 \text{ for a slight level} \\ 0.777 \text{ for an average level} \\ 0.437 \text{ for a severe level} \end{Bmatrix}$$

$$\text{with } \sigma_Y = 215 \text{ N/mm}^2 \quad (8.3.15a)$$

(b) Yield stress of the HAZ material for aluminum alloy 5383-H116

$$\frac{\sigma_{YHAZ}}{\sigma_Y} = \begin{Bmatrix} 0.820 \text{ for a slight level} \\ 0.774 \text{ for an average level} \\ 0.640 \text{ for a severe level} \end{Bmatrix}$$

$$\text{with } \sigma_Y = 220 \text{ N/mm}^2 \quad (8.3.15b)$$

(c) Yield stress of the HAZ material for aluminum alloy 5383-H112

$$\frac{\sigma_{YHAZ}}{\sigma_Y} = 0.891 \text{ for an average level}$$

$$\text{with } \sigma_Y = 190 \text{ N/mm}^2 \quad (8.3.15c)$$

(d) Yield stress of the HAZ material for aluminum alloy 6082-T6

$$\frac{\sigma_{YHAZ}}{\sigma_Y} = 0.703 \text{ for an average level}$$

$$\text{with } \sigma_Y = 240 \text{ N/mm}^2 \quad (8.3.15d)$$

The compressive residual stresses at the plate part and stiffener web may be determined regardless of the aluminum alloy type, as follows (Paik et al., 2006).

$$\sigma_{rcx} = \begin{Bmatrix} -0.110\sigma_{Yp} \text{ for a slight level} \\ -0.161\sigma_{Yp} \text{ for an average level} \\ -0.216\sigma_{Yp} \text{ for a severe level} \end{Bmatrix}$$

$$\text{in the plate part} \quad (8.3.16a)$$

$$\sigma_{rcx} = \begin{Bmatrix} -0.078\sigma_{Ys} \text{ for a slight level} \\ -0.137\sigma_{Ys} \text{ for an average level} \\ -0.195\sigma_{Ys} \text{ for a severe level} \end{Bmatrix}$$

$$\text{in the stiffener web} \quad (8.3.16b)$$

where σ_{Yp} = yield strength of the plate part, σ_{Ys} = yield strength of the stiffener web.

Equation (8.3.13b) may also be applied for the features of the residual stresses in the y direction, as necessary.

8.3.7 Load Applications

Ship and ship-shaped offshore structures are likely to be subjected to complex load applications. For example, the outer bottom stiffened-plate structures of vessels may be subjected to combinations of longitudinal compression, transverse compression, and lateral pressure, as shown in Fig. 8.15.

Order of Load Component Application

In current nonlinear finite element method computation practice, lateral pressure is usually applied first. Then, keeping the lateral pressure constant, a combination of biaxial compressive loads is applied. It is interesting to note that the shape and magnitude of the initial distortions in the plate panels can be markedly changed by the lateral pressure.

Figure 8.16 provides examples of steel panels under longitudinal and transverse compression before and after lateral pressure. The pressure causes an effective "clamping" of the plating and changes the deflected shape away from the buckling mode shape. This may cause the buckling strength value of the in-plane compression to be larger than if the pressure was small or absent. Therefore, in panels that may receive in-plane compression with either a large or a small lateral pressure (such as underwater panels in a tanker or bulker) the ultimate strength should be calculated for both full load and ballasted conditions, and the lower value should be taken as the true ultimate strength.

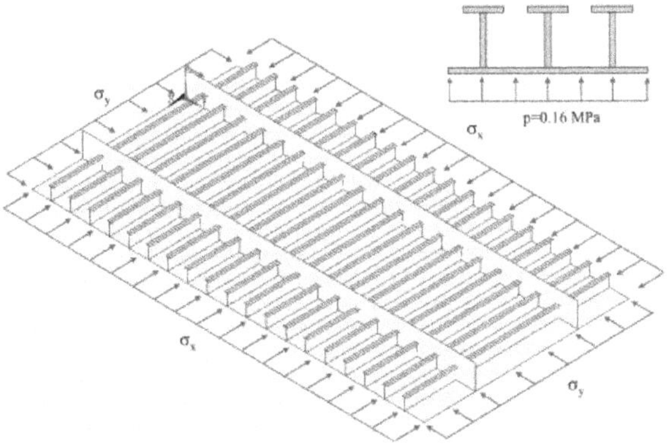

Figure 8.15 Illustrative example of a stiffened panel under combined biaxial compression and lateral pressure (p = 0.16 MPa).

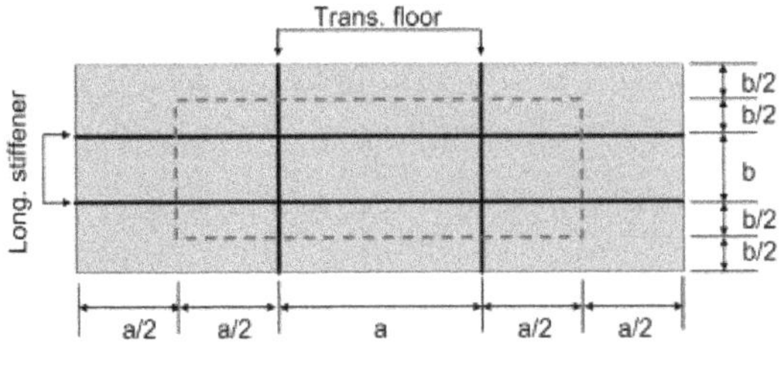

Figure 8.16a Extent of the analysis for the two (½+1+½)-bay plate model.

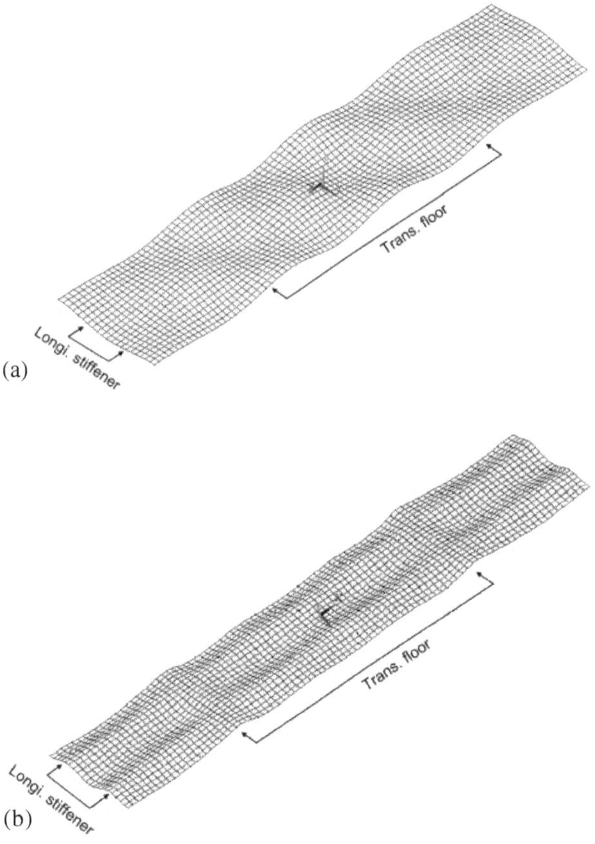

Figure 8.16b The plate initial deflection shapes under predominantly longitudinal compression for the two-bay plate model: (a) Before lateral pressure and (b) after lateral pressure (amplification factor of 30).

Effect of Load Path

In linear structural mechanics under a combination of multiple load components, the principle of linear superposition of structural responses by individual load components is satisfied and the final status of structural response is identical regardless of the load paths. This principle is often adopted even for the nonlinear structural mechanics problems with the focus on buckling or ultimate strength, where the load effects or resulting deformations are not large with small strains

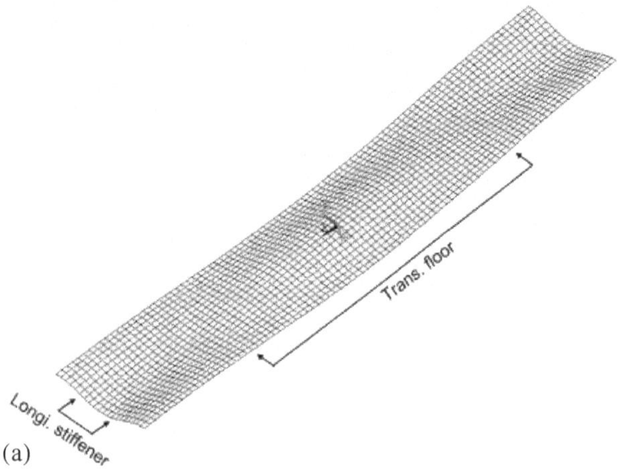

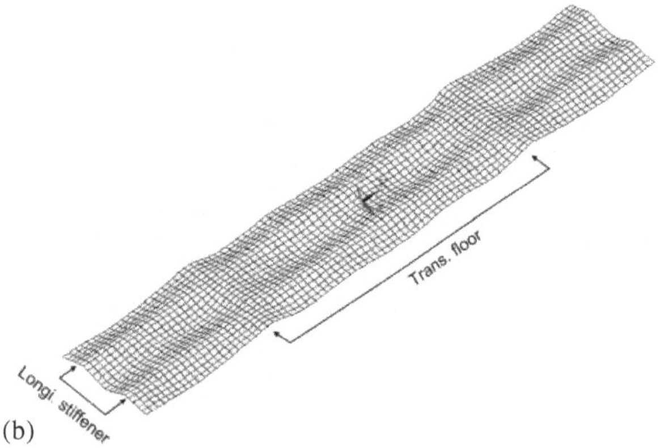

Figure 8.16c The plate initial deflection shapes under predominantly transverse compression for the two-bay plate model: (a) Before lateral pressure and (b) after lateral pressure (amplification factor of 30).

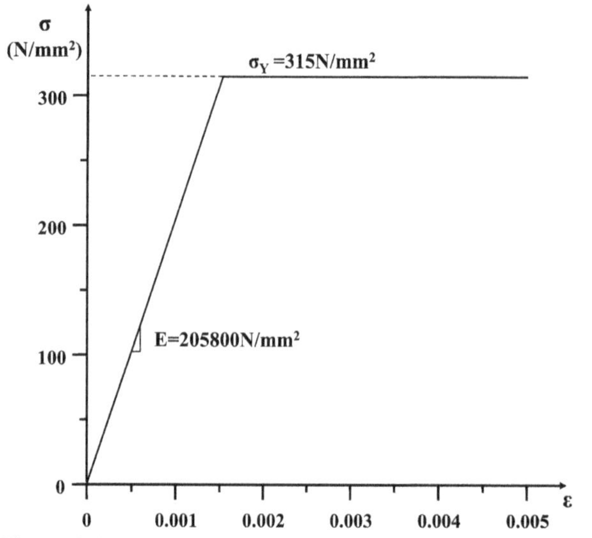

Figure 8.17b Elastic-perfectly plastic model of high-tensile steel considered.

until buckling or ultimate strength is reached. In contrast, the problems of structural crashworthiness in accidental situations such as collisions and grounding exhibit large strains associated with crushing and rupture, so the principle of linear superposition is no longer applicable.

An illustrative example of the ultimate strength behavior for a simply supported steel plate under biaxial compressive loads is now considered, as shown in Fig. 8.17a. While the geometrical and material properties of the plate are indicated in Fig. 8.17a, the plate slenderness coefficient equals $\beta = 1.76$. This plate was actually extracted from a VLCC (very large crude oil carrier) class double-hull oil tanker structure, and its thickness is moderate or medium (i.e., neither very thick nor very thin).

The plate edges are presumed to move in plane but keeping them straight. This edge condition likely represents the situation of a plate in continuous stiffened plate structures. For convenience of this discussion, it is assumed that the welding residual stresses do not exist. The plate initial deflection is assumed with buckling mode shape as follows.

$$w_o = w_{opl} \sin \frac{m\pi x}{a} \sin \frac{\pi y}{b}$$

where m = buckling half-wave number in the x direction, which can be determined from (8.3.10), w_{opl} = maximum initial deflection, which is assumed to be $w_{opl} = b/200$ in the present example.

An elastic-perfectly plastic model of material without considering the strain-hardening effect is adopted for the present nonlinear finite element analysis, as shown in Fig. 8.17b. Figures 8.18a and 8.18b present the finite element mesh models of the plate, including initial deflection, under predominantly longitudinal axial compression and predominantly transverse axial compression,

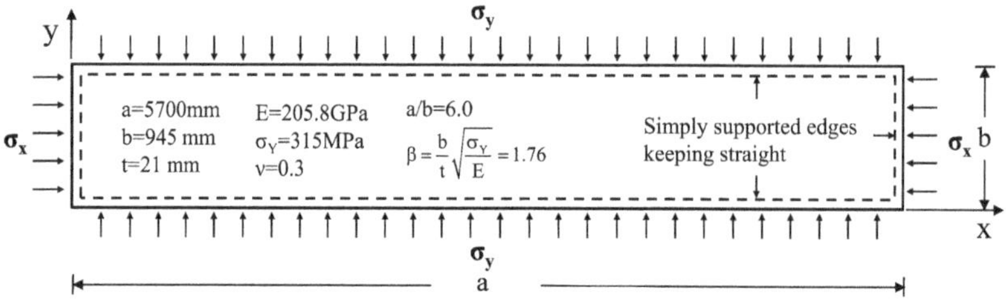

Figure 8.17a A simply supported steel plate under biaxial compressive loads.

respectively. Figure 8.18 shows different initial deflection shapes, depending on the load combination, where the buckling half-wave number is $m = 6$ for predominantly longitudinal axial compression but is $m = 1$ as the transverse axial compressive load becomes predominant.

Four types of load combination, namely pure σ_x with $\sigma_y = 0$, $\sigma_y/\sigma_x = 1$, $\sigma_y/\sigma_x = 0.5$, and pure σ_y with $\sigma_x = 0$ are considered. Figure 8.19 shows the ultimate strength interaction relationships of the plate between σ_x and σ_y, obtained by the ANSYS (2009) nonlinear finite element method and ALPS/ULSAP (2009).

Figure 8.20 presents the relationships of the applied average-stresses versus the resulting average-strains, under purely longitudinal or transverse axial-compression, obtained by ANSYS computations, where $\varepsilon_x = u/a$ and $\varepsilon_y = v/b$ with average axial-compressive-displacements u in the x direction and v in the y direction (compression is taken as positive and tension is taken as negative). In load paths OA or OF under a single load component of σ_x, the $\sigma_x - \varepsilon_x$ relationship is almost linear, as shown in Fig. 8.20, representing that the deflections are not large. On the other hand, in load paths OG or OB, the $\sigma_y - \varepsilon_y$ relationship is nonlinear, as shown in Fig. 8.20, representing that large deflections but with small strains must have occurred due to buckling and/or plasticity.

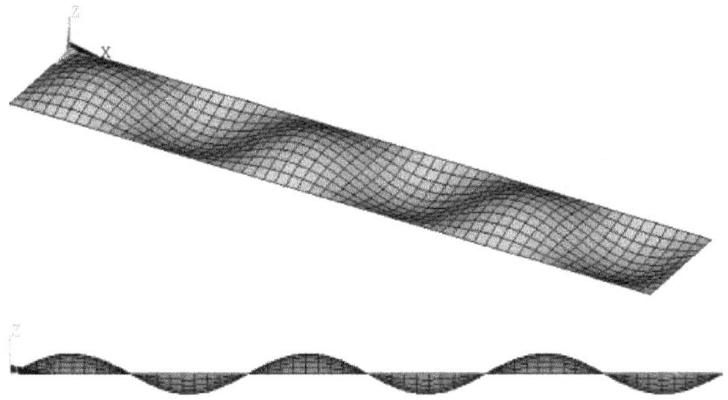

Figure 8.18a Finite element analysis model of the plate with initial deflection (amplified by 150 times), under predominantly longitudinal axial compressive loads.

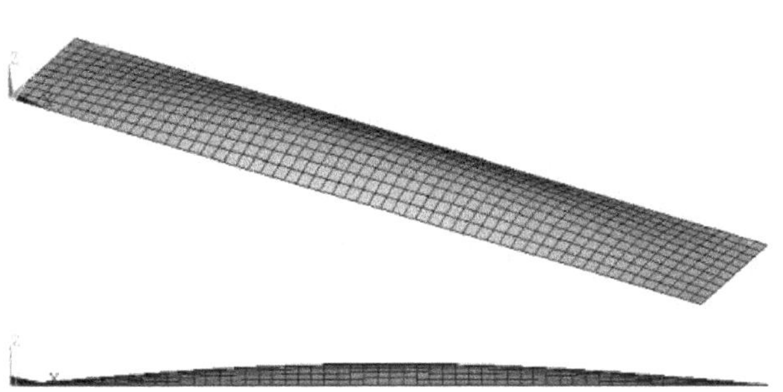

Figure 8.18b Finite element analysis model of the plate with initial deflection (amplified by 150 times), under predominantly transverse axial compressive loads.

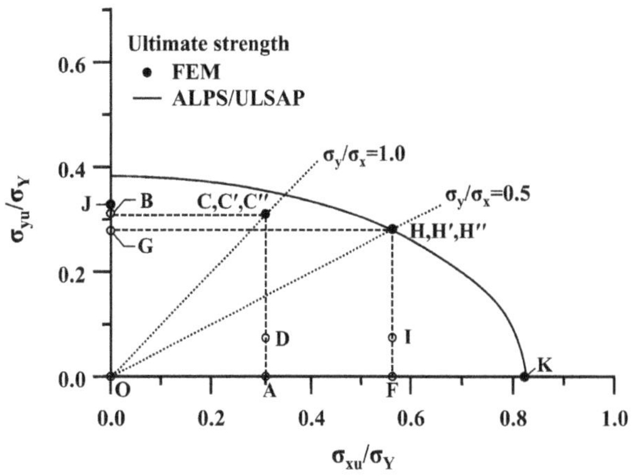

Figure 8.19 Ultimate strength interaction relationships of the plate between σ_x and σ_y.

Figures 8.21a to 8.21c show the membrane stress distributions of the plate at point F of load path OF, at point G of load path OG, and at point H of load path OH with $\sigma_y/\sigma_x = 0.5$, respectively. It is observed from Fig. 8.21c that the membrane stress distribution in the compressive-load direction is nonuniform under a combination of σ_x and σ_y due to the existence of lateral deflection arising from initial deflection and buckling. Even under a single load component, e.g., pure σ_x or pure σ_y, as shown in Figs. 8.21a and 8.21b, the membrane stress distribution along the unloaded edges is nonuniform because of the straight edge condition, although the average stress must be zero because no forces are applied, where the compressive stress develops around the plate corner and the tensile stress develops in the middle of the plate edge.

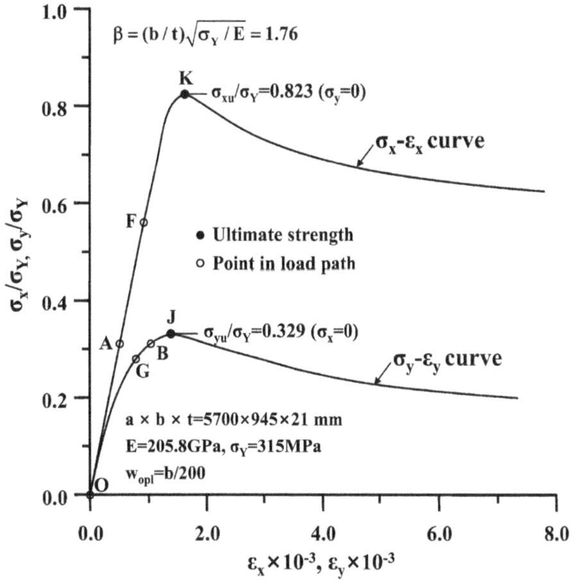

Figure 8.20 Ultimate strength behavior of the plate under purely longitudinal or transverse axial compressive loads.

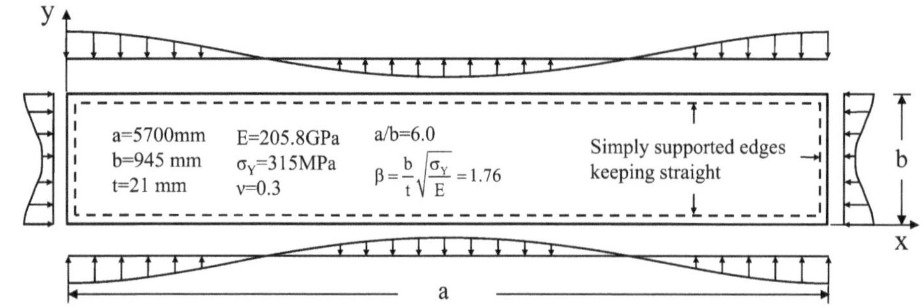

FIgure 8.21a Membrane stress distribution of the plate at point F of load path OF, under purely longitudinal axial compressive loads.

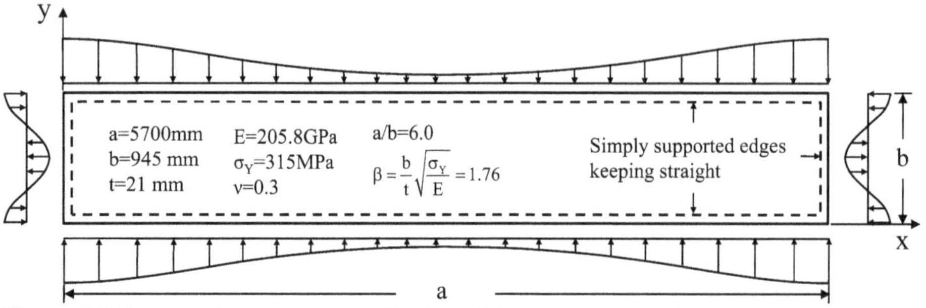

Figure 8.21b Membrane stress distribution of the plate at point G of load path OG, under purely transverse axial compressive loads.

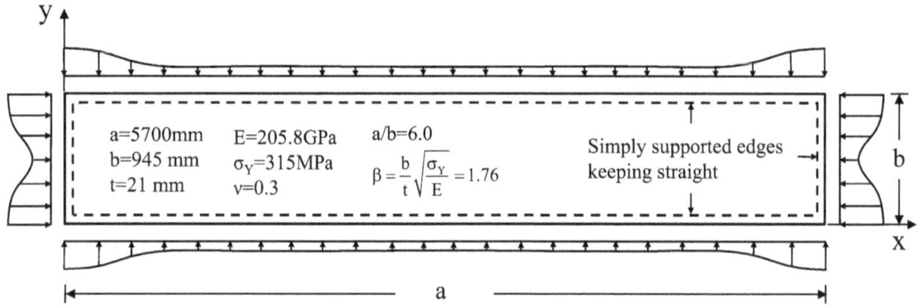

Figure 8.21c Membrane stress distribution of the plate at point H of load path OH, under a combination of longitudinal and transverse axial compressive loads with $\sigma_y/\sigma_x = 0.5$.

Figures 8.22a to 8.22d show that the load path affects the plate's ultimate strength behavior before or after the plate reaches its ultimate limit state, but the effect of that path is negligible in terms of the plate's ultimate strength value. For simplicity of the buckling and ultimate strength computations, therefore, the constant loading ratio approach (i.e., by a simultaneous application of the load components) is often adopted. However, it is noted that this rule cannot be applied for structural crashworthiness simulations that involve large strains due to crushing and rupture in accidental situations. In this case, it is of significant importance to accurately define the load path, as well as the load characteristics.

8.3.8 Verification of Structural Modeling Techniques

Before analysis of the target structure begins, it is necessary to verify that the nonlinear finite element method modeling is adequate. This can be accomplished through comparison with experimental results and/or existing theoretical and numerical computations. For this purpose, useful databases of experimental results are available for steel-stiffened plate structures (Smith, 1976) and aluminum-stiffened plate structures (Paik et al., 2008; Paik 2009). A database of nonlinear finite element method solutions obtained by different computer codes is also available in the report of ISSC (International

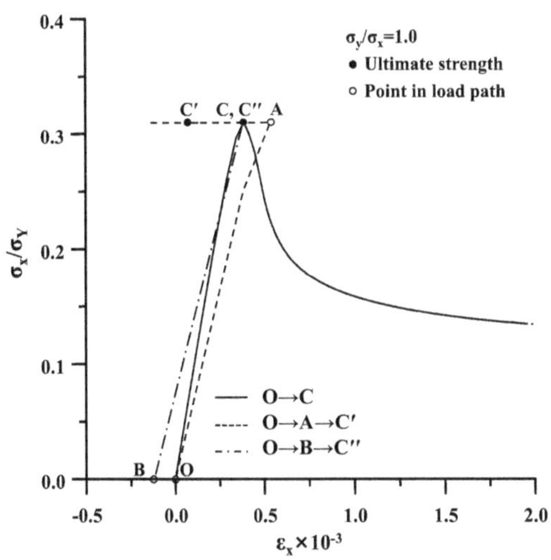

Figure 8.22a Ultimate strength behavior ($\sigma_x - \varepsilon_x$ curve) of the plate with the loading path varied for a combination of longitudinal axial compression and transverse axial compression, $\sigma_y/\sigma_x = 1$.

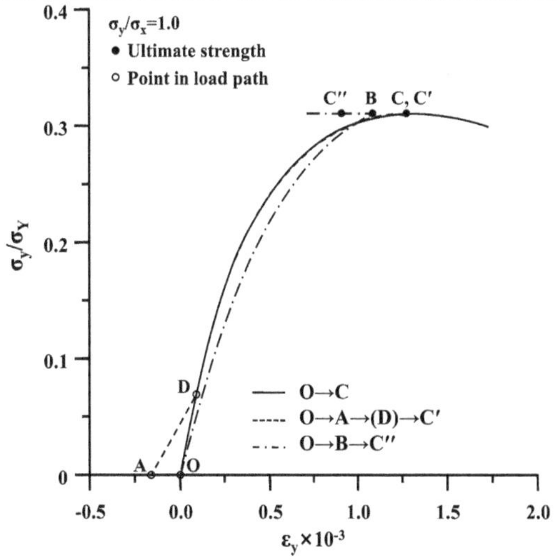

Figure 8.22b Ultimate strength behavior ($\sigma_y - \varepsilon_y$ curve) of the plate with the loading path varied for a combination of longitudinal axial compression and transverse axial compression, $\sigma_y/\sigma_x = 1$.

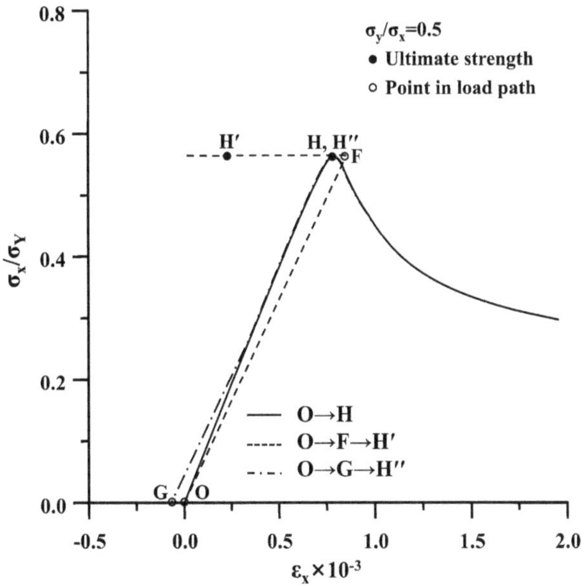

Figure 8.22c Ultimate strength behavior ($\sigma_x - \varepsilon_x$ curve) of the plate with the loading path varied for a combination of longitudinal axial compression and transverse axial compression, $\sigma_y/\sigma_x = 0.5$.

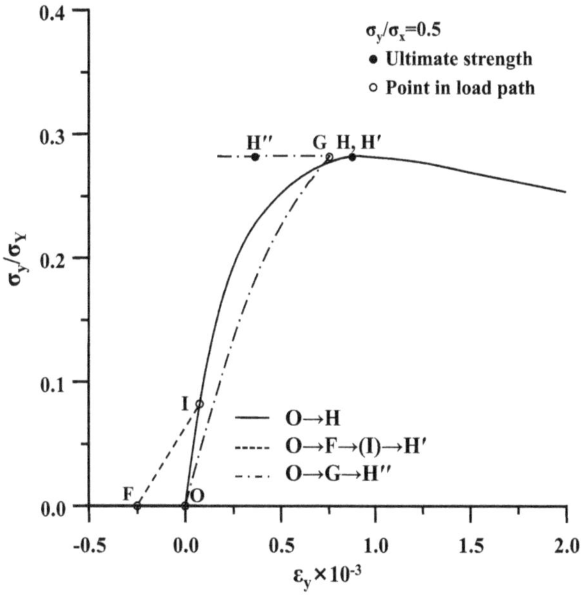

Figure 8.22d Ultimate strength behavior ($\sigma_y - \varepsilon_y$ curve) of the plate with the loading path varied for a combination of longitudinal axial compression and transverse axial compression, $\sigma_y/\sigma_x = 0.5$.

Ship and Offshore Structures Congress), together with useful guidance for nonlinear finite element method modeling (Paik et al., 2009).

REFERENCES

ALPS/ULSAP. (2009). *Ultimate limit state assessment of plate panels*. Stevensville, MD: DRS C3 Systems. Available online at http://www.orca3d.com/ maestro.

ANSYS. (2009). *Nonlinear finite element method analysis*, Version 11.0. Canonsburg, PA: Author.

Bathe, K. J. (1982). *Finite element procedures in engineering analysis*. Englewood Cliffs, New Jersey: Prentice Hall.

Cook, R. D., Malkus, D. S., and Plesha, M. E. (1989). *Concepts and applications of finite element analysis*. New York: Wiley.

Cowper, G. R., and Symonds, P. S. (1957). *Strain-hardening and strain-rate effects in the impact loading of cantilever beams*. Technical Report No. 23. Providence, RI: Division of Applied Mathematics, Brown University.

Crisfield, M. A. (1981). A fast incremental/iterative solution procedure that handles "snap-through." *Computers & Structures*, Vol. 13, 55–62.

Masubuchi, K. (1980). *Analysis of welded structures: Residual stresses, distortion, and their consequences*. Oxford, U.K.: Pergamon Press.

Mazzolani, F. M. (1985). *Aluminum alloy structures*. London: Pitman Publishing Ltd.

Owen, D. R. J., and Hinton, E. (1980). *Finite elements in plasticity*. Swansea, U.K.: Pineridge Press Ltd.

Paik, J. K. (2007a). Practical techniques for finite element modeling to simulate structural crashworthiness in ship collisions and grounding (Part I: Theory). *Ships and Offshore Structures*, Vol. 2, Issue 1, 69–80.

Paik, J. K. (2007b). Practical techniques for finite element modeling to simulate structural crashworthiness in ship collisions and grounding (Part II: Verification). *Ships and Offshore Structures*, Vol. 2, Issue 1, 81–85.

Paik, J. K. (2007c). Characteristics of welding induced initial deflections in welded aluminum plates. *Thin-Walled Structures*, Vol. 45, 493–501.

Paik, J. K. (2009). *Buckling collapse testing of friction stir welded aluminum stiffened plate structures*. Report SSC-456. Washington, DC: Ship Structure Committee.

Paik, J. K., Amdahl, J., Barltrop, N., Donner, E. R., Gu, Y., Ito, H., et al. (2003). Collision and grounding. *Report of ISSC Committee V.3, 15th International Ship and Offshore Structures Congress*, San Diego, CA, August.

Paik, J. K., Branner, K., Choo, Y. S., Czujko, J., Fujikubo, M., Gordo, J. M., et al. (2009). Ultimate strength. *Report of Technical Committee III.1, 17th International Ship and Offshore Structures Congress*, Seoul, Korea, August.

Paik, J. K., and Hughes, O. F. (2007). Ship structures. In Melchers, R. E., and Hough, R. (Eds.). *Modeling complex engineering structures*. Reston, VA: The American Society of Civil Engineers.

Paik, J. K., Kim, B. J., and Seo, J. K. (2008a). Methods for ultimate limit state assessment of ships and ship-shaped offshore structures: Part I—Unstiffened plates. *Ocean Engineering*, Vol. 35, 261–270.

Paik, J. K., Kim, B. J., and Seo, J. K. (2008b). Methods for ultimate limit state assessment of ships and ship-shaped offshore structures: Part II—Stiffened panels. *Ocean Engineering*, Vol. 35, 271–280.

Paik, J. K., Kim, B. J., and Seo, J. K. (2008c). Methods for ultimate limit state assessment of ships and ship-shaped offshore structures: Part III—Hull girders. *Ocean Engineering*, Vol. 35, 281–286.

Paik, J. K., and Sohn, J. M. (2009). Effects of welding residual stresses on high tensile steel plate ultimate strength: Nonlinear finite element method investigations. OMAE2009-79297. *Proc. of 28th International Conference on Offshore Mechanics and Arctic Engineering*, Honolulu, Hawaii, May 31–June 5.

Paik, J. K., and Thayamballi, A. K. (2003). *Ultimate limit state design of steel-plated structures*. Chichester, U.K.: Wiley.

Paik, J. K., and Thayamballi, A. K. (2007). *Ship-shaped offshore installations: Design, building, and operation*. Cambridge, U.K.: Cambridge University Press.

Paik, J. K., Thayamballi, A. K., Ryu, J. Y., Jang, J. H., Seo, J. K., Park, S. W., et al. (2006). The statistics of weld induced initial imperfections in aluminum stiffened plate structures for marine applications. *International Journal of Maritime Engineering*, Vol. 148, Part A1, 1–44.

Paik, J. K., Thayamballi, A. K., Ryu, J. Y., Jang, J. H., Seo, J. K., Park, S. W., et al. (2008). *Mechanical collapse testing on aluminum stiffened panels for marine applications*. Report SSC-451. Washington, DC: Ship Structure Committee.

Ramberg, W., and Osgood, W. R. (1943). *Description of stress strain curves by three parameters*. Technical Note No. 902. Hampton, VA: National Advisory Committee on Aeronautics.

Smith, C. S. (1976). Compressive strength of welded steel grillages. *RINA Trans., Vol.* 118, 325–359.

Smith, C. S., Davidson, P. C., Chapman, J. C., and Dowling, P. J. (1988). Strength and stiffness of ships' plating under in-plane compression and tension. *RINA Trans.*, Vol. 130, 277–296.

Steinhardt, O. (1971). Aluminum constructions in civil engineering. *Aluminum*, Vol. 47, 131–139, 254–261.

White, R. E. (1985). *An introduction to the finite element method with applications to nonlinear problems*. London, U.K.: Wiley.

Wierzbicki, T., and Abramowicz, W. (1983). On the crushing mechanics of thin-walled structures. *J. of Applied Mechanics*, Vol. 50, 727–734.

Zienkiewicz, O. C. (1977). *The finite element method*, (3rd ed.). London: McGraw-Hill.

NINE

PLATE BENDING

Owen Hughes
Professor, Virginia Tech
Blacksburg, VA, USA

John B. Caldwell
Emeritus Professor (University of Newcastle, UK)
Windermere, Cumbria, UK

This chapter examines the response of plating to four types of lateral load:

1. Static uniform pressure (9.4.1)
2. Static concentrated load, for which there are three options
   a. Multiple location, using an equivalent pressure (9.4.14)
   b. Single location (center of the plate) (9.4.16)
   c. A general method that combines the above two methods
3. Quasistatic pressure, such as slamming (9.5.1,2)
4. Dynamic pressure (very short duration) (9.5.9)

A small program called PLATE is available at http://filebox.vt.edu/users/hugheso. PLATE includes all four of these load types and the corresponding equations. It deals with the three most important variables—load, plate thickness and permanent set—such that given any two, it will solve for the third. The methods and equations in PLATE are the same as those in MAESTRO, but PLATE deals with just one plate instead of a large 3D structural model.

Each method is for a specific type of load. In many cases a given piece of plating may be exposed to two or three or even all four types of load. If the methods were used in a hand solution to determine plate thickness, it would be necessary to use all relevant methods in order to determine the largest required thickness, and this would be arduous. But, in the rationally-based, computer-aided structural optimization process of Fig. 1.4, the existence of multiple types of lateral load simply means that in regard to plate thickness there are two or three or four constraints instead of one; there is no need to calculate multiple values of thickness. Indeed, it may turn out that none of these constraints determine the thickness, but rather one of the constraints arising from in-plane strength requirements.

9.1 SMALL DEFLECTION THEORY

9.1.1 "Long" Plates (Cylindrical Bending)

Unlike a beam, in which bending occurs only along the length, the bending in a plate usually occurs in two orthogonal directions. An equation relating the deflections to the loading can be developed for the plate, as for the beam. To show the similarity (and the differences) let us begin with the case of a plate that is bent about one axis only (cylindrical bending) as occurs for long plates ($a >> b$). An elemental strip of such plating of width da is shown in Fig. 9.1.

If this strip were an isolated beam its transverse section would deform as shown by the dashed lines, due to Poisson's ratio effects. This is termed "anticlastic" curvature, and it may be shown that this radius of curvature is $1/\nu$ times the primary or bending radius of curvature. In plating, however, this transverse deformation does not occur because such a deformation would require that the plate take on a saddle shape, which would mean considerable stretching of the neutral surface and would require enormous strain energy. The prevention of this transverse strain ($\varepsilon_y = 0$) gives rise to a transverse stress $\sigma_y = \nu\sigma_x$, as may be seen from the strain equations:

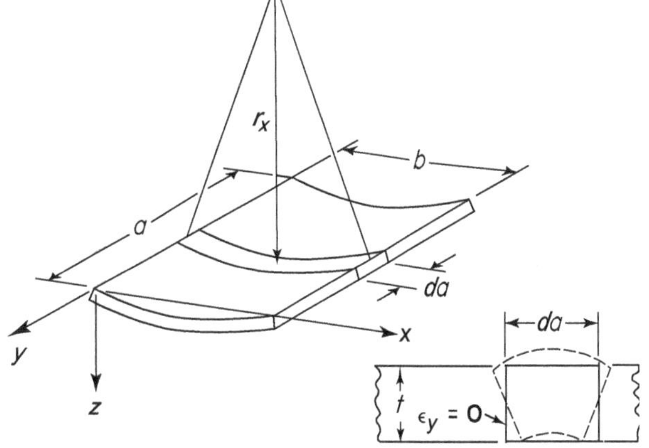

Figure 9.1 Plate in cylindrical bending.

$$\varepsilon_x = \frac{\sigma_x}{E} - \nu\frac{\sigma_y}{E}$$

$$\varepsilon_y = \frac{\sigma_y}{E} - \nu\frac{\sigma_x}{E} = 0$$

From the second equation we have $\sigma_y = \nu\sigma_x$ and hence the first equation becomes:

$$\varepsilon_x = \frac{\sigma_x(1-\nu^2)}{E}$$

from which

$$\sigma_x = \frac{E}{1-\nu^2}\varepsilon_x \tag{9.1.1}$$

The latter equation, when compared to the case of a beam ($\sigma_x = E\varepsilon_x$), suggests the definition of a quantity $E' = E/(1-\nu^2)$, which could be regarded as an "effective" modulus of elasticity. Obviously, this effective modulus is always greater than E, and it may be concluded that a plate is always "stiffer" than a row of beams. The effective modulus is a useful parameter because for a long prismatically loaded plate the effect of this extra stiffness may be fully accounted for by using E' in place of E in all of the various beam deflection formulas (the width of the beam being taken as unity). Thus, for example, the expression for maximum deflection of the long, simply supported plate of Fig. 9.1 due to a lateral pressure p is

$$w_{max} = \frac{5pb^4}{384E'I} = \frac{5pb^4(1-\nu^2)}{32Et^3} \tag{9.1.2}$$

and for the clamped case

$$w_{max} = \frac{pb^4}{384E'I} = \frac{pb^4(1-\nu^2)}{32Et^3} \tag{9.1.3}$$

As with beams, the moment-curvature relation may be obtained by imposing equilibrium of moments over the cross section of the strip of plating (again taking the width as unity).

External bending moment = moment of stress forces.

$$M = \int_{-t/2}^{t/2} \sigma_x z\,dz = \int_{-t/2}^{t/2} \frac{E}{1-\nu^2}\varepsilon_x z\,dz$$

The bending strain is $\varepsilon_x = z/r_x$ and hence

$$M = \int_{-t/2}^{t/2} \frac{E}{1-\nu^2}\left(\frac{z^2}{r_x}\right)dz = \frac{Et^3}{12(1-\nu^2)}\left(\frac{1}{r_x}\right)$$

$$= \frac{D}{r_x} \tag{9.1.4}$$

where

$$D = \frac{Et^3}{12(1-\nu^2)} \tag{9.1.5}$$

Thus for a plate the constant of proportionality between moment and curvature is D; this is referred to as the *flexural rigidity* of the plate, and is analogous to the quantity EI in beam theory. This may also be seen by substituting E' for E and noting that for a strip of plating of unit width, $I = t^3/12$.

The radius of curvature of the plate can be approximated in terms of the deflection w of the plate thus:

$$-\frac{\partial^2 w}{\partial x^2} \cong \frac{1}{r_x} \tag{9.1.6}$$

so that from (9.1.4) the relation between the bending moment and the deflection is

$$-D\frac{\partial^2 w}{\partial x^2} = M$$

and this corresponds exactly to the expression relating bending moment and deflection for a beam, with D substituted for EI.

In a beam, once the maximum bending moment has been determined, the corresponding maximum stress is given by $\sigma = Mc/I$, and this will also be true for a unit strip in a long plate. The section modulus I/c for a unit strip of plating is

$$\frac{\frac{t^3}{12}}{\frac{t}{2}} = \frac{t^2}{6}$$

For a uniform pressure p the maximum bending moment in the unit strip of plating is proportional to pb^2 and hence the maximum stress is proportional to $pb^2/(t^2/6)$. This is usually expressed in the form

$$\sigma_{max} = kp\left(\frac{b}{t}\right)^2 \tag{9.1.7}$$

and the value of the coefficient k depends on the boundary conditions. For simply supported edges k = ¾ and for clamped edges k = ½. As we shall see, this same form of equation is used for all plates, whether they are long or not, and the coefficient k also accounts for the effect of aspect ratio a/b.

9.1.2 Derivation of the Plate Bending Equation

In this section we derive the basic equation governing the behavior of panels of plating under lateral load. This equation was first derived by Lagrange in 1811; the outline that follows is adapted from Timoshenko (1959) and Jaeger (1964). This theory is only applicable if:

1. Plane cross sections remain plane.
2. The deflections of the plate are small (w_{max} not exceeding t).
3. The maximum stress nowhere exceeds the plate yield stress (i.e., the material remains elastic).

In general, a panel of plating will have curvature in two directions at right angles. Let the radii of curvature in these two directions be r_x and r_y, respectively. It follows that if a small element of length dx and breadth dy is considered, as shown in Fig. 9.2, there will be distributed moments m_x and m_y (moments per unit width) along the edges. The values of these moments can be obtained as follows:

$$\varepsilon_x = \frac{\sigma_x}{E} - \frac{\nu\sigma_y}{E}$$

$$\varepsilon_y = \frac{\sigma_y}{E} - \frac{\nu\sigma_x}{E}$$

Multiplying the second of these equations by ν and adding gives

$$\frac{\sigma_x}{E}(1-\nu^2) = \varepsilon_x + \nu\varepsilon_y$$

Assumptions 1 and 2 stated previously give rise to the strain-curvature relations

$$\varepsilon_x = -z\left(\frac{\partial^2 w}{\partial x^2}\right) \quad \text{and} \quad \varepsilon_y = -z\left(\frac{\partial^2 w}{\partial y^2}\right)$$

Hence,

$$\sigma_x = \frac{E}{1-\nu^2}(-z)\left(\frac{\partial^2 w}{\partial x^2} + \nu\frac{\partial^2 w}{\partial y^2}\right)$$

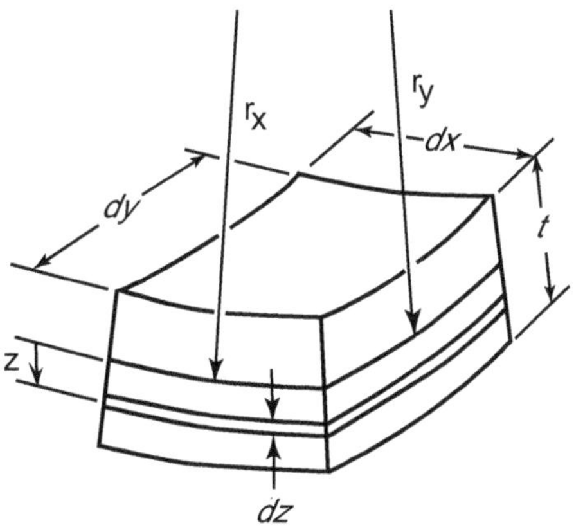

Figure 9.2 Differential element of plating.

Equilibrium of internal and external moments gives

$$m_x = \int_{-t/2}^{t/2} \sigma_x z\, dz$$

$$= -\int_{-t/2}^{t/2} \frac{E}{1-\nu^2}\left(\frac{\partial^2 w}{\partial x^2} + \nu\frac{\partial^2 w}{\partial y^2}\right) z^2 dz$$

$$= -D\left(\frac{\partial^2 w}{\partial x^2} + \nu\frac{\partial^2 w}{\partial y^2}\right) \qquad (9.1.8a)$$

Similarly,

$$m_y = -D\left(\frac{\partial^2 w}{\partial y^2} + \nu\frac{\partial^2 w}{\partial x^2}\right) \qquad (9.1.8b)$$

Suppose now that the intensity of lateral load per unit area on the plate is p; then the load on the element will be $p\, dx\, dy$. This load is carried by the distributed shear forces q acting on the four edges of the element, as shown in Fig. 9.3. In the general case of bending, twisting moments will also be generated on all the four faces, these being denoted by m_{xy} and m_{yx} (again, per unit width).

The complete system of forces and moments on the element is shown in Fig. 9.3. Note that all forces and moments are *per unit width* along the edges of the element.

From equilibrium of vertical forces, it may be seen that

$$\left(q_x + \frac{\partial q_x}{\partial x}dx\right)dy - q_x dy$$

$$+\left(q_x + \frac{\partial q_y}{\partial y}dy\right)dx - q_y dx + p\,dx\,dy = 0$$

or

$$\frac{\partial q_x}{\partial x} + \frac{\partial q_y}{\partial y} + p = 0 \qquad (9.1.9)$$

Taking moments about an axis parallel to the x-axis,

$$\left(m_{xy} + \frac{\partial m_{xy}}{\partial x}dx\right)dy - m_{xy}dy + m_y dx$$

$$-\left(m_y + \frac{\partial m_y}{\partial y}dy\right)dx + q_y dxdy = 0$$

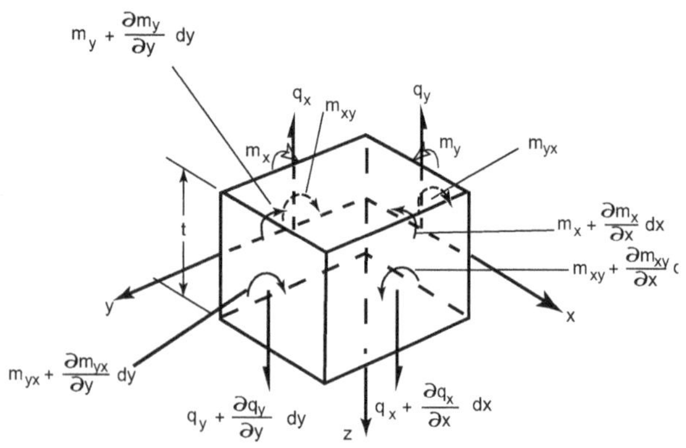

Figure 9.3 Forces and moments in a plate element.

or

$$\frac{\partial m_{xy}}{\partial x} - \frac{\partial m_y}{\partial y} + q_y = 0 \qquad (9.1.10)$$

Similarly, taking moments about an axis parallel to the y-axis leads to

$$\frac{\partial m_{yx}}{\partial y} - \frac{\partial m_x}{\partial x} + q_x = 0 \qquad (9.1.11)$$

Because of the principle of complementary shear stress, it follows that $m_{yx} = -m_{xy}$, so that (9.1.11) becomes

$$-\frac{\partial m_{xy}}{\partial y} + \frac{\partial m_x}{\partial x} + q_x = 0 \qquad (9.1.12)$$

Substituting for q_x and q_y in (9.1.9) from (9.1.10) and (9.1.12)

$$\frac{\partial}{\partial x}\left(-\frac{\partial m_{xy}}{\partial y} + \frac{\partial m_x}{\partial x}\right) + \frac{\partial}{\partial y}\left(\frac{\partial m_y}{\partial y} - \frac{\partial m_{xy}}{\partial_x}\right) + p = 0$$

$$\frac{\partial^2 m_x}{\partial x^2} - 2\frac{\partial^2 m_{xy}}{\partial x \partial y} + \frac{\partial^2 m_y}{\partial y^2} + p = 0 \qquad (9.1.13)$$

It is now necessary to determine the twisting moment m_{xy}, in terms of the deflection. In Fig. 9.4*a* it will be seen that, if a point A in the plate a distance z from the neutral surface is displaced a distance v in the y-direction, the displacement of a nearby point at $x + dx$ will be $v + (\partial v/\partial x)dx$, so that the change in slope of the line AB will be

$$\frac{v + \dfrac{\partial v}{\partial x}dx - v}{dx} = \frac{\partial v}{\partial x}$$

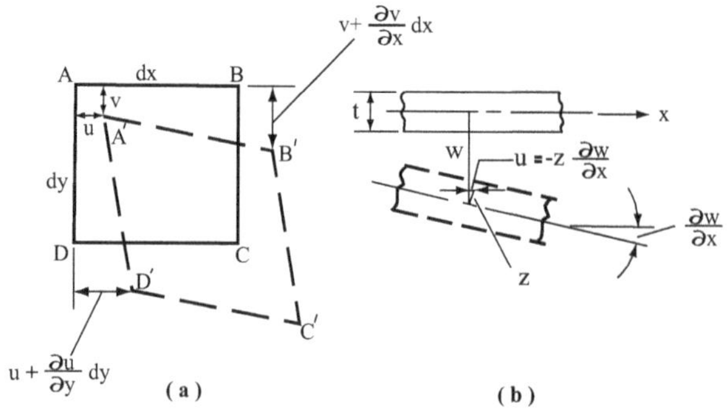

Figure 9.4 Shear strain and plate deflection.

Similarly, the change in slope of the line AD is $\partial u/\partial y$.

The rectangular element ABCD is then changed in shape to a parallelogram A′B′C′D′ and the shear strain is

$$\gamma = \frac{\partial v}{\partial x} + \frac{\partial u}{\partial y}$$

The corresponding shear stress is

$$\tau = G\left(\frac{\partial v}{\partial x} + \frac{\partial u}{\partial y}\right) \qquad (9.1.14)$$

Now from Fig. 9.4(b) it will be seen that

$$u = -z\frac{\partial w}{\partial x}$$

and similarly

$$v = -z\frac{\partial w}{\partial y}$$

Making these substitutions in (9.1.14) gives

$$\tau = -2Gz\frac{\partial^2 w}{\partial x \partial y}$$

The twisting moment per unit width can now be obtained from equilibrium:

$$m_{xy} = -\int_{-t/2}^{t/2} -2G\frac{\partial^2 w}{\partial x \partial y} z^2 dz = \frac{Gt^3}{6}\frac{\partial^2 w}{\partial x \partial y}$$

and substituting

$$G = \frac{E}{2(1+v)}$$

gives

$$m_{xy} = \frac{Et^3}{12(1+\nu)}\frac{\partial^2 w}{\partial x \partial y} = \frac{Et^3(1-\nu)}{12(1+\nu)(1-\nu)}\frac{\partial^2 w}{\partial x \partial y}$$
$$= D(1-\nu)\frac{\partial^2 w}{\partial x \partial y} \qquad (9.1.15)$$

It is now possible to substitute for m_x, m_y, and m_{xy} in (9.1.13) using (9.1.8) and (9.1.15)

$$\frac{\partial^2}{\partial x^2}\left[-D\left(\frac{\partial^2 w}{\partial x^2} + \nu\frac{\partial^2 w}{\partial y^2}\right)\right] - 2\frac{\partial^2}{\partial x \partial y}\left[D(1-\nu)\frac{\partial^2 w}{\partial x \partial y}\right]$$
$$+ \frac{\partial^2}{\partial y^2}\left[-D\left(\frac{\partial^2 w}{\partial y^2} + \nu\frac{\partial^2 w}{\partial x^2}\right)\right] = -p$$

or finally

$$\frac{\partial^4 w}{\partial x^4} + 2\frac{\partial^4 w}{\partial x^2 \partial y^2} + \frac{\partial^4 w}{\partial y^4} = \frac{p}{D} \qquad (9.1.16)$$

This is the equation of equilibrium for the plate. It assumes that the deflection is small and that there is no stretching of the middle plane, that is, that membrane effects are absent. Strictly speaking, this implies that the edges of the plate are free to slide in the plane of the plate. However, if the plate edges are restrained from moving, (9.1.16) will still be applicable provided the plate deflection is sufficiently small for the resulting membrane tensions to be neglected. This is usually the case when the maximum deflection does not exceed t. Equation (9.1.16) conforms to a type of differential equation known as the biharmonic equation and is often abbreviated as $\nabla^4 w = p/D$.

9.1.3 Boundary Conditions

The solution of (9.1.16) must be such as to satisfy the boundary conditions and therefore the basic task is to find an expression for $w(x,y)$ that will satisfy these conditions. The types of restraint around the boundary of a plate can be idealized as follows:

1. Simply supported—edges free to rotate and to move in the plane of the plate.
2. Pinned—edges free to rotate but not free to move in the plane of the plate.
3. Clamped but free to slide—edges not free to rotate but free to move in the plane of the plate.
4. Rigidly clamped—edges not free to rotate or move in the plane of the plate.

The prevention of in-plane movement (no "pulling in" of the edges) of conditions 2 and 4 can only occur if the structure supporting the plate is very rigid in this direction. In most plated frame structures, and particularly in vehicles and freestanding structures, the individual panels of plating receive relatively little restraint against edge pull-in because such restraint would have to come ultimately from the frames at the edges of the overall stiffened panel, and beam bending stiffness is generally insufficient to provide such rigid in-plane support. This has been verified for ship panels by Clarkson's extensive experimental work (1963, pp. 467–484). Therefore conditions 1 and 3 are generally applicable. The former would be appropriate when a load acts on a single panel of plating because neither the stiffeners nor the surrounding panels would provide much rotational restraint. Condition 3 would be appropriate for a distributed pressure loading that extends over several panels.

9.1.4 Solution of Special Cases

SIMPLY SUPPORTED PLATES

The earliest solution to (9.1.16) is that of Navier, who solved the simply supported case by using a Fourier series to represent the load $p(x,y)$. The general expression for the load is

$$p(x,y) = \sum_{m=1}^{\infty}\sum_{n=1}^{\infty} A_{mn} \sin\frac{m\pi y}{a} \sin\frac{n\pi x}{b} \qquad (9.1.17)$$

and the coefficient A_{mn} can be obtained by Fourier analysis for any particular load condition. For example, for the case of a uniform pressure p_o, it may be shown that the coefficient A_{mn} is given by

$$A_{mn} = \frac{16 p_0}{\pi^2 mn} \qquad (9.1.18)$$

It may be readily shown that for the biharmonic equation a sinusoidal load distribution produces a sinusoidal deflection. That is, the general solution to the foregoing case, satisfying both (9.1.16) and the simply supported boundary conditions, is

$$w = \sum_{m=1}^{m=\infty}\sum_{n=1}^{n=\infty} B_{mn} \sin\frac{m\pi y}{a} \sin\frac{n\pi x}{b} \qquad (9.1.19)$$

To find the value of the coefficient in (9.1.19), this, together with (9.1.17) and (9.1.18), is substituted in (9.1.16). This gives, on dropping the common $(\sin m\pi y/a)(\sin n\pi x/b)$ term,

$$B_{mn}\left(\frac{\pi^4 m^4}{a^4}+\frac{2\pi^4 m^2 n^2}{a^2 b^2}+\frac{\pi\ n}{b^4}\right)=\frac{16p_0}{\pi^2 Dmn}$$

and it will thus be seen that

$$B_{mn}=\frac{16p_0}{\pi^6 Dmn\left(\frac{m^2}{a^2}+\frac{n^2}{b^2}\right)^2} \tag{9.1.20}$$

The deflection is then given by

$$w=\sum_{m=1}^{\infty}\sum_{n=1}^{\infty}\frac{16p_0}{\pi^6 Dmn\left(\frac{m^2}{a^2}+\frac{n^2}{b^2}\right)^2}\sin\frac{m\pi y}{a}\sin\frac{n\pi x}{b} \tag{9.1.21}$$

in which m and n, due to the symmetry of the problem, need only take odd values. Values of w_{max} are given in Fig. 9.5 for various values of aspect ratio.

To determine the bending stress in the plate, it is first necessary to calculate the bending moments m_x and m_y. From (9.1.21)

$$\frac{\partial^2 w}{\partial x^2}=-\sum_{m=1}^{m=\infty}\sum_{n=1}^{n=\infty}\frac{16p_0}{\pi^6 Dmn\left(\frac{m^2}{a^2}+\frac{n^2}{b^2}\right)^2}\frac{\pi^2 n^2}{b^2}\sin\frac{m\pi y}{a}\sin\frac{n\pi x}{b}$$

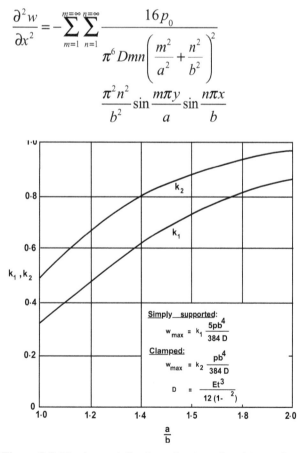

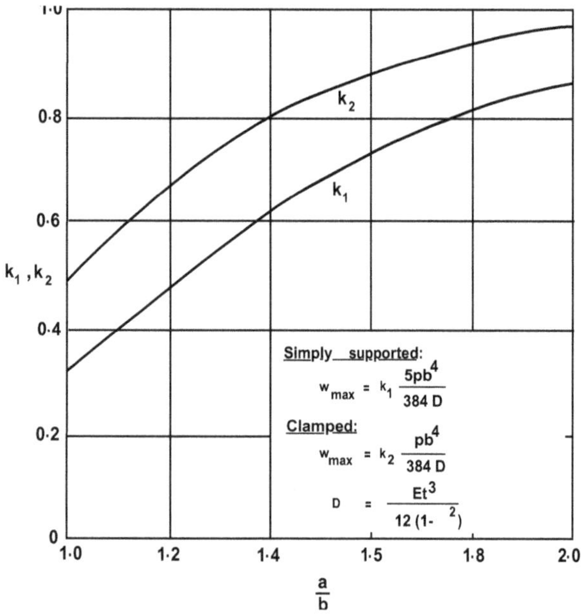

Figure 9.5 Maximum deflection of rectangular plates under uniform pressure (elastic small-deflection theory).

$$=-\sum_{m=1}^{m=\infty}\sum_{n=1}^{n=\infty}\frac{16p_0 n}{\pi^4 Dmb^2\left(\frac{m^2}{a^2}+\frac{n^2}{b^2}\right)^2}\sin\frac{m\pi y}{a}\sin\frac{n\pi x}{b}$$

$$\frac{\partial^2 w}{\partial y^2}=-\sum_{m=1}^{m=\infty}\sum_{n=1}^{n=\infty}\frac{16p_0 m}{\pi^4 Dna^2\left(\frac{m^2}{a^2}+\frac{n^2}{b^2}\right)^2}\sin\frac{m\pi y}{a}\sin\frac{n\pi x}{b}$$

Then, for example,

$$m_x=-D\left(\frac{\partial^2 w}{\partial x^2}+\nu\frac{\partial^2 w}{\partial y^2}\right)$$

$$=\sum_{m=1}^{m=\infty}\sum_{n=1}^{n=\infty}\frac{16p_0}{\pi^4 mn\left(\frac{m^2}{a^2}+\frac{n^2}{b^2}\right)^2}\left(\frac{n^2}{b^2}+\nu\frac{m^2}{a^2}\right)\sin\frac{m\pi y}{a}\sin\frac{n\pi x}{b}$$

The curvature, and hence the bending moment, will always be greater across the shorter span. By convention, the symbol b is used for the shorter dimension (making the aspect ratio a/b always greater than or equal to unity). Thus, in the present case m_x is the larger of the two moments and has its maximum value in the center of the plate. Values of maximum stress for this case and for the case of a clamped plate are given in Fig. 9.6.

CLAMPED PLATES

The solution for a clamped plate is more complicated. The usual methods are the energy (or Ritz) method and the method of Levy. The energy method, while giving only approximate results, has the advantage that the assumed deflected surface need not satisfy the governing differential equation while it must satisfy the boundary conditions. The Levy type solution, as described by Timoshenko (1959) is achieved by the superposition of three loading systems applied to a simply supported plate: (1) uniformly distributed hydrostatic pressure, (2) moments distributed along the short edges, and (3) moments distributed

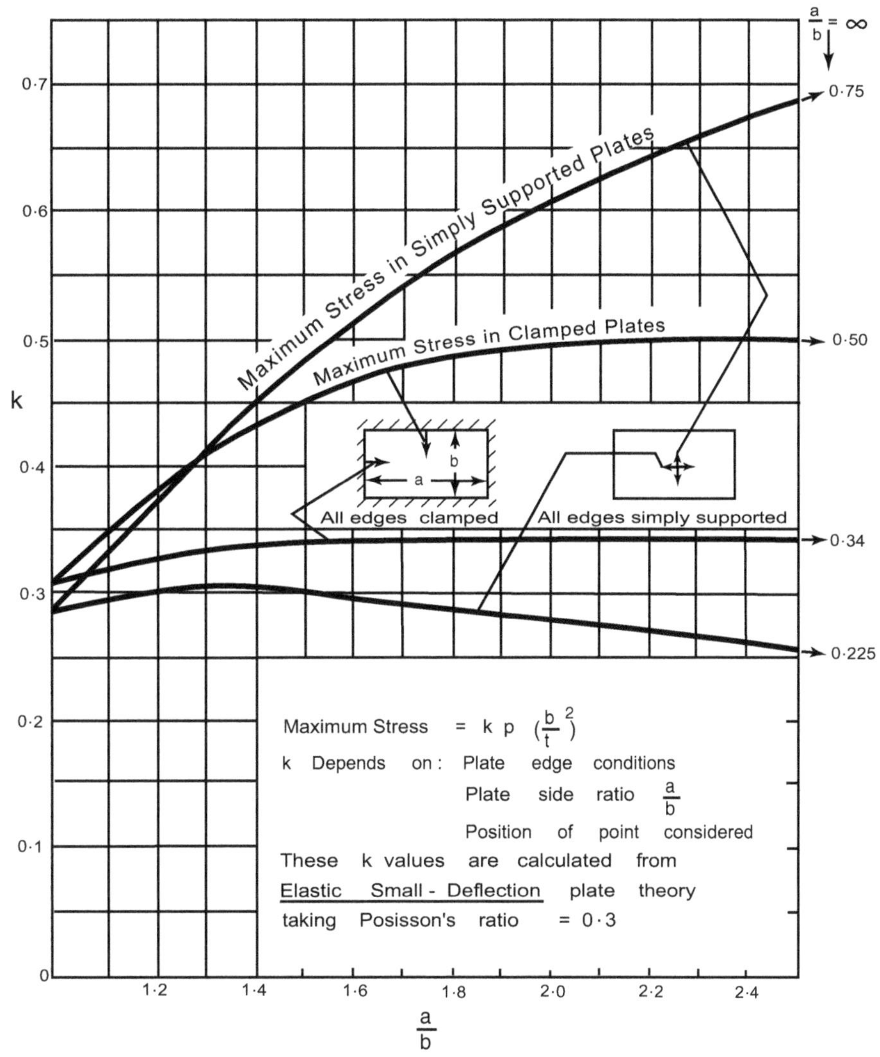

Figure 9.6 Maximum stresses in rectangular plates under uniform lateral pressure.

along the long edges. If w_1, w_2, and w_3 denote the deflections along the appropriate edges for these three cases, the boundary conditions are:

short edges:

$$\left(\frac{\partial w_1}{\partial y}\right)_{y=\pm a/2} + \left(\frac{\partial w_2}{\partial y} + \frac{\partial w_3}{\partial y}\right)_{y=\pm a/2} = 0$$

long edges:

$$\left(\frac{\partial w_1}{\partial x}\right)_{x=\pm b/2} + \left(\frac{\partial w_2}{\partial x} + \frac{\partial w_3}{\partial x}\right)_{y=\pm b/2} = 0$$

where $x = 0$, $y = 0$ is at the center of the plate.

The various deflection components are each represented by trigonometric series, and the foregoing boundary conditions result in two systems of equations each having m unknowns, where m is the number of terms kept in the trigonometric series. The mathematics is quite involved but the answers are well established. The maximum deflection for the case of a uniform load is given in Fig. 9.5. The maximum bending moment occurs at the edge of the plate, and the absolute maximum occurs at the midpoint of the long side. Thus, in terms of the foregoing coordinate system, the greatest bending moment is m_x at $x = \pm b/2$; $y = 0$. The corresponding values of maximum stress at these points are given in Fig. 9.6.

It may be seen from Fig. 9.6 that the effect of aspect ratio is smaller for clamped plates, and beyond about $a/b = 2$ such a plate behaves essentially as a clamped strip and the influence of aspect ratio is negligible.

9.1.5 Elastic Design of Laterally Loaded Plating

If the lateral load is the only significant load, then in many cases the required plate thickness is determined by a maximum allowable inelastic (and therefore permanent) lateral deflection, or "permanent set." But in other cases it is desirable, or even crucial, that the plate should remain flat under the maximum design load, and the design criterion is the avoidance of yield anywhere in the plate. In plate bending the stress is biaxial, and therefore initial yield cannot be predicted simply by equating the largest stress to the yield stress, as measured in a tensile test. Rather, we must apply a suitable yield criterion, such as the Hencky-von Mises criterion.

We take the case of a uniform pressure and assume that this pressure also acts on adjacent panels of plating. Then the appropriate boundary condition is "clamped; free to pull in." As shown in Fig. 9.6, the maximum stress is at the midpoint of the long side, where the plate edge is undergoing cylindrical bending. The figure shows that for any plate with an aspect ratio $a/b \geq 2$, the plate bending coefficient k is ½. Hence the value of this maximum stress is

$$\sigma_x = \frac{1}{2} p \left(\frac{b}{t}\right)^2 \qquad a/b \geq 2$$

Section 9.1.1 also showed that for cylindrical bending the stress in the orthogonal direction is $\sigma_y = \nu\sigma_x$. The third principal stress is zero. The Hencky-von Mises yield criterion states that yielding will occur when $\sigma_{eq} = \sigma_Y$ where

$$\sigma_{ea} = 0.707[(\sigma_1 - \sigma_2)^2 + (\sigma_2 - \sigma_3)^2 + (\sigma_3 - \sigma_1)^2]^{1/2}$$

Substituting for the principal stresses and solving for $(\sigma_x)_Y$, the value of σ_x at which yield occurs, we obtain

$$(\sigma_x)_Y = \frac{\sigma_Y}{\sqrt{1 - \nu + \nu^2}} \tag{9.1.22}$$

For metals ν is of the order of 0.3, and so the right hand side is about $1.125\sigma_Y$. If the σ_y stress was not present, yield would occur when $\sigma_x = \sigma_Y$. Because σ_y is present and always has the same sign as σ_x, yielding does not occur until after σ_x has exceeded σ_Y by about 13%.

9.1.6 Thermal Stress and Strain in Plates

A significant contribution to stresses in ships' plating can sometimes result from temperature gradients across the thickness of a plate, due for example to the presence of cargoes that are hotter or colder than the adjacent sea water. By a simple extension of the linear elastic theory of plate behavior, thermal stresses and strains can be included in the calculations.

For a material having a coefficient of linear expansion α, a temperature change of T will cause a thermal strain αT, where α and T must be in consistent units. Hence the stress-strain temperature equations are

$$\varepsilon_x = \frac{\sigma_x}{E} - \nu\frac{\sigma_y}{E} + \alpha T$$

$$\varepsilon_y = \frac{\sigma_y}{E} - \nu\frac{\sigma_x}{E} + \alpha T$$

Since a change in temperature does not induce shear strain, the relation between shear strain and stress remains unchanged.

Substitution of the above equations in the analysis presented in Section 9.1 enables the effect of temperature changes to be included. Even, for example, in the very simple one-dimensional case where plate bending is not involved, an instructive result can be found. Thus if a flat rectangular plate $a \times b \times t$ is uniformly heated by T, and the edges are fully restrained from axial displacement, then $\varepsilon_x = \varepsilon_y = 0$ everywhere and hence from the above equations, $\sigma_x = \sigma_y = -E\alpha T/(1-\nu)$ so that a uniform biaxial compressive stress, proportional to the temperature change T and also to E, is induced in the plate, regardless of its dimensions. Such biaxial compression may induce buckling of the plating.

If there is a temperature gradient through the thickness of a plate, then bending will be induced. In such cases, T is a function of z, and the integration of stresses across the plate thickness will result in bending moments and corresponding curvatures in the plating.

9.1.7 Strain Energy of Deformed Plating

Since the governing differential equation of plate bending under lateral pressure, $D\nabla^4 w = p$, is difficult to solve exactly, approximate solutions to plate problems can alternatively be found by methods involving considerations of work and energy. Such methods are also particularly useful, as shown later, for estimating natural frequencies of vibration of plating.

The elastic linear theory of Section 9.1.2 can readily be developed to provide expressions for the strain energy of an element of plate $dx \times dy \times t$ when subjected to bending moments m_x, m_y, and twisting moment m_{xy}, per unit width. Thus the work done on, and hence the strain energy stored in, the element by a bending moment $m_x dy$ (see Fig. 9.3) acting on opposite faces is $m_x dy\theta/2$, where θ is the angular deflection (change in slope) caused by m_x.

But from geometry and the small deflection approximation of (9.1.6)

$$\theta = \frac{dx}{r_x} = -\frac{\partial^2 w}{\partial x^2} dx - \frac{1}{2} m_x \frac{\partial^2 w}{\partial x^2} dx\, dy$$

Hence this component of strain energy is

$$\theta = \frac{dx}{r_x} = -\frac{\partial^2 w}{\partial x^2} dx - \frac{1}{2} m_x \frac{\partial^2 w}{\partial x^2} dx\, dy$$

In a similar way, the strain energy components due to m_y and the twisting moment m_{xy} can be expressed in terms of the appropriate moments and curvatures. Also, in (9.1.8) and (9.1.15), m_x, m_y and m_{xy} are expressed in terms of derivatives of w, the plate flexural rigidity D and Poisson's ratio ν. Straightforward development of the above relations and integration over the whole area of the plate, leads to the following expression for the strain energy U of a deformed plate

$$U = \frac{D}{2} \int\int \left\{ \left(\frac{\partial^2 w}{\partial x^2} + \frac{\partial^2 w}{\partial y^2} \right)^2 + 2(1-\nu) \left[\frac{\partial^2 w}{\partial x^2} \frac{\partial^2 w}{\partial y^2} - \left(\frac{\partial^2 w}{\partial x \partial y} \right)^2 \right] \right\} dx\, dy$$

Note that in this analysis the strain energy due to shear actions q_x and q_y has not been included. For the slender plate geometries typical in ships, this is justifiable because of the relative unimportance of such effects.

Note also the very convenient fact that in many practical cases of interest, integration of the term in square brackets in the above equation for U leads to zero.

9.1.8 Plate Vibration

To avoid resonance between plate vibrations and some exciting frequency (for example from propeller or machinery) it is necessary to estimate natural frequencies of plate panels. Assuming free, undamped vibration, considerations of energy require that at any point during the vibration, the sum of kinetic and potential (i.e., strain) energies of a vibrating plate is constant. It follows that if a plate vibration is assumed to be of the form

$$W(x,y,T) = w(x,y)\sin\omega T$$

in which ω is the natural frequency, T is time, and $w(x,y)$ is the mode (or deflected shape) of the vibrating plate, then the maximum kinetic energy V_{max}, occurring when $w = 0$ and velocity v is a maximum, must equal the maximum strain energy, occurring when the amplitude of vibration is a maximum and the velocity is zero.

The kinetic energy of an element $dx \times dy \times t$ of the plate is

$$\frac{1}{2} m v^2 = \frac{1}{2} \rho dx\, dy\, t \left(\frac{\partial W}{\partial T} \right)^2$$

Hence the total kinetic energy of the vibrating plate is

$$V = \frac{1}{2} \rho t \int_0^a \int_0^b \left(\frac{\partial W}{\partial T} \right)^2 dx\, dy$$

Substitution of the above equations for w, U, and V into the equation $U_{max} - V_{max} = 0$ leads to a useful general equation for estimating the natural frequency ω of vibrating rectangular plates:

$$\omega^2 = \frac{D}{\rho t} \frac{\int\int \left(\frac{\partial^2 w}{\partial x^2} + \frac{\partial^2 w}{\partial y^2} \right)^2 dx\, dy}{\int\int (\omega^2) dx\, dy}$$

Note that the frequency ω is in radians per second. The frequency f in cycles per second and the period T_p in seconds are

$$f = \frac{\omega}{2\pi} = \frac{1}{T_p}$$

To illustrate the use of this energy method, consider a simply-supported flat rectangular plate $a \times b \times t$. Both by analogy with the buckling behaviour of plates and by observation of vibrating plates, a plausible expression for the vibration is

$$W(x,y,T) = w_{mn} \sin\frac{m\pi x}{a} \sin\frac{n\pi y}{b} \sin\omega T$$

This implies that the vibration is harmonic, and that the mode of vibration $w(x,y)$ involves m and n half-waves, respectively, in the x and y directions. The expression for $w(x,y)$ must satisfy the appropriate plate boundary conditions. As mentioned in the development of (9.1.19), this double-sine form does indeed satisfy the simply supported edge conditions of zero deflections and bending moments.

Substitution of this expression into the equation for ω^2 is straightforward, if tedious. Double differentiation of w with respect to x or y retains the sine terms, so that evaluation of ω requires integration of the squares of these terms, for which it is helpful to note that

$$\int_0^a \sin^2 \frac{m\pi x}{a} dx = \frac{a}{2}$$

and

$$\int_0^b \sin^2 \frac{n\pi y}{b} dy = \frac{b}{2}$$

Hence it follows that

$$\omega_{m,n} = \pi^2 \left(\frac{m^2}{a^2} + \frac{n^2}{b^2} \right) \sqrt{\frac{D}{\rho t}} \quad (9.1.23)$$

is the natural frequency of free, undamped vibration corresponding to m and n half-waves in the x and y directions. Thus, for example, the fundamental frequency of a square, simply supported plate of side length b is found, by putting $m = n = 1$ and $a = b$, to be

$$= \frac{2\pi^2}{b^2} \sqrt{\frac{D}{\rho t}} = \frac{2\pi^2 t}{b^2} \sqrt{\frac{E}{12\rho(1-\nu^2)}} \quad (9.1.24)$$

Note the relative influences of plate thickness t and side length b on this fundamental frequency.

For simply-supported plates (as also for simply-supported beams) the assumption of a sinusoidal mode of vibration in the above energy method gives the exact solution. (For the solution of this problem by solving the governing differential equation of motion, see Timoshenko [1959].) For other plate boundary conditions, the energy method in general gives frequency predictions slightly higher than exact values. Consider for example the case of a plate with fully clamped edges. For vibration in the fundamental mode, the following expression satisfies the required conditions of zero deflection and slope at the boundaries

$$w(x, y) = w_M \left(x^2 - \frac{a^2}{4} \right)^2 \left(y^2 - \frac{b^2}{4} \right)^2$$

where x and y are measured from the center of the plate. Substituting this in the general equation for ω leads to

$$\omega_{1,1} = 12 \sqrt{\frac{7}{2} \left(\frac{1}{a^4} + \frac{4}{7a^2b^2} + \frac{1}{b^4} \right)} \sqrt{\frac{D}{\rho t}}$$

For a square clamped plate

$$\omega_{1,1} = \frac{36}{b^2} \sqrt{\frac{D}{\rho t}} \quad (9.1.25)$$

compared with the exact value

$$\omega_{1,1} = \frac{35.986}{b^2} \sqrt{\frac{D}{\rho t}} \quad (9.1.26)$$

The following equation gives the first mode frequencies (in cps) for a range of plate edge conditions and aspect ratios, in terms of a frequency coefficient C

$$f = \frac{\omega}{2\pi} = \frac{Ct}{b^2} \sqrt{\frac{E}{\rho}} \quad (9.1.27)$$

in which

C = frequency coefficient, given in Table 9.1
t = plate thickness
b = shorter side of plate
E = Young's modulus
ρ = density of plate
ω = frequency in radians/sec.

Table 9.1 Frequency Coefficient C for Plates

	Aspect Ratio a/b					
Edge Conditions:	1.0	1.5	2.0	2.5	3.0	Infinite
All edges SS	0.951	0.687	0.594	0.552	0.528	0.475
One short edge C1 three edges SS	1.139	0.911	0.835	0.801	0.783	0.743
Short edges C1 long edges SS	1.394	1.207	1.147	1.121	1.108	1.077
All edges C1	1.733	1.301	1.183	1.145	1.117	1.077

C1 = clamped; SS = simply supported

NOTES: (1) This formula is valid only for basic units (N, kg, m or lb, slug, ft) because it uses the conversion 1 N = 1 kg m/sec^2 (or 1 lb = 1 slug ft/sec^2). If other metric units are used then the appropriate factor must be inserted inside the square root: 10^3 for mm and 10^6 for MN. If inches are used, ρ must be in slug/in^3. (2) A value of 0.3 has been assumed for Poisson's ratio; for other values of ν, multiply the result by $[0.91/(1-\nu^2)]^{1/2}$

9.1.9 Plastic Bending of a Beam or Plate Strip

PLASTIC HINGE AND PLASTIC MOMENT

This section presents the simple notion of an "elastic-perfectly plastic" model for ductile material behavior, and uses it to explain how yielding spreads through the rectangular cross section of a beam (namely the unit width plate strip shown in Fig. 9.1) when the bending moment per unit width goes beyond m_Y, the initial yield value, and eventually reaches m_P, the "plastic moment," whereupon it causes a "plastic hinge" in the beam. We will derive equations for calculating m_P and for determining the pressure required to form a plastic hinge. These will be used in Section 9.3.

Plastic theory is based on an idealized "elastic-perfectly plastic" stress-strain curve. As shown in Fig. 9.7, this idealization is quite suitable for steel, with its definite yield point, and it is conservative since it ignores the subsequent strain-hardening of the material.

Let us consider a cross section of a plate strip (or unit width beam) that is subjected to a steadily increasing value of bending moment m. For example, let us consider the cross section at the midlength of a simply supported beam that carries a pressure p, as shown in Fig. 9.8*a*. Within the elastic range the local curvature of the beam ϕ is linearly proportional to m, and the load-deflection curve is also linear as shown in Figs. 9.8*b*,*c*.

Figure 9.9 shows the resulting growth and distribution of the bending stress in the cross section of the plate strip. We assume that there is no axial force and we neglect the effect of shear. At some value of m, say m_Y, the maximum primary bending stress σ_x will reach the value that causes yield. Section 9.1.5 showed that, because of the secondary bending stress $\sigma_y = \nu\sigma_x$, the value of σ_x that causes yield is given by (9.1.22)

$$(\sigma_x)_Y = \frac{\sigma_Y}{\sqrt{1 - \nu + \nu^2}}$$

If the pressure, and therefore the bending moment, is further increased, plasticity will spread throughout the depth of the beam until the section is fully plastic. The local bending

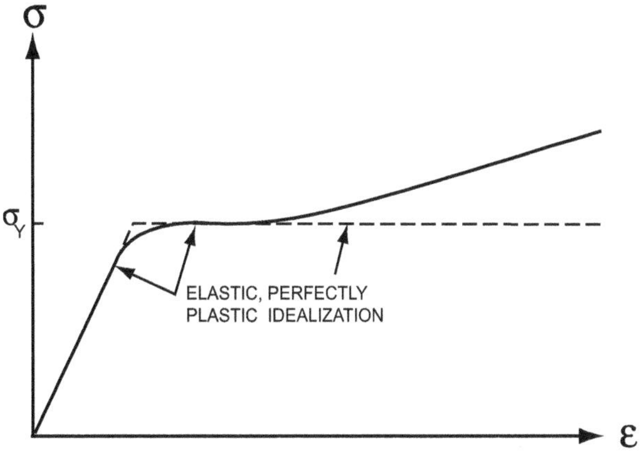

Figure 9.7 Idealized elastic-plastic stress-strain curve.

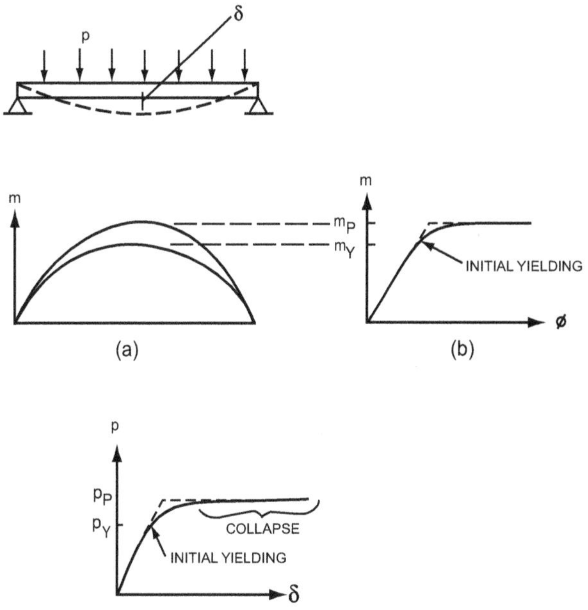

Figure 9.8 Plastic bending of a plate strip.

moment corresponding to this condition is known as the *plastic moment* of the section, m_P. Because all of the fibers have now reached the limit of their load-carrying ability, the beam can absorb no further bending moment at this section. It is as if a hinge had been inserted in the beam at this point, and hence this condition in a beam is referred to as a "plastic hinge." If the beam is merely simply supported as in Fig. 9.8*a*, or if it is a cantilever, then the occurrence of such a hinge makes it completely incapable of carrying any further load and it would collapse. In practice the strain-hardening of the material would delay the collapse slightly, to some value of m slightly greater than m_P, but nevertheless at $m = m_P$ the deflection would already be so large as to constitute effective collapse.

Since the stresses in a plastic hinge are equal and opposite in the upper and lower portions of the cross section and since there is no axial force, equilibrium in the longitudinal direction requires that the areas of these two portions be equal. Therefore, the "plastic neutral axis" (PNA)—that is, the axis or line in the fully plastic section

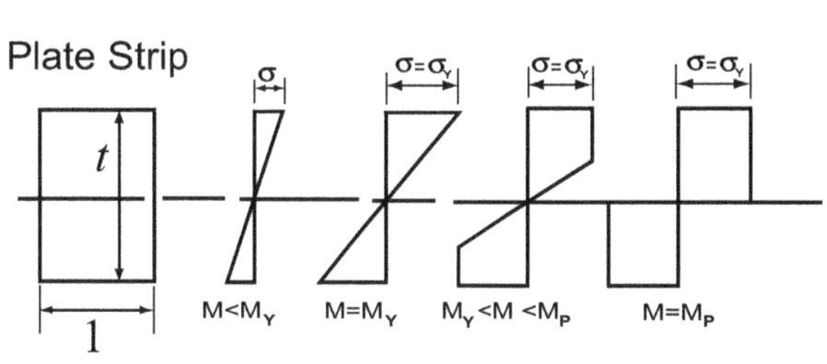

Figure 9.9 Plastic hinge formation.

where the stress reverses sign—is simply the horizontal line that divides the section into two equal areas.

We now consider equilibrium of moments for the fully plastic section, which requires that the external bending moment m_P must be balanced by the moment of the stress forces in the section. In both halves the stress is $(\sigma_x)_Y$ and the moment arm is $t/4$. Therefore

$$m_P = (\sigma_x)_Y \frac{t^2}{4} \qquad (9.1.28)$$

Substituting for $(\sigma_x)_Y$ from (9.1.22) gives

$$m_P = \frac{\sigma_Y}{\sqrt{1-\nu+\nu^2}} \frac{t^2}{4} \qquad (9.1.29)$$

As a further analogy to elastic beam bending we define a "plastic section modulus" Z_P such that

$$m_P = (\sigma_x)_Y Z_P \qquad (9.1.30)$$

From (9.1.28) we obtain

$$Z_P = \frac{t^2}{4} \qquad (9.1.31)$$

and thus we see that Z_P is simply the sum of the first moments of area of the two half-areas of the fully yielded section.

END HINGES IN A CLAMPED PLATE STRIP

In section 9.1.5 we noted that when a uniform pressure p acts over many adjacent panels of plating, the appropriate boundary condition for each panel is "clamped; free to pull in."

Figure 9.10 shows a plate strip of unit width and having a span b (the shorter dimension of the panel). The ends of the plate strip are clamped and the maximum bending moment occurs at these ends:

$$m_e = \frac{pb^2}{12} \qquad (9.1.32)$$

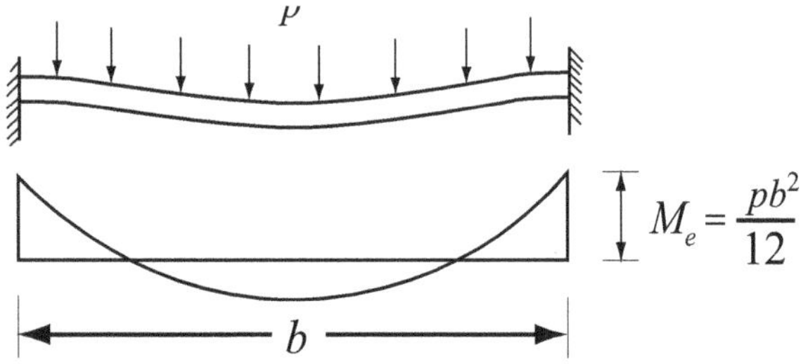

Figure 9.10 Clamped plate strip under uniform load.

By increasing the pressure (and hence increasing m_e) we can cause two successive events to occur:

(1) at a pressure p_Y the end moment reaches m_Y, the value that would cause initial yield at the ends of the plate strip (at its upper and lower surfaces). That is

$$p_Y = \frac{12 m_Y}{b^2} \qquad (9.1.33)$$

(2) at a pressure p_{EH} the end moment reaches m_P, the value that would cause a plastic hinge at the ends of the plate strip. That is

$$p_{EH} = \frac{12 m_P}{b^2} \qquad (9.1.34)$$

and from (9.1.29) this is

$$p_{EH} = \frac{3\sigma_Y}{\sqrt{1-\nu+\nu^2}} \left(\frac{t}{b}\right)^2 \qquad (9.1.35)$$

9.2 COMBINED BENDING AND MEMBRANE STRESSES-ELASTIC RANGE

9.2.1 Large-Deflection Plate Theory

In many cases the use of small-deflection theory for the design of laterally loaded plating leads to thicknesses that are excessive. Small-deflection theory fails to allow for membrane stresses that arise when the deflection becomes large and/or when the edges are prevented from pulling in. With large deflections the plate is no longer a developable surface and the deflection requires in-plane stretching and compression of the plate. If the edges are prevented from pulling in, membrane action becomes significant when the lateral deflection (whether initial deflection or due to load) exceeds approximately half of the plate thickness. If the edges are free to pull in (which is the appropriate assumption for ship plating) membrane action becomes significant somewhat later, typically at a deflection $w > t$. As the deflection increases, an increasing proportion of the load is carried by this membrane action. This situation, in which the lateral load is supported by both bending and membrane action, requires a more comprehensive plate theory, usually referred to as "large-deflection" plate theory. The differential equations for large-deflection plate theory were formulated by von Karman. One of these, the equation relating load to deflection, is a generalization of (9.1.16) and may be derived by considering vertical equilibrium of an element *dx dy*, as was done in deriving (9.1.9). Let us denote the membrane tensions and membrane shearing force per unit width as N_x, N_y and N_{xy}. Taking the *x*-direction, it can be seen from Fig. 9.11 that at side (1) the vertical component of the tension force is $-N_x\,dy\,(\partial w/\partial x)$. The vertical force at side (2) would be the same magnitude (but opposite sign) if the slope $\partial w/\partial x$ were the same, but in general it is not. Therefore we write this component as

$$+N_x\,dy\left[\frac{\partial w}{\partial x}+\frac{\partial}{\partial x}\left(\frac{\partial w}{\partial x}\right)dx\right]$$

and the net vertical force is $N_x\,(\partial^2 w/\partial x^2)\,dx\,dy$. Similarly, in the *y*-direction the net vertical force is $N_y\,(\partial^2 w/\partial y^2)\,dx\,dy$.

The vertical component of the shear force along sides (3) and (4) is

$$+N_{xy}\,dx\left[\frac{\partial w}{\partial x}+\frac{\partial}{\partial y}\left(\frac{\partial w}{\partial x}\right)dy\right]-N_{xy}\,dx\left(\frac{\partial w}{\partial x}\right)$$

and a similar expression is obtained for the other two sides. Collecting results, we find that the new terms to be added to (9.1.9) are

$$\left(N_x\frac{\partial^2 w}{\partial x^2}+2N_{xy}\frac{\partial^2 w}{\partial x\partial y}+N_y\frac{\partial^2 w}{\partial y^2}\right)dx\,dy$$

Thus (9.1.16) becomes

$$\nabla^4 w=\frac{1}{D}\left(p+N_x\frac{\partial^2 w}{\partial x^2}+2N_{xy}\frac{\partial^2 w}{\partial x\partial y}+N_y\frac{\partial^2 w}{\partial y^2}\right) \tag{9.2.1}$$

In general, N_x, N_y, and N_{xy} are functions of *x* and *y* and hence (9.2.1) does not provide sufficient information for solution except in special cases where these membrane forces can be separately calculated. The other equation formulated by von Karman is given in Section 101 of Timoshenko (1959). With these two equations, a complete solution can be achieved. However, except for a few simple cases, precise mathematical solutions to these equations are very difficult. Aalami and Williams (1975) presents solutions to a variety of cases.

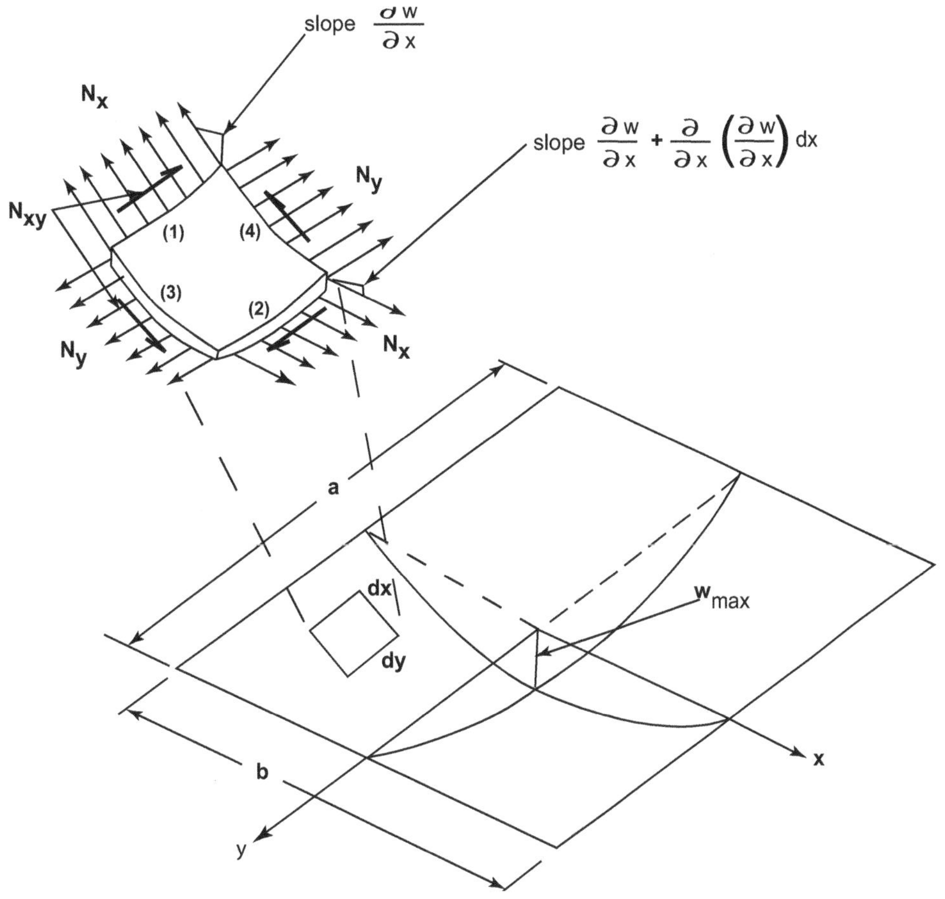

Figure 9.11 Plating with membrane stress.

In ship structures, the principal application of (9.2.1) is for plating that is subjected to an in-plane compressive load N_x, which brings with it the possibility of buckling. In this case, since N_x is an applied load, it is a known quantity. The same may be true of N_y and N_{xy}, or they may continue to be membrane forces arising from large deflections and boundary restraints. In many practical situations N_y and N_{xy} may be taken as zero, and then the solution of (9.2.1) presents little problem. The whole question of plate buckling is dealt with in Chapter 12.

9.2.2 Membrane Tension (Edges Restrained Against Pull-in)

The relative magnitude of membrane effects depends on two things: the degree of lateral deflection or "curvature" of the plate surface, whether present initially or due to the action of the lateral load, and the degree to which the edges are restrained from pulling in. Both factors cause the development of membrane tension in the plate, and if the edges are restrained, then at large deflections ($w > 1.5t$, say) a high proportion of the lateral load is supported by this means rather than by plate bending. On the other hand, so long as a plate is perfectly flat, there are no membrane effects whatever because it is only the normal (or lateral) component of the membrane tension that can carry a portion of the normal (or lateral) pressure load. Likewise, if the plate edges are free to pull in, then no significant membrane tension can arise, and the only membrane effect is the in-plane strain (tensile in the central portion and circumferential compression at the plate edges) that arises because of geometric compatibility. This type of membrane effect only becomes significant for medium to large deflections ($w > t$, say). Note that it cannot arise at all for a very long plate because a cylindrical deflected shape is a developable surface, having no geometric incompatibilities.

As explained in Section 9.1.3, in ship plating there is relatively little restraint against edge pull-in as long as the deflections are not large. For large deflections such restraint does become significant but in many cases the load required to cause such large deflections would have already caused failure in the stiffeners or beams that support the plating. Nevertheless there are some situations in which these large deflections can be permitted, and in these cases the use of membrane tension can give substantial weight savings. Therefore it is important to have at least some idea of the relative magnitude of membrane tension that can arise if the edges are restrained from pulling in. This can be obtained by examining the case of a unit-width strip of laterally loaded plating with edges prevented from approaching. If $w(x)$ is the deflection of the strip, the difference δ between the arc length of the deflected strip and the original straight length is

$$\delta = \int_0^b \left(\sqrt{1 + \left(\frac{dw}{dx}\right)^2} - 1 \right) dx$$

$$\delta \cong \int_0^b \frac{1}{2}\left(\frac{dw}{dx}\right)^2 dx \qquad (9.2.2)$$

In this case the membrane force is unidirectional ($N_x = N_y = 0$). We shall also assume that it is constant over the length. This is a reasonable approximation and becomes more accurate as deflection increases and the plate becomes more of a "pure membrane." If we call this constant membrane tension T, the extension due to tension is $\delta_1 = Tb/AE$ where A, the cross-sectional area per unit width, is equal to t. It follows from (9.2.2) that

$$\delta_1 = \frac{Tb}{tE} = \int_0^b \frac{1}{2}\left(\frac{dw}{dx}\right)^2 dx$$

or

$$T = \frac{Et}{2b}\int_0^b \left(\frac{dw}{dx}\right)^2 dx \qquad (9.2.3)$$

Let us assume the strip to be pinned and, in order to investigate the effect of initial deflection, let us assume that there is an initial deflection w_0. Let us further assume that both w_0 and the deflection w_l due to the load are sinusoidal, so that the total deflection in the loaded condition is

$$w(x) = (w_0 + w_1)\sin\frac{\pi x}{b}$$

From (9.2.2) the total change in length is

$$\delta = \int_0^b \frac{1}{2}(w_0 + w_1)^2 \frac{\pi^2}{b^2}\cos^2\frac{\pi x}{b}\,dx$$

$$= \frac{\pi^2}{4b}(w_0 + w_1)^2$$

Similarly, the change in length due to the initial deflection is

$$\frac{\pi^2 w_0^2}{4b}$$

and the change in length due to the loading is the difference between these two:

$$\delta_1 = \frac{\pi^2(2w_0 w_1 + w_1^2)}{4b}$$

Using this value, Muckle (1967) obtained a solution for w_1 using an energy method as follows:

Strain energy due to *bending*

$$= \int_0^b \frac{E'I}{2}\left(\frac{d^2w}{dx^2}\right)^2 dx$$

$$= \int_0^b \frac{E'I}{2}(w_1)^2 \frac{\pi^4}{b^4}\sin^2\frac{\pi x}{b}dx$$

$$= \frac{\pi^4 E'I(w_1)^2}{4b^3}$$

Strain energy due to *tension*

$$= \frac{1}{2}T\delta_1$$

$$= \frac{1}{2}\left(\frac{AE}{b}\delta_1\right)\delta_1$$

$$= \frac{1}{2}\frac{AE}{b}\delta_1^2$$

In general, for a deflection $w(x)$ due to a load $p(x)$ per unit length, the work done is

$$W = \int_0^b \frac{1}{2}pw\,dx$$

In the present case, p is a constant and the lateral deflection due to the load is $w_1 \sin(\pi x/b)$. Therefore

$$W = \frac{pw_1 b}{\pi}$$

The work done by the load is equal to the total strain energy, so that

$$\frac{pw_1 b}{\pi} = \frac{AE\delta_1^2}{2b} + \frac{\pi^4 E'I(w_1)^2}{4b^3}$$

$$= \frac{\pi^4}{32}\frac{AE}{b^3}(2w_0 w_1 + w_1^2)^2 + \frac{\pi^4 E'I(w_1)^2}{4b^3}$$

Hence

$$\frac{\pi^4 AEw_1^2}{32b^3} + \frac{\pi^4 AEw_0 w_1^2}{8b^3} + \quad 8$$

$$+\left(\frac{\pi^4 AEw_0^2}{8b^3} + \frac{\pi^4 E'I}{4b^3}\right)w_1 - \frac{pb}{\pi} = 0 \quad (9.2.4)$$

For the strip $A = t$, $E' = E/(1 - \nu^2)$ and $I = t^3/12$ per unit width.

On substituting these in (9.2.4) we have

$$w_1^3 + 4w_0 w_1^2 + \ 4$$

$$+\left(4w_0^2 + \frac{2}{3}\frac{t^2}{1-\nu^2}\right)w_1 - \frac{32pb^4}{\pi^5 Et} = 0 \quad (9.2.5)$$

9.2.3 Effect of Initial Deformation

For plates which have no initial deflection $w_0 = 0$ and (9.2.5) becomes

$$w_1^3 + \frac{2}{3}\frac{t^2}{1-\nu^2}w_1 = \frac{32pb^4}{\pi^5 Et} \quad (9.2.6)$$

For the initial stages of loading, the deflection w_1 will be small relative to the thickness, and hence the first term may be neglected. This gives

$$\frac{w_1}{t} = \frac{48(1-\nu^2)p}{\pi^5 E}\left(\frac{b}{t}\right)^4$$

$$= 0.143\frac{p}{E}\left(\frac{b}{t}\right)^4 \quad (9.2.7)$$

and this agrees almost exactly with (9.1.2), which was the result obtained from small-deflection theory, ignoring membrane action:

$$\frac{w_1}{t} = \frac{5(1-\nu^2)p}{32E}\left(\frac{b}{t}\right)^4$$

$$= 0.142\frac{p}{E}\left(\frac{b}{t}\right)^4$$

This illustrates the point made earlier that membrane action requires some deflection, either initial or due to load, and if there is no initial deflection then membrane action does not become significant until the deflection due to load approaches the plate thickness. This is further illustrated in Fig. 9.12, which shows the stresses in a strip of initially flat ($w_0 = 0$) plating 0.75 m wide and 15 mm thick. It will be seen that as the load increases, the bending stress begins to depart from the value obtained when neglecting membrane effects. There is less deflection, and hence less bending (the plate is straighter, due to the tension T) and therefore the magnitude of the bending stress is always lower. However, on the tension side of the plate the membrane stress adds

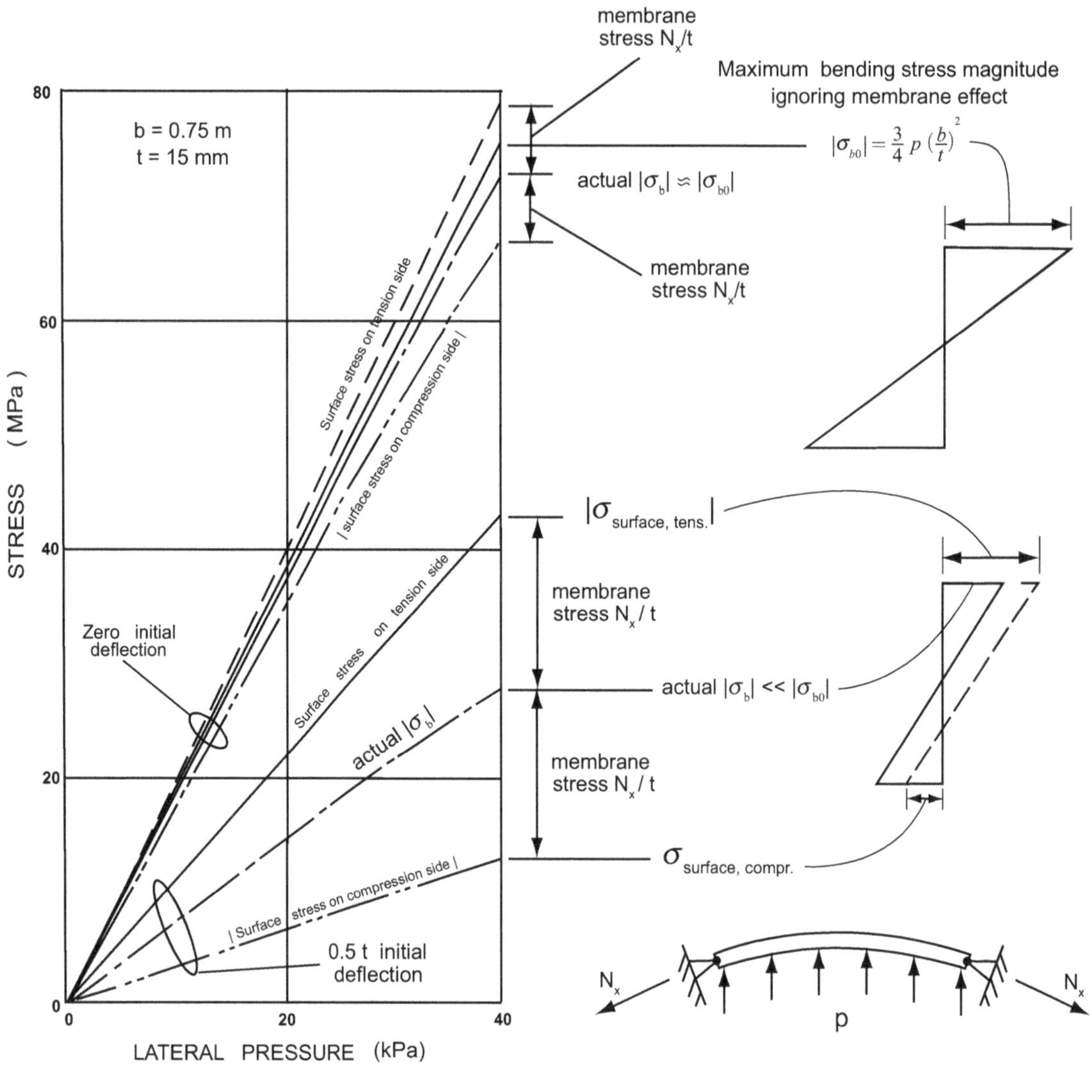

Figure 9.12 Membrane effect for long simply supported plates.

to the bending stress and thus the maximum tensile stress in the plate is slightly greater than the value obtained when membrane effects are neglected. Nevertheless, the overall effect is small, being about 4% for a stress level of 80 MPa.

In contrast to this, the figure also shows the stresses for the case of an initial deflection w_0 = 7.5 mm, that is, half the thickness of the plating. The magnitude of the bending stress $|\sigma_b|$ is greatly reduced and it is obvious that even a small initial deformation allows membrane effects to occur as soon as the load is applied.

In stiffened plating there will always be some degree of inwards dishing of the plate panels bounded by the stiffeners and frames because of the shrinkage of the fillet welds along these boundaries, which pulls the plating inwards. If the plating is relatively thin this causes prominent parallel ridges on the outside of the plating which are reminiscent of a gaunt rib cage, and the effect is commonly referred to as the "hungry horse" look. A similar type of initial deformation may be caused by previous loading that produced "permanent set," that is, a lateral deflection that involved plastic bending strain along the boundaries.

Although such initial deformation may be undesirable from consideration of appearance, fluid resistance and so on, it can have a beneficial influence on the elastic strength of plates, providing that the edges are restrained from pulling in. Defining "elastic strength" as the applied uniform loading that will just cause inelastic behavior, Table 9.2 taken from Clarkson (1958) demonstrates the beneficial effect of initial deformation. The figures quoted are values of the constant C in the equation elastic strength = $C \times$ nondimensional pressure coefficient.

As shown in Table 9.2, if a plate has already acquired some degree of permanent set (and hence "locked-in" or residual stresses) the pressure to cause *further* yielding may be significantly greater than for the *initial* yielding of a plate with a stress-free deflection. However, since the magnitude of such residual stresses will be unknown in practice, it is safer to ignore their effect and work with data for a stress-free plate.

It should be emphasized that initial lateral deformation is beneficial only when the plate edges are at least partly restrained from pulling in, thus allowing the development of in-plane tension. Also, when there is in-plane compression the lateral deformation can be quite detrimental if the lateral pressure causes a dished profile in line with the in-plane load, because the latter acts to increase the deformation, which increases the bending stress and also induces buckling failure. This is examined in Section 12.5.

On the other hand, for transverse bulkheads, platform decks, and other such plating, a larger amount of lateral deformation is permissible and such plating is usually designed on the basis of not exceeding a specified value of permanent set. The choice of a maximum permissible value of permanent set depends upon the application. For example, in the case of a collision bulkhead, a very large value would be chosen. The design of plating on this basis is covered in Section 9.4.

Table 9.2 Beneficial Effects of Initial Deformation

Type of Initial Deformation	Elastic Strength	Source of Increase in Elastic Strength
Flat plate	1.59	-
Initial deflection (stress free) equal to plating thickness	2.58	Membrane action
Initially flat plate dished to a permanent set equal to plating thickness	4.50	Membrane action plus residual stress

9.3 PLATES LOADED BEYOND THE ELASTIC LIMIT

9.3.1 Introduction

Except for the discussion of a "plastic hinge" in Section 9.1.9, the theory presented so far has been elastic theory. However, the pressure applied to a plate that results in the onset of yield does not represent the limit of pressure that the plate can support. The plate may withstand a pressure several times greater than this before it fails in any significant way, or before the deformation becomes unacceptably large. In fact, for continuous plating supported by stiffeners, true "ultimate failure" of the plating almost never occurs because the stiffeners usually have a much lower load capacity than the plating. When the plate deflection becomes very large there does arise some restraint against pull-in, and hence the plate gradually becomes a fully plastic membrane, for which the rupture load is enormous. This extreme level of load and deformation is relevant in some special design applications such as icebreakers and protection against blast or collision, but in general the design of plating in which lateral loads predominate is governed by unserviceability rather than by ultimate failure. In most cases the governing type of unserviceability is a maximum allowable permanent set w_p, rather than a maximum stress level. An exception would be plates subject to cyclic loading where fatigue considerations may impose a limit on the level of the working stress. Also, permanent set does have some effect on ultimate strength in the case of wide or approximately square plates that are subjected to in-plane compression and lateral loading, as would occur in a transversely framed strength deck or bottom. This is dealt with in Chapter 12.

Some examples of serviceability requirements that may dictate the permissible value of w_p are the operation of forklift trucks or other vehicles, the overall flexural stiffness of the deck or panel, the rigidity of support to attached fittings, the robustness of the plating against damage (dents, etc.), and the avoidance of dished-in side plating (the "hungry horse" look) either for stealth or for aesthetics. As this list indicates, it is often a nonstructural consideration that determines the maximum permissible w_p. Because of the variety of serviceability requirements, no single value or expression for the maximum permissible w_p will be universally suitable. Hence the approach that is taken herein is to use a combination of basic elasto-plastic theory and experimental results to derive a semi-empirical mathematical relationship between lateral load and w_p. Some design charts are also provided for hand design.

Theories relating to the behavior of the plate following the onset of yield are called elasto-plastic theories. If the maximum permissible w_p is large, then an accurate analysis requires a combined large deflection and elasto-plastic theory. This is quite complex and so, before proceeding further, it will be helpful to consider which types and levels of theory are generally required for laterally loaded ship plating, as a function of the plate characteristics, the loading and support conditions, and the purpose of the analysis, that is, whether it is for calculating service stresses, or maximum permanent set (and of what magnitude), or possibly ultimate failure, as in the special cases mentioned previously.

To begin, we first consider what are the most basic characteristics of plating and how best to express these. Equations (9.1.7) and (9.2.7) showed that both the bending stress and the lateral deflection depend mainly on the *slenderness ratio* b/t. Ship plating subjected to large lateral loads (e.g., deck and shell plating) is generally within the range $30 < b/t < 80$, whereas superstructure paneling and the stiffened plating in smaller lightweight vessels is usually more slender, having $b/t > 80$. Equation (9.2.7) showed that lateral deflection depends inversely on E. Another fundamental property that characterizes a plate is its yield stress, since this measures its elastic range. Hence in the treatment that is to follow we will see that the most useful nondimensional parameter for plating is the plate *slenderness parameter* β, which is defined as

$$\beta = \frac{b}{t}\sqrt{\frac{\sigma_Y}{E}} \qquad (9.3.1)$$

and which combines all four of the foregoing plate characteristics. As we shall see, the ½ power that is applied to σ_Y/E arises from elasto-plastic theory.

For our purposes in this chapter, plates may be divided into two broad categories; slender plates ($\beta > 2.4$ approximately), and sturdy plates ($\beta < 2.4$). For *sturdy* plates, with a small b/t ratio, the relative deflection w/t will generally be small since it is proportional to $(b/t)^4$. This will apply to both initial deflection and deflection under load. Because of this, and because of the occurrence of edge pull-in, membrane effect in sturdy plates is generally quite small, even after yielding has occurred. Hence small deflection elasto-plastic theory is sufficient for unserviceability analysis. For *slender* plates the deflections will be larger for a given load and hence membrane effects may become significant; in this case, large-deflection theory would be required for unserviceability analysis.

Ship plating that is subjected mainly to lateral loads (e.g., platform decks, tween decks, and shell plating forward) is generally within the range $30 < b/t < 80$, and for mild steel $(\sigma_Y/E)^{½}$ is approximately 0.03, giving β values between 0.9 and 2.4. Therefore large-deflection theory is generally not required for the unserviceability analysis of such plating, but elasto-plastic theory usually is required if the analysis is to be rationally-based and versatile in its application.

9.3.2 Application of Elasto-plastic Theory to Laterally Loaded Plates

This section presents a simplified explanation of the formation and growth of permanent set, taken from Hughes (1981). The purpose is to establish the relative importance of the various parameters and to derive the general form of the relationship between load and permanent set. This general form then provides the basis for the design formulas that are presented in Section 9.4. For the load, we take the case of a uniform pressure and assume that this pressure also acts on adjacent panels of plating. In this case the appropriate boundary condition is "clamped; free to pull in." As shown in Section 9.1.5, yielding first occurs at the boundaries of the plate, at the upper and lower surfaces, and this marks the beginnings of permanent set in the plate. As illustrated in Fig. 9.9, the yielding eventually penetrates through the thickness and forms a "plastic hinge" along each of the plate boundaries. From then on the permanent set increases rapidly in proportion to any further load increase.

In order to determine the basic features of the growth of permanent set, we begin with the case of an infinitely long plate, thus removing the effect of aspect ratio and allowing us to consider a plate strip of unit width. It was shown in Section 9.1.5 that along the clamped edges, the maximum stress is at the plate surface, normal to the edges, and is given by

$$\sigma_x = \frac{1}{2}p\left(\frac{b}{t}\right)^2 \qquad (9.3.2)$$

It was also shown that the other two stresses are $\sigma_y = \nu\sigma_x$ and $\sigma_z = 0$, and that the value of σ_x that causes initial yield is

$$(\sigma_x)_Y = \frac{\sigma_Y}{\sqrt{1-\nu+\nu^2}} \tag{9.3.3}$$

The section modulus for a strip of plating of unit width is $Z = t^2/6$. Therefore, if m_Y denotes the bending moment per unit width corresponding to initial yield, the value of m_Y must be such that $m_Y/Z = \sigma_Y$. That is

$$m_Y = \frac{\sigma_Y}{\sqrt{1-\nu+\nu^2}}\left(\frac{t}{b}\right)^2$$

From (9.3.2) the pressure that causes initial yielding is

$$p_Y = \frac{2\sigma_Y}{\sqrt{1-\nu+\nu^2}}\left(\frac{t}{b}\right)^2 \tag{9.3.4}$$

By defining a nondimensional load parameter

$$Q = \frac{pE}{\sigma_Y^2} \tag{9.3.5}$$

and by making use of the plate slenderness parameter β, (9.3.4) may be put into the form

$$Q_Y = \frac{2}{\sqrt{1-\nu+\nu^2}}\left(\frac{1}{\beta}\right)^2 \tag{9.3.6}$$

As explained in Section 9.1.9, as the pressure load is further increased, the next important event in the plate's response history is the formation of edge plastic hinges. It was shown there that the *plastic moment* m_P (per unit width) required to form these hinges is

$$m_P = \frac{\sigma_Y}{\sqrt{1-\nu+\nu^2}}\frac{t^2}{4} = 1.5m_Y \tag{9.3.7}$$

It was also shown that the "edge hinge" pressure corresponding to M_P is

$$p_{EH} = \frac{3\sigma_Y}{\sqrt{1-\nu+\nu^2}}\left(\frac{t}{b}\right)^2$$

In terms of the nondimensional parameters,

$$Q_{EH} = \frac{3}{\sqrt{1-\nu+\nu^2}}\left(\frac{1}{\beta}\right)^2 \tag{9.3.8}$$

Figure 9.13 shows the theoretical load-deflection curve and the corresponding load-permanent set curve for an infinitely long plate. Complete collapse requires the formation of a third plastic hinge, but we are more interested in events prior to this and we want information about the growth of permanent deflection. The simple "hinge mechanism" analysis gives no information about deflection, and so we must examine the formation of the plastic hinges in more detail. As shown in Fig. 9.13, the commencement and early growth of permanent set is entirely due to the edge hinges, and this continues to be the case until the edge hinges have nearly been completed. We therefore examine the growth of plasticity at the edge and the relation between the end moment m_e and permanent set.

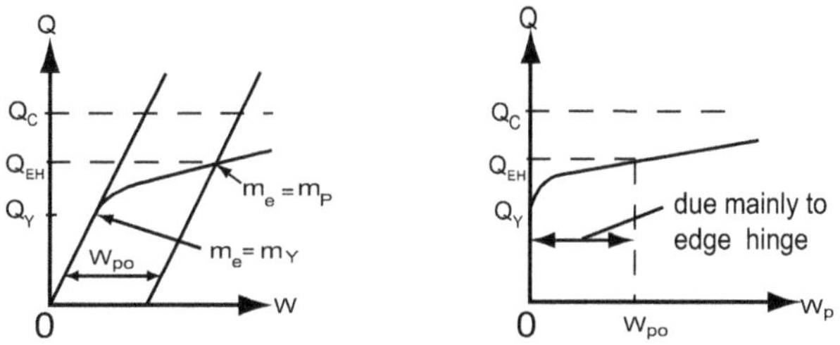

Figure 9.13 Load-deflection and load-parameter set curves for an infinitely long clamped plate.

Figure 9.14*a* is the moment–curvature diagram and we focus attention on the region from $m_e = m_Y$ to $m_e = m_P$. If the bending moment is increased by a differential amount dm_e, the plastic region grows and penetrates into the thickness t. The yielded portion cannot take any further bending moment and hence the additional moment dm_e is resisted only by the inner elastic layer of thickness t_r, as shown in Fig. 9. 15*a*. The increase of curvature $d\phi$ is therefore dm/EI_r, where $I_r = t_r^3/12$. That is, EI_r is the local slope of the m–ϕ curve, as shown in Fig. 9.14*a*. If the moment were removed, the curvature would decrease, approximately by an amount m/EI, leaving some permanent local curvature ϕ_p. The permanent curvature is drawn separately in Fig. 9.14*b*. In the later stages of the hinge formation, the decrease in ϕ (the "spring-back" after load removal) would be somewhat less than m/EI, but the effect is slight and ignoring it is conservative.

Hence we may write

$$\phi_p(m_e) = \int_{m_Y}^{m_e}\left(\frac{1}{EI_r} - \frac{1}{EI}\right)dm_e \tag{9.3.9}$$

We define a nondimensional post-yield end moment $\mathcal{M}_e$, as shown in Fig. 9.14*c*

$$\mathcal{M}_e = \frac{m_e - m_Y}{m_P - m_y} = \frac{m_e - m_Y}{\frac{m_P}{3}} \tag{9.3.10}$$

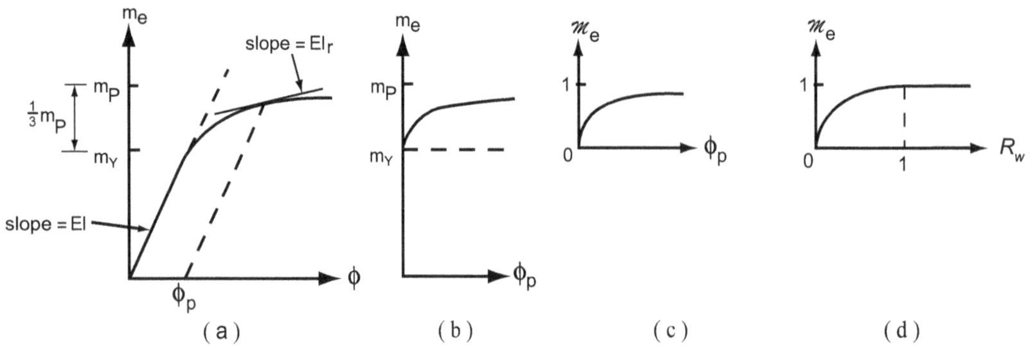

Figure 9.14 Moment-curvature diagrams for an end hinge in an infinitely long clamped plate.

In terms of $\mathcal{M}_e$ (9.3.9) becomes

$$\phi_p(\mathcal{M}_e) = \frac{4m_P}{Et^3}\int_0^{\mathcal{M}_e}\left(\frac{t^3}{t_r^3} - 1\right)d\mathcal{M}_e \quad (9.3.11)$$

after substituting for I and I_r. Figure 9.15*a* shows the stress distribution after yielding has commenced. It may be shown that equilibrium requires

$$m_e = \left(\frac{t_r^2}{6} + \frac{t^2 - t_r^2}{4}\right)(\sigma_x)_Y$$

in which the first term corresponds to the elastic portion. The bending moment at first yield is $m_Y = (\sigma_x)_Y\, t^2/6$. Substituting m_e and m_Y into the definition of $\mathcal{M}_e$ gives

$$\mathcal{M}_e = 1 - \frac{t_r^2}{t^2}$$

Hence

$$\frac{t^3}{t_r^3} = \frac{1}{(1 - \mathcal{M}_e)^{3/2}}$$

and (9.3.11) integrates to

$$\phi_p(\mathcal{M}_e) = \frac{4m_P}{Et^3}\left[\frac{2}{\sqrt{1 - \mathcal{M}_e}} - 2 - \mathcal{M}_e\right] \quad (9.3.12)$$

As shown in Fig. 9.13 the permanent set increases sharply during the formation of the edge hinge. As the load approaches Q_{EH} the plate becomes effectively simply supported with respect to the additional load. Equations (9.1.2) and (9.1.3) derived earlier for a plate strip show that the *rate* of increase in the total deflection would eventually become five times the pre-hinge value. Because of the plastic condition of the plate edges most of this deflection would be locked in and would therefore constitute permanent set. Hence the plate quickly reaches typical permissable values of permanent set and there is no need to consider the formation of a third hinge in the center.

Throughout the range of loading up to $Q = Q_{EH}$, the deflected shape of the plate remains a smooth curve since there is as yet no hinge at the center. If the load is removed after the

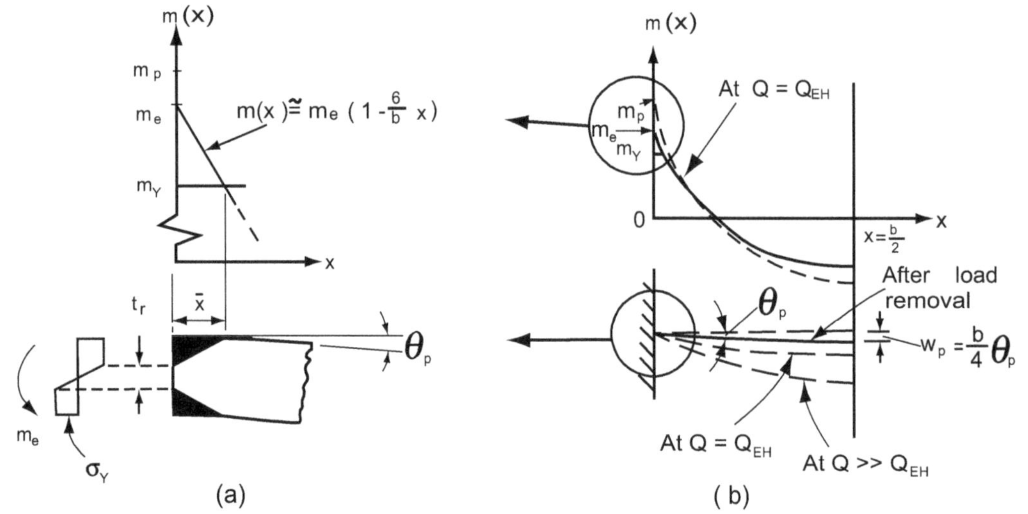

Figure 9.15 Conditions at the edge of a clamped plate.

formation of the edge hinges the plate relaxes to an approximately parabolic shape, with some "locked-in" bending moment and with permanent edge rotation θ_p, as shown in Fig. 9.15*b*. The permanent set w_p is therefore approximately θ_p $b/4$, because the occurrence of "spring-back" in the edge hinges has already been allowed for. In practice, the stored elastic energy in the plate would cause some slight additional spring-back, but the effect is small and is ignored because our goal here is simply to obtain an understanding of the basic nature of the process and the functional relationships among the variables.

The (permanent) angle of rotation of the plate edge θ_p is obtained by integrating the (permanent) curvature of the plate ϕ_p. This curvature occurs within the length of the plastic zone, say $\bar{x}$ (see Fig. 9.15*a*) and this length is dependent on the magnitude of the load, as characterized by the end moment m_e. Also, the bending moment within the plate is a function of position: $m = m\,(x)$. The upper part of Figure 9.15*a* is a plot of $m(x)$, the bending moment per unit width just inside the clamped end, where $m = m_e$. It may be shown that the derivative of m at $x = 0$ is $-6m_e/b$. Since the length of the plastic zone is small relative to the plate width b we take a linear approximation for $m(x)$:

$$m(x) - m_e - \frac{6}{b} m_e x$$

The corresponding linear approximation for $\mathcal{M}(x)$ is

$$\mathcal{M}(x) = \mathcal{M}_e - 6\frac{x}{b}(2 + \mathcal{M}_e) \qquad (9.3.13)$$

The length of the plastic zone is obtained by setting $m(x) = m_Y$, or $\mathcal{M}(x) = 0$, which gives

$$\bar{x} = \frac{b}{6}\mathcal{M}_e / (2 + \mathcal{M}_e) \qquad (9.3.14)$$

Adopting the above estimate for w_p we have

$$w_p \approx \frac{b}{4}\theta_p = \frac{b}{4}\int_0^{\bar{x}} \phi_p\left[\mathcal{M}(x)\right] dx \qquad (9.3.15)$$

in which the brackets indicate functional dependence. Equation 9.3.12 may now be generalized to express the curvature at an arbitrary point within the plastic zone where the bending moment is $\mathcal{M}(x)$:

$$\phi_p(\mathcal{M}) = \frac{4M_P}{Et^3}\left[\frac{2}{\sqrt{1-\mathcal{M}}} - 2 - \mathcal{M}\right. \qquad (9.3.16)$$

To facilitate the integration of (9.3.15) we regard $\mathcal{M}(x)$ as the primary variable, giving it the symbol X:

$$X = \mathcal{M}(x) = \mathcal{M}_e - \frac{6}{b}x(\mathcal{M}_e + 2)$$

from which

$$dx = -\frac{b}{6}dX(\mathcal{M}_e + 2)$$

At $x = 0$, $X = \mathcal{M}_e$, and at $x = \bar{x}$, $X = 0$. With these substitutions, together with (9.3.16), (9.3.15) becomes (after reversing the order of the integration)

$$w_p = \frac{4M_p}{Et^3}\frac{b}{24}\int_0^{\mathcal{M}_e}\left(\frac{2}{\sqrt{1-X}} - 2 - X\right)dX / (\mathcal{M}_e + 2)$$

$$= \frac{M_p b^2}{6Et^3}F(\mathcal{M}_e)$$

where

$$F(\mathcal{M}_e) = \left[4\left(1 - \sqrt{1-\mathcal{M}_e}\right) - 2\mathcal{M}_e - \frac{\mathcal{M}_e^2}{2}\right]\frac{1}{2+\mathcal{M}_e}$$

Substituting for M_P from (9.3.7) and introducing the plate slenderness β from (9.3.1) gives

$$\frac{w_p}{t} = \frac{\beta^2}{24\sqrt{1-\nu+\nu^2}}F(\mathcal{M}_e) \qquad (9.3.17)$$

At the completion of the edge hinge $\mathcal{M}_e = 1$ and the corresponding value of F is 0.5. Hence the value of w_p at this stage of the loading, which was denoted as w_{p0} (see Fig. 9.13) is

$$\frac{w_{p0}}{t} = \frac{\beta^2}{48\sqrt{1-\nu+\nu^2}} \qquad (9.3.18)$$

This expression is for an infinitely long plate. From experiments by Clarkson (1962) and Konieczny and Bogdaniuk (1999), and from finite element analysis, the following expression has been derived for a plate of finite aspect ratio

$$\frac{w_{p0}}{t} = \frac{\beta^2}{48\sqrt{1-\nu+\nu^2}} + (0.36 + 0.33\beta)\frac{b}{a} \qquad (9.3.19)$$

It will be useful to have a nondimensional measure of permanent set. For this purpose we define R_w as the ratio of the permanent set w_p and the edge hinge value w_{p0}. From (9.3.17) and (9.3.18) this is

$$R_w = \frac{w_p}{w_{p0}} = 2F(\mathcal{M}_e)$$
$$= \frac{2}{2+\mathcal{M}_e}\left[4\left(1-\sqrt{1-\mathcal{M}_e}\right)-2\mathcal{M}_e-\frac{\mathcal{M}_e^2}{2}\right]$$

As shown in Fig. 9.14*d*, this equation gives the shape of the curve of end bending moment vs. permanent set during the formation of the end hinge. In other words, it describes the "knee" portion of the complete curve shown in Fig. 9.13. In Fig. 9.14*d* the curve becomes horizontal at $R_w = 1$ (the completion of the end hinge) because we are neglecting plate aspect ratio and the possible beginning of a third (central) plastic hinge.

For our application we require the inverse of the above expression; that is, we need $\mathcal{M}_e$ as a function of R_w, say $\mathcal{M}_e = T(R_w)$. For this purpose we replace the above expression by the following, which can easily be inverted.

$$R_w(\mathcal{M}_e) = \frac{w_p}{w_{p0}} \cong 1-(1-\mathcal{M}_e^3)^{1/3}$$

Figure 9.16 shows the accuracy of the approximation. Upon inverting we obtain the nondimensional end moment $\mathcal{M}_e$ as a function of the nondimensional permanent set R_w. We choose the symbol $T(R_w)$ for this function because it describes the "transition" of the curve from vertical to horizontal.

$$\mathcal{M}_e = T(R_w)$$

in which

$$\begin{cases} T(R_w) = [1-(1-R_w)^3]^{1/3} & \text{for } R_w \geq 1 \\ T(R_w) = 1 & \text{for } R_w \geq 1 \end{cases}$$

$$\text{with } R_w = w_p / w_{p0} \tag{9.3.20}$$

and with w_{p0} as defined in (9.3.18).

The separate definition for $R_w > 1$ is required because the foregoing analysis does not apply to this range. From the definition of $\mathcal{M}_e$ in (9.3.10) it can be seen that $\mathcal{M}_e$ is linearly proportional to the increment of bending moment from m_Y to m_P. Within this same range the load parameter increases from Q_Y to Q_{EH}., and equations (9.3.7) and (9.3.8) show that Q is linearly proportional

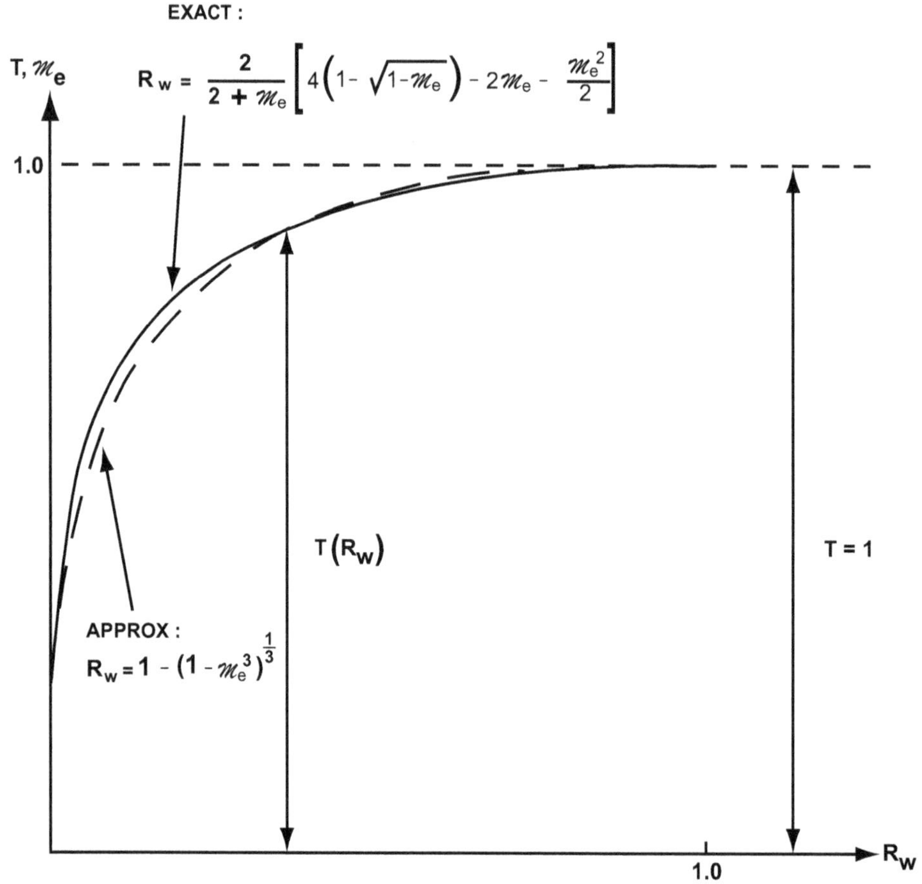

Figure 9.16 Exact and approximate expressions for the permanent set ration $R_w = w_p/w_{po}$.

to $\mathcal{M}_e$ because they both reach a value that is 50% above the yield value. Therefore

$$Q = Q_Y + (Q_{EH} - Q_Y)T(R_w)$$

and since $Q_{EH} = 1.5Q_Y$ this becomes

$$Q = Q_Y + \frac{Q_Y}{2}T(R_w) \qquad (9.3.21)$$

with $T(R_w)$ given by (9.3.20). This expression is plotted in Fig. 9.17 which shows that as the edge hinges begin to form, the load-permanent-set curve turns through 90°, forming a "knee," and the permanent set increases rapidly to the value w_{p0} given by (9.3.18), at which point the curve is horizontal. Beyond this point the foregoing analysis is no longer valid because the central hinge begins to form and this absorbs more load. Nevertheless the foregoing analysis shows that for an infinitely long plate (and hence with no aspect ratio effect) the permanent set begins to increase rapidly before the completion of the edge hinges, and the theoretical asymptote of the load-permanent-set curve is a line passing through the edge hinge load Q_{EH}, rather than the collapse load Q_c. The knee of the curve, that is, the region $w_p < w_{p0}$ (or $R_w < 1$), will be referred to as the "transition zone" and it is defined by the transition function $T(R_w)$ of (9.3.20).

For plates of finite aspect ratio, yielding and hinge formation occur at four edges instead of two, and therefore the load required to cause edge hinges will be larger than for a long plate of the same area. Also, the yielding and hinge formation is a more gradual process, beginning at the midpoint of the long sides and gradually extending along the sides, and this makes the transition zone larger than for a long plate. In addition, some membrane effect will be present due to the requirement of geometric compatibility. Nevertheless, the basic mechanism of permanent set is the same as for a long plate, and experiments (Clarkson, 1962) have verified that the formation of edge hinges marks the upper bound of the loading range within which permanent set begins to increase rapidly.

It has also been observed experimentally that once Q_{EH} is exceeded, the growth of w_p is very nearly linearly proportional to further increases in load. The main reason for this is that once plastic hinges have formed at the boundaries, there is no further change in the basic nature of the boundary conditions with further increase of load. The linearity in the growth of permanent set is illustrated in Fig. 9.18. The figure also shows how the foregoing analysis is applied to plates of finite aspect ratio. The commencement of permanent set occurs at the initial yield load Q_Y, which can be calculated from elastic theory. The subsequent load-permanent-set relationship contains two parts: a curved "transition" portion and a subsequent straight portion. These are defined in terms of two parameters, ΔQ_0 and ΔQ_1, each of which corresponds to a particular increment of load, as shown in Fig. 9.18. The increment ΔQ_0 defines the location of the intercept of the straight portion of the curve and ΔQ_1 is the further increment of load at the end of the transition zone, and this serves to define the slope of the straight portion. In mathematical terms

$$Q = Q_Y + T(R_w)\left[\Delta Q_0\left(\frac{a}{b}, \beta\right) + \Delta Q_1\left(\frac{a}{b}, \beta\right)R_w\right] \qquad (9.3.22)$$

in which the parentheses (but not the square brackets) indicate functional dependence. By definition, $R_w = w_p / w_{p0}$, and w_{p0} is given by (9.3.19). As shown in the next section, experiments indicate that the Q versus R_w relationship remains essentially linear up to relatively large values of permanent set. Ultimately, of course, as

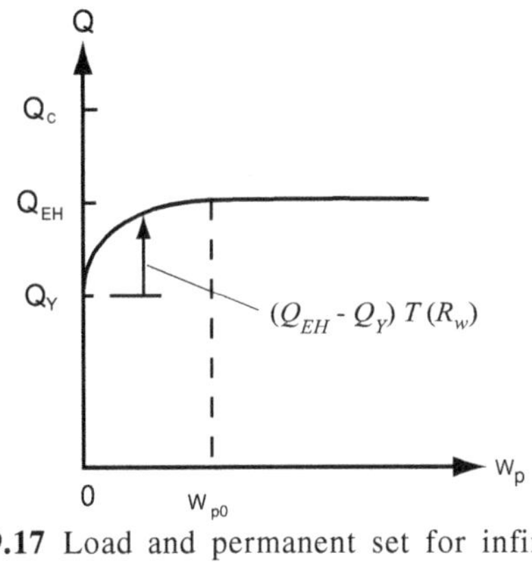

Figure 9.17 Load and permanent set for infinite aspect ratio.

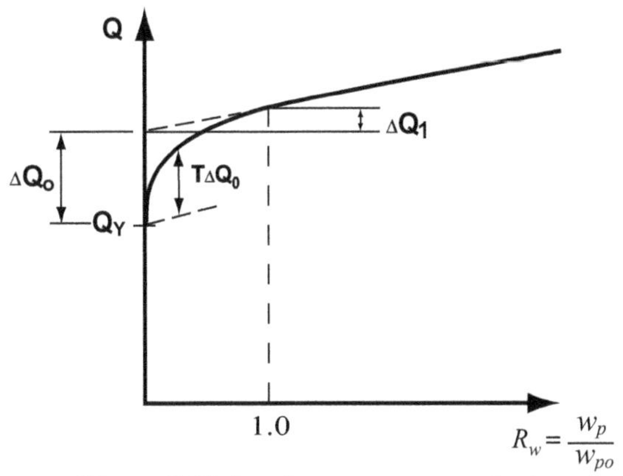

Figure 9.18 Load versus permanent set.

the load increases, plasticity spreads throughout the central region of the plate and some membrane straining begins to occur because of the large deflections. This latter effect would cause the slope of the curve to increase. At this stage small-deflection theory is no longer valid. However, for typical ship plates there is no need to carry the analysis any further because the magnitude of the permanent set has already exceeded typical serviceability limits.

9.4 DESIGN OF PLATING BASED ON ALLOWABLE PERMANENT SET

Because permanent set involves complicated elasto-plastic behavior, there is no analytical, closed-form method available for the direct calculation of the load required to cause a specified level of permanent set. Some approximate analytical solutions have been obtained for special cases, as in the previous section, but an accurate and general solution requires the use of numerical techniques such as incremental finite element analysis or the solution of the nonlinear plate equations by finite differences. These methods involve too much computation for ordinary design applications. For this purpose, the designer requires a rapid and simple method—ideally a single formula—for estimating the load that would produce a given level of permanent set, and since the choice of that level is somewhat arbitrary and does not involve ultimate failure, the calculation does not require the accuracy of these sophisticated numerical techniques.

9.4.1 Plating Subjected to Uniform Pressure

In view of the undesirable complexities that exist on the analytical side and the availability of suitable experimental data obtained by Clarkson (1962), an empirical expression was derived in Hughes (1981) based on the general form of (9.3.22). The expression is

$$Q = Q_Y + T(R_w)\left[\Delta Q_0 + \Delta Q_1 R_w\right] \qquad (9.4.1)$$

in which

Q = the load parameter for the pressure load, p

$$= \frac{pE}{\sigma_Y^2}$$

Note: If, as is usual, the maximum value of pressure is subject to some uncertainty, then the pressure load p should be a *factored* load.

Since the first edition of *Ship Structural Analysis and Design* in 1983, further data have been obtained from experiments and also from nonlinear finite element analysis (Konieczny and Bogdaniuk, 1999). This additional data has made it possible to develop new empirical expressions for Q_Y, ΔQ_0 and ΔQ_1 which are more accurate than previously.

$$Q_Y = \frac{2}{\sqrt{1-\nu+\nu^2}\,\beta^2}\left[1 + 1.46\left(\frac{b}{a}\right)^{1.87}\right]$$

$$-Q_0 = \frac{1 + 3.24\beta^{0.0687}\left(\frac{b}{a}\right)^{1.389}}{\sqrt{1-\nu+\nu^2}\,\beta^2} \qquad (9.4.2)$$

$$-Q_1 = 1.92\frac{\left(\frac{b}{a}\right)^{1.86}}{\beta^{0.94}} \qquad (9.4.3)$$

The first term of ΔQ_0, that is, $1/[(1-\nu+\nu^2)^{½}\beta^2]$, is the increase of load above Q_Y which would cause edge hinges in an infinitely long plate, at Q_{EH}, given by (9.3.8). The second term accounts for the aspect ratio, a/b and for the membrane effect due to compatibility that occurs in plates of finite aspect ratio and that increases with plate slenderness. Equation 9.4.3 gives the further increment in the load parameter at the end of the transition zone, as shown in Fig. 9.18. Figure 9.19 shows that the foregoing empirical expression represents the experimental results of Clarkson and Konieczny and Bogdaniuk with satisfactory accuracy. It therefore provides a simple and rapid means for estimating the load corresponding to a specified level of permanent set.

Clarkson's data is for plating that is initially flat, and therefore in (9.4.1) the value of w_p (within R_w) is that due to the load only. In practice, plates are seldom initially flat; they usually have some initial deformation, the most common being that due to welding. There are two types of initial deformation: that caused by previous loads, which involves plastic deformation and locked-in stresses, and a "stress-free" initial deformation. Both cases are illustrated in Fig. 9.20. Let us first consider a stress-free initial deformation w_{pi} and define the total permanent set as $w_{pt} = w_{pi} + w_p$. The permanent set ratio is then

$$R_w = \frac{w_{pt} - w_{pi}}{w_{p0}} \qquad (9.4.4)$$

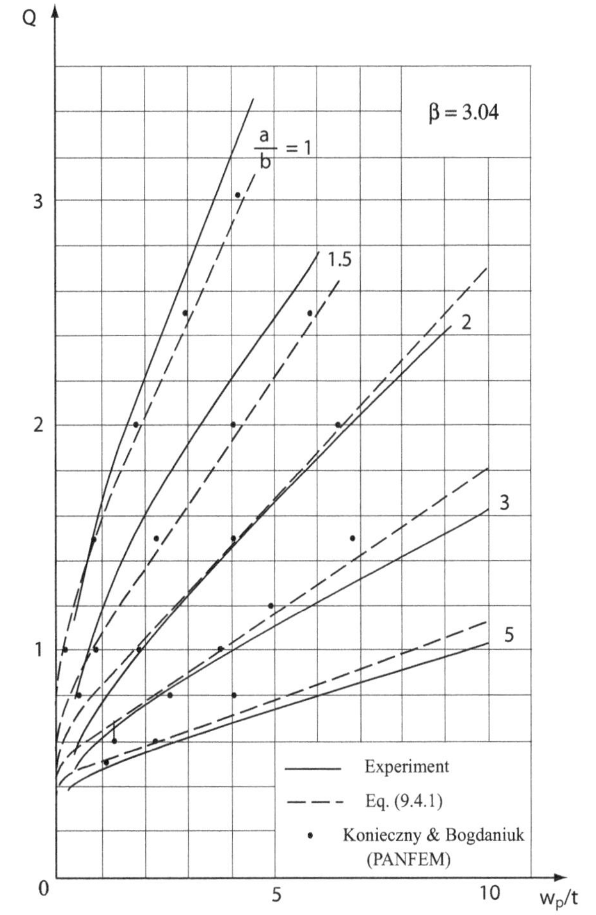

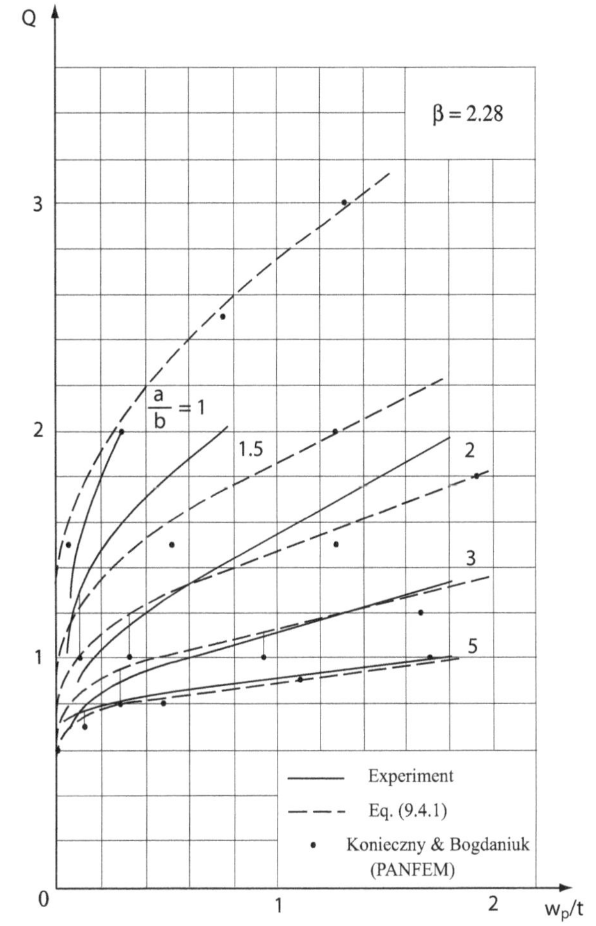

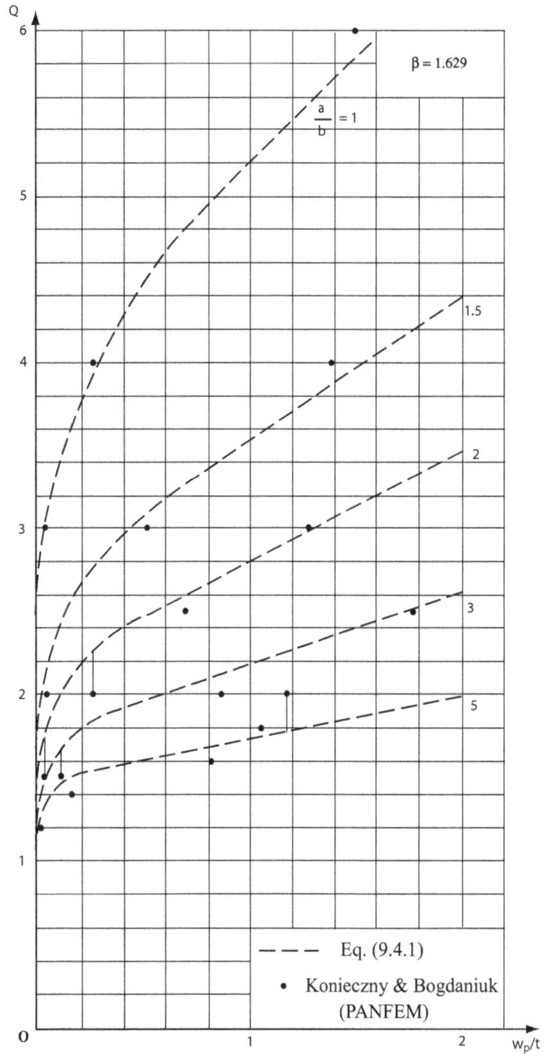

Figure 9.19 Comparison of eq. (9.4.1) with experiments and finite elemental data points. **(a)** $\beta = 3.04$. **(b)** $\beta = 2.28$. **(c)** $\beta = 1.629$.

If the plate edges are free to pull in and if w_{pi} is not large (say, $w_{pi} < t$), then the presence of w_{pi} does not substantially alter the load-deflection curve; it merely shifts it to the right by an amount w_{pi}, and the curve of Q versus w_p/t is likewise shifted rightward by an amount w_{pi}/t. Thus the load-permanent-set relation of (9.4.1) can be used for plates with stress-free initial permanent set w_{pi} by first subtracting w_{pi} from the total allowable permanent set $(w_{pt})_{max}$.

If the initial permanent set w_{pi} is not stress-free but is due to a previous load (curve OAC in Fig. 9.20), the load-deflection curve would commence at $w = w_{pi}$ and would rise linearly until it almost regained its previous point; only then would the permanent set begin to increase further (curve CAB). Thus, if $(w_{pt})_{max}$ denotes the maximum permissible value of total permanent set and if Q_1 denotes the corresponding capability of an initially flat plate (curve OAB), then as shown in

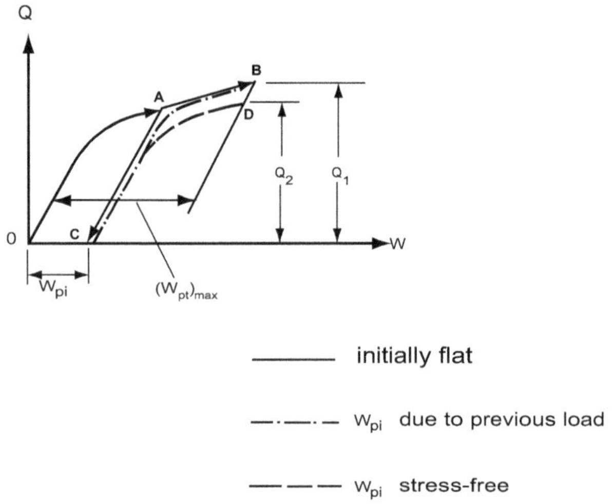

Figure 9.20 Load-deflection curves for various initial-conditions.

the figure an initial permanent set due to a previous load does not reduce the capability of the plate; it is still Q_1. In contrast to this, when the initial permanet set w_{pi} is stress-free the curve of Q versus w is curve CD. As long as w_{pi} is not extreme, this curve has the same shape as for an initially flat plate (curve OAB) but is shifted to the right by an amount w_{pi}. That is, the initial permanent set is not "absorbed" within the subsequent values, but rather is additive. Hence for the same value of maximum permissible permanent set $(w_{pt})_{\max}$, the plate's capability will be Q_2, slightly less than Q_1. Eventually, at very large loads, the curve does reach the "initially flat" curve, but this is usually well beyond the design range.

In the method presented herein, the initial permanent set due to welding is treated as if it were stress-free. In reality, of course, a welded plate is not stress-free, because welding causes large locked-in stresses. Therefore the curve of Q versus w for a welded plate lies somewhere between the "stress-free" curve (CD) and the "initially loaded" curve (CAB). The exact position would be difficult to ascertain and such exactness is not warranted because the prediction of weld-induced deformation, which is considered subsequently, is only approximate. Hence in the method presented here, w_{pi} is simply subtracted from the total allowable permanent set, as in (9.4.4). Within the normal design range the error involved in this approximation is small and it lies on the conservative side.

THE *PLATE* COMPUTER PROGRAM

Equation 9.4.1 provides a relationship between the three principal variables—load, plate thickness (t), and permanent set (w_p). Given any two of them, the equation can provide the third. It thus has three types of use:

1. Design—determine t for a given load and w_p
2. Response analysis—determine w_p for a given load and t
3. Limit analysis—determine the maximum load for a given w_p and t.

Since (9.4.1) is too complex for hand calculation, it has been implemented in a small computer program called PLATE, and a copy is available at http://filebox.vt.edu/users/hugheso. It has also been implemented in MAESTRO, and when structural optimization is being performed, it supplies the constraint against unserviceability due to excessive permanent set.

In order to permit (9.4.1) to be used for hand calculations, it is presented in the form of "design charts" in Figs. 9.21*a–e*. Each figure is for a fixed amount of permanent set, measured in terms of a nondimensional ratio $(w_p/b)(E/\sigma_Y)^{½}$. In the five figures, this ratio is 0.2, 0.4, 0.6, 0.8, and 1.0. If the thickness is to be calculated the steps are:

1. For the specified permissable value of w_p, calculate the non-dimensional ratio and determine which two figures bracket that value. Then, for each figure:
   a. compute Q for the given design load (factored, if appropriate)
   b. for a selected stiffener spacing b (and hence aspect ratio a/b) read the value of β
   c. calculate the thickness.
2. Interpolate between the two thicknesses.

If the result is not satisfactory for some reason, the value of b, or perhaps of σ_Y, may be varied as required.

INITIAL PERMANENT SET DUE TO WELDING

Antoniou (1980, pp. 31–39) has presented the results of a regression analysis of over 2000 values of w_{pi}/t measured on newly built ships over a number of years, showing that the significant parameters are β and the ratio t_w/t, where t_w is the thickness of the stiffener web, and that the following expression gives a satisfactory fit to the data

$$\frac{w_{pi}}{t} = 0.073\beta^{1.65}\left(\frac{t_w}{t}\right)^{0.42} \qquad (\beta \le 2.5) \qquad (9.4.5)$$

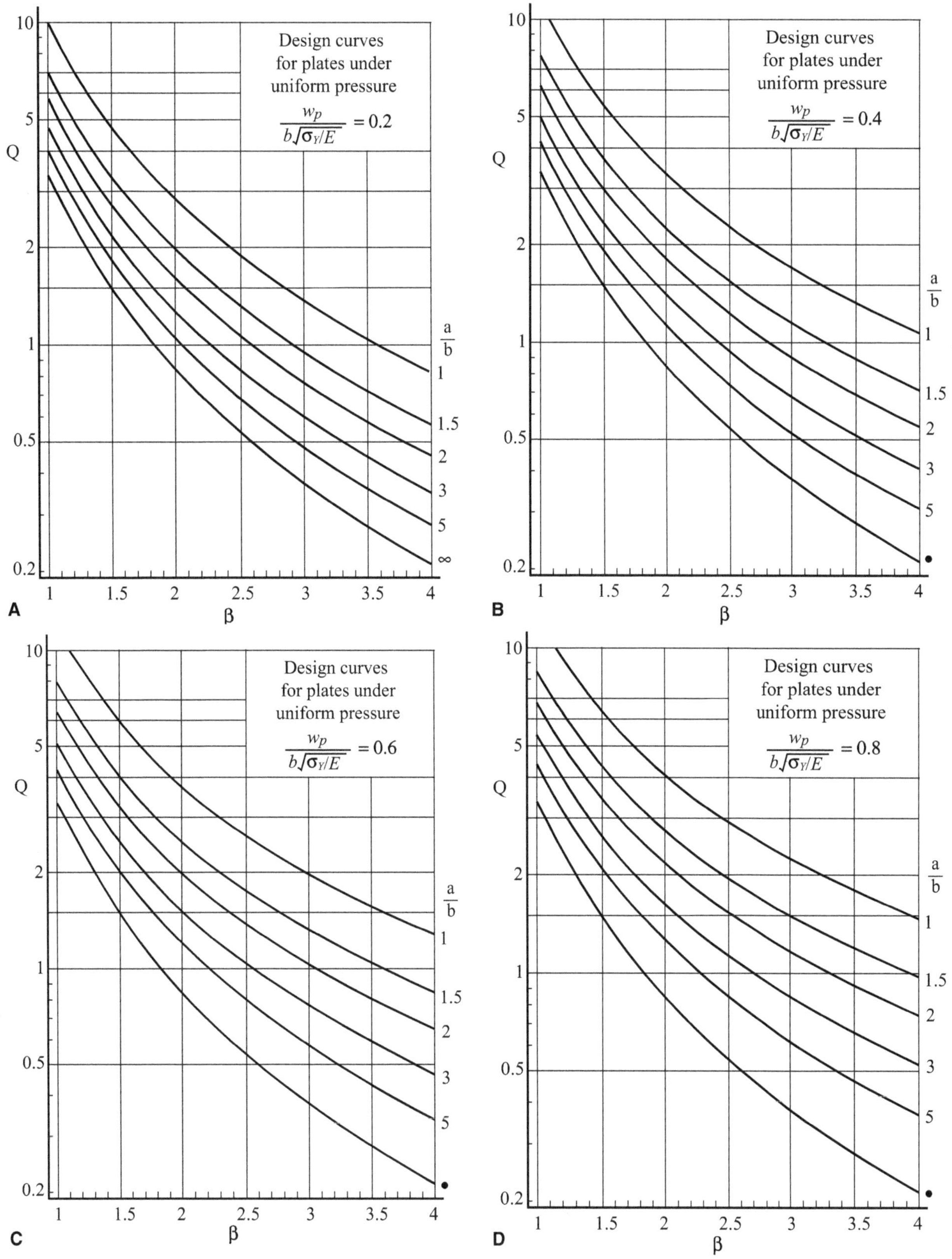

Figure 9.21 Design curves for plates under uniform pressure with an allowable permanent set. *Continued on next page.*

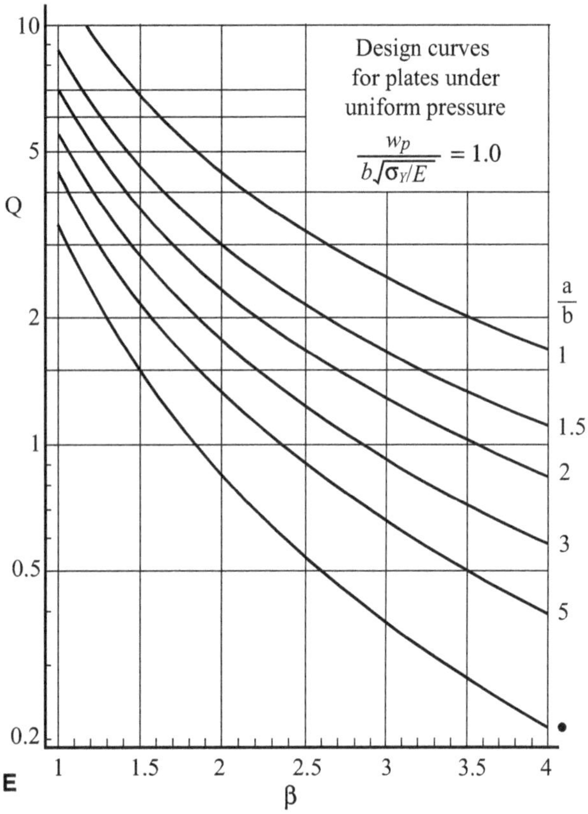

Figure 9.21 *Continued.* Design curves for plates under uniform pressure with an allowable permanent set.

Typical ship plates have $0.6 < t_w/t < 0.9$ and therefore the mean value is approximately

$$\frac{w_{pi}}{t} = 0.065\beta^{1.65}$$

Also, for ship plates the slenderness ratio is usually in the range $1.5 < \beta < 2.5$ and hence the initial welding distortion is usually in the range $0.13 < w_{pi}/t < 0.3$. Similarly, from (9.3.19), the amount of permanent set in the transition zone is usually in the range $0.16 < w_{p0}/t < 0.44$. Thus the initial welding deformation w_{pi} is about equal to the transition amount. This means that if the total allowable permanent set is more than twice the as-welded value (which is often the case) then the allowable permanent set due to the load is beyond the transition zone and (9.4.1) is simplified because the transition function T is unity.

Antoniou (1980, pp. 31–39) also proposed the following formulas for the maximum permissible value of initial weld-induced permanent set

$$\text{for } t \geq 14\ mm{:}\quad \left(\frac{w_{pi}}{t}\right)_{max} = 0.014\frac{b}{t} - 0.32$$
$$\text{for } t < 14\ mm{:}\quad \left(\frac{w_{pi}}{t}\right)_{max} = 0.018\frac{b}{t} - 0.55 \qquad (9.4.6)$$

and showed that this requirement is quite reasonable, being more than two standard deviations above the mean value given by (9.4.5), and is consistent with other formulations and with general shipbuilding practice.

The question remains as to what total level of permanent set is to be considered acceptable. This depends on the particular circumstances and therefore must be left to the designer. For naval vessels, where weight saving is important, a large value would be appropriate, such as $w_{pt} = b/50$. For cargo vessels a lower value would generally be preferred, such as $b/100$. This question will be considered further at the end of Section 9.4.2, after dealing with concentrated loads, because in many cases the design load is of this type rather than a uniform pressure. Also, many areas of ship plating are subjected to both types of loads and both must be investigated because it will not be known *a priori* which type would cause the larger value of permanent set.

The influence of permanent set on the in-plane compressive strength of plating is discussed in Section 12.7. In brief, for longitudinally stiffened panels the permanent set caused by welding and by lateral pressure does not diminish the ultimate compressive strength; it even enhances it slightly. For transversely stiffened panels, even a moderate amount of permanent set causes higher stresses (see Section 12.5) and a lower ultimate compressive strength, and a large permanent set causes a drastic reduction in strength. Since some permanent set is nearly always present, simply due to welding, transverse stiffening should not be used where good in-plane compressive strength is required, and if it is used in any panel that panel should not be counted in the hull girder calculations.

9.4.2 Plating Subjected to Concentrated Loads

Ship plating is also subjected to concentrated loads of various types, such as wheel loads, fenders, pallets, and falling objects. In Hughes (1983) it is shown that for design purposes there are two main types of concentrated loads, depending on the number of different locations where they can occur in the panel: single location and multiple location. Here the word "multiple" is not meant to imply simultaneous, but simply any load that will occur several times over the life of the ship and probably in different locations. The distinction between single and multiple is important because, as will be shown subsequently, these two types require different design methods.

Single location loads are either deliberate, such as the weight of a fixed piece of equipment, or accidental, such as the dropping of a heavy object. Deliberate single location loads do not usually influence plate design because they can either be placed over a stiffener, or additional stiffeners or other supporting structure can be provided. This is nearly always better than having thicker plating. Although the position of accidental loads may be random, for design purposes the worst position—midway between the stiffeners—must be assumed. Note that here the term accidental implies rare: if it is expected to occur several times in the life of a ship and at various locations in a panel (for example, heavy landing of a helicopter) it is classified as multiple location. Obviously, wheel loads and all other moveable loads are classified as multiple location.

Concentrated loads may be either static, quasistatic, or dynamic. Dynamic loads are those that have such a short duration that the inertia and dynamic response of the plating becomes significant. About the only example of a dynamic concentrated load is projectile impact, and in this case it is necessary to investigate the possibilties of puncture and of plate rupture, in addition to large permanent set. These two-limit states are not considered herein. Quasistatic loads are those in which the motion of the load can be accounted for by means of static inertia forces. Ship speeds are such that collision loads fall in this category, but here also it is necessary to investigate puncture and rupture as well as permanent set. Some common examples of quasi-static loads are the heavy landing of an aircraft or helicopter, and wheel loads of vehicles (or any other static weight-related load) when the ship is in heavy seas.

When the load magnitude has a high degree of variability, such as heavy landing of a helicopter, the design load is usually a *design extreme value*, which either corresponds to some probability of occurrence or is associated with some identifiable result (e.g., collapse of the landing gear). Alternatively, in commercial vessels there is sometimes the possibility of controlling and limiting the maximum load by means of operating regulations, such as maximum permissible axle load, wheel load, and tire pressure. If this approach is adopted, the maximum permissible load is derived from the maximum allowable permanent set, and this load (again factored to allow some margin for accidental overload and for change of service of the ship) becomes the design load.

PARAMETERS FOR DESCRIBING CONCENTRATED LOADS

For simplicity, the concentrated load is taken to be rectangular in shape (or footprint) of dimension $e \times f$, with e parallel to b, as shown in Fig. 9.22. The size or extent of the load footprint is measured by the geometric average of its two dimensions

$$e_m = \sqrt{ef} \qquad (9.4.7)$$

This is done because over a period of time a multiple location load will occur in all orientations to the plate and therefore the final cumulative value of permanent set will be independent of the load aspect ratio e/f. Hence both dimensions are given equal weight. The geometric mean is preferable to the arithmetic mean because it relates directly to the area under the load: e_m^2. For wheel loads, e_m is related to the load P and the tire pressure p_t by

$$e_m = \sqrt{p / p_t} \qquad (9.4.8)$$

We also need a measure of just how concentrated (or diffuse) the load footprint is. That is, we need a nondimensional parameter that measures the *relative* size or extent of the load footprint, in comparison to the extent (or span) of the plating. For this purpose we define

$$\lambda = e_m / b \qquad (9.4.9)$$

And finally the load parameter that is most commonly used for concentrated loads is

$$Q_P = \frac{PE}{b^2 \sigma_Y^2} \qquad (9.4.10)$$

9.4.3 Design for Multiple Location Loads

Since a wheel load is the most obvious example of a multiple location load, the discussion herein will be in terms of this type of load, but the results apply to all types of multiple location loads.

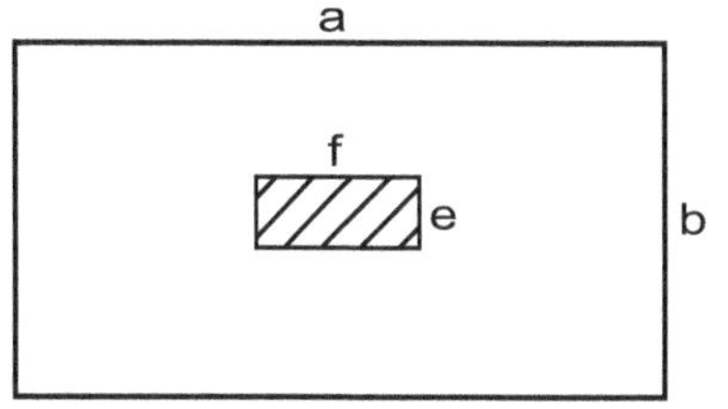

Figure 9.22 Geometry of single location concentrated load.

When a concentrated load that is large enough to produce permanent set moves across a panel from one side to the other, it causes plastic deformation along its entire path and the resulting permanent set is larger than if the same load were applied only at the center. Since wheel loads can occur anywhere in the panel, eventually every part of the panel will have undergone at least some plastic deformation, though it may only be near the plate surfaces. Also, the direction of travel will vary, at least from time to time. There will also be variations in tire diameter and width, and in the wheel arrangement, with twin wheels being especially common. With wide tires or closely spaced twin wheels, the load will occasionally be applied along the edge of a panel, straddling a stiffener, as shown in Fig. 9.23. This will cause some plastic rotation of the panel edges—more than would have been caused by loads crossing the panel—and so the permanent set will again be increased. Although not all wheel loads will equal the full design load, over a period of time this load will occur a number of times, in various locations, direction, and wheel arrangements. As a result the distribution and arrangement of plasticity in the panel gradually reaches a final stationary pattern, and the maximum permanent set reaches a final value corresponding to the design load. It is only this final cumulative value that is of interest to the designer. Moreover, this value is a function only of the panel parameters α and β and the load footprint size parameter λ; it is independent of the load shape and load aspect ratio. In Hughes (1983) it is shown that because of the cumulative nature of the plastic deformation the final distribution of plasticity is basically similar to that caused by a uniform pressure load. This is true not only for wheel loads but also for all multiple location loads, because even if they do not "move" they occur in different locations over the course of time. Because the distribution of plasticity is similar to that for a uniform load, then the relationship between load parameter and permanent set is also similar. Hence for a given design load parameter Q_P, the eventual stationary value of permanent set is essentially the same as that caused by an equivalent uniform pressure p_e, with load parameter Q_e given by

$$Q_e = \frac{p_e E}{\sigma_Y^2} \tag{9.4.11}$$

The value of Q_e may be expressed in terms of the ratio $r = Q_e / Q_P$. That is

$$Q_e = rQ_P \tag{9.4.12}$$

Once we have Q_e we can use all the information we have for pressure loads (9.4.1) and the curves of Fig. 9.21, with Q_e in place of Q.

There are two ways in which the overall plastic deformation caused by concentrated loads can differ from that caused by uniform loads: in geometry (the pattern of the plasticity) and in magnitude (the value of permanent set). Hughes (1983) showed that the geometry is basically similar and that the magnitude of the deformation bears a regular and well-behaved relationship with that caused by a uniform load, and that this relationship depends almost entirely on the load footprint size parameter λ. It is only slightly dependent on α and β because their effect is mainly accounted for by the formula for uniform load.

We can establish the general features of the function $r(\lambda)$ by imagining an experiment in which a pneumatic tire carrying a constant load P is applied over the entire surface of a panel in every direction so as to produce the final cumulative value of permanent set. In order to establish the behavior of r as λ becomes large, let us imagine a very large balloon type of tire as in Fig. 9.24, such that λ is relatively large. As the tire pressure is decreased, the load footprint grows larger, and as this process continues the

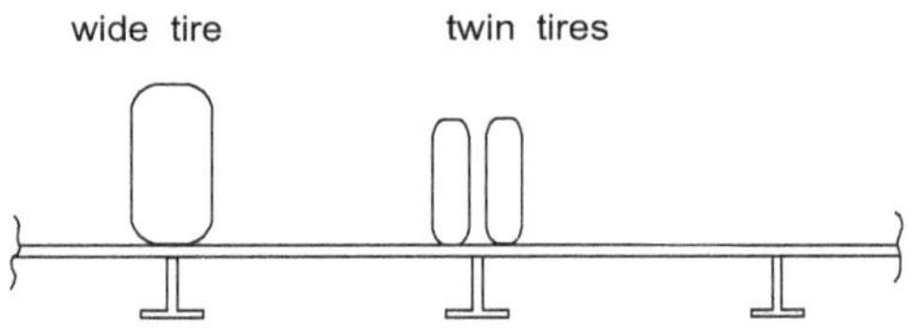

Figure 9.23 Loads causing edge rotation.

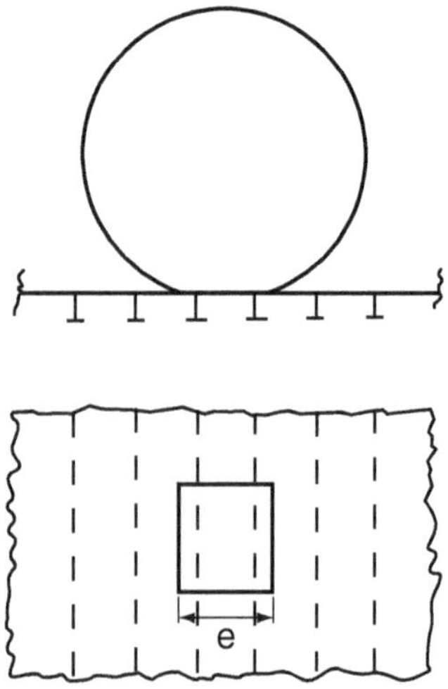

Figure 9.24 Example of relatively large load footprint.

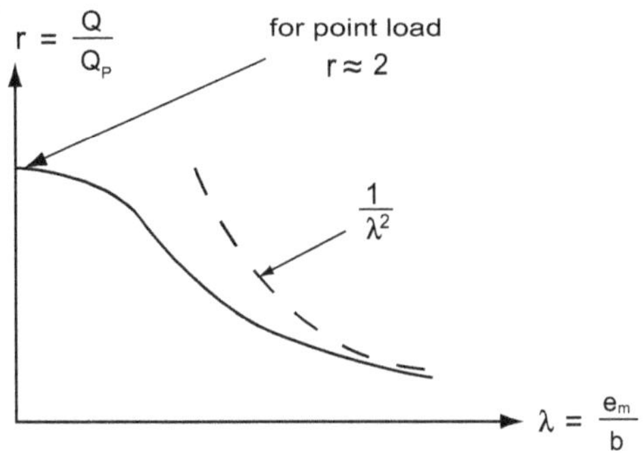

Figure 9.25 Ratio r for converting to an equivalent uniform pressure.

load eventually becomes a uniform pressure load, with the pressure being the tire pressure. From the definition of r we have

$$r = \frac{Q_e}{Q_P} = \frac{p_e E / \sigma_Y^2}{PE / (b^2 \sigma_Y^2)} = \frac{p_e b^2}{P}$$

In the limit we can substitute p_t in place of p_e.

$$r_{\text{limit}} = \frac{p_t b^2}{P}$$

Also, from (9.4.8) the tire pressure is related to e_m by $p_t = P/em^2$ and so the limit becomes

$$r_{\text{limit}} = \frac{b^2}{e_m^2} = \frac{1}{\lambda^2}$$

That is

$$r \to \frac{1}{\lambda^2} \text{ as } \lambda \to \infty \qquad (9.4.13)$$

as illustrated in Fig. 9.25.

To establish the behavior of r as λ approaches zero, let us imagine a series of experiments with relatively small tires, as in Fig. 9.26, and with progressively larger tire pressures. The width of the tire print e_m, given by (9.4.8) will be smaller each time, and therefore λ will be approaching zero. Since P is constant, Q_P is also constant and so the function $r(\lambda)$ will behave in the same manner as Q_e. If the pressure is quadrupled, e_m is halved and so is λ. Since the width of the load is already quite small, halving it does not alter the basic geometry, and since the load P is being kept constant, the final amount of permanent set will only be slightly larger than before. Hence the equivalent pressure and its load parameter Q_e will only increase slightly. In the limit, as the width becomes very small, there is no change in geometry and hence Q_e becomes constant. Therefore as λ approaches zero, the value of r approaches a constant value, as shown in Fig. 9.27. In Hughes (1983) it is shown that this value is approximately 2 and this information, together with experimental data obtained by Sandvik (1974) and calculations by the author using NASTRAN, is used to obtain the following approximate expression for $r(\lambda)$

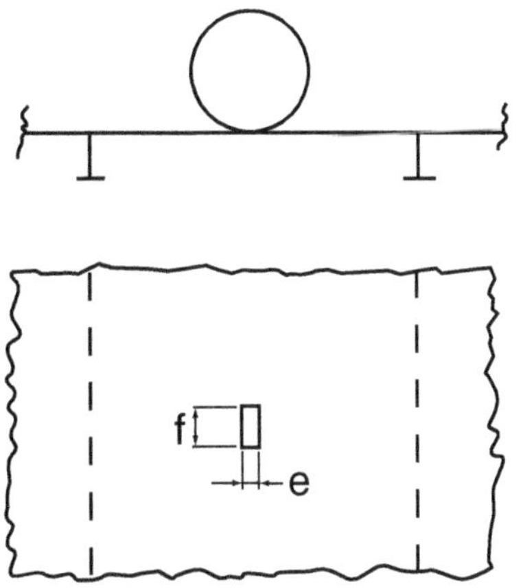

Figure 9.26 Example of relatively small load footprint.

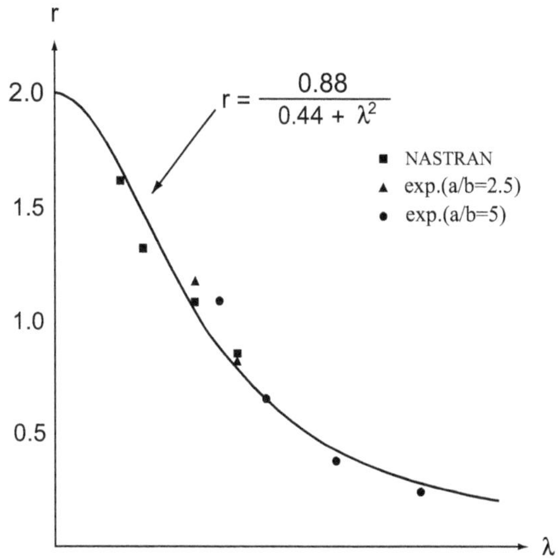

Figure 9.27 Theoretical and experimental values of r.

$$r = \frac{0.88}{0.44 + \lambda^2} \qquad (9.4.14)$$

which is illustrated in Fig. 9.27.

SELECTION OF LOAD WIDTH FOR TWIN WHEELS

With twin wheels the question arises as to what value of load width e should be used in order to best represent the load. This depends on the ratio of the tire width d and the wheel spacing s. It also depends on the relative values of the width of the total load envelope parallel to the axle and

b, the shorter dimension of the panel; that is, it depends on the ratio (see Fig. 9.28)

$$\chi = \frac{s+d}{b} \tag{9.4.15}$$

The shorter panel dimension is used because the plate bending moment is larger in this direction.

To begin let us consider a situation in which χ is small (say less than 0.4) such that both wheels would be either within the same panel or straddling a stiffener but not very far from the stiffener. In Jackson and Frieze (1980) measurements of the contact pressure across the width of a tire showed that the pressure is largest near the tire walls. Therefore if the wheels are closely spaced (small s/d), the effective width e can be taken as the total width: $e = d+s$. However, as the spacing increases, this method of representing the load gradually becomes unrealistic and inaccurate because it artificially increases e_m and hence λ, which in turn causes r, and hence Q_e, to become too small. Thus the error is on the unsafe side. A corollary of this is that making a small or moderate increase in the spacing of twin wheels does not give any substantial decrease in permanent set. Therefore when dealing with moderately spaced twin wheels, the best course is to ignore the spacing and use $e = 2d$.

Of course, as the spacing is further increased there is some benefit, as shown (qualitatively only) in Fig. 9.28. However, in most cases, there is so much variety and uncertainty regarding the spacing and other aspects of the wheel loads that a ship is likely to encounter in its lifetime that it is probably best to adopt the conservative course: $e = 2d$.

For cases in which the wheels are widely spaced relative to b—that is, values of χ approaching or exceeding 1—the worst loading condition will be when the wheels are acting on different panels, and the loading should be represented as one wheel of width $e = d$ carrying a load $P/2$. This is illustrated in Fig. 9.28 showing that with values of χ above 0.8 or so the maximum value of permanent set corresponds to the single wheel $P/2$ condition. The value of permanent set for this condition will not be as large as when the wheels are close together. However it will not be dramatically less than this because although Q_P has been halved so also has λ, and this causes r, and hence Q_e, to increase. The net effect depends on the width of the tire and other factors. The value of χ at which the single wheel condition can be assumed in place of the $e = 2d$ condition depends on many factors and it would be very difficult to formulate an expression that would cover all cases. After examining a number of design situations with helicopter undercarriages, Jackson and Frieze recommended a value of 0.6. Hence the best procedure would seem to be as follows: for $\chi < 0.6$ calculate w_p for both cases (load P with $e = 2d$, and load $P/2$ with $e = d$) and take the larger. For $\chi > 0.6$, take the latter case. This procedure is illustrated in Fig. 9.28.

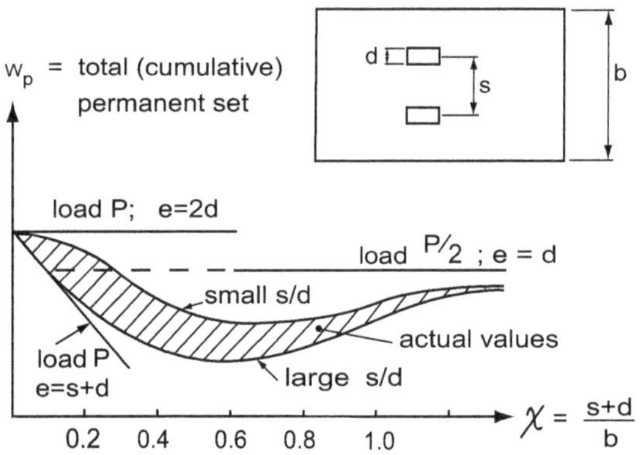

Figure 9.28 Effective width of twin wheels.

PERMISSIBLE LEVELS OF PERMANENT SET

As mentioned earlier, it is not possible to give general rules for deciding on what amount of permanent set is to be specified as the maximum allowable amount. This depends entirely on the particular circumstances, such as the type of ship, the location and purpose of the plating, and the type of loads. Even when these main factors are specified there may be other factors which influence the decision. Consider, for example, the plating in the vehicle deck of a ro-ro ship. Some types of cargo handling vehicles are operated at relatively high speeds and the permissible value of permanent set may be dictated by the need to avoid a driving hazard. A larger value may be permitted in a nontraffic or parking area. Another factor may be the desire to avoid having pools of water on the deck.

9.4.4 Design for Single Location Loads

As noted earlier, concentrated loads that are expected to occur only rarely should be taken as occurring in the center of the panel. With such loads, the pattern of plasticity in the plate is even more complicated than for uniform loads and it would be extremely difficult to derive an expression for permanent set purely from theoretical considerations. Hence for this case also we present an approximate semiempirical design formula obtained in Hughes (1983) by combining some basic theory with some experimental results from Jackson and Frieze (1980). The experiments showed that for this type of load the panel aspect ratio α has little effect on the permanent set and that the only effect of load aspect ratio *e/f* is that loads that are approximately square cause slightly more permanent set than elongated loads. The formula accounts for the latter by means of a simple correction factor Φ.

The range of values of λ covered in the experiment is $0.24 < \lambda < 0.79$. Therefore the design formula may not be accurate for values of λ less than 0.24, and the formula *should not be used* for values of λ greater than 0.8. The formula is

$$Q_P = \Phi\left\{\frac{C_1\lambda}{\beta^2} + \left[C_2 + \frac{C_3}{\beta^k} + C_4\left(\frac{\lambda}{1-\lambda}\right)^l\right]\left(\frac{W_p}{\beta t}\right)^m\right\} \quad (\lambda < 0.8) \tag{9.4.16}$$

in which

$C_1 = 10.45$ $\quad k = 1.6$

$C_2 = 0.34$ $\quad l = 0.8$

$C_3 = 3.56$ $\quad m = 1.1$

$C_4 = 0.23$

The coefficient Φ allows for the load footprint shape $e \times f$. The only effect is that a square footprint ($e = f$) causes slightly more w_p than other shapes. This is achieved by defining

$$\Phi = 1 - 0.8\left(\frac{ef}{e^2 + f^2}\right)^2 \tag{9.4.17}$$

For constant area $e \times f$ let us imagine a change in the aspect ratio of the footprint.

As $e \to 0$ (and $f \to \infty$) the squared term $\to 0$ and $\Phi \to 1$

whereas for $e = f$, the squared term is 0.25 and $\Phi = 0.8$. That is, for a square footprint, the load Q_P only needs to be 80% as large to cause the same w_p. A set of design curves based on (9.4.16) is given in Fig. 9.29.

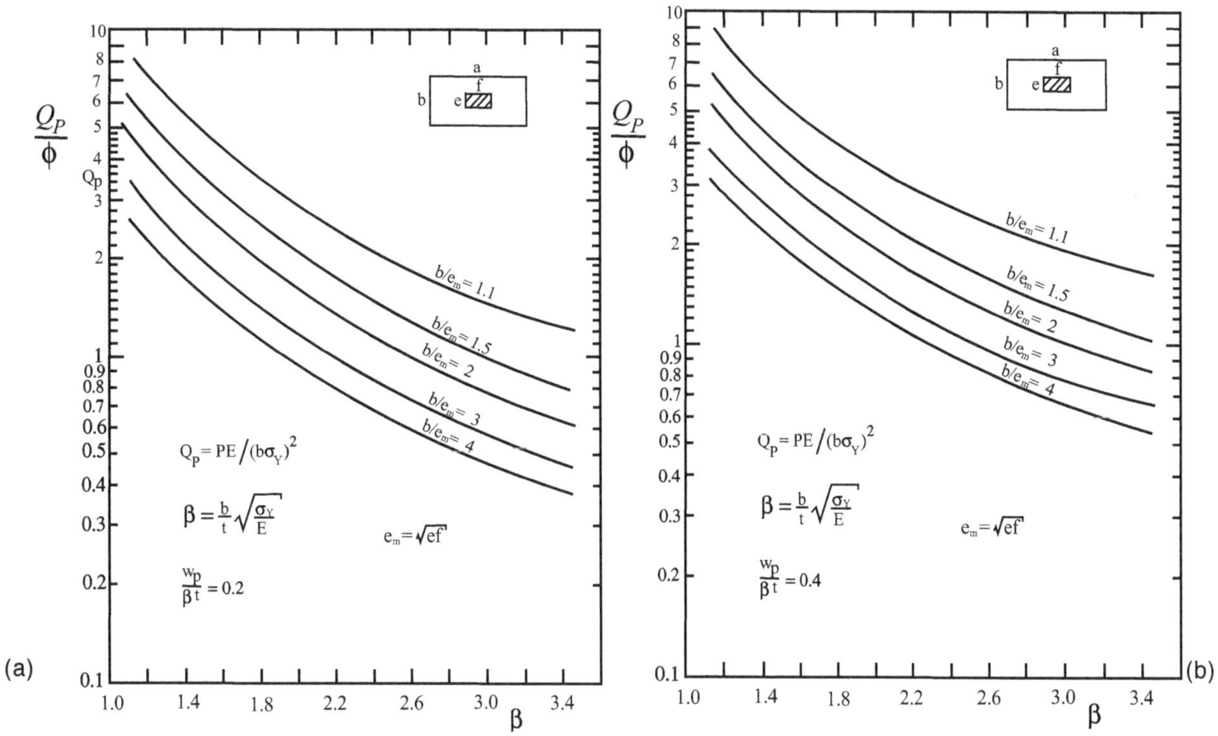

Figure 9.29 Design curves for single location concentrated loads. **(a)** Permanent set ratio = 0.2. **(b)** Permanent set ratio = 0.4. *(Continued on next page.)*

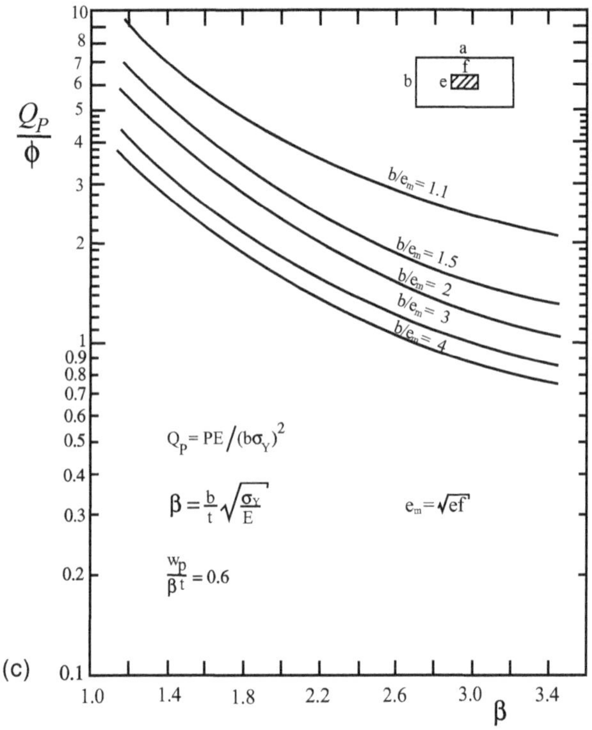

Figure 9.29 *Continued.* (**c**) Permanent set ratio = 0.6.

9.4.5 General Approach for Concentrated Loads

The previous sections have dealt with two types of concentrated loads—single location (SL) and multiple location (ML)—as if they were mutually exclusive. But the experiments that are the basis for the SL formula showed that for loads that are *very* concentrated ($\lambda < 0.3$) the mid-plate value of w_p can be taken as the maximum value; that is, moving the load around does not further increase w_p very much. *Hence for small λ the SL formula is better.*

In contrast, when the load footprint is relatively large—say $\lambda > 0.8$—the load is more like a pressure (even if it does not move) and the ML approach is better because it is based on an equivalent pressure. Also, for $\lambda > 0.8$ the SL formula is not valid. *Hence for large λ the ML formula is better, even if the load occurs only once.* Here the word "formula" means the combination of (9.4.12) and (9.4.14).

Therefore one possible approach is to use a "blend" of the two formulas, with each being given greater weight in the λ–region where it is better. A "weighting equation" that achieves this is

$$Q_P(\lambda) = (Q_P)_{SL}\,\text{sech}^2(8\lambda^3) + (Q_P)_{ML}\tanh^2(8\lambda^3) \tag{9.4.18}$$

This formula is plotted in Fig. 9.30.

9.5 LARGE DEFLECTION ANALYSIS

As mentioned earlier, there are some design applications that require the calculation of the large deflection response and the ultimate strength of laterally loaded plating. Elastic large deflection theory has only limited application for ship steels because loads severe enough to cause large deflections also cause yielding. Therefore, elastic large-deflection theory is not treated here. Aalami and Williams (1975) presents a summary of the theory and a series of nondimensional design curves for various boundary conditions.

Elasto-plastic large-deflection theory is quite complicated because there are three separate sources of nonlinearity:

1. Yielding.
2. Large deflections (compatibility; the membrane effect).
3. Restraint from edge pull-in, which appears and becomes significant as deflections become truly large.

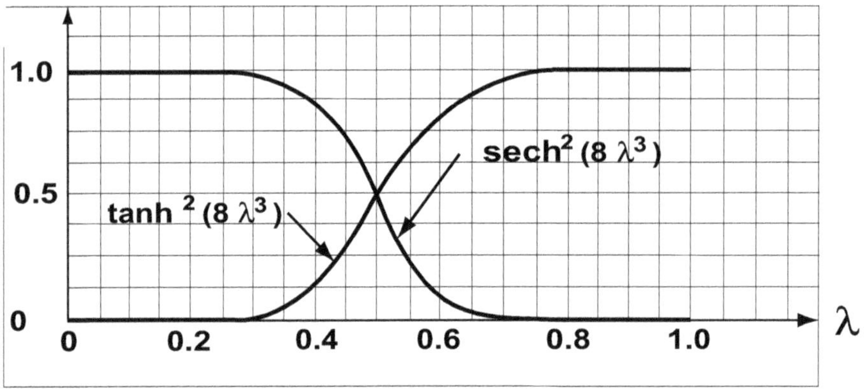

Figure 9.30 Weighting factors for equation (9.4.18) (Blending Algorithm).

Because of the complexity, there is no direct analytical method able to yield accurate results for realistic cases. The calculation generally requires computer-based numerical techniques that are too specialized to be dealt with here. Most general purpose finite element programs have elasto-plastic capability, and some also have large-deflection capability.

9.5.1 Rigid-Plastic Formulas for Static Pressure Loads

There is, however, one approach that yields simple and explicit load-deflection formulas, although these are only approximate. This is the rigid-plastic hinge-line method originally developed in civil engineering for the design of concrete slabs. The basic approach was presented by Wood (1961) and was extended to large deflections by Sawczuk (1964). The principal features of the method, as applied to rectangular plates, are as follows:

1. The edges are assumed to be completely restrained from pulling in, so that at large deflections membrane stresses dominate.
2. Elastic deflections are ignored, and the material is assumed to be rigid-perfectly plastic; that is, zero strain until $\sigma = \sigma_Y$, then unlimited strain with $\sigma = \sigma_Y$.
3. The plate is divided into four rigid regions separated by straight line hinges so as to form a kinematically admissable collapse mechanism, as shown in Fig. 9.31.
4. Yielding is assumed to be governed by the maximum normal stress yield criteria.

With these assumptions, the theory results in a pair of alternative equations, depending on whether the permanent set is less than or greater than the plate thickness.

$$\frac{p}{p_c} = \frac{Q}{Q_c} = 1 + \frac{1}{3}W^2\left[\frac{\zeta_0 + (3-2\zeta_0)^2}{3-\zeta_0}\right] \qquad W \le 1 \qquad (9.5.1)$$

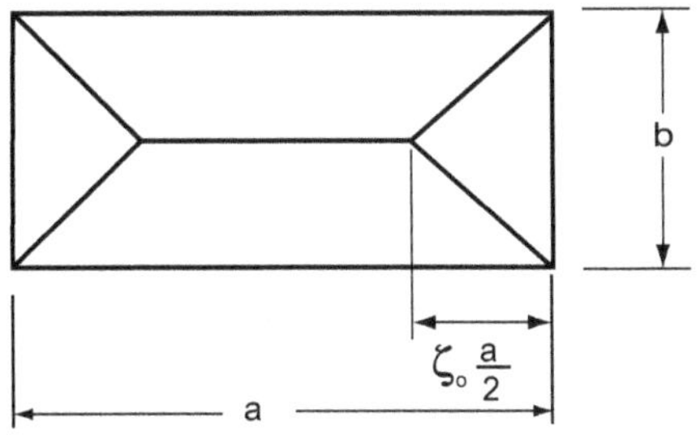

Figure 9.31

$$\frac{p}{p_c} = \frac{Q}{Q_c}2W\left[1 - \frac{\zeta_0(2-\zeta_0)}{3-\zeta_0}\left(1 - \frac{1}{3W^2}\right)\right] \qquad W > 1 \qquad (9.5.2)$$

where

$$\left.\begin{aligned} p_c &= \frac{48M_p}{b^2\left(\sqrt{3+\psi^2}-\psi\right)^2} \\ Q_c &= \frac{p_c E}{\sigma_Y^2} \end{aligned}\right\} \qquad (9.5.3)$$

$$\zeta_0 = \psi\left(\sqrt{3+\psi^2}-\psi\right)$$

$$\psi = \frac{b}{a} \qquad (9.5.4)$$

$$m_P = \frac{\sigma_Y}{\sqrt{1-\nu+\nu^2}}\frac{t^2}{4}$$

and $W = \dfrac{w_p}{t}$

As shown in Fig. 9.31, ζ_0 specifies the location of the interior hinge lines. Equations 9.5.1 and 9.5.2 can be used for either the pressure p or the load parameter $Q = pE/\sigma_Y^2$. Since rigid-plastic theory ignores elastic deflection, Young's modulus E plays no part and so pressure is more appropriate and more convenient. The load parameter is included in order to facilitate comparison with the results of earlier sections. Rigid-plastic theory also ignores the semi-elastic transition phase that occurs at the beginning of permanent set. Instead, the theory postulates a threshold pressure p_c (or a threshold load parameter Q_c) at which permanent set suddenly begins. The threshold pressure is assumed to be the pressure that would cause sufficient hinge lines so as to form a kinetically admissable collapse mechanism, assuming small deflections, as in Fig. 9.31. From the discussion in Section 9.3.2, it is clear that this assumption is too optimistic; that is, the threshold pressure of (9.5.3) is too large, as illustrated in Fig. 9.32. The order of the discrepancy can be ascertained by considering the case of a long plate; if ψ is set to zero in (9.5.3), p_c becomes $16m_P/b^2$, which is the pressure to cause three hinges, whereas in Section 9.3.2 it was shown that a better choice for a threshold load is the pressure to cause edge

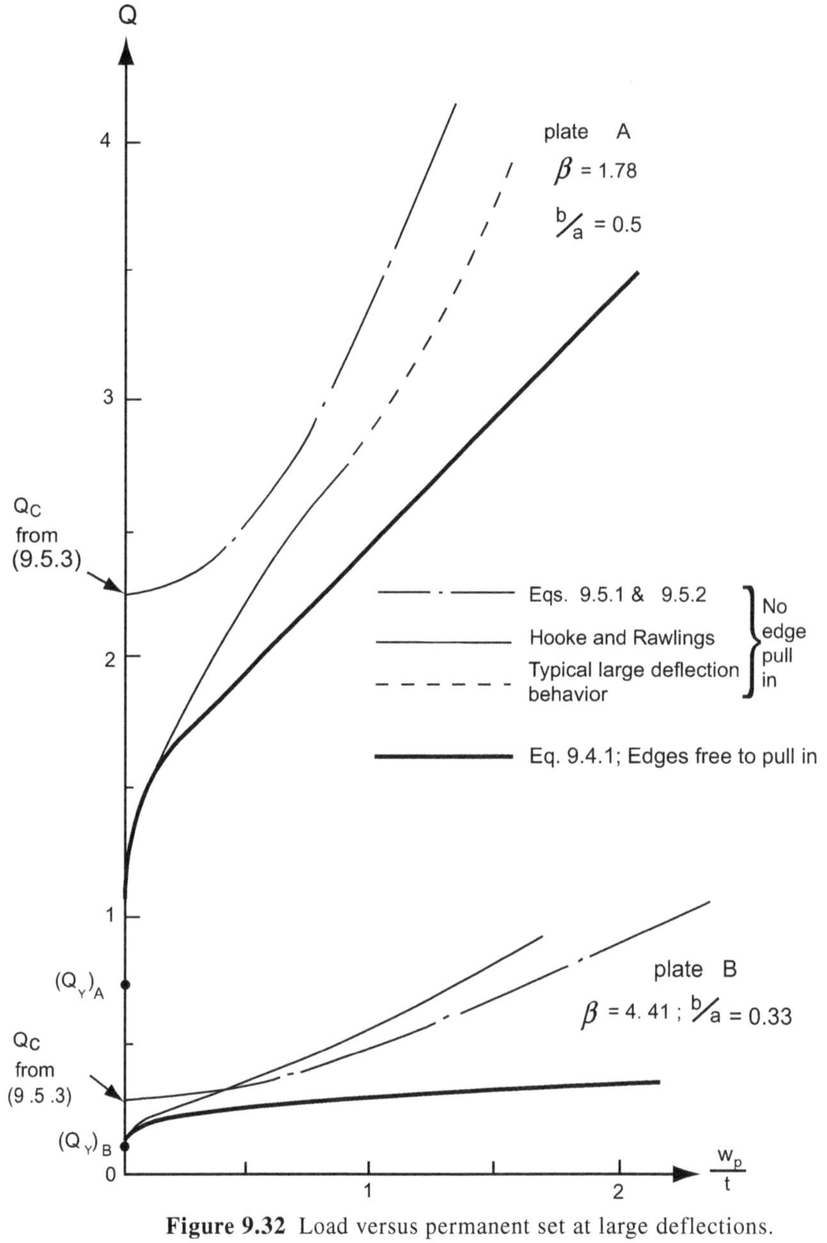

Figure 9.32 Load versus permanent set at large deflections.

hinges only: $12m_P/b^2$. Likewise, the rigid-plastic threshold load parameter, Q_c, is 4/3 of the true value, Q_{EH} (see Fig. 9.13).

The principal limitation of rigid-plastic theory is the assumption that membrane effects dominate. There are two ways in which this can occur; either because the plate edges are physically restrained from pulling in or because the deflections have become so large that in-plane restraint arises from the nonuniform distribution of in-plane strain due to the nonuniform lateral deflection, such that portions of plating with less deflection provide some restraint to portions with greater deflection. The amount of deflection, and hence of in-plane restraint, depends entirely on the plate slenderness β. For slender plates ($\beta > 2.4$) the lateral deflection grows quickly with load, and consequently some in-plane self restraint arises in the plating. Hence it can be expected that rigid-plastic theory will give better results for slender plating.

Hooke and Rawlings (1969) have presented experimental results for laterally loaded plating in which the plate edges were rigidly bolted to a nondeflecting frame. From these results Jones and Walters (1971) have shown that if there is such complete or near-complete restraint, and if the plating is truly slender ($\beta > 2.4$), then the rigid-plastic theory gives good results; for example, plate B in Fig. 9.32. However, the figure also shows that for the relatively sturdy plates which are typical of ship structures the agreement is not as good, especially for $w_p < t$. This important point was omitted in Jones and Walters (1971), in which a comparison was given only for the three most slender plates tested by Hooke and Rawlings (β = 2.76, 4.41, and 5.36) and not for the fourth (plate A, β = 1.78). It is only this fourth plate which has a slenderness that is typical of ship structures.

Moreover, it is important to note that these experimental results are for plates that were physically prevented from pulling in, and this is not a common condition.* In ships, as in any freestanding plated structure, such restraint does not occur (for static loading) until the deflections and the permanent set have become very large. Hence the rigid-plastic theory is not suitable for serviceability-based design (i.e., design based on maximum permanent w_p). In particular, (9.5.1), which is intended for small values of w_p ($w_p < t$), is too optimistic for ship plates of typical slenderness.

In contrast, if there is in-plane restraint and it is maintained throughout the load range, then the rigid-plastic theory gives excellent results for large deflections and large values of w_p. In the experiments of Hooke and Rawlings, one plate (Fig. 9.32*b*) was loaded to tensile facture. Figure 9.33 shows that, apart from some strain hardening near the ultimate load, the rigid-plastic prediction agrees very well with the experimental result.

We can now summarize, with the aid of Fig. 9.33, the respective areas of application, for static loading, of the semiempirical plate formulas of Section 9.4 and the rigid-plastic formulas. In the figure, the heavy dashed lines indicate qualitatively the behavior of typical ship plating. For values of permanent set within the usual design range (say $w_p/\beta t < 3$) the edges are essentially unrestrained from pulling in. Hence the semiempirical formulas are sufficiently accurate, whereas the rigid-plastic formulas are too optimistic and are not appropriate for this type of design.

*In fact, for static loading, complete restraint against edge pull-in is very difficult to achieve at any stage of the loading, even when a deliberate attempt is made to provide it, as in the previously mentioned experiments. In spite of the careful measures which they took to achieve complete restraint, Hooke and Rawlings found that for the sturdier plates ($\beta < 2$) significant edge pull-in occurred and this explains the decreased slope of their experimental curve compared to the rigid-plastic solution for the case of β = 1.78 in Fig. 9.32.

Moreover, even for somewhat larger values of allowable permanent set the empirical formulas have the advantage that their error lies on the conservative side rather than the optimistic side. On the other hand, as the edge slope, which is measured by w_p/b or $w_p/\beta t$, becomes large (say $w_p/\beta t > 3$) membrane restraint grows steadily and takes over more and more of the load, and the load-permanent-set curves begin to approach the "edge restrained" curves (see Fig. 9.33).

Eventually, if the material is sufficiently ductile, the plate becomes a fully plastic membrane and if there are no other sources of failure (stress concentration, welding flaws, etc.) it will finally fail when the membrane stress reaches the ultimate tensile stress. Therefore, rigid-plastic theory is applicable in the ultimate failure range. Specifically, (9.5.2) can be used for estimating the loads which correspond to very large deflections. Unfortunately this equation does not give the value of the ultimate failure load, but only the load-deflection relationship that occurs as the plate becomes a fully plastic membrane. This is because in a fully plastic membrane, as the load is increased the deflection of the membrane increases so as to carry this increased load, while the membrane stress itself remains constant and equal to the yield stress. However, it is possible to obtain an estimate for the ultimate load by first estimating the membrane strain ε_M from the maximum deflection ($\approx w_p$) as in Section 9.2.2, and then equating ε_M to the ultimate tensile strain.

9.5.2 Rapidly Varying Loads: Slamming, Collision

Ochi (1964) has shown that the maximum impact pressure p_s that acts on the bottom of a ship during a slam can be expressed in the form

$$p_s = kV_s^2$$

where k is a constant which must be determined experimentally for a given body section shape, and V_s is the velocity at the moment of impact. In order for slamming to occur, it is necessary for $V_s \geq V_T$ where V_T is the minimum or threshold relative velocity between a wave and a ship's bow necessary for a slam.

If τ, the duration of the pressure pulse, is small relative to T_p, the natural period of vibration of the plate, then the dynamic aspects (kinetic energy of the load and inertia of the plate) must be taken into account. From (9.1.25) the natural period of a clamped elastic square plate is

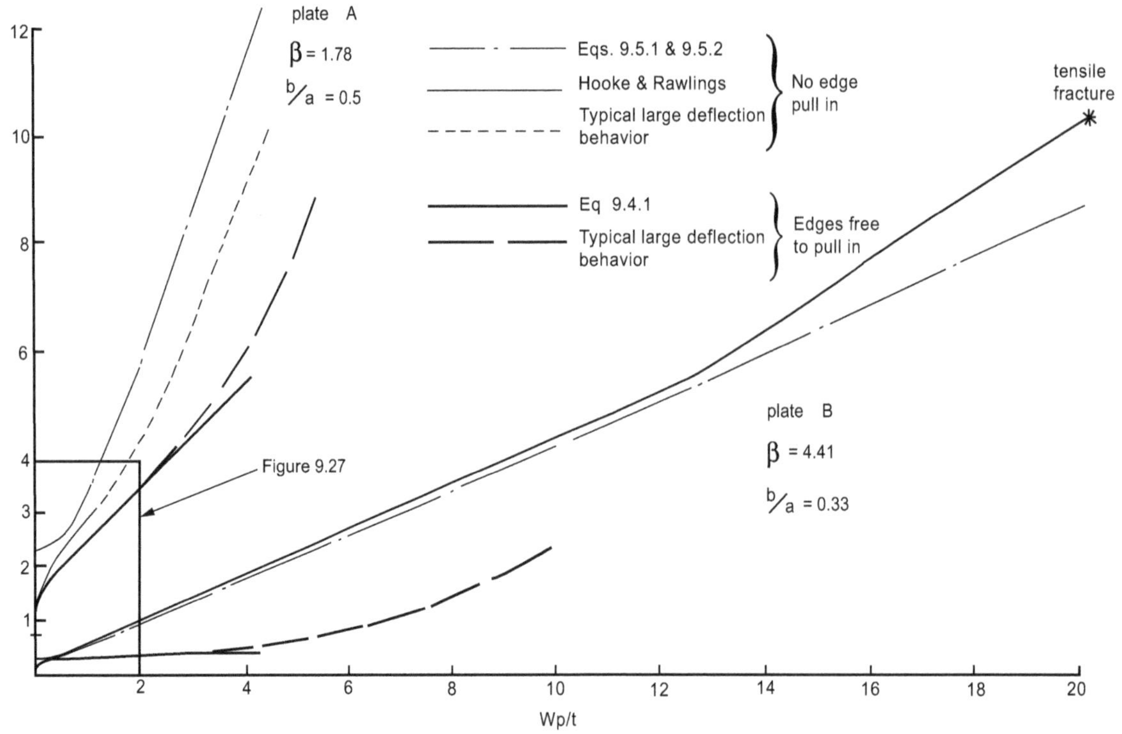

Figure 9.33 Load versus permanent set at large deflections.

$$T_p = \frac{\pi b^2}{9}\sqrt{\frac{3\rho\left(1-\nu^2\right)}{Et^2}}$$

Structural dynamics theory shows that the response is essentially static whenever τ is appreciably larger than T_p. For example, Biggs (1964) has shown that for a one-degree-of-freedom elastic spring-mass system subjected to an impulsive load, the maximum displacement calculated for the same load applied statically agrees with the corresponding maximum dynamic displacement to within approximately 17% when $\tau > 1.75\ T_p$. Analysis of full-scale measurements (Greenspon, 1956; Wheaton et al, 1970) and reduced-scale experiments (Clevenger & Melberg, 1963; Goodwin & Kime, 1963) indicates that slamming loads are of sufficient duration to be treated as static. The same also applies to ship collisions, but not, of course, to blast loads or any other truly high-speed loads. Moreover, treating a varying load as static is conservative because a pressure that is applied for only a brief interval will cause less permanent set than pressure applied over a longer period. At present, there does not seem to be any simple method for estimating or predicting slam duration, and therefore the safest course is to treat slamming as a static load. Jones (1973) has shown that (9.5.1) and (9.5.2) compare well with all four of the aforementioned experimental results. In fact, both equations give better results for moderately dynamic (quasistatic) loads than for loads which are truly static. The contrast is particularly marked for (9.5.1), which neglects the transition phase of static permanent set and which adopts p_c as the threshold pressure. The main reason for the improved accuracy is the fact that with a dynamic load there is more complete restraint against edge pull-in, as is assumed by these equations. The plate can deflect laterally much faster than it can deflect within its own plane, and the result is that the ratio of in-plane stiffness to lateral stiffness is much greater in the dynamic case than in the static case. In addition, there are two particular reasons for the marked improvement in the accuracy of (9.5.1). First, as shown by Symonds (1967), the difference between elastic-plastic and rigid-plastic response is small if τ / T_p is small and if the total dynamic energy is much larger than the maximum amount of elastic strain energy that the system can absorb. Second, the strengthening effect due to the inertia of the plating is greatest in the initial instants of the loading and this effectively raises the threshold load, so that p_c becomes a more accurate estimate of it. Thus, for example, Jones (1973) has shown that for the plates tested in Greenspon (1956) for which there was no permanent set at a peak slamming pressure of 295 psi, (9.5.3) predicts a threshold pressure of 346 psi, whereas in (9.4.1) the threshold load corresponds to $Q_Y + \Delta Q_0$ with T set equal to unity, and this gives a pressure of 245 psi. This is clearly too low since in reality the plating had not yet undergone any permanent set even at a pressure of 295 psi. Therefore, the rigid-plastic approach is appropriate for quasi-static loads such as slamming, and (9.5.1) and (9.5.2) may be used for the design of plating such that some acceptable amount of permanent set w_p would result from a slamming pressure p_s, which itself is chosen on the basis of having an acceptably small risk of occurrence. The latter may be estimated by the methods of Ochi and Motter (1973) and Stavovy and Chuang (1976). To facilitate the calculations (9.5.1) and (9.5.2) are presented here in inverted form:

$$\frac{w_p}{t} = \left[\frac{3(3-\zeta_0)}{\zeta_0 + (3-2\zeta_0)^2}\left(\frac{p_s}{p_c} - \right)\right]^{1/2}$$

$$\frac{w_p}{t} \le 1 \qquad (9.5.5)$$

$$\frac{w_p}{t} = \frac{\dfrac{p_s}{2p_c} + \left\{\left(\dfrac{ps}{2p_c}\right)^2 - \dfrac{4\zeta_0(2-\zeta_0)}{3(3-\zeta_0)}\left[1 - \dfrac{\zeta_0(\frac{2}{3}-\zeta_0)}{\frac{2}{3}-\zeta_0}\right]\right\}^{1/2}}{2\left[1 - \dfrac{\zeta_0(2-\zeta_0)}{3-\zeta_0}\right]} \qquad (9.5.6)$$

when $\dfrac{w_p}{t} > 1$

in which the threshold pressure p_c is given by (9.5.3). If $p_s < p_c$ there is no permanent set and hence, if there is no particular need to keep shell plating thickness to a minimum, a less sophisticated approach may be used in which the design requirement is simply that $p_c \ge p_s$. Equations 9.5.3 and 9.3.7 then give the following explicit equation for the plate thickness, assuming $\nu = 0.3$:

$$t = 0.272b\left(\sqrt{3 + \left(\frac{b}{a}\right)^2} - \frac{b}{a}\right)\sqrt{\frac{p_s}{\sigma_Y}} \qquad (9.5.7)$$

9.5.3 Dynamic Loading and Response

The most useful application of the rigid-plastic approach is in estimating the response of plates to dynamic loads. A principal contributor in this area is Jones, who generalized and extended the hinge line approach to deal with dynamic plastic response (Jones, 1971). This approach takes into consideration the kinetic energy of the load and the inertia of the plate. When inertia terms are retained in the basic equations it is found that a structure can support a load which is larger than the static limit load, provided that it is removed after a small interval of time. In this circumstance the motion of a structure eventually ceases and reaches a permanent deformed state after expending all the external kinetic energy. It was shown in Jones (1971) that the maximum permanent set w_p of a rigid-plastic fully clamped rectangular plate, which is subjected to a uniformly distributed pressure of constant magnitude p_0 [$p_0 \ge p_c$, where p_c is defined by (9.5.3)] for a duration τ is

$$\frac{w_p}{t} = (3-\zeta_0)\left[\frac{\{1 + 2\eta(\eta-1)[1-\cos(\gamma\tau)]\}^{1/2} - 1}{2\{1 + (2-\zeta_0)(1-\zeta_0)\}}\right] \qquad (9.5.8)$$

where $\gamma^2 = \dfrac{96M_p}{\mu t b^2(3-2\zeta_0)}\left(1 - \zeta_0 + \dfrac{1}{2-\zeta_0}\right)$

$$\eta = \frac{p_0}{p_c}$$

and $\mu = \rho\tau$ = mass per unit area (ρ = density)

A limitation of (9.5.8) and (9.5.9) is that they do not allow for the strain-rate sensitivity of steel, which has the effect of reducing the permanent set.

If $p_0 = p_c$, (9.5.8) gives $w_p/t = 0$ because the theory assumes that either the pressure is sufficiently large or the duration is sufficiently small so that inertia plays a part; if the pressure is merely p_c then the theory implies a very small duration and hence the equation states that no permanent set would occur.

IMPULSE LOAD

If a rectangular external pressure-time history has a large magnitude ($p_0 \gg p_c$) and a very short duration τ, then the dynamic loading can be idealized as impulsive. The impulse is the product $p_0\tau$ and is equal to the momentum μV_0, where V_0 is the initial impulsive velocity of the plate. Equation 9.5.8 becomes

$$\frac{w_p}{t} = \frac{(3-\zeta_0)\left[(1+\Gamma)^{1/2}-1\right]}{2\left[1+(\zeta_0-1)(\zeta_0-2)\right]} \qquad (9.5.9)$$

where

$$\Gamma = \frac{\Omega}{6}\left(\frac{b}{a}\right)^2 (3-2\zeta_0)\left(1-\zeta_0+\frac{1}{2-\zeta_0}\right)$$

and

$$\Omega = \frac{\mu V_0^2 a^2}{4 m_P t} = \frac{\rho V_0^2 a^2}{\sigma_Y t^2}$$

DYNAMIC CONCENTRATED LOADS

Since the rigid-plastic approach has been found to be well suited for dynamic uniform pressure loads, it is probably also suitable for dynamic concentrated loads. However it must be borne in mind that this approach examines only permanent set and does not investigate or even allow for the possibility of puncture or tensile fracture, both of which are strong possibilities with loads of this type.

Multiple Location Loads. Here the procedure is the same as for nondynamic multiple location loads except that (9.5.8) is used in place of (9.4.1).

Single Location (Central) Loads. The following relationship between load and permanent set was derived by (Kling, 1980) using rigid-plastic theory and assuming the plate edges to be simply supported.

$$Q_P = Q_{P,T}\left[1+\left(\frac{w_p}{t}\right)^2 \frac{1+2\xi}{3}\right] \qquad \text{for } W < 1$$

$$Q_P = \frac{Q_{P,T}}{2}\left[\frac{3(W-1)3+1}{W}(1-\xi)+1+\frac{W^2}{3}(2+7\xi)\right] \qquad \text{for } W > 1$$

$$(9.5.10)$$

where $Q_{P,T}$ is given by (9.4.16) with $w_p = 0$ (i.e., only the first term) and

$$\xi = \frac{1}{\alpha}\left[\frac{e(b-e)+f(a-f)}{b(b-e)+a(a-f)}\right]$$

and $\quad W = w_p / t$

The load dimensions e and f are shown in Fig. 9.22.

REFERENCES

Aalami, B., and Williams, D. G. (1975). *Elastic design of thin plates, Vol. 1, Plates under transverse load.* London: Crosby Lockwood.

Antoniou, A. C. (1980). On the maximum deflection of plating in newly built ships. *J. Ship Res.*, Vol. 24, Issue 1, 31–39.

Biggs, J. M. (1964). *Introduction to structural dynamics.* New York: McGraw-Hill.

Clarkson, J. (1958). Strength of approximately flat long rectangular plates. *Trans. North East Coast Inst. Eng. Shipbuild.*, Vol. 74, 21–40.

Clarkson, J. (1962). Uniform pressure tests of plates with edges free to slide inwards. *Trans. RINA*, Vol. 104, 67–76.

Clarkson, J. (1963). Tests of flat plated grillages under uniform pressure. *Trans. RINA*, Vol. 105, 467–484.

Clevenger, R. L., and Melberg, L. C. (1963). *Slamming of a ship structural model.* Massachusetts Institute of Technology Engineer's Thesis, May.

Goodwin, J. J., and Kime, J. W. (1963). *DTMB Technical Note SML*, July, 760–761.

Greenspon, J. E. (1956). Sea tests of the U.S.C.G.C. Unimak, Part 3. *DTMB Report 978*, March.

Hooke, R., and Rawlings, B. (1969). An experimental investigation of the behavior of clamped rectangular mild steel plates subjected to uniform transverse pressure. *Proc. Inst. Civil Eng.*, Vol. 42, 75–103.

Hughes, O. (1983). Design of plating under concentrated lateral loads. *J. Ship Res.*, Vol. 27, Issue 4, 252–264.

Jackson, R. L., and Frieze, P. A. (1980). Design of deck structures under wheel loads. *Trans. RINA*, Vol. 122.

Jaeger, L. G. (1964). *Elementary theory of elastic plates.* Oxford: Pergamon Press.

Jones, N. (1971). A theoretical study of the dynamic plastic behavior of beams and plates with finite deflections. *Int'l J. Solids Struct.*, Vol. 7, 1007–1029.

Jones, N. (1973). Slamming damage. *J. Ship Res.*, Vol. 17, Issue 2, 80–86.

Jones, N., and Walters, R. M. (1971). Large deflections of rectangular plates. *J. Ship Res.*, Vol. 15, Issue 2, 164–171.

Kling, M. (1980). Large deflections of rectangular plates subject to concentrated loads. Master's Thesis, Dept. of Ocean Engineering, Massachusetts Institute of Technology, Cambridge, MA.

Konieczny, L., and Bogdaniuk, M. (1999). Design of transversely loaded plating based on allowable permanent set. *Marine Structures*, Vol. 12, 497–519.

Ochi, M. K. (1964). Prediction of occurrence and severity of ship slamming at sea. *5th Symp. on Naval Hydrodynamics*, Office of Naval Research, ACR 112.

Ochi, M. K., and Motter, L. E. (1973). Prediction of slamming characteristics and hull responses for ship design. *Trans. SNAME*, Vol. 81, 144–176.

Sandvik, P. C. (1974). *Deck plates subject to large wheel loads.* Report SK/M. Trondheim, Norway: Univ. of Trondheim.

Sawczuk, A. (1964). On initiation of the membrane action in rigid-plastic plates. *J. Méchanique*, Vol. 3, Issue 1, 15–23.

Stavovy, A., and Chuang, S. L. (1976). Analytical determination of slamming pressures for high speed vehicles in waves. *J. Ship Res.*, Vol. 20, Issue 4, 190–198.

Symonds, P. S. (1967). *Survey of methods of analysis for plastic deformation of structures under dynamic loads*. Report NSRDC/1–67. Providence, RI: Brown University.

Timoshenko, S., and Woinowsky-Krieger, S. (1959). *Theory of plates and shells*, (2nd ed.). New York: McGraw-Hill.

Wheaton, J. W., Kano, C. H., Diarnant, P. T., and Bailer, F. C. (1970). Analysis of slamming data from the S.S. *Wolverine State*. Report SSC-210. Washington, DC: Ship Structure Committee.

Wood, R. H. (19610. *Plastic and elastic design of slabs and plates*. New York: Ronald Press.

DEFORMATION AND STRENGTH CRITERIA FOR STIFFENED PANELS UNDER IMPACT PRESSURE

Jeom Kee Paik
Professor, Pusan National University
Busan, Korea

Owen Hughes
Professor, Virginia Tech
Blacksburg, VA, USA

[Part of this chapter was extracted from J.K. Paik and Y.S. Shin, Structural damage and strength criteria for ship stiffened panels under impact pressure actions arising from sloshing, slamming and green water loading, *Ships and Offshore Structures*, Vol. 1, No. 3, 2006, pp. 249–256. Used with permission from Taylor & Francis.]

10.1 CAUSES OF IMPACT PRESSURE LOADS

The hull structures of ships and ship-shaped offshore installations are likely to be subjected to dynamic (impact) pressure loads that arise from sloshing, slamming, and green seas while in service (Paik and Thayamballi, 2007). The accelerations that arise from a vessel's motions in a seaway produce sloshing loads (i.e., fluctuating pressures on the internal faces of partially filled liquid cargo tanks). The motions of such liquid cargo vessels as oil tankers often produce severe sloshing loads. The tanks of moored, ship-shaped, offshore installations such as FPSOs (floating production, storage, and off-loading units) are continuously loaded and unloaded, and thus sloshing within them is unavoidable. Recently, there has been a trend to reduce the number of tanks and make them larger and wider, which gives them longer natural periods of sloshing. This trend means that the fluid motions in liquid cargo tanks are more sensitive to ocean wave excitations.

The bottom, bow flare and/or overhanging stern regions of vessels are often subjected to the dynamic pressures that arise from slamming. Bottom slamming occurs when a vessel's bottom emerges from the water due to pitching, possibly combined with the occurrence of a wave trough. Bow flare slamming occurs due to the plunging of the upper flared portion of the bow deeper into the water. Large bulk carriers and oil tankers are often exposed to severe bottom slamming, and large container vessels sometimes face severe slamming of a flared bow and an overhanging stern. Deck structures are also subjected to impact pressure loads caused by "green seas on deck" in severe sea states and weather conditions. In particular, the decks of large bulk carriers and moored FPSOs are often exposed to severe green seas.

Under current Classification Society rules, the ship structural design criteria for dynamic pressures are typically based on an equivalent quasi-static pressure, rather than the actual dynamic pressure. This simplification can lead to an overestimation or underestimation of the deformation and strength of stiffened panels.

This chapter presents a refined method for analyzing the deformation and strength of stiffened panels under the impact pressures that arise from sloshing, slamming, or green seas. Closed-form methods for predicting the permanent deflections of the plate panels, which can be a basis of serviceability limit state design (Paik and Thayamballi, 2007), are developed by examining the dynamic pressure characteristics. These methods are verified by comparing them with nonlinear finite element method simulations and experimental results.

10.2 IDEALIZATION OF IMPACT PRESSURE PROFILE AND STRENGTH CRITERIA

For practical design purposes, the behavior of panels under impact pressure can be grouped into three domains, depending on the ratio of the duration of the impact pressure to the natural period of the structure, as follows (NORSOK, 1999).

- Quasi-static domain when $3 \le \tau/T$
- Dynamic domain when $0.3 \le \tau/T < 3$
- Impulsive domain when $\tau/T < 0.3$

In the above, τ is the duration of the impact pressure, and T is the natural period of the structure.

The dynamic pressures that are attributable to sloshing, slamming, and green seas have a very high peak value that lasts for a very short period of time, and they can be characterized by four parameters: (a) the rise time until peak pressure, (b) peak pressure, (c) pressure decay beyond peak pressure, and (d) pressure duration, as illustrated in Fig. 10.1.

When the rise time and duration of the impact pressure are very short, the impact pressure can be approximated as a constant equivalent pressure, p_e, acting over a duration τ, as shown in Fig. 10.1. This approximation is based on equal impulse, that is, the product $p_e\tau$ is required to be equal to the area under the curve of pressure-time history (Paik et al., 2004).

$$I = \int p(\tau)dt = p_e\tau \qquad (10.2.1)$$

where I = impulse of the impact pressure, t = time, p_e = equivalent (design) peak pressure, and τ = duration of p_e.

Taking p_e as of the same order as p_o would imply an instantaneous impact pressure (zero rise time) and would result in too large structural damage evaluation. Therefore p_e is often defined by multiplying p_o with an appropriate knockdown factor. Once impulse, I, and the equivalent peak pressure value, p_e, are defined, duration (τ) can be determined from (10.2.1). Sections 4.6.1 and 4.6.2 give some further information about dynamic pressures due to slamming and sloshing.

In predicting the structural damage (permanent deflection) due to impact pressure, p_e and τ are dealt with as parameters of influence in the design formulations of permanent deflection w_p. For protection against impact pressures, the permanent deflection of the plate panels must be smaller than a prescribed allowable deflection value, namely,

$$w_p \le w_{pa} \qquad (10.2.2)$$

where w_p = permanent deflection, and w_{pa} = allowable value of permanent deflection, which may be taken as a few times the plate thickness.

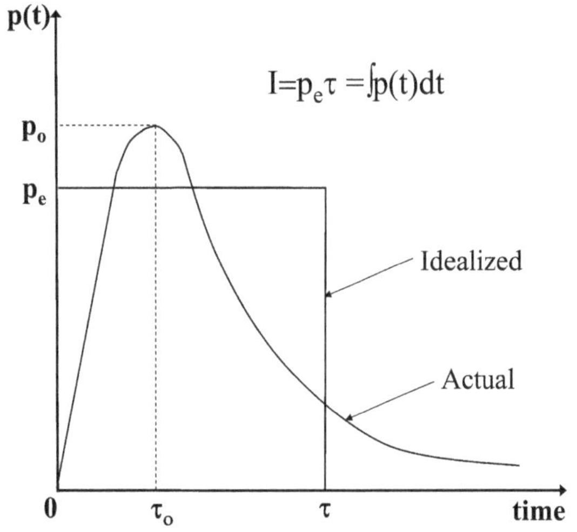

Figure 10.1 An actual profile, and its idealization, of impact pressure in terms of pressure pulse versus time history (Paik and Thayamballi, 2007).

10.3 DESIGN FORMULATIONS FOR PERMANENT PANEL DEFLECTION

The structural damage of stiffened panels under impact pressures is evaluated in terms of the permanent plastic deflection of plate panels at three levels: the plate level between support members; the plate-stiffener combination level representing the support members and the attached plate; and the grillage level as an entire cross-stiffened panel. See Fig. 10.2 (Paik and Thayamballi, 2003). Existing closed-form formulations in the literature, which are derived under a quasi-static pressure loading condition, are employed and expanded to take into account the effects of the strain rate in association with the impact loads.

10.3.1 Plates Between Support Members

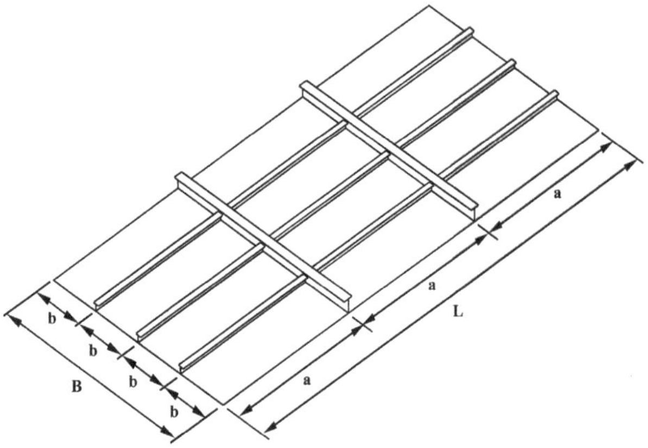

Figure 10.2 Nomenclature: a stiffened panel.

Jones (1997) proposes an approach to computing the permanent deflection of beams and rectangular plates loaded by a pressure pulse while taking into account the large deflection effect. According to the Jones approach, if bending moments and membrane forces are developed within the plate as a result of the axial restraint of the supports, then lateral deflection w obeys the following equation in association with a plastic collapse mechanism.

$$\int_A (p - \mu\ddot{w})\dot{w}dA = \sum_{m=1}^{r} \int_{\ell_m} (M + Nw)\dot{\theta}_m d\ell_m \quad (10.3.1)$$

where p = lateral pressure; w = permanent plastic deflection of plates; μ = mass of the plate per unit area; r = number of hinge lines; ℓ_m = length of the hinge line; θ_m = relative angular rotation across the hinge line; N and M = membrane forces and bending moments acting along the hinge lines, respectively; $\dot{w}$= velocity profile; $\ddot{w}$= acceleration profile; $\dot{\theta}_m$ = rotation rate at the mth discrete location (hinge); $\int_A dA$ = area integration; $\int_{\ell_m} d\ell_m$ = length integration along the hinge line ℓ_m.

Under dynamic pressure loads, plates are likely to deform between the support members, which are generally designed to provide enough support to the plating and not to fail before the plating fails. Since all the individual plates in a continuous stiffened plate structure deflect in the same direction as the pressure, there is almost no rotation of the plating along the support members, so each plate is assumed to be clamped along its four edges. It is also assumed that the material obeys the Tresca-type yield criterion and that shear forces do not affect yielding.

Figure 10.3 shows the pattern of plastic hinge lines that is assumed to constitute the collapse mechanism of the plate. The material is assumed to be rigid—perfectly plastic, and the loaded plate is divided into a number of rigid sections separated by straight line hinges, as depicted in Fig. 10.3. In this case, the bound solutions of the maximum permanent plastic deflection w_p at the center of the plate under dynamic pressures without considering the effect of strain-rate sensitivity of material are given by Chen (1993), as follows.

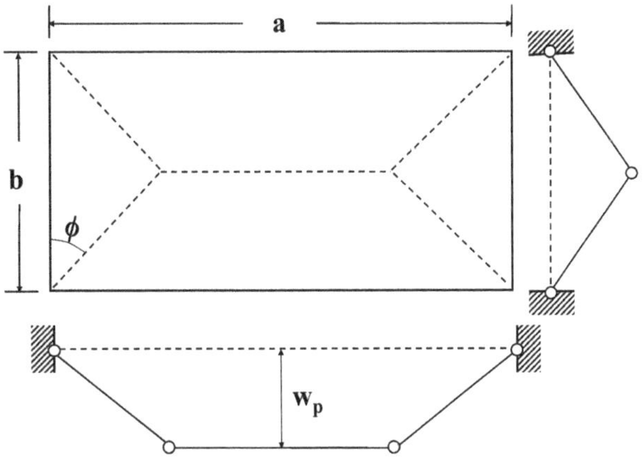

Figure 10.3 Presumed collapse mode of the plate clamped at all (four) edges.

$$\frac{w_p}{t} = \sqrt{2\frac{\alpha}{A_2}\lambda + \left(\frac{A_1}{A_2}\right)^2 - \frac{A_0}{A_2}} - \frac{A_1}{A_2}$$

for the lower bound solution (10.3.2a)

$$\frac{w_p}{t} = \sqrt{2\sqrt{2}\frac{\alpha}{A_2}\lambda + \left(\frac{A_1}{A_2}\right)^2 - \frac{A_0}{A_2}} - \frac{A_1}{A_2}$$

for the upper bound solution (10.3.2b)

where a = plate length, b = plate breadth, t = plate thickness,

$$\lambda = \frac{\mu V_0^2 b^2}{4M_p t}, \; V_0 = \frac{p_e \tau}{\mu}, \; M_p = \frac{\sigma_Y t^2}{4}, \; \mu = \rho t,$$

$$\tan\phi = \sqrt{3 + \alpha^2} - \alpha, \; \beta = \frac{a}{b}, \; \alpha = \frac{1}{\beta} = \frac{b}{a},$$

$$A_0 = \frac{3}{2\sin\phi\cos\phi} + \frac{1}{\alpha} - \tan\phi,$$

$$A_1 = \frac{1}{3\sin\phi\cos\phi} + \frac{2}{\tan\phi} + \frac{2}{\alpha},$$

$$A_2 = 4\left(\frac{1}{\sin\phi\cos\phi} + \frac{1}{\tan\phi} + \frac{4}{\alpha} - 3\tan\phi\right),$$

σ_Y = material yield stress under static load (without allowance for strain rate), p_e = design peak pressure, and ρ = density, which is 7850 N sec^2/m^4 for steel and 2699 N sec^2/m^4 for aluminum alloy.

For a square plate with $\beta = 1$, $A_0 = 3$, $A_1 = \frac{14}{3}$, and $A_2 = 16$, and the bound solutions of w_p/t are therefore obtained as follows.

$$\frac{w_p}{t} = \sqrt{0.1250\lambda - 0.1024} - 0.2917$$

for the lower bound solution (10.3.3a)

$$\frac{w_p}{t} = \sqrt{0.1768\lambda - 0.1024} - 0.2917$$

for the upper bound solution (10.3.3b)

Jones (1971) and Jones and Baeder (1972) obtained experimental results of the central permanent deflection for square plates of aluminum 6061-T6 alloy and mild steel. Figure 10.4a compares the theoretical solutions (10.3.3) and the experimental results for square plates with $\beta = 1$. Jones et al. (1970) also obtained the experimental results of the maximum permanent deflection for rectangular plates with $\beta = 1.686$ subjected to dynamic pressure. In this case, the bound solutions of (10.3.2) are given as follows.

$$\frac{w_p}{t} = \sqrt{0.0504\lambda - 0.0914} - 0.2409$$

for the lower bound solution (10.3.4a)

$$\frac{w_p}{t} = \sqrt{0.0713\lambda - 0.0914} - 0.2409$$

for the upper bound solution (10.3.4b)

Figure 10.4b compares the theoretical solutions and the experimental results for rectangular plates with $\beta = 1.686$ subjected to dynamic pressure. It is observed from Figs. 10.4a and 10.4b that the lower bound solution is in better agreement with the experimental results than is the upper bound solution.

However, for impact or high-speed loading, the effect of the strain-rate sensitivity of material cannot be neglected. To take into account the strain-rate effect, the static yield stress σ_Y in (10.3.2) should be replaced by the dynamic yield stress σ_{Yd}, which is given as

$$\sigma_{Yd} = \left\{1 + \left(\frac{\dot{\varepsilon}}{C}\right)^{\frac{1}{q}}\right\}\sigma_Y \quad (10.3.5)$$

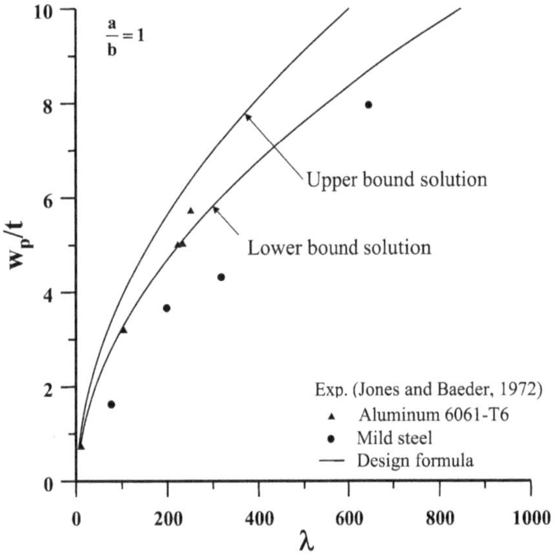

Figure 10.4a The maximum permanent deflection of square plates (with $\beta = 1$) subjected to dynamic pressure.

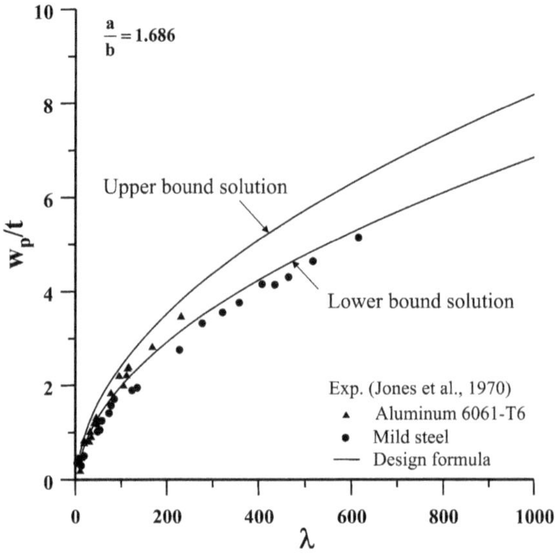

Figure 10.4b The maximum permanent deflection of rectangular plates with ($\beta = 1.686$) subjected to dynamic pressure.

where $\dot{\varepsilon} = \dfrac{V_0}{2w_p}$ = strain rate. C and q are coefficients of the Cowper-Symonds equation (Cowper and Symonds, 1957), which are given by $C = 40.4$/sec and $q = 5$ for mild steel, $C = 3200$/sec and $q = 5$ for high-tensile steel, and $C = 6500$/sec and $q = 4$ for aluminum alloys (Paik and Thayamballi, 2003, 2007).

When the strain-rate effect is accounted for, the permanent deflection w_p/t becomes a function of the strain rate $\dot{\varepsilon}$, which itself is a function of w_p/t. In this case, equation (10.3.2) becomes a nonlinear function with regard to the permanent deflection, as follows.

$$\frac{w_p}{t} = f(\lambda, \dot{\varepsilon}, \cdots) = f\left(\lambda, \frac{w_p}{t}, \cdots\right) \quad (10.3.6)$$

where $\lambda = \dfrac{p_e^2 \tau^2 b^2}{\sigma_{Yd} \rho t^4}$.

The permanent plastic deflection w_p under impact pressure should not be greater than such deflection when the pressure duration is equal to the natural period of the panel (Paik et al., 2004). That is,

$$w_p \le w_p^* \quad (10.3.7a)$$

where $w_p^* = w_p$ at $\tau = T$ with T = natural period.

The natural period of steel plates is approximately calculated as follows (Korean Register, 1997).

$$T = \frac{1}{f_n} \quad (10.3.7b)$$

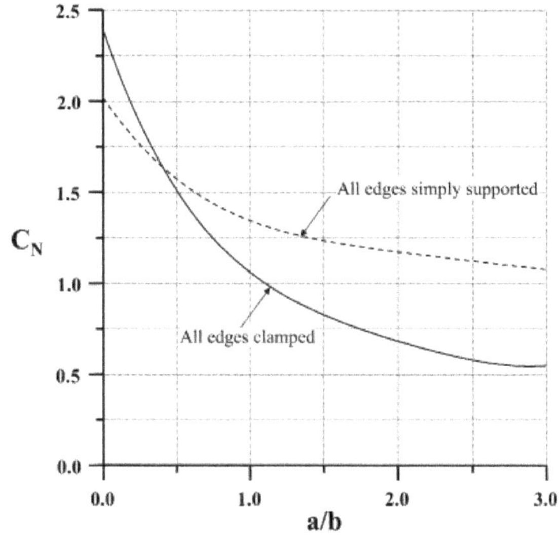

Figure 10.5 Coefficient C_N for determining the natural period of a rectangular plate (Korean Register, 1997).

where $f_n = \dfrac{\lambda_n}{2\pi b^2}\sqrt{\dfrac{D}{\rho t}}$ for square plates, with λ_n = 19.74 for simply supported edges and λ_n = 35.98 for clamped edges; $f_n = \dfrac{C_N \pi}{2b^2}\sqrt{\dfrac{D}{\rho t}}$ for rectangular plates with C_N, as defined in Fig. 10.5, $D = \dfrac{Et^3}{12(1 - \nu^2)}$, E = elastic modulus, ν = Poisson's ratio, and a, b, t, and ρ are as defined in (10.3.2). The constant C_N can be given as a continuous function as follows.

Simply supported plate edges

$$C_N = \begin{cases} 0.021\left(\dfrac{a}{b}\right)^4 - 0.196\left(\dfrac{a}{b}\right)^3 + 0.691\left(\dfrac{a}{b}\right)^2 - 1.190\left(\dfrac{a}{b}\right) \\ + 2.019 \text{ for } 0 < \dfrac{a}{b} \le 3 \\ 1.077 \text{ for } \dfrac{a}{b} > 3 \end{cases} \qquad (10.3.7c)$$

Clamped plate edges

$$C_N = \begin{cases} 0.054\left(\dfrac{a}{b}\right)^4 - 0.443\left(\dfrac{a}{b}\right)^3 + 1.430\left(\dfrac{a}{b}\right)^2 - 2.376\left(\dfrac{a}{b}\right) \\ + 2.395 \text{ for } 0 < \dfrac{a}{b} \le 3 \\ 0.55 \text{ for } \dfrac{a}{b} > 3 \end{cases} \qquad (10.3.7d)$$

10.3.2 Plate-stiffener Combinations

When the pressure pulse is relatively small, the transverse frames in a grillage (i.e., a stiffened panel with both longitudinal and transverse support members) may not fail until the uniaxially stiffened panel between the transverse frames fails. In this case, the uniaxially stiffened panel between two adjacent transverse frames may be modeled as a plate-stiffener combination clamped at both ends, as shown in Figs. 10.6 and 10.7.

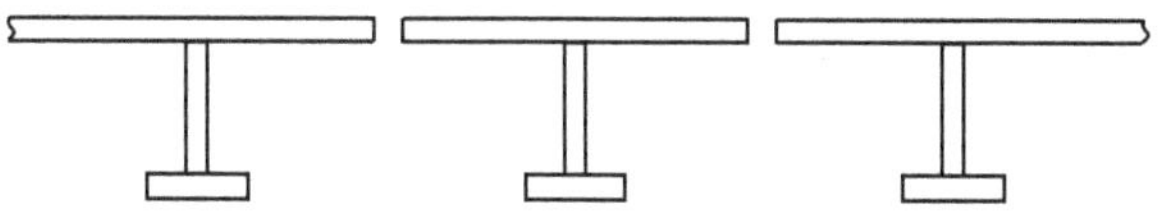

Figure 10.6 An idealization of a stiffened panel as plate-stiffener combinations between the transverse support members.

Jones (1997) derived the following closed-form expression for the permanent plastic deflection of a beam under impact pressure when the strain-rate effect is not accounted for (symbols that are not defined below are defined in (10.3.2) and Fig. 10.7).

$$\frac{w_p}{t_{eq}} = \frac{1}{2}\left[\left(1 + \frac{3\lambda}{4}\right)^{1/2} - 1\right] \qquad (10.3.8)$$

where $\lambda = \dfrac{\mu V_0^2 a^2}{16 M_p t_{eq}}$, $M_p = \dfrac{\sigma_o b t_{eq}^2}{4}$, σ_o = static flow stress, taking account of the strain-hardening effect, which may be taken as $\sigma_o = \dfrac{\sigma_Y + \sigma_T}{2}$, σ_T = ultimate tensile stress of the material under static load, and t_{eq} = equivalent thickness, which is given by $t_{eq} = \dfrac{bt + h_w t_w + b_f t_f}{b}$.

When the strain-rate effect is accounted for, (10.3.8) becomes a nonlinear function of the strain-rate effect because the static flow stress σ_o in (10.3.8) must be replaced by the dynamic flow stress σ_{od}, which is given by

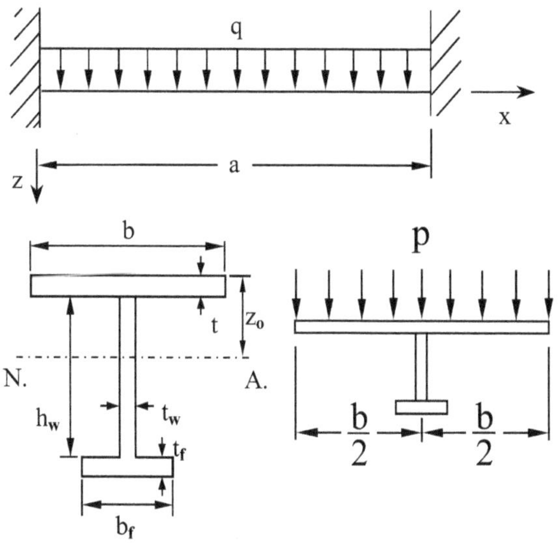

Figure 10.7 A plate-stiffener combination model clamped at both ends, representing a uniaxially stiffened panel between transverse frames and under impact lateral line load, $q = pb$.

$$\sigma_{od} = \left\{1 + \left(\frac{\dot{\varepsilon}}{C}\right)^{\frac{1}{q}}\right\}\sigma_o \qquad (10.3.9)$$

where C and q are as defined in (10.3.5).

10.3.3 Grillages

When the pressure pulse is very large and/or the transverse frames are relatively weak, these frames may fail together with both the longitudinal stiffeners and the plating. In this case, the cross-stiffened panel or grillage shown in Fig. 10.2 may be idealized as an orthotropic plate.

The permanent plastic deflection of the panel may be approximately calculated from the design formulae of the plating, i.e., equation (10.3.2) or (10.3.6), but with the equivalent plate thickness of an orthotropic plate, which is given by

$$t_{eq} = \frac{V}{LB} \qquad (10.3.10)$$

where V = total volume of the grillage, L = length of grillage, B = breadth of grillage. The related parameters, including impact energy parameter λ, are then rewritten as follows.

Without strain-rate effect

$$\lambda = \frac{\mu V_0^2 B^2}{4M_p t_{eq}},\ \alpha = \frac{B}{L} \qquad (10.3.11a)$$

With strain-rate effect

$$\lambda = \frac{p_e^2 \tau^2 B^2}{\sigma_{Yd} \rho t_{eq}^4},\ \sigma_{Yd} = \left\{1 + \left(\frac{\dot{\varepsilon}}{C}\right)^{\frac{1}{q}}\right\}\sigma_Y,\ \dot{\varepsilon} = \frac{V_0}{2w_p} \qquad (10.3.11b)$$

In both cases V_o is as defined in (10.3.2).

Note that the permanent plastic deflection, w_p, under impact pressure should not be greater than such deflection when the pressure duration is equal to the natural period of the panel, as indicated in (10.3.7a).

10.4 APPLICATION EXAMPLES

The impact pressures that arise from sloshing, slamming, and green seas can cause severe structural damage, so the avoidance of such damage is one of the most important tasks undertaken by structural designers. The foregoing theory with the lower bound solutions taking into account the effect of strain-rate sensitivity of material has been implemented in the ALPS/ULSAP program (ALPS/ULSAP, 2009). The nonlinear function of the permanent plastic deflection was solved using the bisection method, and the number of iterations required was less than 15.

This section demonstrates application examples for predicting the permanent plastic deflection of ship's stiffened panels under impact pressure through a comparison with LS-DYNA (2007) simulations and existing experimental results. ALPS/ULSAP (2009) is able to predict the permanent plastic deflections of the plating (between stiffeners), interframe stiffened panels (between transverse frames), and grillages under impact pressure once the impact peak pressure value and its duration are prescribed.

Figure 10.8 compares the permanent plastic deflections of steel plating under impact pressure (with varying peak pressures and/or pressure durations) obtained with the LS-DYNA simulations in experiments performed by Jones and Baeder (1972) and using the current design formula via ALPS/ULSAP. In this comparison, p_c is the collapse strength of plates under quasi-static pressures, which is given as follows (Jones, 1975).

Simply supported plate edges

$$p_c = \begin{cases} \dfrac{8M_p}{b^2}(1 + \alpha + \alpha^2) \text{ for the lower bound solution} \\ \dfrac{24M_p}{b^2}\dfrac{1}{(\sqrt{3+\alpha^2} - \alpha)^2} \text{ for the upper bound solution} \end{cases} \qquad (10.3.12a)$$

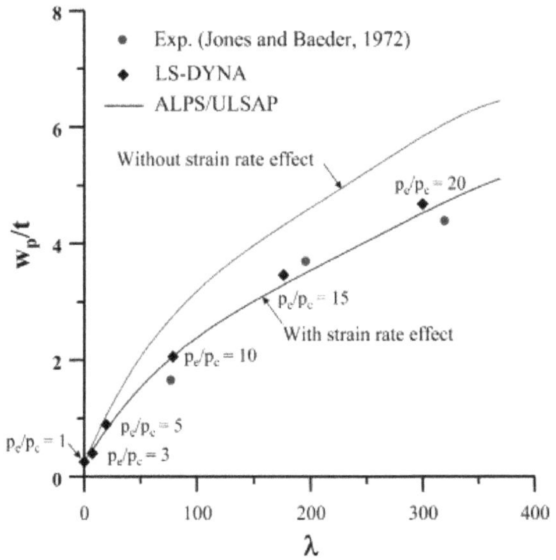

Figure 10.8 Comparison among LS-DYNA simulations, experiments, and the design formula for a mild steel square plate.

Clamped plate edges

$$p_c = \begin{cases} \dfrac{16M_p}{b^2}(1+\alpha^2) \text{ for the lower bound solution} \\ \dfrac{48M_p}{b^2}\dfrac{1}{(\sqrt{3+\alpha^2}-\alpha)^2} \text{ for the upper bound solution} \end{cases} \quad (10.3.12b)$$

where $M_p = \dfrac{\sigma_Y t^2}{4}$ and $\alpha = \dfrac{b}{a}$. If M_p is substituted into the upper bound version of (10.3.12a) the result is

$$p_c = \frac{6t^2\sigma_Y}{b^2}\frac{1}{(\sqrt{3+\alpha^2}-\alpha)^2} \quad (10.3.13)$$

In Fig. 10.8, p_c is calculated using (10.3.13). It can be seen that the permanent deflection for steel plating obtained with the current design formula is in good agreement with those obtained in the LS-DYNA simulations and experimental results.

We now consider a grillage (cross-stiffened panel) under impact pressure taken from the midship section cargo hold of a 300k dwt, ultra-large crude oil carrier (ULCC). The structural dimensions, with the nomenclature indicated in Fig. 10.2, are L = 15300 mm, B = 3760 mm and t = 16 mm. The number of longitudinal stiffeners is 3, and their type is T-bar with a 520-mm × 12-mm web and 150-mm ×20-mm flange. The number of transverse frames is 2, and their type is T-bar with a 2730-mm × 18-mm web and 450-mm × 45-mm flange. The material yield stress is σ_Y = 315 N/mm², and the elastic modulus is E = 205800 N/mm². The mass density is ρ = 7850 kg/m³.

Figure 10.9 presents a comparison of the permanent set deflection predictions obtained by the ALPS/ULSAP and LS-DYNA nonlinear finite element analysis on plating, longitudinally stiffened panels, and grillages (cross-stiffened panels) under impact pressure, with varying peak impact pressure values and impact pressure durations. It is evident from Fig. 10.9 that the ALPS/ULSAP and LS-DYNA solutions are in good agreement for a wide range of impact pressures and durations. Figure 10.10 compares the design formulae with the RADIOSS computations of a mild steel square plate obtained by Saitoh et al. (1995) for underwater explosions. (The recent version of RADIOSS code (2009) is available at http://www.altair.com.) It is believed the design formulae presented in this chapter are also suitable for explosive action cases.

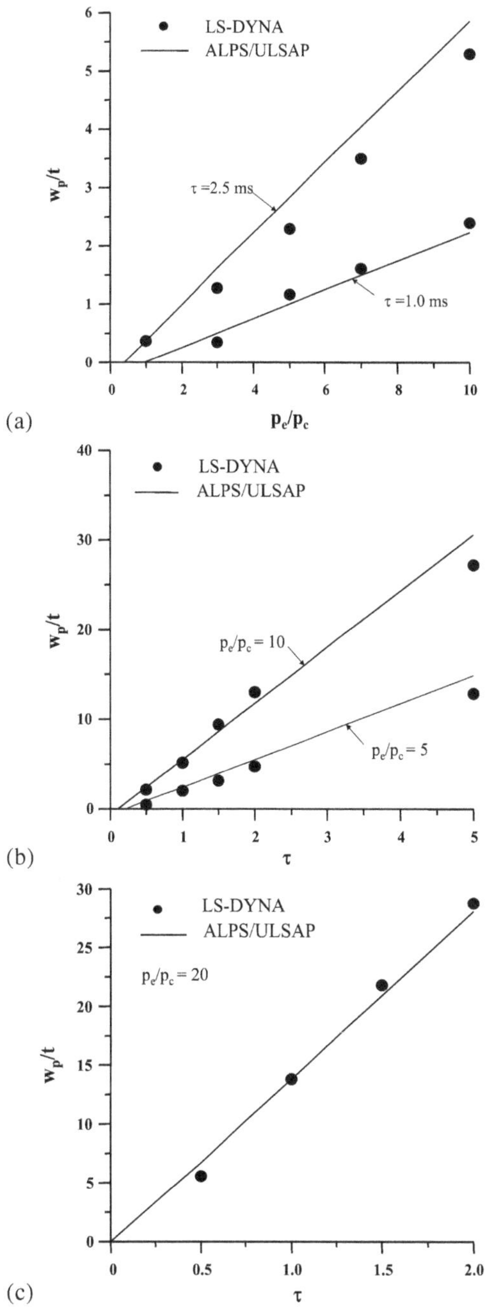

Figure 10.9 Comparison of permanent deflections for (a) the plating between stiffeners with p_c = 0.377 N/mm², (b) longitudinally stiffened panels between transverse frames with p_c = 0.65 N/mm², and (c) cross-stiffened panels with p_c = 0.65 N/mm².

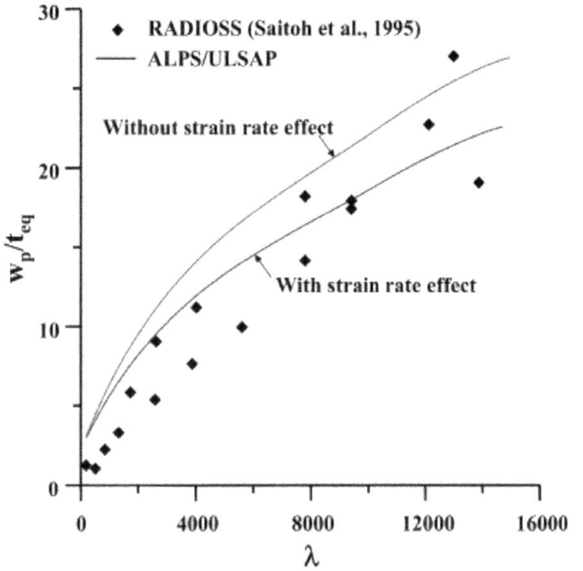

Figure 10.10 Comparison between the design formulae and the nonlinear finite element method computations of a mild steel square plate for underwater explosions.

REFERENCES

ALPS/ULSAP. (2009). *Ultimate limit state assessment of plate panels.* Stevensville, MD: DRS C3 Systems. Available online at http://www.orca3d.com/ maestro.

Chen, W. (1993). A new bound solution for quadrangular plates subjected to impulsive loads. *Proc. of the 3rd International Offshore and Polar Engineering Conference*, Vol. IV, Singapore, June 6–11, 701–708.

Cowper, G. R., and Symonds, P. S. (1957). *Strain-hardening and strain-rate effects in the impact loading of cantilever beams*. Technical Report No. 23. Providence, RI: Division of Applied Mathematics, Brown University.

Jones, N. (1971). A theoretical of the dynamic plastic behavior of beams and plates with finite deflections. *International Journal of Solids and Structures*, Vol. 7, 1007–1029.

Jones, N. (1975). Plastic behavior of beams and plates. In Evans, J. H. (Ed.). *Ship structural design concept.* Centreville, MD: Cornell Maritime Press, 747–778.

Jones, N. (1997). *Structural impact.* Cambridge, U.K.: Cambridge University Press.

Jones, N., and Baeder, R. A. (1972). An experimental study of the dynamic plastic behavior of rectangular plates. *Proc. of Symposium on Plastic Analysis of Structures, Ministry of Education, Polytechnic Institute of Jassy, Civil Engineering Faculty,* Vol. 1, Jassy, Romania, 476–497.

Jones, N., Uran, T. O., and Tekin, S. A. (1970). The dynamic plastic behavior of fully clamped rectangular plates. *International Journal of Solids and Structures*, Vol. 6, 1499–1512.

Korean Register (KR). (1997). *Control of ship vibration and noise*. Daejon, Korea: Korean Register of Shipping.

LS-DYNA. (2007). *Nonlinear dynamic analysis of structures*. Livermore, CA: Livermore Software Technology Corporation. Available online at http://www.lstc.com.

NORSOK. (1999). Actions and action effects. *Norwegian Standards*, NORSOK N003, Norway.

Paik, J. K., Lee, J. M., Shin, Y. S., and Wang, G. (2004). Design principles and criteria for ship structures under impact pressure actions arising from sloshing, slamming and green water. *Trans SNAME*, Vol. 112, 292–313.

Paik, J. K., and Thayamballi, A. K. (2003). *Ultimate limit state design of steel-plated structures*. Chichester, U.K.: Wiley.

Paik, J. K., and Thayamballi, A. K. (2007). *Ship-shaped offshore installations: Design, building, and operation*. Cambridge, U.K.: Cambridge University Press.

RADIOSS. (2009). *Nonlinear dynamic analysis of structures*. Troy, MI: Altair Engineering. Available online at http://www.altair.com.

Saitoh, T., Yoshikawa, T., and Yao, H. (1995). Estimation of deflection of steel panel under impulsive loading. *The Japan Society of Mechanical Engineers*, Vol. 61, 2241–2246.

ELEVEN

BUCKLING AND ULTIMATE STRENGTH OF COLUMNS

Owen Hughes
Professor, Virginia Tech
Blacksburg, VA, USA

11.1 REVIEW OF BASIC THEORY

Ideal Columns

The "ultimate load" P_{ult} is the maximum load that a column can carry, and depends on initial eccentricity of the column, eccentricity of the load, transverse loads, end conditions, local or lateral buckling, inelastic action, and residual stresses. The Euler buckling load P_E, on the other hand, is an idealized quantity which does not take any of the foregoing factors into account (except for end rotational restraint, which can be accounted for by using an effective length L_e):

$$P_E = \frac{\pi^2 EI}{L_e^2} \qquad (11.1.1)$$

The Euler buckling load is the load for which an ideal column will first have an equilibrium deflected shape. Mathematically, it is the eigenvalue in the solution to Euler's differential equation

$$\frac{d^2w}{dx^2} + \frac{Pw}{EI} = 0$$

Due to the factors just mentioned, the ultimate load of a practical column will be less than the Euler buckling load. In fact, strictly speaking, buckling—the sudden transition to a deflected shape—only occurs in the case of "ideal" columns (columns with no residual stress or eccentricity). Moreover, even in ideal columns the buckling load will be less than the Euler load if the compressive stress in the column exceeds the proportional limit stress, because the diminished slope of the stress-strain curve represents an effective weakening of the column. Shanley [1] has shown that for an ideal column (no eccentricity or residual stress) buckling will occur at the *tangent modulus* load, given by

$$P_t = \frac{\pi^2 E_t I}{L_e^2}$$

in which E_t is the slope of the stress-strain curve corresponding to the level of the compressive stress P_t/A in the column. Since E_t depends on P_t the calculation of P_t is generally an iterative process. Figure 11.1 shows the load–deflection behavior for practical columns, containing residual stress and eccentricity. For such columns it is more convenient to deal in terms of stress rather than load because this allows the effects of yielding and residual stress to be included. The Euler buckling stress is

$$\sigma_E = \frac{P_E}{A} = \frac{\pi^2 E}{\left(\frac{L_e}{\rho}\right)^2} \qquad (11.1.2)$$

where ρ is the radius of gyration ($I = \rho^2 A$) and L_e/ρ is the *slenderness ratio*. Similarly, the *ultimate strength* of a column is defined as the average applied stress at collapse:

$$\sigma_{ult} = \frac{P_{ult}}{A}$$

For ideal columns the ultimate strength would correspond to the tangent modulus load, that is,

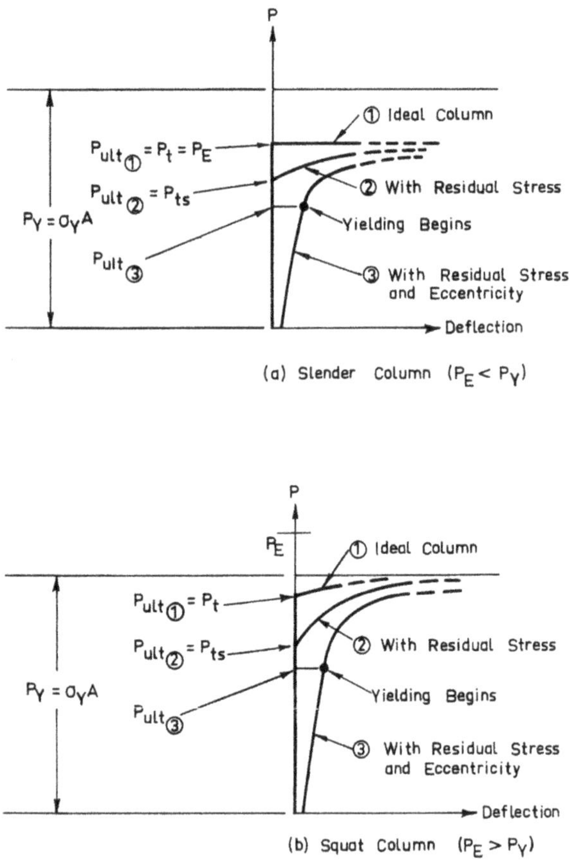

Figure 11.1 Typical load deflection diagrams.

$$(\sigma_{\text{ult}})_{\text{ideal}} = \frac{\pi^2 E_t}{\left(\dfrac{L_e}{\rho}\right)^2} \qquad (11.1.3)$$

The ideal ultimate strength curve, i.e. the relationship between $(\sigma_{\text{ult}})_{\text{ideal}}$ and L_e/ρ, is shown in Fig. 11.2, together with a typical stress-strain curve and the corresponding tangent modulus curve. The figure also shows typical experimental values for essentially straight rolled sections and welded sections. The marked discrepancy is almost entirely due to the compressive residual stress σ_r which exists in rolled, welded, and even cold-worked structural shapes, and this will now be discussed.

Residual Stress

Rolled Sections

In rolled sections the uneven cooling between flange root and flange tip produces tensile residual stresses in the former and compressive residual stresses in the latter, as shown in Fig. 11.3. The effect depends mainly on geometry, and is most pronounced in wide flange sections, in which the residual compressive stress in the flange tips can be as high as 80 MPa for mild steel. The effect is diminished for steels having a higher yield strength; for instance, for $\sigma_Y = 350$ MPa typical values of σ_r range from 20 to 50 MPa. Residual stress due to rolling can usually be eliminated by using quenched and tempered steels or by annealing, but this is costly and often impractical.

The parts of the cross section that have a compressive residual stress will commence yielding when the average applied stress has reached a value of $\sigma_Y - \sigma_r$, and this greatly reduces σ_{ult}. The effect is particularly detrimental because it occurs in the flange tips. The material in the column is no longer homogeneous, and hence the simple tangent modulus approach of (11.1.3) is no longer valid. However, a comparatively simple solution can be achieved for the buckling strength in the primary direction (i.e., bending about the xx-axis in Fig. 11.3) by assuming an elastic-perfectly plastic stress-strain relationship. In this case the progressive loss of bending stiffness, due to progressive yielding of the flanges, is linearly proportional to the extent of the yielded zone. This linearity permits the use of an average value of tangent modulus (averaged over the entire cross section). This

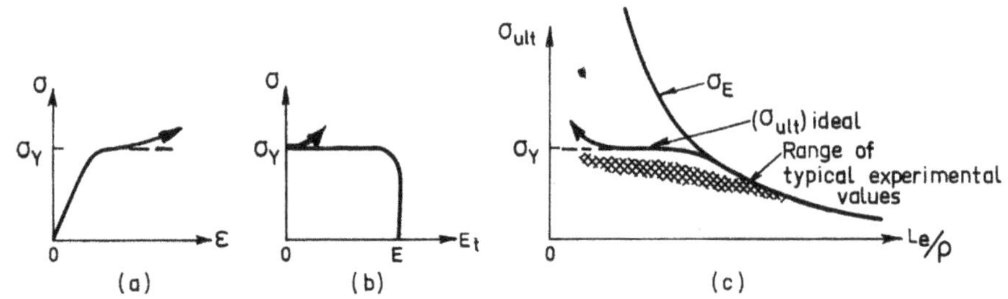

Figure 11.2 Tangent modulus values of ultimate strength for ideal columns.

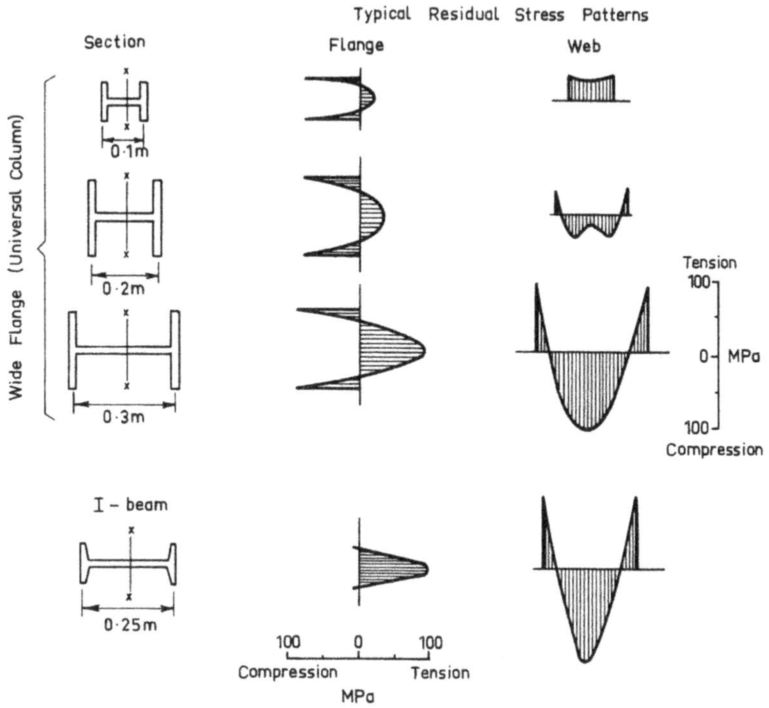

Figure 11.3 Typical residual stresses in rolled sections.

average value is referred to as the *structural tangent modulus*, E_{ts}, and is obtained from stub column tests. A typical stub column stress-strain curve and corresponding structural tangent modulus diagram are shown in Fig. 11.4. For materials that have a definite yield plateau, one of the most useful representations of the structural tangent modulus is the Ostenfeld-Bleich [2] parabola:

$$\frac{E_{ts}}{E} = \frac{\sigma_{av}(\sigma_Y - \sigma_{av})}{\sigma_{spl}(\sigma_Y - \sigma_{spl})} \quad \text{for } (\sigma_{spl} < \sigma_{av} < \sigma_Y) \tag{11.1.4}$$

where σ_{spl} is the *structural proportional limit*. Note that (11.1.4) is only valid in the range $\sigma_{spl} < \sigma_{av} < \sigma_Y$.

Substituting E_{ts} from (11.1.4) in place of E_t in (11.1.3) gives

$$\frac{\sigma_{ult}}{\sigma_Y} = 1 - \frac{\sigma_{spl}}{\sigma_Y}\left[1 - \frac{\sigma_{spl}}{\sigma_Y}\right]\lambda^2 \tag{11.1.5}$$

where
$$\lambda = \frac{L_e}{\rho\pi}\sqrt{\frac{\sigma_Y}{E}} \tag{11.1.6}$$

For rolled, "universal column," (wide flange) sections, a typical value of σ_{spl} is $\frac{1}{2}\sigma_Y$. In this case (11.1.5)

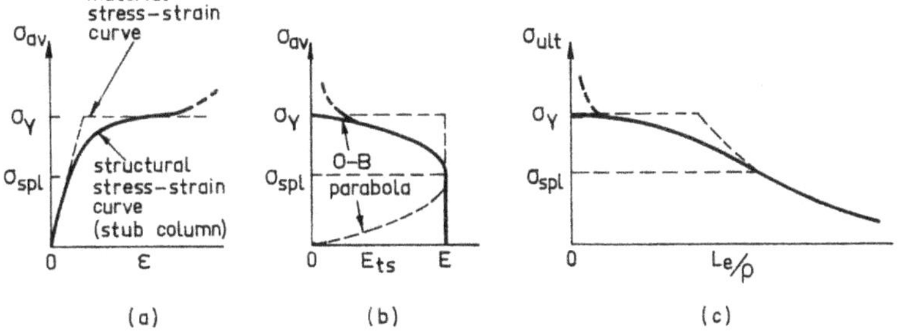

Figure 11.4 Column strength curve based on structural tangent modulus.

becomes

$$\frac{\sigma_{\text{ult}}}{\sigma_Y} = 1 - \frac{\lambda^2}{4} \qquad (11.1.7)$$

This equation is known as the Johnson parabola and is illustrated in Fig. 11.5. Beyond $\lambda = (2)^{1/2}$ the Euler curve applies ($\sigma_{\text{ult}}/\sigma_Y = 1/\lambda^2$). The Johnson parabola has been recommended by the Column Research Council [3] and by the A.I.S.C. [4] for calculating the ultimate strength of a "basic column" (essentially straight; pinned end) for the case of rolled steel sections. This "basic column" value is then used as the basis for design curves for practical, rolled steel columns.

WELDED SECTIONS

The distribution of residual stress due to welding is generally quite different from that due to hot rolling, even though both are caused by time-dependent temperature gradients. During welding the metal is in a perfectly plastic state. At any time during and after the instant of welding, some plastic fibers in the cross section will be cooling and other elastic fibers will be heating. The action of the loads on the elastic fibers due to changes in temperature is not counteracted by the plastic fibers until the plastic fibers have cooled sufficiently to become elastic. The interaction between the different fibers in the cross section results in a locked-in tensile stress in and near the weld which is approximately equal to the yield stress of the material. A typical residual stress distribution is shown in Fig. 11.6 together with an idealized distribution. The extent of the tension yield zone is generally from three to six thicknesses out from the weld on each side, and depends mainly on the total heat input. Other principal factors are the cross-sectional area of the weld deposit, the type of welding, and the welding sequence. This locked-in tension gives rise to a compressive residual stress σ_r in the remaining areas of the section. If ηt denotes the width of the tension yield zone and b the total flange width, then from equilibrium

$$\frac{\sigma_r}{\sigma_Y} = \frac{2\eta}{\dfrac{b}{t} - 2\eta} \qquad (11.1.8)$$

Obviously, for narrow thick sections residual stress will be high, and this will seriously diminish the

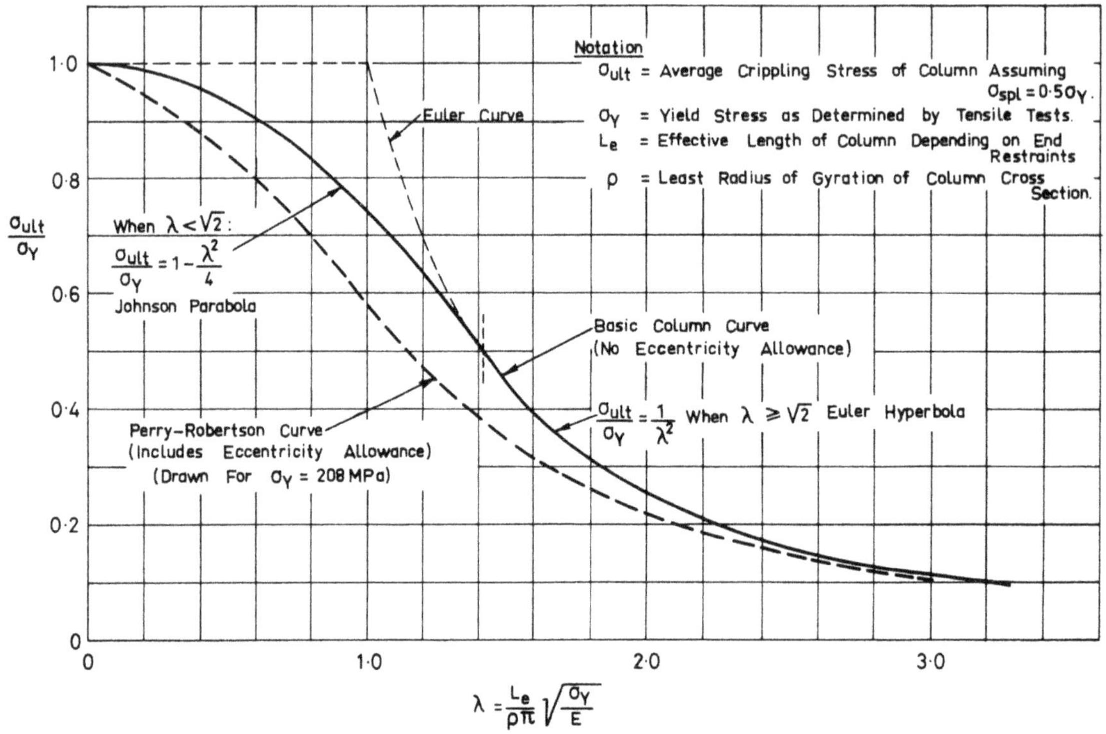

Figure 11.5 Basic column curve (Johnson parabola) and Perry-Robertson curve.

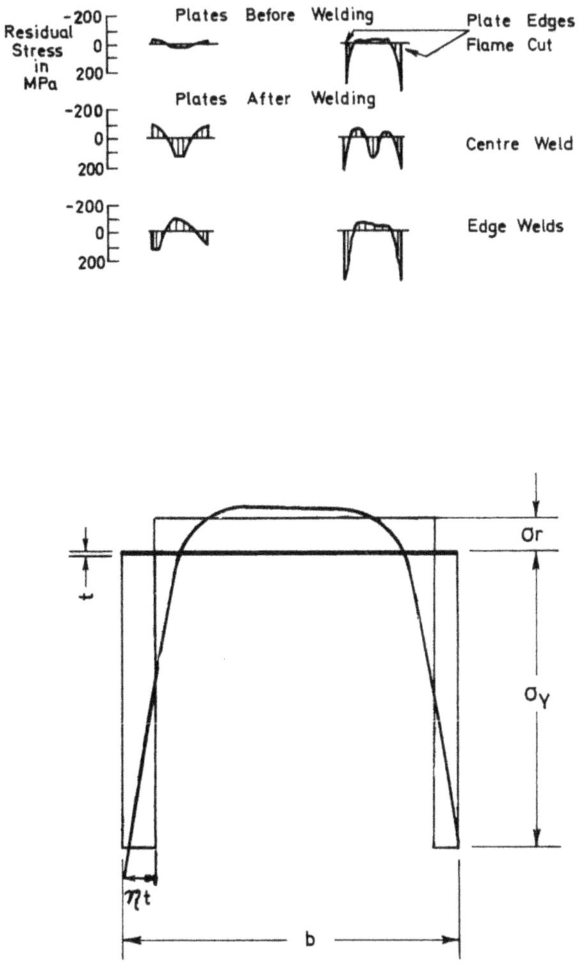

Figure 11.6 Typical residual stresses in welded sections.

strength of the section. It should be noted, however, that the effect will be much worse for open sections than for box sections, since in the latter the tensile stress in the corners delays the final collapse of this region, which is at the extremity of the section. Another point worth noting is that flame-cutting, which is usual ship fabrication practice, creates conditions similar to welding and thus causes high tensile stresses at the edges of flanges, an effect which has been shown to be moderately beneficial [5]. Finally, it should be noted that the effect of residual stress is somewhat diminished for higher yield steels due to a narrower tension zone.

Another important feature of welded sections is that because of the welding distortions the eccentricity is generally larger than for rolled shapes, and this can also seriously diminish the ultimate strength of such sections. This point will be taken up in the next section.

For typical welded universal column (wide flange) sections, the average residual stress is of the order of $\frac{1}{4}\sigma_Y$. Hence $\sigma_{spl} = \sigma_Y - \sigma_r = \frac{3}{4}\sigma_Y$. For stockier sections (such as an I section) and when the bending is in the "strong direction," σ_r is typically $\frac{1}{2}\sigma_Y$, giving $\sigma_{spl} = \frac{1}{2}\sigma_Y$. This again gives the Johnson parabola of (11.1.7), but it should be emphasized that this equation does not allow for eccentricity. This effect will now be discussed.

Eccentricity: Magnification Factor

In practice, columns are seldom perfectly straight, nor is the loading perfectly axial. Figure 11.7 shows a pinned column with an initial deflection $\delta(x)$ and an additional deflection $w(x)$ due to the axial load. For small deflections the governing equation is

$$\frac{d^2w}{dx^2} + \frac{P}{EI}(\delta + w) = 0 \qquad (11.1.9)$$

in which the combination of the total deflection and the axial force gives rise to a bending moment $P(\delta + w)$, which begins to act as soon as the load P is applied. This causes the deflection to increase further, and the deflection continues to grow and be magnified as long as P increases. Thus there is no static equilibrium configuration and no sudden buckling. The solution of (11.1.9) will depend on the nature of δ. Let δ be represented by a Fourier series

$$\delta = \sum_{n=1}^{\infty} \delta_n \sin\frac{n\pi x}{L}$$

and assume that the additional deflection due to the bending is also a Fourier series.

$$w = \sum_{n=1}^{\infty} w_n \sin\frac{n\pi x}{L}$$

Then $$\frac{d^2w}{dx^2} = \sum_{n=1}^{\infty} -\frac{n^2\pi^2}{L^2} w_n \sin\frac{n\pi x}{L}$$

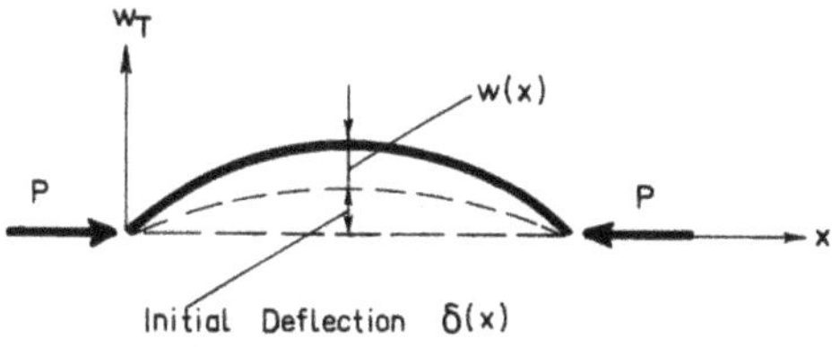

Figure 11.7 Eccentric column.

and from (11.1.9)

$$\sum_{n=1}^{\infty}\left[-\frac{n^2\pi^2}{L^2}w_n + \frac{P}{EI}(w_n + \delta_n)\right]\sin\frac{n\pi x}{L} = 0$$

This must be true for any value of n as well as for the complete summation and therefore, since $\sin n\pi x/L$ cannot be zero, we must have

$$-\frac{n^2\pi^2}{L^2}(w_n) + \frac{P}{EI}(w_n + \delta_n) = 0$$

and therefore

$$w_n = \frac{\delta_n}{\dfrac{n^2 P_E}{P} - 1}, \quad \text{where } n = 1, 2, 3 \ldots \quad (11.1.10)$$

The condition when $(P_E/P) \to 1$ is of particular interest. An examination of (11.1.10) for various values of n gives the following:

n	w_n/δ_n
1	∞
2	$\frac{1}{3}$
3	$\frac{1}{8}$
⋮	⋮
∞	0

from which it is clear that the dominant term is $n = 1$.

Hence for any configuration of initial deflection δ, the extra deflection which P induces is given by

$$w = \frac{\delta}{\dfrac{P_E}{P} - 1}$$

and the total or "magnified" deflection is

$$w_T = \delta + w = \frac{P_E}{P_E - P}\delta = \phi\delta \quad (11.1.11)$$

The factor

$$\phi = \frac{P_E}{P_E - P} \quad (11.1.12)$$

is called the *magnification factor*.

The foregoing load–deflection relationship for an elastic column is illustrated in Fig. 11.8. As the axial

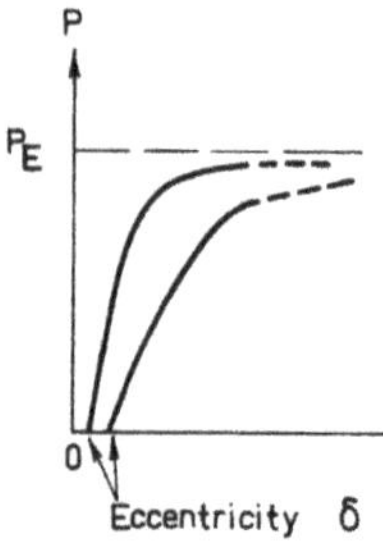

Figure 11.8 Load–deflection curves for eccentric columns.

load approaches the Euler buckling load the deflection becomes very large—regardless of the value of the initial deflection. This implies that for an elastic column the maximum value of axial compressive load is close to the Euler buckling load, and hence the latter represents an upper bound for such a column. Strictly speaking, (11.1.12) is exact only for a sinusoidal initial deflection. However, the most serious shape for an initial deflection is a single half wave shape such as this, and the value of ϕ for other half wave shapes differs only negligibly from the value given by (11.1.12). Hence this equation can be used for nearly all types of initial eccentricity.

Eccentricity effects may also arise due to eccentricity of load, as shown in Fig. 11.9. In this case it may be shown (see, for example, Ref. 6) that the magnification factor is given by

$$\phi = \sec\left(\frac{\pi}{2}\sqrt{\frac{P}{P_E}}\right) \quad (11.1.13)$$

The two types of eccentricity (load application and column geometry) can be combined linearly (superimposed) provided that the correct magnification factor is used for each. By plotting (11.1.12) and (11.1.13) it may be demonstrated that for $P/P_E < 0.5$ they give values of ϕ that never differ from each other by more

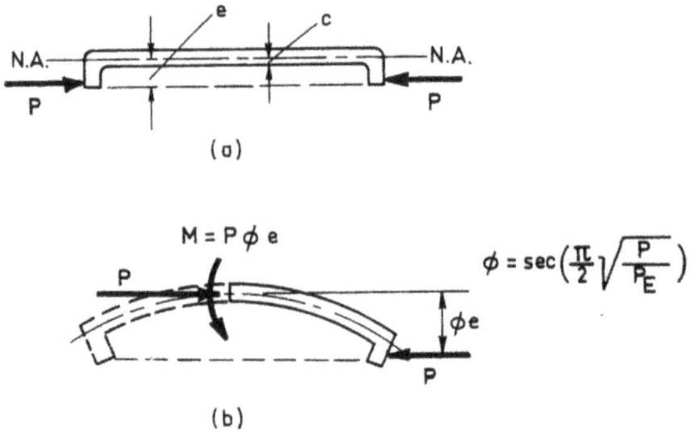

Figure 11.9 Column with eccentric load.

than 12%. Therefore, since it is simpler, (11.1.12) is usually used for eccentric loads as well as for initial deflections. This has the further advantage that the two effects may be combined directly, the total eccentricity $\Delta = \delta + e$ being used in place of δ in (11.1.11). The combination is illustrated in Fig. 11.10. The figure also shows that the maximum compressive stress in such a column is

$$\sigma_{\max} = \frac{P}{A} + \frac{P\phi\Delta}{Z}$$

$$= \frac{P}{A}\left(1 + \phi\frac{\Delta}{r_c}\right) \qquad (11.1.14)$$

in which Z is section modulus (on the compression side) and

$$r_c = \text{core radius} = \frac{Z}{A} = \frac{\rho^2}{c} \qquad (11.1.15)$$

The ratio Δ/r_c is the *eccentricity ratio*. If the eccentricity of the load is large it will cause significant induced bending stresses, and these will tend to eclipse the residual stresses. The choice of a typical value for Δ for design purposes has long been a vexed question. Most authorities have related eccentricity to the slenderness ratio L/ρ, on the basis that a more slender column would be expected to have a larger eccentricity. If the relationship is assumed to be linear then

$$\frac{\Delta}{r_c} = \alpha\frac{L}{\rho} \qquad (11.1.16)$$

On the basis of a series of tests of typical columns (with varying degrees of residual stress and eccentricity) Robertson [7] proposed a mean value of 0.003 for α, since this gave a lower bound to the test results.

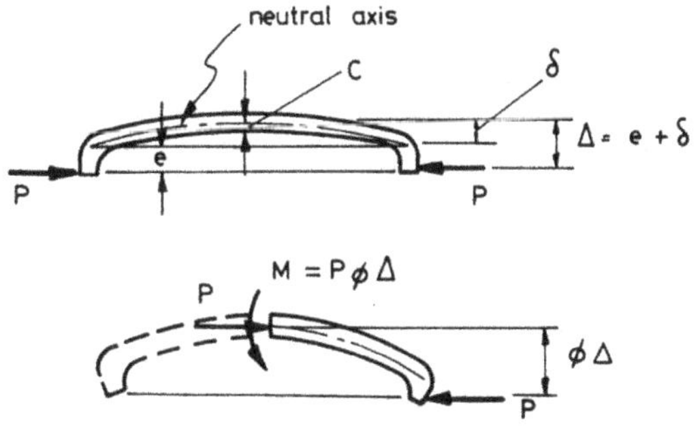

Figure 11.10 Combined eccentricity.

This approach of allowing for both eccentricity and residual stress by means of some effective eccentricity has formed the basis of many structural engineering design codes for some time.

11.2 COLUMN DESIGN FORMULAS

Perry-Robertson Formula

One of the simplest formulas for the ultimate load is the Perry-Robertson formula which adopts Robertson's value for α and assumes that the column will collapse when the maximum compressive stress reaches the yield stress. This gives, from (11.1.14) and (11.1.16)

$$\sigma_Y = \frac{P_{\text{ult}}}{A}\left(1 + \frac{\alpha\left(\frac{L}{\rho}\right)P_E}{P_E - P_{\text{ult}}}\right) \qquad (11.2.1)$$

or, in terms of stress

$$\sigma_Y = \sigma_{\text{ult}}\left(1 + \frac{\frac{\alpha L}{\rho}}{1 - \frac{\sigma_{\text{ult}}}{\sigma_E}}\right) \qquad (11.2.2)$$

with σ_E given by (11.1.2). This equation is a quadratic, and may be rewritten in a nondimensional, factored form

$$(1 - R)(1 - \lambda^2 R) = \eta R$$

in which R is the *strength ratio* of the column:

$$R = \frac{\sigma_{\text{ult}}}{\sigma_Y}$$

η is the *eccentricity ratio*, defined in (11.1.16):

$$\eta = \frac{\alpha L}{\rho}$$

and λ is the *column slenderness parameter*:

$$\lambda = \sqrt{\frac{\sigma_Y}{\sigma_E}} = \frac{L}{\pi\rho}\sqrt{\frac{\sigma_Y}{E}}$$

The solution is

$$R = \frac{1}{2}\left(1 + \frac{1+\eta}{\lambda^2}\right) - \sqrt{\frac{1}{4}\left(1 + \frac{1+\eta}{\lambda^2}\right)^2 - \frac{1}{\lambda^2}} \tag{11.2.3}$$

The Perry-Robertson formula, that is, the specific case of $\alpha = 0.003$, is plotted for mild steel ($E/\sigma_Y = 1000$) in Fig. 11.5, which shows that it lies significantly below the Johnson parabola, indicating the extent to which eccentricity degrades the column strength, quite apart from the effects of residual stress. For a perfectly straight column, having $\eta = 0$, (11.2.3) reduces to the Euler curve, $R = 1/\lambda^2$, providing that $\lambda > 1$.

Accurate Design Curves

Extensive experimental research [8] has shown that different types of sections have significantly different column collapse curves, partly due to geometry and partly to residual stresses. The results have shown that a single design curve such as the Perry-Robertson formula cannot accurately represent all types of sections. Instead, it is necessary to have different sets of curves for each type of section, each curve corresponding to a particular grade of steel. For instance, a Universal Column section has a quite different set of curves to a thin-wall tube, and should be treated accordingly. The newer codes have incorporated this approach. As shown by Dwight [9] the Perry-Robertson formula can still be used as a basis for these curves, with different values of α being used for different types of sections. Three such curves, corresponding to $\alpha = 0.0020$, 0.0035, and 0.0055, are given in Figs. 11.11, 11.12, and 11.13, and for rolled sections the curve to be chosen for each type of section is indicated in Table 11.1. In order to allow for strain hardening, the curves have a horizontal plateau at values of slenderness ratio less than some threshold value $(L/\rho)_0$. To achieve this the value of the eccentricity ratio, Δ/r_c, is defined as

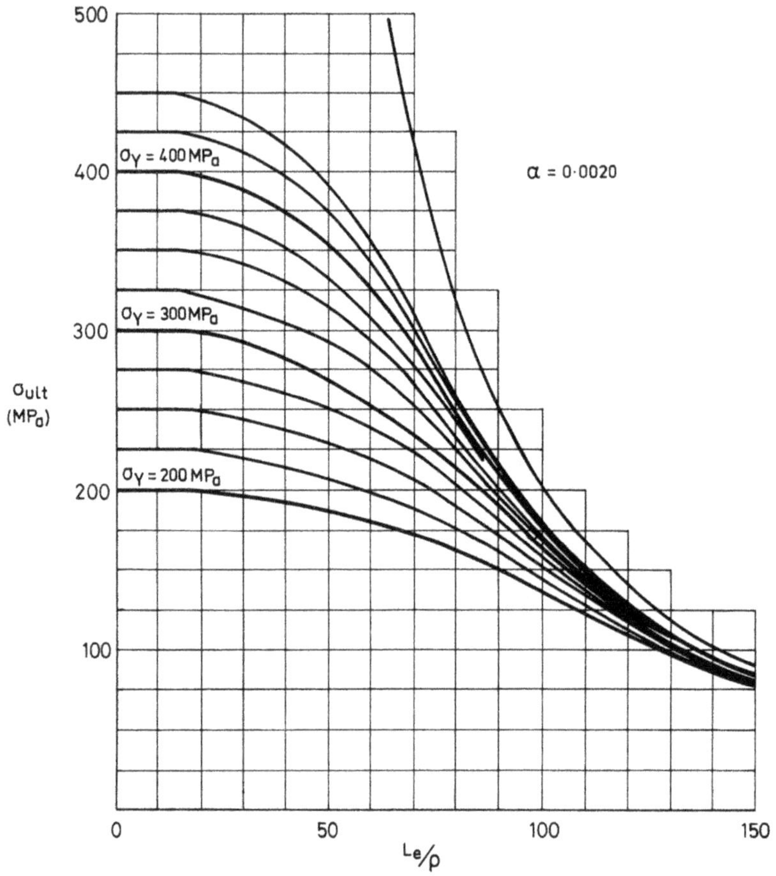

Figure 11.11 Column design curves for $\alpha = 0.002$.

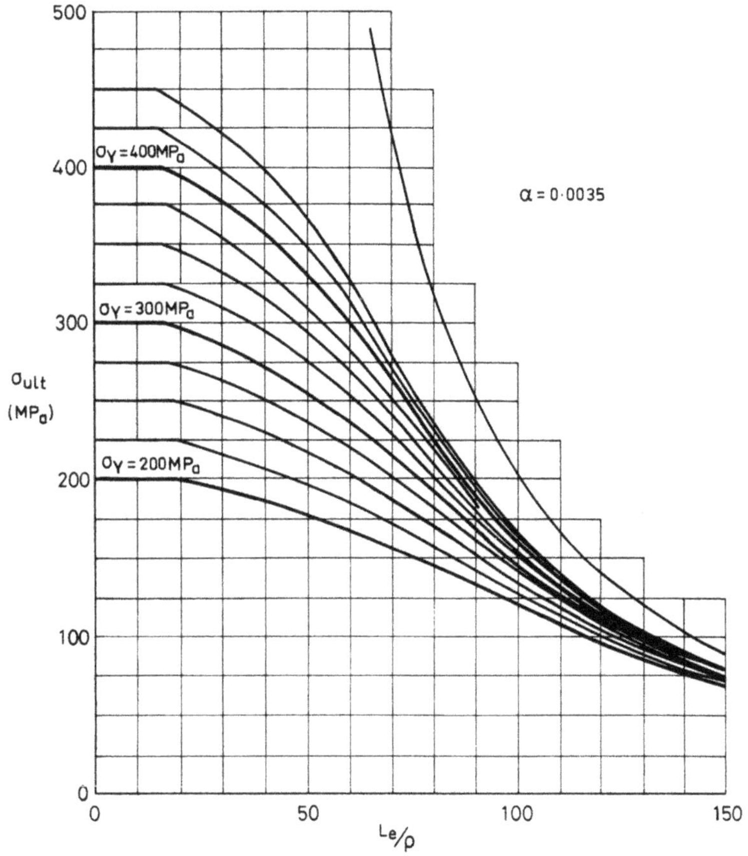

Figure 11.12 Column design curves for $\alpha = 0.0035$.

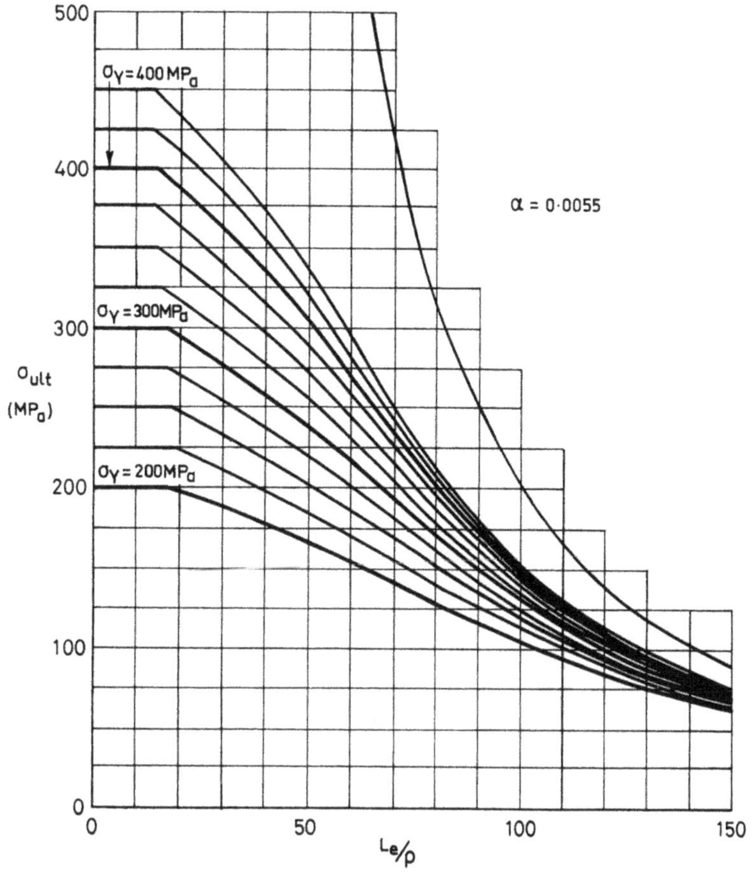

Figure 11.13 Column design curves for $\alpha = 0.0055$.

follows:

$$\left.\begin{array}{ll}\text{for } \dfrac{L}{\rho} < \left(\dfrac{L}{\rho}\right)_{0'} & \dfrac{\Delta}{r_c} = 0 \\ \text{for } \dfrac{L}{\rho} > \left(\dfrac{L}{\rho}\right)_{0'} & \dfrac{\Delta}{r_c} = \alpha\left[\dfrac{L}{\rho} - \left(\dfrac{L}{\rho}\right)_0\right]\end{array}\right\} \quad (11.2.4)$$

where $(L/\rho)_0$ is given by:

$$\left(\frac{L}{\rho}\right)_0 = 0.2\pi\sqrt{\frac{E}{\sigma_Y}} \quad (11.2.5)$$

For welded columns the residual stress σ_r in the flanges is generally larger than for rolled sections, and a further allowance must be made. This may be done to a satisfactory degree of accuracy by taking a reduced yield stress equal to about 95% of true yield stress.

TABLE 11.1 CURVE SELECTION TABLE FOR ROLLED AND WELDED SECTIONS (Note: For welded I- and box-sections the yield stress should be reduced by 5% to allow for welding residual stresses.)

Section	Axis of Buckling	α
Universal column	*xx*	0.0035
Universal column	*yy*	0.0055
Universal beam	*xx*	0.0020
Universal beam	*yy*	0.0035
UC or UB with cover-plates	*xx*	0.0035
UC or UB with cover-plates	*yy*	0.0020
Channel	*xx*	0.0055
Channel	*yy*	0.0055
Tee	*xx*	0.0055
Tee	*yy*	0.0055
Angle	any	0.0055
Round tube	any	0.0020
Rectangular hollow	*xx*	0.0020
Rectangular hollow	*yy*	0.0020
Welded I-sections	*xx*	0.0035
Welded I-sections	*yy*	0.0055
Welded box-sections		0.0035

11.3 EFFECT OF LATERAL LOAD: BEAM COLUMNS

Use of Magnification Factor

It may be shown that for a pinned column subjected to a uniform lateral load q and an axial load P, the exact solution for the maximum deflection is

$$w_{max} = \frac{5qL^4}{384EI}\left[\frac{24}{5\xi^4}\left(\sec \xi - 1 - \frac{\xi^2}{2}\right)\right] \quad (11.3.1)$$

in which $\xi = \left(\dfrac{L}{2}\right)\sqrt{\left(\dfrac{P}{EI}\right)}$

The first factor, $5qL^4/(384EI)$, is the central deflection of a laterally loaded member with pinned ends and with no axial loading. Hence the second factor gives the effect of the axial load in magnifying the central deflection; it is the magnification factor for the deflection for this particular loading and end condition. It may also be shown that the maximum bending moment is

$$M_{max} = \frac{qL^2}{8}\left[\frac{2(1 - \sec \xi)}{\xi^2}\right] \quad (11.3.2)$$

Once again the first factor, $qL^2/8$, is the central bending moment without the axial load, and the second factor is the magnification factor for the bending moment for this case. It is evident that the magnification factor for bending moment is different from that for deflection, and it may be shown that further differences occur for each combination of load and end condition.

Hence, in the interest of simplicity, the usual approach in dealing with laterally loaded columns (or "beam columns") is to use the simpler magnification factor which was derived for sinusoidal initial eccentricity:

$$\phi = \frac{P_E}{P_E - P} \quad (11.3.3)$$

In a beam column the maximum bending moment is the sum of the bending moment due to the lateral load, M_0, plus that due to the eccentricity, which in this case includes the deflection δ_0 caused by the lateral load. That is

$$M_{max} = M_0 + P\phi(\delta_0 + \Delta) \quad (11.3.4)$$

The maximum compressive stress is then

$$\sigma_{max} = \frac{P}{A} + \frac{M_{max}}{Z}$$

To demonstrate the accuracy of this approach let us take the case of a beam column carrying a uniform lateral load q and an axial load $P = 0.5P_E$. We let

$\Delta = 0$ in order to be able to compare with the exact solution given by (11.3.2), which for this case is $M_{max} = 2.030M_0$. In (11.3.4) the value of δ_0 is $5qL^4/(384EI)$ and $\phi = 2$. Substituting these values gives

$$\begin{aligned} M_{max} &= M_0 + 0.5\left(\frac{\pi^2 EI}{L^2}\right)2\frac{5qL^4}{384EI} \\ &= M_0\left(1 + \frac{5\pi^2}{48}\right) \\ &= 2.028M_0 \end{aligned}$$

which agrees with the exact result to three significant figures.

Because of the relatively large bending stress caused by the lateral load, the total bending stresses will be large and will tend to eclipse the residual stress. Hence we again make the simplifying (and only slightly conservative) assumption that the column will collapse when the maximum compressive stress reaches the yield stress. Therefore

$$\sigma_Y = \frac{P_{ult}}{A} + \frac{M_0}{Z} + \frac{P_{ult}(\delta_0 + \Delta)}{\left(1 - \dfrac{P_{ult}}{P_E}\right)Z} \qquad (11.3.5)$$

As before, this equation can be expressed in terms of nondimensional parameters:

$$R = \frac{\sigma_{ult}}{\sigma_Y}$$

$$\lambda = \sqrt{\frac{\sigma_Y}{\sigma_E}} = \frac{L}{\pi\rho}\sqrt{\frac{\sigma_Y}{E}}$$

$$\eta = \frac{(\delta_0 + \Delta)A}{Z}$$

$$\mu = \frac{\dfrac{M_0}{Z}}{\sigma_Y}$$

The resulting equation is the quadratic

$$(1 - R - \mu)(1 - \lambda^2 R) = \eta R \qquad (11.3.6)$$

for which the solution is

$$R = \frac{1}{2}\left(1 - \mu + \frac{1 + \eta}{\lambda^2}\right) - \sqrt{\frac{1}{4}\left(1 - \mu + \frac{1 + \eta}{\lambda^2}\right)^2 - \frac{1 - \mu}{\lambda^2}} \qquad (11.3.7)$$

This expression is plotted in Fig. 11.14 for selected values of η and μ. For a pinned beam column subjected to a specified lateral load, the axial load to cause collapse may be obtained by first calculating M_0 and δ_0, and then η and μ, and then interpolating for R in the figure.

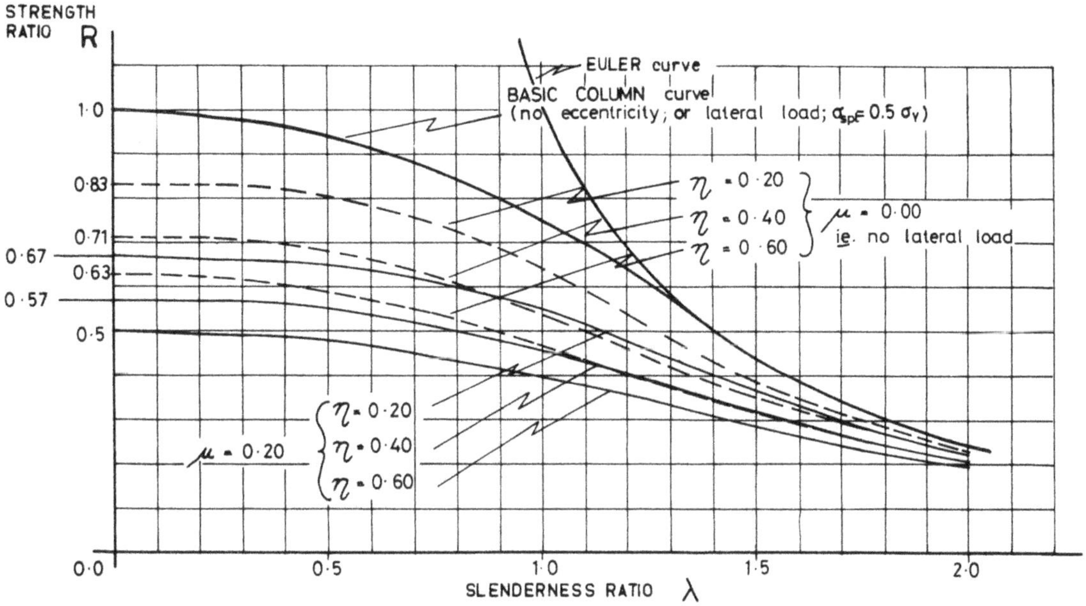

Figure 11.14a

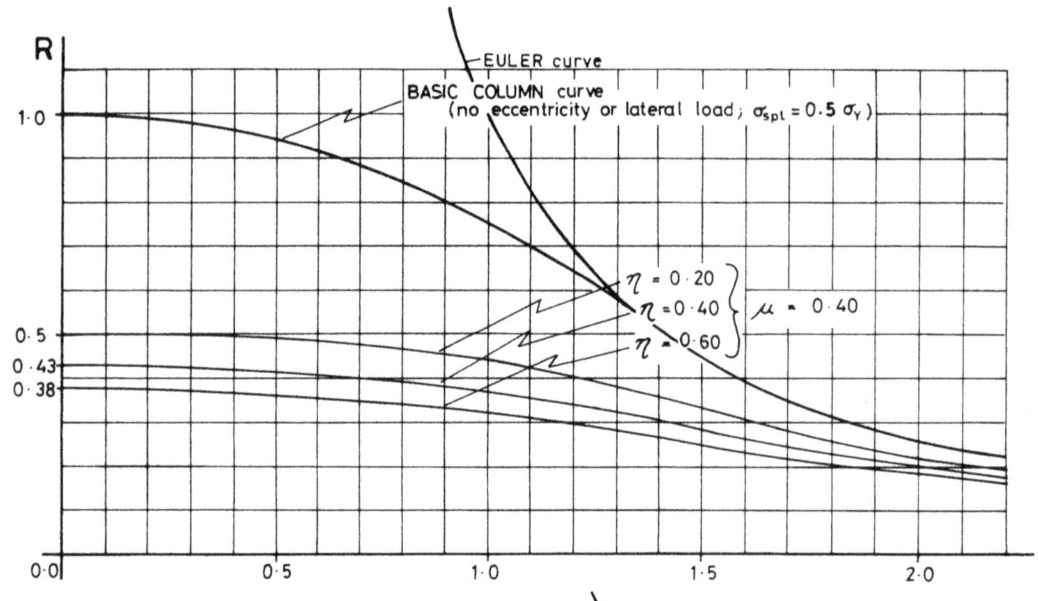

Figure 11.14b

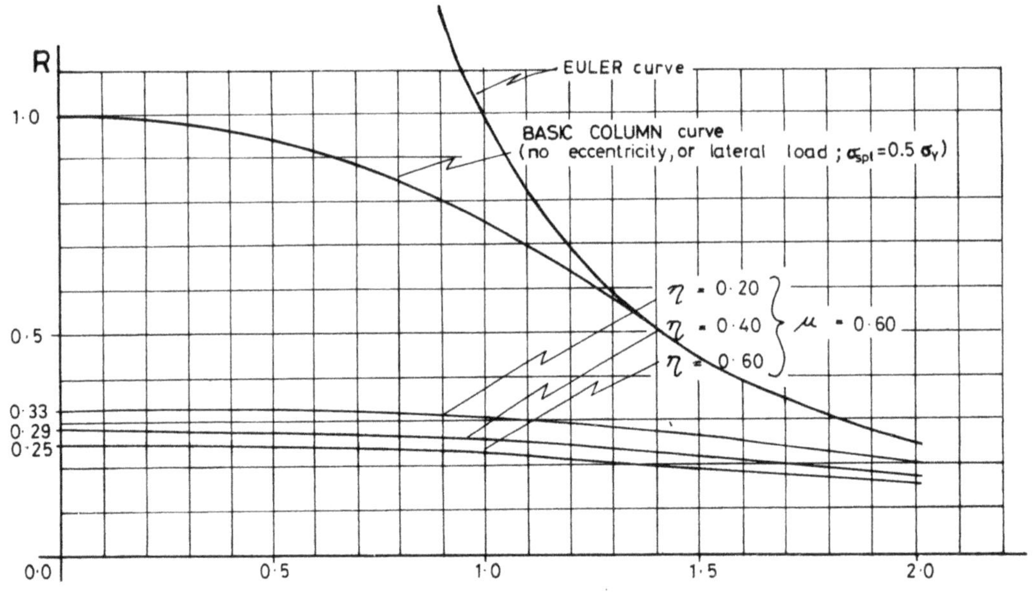

Figure 11.14c

Clamped Beam Column

If the ends of the beam column are rotationally restrained such that there is an end moment M_e, then the stress should be calculated at the end as well as at the center, because it is possible that yield in a compression flange may occur first at the ends. This is particularly likely with side frames, deck beams, and so on, in which one of the flanges is an effective width of plating. The other flange (i.e., the beam flange) is further from the neutral axis and therefore has larger stress, and this flange becomes the compression flange at the ends of the beam (assuming that the lateral load acts on the opposite side of the plating to the beam).

If the load is symmetric and if the ends are clamped then the beam column is equivalent to three pinned columns: the central portion, of length υL between the points of zero bending moment, and the two end portions, each of effective length $(1 - \upsilon)L$. The value of υ is 0.577 for a uniform load and 0.5 for a central point load. Figure 11.15 illustrates this for a uniformly loaded clamped beam column. The central portion is of length $0.577L$ and is subjected to a parabolic bending moment and the end portions are each equivalent to a pinned beam column of length $(1 - \upsilon)L = 0.423L$ and are subjected to an approximately triangular bending moment. Earlier in Section 11.3 it was mentioned that the magnification factor ϕ is relatively insensitive to the shape of the eccentric beam, and that the expression for ϕ for sinusoidal eccentric bending can also be used for a constant eccentricity. The same is true for the three portions of a clamped beam column. As shown in Fig. 11.15 the eccentricity of each end portion is equal to the deflection due to the lateral load (only) at the point of zero bending moment. This may be expressed as a fraction of δ_0, the maximum deflection of the beam due to the lateral load.

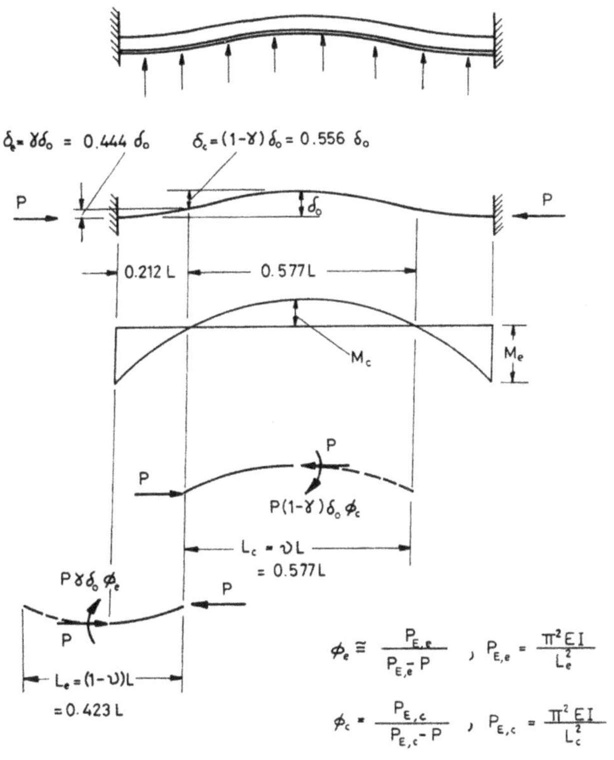

Figure 11.15 Clamped beam-column under uniform load.

$$\delta_e = \gamma\delta_0 \qquad (11.3.8)$$

The value of γ is 0.444 for a uniform load and 0.5 for a central point load. Although the eccentricity caused by the lateral load is not sinusoidal, the maximum eccentricity-induced bending moment in the end portion is approximately $P\phi_e\delta_e$, where the magnification factor ϕ_e is given by

$$\phi_e = \frac{P_{E,e}}{P_{E,e} - P} \qquad (11.3.9)$$

in which

$$P_{E,e} = \frac{\pi^2 EI}{L_e^2}$$

and, as shown in Fig. 11.15

$$L_e = (1 - \upsilon)L$$

Similarly, the maximum eccentricity-induced bending moment in the center portion is approximately $P\phi_c\delta_c$, where δ_c and ϕ_c are the eccentricity and magnification factor of the center portion, given by

$$\delta_c = (1 - \gamma)\delta_0 \qquad (11.3.10)$$

and

$$\phi_c = \frac{P_{E,c}}{P_{E,c} - P} \qquad (11.3.11)$$

in which

$$P_{E,c} = \frac{\pi^2 EI}{L_c^2}$$

and

$$L_c = \upsilon L$$

Thus the maximum values of the extra, eccentricity-induced bending moments at the ends and at the center are $P\phi_e\gamma\delta_0$ and $P\phi_c(1 - \gamma)\delta_0$ respectively. Therefore, at the ends of a clamped beam column the total bending moment under the combined lateral and axial loads is $M_e + P[(1 - \upsilon)\Delta + \gamma\delta_0]\phi_e$, where M_e is the end moment due to the lateral load (acting alone), while at the center of a clamped beam column the total bending moment is $M_c + P[v\Delta + (1 - \gamma)\delta_0]\phi_c$ where M_c is the central bending moment due to the lateral load, acting alone. The two portions of the beam are analyzed separately, and the effective length of each portion must be used in computing λ: vL for the middle portion and $(1 - v)L$ for the end portion.

This approach is much simpler than the exact beam column theory and yet retains sufficient accuracy for engineering use. The exact theory gives the following expression for ϕ_e for a uniform load:

$$\phi_e = \frac{3(\tan \xi - \xi)}{\xi^2 \tan \xi} \qquad (11.3.12)$$

in which

$$\xi = \frac{\pi}{2}\sqrt{\frac{P}{P_E}} = \frac{L}{2}\sqrt{\frac{P}{EI}}$$

In order to assess the accuracy of the simplified approach let us take the case in which $P = 0.5P_E$. The end portion is of length $(1 - v)L = 0.423L$ and therefore

$$P_{E,e} = \frac{\pi^2 EI}{(0.423L)^2}$$

Hence

$$\frac{P}{P_{E,e}} = \frac{0.5P_E}{P_{E,e}} = 0.5(0.423)^2 = 0.0895$$

from which

$$\phi_e = 1.098$$

For $P = 0.5P_E$ the value of ξ in (11.3.12) is $\xi = 1.111$, from which $\phi_e = 1.090$. Thus the discrepancy is less than 1%.

The foregoing theory does not apply to intermediate degrees of end restraint. However, the idealized conditions of simply supported ends and clamped ends constitute the limits of possible variations and hence these two solutions provide useful information for the structural evaluation or design of the member.

The principal advantage of this approach is that it permits the solution for a clamped beam column to be obtained from the same direct and explicit formula as for simply supported beam columns, that is, (11.3.7) or, alternatively, Fig. 11.14. At the end of the beam column the eccentricity ratio and bending stress ratio to be used in (11.3.7) are, respectively,

$$\eta = \frac{[(1 - v)\Delta + \gamma\delta_0]A}{Z_e} \qquad (11.3.13)$$

$$\mu = \frac{\dfrac{M_e}{Z_e}}{\sigma_Y} \qquad (11.3.14)$$

where Z_e is the section modulus for whichever flange is in compression at the ends. Similarly, at the center of a clamped beam column these ratios are

$$\eta = \frac{[v\Delta + (1 - \gamma)\delta_0]A}{Z_c} \qquad (11.3.15)$$

$$\mu = \frac{\dfrac{M_c}{Z_c}}{\sigma_Y} \qquad (11.3.16)$$

where Z_c is the section modulus to the compression flange at the center.

REFERENCES

1. F. R. Shanley, "Inelastic Column Theory," *J. Aero. Sci.*, **14**, May 1947, pp. 261–267.
2. F. Bleich, "Buckling Strength of Metal Structures," McGraw-Hill, 1952.
3. Column Research Council, "Guide to Design Criteria for Metal Compression Members," Crosby Lockwood, London, 1960.
4. American Insititute of Steel Construction, "Specification for the Design, Fabrication, and Erection of Structural Steel for Building," New York, A.I.S.C., 1963.
5. L. Tall, "Recent Developments in the Study of Column Behaviour," *J. Inst. Eng., Aust.*, December 1964.
6. S. Timoshenko and J. Gere, *Theory of Elastic Stability*, McGraw-Hill.
7. A. Robertson, "The Strength of Struts," *Inst. Civ. Eng.*, Selected Paper No. 28, 1925.
8. L. S. Beedle, "Ductility as a Basis for Steel Design," *Conf. on Eng. Plasticity*, Cambridge, 1968, Cambridge University Press.
9. J. B. Dwight, "Use of Perry Formula to Represent the New European Strut Curves," Cambridge University, Dept. of Engineering Report TR.30, 1972.

ELASTIC BUCKLING OF PLATES

Jeom Kee Paik
Professor, Pusan National University
Busan, Korea

12.1 FUNDAMENTALS

This chapter presents analytical solutions to and/or the empirical expressions of the elastic buckling strength of a plate, which is a basic element of a continuous stiffened-plate structure. The plate is surrounded by such support members as longitudinal stiffeners (or girders) and transverse frames (or stiffeners), thus implying that the rotational restraints at the plate edges are neither zero nor infinite. In ships and offshore structures, such plates are likely to be subjected to both in-plane and out-of-plane loads. In-plane loads include longitudinal axial compression/tension, transverse axial compression/tension, edge shear, longitudinal in-plane bending, and transverse in-plane bending. Out-of-plane loads include lateral pressures that are due to cargo and/or water pressure. It should be noted that buckling does not occur in plates under axial tension or out-of-plane actions alone, but rather it occurs through the application of compressive loads.

The approaches to determining the elastic buckling of plates can be categorized into two types. The first type of approach is to search for the bifurcation point at which the plate begins to buckle because the flat form of equilibrium becomes unstable. In the general plate under normal working forces N_x and N_y and shear working force N_{xy}, equation (9.2.1) is availablc. The deflection function w of the plate may be approximately expressed and involve unknown coefficients, but must satisfy both the boundary condition and the general biharmonic equation.

$$w = a_1 f_1(x, y) + a_2 f_2(x, y) + \cdots + a_i f_i(x, y) + \cdots \qquad (12.1.1)$$

where $a_1, a_2, \ldots, a_i, \ldots$ are the unknown coefficients, and $f_1(x, y), f_2(x, y), \ldots, f_i(x, y), \ldots$ are the functions that satisfy the boundary conditions.

We assume that the plate is free to move inward under the in-plane loads (this should always be assumed for compressive loads) and hence there is no strain in the mid-plane of the plate. Under these conditions the strain energy of deformation is due to bending only and is given by (Timoshenko and Gere, 1961)

$$U = \frac{D}{2}\int_0^a\int_0^b\left\{\left(\frac{\partial^2 w}{\partial x^2}+\frac{\partial^2 w}{\partial y^2}\right)^2 - 2(1-\nu)\left[\frac{\partial^2 w}{\partial x^2}\frac{\partial^2 w}{\partial y^2}-\left(\frac{\partial^2 w}{\partial x \partial y}\right)^2\right]\right\}dxdy \qquad (12.1.2)$$

where $D = \dfrac{Et^3}{12(1-\nu^2)}$ = plate bending flexibility, E = Young's modulus, ν = Poisson's ratio.

Likewise, the work done by the applied forces is given by

$$W = -\frac{1}{2}\int_0^a\int_0^b\left[N_x\left(\frac{\partial w}{\partial x}\right)^2 + N_y\left(\frac{\partial w}{\partial y}\right)^2 + 2N_{xy}\frac{\partial w}{\partial x}\frac{\partial w}{\partial y}\right]dxdy \qquad (12.1.3)$$

The principle of virtual work requires that

$$W = U \qquad (12.1.4)$$

Substitution of (12.1.1) into (12.1.4) will yield the elastic buckling stress of the plate. The first type of approach can also be applied by the method of variation. Considering that the following equations are available for normal forces N_x and N_y and shear force N_{xy} with common factor γ, i.e.,

$$N_x = \gamma N_x', \quad N_y = \gamma N_y', \quad N_{xy} = \gamma N_{xy}' \qquad (12.1.5)$$

The buckling point can be determined by increasing γ simultaneously. The critical value of this factor is obtained from

$$\gamma = \frac{U}{W} \qquad (12.1.6)$$

In our calculation of γ, the variation of (12.1.6) should be zero.

$$\delta\gamma = \frac{W\delta U - U\delta W}{W^2} \quad \text{or} \quad \frac{1}{W}(\delta U - \gamma\delta W) = 0$$

$$\text{or} \quad \delta U - \gamma\delta W = 0 \quad (12.1.7)$$

Substituting (12.1.1) into (12.1.7), a set of homogeneous linear equations with regard to the unknown coefficients can then be obtained.

$$\frac{\partial U}{\partial a_1} - \gamma\frac{\partial W}{\partial a_1} = 0, \frac{\partial U}{\partial a_2} - \gamma\frac{\partial W}{\partial a_2} = 0, \ldots,$$

$$\frac{\partial U}{\partial a_i} - \gamma\frac{\partial W}{\partial a_i} = 0, \ldots. \quad (12.1.8)$$

Solutions of (12.1.8) that are different from zero can be obtained only if the determinant of these equations is zero. Once the unknown coefficients are determined, the critical value of γ can be obtained from (12.1.6).

The second type of approach to searching for the buckling point is to investigate the post-buckling strength behavior of the plate by solving the following equilibrium and compatibility equations.

$$D\left(\frac{\partial^4 w}{\partial x^4} + 2\frac{\partial^4 w}{\partial x^2 \partial y^2} + \frac{\partial^4 w}{\partial y^4}\right) - t\left[\frac{\partial^2 F}{\partial y^2}\frac{\partial^2 w}{\partial x^2} - 2\frac{\partial^2 F}{\partial x \partial y}\frac{\partial^2 w}{\partial x \partial y} + \frac{\partial^2 F}{\partial x^2}\frac{\partial^2 w}{\partial y^2} + \frac{p}{t}\right] = 0 \quad (12.1.9a)$$

$$\frac{\partial^4 F}{\partial x^4} + 2\frac{\partial^4 F}{\partial x^2 \partial y^2} + \frac{\partial^4 F}{\partial y^4} - E\left[\left(\frac{\partial^2 w}{\partial x \partial y}\right)^2 - \frac{\partial^2 w}{\partial x^2}\frac{\partial^2 w}{\partial y^2}\right] = 0 \quad (12.1.9b)$$

where F is the Airy stress function and p is lateral pressure. With Airy's stress function F and plate deflection w known, the membrane stresses at the mid-thickness of the plate can be calculated as follows.

$$\sigma_x = \frac{N_x}{t} = \frac{\partial^2 F}{\partial y^2}, \sigma_y = \frac{N_y}{t} = \frac{\partial^2 F}{\partial x^2}, \tau = \frac{N_{xy}}{t} = -\frac{\partial^2 F}{\partial x \partial y} \quad (12.1.10)$$

The buckling point can then be determined from (12.1.10) when the plate deflection is zero immediately before the plate begins to buckle.

For convenience of plate buckling analysis in this chapter, the coordinate system for the plate uses x in the long direction and y in the short direction. The dimensions of the plate are a in length (i.e., in the x or longer direction), b in breadth (i.e., in the y or shorter direction), and t in thickness. The plate aspect ratio, a/b, is then always ≥ 1. Compressive stress is taken as positive and tensile stress as negative.

12.2 SIMPLY SUPPORTED EDGES

Neglecting the rotational restraints that are due to the torsional rigidities of the support members at the plate edges, the boundary condition becomes simply supported. This boundary condition may be relevant when the torsional rigidity of the support members is small relative to plate bending flexibility.

12.2.1 Uniaxial Compression in the *x*-Direction

The elastic buckling stress of a simply supported plate under uniaxial compression in the x-direction is now considered, as shown in Fig. 12.1, where the rotational restraints at all (four) plate edges are assumed to be zero. In this case, $N_x = -\sigma_x t$ and $p = N_y = N_{xy} = 0$ are applicable.

Since the edges are simply supported, the deflected shape can be expressed in the following form.

$$w = \sum_m \sum_n w_{mn} = \sum_m \sum_n C_{mn} \sin\frac{m\pi x}{a} \sin\frac{n\pi y}{b} \quad (12.2.1)$$

Substitution of (12.2.1) into (12.1.2) yields the strain energy of the plate as follows.

$$U = \frac{\pi^4 ab}{8} D \sum_m \sum_m C_{mn}^2 \left(\frac{m^2}{a^2} + \frac{n^2}{b^2}\right)^2 \quad (12.2.2)$$

Also, the work done by the in-plane compressive stress is obtained from (12.1.3) with $N_y = N_{xy} = 0$ as follows.

$$W = \frac{\pi^2 b \sigma_x t}{8a} \sum_m \sum_n C_{mn}^2 m^2 \quad (12.2.3)$$

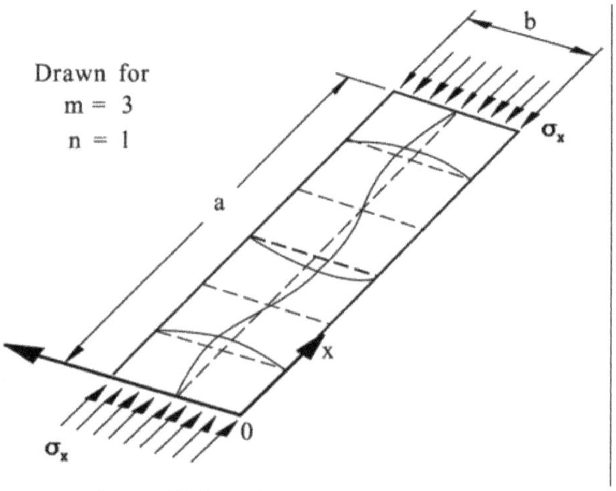

Figure 12.1 Buckled shape of a long plate.

From (12.1.4), the equilibrium value of σ_x is obtained as follows.

$$\sigma_x = \frac{\pi^2 a^2 D \sum_m \sum_n C_{mn}^2 \left(\frac{m^2}{a^2} + \frac{n^2}{b^2}\right)^2}{t \sum_m \sum_n C_{mn}^2 m^2} \qquad (12.2.4)$$

The values of C_{mn} which give the minimum value of σ_x may be determined from the fact that an expression of the form $(c_1 + c_2 + c_3 \cdots)/(d_1 + d_2 + d_3 \cdots)$ has some intermediate value between the maximum and the minimum of the fractions c_1/d_1, c_2/d_2, c_3/d_3, ... Thus there exists one fraction, c_j/d_j, which is less than any other fraction, c_i/d_i, and also less than any other sum or partial sum of fractions of the form shown. Therefore the minimum value of σ_x will be obtained by taking only one term of (12.2.1), say C_{mn}. Then the minimum value of σ_x corresponding to the buckling stress is given by

$$\sigma_x \equiv \sigma_{xE,1} = \frac{\pi^2 a^2 D \left(\frac{m^2}{a^2} + \frac{n^2}{b^2}\right)^2}{tm^2} \qquad (12.2.5)$$

The parameters m and n indicate the number of half-waves in each direction in the buckled shape. Both must be integers, and it can be seen that the value of n that gives the smallest value of σ_x is $n = 1$. Hence the plate will buckle into only one half-wave transversely, and the resulting buckling stress is

$$\sigma_{xE,1} = \frac{\pi^2 D}{a^2 t}\left[m + \frac{1}{m}\left(\frac{a}{b}\right)^2\right]^2 \qquad (12.2.6)$$

This equation was derived by G.H. Bryan in 1891. The plate buckling stress $\sigma_{E,1}$ under a single type of load is usually written in a more general form in terms of a buckling coefficient k and the plate width b as follows.

$$\sigma_{E,1} = k\frac{\pi^2 D}{b^2 t} = k\frac{\pi^2 E}{12(1-\nu^2)}\left(\frac{t}{b}\right)^2 \qquad (12.2.7)$$

The expression for the buckling coefficient k depends on the type of boundary support as well as the loading type. From a comparison of (12.2.6) and (12.2.7) it is evident that for simply supported plates under uniaxial compression in the x-direction k is given by

$$k = \left(\frac{mb}{a} + \frac{a}{mb}\right)^2 \qquad (12.2.8)$$

The buckling half-wave number m can be determined as an integer satisfying the following equation.

$$\frac{mb}{a} + \frac{a}{mb} \le \frac{(m+1)b}{a} + \frac{a}{(m+1)b}$$

$$\text{or } \frac{a}{b} \le \sqrt{m(m+1)} \qquad (12.2.9)$$

In Fig. 12.2, the buckling coefficient k is plotted against aspect ratio a/b for various values of m. This figure shows that the lowest (and therefore truly critical) value of σ_x will occur for different values of m, depending on the aspect ratio. It is also seen that k approaches 4 as a/b increases.

12.2.2 Other Types of Single Load

The elastic buckling stress of a long plate, i.e., one with $a/b \ge 1$, is calculated from (12.2.7), but with different values of buckling coefficient depending on types of loads. Table 12.1 indicates the buckling stress σ_E and the associated buckling coefficient k for various single types of loads.

12.2.3 Combined Biaxial Compression/Tension

The buckling stress of the plate under a combination of σ_x and σ_y is now considered with $\tau = p = 0$. The buckling mode deflection function of the plate simply supported at all (four) edges is assumed as follows.

$$w = A_m \sin\frac{m\pi x}{a}\sin\frac{\pi y}{b} \qquad (12.2.10)$$

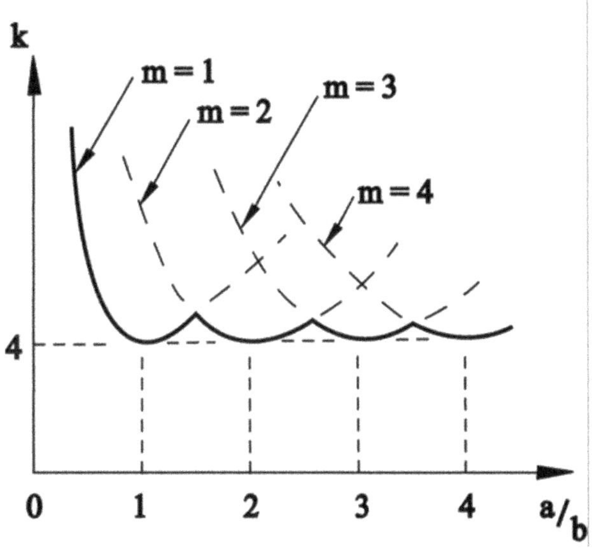

Figure 12.2 Buckling coefficient as a function of the plate aspect ratio.

Table 12.1 Buckling Coefficients for a Simply Supported Plate Under a Single Component of Normal or Shear Stress for $a/b \geq 1$

Load type	σ_E	k
Uniaxial compression in the x-direction, $\sigma_x = N_x/t$	$\sigma_{xE,1}$	$k_x = \left[\dfrac{a}{m_o b} + \dfrac{m_o b}{a}\right]^2$, in which m_o is the buckling half-wave number for the plate in the x-direction, which is the minimum integer satisfying $a/b \leq \sqrt{m_o(m_o + 1)}$. For practical use, the buckling coefficient is often approximated to $k_x = 4$.
Uniaxial compression in the y-direction, $\sigma_y = N_y/t$	$\sigma_{yE,1}$	$k_y = \left[1 + \left(\dfrac{b}{a}\right)\right]^2$
Uniform edge shear, $\tau = N_{xy}/t$	$\tau_{E,1}$	$k_\tau = 4\left(\dfrac{b}{a}\right)^2 + 5.34$
Pure in-plane bending in the x-direction, σ_{bx}	$\sigma_{bxE,1}$	$k_{bx} = 23.9$
Pure in-plane bending in the y-direction, σ_{by}	$\sigma_{byE,1}$	$k_{by} = \begin{cases} 23.9 & \text{for } 1 \leq \dfrac{a}{b} \leq 1.5 \\ 15.87 + 1.87\left(\dfrac{a}{b}\right)^2 + 8.6\left(\dfrac{b}{a}\right)^2 & \text{for } \dfrac{a}{b} > 1.5 \end{cases}$

Notes: (1) Subscript '1' represents buckling under a single load component.
(2) Pure in-plane bending means that the maximum (edge) stresses are $\pm\,\sigma_b$.

where A_m is the unknown amplitude of the deflection function, and m is the buckling half-wave number in the x-direction. Substitution of (12.2.10) into (12.1.9b) yields

$$\frac{\partial^4 F}{\partial x^4} + 2\frac{\partial^4 F}{\partial x^2 \partial y^2} + \frac{\partial^4 F}{\partial y^4}$$
$$= -\frac{m^2\pi^4 E A_m^2}{2a^2b^2}\left(\cos\frac{2m\pi x}{a} + \cos\frac{2\pi y}{b}\right) \qquad (12.2.11)$$

The solution of (12.2.11) with regard to the Airy stress function F is obtained as follows.

$$F = \sigma_x \frac{y^2}{2} + \sigma_y \frac{x^2}{2}$$
$$+ \frac{EA_m^2}{32}\left(\frac{a^2}{m^2b^2}\cos\frac{2m\pi x}{a} + \frac{m^2b^2}{a^2}\cos\frac{2\pi y}{b}\right) \qquad (12.2.12)$$

By substituting (12.2.10) and (12.2.12) into (12.1.9a) and applying the Galerkin method with $p = 0$, the following equation is obtained.

$$\int_0^a\int_0^b \left\{D\left(\frac{\partial^4 w}{\partial x^4} + 2\frac{\partial^4 w}{\partial x^2\partial y^2} + \frac{\partial^4 w}{\partial y^4}\right)\right.$$
$$\left. -t\left(\frac{\partial^2 F}{\partial y^2}\frac{\partial^2 w}{\partial x^2} - 2\frac{\partial^2 F}{\partial x\partial y}\frac{\partial^2 w}{\partial x\partial y} + \frac{\partial^2 F}{\partial x^2}\frac{\partial^2 w}{\partial y^2}\right)\right\}$$
$$\times \sin\frac{m\pi x}{a}\sin\frac{\pi y}{b}\,dxdy = 0 \qquad (12.2.13)$$

By performing the integration of (12.2.13) over the entire plate, a third-order equation with regard to the unknown variable A_m is obtained as follows.

$$A_m\left[\frac{\pi^2 E}{16}\left(\frac{m^4}{a^4} + \frac{1}{b^4}\right)A_m^2 + \frac{m^2}{a^2}\sigma_x\right.$$
$$\left. + \frac{1}{b^2}\sigma_y - \frac{\pi^2 D}{t}\left(\frac{m^2}{a^2} + \frac{1}{b^2}\right)^2\right] = 0 \qquad (12.2.14)$$

Non-zero solution of A_m is readily given by

$$A_m = \left\{-\frac{16}{\pi^2 E(m^4/a^4 + 1/b^4)}\left[\frac{m^2}{a^2}\sigma_x + \frac{1}{b^2}\sigma_y\right.\right.$$
$$\left.\left. -\frac{\pi^2 D}{t}\left(\frac{m^2}{a^2} + \frac{1}{b^2}\right)^2\right]\right\}^{1/2} \qquad (12.2.15)$$

In the post-buckling regime, the plate deflection is figured out of (12.2.10) with (12.2.15). Also, the membrane stresses at the mid-thickness of the plate can be calculated from (12.1.10) with (12.2.12).

Since the deflection immediately before the plate buckles must be zero, i.e., $A_m = 0$, the following equation representing the buckling condition is obtained.

$$\frac{m^2}{a^2}\sigma_x + \frac{1}{b^2}\sigma_y - \frac{\pi^2 D}{t}\left(\frac{m^2}{a^2} + \frac{1}{b^2}\right)^2 = 0 \qquad (12.2.16)$$

If the loading ratio between σ_x and σ_y is kept constant, the buckling stress of the plate can be obtained as follows.

$$\sigma_{xE} = \frac{\pi^2 D}{b^2 t}\,\frac{(1 + m^2b^2/a^2)^2}{c + m^2b^2/a^2} \qquad (12.2.17a)$$

where $c = \sigma_y/\sigma_x$, σ_{xE} = buckling stress component in the x-direction, $\sigma_{yE} = c\sigma_{xE}$ = buckling stress component in the y-direction. The buckling half-wave number m is determined as the minimum integer that satisfies the following condition (Paik and Thayamballi, 2003).

$$\frac{(m^2/a^2 + 1/b^2)^2}{m^2/a^2 + c/b^2} \le \frac{[(m+1)^2/a^2 + 1/b^2]^2}{(m+1)^2/a^2 + c/b^2} \qquad (12.2.17b)$$

When only longitudinal axial compression is applied, i.e., $c = 0$, (12.2.17b) is simplified to (12.2.9).

Figure 12.3 shows the elastic buckling strength interaction of a long plate with $a/b = 5$ between the biaxial compression. When the longitudinal axial compression is predominant, the buckling half wave number m is 5, but it decreases as the transverse axial compression increases, and, eventually, one half wave appears when the transverse axial compression is predominant.

An empirical expression of the plate buckling strength interaction relationship between biaxial compression is given as follows.

$$\left(\frac{\sigma_{xE}}{\sigma_{xE,1}}\right)^{\alpha_1} + \left(\frac{\sigma_{yE}}{\sigma_{yE,1}}\right)^{\alpha_2} = 1 \qquad (12.2.18)$$

where α_1 and α_2 are constants that are a function of the plate aspect ratio. Based on the computed results, these constants may be determined empirically as follows.

$$\alpha_1 = \alpha_1 = 1 \quad \text{for } 1 \le \frac{a}{b} \le \sqrt{2},$$

$$\left.\begin{aligned} \alpha_1 &= 0.0293\left(\frac{a}{b}\right)^3 - 0.3364\left(\frac{a}{b}\right)^2 + 1.5854\left(\frac{a}{b}\right) - 1.0596 \\ \alpha_2 &= 0.0049\left(\frac{a}{b}\right)^3 - 0.1183\left(\frac{a}{b}\right)^2 + 0.6153\left(\frac{a}{b}\right) + 0.8522 \end{aligned}\right\}$$

$$\text{for } \frac{a}{b} > \sqrt{2}.$$

12.2.4 Combined Longitudinal Axial Compression and Longitudinal In-plane Bending

The buckling stress of the plate under combined longitudinal axial compression and longitudinal in-plane bending can be obtained as follows.

$$\sigma_{xE} = \frac{8.4}{\psi + 1.1}\,\frac{\pi^2 D}{b^2 t},\ \sigma_{bxE} = \frac{1-\psi}{1+\psi}\,\sigma_{xE} \text{ for } 0 < \psi < 1 \qquad (12.2.19a)$$

$$\sigma_{xE} = (10\psi^2 - 6.4\psi + 7.6)\,\frac{\pi^2 D}{b^2 t},$$

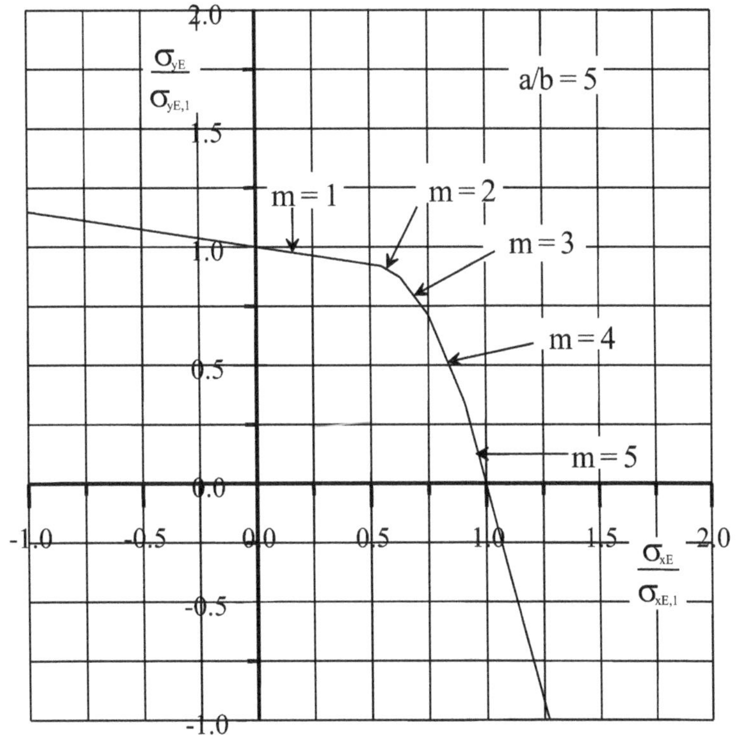

Figure 12.3 Elastic buckling strength interaction of a long plate between biaxial compression/tension.

$$\sigma_{bxE} = \frac{1-\psi}{1+\psi}\sigma_{xE} \text{ for } -1 < \psi < 0 \quad (12.2.19b)$$

where ψ = as defined in Figs.12.4a and 12.4b.

A continuous formula of the buckling strength interaction relationship is often useful for practical design purpose. The following is an empirical formula of the buckling strength interaction between longitudianl axial compression and longitudinal in-plane bending.

$$\frac{\sigma_{xE}}{\sigma_{xE,1}} + \left(\frac{\sigma_{bxE}}{\sigma_{bxE,1}}\right)^c = 1 \quad (12.2.19c)$$

where $c = 1.75 \sim 2$.

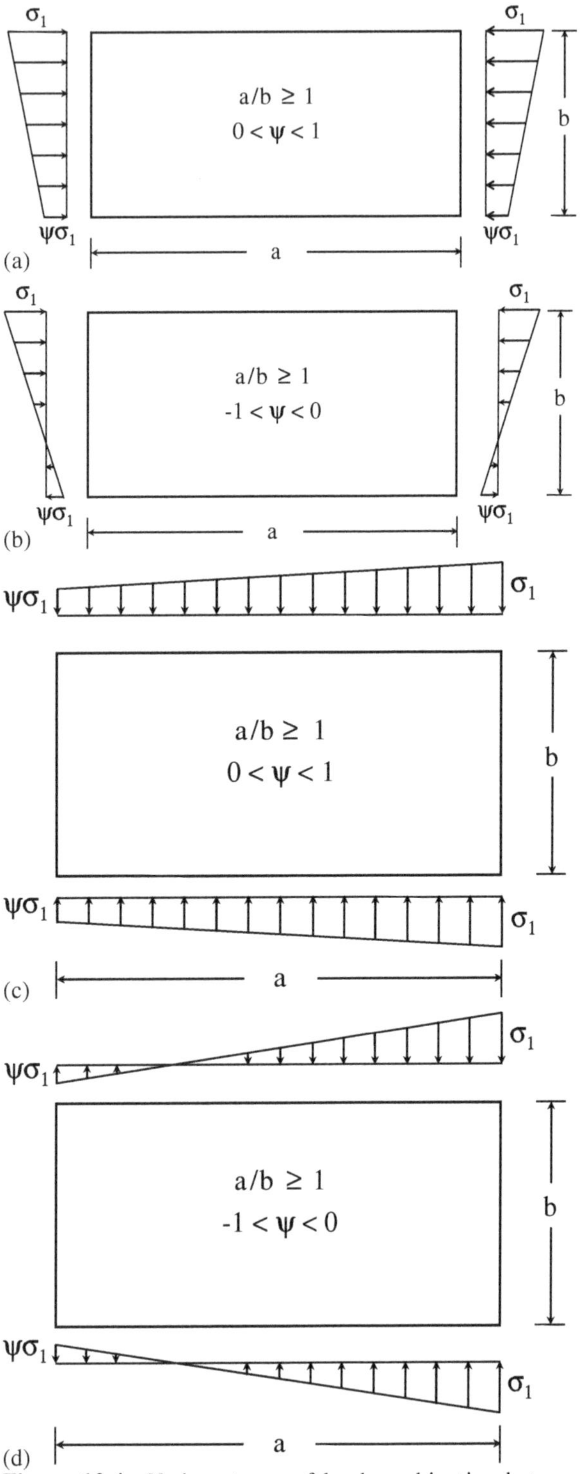

Figure 12.4 Various types of load combination between axial compression and in-plane bending.

12.2.5 Combined Transverse Axial Compression and Longitudinal In-plane Bending

The buckling strength interaction relationship between transverse axial compression and longitudinal in-plane bending is given by

$$\left(\frac{\sigma_{yE}}{\sigma_{yE,1}}\right)^{\alpha_3} + \left(\frac{\sigma_{bxE}}{\sigma_{bxE,1}}\right)^{\alpha_4} = 1 \quad (12.2.20)$$

where

$$\alpha_3 = \alpha_4 = 1.50\left(\frac{a}{b}\right) - 0.30 \text{ for } 1 \le \frac{a}{b} \le 1.6,$$

$$\left.\begin{aligned} \alpha_3 &= -0.625\left(\frac{a}{b}\right) + 3.10 \\ \alpha_4 &= 6.25\left(\frac{a}{b}\right) - 7.90 \end{aligned}\right\} \text{for } 1.6 < \frac{a}{b} \le 3.2,$$

$$\left.\begin{aligned} \alpha_3 &= 1.10 \\ \alpha_4 &= 12.10 \end{aligned}\right\} \text{for } 3.2 < \frac{a}{b}.$$

12.2.6 Combined Longitudinal Axial Compression and Transverse In-plane Bending

The buckling strength interaction relationship between longitudinal axial compression and transverse in-plane bending is given by

$$\left(\frac{\sigma_{xE}}{\sigma_{xE,1}}\right)^{\alpha_5} + \left(\frac{\sigma_{byE}}{\sigma_{byE,1}}\right)^{\alpha_6} = 1 \quad (12.2.21)$$

where

$$\left.\begin{aligned} \alpha_5 &= 0.930\left(\frac{a}{b}\right)^2 - 2.890\left(\frac{a}{b}\right) + 3.160 \\ \alpha_6 &= 1.20 \end{aligned}\right\} \text{for } 1 < \frac{a}{b} \le 2,$$

$$\left.\begin{aligned} \alpha_5 &= 0.066\left(\frac{a}{b}\right)^2 - 0.246\left(\frac{a}{b}\right) + 1.328 \\ \alpha_6 &= 1.20 \end{aligned}\right\} \text{for } 2 < \frac{a}{b} \le 5,$$

$$\left.\begin{aligned}\alpha_5 &= 1.117\left(\frac{a}{b}\right) - 3.837\\ \alpha_6 &= -0.167\left(\frac{a}{b}\right) + 2.035\end{aligned}\right\}\text{for } 5 < \frac{a}{b} \le 8,$$

$$\left.\begin{aligned}\alpha_5 &= 5.10\\ \alpha_6 &= 0.70\end{aligned}\right\}\text{for } 8 < \frac{a}{b}.$$

12.2.7 Combined Transverse Axial Compression and Transverse In-plane Bending

The buckling stress of the plate between transverse axial compression and transverse in-plane bending can be calculated as follows.

$$\sigma_{yE} = \left(\frac{a}{b} + \frac{b}{a}\right)^2 \frac{2.1}{\psi + 1.1}\,\frac{\pi^2 D}{b^2 t},$$

$$\sigma_{byE} = \frac{1-\psi}{1+\psi}\,\sigma_{yE} \text{ for } 0 < \psi < 1, \qquad (12.2.22a)$$

$$\sigma_{yE} = \left[1.91(1+\psi)\left(\frac{a}{b} + \frac{b}{a}\right)^2 - \psi\left\{8.6\left(\frac{b}{a}\right)^2 + 1.87\left(\frac{a}{b}\right)^2 + 15.87\right\} + 10\psi(1+\psi)\right]\frac{\pi^2 D}{b^2 t},$$

$$\sigma_{byE} = \frac{1-\psi}{1+\psi}\,\sigma_{yE} \text{ for } -1 < \psi < 0, \qquad (12.2.22b)$$

where ψ = as defined in Figs.12.4c and 12.4d.

An empirical expression of the continuous buckling strength interaction between transverse axial compression and transverse in-plane bending is given by

$$\left(\frac{\sigma_{yE}}{\sigma_{yE,1}}\right)^{\alpha_7} + \left(\frac{\sigma_{byE}}{\sigma_{byE,1}}\right)^{\alpha_8} = 1 \qquad (12.2.22c)$$

where

$$\left.\begin{aligned}\alpha_7 &= 1.0\\ \alpha_8 &= \frac{1}{6.5}\left(14.0 - \frac{a}{b}\right)\end{aligned}\right\}\text{for } 1 \le \frac{a}{b} \le 7.5,$$

$$\alpha_7 = \alpha_8 = 1.0 \text{ for } 7.5 < \frac{a}{b}.$$

12.2.8 Combined Biaxial In-plane Bending

The buckling strength interaction relationship between biaxial in-plane bending is given by

$$\left(\frac{\sigma_{bxE}}{\sigma_{bxE,1}}\right)^{\alpha_9} + \left(\frac{\sigma_{byE}}{\sigma_{byE,1}}\right)^{\alpha_{10}} = 1 \qquad (12.2.23)$$

where

$$\left.\begin{aligned}\alpha_9 &= 0.050\left(\frac{a}{b}\right) + 1.080\\ \alpha_{10} &= 0.268\left(\frac{a}{b}\right) - 1.248\left(\frac{b}{a}\right) + 2.112\end{aligned}\right\}\text{for } 1 \le \frac{a}{b} \le 3,$$

$$\left.\begin{aligned}\alpha_9 &= 0.146\left(\frac{a}{b}\right)^2 - 0.533\left(\frac{a}{b}\right) + 1.515\\ \alpha_{10} &= 0.268\left(\frac{a}{b}\right) - 1.248\left(\frac{b}{a}\right) + 2.112\end{aligned}\right\}\text{for } 3 < \frac{a}{b} \le 5,$$

$$\left.\begin{aligned}\alpha_9 &= 3.20\left(\frac{a}{b}\right) - 13.50\\ \alpha_{10} &= -0.70\left(\frac{a}{b}\right) + 6.70\end{aligned}\right\}\text{for } 5 < \frac{a}{b} \le 8,$$

$$\left.\begin{aligned}\alpha_9 &= 12.10\\ \alpha_{10} &= 1.10\end{aligned}\right\}\text{for } 8 < \frac{a}{b}.$$

12.2.9 Combined Longitudinal Axial Compression and Edge Shear

The buckling strength interaction relationship between longitudinal axial compression and edge shear is given by

$$\frac{\sigma_{xE}}{\sigma_{xE,1}} + \left(\frac{\tau_E}{\tau_{E,1}}\right)^{\alpha_{11}} = 1 \qquad (12.2.24)$$

where

$$\alpha_{11} = \begin{cases} -0.160\left(\frac{a}{b}\right)^2 + 1.080\left(\frac{a}{b}\right) + 1.082 & \text{for } 1 \le \frac{a}{b} \le 3.2.\\ 2.90 & \text{for } \frac{a}{b} > 3.2\end{cases}$$

12.2.10 Combined Transverse Axial Compression and Edge Shear

The buckling strength interaction relationship between transverse axial compression and edge shear is given by

$$\frac{\sigma_{yE}}{\sigma_{yE,1}} + \left(\frac{\tau_E}{\tau_{E,1}}\right)^{\alpha_{12}} = 1 \qquad (12.2.25)$$

where

$$\alpha_{12} = \begin{cases} 0.10\left(\frac{a}{b}\right) + 1.90 & \text{for } 1 \le \frac{a}{b} \le 2 \\ 0.70\left(\frac{a}{b}\right) + 0.70 & \text{for } 2 \le \frac{a}{b} \le 6. \\ 4.90 & \text{for } 6 < \frac{a}{b} \end{cases}$$

12.2.11 Combined Longitudinal In-plane Bending and Edge Shear

The buckling strength interaction relationship between longitudinal in-plane bending and edge shear is given by

$$\left(\frac{\sigma_{bxE}}{\sigma_{bxE,1}}\right)^2 + \left(\frac{\tau_E}{\tau_{E,1}}\right)^2 = 1 \qquad (12.2.26)$$

12.2.12 Combined Transverse In-plane Bending and Edge Shear

The buckling strength interaction relationship between transverse in-plane bending and edge shear is given by

$$\left(\frac{\sigma_{byE}}{\sigma_{byE,1}}\right)^2 + \left(\frac{\tau_E}{\tau_{E,1}}\right)^2 = 1 \qquad (12.2.27)$$

12.2.13 Combined In-plane Load Components of All Types

The buckling strength interaction relationship among all types of loads is given by

$$\left[\frac{\sigma_{xE}}{C_1 C_4 \sigma_{xE,1}\left\{1 - \left(\frac{\tau_E}{C_3 C_6 \tau_{E,1}}\right)^{\alpha_{11}}\right\}}\right]^{\alpha_1} +$$

$$\left[\frac{\sigma_{yE}}{C_2 C_5 \sigma_{yE,1}\left\{1 - \left(\frac{\tau_E}{C_3 C_6 \tau_{E,1}}\right)^{\alpha_{12}}\right\}}\right]^{\alpha_2} = 1 \qquad (12.2.28)$$

where

$$C_1 = 1 - \left(\frac{\sigma_{bxE}}{C_7 \sigma_{bxE,1}}\right)^2, \; C_2 = \left\{1 - \left(\frac{\sigma_{bxE}}{C_7 \sigma_{bxE,1}}\right)^{\alpha_4}\right\}^{1/\alpha_3},$$

$$C_3 = \left\{1 - \left(\frac{\sigma_{bxE}}{C_7 \sigma_{bxE,1}}\right)^2\right\}^{0.5}, \; C_4 = \left\{1 - \left(\frac{\sigma_{byE}}{\sigma_{byE,1}}\right)^{\alpha_6}\right\}^{1/\alpha_5},$$

$$C_5 = \left\{1 - \left(\frac{\sigma_{byE}}{\sigma_{byE,1}}\right)^{\alpha_8}\right\}^{1/\alpha_7}, \; C_6 = \left\{1 - \left(\frac{\sigma_{byE}}{\sigma_{byE,1}}\right)^2\right\}^{0.5},$$

$$C_7 = \left\{1 - \left(\frac{\sigma_{byE}}{\sigma_{byE,1}}\right)^{\alpha_{10}}\right\}^{1/\alpha_9}.$$

12.3 CLAMPED EDGES

12.3.1 Single Types of Loads

Equation (12.2.7) is still used to compute buckling strength, but the buckling coefficient is now calculated from Table 12.2. Figure 12.5 shows the buckling coefficient k of the plate under uniaxial compression, plotted against the aspect ratio for different conditions of plate edges.

12.3.2 Combined Loads

The plate buckling interaction relationships between combined loads with clamped edges are the same as those with simply supported edges, whereas the plate buckling strength under a single type of load must be computed with the clamped edge condition.

12.4 PARTIALLY ROTATION-RESTRAINED EDGES

Figure 12.6 shows a plate surrounded by support members, e.g., longitudinal stiffeners and transverse frames. It is assumed that the structural geometry and dimensions of these support members are the same in the same direction. The support members' degree of rotational restraints can be defined using a parameter that is a ratio of the torsional rigidity of the support member to the bending flexibility of the plate, as follows.

$$\zeta_L = C_L \frac{GJ_L}{bD}, \quad \zeta_S = C_S \frac{GJ_S}{aD} \qquad (12.4.1)$$

where ζ_L, ζ_S = rotational restraint parameters for the longitudinal or transverse support member, $J_L = \frac{h_{wx}t_{wx}^3 + b_{fx}t_{fx}^3}{3}$, $J_S = \frac{h_{wy}t_{wy}^3 + b_{fy}t_{fy}^3}{3}$, and $G = \frac{E}{2(1+\nu)}$.

$C_L = C_S = 1.0$ is usually applied, but in some cases, the support members may not be fully effective because they may distort sideways prior to plate buckling. Subsequently, their rotational restraints along the plate edges may decrease. C_L and C_S in (12.4.1) are constants that take this effect into account:

$$C_L = \frac{J_L}{J_{PL}} \le 1.0, \quad C_S = \frac{J_S}{J_{PS}} \le 1.0 \qquad (12.4.2)$$

where $J_{PL} = \frac{bt^3}{3}$, $J_{PS} = \frac{at^3}{3}$.

Table 12.2 Elastic Buckling Coefficients of Clamped Plates Under Single Types of Loads for $\frac{a}{b} \geq 1$

Load Type	σ_E	B.C.	k
Uniaxial compression in the x-direction, $\sigma_x = N_x/t$	$\sigma_{xE,1}$	TSLC	$k_x = \begin{cases} 7.39\left(\frac{a}{b}\right)^2 - 19.6\left(\frac{a}{b}\right)^2 + 20 & \text{for } 1.0 \leq \frac{a}{b} \leq 1.33 \\ 6.98 & \text{for } 1.33 < \frac{a}{b} \end{cases}$
		TCLS	$k_x = \begin{cases} -0.95\left(\frac{a}{b}\right)^3 + 6.4\left(\frac{a}{b}\right)^2 - 14.86\left(\frac{a}{b}\right) + 16.34 & \text{for } 1.0 \leq \frac{a}{b} \leq 2.0 \\ 0.2\left(\frac{a}{b}\right)^2 - 1.4\left(\frac{a}{b}\right) + 6.64 & \text{for } 2.0 \leq \frac{a}{b} \leq 3.0 \\ -0.05\left(\frac{a}{b}\right) + 4.4 & \text{for } 3.0 \leq \frac{a}{b} < 8.0 \\ 4.0 & \text{for } 8.0 \leq \frac{a}{b} \end{cases}$
		AC	$k_x = \begin{cases} -1.23\left(\frac{a}{b}\right)^3 + 7.9\left(\frac{a}{b}\right)^2 - 17.65\left(\frac{a}{b}\right) + 21.35 & \text{for } 1.0 \leq \frac{a}{b} \leq 2.0 \\ 0.2\left(\frac{a}{b}\right)^2 - 1.62\left(\frac{a}{b}\right) + 10.35 & \text{for } 2.0 \leq \frac{a}{b} \leq 3.0 \\ -0.062\left(\frac{a}{b}\right) + 7.476 & \text{for } 3.0 \leq \frac{a}{b} < 8.0 \\ 6.98 & \text{for } 8.0 \leq \frac{a}{b} \end{cases}$
Uniaxial compression in the y-direction, $\sigma_y = N_y/t$	$\sigma_{yE,1}$	TSLC	$k_y = \left[1.0 + \left(\frac{b}{a}\right)^2\right]^2 + 3.01 \quad \text{for } 0.0 < \frac{b}{a} \leq 1.0$
		TCLS	$k_y = \begin{cases} \left[1.0 + \left(\frac{b}{a}\right)^2\right]^2 + 0.12 & \text{for } 0.0 < \frac{b}{a} < 0.34 \\ \left[0.95 + 1.89\left(\frac{b}{a}\right)^2\right]^2 & \text{for } 0.34 \leq \frac{b}{a} \leq 0.96 \\ 13.98\left(\frac{b}{a}\right) - 6.20 & \text{for } 0.96 < \frac{b}{a} \leq 1.0 \end{cases}$
		AC	$k_y = \begin{cases} \left[1.0 + \left(\frac{b}{a}\right)^2\right]^2 + 4.8 & \text{for } 0.0 < \frac{b}{a} < 0.8 \\ \left[1.92 + 1.305\left(\frac{b}{a}\right)^2\right]^2 & \text{for } 0.8 \leq \frac{b}{a} \leq 1.0 \end{cases}$

(continued)

12.4.1 Partially Rotation-Restrained at the Longitudinal Edges and Simply Supported at the Transverse Edges

Equation (12.2.7) is used to compute the plate buckling strength, but buckling coefficient k in (12.2.7) is given as a function of the rotational restraint parameter defined in (12.4.1).

Longitudinal Axial Compression

In this case, the buckling strength coefficient is given by (Paik and Thayamballi, 2003).

$$k_x = \begin{cases} 0.396\zeta_L^3 - 1.974\zeta_L^2 + 3.565\zeta_L + 4.0 & \text{for } 0 \leq \zeta_L \leq 2 \\ 6.951 - \dfrac{0.881}{\zeta_L - 0.4} & \text{for } 2 \leq \zeta_L < 20 \\ 7.025 & \text{for } 20 \leq \zeta_L \end{cases} \qquad (12.4.3)$$

Transverse Axial Compression

In this case, the buckling coefficient is given by

$$k_y = e_1\zeta_L^2 + e_2\zeta_L + e_3 \qquad (12.4.4)$$

Table 12.2 Elastic Buckling Coefficients of Clamped Plates Under Single Types of Loads for $\frac{a}{b} \geq 1$ *(Continued)*

Load Type	σ_E	B.C.	k
Uniform edge shear, $\tau = N_{xy}/t$	$\tau_{E,1}$	TSLC	$k_\tau = 2.4\left(\frac{b}{a}\right)^2 + 1.08\left(\frac{b}{a}\right) + 9.0$ for $0.0 < b/a \leq 1.0$
		TCLS	$k_\tau = \begin{cases} 2.25\left(\frac{b}{a}\right)^2 + 1.95\left(\frac{b}{a}\right) + 5.35 & \text{for } 0.0 < \frac{b}{a} \leq 0.4 \\ 22.92\left(\frac{b}{a}\right)^3 - 33.0\left(\frac{b}{a}\right)^2 + 20.43\left(\frac{b}{a}\right) + 2.13 & \text{for } 0.4 < \frac{b}{a} \leq 1.0 \end{cases}$
		AC	$k_\tau = 5.4\left(\frac{b}{a}\right)^2 + 0.6\left(\frac{b}{a}\right) + 9.0$ for $0.0 < \frac{b}{a} \leq 1.0$

Notes: B.C. = boundary condition, TSLC = transverse (y) edges simply supported and longitudinal (x) edges clamped, TCLS = transverse (y) edges clamped and longitudinal (x) edges simply supported, and AC = all edges clamped.

where

$$e_1 = \begin{cases} 1.322(b/a)^4 - 1.919(b/a)^3 + 0.021(b/a)^2 \\ \quad + 0.032(b/a) \quad \text{for } 0 \leq \zeta_L < 2 \\ -0.463(b/a)^4 + 1.023(b/a)^3 - 0.649(b/a)^2, \\ \quad + 0.073(b/a) \quad \text{for } 2 \leq \zeta_L < 8 \\ 0.0 \quad \text{for } 8 \leq \zeta_L \end{cases}$$

$$e_2 = \begin{cases} -0.179(b/a)^4 - 3.098(b/a)^3 + 5.648(b/a)^2 \\ \quad - 0.199(b/a) \quad \text{for } 0 \leq \zeta_L < 2 \\ 5.432(b/a)^4 - 11.324(b/a)^3 + 6.189(b/a)^2 \\ \quad - 0.068(b/a) \quad \text{for } 2 \leq \zeta_L < 8 \\ -1.047(b/a)^4 + 2.624(b/a)^3 - 2.215(b/a)^2 \\ \quad + 0.646(b/a) \quad \text{for } 8 \leq \zeta_L < 20 \\ 0.0 \quad \text{for } 20 \leq \zeta_L \end{cases},$$

$$e_3 = \begin{cases} 0.994(b/a)^4 + 0.011(b/a)^3 + 1.991(b/a)^2 \\ \quad + 0.003(b/a) + 1.0 \quad \text{for } 0 \leq \zeta_L < 2 \\ -3.131(b/a)^4 + 4.753(b/a)^3 + 3.587(b/a)^2 \\ \quad - 0.433(b/a) + 1.0 \quad \text{for } 2 \leq \zeta_L < 8 \\ 20.111(b/a)^4 - 43.697(b/a)^3 + 30.941(b/a)^2 \\ \quad - 1.836(b/a) + 1.0 \quad \text{for } 8 \leq \zeta_L < 20 \\ 0.751(b/a)^4 - 0.047(b/a)^3 + 2.053(b/a)^2 \\ \quad - 0.015(b/a) + 4.0 \quad \text{for } 20 \leq \zeta_L \end{cases}.$$

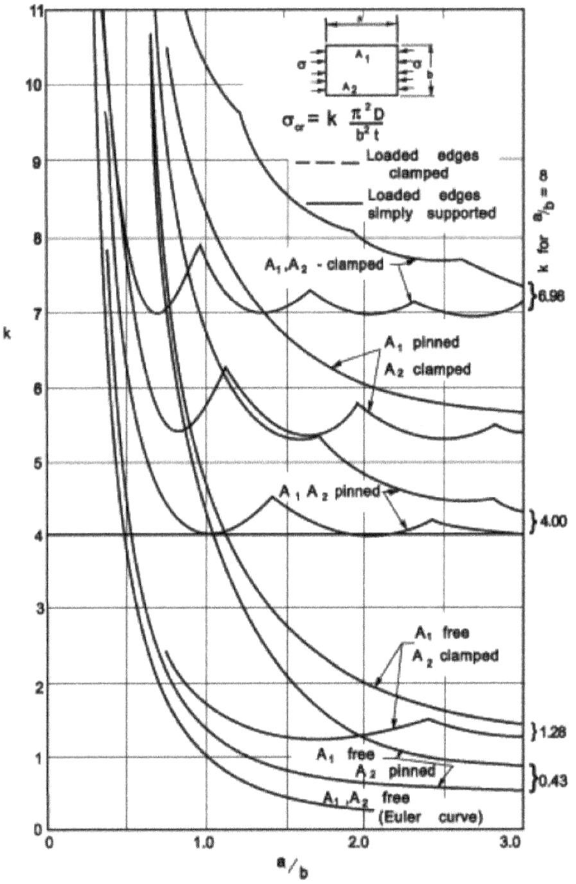

Figure 12.5 Buckling coefficient k of a plate under uniaxial compression for different conditions of plate edges.

Figures 12.7a and 12.7b indicate the variations of buckling coefficient k_x under longitudinal axial

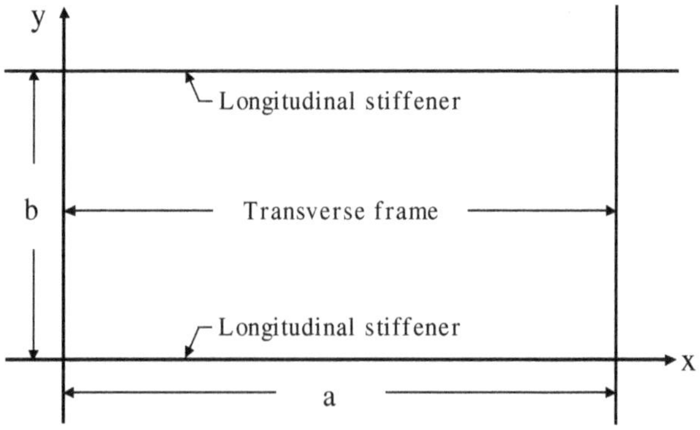

Figure 12.6a Plate surrounded by longitudinal and transverse support members.

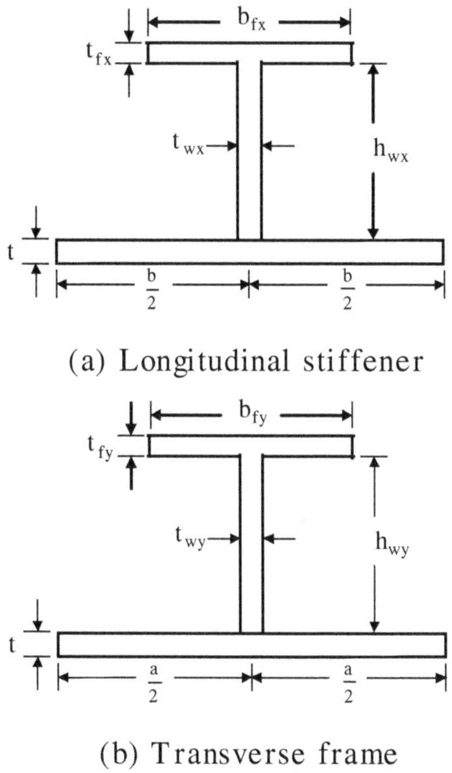

Figure 12.6b Dimensions of longitudinal (a) or transverse (b) support members.

compression and those of buckling coefficient k_y under transverse axial compression, respectively, for a plate that is rotationally restrained at its longitudinal edges and simply supported at its transverse edges. The exact solutions in these figures are those obtained by solving the plate governing differential equations (Paik and Thayamballi, 2000, 2003), and the approximate formula solutions are those of (12.4.3) or (12.4.4).

$\zeta_L = 0$ indicates the simply supported condition with zero rotational restraint, and $\zeta_L = \infty$ indicates the clamped condition with infinite rotational restraints. It is evident from Figs. 12.7a and 12.7b that the buckling coefficients (and thus the buckling strengths) of the plates increase significantly with an increase in the rotational restraints of the support members. It is found that, under longitudinal axial compression, the buckling strength of a plate with ζ_L of 20 has almost the same value as a plate clamped at its longitudinal edges and simply supported at its transverse edges. However, ζ_L must be a very large value, e.g., 500, to be equivalent to the buckling strength of a plate that is under transverse axial compression, clamped at its longitudinal edges, and simply supported at its transverse edges.

12.4.2 Simply Supported at the Longitudinal Edges and Partially Rotation-Restrained at the Transverse Edges

Longitudinal Axial Compression

In this case, the buckling coefficient is given by (Paik and Thayamballi, 2003).

$$k_x = d_1\zeta_S^4 + d_2\zeta_S^3 + d_3\zeta_S^2 + d_4\zeta_S + d_5 \qquad (12.4.5)$$

where

$$d_1 = \begin{cases} -1.010(a/b)^4 + 12.827(a/b)^3 - 52.553(a/b)^2 \\ \quad + 67.072(a/b) - 27.585 \quad \text{for } 0 \le \zeta_S < 0.4 \\ 0.047(a/b)^4 - 0.586(a/b)^3 + 2.576(a/b)^2 \\ \quad - 4.410(a/b) + 1.748 \quad \text{for } 0.4 \le \zeta_S < 0.8 \\ -0.017(a/b)^2 + 0.099(a/b) - 0.150 \\ \quad \text{for } 0.8 \le \zeta_S < 2 \\ 0.0 \quad \text{for } 2 \le \zeta_S \end{cases},$$

$$d_2 = \begin{cases} 0.881(a/b)^4 - 10.851(a/b)^3 + 41.688(a/b)^2 \\ \quad - 43.150(a/b) + 14.615 \quad \text{for } 0 \le \zeta_S < 0.4 \\ -0.123(a/b)^4 + 1.549(a/b)^3 - 6.788(a/b)^2 \\ \quad + 11.299(a/b) - 3.662 \quad \text{for } 0.4 \le \zeta_S < 0.8 \\ 0.138(a/b)^2 - 0.793(a/b) + 1.171 \\ \quad \text{for } 0.8 \le \zeta_S < 2 \\ 0.0 \quad \text{for } 2 \le \zeta_S \end{cases},$$

$$d_3 = \begin{cases} -0.190(a/b)^4 + 2.093(a/b)^3 - 5.891(a/b)^2 \\ \quad - 2.096(a/b) + 1.792 \quad \text{for } 0 \le \zeta_S < 0.4 \\ 0.114(a/b)^4 - 1.412(a/b)^3 + 5.933(a/b)^2 \\ \quad - 8.638(a/b) + 0.224 \quad \text{for } 0.4 \le \zeta_S < 0.8 \\ -0.457(a/b)^2 + 2.571(a/b) - 3.712 \\ \quad \text{for } 0.8 \le \zeta_S < 2 \\ 0.0 \quad \text{for } 2 \le \zeta_S \end{cases},$$

$$d_4 = \begin{cases} 0.004(a/b)^4 - 0.007(a/b)^3 - 0.243(a/b)^2 \\ \quad + 0.630(a/b) + 3.617 \quad \text{for } 0 \le \zeta_S < 0.4 \\ -0.021(a/b)^4 + 0.184(a/b)^3 - 0.126(a/b)^2 \\ \quad - 2.625(a/b) + 6.457 \quad \text{for } 0.4 \le \zeta_S < 0.8 \\ 0.822(a/b)^2 - 4.516(a/b) + 6.304 \\ \quad \text{for } 0.8 \le \zeta_S < 2 \\ -0.106(a/b) + 0.176 \quad \text{for } 2 \le \zeta_s < 20 \\ 0.0 \quad \text{for } 20 \le \zeta_S \end{cases},$$

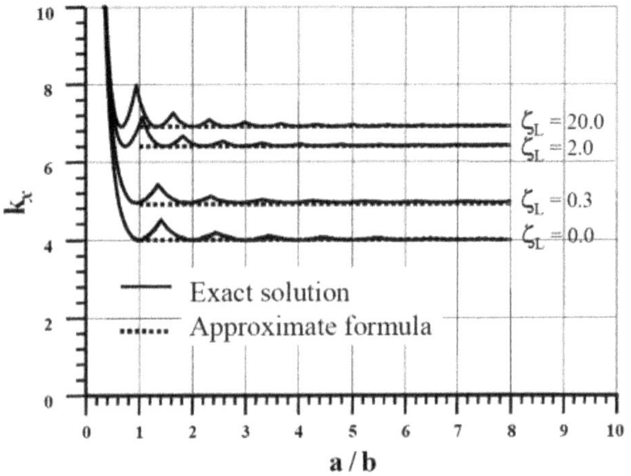

Figure 12.7a Variation of buckling coefficient k_x for a plate under longitudinal axial compression, rotationally restrained at its longitudinal edges, and simply supported at its transverse edges.

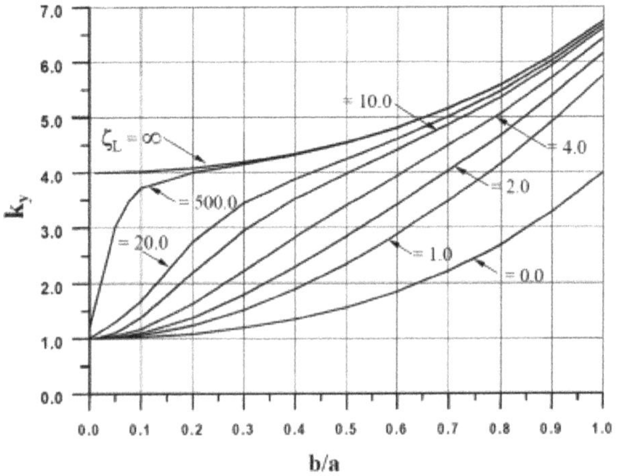

Figure 12.7b Variation of buckling coefficient k_y for a plate under transverse axial compression, rotationally restrained at its longitudinal edges, and simply supported at its transverse edges.

$$d_5 = \begin{cases} 4.0 \quad \text{for } 0 \le \zeta_S < 0.4 \\ -0.001(a/b)^4 + 0.033(a/b)^3 - 0.241(a/b)^2 \\ \quad + 0.684(a/b) + 3.539 \quad \text{for } 0.4 \le \zeta_S < 0.8 \\ -0.148(a/b)^2 - 0.596(a/b) + 3.847 \\ \quad \text{for } 0.8 \le \zeta_S < 2 \\ -1.822(a/b) + 7.850 \quad \text{for } 2 \le \zeta_s < 20 \\ 0.041(a/b)^4 - 0.602(a/b)^3 + 3.303(a/b)^2 \\ \quad - 8.176(a/b) + 12.144 \quad \text{for } 20 \le \zeta_S \end{cases}.$$

When calculating k_x in (12.4.5), the following conditions must be satisfied in order for the approximations to hold. (1) If $4.0 < a/b \le 4.5$ and $\zeta_S \ge 0.2$, then $\zeta_S = 0.2$; (2) if $a/b > 4.5$ and $\zeta_S \ge 0.1$, then $\zeta_S = 0.1$; (3) if $a/b \ge 2.2$ and $\zeta_S \ge 0.4$, then $\zeta_S = 0.4$; (4) if $a/b \ge 1.5$ and $\zeta_S \ge 1.4$, then $\zeta_S = 1.4$; (5) if $8 \le a/b \le 20$, then $\zeta_S = 8$; and (6) if $a/b \ge 5$, then $a/b = 5$.

Transverse Axial Compression

In this case, the buckling coefficient is given by (Paik and Thayamballi, 2003).

$$k_y = f_1\zeta_S^2 + f_2\zeta_S + f_3 \qquad (12.4.6)$$

where

$$f_1 = \begin{cases} 0.543(b/a)^4 - 1.297(b/a)^3 + 0.192(b/a)^2 \\ \quad - 0.016(b/a) \quad \text{for } 0 \le \zeta_S < 2 \\ -0.347(b/a)^4 + 0.403(b/a)^3 - 0.147(b/a)^2 \\ \quad + 0.016(b/a) \quad \text{for } 2 \le \zeta_S < 6 \\ 0.0 \quad \text{for } 6 \le \zeta_S \end{cases},$$

$$f_2 = \begin{cases} -1.094(b/a)^4 + 4.401(b/a)^3 - 0.751(b/a)^2 \\ \quad + 0.068(b/a) \quad \text{for } 0 \le \zeta_S < 2 \\ 2.139(b/a)^4 - 1.761(b/a)^3 + 0.419(b/a)^2 \\ \quad - 0.030(b/a) \quad \text{for } 2 \le \zeta_S < 6 \\ -0.199(b/a)^4 + 0.308(b/a)^3 - 0.118(b/a)^2 \\ \quad + 0.013(b/a) \quad \text{for } 6 \le \zeta_S < 20 \\ 0.0 \quad \text{for } 20 \le \zeta_S \end{cases},$$

$$f_3 = \begin{cases} 0.994(b/a)^4 + 0.011(b/a)^3 + 1.991(b/a)^2 \\ \quad + 0.003(b/a) + 1.0 \quad \text{for } 0 \le \zeta_S < 2 \\ -2.031(b/a)^4 + 5.765(b/a)^3 + 0.870(b/a)^2 \\ \quad + 0.102(b/a) + 1.0 \quad \text{for } 2 \le \zeta_S < 6 \\ -0.289(b/a)^4 + 7.507(b/a)^3 - 1.029(b/a)^2 \\ \quad + 0.398(b/a) + 1.0 \quad \text{for } 6 \le \zeta_S < 20 \\ -6.278(b/a)^4 + 17.135(b/a)^3 - 5.026(b/a)^2 \\ \quad + 0.860(b/a) + 1.0 \quad \text{for } 20 \le \zeta_S \end{cases}.$$

Figure 12.8 shows the variation of buckling coefficients k_x and k_y for a plate simply supported at its longitudinal edges and rotationally restrained at its transverse edges. It is found that the effect of rotational restraints becomes more severe for shorter plates and more moderate for longer plates.

12.4.3 Partially Rotation-Restrained at All Edges

Longitudinal Axial Compression

The elastic buckling strength of a plate rotationally restrained at all of its edges, and under longitudinal axial compression, can be calculated by (12.2.7),

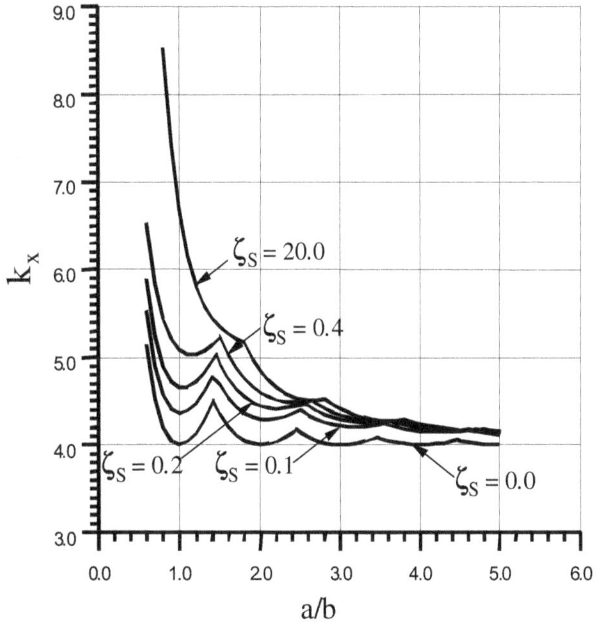

Figure 12.8a Variation of buckling coefficient k_x for a plate under longitudinal axial compression, simply supported at its longitudinal edges, and rotationally restrained at its transverse edges.

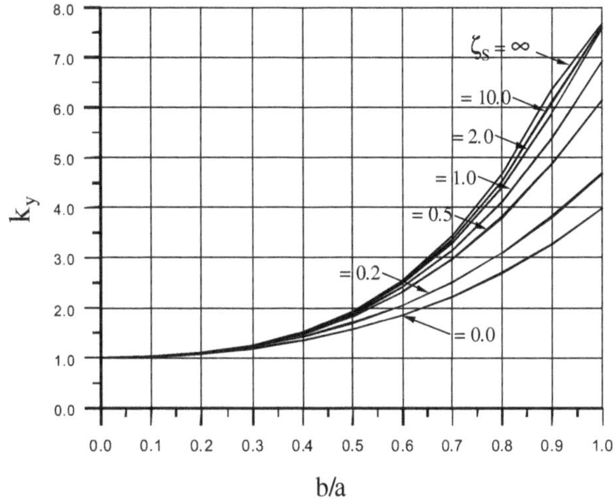

Figure 12.8b Variation of buckling coefficient k_y for a plate under transverse axial compression, simply supported at its longitudinal edges, and rotationally restrained at its transverse edges.

and buckling coefficient k_x is approximately given by (Paik and Thayamballi, 2003)

$$k_x = k_{x1} + k_{x2} - k_{x0} \tag{12.4.7}$$

where $k_{x1} = k_x$, as defined in (12.4.3), is equivalent to a plate rotationally restrained at its longitudinal edges and simply supported at its transverse edges, $k_{x2} = k_x$, as defined in (12.4.5), is equivalent to a plate simply supported at its longitudinal edges and rotationally restrained at its transverse edges, and $k_{x0} = k_x$, as defined in Table 12.1, is equivalent to a plate simply supported at all of its edges.

Transverse Axial Compression

The elastic buckling strength of a plate rotationally restrained at all of its edges, and under transverse axial compression, can also be calculated by (12.2.7), although the buckling coefficient k_y is now given by (Paik and Thayamballi, 2003)

$$k_y = k_{y1} + k_{y2} - k_{y0} \tag{12.4.7}$$

where $k_{y1} = k_y$, as defined in (12.4.4), is equivalent to a plate rotationally restrained at its longitudinal edges and simply supported at its transverse edges, $k_{y2} = k_y$, as defined in (12.4.6), is equivalent to a plate simply supported at its longitudinal edges and rotationally restrained at its transverse edges, and $k_{y0} = k_y$, as defined in Table 12.1, is equivalent to a plate simply supported at all of its edges.

12.5 EFFECT OF WELDING RESIDUAL STRESSES

Figure 12.9 presents an idealized distribution of welding residual stresses in which tensile and compressive residual stress blocks are developed in both the longitudinal (x) and transverse (y) directions, as welding must be undertaken along the plate edges in these two directions.

These welding residual stresses decrease the elastic plate buckling strength. Under longitudinal axial compression, this strength can be computed as follows.

$$\sigma_{xE,1} = k_x \frac{\pi^2 E}{12(1-\nu^2)}\left(\frac{t}{b}\right)^2 - \sigma_{rex} \tag{12.5.1}$$

where $\sigma_{rex} = \sigma_{rcx} + \dfrac{2}{b}(\sigma_{rtx} - \sigma_{rcx})\left(b_t - \dfrac{b}{2\pi}\sin\dfrac{2\pi b_t}{b}\right)$, σ_{rcx} = compressive residual stress in the x-direction, σ_{rtx} = tensile residual stress in the x-direction, and b_t =

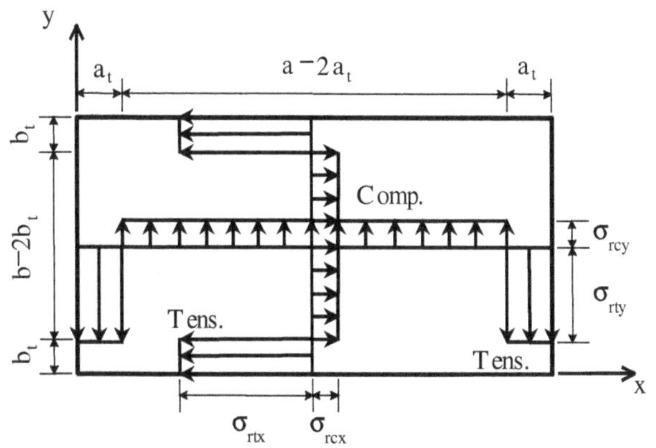

Figure 12.9 Idealized distribution of the welding residual stresses in a plate.

$\frac{\sigma_{rcx}}{\sigma_{rcx} - \sigma_{rtx}} \frac{b}{2}$ = breadth of the tensile residual stress block in the x-direction.

In a similar manner, the effect of welding residual stresses on the plate buckling strength under transverse axial compression can be calculated by

$$\sigma_{yE,1} = k_y \frac{\pi^2 E}{12(1 - \nu^2)} \left(\frac{t}{b} \right)^2 - \sigma_{rey} \qquad (12.5.2)$$

where $\sigma_{rey} = \sigma_{rcy} + \frac{2}{a} (\sigma_{rty} - \sigma_{rcy}) \left(a_t - \frac{a}{2\pi} \sin \frac{2\pi a_t}{a} \right)$, σ_{rcy} = compressive residual stress in the y-direction, σ_{rty} = tensile residual stress in the y-direction, and $a_t = \frac{\sigma_{rcy}}{\sigma_{rcy} - \sigma_{rty}} \frac{a}{2}$ = length of the tensile residual stress in the y-direction. The tensile residual stresses of mild steel plates are often taken as $\sigma_{rtx} = \sigma_{rty} = \sigma_Y$, where σ_Y is the yield stress of the material.

12.6 EFFECT OF LATERAL PRESSURE

In the plate elements of ships and ship-shaped offshore structures, the lateral pressures that arise from cargo and/or water are usually applied first, and then additional in-plane loads are applied. This implies that any lateral deflection may be caused by lateral pressure before the in-plane loads are applied. As a result, a clear buckling phenomenon may not appear in the actual plates until the ultimate strength has been reached when a very large amount of lateral pressure is applied. However, as far as elastic buckling is concerned, the plate will eventually buckle even when a relatively small or large amount of lateral pressure is applied. Fig. 12.10 shows a schematic of the plate buckling pattern with and without lateral pressure. It should be noted that square plates with $a/b \approx 1$ do not show a clear buckling phenomenon when lateral pressure is applied because the plate would already have deflected before the in-plane loads were applied.

The elastic buckling strength of a plate is normally increased by lateral pressure because more external work in association with the in-plane loads is required to determine the original plate buckling pattern that has been disturbed by the lateral pressure. This increase is also partly due to the rotational restraints at the plate edges becoming greater because of these actions, as can be clearly seen in Fig. 12.10c, in which the plate is likely to deflect in the same direction, and, subsequently, a large degree of rotational restraint develops at the plate edges.

The increase of elastic buckling strength due to lateral pressure can be estimated for longitudinal or transverse compression respectively, as follows (Fujikubo et al., 1998).

$$C_{px} = 1 + \frac{1}{576} \left(\frac{pb^4}{Et^4} \right)^{1.6} \quad \text{for } \frac{a}{b} \geq 2 \qquad (12.6.1a)$$

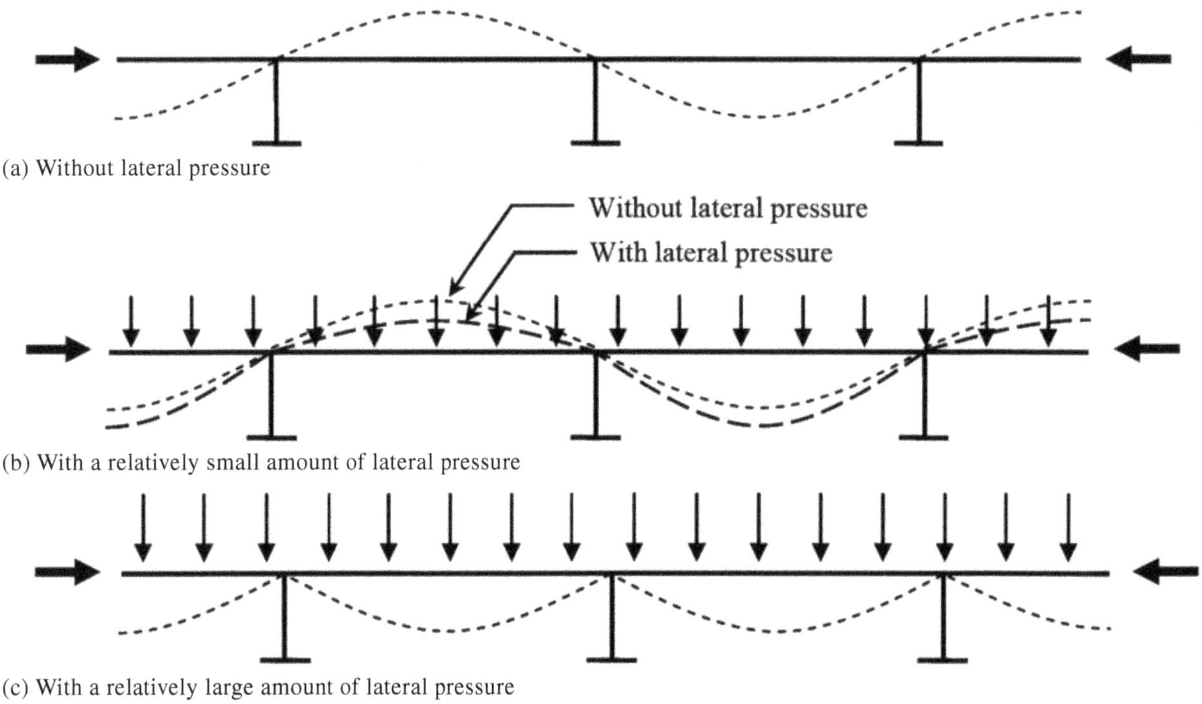

Figure 12.10 Schematic of the axial compressive buckling pattern of a plate with and without lateral pressures.

$$C_{py} = 1 + \frac{1}{160}\left(\frac{b}{a}\right)^{0.95}\left(\frac{pb^4}{Et^4}\right)^{1.75} \quad \text{for } \frac{a}{b} \geq 2 \tag{12.6.1b}$$

where p = lateral pressure.

The elastic buckling strengths of plates under longitudinal or transverse axial compression taking into account the effect of lateral pressure can then be obtained as follows.

$$\sigma_{xE,1} = C_{px}\, k_x \frac{\pi^2 E}{12(1-\nu^2)}\left(\frac{t}{b}\right)^2 \tag{12.6.2}$$

$$\sigma_{yE,1} = C_{py}\, k_y \frac{\pi^2 E}{12(1-\nu^2)}\left(\frac{t}{b}\right)^2 \tag{12.6.3}$$

Note that the elastic buckling strengths of plates computed from (12.6.2) or (12.6.3) should not be greater than those of plates clamped at all edges.

12.7 EFFECT OF OPENINGS

Openings (or cut-outs) are occasionally located in plate elements to make way for access or to lighten the structure, as shown in Fig. 12.11. Such perforations of course decrease buckling strength. Where it is significant, therefore, the effects of such an opening must be taken into account in buckling strength calculations. However, it must be cautioned that the plasticity correction approach of elastic plate buckling strength using the Johnson-Ostenfeld formula may cause a significant overestimate of elastic-plastic buckling or the ultimate strength of plates, particularly when the size of the opening and/or the plate thickness is relatively large, in contrast to plates without opening.

The following are empirical formulations of the elastic buckling strength of plates with a circular opening, i.e., those with $a_c = b_c = d_c$, located at their center.

12.7.1 Longitudinal Axial Compression

$$\sigma_{xE,1} = R_{xE} k_{xo} \frac{\pi^2 E}{12(1-\nu^2)}\left(\frac{t}{b}\right)^2 \tag{12.7.1}$$

where

k_{xo} = elastic buckling coefficient of the plate without opening,

$$R_{xE} = \alpha_{E1}\left(\frac{d_c}{b}\right)^3 + \alpha_{E2}\left(\frac{d_c}{b}\right)^2 + \alpha_{E3}\frac{d_c}{b} + 1,$$

$$\alpha_{E1} = \begin{cases} 0.002\left(\frac{a}{b}\right)^{8.238} & \text{for } 1 \leq \frac{a}{b} < 2 \\ -1.542\left(\frac{a}{b}\right)^2 + 7.232\frac{a}{b} - 7.666 & \text{for } 2 \leq \frac{a}{b} < 3 \\ -0.052\left(\frac{a}{b}\right)^2 + 0.526\frac{a}{b} - 0.964 & \text{for } 3 \leq \frac{a}{b} < 6 \end{cases},$$

$$\alpha_{E2} = \begin{cases} 0.655 + \frac{1}{4.123(a/b) - 8.922} & \text{for } 1 \leq \frac{a}{b} < 2 \\ 1.767\left(\frac{a}{b}\right)^2 - 7.937\frac{a}{b} + 7.982 & \text{for } 2 \leq \frac{a}{b} < 3 \\ 0.071\left(\frac{a}{b}\right)^2 - 0.732\frac{a}{b} + 1.631 & \text{for } 3 \leq \frac{a}{b} < 6 \end{cases},$$

$$\alpha_{E3} = \begin{cases} -0.945 + \frac{1}{-5.661(a/b) + 12.342} & \text{for } 1 \leq \frac{a}{b} < 2 \\ -0.248\left(\frac{a}{b}\right)^2 + 0.796\frac{a}{b} - 0.565 & \text{for } 2 \leq \frac{a}{b} < 3 \\ -0.020\left(\frac{a}{b}\right)^2 + 0.199\frac{a}{b} - 0.826 & \text{for } 3 \leq \frac{a}{b} < 6 \end{cases}.$$

12.7.2 Transverse Axial Compression

$$\sigma_{yE,1} = R_{yE} k_{yo} \frac{\pi^2 E}{12(1-\nu^2)}\left(\frac{t}{b}\right)^2 \tag{12.7.2}$$

where

k_{yo} = elastic buckling coefficient of the plate

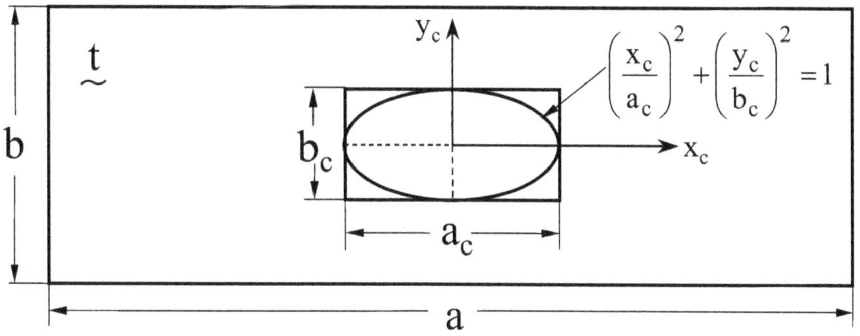

Figure 12.11 Opening in a plate.

without opening,

$$R_{yE} = \alpha_{E4}\left(\frac{d_c}{b}\right)^2 + \alpha_{E5}\frac{d_c}{b} + 1,$$

$$\alpha_{E4} = \begin{cases} 0.034\left(\dfrac{a}{b}\right)^2 - 0.327\dfrac{a}{b} + 0.768 & \text{for } 1 < \dfrac{a}{b} \le 4 \\ 0.004 & \text{for } 0.4 \le \dfrac{a}{b} \le 6 \end{cases},$$

$$\alpha_{E5} = -0.008 - \frac{1}{0.976(a/b) + 0.302} \quad \text{for } 1 \le \frac{a}{b} \le 6.$$

12.7.3 Edge Shear

$$\tau_{E,1} = R_{\tau E} k_{\tau o} \frac{\pi^2 E}{12(1 - \nu^2)}\left(\frac{t}{b}\right)^2 \qquad (12.7.3)$$

where

$k_{\tau o}$ = elastic buckling coefficient of the plate without opening,

$$R_{\tau E} = \alpha_{E6}\left(\frac{d_c}{b}\right)^3 + \alpha_{E7}\left(\frac{d_c}{b}\right)^2 + \alpha_{E8}\frac{d_c}{b} + 1,$$

$$\alpha_{E6} = \begin{cases} 0.094\left(\dfrac{a}{b}\right)^2 + 0.035\dfrac{a}{b} + 1.551 & \text{for } 1 \le \dfrac{a}{b} < 3 \\ 2.502 & \text{for } 3 \le \dfrac{a}{b} \le 6 \end{cases},$$

$$\alpha_{E7} = \begin{cases} -0.039\left(\dfrac{a}{b}\right)^2 - 0.807\dfrac{a}{b} - 0.405 & \text{for } 1 \le \dfrac{a}{b} < 3 \\ -3.177 & \text{for } 3 \le \dfrac{a}{b} \le 6 \end{cases},$$

$$\alpha_{E8} = \begin{cases} -0.053\left(\dfrac{a}{b}\right)^2 + 0.785\dfrac{a}{b} - 1.875 & \text{for } 1 \le \dfrac{a}{b} < 3 \\ 0.003 & \text{for } 3 \le \dfrac{a}{b} \le 6 \end{cases}.$$

For elliptical or rectangular types of openings, the elastic buckling strength of the perforated plates can be obtained using the following values of R_{xE} in (12.7.1) for longitudinal axial compression and those of R_{yE} in (12.7.2) for transverse axial compression, namely,

$$R_{xE} = \alpha_{E1}\left(\frac{b_c}{b}\right)^3 + \alpha_{E2}\left(\frac{b_c}{b}\right)^2 + \alpha_{E3}\frac{b_c}{b} + 1 \qquad (12.7.4)$$

$$R_{yE} = \alpha_{E4}\left(\frac{a_c}{a}\right)^2 + \alpha_{E5}\frac{a_c}{a} + 1 \qquad (12.7.5)$$

REFERENCES

Fujikubo, M., Yao, T., Varghese, B., Zha, Y., and Yamamura, K. (1998). Elastic local buckling strength of stiffened plates considering plate/stiffener interaction and lateral pressure. *Proc. of the International Offshore and Polar Engineering Conference*, Vol. IV, Montreal, Canada, 292–299.

Paik, J. K., and Thayamballi, A. K. (2000). Buckling strength of steel plating with elastically restrained edges. *Thin-Walled Structures*, Vol. 37, 27–55.

Paik, J. K., and Thayamballi, A. K. (2003). *Ultimate limit state design of steel-plated structures*. Chichester, U.K.: Wiley.

Timoshenko, S., and Gere, J. M. (1961). *Theory of elastic stability*. New York: McGraw-Hill.

CHAPTER

THIRTEEN

LARGE DEFLECTION BEHAVIOR AND ULTIMATE STRENGTH OF PLATES

Jeom Kee Paik
Professor, Pusan National University
Busan, Korea

13.1 FUNDAMENTALS OF ULTIMATE PLATE STRENGTH BEHAVIOR

It is not possible to determine the true margin of structural safety for structural components and system structures under extreme loads if the ultimate strength remains unknown. One of the primary failure modes of stiffened panels is the buckling and plastic collapse of the plates surrounded by support members, and thus an evaluation of the buckling and plastic collapse behavior of plates is essential to identifying the failure of ship structures (Paik and Thayamballi, 2003, 2007; ISO, 2007).

Figure 13.1 shows a schematic of plate behavior subject to predominantly axial compressive loads. Such behavior always involves a large degree of deflection (geometric nonlinearity) and/or plasticity (material nonlinearity) before and after the ultimate strength has been reached. Plate behavior depends on a variety of influential factors, namely the plate's geometric and material properties, loading characteristics, initial imperfections (e.g., initial deflections and residual stresses), boundary conditions, and the existing local damage related to corrosion, fatigue crack, and denting.

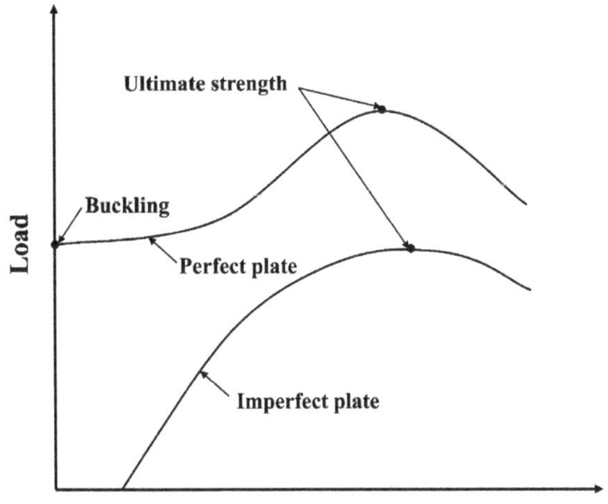

Figure 13.1 Schematic of the collapse behavior of ship plates subject to predominantly axial compressive loads.

A clear buckling phenomenon may be seen in perfectly flat plates without initial deflection as the type of bifurcation described in Chapter 12. Depending on the geometric and material properties, together with the loading and boundary conditions, buckling may occur in an entirely elastic, elastic–plastic, or fully plastic regime. When the plate buckles in the elastic or elastic–plastic regime, it retains some residual load-carrying capacity until the ultimate strength has been reached; although it will collapse immediately after the inception of plastic buckling. This is in contrast to columns in which buckling, even in the elastic regime, corresponds to the ultimate strength, which indicates that no residual load-carrying capacity exists after buckling.

Buckling strength therefore serves as a good indicator in the ultimate strength computations of perfect plates. Even for imperfect plates, buckling strength is often used as an indicator of ultimate strength, although such plates may not exhibit a clear buckling phenomenon because plate deflection exists from the very beginning of compressive loading, similar to plates subject to predominantly lateral pressure loads. The buckled or deflected plates eventually reach their ultimate strength through the progressive expansion of plasticity.

Various methods of computing the ultimate strength of structural components or entire structural systems can be found in the literature. Some methods are simple, and others are more sophisticated. However, all of these methods basically involve both geometric and material nonlinearities, with the former being associated with buckling and large deformation and the latter due to plasticity. The factors that affect ultimate strength behavior are as follows (ISSC, 2009).

- Geometrical factors associated with buckling, large deflection, crushing, or folding
- Material factors associated with yielding/plasticity, ductile/brittle fracture, rupture, or cracking damage
- Fabrication-related initial imperfections such as initial distortion, residual stress, and softening
- Temperature factors, such as low temperatures associated with operation in cold waters or low-temperature cargo and high temperatures due to fire and explosions
- Dynamic factors (strain rate sensitivity, inertia effect) associated with freak/rogue/abnormal waves and impact pressure actions arising from sloshing, slamming, or green water, overpressure actions arising from explosions, and impacts due to collisions, grounding, or dropped objects
- Age-related deterioration such as corrosion and fatigue cracking
- Human factors relating to unusual operations in terms of ship speed (compared to maximum permitted speed or acceleration), ship heading, loading conditions, etc.

This chapter presents theoretical approaches to the computation of the large deflection behavior and ultimate strength of plates used for ships and ship-shaped offshore structures.

13.2 BASIC IDEALIZATIONS OF PLATES

13.2.1 Geometric and Material Properties

The geometry of the plates found in ships and ship-shaped offshore structures is usually rectangular, and the material used is mild- or high-tensile steel. Note that the use of aluminum alloys is increasing in the design and fabrication of weight-critical structures such as high-speed vessels.

Figure 13.2 shows a schematic of a typical stiffened plate structure. The responses of such a structure can be classified into three levels: the entire structure level, the stiffened panel level, and the bare plate element level. This chapter deals with the last level (i.e., the bare plate element level), as shown in Fig. 13.3.

The coordinates of the plate are taken as the x axis in the longitudinal direction and the y axis in the transverse direction. The length and breadth of the plating are a (along the x axis) and b (along the y axis), respectively. The long edges are not necessarily taken as the x axis, in contrast to the definition given in Chapter 12, such that the plate aspect ratio (i.e., a/b) will be greater than 1 for a long plate and

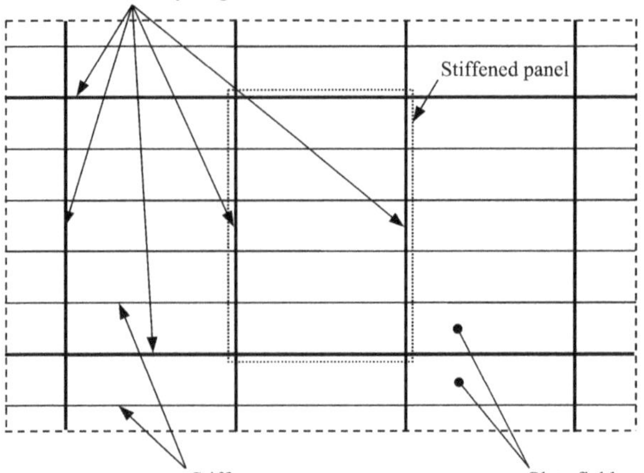

Figure 13.2 Typical stiffened plate structure.

less than 1 for a wide plate. This rule may be different from that given by the traditional classification societies in which the long edge is always taken as the x axis, such that a/b is always greater than 1.

One of the benefits of the coordinate rule used in this chapter is that the computerization of the design equations is more general for calculating the large deflection and ultimate strength behavior of a large plated structure that is composed of a number of plate elements, some of which are long and some wide. The thickness of the plate is t. Young's modulus and Poisson's ratio are E and ν, respectively. By definition, the elastic shear modulus G is $G = E/[2(1 + \nu)]$. The yield stress of the material is σ_Y. The plate bending rigidity D is defined by $D = Et^3/[12(1 - \nu^2)]$, and the plate slenderness coefficient β is defined by $\beta = (b/t)\sqrt{\sigma_Y/E}$.

This chapter deals with theoretical solution methods and applies the elastic–perfectly plastic material model and neglects the effect of strain hardening.

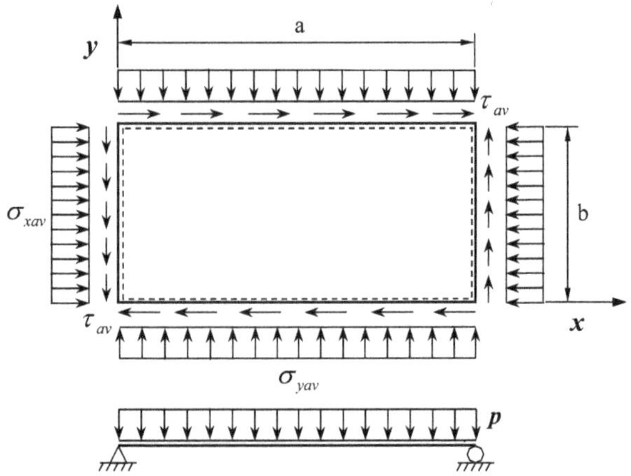

Figure 13.3 Simply supported rectangular plate subject to biaxial compression/tension, edge shear, and lateral pressure loads.

13.2.2 Plate Edge Conditions

In maritime engineering practice, it is often assumed that the boundary condition for plates is that they are simply supported at their edges, although the edges are surrounded by support members and thus are neither simply supported nor clamped, as described in Chapter 12.

The plate is supported at its four edges by beam members (e.g., longitudinal stiffeners and transverse frames). The bending rigidities of the boundary support members are usually quite large compared to that of the plate itself. This implies that the displacements of the support members normal to the plane of the plating are very small even up to plate collapse, and thus it is presumed that all (four) of the plate edges remain in-plane. The rotational restraints along the plate edges depend on the torsional rigidities of the support members, and these are neither zero nor infinite.

When predominantly in-plane compressive loads are applied on a continuous plated structure, the buckling pattern of the plate elements is expected to be unsymmetrical (i.e., one plate element will tend to buckle up, and the adjacent plate element will tend to deflect down). In this case, the rotational restraints along the plate edges are considered to be small. When the plated structure is subjected predominantly to laterally distributed loads, however, its deflection pattern tends to be symmetrical, at least when the pressure is sufficiently large (i.e., each adjacent plate element may deflect in the direction of lateral pressure loading). The rotational restraints at the plate edges may then become large, such that they correspond to a clamped condition at the beginning of loading. However, if plasticity occurs earlier along the edges at which the large bending moments are developed, then the rotational restraints at the yielded edges will be reduced as the applied loads increase.

In a continuous plated structure, the edges of the individual plate elements are considered to remain almost straight because of the relative structural response to the adjacent plate elements even after plate deflection. In this regard, an idealized condition (i.e., one with zero rotational restraints along the plate edges) has been widely used for practical analytical purpose in maritime engineering practice. In this chapter, therefore, it is also assumed that the plate edges are simply supported, with zero deflection and zero rotational restraints along the four edges, with all of the edges kept straight. In most practical situations, this approximation will lead to slightly pessimistic but adequate results. However, for comparison purposes, several theoretical approaches to identifying the effects of edge rotational restraints or clamped condition on the large deflection behavior of plates are also introduced.

13.2.3 Loading Conditions

Ship plates are likely to be subjected to combined in-plane and lateral pressure loads. The former include biaxial compression and/or tension and edge shear, which are mainly induced by overall hull girder bending and/or vessel torsion, whereas the latter are due to water pressure and/or cargo. The extrema of such load components may not occur simultaneously, but several load components usually exist and interact with one another. Hence, for more advanced ship structural designs, it is of crucial importance to understand the ultimate strength characteristics of ship plates subject to combined loads.

The potential load components acting on plate elements are generally of four types (or six load components): biaxial loads (i.e., compression or tension), edge shear, biaxial in-plane bending, and lateral pressure. When the plate size is relatively small compared to the entire stiffened plate structure, the influence of in-plane bending may be negligible, an insight that may also be applicable to the elastic buckling strength calculations described in Chapter 12.

In this chapter, we thus deal with three types of loads (or four load components): longitudinal compression or tension (denoted by σ_{xav}), transverse compression or tension (denoted by σ_{yav}), edge shear (denoted by $\tau = \tau_{av}$), and lateral pressure loads (denoted by p), as shown in Fig. 13.3. For the sake of simplicity, the in-plane bending effects are not considered here.

In actual ship structures, lateral pressure loads arise from water pressure and/or cargo weight. The still-water magnitude of water pressure depends on the vessel draft, and the still-water value of cargo pressure is determined by the amount and density of the cargo that is loaded. These still-water pressure values will be augmented by vessel motion in association with wave actions. Larger in-plane loads are typically caused by longitudinal hull girder bending, both in still-water and in waves at sea.

13.2.4 Initial Imperfections

Initial plate imperfections in the form of initial deflections and residual stresses are primarily caused by welding during the fabrication process. Because these initial imperfections can significantly affect (reduce) the strength performance, they should be included in the strength calculations as parameters of influence. In welded aluminum plates, the softening of the material in the heat-affected zone is also a parameter of influence. Section 8.3.6 presents the modeling of welding-induced initial imperfections, which is adopted for the ultimate plate strength calculations in this chapter.

The initial plate deflection can generally be expressed as follows.

$$w_o = \sum_{m=1}^{M}\sum_{n=1}^{N} A_{omn} \sin\frac{m\pi x}{a}\sin\frac{n\pi y}{b} \quad (13.2.1)$$

where w_o is initial plate deflection, m is half-wave number in the x direction, n is half-wave number in the y direction, A_{omn} is known amplitude of initial deflection, and M, N is maximum number of half-waves in the x or y directions, respectively.

In the shorter direction, the initial plate deflection can be fairly well expressed with one half-wave. For example, if the x coordinate is taken such that $a/b \geq 1$, then $n = 1$ in equation (13.2.1) becomes

$$w_o = \sum_{m=1}^{M} A_{om} \sin\frac{m\pi x}{a}\sin\frac{\pi y}{b} \quad (13.2.2)$$

where A_{om} corresponds to A_{om1} with m half-waves in the x direction and one half-wave in the y direction, as is referred to in equation (13.2.1).

The distribution of the welding residual stress can be idealized, as shown in Fig. 8.13 of Chapter 8, which depicts tensile and compressive stress blocks. The tensile residual stress in the heat-affected zone develops with a magnitude of σ_{rt}. The equilibrium between the tensile and compressive stress determines the breadths of the tensile residual stress blocks, as indicated in equation (8.3.12) of Chapter 8. The welding residual stress distributions can then be formulated with the nomenclature described in Fig. 8.13, as follows.

$$\sigma_{rx} = \begin{cases} \sigma_{rtx} & for\ 0 \leq y < b_t \\ \sigma_{rcx} & for\ b_t \leq y < b - b_t \\ \sigma_{rtx} & for\ b - b_t \leq y \leq b \end{cases} \quad (13.2.3a)$$

$$\sigma_{ry} = \begin{cases} \sigma_{rty} & \text{for}\ 0 \leq x < a_t \\ \sigma_{rcy} & \text{for}\ a_t \leq x < a - a_t \\ \sigma_{rty} & \text{for}\ a - a_t \leq x \leq a \end{cases} \quad (13.2.3b)$$

where σ_{rx}, σ_{ry} is welding residual stress distributions in the x or y directions, respectively.

13.3 NONLINEAR GOVERNING DIFFERENTIAL EQUATIONS FOR PLATES

The elastic large deflection behavior of plates can be identified by solving the following nonlinear governing differential equations, which comprise equilibrium and compatibility equations (Timoshenko and Woinowsky-Krieger, 1959).

$$D\left(\frac{\partial^4 w}{\partial x^4} + 2\frac{\partial^4 w}{\partial x^2 \partial y^2} + \frac{\partial^4 w}{\partial y^4}\right) - t\left[\frac{\partial^2 F}{\partial y^2}\frac{\partial^2 (w+w_o)}{\partial x^2} - 2\frac{\partial^2 F}{\partial x \partial y}\frac{\partial^2 (w+w_o)}{\partial x \partial y} + \frac{\partial^2 F}{\partial x^2}\frac{\partial^2 (w+w_o)}{\partial y^2} + \frac{p}{t}\right] = 0 \quad (13.3.1)$$

$$\frac{\partial^4 F}{\partial x^4} + 2\frac{\partial^4 F}{\partial x^2 \partial y^2} + \frac{\partial^4 F}{\partial y^4} - E\left[\left(\frac{\partial^2 w}{\partial x \partial y}\right)^2 - \frac{\partial^2 w}{\partial x^2}\frac{\partial^2 w}{\partial y^2} + 2\frac{\partial^2 w_o}{\partial x \partial y}\frac{\partial^2 w}{\partial x \partial y} - \frac{\partial^2 w_o}{\partial x^2}\frac{\partial^2 w}{\partial y^2} - \frac{\partial^2 w}{\partial x^2}\frac{\partial^2 w_o}{\partial y^2}\right] = 0 \quad (13.3.2)$$

where w is added deflection caused by applied loads, w_o is initial deflection caused by welding and other reasons prior to the application of external loads, and F is stress function.

Once the deflection w and the stress function F have been identified, the membrane stresses inside the plate can be obtained as follows.

$$\sigma_x = \frac{\partial^2 F}{\partial y^2} - \frac{Ez}{1-\nu^2}\left[\frac{\partial^2 w}{\partial x^2} + \nu\frac{\partial^2 w}{\partial y^2}\right] \quad (13.3.3a)$$

$$\sigma_y = \frac{\partial^2 F}{\partial x^2} - \frac{Ez}{1-\nu^2}\left[\frac{\partial^2 w}{\partial y^2} + \nu\frac{\partial^2 w}{\partial x^2}\right] \quad (13.3.3b)$$

$$\tau = \tau_{xy} = -\frac{\partial^2 F}{\partial x \partial y} - \frac{Ez}{2(1+\nu)}\frac{\partial^2 w}{\partial x \partial y} \quad (13.3.3c)$$

where z is the coordinate in the plate thickness direction. At the plate midthickness, $z = 0$.

13.4 ANALYTICAL METHODS FOR ULTIMATE PLATE STRENGTH CALCULATIONS

Within the framework of structural design and strength assessment, it is more convenient to use analytical or closed-form formulations than numerical methods. This section presents analytical approaches that can produce closed-form design formulations for ultimate plate strength computations.

13.4.1 The Johnson-Ostenfeld Formula Method

In current maritime engineering practice, the so-called critical buckling strength is often regarded as a pseudo-ultimate strength and is estimated by a plasticity correction of the elastic buckling strength using the Johnson-Ostenfeld formula, as follows.

$$\sigma_{cr} = \begin{cases} \sigma_E & \text{for}\ \sigma_E \leq 0.5\sigma_F \\ \sigma_F\left(1 - \dfrac{\sigma_F}{4\sigma_E}\right) & \text{for}\ \sigma_E > 0.5\sigma_F \end{cases} \quad (13.4.1)$$

where σ_E is elastic buckling stress; σ_{cr} is critical (elastic-plastic) buckling stress; σ_F is reference yield stress, $\sigma_F = \sigma_Y$ for compressive stress, and $\sigma_F = \tau_Y = \sigma_Y/\sqrt{3}$ for shear stress; and σ_Y is material yield stress. In using equation (13.4.1), the sign of the compressive stress is taken as positive.

The elastic buckling strength formulations of plates are presented in detail in Chapter 12. For thick or moderately thick plates, the critical buckling stress σ_{cr} estimated by equation (13.4.1) gives a good indication of the ultimate strength, albeit somewhat on the pessimistic side. However, it equals the elastic buckling stress for thin plates by nature of equation (13.4.1), which cannot account for the reserve strength after buckling. Equation (13.4.1) also significantly overestimates the load-carrying capacity (ultimate strength) for perforated plates (i.e., with openings), particularly when the opening size and/or plate thickness is large (Paik and Thayamballi, 2003). Equation (13.4.1) is therefore not relevant for ultimate plate strength computations in such cases.

13.4.2 Rigid-Plastic Theory Method

In rigid-plastic theory, the kinematically admissible collapse mechanisms of the plate are presumed on the basis of prior insights. The collapse (ultimate) strength formula is then derived by applying the principle of minimum potential energy. Figure 10.3 in Chapter 10 provides an example of the plastic collapse mechanism of a plate in association with the plate's collapse strength computations. In the following discussion, useful formulations of the rigid-plastic theory approach are presented for plates subject to lateral pressure or axial compressive loads.

13.4.2.1 Lateral Pressure Loads

The collapse strength of rectangular plates subject to uniformly distributed lateral pressure loads is obtained by employing the rigid-plastic theory approach in between the lower and upper bounds, as follows (Jones, 1975).

$$\frac{8M_p}{b^2}(1+\phi+\phi^2) \le p_u \le \frac{24\,M_p}{b^2}\frac{1}{\left(\sqrt{3+\phi^2}-\phi\right)^2}$$

for simply supported plates (13.4.2a)

$$\frac{16\,M_p}{b^2}\left(1+\phi^2\right) \le p_u \le \frac{48\,M_p}{b^2}\frac{1}{\left(\sqrt{3+\phi^2}-\phi\right)^2}$$

for clamped plates (13.4.2b)

where p_u is ultimate lateral pressure loads, $M_p = \sigma_Y t^2/4$ is the plastic bending moment per unit breadth that the plate cross-section may carry, and $\phi = b/a$.

Equations (13.4.2a) and (13.4.2b) are derived using the upper and lower bound theorems for plates made of rigid-plastic material, which obey the Tresca yield criterion. The effect of shear on yielding has been neglected, assuming that the plates are thin. Interestingly, equations (13.4.2a) and (13.4.2b) can be simplified for a square plate with $\phi = 1$, as follows.

$$\frac{24\,M_p}{b^2} \le p_u \le \frac{24\,M_p}{b^2} \text{ for simply supported plates} \quad (13.4.3a)$$

$$\frac{32\,M_p}{b^2} \le p_u \le \frac{48\,M_p}{b^2} \text{ for clamped plates} \quad (13.4.3b)$$

13.4.2.2 Axial Compression

Paik and Pedersen (1996) derived the ultimate strength formulations of long plates subject to uniaxial compression by applying combined elastic large deflection analysis with rigid-plastic large deflection theory. The Paik-Pedersen method has the benefit of taking into account the initial deflection of a complex shape, which is often found in the plates of ships and offshore structures. This method also considers the effect of welding residual stress. The elastic large deflection behavior is identified by solving equations (13.3.1) and (13.3.2) when the plate is assumed to be simply supported at all (four) edges. The rigid-plastic analysis is based on the collapse mechanism, which also takes account of the large deformation effect. Both the elastic large deflection and rigid-plastic analyses are performed for each of the individual collapse modes or half-waves to find their intersection values, which are regarded as the corresponding ultimate strength values, while the number of half-waves may be considered in the range of 1 to $2m$, where m is the buckling half-wave number of the plate. The real value of the ultimate strength is determined as the minimum value among those so computed. Figure 13.4 shows a schematic of the Paik-Pedersen method for ultimate strength calculations of plates.

ELASTIC LARGE DEFLECTION ANALYSIS

The elastic large deflection behavior of an imperfect plate can be identified by solving the equilibrium equation (13.3.1) and the compatibility equation (13.3.2). For convenience, the elastic large deflection analysis is undertaken with the initial and added deflection functions having only one specific half-wave component, and this analysis is subsequently performed for each of the half-waves considered. Therefore, when the coordinate is taken such that $a/b \ge 1$, the following deflection functions can be given.

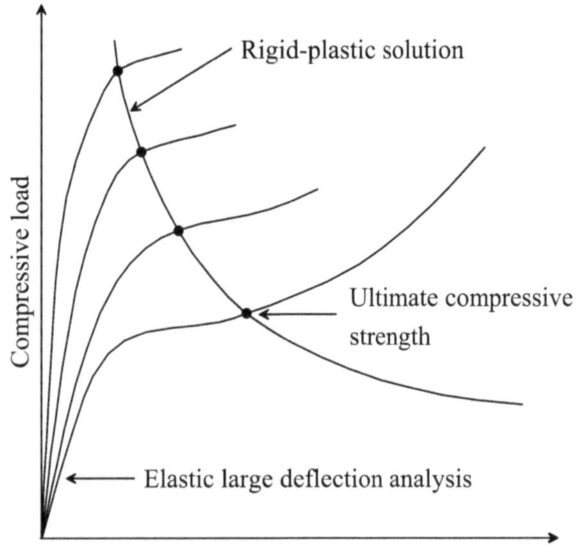

Figure 13.4 Schematic of the Paik-Pedersen method for ultimate strength calculations of plates under axial compressive loads, with multiple solutions of the elastic large deflection analysis corresponding to different values of half-wave i.

$$w_o = A_{oi} \sin\frac{i\pi x}{a} \sin\frac{\pi y}{b} \quad (13.4.4a)$$

$$w = A_i \sin\frac{i\pi x}{a} \sin\frac{\pi y}{b} \quad (13.4.4b)$$

where i is usually considered to be from 1 to $2m$ in which m is the buckling half-wave number to be taken as an integer that satisfies $a/b \le \sqrt{m(m+1)}$.

Regarding the welding residual stress as the initial stress, the stress function F can be obtained from equation (13.3.2) after substitution of equations (13.3.4a) and (13.3.4b), as follows.

$$F = (\sigma_{xav} + \sigma_{rx})\frac{y^2}{2} + \frac{EA_i(A_i + 2A_{oi})}{32}\left(\frac{a^2}{i^2b^2}\cos\frac{2i\pi x}{a} + \frac{i^2b^2}{a^2}\cos\frac{2\pi y}{b}\right) \quad (13.4.5)$$

where σ_{xav} is average (applied) compressive stress = P_x/bt, and P_x is the axial compressive load in the x direction.

The Galerkin method is applied to determine the unknown amplitude A_i of added deflection with equation (13.3.1), as follows.

$$\int_0^a\int_0^b \left[D\left(\frac{\partial^4 w}{\partial x^4} + 2\frac{\partial^4 w}{\partial x^2 \partial y^2} + \frac{\partial^4 w}{\partial y^4}\right) - t\left\{\frac{\partial^2 F}{\partial y^2}\frac{\partial^2 (w+w_o)}{\partial x^2} - 2\frac{\partial^2 F}{\partial x \partial y}\frac{\partial^2 (w+w_o)}{\partial x \partial y} + \frac{\partial^2 F}{\partial x^2}\frac{\partial^2 (w+w_o)}{\partial y^2}\right\}\right] \sin\frac{i\pi x}{a}\sin\frac{\pi y}{b} = 0 \quad (13.4.6)$$

If equations (13.4.4a), (13.4.4b), and (13.4.5) are substituted into equation (13.4.6), and the integration is performed over the plate, then the following third-order equation with regard to the unknown amplitude A_i is obtained.

$$C_1 A_i^3 + C_2 A_i^2 + C_3 A_i + C_4 = 0 \quad (13.4.7)$$

where, $C_1 = \frac{\pi^2 E}{16}\left(\frac{i^4 b}{a^3} + \frac{a}{b^3}\right)$, $C_2 = \frac{3\pi^2 EA_{oi}}{16}\left(\frac{i^4 b}{a^3} + \frac{a}{b^3}\right)$,

$$C_3 = \frac{\pi^2 EA_{oi}^2}{8}\left(\frac{i^4 b}{a^3} + \frac{a}{b^3}\right) + \frac{i^2 b}{a}(\sigma_{xav} + \sigma_{rex}) + \frac{\pi^2 D}{t}\frac{i^2}{ab}\left(\frac{ib}{a} + \frac{a}{ib}\right)^2,$$

and $C_4 = A_{oi}\frac{i^2 b}{a}(\sigma_{xav} + \sigma_{rex})$,

$$\sigma_{rex} = \sigma_{rcx} + \frac{2}{b}(\sigma_{rtx} - \sigma_{rcx})\left(b_t - \frac{b}{2\pi}\sin\frac{2\pi b_t}{b}\right).$$

The following is the FORTRAN language program CARDANO used to solve equation (13.4.7) with regard to the unknown value A_i, when coefficients C_1 to C_4 are predefined.

```
      SUBROUTINE CARDANO(C1,C2,C3,C4,W)
      IMPLICIT REAL*8(A-H,O-Z)
C
C***  C1*W**3+C2*W**2+C3*W+C4=0
C***  INPUT: C1,C2,C3,C4
C***  OUTPUT: W
C PROGRAMMED BY PROF. J.K. PAIK
C © J.K. PAIK. ALL RIGHTS RESERVED
C
   S1=C2/C1
   S2=C3/C1
   S3=C4/C1
   P=S2/3.0-S1**2/9.0
   Q=S3-S1*S2/3.0+2.0*S1**3/27.0
   Z=Q**2+4.0*P**3
   IF(Z.GE.0.0) THEN
   AZ=(-Q+SQRT(Z))*0.5
   BZ=(-Q-SQRT(Z))*0.5
   AM=ABS(AZ)
   BM=ABS(BZ)
   IF(AM.LT.1.0E-10) THEN
   CA=0.0
   ELSE
   CA=AZ/AM
   END IF
   IF(BM.LT.1.0E-10) THEN
   CB=0.0
   ELSE
   CB=BZ/BM
   END IF
   W=CA*AM**(1.0/3.0)+CB*BM**(1.0/3.0)-
   S1/3.0
   ELSE
   TH=ATAN(SQRT(-Z)/(-Q))
```

```
W=2.0*(-P)**0.5*COS(TH/3.0)-S1/3.0
END IF
RETURN
END
```

Once the solution of equation (13.4.7) has been obtained for a specific half-wave, the elastic large deflection behavior of the plate can be identified. This analysis should be undertaken for all of the half-wave components considered.

Rigid-Plastic Analysis

When a rectangular plate is subjected to a uniform virtual-displacement δu in the x direction, the principle of virtual work emerges from the following equilibrium equation between the external virtual-work and the internal virtual-energy dissipation for a kinematically admissible collapse mechanism.

$$\sigma_{xav} bt\delta u = -\sum_{n=1}^{r} \int_{L_n} N\delta U dL_n + \sum_{n=1}^{s} (M + wN)\delta\theta\, dL_n \quad (13.4.8)$$

where, L_n is the length of the nth plastic hinge, M is the moment per unit length along the plastic hinge line, N is the axial force per unit length along the plastic hinge line, r is the number of inclined hinge lines, s is the number of horizontal or vertical hinge lines, U is the axial displacement along the plastic hinge line, u is the axial displacement in the x direction, w is the lateral deflection of the plate, and θ is the rotation along the plastic hinge line.

In equation (13.4.8), the prefix δ denotes the virtual variable. The left and right terms of equation (13.4.8) indicate the external virtual-work and the internal virtual-energy dissipation, respectively. The first and second terms on the right-hand side represent the energy contributions due to virtual-axial displacement and virtual rotation along the plastic hinge lines, respectively. The plate is considered to have three different types of collapse mechanisms, depending on the plate aspect ratio and the deflection shape, among other factors.

(a) Mode I for $\frac{a}{ib} > 1$

Figure 13.5(a) depicts Mode I, where the angle between hinge lines I and II is defined by α. In this mode, the virtual deflections, virtual rotations, and virtual axial-displacements along plastic hinge lines I and II are given as follows.

$$w^I = A_i\left(1 - \frac{2\sin\alpha}{b} L_n\right),\ w^{II} = A_i,\ \delta\theta^I = \frac{4A_i \sin^2\alpha}{b\cos\alpha},$$

$$\delta\theta^{II} = \frac{4A_i}{b},\ \delta u^I = \delta u \sin\alpha,\ \delta U^{II} = 0 \quad (13.4.9a)$$

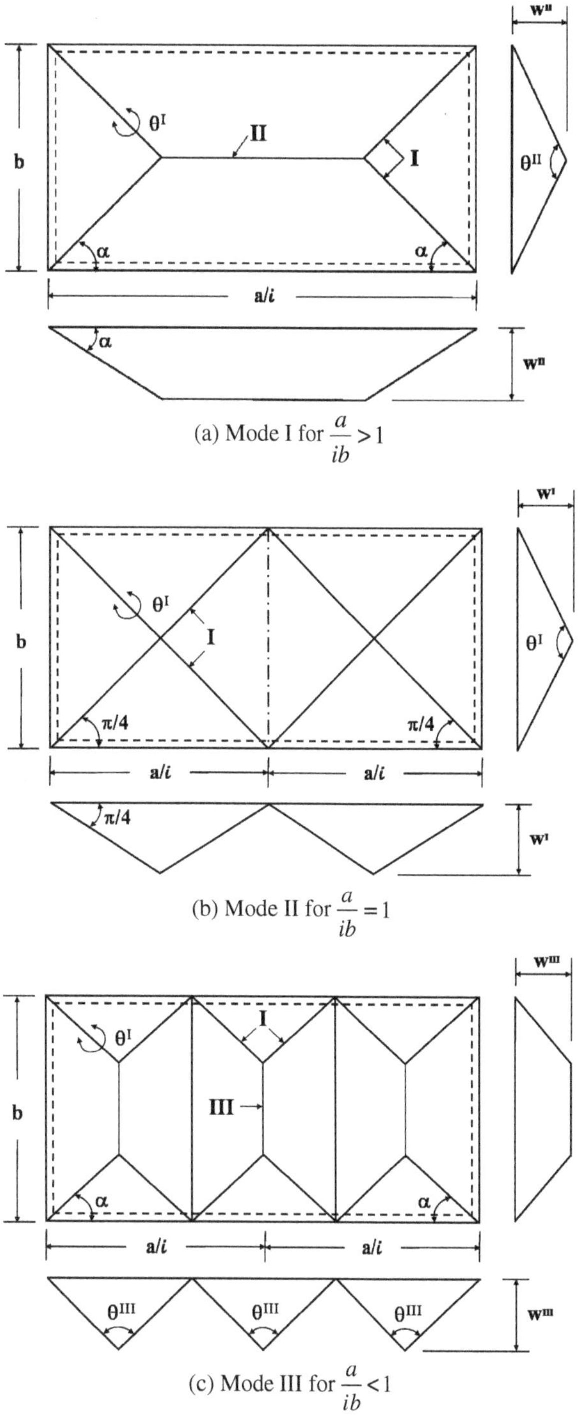

Figure 13.5 Collapse mechanisms for a plate subject to axial compressive loads.

The axial force and bending moment per unit length along the hinge lines are calculated as follows.

$$N^I = \frac{\sigma_{xav} t}{2}(\cos 2\alpha - 1),\ N^{II} = 0,$$

$$M^I = \frac{4\left(1 - p_x^2\right) M_p}{\sqrt{16 - 3p_x^2(\cos 2\alpha + 1)^2 - 12 p_x^2 \sin^2 2\alpha}},$$

$$M^{II} = \frac{2\left(1-p_x^2\right)M_p}{\sqrt{4-3p_x^2}} \tag{13.4.9b}$$

where $p_x = \frac{\sigma_{xav}}{\sigma_Y}$, $M_p = \frac{t^2}{4}\sigma_Y$ is the plastic moment along the plastic hinge lines and σ_Y is the material yield stress.

The substitution of equations (13.4.9a) and (13.4.9b) into equation (13.4.8) yields the following.

$$\sigma_{xav} bt\delta u = -4\int_0^{\frac{b}{2\sin\alpha}} \left(N^I \delta U^I - M^I \delta\theta_I - w^I N^I \delta\theta^I\right) dL_n$$

$$-\int_0^{\frac{a}{i}-b\cot\alpha} \left(N^{II} \delta U^{II} - M^{II} \delta\theta^{II} - w^{II} N^{II} \delta\theta^{II}\right) dL_n$$

$$= -\sigma_{xav} bt\delta u\left(\cos 2\alpha - 1\right) + 8A_i M^I \tan\alpha$$

$$+4A_i M^{II}\frac{1}{b}\left(\frac{a}{i} - b\cot\alpha\right) + 2A_i^2 \sigma_{xav} t\left(\cos 2\alpha - 1\right)\tan\alpha \tag{13.4.10}$$

The exact angle between hinge lines I and II for the corresponding collapse mechanism can be determined by minimizing the total potential energy with regard to the angle α. For the sake of simplicity, however, $\alpha = \pi/4$ is assumed. The axial compressive stress versus the maximum deflection relation for the Mode I collapse mechanism is obtained as follows.

$$W_i = \frac{A_i}{t} = \frac{1-p_x^2}{p_x}\left[\frac{4}{\sqrt{16-15p_x^2}} + \frac{1}{\sqrt{4-3p_x^2}}\left(\frac{a}{ib} - 1\right)\right] \tag{13.4.11}$$

(b) Mode II for $\frac{a}{ib} = 1$

Figure 13.5(b) depicts the collapse mechanism of Mode II, where the angle between the hinge lines is $\pi/4$, and there are no hinge lines parallel to the axial load direction. The virtual deflections, virtual rotations, and virtual in-plane deformations along hinge line I are determined as follows.

$$w^I = A_i\left(1 - \frac{\sqrt{2}L_n}{b}\right),\quad \delta\theta^I = \frac{2\sqrt{2}A_i}{b},\quad \delta U^I = \frac{\sqrt{2}\delta u}{2} \tag{13.4.12a}$$

The axial forces and bending moments along hinge line I are determined as follows.

$$N^I = -\frac{\sigma_{xav} t}{2},\quad M^I = \frac{4\left(1-p_x^2\right)M_p}{\sqrt{16-15p_x^2}} \tag{13.4.12b}$$

The substitution of equations (13.4.12a) and (13.4.12b) into equation (13.4.8) yields the following.

$$\sigma_{xav} bt\delta u = -4\int_0^{\frac{\sqrt{2}b}{2}} \left(N^I \delta U^I - M^I \delta\theta^I - w^I N^I \delta\theta^I\right) dL_n$$

$$= \sigma_{xav} bt\delta u + 8A_i M^I - 2\sigma_{xav} t A_i^2 \tag{13.4.13}$$

The axial force versus plate deflection for Mode II is obtained as follows.

$$W_i = \frac{A_i}{t} = \frac{4\left(1-p_x^2\right)}{p_x\sqrt{16-15p_x^2}} \tag{13.4.14}$$

(c) Mode III for $\frac{a}{ib} < 1$

Figure 13.5(c) depicts the collapse mechanism of Mode III. Virtual deflections, virtual rotations, and virtual in-plane deformations along the hinge lines are determined as follows.

$$w^I = A_i\left(1 - \frac{2i\cos\alpha}{a}L_n\right),\quad w^{III} = A_i,\quad \delta\theta^I = \frac{4A_i\cos^2\alpha}{a\sin\alpha}$$

$$\delta\theta^{III} = \frac{4i}{a}A_i,\quad U^I = \delta u\sin\alpha,\quad U^{III} = \delta u \tag{13.4.15a}$$

The axial force and the bending moment along the hinge lines are determined as follows.

$$N^I = \frac{\sigma_{xav} t}{2}\left(\cos 2\alpha - 1\right),\quad N^{III} = -\sigma_{xav} t,$$

$$M^I = \frac{4\left(1-p_x^2\right)M_p}{\sqrt{16-3p_x^2\left(\cos 2\alpha + 1\right)^2 - 12p_x^2\sin^2 2\alpha}}$$

$$M^{III} = \left(1-p_x^2\right)M_p \tag{13.4.15b}$$

The substitution of equations (13.4.15a) and (13.4.15b) into equation (13.4.8) yields the following.

$$\sigma_{xav} bt\delta u = -4\int_0^{\frac{a}{2i\cos\alpha}} \left(N^I \delta U^I - M^I \delta\theta^I - w^I N^I \delta\theta^I\right) dL_n$$

$$-\int_0^{b-\frac{a}{i\tan\alpha}} \left(N^{III} \delta U^{III} - M^{III} \delta\theta^{III} - w^{III} N^{III} \delta\theta^{III}\right) dL_N$$

$$= \sigma_{xav} bt\delta u - \frac{\sigma_{xav} at\delta u}{i}\tan\alpha\cos 2\alpha$$

$$+8A_i M^I\cot\alpha + 4A_i M^{III}\left(\frac{ib}{a} - \tan\alpha\right)$$

$$+2A_i^2\sigma_{xav} t\left[\left(\cos 2\alpha - 1\right)\cot\alpha + 2\tan\alpha - \frac{2ib}{a}\right] \tag{13.4.16}$$

By assuming that $\alpha = \pi/4$, the axial force versus the plate deflection relation for Mode III can be obtained as follows.

$$W_i = \frac{A_i}{t} = \frac{a}{2ib-1}\frac{i-p_x^2}{p_x}\left(\frac{4}{\sqrt{16-15p_x^2}} + \frac{ib}{2a} - \frac{1}{2}\right) \tag{13.4.17}$$

Combined Elastic Large Deflection Analysis and Rigid-Plastic Analysis

The ultimate strength is determined as the intersection between the elastic large deflection behavior and the rigid-plastic large deflection behavior, as illustrated in Fig. 13.4. This calculation is undertaken for each of the possible collapse modes or half-waves varying from $i = 1$ to $2m$, where m is determined from equation (12.2.9). The minimum value among the ultimate strengths so obtained is regarded as the real ultimate strength of the plate.

13.4.3 Membrane Stress-Based Method (Plate Edge-Oriented Plastic Hinge Approach)

The membrane stress inside a deflected or buckled plate is nonuniform. Figure 13.6 shows a typical example of the axial membrane stress distribution inside a plate subject to uniaxial compressive loading before and after buckling occurs, where for simplicity, the case of just one bulge in the middle of the plate ($m = n = 1$) is drawn, but later when we present the theory in Section 13.5, equation (13.5.5) allows m and n to have any value. The membrane stress distribution in the loading (x) direction becomes nonuniform as the plate starts to deflect (e.g., due to buckling). The y direction also becomes nonuniform as long as the unloaded plate edges remain straight, although no membrane stresses will develop in the y direction if the unloaded plate edges move freely in plane. It is noted that for a plate that is part of a stiffened panel, the unloaded edges are likely to remain straight.

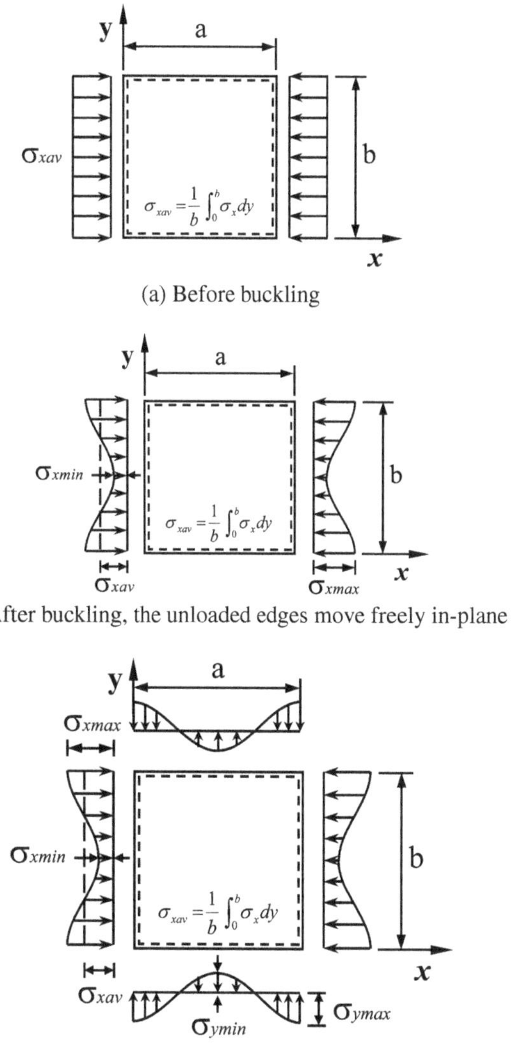

Figure 13.6 Membrane stress distribution inside a plate subject to uniaxial compressive loads for the case of one bulge in the middle of the plate ($m = n = 1$).

The maximum compressive membrane stresses are developed around the plate corners, and the minimum (tensile) membrane stresses occur in the middle of the plate, where a membrane tension field is formed by the plate deflection because the plate edges remain straight. A similar nonlinear distribution of membrane stresses may appear inside a deflected plate subject to combined axial compression and lateral pressure loads. Edge shear loading may render the membrane stress distribution pattern more complex than that under biaxial and lateral pressure load conditions, but as long as the edge shear is a secondary load component, the basic membrane stress distribution pattern inside the plate may be similar to Fig.13.6c.

With an increase in plate deflection, the membrane stress is redistributed as in Fig.13.6c, but although it is lower in the midwidth of the plate, the stress in the upper and/or lower faces in the midwidth of the plate will initially yield through bending action. However, as long as it is possible to redistribute the stress to the straight plate boundaries through membrane action, the plate will not collapse. Collapse will occur when the most stressed boundary locations yield because the plate can no longer keep the boundaries straight, thus resulting in a rapid increase in lateral deflection, which corresponds to the ultimate limit state or ultimate strength (Paik and Thayamballi, 2003).

Because of the nature of the combined membrane axial stresses in the x and y directions, there are three possible locations for initial yield at the edges, namely, the plate corners, longitudinal midedges, and transverse midedges, as shown in Fig. 13.7. The stress at the two midedge locations (i.e., that at each longitudinal or transverse midedge) can be expected to be the same as long as the longitudinal or transverse axial stresses are uniformly applied (i.e., without in-plane bending). Depending

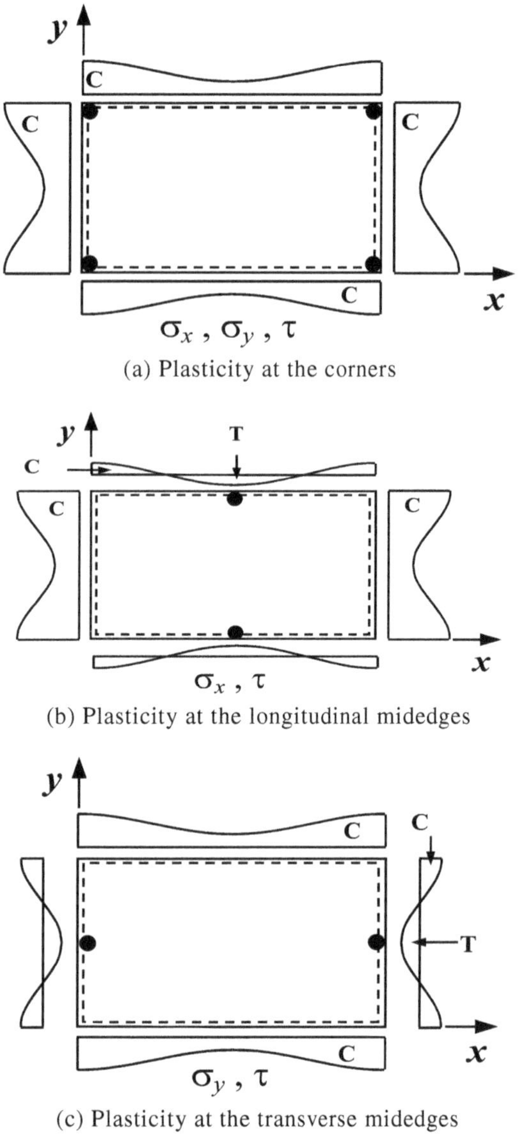

Figure 13.7 Three possible locations for the initial plastic yield at plate edges subject to combined loads (•, expected plasticity location; T, tension; C, compression).

on the predominant half-wave mode in the length direction, the location of possible plasticity may vary at the long edges because the location of the minimum membrane stresses may differ, whereas it is always at the midedges in the short direction. In this regard, the membrane stress-based method can be termed the plate edge-oriented plastic hinge approach.

The occurrence of plasticity can be assessed using the von Mises yield criterion. The following three resulting ultimate strength criteria for the most probable yield locations will be found once the maximum or minimum membrane stresses are defined, as shown schematically in Fig. 13.6.

(a) Plasticity at plate corners

$$\sigma_{eq1} = \sqrt{\sigma_{x\max}^2 - \sigma_{x\max}\sigma_{y\max} + \sigma_{y\max}^2 + 3\tau^2} = \sigma_Y \quad (13.4.18a)$$

(b) Plasticity at longitudinal midedges

$$\sigma_{eq2} = \sqrt{\sigma_{x\max}^2 - \sigma_{x\max}\sigma_{y\min} + \sigma_{y\min}^2 + 3\tau^2} = \sigma_Y \quad (13.4.18b)$$

(c) Plasticity at transverse midedges

$$\sigma_{eq3} = \sqrt{\sigma_{x\min}^2 - \sigma_{x\min}\sigma_{y\max} + \sigma_{y\max}^2 + 3\tau^2} = \sigma_Y \quad (13.4.18c)$$

As the applied loads increase, the plate will collapse if any one of the three foregoing equivalent stresses, namely, σ_{eq1}, σ_{eq2}, or σ_{eq3}, reaches the yield stress of the material. The minimum values of the applied load components among those that satisfy the three equations must then be the real ultimate strength of the plate.

The remaining task in this approach is to identify the maximum and minimum membrane stresses included in equation (13.4.18), which are formulated as a function of the various parameters of influence addressed in Section 13.5.

13.4.4 Effective Width or Breadth Method

The membrane stress-based method described in Section 13.4.3 is the preferred method, but for historical reasons the effective width or breadth method is presented here.

A plate that is buckled or deflected by predominantly axial compressive loads with or without lateral pressure loads is often modeled as an equivalent flat plate, but with reduced plate breadth or effectiveness. Two different terms are sometimes employed in the evaluation of plate effectiveness (Paik, 2008c), depending on the cause of plate deflection. When the plate deflects by buckling under predominantly axial compressive loads, the term "effective width" is used, whereas when out-of-plane or lateral actions such as lateral pressure loads are the dominant load components causing lateral deflection, the term "effective breadth" is used in evaluating the effectiveness of the deflected plate in association with the shear-lag effect.

By definition, the effective width in the y direction or the effective length in the x direction can be expressed as a ratio of the average axial compressive stress to the corresponding maximum (compressive) axial stress, as follows.

$$\frac{b_e}{b} = \frac{\sigma_{xav}}{\sigma_{x\max}} \quad (13.4.19a)$$

$$\frac{a_e}{a} = \frac{\sigma_{yav}}{\sigma_{y\max}} \quad (13.4.19b)$$

where b_e is effective plate width and a_e is effective plate length.

Equations (13.4.19a) and (13.4.19b) are valid only when σ_{xav} and σ_{yav} are the non-zero axial compressive stresses, respectively. The denominators $\sigma_{x\,\max}$ and $\sigma_{y\,\max}$ are functions of the applied load components (e.g., biaxial loads, lateral pressure, edge shear) and initial imperfections, as well as the geometric and material properties (Paik and Thayamballi, 2003; Paik, 2008c). In other words, the effective width or length of a plate does not take a constant value, but varies with the applied loads among the other parameters of influence.

The effective width or length formula of a plate that has just reached its ultimate strength is often used to predict that strength, as follows.

$$\sigma_{xu} = \frac{b_{eu}}{b}\sigma_Y \qquad (13.4.20a)$$

$$\sigma_{yu} = \frac{a_{eu}}{a}\sigma_Y \qquad (13.4.20b)$$

where σ_{xu}, σ_{yu} is the ultimate compressive stresses in the x or y directions, and b_{eu} and a_{eu} are the effective width or length at the ultimate strength, respectively.

Faulkner (1975) suggested the following empirical formulation of the effective width of a plate subject to axial compressive loads in the x direction.

$$\frac{b_{eu}}{b} = \begin{cases} 1.0 & \text{for } \beta < C_3 \\ \dfrac{C_1}{\beta} - \dfrac{C_2}{\beta^2} & \text{for } \beta \ge C_3 \end{cases} \qquad (13.4.21)$$

where $\beta = \frac{b}{t}\sqrt{\frac{\sigma_Y}{E}}$ is the plate slenderness coefficient, $C_1 = 2$, $C_2 = 1$, and $C_3 = 1$ for simply supported plates, and $C_1 = 2.25$, $C_2 = 1.25$, and $C_3 = 1.25$ for clamped plates.

Figure 13.8 shows illustrative examples of the effective width of a simply supported plate subject to uniaxial compression with or without lateral pressure loads. When either welding residual stress or lateral pressure exists, it appears that the effective plate width increases from zero as the axial compressive loads increase until the membrane axial compressive stresses that are results of these loads become dominant. This is because of the definition of the effective width of a plate, as indicated in equation (13.4.19a), but does not necessarily mean that the plate effectiveness is zero or increasing.

13.4.5 Effective Shear Modulus Method

A plate that buckles by predominantly edge shear loads can be modeled as an equivalent flat plate, but with a reduced shear modulus (Paik, 1995). The

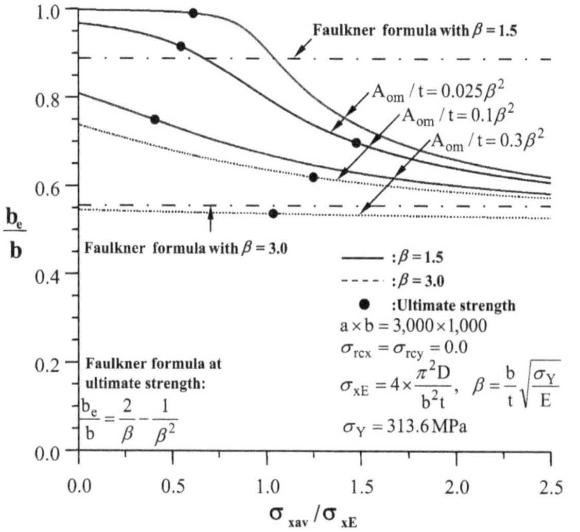

(a) Effect of initial deflection without welding residual stresses

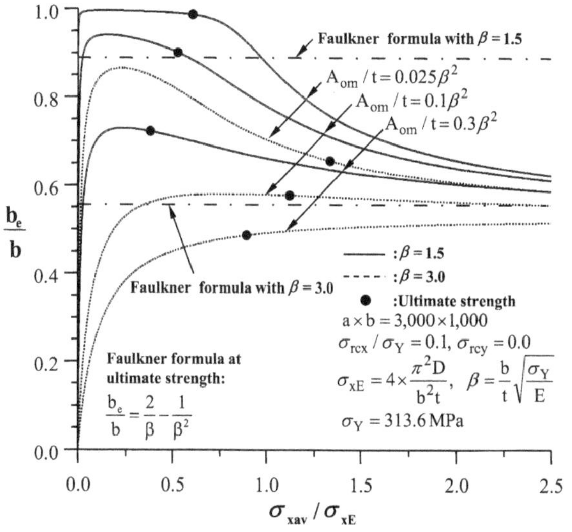

(b) Effect of initial deflection with welding residual stresses

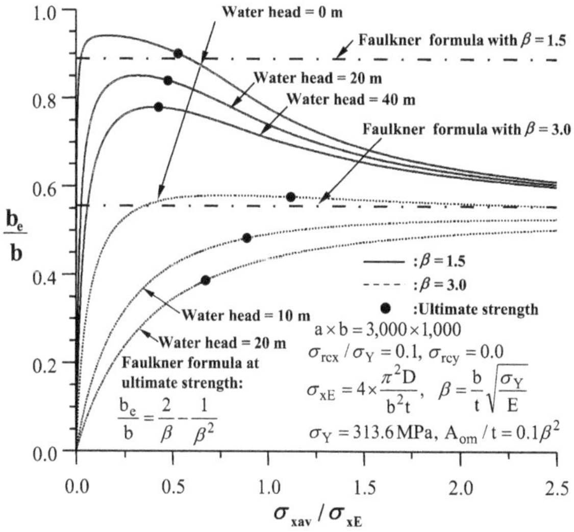

(c) Effect of lateral pressure

Figure 13.8 Variation in the effective width of a simply supported plate subject to uniaxial compression (σ_{xE} = elastic compressive buckling stress).

membrane shear strain component γ_m of the buckled plate must in this case be evaluated as follows.

$$\gamma_m = \frac{\partial u}{\partial y} + \frac{\partial v}{\partial x} = \frac{\tau_{av}}{G} - \left[\frac{\partial w}{\partial x}\frac{\partial w}{\partial y} + \frac{\partial w}{\partial x}\frac{\partial w_o}{\partial y} + \frac{\partial w_o}{\partial x}\frac{\partial w}{\partial y}\right] \quad (13.4.22)$$

where γ_m is the membrane shear strain, τ_{av} is the average edge shear stress, and u, v are the axial displacements in the x or y directions, respectively.

The mean membrane shear strain γ_{av} can be obtained as an average of the so-computed shear strains over the entire plate, as follows.

$$\gamma_{av} = \frac{1}{ab}\int_0^a\int_0^b \gamma_m \, dxdy \quad (13.4.23)$$

Because the shear stress at the plate edges may equal the average shear stress (i.e., $\tau = \tau_{av}$), the effective shear modulus G_e, which represents the effectiveness of the plate buckled in edge shear, can be defined as follows.

$$G_e = \frac{\tau_{av}}{\gamma_{av}} \quad (13.4.24)$$

The effective shear modulus can be evaluated by numerical computations (Paik, 1995) and is a function of the applied shear forces. The ultimate shear strength of the plate is then obtained as follows.

$$\tau_u = \frac{G_{eu}}{G}\tau_Y \quad (13.4.25)$$

where τ_u is the ultimate shear stress, $\tau_Y = \sigma_Y/\sqrt{3}$ is the shear yield stress, and G_{eu} is the effective shear modulus at ultimate strength.

An empirical formulation of the effective shear modulus at ultimate strength for a simply supported plate subject to edge shear, and with an average level of initial deflection, is given by the following.

$$\frac{G_{eu}}{G} = \begin{cases} 1.324\left(\frac{\tau_E}{\tau_Y}\right) & \text{for } 0 < \frac{\tau_E}{\tau_Y} \le 0.5 \\ 0.039\left(\frac{\tau_E}{\tau_Y}\right)^3 - 0.274\left(\frac{\tau_E}{\tau_Y}\right)^2 + 0.676\left(\frac{\tau_E}{\tau_Y}\right) + 0.388 & \text{for } 0.5 < \frac{\tau_E}{\tau_Y} \le 2.0 \\ 0.956 & \text{for } \frac{\tau_E}{\tau_Y} > 2.0 \end{cases} \quad (13.4.26)$$

where τ_E is the elastic shear buckling stress, which is given by $\tau_E = k_\tau \frac{\pi^2 E}{12(1-\nu^2)}\left(\frac{t}{b}\right)^2$, with $k_\tau \approx 4\left(\frac{b}{a}\right)^2 + 5.34$ for $\frac{a}{b} \ge 1$ or $k_\tau \approx 5.34\left(\frac{b}{a}\right)^2 + 4.0$ for $\frac{a}{b} < 1$. Figure 13.9 shows the ultimate strength of a simply supported plate subject to edge shear obtained from equation (13.4.25) in comparison with nonlinear finite element method (FEM) solutions.

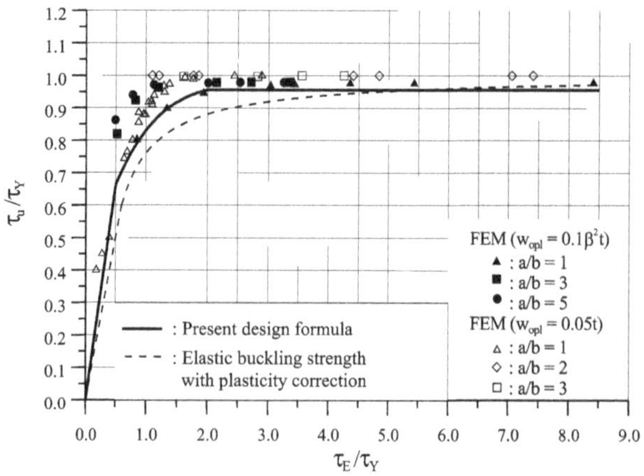

Figure 13.9 Ultimate strength of a simply supported plate under edge shear conditions.

Figure 13.10 shows the effect of the plate aspect ratio on the plate ultimate shear strength. As the plate aspect ratio increases, the plate ultimate shear strength tends to decrease. As apparent from Fig. 13.10, the ultimate shear strength depends weakly on the plate aspect ratio, especially for relatively thick plates. In this regard, one may neglect the effect of the plate aspect ratio on the plate ultimate shear strength for the sake of simplicity.

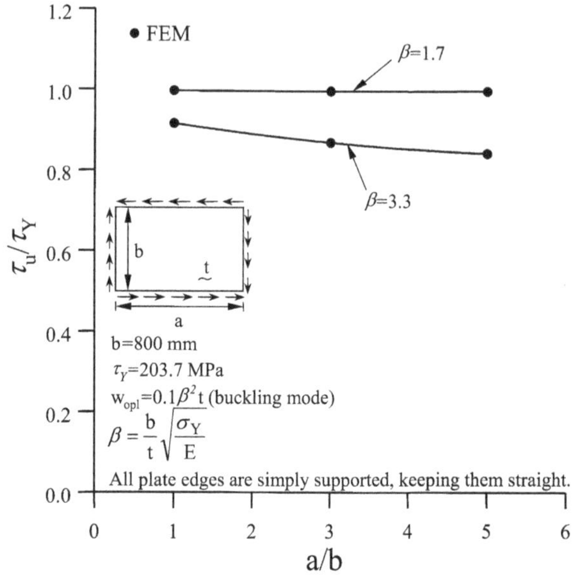

Figure 13.10 Effect of the aspect ratio on the plate ultimate shear strength.

13.5 ELASTIC LARGE DEFLECTION BEHAVIOR: DEFINITION OF MAXIMUM AND MINIMUM MEMBRANE STRESSES

The membrane stress-based method, or plate edge-oriented plastic hinge approach, described in Section 13.4.3 is useful for ultimate plate strength computations because it is able to take into account the effects of combined loads and initial imperfections as parameters of influence. The remaining task in terms of applying the plate edge-oriented plastic hinge approach is to define the maximum and minimum membrane stresses that should be substituted into equation (13.4.18).

In the following, these stresses are derived for plates subject to different types of applied loads by solving nonlinear governing differential equations.

13.5.1 Simply Supported Plates

The maximum and minimum membrane stresses of a plate simply supported at all edges are derived here. The effects of initial imperfections are taken into account, and it is presumed that the direction of the initial deflection corresponds to that of the lateral deflection as a result of pressure loads, which will result in a somewhat pessimistic evaluation of plate strength.

13.5.1.1 Lateral Pressure Loads

In this case, it is presumed that the initial and added deflection functions can be approximately expressed as follows.

$$w_o = A_{o1} \sin\frac{\pi x}{a} \sin\frac{\pi y}{b} \tag{13.5.1a}$$

$$w = A_1 \sin\frac{\pi x}{a} \sin\frac{\pi y}{b} \tag{13.5.1b}$$

where A_{o1} is the known amplitude of initial deflection and A_1 is the unknown amplitude of added deflection.

With deflection functions (13.5.1a) and (13.5.1b), the nonlinear governing differential equations (13.3.1) and (13.3.2) can be solved by the Galerkin method. The stress function F is then obtained as follows.

$$F = \sigma_{rx}\frac{y^2}{2} + \sigma_{ry}\frac{x^2}{2} + \frac{EA_1(A_1+2A_{o1})}{32}\left(\frac{a^2}{b^2}\cos\frac{2\pi x}{a} + \frac{b^2}{a^2}\cos\frac{2\pi y}{b}\right) \tag{13.5.2}$$

The following third-order equation for the unknown amplitude of added deflection A_1 is obtained in a similar way to equation (13.4.6), as follows.

$$C_1 A_1^3 + C_2 A_1^2 + C_3 A_1 + C_4 = 0 \tag{13.5.3}$$

where, $C_1 = \dfrac{\pi^2 E}{16}\left(\dfrac{b}{a^3} + \dfrac{a}{b^3}\right)$, $C_2 = \dfrac{3\pi^2 EA_{o1}}{16}\left(\dfrac{b}{a^3} + \dfrac{a}{b^3}\right)$,

$$C_3 = \frac{\pi^2 EA_{o1}^2}{8}\left(\frac{b}{a^3} + \frac{a}{b^3}\right) + \frac{b}{a}\sigma_{rex} + \frac{a}{b}\sigma_{rey} + \frac{\pi^2 D}{t}\frac{1}{ab}\left(\frac{b}{a} + \frac{a}{b}\right)^2,$$

$$C_4 = A_{o1}\left(\frac{b}{a}\sigma_{rex} + \frac{a}{b}\sigma_{rey}\right) - \frac{16ab}{\pi^4 t}p,$$

$$\sigma_{rex} = \sigma_{rcx} + \frac{2}{b}(\sigma_{rtx} - \sigma_{rcx})\left(b_t - \frac{b}{2\pi}\sin\frac{2\pi b_t}{b}\right), \text{ and}$$

$$\sigma_{rey} = \sigma_{rcy} + \frac{2}{b}(\sigma_{rty} - \sigma_{rcy})\left(a_t - \frac{a}{2\pi}\sin\frac{2\pi a_t}{a}\right).$$

Once A_1 has been determined as a solution of equation (13.5.3), the membrane stresses inside the plate at the midthickness can be obtained from equations (13.3.3a) and (13.3.3b). The maximum and minimum membrane stresses in the x and y directions are obtained as follows.

$$\sigma_{x\max} = \left.\frac{\partial^2 F}{\partial y^2}\right|_{x=0,\,y=b_t} = \sigma_{rtx} - \frac{E\pi^2 A_1(A_1+2A_{o1})}{8a^2}\cos\frac{2\pi b_t}{b} \tag{13.5.4a}$$

$$\sigma_{x\min} = \left.\frac{\partial^2 F}{\partial y^2}\right|_{x=0,\,y=b/2} = \sigma_{rcx} + \frac{E\pi^2 A_1(A_1+2A_{o1})}{8a^2} \tag{13.5.4b}$$

$$\sigma_{y\max} = \left.\frac{\partial^2 F}{\partial x^2}\right|_{x=a_t,\,y=0} = \sigma_{rty} - \frac{E\pi^2 A_1(A_1+2A_{o1})}{8b^2}\cos\frac{2\pi a_t}{a} \tag{13.5.4c}$$

$$\sigma_{y\min} = \left.\frac{\partial^2 F}{\partial x^2}\right|_{x=a/2,\,y=0} = \sigma_{rcy} + \frac{E\pi^2 A_1(A_1+2A_{o1})}{8b^2} \tag{13.5.4d}$$

13.5.1.2 Combined Biaxial Loads

In this case, the initial and added deflection functions with a single deflection component may be approximately adopted as follows.

$$w_o = A_{omn} \sin\frac{m\pi x}{a} \sin\frac{n\pi y}{b} \tag{13.5.5a}$$

$$v = A_{mn} \sin\frac{m\pi x}{a} \sin\frac{n\pi y}{b} \tag{13.5.5b}$$

where m and n are the buckling half-wave numbers in the x and y directions, respectively. In the short direction of the plate, the buckling half-wave number must be taken as 1. For example, if the x coordinate is taken such that $a/b \geq 1$, then $n = 1$ is identified, and m is determined as a minimum integer that satisfies the following equation (Paik and Thayamballi, 2003).

(a) When σ_{xav} and σ_{yav} are both non-zero compressive (negative):

$$\frac{(m^2/a^2 + 1/b^2)^2}{m^2/a^2 + c/b^2} \leq \frac{\left[(m+1)^2/a^2 + 1/b^2\right]^2}{(m+1)^2/a^2 + c/b^2} \tag{13.5.6a}$$

where $c = \sigma_{yav}/\sigma_{xav}$.

(b) When σ_{xav} is tensile (positive) or zero, no matter what value of σ_{yav}:

$$m = 1 \tag{13.5.6b}$$

(c) When σ_{xav} is compressive and σ_{yav} is tensile or zero:

$$\frac{a}{b} \le \sqrt{m(m+1)} \tag{13.5.6c}$$

In a similar way, the applicable stress function F can be expressed as follows.

$$F = (\sigma_{xav} + \sigma_{rx})\frac{y^2}{2} + (\sigma_{yav} + \sigma_{ry})\frac{x^2}{2} + \frac{EA_{mn}(A_{mn} + 2A_{omn})}{32}\left(\frac{n^2a^2}{m^2b^2}\cos\frac{2m\pi x}{a} + \frac{m^2b^2}{n^2a^2}\cos\frac{2n\pi y}{b}\right) \tag{13.5.7}$$

By applying the Galerkin method, the following third-order equation with regard to the unknown amplitude A_{mn} is obtained as follows.

$$C_1A_{mn}^{\ 3} + C_2A_{mn}^{\ 2} + C_3A_{mn} + C_4 = 0 \tag{13.5.8}$$

where $C_1 = \frac{\pi^2E}{16}\left(\frac{m^4b}{a^3} + \frac{n^4a}{b^3}\right)$, $C_2 = \frac{3\pi^2EA_{omn}}{16}\left(\frac{m^4b}{a^3} + \frac{n^4a}{b^3}\right)$,

$$C_3 = \frac{\pi^2EA_{omn}^{\ 2}}{8}\left(\frac{m^4b}{a^3} + \frac{n^4a}{b^3}\right) + \frac{m^2b}{a}(\sigma_{xav} + \sigma_{rex}) + \frac{n^2a}{b}(\sigma_{yav} + \sigma_{rey}) + \frac{\pi^2D}{t}\frac{m^2n^2}{ab}\left(\frac{mb}{na} + \frac{na}{mb}\right)^2,$$

$$C_4 = A_{omn}\left[\frac{m^2b}{a}(\sigma_{xav} + \sigma_{rex}) + \frac{n^2a}{b}(\sigma_{yav} + \sigma_{rey})\right],$$

$$\sigma_{rex} = \sigma_{rcx} + \frac{2}{b}(\sigma_{rtx} - \sigma_{rcx})\left(b_t - \frac{b}{2n\pi}\sin\frac{2n\pi b_t}{b}\right), \text{ and}$$

$$\sigma_{rey} = \sigma_{rcy} + \frac{2}{b}(\sigma_{rty} - \sigma_{rcy})\left(a_t - \frac{a}{2m\pi}\sin\frac{2m\pi a_t}{a}\right)$$

The unknown deflection component (amplitude) A_{mn} can be obtained as a solution of equation (13.5.8). The maximum and minimum membrane stresses inside the plate are determined as follows.

$$\sigma_{x\max} = \left.\frac{\partial^2F}{\partial y^2}\right|_{x=0,\,y=b_t} = \sigma_{xav} + \sigma_{rtx} - \frac{E\pi^2m^2A_{mn}(A_{mn} + 2A_{omn})}{8a^2}\cos\frac{2n\pi b_t}{b} \tag{13.5.9a}$$

$$\sigma_{x\min} = \left.\frac{\partial^2F}{\partial y^2}\right|_{x=0,\,y=b/2} = \sigma_{xav} + \sigma_{rcx} + \frac{E\pi^2m^2A_{mn}(A_{mn} + 2A_{omn})}{8a^2} \tag{13.5.9b}$$

$$\sigma_{y\max} = \left.\frac{\partial^2F}{\partial x^2}\right|_{x=a_t,\,y=0} = \sigma_{yav} + \sigma_{rty} - \frac{E\pi^2n^2A_{mn}(A_{mn} + 2A_{omn})}{8b^2}\cos\frac{2m\pi a_t}{a} \tag{13.5.9c}$$

$$\sigma_{y\min} = \left.\frac{\partial^2F}{\partial x^2}\right|_{x=a/2,\,y=0} = \sigma_{yav} + \sigma_{rcy} + \frac{E\pi^2n^2A_{mn}(A_{mn} + 2A_{omn})}{8b^2} \tag{13.5.9d}$$

For the particular case of perfect plates without initial imperfections, equation (13.5.8) is simplified, as follows, because $C_2 = C_4 = 0$.

$$A_{mn}\left(C_1A_{mn}^2 + C_3\right) = 0 \tag{13.5.10}$$

where $C_1 = \frac{\pi^2E}{16}\left(\frac{m^4b}{a^3} + \frac{n^4a}{b^3}\right)$ and

$$C_3 = \frac{m^2b}{a}\sigma_{xav} + \frac{n^2a}{b}\sigma_{yav} + \frac{\pi^2D}{t}\frac{m^2n^2}{ab}\left(\frac{mb}{na} + \frac{na}{mb}\right)^2$$

The non zero solution of A_{mn} is defined from equation (13.5.10) as follows.

$$A_{mn} = \sqrt{-\frac{C_3}{C_1}} \tag{13.5.11}$$

The elastic buckling condition of a plate subject to biaxial compression is determined when $A_{mn} = 0$ or $C_3 = 0$ in equation (13.5.11), which is identical to equation (12.2.15) in Chapter 12, because the plate deflection must be zero immediately before buckling inception. Equation (12.2.16) of Chapter 12 gives the plate buckling condition.

Figure 13.11 shows the variation in the maximum and minimum membrane stresses in a square plate subject to longitudinal axial compression in comparison with the SPINE semianalytical method solution (Paik et al., 2001).

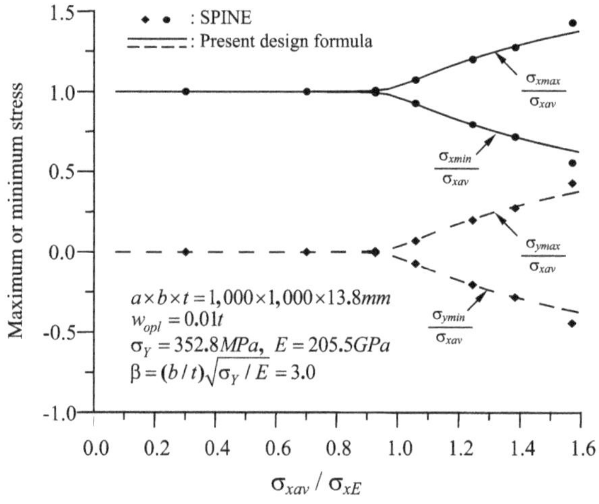

Figure 13.11 Variation in the maximum and minimum membrane stresses at the plate edges under longitudinal compressive load conditions.

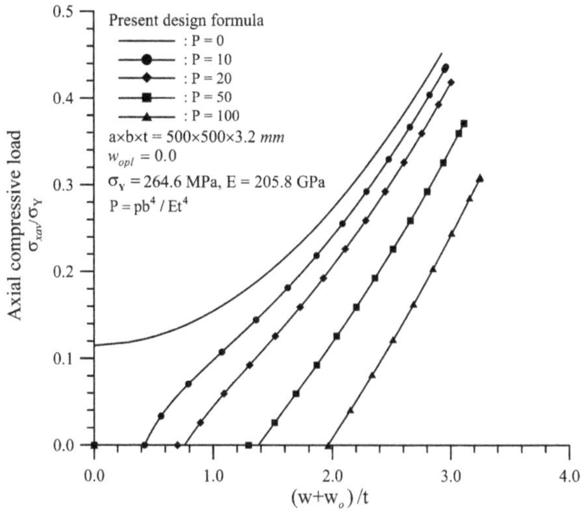

Figure 13.12 Axial compressive load versus deflection curves for a simply supported square plate element subject to combined longitudinal axial compression and lateral pressure.

Figure 13.12 shows an example of the relationship between longitudinal compressive loads and maximum deflections for a simply supported square plate with varying lateral pressure load magnitudes, as obtained by solving the elastic large deflection governing equations. It can be seen from this figure that, because of pressure loads, the lateral deflection increases from the beginning of axial compressive loading; therefore, no bifurcation (buckling) point can be defined because lateral pressure loads are applied. It should be noted, however, that this observation is true only for a square or near-square plate. For a long plate, a bifurcation point in longitudinal compression may appear even in the presence of lateral pressure loads as long as the magnitude of these loads is not large. In this case, the value of the elastic bifurcation load is, however, normally greater than that without lateral pressure loads (Okada, et al., 1979).

13.5.1.3 Interaction Effect Between Biaxial Loads and Lateral Pressure

The elastic large deflection behavior of plates under combined biaxial loads and lateral pressure is significantly affected by the amount of lateral pressure load, among other factors. In fact, it is not possible to analyze the large deflection plate behavior with the deflection functions of equations (13.5.5a) and (13.5.5b), which have a single deflection component. A greater number of deflection components must be included in these functions to make it possible. For the sake of simplicity, however, the contribution made by the lateral pressure loads to the nonlinear membrane stresses inside the plate is approximately accounted for here, where the membrane stresses arising from only the deflection components of $m = 1$ and $n = 1$ are linearly superposed to those arising from the biaxial in-plane loads. In this case, the coefficient C_4 of equation (13.5.8) is redefined as follows.

$$C_4 = A_{omn}\left[\frac{m^2 b}{a}(\sigma_{xav} + \sigma_{rex}) + \frac{n^2 a}{b}(\sigma_{yav} + \sigma_{rey})\right] - \frac{16ab}{\pi^4 t}p \tag{13.5.12}$$

When lateral pressure loads are applied together with predominantly biaxial loads, the third-order equation of equation (13.5.3) should be solved with the coefficient C_4, as defined in equation (13.5.12).

13.5.1.4 Effect of Edge Shear

Edge shear loading can affect the membrane normal stress distribution of a plate subject to predominantly biaxial loads. To analyze the large deflection behavior of a plate subject to edge shear loads, which is very complex in terms of geometrical shape, the deflection function must include a large number of deflection components. However, analytically solving the nonlinear governing differential equations with such deflection functions is no straightforward matter. In this regard, the deflection function with a single component is retained, but with the effects of edge shear empirically accounted for.

The maximum and minimum membrane stresses inside the plate under biaxial and lateral pressure load conditions are given by taking into account the effect of edge shear, as follows (Ueda, et al., 1984).

$$\sigma_{x\max} = \left.\frac{\partial^2 F}{\partial y^2}\right|_{x=0,\,y=b_t} = \sigma_{xav} + \sigma_{rtx} - \frac{E\pi^2 m^2 A_{mn}\left(A_{mn} + 2A_{omn}\right)}{8a^2}$$
$$\cos\frac{2n\pi b_t}{b}\left[1.3\left(\frac{\tau_{av}}{\tau_E}\right)^c + 1\right] + 1.62\sigma_{xE}\left(\frac{\tau_{av}}{\tau_E}\right)^{2.4} \tag{13.5.13a}$$

$$\sigma_{x\min} = \left.\frac{\partial^2 F}{\partial y^2}\right|_{x=0,\,y=b/2} = \sigma_{xav} + \sigma_{rcx} + \frac{E\pi^2 m^2 A_{mn}\left(A_{mn} + 2A_{omn}\right)}{8a^2}$$
$$\left(0.3\frac{\tau_{av}}{\tau_E} + 1\right) - 1.3\sigma_{xE}\left(\frac{\tau_{av}}{\tau_E}\right)^{2.1} \tag{13.5.13b}$$

$$\sigma_{y\max} = \left.\frac{\partial^2 F}{\partial x^2}\right|_{x=a_t,\,y=0} = \sigma_{yav} + \sigma_{rty} - \frac{E\pi^2 n^2 A_{mn}\left(A_{mn} + 2A_{omn}\right)}{8b^2}$$
$$\cos\frac{2m\pi a_t}{a}\left[1.3\left(\frac{\tau_{av}}{\tau_E}\right)^c + 1\right] + 1.62\sigma_{yE}\left(\frac{\tau_{av}}{\tau_E}\right)^{2.4} \tag{13.5.13c}$$

$$\sigma_{y\min} = \left.\frac{\partial^2 F}{\partial x^2}\right|_{x=a/2,\, y=0} = \sigma_{yav} + \sigma_{rcy} + \frac{E\pi^2 n^2 A_{mn}\left(A_{mn} + 2A_{omn}\right)}{8b^2}$$

$$\left(0.3\frac{\tau_{av}}{\tau_E} + 1\right) - 1.3\sigma_{yE}\left(\frac{\tau_{av}}{\tau_E}\right)^{2.1} \tag{13.5.13d}$$

where σ_{xE} is the elastic buckling stress subject to axial compression in the x direction, as defined in Table 12.1 of Chapter 12, σ_{yE} is the elastic buckling stress subject to axial compression in the y direction, as defined in Table 12.1 of Chapter 12, τ_E is the elastic buckling stress subject to edge shear, as defined in Table 12.1 of Chapter 12, and $c = 1.5$ for $\tau_{av} \leq \tau_E$ and $c = 1$ for $\tau_{av} > \tau_E$.

Figure 13.13 shows the variation in the maximum and minimum membrane stresses at the plate edges under combined longitudinal compression and edge shear conditions. It is seen from this figure that edge shear amplifies the maximum and minimum membrane stresses at the plate edges.

13.5.2 Plates with Partially Rotation-Restrained Edges

Plate edges are supported by beams such as longitudinal stiffeners and transverse frames. The rotational restraints at the plate edges are therefore neither zero nor infinite, but rather depend on the torsional rigidities of the support members, as discussed in Section 12.4 of Chapter 12. The degree of rotational restraint at the edges of a plate subject to lateral pressure loads also tends to increase because the individual plate elements deflect in the same direction.

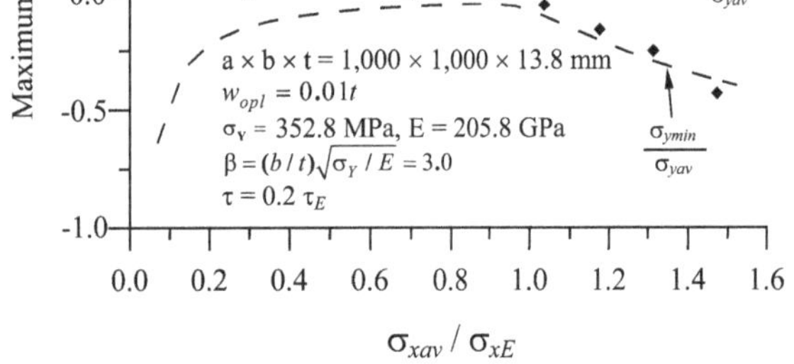

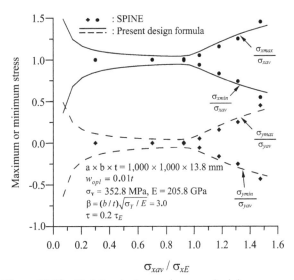

Figure 13.13 Variation in the maximum and minimum membrane stresses at the plate edges under combined longitudinal compression and edge shear conditions.

The large deflection behavior of such plates will depend on the degree of rotational restraint, among other factors. Dealing with the effects of rotational restraints on plate behavior is not easy. For the sake of simplicity, the third-order equation of equation (13.5.3) is used with the coefficients of equation (13.5.8), except for coefficient C_3, which is redefined as follows.

$$C_3 = \frac{\pi^2 E A_{omn}{}^2}{8}\left(\frac{m^4 b}{a^3} + \frac{n^4 a}{b^3}\right) + \frac{m^2 b}{a}(\sigma_{xav} + \sigma_{rex}) + \frac{n^2 a}{b}(\sigma_{yav} + \sigma_{rey})$$

$$+ \frac{\pi^2 D}{t}\frac{m^2 n^2}{ab}\left(\frac{mb}{na} + \frac{na}{mb}\right)^2 \sqrt{\frac{k_x k_y}{k_{xo} k_{yo}}}\sqrt{C_{px} C_{py}} \tag{13.5.14}$$

where k_x and k_y are the buckling coefficients considering the effect of rotational restraints, as defined in Section 12.4 of Chapter 12, k_{xo} and k_{yo} are the buckling coefficients of simply supported plates with $k_{xo} = k_x$ and $k_{yo} = k_y$, as defined in Table 12.1 of Chapter 12, C_{px} is as defined in equation (12.4.10a), and C_{py} is as defined in equation (12.4.10b).

It should be noted that the lateral deflection of a plate obtained using equation (13.5.14) should not be smaller than that of a plate clamped at all (four) edges.

As an illustrative example to validate the theory of equation (13.5.14), Figure 13.14 shows the elastic large deflection behavior of a plate surrounded by support members by comparison with two extreme cases (i.e., simply supported edges [zero rotational restraints] and clamped edges [infinite rotational restraints]). It is noted that the degree of rotational restraints at long edges is different from that at short edges because the size and geometry of support members at long edges are different from those at short edges. ζ_L and ζ_S are the rotational restraint parameters as defined in equation (12.4.1) of Chapter 12. The plate behavior of simply supported plates or clamped plates was obtained from the theory described in Sections 13.5.1 or 13.5.3, respectively. In addition, the nonlinear finite element method solutions are also compared with theoretical results.

From Figs. 13.14a and 13.14b, which represent the elastic large deflection behavior of a plate under σ_{xav} or σ_{yav}, respectively, it can be seen that the plate behavior with partially rotation-restrained edges is in between the two extreme cases (i.e., with simply supported edges and clamped edges), as would be expected. From Fig.13.14c for the case of lateral pressure loads, it is observed that the behavior of the plate with partially rotation-restrained edges is similar to that of simply supported plates when the amount of lateral pressure loads is relatively small. With fur-

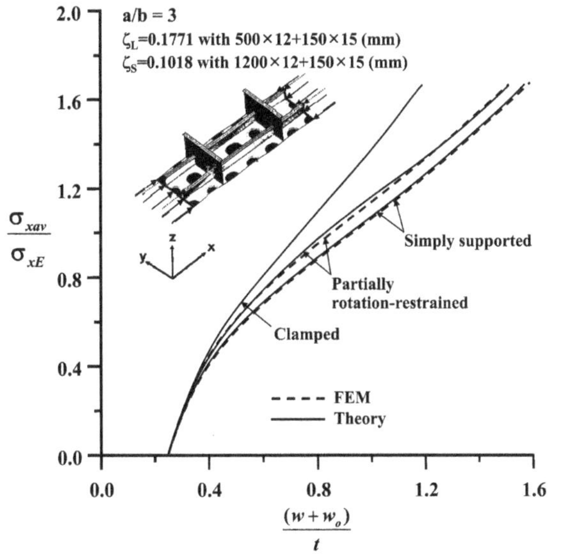

Figure 13.14a Elastic large deflection behavior of a plate under longitudinal axial compression (σ_{xE} is the elastic compressive buckling stress in the x direction for a simply supported plate).

ther increase in lateral pressure loads, however, the plate behavior approaches that of clamped plates.

13.5.3 Clamped Plates

The clamped edge condition is sometimes adopted when the torsional rigidity of the support members is very large compared to the bending rigidity of the plate itself and/or when lateral pressure loads are dominant, and, subsequently, the rotational restraints at the plate edges tend to become large.

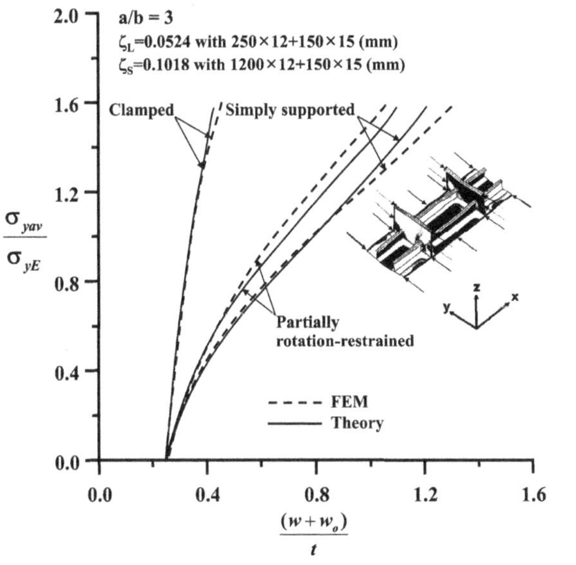

Figure 13.14b Elastic large deflection behavior of a plate under transverse axial compression (σ_{yE} is the elastic compressive buckling stress in the y direction for a simply supported plate).

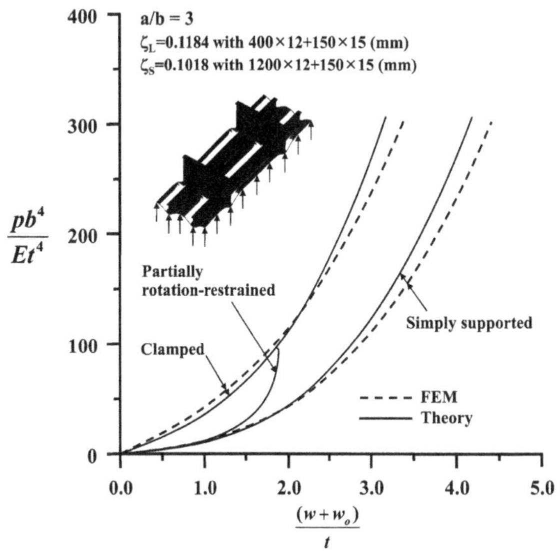

Figure 13.14c Elastic large deflection behavior of a plate under lateral pressure loads.

13.5.3.1 Lateral Pressure Loads

In this case, the initial and added deflection functions may be assumed as follows.

$$w_o = \frac{1}{4}A_{o1}\left(1-\cos\frac{2\pi x}{a}\right)\left(1-\cos\frac{2\pi y}{b}\right) \quad (13.5.15a)$$

$$w = \frac{1}{4}A_1\left(1-\cos\frac{2\pi x}{a}\right)\left(1-\cos\frac{2\pi y}{b}\right) \quad (13.5.15b)$$

where A_{o1} is the known component of initial deflection, and A_1 is the unknown component of added deflection.

The substitution of equations (13.5.15a) and (13.5.15b) into equation (13.3.2) yields the following.

$$\frac{\partial^4 F}{\partial x^4}+2\frac{\partial^4 F}{\partial x^2 \partial y^2}+\frac{\partial^4 F}{\partial y^4} = \frac{4\pi^4 EA_1(A_1+2A_{o1})}{2a^2b^2}\left(1+\cos\frac{2\pi x}{a}+\cos\frac{2\pi y}{b}\right)\sin^2\frac{\pi x}{a}\sin^2\frac{\pi y}{b} \quad (13.5.16)$$

The stress function F is obtained as follows.

$$F = \sigma_{rx}\frac{y^2}{2}+\sigma_{ry}\frac{x^2}{2} + \frac{EA_1(A_1+2A_{o1})}{512a^2b^2}\left[16a^4\cos\frac{2\pi x}{a}-a^4\cos\frac{4\pi x}{a}+b^4\left\{\left(16\cos\frac{2\pi y}{b}-\cos\frac{4\pi y}{b}\right)\right.\right.$$
$$+8a^4\left\{\frac{\cos\left(\frac{2\pi x}{a}-\frac{4\pi y}{b}\right)}{(b^2+4a^2)^2}-\frac{2\cos\left(\frac{2\pi x}{a}-\frac{2\pi y}{b}\right)}{(b^2+a^2)^2}+\frac{\cos\left(\frac{4\pi x}{a}-\frac{2\pi y}{b}\right)}{(4b^2+a^2)^2}\right.$$
$$\left.\left.\left.-\frac{2\cos\left(\frac{2\pi x}{a}+\frac{2\pi y}{b}\right)}{(b^2+a^2)^2}+\frac{2\cos\left(\frac{4\pi x}{a}+\frac{2\pi y}{b}\right)}{(4b^2+a^2)^2}+\frac{\cos\left(\frac{2\pi x}{a}+\frac{4\pi y}{b}\right)}{(b^2+4a^2)^2}\right\}\right\}\right] \quad (13.5.17)$$

The Galerkin method is applied to equation (13.3.1) as follows.

$$\int_0^b \int_0^a \left[D\left(\frac{\partial^4 w}{\partial x^4} + 2\frac{\partial^4 w}{\partial x^2 \partial y^2} + \frac{\partial^4 w}{\partial y^4} \right) - t\left\{ \frac{\partial^2 F}{\partial y^2}\frac{\partial^2 (w+w_o)}{\partial x^2} - 2\frac{\partial^2 F}{\partial x \partial y}\frac{\partial^2 (w+w_o)}{\partial x \partial y} + \frac{\partial^2 F}{\partial x^2}\frac{\partial^2 (w+w_o)}{\partial y^2} + \frac{p}{t} \right\} \right] \left(1-\cos\frac{2\pi x}{a}\right)\left(1-\cos\frac{2\pi y}{b}\right) dxdy = 0 \tag{13.5.18}$$

The substitution of equations (13.5.15a), (13.5.15b), and (13.5.17) into equation (13.5.18), and performing the integration over the entire plate, gives the following equation with regard to the unknown component A_1 of equation (13.3.15b).

$$C_1 A_1^3 + C_2 A_1^2 + C_3 A_1 + C_4 = 0 \tag{13.5.19}$$

where $C_1 = \frac{\pi^2 E}{256 a^3 b^3} K$, $C_2 = \frac{3\pi^2 E A_{01}}{256 a^3 b^3} K$,

$$C_3 = \frac{\pi^2 E A_{01}^2}{128 a^3 b^3} K + \frac{3b}{4a}\sigma_{rex} + \frac{3a}{4b}\sigma_{rey} + \frac{\pi^2 D}{t}\left(\frac{3b}{a^3} + \frac{3a}{b^3} + \frac{2}{ab} \right),$$

$$C_4 = \frac{3A_{01}}{4ab}\left[b^2\sigma_{rex} + a^2\sigma_{rey} \right] - \frac{ab}{\pi^2 t} p,$$

$$\sigma_{rex} = \sigma_{rcx} + \frac{2}{b}(\sigma_{rtx} - \sigma_{rcx})\left(b_t - \frac{b}{2\pi}\sin\frac{2\pi b_t}{b} \right),$$

$$\sigma_{rey} = \sigma_{rcy} + \frac{2}{b}(\sigma_{rty} - \sigma_{rcy})\left(a_t - \frac{a}{2\pi}\sin\frac{2\pi a_t}{a} \right),$$

$$K = \frac{H}{(4a^6 + 21a^4 b^2 + 21 a^2 b^4 + 4b^6)^2}, \text{ and}$$

$$H = 272a^{16} + 2856 a^{14} b^2 + 11273 a^{12} b^4 + 23146 a^{10} b^6 + 23146 a^6 b^{10} + 31506 a^8 b^8 + 11273 a^4 b^{12} + 2856 a^2 b^{14} + 272 b^{16}.$$

Once the unknown deflection component A_1 is obtained as a solution of equation (13.5.19), the maximum and minimum membrane stresses can be determined by equation (13.3.3) with the stress function F of equation (13.5.17), where a similar definition of equation (13.5.4) is used for the plate locations. An illustrative example of this theory is presented in Fig.13.14c.

13.5.3.2 Combined Biaxial Loads

In this case, the initial and added deflection functions may be assumed as follows.

$$w_o = \frac{1}{4} A_{omn}\left(1-\cos\frac{2m\pi x}{a}\right)\left(1-\cos\frac{2n\pi y}{b}\right) \tag{13.5.20a}$$

$$w = \frac{1}{4} A_{mn}\left(1-\cos\frac{2m\pi x}{a}\right)\left(1-\cos\frac{2n\pi y}{b}\right) \tag{13.5.20b}$$

where m and n are the buckling half-wave numbers in the x and y directions, respectively. When the x coordinate is taken such that $a/b \geq 1$, $n = 1$ is applied. The theoretical description below follows this definition.

The substitution of equations (13.5.20a) and (13.5.20b) into equation (13.3.2) yields the following.

$$\frac{\partial^4 F}{\partial x^4} + 2\frac{\partial^4 F}{\partial x^2 \partial y^2} + \frac{\partial^4 F}{\partial y^4} = \frac{4m^2 n^2 \pi^4 E A_{mn}(A_{mn} + 2A_{omn})}{2a^2 b^2}\left(1 + \cos\frac{2m\pi x}{a} + \cos\frac{2n\pi y}{b}\right) \sin^2\frac{m\pi x}{a}\sin^2\frac{n\pi y}{b} \tag{13.5.21}$$

The stress function F is obtained as follows.

$$\begin{aligned} F =& (\sigma_{xav} + \sigma_{rx})\frac{y^2}{2} + (\sigma_{yav} + \sigma_{ry})\frac{x^2}{2} \\ &+ \frac{E A_{mn}(A_{mn} + 2A_{omn})}{512 m^2 n^2 a^2 b^2}\Bigg[16 n^4 a^4 \cos\frac{2m\pi x}{a} - n^4 a^4 \cos\frac{4m\pi x}{a} + m^4 b^4 \Bigg\{ \left(16\cos\frac{2n\pi y}{b} - \cos\frac{4n\pi y}{b} \right) \\ &+ 8n^4 a^4 \Bigg\{ \frac{\cos\left(\frac{2m\pi x}{a} - \frac{4n\pi y}{b}\right)}{(m^2 b^2 + 4n^2 a^2)^2} - \frac{2\cos\left(\frac{2m\pi x}{a} - \frac{2n\pi y}{b}\right)}{(m^2 b^2 + n^2 a^2)^2} + \frac{\cos\left(\frac{4m\pi x}{a} - \frac{2n\pi y}{b}\right)}{(4m^2 b^2 + n^2 a^2)^2} \\ &- \frac{2\cos\left(\frac{2m\pi x}{a} + \frac{2n\pi y}{b}\right)}{(m^2 b^2 + n^2 a^2)^2} + \frac{2\cos\left(\frac{4m\pi x}{a} + \frac{2n\pi y}{b}\right)}{(4m^2 b^2 + n^2 a^2)^2} + \frac{\cos\left(\frac{2m\pi x}{a} + \frac{4n\pi y}{b}\right)}{(m^2 b^2 + 4n^2 a^2)^2} \Bigg\}\Bigg\}\Bigg] \end{aligned} \tag{13.5.22}$$

The Galerkin method is applied to equation (13.3.1), with $p = 0$, as follows.

$$\int_0^b \int_0^a \left[D\left(\frac{\partial^4 w}{\partial x^4} + 2\frac{\partial^4 w}{\partial x^2 \partial y^2} + \frac{\partial^4 w}{\partial y^4} \right) - t\left\{ \frac{\partial^2 F}{\partial y^2}\frac{\partial^2 (w+w_o)}{\partial x^2} - 2\frac{\partial^2 F}{\partial x \partial y}\frac{\partial^2 (w+w_o)}{\partial x \partial y} + \frac{\partial^2 F}{\partial x^2}\frac{\partial^2 (w+w_o)}{\partial y^2} \right\} \right] \left(1-\cos\frac{2m\pi x}{a}\right)\left(1-\cos\frac{2n\pi y}{b}\right) dxdy = 0 \tag{13.5.23}$$

The following third-order equation with regard to the unknown deflection component A_{mn} is obtained once the integration of equation (13.5.23) has been completed.

$$C_1 A_{mn}^3 + C_2 A_{mn}^2 + C_3 A_{mn} + C_4 = 0 \tag{13.5.24}$$

where $C_1 = \frac{\pi^2 E}{256 a^3 b^3} K$, $C_2 = \frac{3\pi^2 E A_{omn}}{256 a^3 b^3} K$,

$$C_3 = \frac{\pi^2 E A_{omn}^2}{128 a^3 b^3} K + \frac{3m^2 b}{4a}(\sigma_{xav} + \sigma_{rex}) + \frac{3n^2 a}{4b}(\sigma_{yav} + \sigma_{rey}) + \frac{\pi^2 D}{t}\left(\frac{3m^4 b}{a^3} + \frac{3n^4 a}{b^3} + \frac{2m^2 n^2}{ab} \right),$$

$$C_4 = \frac{3A_{omn}}{4ab}\left\{m^2b^2(\sigma_{xav}+\sigma_{rex})+n^2a^2(\sigma_{yav}+\sigma_{rey})\right\},$$

$$\sigma_{rex} = \sigma_{rcx}+\frac{2}{b}(\sigma_{rtx}-\sigma_{rcx})\left(b_t-\frac{b}{2n\pi}\sin\frac{2n\pi b_t}{b}\right),$$

$$\sigma_{rey} = \sigma_{rcy}+\frac{2}{b}(\sigma_{rty}-\sigma_{rcy})\left(a_t-\frac{a}{2m\pi}\sin\frac{2m\pi a_t}{a}\right),$$

$$K = \frac{H}{(4n^6a^6+21m^2n^4a^4b^2+21m^4n^2a^2b^4+4m^6b^6)^2}, \text{ and}$$

$$H = 272n^{16}a^{16}+2856m^2n^{14}a^{14}b^2+11273m^4n^{12}a^{12}b^4 +23146m^6n^{10}a^{10}b^6+23146m^{10}n^6a^6b^{10} +31506m^8n^8a^8b^8+11273m^{12}n^4a^4b^{12} +2856m^{14}n^2a^2b^{14}+272m^{16}b^{16}.$$

Once the unknown deflection component A_{mn} is obtained as a solution of equation (13.5.24), the maximum and minimum membrane stresses can be determined from equation (13.3.3), with the stress function F of equation (13.5.22), where a similar definition of equation (13.5.4) is used for the plate locations. An illustrative example of this theory is presented in Figs. 13.14a and 13.14b.

For the particular case of no initial imperfections, $C_2 = C_4 = 0$ in equation (13.5.24) is identified. A similar procedure to equation (13.5.11) is then applied to obtain the buckling condition of clamped plates subject to biaxial loads, as follows.

$$C_3 = \frac{3m^2b}{4a}\sigma_{xav}+\frac{3n^2a}{4b}\sigma_{yav}+\frac{\pi^2D}{t}\left(\frac{3m^4b}{a^3}+\frac{3n^4a}{b^3}+\frac{2m^2n^2}{ab}\right)=0 \quad (13.5.25)$$

For uniaxial compression σ_{xav} with $\sigma_{yav} = 0$, the elastic buckling stress σ_{xE} of clamped plates is obtained from equation (13.5.25), as follows.

$$\sigma_{xE} = -\frac{\pi^2D}{t}\frac{4a}{3m^2b}\left(\frac{3m^4b}{a^3}+\frac{3n^4a}{b^3}+\frac{2m^2n^2}{ab}\right) \quad (13.5.26a)$$

where $n = 1$ is taken for $a/b \geq 1$. The buckling half-wave number m in the x (plate length) direction is determined as a minimum integer that satisfies the following condition.

$$\frac{1}{m^2}\left(\frac{3m^4b}{a^3}+\frac{3a}{b^3}+\frac{2m^2}{ab}\right) \leq \frac{1}{(m+1)^2}\left\{\frac{3(m+1)^4b}{a^3}+\frac{3a}{b^3}+\frac{2(m+1)^2}{ab}\right\} \quad (13.5.26b)$$

For uniaxial compression σ_{yav} with $\sigma_{xav} = 0$, the elastic buckling stress σ_{yE} of clamped plates is obtained from equation (13.5.25), as follows.

$$\sigma_{yE} = -\frac{\pi^2D}{t}\frac{4b}{3n^2a}\left(\frac{3m^4b}{a^3}+\frac{3n^4a}{b^3}+\frac{2m^2n^2}{ab}\right) \quad (13.5.27)$$

where $m = n = 1$ is taken for $a/b \geq 1$. Equations (13.5.26) and (13.5.27) can be compared with the corresponding buckling stresses indicated in Table 12.2 of Chapter 12.

For biaxial compression with a constant loading ratio of $c = \sigma_{yav}/\sigma_{xav}$, the buckling stresses σ_{xE} and σ_{yE} are defined from equation (13.5.25), as follows.

$$\sigma_{xE} = -\frac{\pi^2D}{t}\frac{4ab}{3m^2b^2+3cn^2a^2}\left(\frac{3m^4b}{a^3}+\frac{3n^4a}{b^3}+\frac{2m^2n^2}{ab}\right) \quad (13.5.28a)$$

$$\sigma_{yE} = c\sigma_{xE} = -c\frac{\pi^2D}{t}\frac{4ab}{3m^2b^2+3cn^2a^2}\left(\frac{3m^4b}{a^3}+\frac{3n^4a}{b^3}+\frac{2m^2n^2}{ab}\right) \quad (13.5.28b)$$

where $n = 1$ is taken for $a/b \geq 1$, and m is determined as a minimum integer that satisfies the following condition.

$$\frac{1}{m^2b^2+ca^2}\left(\frac{3m^4b}{a^3}+\frac{3a}{b^3}+\frac{2m^2}{ab}\right) \leq \frac{1}{(m+1)^2b^2+ca^2}\left\{\frac{3(m+1)^4b}{a^3}+\frac{3a}{b^3}+\frac{2(m+1)^2}{ab}\right\} \quad (13.5.28c)$$

13.5.3.3 Interaction Effect Between Biaxial Loads and Lateral Pressure

A simplification similar to simply supported plates subject to combined biaxial loads and lateral pressure is applied as indicated in equation (13.5.12), namely,

$$C_4 = \frac{3A_{omn}}{4ab}\left[m^2b^2(\sigma_{xav}+\sigma_{rex})+n^2a^2(\sigma_{yav}+\sigma_{rey})\right]-\frac{ab}{\pi^2t}p \quad (13.5.29)$$

Equation (13.5.24) is applied for clamped plates subject to combined biaxial and lateral pressure loads, but with the coefficient C_4 of equation (13.5.29).

13.6 ULTIMATE STRENGTH FORMULATIONS

The aforementioned membrane stress-based method can be applied to the ultimate strength computations of plates under combined in-plane and lateral pressure loads. In cases in which it is difficult to identify the membrane stresses theoretically (e.g., under edge shear loading or a combination of more load components), the ultimate strength interaction relationships are often utilized, whereas the ultimate strength under single or multiple load components may be determined by the application of different methods, such as the membrane stress-based method, the rigid-plastic theory method, or even an empirical formula method. In the following, a number of useful ultimate plate strength relationships are introduced.

The membrane stress-based method of calculating ultimate plate strength that has been described in this chapter has been implemented in the ALPS/ULSAP computer program (2009). A number of application examples of the large deflection behavior and ultimate strength of plates obtained from this method are now illustrated through comparison with nonlinear FEM solutions and SPINE semianalytical method solutions (Paik et al., 2001; ALPS/SPINE 2008). Only simply supported plates are dealt with here, but plates with partially or fully rotation-restrained edges can also be considered.

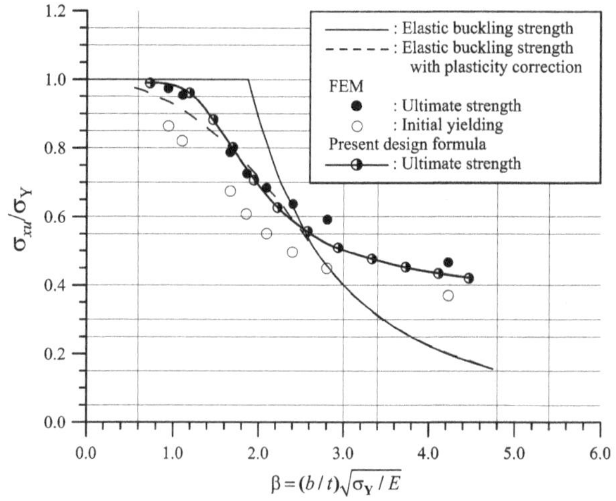

Figure 13.16 Variation in the ultimate longitudinal compressive strength of a square plate shown as a function of the plate slenderness coefficient when $a/b = 1$.

13.6.1 Uniaxial Loads

This is one of the simplest but the most important load cases. The membrane stress-based method can be applied where the maximum and minimum membrane stresses of the plate under uniaxial loads are calculated, as described in Section 13.5, and substituted into the three ultimate strength conditions (i.e., equation [13.4.18a] to equation [13.4.18c]). Each of these conditional equations is solved with regard to the applied loads, and the ultimate plate strength can then be obtained as a minimum value of the applied loads among the three solutions.

Figure 13.15 compares the formula solutions with the mechanical collapse tests and the nonlinear finite element method solutions for long plates with different plate aspect ratios and under longitudinal axial compressive loads. While the formula deals with initial imperfections as direct parameters of influence, the mechanical collapse tests involve various uncertain levels of both initial deflections and residual stresses. Details of the test data are given in Ellinas et al. (1984). In the finite element analysis, two types of the unloaded plate edge condition are applied: (1) the unloaded plate edges move freely in plane, and (2) they are kept straight. An "average" level of initial deflections is assumed, and the welding residual stresses are not included. It is seen from this figure that the finite element method solutions with the edge condition (1) are smaller than those with the edge condition (2), as would be expected.

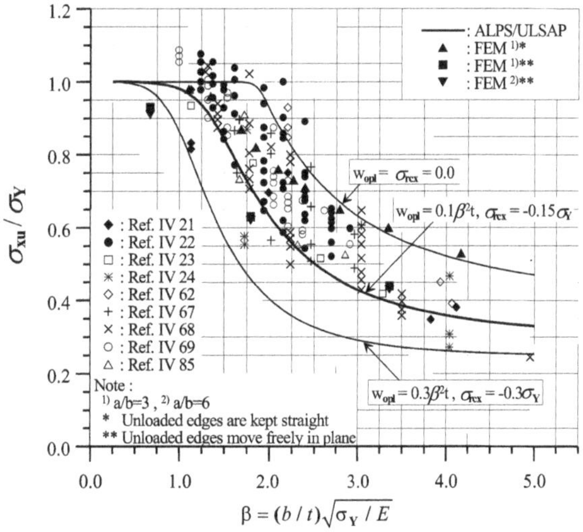

Figure 13.15 Variations in the ultimate strength of steel plates under axial compression as a function of the plate slenderness coefficient; reference numbers for test data are extracted from Ellinas et al. (1984).

Figure 13.16 shows the ultimate longitudinal compressive strength of a square plate plotted as a function of the plate slenderness coefficient. Figure 13.17 shows the effect of initial deflection on the plate ultimate strength. When the magnitude of initial deflection is large, the present design formula tends to underestimate the ultimate strength compared to the finite element method solutions. Figure 13.18 shows the ultimate strength of a long plate of $a/b = 3$ under longitudinal compression in the x direction or transverse direction in the y direction.

13.6.2 Combined Axial and Lateral Pressure Loads

In this case, the application of the membrane stress-based method alone can also solve the problem. Figure 13.19 compares the membrane stress-based method results (from ALPS/ULSAP) against corresponding mechanical collapse test results from Yamamoto et al. (1970) and SPINE semianalytical solutions for long plating of $a/b = 3$ under combined

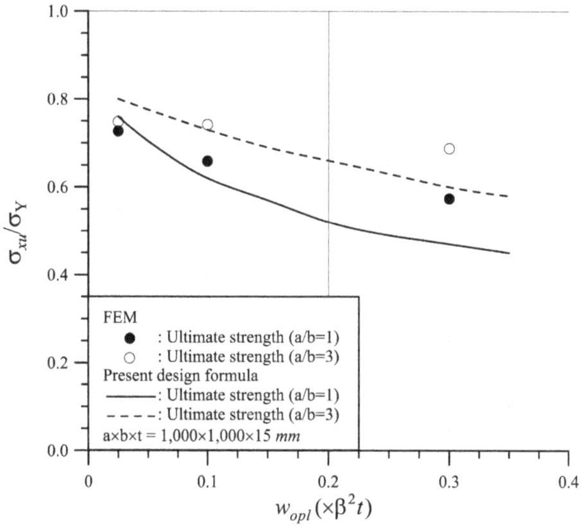

Figure 13.17 Variation in the ultimate longitudinal compressive strength of square and long plates as a function of the magnitude of the maximum initial deflection.

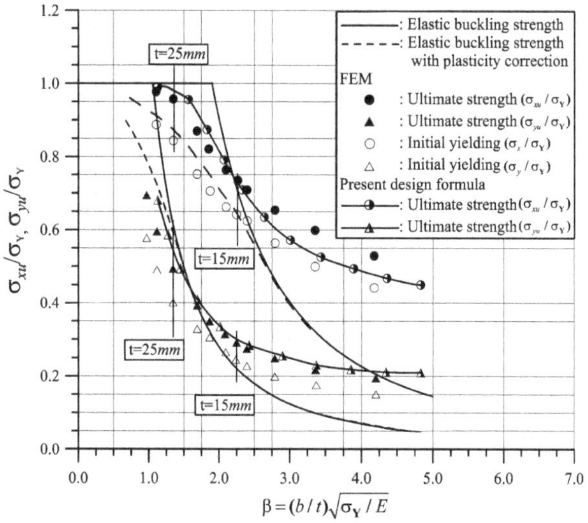

Figure 13.18 Variation in the ultimate longitudinal or transverse compressive strength of a long plate shown as a function of the plate slenderness coefficient when $a/b = 3$.

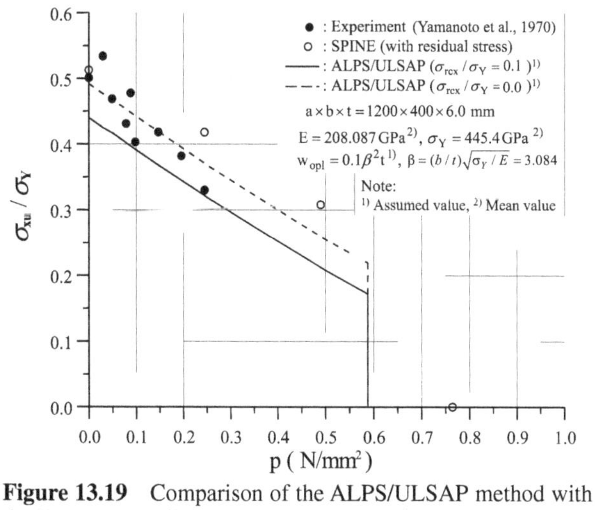

Figure 13.19 Comparison of the ALPS/ULSAP method with the Yamamoto collapse test results for plating under combined longitudinal axial compression and lateral pressure loads.

longitudinal axial compression and lateral pressure loads. The model uncertainties for the ALPS/ULSAP method on the basis of the Yamamoto testing are mean of 0.967 and coefficient of variation of 0.064.

13.6.3 Combined Edge Shear and Lateral Pressure Loads

In this case, the following equation may be applied (Paik and Thayamballi, 2003).

$$\left(\frac{\tau_{av}}{\tau_{uo}}\right)^{1.5} + \left(\frac{p}{p_{uo}}\right)^{1.2} = 1 \qquad (13.6.1)$$

where τ_{uo} is the ultimate strength of a plate subject to edge shear alone, as defined in equation (13.4.25), and p_{uo} is the ultimate strength of a plate under lateral pressure load alone, as defined by the rigid-plastic theory method or membrane stress-based method.

Figure 13.20 shows the ultimate strength interaction relationship for a plate under combined edge shear and lateral pressure obtained using the SPINE semianalytical method. It is seen from this figure that the ultimate strength interaction between edge shear and lateral pressure cannot be ignored. It is confirmed that the design formula of equation (13.6.1) represents the ultimate strength relationship between edge shear and lateral pressure well.

13.6.4 Combined Biaxial, Edge Shear, and Lateral Pressure Loads

In this case, the following relationship may be applicable.

$$\left(\frac{\sigma_{xav}}{\sigma_{xu}^{*}}\right)^{c_1} + \kappa\left(\frac{\sigma_{xav}}{\sigma_{xu}^{*}}\right)\left(\frac{\sigma_{yav}}{\sigma_{yu}^{*}}\right) + \left(\frac{\sigma_{yav}}{\sigma_{yu}^{*}}\right)^{c_2} + \left(\frac{\tau_{av}}{\tau_{u}^{*}}\right)^{c_3} = 1$$

(13.6.2)

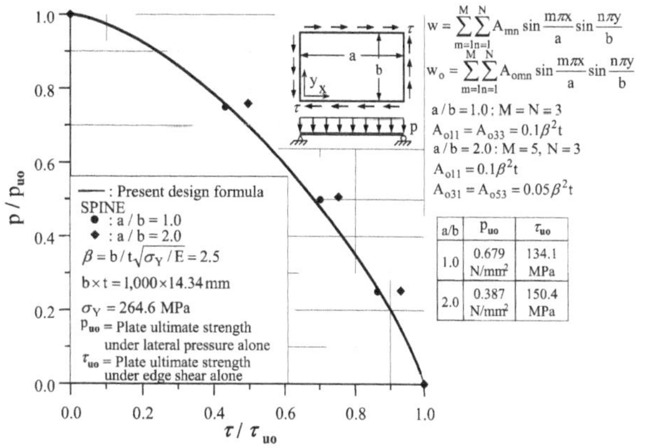

a/b	p_{uo}	τ_{uo}
1.0	0.679 N/mm²	134.1 MPa
2.0	0.387 N/mm²	150.4 MPa

Figure 13.20 Ultimate strength interaction relationship for a simply supported plate subjected to edge shear and lateral pressure.

where σ^*_{xu} is the ultimate strength of the plate under uniaxial load σ_{xav} with lateral pressure load p, σ^*_{yu} is the ultimate strength of the plate under uniaxial load σ_{yav} with lateral pressure load p, and τ^*_u is the ultimate strength of the plate under edge shear τ_{av} with lateral pressure load p.

In equation (13.6.2), c_1, c_2, c_3, and κ are constants that may be defined empirically. For example, the following values may be used (Paik and Thayamballi, 2003).

$$c_1 = c_2 = c_3 = 2 \qquad (13.6.3a)$$

$$\kappa = \begin{cases} 0 \text{ when both } \sigma_{xav} \text{ and } \sigma_{yav} \text{ are compressive (negative)} \\ -1 \text{ when either } \sigma_{xav} \text{ or } \sigma_{yav} \text{ or both are tensile (positive)} \end{cases} \qquad (13.6.3b)$$

For the particular case in which biaxial loads are applied with a constant loading ratio of $c = \sigma_{yav}/\sigma_{xav}$, with or without lateral pressure loads, the ultimate strength of the plate subject to combined biaxial and lateral pressure loads, but without edge shear, can be determined by the membrane stress-based method. In this case, the expression of the ultimate plate strength interaction relationship can be simplified to the following.

$$\left(\frac{\sigma_{xav}}{\sigma^{**}_{xu}}\right)^2 + \left(\frac{\tau_{av}}{\tau^*_u}\right)^2 = 1 \text{ or } \left(\frac{c\sigma_{xav}}{\sigma^{**}_{yu}}\right)^2 + \left(\frac{\tau_{av}}{\tau^*_u}\right)^2 = 1 \qquad (13.6.4)$$

where σ^{**}_{xu} is the ultimate strength axial component of the plate in the x direction under biaxial and lateral pressure loads, $\sigma^{**}_{yu} = c\sigma^{**}_{xu}$ is the ultimate strength axial component of the plate in the y direction under biaxial and lateral pressure loads, and τ^*_u is as defined in equation (13.6.2).

Figures 13.21 to 13.25 show some selected examples of theoretical ultimate strength computations by comparison with nonlinear FEM solutions or SPINE semianalytical method solutions. In these examples, the yield stress of the material is σ_Y = 274.4 MPa, and the plate breadth is b = 1000 mm.

Figures 13.21a and 13.21b display the ultimate strength interaction relationships between biaxial compressive loads for a square plate with plate thicknesses of 15 mm (or β = 2.254) and 25 mm (or β = 1.352), respectively. The maximum initial deflection level is varied.

Figures 13.22a and 13.22b show the ultimate strength interaction relationships between biaxial compressive loads for a long plate of a/b = 3 with plate thicknesses of 15 mm (or β = 2.254) and 25 mm (or β = 1.352), respectively. The maximum initial deflection level is also varied.

Figure 13.23 depicts the ultimate strength interaction relationship between longitudinal compres-

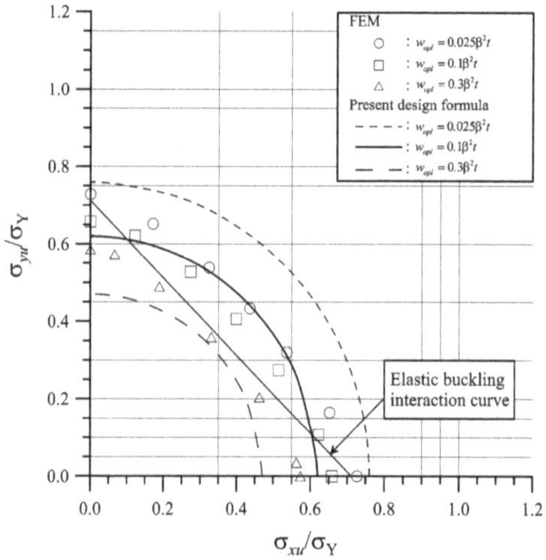

Figure 13.21a Ultimate strength interaction relationship between biaxial compressive loads for a thin plate, when a/b = 1 and t = 15 mm.

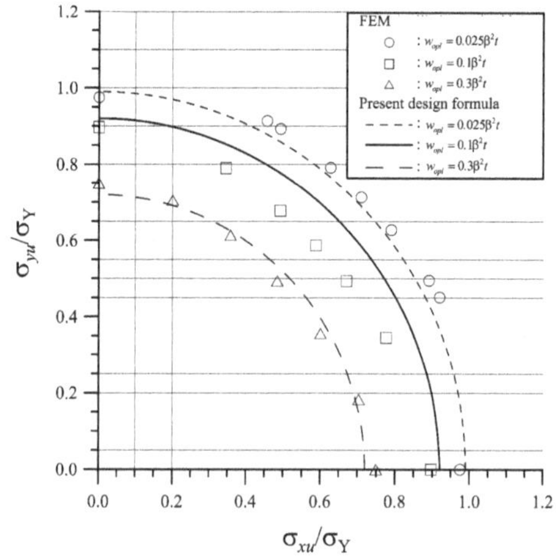

Figure 13.21b Ultimate strength interaction relationship between biaxial compressive loads for a thick plate, when a/b = 3 and t = 25 mm.

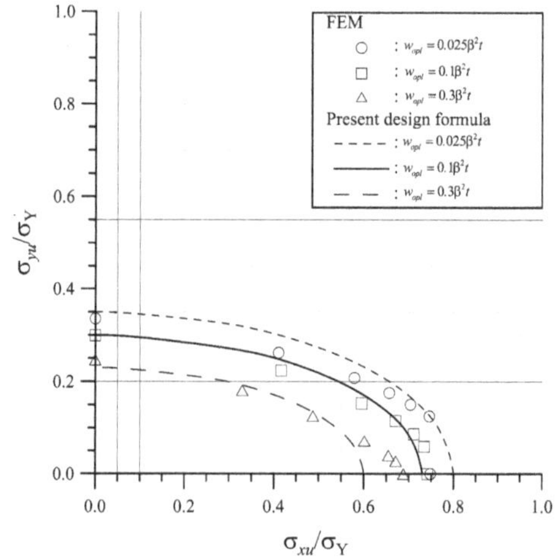

Figure 13.22a Ultimate strength interaction relationship between biaxial compressive loads for a thin plate, when a/b = 3 and t = 15 mm.

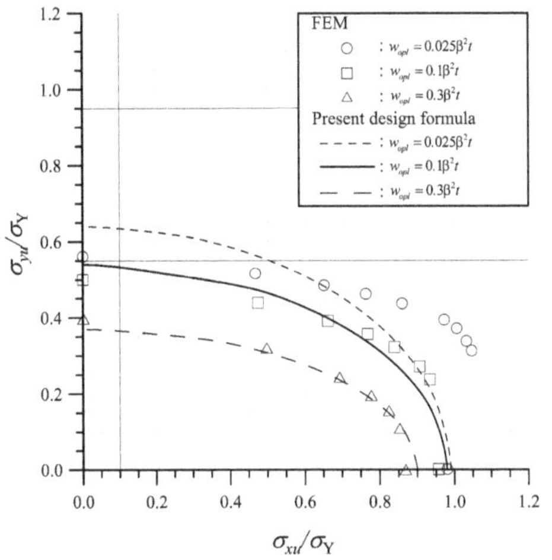

Figure 13.22b Ultimate strength interaction relationship between biaxial compressive loads for a thick plate, whena/b = 3 and t = 25 mm.

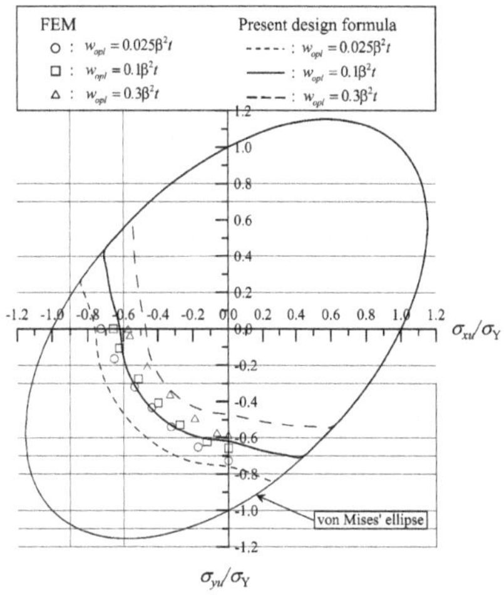

Figure 13.24a Ultimate strength interaction relationship between biaxial compression or tension for a thin plate, when a/b = 1 and t = 15 mm.

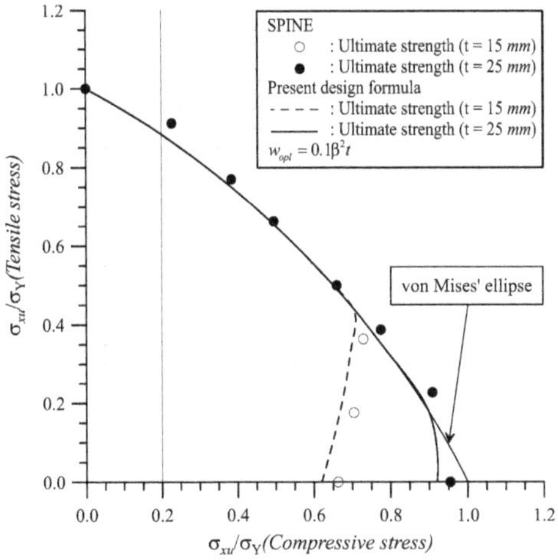

Figure 13.23 Ultimate strength interaction relationship between longitudinal compression and transverse tension, when a/b = 1.

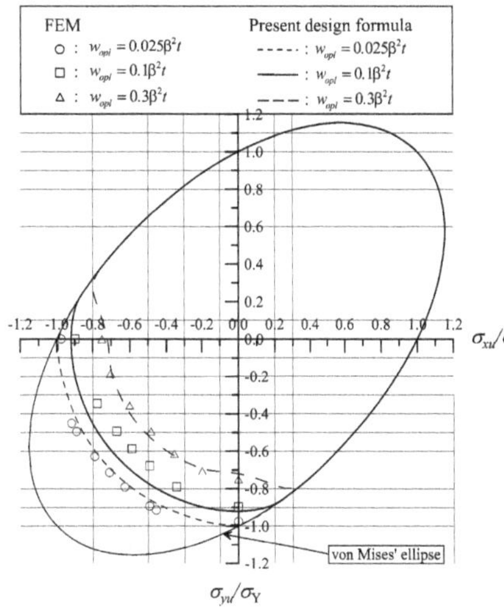

Figure 13.24b Ultimate strength interaction relationship between biaxial compression or tension for a thick plate, when a/b = 1 and t = 25 mm.

sion and transverse tension for a square plate. A very small initial deflection (i.e., w_{opl} = 0.01t) is presumed. When transverse tensile loads are predominant, the plate can sustain external loads until the plate cross-section in the transverse direction yields almost completely.

Figures 13.24a and 13.24b show the ultimate strength interaction relationships between all possible combinations of longitudinal compression or tension and transverse compression or tension for a simply supported square plate with plate thicknesses of 15 mm and 25 mm, respectively. The maximum initial deflection level is varied. The compressive stress is taken as negative in these figures, whereas the tensile stress is taken as positive. A similar comparison but for a simply supported plate of a/b = 3 is shown in Figs. 13.25a and 13.25b.

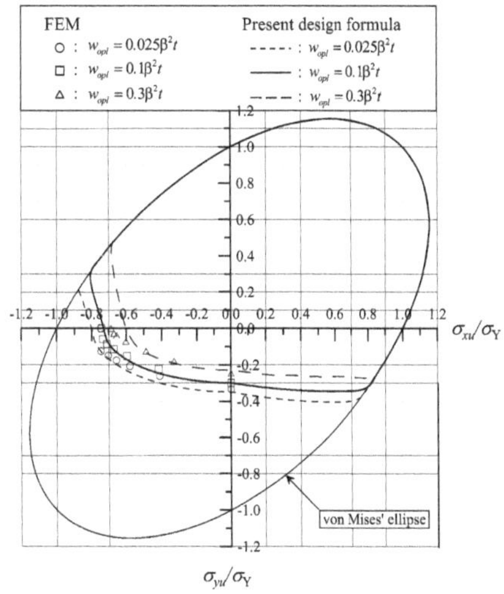

Figure 13.25a Ultimate strength interaction relationship between biaxial compression or tension for a thin plate, when a/b = 3 and t = 15 mm.

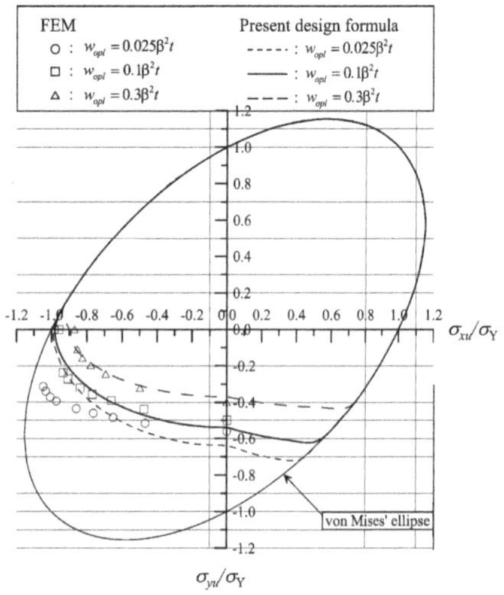

Figure 13.25b Ultimate strength interaction relationship between biaxial compression or tension for a thick plate, when $a/b = 3$ and $t = 25$ mm.

Figure 13.26 shows the ultimate strength interaction curves of a square plate between longitudinal compression and edge shear. Figures 13.27 and 13.28 show the ultimate strength interactions for a simply supported plate of $a/b = 3$ under edge shear together with longitudinal compression or transverse compression, respectively.

13.7 EFFECT OF OPENINGS

Openings decrease the ultimate strength of plates. When the opening size is relatively large, its effect on ultimate plate strength cannot be neglected and must be taken into account in strength calculations and design. As pointed out in Section 12.7 of Chapter 12, the application of the Johnson-Ostenfeld formula approach is inadequate for the ultimate strength calculation of perforated plates, particularly when the opening size and/or plate thickness is large.

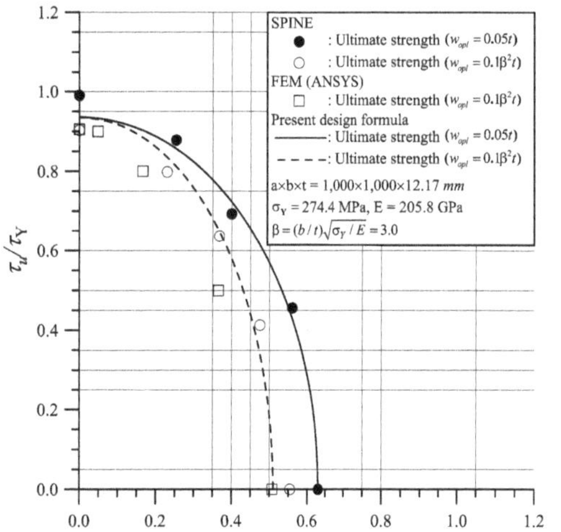

Figure 13.26 Ultimate strength interaction relationship between longitudinal compression and edge shear, when $a/b = 1$.

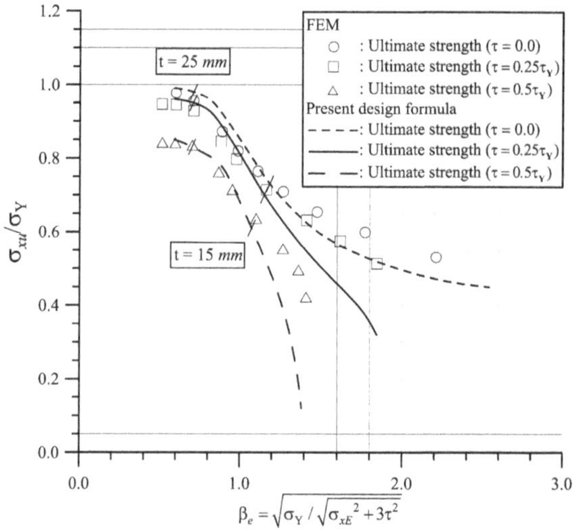

Figure 13.27 Variation in the ultimate longitudinal compressive strength of a long plate subject to combined longitudinal compression and edge shear, when $a/b = 3$.

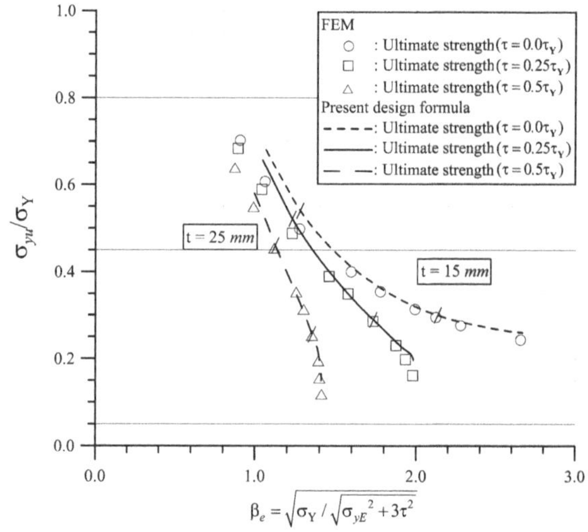

Figure 13.28 Variation in the ultimate transverse compressive strength of a long plate subject to combined transverse compression and edge shear, when $a/b = 3$.

In the following, a number of empirical formulations for calculating the ultimate strength of perforated plates subject to single load components are presented in the form of ultimate strength reduction (Paik and Thayamballi, 2003), with the opening nomenclature defined in Fig. 12.11. Under combined load conditions, the ultimate strength interaction relationship of perforated plates is considered to take the same expression as that of intact plates, although the ultimate strengths of plates subject to individual single load components will differ.

13.7.1 Circular Openings

The ultimate strength formulations of a plate with a circular opening located at its center (i.e., $a_c = b_c = d_c$) are presented here.

13.7.1.1 Longitudinal Axial Compression

$$\sigma_{xu} = \sigma_{xuo}\left[1.0 - 0.365\frac{d_c}{b} - 0.7\left(\frac{d_c}{b}\right)^2\right] \quad (13.7.1)$$

where σ_{xu} and σ_{xuo} are the ultimate strengths with and without an opening, under longitudinal axial compression, respectively.

13.7.1.2 Transverse Axial Compression

$$\sigma_{yu} = \sigma_{yuo}\left[1.0 + C_1\frac{d_c}{b} + C_2\left(\frac{d_c}{b}\right)^2\right] \quad (13.7.2)$$

where σ_{yu} and σ_{yuo} are the ultimate strengths with and without an opening, under transverse axial compression, respectively. The coefficients are defined as follows.

$$C_1 = \begin{cases} -0.048\left(\frac{a}{b}\right)^2 + 0.252\frac{a}{b} - 0.386 & \text{for } 1 \le \frac{a}{b} \le 3 \\ -0.062 & \text{for } 3 < \frac{a}{b} \le 6 \end{cases}$$

and

$$C_2 = \begin{cases} -0.177\left(\frac{a}{b}\right)^2 + 1.088\frac{a}{b} - 1.671 & \text{for } 1 \le \frac{a}{b} \le 3 \\ 0.0 & \text{for } 3 < \frac{a}{b} \le 6. \end{cases}$$

13.7.1.3 Edge Shear

$$\tau_u = \tau_{uo}\left[1.0 + C_1\frac{d_c}{b} + C_2\left(\frac{d_c}{b}\right)^2\right] \quad (13.7.3)$$

where τ_u and τ_{uo} are the ultimate strengths with and without an opening, under transverse axial compression, respectively. The coefficients are defined as follows.

$$C_1 = -0.025\left(\frac{a}{b}\right)^2 + 0.309\frac{a}{b} - 0.787,$$

$$C_2 = -0.009\left(\frac{a}{b}\right)^2 - 0.068\frac{a}{b} - 0.415.$$

13.7.2 Elliptical or Rectangular Openings

The ultimate strength formulations of a plate with an elliptical or rectangular opening located at its center are presented here.

13.7.2.1 Longitudinal Axial Compression

$$\sigma_{xu} = \sigma_{xuo}\left[1.0 - 0.365\frac{b_c}{b} - 0.7\left(\frac{b_c}{b}\right)^2\right] \quad (13.7.4)$$

13.7.2.2 Transverse Axial Compression

$$\sigma_{yu} = \sigma_{yuo}\left[1.0 + C_1\frac{a_c}{b} + C_2\left(\frac{a_c}{b}\right)^2\right] \quad (13.7.5)$$

where coefficients C_1 and C_2 are as defined in equation (13.7.2).

13.7.2.3 Edge Shear

$$\tau_u = \tau_{uo}\left[1.0 + C_1\frac{a_c + b_c}{2b} + C_2\left(\frac{a_c + b_c}{2b}\right)^2\right] \quad (13.7.6)$$

where coefficients C_1 and C_2 are as defined in equation (13.7.3).

13.8 EFFECT OF CORROSION WASTAGE

The typical types of corrosion wastage are general (or uniform) corrosion and localized (or pit) corrosion. When the extent of the corrosion is substantial, the ultimate plate strength is significantly reduced. Therefore, the effect of corrosion wastage must be taken into account in the ultimate strength calculations of aged plates (Paik and Melchers, 2008).

The effects of general corrosion can be readily dealt with because the plate thickness decreases uniformly, and the ultimate strength of corroded plates can thus be calculated by simply using the reduced (corroded) plate thickness in all of the ultimate strength formulations.

However, the effects of localized corrosion are more difficult to deal with. The severity or extent of the corrosion is often represented by an index termed "the degree of pit corrosion intensity" (DOP), which is defined on a volumetric basis, as follows.

$$\text{DOP} = \eta = \frac{1}{abt}\sum_{i=1}^{n} V_{pi} \times 100(\%) \quad (13.8.1)$$

where V_{pi} is the volume of the ith pit, which may be determined as $V_{pi} = \frac{\pi d_{di} d_{ri}^2}{4}$, d_{di} is the depth of the ith pit and d_{ri} is the diameter of the ith pit, and n is the total number of pits.

The following presents empirical formulations for calculating the ultimate strength of plates with pit corrosion. Because corrosion wastage is time-variant, the resulting ultimate strength formula must also be a function of time (i.e., the age of the structure).

13.8.1 Axial Tension

The following formulations may be used under conditions of axial tension (Paik et al., 2003a).

$$\sigma_{xu} = \sigma_{xuo} \frac{A_{xo} - A_{xr}}{A_{xo}} \quad (13.8.2a)$$

$$\sigma_{yu} = \sigma_{yuo} \frac{A_{yo} - A_{yr}}{A_{yo}} \quad (13.8.2b)$$

where A_{xo} and A_{yo} are the total cross-sectional areas of the original (intact) plate without pits in the x or y directions, A_{xr} and A_{yr} are the total cross-sectional areas associated with all pits at the cross-section of the largest number of pits in the x or y directions, σ_{xuo} and σ_{yuo} are the ultimate strengths of intact plates subject to axial loads in the x or y directions, and σ_{xu} and σ_{yu} are the ultimate strengths of pit-corroded plates subject to axial loads in the x or y directions.

13.8.2 Axial Compression

Paik et al. (2003a) proposed the following ultimate strength formula for pit-corroded plates subject to axial compression.

$$\sigma_{xu} = \sigma_{xuo} \left(\frac{A_{xo} - A_{xr}}{A_{xo}} \right)^{0.73} \quad (13.8.3a)$$

$$\sigma_{yu} = \sigma_{yuo} \left(\frac{A_{yo} - A_{yr}}{A_{yo}} \right)^{0.73} \quad (13.8.3b)$$

where the nomenclature is as indicated in equation (13.8.2).

13.8.3 Edge Shear

Paik et al. (2004) proposed the following ultimate strength formula for pit-corroded plates subject to edge shear.

$$\tau_u = \begin{cases} \tau_{uo} & \text{for } \eta \le 1.0 \\ \tau_{uo} \{1.0 - 0.18 \ell n(\eta)\} & \text{for } \eta > 1.0 \end{cases} \quad (13.8.4)$$

where τ_{uo} and τ_u are the ultimate shear strengths without and with pit corrosion wastage, respectively.

13.9 EFFECT OF CRACKING DAMAGE

Cracking damage is another important type of age-related degradation, and one that certainly reduces the ultimate strength of plates (Paik and Melchers, 2008). In axial tension, premised cracks can propagate, and the effects of those cracks on the resulting ultimate strength can be significant. Even in axial compression, buckling results in lateral deflection, and, subsequently, the premised cracks may play a role in decreasing the maximum load-carrying capacity and structural stiffness. Cracking damage caused by fatigue loads is also time-variant in nature, and thus the ultimate strength of cracked plates must also vary with time.

In the following, several empirical formulations for calculating the ultimate strength of cracked plates subject to axial tension or compression are presented. The cracking damage is positioned approximately in the direction normal to the axial loading direction. It is found that the ultimate strength of plates with crack damage located parallel to the axial loading direction is also significantly reduced (Paik, 2008a, 2008b).

13.9.1 Axial Tension

Paik et al. (2005) suggested the following ultimate strength formula for cracked plates subject to axial tension.

$$\sigma_{xu} = \frac{A_{xo} - A_{xc}}{A_{xo}} \sigma_Y \quad (13.9.1a)$$

$$\sigma_{yu} = \frac{A_{yo} - A_{yc}}{A_{yo}} \sigma_Y \quad (13.9.1b)$$

where A_{xo} and A_{yo} are the cross-sectional areas of intact plates in the x or y directions, A_{xc} and A_{yc} are the cross-sectional areas associated with crack damage projected to the x or y directions, and σ_{xu} and σ_{yu} are the ultimate strengths of cracked plates in the x or y directions.

13.9.2 Axial Compression

Paik et al. (2005) suggested the following ultimate strength formula of cracked plates subject to axial compression, which takes the same expression as the formula for axial tension.

$$\sigma_{xu} = \frac{A_{xo} - A_{xc}}{A_{xo}} \sigma_{xuo} \quad (13.9.2a)$$

$$\sigma_{yu} = \frac{A_{yo} - A_{yc}}{A_{yo}} \sigma_{yuo} \quad (13.9.2b)$$

where σ_{xuo} and σ_{yuo} are the ultimate compressive strengths of intact plates in the x or y directions, together with the other notations indicated in equation (13.9.1).

13.10 EFFECT OF LOCAL DENTS

Local dent damage can occur in plate panels. For example, the inner bottom plates of the cargo holds of bulk carriers can suffer local dents through the mishandled loading or unloading of cargo. Local dents may occur in these plates when they are struck during the loading of iron ore. In the unloading of

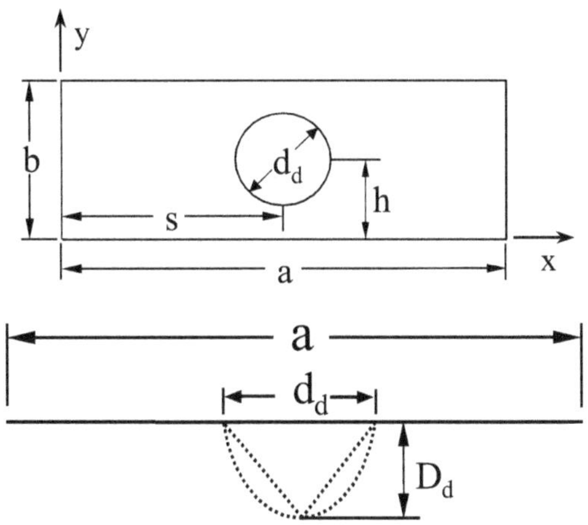

Figure 13.29 Geometric parameters of a local dent.

bulk cargo such as iron ore or coal, the excavator hits the inner bottom plates mechanically. The deck plates of offshore platforms may be subjected to the impact of objects dropped from a crane.

Such mechanical damage may involve denting, cracking, residual stresses or strains associated with plastic deformation, and coating damage (Paik et al., 2003b; Paik, 2005; Paik and Melchers, 2008). In this section, the effects of local dents on ultimate plate strength are described with the nomenclature indicated in Fig. 13.29, where d_d is the diameter of the local dent, D_d is the depth of the local dent, and h and s are the distances from the plate origin to the center of the local dent in the y and x directions, respectively.

13.10.1 Axial Compression

Paik et al. (2003) proposed the following ultimate strength formula for dented plates subject to longitudinal axial compression, with $s = \frac{a}{2}$.

$$\sigma_{xu} = \sigma_{xuo}\left[C_1 \ell n\left(\frac{D_d}{t}\right) + C_2\right]C_3 \qquad (13.10.1)$$

where σ_{xuo} and σ_{xu} are the ultimate longitudinal axial compressive strengths of plates without and with local dents, respectively. The coefficients are defined as follows.

$$C_1 = -0.042\left(\frac{d_d}{b}\right)^2 - 0.105\frac{d_d}{b} + 0.015,$$

$$C_2 = -0.138\left(\frac{d_d}{b}\right)^2 - 0.302\frac{d_d}{b} + 1.042, \text{ and}$$

$$C_3 = \begin{cases} -1.44\left(\frac{h}{b}\right)^2 + 1.74\frac{h}{b} + 0.49 & \text{for } h \le \frac{b}{2} \\ -1.44\left(\frac{b-h}{b}\right)^2 + 1.74\frac{b-h}{b} + 0.49 & \text{for } h > \frac{b}{2}. \end{cases}$$

13.10.2 Edge Shear

Paik (2005) proposed the following ultimate strength formula for dented plates subject to edge shear.

$$\tau_u = \begin{cases} \tau_{uo}\left[1.0 + C_1\left(\frac{D_d}{t}\right)^2 - C_2\frac{D_d}{t}\right] & \text{for } 1 < \frac{D_d}{t} \le 10 \\ \tau_{uo}\left(1.0 + 100C_1 - 10C_2\right) & \text{for } \frac{D_d}{t} > 10 \end{cases} \qquad (13.10.2)$$

where τ_{uo} and τ_u are the ultimate edge shear strengths of plates without and with local dents, respectively. The coefficients are defined as follows.

$$C_1 = 0.0129\left(\frac{d_d}{b}\right)^{0.26} - 0.0076 \text{ and } C_2 = 0.1888\left(\frac{d_d}{b}\right)^{0.49} - 0.07.$$

REFERENCES

ALPS/SPINE. (2008). Elastic-plastic large deflection analysis of stiffened panels by incremental energy method. Stevensville, MD: DRS C3 Systems. Available online at http://www.orca3d.com/maestro.

ALPS/ULSAP. (2009). Ultimate limit state assessment of plate panels. Stevensville, MD: DRS C3 Systems. Available online at http://www.orca3d.com/maestro.

Ellinas, C. P., Supple, W. J., and Walker, A. C. (1984). *Buckling of offshore structures: A state-of-the-art review*. Houston: Gulf Publishing Company.

Faulkner, D. (1975). A review of effective plating for use in the analysis of stiffened plating in bending and compression. *J. of Ship Research*, Vol. 19, Issue 1, 1–17.

International Organization for Standardization (2007). ISO 18072-1. Ships and marine technology—ship structures—part 1: General requirements for their limit state assessment. *Ships and Marine Technology*, Geneva, Switzerland, November.

International Ship and Offshore Structure Congress (ISSC). (2009). Ultimate strength. *Report of Technical Committee III.1, International Ship and Offshore Structures Congress*, Seoul, Korea, August.

Jones, N. (1975). Plastic behavior of beams and plates. Chapter 23 in *Ship Structural Design Concepts*. Cornell Maritime Press, Cambridge, MD, 747–778.

Okada, H., Oshima, K., and Fukumoto, Y. (1979). Compressive strength of long rectangular plates under hydrostatic pressure [in Japanese]. *J. of the Society of Naval Architects of Japan*, Vol. 146, 270–280.

Paik, J. K. (1995). A new concept of the effective shear modulus for a plate buckled in shear. *J. of Ship Research*, Vol. 39, Issue 1, 70–75.

Paik, J. K. (2005). Ultimate strength of dented steel plates under edge shear loads. *Thin-Walled Structures*, Vol. 43, 1475–1492.

Paik, J. K. (2008a). Residual ultimate strength of steel plates with longitudinal cracks under axial compression experiments. *Ocean Engineering*, Vol. 35, 1775–1783.

Paik, J. K. (2008b). Residual ultimate strength of steel plates with longitudinal cracks under axial compression-nonlinear finite element method investigations. *Ocean Engineering*, Vol. 36, 266–276.

Paik, J. K. (2008c). Some recent advances in the concepts of plate-effectiveness evaluation. *Thin-Walled Structures*, Vol. 46, 1035–1046.

Paik, J. K., Lee, J. M., and Ko, M. J. (2003a). Ultimate compressive strength of plate elements with pit corrosion wastage. *J. of Engineering for the Maritime Environment*, Vol. 217, Issue M4, 185–200.

Paik, J. K., Lee, J. M., and Ko, M. J. (2004). Ultimate shear strength of plate elements with pit corrosion wastage. *Thin-Walled Structures*, Vol. 42, Issue 8, 1161–1176.

Paik, J. K., Lee, J. M., and Lee, D. H. (2003b). Ultimate strength of dented steel plates under axial compressive loads. *International Journal of Mechanical Sciences*, Vol. 45, 433–448.

Paik, J. K., and Melchers, R. E. (2008). *Condition assessment of aged structures.* New York: CRC Press.

Paik, J. K., and Pedersen, P. T. (1996). A simplified method for predicting ultimate compressive strength of ship panels. *Shipbuilding Progress*, Vol. 43, Issue 434, 139–157.

Paik, J. K., Satish Kumar, Y. V., and Lee, J. M. (2005). Ultimate strength of cracked plate elements under axial compression or tension. *Thin-Walled Structures*, Vol. 43, 237–272.

Paik, J. K., and Thayamballi, A. K. (2003). *Ultimate limit state design of steel-plated structures.* Chichester, U.K.: Wiley.

Paik, J. K., and Thayamballi, A. K. (2007). *Ship-shaped offshore installations: Design, building, and operation.* Cambridge, U.K.: Cambridge University Press.

Paik, J. K., Thayamballi, A. K., Lee, S. K., and Kang, S. J. (2001). A semi-analytical method for the elastic-plastic large deflection analysis of welded steel or aluminum plating under combined in-plane and lateral pressure loads. *Thin-Walled Structures*, Vol. 39, 125–152.

Timoshenko, S. P., and Woinowsky-Krieger, S. (1959). *Theory of plates and shells.* New York: McGraw-Hill.

Ueda, Y., Rashed, S. M. H., and Paik, J. K. (1984). Buckling and ultimate strength interactions of plates and stiffened plates under combined loads (1st report)–In-plane biaxial and shearing forces [in Japanese]. *J. of the Society of Naval Architects of Japan*, Vol. 156, 377–387.

Yamamoto, Y., Matsubara, N., and Murakami, T. (1970). Buckling strength of rectangular plates subjected to edge thrusts and lateral pressure (2nd Report) [in Japanese]. *J. of the Society of Naval Architects of Japan*, Vol. 127, 171–179.

FOURTEEN

ELASTIC BUCKLING OF STIFFENED PANELS

Owen Hughes
Professor, Virginia Tech
Blacksburg, VA, USA

This chapter deals with the elastic buckling of flat rectangular panels that are stiffened in one or both directions, under various combinations of loading. It is not possible, in one chapter, to present solutions for all types of loading, stiffener arrangements and geometry, and boundary conditions. This type of information is available in structural stability handbooks. The Column Research Committee of Japan (1971), for example, contains an extensive collection of solutions to specific cases. The purpose of this chapter is to explain the principal features of stiffened panel buckling, to discuss its applications and limitations with respect to ship panels, and to present solutions for the most common cases.

Stiffened panels can buckle in essentially two different ways. In overall buckling, the stiffeners buckle along with the plating; in local buckling, either the stiffeners buckle prematurely because of inadequate rigidity or stability, or the plate panels buckle between the stiffeners, thus shedding extra load into the stiffeners so that eventually the stiffeners buckle in the manner of columns. For most ship panels, the proportions are such that the buckling—of either type—is inelastic, and therefore, the word "failure" is more properly used instead of "buckling": overall panel failure, local stiffener failure, local plate failure. Nevertheless, an elastic buckling analysis gives a good indication of the likely modes of failure, and also provides a foundation for the more complex question of the inelastic buckling and ultimate strength of stiffened panels, which is covered in the next chapter.

As noted earlier when dealing with unstiffened plates, the calculation of elastic buckling stress should usually assume simple support, irrespective of the presence of a lateral load, because in some cases such a load may be absent or may not be large enough to provide rotational restraint; apart from this possible influence on boundary conditions, a lateral load has little effect on elastic buckling. Therefore, throughout this chapter, unless stated otherwise, it is assumed that the edges of the panel are simply supported. It is also assumed that individual elements of the stiffeners are not subject to instability.

The way in which a stiffened panel will buckle when subjected to in-plane longitudinal compressive forces depends mainly on the stiffeners. The two principal requirements are that they have sufficient torsional stability so that they do not buckle prematurely (i.e., before the plating) and that they have sufficient lateral rigidity so that the possibility of overall buckling is either eliminated or made sufficiently unlikely. For all practical purposes, stiffener buckling is synonymous with overall buckling, because if the stiffeners buckle the plating is left with almost no lateral rigidity whatever. Since overall buckling involves the entire panel, it is usually regarded as collapse rather than as unserviceability. Moreover, it is a quite sudden mode of collapse and therefore more undesirable than other modes. Thus, the first and most basic principle in regard to stiffeners is that they should be at least as strong as the plating; that is, they should be sufficiently rigid and stable so that neither overall buckling nor local stiffener buckling occurs before local plate buckling. Most ship panels must carry substantial lateral loads and this requirement usually produces stiffeners that are already larger and more rigid than the minimum sizes required by consideration of elastic overall buckling. Therefore, in most cases an inelastic failure analysis is required. However, if the panel is *slender* (this term will be quantified later) as may occur with small lateral load and in lightweight construction (small, closely spaced stiffeners), then elastic overall buckling becomes a possible collapse mode and an elastic buckling analysis becomes essential. Generally speaking, it is not possible, without a specific analysis, to know with certainty just what failure modes—elastic or inelastic—will be the governing requirements that determine the plate thickness and the stiffener sizes and spacing. The best approach is to first perform an elastic buckling analysis because 1) it is relatively simple, consisting mostly of explicit formulas; 2) for slender panels, elastic buckling of one type or another may be possible and may be one of the governing failure modes; and 3) the elastic analysis indicates whether or not an inelastic analysis is required, and some of the elastic buckling parameters are needed for the inelastic analysis.

14.1 LONGITUDINALLY STIFFENED PANELS

Figure 14.1 illustrates overall buckling and the two types of local buckling for longitudinally stiffened panels.* We first investigate the minimum lateral rigidity to ensure that overall buckling does not precede plate buckling, and then the required torsional rigidity to prevent local stiffener hackling.

14.1.1 Overall Buckling versus Plate Buckling

14.1.1.1 Minimum Flexural Rigidity to Avoid Overall Buckling

The minimum rigidity of longitudinal stiffeners necessary to ensure that overall buckling does not precede plate buckling has been investigated by various authors (Cox & Riddel, 1949; Seide, 1953; Timoshenko & Gere, 1961) for a panel containing one, two, or three equally spaced longitudinal stiffeners. For the first two cases, Bleich (1952) presented approximate formulas, which are reproduced here. As before, the symbol I_x denotes the moment of inertia of a section comprised of a stiffener together with a width b of plate. The minimum required rigidity is expressed in terms of a parameter γ_x, which is the ratio of the flexural rigidity of the combined section to the flexural rigidity of the plating

$$\gamma_x = \frac{EI_x}{Db} = \frac{12(1-\nu^2)I_x}{bt^3} \tag{14.1.1}$$

The other parameters are the panel aspect ratio Π

$$\Pi = \frac{L}{B} = \frac{a}{B} \tag{14.1.2}$$

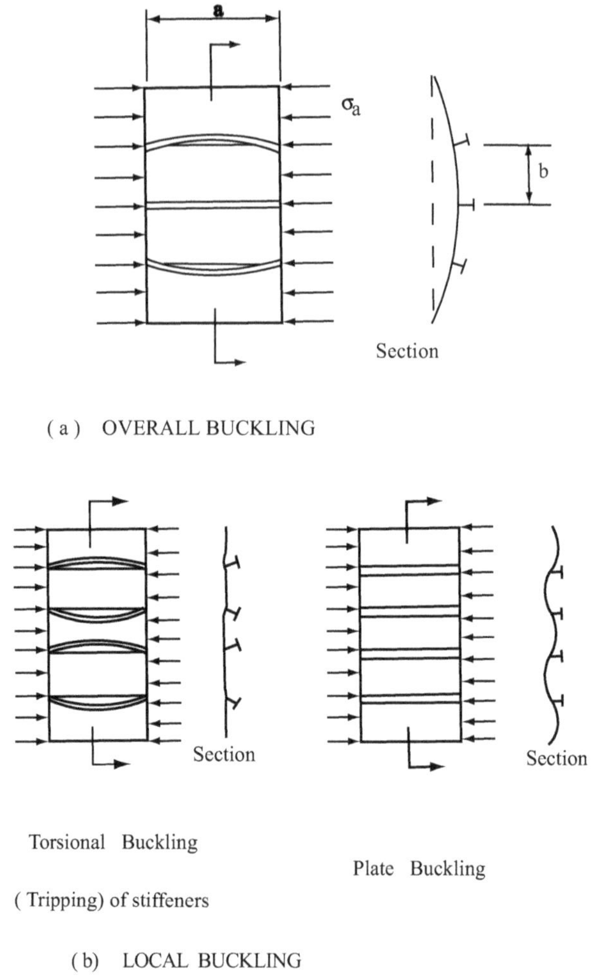

Figure 14.1 Buckling of longitudinally stiffened panels.

in which L is the panel length (which in this case equals the plate length a) and the area ratio δ_x

$$\delta_x = \frac{A_x}{bt} \tag{14.1.3}$$

in which A_x is the cross-sectional area of the stiffener (only).

In terms of these parameters, the minimum value of γ_x to ensure that stiffener buckling does not precede plate buckling is given by the following formulas:

(a) Panel with one central longitudinal stiffener: either

$$(\gamma_x)_{MIN} = 22.8\Pi + (2.5 + 16\delta_x)\Pi - 10.8\sqrt{\Pi}$$

or

$$(\gamma_x)_{MIN} = 48.8 + 112\delta_x(1 + 0.5\delta_x) \tag{14.1.4}$$

whichever is less.

*The term "overall buckling" is used with reference to the type of panel being considered. Thus, it is important to distinguish overall buckling of a longitudinally stiffened panel from overall buckling of a cross-stiffened panel, which is dealt with in Section 14.5. To assist in making this distinction, the term "gross panel buckling" will be used for the latter.

(b) Panel with two equally spaced longitudinal stiffeners:
either

$$(\gamma_x)_{MIN} = 43.5\sqrt{\Pi^3} + 36\Pi^2\delta_x$$

or

$$(\gamma_x)_{MIN} = 288 + 610\delta_x + 325\delta_x^2 \qquad (14.1.5)$$

whichever is less.

NOTE: The range of validity for the above formulas is $0 < \delta_x < 0.2$.

Cox and Riddel (1949) discussed in detail plates with one, two, or three stiffeners, and their analysis is capable of extension to four, five, or more stiffeners. They investigated the smallest size of stiffeners necessary to prevent overall buckling of a flat panel before buckling of the plate between stiffeners. The effect of torsional stiffness of the stiffeners is included. The analysis was done using a strain energy method and the solution was given in closed form.

A more general solution, valid for any number of stiffeners, has been presented by Klitchieff (1951)

$$(\gamma_x)_{MIN} = \delta_x(1 + N_B^2\Pi^2)^2 + \frac{4}{\pi}\Pi(1 + N_B^2\Pi^2)\sqrt{2 + N_B^2\Pi^2}$$

where N_B = number of panels = 1 + number of longitudinal stiffeners.

A more accurate expression will be presented in Section 14.6, based on the orthotropic plate theory given in Section 14.5.2.

14.1.1.2 Calculation of Overall Buckling Stress

An alternative approach is to calculate the overall buckling stress $(\sigma_a)_{cr}$ and compare it to the plate buckling stress to ensure that it is larger. One method of calculating $(\sigma_a)_{cr}$ is to regard each stiffener and its associated width of plating as a column, having some equivalent slenderness ratio $(L/\rho)_{eq}$. The elastic buckling stress $(\sigma_a)_{cr}$ is then obtained from the Euler column formula and the requirement that overall buckling not precede plate buckling then appears in a more explicit form:

$$(\sigma_a)_{cr} \geq \sigma_o$$

where

$$\sigma_o = 3.62\,E\left(\frac{t}{b}\right)^2$$

In short panels, this approach is virtually exact; the stiffeners are in fact a row of parallel columns that are essentially identical and, as far as buckling is concerned, independent. The equivalent slenderness ratio of each column is the actual slenderness of the section

$$\left(\frac{L}{\rho}\right)_{eq} = \frac{a}{\rho} = \frac{a}{\sqrt{\dfrac{I_x}{(A_x + bt)}}} \qquad (14.1.6a)$$

or, in terms of nondimensional parameters,

$$\left(\frac{L}{\rho}\right)_{eq} = \frac{a}{t}\sqrt{\frac{12(1-\nu)^2(1+\delta)^2}{\gamma_x}} \qquad (14.1.6b)$$

In long panels, the stiffeners receive some lateral restraint from the sides of the panel and this may cause them to buckle in more than one half-wave. In this case, the equivalent slenderness ratio is smaller than the value given by equations (14.1.6). The effect occurs for large values of the panel aspect ratio Π and for small values of stiffener rigidity relative to the plating (i.e., small values of γ_x). From the work of Sharp (1966), it is possible to derive an aspect ratio coefficient C_Π which accounts for this effect as follows

$$\left(\frac{L}{\rho}\right)_{eq} = C_\Pi\frac{a}{\rho} = \frac{C_\Pi a}{\sqrt{\dfrac{I_x}{(A_x + bt)}}} \qquad (14.1.7a)$$

in which C_Π, is given by either

$$C_\Pi = \frac{1}{\Pi}\sqrt{\frac{\gamma_x}{2(1 + \sqrt{1 + \gamma_x})}}$$

or

$$C_\Pi = 1 \qquad (14.1.7b)$$

whichever is less. The resulting value of $(L/\rho)_{eq}$ *is* then used in the standard column buckling formula

$$(\sigma_a)_{cr} = \frac{\pi^2 E}{\left(\dfrac{L}{\rho}\right)_{eq}^2} \qquad (14.1.8)$$

As is stands, the purpose of this equation is simply to verify that the critical stress for the combined stiffener and plate is greater than the plate buckling stress σ_0, and for this purpose the plate flange should be taken at its full width b because a value less than b would be begging the question, since it could only be less if the plate had already buckled. Of course, the above value of $(\sigma_a)_{cr}$ does not constitute a true overall buckling stress if it exceeds the yield stress. For convenience, we shall use the adjective "slender" to describe a panel in which the overall buckling stress calculated from elastic theory is less than the yield stress. That is, we say that a panel is slender if

$$\frac{\pi^2 E}{\left(\frac{L}{\rho}\right)^2_{eq}} < \sigma_Y$$

and so in terms of the column slenderness parame-ter $\lambda = (L/\pi\rho)(\sigma_Y/E)^{1/2}$, which was introduced in Chapter 11, the foregoing criterion becomes simply $\lambda_{eq} > 1$.

Since slender panels are normally designed such that plate buckling precedes overall buckling, when the latter occurs, the plate flange of the stiffener will not be fully effective over the width b. Instead, it is necessary to take some reduced effective width, b_e. Note that this is not the same as the reduced effective *breadth* that was used in Chapter 3 in order to correct for shear lag effects. In that case, the lack of effectiveness was because of in-plane deformation of the plating caused by shear; in the present case, it is because of out-of-plane deformation caused by buckling.

The effective width caused by buckling has long been a vexed question, mainly because in most cases it was being discussed and applied in the difficult context of the ultimate strength of panels that were not slender and therefore did not buckle elastically. For elastic or near-elastic buckling, a satisfactory formula was derived by von Karman, Sechler, and, Donnell (1932) as early as 1924 (the reference quoted is a later paper, in English). Their approach was characteristically simple and practical. They idealized the state of stress within the buckled plate by assuming that, because of buckling, the center portion has no compressive stress, while the edge portions of the plate remain fully effective and carry a uniform stress σ, as shown in Fig. 14.2. In other words, the buckled center portion is discounted completely and the original plate of width b is replaced by a narrower unbuckled plate of effective width b_e. From statics, it is clear that σ_e and σ_a are related by

$$\sigma_e = \frac{b}{b_e}\sigma_a \tag{14.1.9}$$

To simulate the progressive growth of the buckling, it is further assumed that the (as yet unbuckled) effective plate is always on the verge of further buckling; that is, the effective width is taken to be the width at which the equivalent plate would buckle at an applied stress of σ_e. This implies that

$$\sigma_e = k\frac{\pi^2 D}{b_e^2 t}$$

For the original plate

$$(\sigma_a)_{cr} = k\frac{\pi^2 D}{b^2 t}$$

and if it is assumed that k *is* the same in both cases then

$$\frac{b_e}{b} = \sqrt{\frac{(\sigma_a)_{cr}}{\sigma_e}} \tag{14.1.10}$$

The latter assumption is not strictly correct because although the boundary conditions can be regarded as similar in both cases, the aspect ratios are different. However, it was shown in Section 12.1 that for aspect ratios greater than 1.0, k can be taken as 4.0. Substitution of this value, together with $\nu = 0.3$, into the expression for $(\sigma_a)_{cr}$ converts equation (14.1.10) into

$$\frac{b_e}{b} = 1.9\frac{t}{b}\sqrt{\frac{E}{\sigma_e}} \tag{14.1.11}$$

The effective width would reach its smallest possible value when σ_e reached yield. In this case, the right-hand side becomes a simple inverse function of the plate slenderness parameter β

$$\left(\frac{b_e}{b}\right)_{\min} = \frac{1.9}{\beta} \tag{14.1.12}$$

However, if the panel is truly slender, overall panel buckling would in most cases occur before σ_e reaches σ_Y and so we use the more general expression for b_e given by equation (14.1.11).

Now that we have an expression for b_e, we can proceed to obtain an expression for the ultimate or collapse load of slender stiffened panels, that is panels for which collapse is by overall elastic buckling. The effective width is used in calculating the equivalent slenderness ratio from equation (14.1.7a); that is, b_e is used as the plate flange width in calculating I_x, and ρ. We shall denote these values as I_{xe}, and ρ_e and the resulting value of equivalent slenderness

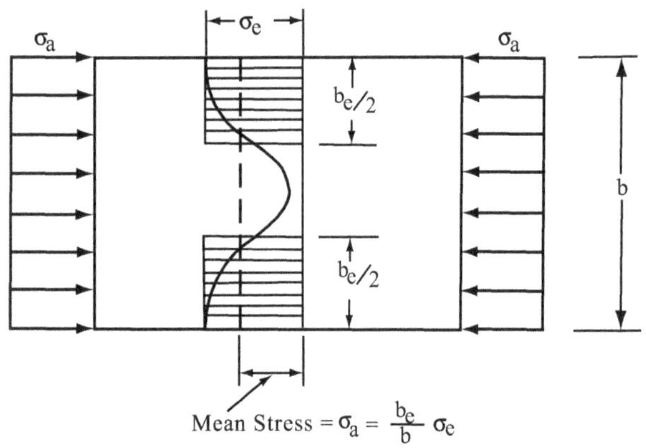

Figure 14.2 Postbuckling stress distribution; effective width.

ratio as $(L/\rho_e)_{eq}$*. The axial stress in the stiffener "columns" is σ_e, and the critical value of this stress is given by equation (14.1.8)

$$(\sigma_e)_{cr} = \frac{\pi^2 E}{\left(\frac{L}{\rho_e}\right)^2_{eq}} \qquad (14.1.13)$$

Note that this equation refers to σ_e rather than to σ_a; the axial stress in the stiffener is larger than the external applied stress σ_a because of the reduced width of the plate. The quantity of interest is the value of σ_a corresponding to $(\sigma_e)_{cr}$. From statics, the two are related by $\sigma_a(bt + A_x) = \sigma_e(b_e t + A_x)$, and from this together with equation (14.1.13) we have

$$(\sigma_a)_{cr} = \left(\frac{b_e t + A_x}{bt + A_x}\right) \frac{\pi^2 E}{\left(\frac{L}{\rho_e}\right)^2_{eq}} \qquad (14.1.14)$$

Because of the presence of σ_e in equation (14.1.11), the foregoing sequence of calculations must be performed iteratively. A suitable procedure would be:

1. Assume some initial value of b_e (e.g., $0.8b$).
2. Calculate I_{xe}, and then evaluate $(L/\rho_e)_{eq}$ from equation (14.1.7a), using b_e in place of b.
3. Calculate $(\sigma_e)_{cr}$ from equation (14.1.13). Check that $(\sigma_e)_{cr} > \sigma_o$.
4. Using this value of σ_e, recalculate b_e from equation (14.1.11).
5. Repeat from step 2 until b_e has converged.
6. Calculate $(\sigma_e)_{cr}$ from equation (14.1.14).

In computer-aided design, steps 1 to 5 would simply become an inner loop of the overall procedure. For short panels ($\Pi < 1$) in which $C_\Pi = 1$, steps 1 to 5 can be condensed into a single nonlinear equation for b_e by expressing ρ_e in terms of the stiffener web and flange areas A_w and A_f. The resulting equation is

$$\frac{\pi d}{1.9at}\, b_e \sqrt{\frac{A_w}{12}(A_w + 4A_f) + \left(\frac{A_w}{3} + A_f\right) b_e t} = A_w + A_f + b_e t$$

The preceding analysis only applies to slender panels, that is, panels for which the overall buckling stress is, at its highest, less than the yield stress. The phrase "at its highest" is required because the value of $(\sigma_e)_{cr}$ depends on ρ_e, which in turn depends on the value of b_e. As shown in Fig. 14.3a, if the plate flange area bt is substantially greater than the stiffener area A_x (say, $bt > 2A_x$) *as is usually the case,* then the progressive decrease in effective width causes an increase in ρ_e, making the panel *sturdier* (because L/ρ_e is smaller). If the plate area is less than $2A_x$, then the effect can be neutral or even opposite, but this is not common. Because of this somewhat paradoxical relationship between ρ_e and b_e, the lowest (hypothetical) value of $(\sigma_e)_{cr}$ corresponds to $b_e = b$; that is, plating fully effective. This is illustrated in Fig. 14.3b, which also shows that the same paradoxical relationship usually holds for $(\sigma_a)_{cr}$, but the effect is somewhat diminished because of the multiplication by the area ratio in equation (14.1.14). The full range of possible situations is shown in Fig. 14.4. As long as the value of $(\sigma_a)_{cr}$ corresponding to $b = b_e$ is greater than the plate buckling stress σ_0, then it remains merely hypothetical; it would not be the actual collapse stress because the plating would buckle first. But, if this value of $(\sigma_a)_{cr}$ were less than σ_0, then the panel would collapse by overall elastic buckling at an applied stress equal to $(\sigma_a)_{cr}$ with no prior plate buckling. As mentioned earlier, this is a highly undesirable mode of collapse and for this reason one of the basic steps in panel design is to ensure that $(\sigma_a)_{cr}$ for fully effective plating is greater than the plate buckling stress. We see now that if $bt > 2A_x$ (approximately), then in making a preliminary check as to whether or not a panel is slender (by calculating λ_{eq} and seeing if it is greater than 1.0) it is best to use b_e.

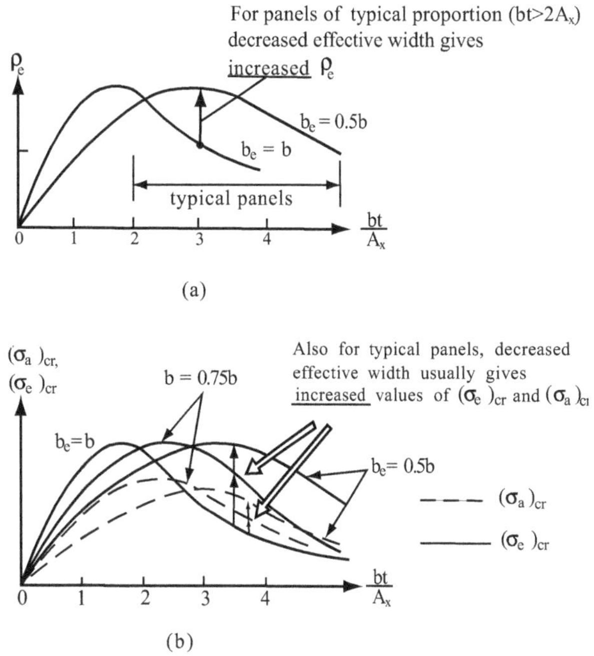

Figure 14.3 Flexural properties of panels as a function of effective width for various plating/stiffener area ratios.

*C_Π is for the original panel and does not need updating.

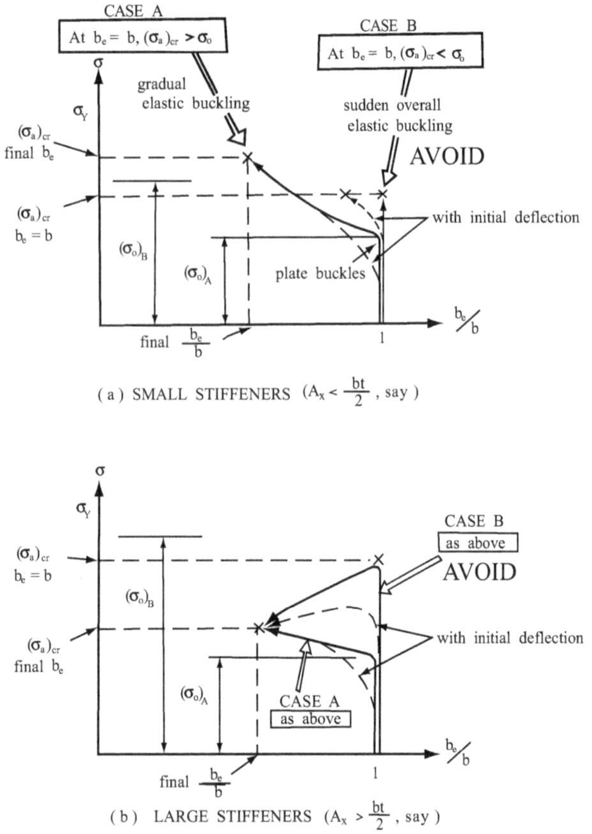

Figure 14.4 Possible modes of elastic buckling of stiffened panels.

14.1.2 Local Buckling of Stiffener (Tripping)

As shown in Fig. 14.5, a stiffener may buckle by twisting about its line of attachment to the plating. This is commonly referred to as tripping. The plate may rotate somewhat to accommodate the stiffener rotation, and the direction of rotation usually alternates as shown because this involves less elastic strain energy in the plating. However, this is not plate buckling. Tripping and plate buckling do interact but they can occur in either order, depending on the stiffener and plating proportions. As noted earlier, tripping

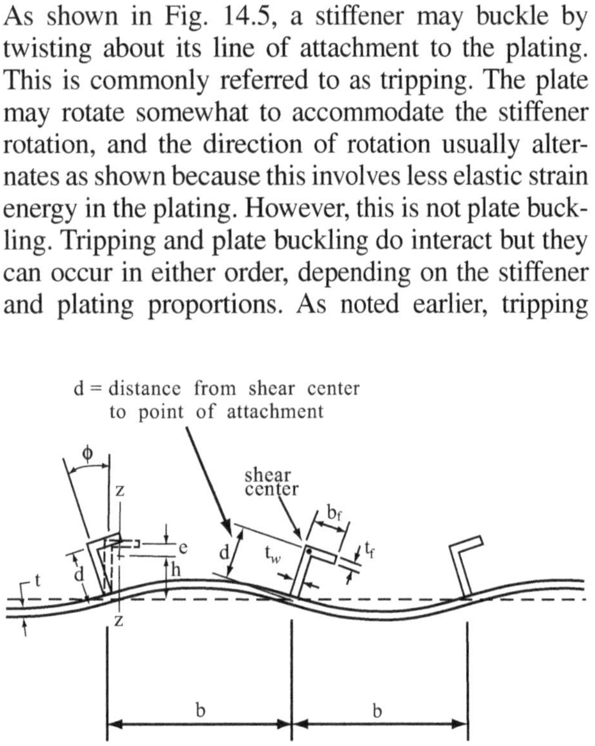

Figure 14.5 Stiffener tripping.

failure is regarded as collapse because once tripping occurs the plating is left with no stiffening and so overall buckling follows immediately. Also, *elastic* tripping is a quite sudden phenomenon and hence it is a most undesirable mode of collapse, akin to elastic overall panel buckling. Since the open sections that are used as stiffeners in ship panels have relatively little torsional rigidity, such panels can be susceptible to tripping and so it is very important to consider this mode of buckling and to provide an adequate margin of safety. With stiffeners of closed cross-section, tripping cannot occur and the use of such stiffeners produces panels that have a larger strength-to-weight ratio. However, their use also increases the cost of panel fabrication and there may be difficulties regarding inspection and control of corrosion.

14.1.2.1 Types of Stiffener Buckling and Methods of Analysis

The theory presented in this section is for a stiffener subjected to axial compression. Torsional buckling of a stiffener may also be caused by a bending moment that puts the flange in compression; this is referred to as flexural-torsional buckling. It is decidedly more complicated than buckling under a purely axial load because of the strong influence of eccentricity and the inherently nonlinear coupling between the flexural and torsional response of the stiffener. There are two quite different methods for dealing with nonlinear elastic buckling: 1) "folded plate" analysis based on finite difference methods (this approach is summarized and applied to ship structural components in Smith [1968]); and 2) nonlinear frame (or finite element) analysis, which accounts for nonlinearity by means of a "geometric stiffness matrix." The first method is simpler but is restricted to certain boundary conditions and to simple forms of structural geometry; also, it is basically limited to elastic buckling of bifurcation type. The second method involves more computation and is necessarily computer-based, but is nevertheless quite economical. It is much more general and it can deal with other forms of nonlinearity such as yielding, large deflections, plastic hinge formation, and the development of a "mechanism" in the overall structure. Therefore, only this second method is considered in this text, in Chapter 15. The present chapter concentrates on the other forms of elastic panel buckling, for which analytical methods, and often explicit expressions, are available.

14.1.2.2 Stiffener Buckling Due to Axial Compression

Under this type of load, a stiffener acts essentially as a column, but its tripping or torsional buckling differs from that of a column in three ways: first, because the rotation occurs about an enforced axis—the line of attachment to the plating; second, because the plate offers some restraint against this rotation; and third, because it is not necessarily rigid body rotation (if the plating is sturdy there will be some distortion of the stiffener due to web bending, as shown in Fig. 14.6). The analysis that follows takes account of all these factors and it applies to both symmetric and unsymmetric stiffeners.†

The governing differential equation for the torsional buckling of a column about an enforced axis of rotation under the action of an applied axial stress is presented in Timoshenko and Gere (1961). In applying this to stiffeners, we may take advantage of the fact, known from the basic theory of the torsion of thin-walled sections, that for all cross-sectional shapes that are composed of thin rectangles which meet at a common point, the shear center is at this point and the warping constant is zero. In this case, it may be shown that the governing differential equation for the rotation ϕ is

$$EI_{sz}d^2\frac{d^4\phi}{dx^4} - (GJ - \sigma_a I_{sp})\frac{d^2\phi}{dx^2} + K_\phi\phi = 0$$

in which

I_{sz} = moment of inertia of the stiffener (only) about an axis through the centroid of the stiffener and parallel to the web

d = stiffener web height + $(t + t_f)/2$

I_{sp} = polar moment of inertia of the stiffener about the center of rotation; it may be shown that for thin-wall open sections $I_{sp} \cong d^2(A_f + A_w/3)$, where A_f is the area of the flange and A_w is the area of the web

K_ϕ = distributed rotational restraint which the plating exerts on the stiffener.

If the ends of the stiffener are regarded as simply supported, the solution for $\phi(x)$ is a buckled shape in which the rotation ϕ varies sinusoidally in m half-waves over the length a. The elastic tripping stress, that is, the value of the applied in-plane stress σ_a, that would cause tripping according to elastic theory, will be denoted as $\sigma_{a,T}$. From the foregoing equation, it may be seen that $\sigma_{a,T}$ is the minimum value of σ_a that satisfies the following, in which m is a positive integer

$$EI_{sz}d^2\frac{m^4\pi^4}{a^4} + (GJ - \sigma_a I_{sp})\frac{m^2\pi^2}{a^2} + K_\Phi(\sigma_a, m) = 0 \qquad (14.1.15)$$

†Note, however, that in regard to flexural-torsional buckling, unsymmetric stiffeners (angles, etc.) have a lower buckling load than symmetric stiffeners (tees) because they begin to rotate as soon as a lateral load is applied.

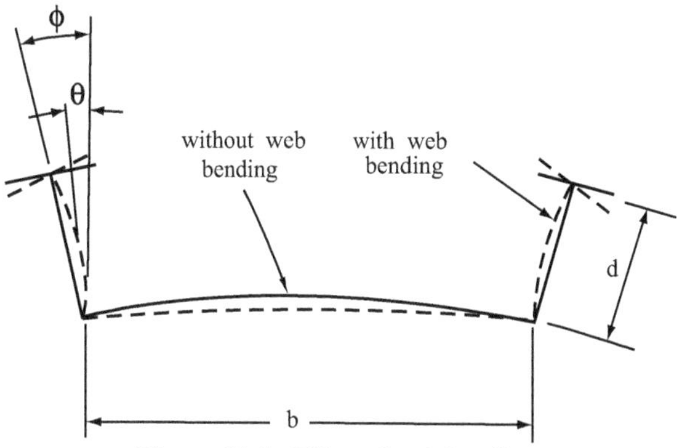

Figure 14.6 Effect of web bending.

As with plate buckling, the critical mode of tripping corresponds to whichever integer value of m gives the minimum value of σ_a in the foregoing expression. In this expression, K_ϕ is written as a function of σ_a and m because it is strongly dependent on both of these quantities. It is dependent on σ_a because the onset of plate buckling can diminish or eliminate K_ϕ, or even make it negative; and it depends on m because this diminishing of K_ϕ depends on whether m matches or approaches the number of half-waves in the plate's own critical buckling mode. Conversely, the value of m depends on the value of K_ϕ; if K_ϕ is small relative to EI_{sz}, the critical mode will be $m = 1$, but for larger values of K_ϕ, m may be larger.

In the absence of other factors, the rotational restraint offered by the plating comes directly from the plate's flexural rigidity which causes, in response to the rotation ϕ of the stiffener, a total distributed restraining moment $M_R = 2M$ along the line of the stiffener attachment, as shown in Fig. 14.7. If the individual plate panels are long, that is, if $a \gg b$, then we may ignore aspect ratio effects, and by considering a unit strip of plating across the span b it may be shown that $\phi = \frac{1}{2}Mb/D$. Therefore, the rotational restraint coefficient, or rotational stiffness, is

$$K_\phi = \frac{M_R}{\Phi} = \frac{4D}{b}$$

However, this assumes that the buckled displacement of the stiffener is entirely due to rigid body rotation. This is only accurate if the flexural rigidity of the stiffener web is much larger than that of the plate. In practice, some of the sideways displacement of the stiffener flange occurs because of bending of the web, and this effect becomes important if the plating is sturdy or if the stiffener web is slender. Sharp (1966) has presented an expression that accounts for this effect. Figure 14.6 shows the deflected shape when web bending does and does not occur, for the same amount of maximum sideways displacement of the stiffener flange. The rotational restraining moment M_R exerted by the plate (Fig. 14.7) is proportional to the angle of rotation of the plate along the line of attachment, and this angle is denoted as θ and ϕ in the two cases, respectively. It may readily be shown that the two angles are related by

$$\theta = C_r \phi$$

in which

$$C_r = \frac{1}{1 + \frac{2}{3}\left(\frac{t}{t_w}\right)^3 \frac{d}{b}}$$

That is, C_r is the factor by which the plate rotational restraint is reduced because of web bending. This correction does not account for all of the effects of web bending; for that, it would be necessary to take web bending into account from the very beginning, in the derivation of the governing differential equation. Nevertheless, on comparing the results of the foregoing treatment with a more detailed computer-based solution, Sharp (1966) found good agreement, particularly if the 2/3 factor was changed to 0.4. Also, the effect of plate aspect ratio may be accounted for by applying another correction factor $C_\alpha = (1 + m^2/\alpha^2)$. We then have

$$K_\phi = \frac{4D}{b} C_r C_\alpha \tag{14.1.16}$$

with

$$C_r = \frac{1}{1 + 0.4\left(\frac{t}{t_w}\right)^3 \frac{d}{b}} \tag{14.1.17}$$

and

$$C_\alpha = 1 + \frac{m^2}{\alpha^2} \quad (\text{for } \sigma_a = 0)$$

The qualification regarding σ_a is added because we have not yet allowed for the decrease in plate stiffness which occurs as the applied stress σ_a approaches the plate buckling stress σ_0 [which, for $\alpha > 1$, is $\sigma_0 = 3.62\, E(t/b)^2$]. Kroll (1943) has calculated a systematic set of values of a plate stiffness factor that accounts for both aspect ratio and buckling effects, and these values can be used for C_α in

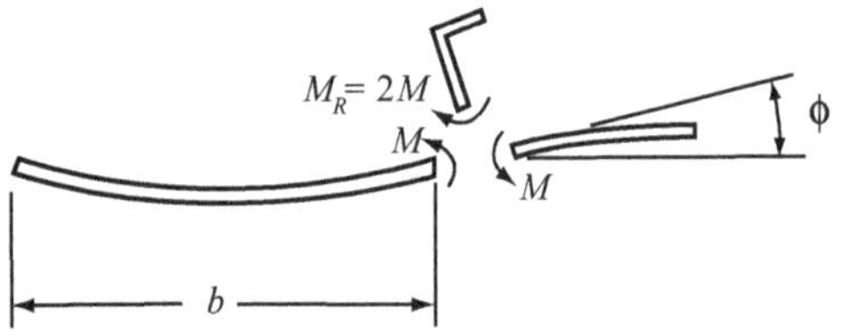

Figure 14.7 Restraining moment exerted by plating.

place of $(1 + m^2/\alpha^2)$. However, Kroll's results are in tabular form, whereas for purposes of computer-aided design it is preferable to have an analytical expression for C_α as a function of σ_a, m, and α, even at the cost of some accuracy. The following expression is an adequate representation of the tabulated values, and the small discrepancies that are introduced occur on the conservative side.

$$C_\alpha = 1 - \left(\frac{2\sigma_a}{\sigma_0} - 1\right)\frac{m^2}{\alpha^2} \tag{14.1.18}$$

From this expression, it may be seen that for $\sigma_a = \sigma_0$, the factor C_α, and hence also the plate rotational restraint K_ϕ, is proportional to $1 - (m/\alpha)^2$; that is, the restraint disappears when $m = \alpha$. This reflects the fact that if the plate panel between the stiffeners is long, or at least square, it will buckle into a number (approximately α) of square subpanels. In most cases, the number will exceed m, the number of half-waves of the stiffener tripping mode, and in such cases the plate buckling has little deleterious effect on K_ϕ. However, if the plate buckling pattern matches that of the stiffener ($\alpha = m$) then as σ_a approaches σ_0, the plate loses its ability to provide any rotational restraint. For stiffened panels of usual proportions, tripping occurs in a single half-wave, $m = 1$, and hence it is mainly square or short panels in which this loss of stiffness can occur. Applying the correction factors C_r and C_a gives

$$K_\phi = \frac{4D}{b}\left[\frac{1}{1 + 0.4\left(\frac{t}{t_w}\right)^3 \frac{d}{b}}\right]\left[1 - \left(\frac{2\sigma_a}{\sigma_0} - 1\right)\frac{m^2}{\alpha^2}\right] \tag{14.1.19}$$

We have now seen that tripping involves three variables that are interrelated in a rather complex fashion: σ_a, m, and K_ϕ. The critical value of σ_a given by equation (14.1.15) depends on K_ϕ and m, the value of m depends on the magnitude of K_ϕ relative to EI_{sz}, and equation (14.1.19) shows that K_ϕ depends on both σ_a and m.

As mentioned previously, in stiffened panels of average proportions the critical tripping mode is usually the $m = 1$ mode, but of course this cannot be simply assumed; the correct value of m must be ascertained in each case. The only exception to this is the calculation of a lower bound solution in which the plating rotational restraint is deliberately ignored, in which case the critical mode is always $m = 1$. Also, in this case the solution for $\sigma_{a,T}$ is much simpler, as will be shown. Therefore, in the analysis of tripping we distinguish three different cases: $m \geq 2$, $m = 1$, and the lower bound case. Accordingly, it will be helpful and even necessary to have some method for predicting the likely value of m, even if only approximate, instead of having to use a purely trial-and-error approach. An estimate can be obtained by substituting equation (14.1.19) for K_ϕ in equation (14.1.15) and solving the latter for σ_a. The result is

$$\sigma_{a,T} = \underset{m=1,2,\ldots}{\mathit{Minimum}} \left\{ \frac{1}{I_{sp} + \dfrac{2C_r b^3 t}{\pi^4}} \left[GJ + \frac{m^2\pi^2}{a^2} EI_{sz} d^2 + \frac{4DC_r}{\pi^2 b}\left(\frac{a^2}{m^2} + b^2\right)\right] \right\} \tag{14.1.20}$$

in which C_r is defined in equation (14.1.17). We then temporarily regard m as a continuous variable, differentiate equation (14.1.20) with respect to m, and set this equal to zero. The result is

$$m \cong \frac{a}{\pi}\sqrt[4]{\frac{4DC_r}{EI_{sz} d^2 b}} \tag{14.1.21}$$

After obtaining this estimate, try the two integer values above and below it to see which value gives the lowest value of $\sigma_{a,T}$ in equation (14.1.20).

14.2 TRANSVERSELY STIFFENED PANELS

The panel geometry for this case is shown in Fig. 14.8. The minimum size of transverse stiffeners for plates loaded in uniaxial compression has been defined by Timoshenko and Gere (1961) for one, two, or three equally spaced stiffeners, and by Klitchieff (1949) for any number of stiffeners. The stiffeners as sized provide a nodal line for the buckled plate and thus prohibit overall buckling of the stiffened panel. The strength of the stiffened panel is then determined by the buckling strength of the plate between stiffeners. The required minimum value of γ_y given by Klitchieff is

$$(\gamma_y)_{MIN} = \frac{(4N_L^2 - 1)\left[(N_L^2 - 1)^2 - 2(N_L^2 + 1)\kappa + \kappa^2\right]}{2(5N_L^2 + 1 - \kappa)\Pi^4} \tag{14.2.1}$$

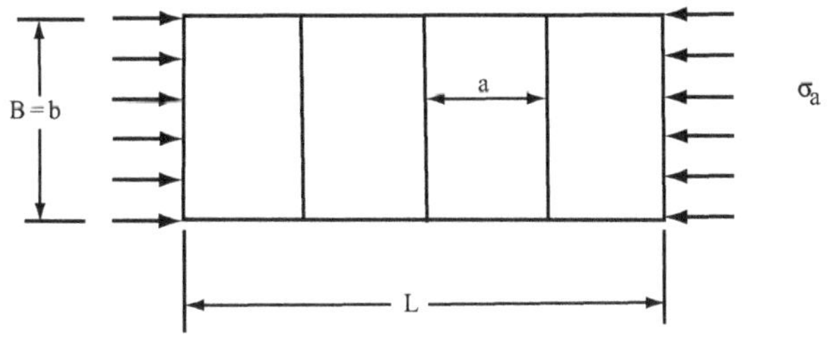

$N_L = \dfrac{L}{a}$ = number of subpanels

$\gamma_y = \dfrac{EI_y}{Da}$

Figure 14.8 Transversely stiffened panel.

14.3 BUCKLING AS A RESULT OF SHEAR

Figure 14.9 shows various types of stiffened panels subjected to a uniform shear loading. If s and l denote the stiffener spacing and stiffener length, then $s = b$ or a, and $l = a$ or b, depending on whether the panel is (1) longitudinally or (2) transversely stiffened, respectively. For the present, we will assume that $a > b$, as in Fig. 14.9a. If the stiffeners have sufficient rigidity to remain straight until the plate elements buckle, then the critical shear stress is $\tau_{cr} = K_s E(t/b)^2$, where K_s is given in Fig. 12.11b as a function of the aspect ratio and the degree of restraint at the edges. If the stiffeners do not have sufficient rigidity then the panel will undergo overall buckling at a smaller value of applied shear stress. The expression for τ_{cr} can still have the foregoing form provid-

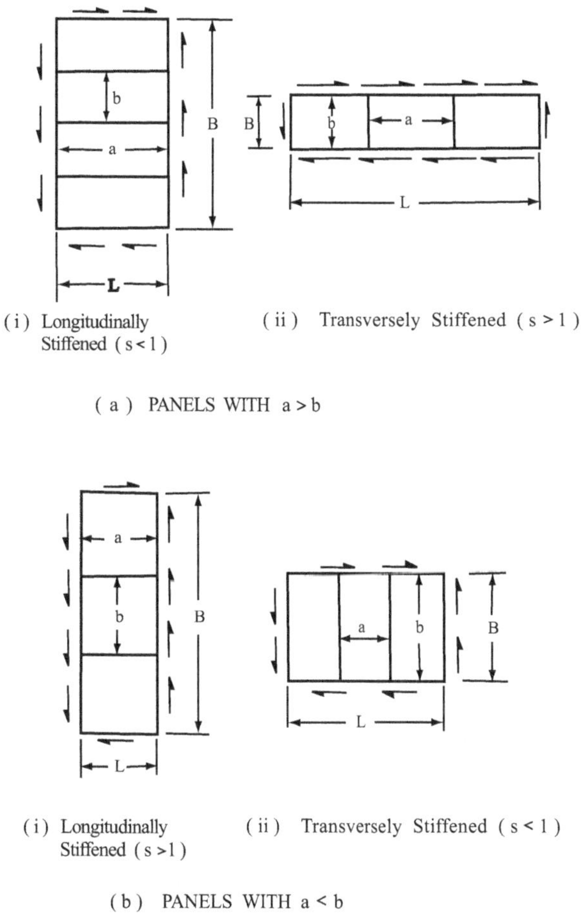

Figure 14.9 Variations in subpanel aspect ratio.

ing that we obtain a general expression for K_s that accounts for the rigidity of the stiffeners.

When the stiffeners are relatively weak flexurally and closely spaced, the buckle pattern becomes independent of stiffener spacing, and the analysis follows that for orthotropic plates in shear, in which the general form of the buckling stress equation is $\tau_{cr} = k_s(\pi^2 D/l^2 t)$. If we define the bending stiffness parameter γ_s as either EI_x/Db or EI_y/Da for (1) or (2), respectively, then the relationship between γ_s and the plate buckling coefficient k_s is given in the accompanying table from Stein and Fralich (1949).

γ_s	0	2	5	20	50	100	200
k_s	5.34	10.34	16.07	37.14	68.99	112.2	184.6

The problem was investigated in more detail by Timoshenko and Gere (1961) for the cases of one and two stiffeners, using an energy approach. For one stiffener (say in the transverse or y-direction, such that $a = L/2$) the critical value of shear stress is

$$\tau_{cr} = \frac{9\pi^2 D(1+\Pi^2)^2}{32BLt}\sqrt{1+\frac{2\gamma_y \Pi^4}{(1+\Pi^2)^2}}$$

where $\gamma_s = \gamma_y = EI_y/Da$ and $\Pi = L/B$. For a single longitudinal stiffener, $\gamma_s = \gamma_x = EI_x/Db$ and $1/\Pi$ replaces Π.

Rockey and Cook (1962) presented a solution for the opposite case: when the number of stiffeners is large. From these results and from some further numerical solutions, it has been possible to obtain an empirical expression for K_s that covers the full range of panel geometry. The values from the expression do not depart from the numerical results by more than 8% and the discrepancy is on the conservative side. For the cases shown in Fig. 14.9a, in which the subpanel aspect ratio a/b is greater than 1.0, the complete expression for τ_{cr} is

$$\tau_{cr} = K_s E\left(\frac{t}{b}\right)^2 \quad (a \geq b) \qquad (14.3.1)$$

in which

$$K_s = 4.5\left[\left(\frac{b}{a}\right)^2 + \frac{1}{N^2} + \left(\frac{N^2-1}{N^2}\right)\left(\frac{\omega}{1+\omega}\right)^r\right]$$

in which N = number of subpanels

$$= \begin{cases} B/b \text{ for longitudinal stiffening} \\ L/a \text{ for transverse stiffening} \end{cases}$$

$$\omega = \frac{I_{se}}{lt^3}$$

I_{se} = moment of inertia of a section consisting of the stiffener and a plate flange of effective width $s/2$.

$$r = \begin{cases} 1 - 0.75\,(s/l) \text{ for longitudinal stiffening} \\ 0.25 \text{ for transverse stiffening} \end{cases}$$

The numerical solutions showed that when the stiffener spacing exceeds the stiffener length (i.e., when $s/l > 1$) the effect of stiffener size remains the same as for a square subpanel ($s = l$). Since we are dealing with the case when $a \geq b$, transversely stiffened panels will have $s/l \geq 1$, and hence for such panels, r is set to the value corresponding to $s/l = 1$, namely 0.25.

For the panel geometry shown in Fig. 14.9b, in which the subpanel aspect ratio is less than 1.0, a and b must be interchanged in the foregoing equations because a shear load has no directionality. Therefore, in the analysis of shear buckling, the subpanel aspect ratio is always taken as being greater than or equal to 1.0. Therefore, whenever $a < b$, the equation for the shear buckling stress is

$$\tau_{cr} = K_s E\left(\frac{t}{a}\right)^2 \quad (a < b) \qquad (14.3.2)$$

in which

$$K_s = 4.5\left[\left(\frac{a}{b}\right)^2 + \frac{1}{N^2} + \left(\frac{N^2-1}{N^2}\right)\left(\frac{\omega}{1+\omega}\right)^r\right]$$

All other quantities are as defined previously except that the two expressions for r must be reversed:

$$r = \begin{cases} 0.25 \text{ for longitudinal stiffening} \\ 1 - 0.75\,(s/l) \text{ for transverse stiffening} \end{cases}$$

14.4 BUCKLING OF LONGITUDINALLY STIFFENED PANELS UNDER COMBINED COMPRESSION AND SHEAR

Analytical and experimental results on the buckling behavior of stiffened plates under combined compression and shear are relatively scarce. Recourse is therefore usually needed for unstiffened plates supplemented with whatever data are available for the type of longitudinally stiffened panels used in ships, such as that shown in Fig. 14.10. The case of unstiffened rectangular plates under combined compressive and shear stresses was dealt with in Section 12.4, which for $\alpha \geq 1$ gave the parabolic interaction formula of equation (12.4.5) which is

$$R_x + R_s^2 = 1 \quad (\alpha \geq 1) \qquad (14.4.1)$$

where $R_x = \dfrac{\sigma_a}{(\sigma_a)_{cr}}$

= ratio of compressive stress when buckling occurs in combined shear and direct stress to compressive stress when buckling occurs in pure compression

and $R_s = \dfrac{\tau}{\tau_{cr}}$

= ratio of shear stress when buckling occurs in combined shear and direct stress to shear stress when buckling occurs in pure shear

Harris and Pifko (1969) obtained a finite element solution for the buckling of infinitely wide longitudinally stiffened panels under compression and shearing stresses. The stiffeners were assumed to have both bending and torsional stiffness, and the grid refinement used was judged to be adequate to ensure accurate results. The results are compared to the parabolic expression of equation (14.4.1) in Fig. 14.11. Except for the case of assumed large torsional stiffness ratio ($GJ/bD = 10^6$), the analytical points follow the parabolic relationship very well.

From the work of Section 14.1, we have equation (14.1.14) which, together with equation (14.1.7), gives $(\sigma_a)_{cr}$ for longitudinally stiffened panels of any panel aspect ratio Π, stiffener spacing b, and stiffener rigidity γ_x. Likewise, equations (14.3.1) and (14.3.2) give the value of τ_{cr} for uniaxially stiffened panels of any value of Π, b, and ω (which is an alternative measure of stiffener rigidity). Therefore, these general expressions can be used to calculate the denominators of the strength ratios in the interaction formula given in equation (14.4.1).

In fact, we can go a few steps further and make some allowance for the presence of a transverse compressive stress σ_{ay} and in-plane bending stress σ_b. First, in regard to transverse compression, equation (14.2.1) gives the minimum stiffener rigidity at

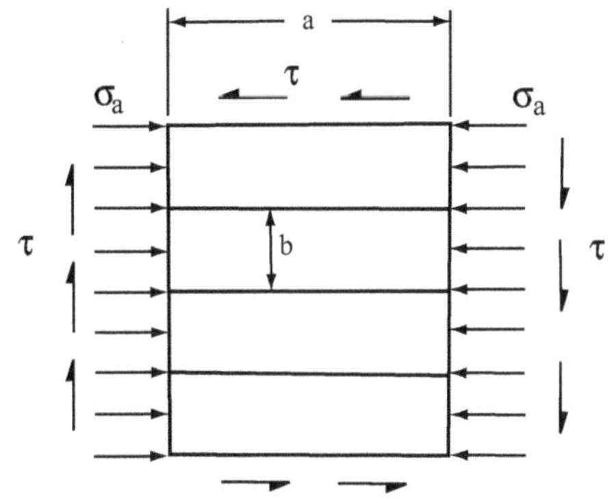

Figure 14.10 Longitudinally stiffened panel under combined compression and shear stresses.

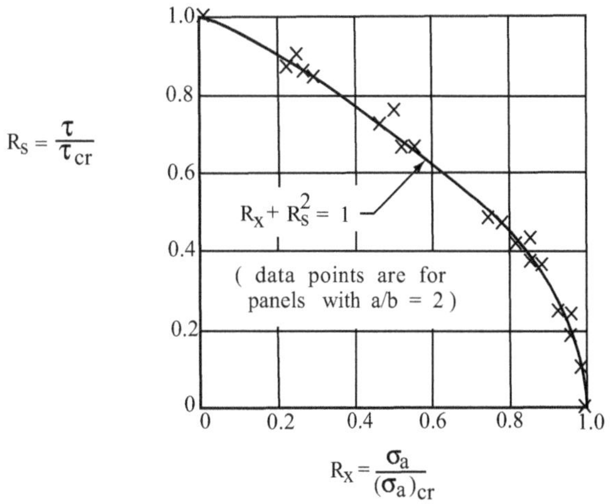

Figure 14.11 Interaction relation for longitudinally stiffened panel under combined compression and shearing stresses.

which plate buckling occurs before stiffener buckling under transverse compression. A numerical parametric study of this equation indicates that for most proportions of stiffened panels, the stiffeners exceed this rigidity. This is nearly always true if the panel is intended to carry a primary longitudinal compressive stress, σ_{ax}, and it is virtually guaranteed if the panel must withstand a significant lateral pressure load. Therefore, the main effect of a transverse compressive stress is to promote plate buckling between the stiffeners. That is, the interaction of σ_{ay} and τ is mainly at the plate level and the interaction equation is

$$R_x + \frac{0.625\left(1 + \dfrac{0.6}{\alpha}\right)R_y}{1 + R_x} + R_s^2 = 1 \quad (\alpha \geq 1) \qquad (14.4.2)$$

in which $R_y = \dfrac{\sigma_{ay}}{(\sigma_{ay})_{cr}}$.

Since σ_{ay} interacts at the plate level, the value of $(\sigma_a)_{cr}$ is obtained from equation (12.2.7) with k given in the second row of Table 12.1.

This interaction formula is a generalization of equation (14.4.1), which Harris and Pifko (1969) verified for stiffened panels, and it reduces to equation (14.4.1) when $R_y = 0$.

The full version of equation (12.4.7) includes the effect of in-plane bending stress, but this loading cannot be associated exclusively with plate buckling; in its compressive portion, it is similar to σ_{ax}, the only difference being that it starts at a maximum value σ_b at one edge and attenuates linearly across the panel. In a uniaxially stiffened panel, each stiffener is essentially an independent column. The column that has the largest compressive stress will buckle first; this will throw more load onto the others, causing the next most heavily loaded stiffener to buckle and continuing in a rapid chain reaction that can scarcely be distinguished from simultaneous buckling of all stiffeners. Therefore, the safest and easiest way of handling an in-plane bending stress of maximum value σ_b is to simply add σ_b to σ_{ax}. The interaction formula of equation (14.4.2) then becomes

$$R_{x+b} + \frac{0.625\left(1+\frac{0.6}{\alpha}\right)R_y}{1+R_{x+b}} + R_s^2 = 1 \qquad (\alpha \geq 1) \qquad (14.4.3)$$

in which

$$R_{x+b} = \frac{\sigma_{ax}+\sigma_b}{(\sigma_{ax})_{cr}}$$

$$R_y = \frac{\sigma_{ay}}{(\sigma_{ay})_{cr}}$$

$$R_s = \frac{\tau}{\tau_{cr}}$$

The value of $(\sigma_{ax})_{cr}$ is obtained from equations (14.1.14) and (14.1.7),‡ τ_{cr} is given by either equation (14.3.1) or equation (14.3.2), depending on a/b, and $(\sigma_{ay})_{cr}$ is given by

$$(\sigma_{ay})_{cr} = \overline{K}E\left(\frac{t}{b}\right)^2 \qquad (14.4.4)$$

in which

$$\overline{K} + 0.905\left[1+\frac{1}{\alpha^2}\right]^2$$

The foregoing approach lacks experimental confirmation but seems reasonable and should be sufficiently accurate for panels of normal proportions, as long as R_{ay} and σ_b/σ_{ax} are not large (each less than, say, 0.3) as is usually the case.

14.5 BUCKLING OF CROSS-STIFFENED PANELS

When the panel is large, it will generally be necessary to provide stiffening in both directions. In general, the relative proportions of the stiffeners will depend on the overall panel aspect ratio, the direction and size of the primary in-plane compressive load σ_{ax}, and the size of the secondary compressive load σ_{ay}, if there is one. As mentioned earlier, if there is a lateral load, it is likely that the scantlings that are required to provide adequate bending strength will preclude elastic buckling, and the ultimate compressive strength would be calculated by the methods to be presented in Chapter 15. But, as also mentioned earlier, the elastic analysis provides a relatively simple method for testing whether this is true.

In cross-stiffened panels that are designed to carry in-plane loads, there is usually a hierarchy of structural elements. In the first place, because of its larger cross-sectional area, the plating takes most of the in-plane compressive load. Next are the stiffeners, which carry most of the lateral load and which also are required to stiffen and stabilize the plating so that it can carry the in-plane load. Therefore, wherever practicable, the stiffeners should be oriented parallel to the primary in-plane load, which is nearly always the longitudinal direction; such stiffeners are commonly referred to as "longitudinals." Finally, there are the transverse members, whose main role is to provide nondeflecting intermediate supports for the longitudinals; the latter act as columns and the most efficient way of strengthening a column is to give it intermediate support, thus shortening its span. In general, the transverse members are not expected to provide rotational restraint because this is much more difficult and the benefits are not commensurate. Because they must provide (nearly) undeflecting support to the longitudinals, the transverse members must usually have a larger rigidity; that is, $I_y > I_x$. Therefore, they usually have deeper webs than the longitudinals; for this reason, they are often referred to as "web frames" or "deep webs." Another reason for making them deeper is to simplify the construction by permitting one set of members to pierce the webs of the other set, leaving the flanges of both undisturbed.

If the transverse members are not rigid enough, the panel may undergo "gross panel" buckling, in which the transverses buckle with the longitudinals. This is also referred to as "grillage" buckling to indicate that both sets of stiffeners are involved. If, on the other hand, the transverses are sufficiently rigid, then the stiffened panels between them are ordinary, simply supported, longitudinally stiffened panels, and can be analyzed by the methods of the previous sections. Thus, the provision of cross-stiffening introduces two new quantities to be calculated:

1. The minimum rigidity of the transverses to provide nondeflecting support to the longitudinals.
2. Alternatively, if large transverse stiffeners are undesirable or impractical, the value of applied stress at which gross panel buckling would occur.

14.5.1 Minimum Transverse Rigidity to Prevent Gross Panel Buckling (Uniaxial Load)

Johnston (1976) derived an expression for estimating the required stiffness of transverse stiffeners from a consideration of the buckling of columns

‡Providing that plate buckling occurs first (i.e., that the plate buckling stress is less than $(\sigma_a)_{cr}$ calculated from equation (14.1.8) with $b_e = b$ (case A in Fig. 14.4). If not, then equation (14.1.8) with $b_e = b$ constitutes the value of $(\sigma_{ax})_{cr}$.

having elastic springs at evenly spaced intermediate points which restrain the lateral deflection at those points. Johnston's results are based on the work of Timoshenko and Gere (1961), who showed that for a single column having multiple spring supports, the required spring constant K_y for the supports to remain undeflected and to constitute nodal points is given by:

$$K_y = \frac{NP}{CL} \tag{14.5.1}$$

where

$$P = \frac{N^2 \pi^2 EI_x}{L^2}$$

and EI_x = flexural stiffness of column
N = number of spans
L = total length of column

C is a parameter that depends on N, and that decreases from 0.5 for $N = 2$ to 0.25 for infinitely large N. It is given approximately by

$$C = 0.25 + \frac{2}{N^3} \tag{14.5.2}$$

In the case of a cross-stiffened panel (Fig. 14.12), the longitudinal stiffeners act as columns that are elastically restrained by the transverse stiffeners. Assuming that the loading from the longitudinal stiffener to the transverse stiffener is proportional to the deflection of the latter, the spring constant for each column support can be estimated. For example, for the limit case of an infinite number of small longitudinals, the deflected shape of the transverses is a half sine wave and the spring constant per unit width is

$$k_y = \frac{\pi^4 EI_y}{B^4} \tag{14.5.3}$$

For this case the spring constant K_y, which is required at each stiffener, must also be converted to a per-unit-width value by dividing by the stiffener spacing b. Thus, from equation (14.5.1), the required stiffness per unit width is

$$k_y = \frac{K_y}{b} = \frac{NP}{bCL} \tag{14.5.4}$$

Equating the two values of k_y and inserting the value given for P yields

$$\frac{bEI_y}{BEI_x} = \frac{N^3}{\pi^2 C\left(\frac{L}{B}\right)^3} = \frac{B^3}{\pi^2 Ca^3}$$

Therefore, in panels containing a large number of longitudinal stiffeners, the minimum rigidity $\gamma_y(= EI_y/Da)$ of the transverse members to prevent gross panel buckling is

$$\frac{\gamma_y}{\gamma_x} = \frac{B^4}{\pi^2 Ca^4} \tag{14.5.5}$$

in which C is given by equation (14.5.2). Timoshenko and Gere (1961) presented solutions for several particular cases involving a specific number of longitudinals, and these are given in Table 14.1. Johnston (1976) showed that a general expression that covers all of these cases with satisfactory accuracy is

$$\frac{\gamma_y}{\gamma_x} \quad \frac{B^4}{\pi^2 Ca^4}\left(1 + \frac{1}{p}\right) \tag{14.5.6}$$

where p is the number of longitudinal stiffeners, and the other parameters are as defined previously (see also Fig. 14.12). As long as the relative rigidity of the transverse stiffener exceeds the value given by equation (14.5.6), the compressive strength of the cross-stiffened panel is determined by the compres-

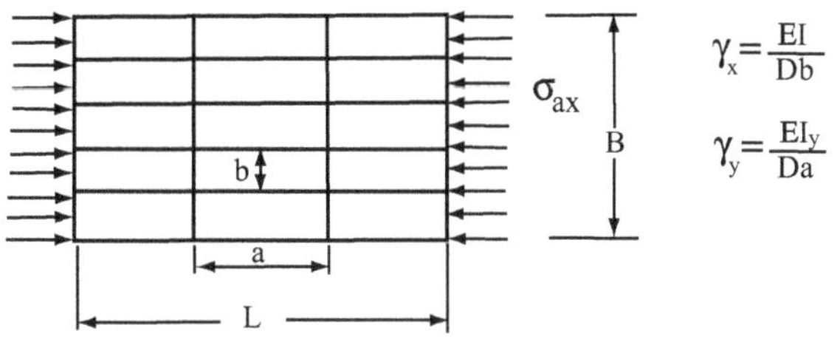

$N_B = \frac{B}{b}$ = no. of panels transversely
$p = N_B - 1$ = no. of longitudinal stiffeners
$N = \frac{L}{a}$ = no. of panels longitudinally
$q = N_L - 1$ = no. of transverse stiffeners

Figure 14.12 Cross-stiffened panel.

Table 14.1

Number of Longitudinal Stiffeners	**Minimum Transverse Rigidity Ratio γ_y/γ_x to prevent Gross Panel Buckling**
1	$\frac{0.206}{C}\left(\frac{B}{a}\right)^4$
2	$\frac{0.152}{C}\left(\frac{B}{a}\right)^4$
4	$\frac{0.133}{C}\left(\frac{B}{a}\right)^4$

sive strength of the longitudinally stiffened panels between transverse stiffeners.

14.5.2 Gross Panel Buckling—Orthotropic Plate Approach

Panels that are stiffened in two orthogonal directions may be idealized as orthotropic plates by "smearing" the bending rigidity of the stiffeners over the region of the plating. This approach is satisfactory only when the stiffeners are uniform and closely spaced in each direction. As noted in the introduction, this approach can also be used for panels that are stiffened in only one direction, but in this case the stiffeners are much more independent and, unless they are very closely spaced, the discrete beam approach is preferable. The equations and solutions for unidirectional stiffened panels can be obtained from those presented in this section simply by setting D_x or D_Y equal to D.

Much work has been done regarding the elastic buckling of stiffened panels, particularly in the aircraft field. Solutions have been achieved for a wide variety of cases and at all three levels of accuracy: linear small-deflection theory, corrected small-deflection theory, and large-deflection theory. However, as explained at the beginning of this chapter, because of the lateral loads and the general sturdiness of ship panels, the buckling of such panels is usually inelastic. Elastic gross panel buckling can occur only in lightweight panels (thin plating; slender, closely spaced stiffeners). For panels of this type, elastic gross panel buckling should be investigated as accurately as possible because it constitutes collapse (and usually a quite undesirable type of collapse). For best results, a large deflection method that accounts for the interaction between lateral and in-plane loads should be used. For this purpose, Mansour (1976) presented a comprehensive and convenient set of charts which give deflection, bending moment, buckling stress, and effective width. However, since this type of ultimate strength analysis is only relevant for lightweight panels, it will not be discussed further, either here or in Chapter 15, which deals with the ultimate strength of stiffened panels.

Also, the present chapter deals only with singly plated panels because for double wall panels the elastic buckling stress far exceeds the yield stress and therefore has little meaning, even as a reference parameter.

Because the elastic gross panel buckling stress for typical ship panels exceeds the yield stress and is therefore not the actual collapse stress but merely a parameter that represents the panel characteristics, there is no need for great accuracy in its calculation. Hence, it is usually sufficient to use small-deflection theory. As noted earlier, most ship panels are sufficiently sturdy that by the time a lateral load is large enough to produce any significant lateral deflection, it would already have produced extensive yielding in the panel, thus rendering an elastic analysis invalid. However, if the panel is slender enough to undergo a significant elastic lateral deflection, either because of panel eccentricity or a large lateral load, then the corrected small-deflection approach is sufficient. Mansour (1967), Falconer and Chapman (1953), and Chapman and Slatford (1959) have presented suitable methods.

In all of its basic aspects, the buckling of orthotropic plates is similar to that of isotropic plates. For a uniform compressive distributed load N_x (positive if compressive), the governing differential equation of small deflection orthotropic plate theory is

$$D_x \frac{\partial^4 w}{\partial x^4} + 2H \frac{\partial^4 w}{\partial x^2 \partial y^2} + D_y \frac{\partial^4 w}{\partial y^4} = -N_x \frac{\partial^2 w}{\partial x^2} \tag{14.5.7}$$

It is often more appropriate or more convenient to express the load as an applied stress σ_{ax} (e.g., the hull girder stress). The relationship between the two is $N_x = \sigma_{ax}\ (t + A_x/b)$, where A_x *is* the area of each longitudinal stiffener.

In the foregoing equation, D_x and D_y are the bending rigidities and H is the torsional rigidity of the orthotropic plate. Orthotropic plate solutions are most conveniently expressed in terms of two nondimensional parameters

$$\Pi_{\text{orth}} = \frac{L}{B}\left(\frac{D_y}{D_x}\right)^{1/4} = \text{orthotropic aspect ratio}$$

$$\eta = \frac{H}{\sqrt{D_x D_y}} = \text{torsional stiffness parameter} \tag{14.5.8}$$

For simply supported edges, the solution to equation (14.5.7) is

$$(N_x)_{cr,gp} = k_0 \frac{\pi^2 \sqrt{D_x D_y}}{B^2} \tag{14.5.9}$$

where

$$k_0 = \frac{m^2}{\Pi_{\text{orth}}^2} + 2\eta + \frac{\Pi_{\text{orth}}^2}{m^2} \tag{14.5.10}$$

and m is the number of half-waves in the x-direction. Note that all of the formulas presented in this section are based on the assumption that the panel buckles into only one half-wave in the y-direction.

In terms of an applied stress σ_{ax}, the solution is

$$(\sigma_{ax})_{cr,gp} = k_0 \frac{\pi^2 \sqrt{D_x D_y}}{B^2\left(t + \frac{A_x}{b}\right)} \tag{14.5.11}$$

The close parallel with the isotropic case can be seen by comparing these equations with equations (12.2.7) and (12.2.8). If η is set to unity, equation (14.5.10) becomes

$$k_0 = \left(\frac{m}{\Pi_{orth}} + \frac{\Pi_{orth}}{m^2} \right)^2$$

which is the same as equation (12.2.8), with the orthotropic aspect ratio Π_{orth} in place of the isotropic aspect ratio a/b.

The primary buckling load corresponds to whichever value of m gives the smallest value of k_0 in equation (14.5.10). We can estimate this by temporarily regarding m as a continuous variable, differentiating equation (14.5.10) with respect to m and setting this equal to zero. The result is $m = \Pi_{orth}$; that is, the number of half-waves in the x-direction is the integer nearest to Π_{orth}, the orthotropic aspect ratio of the panel. This also is completely analogous to the situation with isotropic plates, for which the primary buckling mode is $m = \alpha$ (buckling into square subpanels). Thus, if $D_x \cong D_y$, a cross-stiffened panel will also tend to buckle into square subpanels, but if the two flexural rigidities are unequal, the primary buckling mode will be some other pattern. Obviously, if $\rho < 1$, (a short panel and/or relatively rigid longitudinal stiffeners) the primary mode will be $m = 1$.

In orthotropic plate theory, the bending rigidity of the stiffeners in both directions is smeared into the plating. This implicitly assumes that in the buckled deformation pattern, all of the stiffeners are participating; that is, they are all undergoing flexure. If the primary mode is such that $m > 1$ and if N_L, the number of subpanels in the x-direction, is a multiple of m (i.e., if $N_L = m$, $2m$, $3m$, etc., where m is the nearest integer to Π_{orth}), then $m - 1$ of the transverse stiffeners are located at or near node lines and will not undergo any significant flexure. The extreme case is $N_L = m$ because then *all* of the transverse stiffeners occur at node lines. In any of these cases, the elastic buckling stress predicted by orthotropic plate theory will be larger than the true value, and the error is important because it lies on the optimistic (nonconservative) side. Therefore, when using the buckling formulas, which are presented below, it is very important to check whether N_L is a multiple of either of the integers m_1, m_2 above and below Π_{orth}. If so (if, for example, $N_L = jm_1$) then the subpanel of length L/m_1 should also be investigated, again using the formulas presented in this section. That is, Π_{orth} should be calculated for the reduced length, and the elastic buckling stress evaluated for this value of Π_{orth}. The true buckling stress is then the lower of the two calculated values.

For singly plated panels, the neutral axis in each direction is close to the plating and therefore, as shown by Schultz (1962), the effect of Poisson's ratio may be ignored. Therefore, the flexural rigidities are

$$D_x = \frac{EI_x}{b} = Ei_x \quad \text{and} \quad D_y = \frac{EI_y}{a} = Ei_y \qquad (14.5.12)$$

where I_x and I_y are the moments of inertia of the stiffeners, including a plate flange of width b and a, respectively. The full width is used because we want to guard against overall buckling occurring before local plate buckling and therefore we want the value of $(\sigma_{ax})_{cr}$ with the plating fully effective.

The torsional rigidity is

$$H = D_{xy} = \frac{1}{6}Gt^3 + \frac{GJ_x}{b} + \frac{GJ_y}{a} \qquad (14.5.13)$$

where J_x and J_y are the St. Venant torsion constants of the stiffeners, which for open sections is given by

$$J = \frac{1}{3}\sum_i l_i t_i^3$$

in which t_i and l_i are the thickness and width of the stiffener flange and web.

The following two cases are dealt with in this section:

1. Uniaxial load: ends simply supported, sides elastically restrained
2. Uniaxial load: sides clamped, ends simply supported or clamped

Case 1 is based on work by Schultz (1962); case 2 is based on work by Wittrick (1952).

14.5.2.1 Uniaxial Load: Ends Simply Supported, Sides Elastically Restrained

The buckling stress is given by

$$(N_x)_{cr,gp} = (k + 2\eta)\frac{\pi^2 E\sqrt{i_x i_y}}{B^2} \qquad (14.5.14)$$

The coefficient k is given by Bleich (1952) in the form of curves. For computer-aided design, a mathematical expression is more suitable, and for this purpose the following expression, which gives a satisfactory representation of these curves, has been developed.

$$k = \frac{\Pi_{orth}^2}{0.4 + \Pi_{orth}^2}\left(2 + \frac{0.5}{0.2 + C_{fy}}\right) \qquad (14.5.15)$$

In this expression, C_{fy} is a flexibility coefficient that accounts for the rotational restraint along the sides of the panel resulting from adjacent structure. If $\overline{K}_y$ denotes the rotational stiffness of the adjacent structure (such that a distributed edge moment m_y acting on the adjacent structure would cause it to rotate an angle of $m_y/\overline{K}_y$), then the flexibility coefficient is

$$C_{fy} = \frac{2D_y}{B\overline{K}_y}$$

For clamped edges, $C_{fy} = 0$ and for simply supported edges $C_{fy} = \infty$. When the two sides are of differing flexibility, sufficient accuracy is achieved by averaging the two values of k that correspond to the values of C_{fy}.

In ship panels, the adjacent structure along each side of the panel is usually another panel. For this case, Schultz (1962) derived the following expression for C_{fy}

$$C_{fy} = \frac{D_y}{\pi \overline{D}_y}\left(\frac{\overline{D}_x}{\overline{D}_y}\right)^{1/4} \xi r \qquad (14.5.16)$$

where the barred symbol refers to the restraining panel. The effect of an in-plane load $\overline{N}_x$ acting on the restraining panel is reflected by r, which for a general, nonuniform compressive load is given by

$$r = \frac{1}{1-\left(\frac{1+\psi}{2}\right)\left(\frac{1+\eta}{1+\overline{\eta}}\right)\frac{D_x D_y}{\sqrt{\overline{D}_x \overline{D}_y}}\left(\frac{\overline{B}}{B}\right)^2} \qquad (14.5.17)$$

in which all barred symbols refer to the restraining panel. The parameter ψ describes the distribution of $\overline{N}_x$, as shown in Fig. 14.13. The factor ξ accounts for the torsional stiffness parameter $\overline{\eta}[=\overline{H}/(\overline{D}_x \overline{D}_y)^{1/2}]$ of the restraining panel. Its value may be obtained from Table 14.2, in which

$$\varepsilon = \frac{\pi \overline{B}}{B}\left(\frac{\overline{D}_y}{\overline{D}_x}\right)^{1/4} \qquad (14.5.18)$$

For a simply supported panel, equation (14.5.14) reduces to

$$(N_x)_{cr,gp} = \left(\frac{2\Pi^2_{\text{orth}}}{0.4+\Pi^2_{\text{orth}}} + 2\eta\right)\frac{\pi^2 E\sqrt{i_x i}}{B^2} \qquad (14.5.19)$$

14.5.2.2 Uniaxial Load: Edges Simply Supported or Clamped

Wittrick (1952) derived an approximate expression for $(N_x)_{cr,gp}$ utilizing the fact that the principal term, k, in the orthotropic buckling coefficient k_0 and the virtual aspect ratio ρ have exactly the same relationship as the buckling coefficient k and the ordinary aspect ratio $\alpha = a/b$ of isotropic plates. Therefore, eq. (12.2.7) can be used for an orthotropic panel, with the value of ρ being used in place of α. The value of k is obtained from Table 12.1 for simple support and Table 12.2 for clamped support.

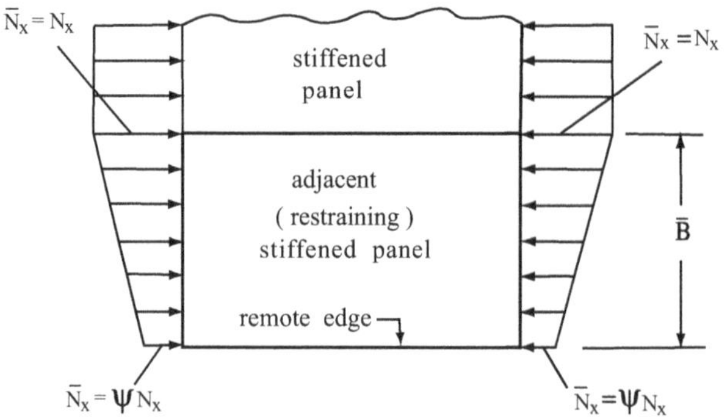

Figure 14.13 Load distribution in adjacent panels.

14.6 MINIMUM RIGIDITY RATIO TO AVOID HAVING OVERALL PANEL BUCKLING AS THE PRIMARY BUCKLING MODE

Section 14.1.1 gave some approximate formulas for $(\gamma_x)_{MIN}$, the minimum rigidity ratio such that the stiffeners are sufficiently rigid that they prevent overall panel buckling from occurring before local plate buckling. Now that we have equation 14.5.11 for overall panel buckling based on orthotropic plate theory, we can obtain a more accurate formula for this minimum ratio. We will obtain this by equating the local plate buckling stress from Chapter 12 and the overall panel buckling stress from (14.5.11). Before this, we will present a recently "re-discovered" reduction factor allowing for shear deflection of the stiffener web.

14.6.1 Reduction Factor for Shear Deflection of the Stiffener Web

This reduction factor arises from a long forgotten phenomenon that was described and quantified in the original 1936 edition of Timoshenko's famous textbook *Theory of Elastic Stability*. His presentation was for a typical simply supported column, but as we shall see, it is even more relevant for a stiffened panel, which is essentially a row of identical columns, with the plating acting as a second, and very wide, flange. As with any column, the "stiffener-columns" have some initial eccentricity, which means they begin to deflect as soon as an axial load P is applied, and the deflection increases with the load. The deflection causes

Table 14.2 Effect of Orthotropic Parameters and Remote Edge Support of Restraining Panel

Remote edge simply supported	
$\bar{\eta} > 1$	
$\xi = \dfrac{1}{\sqrt{\dfrac{\bar{H}^2}{\bar{D}_y^2} - \dfrac{\bar{D}_x}{\bar{D}_y}}}(\rho_1^* \coth \varepsilon\rho_1^* - \rho_2^* \coth \varepsilon\rho_2^*)$	$\rho_1^* = \sqrt{\dfrac{\bar{H}}{\bar{D}_y} + \sqrt{\dfrac{\bar{H}^2}{\bar{D}_y^2} - \dfrac{\bar{D}_x}{\bar{D}_y}}}$ $\rho_2^* = \sqrt{\dfrac{\bar{H}}{\bar{D}_y} - \sqrt{\dfrac{\bar{H}^2}{\bar{D}_y^2} - \dfrac{\bar{D}}{\bar{D}}}}$
$\bar{\eta} = 1$	
$\xi = \dfrac{\cosh \varepsilon\rho^* \sinh \varepsilon\rho^* - \varepsilon\rho^*}{\rho^* \sinh^2 \varepsilon\rho^*}$	$\rho^* = \sqrt[4]{\dfrac{\bar{D}_x}{\bar{D}_y}}$
$\bar{\eta} < 1$	
$\xi = \dfrac{\rho_2^* \sinh \varepsilon\rho_1^* \cosh \varepsilon\rho_1^* - \rho_1^* \sin \varepsilon\rho_2^* \cos \varepsilon\rho_2^*}{\rho_1^*\rho_2^*(\sinh \varepsilon\rho_1^* + \sin^2 \varepsilon\rho_2^*)}$	$\rho_1^* = \sqrt{\dfrac{1}{2}\sqrt{\dfrac{\bar{D}_x}{\bar{D}_y}} + \dfrac{\bar{H}}{\bar{D}_y}}$ $\rho_2^* = \sqrt{\dfrac{1}{2}\sqrt{\dfrac{\bar{D}_x}{\bar{D}_y}} - \dfrac{\bar{H}}{\bar{D}_y}}$
Remote edge clamped	
$\bar{\eta} > 1$	
$\xi = \dfrac{2\rho_1^*\rho_2^* \cosh \varepsilon\rho_2^* - \sinh \varepsilon\rho_1^* \sinh \varepsilon\rho_2^*(\rho_2^{*2} + \rho_2^{*2}) - 2\rho_1^*\rho_2^*}{\sqrt{\dfrac{\bar{H}^2}{\bar{D}^2} - \dfrac{\bar{D}_x}{\bar{D}_y}}(\rho_2^* \sinh \varepsilon\rho_1^* \cosh \varepsilon\rho_2^* - \rho_1^* \sinh \varepsilon\rho_2^* \cosh \varepsilon\rho_1^*}$	$\rho_1^* = \sqrt{\dfrac{\bar{H}}{\bar{D}_y} + \sqrt{\dfrac{\bar{H}^2}{\bar{D}_y^2} - \dfrac{\bar{D}_x}{\bar{D}_y}}}$ $\rho_2^* = \sqrt{\dfrac{\bar{H}}{\bar{D}_y} - \sqrt{\dfrac{\bar{H}^2}{\bar{D}_y^2} - \dfrac{\bar{D}}{\bar{D}}}}$
$\bar{\eta} = 1$	
$\xi = \dfrac{1}{\rho^*}\dfrac{\sinh^2 \varepsilon\rho^* - (\varepsilon\rho^*)^2}{\sinh \varepsilon\rho^* \cosh \varepsilon\rho^* - \varepsilon\rho^*}$	$\rho^* = \sqrt[4]{\dfrac{\bar{D}_x}{\bar{D}_y}}$
$\bar{\eta} < 1$	
$\xi = \dfrac{1}{\rho_1^*\rho_2^*}\dfrac{\rho_2^{*2} \sinh^2 \varepsilon\rho_1^* - \rho_1^{*2} \sin \varepsilon\rho_2^*}{\rho_2^* \sinh \varepsilon\rho_1^* \cosh \varepsilon\rho_1^* - \rho_1^* \sin \varepsilon\rho_2^* \cos \varepsilon\rho_2^*}$	$\rho_1^* = \sqrt{\dfrac{1}{2}\sqrt{\dfrac{\bar{D}_x}{\bar{D}_y}} + \dfrac{\bar{H}}{\bar{D}_y}}$ $\rho_2^* = \sqrt{\dfrac{1}{2}\sqrt{\dfrac{\bar{D}_x}{\bar{D}_y}} - \dfrac{\bar{H}}{\bar{D}_y}}$

each column to have a slightly "bowed" shape, such that at the ends there is a small angle θ between the neutral axis and the load P, as shown in Fig. 14.14. Therefore within the column there is a component of P that acts transversely and thus constitutes a shear force $Q = P \tan \theta$ in the web of the column. Timoshenko showed that the resulting shear deflection slightly reduces the overall Euler buckling stress of the column. For typical columns the effect is negligible, and this topic was dropped

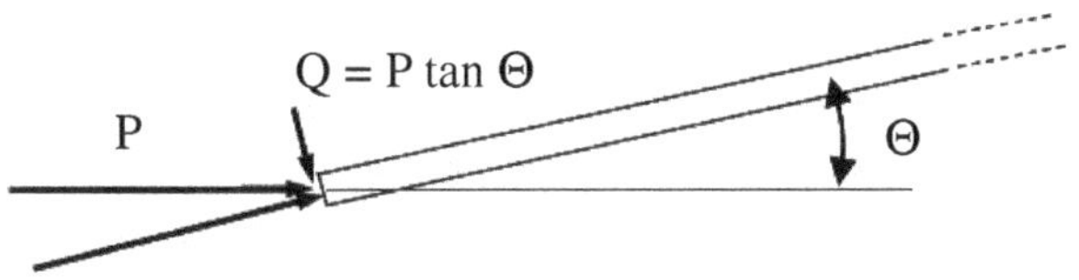

Figure 14.14 Transverse shear force in the web of a column or stiffener

in subsequent editions of Timoshenko's book. It remained largely forgotten until (Hughes et al, 2004) showed that for stiffened panels it can be significant. In a stiffened panel most of the cross-sectional area is in the plating, and the sectional area of the web A_w is relatively small. Therefore the shear stress due to Q is much larger than in a typical column. This causes a larger shear deflection, which in turn means that the reduction in the Euler buckling stress is not negligible.

Let F_Q denote the factor by which the Euler buckling stress is reduced by the shear deflection caused by Q. That is

$$\sigma'_E = F_Q\,\sigma_E$$

Timoshenko showed that F_Q is given by

$$F_Q = \left(\frac{A_w G}{A_w\,G + A_T\,\sigma_E}\right) \qquad (14.6.1)$$

where A_w is the area of the web, A_T is the total sectional area (including the plating) and G is the shear modulus.

We can see whether this factor is significant by considering typical proportions of columns and stiffener-columns. For steel $G = E/2(1+\nu) = E/2.6$. The above factor becomes

$$\frac{1}{1 + \frac{A_T}{A_w}\frac{2.6\pi^2 \rho^2}{L^2}}$$

where ρ is the radius of gyration.

For an ordinary column and a stiffener-column A_T/A_w is typically 3 and 6 respectively. From eq. (3.12.1) for typical proportions ρ is $0.6d$ and $1.0d$ respectively, where d is the flange-to-flange web height (see Fig. 3.50). For both types of members d/L is typically 0.03. With these values the reduction factor becomes 0.96 and 0.88 respectively. Thus in an ordinary column the reduction is merely 4% whereas in a stiffener-column it is 12%. For the 55 panels studied in (Hughes et al, 2004) the reduction ranged from 2% to 29%.

14.6.2 Improved Expression for the Minimum Rigidity Ratio

For overall buckling of longitudinally stiffened panels of normal proportions the number of buckling half waves, m, is 1, and we will use that value. In (14.5.11) D_y becomes D and we replace the subscript "gp" by "overall". We can obtain a more concise form of (14.5.11) by dividing it by Π^2_{orth}, multiplying (14.5.10) by Π^2_{orth} and combining them. We also apply the reduction factor F_Q. The final result is

$$(\sigma_{ax})_{cr,\ overall} = \{1 + 2\eta\Pi^2_{\text{orth}} + \Pi^4_{\text{orth}}\}\ F_Q \frac{\pi^2 D_x}{L^2\left(t + \frac{A_x}{b}\right)} \qquad (14.6.2)$$

Equation 12.2.7 gave the elastic buckling stress for local plate buckling:

$$(\sigma_{ax})_{cr,\ overall} = k_x \frac{\pi^2 D}{b^2 t} \qquad (14.6.3)$$

The buckling coefficient k_x, given by (12.4.3), allows for the rotational restraint exerted by the stiffeners on the plating, in terms of a parameter ζ_L

$$\zeta_L = C_L \frac{GJ}{Db} \qquad (14.6.4)$$

This rotational restraint is reduced if there is local out-of-plane bending of the stiffener web, and the coefficient C_L, given by (12.4.2), allows for this.

We can now obtain the desired minimum rigidity ratio by equating the overall panel buckling stress of (14.6.2) to the local plate buckling stress of (14.6.3). Since we are using orthotropic theory for the former, we note that the rigidity ratio defined in (14.1.1) can also be expressed in terms of D_x and D

$$\gamma_x = \frac{D_x}{D} \qquad (14.6.5)$$

The result of equating (14.6.2) and (14.6.3) and solving for D_x/D is

$$(\gamma_x)_{\text{MIN}} = \left(\frac{L}{b}\right)^2 \frac{k_x(1+\delta_x)}{\{1 + 2\eta\Pi^2_{\text{orth}} + \Pi^4_{\text{orth}}\}\ F_Q} \qquad (14.6.6)$$

in which δ_x is the ratio of stiffener area to plate area, defined in (14.1.3).

14.6.3 Validation

In (Hughes et al, 2004) Table 3 compares values of $(\gamma_x)_{\text{MIN}}$ with ABAQUS eigenvalue results for 55 typical welded steel panels. It also gives the values from the Klitchieff equation at the end of Section 14.1.1 (page 14-3). The new expression has an average error of 4.4% and COV = 0.073, whereas the Klitchieff expression has an average error of 26.1% and COV = 0.148.

14.7 REFERENCES

Bleich, F. (1952). *Buckling strength of metal structures*. New York: McGraw-Hill, New York.

Chapman, J. C., and Slatford, J. E. (1959). Design of stiffened plating in compression. *The Engineer*, Vol. 207, 292–294.

Column Research Committee of Japan. (1971). *Handbook of structural stability*. Tokyo: Corona Publishing Co., Ltd.

Cox, H. L., and Riddel, J. R. (1949). Buckling of a longitudinally stiffened flat panel. *Aeronaut. Q.*, Vol. 1, 225–244.

Falconer, B. H., and Chapman, J. C. (1953). Compressive buckling of stiffened plates. *The Engineer*, Vol. 195, 789–791, 822–825.

Faulkner, D. (1973). *The overall compression buckling of partially constrained ship grillages*. M.I.T. Sea Grant Publication No. MITSG-73-10.

Harris, H. G., and Pifko, A. B. (1969). Elastic-plastic buckling of stiffened rectangular plates. *Proc. Symp. Appl. Finite Element Meth. Civil Eng.*, Vanderbilt University, November.

Hughes, O. F., Ghosh, B. and Chen, Y. (2004). Improved prediction of simultaneous local and overall buckling of stiffened panels. *Thin-Walled Structures*, Vol. 42, 827–856.

Johnston, B. G. (Ed.). (1976). *Guide to stability design criteria for metal structures* (3rd ed.). New York: Wiley.

Klitchieff, J. M. (1949). On the stability of plates reinforced by ribs. J. *Appl. Mech.*, Vol. 16, Issue 1, 74–76.

Klitchieff, J. M. (1951). On the stability of plates reinforced by longitudinal ribs. *J. Appl. Mech.*, Vol. 18, Issue 4, 364–366.

Kroll, W. D. (1943). Tables of stiffness and carry-over factor for flat rectangular plates under compression. *NACA Advanced Report*, No. 3K27.

Mansour, A. (1967). Ship bottom structure under uniform lateral and inplane loads. *Schiff und Hafen*, Vol. 5, Issue 19, 323–339.

Mansour, A. (1976). Charts for buckling and post-buckling analyses of stiffened plates under combined loading. *SNAME T & R Bull.*, 2–22.

Rockey, K. C., and Cook, I. T. (1962). Shear buckling of clamped and simply-supported infinitely long plates reinforced by transverse stiffeners. *Aeronaut. Q.*, Vol. 8, 41.

Schultz, H. G. (1962). Sum stabilitatsproblem elastisch eingespannter orthotroper platten. *Schiff und Hafen*, Vol. 14, Issue 617, 479–486, 569–576.

Seide, P. (1953). *The effect of longitudinal stiffeners located on one side of a plate on the compressive buckling stress of the plate-stiffener combination.* Washington, DC: NACA TN 2873.

Sharp, M. L. (1966). Longitudinal stiffeners for compression members. *ASCE J. Struct. Div.*, Vol. 92, ST5.

Smith, C. S. (1968). Bending, buckling and vibration of orthotropic plate-beam structures. *J. Ship. Res.*, Vol. 12, Issue 4, 249–268.

Stein, M., and Fralich, R. W. (1949). Critical shear stress of infinitely long, simply supported plate with transverse stiffeners. *NACA Tech. Note 1851.*

Timoshenko, S. P. (1936). *Theory of Elastic Stability.* New York: McGraw-Hill, New York.

Timoshenko, S. P., and Gere, J. M. (1961). *Theory of elastic stability* (2nd ed.). New York: McGraw-Hill.

von Karman, T., Sechler, E., and Donnell, L. H. (1932). The strength of thin plates in compression. *Trans. ASME*, Vol. 54, 53.

Wittrick, W. H. (1952). Correlation between some stability problems for orthotropic and isotropic plates under bi-axial and uni-axial direct stress. *Aeronaut. Q.*, Vol. 4, 83–92.

CHAPTER

FIFTEEN

LARGE DEFLECTION BEHAVIOR AND ULTIMATE STRENGTH OF STIFFENED PANELS

Jeom Kee Paik
Professor, Pusan National University
Busan, Korea

15.1 FUNDAMENTALS OF THE ULTIMATE STRENGTH BEHAVIOR OF STIFFENED PANELS

The overall failure of ship structures is mainly governed by the buckling and plastic collapse of the stiffened panels in the deck, bottom, and sometimes the side shell. Therefore, the accurate and efficient calculation of the collapse strength of stiffened panels is an important task in the design and safety assessment of ship structures.

A stiffened panel is an assembly of plate elements and support members (e.g., longitudinal stiffeners), as shown in Fig. 13.2 of Chapter 13. The interaction between the plate elements and support members in terms of their geometrical and material properties and other factors such as loading condition and initial imperfections plays an important role in the ultimate strength, buckling, and plastic collapse patterns of stiffened panels.

The possible collapse modes of a stiffened panel can be categorized into the following six types (Paik and Thayamballi, 2003).

1. Collapse mode I: Overall collapse of the plating and stiffeners as a unit; see Fig. 15.1a
2. Collapse mode II: Biaxial compressive collapse without failure of the stiffeners; see Fig. 15.1b
3. Collapse mode III: Beam-column type collapse; see Fig. 15.1c
4. Collapse mode IV: Local buckling of the stiffener web (after the inception of the buckling collapse of the plating between the stiffeners); see Fig. 15.1d
5. Collapse mode V: Flexural–torsional buckling or tripping of the stiffeners; see Fig. 15.1e
6. Collapse mode VI: Gross yielding

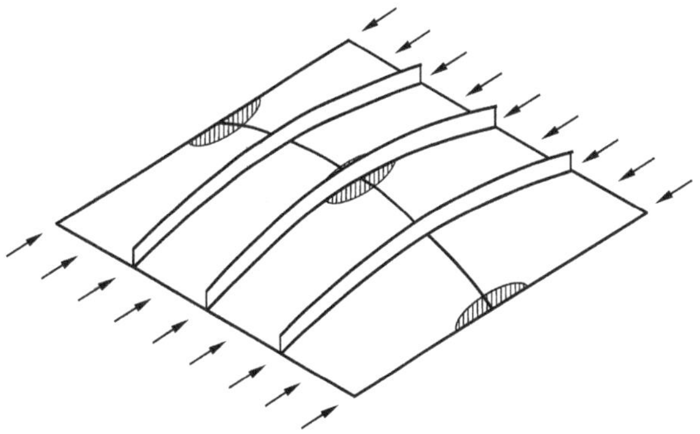

Figure 15.1a Collapse mode I: Overall collapse of the plating and stiffeners as a unit (shaded areas represent yielded regions).

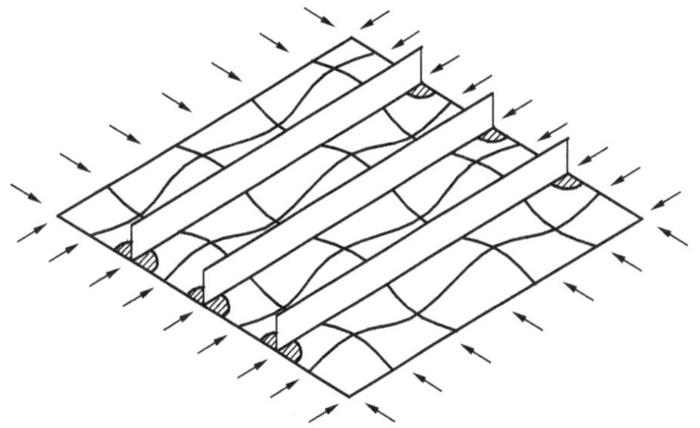

Figure 15.1b Collapse mode II: Biaxial compressive collapse (shaded areas represent yielded regions).

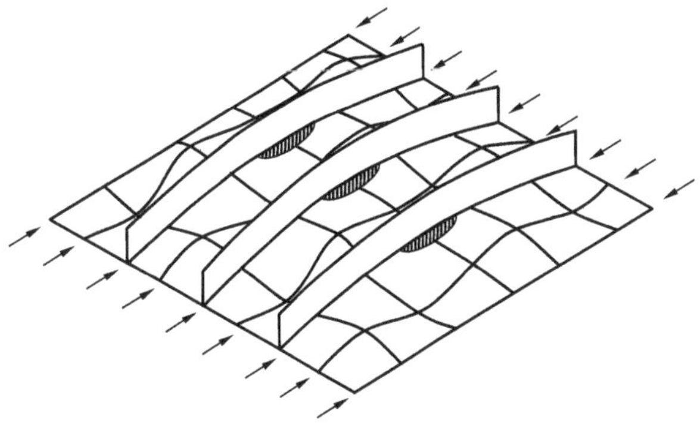

Figure 15.1c Collapse mode III: Beam-column type collapse (shaded areas represent yielded regions).

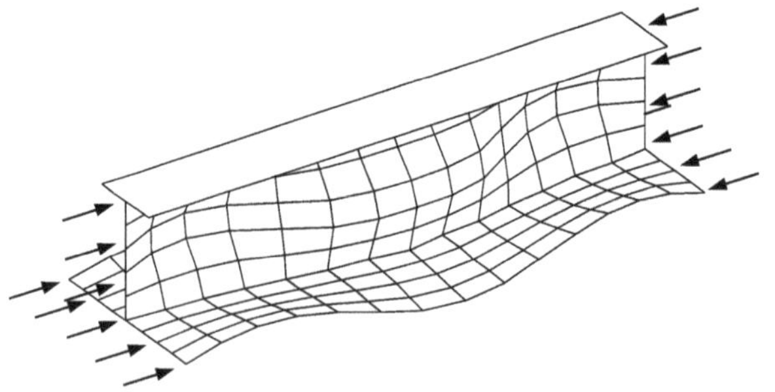

Figure 15.1d Collapse Mode IV: Local buckling of the stiffener web (after the buckling collapse of the plating between the stiffeners).

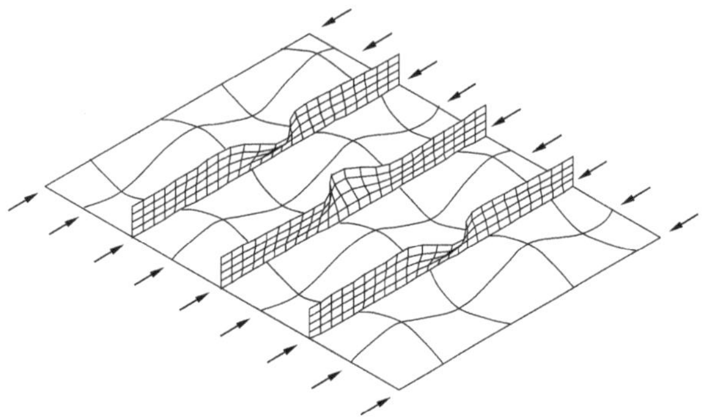

Figure 15.1e Collapse mode V: Flexural–torsional buckling of the stiffeners (after the buckling collapse of the plating between the stiffeners).

This classification of the collapse modes is applicable for any load combination, such as uniaxial compressive loads and combined in-plane loads with or without lateral pressure loads. Collapse mode I represents the overall collapse after overall buckling. In this mode, the stiffeners buckle with the plating as a unit, and overall buckling often occurs under an elastic regime. This collapse mode typically occurs when the stiffeners are relatively weak compared with the plating.

Collapse mode II occurs when the panel is predominantly subjected to biaxial compressive loads, causing the panel to collapse because of yielding along the plate–stiffener intersection at panel edges, with no distinct stiffener failure. This mode assumes that the stiffeners do not fail first in contrast to collapse modes III, IV, and V. Depending on the sturdiness of the plating, there are two possibilities:

1. If it is sturdy, then the plating between stiffeners will reach plasticity at its corners. During this process, the load in the plating is transferred to the stiffeners. At some point, the stiffeners may collapse by yielding or by column or torsional buckling, and these possibilities are covered by collapse modes III, IV, and V. Alternatively, the stiffeners may carry a small additional load before collapsing. For simplicity, this small additional load is often neglected.

2. If the plating is not sturdy and buckles at a relatively low load, then the panel strength comes mainly from the stiffeners, and this is covered by collapse modes III, IV, and V.

When the dimensions of the stiffeners are intermediate, that is, when they are neither weak nor strong, the stiffened panel is likely to behave as a plate–stiffener combination that is representative of the entire panel, and thus reaches ultimate strength by collapse mode III, which is a beam-column type collapse.

When the height to thickness ratio of the stiffener web is large, local buckling is likely to take place in the web. Collapse mode IV is the collapse pattern that occurs when the stiffener web buckles together with the inception of failure in the plating between the stiffeners.

When the stiffener flange is of a type that is unable to remain straight, the stiffeners twist sideways, which is a phenomenon termed flexural–torsional buckling or tripping. Collapse mode V represents the failure pattern in which the panel collapses because of the lateral–torsional buckling or tripping of the stiffeners.

The stiffened panel reaches ultimate strength in collapse mode VI when the panel is stocky or is predominantly subjected to axial tensile loading so that neither local nor overall buckling occurs until the panel cross-section yields either entirely or to a large extent.

Although these collapse modes are illustrated separately, some modes may interact and occur simultaneously. For the sake of simplicity, however, a stiffened panel will reach ultimate strength by the dominant collapse mode that occurs first among the six types of collapse patterns. Hence, the ultimate strengths of the panel are calculated separately for each of the six collapse patterns, and the smallest value among the computed strengths is taken as the real ultimate strength of the panel. This chapter presents the ultimate strength formulations of stiffened panels for each of the six types of collapse patterns.

15.2 BASIC IDEALIZATIONS

15.2.1 Geometric and Material Properties

Figure 15.2 shows a typical stiffened panel within a continuous-stiffened plate structure under a combined load. The length and breadth of the stiffened panel are denoted by a and B, respectively. The panel has identical stiffeners in terms of geometry and material, with the same spacing. The number of stiffeners is n_s, and thus the stiffener spacing is $b = B/(n_s + 1)$. The stiffeners are arranged in the longitu-

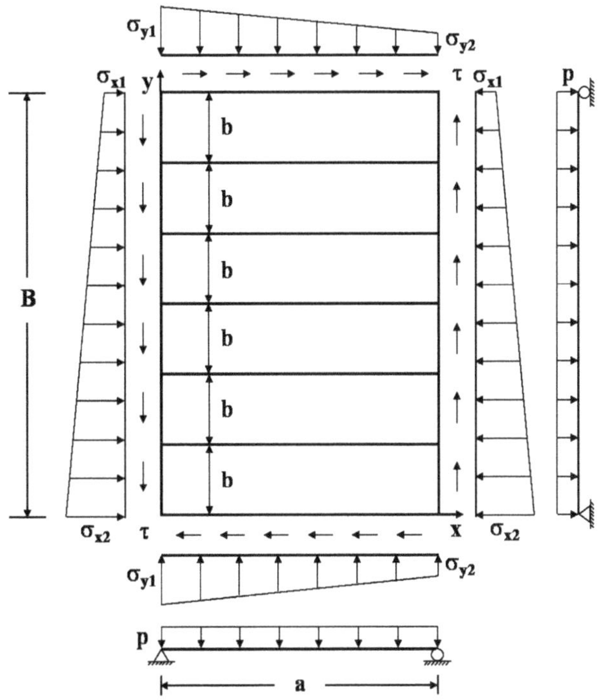

Figure 15.2 Stiffened panel under a combined in-plane and lateral pressure load.

dinal (x) direction and are attached to one side of the panel, that is, they are placed on the positive side in the z direction.

The thickness of the plating between the stiffeners is t. In a ship-stiffened panel, the thicknesses of the individual plates between the stiffeners are sometimes not the same. In this case, the plate thickness t used in the ultimate strength formulations of panels is approximated as follows.

$$t = \frac{b}{B}\sum_{i=1}^{n_s+1} t_i \tag{15.2.1}$$

where t_i = thickness of the ith plate.

Figure 15.3 gives the cross-sectional nomenclature for the stiffener geometry, in which the stiffener height h_w is defined by excluding the flange thickness t_f. The plating and stiffeners are of the same material, although their yield stresses may differ. The plating and stiffeners have the same value of either the Young's modulus or Poisson's ratio, defined by E and ν, respectively, and the elastic shear modulus is defined by $G = \dfrac{E}{2(1+\nu)}$.

The material yield stress is σ_{Yp} for the plating and σ_{Ys} for the stiffeners. The slenderness coefficient and the flexural (bending) rigidity of the plating between the stiffeners are given by $\beta = \dfrac{b}{t}\sqrt{\dfrac{\sigma_{Yp}}{E}}$ and $D = \dfrac{Et^3}{12(1-\nu^2)}$, respectively.

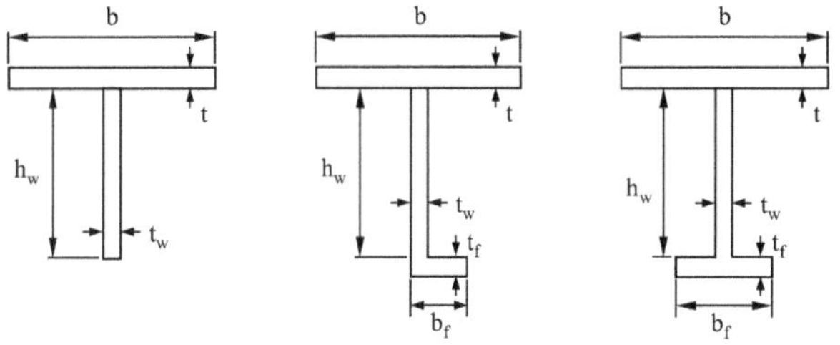

Figure 15.3 Typical cross-section types of the stiffeners and their nomenclature.

In ship-stiffened panels, the material yield stress of the plating is sometimes different from that of the stiffeners. For example, the plating may be made of mild steel, whereas the stiffeners are made of high tensile steel. In such cases, an equivalent yield stress σ_{Yeq} can be defined to represent the yield stress of the entire panel, as follows.

$$\sigma_{Yeq} = \frac{Bt\sigma_{Yp} + n_s(h_w t_w + b_f t_f)\sigma_{Ys}}{Bt + n_s(h_w t_w + b_f t_f)} \tag{15.2.2a}$$

Alternatively, when a single stiffener with its attached "plate flange" is being analyzed (e.g., for column buckling), the equivalent yield stress is defined as follows.

$$\sigma_{Yeq} = \frac{bt\sigma_{Yp} + (h_w t_w + b_f t_f)\sigma_{Ys}}{bt + h_w t_w + b_f t_f} \tag{15.2.2b}$$

15.2.2 Panel Edge Conditions

As shown in Fig. 13.2 of Chapter 13, the stiffened panels under consideration are surrounded by strong support members such as longitudinal girders and transverse frames, the bending rigidities of which are normally quite large compared to the rigidity of the panel itself. In a similar way to the boundary condition of plate elements located between longitudinal stiffeners and transverse frames described in Chapter 13, the rotational restraints along the panel edges depend on the relative values of the torsional rigidities of the support members to the flexural rigidity of the panel, and these are neither zero nor infinite.

For the sake of simplicity, however, it is often assumed that the stiffened panel edges are simply supported, with zero deflection and zero rotational restraints along the four edges and with all edges kept straight. In maritime engineering practice, this approximation is considered adequate. However, as shown in Chapter 12, in calculating the local buckling of either the plating between the stiffeners or the stiffener web, the influence of rotational restraints along the junctions of the plate and stiffener or stiffener web-flange may need to be accounted for. These are thus applicable for failure evaluation of collapse modes IV and V.

15.2.3 Loading Conditions

As shown in Fig. 15.2, the potential load components acting on a stiffened panel generally comprise the following six types.

1. Longitudinal axial load in the x direction
2. Transverse axial load in the y direction
3. Edge shear stress
4. Longitudinal in-plane bending moment in the x direction
5. Transverse in-plane bending moment in the y direction
6. Lateral pressure

To develop the ultimate strength formulations for the panel, this chapter simplifies some of the applied load components depending on the collapse mode. These are described in the following using the nomenclature of Fig. 15.2, where σ_{xav} is the average axial stress in the x direction, σ_{yav} is the average axial stress in the y direction, τ_{av} is the average edge shear stress, and p is the lateral pressure. Note that in the x direction, σ_{x2} is always the larger edge stress than σ_{x1}.

15.2.3.1 Collapse Modes I and VI

The effect of in-plane bending moments is neglected over the stiffened panel, and the following four load components are defined in this case.

$$\sigma_{xav} = \frac{\sigma_{x1} + \sigma_{x2}}{2}, \ \sigma_{yav} = \frac{\sigma_{y1} + \sigma_{y2}}{2}, \ \tau_{av}, p \qquad (15.2.3)$$

15.2.3.2 Collapse Mode II

The most highly stressed plating between the stiffeners is considered to determine the ultimate strength of the panel.

$$\sigma_{xav} = \sigma_{x2} - \frac{b}{2B}(\sigma_{x2} - \sigma_{x1}), \ \sigma_{yav} = \frac{\sigma_{y1} + \sigma_{y2}}{2}, \ \tau_{av}, p \qquad (15.2.4)$$

15.2.3.3 Collapse Modes III, IV, and V

The most highly stressed stiffener is considered to determine the ultimate strength of the panel.

$$\sigma_{xav} = \sigma_{x2} - \frac{b}{B}(\sigma_{x2} - \sigma_{x1}), \ \sigma_{yav} = \frac{\sigma_{y1} + \sigma_{y2}}{2}, \ \tau_{av}, p \qquad (15.2.5)$$

15.2.4 Initial Imperfections

The configuration of initial imperfections for the plate part is defined in the same way as described in Section 13.2.4 of Chapter 13 or Section 8.3.6 of Chapter 8. The initial distortions of the stiffeners are classified into two types depending on their deflected direction, that is, the z or y directions. The former corresponds to a column type of initial distortion in the direction of the stiffener height, and the latter type corresponds to a torsional initial distortion in which the stiffener flange is distorted sideways.

It is recognized that the welding-induced residual stresses in ship structures under operation in waves are well-released because of hull girder low cycle fatigue or cyclic loading such as hogging and sagging bending in conjunction with hull girder wave actions. In this regard, the effect of welding-induced residual stresses is sometimes overlooked either entirely or to some extent in the buckling and ultimate strength calculations of ship structures.

15.3 ULTIMATE STRENGTH FORMULATIONS FOR THE PANEL FOR COLLAPSE MODE I

When a stiffened panel reaches its ultimate strength by collapse mode I, it can be reasonably modeled as an orthotropic plate to analyze its large deflection behavior and ultimate strength if there are numerous small stiffeners. The orthotropic plate is considered to have an initial deflection, but the influence of welding-induced residual stresses is usually ignored because in a panel with numerous small stiffeners, the tensile and compressive residual stresses effectively cancel one another.

15.3.1 Nonlinear Governing Differential Equations for Orthotropic Plates

For the purpose of computing the ultimate strength of the panel associated with collapse mode I, the membrane stress-based method for plates described in Chapter 12 can be applied. The maximum and minimum membrane stresses can be obtained by solving the nonlinear governing differential equations for orthotropic plates, taking into account the effect of large deflection behavior.

In the orthotropic plate theory method, the stiffeners are in a sense smeared into the plating, given that they are relatively numerous and small and thus deflect together with the plating and remain stable through the range of orthotropic plate behavior.

The nonlinear governing differential equations for large deflection orthotropic plate theory comprise the equilibrium equation and the compatibility equation, as follows (Paik and Thayamballi, 2003).

$$D_x \frac{\partial^4 w}{\partial x^4} + 2H \frac{\partial^4 w}{\partial x^2 \partial y^2} + D_y \frac{\partial^4 w}{\partial y^4}$$

$$-t\left[\frac{\partial^2 F}{\partial y^2}\frac{\partial^2 (w+w_o)}{\partial x^2} - 2\frac{\partial^2 F}{\partial x \partial y}\frac{\partial^2 (w+w_o)}{\partial x\, \partial y}\right.$$

$$\left.+\frac{\partial^2 F}{\partial x^2}\frac{\partial^2 (w+w_o)}{\partial y^2} + \frac{p}{t}\right] = 0 \qquad (15.3.1a)$$

$$\frac{1}{E_y}\frac{\partial^4 F}{\partial x^4} + \left(\frac{1}{G_{xy}} - 2\frac{\nu_x}{E_x}\right)\frac{\partial^4 F}{\partial x^2 \partial y^2} + \frac{1}{E_x}\frac{\partial^4 F}{\partial y^4}$$

$$-\left[\left(\frac{\partial^2 w}{\partial x \partial y}\right)^2 - \frac{\partial^2 w}{\partial x^2}\frac{\partial^2 w}{\partial y^2} + 2\frac{\partial^2 w_o}{\partial x \partial y}\frac{\partial^2 w}{\partial x \partial y}\right.$$

$$\left.-\frac{\partial^2 w_o}{\partial x^2}\frac{\partial^2 w}{\partial y^2} - \frac{\partial^2 w}{\partial x^2}\frac{\partial^2 w_o}{\partial y^2}\right] = 0 \qquad (15.3.1b)$$

where w_o and w are the initial and added deflection functions, respectively, for the orthotropic plate and F is the stress function. E_x and E_y are the elastic moduli of the orthotropic plate in the x and y directions, respectively, and G_{xy} is the elastic shear modulus of the orthotropic plate, which is approximately defined as follows.

$$G_{xy} = \frac{E_x E_y}{E_x + \left(1 + 2\sqrt{\nu_x \nu_y}\right) E_y} \approx \frac{\sqrt{E_x E_y}}{2\left(1 + \sqrt{\nu_x \nu_y}\right)} \qquad (15.3.2)$$

D_x and D_y in equation (15.3.1) are the flexural rigidities of the orthotropic plate in the x and y directions, respectively. H is the effective torsional rigidity of the orthotropic plate. Once the stress function F and the added deflection w have been determined, the membrane stresses inside the orthotropic plate can be calculated using the following equations.

$$\sigma_x = \frac{\partial^2 F}{\partial y^2} - \frac{E_x z}{1 - \nu_x \nu_y}\left(\frac{\partial^2 w}{\partial x^2} + \nu_y \frac{\partial^2 w}{\partial y^2}\right) \qquad (15.3.3a)$$

$$\sigma_y = \frac{\partial^2 F}{\partial x^2} - \frac{E_y z}{1 - \nu_x \nu_y}\left(\frac{\partial^2 w}{\partial y^2} + \nu_x \frac{\partial^2 w}{\partial x^2}\right) \qquad (15.3.3b)$$

$$\tau = -\frac{\partial^2 F}{\partial x\, \partial y} - 2G_{xy} z \frac{\partial^2 w}{\partial x \partial y} \qquad (15.3.3c)$$

where σ_x and σ_y are the axial stresses in the x and y directions, respectively, τ is the shear stress, and z is the axis in the plate thickness direction with $z = 0$ at the midthickness.

The reliability of orthotropic plate analysis depends significantly on various elastic constants that must be determined when a stiffened panel is replaced by an equivalent orthotropic plate. The large deflection orthotropic plate theory constants developed by Paik et al. (2001) is useful in this respect, and are thus introduced here.

For an isotropic plate, there are two independent elastic constants: the Young's modulus E and Poisson's ratio ν. For an orthotropic plate, the four elastic constants E_x, E_y, ν_x, and ν_y, are required to describe the orthotropic stress–strain relationship.

In real stiffened panels, the anisotropy in the two mutually perpendicular directions arises from different geometric properties rather than different properties of the material, which are inherently isotropic. Because the stiffened panel considered in this chapter has stiffeners in the x direction only, the corresponding orthotropic constants of the elastic moduli can be approximately given by

$$E_x = E\left(1 + \frac{n_s A_s}{Bt}\right),\ E_y = E \qquad (15.3.4)$$

where $A_s = h_w t_w + b_f t_f$.

The flexural and torsional rigidities of the orthotropic plate can be approximately expressed by

$$D_x = \frac{Et^3}{12(1 - \nu_x \nu_y)} + \frac{Etz_o^2}{1 - \nu_x \nu_y} + \frac{EI}{b} \qquad (15.3.5a)$$

$$D_y = \frac{Et^3}{12(1 - \nu_x \nu_y)} \qquad (15.3.5b)$$

$$H = \frac{1}{2}\left(\nu_y D_x + \nu_x D_y + G_{xy}\frac{t^3}{3}\right) \qquad (15.3.5c)$$

where $I = \frac{t_w h_w^3}{12} + t_w h_w \left(\frac{h_w}{2} + \frac{t}{2} - z_o\right)^2 + \frac{b_f t_f^3}{12}$

$$+ b_f t_f \left(\frac{t_f}{2} + h_w + \frac{t}{2} - z_o\right)^2$$

and $z_o = \frac{0.5 h_w t_w (h_w + t) + b_f t_f (0.5 t_f + h_w + 0.5t)}{bt + h_w t_w + b_f t_f}$.

For an isotropic plate, the flexural rigidities will simplify to the following well-known expression.

$$D_x = D_y = H = D = \frac{Et^3}{12(1 - \nu^2)} \qquad (15.3.6)$$

To determine the various elastic constants indicated above, it is necessary to predefine the Poisson's ratios ν_x and ν_y that are results of orthotropy. These are not material properties, but rather elastic constants that correspond to the given geometrical configuration. Based on Betti's reciprocity theorem, the following two requirements are then pertinent.

$$\nu_x E_y = \nu_y E_x,\ \nu_x D_y = \nu_y D_x \qquad (15.3.7)$$

The substitution of equations (15.3.4), (15.3.5a), and (15.3.5b) into equation (15.3.7) gives

$$\frac{EI}{b}\frac{E_y}{E_x}\nu_x^3 - \left(\frac{Et^3}{12} + Etz_o^2 + \frac{EI}{b}\right)\nu_x = 0 \quad (15.3.8)$$

A non-zero solution to equation (13.5.8) together with equation (13.5.7) gives the effective Poisson's ratios in the x and y directions.

$$\nu_x = c\left[\frac{\frac{Et^3}{12} + Etz_o^2 + \frac{EI}{b}}{\frac{EI}{b}\left(\frac{E_y}{E_x}\right)}\right]^{0.5}, \quad \nu_y = \frac{D_y}{D_x}\nu_x = \frac{E_y}{E_x}\nu_x \quad (15.3.9)$$

where c is a correction factor to correlate the Poisson's ratios with $\nu_x = \nu_y = \nu$ for an isotropic plate that can be approximately taken as $c = \nu/0.86$.

15.3.2 Combined Longitudinal Axial Load and Lateral Pressure

The membrane stress-based method described in Chapter 13 can be used to calculate the ultimate strength of the panel σ_{xu} when the stiffened panel reaches the ultimate strength by collapse mode I. The maximum and minimum membrane stresses inside the orthotropic plate can be computed by solving the governing differential equations (15.3.1a) and (15.3.1b), which are then substituted into the three failure-conditional equations (13.4.18a), (13.4.18b), and (13.4.18c) by replacing σ_Y with σ_{Yeq} to give the ultimate strength formulations for the panel for collapse mode I.

To calculate the maximum and minimum membrane stresses by solving the governing differential equations (15.3.1a) and (15.3.1b), the initial and added deflection functions of the orthotropic plate are presumed by including the following single deflection component.

$$w_o = A_{om}\sin\frac{m\pi x}{a}\sin\frac{\pi y}{B} \quad (15.3.10a)$$

$$w = A_m\sin\frac{m\pi x}{a}\sin\frac{\pi y}{B} \quad (15.3.10b)$$

where it is assumed that one half-wave mode is dominant in the unloaded (y) direction. m is the half-wave number in buckling when σ_{xav} is compressive, although $m = 1$ is applied when σ_{xav} is predominantly tensile. A_{om} is the amplitude of the initial deflection of the buckling mode, and A_m is the unknown amplitude of the added deflection function.

Considering the basic idealizations of the stiffened panel described previously, the initial and added deflection functions of equations (15.3.10a) and (15.3.10b) are substituted into equation (15.3.1b) to give the following equation.

$$\frac{1}{E_y}\frac{\partial^4 F}{\partial x^4} + \left(\frac{1}{G_{xy}} - 2\frac{\nu_x}{E_x}\right)\frac{\partial^4 F}{\partial x^2\partial y^2} + \frac{1}{E_x}\frac{\partial^4 F}{\partial y^4} = \frac{m^2\pi^4}{2a^2B^2}A_m\left(A_m + 2A_{om}\right)\left(\cos\frac{2m\pi x}{a} + \cos\frac{2\pi y}{B}\right) \quad (15.3.11)$$

The applicable stress function is then obtained by solving equation (15.3.11) as follows.

$$F = \sigma_{xav}\frac{y^2}{2} + \frac{A_m\left(A_m + 2A_{om}\right)}{32}\left(E_y\frac{a^2}{m^2B^2}\cos\frac{2m\pi x}{a} + E_x\frac{m^2B^2}{a^2}\cos\frac{2\pi y}{B}\right) \quad (15.3.12)$$

The Galerkin method is applied to find the unknown amplitude A_m by substituting equations (15.3.10a), (15.3.10b), and (15.3.12) into equation (15.3.1a) as follows.

$$\int_0^a\int_0^B\left[D_x\frac{\partial^4 w}{\partial x^4} + 2H\frac{\partial^4 w}{\partial x^2\partial y^2} + D_y\frac{\partial^4 w}{\partial y^4} - t\left\{\frac{\partial^2 F}{\partial y^2}\frac{\partial^2(w+w_o)}{\partial x^2} - 2\frac{\partial^2 F}{\partial x\partial y}\frac{\partial^2(w+w_o)}{\partial x\partial y} + \frac{\partial^2 F}{\partial x^2}\frac{\partial^2(w+w_o)}{\partial y^2} + \frac{p}{t}\right\}\right]\sin\frac{m\pi x}{a}\sin\frac{\pi y}{B}dxdy = 0 \quad (15.3.13)$$

By performing the integration of equation (15.3.13) over the entire panel, the following third-order equation with respect to A_m is obtained.

$$C_1A_m^3 + C_2A_m^2 + C_3A_m + C_4 = 0 \quad (15.3.14)$$

where $C_1 = \dfrac{\pi^2}{16}\left(E_x\dfrac{m^4B}{a^3} + E_y\dfrac{a}{B^3}\right)$,

$$C_2 = \frac{3\pi^2A_{om}}{16}\left(E_x\frac{m^4B}{a^3} + E_y\frac{a}{B^3}\right),$$

$$C_3 = \frac{\pi^2A_{om}^2}{8}\left(E_x\frac{m^4B}{a^3} + E_y\frac{a}{B^3}\right) + \frac{m^2B}{a}\sigma_{xav} + \frac{\pi^2}{t}\left(D_x\frac{m^4B}{a^3} + 2H\frac{m^2}{aB} + D_y\frac{a}{B^3}\right)$$

and $C_4 = A_{om}\dfrac{m^2B}{a}\sigma_{xav} - \dfrac{16aB}{\pi^4 t}p$.

It is noted that the interaction between the axial compressive load and lateral pressure is approximately taken into account, as indicated in equation (13.5.12) of Chapter 13. The solution to equation (15.3.14) with regard to A_m can be obtained by the Cardano method described in Section 13.4.2 of Chapter 13. Once A_m and the related stress function

F have been identified, the maximum and minimum membrane stresses can be computed from equation (15.3.3) at the midthickness ($z = 0$) as follows.

$$\sigma_{x\max} = \sigma_x\Big|_{x=0,\,y=0} = \sigma_{xav} - \frac{m^2\pi^2 E_x A_m (A_m + 2A_{om})}{8a^2} \tag{15.3.15a}$$

$$\sigma_{x\min} = \sigma_y\Big|_{x=0,\,y=B/2} = \sigma_{xav} + \frac{m^2\pi^2 E_x A_m (A_m + 2A_{om})}{8a^2} \tag{15.3.15b}$$

$$\sigma_{y\max} = \sigma_y\Big|_{x=0,\,y=0} = -\frac{\pi^2 E_y A_m (A_m + 2A_{om})}{8B^2} \tag{15.3.15c}$$

$$\sigma_{y\min} = \sigma_y\Big|_{x=a/2,\,B=0} = -\frac{\pi^2 E_y A_m (A_m + 2A_{om})}{8B^2} \tag{15.3.15d}$$

As the applied load increases, the panel will collapse when any one of the following three failure-conditional equations is satisfied.

$$\sigma_{x\max}^2 - \sigma_{x\max}\sigma_{y\max} + \sigma_{y\max}^2 = \sigma_{Yeq}^2 \tag{15.3.16a}$$

$$\sigma_{x\max}^2 - \sigma_{x\max}\sigma_{y\min} + \sigma_{y\min}^2 = \sigma_{Yeq}^2 \tag{15.3.16b}$$

$$\sigma_{x\min}^2 - \sigma_{x\min}\sigma_{y\max} + \sigma_{y\max}^2 = \sigma_{Yeq}^2 \tag{15.3.16c}$$

In this formulation, the buckling half-wave number m remains undetermined. A clear buckling phenomenon occurs only in a perfect plate, that is, one without initial deflections induced by welding or lateral pressure loads. In this regard, equation (15.3.14) is simplified for a perfect plate taking $A_{om} = 0$ and $p = 0$ into consideration, as follows.

$$C_1 A_m^3 + C_3 A_m = 0 \tag{15.3.17}$$

where $C_1 = \dfrac{\pi^2}{16}\left(E_x \dfrac{m^4 B}{a^3} + E_y \dfrac{a}{B^3}\right)$ and

$$C_3 = \frac{m^2 B}{a}\sigma_{xav} + \frac{\pi^2}{t}\left(D_x \frac{m^4 B}{a^3} + 2H\frac{m^2}{aB} + D_y \frac{a}{B^3}\right).$$

A non-zero solution to equation (15.3.17) with regard to A_m is given by

$$A_m = \sqrt{-\frac{C_3}{C_1}} \tag{15.3.18}$$

The deflection amplitude A_m must be zero for a perfect plate until immediately before buckling takes place, although it will rapidly increase as the axial shortening increases immediately after buckling. This means that the following equation must be available at the instant of buckling.

$$A_m = \sqrt{-\frac{C_3}{C_1}} = 0 \quad \text{or} \quad C_3 = 0 \tag{15.3.19}$$

Equation (15.3.19) corresponds to the buckling condition of the orthotropic plate under a longitudinal axial compressive load. From the definition of C_3, this condition is written as follows.

$$\frac{m^2 B}{a}\sigma_{xav} + \frac{\pi^2}{t}\left(D_x \frac{m^4 B}{a^3} + 2H\frac{m^2}{aB} + D_y \frac{a}{B^3}\right) = 0 \tag{15.3.20}$$

Denoting σ_{xav} with σ_{xEO} when equation (15.3.20) is satisfied gives the elastic overall buckling stress σ_{xEO} of a stiffened panel under σ_{xav} in compression.

$$\sigma_{xEO} = -\frac{\pi^2}{t}\left(D_x \frac{m^2}{a^2} + 2H\frac{1}{B^2} + D_y \frac{a^2}{m^2 B^4}\right) \tag{15.3.21}$$

The buckling load must be identical at the transition of buckling half-waves, and thus the buckling half-wave number m can be determined as a minimum integer that satisfies the following condition.

$$D_x \frac{m^2}{a^2} + 2H\frac{1}{B^2} + D_y \frac{a^2}{m^2 B^4} \le D_x \frac{(m+1)^2}{a^2} + 2H\frac{1}{B^2} + D_y \frac{a^2}{(m+1)^2 B^4} \tag{15.3.22a}$$

or more simply

$$\frac{a}{B} \le \sqrt[4]{\frac{D_x}{D_y} m^2 (m+1)^2} \tag{15.3.22b}$$

It is evident from equation (15.3.22b) that the buckling mode depends on both the plate aspect ratio and the structural orthotropy. For an isotropic plate under σ_{xav} in compression, equation (15.3.22b) simplifies to the well-known condition corresponding to equation (13.5.6c) of Chapter 13 because $D_x = D_y = D$, which depends on the plate aspect ratio only.

$$\frac{a}{B} \le \sqrt{m(m+1)} \tag{15.3.22c}$$

15.3.3 Combined Transverse Axial Load and Lateral Pressure

The membrane stress-based method is also applicable to calculate the transverse ultimate strength of the panel σ_{yu}. Under a transverse axial load σ_{yav} and lateral pressure load p, the initial and added deflection functions of the orthotropic plate can be assumed with a single deflection component as follows.

$$w_o = A_{on}\sin\frac{\pi x}{a}\sin\frac{n\pi y}{B} \qquad (15.3.23a)$$

$$w = A_n\sin\frac{\pi x}{a}\sin\frac{n\pi y}{B} \qquad (15.3.23b)$$

where A_{on} is the initial deflection amplitude, A_n is the unknown amplitude of the added deflection function, and n is the buckling half-wave number in the y direction.

A similar procedure to that given in Section 15.3.2 can be applied until the following third-order equation with regard to A_n is obtained.

$$C_1A_n^3 + C_2A_n^2 + C_3A_n + C_4 = 0 \qquad (15.3.24)$$

where $C_1 = \frac{\pi^2}{16}\left(E_x\frac{B}{a^3} + E_y\frac{n^4a}{B^3}\right)$,

$$C_2 = \frac{3\pi^2A_{on}}{16}\left(E_x\frac{B}{a^3} + E_y\frac{n^4a}{B^3}\right),$$

$$C_3 = \frac{\pi^2A_{on}^2}{8}\left(E_x\frac{B}{a^3} + E_y\frac{n^4a}{B^3}\right) + \frac{n^2a}{B}\sigma_{yav}$$

$$+\frac{\pi^2}{t}\left(D_x\frac{B}{a^3} + 2H\frac{n^2}{aB} + D_y\frac{n^4a}{B^3}\right), \text{ and}$$

$$C_4 = A_{on}\frac{n^2a}{B}\sigma_{yav} - \frac{16aB}{\pi^4t}p.$$

The Cardano method is used to solve equation (15.3.24). The maximum and minimum membrane stresses can then be obtained from equation (15.3.3) at the midthickness ($z = 0$), as follows.

$$\sigma_{x\max} = \sigma_x\big|_{x=0,y=0} = -\frac{\pi^2E_xA_n(A_n + 2A_{on})}{8a^2} \qquad (15.3.25a)$$

$$\sigma_{x\min} = \sigma_x\big|_{x=0,y=B/2} = \frac{\pi^2E_xA_n(A_n + 2A_{on})}{8L^2} \qquad (15.3.25b)$$

$$\sigma_{y\max} = \sigma_y\big|_{x=0,y=0} = \sigma_{yav} - \frac{n^2\pi^2E_yA_n(A_n + 2A_{on})}{8B^2} \qquad (15.3.25c)$$

$$\sigma_{y\min} = \sigma_y\big|_{x=a/2,y=0} = \sigma_{yav} + \frac{n^2\pi^2E_yA_n(A_n + 2A_{on})}{8B^2} \qquad (15.3.25d)$$

The three failure-conditional equations (15.3.16a), (15.3.16b), and (15.3.16c) are employed to check the ultimate strength of an orthotropic plate under σ_{yav} and p, but using the maximum and minimum stresses defined in equation (15.3.25).

The buckling half-wave number n, which has been left undetermined, is now calculated in the following. For a perfect plate under σ_{yav} alone, $A_{om} = 0$ and $p = 0$ must be taken into account in equation (15.3.24). Thus, equation (15.3.24) becomes

$$A_n\left(C_1A_n^2 + C_3\right) = 0 \qquad (15.3.26)$$

where $C_1 = \frac{\pi^2}{16}\left(E_x\frac{B}{a^3} + E_y\frac{n^4a}{B^3}\right)$ and

$$C_3 = \frac{n^2a}{B}\sigma_{yav} + \frac{\pi^2}{t}\left(D_x\frac{B}{a^3} + 2H\frac{n^2}{aB} + D_y\frac{n^4a}{B^3}\right).$$

Immediately before buckling, the plate deflection must still be zero. Hence, a non-zero solution to equation (15.3.26) gives the plate buckling condition as follows.

$$A_n = \sqrt{-\frac{C_3}{C_1}} = 0 \quad \text{or} \quad C_3 = 0 \qquad (15.3.27a)$$

Equation (15.3.27a) can then be rewritten as follows.

$$\frac{n^2a}{B}\sigma_{yav} + \frac{\pi^2}{t}\left(D_x\frac{B}{a^3} + 2H\frac{n^2}{aB} + D_y\frac{n^4a}{B^3}\right) = 0 \qquad (15.3.27b)$$

Using σ_{yEO} to denote the value of σ_{yav} when equation (15.3.27b) is satisfied, the elastic overall buckling stress under axial compression in the y direction is

$$\sigma_{yEO} = -\frac{\pi^2}{t}\left(D_x\frac{B^2}{n^2a^4} + 2H\frac{1}{a^2} + D_y\frac{n^2}{B^2}\right) \qquad (15.3.28)$$

In a similar way to equation (15.3.22), the buckling half-wave number n is defined as a minimum integer that satisfies the following condition.

$$D_x\frac{B^2}{n^2a^4} + 2H\frac{1}{a^2} + D_y\frac{n^2}{B^2} \le D_x\frac{B^2}{(n+1)^2a^4} + 2H\frac{1}{a^2} + D_y\frac{(n+1)^2}{B^2} \qquad (15.3.29a)$$

or more simply

$$\frac{B}{a} \le \sqrt[4]{\frac{D_y}{D_x}n^2(n+1)^2} \qquad (15.3.29b)$$

Again, the buckling half-wave number of the orthotropic plate depends on both the aspect ratio and the structural orthotropy. However, for an isotropic plate, the buckling half-wave number is determined only by the plate aspect ratio because $D_x = D_y = D$.

$$\frac{B}{a} \le \sqrt{n(n+1)} \qquad (15.3.29c)$$

15.3.4 Combined Edge Shear and Lateral Pressure

Empirical formulations for the ultimate strength τ_u of an orthotropic plate are used for the edge shear loading condition because it is not straightforward enough to analytically identify the membrane stress distribution inside a plate buckled predominantly by edge shear. However, the membrane stress-based method can be applied to calculate the ultimate strength of an orthotropic plate under lateral pressure. An empirical formulation can then be used to determine the ultimate strength relationship between the combined edge shear and lateral pressure loads.

15.3.4.1 Edge Shear

Bleich (1952) suggested the following elastic shear buckling stress formula for the stiffened panel shown in Fig. 15.2, which is available for $1 \le \alpha \le 5$.

$$\tau_E = k_\tau \frac{\pi^2 E}{12(1-\nu)^2}\left(\frac{t}{a}\right)^2 \qquad (15.3.30)$$

where $\alpha = \frac{a}{b}$, $\gamma = \frac{EI}{bD}$, $D = \frac{Et^3}{12(1-\nu^2)}$,

$$I = \frac{bt^3}{12} + bt\left(z_o - \frac{t}{2}\right)^2 + \frac{h_w^3 t_2}{12} + h_w t_w\left(z_o - t - \frac{h_w}{2}\right)^2 + \frac{b_f t_f^3}{12} + b_f t_f\left(t + h_w + \frac{t_f}{2} - z_o\right)^2,$$

$$z_o \frac{0.5bt^2 + h_w t_w(t + 0.5h_w) + b_f t_f(t + h_w + 0.5t_f)}{bt + h_w t_w + b_f t_f},$$

and
$$k_\tau = \begin{cases} 5.34 + (5.5\alpha^2 - 0.6)\sqrt[3]{\dfrac{\gamma}{4(7\alpha^2-5)}} \\ \quad \text{for } 0 \le \gamma \le 4(7\alpha^2 - 5) \\ 4.74 + 5.5\alpha^2 \text{ for } \gamma > 4(7\alpha^2 - 5) \end{cases}.$$

Once the elastic shear buckling stress has been determined, the following equation can be used to calculate the ultimate strength of the stiffened panel subject to edge shear alone (Paik and Thayamballi, 2003).

$$\frac{\tau_u}{\tau_{Yeq}} = \begin{cases} 1.324\left(\dfrac{\tau_E}{\tau_{Yeq}}\right) \text{ for } 0 < \dfrac{\tau_E}{\tau_{Yeq}} \le 0.5 \\ 0.039\left(\dfrac{\tau_E}{\tau_{Yeq}}\right)^3 - 0.274\left(\dfrac{\tau_E}{\tau_{Yeq}}\right)^2 + 0.676\left(\dfrac{\tau_E}{\tau_{Yeq}}\right) \\ +0.388 \text{ for } 0.5 < \dfrac{\tau_E}{\tau_{Yeq}} \le 2.0 \\ 0.956 \text{ for } \dfrac{\tau_E}{\tau_{Yeq}} > 2.0 \end{cases} \qquad (15.3.31a)$$

where τ_u is the ultimate shear stress and $\tau_{Yeq} = \frac{\sigma_{Yeq}}{\sqrt{3}}$.

Alternatively, the Johnson-Ostenfeld formula can be applied to calculate the pseudoultimate strength of the panel.

$$\frac{\tau_u}{\tau_{Yeq}} = \begin{cases} \dfrac{\tau_E}{\tau_{Yeq}} \text{ for } \dfrac{\tau_E}{\tau_{Yeq}} \le 0.5 \\ 1 - \dfrac{\tau_{Yeq}}{4\tau_E} \text{ for } \dfrac{\tau_E}{\tau_{Yeq}} > 0.5 \end{cases} \qquad (15.3.31b)$$

A comparison between the Paik-Thayamballi formula and the Johnson-Ostenfeld formula is given in Fig. 13.9 of Chapter 13, while the former is recommended to be used.

15.3.4.2 Lateral Pressure

The membrane stress-based method can be used to calculate the ultimate strength of an orthotropic plate under lateral pressure. In this case, the plate deflection amplitude can be determined as the solution to the following equation from solution (15.3.14), but taking $m = 1$ and $\sigma_{xav} = 0$ into consideration.

$$C_1 A_1^3 + C_2 A_1^2 + C_3 A_1 + C_4 = 0 \qquad (15.3.32)$$

where $C_1 = \frac{\pi^2}{16}\left(E_x \frac{B}{a^3} + E_y \frac{a}{B^3}\right)$,

$$C_2 = \frac{3\pi^2 A_{o1}}{16}\left(E_x \frac{B}{a^3} + E_y \frac{a}{B^3}\right),$$

$$C_3 = \frac{\pi^2 A_{o1}^2}{8}\left(E_x \frac{B}{a^3} + E_y \frac{a}{B^3}\right) + \frac{\pi^2}{t}\left(D_x \frac{B}{a^3} + 2H\frac{1}{aB} + D_y \frac{a}{B^3}\right),$$

and $C_4 = -\frac{16aB}{\pi^4 t} p$.

The maximum and minimum membrane stresses inside the orthotropic plate can then be obtained as follows.

$$\sigma_{x\max} = \sigma_x\big|_{x=0,y=0} = -\frac{\pi^2 E_x A_1(A_1 + 2A_{o1})}{8a^2} \qquad (15.3.33a)$$

$$\sigma_{x\min} = \sigma_y\big|_{x=0,y=B/2} = \frac{\pi^2 E_x A_1(A_1 + 2A_{o1})}{8a^2} \qquad (15.3.33b)$$

$$\sigma_{y\max} = \sigma_y\big|_{x=0,y=0} = -\frac{\pi^2 E_y A_1(A_1 + 2A_{o1})}{8B^2} \qquad (15.3.33c)$$

$$\sigma_{y\min} = \sigma_y\big|_{x=a/2,B=0} = \frac{\pi^2 E_y A_1(A_1 + 2A_{o1})}{8B^2} \qquad (15.3.33d)$$

The maximum and minimum membrane stresses calculated from equation (15.3.33) are substituted into the three failure-conditional equations (15.3.16a), (15.3.16b), and (15.3.16c), and the ultimate lateral pressure load is determined as the lowest value of the three solutions.

15.3.4.3 Interaction Relationship Between Edge Shear and Lateral Pressure

The same expression of the ultimate strength interaction relationship for an isotropic plate described in Section 13.6.3 of Chapter 13 can be applied as follows (Paik and Thayamballi, 2003).

$$\left(\frac{\tau_{av}}{\tau_{uo}}\right)^{1.5} + \left(\frac{p}{p_{uo}}\right)^{1.2} = 1 \qquad (15.3.34)$$

where τ_{uo} is the ultimate strength of the orthotropic plate subject to edge shear alone as defined in equation (15.3.31), and p_{uo} is the ultimate strength of the orthotropic plate subject to lateral pressure load alone as defined by the membrane stress-based method already described.

The ultimate shear strength of the panel taking account of lateral pressure as a secondary load component can then be derived from equation (15.3.34) as follows.

$$\tau_u = \tau_{uo}\left[1 - \left(\frac{p}{p_{uo}}\right)^{1.2}\right]^{\frac{1}{1.5}} \qquad (15.3.35)$$

15.3.5 Combined Biaxial Load, Edge Shear, and Lateral Pressure

Under the combination of σ_{xav}, σ_{yav}, τ_{uo}, and p defined in Section 15.2.3, the following ultimate strength interaction equation is applicable for the collapse mode I failure of a stiffened panel using the collapse mode I ultimate strength components obtained thus far.

$$\left(\frac{\sigma_{xav}}{\sigma^I_{xu}}\right)^{c_1} - \kappa\left(\frac{\sigma_{xav}}{\sigma^I_{xu}}\right)\left(\frac{\sigma_{yav}}{\sigma^I_{yu}}\right) + \left(\frac{\sigma_{yav}}{\sigma^I_{yu}}\right)^{c_2} + \left(\frac{\tau_{av}}{\tau^I_u}\right)^{c_3} = 1 \qquad (15.3.36)$$

where κ is an interaction constant that can be taken as $\kappa = 0$ when both σ_{xav} and σ_{yav} are compressive (negative) and $\kappa = 1$ when either σ_{xav}, σ_{yav}, or both are tensile (positive). The coefficients $c_1 \sim c_3$ are often taken as $c_1 = c_2 = c_3 = 2$ following the envelope of the von Mises yield condition (Paik and Thayamballi, 2003).

The superscript I in equation (15.3.36) denotes collapse mode I failure. The ultimate strengths of the individual load components in equation (15.3.36) are calculated by taking into account the effect of lateral pressure loads, although the lateral pressure p is not dealt with explicitly as a parameter in the equation.

15.4 ULTIMATE STRENGTH FORMULATIONS FOR THE PANEL FOR COLLAPSE MODE II

Collapse mode II is solely associated with the plating between the stiffeners, and specifically with the most highly stressed plating. The stiffened panel reaches ultimate strength by collapse mode II if the most highly stressed plating between the stiffeners, as defined in Section 15.2.3, has plasticity at its corners.

15.4.1 Combined Longitudinal Axial Load and Lateral Pressure

In this case, the ultimate compressive strength σ_{xu} is calculated by the membrane stress-based method for the most highly stressed plating between the stiffeners. The panel collapses if the following equation is satisfied in association with the plasticity at the corners of the plate between the stiffeners.

$$\sigma^2_{x\max} - \sigma_{x\max}\sigma_{y\max} + \sigma^2_{y\max} = \sigma^2_{Yp} \qquad (15.4.1)$$

where $\sigma_{x\max}$ and $\sigma_{y\max}$ are the maximum membrane stresses in the x and y directions, respectively, under σ_{xav} and p, as defined in Section 15.2.3. The maximum and minimum membrane stresses are obtained from equation (13.5.9) of Chapter 13 as follows.

$$\sigma_{x\max} = \sigma_{xav} + \sigma_{rtx} - \frac{E\pi^2 m^2 A_{mn}(A_{mn} + 2A_{omn})}{8a^2}\cos\frac{2n\pi b_t}{b} \qquad (15.4.2a)$$

$$\sigma_{y\max} = \sigma_{rty} - \frac{E\pi^2 n^2 A_{mn}(A_{mn} + 2A_{omn})}{8b^2}\cos\frac{2m\pi a_t}{a} \qquad (15.4.2b)$$

where the nomenclature is as defined in Section 13.5.1 of Chapter 13. A_m is the solution to equation (13.5.8) under σ_{xav} taking $\sigma_{yav} = 0$ into consideration, and C_3 must be defined by equation (13.5.12) to take into account the effect of the lateral pressure p.

15.4.2 Combined Transverse Axial Load and Lateral Pressure

The membrane stress-based method can also be applied to calculate the ultimate compressive strength σ_{yu}. Equation (15.4.1) is employed to check the collapse mode II failure under σ_{yav} and p as defined in Section 15.2.3 using the following maximum membrane stresses obtained from equation (13.5.9) of Chapter 13.

$$\sigma_{x\max} = \sigma_{rtx} - \frac{E\pi^2 m^2 A_{mn}(A_{mn} + 2A_{omn})}{8a^2}\cos\frac{2n\pi b_t}{b} \qquad (15.4.3a)$$

$$\sigma_{y\max} = \sigma_{yav} + \sigma_{rty} - \frac{E\pi^2 n^2 A_{mn}\left(A_{mn} + 2A_{omn}\right)}{8b^2}\cos\frac{2m\pi a_t}{a} \qquad (15.4.3b)$$

where the nomenclature is as defined in Section 13.5.1 of Chapter 13. A_m is the solution to equation (13.5.8) under σ_{yav} taking $\sigma_{xav} = 0$ into consideration, and C_3 must be defined by equation (13.5.12) to take into account the effect of the lateral pressure p.

15.4.3 Combined Edge Shear and Lateral Pressure

Equation (15.3.34) can be applied to the interaction between the edge shear and the lateral pressure, and the ultimate strength of the plating subject to edge shear alone can be obtained from equation (13.4.25) of Chapter 13. The ultimate strength of the plating subject to lateral pressure alone is determined as the solution to the failure-conditional equation (15.4.1), which corresponds to the condition of plasticity at the corners of plating, but using $\sigma_{x\max}$ and $\sigma_{y\max}$ obtained from equations (13.5.4a) and (13.5.4c).

15.4.4 Combined Biaxial Load, Edge Shear, and Lateral Pressure

Under the combination of σ_{xav}, σ_{yav}, τ_{av}, and p, as defined in Section 15.2.3, the following ultimate strength interaction equation can be applied to the collapse mode II failure of a stiffened panel, using the ultimate strength components for collapse mode II obtained thus far.

$$\left(\frac{\sigma_{xav}}{\sigma_{xu}^{II}}\right)^{c_1} - \kappa\left(\frac{\sigma_{xav}}{\sigma_{xu}^{II}}\right)\left(\frac{\sigma_{yav}}{\sigma_{yu}^{II}}\right) + \left(\frac{\sigma_{yav}}{\sigma_{yu}^{II}}\right)^{c_2} + \left(\frac{\tau_{av}}{\tau_u^{II}}\right)^{c_3} = 1 \qquad (15.4.4)$$

where the superscript *II* denotes collapse mode II failure, and the ultimate strengths of the individual load components have been calculated for collapse mode II failure taking into account the effect of lateral pressure. The coefficients κ, c_1, c_2, and c_3 are as defined in equation (15.3.36).

15.5 ULTIMATE STRENGTH FORMULATIONS FOR THE PANEL FOR COLLAPSE MODE III

To calculate the ultimate strength of a stiffened panel, the plate–stiffener combination is used to represent the entire stiffened panel, as illustrated in Fig. 15.4. The stiffened panel reaches ultimate strength by collapse mode III if the most highly stressed stiffener together with the attached plating collapses as a beam-column. This section presents the ultimate strength formulations

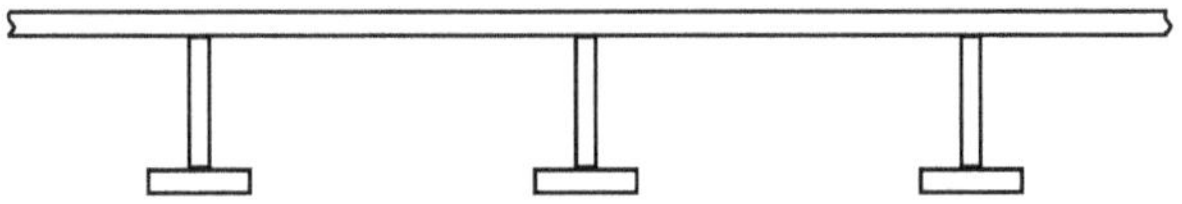

(a) Continuous stiffened plate structure

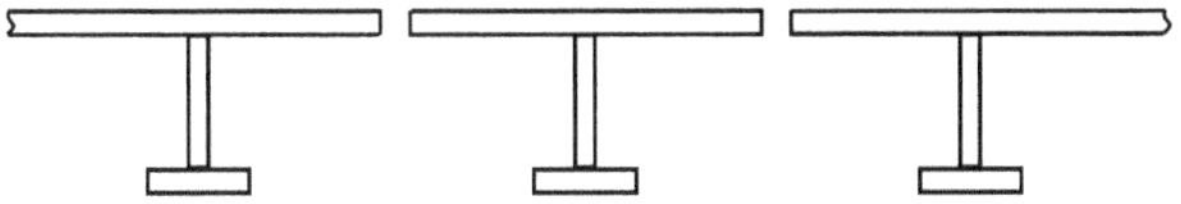

(b) Plate–stiffener combination model

Figure 15.4 Plate–stiffener combination model representing the entire stiffened panel.

for collapse mode III failure under combined in-plane and lateral pressure loads, as defined in Section 15.2.3.

15.5.1 Combined Longitudinal Axial Load and Lateral Pressure

The ultimate compressive strength σ_{xu} of the stiffened panel for collapse mode III failure is calculated for a plate–stiffener combination, which represents the entire panel.

15.5.1.1 Axial Compression

Three methods to calculate the ultimate strength of the stiffened panel under axial compression alone are introduced: the Johnson-Ostenfeld formula method, the Perry-Robertson formula method, and the Paik-Thayamballi empirical formula method. Each of these three methods has its own unique advantage in terms of practical applications, while the software ALPS/ULSAP (2009) developed by the theory described in this chapter employs the Perry-Robertson formula method for evaluating collapse mode III failure.

THE JOHNSON-OSTENFELD FORMULA METHOD

A pseudoultimate strength σ_{xu} can be obtained by the Johnson-Ostenfeld formula as a plasticity correction of the elastic buckling stress, as follows.

$$\sigma_{xu} = \begin{cases} \sigma_{xE} & \text{for } \sigma_{xE} \le 0.5\sigma_{Yeq} \\ \sigma_{Yeq}\left(1 - \dfrac{\sigma_{Yeq}}{4\sigma_{xE}}\right) & \text{for } \sigma_{xE} > 0.5\sigma_{Yeq} \end{cases} \qquad (15.5.1)$$

where σ_{xE} is the elastic buckling stress and σ_{Yeq} is the equivalent yield stress as defined in equation (15.2.2b).

In equation (15.5.1), the elastic buckling stress can be obtained for the plate–stiffener combination as a column simply supported at both ends, as follows.

$$\sigma_{xE} = \frac{\pi^2 EI_e}{a^2 A_s} \qquad (15.5.2)$$

where $A_s = bt + h_w t_w + b_f t_f$,

$$z_o = \frac{0.5 b_e t^2 + h_w t_w (t + 0.5 h_w) + b_f t_f (t + h_w + 0.5 t_f)}{b_e t + h_w t_w + b_f t_f},$$

$$I_e = \frac{b_e t^3}{12} + b_e t\left(z_o - \frac{t}{2}\right)^2 + \frac{h_w^3 t_w}{12} + h_w t_w \left(z_o - t - \frac{h_w}{2}\right)^2 + \frac{b_f t_f^3}{12} + b_f t_f \left(t + h_w + \frac{t_f}{2} - z_o\right)^2$$ is the effective moment of inertia, and b_e is the effective width of the attached plating at the ultimate limit state.

It is noted that the effective width of the attaching plating does not take a constant value, but rather it is a function of applied loads as defined by equation (13.4.19a). The plate effective width at the ultimate limit state can be obtained by equation (13.4.20a), taking into account the effect of initial imperfections, or alternatively by equation (13.4.21) for the sake of simplicity. The software ALPS/ULSAP (2009) employs equation (13.4.19) to determine the plate effective width at the ultimate limit state.

THE PERRY-ROBERTSON FORMULA METHOD

As described in Chapter 11 and in Paik and Thayamballi (2003), this method considers that a column collapses when the maximum compressive stress at the outermost fiber of the column cross-section reaches the yield stress. Here, it is being used for a plate–stiffener combination in association with collapse mode III failure, and so it is briefly summarized using that notation.

The maximum bending moment at the midspan of an initially deflected simply supported column (plate–stiffener combination) is

$$M_{max} = P w_{max} \tag{15.5.3}$$

where $w_{max} = \frac{w_{oc}}{1 - P/P_E}$ is the maximum deflection at the midspan of the column, w_{oc} is the column type initial distortion of the stiffener, $P = A_s \sigma_{xav}$, $A_s = bt + h_w t_w + b_f t_f$, $P_E = A_s \sigma_{xE}$, and σ_{xE} is as defined in equation (15.5.2).

The maximum compressive stress at the outermost fiber of the cross-section can then be obtained from the sum of the axial stress and bending stress.

$$\sigma_{max} = \frac{P}{A_s} + \frac{M_{max}}{I_e} z_c = \frac{P}{A_s} + \frac{z_c}{I_e} \frac{P w_{oc}}{1 - P/P_E}$$

$$= \sigma_{xav} + \frac{A_s w_{oc} z_c}{I_e} \frac{\sigma_{xav}}{1 - \sigma_{xav}/\sigma_{xE}} \tag{15.5.4}$$

where z_c is the distance from the elastic neutral axis to the outermost fiber on the compressed side and I_e is as defined in equation (15.5.2).

The failure condition by the Perry-Robertson method is then

$$\sigma_{max} = \sigma_{Yeq} \tag{15.5.5}$$

where σ_{Yeq} is as defined in equation (15.2.2b).

By substituting equation (15.5.4) into equation (15.5.5) and by replacing σ_{xav} with σ_{xu}, the following second-order equation with regard to σ_{xu} is obtained.

$$\sigma_{xu}^2 - \left[(1+\eta)\sigma_{xE} + \sigma_{Yeq}\right]\sigma_{xu} + \sigma_{Yeq}\sigma_{xE} = 0 \tag{15.5.6}$$

where $\eta = \frac{A_s w_{oc} z_c}{I_e} = \frac{w_{oc} z_c}{r_e^2}$ and $r_e = \sqrt{\frac{I_e}{A_s}}$.

The minimum solution to equation (15.5.6) is the real ultimate compressive strength and is given by

$$\frac{\sigma_{xu}}{\sigma_{Yeq}} = \frac{1}{2}\left(1 + \frac{1+\eta}{\lambda_e^2}\right) - \left[\frac{1}{4}\left(1 + \frac{1+\eta}{\lambda_e^2}\right)^2 - \frac{1}{\lambda_e^2}\right]^{0.5} \tag{15.5.7}$$

where $\lambda_e = \frac{a}{\pi r_e}\sqrt{\frac{\sigma_{Yeq}}{E}} = \sqrt{\frac{\sigma_{Yeq}}{\sigma_{xE}}}$.

A straight column does not have initial distortion, that is, $w_{oc} = 0$, and consequently the constant η becomes $\eta = 0$. This reduces equation (15.5.7) to the Euler formula applicable for $\lambda_e \geq 1$, that is,

$$\frac{\sigma_{xu}}{\sigma_{Yeq}} = \frac{1}{\lambda_e^2} \tag{15.5.8}$$

THE PAIK-THAYAMBALLI EMPIRICAL FORMULA METHOD

Empirical formulae that can be developed by curve-fitting experimental and/or numerical data are sometimes useful for predicting the ultimate strength of a plate–stiffener combination under axial compression alone without taking into account the effect of lateral pressure, although they are usually limited to the applicable range that has been considered in the data.

Paik and Thayamballi (1997) suggested the following empirical formula for the ultimate compressive strength of plate–stiffener combinations, which was developed using extensive experimental data collected from across the world (Paik, 2007). The Paik-Thayamballi formula is expressed as a function of two parameters: the column slenderness coefficient λ and the attached plate slenderness coefficient β.

$$\frac{\sigma_{xu}}{\sigma_{Yeq}} = \frac{1}{\sqrt{0.995 + 0.936\lambda^2 + 0.170\beta^2 + 0.188\lambda^2\beta^2 - 0.067\lambda^4}} \tag{15.5.9}$$

where $\lambda = \frac{a}{\pi r}\sqrt{\frac{\sigma_{Yeq}}{E}}$, $\beta = \frac{b}{t}\sqrt{\frac{\sigma_{Yp}}{E}}$, $r = \sqrt{\frac{I}{A_s}}$, I is the moment of inertia as defined in equation (15.3.30) for a single stiffener together with "fully effective plating", A_s is as defined in equation (15.5.3), and σ_{Yeq} is as defined in equation (15.2.2b). If σ_{xu} calculated from equation (15.5.9) is greater than σ_{Yeq}/λ^2, then $\sigma_{xu} = \sigma_{Yeq}/\lambda^2$ must be adopted by the definition of column buckling.

The Paik-Thayamballi formula implicitly includes the possible effects of stiffener web buckling (collapse mode IV) or tripping (collapse mode V) as well as beam-column type collapse (collapse mode III), because it was developed based on an experimental database that covered all such collapse modes in mechanical collapse tests (Paik, 2007). An average level of initial imperfections (initial deflection and welding residual stress) was considered in the curve-fitting of the data.

15.5.1.2 Effect of Lateral Pressure—Modified Perry-Robertson Formula Method

The concept of the original Perry-Robertson formula can still be applied to take into account the effect of lateral pressure, with the difference that the plate–stiffener combination reaches ultimate strength if the outermost fiber on either the stiffener flange side or the attached plate side yields. For convenience, the former is denoted as stiffener-induced failure (SIF) and the latter as plate-induced failure (PIF). This approach is often called the modified Perry-Robertson formula method (Paik and Thayamballi, 2003).

The maximum bending moment of the plate–stiffener combination under axial compression σ_{xav} and lateral pressure p can now be calculated as the sum of the maximum bending moments caused by both the initial distortion of the stiffener and lateral pressure, with the nomenclature indicated in equations (15.5.3) and (15.5.4), as follows.

$$M_{\max} = \frac{P}{1 - P/P_E}\left(w_{oc} + w_{q\max}\right) + M_{q\max} \qquad (15.5.10)$$

where $w_{q\max}$ is the maximum deflection resulting from the "lateral line load" defined by $q = bp$ at the midspan of the plate–stiffener combination, which can be approximately calculated by linear structural mechanics as $w_{q\max} = \frac{5qa^4}{384EI_e}$, and $M_{q\max}$ is the maximum bending moment resulting from the lateral line load $q = bp$ at the midspan of the plate–stiffener combination, which is given as $M_{q\max} = \frac{qa^2}{8}$.

It is noted that lateral pressure loads acting on the plate part of the stiffened panel in reality are condensed into the lateral line load, which is applied along the line of the plate–stiffener intersection because the attached plating (flange) in a separate plate–stiffener combination or a beam-column cannot theoretically resist lateral pressure loads.

The maximum stress at the outermost fiber of the plate–stiffener combination is then obtained as follows.

$$\sigma_{\max} = \frac{P}{A_s} + \frac{M_{\max}}{I_e} z_c$$
$$= \frac{P}{A_s} + \frac{qa^2}{8}\frac{z_c}{I_e} + \frac{P}{1 - P_E}\left(\frac{5qa^4}{384EI_e} + w_{oc}\right)\frac{z_c}{I_e} \qquad (15.5.11)$$

Considering the failure condition of equation (15.5.5) and replacing $\sigma_{xav} = P/A_s$ with σ_{xu}, the following second-order equation with regard to σ_{xu} is derived.

$$\lambda_e^2 R^2 - \left(\lambda_e^2 - \mu\lambda_e^2 + 1 + \eta\right)R + \left(1 + \mu\right) = 0 \qquad (15.5.12)$$

where $R = \frac{\sigma_{xu}}{\sigma_{Yeq}}$, $\lambda_e = \sqrt{\frac{\sigma_{Yeq}}{\sigma_{xE}}} = \frac{a}{\pi r_e}\sqrt{\frac{\sigma_{Yeq}}{E}}$,

$\eta = \frac{Az_c}{I_e}\left(\frac{5qa^4}{384EI_e} + w_{oc}\right)$, and $\mu = \frac{1}{\sigma_{Yeq}}\frac{qa^2}{8}\frac{z_c}{I_e}$.

The minimum solution to equation (15.5.12) with regard to R gives the real ultimate strength of the plate–stiffener combination under σ_{xav} and p, which is given by

$$\frac{\sigma_{xu}}{\sigma_{Yeq}} = \frac{1}{2}\left(1 - \mu + \frac{1+\eta}{\lambda_e^2}\right) - \left[\frac{1}{4}\left(1 - \mu + \frac{1+\eta}{\lambda_e^2}\right)^2 - \frac{1-\mu}{\lambda_e^2}\right]^{0.5} \qquad (15.5.13)$$

15.5.1.3 Stiffener-Induced Failure versus Plate-Induced Failure

The direction of the beam-column deflection is governed by the direction of the initial deflection and/or the direction of lateral pressure loads. Because the directional nature of the initial deflection or lateral pressure is somewhat uncertain, the failure mode of the plate–stiffener combination model can be either PIF or SIF. For this reason, the ultimate strength of the Perry-Robertson formula method is determined as the minimum value of the two strengths.

In a continuous stiffened-plate structure, SIF triggers the collapse of the entire panel. The original idea of the Perry-Robertson formula method assumes that SIF occurs if the tip of the stiffener yields. However, this assumption may in some cases be too pessimistic in terms of the collapse strength predictions of a stiffened panel. Rather, plasticity may grow into the stiffener web as long as lateral–torsional buckling or stiffener web buckling does not occur, such that the stiffener may resist further loading even after the first yielding at the outermost fiber (Hughes and Ma, 1996).

In maritime engineering practice, often only the PIF-based Perry-Robertson formula method, which excludes SIF, is adopted to predict the ultimate strength of a plate–stiffener combination representing a continuous stiffened panel, where a lower limit must be the ultimate strength of the entire panel without the stiffeners (Paik and Thayamballi, 2003). The software ALPS/ULSAP (2009) also applies this practice for predicting the collapse mode III failure.

15.5.2 Combined Transverse Axial Load and Lateral Pressure

The membrane stress-based method can be applied to calculate the ultimate strength of the panel σ_{yu}. The stiffened panel reaches ultimate strength when the most highly stressed plating between the stiffeners collapses.

Equation (13.5.8) in Chapter 13 can be used to calculate the maximum deflection of the plate, but using C_4 as defined in equation (13.5.12) and $\sigma_{xav} = 0$. Further, $m = n = 1$ is applied in equation (13.5.8) because $a/b \geq 1$.

$$C_1 A_1^3 + C_2 A_1^2 + C_3 A_1 + C_4 = 0 \qquad (15.5.14)$$

where $C_1 = \frac{\pi^2 E}{16}\left(\frac{b}{a^3} + \frac{a}{b^3}\right)$, $C_2 = \frac{3\pi^2 E A_{o1}}{16}\left(\frac{b}{a^3} + \frac{a}{b^3}\right)$,

$$C_3 = \frac{\pi^2 E A_{o1}^2}{8}\left(\frac{b}{a^3} + \frac{a}{b^3}\right) + \frac{b}{a}\sigma_{rex} + \frac{a}{b}(\sigma_{yav} + \sigma_{rey}) + \frac{\pi^2 D}{t}\frac{1}{ab}\left(\frac{b}{a} + \frac{a}{b}\right)^2$$

$$C_4 = A_{o1}\left[\frac{b}{a}\sigma_{rex} + \frac{a}{b}(\sigma_{yav} + \sigma_{rey})\right] - \frac{16ab}{\pi^4 t}p,$$

$$\sigma_{rex} = \sigma_{rcx} + \frac{2}{b}(\sigma_{rtx} - \sigma_{rcx})\left(b_t - \frac{b}{2\pi}\sin\frac{2\pi b_t}{b}\right), \text{ and}$$

$$\sigma_{rey} = \sigma_{rcy} + \frac{2}{b}(\sigma_{rty} - \sigma_{rcy})\left(a_t - \frac{a}{2\pi}\sin\frac{2\pi a_t}{a}\right),$$

with the nomenclature of welding-induced residual stresses as defined in Fig. 8.13 of Chapter 8.

Once A_1 in equation (15.5.14) is determined by the Cardano method described in Chapter 13, the maximum and minimum membrane stresses inside the plate can be obtained as follows.

$$\sigma_{x\max} = \sigma_{rtx} - \frac{E\pi^2 A_1(A_1 + 2A_{o1})}{8a^2}\cos\frac{2\pi b_t}{b} \qquad (15.5.15a)$$

$$\sigma_{x\min} = \sigma_{rcx} + \frac{E\pi^2 A_1(A_1 + 2A_{o1})}{8a^2} \qquad (15.5.15b)$$

$$\sigma_{y\max} = \sigma_{yav} + \sigma_{rty} - \frac{E\pi^2 A_1(A_1 + 2A_{o1})}{8b^2}\cos\frac{2\pi a_t}{a} \qquad (15.5.15c)$$

$$\sigma_{y\min} = \sigma_{yav} + \sigma_{rcy} + \frac{E\pi^2 A_1(A_1 + 2A_{o1})}{8b^2} \qquad (15.5.15d)$$

Now that the maximum and minimum membrane stresses are known, the ultimate compressive strength of the plating between the stiffeners can be determined as the lowest value among the solutions of the following three failure-conditional equations.

$$\sigma_{x\max}^2 - \sigma_{x\max}\sigma_{y\max} + \sigma_{y\max}^2 = \sigma_{Yp}^2 \qquad (15.5.16a)$$

$$\sigma_{x\max}^2 - \sigma_{x\max}\sigma_{y\min} + \sigma_{y\min}^2 = \sigma_{Yp}^2 \qquad (15.5.16b)$$

$$\sigma_{x\min}^2 - \sigma_{x\min}\sigma_{y\max} + \sigma_{y\max}^2 = \sigma_{Yp}^2 \qquad (15.5.16c)$$

15.5.3 Combined Edge Shear and Lateral Pressure

Equation (15.3.34) can be applied, and the ultimate strength of the panel τ_u under edge shear is calculated from equation (13.4.25) of Chapter 13. The ultimate strength p_{uo} of the plating under lateral pressure alone is calculated as the lowest value of the three failure-conditional equations (15.5.16a), (15.5.16b), and (15.5.16c) but using the maximum and minimum membrane stresses defined in equations, (13.5.4a) and (13.5.4c).

15.5.4 Combined Biaxial Load, Edge Shear, and Lateral Pressure

Under the combination of σ_{xav}, σ_{yav}, τ_{av}, and p, as defined in Section 15.2.3, the following ultimate strength interaction equation can be applied for the collapse mode III failure of a stiffened panel, using the collapse mode III ultimate strength components obtained thus far.

$$\left(\frac{\sigma_{xav}}{\sigma_{xu}^{III}}\right)^{c_1} - \kappa\left(\frac{\sigma_{xav}}{\sigma_{xu}^{III}}\right)\left(\frac{\sigma_{yav}}{\sigma_{yu}^{III}}\right) + \left(\frac{\sigma_{yav}}{\sigma_{yu}^{III}}\right)^{c_2} + \left(\frac{\tau_{av}}{\tau_u^{III}}\right)^{c_3} = 1 \qquad (15.5.17)$$

where the superscript *III* denotes collapse mode III failure, and the ultimate strengths of the individual load components have been calculated for collapse mode III failure taking into account the effect of lateral pressure. The coefficients κ, c_1, c_2, and c_3 are as defined in equation (15.3.36).

15.6 ULTIMATE STRENGTH FORMULATIONS FOR THE PANEL FOR COLLAPSE MODE IV

The stiffened panel reaches ultimate strength by collapse mode IV if the most highly stressed stiffener together with its attached plating collapses by the buckling of the stiffener web. This section presents the ultimate strength formulations for collapse mode IV failure under combined in-plane and lateral pressure loads, as defined in Section 15.2.3.

15.6.1 Combined Longitudinal Axial Load and Lateral Pressure

The ultimate strength of the panel for collapse mode IV failure is calculated as the sum of the ultimate plate strength and stiffener web buckling strength, as follows.

$$\sigma_{xu} = \frac{\sigma_{xu}^{P} bt + \sigma_{u}^{W}\left(h_w t_w + b_f t_f\right)}{bt + h_w t_w + b_f t_f} \tag{15.6.1}$$

where σ_{xu}^{P} is the ultimate strength of the plating between the stiffeners and σ_u^W is the ultimate strength of the stiffener because of web buckling.

15.6.1.1 Ultimate Strength of the Plating Between the Stiffeners

The ultimate strength σ_{xu}^{P} in equation (15.6.1) of the plating between the stiffeners under σ_{xav} and p can be obtained by using the membrane stress-based method, and is taken as the lowest value among the three solutions to the three failure-conditional equations (15.5.16a), (15.5.16b), and (15.5.16c). In this case, the maximum and minimum membrane stresses are determined from equation (13.5.9) of Chapter 13 taking $n = 1$ and $\sigma_{yav} = 0$ into consideration, as follows.

$$\sigma_{x\max} = \sigma_{xav} + \sigma_{rtx} - \frac{E\pi^2 m^2 A_m\left(A_m + 2A_{om}\right)}{8a^2}\cos\frac{2\pi b_t}{b} \tag{15.6.2a}$$

$$\sigma_{x\min} = \sigma_{xav} + \sigma_{rcx} + \frac{E\pi^2 m^2 A_m\left(A_m + 2A_{om}\right)}{8a^2} \tag{15.6.2b}$$

$$\sigma_{y\max} = \sigma_{rty} - \frac{E\pi^2 A_m\left(A_m + 2A_{om}\right)}{8b^2}\cos\frac{2m\pi a_t}{a} \tag{15.6.2c}$$

$$\sigma_{y\min} = \sigma_{rcy} + \frac{E\pi^2 A_m\left(A_m + 2A_{om}\right)}{8b^2} \tag{15.6.2d}$$

where the nomenclature is as defined in Section 13.5.1 of Chapter 13.

The maximum deflection A_m in equation (15.6.2) is the solution to the following third-order equation.

$$C_1 A_m^3 + C_2 A_m^2 + C_3 A_m + C_4 = 0 \tag{15.6.3}$$

where $C_1 = \frac{\pi^2 E}{16}\left(\frac{m^4 b}{a^3} + \frac{a}{b^3}\right)$, $C_2 = \frac{3\pi^2 E A_{om}}{16}\left(\frac{m^4 b}{a^3} + \frac{a}{b^3}\right)$,

$$C_3 = \frac{\pi^2 E A_{om}^2}{8}\left(\frac{m^4 b}{a^3} + \frac{a}{b^3}\right) + \frac{m^2 b}{a}\left(\sigma_{xav} + \sigma_{rex}\right) + \frac{a}{b}\sigma_{rey} + \frac{\pi^2 D}{t}\frac{m^2}{ab}\left(\frac{mb}{a} + \frac{a}{mb}\right)^2,$$

$$C_4 = A_{om}\left(\frac{m^2 b}{a}\left(\sigma_{xav} + \sigma_{rex}\right) + \frac{a}{b}\sigma_{rey}\right) - \frac{16ab}{\pi^4 t}p$$

$$\sigma_{rex} = \sigma_{rcx} + \frac{2}{b}\left(\sigma_{rtx} - \sigma_{rcx}\right)\left(b_t - \frac{b}{2\pi}\sin\frac{2\pi b_t}{b}\right), \text{ and}$$

$$\sigma_{rey} = \sigma_{rcy} + \frac{2}{b}\left(\sigma_{rty} - \sigma_{rcy}\right)\left(a_t - \frac{a}{2m\pi}\sin\frac{2m\pi a_t}{a}\right),$$

with the nomenclature of welding-induced residual stresses as defined in Fig. 8.13 of Chapter 8.

15.6.1.2 Ultimate Strength of the Stiffener as a Result of Web Buckling

The Johnson-Ostenfeld formula is used to calculate the ultimate strength σ_u^W of a stiffener subject to web buckling. The elastic buckling strength of the stiffener web is calculated by the following equation.

$$\sigma_E^W = -k_w \frac{\pi^2 E}{12\left(1-\nu^2\right)}\left(\frac{t_w}{h_w}\right)^2 \tag{15.6.4}$$

where the compressive stress takes the negative sign.

In equation (15.6.4), k_w is the elastic web buckling strength coefficient, which is given by (Paik and Thayamballi, 2003)

$$k_w = \begin{cases} C_1\zeta_p + C_2 & \text{for } 0 \le \zeta_p \le \eta_w \\ C_3 - \dfrac{1}{C_4\zeta_p + C_5} & \text{for } \eta_w < \zeta_p \le 60 \\ C_3 - \dfrac{1}{60C_4 + C_5} & \text{for } 60 < \zeta_p \end{cases} \tag{15.6.5}$$

where $\zeta_p = \frac{GJ_p}{h_w D_w}$, $\zeta_f = \frac{GJ_f}{h_w D_w}$, $J_p = \frac{b_e t^3}{3}$ is the torsion constant of the attached effective plating, $J_f = \frac{b_f t_f^3}{3}$ is the torsion constant of the stiffener flange, $D_w = \frac{E t_w^3}{12(1-\nu^2)}$ is the bending rigidity of the stiffener web, b_e is the effective width of the plating at the ultimate limit state as defined in equation (15.5.2), $G = \frac{E}{2(1+\nu)}$, $\eta_w = 0.444\zeta_f^2 + 3.333\zeta_f + 1.0$, $C_1 = -0.001\zeta_f + 0.303$, $C_2 = 0.308\zeta_f + 0.4267$,

$$C_3 = \begin{cases} -4.350\zeta_f^2 + 3.965\zeta_f + 1.277 & \text{for } 0 \le \zeta_f \le 0.2 \\ -0.427\zeta_f^2 + 2.267\zeta_f + 1.460 & \text{for } 0.2 < \zeta_f \le 1.5 \\ -0.133\zeta_f^2 + 1.567\zeta_f + 1.850 & \text{for } 1.5 < \zeta_f \le 3.0 \\ 5.354 & \text{for } 3.0 < \zeta_f \end{cases},$$

$$C_4 = \begin{cases} -6.70\zeta_f^2 + 1.40 & \text{for } 0 \le \zeta_f \le 0.1 \\ \dfrac{1}{5.10\zeta_f + 0.860} & \text{for } 0.1 < \zeta_f \le 1.0 \\ \dfrac{1}{4.0\zeta_f + 1.814} & \text{for } 1.0 < \zeta_f \le 3.0 \\ 0.0724 & \text{for } 3.0 < \zeta_f \text{, and} \end{cases},$$

$$C_5 = \begin{cases} -1.135\zeta_f + 0.428 & \text{for } 0 \le \zeta_f \le 0.2 \\ -0.299\zeta_f^3 + 0.803\zeta_f^2 - 0.783\zeta_f + 0.328 & \text{for } 0.2 < \zeta_f \le 1.0 \\ -0.016\zeta_f^3 + 0.117\zeta_f^2 - 0.285\zeta_f + 0.235 & \text{for } 1.0 < \zeta_f \le 3.0 \\ 0.001 & \text{for } 3.0 < \zeta_f \end{cases}.$$

For flat-bar stiffeners, k_w in equation (15.6.5) becomes much simpler because $\zeta_f = 0$, as follows.

$$k_w = \begin{cases} 0.303\zeta_p + 0.427 & \text{for } 0 \le \zeta_p \le 1 \\ 1.277 - \dfrac{1}{1.40\zeta_p + 0.428} & \text{for } 1 < \zeta_p \le 60 \\ 1.2652 & \text{for } 60 < \zeta_p \end{cases} \quad (15.6.6)$$

σ_u^W in equation (15.6.1) is then obtained by the Johnson-Ostenfeld formula as a plasticity correction of the elastic buckling stress.

$$\sigma_u^W = \begin{cases} \sigma_E^W & \text{for } \sigma_E^W \le 0.5\sigma_{Ys} \\ \sigma_{Ys}\left(1 - \dfrac{\sigma_{Ys}}{4\sigma_E^W}\right) & \text{for } \sigma_E^W > 0.5\sigma_{Ys} \end{cases} \quad (15.6.7)$$

15.6.2 Combined Transverse Axial Load and Lateral Pressure

For collapse mode IV failure, the ultimate strength σ_{yu} can be calculated using the same method as that described in Section 15.5.2.

15.6.3 Combined Edge Shear and Lateral Pressure

The ultimate strength of the panel τ_u for collapse mode IV failure is calculated using the same method as that described in Section 15.5.3.

15.6.4 Combined Biaxial Load, Edge Shear, and Lateral Pressure

Under the combination of σ_{xav}, σ_{yav}, τ_{av}, and p, as defined in Section 15.2.3, the following ultimate strength interaction equation is applicable for the collapse mode IV failure of a stiffened panel, using the collapse mode IV ultimate strength components obtained thus far.

$$\left(\frac{\sigma_{xav}}{\sigma_{xu}^{IV}}\right)^{c_1} - \kappa\left(\frac{\sigma_{xav}}{\sigma_{xu}^{IV}}\right)\left(\frac{\sigma_{yav}}{\sigma_{yu}^{IV}}\right) + \left(\frac{\sigma_{yav}}{\sigma_{yu}^{IV}}\right)^{c_2} + \left(\frac{\tau_{av}}{\tau_u^{IV}}\right)^{c_3} = 1 \quad (15.6.8)$$

where the superscript *IV* denotes collapse mode IV failure, and the ultimate strengths of the individual load components have been calculated for collapse mode IV failure taking into account the effect of lateral pressure. The coefficients κ, c_1, c_2, and c_3 are as defined in equation (15.3.36).

15.7 ULTIMATE STRENGTH FORMULATIONS FOR THE PANEL FOR COLLAPSE MODE V

The stiffened panel reaches ultimate strength by collapse mode V if the most highly stressed stiffener together with the attached plating collapses by flexural–torsional buckling or tripping. This section presents the ultimate strength formulations for collapse mode V failure under combined in-plane and lateral pressure loads, as defined in Section 15.2.3.

15.7.1 Combined Longitudinal Axial Load and Lateral Pressure

The ultimate strength of the panel σ_{xu} for collapse mode V failure is calculated as the sum of the ultimate plate strength and stiffener flexural–torsional buckling strength.

$$\sigma_{xu} = \frac{\sigma_{xu}^P bt + \sigma_u^T (h_w t_w + b_f t_f)}{bt + h_w t_w + b_f t_f} \quad (15.7.1)$$

where σ_{xu}^P is the ultimate strength of the plating between the stiffeners and σ_u^T is the ultimate strength of the stiffener resulting from lateral–torsional buckling or tripping.

15.7.1.1 Ultimate Strength of the Plating Between the Stiffeners

The ultimate strength σ^P_{xu} in equation (15.7.1) of the plating between the stiffeners under σ_{xav} and p is obtained using the same method as that described in Section 15.6.1.

15.7.1.2 Ultimate Strength of the Stiffener Resulting From Flexural–Torsional Buckling

The Johnson-Ostenfeld formula is used to calculate the ultimate strength σ^T_u of a stiffener subject to flexural–torsional buckling. The elastic flexural–torsional buckling strength formulae for stiffeners differ depending on the types of stiffeners (Paik and Thayamballi, 2003).

FLAT-BAR STIFFENERS

In this case, it is assumed that the elastic flexural–torsional buckling strength equals the elastic stiffener web buckling strength.

$$\sigma^T_E = \sigma^W_E \qquad (15.7.2)$$

where σ^W_E is as defined in equation (15.6.4).

ASYMMETRIC ANGLE STIFFENERS

$$\sigma^T_E = (-1)\min_{m=1,2,3\cdots}.\left|\frac{C_2+\sqrt{C_2^2-4C_1C_3}}{2C_1}\right| \qquad (15.7.3)$$

where

$C_1 = (b_et + h_wt_w + b_ft_f)I_p - S_f^2$,

$$C_2 = -I_p\left[EI\left(\frac{m\pi}{a}\right)^2 - \frac{qa^2}{12}\frac{S_1}{I_y}\left(1-\frac{3}{m^2\pi^2}\right)\right],$$

$$-\left(b_et+h_wt_w+b_ft_f\right)\left[G\left(J_w+J_f\right)\right.$$

$$\left.+EI_zh_w^2\left(\frac{m\pi}{a}\right)^2 - \frac{qa^2}{12}\frac{S_2}{I_y}\left(1-\frac{3}{m^2\pi^2}\right)\right]$$

$$+2S_f\left[EI_{zy}h_w\left(\frac{m\pi}{a}\right)^2 - \frac{qa^2}{12}\frac{S_3}{I_y}\left(1-\frac{3}{m^2\pi^2}\right)\right],$$

$$C_3 = \left[EI_y\left(\frac{m\pi}{a}\right)^2 - \frac{qa^2}{12}\frac{S_1}{I_y}\left(1-\frac{3}{m^2\pi^2}\right)\right]\left[G\left(J_w+J_f\right)\right.$$

$$\left.+EI_zh_w^2\left(\frac{m\pi}{a}\right)^2 - \frac{qa^2}{12}\frac{S_2}{I_y}\left(1-\frac{3}{m^2\pi^2}\right)\right]$$

$$-\left[EI_{zy}h_w\left(\frac{m\pi}{a}\right)^2 - \frac{qa^2}{12}\frac{S_3}{I_y}\left(1-\frac{3}{m^2\pi^2}\right)\right]^2,$$

$$S_f = -\frac{t_fb_f^2}{2},$$

$$S_1 = -\left(z_p-h_w\right)t_fb_f - b_etz_p - h_wt_w\left(z_p-\frac{h_w}{2}\right),$$

$$S_2 = -\left(z_p-h_w\right)t_f\left(h_w^2b_f+\frac{b_f^3}{3}\right) - h_w^3t_w\left[\frac{1}{3}z_p-\frac{h_w}{4}\right],$$

$$S_3 = \left(z_p-h_w\right)\frac{b_f^2t_f}{2},$$

$$I_y = \frac{b_et^3}{12} + b_etz_p^2 + \frac{t_wh_w^3}{12} + h_wt_w\left(z_p-\frac{t}{2}-\frac{h_w}{2}\right)^2$$

$$+\frac{b_ft_f^3}{12} + b_ft_f\left(z_p-\frac{t}{2}-h_w-\frac{t_f}{2}\right)^2,$$

$$I_z = b_ety_o^2 + h_wt_wy_o^2 + b_ft_f\left(y_o^2-b_fy_o+\frac{b_f^2}{3}\right),$$

$$I_{zy} = b_etz_py_o + h_wt_w\left(z_p-\frac{t}{2}-\frac{h_w}{2}\right)y_o$$

$$+b_ft_f\left(z_p-\frac{t}{2}-h_w-\frac{t_f}{2}\right)\left(y_o-\frac{b_f}{2}\right),$$

I_p = polar moment of inertia of the stiffener about the toe, which is given by

$$I_p = \frac{t_wh_w^3}{3} + \frac{t_w^3h_w}{3} + \frac{b_f^3t_f}{3} + \frac{b_ft_f^3}{3} + b_ft_fh_w^2,$$

$$z_p = \frac{0.5h_wt_w\left(t+h_w\right)+b_ft_f\left(0.5t+h_w+0.5t_f\right)}{b_et+h_wt_w+b_ft_f},$$

$$y_o = \frac{b_f^2t_f}{2\left(b_et+h_wt_w+b_ft_f\right)},$$

b_e = effective width of the attached plating at the ultimate limit state,
q = equivalent line pressure ($q = pb$),
m = tripping half-wave number of the stiffener,
J_w = torsion constant for the web, which is given by

$$J_w = \frac{1}{3}t_w^3h_w\left(1-\frac{192}{\pi^5}\frac{t_w}{h_w}\sum_{n=1,3,5}^{\infty}\frac{1}{n^5}\tanh\frac{n\pi h_w}{2t_w}\right),$$

J_f = torsion constant for the flange, which is given by

$$J_f = \frac{1}{3}t_f^3b_f\left(1-\frac{192}{\pi^5}\frac{t_f}{b_f}\sum_{n=1,3,5}^{\infty}\frac{1}{n^5}\tanh\frac{n\pi b_f}{2t_f}\right).$$

SYMMETRIC T-STIFFENERS

$$\sigma_E^T = (-1) \min_{m=1,2,3\cdots} \left| -\frac{a^2 G (J_w + J_f) + E I_f h_w^2 m^2 \pi^2}{I_p a^2} + \frac{q a^2}{12} \frac{S_4}{I_y I_p} \left(1 - \frac{3}{m^2 \pi^2}\right) \right| \quad (15.7.4)$$

where $S_4 = -(z_p - h_w) t_f \left(h_w^2 b_f + \frac{b_f^3}{12} \right) - h_w^3 t_w \left[\frac{1}{3} z_p - \frac{h_w}{4} \right]$,

$I_p = \frac{t_w h_w^3}{3} + \frac{t_w^3 h_w}{12} + \frac{b_f t_f^3}{3} + \frac{b_f^3 t_f}{12} + b_f t_f h_w^2$, $I_f = \frac{b_f^3 t_f}{12}$, and

I_y, I_p, z_p is as defined in equation (15.7.3).

σ_u^T in equation (15.7.1) is then obtained by the Johnson-Ostenfeld formula as a plasticity correction of the elastic buckling stress.

$$\sigma_u^T = \begin{cases} \sigma_E^T & \text{for } \sigma_E^T \le 0.5\sigma_{Ys} \\ \sigma_{Ys}\left(1 - \frac{\sigma_{Ys}}{4\sigma_E^T}\right) & \text{for } \sigma_E^T > 0.5\sigma_{Ys} \end{cases} \quad (15.7.5)$$

15.7.2 Combined Transverse Axial Load and Lateral Pressure

For collapse mode V failure, the ultimate strength σ_{yu} can be calculated using the same method as that described in Section 15.5.2.

15.7.3 Combined Edge Shear and Lateral Pressure

The ultimate strength of the panel τ_u for collapse mode V failure can be calculated using the same method as that described in Section 15.5.3.

15.7.4 Combined Biaxial Load, Edge Shear, and Lateral Pressure

Under the combination of σ_{xav}, σ_{yav}, τ_{av}, and p, as defined in Section 15.2.3, the following ultimate strength interaction equation can be applied for the collapse mode V failure of a stiffened panel, using the collapse mode V ultimate strength components obtained thus far.

$$\left(\frac{\sigma_{xav}}{\sigma_{xu}^V}\right)^{c_1} - \kappa \left(\frac{\sigma_{xav}}{\sigma_{xu}^V}\right)\left(\frac{\sigma_{yav}}{\sigma_{yu}^V}\right) + \left(\frac{\sigma_{yav}}{\sigma_{yu}^V}\right)^{c_2} + \left(\frac{\tau_{av}}{\tau_u^V}\right)^{c_3} = 1 \quad (15.7.6)$$

where the superscript V denotes the collapse mode V failure, and the ultimate strengths of the individual load components have been calculated for collapse mode IV failure taking into account the effect of lateral pressure. The coefficients κ, c_1, c_2, and c_3 are as defined in equation (15.3.36).

15.8 ULTIMATE STRENGTH FORMULATIONS FOR THE PANEL FOR COLLAPSE MODE VI

The stiffened panel reaches ultimate strength by collapse mode VI if plasticity occurs over the en-tire panel and no local buckling occurs. This section presents the ultimate strength formulations for collapse mode VI failure under combined in-plane and lateral pressure loads, as defined in Section 15.2.3.

15.8.1 Combined Longitudinal Axial Load and Lateral Pressure

The ultimate strength of the panel σ_{xu} for collapse mode VI failure is calculated as follows.

$$\sigma_{xu} = \frac{\sigma_{xu}^P bt + \sigma_{Ys}(h_w t_w + b_f t_f)}{bt + h_w t_w + b_f t_f} \quad (15.8.1)$$

where σ_{xu}^P is the ultimate strength of the plating between the stiffeners for collapse mode VI failure under σ_{xav} and p.

To calculate σ_{xu}^P in equation (15.8.1) taking into account the effect of lateral pressure, the following interaction relation between σ_{xav} and p is sometimes applied when no local buckling is considered.

$$\left(\frac{\sigma_{xav}}{\sigma_{Yp}}\right)^{c_1} + \left(\frac{p}{p_{uo}}\right)^{c_2} = 1 \quad (15.8.2)$$

where p_{uo} is the ultimate strength of the plating under lateral pressure alone, which can be calculated as described in Section 15.3.4.2. c_1 and c_2 are constants that are often taken as $c_1 = 2$ and $c_2 = 1$. σ_{xu}^P in equation (15.8.1) is then calculated as follows.

$$\sigma_{xu}^P = \sigma_{Yp}\left[1 - \left(\frac{p}{p_{uo}}\right)^{c_2}\right]^{\frac{1}{c_1}} \quad (15.8.3)$$

15.8.2 Combined Transverse Axial Load and Lateral Pressure

In this case, the following ultimate strength interaction relationship is applied.

$$\left(\frac{\sigma_{yav}}{\sigma_{Yp}}\right)^{c_1} + \left(\frac{p}{p_{uo}}\right)^{c_2} = 1 \quad (15.8.4)$$

where p_{uo}, c_1 and c_2 are as defined in equation (15.8.2). σ_{yu} is then calculated from equation (15.8.4) as follows.

$$\sigma_{yu} = \sigma_{Yp}\left[1-\left(\frac{p}{p_{uo}}\right)^{c_2}\right]^{\frac{1}{c_1}} \quad (15.8.5)$$

15.8.3 Combined Edge Shear and Lateral Pressure

In this case, the following ultimate strength interaction relationship, which is similar to equation (15.3.34), is applied.

$$\left(\frac{\tau_{av}}{\tau_{Yp}}\right)^{1.5} + \left(\frac{p}{p_{uo}}\right)^{1.2} = 1 \quad (15.8.6)$$

where $\tau_{Yp} = \frac{\sigma_{Yp}}{\sqrt{3}}$.

The ultimate shear stress τ_u is then obtained from equation (15.8.6) as follows.

$$\tau_u = \tau_{Yp}\left[1-\left(\frac{p}{p_{uo}}\right)^{1.2}\right]^{\frac{1}{1.5}} \quad (15.8.7)$$

15.8.4 Combined Biaxial Load, Edge Shear, and Lateral Pressure

Under the combination of σ_{xav}, σ_{yav}, τ_{av}, and p, as defined in Section 15.2.3, the following ultimate strength interaction equation can be applied to derive the collapse mode VI failure of a stiffened panel using the collapse mode VI ultimate strength components obtained thus far.

$$\left(\frac{\sigma_{xav}}{\sigma_{xu}^{VI}}\right)^{c_1} - \kappa\left(\frac{\sigma_{xav}}{\sigma_{xu}^{VI}}\right)\left(\frac{\sigma_{yav}}{\sigma_{yu}^{VI}}\right) + \left(\frac{\sigma_{yav}}{\sigma_{yu}^{VI}}\right)^{c_2} + \left(\frac{\tau_{av}}{\tau_{u}^{VI}}\right)^{c_3} = 1 \quad (15.8.8)$$

where the superscript *VI* denotes collapse mode VI failure, and the ultimate strengths of the individual load components have been calculated for collapse mode IV failure taking into account the effect of lateral pressure. The coefficients κ, c_1, c_2, and c_3 are as defined in equation (15.3.36).

15.9 APPLIED EXAMPLES

The theory described in this chapter has been implemented in the software ALPS/ULSAP (2009). Some application examples of ALPS/ULSAP are presented and compared with experimental results or solutions derived using nonlinear finite element method. It is noted that the theory described in this chapter can readily be expanded to orthogonally-stiffened panels (i.e., with stiffeners in both directions) (Paik and Thayamballi, 2003). In fact, the software ALPS/ULSAP is able to compute the ultimate strength of unstiffened plates and stiffened-plate structures with stiffeners or support members in either longitudinal or transverse direction or both directions, although the theory described in this chapter is addressed only for longitudinally-stiffened panels.

15.9.1 Smith's Test Database

Smith (1976) carried out a series of collapse tests using 11 full-scale welded steel-stiffened plate structures representing typical warship deck structures under axial compression, or bottom structures under combined axial compression and lateral pressure.

Figure 15.5 shows a typical configuration for Smith's test structures. The overall dimensions of each structure are $L = 6096$ mm long by $B = 3048$ mm wide, excluding the panel ends, which are bolted to the test frames along the edges. Except for test structure numbers 4a and 4b, which have large girders and small stiffeners in the longitudinal direction, all of the test structures have identical T-type longitudinal stiffeners and identical T-type transverse frames.

The test structures include four pairs of nominally identical stiffened panels (model numbers 1a, 1b, 2a, 2b, 3a, 3b, 4a, and 4b) representing the configuration of a ship's base, together with two stiffened panels (model numbers 5 and 7) representing frigate strength decks, and one stiffened panel (model number 6) corresponding to a light superstructure deck.

Table 15.1 shows the geometric properties of the longitudinal stiffeners and transverse frames together with the material yield stresses for the plating and stiffeners. Table 15.2 presents the other geometric characteristics of each of the test struc-

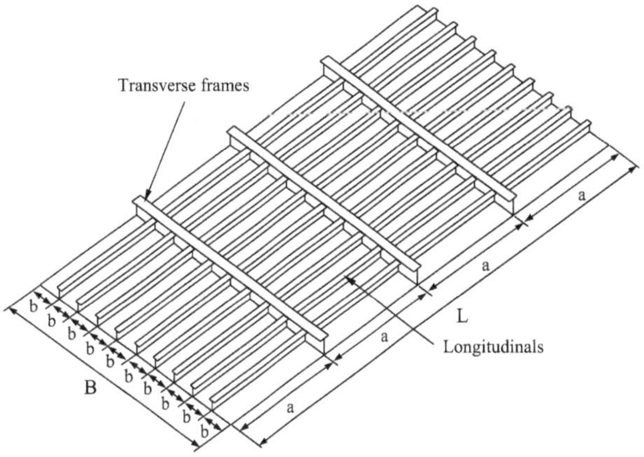

Figure 15.5 Schematic of one of Smith's test structures.

Table 15.1 Geometric Properties and Material Yield Stresses of Smith's Test Structures

Structure Number	a (mm)	B (mm)	t (mm)	n_{sx}	h_{wx} (mm)	t_{wx} (mm)	b_{fx} (mm)	t_{fx} (mm)	n_{sy}	h_{wy} (mm)	t_{wy} (mm)	b_{fy} (mm)	t_{fy} (mm)	σ_{Yp} (MPa)	σ_{Ys} (MPa)	σ_{Yeq} (MPa)
1a	6096	3048.0	8.00	4	153.67	7.21	78.99	14.22	4	257.56	9.37	125.48	18.29	249.1	253.7	250.4
1b	6096	3048.0	7.87	4	152.40	7.11	76.20	14.22	4	254.00	9.14	127.00	18.29	252.2	252.4	252.3
2a	6096	3048.0	7.72	9	115.57	5.44	45.97	9.53	3	204.98	8.31	102.62	16.26	261.3	268.9	263.1
2b	6096	3048.0	7.37	9	114.30	5.38	44.70	9.53	3	203.71	8.33	102.62	16.26	259.7	274.9	263.3
3a	6096	3048.0	6.38	9	77.72	4.52	25.91	6.35	3	156.21	6.81	78.99	14.22	250.6	227.9	246.8
3b	6096	3048.0	6.40	9	77.22	4.65	27.94	6.35	3	153.92	6.88	79.25	14.22	252.2	223.3	247.3
4a	1219.2	1016.0	6.43	3	76.71	4.85	27.69	6.35	—	—	—	—	—	259.7	223.9	252.5
4b	1219.2	1016.0	6.40	3	76.96	4.55	26.16	6.35	—	—	—	—	—	264.3	227.9	257.3
5	6096	3048.0	6.43	4	116.08	5.33	46.23	9.53	3	154.18	6.76	77.22	14.22	247.6	230.9	244.9
6	6096	3048.0	6.32	4	76.20	4.55	27.43	6.35	4	114.55	5.36	46.23	9.53	256.7	241.5	255.2
7	6096	3048.0	6.30	4	115.06	5.16	45.21	9.53	3	153.92	6.65	78.74	14.22	290.1	305.3	303.3

Notes:
1. Test structure numbers 4a and 4b represent longitudinally-stiffened panels between two adjacent longitudinal girders and two adjacent transverse frames: $\nu = 0.3$ and $E = 205.8$GPa.
2. The subscripts x and y denote the x and y directions, respectively.

tures. The initial deflection of the plating and the initial distortions of longitudinal stiffeners and transverse frames were measured in these tests. A high degree of variability associated with the plate initial deflection measurements was reported, with coefficients of variation (COV) for w_{opl} and w_{oc} in the range of 0.22~0.63 and 0.29~1.04, respectively. Specifically, it was observed that the plating and stiffener imperfections for model number 3b were abnormally large, with an "unfavorable" relative stiffener distortion. Structure number 6, which represents a light superstructure deck, also had a "serious level" of initial imperfections, which would not be typical in a real structure. The welding-induced residual stresses of the plating were also measured for selected test structures. The corresponding COV for the compressive residual stress σ_{rcx} in the longitudinal direction was in the range of 0.12~0.52. The residual stresses of the longitudinal stiffeners or transverse frames were neither measured nor reported.

Table 15.3 summarizes the initial imperfections of the plating and stiffeners for each of the test structures on the basis of the measurements and insights

Table 15.2 Other Geometric Properties of Smith's Test Structures

Structure Number	$\frac{b}{t}$	β	$\frac{a}{r}$	λ	$\frac{A_{sx}}{bt}$
1a	76.2	2.67	21	0.24	0.42
1b	77.4	2.72	21	0.23	0.43
2a	39.5	1.42	36.5	0.42	0.40
2b	41.4	1.48	36	0.42	0.42
3a	47.8	1.68	66	0.70	0.24
3b	47.6	1.68	66	0.70	0.24
4a	39.5	1.41	50	0.54	0.28
4b	39.7	1.43	50	0.53	0.28
5	94.9	3.31	42	0.45	0.24
6	96.4	3.42	68	0.75	0.12
7	96.8	3.65	42	0.52	0.24

Note: $\beta = \frac{b}{t}\sqrt{\frac{\sigma_{Yp}}{E}}$, $\lambda = \frac{a}{\pi r}\sqrt{\frac{\sigma_{Yeq}}{E}}$, $r = \sqrt{\frac{I}{A_{sx}}}$, I = as defined in equation (15.3.30), and $A_{sx} = bt + h_{wx}t_{wx} + b_{fx}t_{fx}$.

Table 15.3 Initial Imperfections of the Plating, Longitudinal Stiffeners, and Transverse Frames for Smith's Test Structures

Structure Number	$\frac{w_{opl}}{b}$	$\frac{w_{ocx}}{a}$	$\frac{w_{oxy}}{w_{ocx}}$	$\frac{\sigma_{rcx}}{\sigma_{Yp}}$	$\frac{\sigma_{rcy}}{\sigma_{rcx}}$	$\frac{A_{om}}{w_{opl}}$
1a	0.0060	0.0007	0.7	—	—	0.1
1b	0.0077	0.0011	—	—	—	0.1
2a	0.0044	0.0025	—	0.48	0.10	0.1
2b	0.0060	0.0010	—	0.33	0.10	1.0
3a	0.0093	0.0028	0.2	0.38	0.10	0.7
3b	0.0150	0.0019	-0.8	0.43	0.10	1.0
4a	0.0081	0.0023	0.5	0.38	0.10	0.8
4b	0.0063	0.0008	0.5	0.41	0.10	0.7
5	0.0100	0.0008	-0.4	0.16	0.10	0.1
6	0.0125	0.0020	0.4	0.31	0.10	1.0
7	0.0094	0.0007	—	0.08	0.10	0.1

Notes:
1. A_{om} = buckling mode initial deflection
2. w_{opl} = maximum initial deflection of the plating
3. w_{ocx} = maximum column-type initial distortion of longitudinal stiffeners
4. w_{ocy} = maximum column-type initial distortion of transverse frames
5. σ_{rcs}, σ_{rcy} = compressive residual stresses in the longitudinal (x) or transverse (y) directions, respectively

Sources: Smith, C. S. (1976). Compressive strength of welded steel ship grillages. *Trans. RINA*, Vol. 118, 325–359; Smith, C. S., Anderson, N., Chapman, J. C., Davidson, P. C., and Dowling, P. J. (1992). Strength of stiffened plating under combined compression and lateral pressure. *Trans. RINA*, Vol. 134, 131–147.

provided in Smith (1976) and Smith et al. (1992). Based on the measured initial deflection patterns of the plating, Table 15.3 also represents the buckling mode initial deflection component for each of the test structures.

Smith et al. (1992) later computed the ultimate strengths of Smith's test structures using the nonlinear finite element method, in which they used the "two bay beam-column model" to represent the test structures, as shown in Fig. 8.6(c) in Chapter 8. They performed two types of nonlinear finite element method computations with different levels of initial deflections: finite element analysis (FEA)-1 with average initial imperfections and FEA-2 with actual initial imperfections.

Table 15.4 compares the ALPS/ULSAP solutions with Smith's mechanical test results and nonlinear finite element method solutions. Figure 15.6 represents the correlation of the theoretical solutions and the experimental results. The collapse modes predicted by ALPS/ULSAP and the experiments are also indicated in Table 15.4. The collapse of most of the test structures involves the lateral–torsional buckling of the longitudinal stiffeners (i.e., collapse mode V), as was observed in the experiments. The ALPS/ULSAP predicts the panel collapse modes reasonably well, and compares fairly well with the more refined ultimate strength data in most cases.

15.9.2 Effect of the Stiffener Dimensions

Figure 15.7 shows the variation in the ultimate strength of the longitudinally-stiffened panel under axial compression as a function of the ratio of stiffener web height to web thickness or the h_w/t_w ratio. The panel has three T-type stiffeners. The results are compared with the nonlinear finite element method solutions obtained using the two-bay–stiffened panel model shown in Fig. 8.6e of Chapter 8.

It is evident from Fig. 15.7 that as the height of stiffener web increases, the ultimate strength of the panel also increases but shows different collapse modes. This means that any ultimate panel strength calculation methods with the focus on any specific collapse mode failure cannot be applied for a wide range of the h_w/t_w ratio, but that the six collapse modes defined in this chapter must be considered altogether.

When the stiffeners are relatively small, the stiffeners buckle together with the plating in the collapse mode I failure. When the stiffeners become stiff, the plating between the stiffeners buckles while the stiffeners remain straight, and ultimate strength is eventually reached with the collapse mode III failure in accordance with beam-column type collapse. If the height of the stiffener web exceeds a critical value, however, the ultimate strength tends to flatten out. This is because the stiffener web buckles locally or twists sideways when its height is large.

Table 15.4(a) Comparison of Smith's Finite Element Analysis With the Experimental Data for the Ultimate Strength of the Test Structures

Structure Number	***p* (N/mm²)**	$\left(\frac{\sigma_{xu}}{\sigma_{Yeq}}\right)_{Exp.}$	$\left(\frac{\sigma_{xu}}{\sigma_{Yeq}}\right)_{FEA-1}$	$\left(\frac{\sigma_{xu}}{\sigma_{Yeq}}\right)_{FEA-2}$	$\frac{(\sigma_{xu})_{FEA-1}}{(\sigma_{xu})_{Exp.}}$	$\frac{(\sigma_{xu})_{FEA-2}}{(\sigma_{xu})_{Exp.}}$
1a	0	0.76	0.65	0.69	0.855	0.908
1b	0.103 (15 psi)	0.73	0.57	0.57	0.781	0.781
2a	0.048 (7 psi)	0.91	0.81	0.81	0.890	0.890
2b	0	0.83	0.82	0.82	0.988	0.988
3a	0.021 (3 psi)	0.69	0.69	0.63	1.000	0.913
3b	0	0.61	0.71	0.60	1.164	0.984
4a	0	0.82	0.80	0.75	0.976	0.915
4b	0.055 (8 psi)	0.83	0.73	0.76	0.880	0.916
5	0	0.72	0.51	0.55	0.708	0.764
6	0	0.49	—	—	—	-
7	0	0.65	0.49	0.53	0.754	0.815
				Mean	0.900	0.887
				COV	0.152	0.087

Notes: psi, pounds per square inch.

continued

Table 15.4(b) Comparison of the ALPS/ULSAP Results With the Experimental Data and Finite Element Analysis for the Ultimate Strength of the Test Structures *Continued*

Structure Number	p (N/mm²)	$\left(\frac{\sigma_{xu}}{\sigma_{Yeq}}\right)_{ULSAP}$	$\frac{(\sigma_{xu})_{ULSAP}}{(\sigma_{xu})_{Exp.}}$	$\frac{(\sigma_{xu})_{ULSAP}}{(\sigma_{xu})_{FEA\text{-}1}}$	$\frac{(\sigma_{xu})_{ULSAP}}{(\sigma_{xu})_{FEA\text{-}2}}$	Collapse Mode: Experimental Data	Collapse Mode: ULSAP
1a	0	0.76	1.000	1.169	1.101	V	V
1b	0.103	0.62	0.849	1.088	1.088	V	V
2a	0.048	0.79	0.868	0.975	0.975	III+V	I
		(0.86)	(0.945)	(1.062)	(1.062)		III
		(0.90)	(0.989)	(1.111)	(1.111)		V
2b	0	0.79	0.952	0.963	0.963	III+V	V
3a	0.021	0.69	1.000	1.000	1.095	III+V	V
3b	0	0.58	0.951	0.817	0.967	III+V	V
4a	0	0.80	0.976	1.000	1.067	III+V	V
4b	0.055	0.81	0.976	1.110	1.066	III+V	V
5	0	0.52	0.722	1.020	0.945	III+V	V
6	0	0.37	0.755	—	—	I+V	V
7	0	0.52	0.800	1.061	0.981	III+V	V
		Mean	0.895	1.020	1.025		
		COV	0.113	0.095	0.062		

Note: I+V or III+V indicate that the structures collapsed in mode I or III together with mode V. The values in parentheses are given for comparison when the ALPS/ULSAP method predicted a different collapse mode from that observed in the experiment.

15.9.3 Stiffened Panel Under Combined Biaxial Load and Lateral Pressure

The ultimate strength of a stiffened panel at the bottom of a double-hulled oil tanker is calculated under any combination of biaxial load and lateral pressure. Three methods are applied to calculate the ultimate strength of the stiffened panel: ALPS/ULSAP (2009), DNV PULS (2008), and the ANSYS nonlinear finite element method (2009). Part of the study results is extracted from Paik et al. (2008).

Figure 15.8 shows the geometrical configuration of the stiffened panel, and Table 15.5 indicates the panel's geometric properties. The spacing of the bottom longitudinal stiffeners and transverse floors of the standard bottom panel is 815 mm and 4300 mm, respectively. Using the nomenclature in Fig. 15.3, the dimension of the stiffeners is $h_w \times t_w$ = 463 × 8 mm and $b_f \times t_f$ = 172 × 17 mm. The stiffened panel is subject to biaxial compression and lateral pressure, and the magnitude of the lateral pressure applied to the structure is p = 0.16 N/mm².

The initial deflection of the plating is assumed to be $w_{opl} = \frac{b}{200}$ and the column-type initial distortion of the stiffener is assumed to be $w_{oc} = \frac{a}{1000}$. Figures 15.9 and 15.10 show the initial distortion shapes of the stiffened panel under predominantly longitudinal or transverse compressions, respectively. For

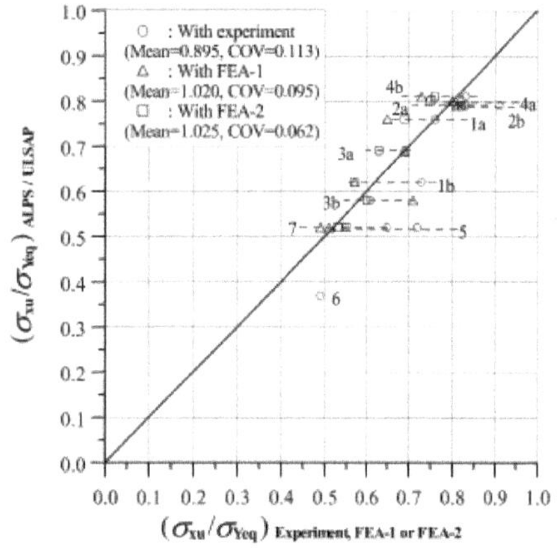

Figure 15.6 Correlation of the ALPS/ULSAP results with the experimental data and FEA solutions for Smith's test structures.

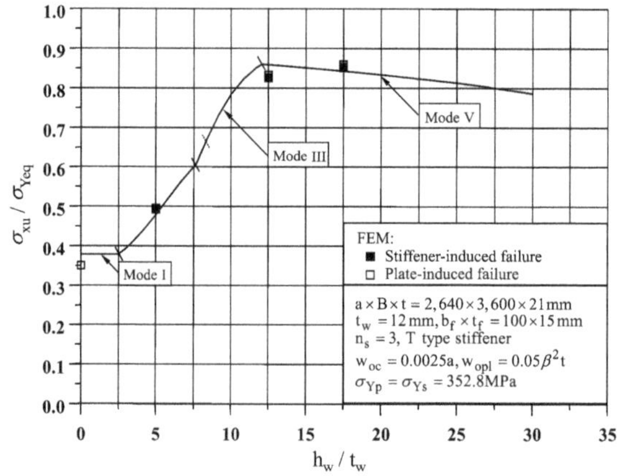

Figure 15.7 Effect of the stiffener dimensions (h_w/t_w ratio) on the ultimate strength of a stiffened panel under axial compression.

Table 15.5 Geometric Properties of the Stiffened Panel

a (mm)	B (mm)	b (mm)	n_s	t (mm)	h_w (mm)	t_w (mm)	b_f (mm)	t_f (mm)
4300	16,300	815	19	17.80	463	8	172	17

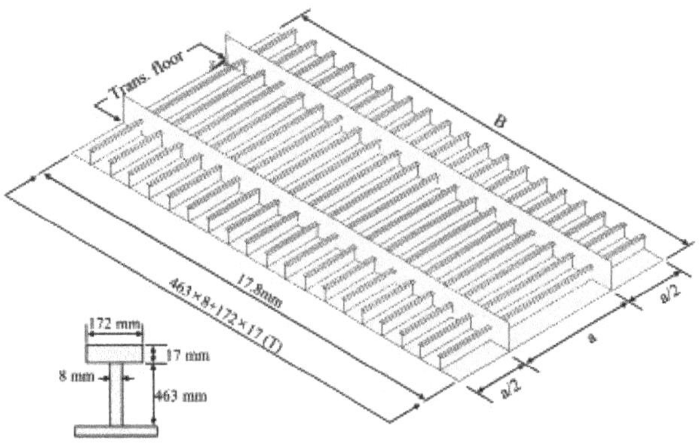

Figure 15.8 Stiffened plate structures at the base of a ship.

predominantly longitudinal compressive loads, the buckling half-wave number (m) in the longitudinal direction is 5, whereas for predominantly transverse compressive loads m becomes 1. In the nonlinear FEA, the axial compressive load is applied after the lateral pressure loading, and it is evident from Figs. 15.9 and 15.10 that the initial distortion shapes change after the lateral pressure load is applied.

Figure 15.11 shows the ANSYS nonlinear FEA results in terms of the ultimate strength behavior of the panel for various biaxial compressive loading ratios with and without lateral pressure. It is evident from the figure that lateral pressure significantly reduces the ultimate strength. When longitudinal compressive loads are predominant, the effect of the two boundary conditions—simply supported or clamped—at the longitudinal edges (i.e., along the bottom girders) is negligible. However, the boundary conditions at the longitudinal edges play a significant role when transverse axial compressive loads are predominant.

Table 15.6 summarizes the ultimate strength computations obtained by ANSYS FEA, DNV PULS, and ALPS/ULSAP. Neither the DNV PULS nor the ALPS/ULSAP method provides any specific implementation for the boundary conditions along the longitudinal edges, but the ANSYS FEA method studies the effect of longitudinal edge conditions in terms of ultimate strength. The ultimate strength of the panel with the clamped boundary condition at longitudinal edges is larger than that for its simply supported counterpart by 9.4% when uniaxial com-

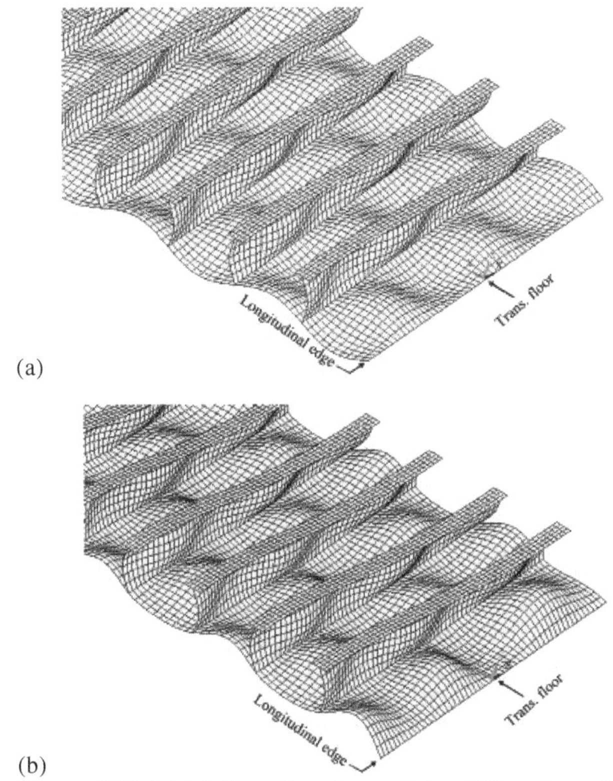

Figure 15.9 Initial distortion shape of the stiffened panel (amplification factor of 30): (a) under predominantly longitudinal compression (with m = 5) before lateral pressure loading, and (b) under predominantly longitudinal compression after lateral pressure loading.

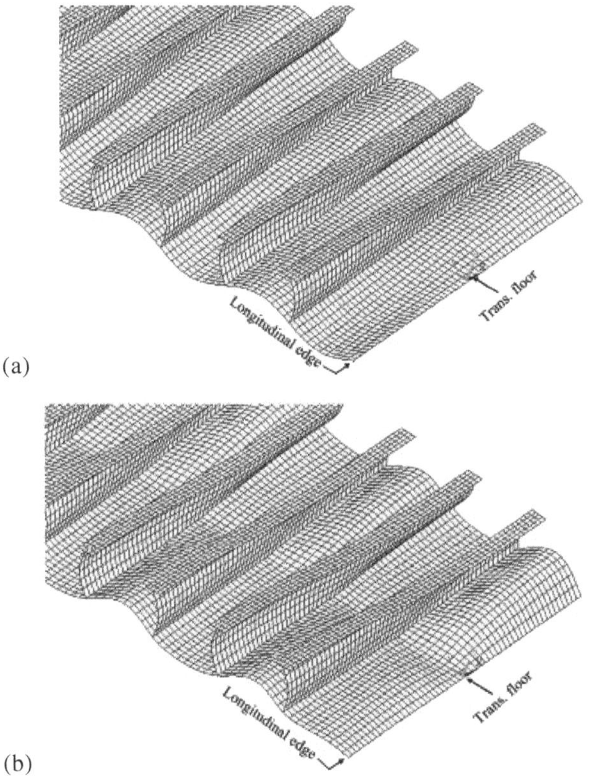

Figure 15.10 Initial distortion shape of the bottom stiffened panel (amplification factor of 30): (a) under predominantly transverse compression (m = 1) before lateral pressure loading, and (b) under predominantly longitudinal compression after lateral pressure loading.

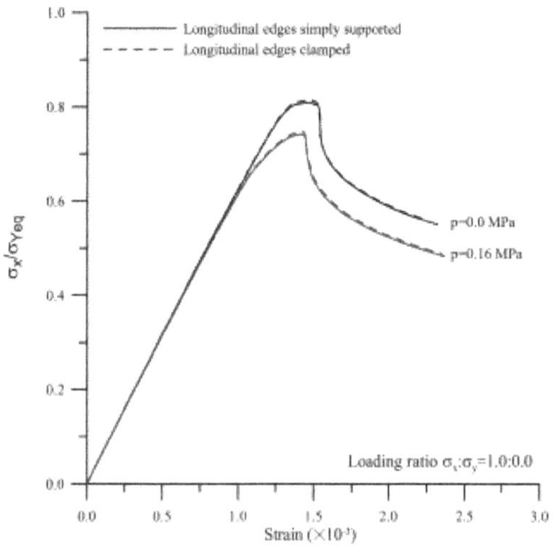

(a) Loading ratio σ_{xav}:σ_{yav} = 1.0:0.0

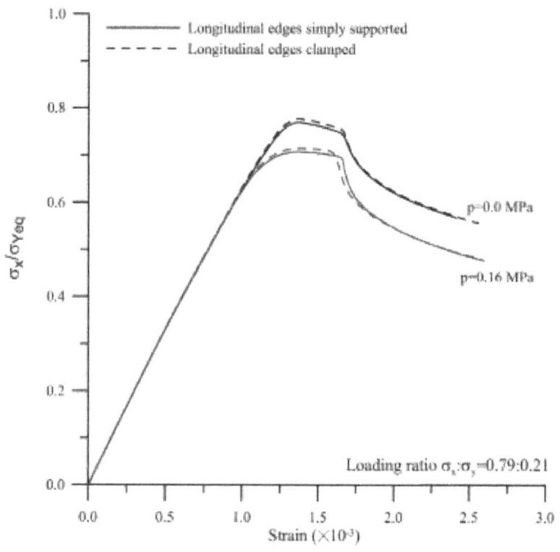

(b) Loading ratio σ_{xav}:σ_{yav} = 0.79:0.21

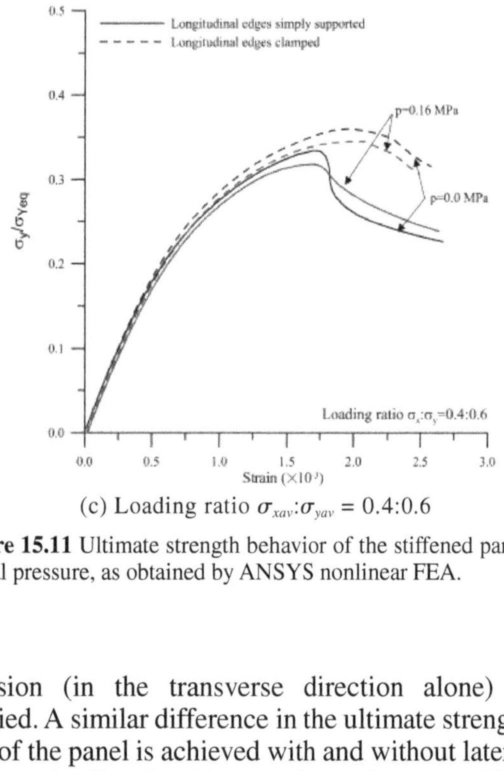

(c) Loading ratio σ_{xav}:σ_{yav} = 0.4:0.6

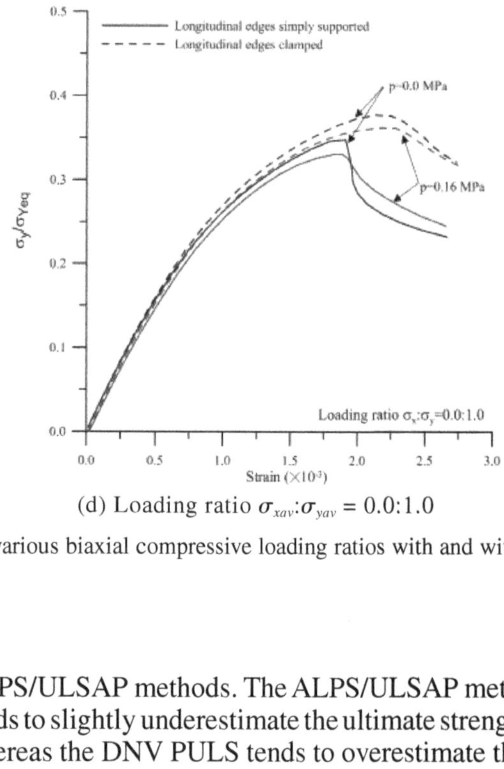

(d) Loading ratio σ_{xav}:σ_{yav} = 0.0:1.0

Figure 15.11 Ultimate strength behavior of the stiffened panel under various biaxial compressive loading ratios with and without lateral pressure, as obtained by ANSYS nonlinear FEA.

pression (in the transverse direction alone) is applied. A similar difference in the ultimate strength rate of the panel is achieved with and without lateral pressure loading for this specific study case, with p = 0.16 N/mm^2.

Figures 15.12 and 15.13 show the ultimate strength relationships of the stiffened panel under biaxial compressive loads with and without lateral pressure loads for the two boundary conditions (simply supported or clamped) at longitudinal edges as obtained by the ANSYS FEA, DNV PULS, and ALPS/ULSAP methods. The ALPS/ULSAP method tends to slightly underestimate the ultimate strengths, whereas the DNV PULS tends to overestimate them compared with the nonlinear FEA. A comparison of the results with the more refined nonlinear FEA, however, shows that both the DNV PULS and ALPS/ULSAP methods are useful for practical design purposes. It is also interesting to note that the DNV PULS solutions are closer to the ANSYS FEA results obtained for the clamped condition at longitudinal edges when lateral pressure loads are applied.

Table 15.6 Ultimate Strength Computations for a Stiffened Panel With a Varying Biaxial Compressive Loading Ratio With and Without Lateral Pressure Loads

Loading ratio	p (N/mm^2)	BC	ANSYS		ALPS/ULSAP		DNV PULS	
σ_x:σ_y			σ_{xu}/σ_{Yeq}	σ_{yu}/σ_{Yeq}	σ_{xu}/σ_{Yeq}	σ_{yu}/σ_{Yeq}	σ_{xu}/σ_{Yeq}	σ_{yu}/σ_{Yeq}
1.0:0.0	0	LS	0.8096	0	0.7920	0	0.8742	0
		LC	0.8147	0				
	0.16	LS	0.7421	0	0.7650	0	0.7905	0
		LC	0.7479	0				
0.79:0.21	0	LS	0.7698	0.2090	0.6882	0.1868	0.8414	0.2284
		LC	0.7772	0.2109				
	0.16	LS	0.7070	0.1919	0.6328	0.1718	0.7766	0.2108
		LC	0.7146	0.1940				
0.4:0.6	0	LS	0.2290	0.3341	0.2410	0.3615	0.2468	0.3702
		LC	0.2397	0.3596				
	0.16	LS	0.2120	0.3180	0.1969	0.2953	0.2319	0.3478
		LC	0.2300	0.3450				
0.0:1.0	0	LS	0	0.3478	0	0.3795	0	0.3766
		LC	0	0.3770				
	0.16	LS	0	0.3305	0	0.3056	0	0.3497
		LC	0	0.3616				

Note: BC, boundary condition; LC, longitudinal edges clamped; LS, longitudinal edges simply supported.

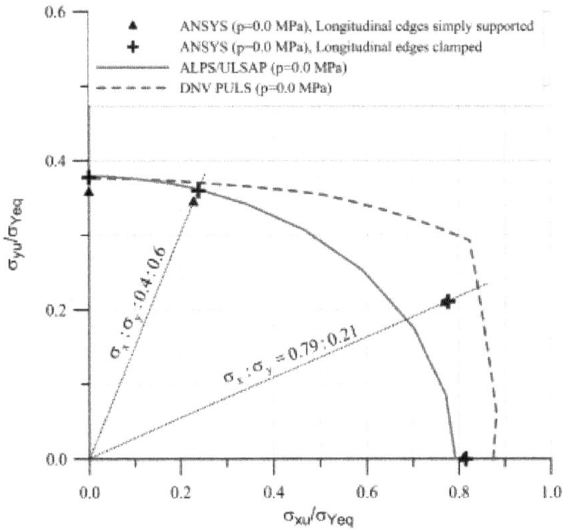

Figure 15.12 Ultimate strength interaction relationships of the stiffened panel under a biaxial compressive load with and without lateral pressure.

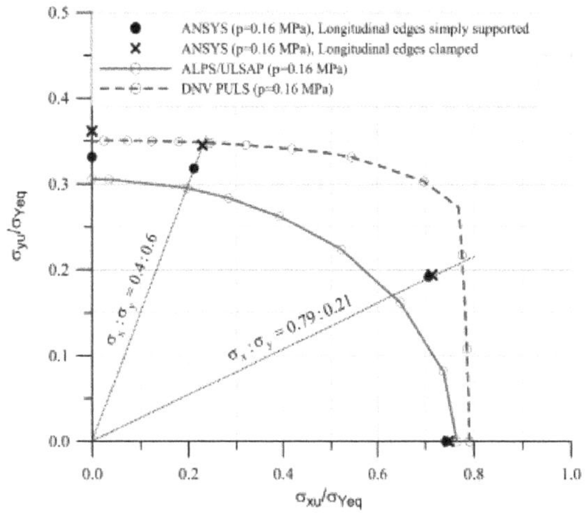

Figure 15.13 Ultimate strength interaction relationship of the stiffened panel under biaxial compressive loads with lateral pressure loads.

REFERENCES

ALPS/ULSAP. (2009). Ultimate limit state assessment of plate panels. Stevensville, MD: DRS C3 Systems. Available online at: http://www.orca3d.com/maestro.

ANSYS. (2009). Version 11.0, ANSYS Inc., Canonsburg, PA, USA.

Bleich, F. (1952). *Buckling strength of metal structures*. New York: McGraw-Hill.

DNV PULS. (2008). *User's manual (version 2.05)*. Technical Report No. 2004-0406. Oslo, Norway: Det Norske Veritas.

Hughes, O. F., and Ma, M. (1996). Inelastic analysis of panel collapse by stiffener buckling. *Computers & Structures*, Vol. 61, Issue 1, 107–117.

Paik, J. K. (2007). The effective use of experimental and numerical data for validating simplified expressions of stiffened steel panel ultimate compressive strength. *Marine Technology*, Vol. 44, Issue 2, 93–105.

Paik, J. K., Kim, B. J., and Seo, J. K. (2008). Methods for ultimate limit state assessment of ships and ship-shaped offshore structures: Part II stiffened panels. *Ocean Engineering*, Vol. 35, 271–280.

Paik, J. K., and Thayamballi, A. K. (1997). An empirical formulation for predicting the ultimate compressive strength of stiffened panels. *Proc. of International Offshore and Polar Engineering Conference*, Honolulu, Vol. IV, 328–338.

Paik, J. K., and Thayamballi, A. K. (2003). *Ultimate limit state design of steel-plated structures*. Chichester, United Kingdom: Wiley.

Paik, J. K., Thayamballi, A. K., and Kim, B. J. (2001). Large deflection orthotropic plate approach to develop ultimate strength formulations for stiffened panels under combined biaxial compression/tension and lateral pressure. *Thin-Walled Structures*, Vol. 39, 215–246.

Smith, C. S. (1976). Compressive strength of welded steel ship grillages. *Trans. RINA*, Vol. 118, 325–359.

Smith, C. S., Anderson, N., Chapman, J. C., Davidson, P. C., and Dowling, P. J. (1992). Strength of stiffened plating under combined compression and lateral pressure. *Trans. RINA*, Vol. 134, 131–147.

CHAPTER

SIXTEEN

ULTIMATE STRENGTH OF SHIP HULLS

Jeom Kee Paik
Professor, Pusan National University
Busan, Korea

16.1 INTRODUCTION

As applied hull girder loads increase, the most highly stressed structural components of a ship's hull buckle in compression or yield in tension. A ship can withstand further hull girder loading even after the buckling or yielding of a few structural components. However, the structural effectiveness of the hull decreases because of local failures, and eventually the overall hull structure reaches the ultimate limit state as the redundancy of the ship's hull becomes exhausted because of the progressive structural failures under applied hull girder loads.

Hull collapse is more likely to occur in ships suffering age-related degradation, such as corrosion wastage and fatigue cracking damage, or in those with in-service or accidental damage associated with accidental events such as collision, grounding, fire, or explosion. Although the strength performance of ship structures is not necessarily insufficient for their designed loads, which are determined for the most unfavorable environmental conditions, a ship's hull can break because of accidental flooding or unintended water ingress into the ship as this causes the hull girder loads to increase to the extent that the hull cannot sustain them.

The collapse of a ship's hull is the most catastrophic failure event because it is almost always entails the complete loss of the ship. A ship's hull can collapse if its maximum load-carrying capacity (or ultimate hull girder strength) is insufficient to sustain the corresponding hull girder loads applied. The most typical consequence of hull girder collapse is the breaking of the hull into two parts as a result of the action of extreme vertical bending moments that exceed the ultimate hull girder strength.

The prevention of hull collapse is the most important task in the design and safety assessment of ship structures. Thus, an accurate and efficient method for computing the ultimate hull girder strength is always required in robust ship structural design.

Methodologies for the computation of the ultimate hull girder strength are classified into five types: the simple-beam theory method, the presumed stress distribution-based method, the nonlinear finite element method, the idealized structural unit method (ISUM), and the intelligent supersize finite element method (ISFEM). The first two are derived by closed-form formulations that are easy to apply, whereas the latter three allow a progressive hull collapse analysis that is more sophisticated and gives more refined solutions. This chapter presents the procedures for applying these five methods to calculate the collapse strength of a ship's hull.

16.2 LESSONS LEARNED FROM PAST HULL COLLAPSE ACCIDENTS

This section presents examples of the total losses of ships associated with hull collapse accidents, and the lessons that have been learned from these events.

16.2.1 The *Titanic* Accident—Passenger Ship

One of the best-known accidents in the history of shipping is the sinking of the *Titanic Liverpool*, the full name of the *Titanic*. The ship's overall length and breadth were 269.1 m and 28.2 m, respectively, the maximum number of passengers and crew that she could carry was 2300, and her maximum operating speed was 24 knots. Figure 16.1 shows the dignified appearance of the *Titanic Liverpool*.

On April 10, 1912, she left the port of Southampton, England, on her maiden voyage to New York City. Two thousand and two hundred passengers and crew were on the voyage. Four days into her journey, at

Figure 16.1 The *Titanic Liverpool* before the accident.

11:40 PM on the night of April 14, she struck an iceberg on the port-side bow. The ship's speed at the moment of collision was reportedly 23 knots, which is an amazing speed for passenger ships even today. The consequence of the collision with the iceberg was catastrophic, because the bow structure was fractured by the collision and icy water soon poured through the ship. Because of the accidental flooding, the ship was subject to a large sagging bending moment, and when five watertight subdivisions and one boiler room were flooded, her back broke entirely in two. She sank at 2:20 AM on April 15. As the survivors reported after the accident, a part of the ship over 75 m in length rose into the sky and reached a 65- or 70-degree angle before sinking. The ship took 2 hours and 40 minutes to sink completely following the collision with the iceberg. Figure 16.2 shows a digitized image of the *Titanic Liverpool* on the seabed based on photographs taken from a deep sea surveying vehicle.

Several lessons can be learned from this accident from the viewpoint of structural mechanics and design.

First, steel tends to become brittle at low temperatures. Although even modern steel products are no exception to this rule, it is suspected that the steel material used to build the *Titanic*'s structure had insufficient fracture toughness at low temperatures. In other words, the hull structure of the *Titanic* must have been prone to brittle failure resulting from local impacts.

Second, the impact velocity at the moment of collision with the iceberg was reportedly 23 knots (or 11.8 m/s), which probably caused a large amount of initial kinetic energy and subsequently made large holes that allowed a significant amount of water to enter the ship.

Third, accidental flooding can change the hull girder load distribution and amplify the maximum hull girder bending moments.

Fourth, because the ship was sagging, the deck structures were subjected to large axial compressive loads and must have buckled and collapsed. To prevent hull breakage, ultimate limit state design meth-

Figure 16.2 The back of the *Titanic* broke in two because of accidental flooding after collision with an iceberg.

ods that consider buckling and plastic collapse should be applied in the design of ship structures.

16.2.2 The *Energy Concentration* Accident—Single-Hulled Oil Tanker

The *Energy Concentration* was a single-hulled crude oil carrier of 312.73 m in length between the perpendiculars, 48.24 m in the beam, and 25.20 m in depth. Her gross tonnage was 98,894 tons. On July 21, 1980, the back of the ship broke at the Europort in Rotterdam during the unloading of cargo oil. Figure 16.3 shows the *Energy Concentration* after the breakage of her hull girder. Because of the shallow depth of the harbor, she did not disappear beneath the water after breaking into two, but her midsection reportedly touched the bottom of the pier. Evidently, total loss was the outcome.

Again, several lessons can be learned from this accident from the point of view of ship structural design. First, the poorly executed unloading of cargo can amplify the maximum hull girder bending moments to the extent that they exceed the maximum load-carrying capacity of the hull structures. Second, deck panels or bottom panels should be designed using ultimate limit state design methods so that the ultimate hull girder strength is able to withstand unintended scenarios of cargo loading and unloading that cause uncertainties in the design load calculations and subsequently affect the structural design process. See Rutherford and Caldwell (1990), who investigated the ultimate hull strength of the *Energy Concentration*, for more details.

16.2.3 The *M.V. Derbyshire* Accident—Double Side-Hulled Bulk Carrier

The *M.V. Derbyshire* was a double side-hulled Capesize bulk carrier of 281.94 m in length between the perpendiculars, 44.2 m in the beam, and 25 m in depth. Her maximum deadweight was 173,218 tons. She was 5 years old at the time of the accident, and was believed to have suffered almost no age-related degradation such as corrosion wastage. Another distinct characteristic of the ship is that she had a double-sided hull arrangement that aimed to prevent unintended water ingress into the cargo holds from the failure of the side shell structures.

On September 9, 1980, she sank in the northwest Pacific, some 400 miles south of Shikoku Island, Japan, during typhoon Orchid while on a voyage from Canada to Japan carrying fine iron ore concentrates. On her last voyage from the Sept Isles, Canada, to Yokohama, Japan, she was carrying about 158,000 tons of fine ore concentrates distributed across seven of her nine holds. Her estimated displacement as she approached Japan was about 194,000 tons, indicating a mean draught of approximately 17 m.

Just before sinking, she was within the most dangerous ambit of typhoon Orchid, and the significant wave height soon before her sinking was reportedly 14 m. There was no distress signal, and only two sightings of oil up-welling were seen some days later to indicate the position of the sunken craft. A damaged lifeboat from the ship was sighted, but this was not recovered and subsequently sank. This and the absence of a distress signal were taken to imply that she sank very quickly.

The lessons that can be learned from this accident from the point of view of ship structural design are as follows. First, abnormal waves not expected in the structural design can occur and amplify the maximum hull girder loads, which may reach or even exceed the corresponding design values. Second, unintended water ingress into cargo holds, which may occur because of hatch cover failure, can further amplify the hull girder loads. Third, the allowable working stress design approach that was applied in the structural design of the *M.V. Derbyshire* cannot deal with this issue, and thus the ultimate limit state design method should be employed to prevent hull girder collapse accidents.

Readers are referred to the papers of Paik and Faulkner (2003) and Paik et al. (2008), who investigated the sinking of the *M.V. Derbyshire* with a focus on hull girder collapse. Paik and Thayamballi (1998) also established some credible scenarios for the sinking of bulk carriers, a ship type for which total loss very frequently occurred in the 1980s and early 1990s. Figure 16.4 shows the total loss scenarios of bulk carriers developed by Paik and Thayamballi (1998).

16.2.4 Anonymous Capesize Bulk Carrier

A similar type of hull collapse accident to that of the *Energy Concentration* occurred in a Capesize bulk carrier during the unloading of cargo iron ore. Figure 16.5 shows the sagging ship with a broken back. At the time of the accident, the central cargo hold of the ship was still full, but the bow and aft holds were empty. The ship was 23 years old, implying that she must have suffered age-related degradation such as corrosion wastage and fatigue cracking damage.

Several lessons can be learned from this accident from the point of view of ship structural design. First, as was found with the *Energy Concentration*, the poor execution of cargo unloading can amplify the hull girder bending moments. Second, age-related degradation such as corrosion wastage and fatigue cracking damage can reduce the hull girder strength performance. Third, ultimate limit state-based methods can

Figure 16.3 The back of the *Energy Concentration* broke because of poorly executed cargo oil unloading.

better deal with the issue of hull collapse in ship structural design, as described in Fig. 16.4.

16.2.5 The *Erika* Accident—Single-Hulled Oil Tanker

The 24-year-old single-hulled oil tanker *Erika* broke up in the Bay of Biscay on December 12, 1999, causing the spillage of some 7000 to 10,000 tons of oil. Immediately before the accident, she was faced with structural problems in very rough sea conditions, which were reportedly a westerly wind of force 8 to 9 with a 6 m swell. Figure 16.6 shows the *Erika* as she sank.

Several lessons can be learned from this accident from the point of view of ship structural design. First, rough sea conditions can amplify hull girder loads to the extent that they reach or even exceed the corresponding design values. Second, age-related degradation such as corrosion wastage and fatigue cracking damage can decrease the hull girder strength. Third, either an increase in applied hull girder loads or a decrease in hull girder strength or both can result in the collapse of the hull girder. Figure 16.4 can also be applied to explain this scenario.

16.2.6 The *Prestige* Accident—Single-Hulled Oil Tanker

A similar accident to that which befell the *Erika* happened to the 26-year-old single-hulled oil tanker *Prestige* in heavy weather conditions on November 13, 2002. Figure 16.7 shows the *Prestige* accident as her back broke. The ship, which was carrying 77,000 tons of heavy fuel oil loaded in St Petersburg, Russia, and Ventspils, Latvia, was heading to Singapore via

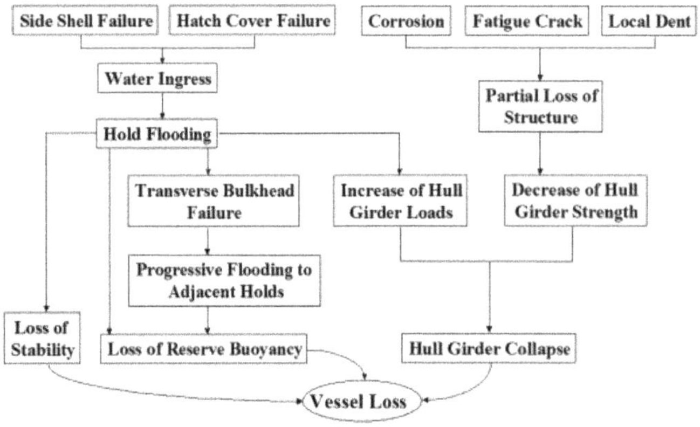

Figure 16.4 Total loss scenarios of bulk carriers developed by Paik and Thayamballi (1998).

Figure 16.5 The back of the 23-year-old Capesize bulk carrier broke during the unloading of cargo at port.

Gibraltar. The findings and lessons of this accident are similar to those of the *Erika* accident.

16.2.7 The *M.S.C. Napoli* Accident—4419-Twenty Equivalent Unit Container Vessel

On January 18, 2007, the British container ship *M.S.C. Napoli*, which could carry up to 4419 twenty equivalent unit containers, was en route from Antwerp to Lisbon when it was caught in a storm at the entry to the Channel and suffered a leak and failure of the steering system. She was transporting 2394 containers on this voyage that held nearly 42,000 tons of merchandise, of which some 1700 tons were classed as hazardous substances such as explosives, flammable gases, liquids and solids, oxidants, toxic substances, and corrosive materials. In her bunkers, she held over 3000 tons of heavy fuel oil.

In contrast to the foregoing hull collapse accidents, the back of the *M.S.C. Napoli* did not break, as shown in Fig. 16.8, but the bulkhead structures in between the engine room and the aft cargo hold buckled, requiring the performance of an emergency risk assessment to establish schemes for the evacuation and treatment of the hazardous substances. The

Figure 16.6 The back of the 24-year-old *Erika* broke in rough weather conditions.

Figure 16.7 The back of the 26-year-old *Prestige* broke in rough weather conditions.

ultimate strength and hull collapse performance of the ship were certainly the main issues.

Three salient lessons can be learned from this accident from the point of view of structural design (Marine Accident Investigation Branch, 2008; Ko et al., 2010). First, the bulkhead structure of the container ship between the engine room and the cargo hold was subject to a large vertical shearing force. Second, the horizontally-framed transverse bulkhead in front of the aft cargo holds is prone to buckle under shearing forces. Third, the buckling and ultimate strength performance must be checked to optimize the structural design of the transverse bulkheads.

16.2.8 The *Sea Prince* Accident—Single-Hulled Oil Tanker

On July 23, 1995, the single-hulled oil tanker *Sea Prince* grounded as she attempted to leave the port of Yosu in South Korea to a safety bay to shelter from an incoming typhoon, causing the spillage of 5000 tons of oil of the 85,000 tons that had been loaded. The engine room caught fire, as shown in Fig. 16.9, and presumably the bottom structures suffered grounding damage. The remaining oil in the

Figure 16.8 The *M.S.C. Napoli* accident.

Figure 16.9 The *Sea Prince* after grounding and a fire in her engine room.

cargo tanks was transferred to barges when the fire was eventually put out on July 24. When the weather conditions further improved on July 25, lighterage and recovery operations were started and continued for 19 days. The ship was eventually refloated and towed out of Korean waters, but sank during towing. It is believed that the ship sank possibly because of hull girder collapse initiated by the failure of the damaged bottom structures.

The structural design lessons that can be learned from this accident are first that accidents such as grounding or collision can cause structural damage to the bottom or side structures, and second that hull girder strength can be decreased because of accidental damage, which means that the residual strength of damaged hulls may be lower than the applied hull girder loads, causing hull girder collapse.

16.3 SIMPLE-BEAM THEORY METHOD

The simple-beam theory method gives the first-failure hull girder strength rather than the ultimate hull girder strength. This method is very easy to apply, and is thus considered useful at the very early stages in the design of hull structures. However, the simple-beam theory method does not take into account the effect of local failures of structural components, except for compressed flanges, or the interacting effect between local and global system failures.

This section presents the details of the simple-beam theory method. Some examples of the application of the method to calculate the hull strength using the simple-beam theory method are presented later in Section 16.10.

16.3.1 First-Failure Vertical Bending Moments

According to the simple-beam theory, the bending stress at the cross-section for a beam subject to a bending moment is calculated as follows.

$$\sigma = \frac{M}{I} z \tag{16.3.1}$$

where σ is the bending stress, M is the applied bending moment, I is the moment of inertia, and z is the distance from the neutral axis position of the beam cross-section to the location of the bending stress calculation in the direction of the depth of the beam.

The maximum bending stress will develop at the outmost fiber of the cross-section of the beam, and can thus be obtained from equation (16.3.1) as follows.

$$\sigma_{\max} = \frac{M}{S} \tag{16.3.2}$$

where $\sigma_{\max}$ is the maximum bending stress at the outmost fiber of the beam's cross-section and S is the section modulus.

In equation (16.3.2), the section modulus S for the cross-section of a ship's hull has two components, as follows.

$$S_d = \frac{I}{z_d} \text{ at the deck and } S_b = \frac{I}{z_b} \text{ at the bottom} \tag{16.3.3}$$

where S_d and S_b are the vertical section moduli at the deck or bottom, respectively, and z_d and z_b are the distances from the neutral axis position of the hull cross-section to the deck or bottom, respectively.

z_d and z_b in equation (16.3.3) can be obtained as follows.

$$z_b = g - \frac{t_b}{2},\ z_d = D - g - \frac{t_d}{2} \tag{16.3.4}$$

where D is the depth of the ship, t_b is representative thickness of the bottom plate, t_d is representative thickness of the deck plate, and g is the distance from the baseline of the ship to the neutral axis position.

In equation (16.3.4), g can be calculated as follows.

$$g = \frac{\sum_{j=1}^{n} a_j z_j}{\sum_{j=1}^{n} a_j} \tag{16.3.5}$$

where a_j is the cross-sectional area of the jth member (portion), z_j is the distance from the baseline to the neutral axis of the jth member (portion), and n is the total number of members to be included in the section modulus calculation.

The moment of inertia I for the ship cross-section in equation (16.3.3) can now be calculated as follows.

$$I = \sum_{j=1}^{n} (a_j z_j^2 + i_j) - Ag^2 \quad (16.3.6)$$

where $A = \sum_{j=1}^{n} a_j$ is the total area of the hull cross-section, a_j, z_j, g, and n are as defined in equation (16.3.5), and i_j is the moment of inertia for the jth member (portion) about its own neutral axis. The moment of inertia i and neutral axis position z_o of the inclined and curved plating shown in Fig. 16.10 are approximately given by

$$i = \frac{1}{12} ad^2, \; z_o = \frac{d}{2} \text{ for the inclined plating} \quad (16.3.7a)$$

$$i = \left(\frac{1}{2} - \frac{4}{\pi^2}\right) ar^2, \; z_o = \frac{(\pi - 2)r}{\pi} \text{ for the curved plating} \quad (16.3.7b)$$

where a is the cross-sectional area of the inclined or curved plating, d is the projected depth of the inclined plating as defined in Fig. 16.10, and r is the radius of the curvature of the curved plating as defined in Fig. 16.10.

In the simple-beam theory method, a ship's hull will reach the first-failure (collapse) state when the maximum bending stress on the compressed side reaches the ultimate compressive strength of the compressed flange, which is the deck panel with sagging and the outer bottom panel with hogging.

The first-failure bending moments for a ship's hull subject to a vertical bending moment are calculated as follows.

$$M^v_{fs} = S_d \sigma_{ud}, \; M^v_{fh} = S_b \sigma_{ub} \quad (16.3.8)$$

where M^v_{fs} and M^v_{fh} are the first-failure vertical bending moments for sagging and hogging, respectively, and σ_{ud} and σ_{ub} are the ultimate compressive stresses for the deck panel and the outer bottom panel, respectively.

The ship's hull can usually sustain further hull girder loading even after the first-failure status is reached because the structural failures can grow into the vertically positioned structures such as longitudinal bulkheads and side shell structures until the ship's hull reaches the ultimate limit state.

16.3.2 First-Failure Horizontal Bending Moments

The first-failure hull strength for a horizontal bending moment can be calculated in a similar way to the vertical bending moment by simply rotating the reference axis. The first-failure bending moments for a ship's hull subject to a horizontal bending moment can be calculated as follows.

$$M^h_{fs} = S_p \sigma_{up}, \; M^h_{fh} = S_s \sigma_{us} \quad (16.3.9)$$

where M^h_{fs} and M^h_{fh} are the first-failure horizontal bending moments for sagging and hogging, respectively, S_p and S_s are the horizontal section moduli at the port side and starboard side panels, respectively, and σ_{up} and σ_{us} are the ultimate compressive stresses for the port side and the starboard side panel, respectively.

16.4 PRESUMED STRESS DISTRIBUTION-BASED METHOD

In the presumed stress distribution-based method, the stress distribution at the ultimate limit state of a ship's hull is presumed over the hull cross-section based on theoretical, numerical, or experimental investigations, and the presumed stresses are then integrated across the hull cross-section to calculate the corresponding ultimate hull girder strength. This

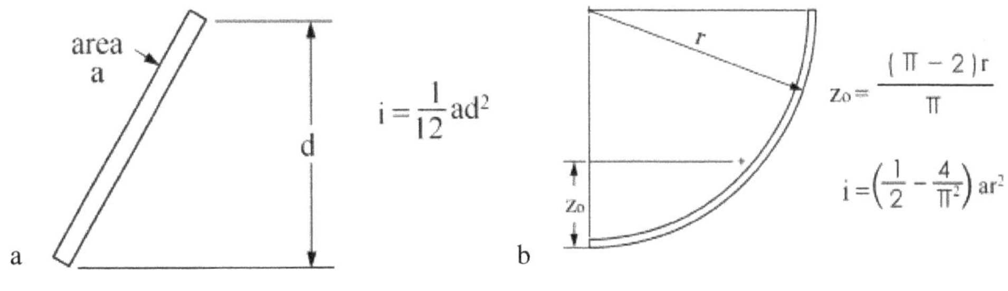

Figure 16.10 Nomenclature for the inclined and curved plating.

method takes into account the effect of local structural failures more precisely than the simple-beam theory method.

16.4.1 Ultimate Vertical Bending Moments

The pioneer of the presumed stress distribution-based method for calculating the ultimate vertical bending moments of a ship's hull was Caldwell (1965). He presumed a bending stress distribution over the hull cross-section at the ultimate limit state under vertical bending moments, as shown in Fig. 16.11, in which all of the materials in compression have reached their ultimate strength with buckling and all of the materials in tension have yielded. He then calculated the ultimate bending moments by integrating the presumed bending stresses over the hull cross-section.

The stress distribution presumed by Caldwell, however, does not represent the ultimate limit states of modern ship structures, resulting in overestimated calculations of the ultimate hull girder strength.

Based on experimental studies of large-scale ship's hull models (e.g., Dow, 1991) and numerical studies of full-scale ships (e.g., Rutherford and Caldwell, 1990; Paik et al., 1996), it is recognized that the overall collapse of a ship's hull under a vertical bending moment is governed by the collapse of the compressed flange, but there is still some reserve strength after the compressed flange has collapsed. This is because after the compressed flange buckles, the neutral axis of the hull cross-section moves toward the tensioned flange, and a further increase in the applied bending moment is sustained until the tensioned flange yields. At later stages of this process, vertical structures (e.g., longitudinal bulkheads or side shell structures) around the compressed flange and the tensioned flange may also fail. However, in the vicinity of the final neutral axis, the vertical structures usually remain in a linear elastic state until the overall collapse of the hull girder occurs. Depending on the geometric and material properties of the hull's cross-section, these parts may of course fail, which corresponds with Caldwell's presumption.

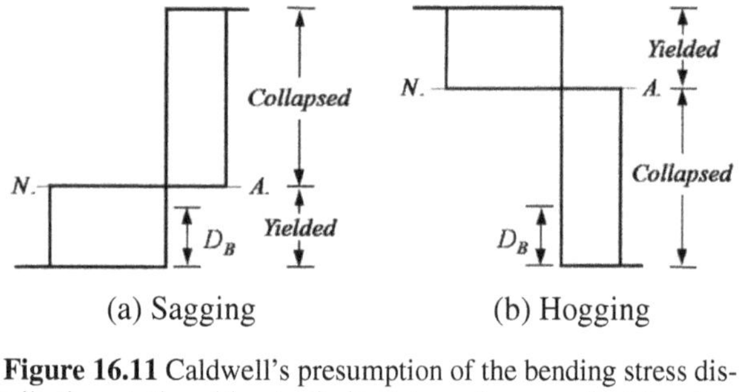

Figure 16.11 Caldwell's presumption of the bending stress distribution at the ultimate limit state under a vertical bending moment for a simplified cross-section of a ship's hull under sagging or hogging (N.A., neutral axis).

Figure 16.12 shows a typical example of the bending stresses across the hull cross-section of a single-hulled oil tanker at the ultimate limit state under a vertical hogging bending moment obtained through numerical investigations (Paik et al., 1996). It is evident from Fig. 16.12 that the compressed flange (the bottom panel) collapses and the tensioned flange (the deck panel) yields until the ultimate strength has been reached, but the vertical structures in the vicinity of the neutral axis (N.A.) are still intact (linear elastic). This means that the approach based on Caldwell's presumption of the bending stress distribution can greatly overestimate the strength of a ship's hull against collapse.

Figure 16.13 shows the bending stress distribution across the cross-section of a ship's hull at the ultimate limit state under sagging or hogging bending moments presumed by Paik and Mansour (1995), in which the bending stress distribution is grouped into four regions to represent a more realistic configuration than that given by Caldwell (1965).

In a sagging condition, regions 1 and 2 are under tension and regions 3 and 4 are under compression. Region 1 represents the outer bottom panels, which have yielded to reach a yield stress σ^Y_x, and region 4

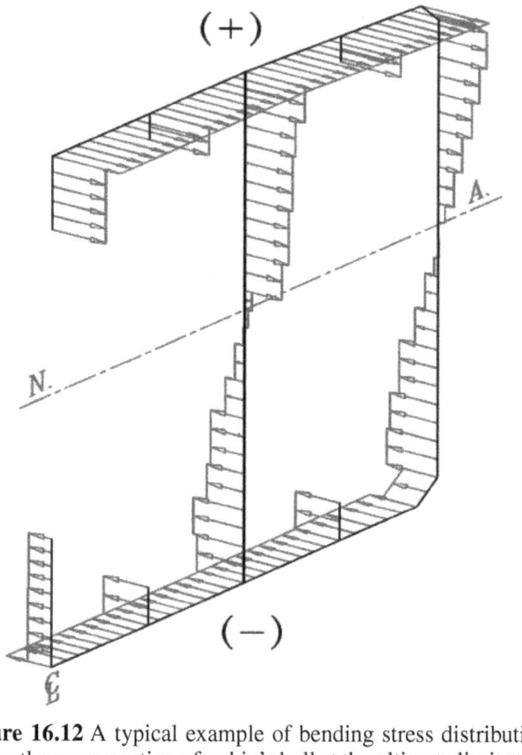

Figure 16.12 A typical example of bending stress distribution across the cross-section of a ship's hull at the ultimate limit state under a hogging bending moment (+, tension; –, compression), obtained by numerical investigations (Paik et al., 1996).

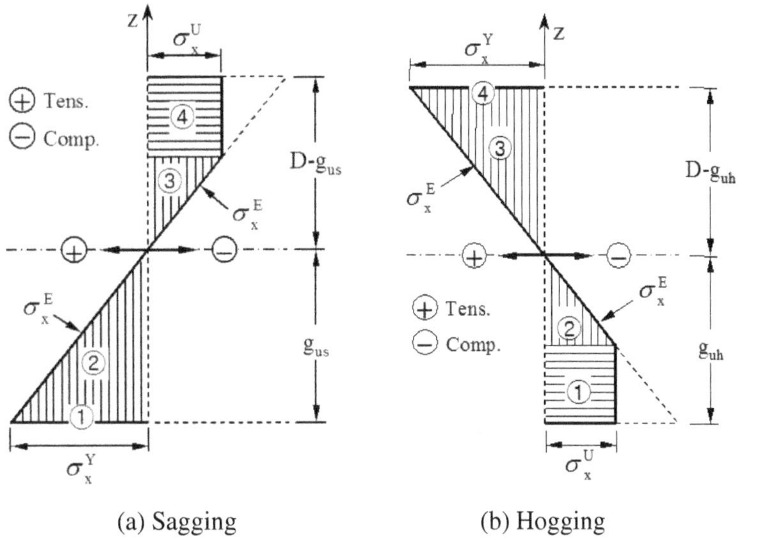

Figure 16.13 Paik and Mansour's (1995) presumption of the bending stress distribution across the cross-section of a ship's hull at the ultimate limit state under sagging or hogging (+, tension; –, compression) (the superscripts *U*, *Y*, *E* denote the ultimate strength, yielding, and elastic region, respectively).

represents the upper deck panels and the upper part of the vertical structures, which have buckled and collapsed to reach an ultimate stress σ^U_x. Regions 2 and 3, however, remain in a linear elastic or unfailed state, reaching an elastic stress σ^E_x.

In a hogging condition, regions 1 and 2 are under compression and regions 3 and 4 are under tension. Region 1, which represents the outer bottom panels and the lower part of the vertical structures, has buckled and collapsed to reach an ultimate stress σ^U_x, and region 4, which represents the upper deck panels, has yielded to reach a yield stress σ^Y_x. Regions 2 and 3 remain in the linear elastic regime, reaching an elastic stress σ^E_x.

The height of region 4 (the upper part of the vertical structures) in a sagging condition or the height of region 1 (the lower part of the vertical structures) in a hogging condition following buckling and collapse can be assigned depending on the geometrical and material properties of the ship's hull structure.

Under a vertical bending moment, the summation of axial forces over the entire cross-section of the hull becomes zero, as follows.

$$\int \sigma_x dA = 0 \qquad (16.4.1)$$

where $\int dA$ is the integration across the entire cross-section of the hull.

By solving equation (16.4.1), the height of region 4 in a sagging condition or the height of region 1 in a hogging condition can be defined. The distance g_u from the ship's baseline (reference position) to the horizontal neutral axis of the cross-section of the ship's hull at the ultimate limit state can then be obtained as follows.

$$g_u = \frac{\sum_{i=1}^{n} |\sigma_{xi}| a_i z_i}{\sum_{i=1}^{n} |\sigma_{xi}| a_i} \qquad (16.4.2)$$

where z_i is the distance from the ship's baseline (reference position) to the horizontal neutral axis of the ith structural component, σ_{xi} is the longitudinal stress of the ith structural component following the presumed stress distribution, a_i is the cross-sectional area of the ith structural component, and n is the total number of structural components. g_u is denoted by g_{us} in a sagging condition and by g_{uh} in a hogging condition.

The ultimate vertical bending moment M^v_u is then calculated as the first moment of the bending stresses about the neutral axis position, as follows.

$$M^v_u = \sum_{i=1}^{n} \sigma_{xi} a_i (z_i - g_u) \qquad (16.4.3)$$

where n is the total number of structural components and g_u is as defined in equation (16.4.2). M^v_u is denoted by M^v_{us} (negative value) for a sagging condition and by M^v_{uh} (positive value) for a hogging condition.

16.4.2 Ultimate Horizontal Bending Moments

Under horizontal bending moments, the distance g_u from the ship's reference position (the outermost point on the starboard side) to the vertical neutral axis of the ship's hull cross-section can be obtained as follows.

$$g_u = \frac{\sum_{i=1}^{n} |\sigma_{xi}| a_i y_i}{\sum_{i=1}^{n} |\sigma_{xi}| a_i} \tag{16.4.4}$$

where y_i is the distance from the ship's outermost point on the starboard side (reference position) to the vertical neutral axis of the ith structural component, σ_{xi} is the longitudinal stress of the ith structural component following the presumed stress distribution, and a_i is the cross-sectional area of the ith structural component. g_u is denoted by g_{us} for a sagging condition and by g_{uh} for a hogging condition.

The ultimate horizontal bending moment M_u^h is then calculated as the first moment of the bending stresses about the neutral axis position, as follows.

$$M_u^h = \sum_{i=1}^{n} \sigma_{xi} a_i (y_i - g_u) \tag{16.4.5}$$

where n is the total number of structural components and g_u is as defined in equation (16.4.4). M_u^h is denoted by M_{us}^h (negative value) for a sagging condition and by M_{uh}^h (positive value) for a hogging condition.

16.4.3 Ultimate Vertical Shearing Forces

Under vertical shearing forces, it is considered that horizontally positioned structures such as the deck panels and inner or outer bottom panels do not provide any resistance against the applied loads, and that it is the vertically positioned structures such as longitudinal bulkheads and side shell structures that sustain such loads. In this regard, a shear stress distribution can be presumed such that all of the vertically positioned structures have reached the ultimate stress τ_u, but the stress of all of the horizontally positioned structures is assumed to be zero.

The ultimate vertical shearing force F_u^v can then be calculated based on the presumed stress distribution, as follows.

$$F_u^v = \sum_{i=1}^{n} a_{vi} \tau_{ui} \tag{16.4.6}$$

where τ_{ui} is the ultimate shear stress of the ith vertically positioned structural component, a_{vi} is the cross-sectional area of the ith vertically positioned structural component, and n is the total number of the vertically positioned structural components.

16.4.4 Ultimate Horizontal Shearing Forces

Under horizontal shearing forces, it is considered that only the horizontally positioned structures now contribute to sustaining the applied loads. The stress distribution at the ultimate limit state for the cross-section of a ship's hull under horizontal shearing forces is then presumed such that the horizontally positioned structures have reached the ultimate shear stress and the stress of the vertically positioned structures is zero. The ultimate horizontal shearing forces can be obtained by the integration of the stress distribution, as follows.

$$F_u^h = \sum_{i=1}^{n} a_{hi} \tau_{ui} \tag{16.4.7}$$

where τ_{ui} is the ultimate shear stress of the ith horizontally positioned structural component, a_{hi} is the cross-sectional area of the ith horizontally positioned structural component, and n is the total number of horizontally positioned structural components.

16.5 NONLINEAR FINITE ELEMENT METHOD

The simple-beam theory method and the presumed stress distribution-based method described previously cannot take into account the effect of progressive failures of structural components until the ship's hull reaches the ultimate limit state and beyond, although they do give approximate solutions for the first-failure or ultimate hull girder strength itself.

The nonlinear finite element method (FEM) described in Chapter 8 gives a much more refined computation of the progressive collapse behavior of a ship's hull, as it takes into account the effect of interactions between local failures of individual structural components and the overall failure of the hull system structure. It is, however, important to realize that the resulting computations may be totally wrong if the FEM modeling technique applied is inadequate.

The structural modeling techniques described in Chapter 8 are applicable to the analysis of progressive hull collapse in terms of modeling various aspects such as mesh size and initial imperfections. Six types of modeling can be considered in determining the extent of progressive hull collapse: (1) the entire hull model, (2) the three cargo hold model, (3) the two cargo hold model, (4) the one cargo hold model, (5) the two-bay sliced hull model, and (6) the one-bay sliced hull model, as shown in Fig. 16.14.

The computational accuracy may worsen from (1) to (6), but the computational efficiency improves. In reality, the application of the conventional nonlinear FEM to (1) the entire hull model is usually impractical because of the great computational effort required. When a vertical or horizontal bending moment is a predominant hull-girder load compo-

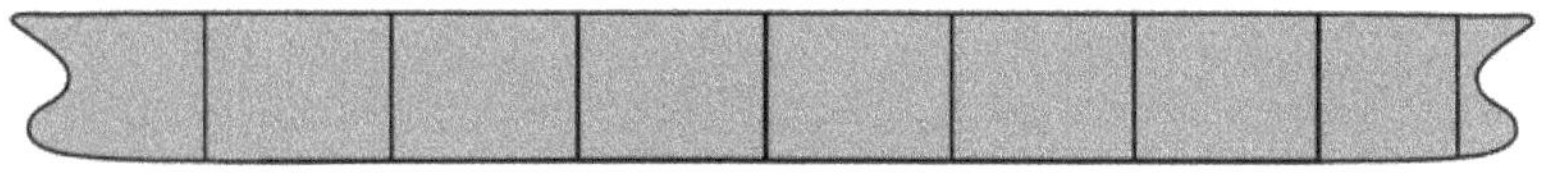

(a) The entire hull model

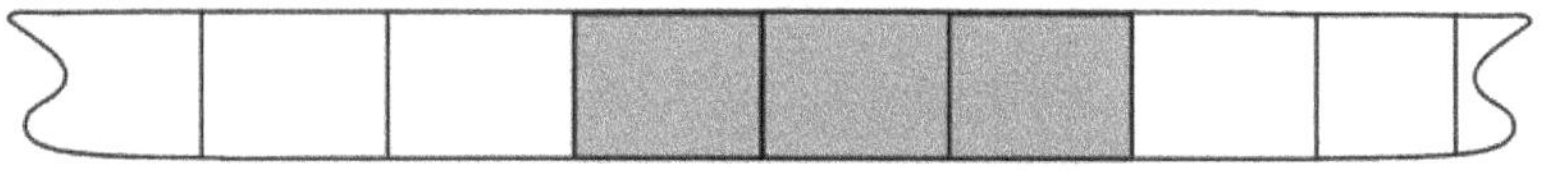

(b) The three cargo hold model

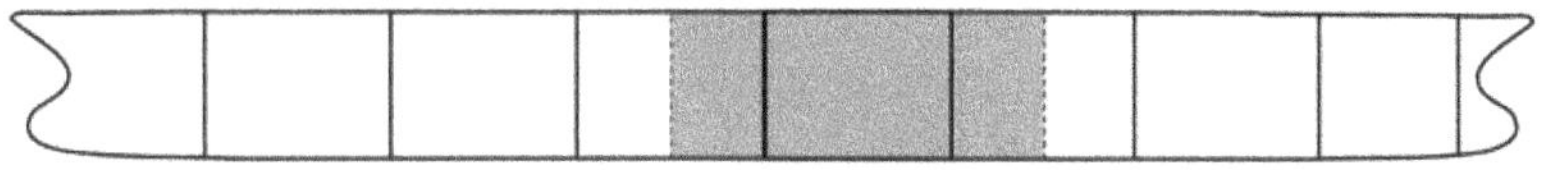

(c) The two cargo hold model

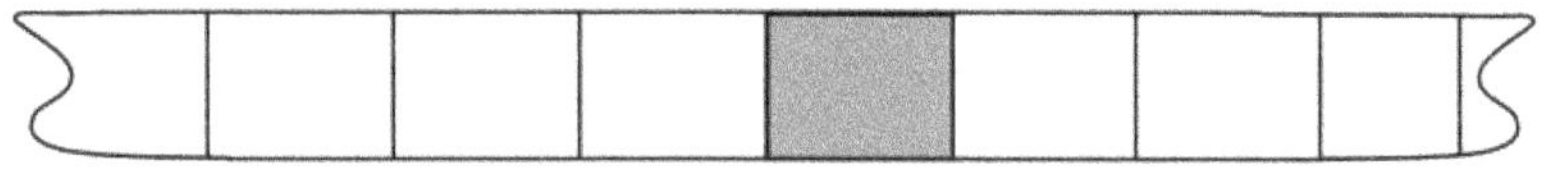

(d) The one cargo hold model

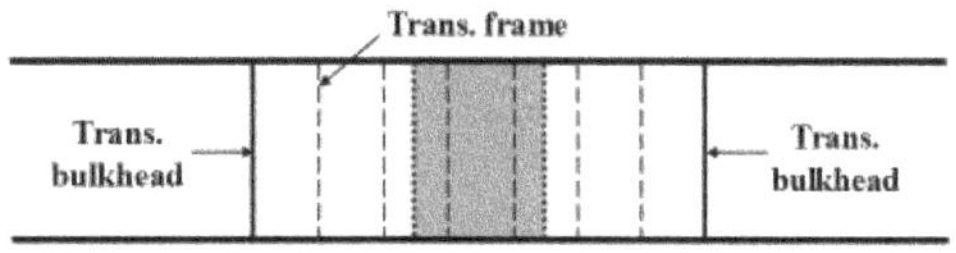

(e) The two-bay sliced hull cross-section model

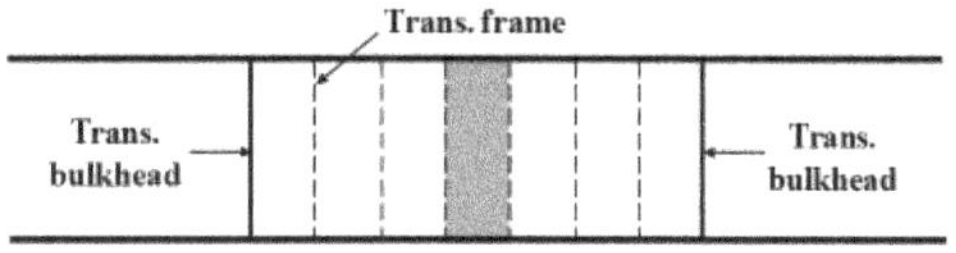

(f) The one-bay sliced hull cross-section model

Figure 16.14 Model types for progressive hull collapse analysis by the nonlinear finite element method.

nent and the transverse frames are strong enough not to fail before the stiffened panels between the two adjacent transverse frames reach the ultimate limit state, (6) the one-bay sliced hull model is often adopted, as it is considered that the resulting computations are good enough.

To take into account the effect of rotational restraints on the transverse frames, it is recommended to adopt (5) the two-bay sliced hull model, which is composed of half a bay panel, one bay panel, and half a bay panel with two transverse frames. When vertical or horizontal shearing forces are applied, with or without vertical or horizontal bending moments, however, the transverse frames can fail or at least deform significantly before the stiffened panels between the adjacent transverse frames reach the ultimate limit state, and thus at least (4) the one cargo hold model must be applied in this case.

To take into account the effect of rotational restraints at the transverse bulkheads, the use of (2) the three cargo hold model or (3) the two cargo hold model composed of half a cargo hold, one cargo hold, and half a cargo hold with two transverse bulkheads is recommended.

General purpose FEM software typically employs an incremental technique to assign the applied hull girder loads until the ultimate hull girder strength is reached and beyond, but it is not always convenient to deal with the progressive hull collapse analysis under hull girder loads in this way. Instead, the use of purpose-built software is often desirable.

When vertical or horizontal bending moments are applied, the hull cross-section is considered to remain plane, and thus the hull girder loads should be applied with reference to the neutral axis of the hull cross-section at each incremental loading step. When the compressed flanges of hull girder structures under bending moments, such as the deck panels in a vertical sagging condition or the bottom panels in a vertical hogging condition, start to fail by buckling and plastic collapse, the neutral axis position of the hull cross-section is subsequently changed, for example moving downward in a sagging condition and upward in a hogging condition.

Figure 16.15 illustrates—and the text that follows describes—a technique for managing changes in the neutral axis position of the hull cross-section using general purpose FEM software and the one-bay

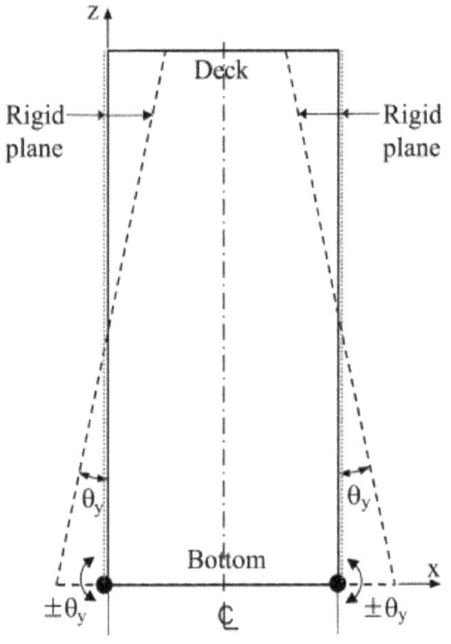

Figure 16.15 Management of changes in the neutral axis position of the hull cross-section under vertical bending moments in the one-bay sliced hull model using general purpose nonlinear FEM software.

sliced hull model between two transverse frames. It is an easy-to-apply procedure for dealing with changes in the neutral axial position under vertical bending moments.

- Step 1: Keeping the hull cross-section plane, apply the rotation angle θ_y incrementally with regard to the two corners at either the outer bottom or upper deck of the hull model. This can be achieved by applying the displacement control technique to the nodal points. Negative rotation generates a sagging bending moment and positive rotation generates a hogging bending moment. The nonlinear finite element analysis is continued until the hull structure reaches the ultimate strength and beyond.
- Step 2: In the postprocessing of the computed nonlinear FEM results, the vertical bending moment versus the curvature curve is identified as the rotation angle θ_y increases. The vertical bending moment is calculated as the first moment of the longitudinal stresses in individual finite elements in terms of the updated neutral axis position, which is obtained in a similar way to equation (16.4.1) by

$$g = \frac{\sum_{i=1}^{n} |\sigma_{xi}| a_i z_i}{\sum_{i=1}^{n} |\sigma_{xi}| a_i} \tag{16.5.1}$$

where g is the distance from the baseline of the ship to the neutral axis of the hull cross-section, σ_{xi} is the longitudinal stress in the ith finite element, a_i is the cross-sectional area of the ith finite element, z_i is the distance from the baseline of the ship to the neutral axis of the ith finite element, and n is the total number of finite elements to be considered for the vertical bending moment calculations.

For convenience of the calculations in equation (16.5.1), it is desirable to employ the rectangular type of finite elements to handle equation (16.5.1) for individual finite elements.

The vertical bending moment of the hull cross-section can thus be obtained as follows.

$$M = \sum_{i=1}^{n} \sigma_{xi} a_i (z_i - g) \tag{16.5.2}$$

where z_i is measured from the baseline of the ship upward and compressive stress takes a negative sign and tensile stress a positive sign. Equation (16.5.2) thus gives a negative value for a sagging moment and a positive value for a hogging moment.

When the rotation angle θ_y is known, the corresponding bending curvature $1/R$ of the hull cross-section can be obtained as follows.

$$\frac{1}{R} = \frac{\theta_y}{L} \tag{16.5.3}$$

where L is the length of the hull model.

The bending moment versus curvature curves can be identified for various cross-sections based on the computed finite element analysis results, but of most interest is the cross-section at which the maximum bending moments are applied because it is here that local structures are prone to failure because of the resulting high stresses.

16.6 IDEALIZED STRUCTURAL UNIT METHOD

The conventional nonlinear FEM introduced in Section 16.5 provides more refined computations as long as its modeling process is relevant, as described in Chapter 8. However, it requires a great deal of computational effort as the extent of the analysis increases. This is because the conventional nonlinear FEM algorithm involves a large number of physical unknown values for the nodal points of the finite elements, and the iterative computations needed to solve the nonlinear stiffness equations take a large amount of computation time.

The ISUM resolves the issue of computation time, while keeping the computational accuracy at a reasonable level. In this method, a large-sized structural member is modeled as a structural unit. For example, the stiffened-plate structure shown in Fig. 15.4a in Chapter 15 or Fig. 16.16a can be modeled

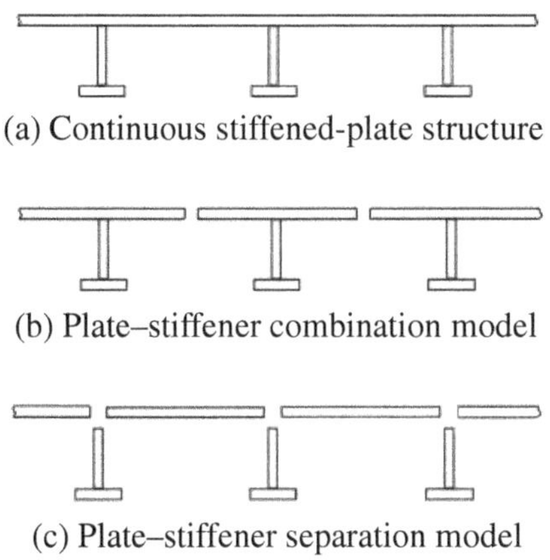

Figure 16.16 Idealization of a continuous stiffened-plate structure.

in ISUM as an assembly of plate–stiffener combinations or beam-column members, as shown in Fig. 15.4b or Fig. 16.16b.

The ultimate strength behavior in terms of the elastic large deflection and plasticity of the structural units is formulated as a closed-form expression. The necessary parameters of influence, such as initial imperfections, on the ultimate strength behavior can be taken into account in the process of formulating the structural behavior. Theoretical, numerical, and even experimental results that have been obtained beforehand are often used to formulate the structural behavior.

Historically, the idea of this method was initiated by Ueda and Rashed (1974) who named this method the idealized structural unit method. The first effort of Ueda and Rashed was to formulate the ultimate strength behavior of the deep girder unit for the purpose of analyzing the ultimate transverse strength of a ship. In an almost parallel development, Smith (1977) considered the same idea, which is sometimes called the Smith method, formulating the nonlinear structural behavior of the plate–stiffener combinations shown in Fig. 15.4b or Fig. 16.16b for analyzing the ultimate longitudinal strength of a ship.

The rules of the International Association of Classification Societies (IACS) (2008) provide details of the idealized structural unit method (or Smith method) to calculate the progressive collapse behavior of a ship's hull under a vertical sagging bending moment, in which the one-bay sliced hull model between two transverse frames shown in Fig. 16.14f is applied as an assembly of plate–stiffener combinations, but the effect of initial imperfections is not taken into consideration.

A unique feature of the idealized structural unit method is that the nonlinear behavior of the structural units is formulated in an explicit fashion in terms of the force versus displacement relationship or the stress versus strain relationship. The analysis of the progressive collapse of a system structure using this method is then calculated as the sum of the nonlinear behavior of individual structural units with an increasing applied force.

This method is very useful for saving computational effort while maintaining accuracy. However, it also has several disadvantages. First, the formulation of the nonlinear structural behavior of the structural units is difficult work that depends on the skill of the developer, and thus a large number of different formulations have been suggested by different analysts for the same problems. Second, formulating the highly complicated nonlinearities in a closed-form expression that arise from aspects, such as initial imperfections, loading conditions, age-related degradation, and in-service damage, is not straightforward. Third, it is difficult to deal with the unloading behavior of individual structural components and their system structure in the postultimate strength regime. Finally, this method cannot deal with the interacting effect of local failures of individual structural components and the overall failure of the system structure. These disadvantages are not, in contrast, issues in the conventional nonlinear finite element method.

16.7 INTELLIGENT SUPERSIZE FINITE ELEMENT METHOD

To resolve the issues of the enormous computational effort required with the conventional nonlinear FEM and the difficulty with the closed-form formulations of the ISUM, Paik (2006) suggested the concept of the ISFEM.

In fact, this method was once classified as one approach of ISUM, because in this method, large-sized structural components are modeled as supersize finite elements in a similar way to that employed in the ISUM, but here a variety of structural modeling techniques is possible by using an assembly of multiple structural elements of different types. A further difference between the ISUM and ISFEM is that the former attempts to theoretically formulate the nonlinear behavior of the structural units in a closed-form expression in terms of the force versus displacement relationship (or the stress versus strain relationship), whereas the theory of the latter is derived using the same framework of the conventional nonlinear FEM.

The ISFEM can readily take into account the interacting effects of local failures of individual structural components and the overall failure of the system structure. The theoretical formulations of the method are solid and systematic because they follow the approach of the conventional FEM.

In the analysis of progressive hull collapse using ISFEM, the stiffened-plate structure can be modeled as an assembly of plate elements and stiffeners, as shown in Fig. 16.16c, using two types of supersize finite elements. The supersize plate element is formulated with four nodal points involving six degrees of freedom per nodal point, as shown in Fig. 16.17, and the supersize stiffener element is formulated with two nodal points involving six degrees of freedom per nodal point, as shown in Fig. 16.18.

The theories of the two supersize finite elements are derived using exactly the same procedures as for the conventional rectangular-type plate-shell finite element and beam-type finite element, taking into account the geometrical and material nonlinearities. However, the stress–strain relationships for the supersize finite elements must be different from those of the conventional finite elements, because the supersize finite elements must take into account the size effect in the geometry of all possible failure modes. For example, progressive collapse analysis theory must involve buckling, plastic collapse, and ductile fracture, whereas the structural crashworthiness analysis theory associated with accidental events should involve crushing and rupture together with the strain-rate effect in addition to buckling and plastic collapse. This approach is called the intelligent method because the supersize element itself takes care of such nonlinear structural behavior, in contrast to the conventional nonlinear FEM.

The theories of the ISFEM have been implemented in various computer software packages, including ALPS/GENERAL (2010), ALPS/HULL (2010), and ALPS/SCOL (2010). ALPS/GENERAL analyzes the progressive collapse behavior of general-type plated structures up to and beyond the ultimate strength. ALPS/HULL is a purpose-built software for analyzing the progressive collapse of ship hulls under any combination of vertical bending, horizontal bending, vertical shearing force, horizontal shearing force, and torsion. ALPS/SCOL is a purpose-built software for analyzing structural crashworthiness involving crushing and rupture in association with collisions and grounding.

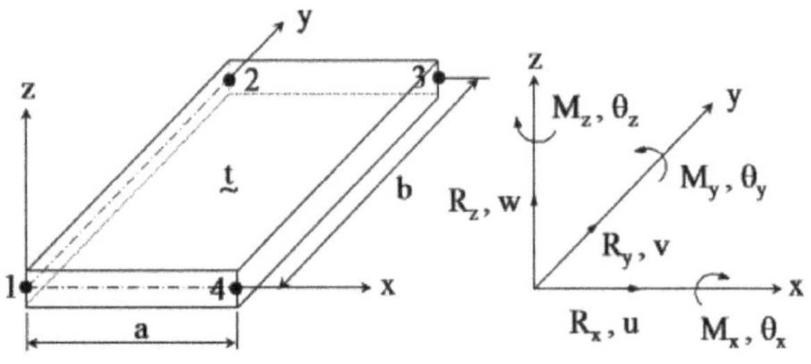

Figure 16.17 Local coordinate system, nodal forces, and displacements for the supersize plate element.

16.8 ULTIMATE HULL COLLAPSE STRENGTH UNDER COMBINED HULL GIRDER LOADS

A ship's hull can be subjected to multiple combinations of hull girder loads and thus the interacting effect of hull girder load components should be considered. The effect of lateral pressure loads should also be taken into consideration where they are applied (Paik et al., 2009).

The following ultimate collapse strength interaction relationship for any combination of vertical bending, horizontal bending, and vertical shearing force is suggested (Paik and Thayamballi, 2003, 2007).

$$\left(\frac{M_v}{M_u^v F_1}\right)^{c_1} + \left(\frac{M_h}{M_u^h F_2}\right)^{c_2} = 1 \qquad (16.8.1)$$

where $F_1 = \{1-(F^v/F^v_u)^{c_4}\}^{1/c_3}$; $F_2 = \{1-(F^v/F^v_u)^{c_6}\}^{1/c_5}$; M_u^v, M_u^h and F_u^v are the ultimate hull girder strength under a vertical bending moment (hogging or sagging) alone, a horizontal bending moment (hogging or sagging) alone, and a shearing force alone, respectively; and M_v, M_h, and F^v are the applied vertical bending moment, applied horizontal moment, and applied vertical shearing force, respectively.

In equation (16.8.1), the coefficients c_1 to c_6 represent the effect of different load combinations for the following two load components.

$$\left(\frac{M_v}{M_u^v}\right)^{c_1} + \left(\frac{M_h}{M_u^h}\right)^{c_2} = 1 \qquad (16.8.2)$$

$$\left(\frac{M_v}{M_u^v}\right)^{c_3} + \left(\frac{F^v}{F_u^v}\right)^{c_4} = 1 \qquad (16.8.3)$$

$$\left(\frac{M_h}{M_u^h}\right)^{c_5} + \left(\frac{F^v}{F_u^v}\right)^{c_6} = 1 \qquad (16.8.4)$$

Regardless of the vessel type and direction of bending (i.e., hogging or sagging), Paik et al. (1996) suggested using $c_1 = 1.85$, $c_2 = 1.0$, $c_3 = 2.0$, $c_4 = 5.0$, $c_5 = 2.5$, and $c_6 = 5.5$. Gordo and Guedes Soares (1997) suggested using $c_1 = c_2 = 1.50$~1.66 for trading tankers and container ships. Ozguc et al. (2005) suggested using $c_1 = 2.0$ and $c_2 = 1.45$ for hogging and $c_2 = 1.35$ for sagging.

Figure 16.19 compares the ultimate hull girder strength interactions for oil tanker hulls under vertical and horizontal bending moments using the three suggested sets of coefficients. It can be seen from Fig. 16.19 that the Paik formula gives the most pessimistic solutions, but is based on numerical solu-

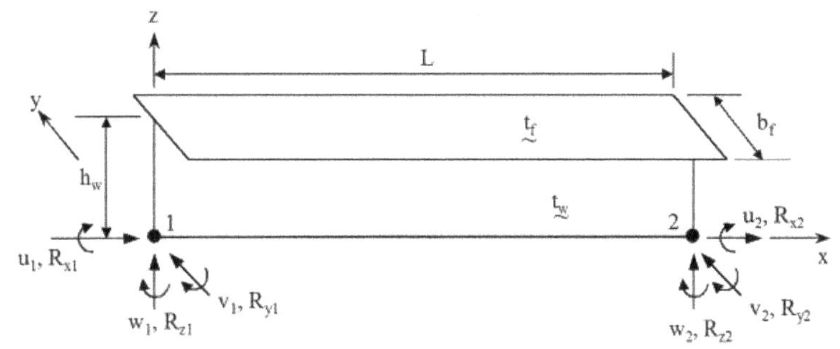

Figure 16.18 Local coordinate system, nodal forces, and displacements for the supersize stiffener element.

tions of progressive hull collapse analyses that take into account the effect of an average level of initial imperfections (initial deflections and welding residual stresses) in all of the structural components of the various types of ship hulls.

16.9 MODELING OF AGE-RELATED DEGRADATION AND IN-SERVICE DAMAGE

It is important to take into account the effects of initial imperfections, age-related degradation, and in-service damage in terms of computing the ultimate hull girder strength where they are involved. Related to this, it is interesting to note that aged members are prone to fracture under tensile loads (Paik, 1994a, 1994b; Drouin, 2006). The methods described in the previous sections allow the effects of initial imperfections, age-related degradation, and in-service and accidental damage to be approximately taken into consideration.

Typical types of age-related degradation are corrosion wastage, fatigue cracking damage, and local denting. Corrosion wastage is classified into two types: general (uniform) corrosion and pit corrosion. In ultimate hull collapse strength computations, general corrosion can be treated simply by deducing the plate thickness uniformly associated with corrosion wastage. However, it is not straightforward to deal with localized pit corrosion in ultimate strength computations. The parameter of the degree of pit corrosion intensity (DOP) defined in equation (13.8.1) in Chapter 13 can be employed to predict the ultimate strength of the pit corroded plates, as described in Section 13.8.

If a ship's hull has been inclined because of unintended accidental flooding, then calculations of the ultimate hull collapse strength must be undertaken for the hull's geometrical characteristics in the inclined position. To calculate the residual strength of a ship's hull following structural damage resulting from accidents such as collisions or grounding, a simplified modeling technique is that the area of the corresponding damage is removed from the structure, such that the structural effectiveness of the damaged area is taken as zero.

It must be noted that premised structural damage can expand until a ship's hull reaches the ultimate limit state as the hull girder loads increase. For example, existing fatigue cracking damage in structural components can propagate under tension or even compression arising from the action of the hull girder loads. The expanding effect of premised damage cannot be neglected where it significantly depends on the types of damage and loading.

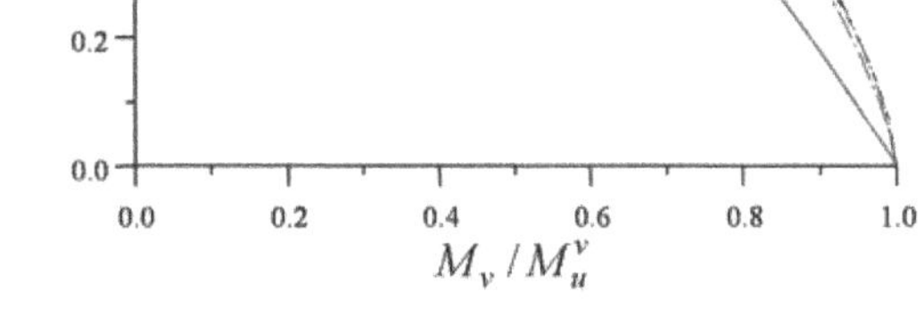

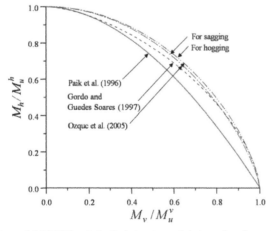

Figure 16.19 Ultimate hull girder strength interaction formulations for oil tanker hulls under vertical and horizontal bending moments.

16.10 APPLIED EXAMPLES

The five methods introduced in this chapter are now applied to predict the ultimate collapse strength of a one-third–scale frigate hull under a sagging bending moment tested by Dow (1991). Figure 16.20 shows the midship cross-section of the hull. Table 16.1 indicates the coordinates of the plate–stiffener intersections in the hull, together with the structural dimensions of the plating and stiffeners. The struc-

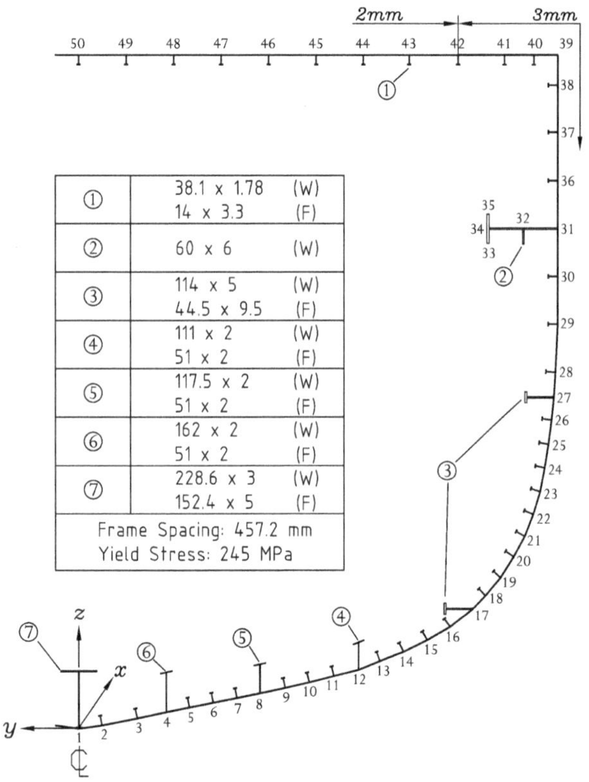

Figure 16.20 Midship cross-section of the one-third–scale frigate hull model tested by Dow (1991).

tures are made of mild steel with a yield stress of 245 MPa for both the plating and stiffeners. The length of the test hull is 18 m where the spacing of the transverse frames is 457.2 mm. The depth and breadth of the hull model are 2.8 m and 4.1 m, respectively.

The material properties of the structural members are defined as follows.

E = 207 GPa (Young's modulus)
σ_Y = 245 MPa (yield stress)
σ_T = 408 MPa (ultimate tensile stress)
ν = 0.3 (Poisson's ratio)

Measurements of the initial deflection and residual stresses because of fabrication-related initial imperfections were reported by Dow (1991). However, for the sake of simplicity, an average level of the initial imperfections of the plating and the stiffeners in the test model is assumed in the present computations as follows.

$$w_{opl} = 0.1t,\ \sigma_{rcx} = -0.1\ \sigma_Y,\ w_{oc} = w_{os} = 0.0015a$$

where w_{opl} is the maximum initial deflection of the plating, σ_{rcx} is the compressive residual stress of the plating in the longitudinal (x) direction, w_{oc} is the column type initial distortion of the stiffeners, w_{os} is the sideways initial distortion of the stiffeners, a is the length of the stiffener between the transverse frames, and t is the plate thickness.

Figure 16.21 shows the sagging bending moment versus curvature curves obtained from the test. The ultimate sagging moment of the hull model obtained from the test is –9.95 MNm.

16.10.1 Simple-Beam Theory Method

The hull cross-sectional properties, including the section modulus S_d at the deck relating to equation (16.3.3), are calculated first. The computed values of some important parameters are given for a fully effective hull cross-section using the nomenclature defined in Section 16.3, as follows.

$$g = 1417.8\text{mm}, A = 56{,}970\text{mm}^2, I = 0.060914\text{m}^4, S_d = 0.044069\text{m}^3$$

where g is the distance from the ship's baseline to the neutral axis, A is the total area of the hull cross-section, I is the moment of inertia of the hull cross-section, and S_d is the section modulus of the hull cross-section at the deck.

Under axial compression, the plating between the stiffeners will not be fully effective. The effective width b_e of the plating between the stiffeners may be approximately evaluated by equation (13.4.21) in Chapter 13, where the simply supported edge condition can be applied. The properties of the parameters

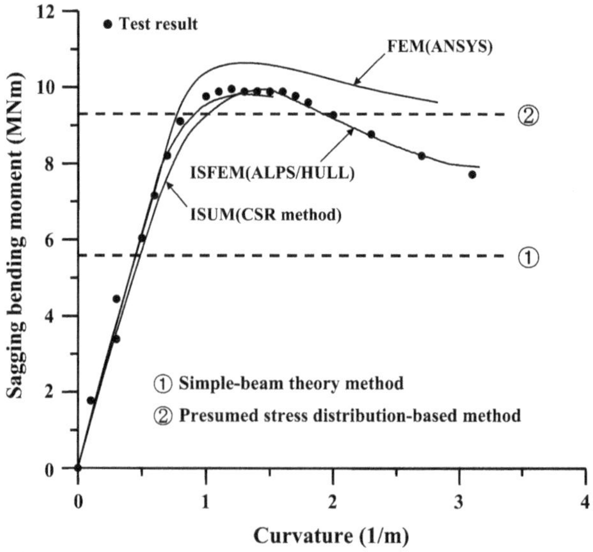

Figure 16.21 Sagging bending moment versus curvature curves or the ultimate sagging bending moments of the one-third–scale frigate hull model, obtained from the test and the five methods considered.

Table 16.1 Coordinates of the Plate–Stiffener Intersections Together with the Structural Dimensions of the Plating and Stiffeners in the One-Third–Scale Frigate Hull Model Tested by Dow (1991)

No.	*x*	*y*(mm)	*z*(mm)	Portion	Plate(mm)	No.	Web(mm)	Flange(mm)
1	0.0	0.0	0.0	1–2	99.2 × 3	1	228.6 × 3	152.4 × 5
2	0.0	−98.4	12.9	2–3	153.7 × 3	2	38.1 × 1.78	14 × 3.3
3	0.0	−249.3	41.9	3–4	127.2 × 3	3	38.1 × 1.78	14 × 3.3
4	0.0	−373.9	67.7	4–5	100.3 × 3	4	162 × 2	51 × 2
5	0.0	−472.3	87.1	5–6	103.5 × 3	5	38.1 × 1.78	14 × 3.3
6	0.0	−574.0	106.5	6–7	103.5 × 3	6	38.1 × 1.78	14 × 3.3
7	0.0	−675.7	125.8	7–8	100.3 × 3	7	38.1 × 1.78	14 × 3.3
8	0.0	−774.1	145.2	8–9	110.5 × 3	8	117.5 × 2	51 × 2
9	0.0	−882.3	167.7	9–10	104.2 × 3	9	38.1 × 1.78	14 × 3.3
10	0.0	−984.0	190.3	10–11	108.1 × 3	10	38.1 × 1.78	14 × 3.3
11	0.0	−1089.0	216.1	11–12	111.2 × 3	11	38.1 × 1.78	14 × 3.3
12	0.0	−1197.0	241.9	12–13	101.5 × 3	12	111 × 2	51 × 2
13	0.0	−1292.0	277.4	13–14	108.8 × 3	13	38.1 × 1.78	14 × 3.3
14	0.0	−1394.0	316.1	14–15	109.6 × 3	14	38.1 × 1.78	14 × 3.3
15	0.0	−1492.0	364.5	15–16	109.8 × 3	15	38.1 × 1.78	14 × 3.3
16	0.0	−1588.0	419.4	16–17	123.2 × 3	16	38.1 × 1.78	14 × 3.3
17	0.0	−1686.0	493.5	17–18	78.3 × 3	17	114 × 5	44.5 × 9.5
18	0.0	−1742.0	548.4	18–19	99.0 × 3	18	38.1 × 1.78	14 × 3.3
19	0.0	−1807.0	622.6	19–20	103.4 × 3	19	38.1 × 1.78	14 × 3.3
20	0.0	−1863.0	709.7	20–21	95.6 × 3	20	38.1 × 1.78	14 × 3.3
21	0.0	−1909.0	793.5	21–22	97.3 × 3	21	38.1 × 1.78	14 × 3.3
22	0.0	−1945.0	883.9	22–23	98.1 × 3	22	38.1 × 1.78	14 × 3.3
23	0.0	−1975.0	977.4	23–24	101.9 × 3	23	38.1 × 1.78	14 × 3.3
24	0.0	−1994.0	1077.4	24–25	98.2 × 3	24	38.1 × 1.78	14 × 3.3
25	0.0	−2011.0	1174.2	25–26	100.9 × 3	25	38.1 × 1.78	14 × 3.3
26	0.0	−2024.0	1274.2	26–27	94.0 × 3	26	38.1 × 1.78	14 × 3.3
27	0.0	−2034.0	1367.7	27–28	103.5 × 3	27	114 × 5	44.5 × 9.5
28	0.0	−2040.0	1471.0	28–29	200.2 × 3	28	38.1 × 1.78	14 × 3.3
29	0.0	−2050.0	1671.0	29–30	196.7 × 3	29	38.1 × 1.78	14 × 3.3
30	0.0	−2050.0	1867.7	30–31	196.8 × 3	30	38.1 × 1.78	14 × 3.3
31	0.0	−2050.0	2064.5	31–32	146 × 6	31	–	–
32	0.0	−1904.0	2064.5	32–33	146 × 6	32	60 × 6	–
33	0.0	−1758.0	2004.5	33–34	60 × 10	33	–	–
34	0.0	−1758.0	2064.5	34–35	60 × 10	34	–	–
35	0.0	−1758.0	2124.4	31–36	200 × 3	35	–	–
36	0.0	−2050.0	2264.5	36–37	200 × 3	36	38.1 × 1.78	14 × 3.3
37	0.0	−2050.0	2464.5	37–38	193.6 × 3	37	38.1 × 1.78	14 × 3.3
38	0.0	−2050.0	2658.1	38–39	141.9 × 3	38	38.1 × 1.78	14 × 3.3
39	0.0	−2050.0	2800.0	39–40	101.7 × 3	39	–	–
40	0.0	−1948.3	2800.0	40–41	124 × 3	40	38.1 × 1.78	14 × 3.3
41	0.0	−1824.3	2800.0	41–42	202.7 × 3	41	38.1 × 1.78	14 × 3.3
42	0.0	−1621.6	2800.0	42–43	202.7 × 2	42	38.1 × 1.78	14 × 3.3
43	0.0	−1418.9	2800.0	43–44	202.7 × 2	43	38.1 × 1.78	14 × 3.3
44	0.0	−1216.2	2800.0	44–45	202.7 × 2	44	38.1 × 1.78	14 × 3.3
45	0.0	−1013.5	2800.0	45–46	202.7 × 2	45	38.1 × 1.78	14 × 3.3
46	0.0	−810.8	2800.0	46–47	202.7 × 2	46	38.1 × 1.78	14 × 3.3
47	0.0	−608.1	2800.0	47–48	202.7 × 2	47	38.1 × 1.78	14 × 3.3
48	0.0	−405.4	2800.0	48–49	202.7 × 2	48	38.1 × 1.78	14 × 3.3
49	0.0	−202.7	2800.0	49–50	202.7 × 2	49	38.1 × 1.78	14 × 3.3
50	0.0	0.0	2800.0	–	–	50	38.1 × 1.78	14 × 3.3

for the effective hull cross-section under a sagging condition are then obtained as follows.

$$g = 1281.3\text{mm}, A = 50{,}705\text{mm}^2, I = 0.051189\text{m}^4, S_d = 0.033705\text{m}^3$$

The fully plastic bending moment denoted by M_P, in which both the compressed and tensioned regions have fully yielded, is derived as follows.

$$M_P = 13.115\text{MNm}$$

For convenience, the ultimate compressive strength σ_{ud} of the deck panel in equation (16.3.8) is calculated by the Paik-Thayamballi empirical formula method indicated in equation (15.5.9) in Chapter 15, taking into account the effect of an average level of initial imperfections rather than the measured data as defined above. A representative of the plate–stiffener combinations in the deck panel, as shown in Fig. 16.16b, that is, with $b \times t = 202.7 \times 2$ (mm), $h_w \times t_w = 38.1 \times 1.78$ (mm), and $b_f \times t_f = 14 \times 3.3$ (mm), has the following values of column slenderness coefficient λ and attached plate slenderness coefficient β.

$$\lambda = 0.177, \beta = 3.497$$

Thus, the ultimate compressive strength σ_{ud} of the deck panel can be obtained from equation (15.5.9) as follows.

$$\sigma_{ud} = -137.5\text{MPa}$$

The first-failure sagging bending moment of the hull model using the simple-beam theory method is then obtained from equation (16.3.8) for a fully effective hull cross-section as follows.

$$M^v_{fs} = S_d \times \sigma_{ud} = -0.044069\text{m}^3 \times 137.5\text{MPa} = -5.59\text{MNm}$$

The first-failure hull strength computed by the simple-beam theory method is 56.2% of the test result. This deviation in the computed ultimate hull strength prediction arises from the concept of the simple-beam theory method itself that does not take into account the expansion of local failures into the vertical structures in addition to the compressed flange. The simple-beam theory method thus significantly underestimates the ultimate hull collapse strength.

For comparison, an alternative hull strength is calculated using the yield stress rather than the ultimate compressive stress in the deck, as follows.

$$M^v_{us} = S_d \times \sigma_Y = -0.044069\text{m}^3 \times 245\text{MPa} = -10.80\text{MNm}$$

which is 108.5% of the test result.

16.10.2 Presumed Stress Distribution-Based Method

The longitudinal stress distribution shown in Fig. 16.13a is considered under a sagging bending moment. The depth of collapsed region 4 in the upper structure under compression can be determined from the condition of equation (16.4.1), and is found to be 1806.8 mm.

The hull is modeled as an assembly of the plate–stiffener combination models, as indicated in Fig. 16.16b. The ultimate strengths of the individual stiffeners with the attached plating are predicted by the Paik-Thayamballi empirical formula method given in (15.5.9) in Chapter 15. The neutral axis position of the hull model under a sagging condition is then obtained from equation (16.4.2), as follows.

$$g_u = 1307.7\text{mm}$$

The ultimate sagging moment of the hull model is then calculated from (16.4.3) as follows.

$$M^v_{us} = -9.30\text{MNm}$$

The ultimate sagging moment computed by the presumed stress distribution-based method is 93.5% of the test result. It is noted that this deviation arises from the ultimate strength predictions of individual structural components as well as the feature of the method itself.

16.10.3 Nonlinear Finite Element Method

The nonlinear FEM is now applied using ANSYS (2009) computer software. The one-bay sliced hull model shown in Fig. 16.14f is adopted. Figure 16.22 shows the ANSYS nonlinear FEM model of the tested structure, where 12 plate-shell elements are used in the transverse direction of the plating between the longitudinal stiffeners, 6 plate-shell elements are used for the stiffener web, and 2 plate-shell elements are used for each of the T-type stiffener flanges (i.e., one element per each side of the flange with regard to the intersection with the stiffener web). The element aspect ratio is about 1.0 for the plating, 2.4 for the stiffener web, and 2.3 for the stiffener flange.

The initial deflection of the plating between the stiffeners and the initial distortions (both column-type and sideways) of the stiffeners defined previously are included in the structural modeling, but the welding residual stress is not considered in the present analysis for the sake of simplicity. Figure 16.23 shows the configuration of the initial imperfections of the plating and the stiffeners applied in

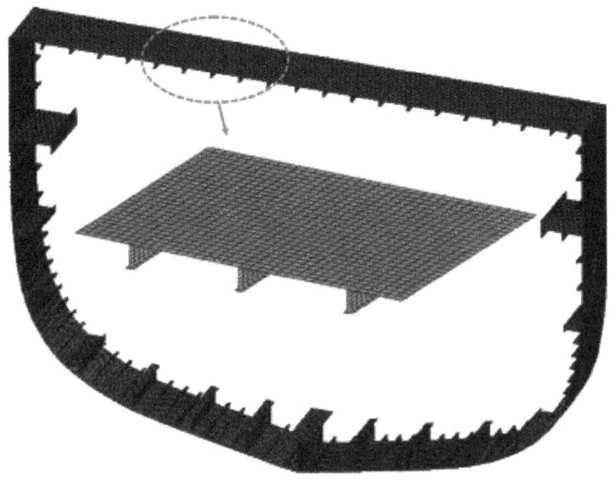

Figure 16.22 ANSYS nonlinear FEM model in the one-bay sliced hull between the transverse frames.

the structural modeling, where the buckling mode initial deflection (i.e., with two half-waves in the length direction of the plating), column-type (one half-wave) initial distortions of the stiffener, and sideways initial distortions of the stiffeners are all combined.

Figure 16.21 shows the sagging bending moment versus curvature curve of the hull model obtained by ANSYS nonlinear FEM analysis. It was found from the computation that the deck stiffened panels have failed with flexural–torsional buckling of the stiffeners. The ultimate bending moment computed by the nonlinear FEM is −10.62 MNm, which is 106.7% of the test result. This overestimation of the ultimate hull strength is partly due to the fact that the welding residual stress is not accounted for in the computation. The changes in the neutral axis position are dealt with as indicated in equation (16.5.1). Figure 16.24 shows the change of the neutral axis with an increase in the sagging bending moment obtained by the nonlinear FEM.

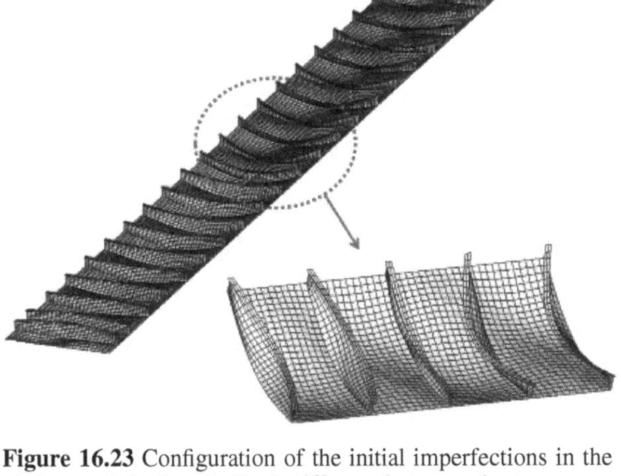

Figure 16.23 Configuration of the initial imperfections in the plating and the longitudinal stiffeners between the transverse frames applied in the ANSYS structural modeling (amplified by 80 times).

16.10.4 Idealized Structural Unit Method

The common structural rule (CSR) method based on an incremental-iterative approach specified by IACS (2008) is applied to predict the ultimate sagging moment. The CSR method has been developed to predict the ultimate collapse strength of double-hulled oil tanker hulls under a sagging bending moment. The hull structure is modeled as an assembly of plate–stiffener combinations, as shown in Fig. 16.16b. The details of the ultimate strength calculations for the plate–stiffener combinations are presented in IACS (2008). The IACS CSR method cannot deal with the initial imperfections as parameters of influence and therefore the effect of the initial imperfections is not considered in this computation.

Figure 16.21 shows the sagging bending moment versus curvature curve of the hull model obtained by the idealized structural unit method (IACS CSR approach). The ultimate sagging moment computed by the idealized structural unit method is −9.83 MNm, which is 98.8% of the test result.

16.10.5 Intelligent Supersize Finite Element Method

The theory of the ISFEM has been applied in the computer software ALPS/HULL (2010), which has been developed to conduct progressive collapse analyses of a ship's hull under a combined hull

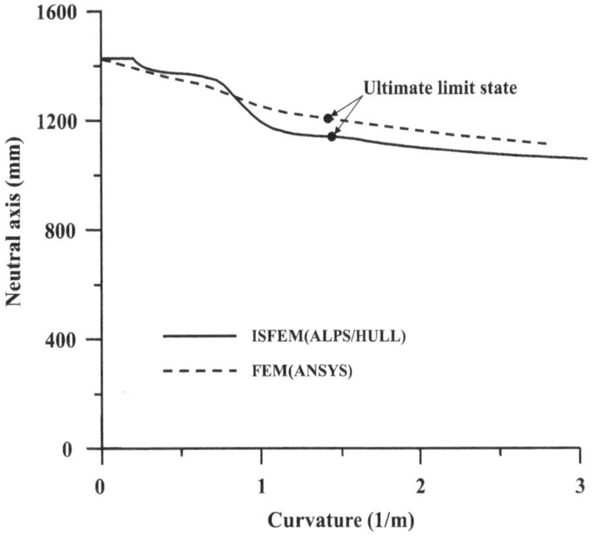

Figure 16.24 Change of the neutral axis of the hull model obtained by the ANSYS nonlinear FEM and the ALPS/HULL ISFEM.

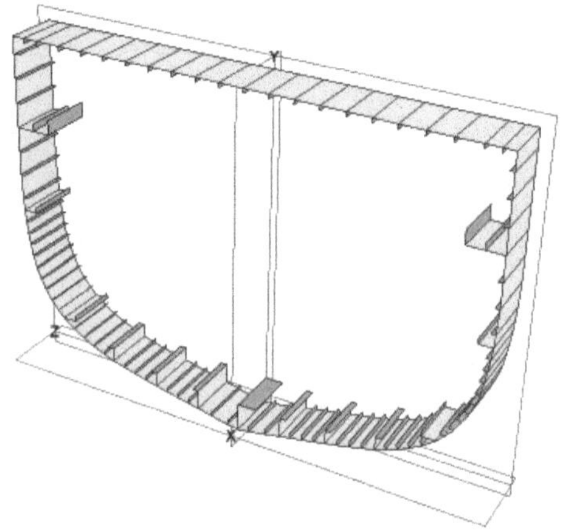

Figure 16.25 ALPS/HULL ISFEM model in the one-bay sliced hull between the transverse frames.

girder load. The one-bay sliced hull model shown in Fig. 16.14f is adopted for convenience. The hull structure is modeled as an assembly of plate–stiffener separations, as shown in Fig. 16.16c. The effects of the initial imperfections defined previously are taken into consideration in this computation: initial deflection of the plating with buckling mode, column-type initial distortions of the stiffeners, sideways initial distortions of the stiffeners, and welding residual stress in the plating. Figure 16.25 illustrates the ALPS/HULL model, although it does not necessarily represent the real geometry or type of the stiffeners but simply shows an overall picture of the structural model, because of using the super-size stiffener elements.

Figure 16.21 shows the sagging bending moment versus curvature curve of the hull model computed by ALPS/HULL. The ultimate sagging moment computed by ALPS/HULL is −9.94 MNm, which is 99.9% of the test result. Figure 16.24 shows the change of the neutral axis with an increase in the sagging bending moment obtained by the ALPS/HULL ISFEM.

16.10.6 Discussion

Figure 16.21 and Table 16.2 compare the ultimate sagging moments computed by the five methods together with the test result. Figure 16.24 compares the changes of the neutral axis obtained by the nonlinear FEM and the ISFEM as the sagging bending curvature increases. The effect of initial imperfections was dealt with in a different way by each of the five methods as addressed in the previous subsections.

Both the nonlinear FEM and the ISFEM take into account the effect of interactions between local failures of individual structural components and the overall failure of the system structure, whereas the other three methods do not consider this.

Both the simple-beam theory method and the presumed stress distribution-based method give closed-form formulations for the hull strength computations. The simple-beam theory method gives the first-failure hull strength at the collapse of the compressed flange, which is usually much smaller than the ultimate hull strength. This is partly because local failures can grow further into the vertical structures in addition to the hull flanges (the deck or bottom), but the simple-beam theory method cannot take into account this effect. In contrast, the presumed stress distribution-based method following the Paik-Mansour presumption of the stress distribution over the hull cross-section gives quite reasonable solutions, taking into account the effect of local failures more precisely than the simple-beam theory method.

The idealized structural unit method is difficult to apply in a wider analysis at a higher level than the

Table 16.2 Comparison of the Ultimate Sagging Bending Moments Obtained by the Five Methods Together with the Test Result

Simple-beam Theory Method		Presumed Stress Distribution-based Method		Nonlinear FEM (ANSYS)*		ISUM (CSR method)†		ISFEM (ALPS/HULL)‡	
M^v_{fs}	M^v_{us}/M_{exp}	M^v_{us}	M^v_{us}/M_{exp}	M^v_{us}	M^v_{us}/M_{exp}	M^v_{us}	M^v_{us}/M_{exp}	M^v_{us}	M^v_{us}/M_{exp}
–5.59	0.562	–9.30	0.935	–10.62	1.067	–9.83	0.988	–9.94	0.999

M^v_{fs} is the first-failure sagging moment (MNm), M^v_{us} is the computed ultimate sagging moment (MNm), M_{exp} is the ultimate sagging moment obtained by the test, = –9.95 MNm.

*ANSYS nonlinear FEM takes into account the effects of initial deflection of the plating and initial distortions of the stiffeners predefined in the foregoing, but residual stress is not accounted for.

†ISUM (CSR method) does not deal with the initial imperfections as parameters of influence.

‡ISFEM (ALPS/HULL) deals with the initial imperfections (initial deflection of the plating, initial distortion of the stiffeners and welding residual stress) as parameters of influence.

two-bay sliced hull model shown in Fig. 16.14(e) because it is very difficult to derive closed-form formulations of the failure behavior of structural components with three-dimensional system coordinates. Neither the nonlinear FEM nor the ISFEM, however, has limitations in terms of the extent of analysis that can be performed, although the latter method requires much less computation time than the former. It is also noted that the derivation of the ISFEM theory is much easier than that of the idealized structural unit method because the former method follows a systematic process similar to that used in the conventional nonlinear FEM.

REFERENCES

ALPS/GENERAL. (2010). Progressive collapse analysis of general-type plated structures. Stevensville, MD: DRS C3 Systems.

ALPS/HULL. (2010). Progressive collapse analysis of ship hulls. Stevensville, MD: DRS C3 Systems.

ALPS/SCOL.(2010). Structural crashworthiness analysis of plated structures. Stevensville, MD: DRS C3 Systems.

ANSYS. (2009). Version 11.0, ANSYS Inc., Canonsburg, PA, USA.

Caldwell, J. B. (1965). Ultimate longitudinal strength. *Trans. RINA*, Vol. 107, 411–430.

Dow, R. S. (1991). Testing and analysis of a 1/3-scale welded steel frigate model. *Proc. of International Conference on Advances in Marine Structures*, Vol. 2, Scotland, May.

Drouin, P. (2006). Brittle fracture in ships – a lingering problem. *Ships and Offshore Structures*, Vol. 1, Issue 3, 229–233.

Gordo, J. M., and Guedes Soares, C. (1997). Interaction equation for the collapse of tankers and containerships under combined bending moments. *Journal of Ship Research*, Vol. 41, Issue 3, 230–240.

International Association of Classification Societies. (2008). Common structural rules for double hull oil tankers. London, England.

Ko, T. J., Paik, J. K., Kim, B. J., et al. (2010). Structural failure assessment of the engine room region for a post-Panamax class containership: Lessons learned from the MSC Napoli accident. Proc. of International Conference on Ocean, Offshore, and Arctic Engineering, OMAE 2010-20996, Shanghai, June.

Marine Accident Investigation Branch. (2008). *Report on the investigation of the structural failure of MSC Napoli, English channel on 18 January 2007*. Report No.9/2008. Southampton, England.

Ozguc, O., Das, P. K., and Barltrop, N. D. P. (2005). Rational interaction design equations for the ultimate longitudinal strength of tankers, bulk carriers, general cargo and container ships under coupled bending moment. Glasgow, UK: Department of Naval Architecture and Marine Engineering, Universities of Glasgow and Strathclyde.

Paik, J. K. (1994a). Hull collapse of an aging bulk carrier under combined longitudinal bending and shearing force. *Trans. RINA*, Vol. 136, 217–228.

Paik, J. K. (1994b). Tensile behavior of local members on ship hull collapse. *Journal of Ship Research*, Vol. 38, Issue 3, 239–244.

Paik, J. K. (2006). *The intelligent super-size finite element method: Theory and practice*. Busan, South Korea: Department of Naval Architecture and Ocean Engineering, Pusan National University.

Paik, J. K., and Faulkner, D. (2003). Reassessment of the *M.V. Derbyshire* sinking with the focus on hull-girder collapse. *Marine Technology*, Vol. 40, Issue 4, 258–269.

Paik, J. K., Kim, D. K., and Kim, M. S. (2009). Ultimate strength performance of Seuezmax tanker structures: Pre-CSR versus CSR designs. *International Journal of Maritime Engineering*, Vol. 151, Issue A2, 1–14.

Paik, J. K., and Mansour, A. E. (1995). A simple formulation for predicting the ultimate strength of ships. *Journal of Marine Science and Technology*, Vol. 1, Issue 1, 52–62.

Paik, J. K., Seo, J. K., and Kim, B. J. (2008). Ultimate limit state assessment of the M.V. *Derbyshire* hull structures. *Journal of Offshore Mechanics and Arctic Engineering*, Vol. 130, 021002-1–021002-9.

Paik, J. K., and Thayamballi, A. K. (1998). The strength and reliability of bulk carrier structures subject to age and accidental flooding. *Trans. SNAME*, Vol. 106, 1–40.

Paik, J. K., and Thayamballi, A. K. (2003). *Ultimate limit state of steel-plated structures*. Chichester, UK: Wiley.

Paik, J. K., and Thayamballi, A. K. (2007). *Ship-shaped offshore installations: Design, building, and operation*. Cambridge, UK: Cambridge University Press.

Paik, J. K., Thayamballi, A. K., and Che, J. S. (1996). Ultimate strength of ship hulls under combined vertical bending, horizontal bending and shearing forces. *Trans. SNAME*, Vol. 104, 31–59.

Rutherford, S. E., and Caldwell, J. B. (1990). Ultimate longitudinal strength of ships: A case study. *Trans. SNAME*, Vol. 98, 441–471.

Smith, C. S. (1977). Influence of local compressive failure on ultimate longitudinal strength of ship's hull. *Proc. of the International Symposium on Practical Design in Shipbuilding*, Tokyo, 73–79.

Ueda, Y., and Rashed, S. M. H. (1974). An ultimate transverse strength analysis of ship structures. *Journal of the Society of Naval Architects of Japan*, Vol. 136, 309–324 (in Japanese).

SEVENTEEN

FATIGUE OF SHIP STRUCTURAL DETAILS

Dominique Béghin
Bureau Veritas, Paris, France (ret)

17.1 GENERAL

17.1.1 Definition of Fatigue

Welded structures made of steel or other metals subjected to cyclic loads lower than those upon which the design is based can initiate microscopic cracks which gradually increase in size until, after a certain number of cycles has been experienced, the cracks have become so large that fracture occurs. This structural failure is known as the phenomenon of fatigue. In other words, fatigue is a process of cycle-by-cycle accumulation of damage in a structure subjected to fluctuating stresses, going through several stages from the initial "crack-free" state to a "failure" state. The most important load effect parameter is the fluctuating component of stress, commonly referred to as stress range.

There are two different types of fatigue:

1. Low-cycle fatigue occurring for a low number of cycles, less than 5 10^3, in the range of plastic deformations.
2. High-cycle fatigue occurring for a high number of cycles in the range of elastic deformations.

Fatigue fractures observed on ship structures are generally of the second type.

For welded structures, the fatigue process includes three main phases:

1. *Initiation of macrocracks.* This phase is characterized by the development of cumulative plastic strains at the tip of microcracks concomitant with changes in material microstructure, leading to the growth and coalescence of existing microscopic weld defects and, finally, to the formation of a macrocrack.
2. *Propagation or crack growth.* In this second phase, the macrocrack grows normal to the direction of the largest principal stress with a propagation rate of about 10^{-6} to 10^{-3} mm per cycle.
3. *Final failure.* The final phase occurs according to one of the following three mechanisms:
   a. Brittle fracture.
   b. Ductile fracture.
   c. Plastic collapse.

These three failure modes occur depending on the toughness of the material, temperature, loading rate, plate thickness, and constraint. For fatigue analyses, the final failure is defined by *S-N* tests or by the maximum tolerable defect size. The phases of high-cycle fatigue failures are:

a. An area of initiation.
b. Minor plastic deformations and beachmarks revealing the stress variations.
c. Final granular fracture.

It is generally accepted, as mentioned for instance by Almar-Naess et al. (1985), that the crack growth represents the predominant part of the fatigue life, taking into account that defects are inherent to welded structures and that the initiation period of a macrocrack growing from these defects is insignificant compared to the crack growth period. This conclusion may however not be valid for welded joints subjected to post-weld improvement techniques.

17.1.2 Main Contributing Factors to Fatigue

There are many factors that affect the fatigue behavior of ship structural details subjected to fluctuating loads:

1. The general configuration and local geometry of details can lead to structural discontinuities and, consequently, produce local stress concentrations.
2. The local configuration and geometry of weld details (e.g., radius at weld toe, weld angle, throat thickness) also produce a local increase in stresses.
3. Weld material defects and internal discontinuities such as undercuts, porosity, slag inclusions, lack of fusion or penetration, solidification cracks, etc., are recognized to reduce fatigue life.
4. Bad workmanship including such problems as misalignment, angular distortions, and insufficient quality of welding, introduces additional stress concentrations.
5. The use of higher tensile steels (HTS) is frequently considered as a significant contributing factor to fatigue since experiments show that fatigue properties of welded structures are practically not improved with the material properties. On the one hand permissible stresses are increased and on the other hand the fatigue damage varies with the cube of the stress range. HTS structures may therefore be more prone to fatigue damage than mild steel structures if no particular measures are taken at the design stage (refer to Section 17.3.6).
6. Cyclic loads and, especially, wave-induced loads are of primary concern in fatigue.
7. Corrosive environment significantly reduces fatigue life unless appropriate measures are taken during the construction and also during the ship's life to protect the structure against corrosion.

To summarize, fatigue cracking generally occurs on welded structural details subjected to fluctuating stresses, due to either incorrect prediction of cyclic loads, improper design, or to bad workmanship. This review of the main factors contributing to fatigue highlights why assessment of the fatigue life of ship structures is a complex task, all the more so as the various contributing factors have generally large uncertainties that are difficult to quantify, in particular when they involve the human factor (e.g., workmanship and quality of welding).

17.1.3 Fatigue Damage in Ship Structures

Fatigue and corrosion are recognized as the main causes of structural damage observed on ships in service. Though fatigue does not generally result in catastrophic failures, its impact on the cost of maintenance of ships is very high. Fatigue cracks develop in areas where nominal stresses are not necessarily high but where there are locally high stress concentrations at structural discontinuities. It is therefore essential to provide designers with appropriate design tools to assess the fatigue life of critical structural details.

In this respect, the International Association of Classification Societies (IACS 1994) and the Tanker Structure Co-operative Forum (TSCF 1997) give valuable information on the critical structural details of bulk carriers and oil tankers that experienced structural failures partly due to fatigue and also provide recommendations for repair. Figures 17.1 and 17.2 show two examples of failures observed on a critical double bottom structural detail:

1. Radiused knuckled joints (Fig. 17.1).
2. Welded knuckled joints (Fig. 17.2).

17.1.4 Fatigue Design Strategies

Strength criteria defined by the classification society rules are generally determined for intact structures. The safety factors do take into account effects of degradations that essentially result from wear and tear, corrosion, and fatigue, bearing in mind that the structural reliability is maintained over the ship's life thanks to a program of periodical class (annual, intermediate, and special) and statutory surveys.

Rules give the necessary information to account for these aspects at the design stage. In particular, the fatigue strength criteria depend on the inspection and maintenance strategy chosen by the owner and also on the consequences that fatigue cracks may have on the environment or on the cost of repairs.

One of the following three strategies may be adopted:

1. *Safe-life design.* Safe-life design is based on a high survival probability assuming that no regular inspection in service is required.
2. *Fail-safe design.* Fail-safe design is based on a moderate survival probability assuming that regular inspection in service is provided. Structural design strategy has always been based on the assumption that ship structures are inspectable, maintainable, and repairable. In recent years, more stringent procedures have been implemented by classification societies for monitoring the hull condition during the ship's life. In particular, they give their field surveyors more precise information on
    - where to look,
    - what to examine, and
    - what to measure and report.

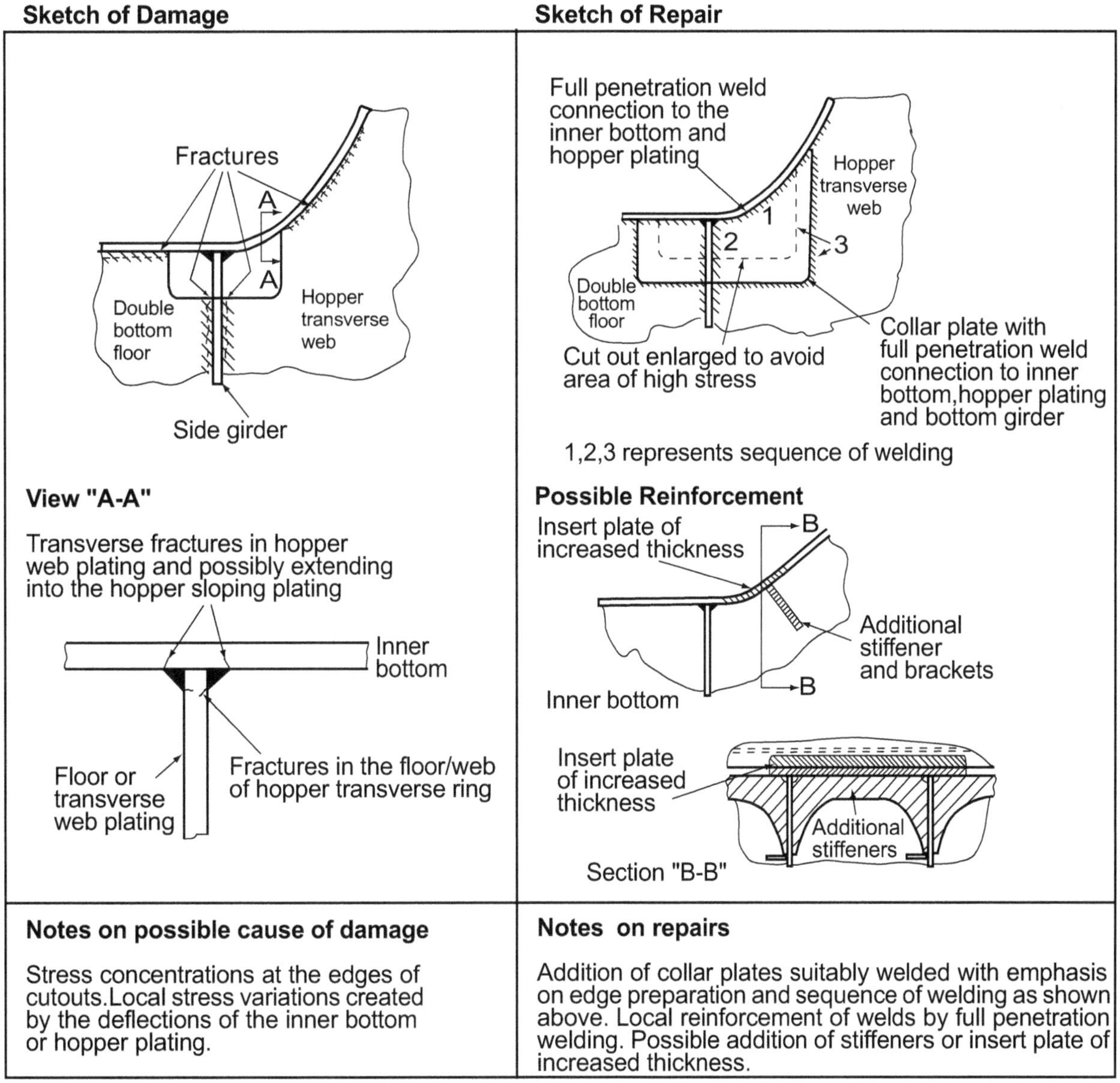

Figure 17.1 Double bottom structure (radiused knuckle).

These instructions are aimed at helping field surveyors make the right decisions on the type and extent of repairs.

3. *Damage tolerant design.* Damage tolerant design is based on a moderate survival probability assuming that the presence of cracks is detected by non-destructive methods and that fracture mechanics is used for calculation of the remaining lifetime until failure.

In general, fatigue is prevented by controlling the cyclic stress amplitude, and in most cases, the most efficient way to control stresses is to increase the local scantlings and/or modify the local geometry so as to reduce the stress concentrations and discontinuities.

17.1.5 General Procedure for Assessment of the Fatigue Strength

As for any other mode of failure, assessment of the fatigue strength of welded ship structures requires the determination of the following.

1. *The demand characterized by the load/stress history diagram over the ship's life gives the distribution of stress variations.* Figure 17.3 shows an example of random or stochastic loading applied to ship structures.

The load history diagram can be determined by either measurements or direct calculations depending on the type of load environment. The diagram

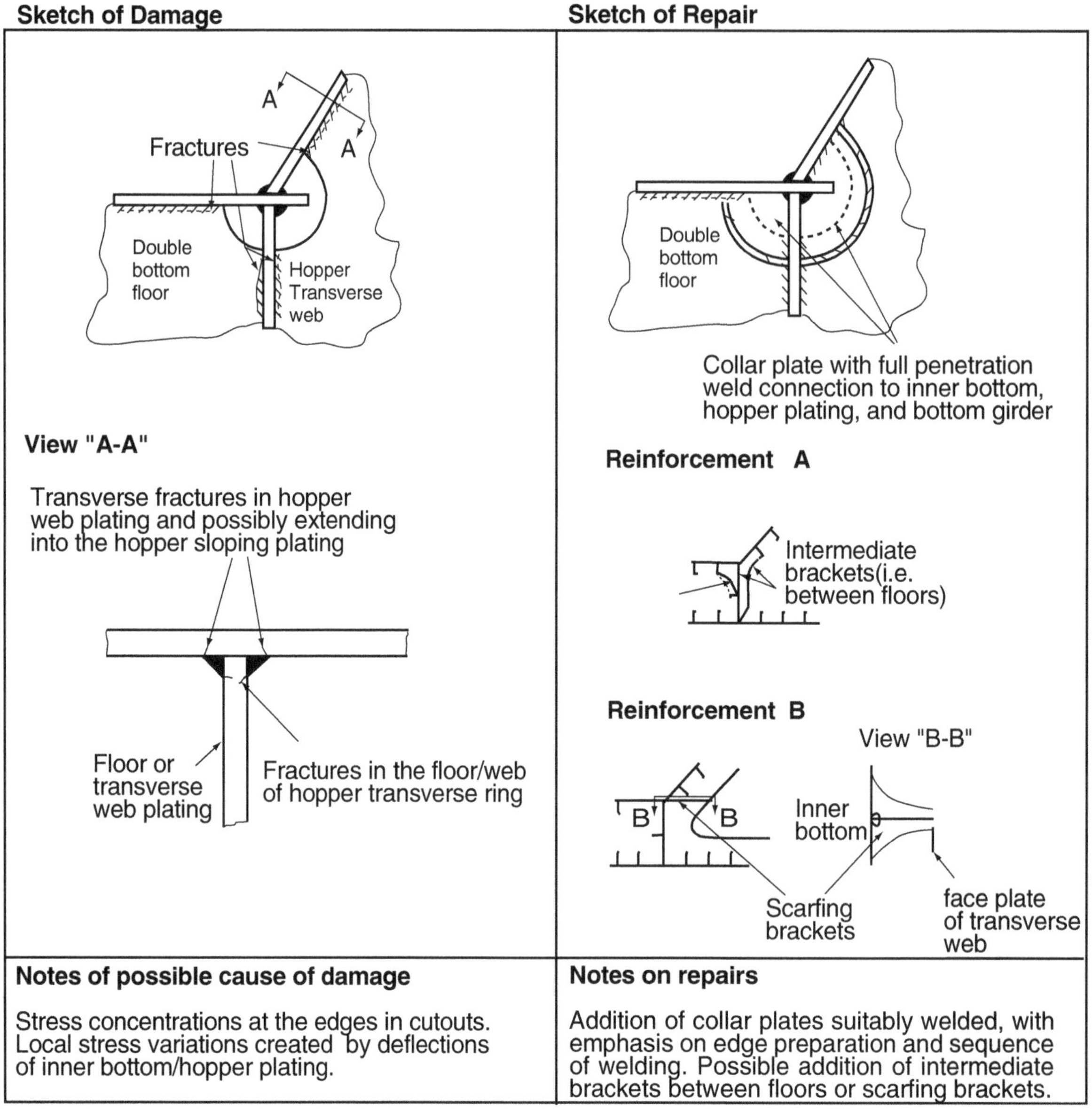

Figure 17.2 Double bottom structure (welded connection).

is determined experimentally when the design is based on an actual local load environment. In this case, the construction of the actual diagram requires the use of statistical counting methods for analysis of load or stress measurements carried out for actual structures or representative models. On the contrary, ships encounter over their life various load environments depending on their route and consequently the load history diagram can only be determined by direct calculations. The method most commonly used in ship design is based on the calculation of the exceedance stress range spectrum, as shown in Figure 17.4, giving the frequency or number of cycles that a given stress range will be exceeded over the ship's life. However with this type of representation, the information on the nature of the phenomenon is lost, in particular the order of occurrence of events and the frequency of stress changes.

2. *The fatigue capacity of the structure, characterized by either S-N curves or crack growth rate curves.* Although fracture mechanics provides a more rational tool than the *S-N* methodology to assess the fatigue strength of welded details, there are many uncertainties in the calculation procedure and intensive research is required before it can be applied successfully in ship design. *S-N* methodology is therefore the most common way to represent the fatigue capacity of welded steel joints and is used by many design codes and classification

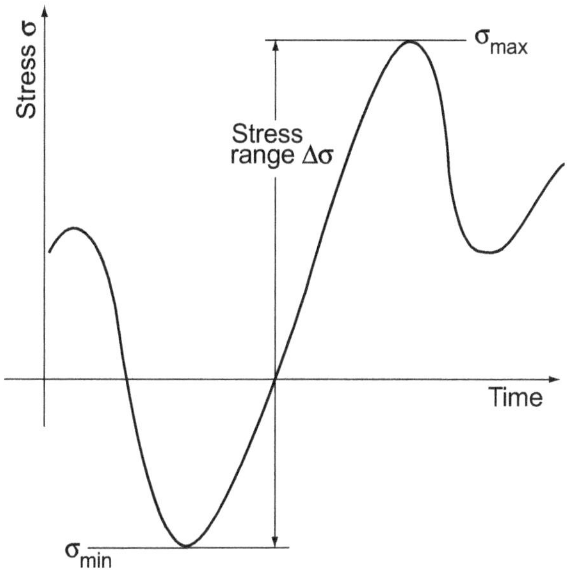

Figure 17.3 Random ship loading.

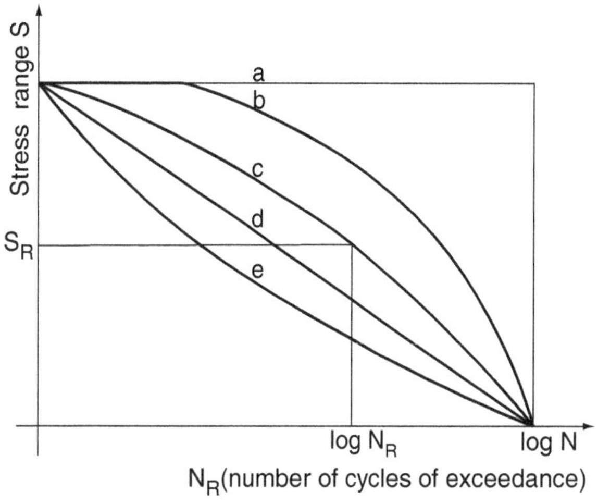

a: constant amplitude spectrum
b: limited maximum stress range
c, d and e: typical stress range spectra for ships

Figure 17.4 Stress range spectra.

society rules. *S-N* curves give the relationship between the fatigue life and the nominal stress range *S* applied to a given sample, the fatigue life being defined as the number of constant amplitude load cycles *N* to failure. However, the use of fracture mechanics whose principles are addressed briefly in Section 17.3.3, may be necessary for particular cases, e.g., for prediction of the remaining life of damage tolerant structures when cracks are detected.

3. *The failure criterion and the selected numerical value above which the structure is considered as having failed.* In the *S-N* methodology, the fatigue capacity is defined by *S-N* curves and assessment of the fatigue strength based on the hypothesis, commonly known as the Miner-Palmgren rule, that fatigue damage accumulates linearly.

The general procedure for assessment of the fatigue strength of structural details which is summarized in Figure 17.5 includes the following three steps:

1. Determination of loads and stresses using either the spectral fatigue analysis (refer to Section 17.2.2), the simplified fatigue analysis (refer to Section 17.2.3) or the equivalent regular wave concept (refer to Section 17.2.4).
2. Definition of the design *S-N* curve for the structural detail considered (refer to Sections 17.3.5 and 17.3.6 for steel joints and 17.3.7 for aluminum joints).
3. Assessment of the fatigue strength and calculation of either the fatigue life (refer to Section 17.4.3) or the probability of failure over the expected ship's life (refer to Section 17.5.2).

17.2 LONG-TERM DISTRIBUTION OF LOADS AND STRESSES

17.2.1 Loads on Ships

Since fatigue is a process of cycle-by-cycle accumulation of damage, assessment of the fatigue life of any structural detail requires the determination of the long-term load history that specifies the distribution of stress variations over a long period of time, with due consideration given to the variations in sea routes, ship speed, and loading conditions. The construction of this long-term load history requires accounting for the loads that can influence the fatigue life.

1. Still water loads (for example, cargo loads that vary from voyage to voyage).
2. Transient loads such as thermal stresses.
3. Wave-induced loads, directly generated by the action of waves.
4. Vibratory loads resulting from main engine or propeller induced vibratory forces.
5. Impact loads such as bottom slamming, bow flare impact (whipping), sloshing, and shipping of green seas.
6. Residual stresses.

Depending on the type of ship, still water stresses can change significantly from voyage to voyage (fully laden or ballast) as shown by Guedes and Moan (1988). With the exception of particular cargo ships, thermal stresses occur in weather-exposed areas and are mainly governed by the diurnal changes

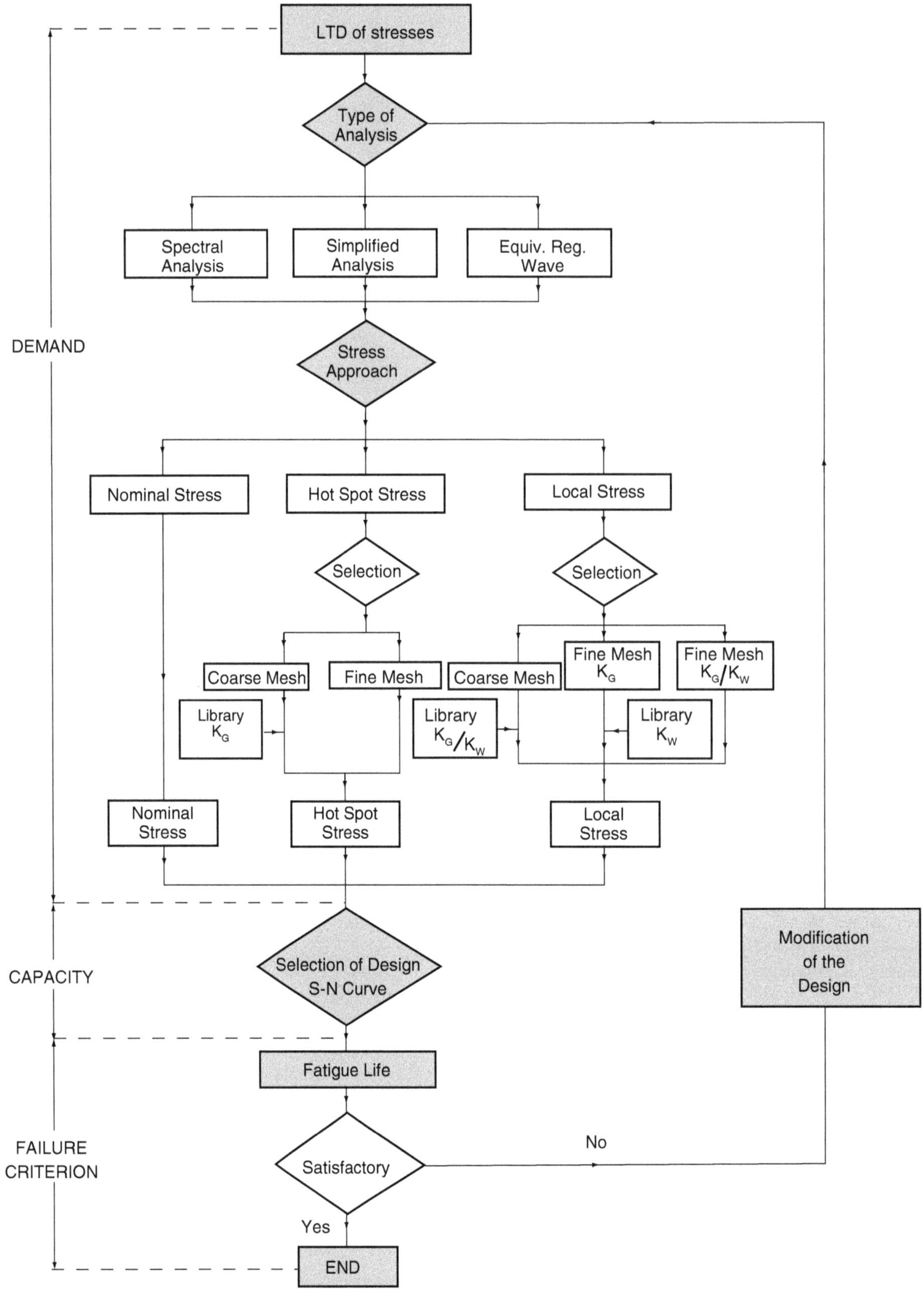

Figure 17.5 Procedure for assessment of the fatigue strength and modification of the design.

in air temperature. Therefore, still water and thermal stresses, which are very low frequency, may be considered as static stresses whose influence is to shift the mean stress.

Wave-induced loads are the most significant contributing factor to fatigue. Among the various methods proposed to calculate these loads and to build up the stress history, the most commonly used in ship design consists of carrying out a direct ship motion and load analysis. Where necessary, the effects of impact and vibratory loads have also to be accounted for (refer to Section 17.5.3).

In ship structures, residual stresses from welding are present and may be separated into two types:

1. *Local stresses.* Local stresses are close to the weld and self-balanced over the cross section of the member. After welding and during the cooling process the interaction between the different fibers results in a locked-in tensile stress in and near the weld, approximately equal to the yield stress of the material. The extent of the tension yield zone is generally from three to six thicknesses out from the weld on each side. This locked-in tension gives rise to a compressive residual stress σ_r in the remaining areas of the section. These stresses may be reduced by either heat treatment or local yielding caused by peak wave loading. High local tensile residual stresses are generally not present in the specimens used for determination of *S-N* curves, and this must be considered when defining the design *S-N* curve (refer to Section 17.3.6).
2. *Regional stresses.* Regional stresses are uniform throughout a member and self-balanced within the structure. These stresses are caused by the procedure of ship construction from prefabricated blocks and are generally small compared to the yield stress. Unlike local stresses, they are not easily reduced by heat treatment or by peak loading.

17.2.2 Spectral Fatigue Analysis

GENERAL

In spectral fatigue analysis, the long-term distribution of stresses is determined by carrying out a direct ship motion and load analysis for:

1. Calculation of the transfer functions of ship motions, load effects, and stresses for regular waves of unit amplitude and for a range of wave periods, heading angles and ship speeds.
2. Determination of the response spectra of stresses for various wave spectra, heading angles and ship speeds (each sea state is represented by a 2D directional wave spectrum defined in terms of two parameters, significant wave height and wave frequency).
3. Determination of the short-term structural response for various sea states, heading angles, and ship speeds.
4. Construction of the long-term distribution of stress range giving the probability $P(S_0)$ of the stress range exceeding a specified value S_0.

Although many different loading conditions occur during a ship's life, for the majority of cargo ships these may be idealized as two standard conditions, fully laden and ballast. Consequently, unless different loading conditions need to be considered depending on the ship type, calculation of the loads and stresses may be limited to these two conditions. For more detailed information on calculation of the ship response to random seas refer to Hughes (1988).

STRESS TRANSFER FUNCTIONS

For calculation of the ship's response to a random sea the usual practice is to represent the irregular sea surface as a linear superposition of a large number of regular waves having different amplitudes, lengths, directions, and random phase differences. The ship's response to regular waves is characterized for each ship motion and load effect by a frequency response function or transfer function. For example, for constant amplitude cosine waves it can be shown that, for each ship motion and load effect X_i, the transfer function $H_{X_i}(\omega,\phi,V)$ is a complex function defined such that its magnitude is equal to the amplitude ratio x_i/x_0 and the ratio of its imaginary part to its real part is equal to the tangent of the phase angle between the output and the regular wave assuming the crest is located at the ship center gravity (x_i and x_0 are the amplitudes of X_i and of the harmonic wave). If

$$H_{X_i}(\omega,\phi,V) = A(\omega) - iB(\omega) \qquad (17.2.1)$$

where $A(\omega)$ and $B(\omega)$ are real functions of ω, then

$$\left|H_{X_i}(\omega,\phi,V)\right| = \sqrt{A^2+B^2} = x_i/x_0 \quad \text{and}$$

$$\frac{\text{imaginary part}}{\text{real part}} = \frac{B}{A} = \tan\theta$$

Using complex exponential notation, we can say that if the input is a harmonic wave of amplitude x_0:

$$x(t) = x_0 \cos\omega t = x_0 \,\mathrm{Re}\left[e^{i\omega t}\right] \qquad (17.2.2)$$

then the corresponding harmonic output will be

$$x_i(t) = x_0 \operatorname{Re}\left[H_{X_i}(\omega)\, e^{i\omega t}\right] \qquad (17.2.3)$$

Transfer functions of ship motions and load effects are obtained by applying the linear potential theory to a series of prismatic girthwise strips of the ship's hull and solving the equations of ship motion. They are typically calculated for wave frequencies ω between 0.3 rad/sec and 1.8 rad/sec, heading angles ϕ between 0° and 360°, various ship speeds V and for the following outputs:

1. Ship motions and accelerations.
2. Vertical bending moment.
3. Horizontal bending moment.
4. Torsional moment (for open deck ships, such as container ships).
5. External sea pressures. However, the strip method does not provide accurate values for external sea pressures, especially in the vicinity of the waterline, and therefore more accurate calculations using 3D hydrodynamic programs are necessary to calculate sea pressures.
6. Internal cargo loads accounting for the inertial forces.

Then, for linear and harmonic responses the resulting stress transfer function at the location considered is obtained as a linear complex combination of the transfer functions for the various contributing load components:

$$H_\sigma(\omega,\phi,V) = \sum_{i=1}^{n} \sigma_{X_i} H_{X_i}(\omega,\phi,V) \qquad (17.2.4)$$

where σ_{X_i} = value of the stress calculated at the location considered for a unit load component X_i
$H_{X_i}(\omega,\phi,V)$ = transfer function for the load component X_i

For nonlinear and nonharmonic responses (e.g., connections of side shell longitudinals to transverse webs in the splash zone) special models are to be developed for each particular case. For instance, let us consider radiused or welded knuckle joints of double bottoms (refer to Figure 17.1 or 17.2 and Section 17.2.5). The load components contributing to the resultant stress at the knuckle are:

1. *The vertical wave-induced bending moment.* If $H_{M_{vw}}(\omega,\phi,V)$ is the transfer function of the vertical wave bending moment M_{vw}, the transfer function of the hull girder bending stress is:

$$H_{\sigma_{vw}}(\omega,\phi,V) = \frac{z_k - z_{na}}{I_v} H_{M_{vw}}(\omega,\phi,V) \qquad (17.2.5)$$

where z_{na} = distance of the neutral axis to the base line
z_k = distance of the knuckle joint to the base line
I_v = moment of inertia of the cross section about the neutral axis

Note that the following sign convention is used: tensile stresses are positive; the sagging wave bending moment $(M_{vw})_S$ is negative.

Due to nonlinear effects there are differences between transfer functions in hogging and sagging for large wave heights. Therefore, for a unit wave bending moment the distribution of wave-induced hull girder bending stresses can be represented by a sinusoid with trough-to-crest amplitude A equal to

$$A = \left|\frac{z_k - z_{na}}{I_v}\right| \left(1 + \left|\frac{(M_{vw})_H}{(M_{vw})_S}\right|\right) \qquad (17.2.6)$$

and the transfer function of the hull girder longitudinal bending stresses is

$$H_{\sigma_{vw}} = \frac{z_k - z_{na}}{2I_v} \left(1 + \left|\frac{(M_{vw})_H}{(M_{vw})_S}\right|\right) H_{M_{vw}}(\omega,\phi,V) \qquad (17.2.7)$$

where $(M_{vw})_S$ = sagging wave bending moment
$(M_{vw})_H$ = hogging wave bending moment

2. *The external sea pressures p_{ext}.* For transverse stresses due to the sea pressure, the assumption of linear superposition is applicable at knuckle considering its position with respect to the waterline. The corresponding transfer function $H_{\sigma_{p_{ext}}}(\omega,\phi,V)$ of transverse stresses is

$$H_{\sigma_{p_{ext}}}(\omega,\phi,V) = \sigma_{p_{ext}} H_{p_{ext}}(\omega,\phi,V) \qquad (17.2.8)$$

where $H_{p_{ext}}(\omega,\phi,V)$ = transfer function of the external sea pressure at knuckle
$\sigma_{p_{ext}}$ = transverse nominal stress at knuckle for a regular wave of unit amplitude

3. *The inertial cargo loads.* For example, from Bureau Veritas (1998) the general expression of inertial pressures calculated at mid-length of cargo tanks is

$$p_{\text{int}} = \rho_c \left[0.5\ \gamma_x \ell_c + \gamma_y\ (y_P - y_A) + \gamma_z\ (z_A - z_P) \right] \tag{17.2.9}$$

where ρ_c = cargo density
γ_x = rule longitudinal ship's acceleration
γ_y = rule transverse ship's acceleration
γ_z = rule vertical ship's acceleration
ℓ_c = length of the cargo tank
y_P , z_P = coordinates of the load point
y_A , z_A = coordinates of the starboard uppermost point of the cargo tank, the ship being in heeled condition $(y_A) < 0$

The distribution of transverse membrane stresses σ_{γ_y} and σ_{γ_z} in the inner hull plating at the knuckle joint is calculated for unit transverse and vertical accelerations γ_y and γ_z. The corresponding stress range can be represented by a sinusoid with trough-to-crest amplitude equal to

$$A = \left| \sigma_{\gamma_y}(y_k) - \sigma_{\gamma_y}(-y_k) \right|$$ for unit transverse acceleration

$$A = 2\left| \sigma_{\gamma_z} \right|$$ for unit vertical acceleration

where y_k = transverse coordinate of the port side knuckle joint under consideration $(y_k) > 0$

Therefore, the transfer function $H_{\sigma_{nom}}(\omega, \phi, V)$ of the transverse nominal stress at the knuckle joint due to inertial pressures is

$$H_{\sigma_{nom}}(\omega,\phi,V) = \frac{\sigma_{\gamma_y}(y_k) - \sigma_{\gamma_y}(-y_k)}{2} H_{\gamma_y}(\omega,\phi,V) + \sigma_{\gamma_z} H_{\gamma_z}(\omega,\phi,V) \tag{17.2.10}$$

where σ_{γ_y} = transverse nominal stresses at knuckle for unit ship's acceleration γ_y
σ_{γ_z} = transverse nominal stresses at knuckle for unit ship's acceleration γ_z
$H_{\gamma_y}(\omega,\phi,V)$ = transfer function of the transverse acceleration
$H_{\gamma_z}(\omega,\phi,V)$ = transfer function of the vertical acceleration

Using the notations of Section 17.2.5 the transfer function $H_{\sigma_G}(\omega,\phi,V)$ of the geometric stress at the knuckle is

Hot spot A

$$H_{\sigma_G} = K_{gy} \left[\frac{\sigma_{\gamma_y}(y_k) - \sigma_{\gamma_y}(-y_k)}{2} H_{\gamma_y} + \sigma_{\gamma_z} H_{\gamma_z} + \sigma_{p_{ext}} H_{p_{ext}} \right] \tag{17.2.11}$$

Hot spot B

$$H_{\sigma_G} = K_{gx} \frac{z_k - z_{na}}{I_v} \left(1 + \left| \frac{(M_{vw})_H}{(M_{vw})_S} \right| \right) H_{M_{vw}} + K_{gxy} \left[\frac{\sigma_{\gamma_y}(y_k) - \sigma_{\gamma_y}(-y_k)}{2} H_{\gamma_y} + \sigma_{\gamma_z} H_{\gamma_z} + \sigma_{p_{ext}} H_{p_{ext}} \right] \tag{17.2.12}$$

RESPONSE SPECTRA

A sea state is made up of a multitude of waves with various amplitudes, frequencies, phases and directions and is represented by its energy spectrum. The relationship between the wave spectrum $S_w(\omega)$ and the amplitude a_i of each component wave is (refer to Hughes (1988))

$$a_i = \sqrt{2S_w(\omega)\,\delta\omega}$$

Bretschneider (1959) was the first to propose that the wave spectrum for a given sea state could be represented in terms of two parameters that were characteristic of that sea state such as average wave height and average period. Various other formulas have been proposed such as those of Pierson-Moskowitz, the International Towing Tank Committee, and the International Ship and Offshore Structures Congress (1985) (ISSC) as recommended by IACS:

$$S_w(\omega) = 0.11 \left(\frac{\overline{H}_S^2}{\omega_S} \right) \left(\frac{\omega}{\omega_S} \right)^{-5} e^{-0.44(\omega/\omega_S)^{-4}} \tag{17.2.13}$$

where $\overline{H_S}$ = significant wave height defined as the average of all of the values above the one-third value $H_{1/3}(H_{1/3} = 1.48\sqrt{M_0}$ if we assume that the peak values of the wave height follow a Rayleigh distribution)
m_0 = area under the short-term wave spectrum
ω_S = mean wave frequency given by
$$\omega_S = \frac{2\pi}{T_S} = \frac{m_1}{m_0} \text{ with } m_k = \int \omega^{\kappa} S_w(\omega)d\omega$$

As already mentioned the ship's response to a random sea is based on the linear superposition of a large number of regular waves of various amplitudes, frequencies, phases, and directions. As a consequence it can be shown that the response spectrum $S_X(\omega,\phi,V)$ for any load effect X is given by

$$S_X(\omega,\phi,V) = \left| H_X(\omega,\phi,V) \right|^2 S_w(\omega) \tag{17.2.14}$$

Therefore, for a given sea state and heading angle ϕ the response spectrum of the stress $S_\sigma(\omega, \phi, V)$ at the location considered is given by

$$S_\sigma(\omega,\phi,V) = |H_\sigma(\omega,\phi,V)|^2 S_w(\omega)$$

$$= S_w(\omega) \sum_{i=1}^{n}\sum_{j=1}^{n} \sigma_{X_i}\sigma_{X_j} H_{\sigma X_i} H^*_{\sigma X_j} \quad (17.2.15)$$

where H_σ = resultant stress transfer function as given by equation (17.2.4)

$H^*_{\sigma X_j}$ = complex conjugate of $H_{\sigma X_j}$

Equation (17.2.13) is applicable to long-crested irregular waves. For short-crested waves, a more complete representation of the sea is given by a 2D directional spectrum $S_w(\omega, \theta)$, which indicates the direction θ as well as the frequencies of the wave components. The most common method for approximating $S_w(\omega, \theta)$ is to use the form

$$S_w(\omega, \theta) = S_w(\omega)\, f(\theta) \quad (17.2.16)$$

and the spreading function $f(\theta)$ is taken as

$$f(\theta) = \frac{2\cos^2\theta}{\pi} \quad \text{for } -\frac{\pi}{2} \le \theta \le \frac{\pi}{2} \quad (17.2.17)$$

$$f(\theta) = 0 \qquad \text{otherwise}$$

In this case, the response spectrum of the wave-induced load component X is given by

$$S_X(\omega, \phi, V) = \frac{2}{\pi} \int_{-\pi/2}^{\pi/2} S_W(\omega) \left|H_X(\omega, \phi - \theta, V)\right|^2 \cos^2\theta \, d\theta \quad (17.2.18)$$

where $H_X(\omega, \phi - \theta, V)$ is the transfer function for the load component X and the response spectrum S_σ of the stress at the location considered is

$$S_\sigma(\omega,\phi,V) = \frac{2}{\pi} \int_{-\pi/2}^{\pi/2} S_w(\omega) \sum_{i=1}^{n}\sum_{j=1}^{n} \sigma_{X_i}\sigma_{X_j} H_{X_i} \times (\omega, \phi - \theta, V)\, H^*_{Xj}(\omega, \phi - \theta, V) \cos^2\theta \, d\theta \quad (17.2.19)$$

Since the ship does not oscillate with the wave frequency ω but rather with the wave encounter frequency $\omega_e = \omega\,(1 + \omega V/g)\cos\phi$, the wave spectrum has to be expressed in terms of the wave encounter frequency and the stress range spectrum $S_\sigma(\omega, \phi, V)$ transformed into the stress range encounter frequency spectrum $S_\sigma(\omega_e, \phi, V)$.

SHORT-TERM RESPONSE

The short-term response which consists in determining the structural response for a given sea state, is based on the following assumptions:

1. Analysis of ocean wave data has shown that for a fully developed, wind-generated, mid-ocean sea state the wave spectrum is relatively narrow-banded. The ship acts as a filter, such that the spectra of ship motions and load effects are even more narrow-banded. Also, like the waves, these various responses have distributions that are Gaussian and stationary in the short term, that is, for a given sea state. Therefore, it can be shown that the peak values of the wave height follow a Rayleigh distribution and that the peak values of ship motions, load effects and stresses also follow a Rayleigh distribution. Moreover, the statistical properties of negative peak values are essentially the same as those of positive values.
2. Linear superposition is applicable. For nonlinear systems special procedures are to be used (e.g., time domain simulation) for determination of the short-term response.

Under these assumptions and noting that $S = 2\sigma$ the probability density function of the stress range response S at any location is given by the Rayleigh distribution:

$$p(S) = \frac{S}{4\mu^2} e^{-\frac{S^2}{8\mu^2}} \quad (17.2.20)$$

where the variance μ^2 is equal for a process with zero mean to the area under the stress response spectrum:

$$\mu^2 = \int_0^\infty S_\sigma(\omega_e,\phi,V)\, d\omega = m_0 \quad (17.2.21)$$

The probability of the stress range S exceeding a specified value S_0 for a given sea state is

$$P(S > S_0) = e^{-\frac{S_0^2}{8m_0}} \quad (17.2.22)$$

and the cumulative probability distribution for a given sea state, heading angle and ship speed is given by

$$P_S(S_0) = 1 - e^{-\frac{S_0^2}{8m_0}} \quad (17.2.23)$$

That is, $P_S(S_0)$ is the probability that a stress range of magnitude S_0 will not be exceeded for a given sea state, heading angle and ship speed.

LONG-TERM RESPONSE

During the 20 years or so of the average working lifetime of a ship a wide range of weather conditions, and hence of sea states, will be encountered. This total time span may be regarded as a large number of short intervals, each of a few hours duration, during which the sea state remains constant. The values of the significant wave height $\bar{H}_S$ computed for such intervals throughout the lifetime of a ship will be characterized by some kind ofdistribution, or probability density function, over the lifetime of the ship. Likewise, the total lifetime response history of the ship may be thought of as a series of short-term episodes. Assuming that the short-term stress range response follows a Rayleigh distribution, the long-term probability $P_S(S>S_0)$ of the resultant stress range exceeding S_0 is obtained by combining:

1. The short-term probability of the stress range exceeding a specified value S_0.
2. The probability $p(\bar{H}_S,T_m)$ encountering each sea state defined by the average significant wave height H_S and the mean period T_m. Wave data considered for this long-term analysis correspond generally to a worldwide service, but may take into account the expected service route of the ship.
3. The probability $p(\phi)$ of occurrence of the heading angle ϕ (for most sea states all headings are equally probable, but for very severe seas ship's masters generally alter the ship's course in order to reduce the severity of the response).
4. The probability $p(V)$ of occurrence of the maximum speed or a reduced speed.

For a given loading condition the long-term probability of the stress range exceeding a specified value S_0 is given by

$$P(S>S_0)=\iiiint e^{-\frac{S_0^2}{8m_0}}p(\bar{H}_S,T_m)p(\phi)\times p(V)\,d\phi\, dV\, dT\, d\bar{H} \qquad (17.2.24)$$

Finally the long-term distribution of the stress range for the structural detail considered is obtained by performing this calculation for various values of S_0.

PROBABILITY DISTRIBUTION FUNCTIONS TO DESCRIBE THE LONG-TERM DISTRIBUTION OF STRESSES

In order to express the fatigue life by a closed-form equation, it is convenient to find a probability distribution that gives the best fit to the calculated long-term distribution of stress ranges, bearing in mind that fatigue is a cumulative process contrary to the other limit states for which it is more important to know the most probable extreme value of the relevant load effect amplitudes over the period of time considered. The Weibull distribution function seems to be the most suitable function to represent the long-term distribution of stresses. Figure 17.6 gives measured values of the cumulative probability distribution of the long-term hull girder stress range for large tankers and dry cargo vessels as obtained by Little et al. (1971) and Hoffman and Lewis (1969). From these and other at-sea measurements carried out by Fain

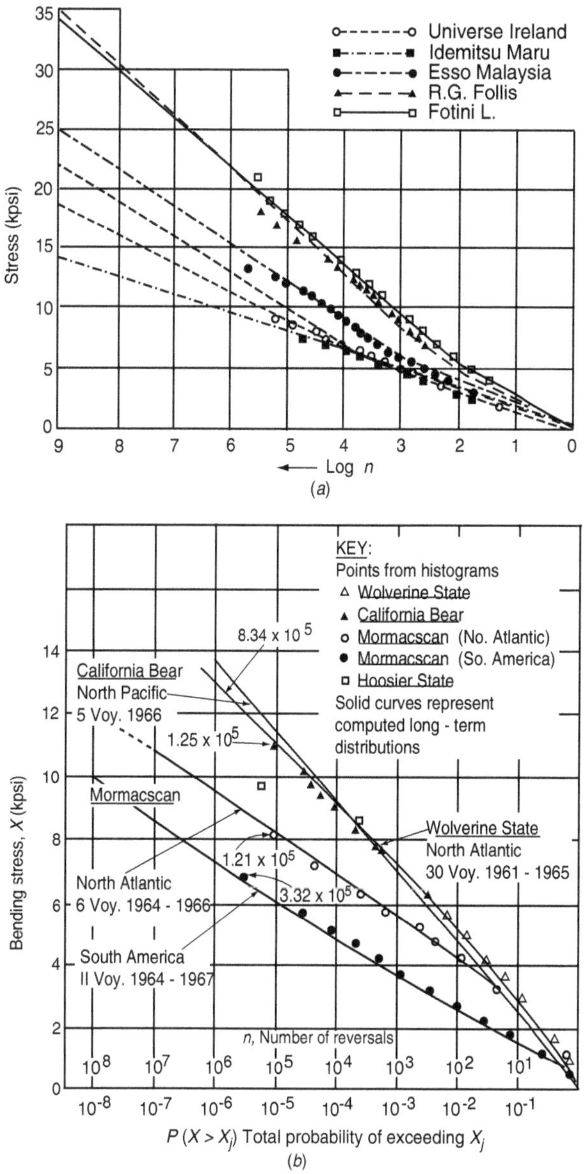

Figure 17.6 Long-term trend in service stresses for (a) large bulk carriers, R. S. Little (1971)and (b) dry cargo vessels, D. Hoffman (1969).

and Booth (1979) and Lewis and Zubaly (1975), it has been found that the long-term distribution of hull girder stresses can be represented satisfactorily by the two-parameter Weibull probability distribution.

For this distribution, the probability density function is

$$p_S(S) = \frac{k}{w}\left(\frac{S}{w}\right)^{k-1} e^{-(S/w)^k} \qquad (17.2.25)$$

in which w and k are the characteristic parameters of the distribution.

This density function is plotted in Fig. 17.7. The cumulative distribution function is the integral of $p_S(S)$, which is:

$$P(S) = 1 - e^{-(S/w)^k} \qquad (17.2.26)$$

where $P(S)$ = probability that a stress range of magnitude S will not be exceeded
w = characteristic value of the distribution
k = Weibull shape parameter

The probability of exceedance of a given stress range S_0 is

$$P(S > S_0) = 1 - P(S_0) = e^{-(S_0/w)^k} \qquad (17.2.27)$$

Structural analyses are generally carried out for loads corresponding to a given probability of exceedance. The number of cycles N_R corresponding to that probability of exceedance is

$$N_R = \frac{1}{P(S > S_R)} = e^{(S_0/w)^k} \qquad (17.2.28)$$

or

$$\ln N_R = (S_R/w)^k \qquad (17.2.29)$$

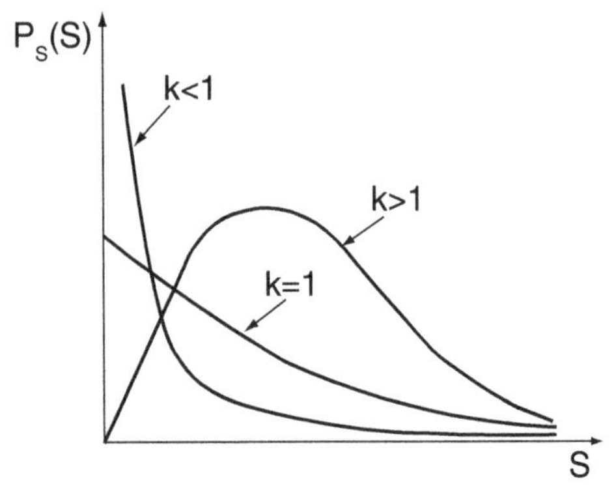

Figure 17.7 Weibull probability density function

and the characteristic value w of the distribution is

$$w = \frac{S_R}{(\ln N_R)^{1/k}} \qquad (17.2.30)$$

Substituting equation (17.2.30) into equation (17.2.26) gives the expression of the cumulative distribution function:

$$P(S) = 1 - e^{-(S/w)^k} = 1 - e^{-\ln N_R (S/S_R)^k} \qquad (17.2.31)$$

The Weibull distribution is plotted in Fig.17.8 for various values of k. Table 17.1 from Munse (1981) summarizes the results of the at-sea measurements illustrated in Fig. 17.6, and shows that for large tankers and bulk carriers k ranges from 0.7 to 1.0, while for dry cargo ships and containerships it ranges from 1 to 1.3. For design applications, k can either be set at an approximate value or it can be estimated more accurately by using the results of direct ship motions and loads analyses. If no direct ship motion and load analysis is carried out, the IACS (1999) suggests that the Weibull shape parameter of the long-term distribution of hull girder bending stresses be approximated by the following formula:

$$k = 1 - 0.35\,\frac{L - 100}{300} \qquad (17.2.32)$$

17.2.3 Simplified fatigue analysis

GENERAL

Direct determination of the long-term distribution of loads and stresses as described in Section 17.2.2 is a time-consuming process and classification societies have developed, as an alternative to direct calculation

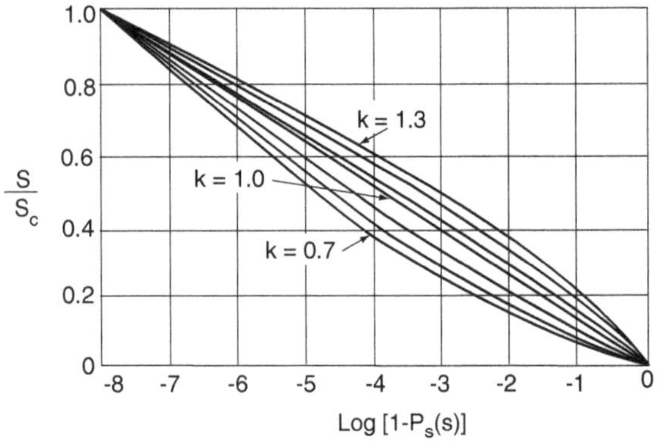

Figure 17.8 Ship loading histories modeled by Weibull distribution

Table 17.1 Ship Loading Histories Compared With Weibull Distributions

Type of ship	Name of ship	Notes	Weibull distribution shape parameter	Stress range at probability of exceedance = 10^{-8} (MPa)
Dry cargo ships	Wolverine state	1,5	1.2	113.8
	California state	1,5	1.0	124.2
	Mormacscan	1,5,7	1.3	82.8
	Mormacscan	1,5,8	1.0	69.0
Large tankers	Idemitsu Maru	2,5	1.0	84.9
	R. G. Follis	2,5	0.8	207.0
	Esso Malaysia	2,5	0.8	150.4
	Universe Ireland	2,3,5	0.7	129.0
Bulk carrier	Fotini L.	2,5	0.9	203.5
SL-7 Container ship		4,6,9	1.2	235.3

Notes:
1. Data from Hoffman and Lewis (1969)
2. Data from Little et al. (1971)
3. Data from Lewis and Zubaly (1975)
4. Data from Fain and Booth (1979)
5. Load history is for wave-induced loading with dynamic effects filtered.
6. Load history is for wave-induced loading with dynamic effects included.
7. Load history based on North-Atlantic voyages.
8. Load history based on South-American voyages.
9. Load history based on data collected from eight SL-7 containerships.

procedures, simplified procedures for calculation of loads and stresses. Different basic loading cases combining the various dynamic effects of the environment on the hull structure are considered for calculation of the stress ranges aiming at covering the most severe conditions. They include the following load components:

1. Hull girder loads (i.e., rule wave bending and torsional moments and shear forces).
2. External sea pressures.
3. Internal inertial and fluctuating loads.

Design loads given in the various classification society rules are generally based on different levels of probability of exceedance and the resultant long-term distribution of stresses (hull girder + local bending stresses) is assumed to be represented by a two-parameter Weibull distribution. If the Weibull shape parameter of the long-term distribution of stresses is known; the choice of the extreme stress range S_R and the associated probability of exceedance does not affect the fatigue damage (refer to Section 17.4.3). Unfortunately, there are many uncertainties on the shape parameter, depending on the type of ship, sailing route and location of the structural detail. As mentioned by the IACS (1997), the smaller the probability of exceedance of the extreme stress range, the greater is the influence of the Weibull shape parameter on the calculated fatigue damage ratio. Table 17.2 gives values of the damage ratios calculated according to equation (17.4.10) for two probabilities of exceedance (10^{-4} and 10^{-8}) and three values

Table 17.2 Influence of fhe Shape Parameter on the Calculated Damage Ratio

Shape Parameter k	$P = 10^{-4}$		$P = 10^{-8}$	
	D_k	$D_k/D_{k}{=}1$	D_k	$D_k/D_{k}{=}1$
1.2	0.124	1.68	0.176	2.38
1	0.074	1	0.074	1
0.8	0.039	0.525	0.023	0.31

of the shape parameter, considering the following input data:

N_t = 5.6×10^7 cycles
K_p = 5.8×10^{12}
S_R = 200 MPa for $p = 10^{-8}$
= 100 MPa for $p = 10^{-4}$

Therefore, loads and stresses should be calculated for moderate probabilities of exceedance (for example, 10^{-4} to 10^{-5}), all the more as most of the fatigue damage occurs in the high life range of *S-N* curves, as shown in Fig. 17.9.

The relative contribution of the local bending stress to the total stress depends on the type and location of the structural detail considered, and therefore the Weibull shape parameter differs from one particular structural detail to another one. As an example, let us examine the long-term distribution of deck longitudinals and side shell longitudinals near the waterline. Based on the following assumptions for the side longitudinal at the waterline:

- hull girder bending stress σ = 50 MPa and $k = 0.9$
- wave induced local stress σ = 100 MPa and $k = 1.1$

The Weibull shape parameter of the resultant stress distribution is 1.02 while it is 0.9 for the deck longitudinals. This problem is addressed in the rules of several Classification Societies that consider different values for the shape parameter, depending on the type and location of the welding connection.

American Bureau of Shipping (2002) provides in Section 5.1.1/5.5 simplified formulas for estimating the Weibull shape parameter :

$$k = 1.4 - 0.036\alpha\sqrt{L} \quad \text{for } 190 < L < 305 \quad (17.2.33)$$

$$k = 1.4 - 0.044\alpha^{0.8}\sqrt{L} \quad \text{for } 305 < L \quad (17.2.34)$$

with $\alpha = 1$ for deck structure, including upper part of side shell and longitudinal bulkheads
$\alpha = 0.93$ for bottom structure, including lower part of side shell and longitudinal bulkheads
$\alpha = 0.86$ for side shell and longitudinal bulkhead structure within mid-depth region
$\alpha = 0.80$ for transverse bulkhead structure

EXAMPLE OF RULE APPLICATION

The following presents briefly the procedure as proposed by Bureau Veritas (1998) in Section 2.2 (refer also to Bureau Veritas (2000)) for determination of the long-term distribution of global and local stresses at the connection of side shell longitudinal stiffeners to transverse webs of oil tankers. Two loading conditions are generally considered, full load and ballast conditions. For each of these two loading conditions four basic cases combining the various dynamic effects of the environment on the longitudinals are considered:

1. *Case 1—Head sea condition.* Static sea pressures with maximum and minimum inertial cargo or ballast loads.

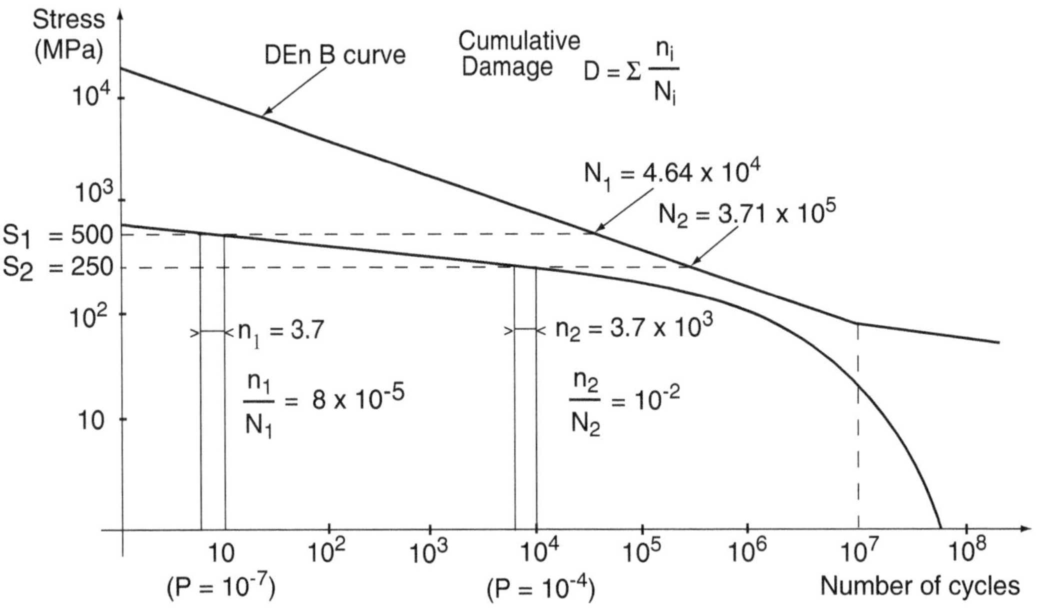

Figure 17.9 Comparative influence of the probability of exceedance of stresses on the fatigue damage ratio.

2. *Case 2—Head sea condition.* Maximum and minimum wave-induced sea pressures with static internal cargo or ballast loads.
3. *Case 3—Oblique sea condition.* Static sea pressures with maximum and minimum inertial cargo or ballast loads.
4. *Case 4—Oblique sea condition.* Maximum and minimum wave-induced sea pressures with static internal cargo or ballast loads.

Note the following in regard to these cases.

1. Tensile stresses are positive and compressive stresses are negative.
2. For side shell longitudinal stiffeners close to neutral axis hull girder longitudinal bending stresses due to the vertical wave bending moment may be disregarded.
3. All formulas are not developed. For more information refer to the Bureau Veritas (1998) and (2000).

For each of these four cases wave-induced global and local loads are calculated as follows:

1. *Case 1.*

a. Global loads:

$M_{vw} = 0.45\,(M_{vw})_S$ for maximum inertial loads

$M_{vw} = 0.45\,(M_{vw})_H$ for minimum inertial loads

where $(M_{vw})_S$ = rule sagging wave bending moment
$(M_{vw})_H$ = rule hogging wave bending moment

b. Local loads. Inertial cargo or ballast pressures are given by

$$p_{iner} = \pm\gamma_z d_c \text{ kN/m}^2$$

where d_c = distance of the load point to the top of compartment
γ_z = vertical axis acceleration of the center of gravity of the compartment

2. *Case 2.*

a. Global loads:

$M_{vw} = 0.625\,(M_{vw})_H$ for ship on crest of wave

$M_{vw} = 0.625\,(M_{vw})_S$ for ship on trough of wave

b. Local loads. Wave-induced sea pressures, in kN/m², are calculated assuming the ship on crest or trough of wave and given by

Ship on crest of wave

$$p_w = 10h_1 e^{-2\pi(T_1-z)/L} \quad \text{for } z \le T_1$$

$$p_w = 10h_1 - 10(z-T_1) \ge 0.15L \quad \text{for } z > T_1$$

Ship on trough of wave $p_w = -10h_1 e^{-2\pi(T_1-z)/L}$ without being taken less than $p_w = 10(z-T_1)$

where L = ship's length
T_1 = draught for the loading case considered
$h_1 = 0.42\,C\,(C_B + 0.7)$ within 0.4L amidships

$$C = 10.75 - \left(\frac{300-L}{100}\right)^{1.5}$$

3. *Case 3.*

a. Global loads:

Maximum inertial loads ($\gamma_z > 0$)

$M_{vw} = 0.30(M_{vw})_S$

$M_{hw} = 0.45(M_{hw})_0$ (port side in tension)

Minimum inertial loads ($\gamma_z < 0$)

$M_{vw} = 0.30(M_{vw})_H$

$M_{hw} = -0.45(M_{hw})_0$ (starboard in tension)

where $(M_{hw})_0$ = rule horizontal wave bending moment $(M_{hw})_0 > 0$

b. Local loads. Maximum and minimum inertial cargo or ballast pressures, in kN/m², are calculated as follows:

$$p_{max} = \rho_c\left[(y_p - y_A)\gamma_y + (z_A - z_P)\gamma_z\right]$$

$$p_{min} = \rho_c\left[(y_A + b_c - y_p) - (z_A - z_p)\gamma_z\right]$$

where γ_y = rule transverse acceleration of the center of gravity of the compartment
y_A, z_A = coordinates of the uppermost point of the cargo or ballast tank, the ship being in heeled condition
y_P, z_P = coordinates of the load point
b_c = breadth of the compartment

4. *Case 4.*

a. Global loads:

$M_{vw} = 0.3(M_{vw})_S$ or $0.3\,(M_{vw})_H$, depending on the location of the side shell longitudinal

$M_{hw} = -0.625\ (M_{hw})_0$ (port side assumed in compression)

b. Local loads. Wave-induced sea pressures, in kN/m², are calculated as follows:

Ship side on crest of wave ($y > 0$)

$$p_w = 5h_1 \frac{y}{y_{WL}} e^{-2\pi(T_1-z)/L} + 10yA_R e^{-\pi(T_1-z)/L} \quad \text{for } z \le T_1$$

$$p_w = 10h_2 - 10(z-T_1) \ge 0.15L \quad \text{for } z > T_1$$

Ship on trough of wave ($y < 0$)

$$p_w = -\left[5h_1 \frac{y}{y_{WL}} e^{-2\pi(T_1-z)/L} + 10yA_R e^{-\pi(T_1-z)/L}\right]$$

without being taken less than $p_w = 10\,(z-T_1)$

where y_{WL} = half breadth of ship measured at waterline
$h_2 = 0.5\, h_1 + y_{WL} A_R$
A_R = roll angle

17.2.4 Equivalent Regular Wave Concept

GENERAL

A different approach, intermediate between the spectral and simplified analyses and based on the so-called equivalent regular wave concept as developed by Bureau Veritas (2000), may be used for determination of the long-term distribution of stresses. Examination of the results of fatigue analyses, carried out for various types of ships, shows that the following wave-induced load effects are the leading factors that govern the fatigue life of structural details:

1. Vertical wave bending moment.
2. Horizontal wave bending moment.
3. Wave torsional moment, where applicable.
4. External sea pressures, especially near the waterline.
5. Internal pressures.

For each of these affects an equivalent regular wave, defined by its wave height, wave length, heading angle and position along the ship length, is determined so that the maximum response for the selected load effect be equal to its value given by the rules for the probability of exceedance considered. The amplitude of the other effects is obtained from a ship motion analysis assuming the ship to be positioned on the equivalent regular wave.

AMPLITUDE OF THE EQUIVALENT REGULAR WAVE

For each relevant load effect, the amplitude of the equivalent regular wave is taken as

$$A = \frac{X_{LT}}{X_{TF}} \qquad (17.2.35)$$

where X_{LT} = long-term response of the selected load effect for the required probability of exceedance
X_{TF} = maximum value of the load effect as obtained from the transfer function for theworst frequency and heading angle

The frequency and heading angle of the equivalent regular wave are those of the regular wave for which X_{TF} is maximum.

POSITION OF THE EQUIVALENT REGULAR WAVE

A ship responds to regular waves with harmonic oscillations and the effects related to ship motions are also harmonic, so that any load effect X can be described by

$$X(t) = X_0 \cos(\omega_e t + \phi_X) \qquad (17.2.36)$$

where ω_e is the encounter frequency and ϕ_X the phase angle of the load effect X.

The maximum value of $X(t)$ is obtained for $\omega_e t + \phi_X = \lambda\pi$, i.e., at time $t_0 = \dfrac{\lambda\pi - \phi_X}{\omega_e}$.

The equation of the wave profile is given by:

$$\xi(x,y,t) = A\cos(kx\cos\alpha + ky\sin\alpha - \omega_e t) \qquad (17.2.37)$$

where $k = 2\pi/\Lambda$ wave number
Λ = wave length
α = wave direction
and the position of the wave crest is such that:

$$k\,x\cos\alpha + k\,y\sin\alpha - \omega_e t = 0$$

that is, $kx\cos\alpha + ky\sin\alpha = \omega_e t = \lambda\pi - \phi_X$ (17.2.38)

Equation (17.2.38) is valid irrespective of the couple (x,y) and, in particular for $y = 0$, that is, in the centerline, which gives $kx\cos\alpha = \lambda\pi - \phi_X$. Therefore, the crest position is given by

$$x = \frac{\lambda\pi - \phi_X}{k\cos\alpha} = \frac{\Lambda(\lambda\pi - \phi_X)}{2\pi\cos\alpha} \qquad (17.2.39)$$

The instantaneous values of the other load effects X_i are calculated for $t = t_0$ and given by $X = X_{0i}\cos(\omega_e t + \phi_{X_i})$ where ϕ_{X_i} is the phase angle of the load effect X_i.

Finally, a 3D FEM stress analysis of the full ship structure is carried out for each relevant load effect and corresponding equivalent regular wave for determination of the stress pattern in the vicinity of the structural detail considered.

17.2.5 Stresses to be Used

DEFINITION OF STRESSES

Assessment of the fatigue strength of structural details requires the determination of the stress at the *hot spot*, that is, where cracks are initiated. Depending on the level of refinement of the method used to calculate the stresses, three kinds of stresses are considered in fatigue analyses, as shown in Fig. 17.10:

1. *Nominal stress.* Nominal stress is a general stress in a structural component taking into account macro-geometric effects but disregarding the stress-raising effects due to structural discontinuities and the presence of welds. Nominal stresses may be calculated either by structural mechanics or by coarse mesh FEM stress analysis and are based on the applied loads and properties of the component.

For beam members such as longitudinal stiffeners the nominal stress is given by

$$\sigma_n = \frac{P}{A} + \frac{Mv}{I} = \sigma_a + \sigma_b \qquad (17.2.40)$$

where σ_a = axial stress (e.g., from hull girder bending)
σ_b = local bending stress

2. *Hot spot stress.* Hot spot stress is the stress at the hot spot, as shown in Fig. 17.10, taking into account the influence of structural discontinuities due to the geometry of the connection but excluding stress concentrations due to the weld profile. The hot spot stress is expressed as follows:

$$\sigma_G = K_G \sigma_n \qquad (17.2.41)$$

in which σ_G = hot spot stress
σ_n = nominal stress
K_G = stress concentration factor due to the geometrical configuration of the connection

The hot spot stress at the surface of plates, as obtained from FEM stress analysis, is given by

$$\sigma_G = \sigma_m + \sigma_b \qquad (17.2.42)$$

in which σ_m = membrane stress
σ_b = bending stress through the plate thickness

3. *Notch stress.* Notch stress is the total stress at a notch, that is, the root of a weld (as in Fig. 17.10), assuming a linear-elastic material behavior. This peak stress, which is limited to a small area, takes into account the stress concentrations due to the effects of structural geometry and the presence of a notch. The notch stress may be expressed as follows:

$$\sigma_{loc} = K_w \sigma_G = K_w K_G \sigma_n = K_t \sigma_n \qquad (17.2.43)$$

where σ_{loc} = notch stress
K_t = total stress concentration factor including the effects associated with the structural and weld geometry
K_w = weld shape concentration factor

DETERMINATION OF STRESSES

Nominal Stresses The following rules may be considered for the calculation of nominal stresses:

1. Nominal stresses are to take into account the overall geometry of the detail, but not the geometry of the hot spot.
2. Where nominal stresses are obtained from a finite element stress analysis, a uniform mesh is to be used with smooth transition and avoidance of abrupt changes in mesh size. The nominal stress at

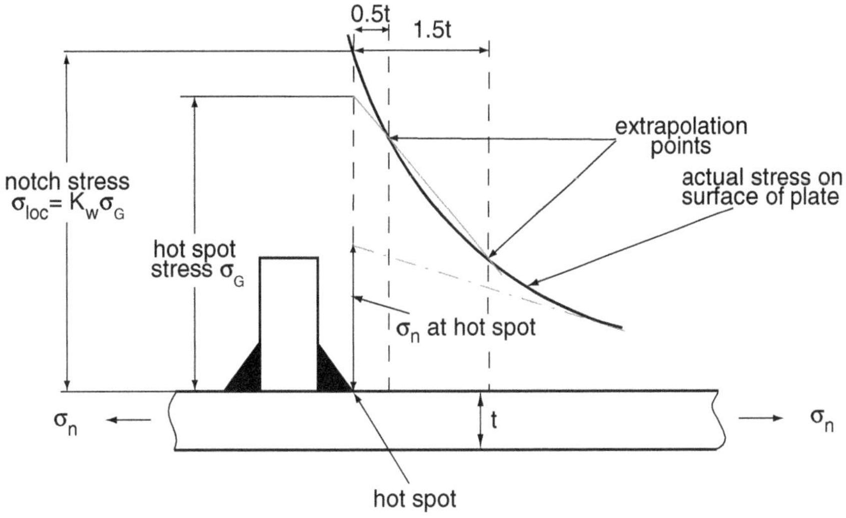

Figure 17.10 Definition of stresses.

the hot spot is determined by extrapolation from the stresses in the region surrounding the hot spot calculated at the Gaussian points of each element.

Hot Spot Stresses The determination of hot spot stresses generally requires 2D or 3D fine mesh stress analyses beyond the 3D coarse mesh structural analysis. In this case, boundary nodal displacements or forces obtained from the 3D coarse mesh model are applied to the fine mesh model as boundary conditions.

In highly stressed areas, in particular in the vicinity of geometrical singularities, the level of stresses depends on the size of elements, due to the high stress gradient. As recommended by the International Institute of Welding (1996) (IIW) and Niemi (1992) the following rules may be considered for the modeling of local structures:

1. *Hot spot stresses are calculated using an idealized welded joint with no misalignment.*
2. *The finite element mesh is to be fine enough near the hot spot such that stresses and stress gradients can be determined at points comparable with the extrapolation points used for strain gauge measurements.*
3. *Plating, webs, and faceplates of primary and secondary members are modeled by 4-node thin shell or 8-node solid elements. In the case of a steep stress gradient 8-node thin shell elements or 20-node solid elements are recommended.*
4. *When thin shell elements are used, the structure is modeled at mid-thickness of the plates. Where considered necessary, the stiffness of the weld intersection should be taken into account (for example, by modeling the welds by inclined shell elements).*
5. *The aspect ratio of elements is not to be greater than 3.*
6. *The size of elements located in the vicinity of the hot spot is to be about one to two times the thickness of the structural member.*
7. *The centroid of the first element adjacent to the weld toe is to be located between the weld toe and 0.4t of the toe, where t is the plate thickness.*
8. *Stresses are to be calculated at the surface of the plate in order to take into account the plate bending moment, where relevant.*

From the results of FEM analyses, three different procedures may be used to determine the hot spot stresses:

1. Stress extrapolation at the structural discontinuity where large stress gradient is expected, such as the connection of longitudinals to transverse webs as in Fig. 17.11. Stress values at a distance of $0.5t$ and $1.5t$ from the weld toe are determined by interpolation of the maximum tensile principal stress ranges at the centers of element faces in the region. Principal stresses to be considered are those forming an angle of less than 45° with the normal to the weld toe, bearing in mind that fatigue cracks tend generally to propagate in a direction normal to the largest tensile principal stresses. Then as illustrated, the hot spot stress is obtained by linear extrapolation to the weld toe.
2. Stress in the element at the hot spot where the geometry does not permit a clear development of a stress gradient.
3. Stress in the free edge for areas where no structural discontinuity exists (e.g., at the smallest radius of a cut-out as in Fig. 17.12). In that case, the hot spot stress may be obtained by placing small diameter bar elements along the free edge.

Hot spot stresses may also be determined by using parametric formulas giving the stress concentration factor K_G of typical structural connections for which nominal stresses can be calculated. In such a case, the hot spot stress is obtained from equation (17.2.41). Classification society rules provide these formulas for K_G for typical structural details such as connections of secondary stiffeners to webs of primary members or knuckle joints, as in Figs. 17.1 and 17.2. The latter type of connection particularly needs to be considered at the design stage because these joints are critical to the ship structural safety as pointed out by Beghin and Cambos (1997).

For instance, Table 17.3 from Bureau Veritas (1998) gives the geometric stress concentration factors in longitudinal and transverse directions for two basic designs of the double bottom structure in way of hopper tanks, as shown in Figs. 17.13 and 17.14. Depending on the location, the hot spot stress is given by

Hot spot A $$\sigma_{hot\ spot} = K_{gy}\,\sigma_{ny} \qquad (17.2.44)$$

Hot spot B $$\sigma_{hot\ spot} = K_{gx}\,\sigma_{nx} + K_{gxy}\,\sigma_{ny} \qquad (17.2.45)$$

where K_{gx} = geometric stress concentration factor for hull girder bending
K_{gxy}, K_{gy} = geometric stress concentration factors for transverse loading
σ_{nx} = nominal hull girder bending stress
σ_{ny} = nominal membrane stress in transverse direction at the hot spot, as obtained from a 3D FEM stress analysis by extrapolation of the stresses at the Gaussian points of each element located in the region of the knuckle

As reported by the TSCF (1997), many cracks have been detected at the connection of side shell

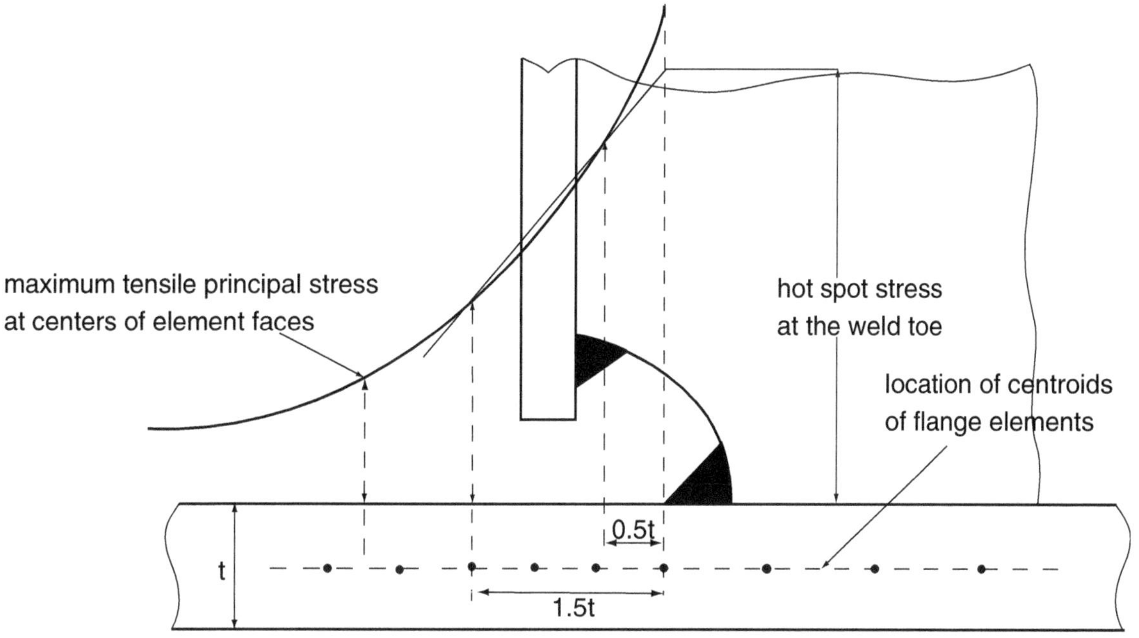

Figure 17.11 Calculation of hot spot stress from finite element analysis.

longitudinal stiffeners to transverse webs of oil tankers. Asymmetry of the stiffener flange was generally a major contributing factor to the fatigue damage. If the geometric stress concentration factors of the connection are known, the hot spot stress is

$$\sigma_G = K_{ga}\,\sigma_{na} + K_{gb}\,\sigma_{nb} \qquad (17.2.46)$$

where K_{ga} = geometric stress concentration factor for axial loading

K_{gb} = geometric stress concentration factor for local bending

σ_{na} = nominal hull girder bending stress

σ_{nb} = nominal local bending stress given by

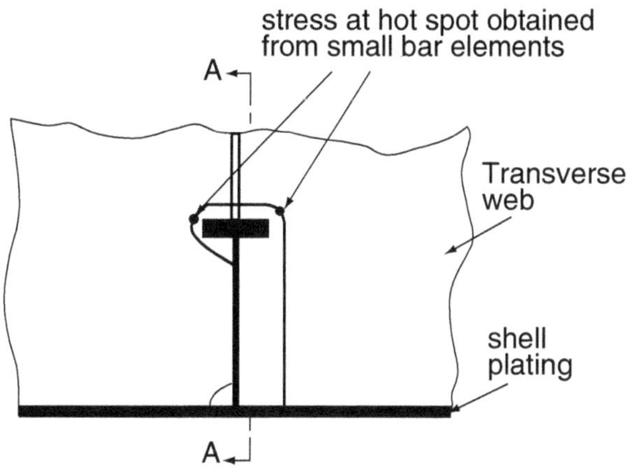

Figure 17.12 Location of hot spot stress in a cut-out.

$$\sigma_{nb} = \sigma_b + \frac{6EI_v\delta}{Z\ell^2} \qquad (17.2.47)$$

$$\sigma_b = \frac{p_{res}\ell^2}{12Z}\left\{1+\left[\frac{t_f(a^2-b^2)}{2Z_B}\right]\left[1-\frac{b}{b_f}\left(1+\frac{Z_B}{Z_A}\right)\right]\right\} \qquad (17.2.48)$$

E = Young's modulus

b_f = flange width

t_f = flange thickness

b = eccentricity of the flange as defined in Fig.17.15

Z = section modulus of the longitudinal stiffener with attached plating

Z_A = section modulus to A of the longitudinal stiffener about z axis without attached plating

Z_B = section modulus to B of the longitudinal stiffener about z axis without attached plating

p_{res} = resultant lateral pressure applied on the plate attached to the longitudinal stiffener

I_v = inertia of the longitudinal stiffener about y axis with attached plating

δ = relative deflection between the transverse bulkhead and adjacent transverse

Table 17.3 Geometric stress concentration Factors of Knuckle joints

Configuration	K_G	Configuration	K_G
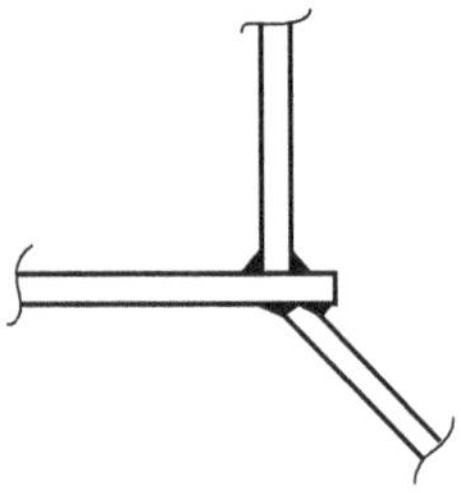	$K_{gx} = 1.30$ $K_{gxy} = 2.00$ $K_{gy} = 3.85$	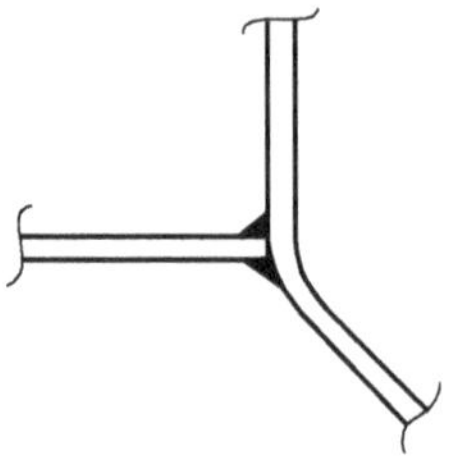	$K_{gx} = 1.30$ $K_{gxy} = 2.25$ $K_{gy} = 3.30$
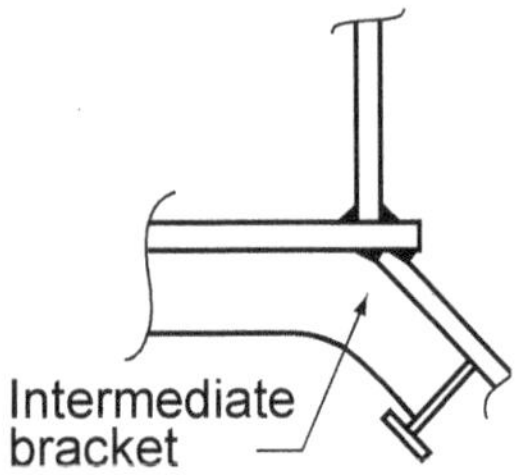	$K_{gx} = 1.30$ $K_{gxy} = 1.75$ $K_{gy} = 3.55$	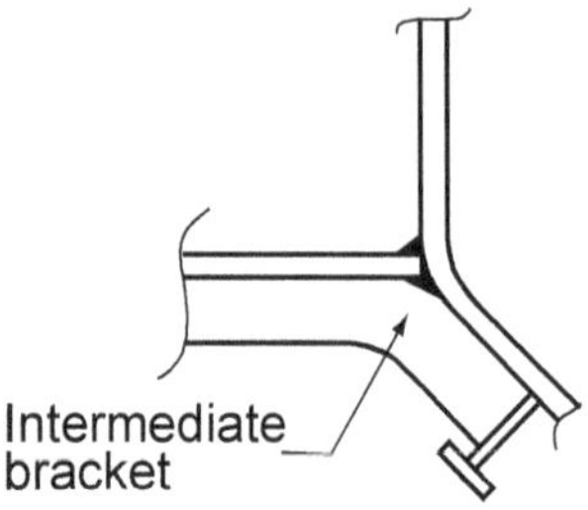	$K_{gx} = 1.30$ $K_{gxy} = 2.05$ $K_{gy} = 3.15$
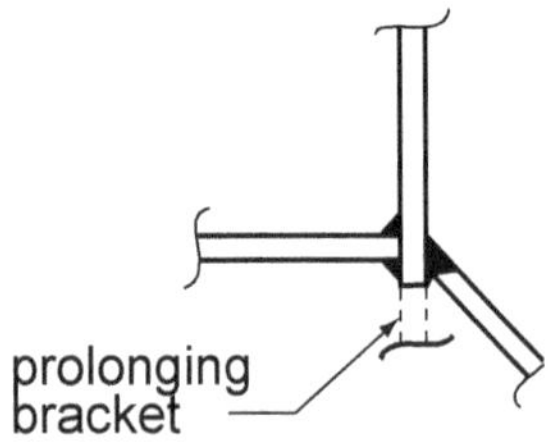	$K_{gx} = 1.30$ $K_{gxy} = 1.50$ $K_{gy} = 2.40$	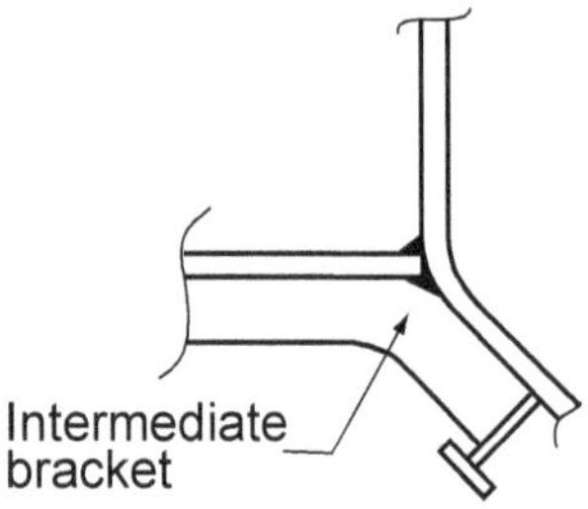	$K_{gx} = 1.30$ $K_{gxy} = 4.50$ $K_{gy} = 3.85$

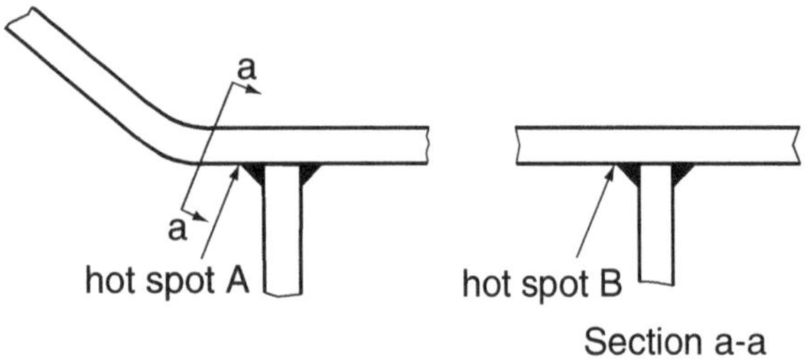

Figure 17.13 Radiused knuckle joint

webs, as obtained from a 3D structural analysis, where applicable

ℓ = span of the longitudinal stiffener

Moreover, the influence of workmanship, such as misalignment or angular distortion, may have to be

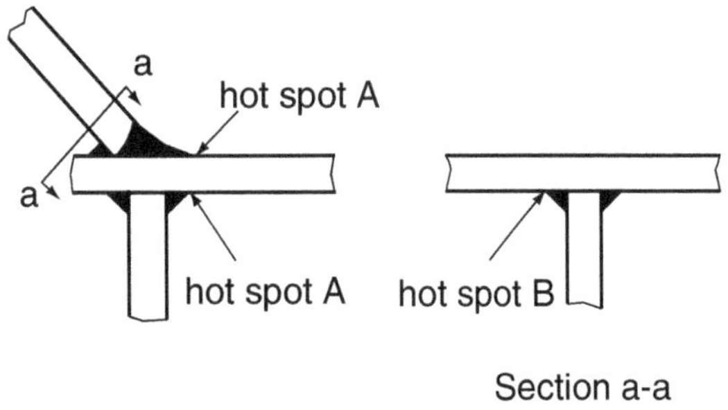

Figure 17.14 Welded knuckle joint.

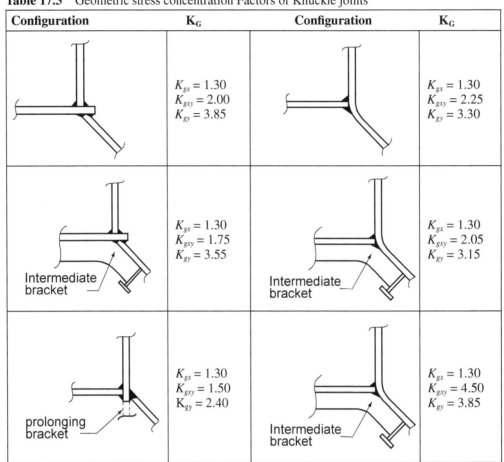

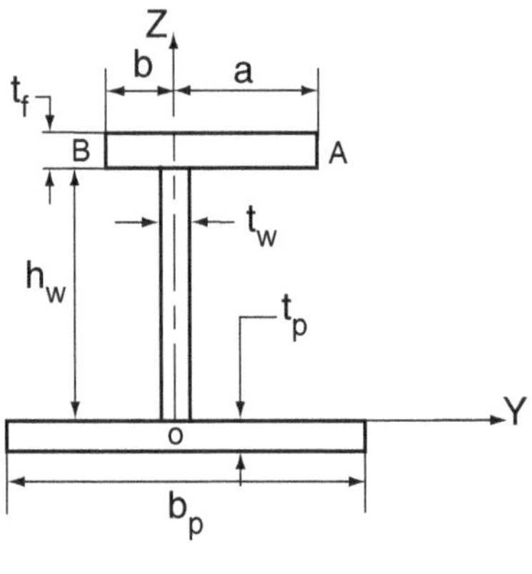

Figure 17.15 Asymmetrical stiffener.

Table 17.4 Fabrication Stress Concentration Factors

<table>
<tr><th>Geometry</th><th>Stress concentration factor K_m</th></tr>
<tr><td>Axial misalignment between flat plates
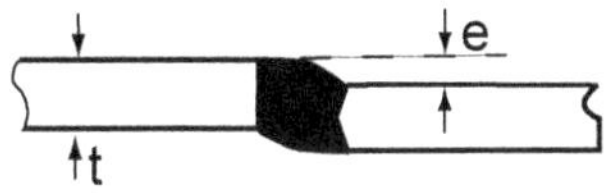
</td><td>$$K_m = 1 + \frac{3(e - e_0)}{t} \geq 1$$
e_0 : permissible misalignment</td></tr>
<tr><td>Axial misalignment between flat plates of different thickness
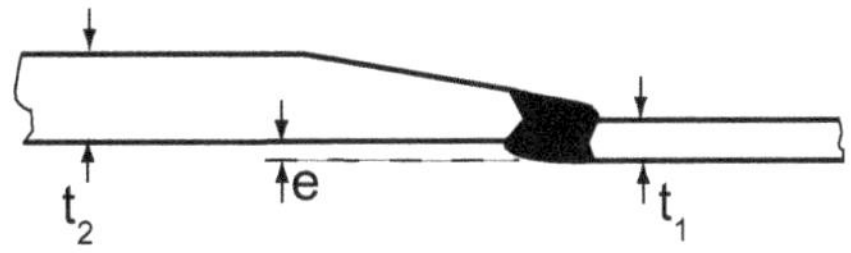
</td><td>$$K_m = 1 + \frac{6(e - e_0)}{t_1} \frac{t_1^{1,5}}{t_1^{1,5} + t_2^{1,5}} \geq 1$$</td></tr>
<tr><td>Angular misalignment between flat plates
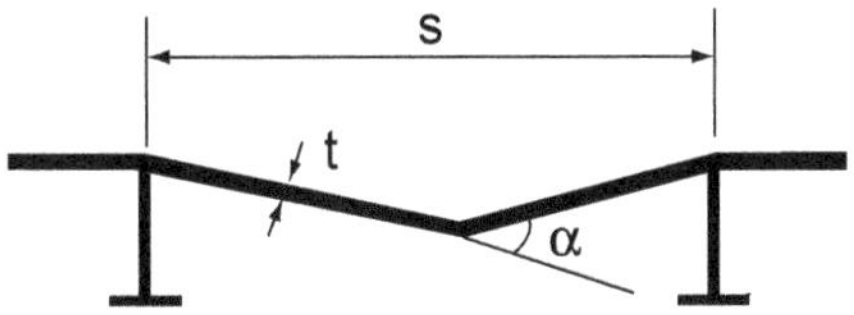
</td><td>$$K_m = 1 + \frac{\lambda}{4} \frac{(\alpha - \alpha_0)s}{t} \geq 1$$
α_0 : permissible angular misalignment
$\lambda = 3$ for fixed ends
$\lambda = 6$ for pinned ends</td></tr>
<tr><td>Axial misalignment in fillet welded cruciform joints
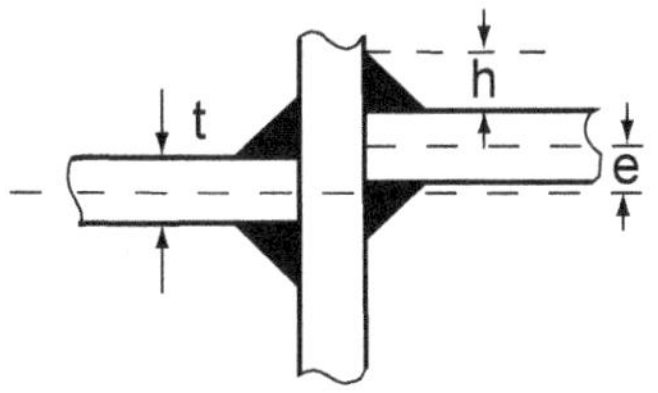
</td><td>$$K_m = 1 + \frac{e - e_0}{t + h} \geq 1$$</td></tr>
</table>

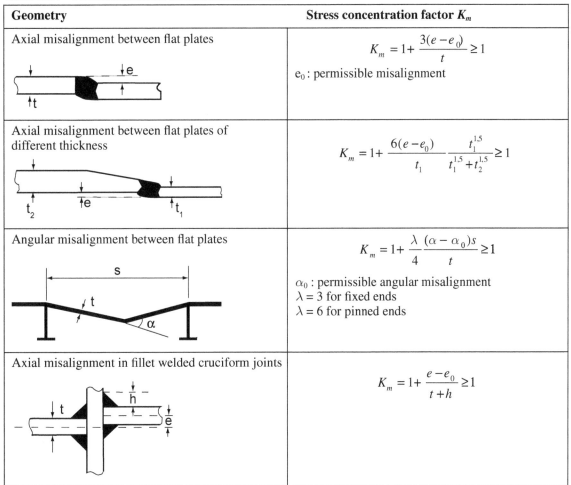

taken into account when calculating the geometric stresses. *S-N* curves are generally assumed to be representative of standard workmanship and welding procedures. However, for particular structural details, it may be necessary to take into account the effects of imperfections and, in particular, of misalignment when determining the hot spot stress. In such a case, the hot spot stress is to be multiplied by an additional stress concentration factor K_m. Table 17.4 gives examples of fabrication stress concentration factors as recommended by the IIW (1996).

Notch Stresses Determination of the peak or notch stress that depends on the weld profile requires a very fine mesh FEM analysis. The actual weld profile is generally not known and the following rules, as recommended by the IIW (1996), may be considered for the modeling of welds:

1. *An effective notch bottom radius of r = 1 mm is to be considered.*
2. *The method is restricted to weld joints that are expected to fail from weld toe or weld root. Other causes of fatigue failure from, for instance, surface roughness or embedded defects are not covered.*
3. *Flank angles of 30° for butt welds and 45° for fillet welds are suggested.*
4. *In cases where a mean geometrical notch root radius can be defined, for instance after certain post weld improvement procedure, this geometrical radius plus 1 mm may be used in the calculation of effective notch stresses.*
5. *The method is limited to thicknesses greater than 5 mm.*

According to equation (17.2.43) the notch stress depends on the weld shape concentration factor K_w which includes only the effects associated with the weld geometry. This coefficient K_w is obtained from diagrams or parametric formulas or from the results of FEM calculations or from measurements. For example, based on studies carried out by Fricke and Petershagen (1992) K_w may be approximated by the following formula:

$$K_w = \lambda\sqrt{\theta/30} \qquad (17.2.49)$$

where λ = coefficient depending on the weld configuration. Table 17.5 gives values of λ, as obtained from Stambaugh et al. (1994) for various typical weld configurations, noting that $K_w = 0.7\ K_f$ with K_f defined as the ratio of the mean fatigue strength at 10^6 cycles of smooth specimens to that of plates in the as-rolled condition mean

θ = weld toe angle, in degrees, taken generally as 30° for butt welds and 45° for fillet welds

Lawrence (1984) suggests a different equation for determination of the weld concentration factor K_w of butt welds:

$$K_w = 1 + 0.27(\tan\theta)^{0.25}(t/\rho)^{0.25} \qquad (17.2.50)$$

where ρ is the notch factor radius at the weld toe. For θ = 30° and $(t/\rho) = 10$ the weld concentration factor is $K_w = 1.42$.

ADVANTAGES AND DISADVANTAGES OF THE THREE APPROACHES

Nominal Stress Approach The nominal stress approach allows the direct use of the experimental *S-N* curves provided that the nominal stress, calculated according to the recommendations given in Section 17.2.5, can be easily defined and the applicable *S-N* curve clearly identified. Unfortunately, due to the complexity of ship structures the latter condition is difficult to fulfill, as it is essentially a matter of judgment.

Hot Spot Stress Approach The hot spot stress approach takes into account the change in geometry of the structure and, therefore, is more suitable when the nominal stress cannot be clearly defined. Moreover, this approach makes it possible to account for fabrication imperfections such as linear or angular misalignments that introduce additional geometric stress concentrations. If the weld configuration corresponds to that of one experimental *S-N* curve, this *S-N* curve which accounts for the weld effects can be used.

The distribution of stresses near the hot spot is highly dependent on the finite element mesh used and the recommendations given by the IIW (1996) and Niemi (1992), based on comparisons of theoretical and experimental results, should be applied, especially for extrapolation of the stresses at the hot spot. Moreover, comparison has to be made between the geometric stress concentration factors K_G of the specimen and actual structural detail, for possible correction of the selected *S-N* curve, as proposed in Section 17.3.5.

Notch Stress Approach The notch stress approach includes the effects of welds. As for the hot spot stress, recommendations given by the IIW (1996) and Niemi (1992) should be applied for determination of the local stress. In this approach, the same *S-N* curve (namely the curve for unwelded steel in the as-rolled condition) may be considered for all the types of structural details. This approach presents the advantage of starting from the unwelded steel and of accounting for all the fatigue degradations resulting from the geometric stress concentrations and weld effects.

If the hot spot stress is known, this approach may also be applied, using a weld shape concentration factor K_w from diagrams or parametric formulas.

17.3 FATIGUE CAPACITY OF WELDED STRUCTURES

17.3.1 Testing Methodologies

The fatigue capacity of welded structures may be assessed according to three different methodologies:

1. *S-N* curves.
2. Fracture mechanics.
3. Prototype testing.

S-N CURVES

S-N curves, which are the most common way to represent the fatigue capacity of welded steel joints, give the relationship between the fatigue life and the nominal stress range *S* applied to a given sample, the fatigue life being defined as the number of constant amplitude load cycles *N* to failure.

Fatigue cracks in welded structures are generally confined to welded joints or flame-cut edges where notches and initial defects are located and *S-N* curves are given for welded joints and flame-cut edges. For cases where cracks can occur in the weld throat, specific *S-N* curves are also provided.

FRACTURE MECHANICS

The fatigue capacity of welded steel joints may also be represented by crack growth rate curves that give

Table 17.5 Weld Shape Concentration Factors

Description	**Weld configuration**	λ
Longitudinally loaded butt weld	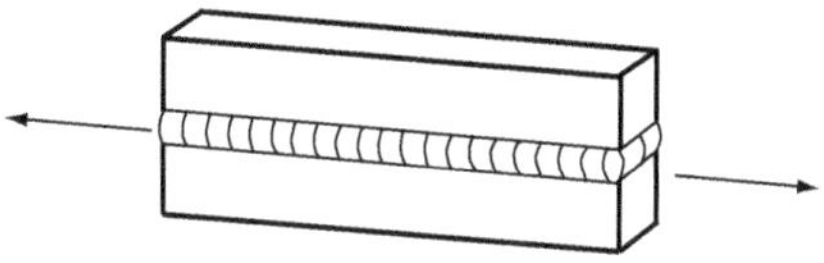	1.45
Longitudinally loaded fillet weld	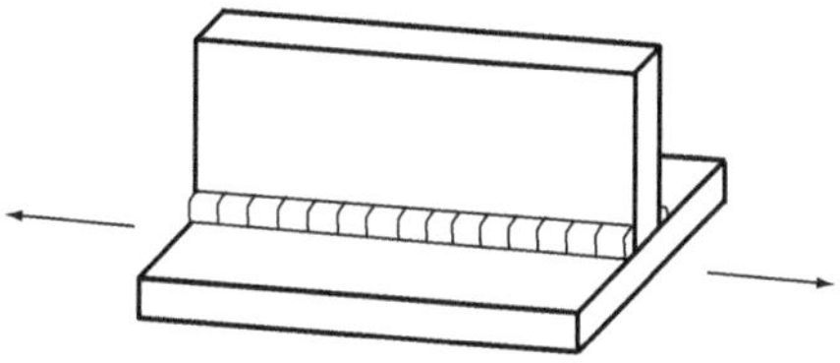	1.25
Transversely loaded butt weld	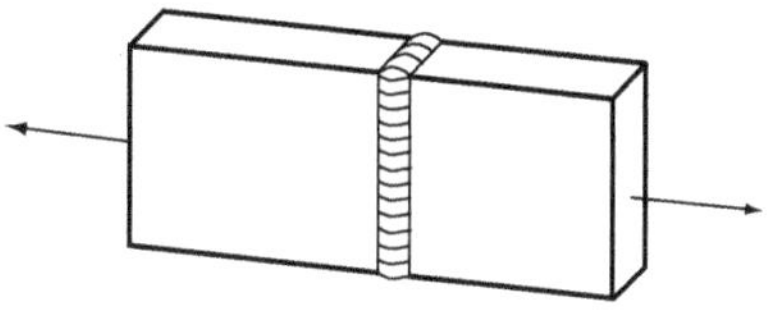	1.7
Transversely loaded fillet weld	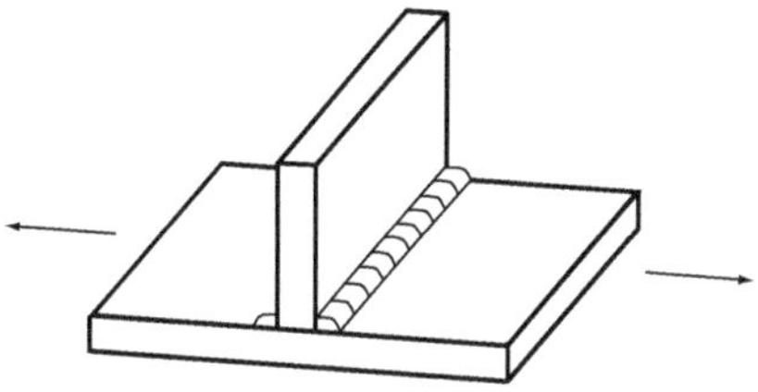	1.55
Axially loaded fillet weld	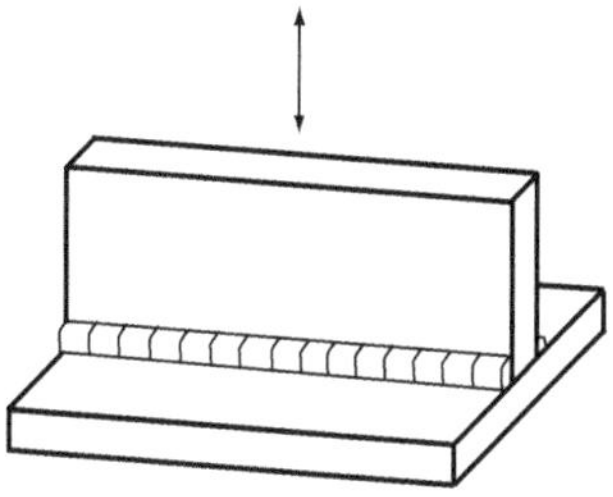	3.15
Lap weld (e.g., for beam brackets)	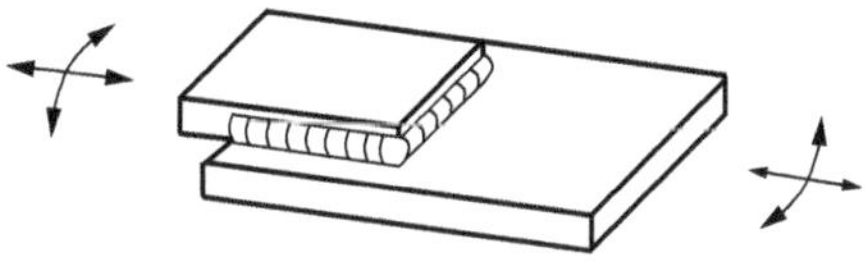	1.65
Cruciform joint (full penetration) ($\theta = 45°$)	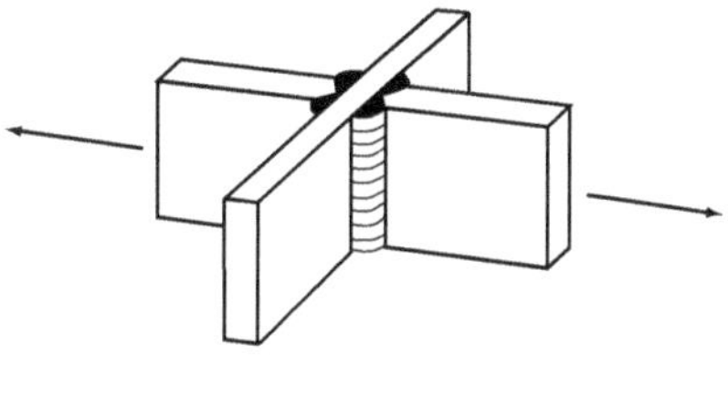	1.5

the relationship between the crack growth rate *da/dN* and the stress intensity factor range ΔK. Experimental curves are given either for the plain material, the heat affected zone, or the weld zone.

PROTOTYPE TESTING

Prototype testing is the most direct way of assessing the fatigue strength for a particular structure. However, prototype testing is expensive and only cost-effective for critical structures or structures which are produced in large numbers. Most marine structures are too big for the available testing machines, and the scaling of defects is uncertain. Therefore this approach is generally not considered feasible for ship structures, except for very particular structural details.

17.3.2 Experimental S-N Curves

GENERAL

Most of the *S-N* curves are determined in laboratories where specimens are subjected to constant amplitude cyclic loadings until failure. The main parameters that influence the fatigue life of specimens are:

1. The stress range $S = \sigma_{max} - \sigma_{min}$, where σ_{min} and σ_{max} are defined in Fig. 17.16.
2. The stress ratio $R = \sigma_{min}/\sigma_{max}$. Fatigue tests are generally performed at a constant stress ratio R lying between 0 and 0.1. Note that the mean stress σ_{mean} is given by:

$$\sigma_{mean} = \frac{1+R}{1-R}\frac{S}{2} \qquad (17.3.1)$$

3. The geometric and weld stress concentrations.
4. The direction of fluctuating stresses.
5. The residual stresses and welding procedures.

S-N curves are generally defined by their mean fatigue life and standard deviation in log N (refer to Section 17.3.4). The mean *S-N* curve, which determines the number of stress cycles of level S for which the sample will fail with a probability of 50%, is given by

$$S^m N = K_{50} \qquad (17.3.2)$$

On a log-log basis, mean *S-N* curves are represented by straight lines, as shown in Fig. 17.17:

$$m \log S + \log N = \log K_{50} \qquad (17.3.3)$$

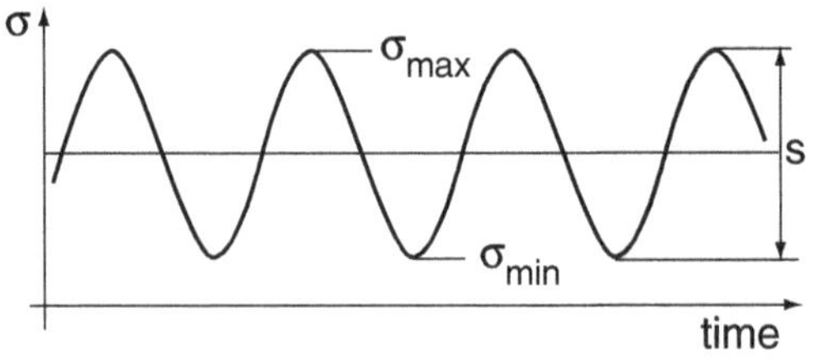

Figure 17.16 Definition of σ_{min} and σ_{max}.

where S = nominal stress range, $S = \sigma_{max} - \sigma_{min}$, ($\sigma = P/A$ for tension tests)
P = load applied on the test specimen
A = cross sectional area of the test specimen
N = number of cycles to failure
m, K_{50} = constants depending on the type of welded connection

Experimental *S-N* curves, which are obtained from constant amplitude tests, show a threshold S_∞ below which the fatigue life is infinite. This level is known as the fatigue limit but, as will be shown in Section 17.3.5, this fatigue limit does not exist for randomly loaded structures.

STANDARD EXPERIMENTAL *S-N* CURVES FOR WELDED JOINTS IN STEEL

Munse (1982) introduced *S-N* curves created specifically for the fatigue design of ship structures. The curves were based on fatigue data accumulated over a period of more than 50 years by the Department of Civil Engineering of the University of Illinois. Although it has been noted that some of the curves were determined with insufficient data or were inconsistent, they have the virtue of representing actual ship structural details. There are many sets of standard *S-N* curves used by designers for assess-

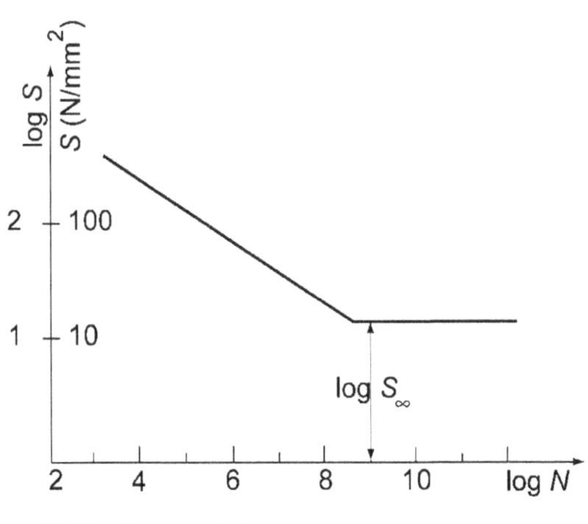

Figure 17.17 Typical *S-N* curve.

ment of the fatigue strength of marine structures. All these curves are established from extensive experimental and theoretical data on the performance of welded connections. Only two of these are presented herein:

1. HSE *S-N* curves.
2. IIW *S-N* curves.

U.K. HSE Basic S-N Curves The U.K. Health and Safety Executive (2001) (HSE) proposed a new set of eight *S-N* curves, identified as B, C, D, E, F, F2, G, and W, corresponding to non-corrosive conditions, superseding DEn (1990). The HSE classification of welded details depends on:

1. The geometrical arrangement of the detail.
2. The direction of the fluctuating stresses.
3. The method of fabrication and inspection of the detail.

and on the type of connection:

- Category 1 Material free from weld.
- Category 2 Continuous weld essentially parallel to the direction of stress.
- Category 3 Transverse butt welds.
- Category 4 Welded attachments on the surface or edge of a stressed member.
- Category 5 Load-carrying fillet and T butt welds.
- Category 6 Details in welded girders.

As shown in Fig. 17.18, each *S-N* curve represents a class of welded details and gives the relationship between the nominal stress range *S* and the number of cycles to failure *N*, (for more information refer to Appendix A). The HSE *S-N* curves are "characteristic" curves (refer to Section 5.2.4) based on statistical analysis of relevant experimental data and defined as the mean *S-N* curve minus two standard deviations. The slope of all curves is $m = 3$ for $N < 10^7$ cycles and $m = 5$ for $N > 10^7$ cycles. These *S-N* curves include the stress concentration factor (K_G) associated with the geometry of the detail shown and the stress concentration factor (K_w) associated with the local weld detail.

The fatigue strength of load-carrying partial penetration or fillet-welded joints where cracks can occur in the weld throat itself is also given. In that case, the *W* curve is to be used in association with the shear stress range across the weld throat.

IIW S-N Curves The International Institute of Welding (IIW 1996) established, for unwelded components and various welded joints in the as-welded

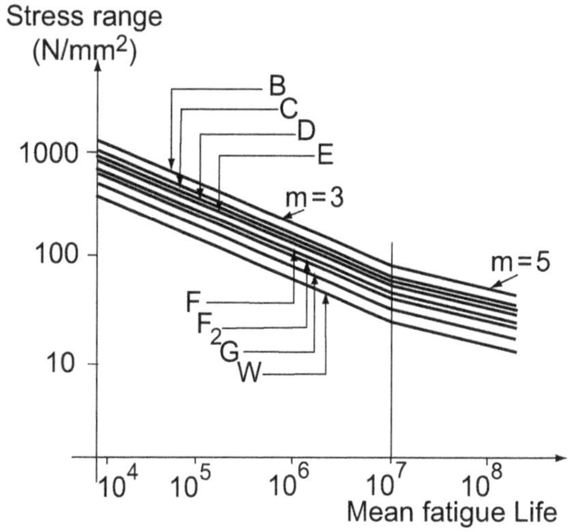

Figure 17.18 HSE *S-N* curves.

condition, a set of *S-N* curves based on constant amplitude tests and nominal stress range. The structural details are classified into seven categories:

- Category 1 Unwelded parts of components.
- Category 2 Transversely loaded butt welds.
- Category 3 Longitudinal load-carrying welded joints.
- Category 4 Cruciform and/or T joints.
- Category 5 Non-load-carrying attachments.
- Category 6 Lap joints.
- Category 7 Reinforcements.

and each *S-N* curve, as shown in Fig. 17.19, represents a class of welded details (for more information refer to Appendix B). The FAT classes refer to a standard quality for welding and inspection defined in Appendix B. As for the HSE curves, the IIW *S-N* curves include the geometric stress concentrations for the detail shown and the local stress concentrations due to the weld geometry.

The IIW *S-N* curves are defined by the fatigue strength of the detail at 2×10^6 cycles, named the fatigue class FAT (refer to Fig. 17.19). They are characteristic curves corresponding to a survival probability of 95% (or 5% failure probability) associated with a 75% confidence interval of the mean and standard deviation of log *K* (12.5% probability of being below or above the extreme value of the confidence interval). For a number of failed specimens greater than 40 it represents two standard deviations below the mean lines. The IIW curves are given for non-corrosive conditions. The slope of all *S-N* curves is $m = 3$, except for unwelded components

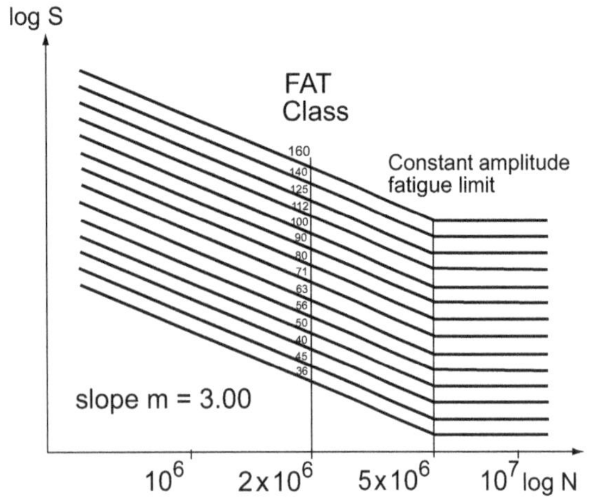

Figure 17.19 IIW *S-N* Curves.

for which $m = 5$ and the constant amplitude fatigue limit occurs for $N = 5 \times 10^6$ cycles. For rolled and extruded products (plates and flats, rolled sections) FAT = 160.

In addition, two *S-N* curves are given for assessment of the fatigue shear strength. The slope is $m = 5$ with no cutoff at $N = 5 \times 10^6$ cycles and the fatigue strength $\Delta\tau$ at $N = 2 \times 10^6$ cycles is

a. Parent metal, full penetration butt welds FAT = 100 MPa
b. Fillet welds, partial penetration butt welds FAT = 80 Mpa

STANDARD EXPERIMENTAL *S-N* CURVES FOR WELDED JOINTS IN ALUMINUM

With the development of high-speed vessels made of aluminum for which the weight of the structure is a governing design parameter, it is essential, from a safety and maintenance point of view, to build such ships with a high degree of soundness with respect to fatigue. As the mechanical and fatigue properties of aluminum are much lower than for steel, great care is required in the design of aluminum ships and especially of structural details.

Consequently, as for welded joints in steel, many tests have been performed in laboratories to determine the *S-N* curves for various types of welded joints in aluminum. There are many sets of experimental *S-N* curves available for assessment of the fatigue strength of aluminum structures, including those proposed by:

1. The British Standards Institute (1991).
2. The European Convention for Constructional Steelwork (1992).
3. The International Institute of Welding (1996).
4. The Eurocode 9 (1999).

In all codes, the classification of the structural details depends on:

a. The direction of fluctuating stresses.
b. The geometry of the detail.
c. The method of fabrication and inspection of the detail.

and on parameters more specific to aluminum structures such as the product form, and the method and quality of fabrication.

ECCS S-N Curves The *S-N* curves proposed by the European Convention for Constructional Steelwork (ECCS) (1992) are based on the experiments carried out by Kosteas (1989) and Urhy (1989). They are characterized by the fatigue strength (FAT) of the detail at $N = 2 \times 10^6$ cycles. The ECCS *S-N* curves are defined for high tensile stress conditions ($R = 0.5$) and represent two standard deviations below the mean line. The slope m is not constant ($3.34 \leq m \leq 7$) and depends on the detail category. For random loading there is a change in slope at $N = 2 \times 10^6$ or 5×10^6 cycles ($m' = m+2$) depending on the structural detail and a cutoff at $N = 10^8$ cycles.

IIW S-N Curves The IIW (1996) established, for unwelded components and various welded joints in the as-welded condition, a set of *S-N* curves based on constant amplitude tests and nominal stress range. Appendix C gives the fatigue strength (FAT) at 2×10^6 cycles of typical welded connections, based on the same assumptions as for steel joints (values of the fatigue strength for similar steel specimens are given for comparison). The slope of all *S-N* curves is $m = 3$, except for unwelded components for which $m = 5$ and the constant amplitude fatigue limit occurs for $N = 5 \times 10^6$ cycles. The IIW curves are given for non-corrosive conditions. In addition, two *S-N* curves are given for assessment of the fatigue shear strength. The slope is $m = 5$ with no cutoff at $N = 5 \times 10^6$ cycles and the fatigue strength $\Delta\tau$ at $N = 2 \times 10^6$ cycles is:

a. Parent metal, full penetration butt welds FAT = 36 = steel / 2.77
b. Fillet welds, partial penetration butt welds FAT = 28 = steel / 2.86

Eurocode S-N Curves The Eurocode *S-N* curves are given for the following details:

a. Non-welded details in wrought and cast alloys.

b. Welded details on the surface of loaded member.
c. Welded details at end connections.
d. Mechanically fastened joints.
e. Adhesively bonded joints.

The curves correspond to non-corrosive conditions and represent two standard deviations below the mean line with a change in slope at $N = 5 \times 10^6$ cycles ($m' = m+2$) and a cutoff at $N = 10^8$ cycles. The slope m is not constant ($3.2 \leq m \leq 7$) and depends on the detail category.

17.3.3 Fracture Mechanics

GENERAL

As pointed out by the HSE (2001), the use of fracture mechanics may be recommended for cases where the standard *S-N* procedure is inappropriate, in particular for:

- Cracked joints difficult to repair.
- Unusual structural details not covered by experimental *S-N* curves.
- Definition of the periodicity of in-service inspections.
- Assessment of the remaining life of cracked joints.

Fracture mechanics aims at establishing a relationship between the rate of growth of a crack and the conditions that contribute to the crack growth (material properties, loads, crack geometry, distribution, and intensity of stresses around a crack).

LINEAR–ELASTIC FRACTURE MECHANICS

The distribution and intensity of stresses and strains in the vicinity of a crack is obtained using either the Linear–Elastic (LEFM) or the Elastic-Plastic (EPFM) theory. LEFM, which is the most currently used approach, is based on the assumption that the plastic zone occurring at the crack tip is too small to significantly modify the stress distribution. There are three different basic modes of cracking, the opening, sliding, and tearing modes. For the opening mode, which is the predominant mode of cracking, the distribution of stresses near the tip of a through thickness crack (refer to Fig. 17.20) is given by

$$\sigma_x = \frac{K}{\sqrt{2\pi r}}\cos\frac{\theta}{2}\left(1-\sin\frac{\theta}{2}\sin\frac{3\theta}{2}\right) \quad (17.3.4)$$

$$\sigma_y = \frac{K}{\sqrt{2\pi r}}\cos\frac{\theta}{2}\left(1+\sin\frac{\theta}{2}\sin\frac{3\theta}{2}\right) \quad (17.3.5)$$

$$\tau_{xy} = \frac{K}{\sqrt{2\pi r}}\cos\frac{\theta}{2}\sin\frac{\theta}{2}\cos\frac{3\theta}{2} \quad (17.3.6)$$

$$\sigma_z = \tau_{yz} = \tau_{zx} = 0 \quad \text{for plane stress}$$

$$\sigma_z = \nu(\sigma_x + \sigma_y) \quad \text{for plane strain} \quad (17.3.7)$$

$$\tau_{yz} = \tau_{zx} = 0 \quad \text{for plane strain}$$

Examination of equations (17.3.4) through (17.3.7) shows that the elastic stresses near the tip of a crack depend on:

a. A constant K, called the stress intensity factor.
b. Polar coordinates r, θ of the element.
c. Poisson's ratio, ν.

and are completely defined, at any position, by the stress intensity factor K.

The stress intensity factor K depends on the loading, external geometry, and crack size and shape and may be expressed by the following general equation:

$$K = \sigma\sqrt{\pi a}\,Y \quad (17.3.8)$$

where σ = nominal or geometric stress, assuming no crack
a = depth of a surface crack or half crack length of a through crack
Y = function of the joint geometry, crack size and shape, and stress gradient at the crack tip and taking into account the stress concentration due to the weld profile

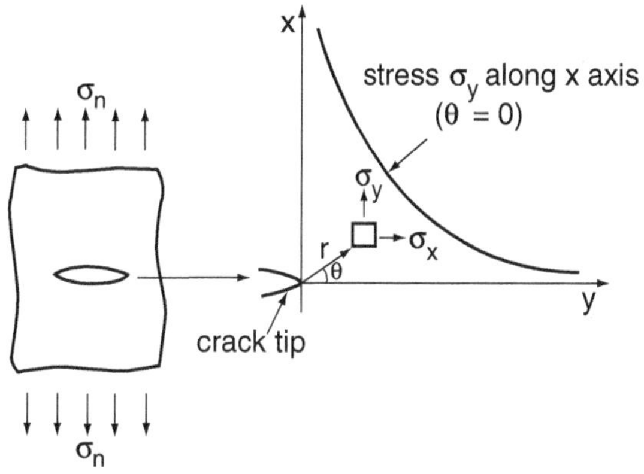

Figure 17.20 Distribution of stresses around the tip of a crack.

The IIW (1996) proposes a more elaborate expression for K with explicit differentiation between membrane and bending stresses:

$$K = \sqrt{\pi a}\left(\sigma_m Y_m M_{k,m} + \sigma_b Y_b M_{k,b}\right) \quad (17.3.9)$$

where σ_m = nominal membrane stress
σ_b = nominal shell bending stress
Y_m = correction function for membrane stress intensity factor
Y_b = correction function for bending stress intensity factor
$M_{k,m}$ = correction factor to account for the local membrane stress concentration due to the weld profile
$M_{k,b}$ = correction factor to account for the local bending stress concentration due to the weld profile

Many proposals (for example, Newman and Raju (1981) and (1983)) are available in the literature for calculation of the correction functions Y_m and Y_b, accounting for various geometrical and loading configurations. Moreover, particular methods of calculation, including the superposition and influence function methods, have been developed for more complex cases not given in the literature. As an example, in the superposition method the actual case is decomposed into basic cases that have known solutions for Y and then combined linearly to obtain the actual solution. They can also be calculated using semi-analytical methods, such as weight functions or the finite element method (FEM). The correction factors M_k can be found from Maddox et al. (1986) and Hobbacher (1994).

CRACK PROPAGATION LAWS

Fracture mechanics enables the prediction of crack propagation by the use of crack growth rate curves, as shown in Fig. 17.21, which give the relationship between the crack growth rate and the stress intensity factor range ΔK ($\Delta K = K_{max} - K_{min}$).

The crack growth diagram is divided into three regions:

1. Region A where the crack growth rate occurs as soon as $\Delta K \geq \Delta K_{th}$, where ΔK_{th} is the threshold value of ΔK. The threshold value depends on numerous factors such as the stress ratio $R = K_{min} / K_{max}$, sequence effect, residual stresses, loading frequency, and environment.
2. Region B where the crack growth rate increases uniformly with ΔK.
3. Region C where the crack growth rate increases rapidly until failure occurs as soon as $K_{max} \geq K_c$ (K_c is the critical stress intensity factor at failure and is a characteristic data of the material).

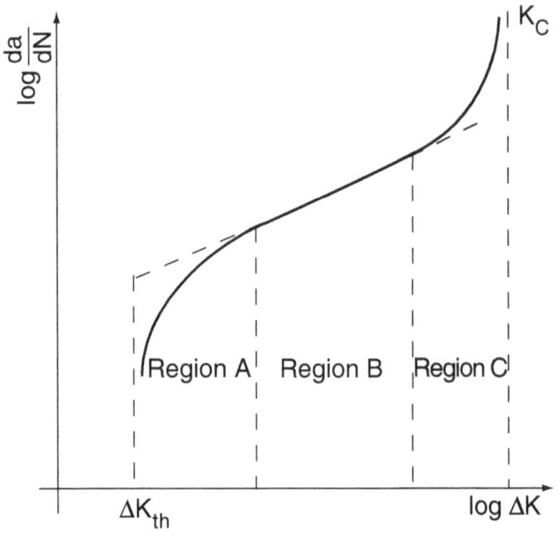

Figure 17.21 Crack growth rate curve.

Many proposals have been made for predicting crack growth. The well-known Paris and Ergodan (1963) equation describes the crack growth rate in the intermediate region:

$$\frac{da}{dN} = C_0 \left(\Delta K\right)^m \qquad (17.3.10)$$

where $\frac{da}{dN}$ = crack growth rate or crack growth per cycle ranging from 10^{-3} to 10^{-6} mm/cycle for most marine load cases of interest
ΔK = stress intensity factor range
C_0 and m = crack growth parameters depending on the material and testing conditions (stress ratio R, environment), and determined experimentally.

The Paris-Ergodan equation may be used to describe crack growth over the whole fatigue life, considering the theoretical crack growth diagram, as shown in Fig. 17.22, and assuming that the limits for integration (ΔK_{th} and K_c) are known. More refined formulas accounting for the stress ratio $R = K_{min}/K_{max}$ and applicable to regions A, B, and C of the crack growth diagram have been proposed, such as the Schütz (1981) crack propagation law:

$$\frac{da}{dN} = \frac{C\left[\left(\Delta K\right)^m - \left(\Delta K_{th}\right)^m\right]}{\left(1 - R\right)K_c - \Delta K} \qquad (17.3.11)$$

For high stress intensity factors compared to the fracture toughness K_0 of the material, the IIW (1996) recommends the use of the following crack propagation law:

$$\frac{da}{dN} = \frac{C_0 \Delta K^m}{(1-R) - \Delta K / K_0} \qquad (17.3.12)$$

Where no specific material data accounting for the actual conditions of the welded joint such as residual and mean stresses and environmental conditions are available, the following characteristic values of the parameters may be considered, as proposed by the IIW (1996) for steel and aluminum in air:

Steel
$C_0 = 9.5 \times 10^{-12}$ and $m = 3$ (C_0 mean value plus two standard deviations)
$\Delta K_{th} = 6 - 4.56R$ MPa $\times$ m$^{0.5}$, without being less than 2 (ΔK_{th} mean value minus two standard deviations) This threshold value is applicable for $R > 0$ irrespective of the environment.

Aluminum
$C_0 = 8.51 \times 10^{-11}$ and $m = 3$ (mean value minus two standard deviations)
$\Delta K_{th} = 2 - 1.5R$ MPa $\times$ m$^{0.5}$ without being less than 0.7 (mean value minus two standard deviations)

Moreover, experiments show that the crack growth rates for steel and aluminum are independent of the material properties and, for stress ratios greater than 0.4, are practically independent of the mean stress.

FATIGUE LIFE

Constant Amplitude Fatigue Life According to the Paris-Ergodan equation, the number of cycles from a given initial crack depth a_0 to the crack depth/length at failure a_f is given by

$$N = \int_{a_0}^{a_f} \frac{da}{C(\Delta K)^m}$$

or

$$N = \frac{1}{C\sqrt{\pi^m}(\Delta\sigma)^m} \int_{a_0}^{a_f} \frac{da}{(Y\sqrt{a})^m} \qquad (17.3.13)$$

Equation (17.3.13) may be rearranged as follows:

$$N(\Delta\sigma)^m = \frac{\frac{1}{\sqrt{\pi^m}} \int_{a_0}^{a_f} \frac{da}{(Y\sqrt{a})^m}}{C} = A \qquad (17.3.14)$$

which is the general expression of *S-N* curves with the constant *A* depending on the material, testing conditions, crack shape, and size and stress gradient. Equation (17.3.14) shows that fracture mechanics may be successfully applied to the determination of *S-N* curves for any particular welded joint.

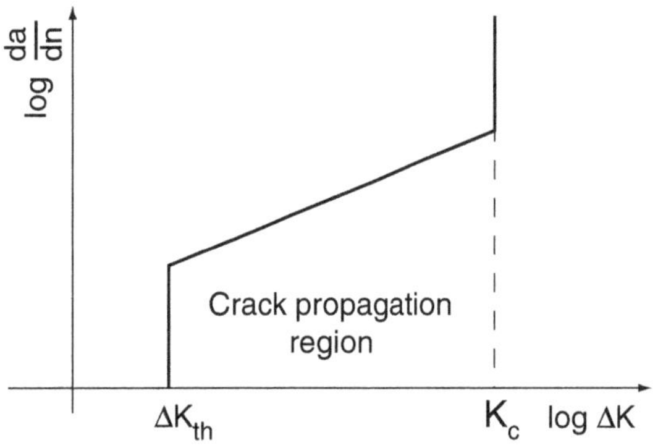

Figure 17.22 Theoretical crack growth diagram.

Variable Amplitude Fatigue Life If the long-term distribution of stresses is converted into a step curve of *n* blocks generally of equal length in log *N*, the crack depth/length increment for the step *i* is

$$\Delta a_i = C(\Delta K_i)^m \Delta N_i \qquad (17.3.15)$$

and the final crack depth/length at the end of the *N* cycles is obtained by summing equation (17.3.15) for the *n* stress blocks:

$$a_N = a_0 + \sum_{i=1}^{n} \Delta a_i \qquad (17.3.16)$$

Equation (17.3.15) is only valid for small values of Δa_i since ΔK_i depends on the crack depth/length, which requires dividing the stress range spectrum into a large number of stress blocks.

The number of cycles to failure may also be calculated according to equation (17.3.13) using an equivalent constant amplitude stress range $(\Delta\sigma)_{eq}$ (refer to Section 17.4.3) giving the same amount of damage:

$$S_{eq} = \left[\int_0^\infty S^\beta p_S(S)dS\right]^{1/\beta} \qquad (17.3.17)$$

where β is an empirical coefficient taking into account the load interaction effects (refer to the next Section). For $\beta = 3$ equation (17.3.17) is identical to equation (17.4.8) and interaction effects are not taken into account.

For the stress range spectrum given in Fig. 17.23, the fatigue damage corresponding to each block of n_i cycles at a stress range S_i and calculated according to equation (17.3.13) is

$$\int_{a_{i-1}}^{a_i} \frac{da}{\left(Y\sqrt{a}\right)^m} = C\sqrt{\pi}^m S_i^m n_i \qquad (17.3.18)$$

and the number of cycles N_i at failure for each block is given by

$$\int_{a_0}^{a_f} \frac{da}{\left(Y\sqrt{a}\right)^m} = C\sqrt{\pi}^m S_i^m N_i \qquad (17.3.19)$$

where a_0 is the initial crack length and a_f the final crack length.

Consequently, we may write:

$$\sum_{i=1}^{k} \frac{n_i}{N_i} = \frac{\sum_{i=1}^{k} \int_{a_{i-1}}^{a_i} \frac{da}{\left(Y\sqrt{a}\right)^m}}{\int_{a_0}^{a_f} \frac{da}{\left(Y\sqrt{a}\right)^m}} = \frac{\int_{a_0}^{a_k} \frac{da}{\left(Y\sqrt{a}\right)^m}}{\int_{a_0}^{a_f} \frac{da}{\left(Y\sqrt{a}\right)^m}} \qquad (17.3.20)$$

which is the expression of the Miner cumulative damage principle:

$a_k < a_f$ no failure if $\sum \frac{n_i}{N_i} < 1$

$a_k < a_f$ failure if $\sum \frac{n_i}{N_i} \geq 1$

FACTORS THAT NECESSITATE CORRECTIONS

As for assessment of the fatigue strength based on the *S-N* approach (refer to Sections 17.3.5 and 17.3.6), the application of fracture mechanics for calculation of the fatigue life needs due consideration of similar correction factors and, in particular, of the following ones:

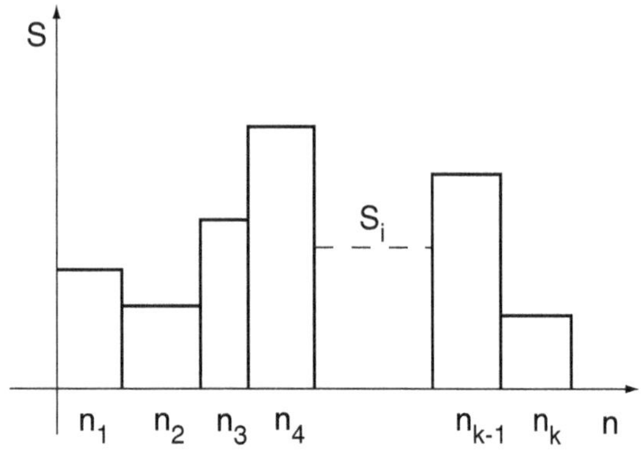

Figure 17.23 Stress range spectrum.

1. *Residual stresses.* Assuming that the residual stresses are of the magnitude of the yield stress (refer to Section 17.3.6), the stresses fluctuate downwards from the yield stress and K_{max} is to be calculated for $\sigma = \sigma_Y$.
2. *Initial crack depth/length.* When determining the initial crack depth/length a_0, attention is to be paid to the accuracy of the method used for measurement of the flaw size, as in some cases the fatigue life is very sensitive to the value of a_i. The HSE (2001) recommends taking an initial flaw size in the range of 0.1 to 0.25 mm.
3. *Crack closure.* A given crack remains closed unless the stress intensity factor is greater than the crack closure stress intensity factor K_{cl}. Therefore, if $K_{min} < K_{el}$ the effective stress range to be considered is given by

$$K_{eff} = K_{max} - K_{el} \qquad (17.3.21)$$

4. *Loading sequence.* As mentioned in Section 17.1.1 fatigue is a process of cycle-by-cycle accumulation of damage and, consequently, the local damage depends not only on the stress level but also on the sequence of events, that is, on the preceding cycles, which is lost in the exceedance stress range spectrum representation. Interaction between the stress cycles introduces retardation or acceleration in the crack growth. To take into account the load history effects, an empirical interaction coefficient which modifies the constant amplitude crack growth can be introduced in equation (17.3.15) at each step of the summing process:

$$\Delta a_i = \beta_i C (\Delta K_i)^m \Delta N_i$$

where β_i is the interaction coefficient applicable to step i.

To deal with this problem, Eurocode 9 (1999) proposes the application of the long-term stress range spectrum in 10 identical sequences with the same stress ranges and R ratios but with one tenth of the number of cycles.

17.3.4 Prototype Testing

GENERAL

For particular structural details, it may be advisable to have recourse to prototype testing with a view to determining more appropriate *S-N* or crack propagation curves than the standard ones given by the codes. In such a case, all measures are to be taken to collect as much information as possible on the initiation and crack propagation process, in particular for improvement of the fatigue design procedures. The testing procedure described in this Section is applicable to the determination of *S-N* curves for small size specimens and may be easily extended to the determination of crack propagation curves. For large scale specimens other procedures are to be applied.

Fatigue tests show a large scattering of the results, requiring a rather large number of specimens. Generally, the number of failed specimens is to be greater than 10, unless the slope of the *S-N* curve is known. In such a case, the *S-N* curve may be determined accurately with only 10 tests:

- 5 at stress level corresponding to $N = 10^4$ cycles
- 5 at stress level corresponding to $N = 5 \times 10^5$ cycles

Application of statistical methods to the test results enables the determination of a characteristic *S-N* curve corresponding to the required survival probability.

FATIGUE TESTING RECOMMENDATIONS

Fatigue tests are generally performed for constant amplitude loadings and the following precautions should be taken:

1. The steel grade used for the test pieces should be the same as that provided for the actual structural detail.
2. Welding procedures should be representative of the actual conditions of welding.
3. The size of test specimens should be such that the level of residual stress is equivalent to that of the actual structure.
4. The stress ratio $R = \sigma_{\min}/\sigma_{\max}$ should remain constant during the experiments. Residual stresses in small-scale specimens are generally smaller than in actual structures. This may be accounted for by taking $R = 0.5$ for the experiments or R between 0 and 0.1 associated with a modification of the experimental characteristic *S-N* curve (refer to Section 17.3.6).
5. Visual examination and non-destructive testing (NDT) should be used for detection and identification of the possible surface defects.
6. The scatter of the results is to be clearly displayed.

Detailed structural analyses of test specimens are to be performed for validation of the calculation procedure of the stress distribution near the hot spot of the actual structural detail. The IIW (1996) recommendations are to be applied for that FEM analysis. In particular, theoretical stresses will have to be computed at locations where stress measurements are carried out during the fatigue testing.

FATIGUE TESTING PROCEDURE

The results of tests are expressed by couples of values (S_i , N_i) where S_i is the applied stress range and N_i the number of cycles to failure (refer to Fig. 17.24). Only failed test specimens are considered to determine the *S-N* curve, except for determination of the fatigue limit. On a log-log basis, the *S-N* curve is represented by a straight line given by

$$\log N = \log K - m \log S \qquad (17.3.22)$$

From the results of tests, many statistical methods exist for determination of the design or characteristic *S-N* curves. Irrespective of the method selected, the following problems have to be solved:

1. Select the probability distributions of the random variables. Fatigue experiments show that for a given stress range *S*, the measured number of cycles *N* to failure is not constant but random, as shown in Fig. 17.24. Based on the analysis of statistical data, it is frequently assumed that *N* or *K*, which is equivalent, may be modeled by the lognormal distribution (refer to Section 17.5.2).
2. Calculate the mean value and the standard deviation of log *K*.
3. Verify that the selected distribution fits with the whole set of data, for example by using the χ^2 test.

The following procedure may be applied for determination of the characteristic *S-N* curve:

1. Fit the couples of variables (S_i N_i) to equation (17.3.22) by linear regression analysis, which enables the determination of the slope *m* and mean value log K_{50} of the constant log *K*. If *m* is known, the mean value log K_{50} of the constant $\log K$ is obtained by the following formula:

$$\log K_{50} = \frac{\sum_i \log K_i}{n} \qquad (17.3.23)$$

where $\log K_i = \log N_i + m \log S_i$

2. Calculate the standard deviation σ_d $(\log K)$ of log *K*:

$$\sigma_d(\log K) = \sqrt{\frac{\sum_i \left[\log K_{50} - \log K_i\right]^2}{n-1}} \qquad (17.3.24)$$

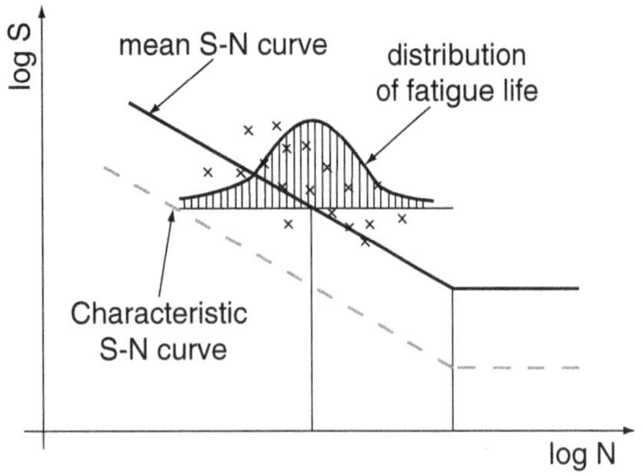

Figure 17.24 Presentation of test results.

For another set of data (S_k N_k), the mean value log K_{50} as well as the standard deviation $\sigma_d(\log K)$ will be different and, therefore, are random variables.

3. For design purposes, a characteristic value of the constant log K corresponding to a specified percentage of the area under the probability curve is considered. Based on that definition, the design or characteristic *S-N* curve is:

$$S^m N = K_P \qquad (17.3.25)$$

where $\log K_P = \log K_{50} - \lambda_p\, \sigma_d(\log K)$ (17.3.26)

$\sigma_d(\log K)$ = standard deviation of log K, as given by equation (17.3.24).

λ_p = coefficient depending on the selected survival probability and number of failed test specimens considered to determine the *S-N* curve. Table 17.6 gives the value of λ_p calculated by the IIW (1996) for a survival probability of 95% (or 5% failure probability) associated with a 75% confidence interval for the mean and standard deviation of log K, assuming that the probability distributions of these random variables follow a Student law (t-distribution) and a chi-square law (χ^2). The coefficient λ_p corresponds to the minimum value of the confidence interval for log K_{50} (12.5% probability of being below that minimum value) and to the maximum value of the confidence interval for $\sigma_d(\log K)$ (12.5% probability of being above that maximum value).

As already mentioned, experimental *S-N* curves, which are determined for constant amplitude loadings, show a threshold level of stress below which N is infinite (refer to Fig. 17.17). Determination of this threshold, known as the fatigue limit, requires a different procedure.

17.3.5 Design *S-N* Curve for Steel Structures

BASIC DESIGN *S-N* CURVE

The design *S-N* curve used to assess the fatigue strength of a given structural detail depends on the type of stress approach considered for determining the long-term distribution of stresses:

1. Nominal stress approach.
2. Hot spot stress approach.
3. Notch stress approach.

Nominal stress approach When the fatigue analysis is based on the nominal stress approach, the HSE or IIW *S-N* curves corresponding to the type of welded connection may be used, subject to appropriate corrections as given hereafter.

Hot spot stress approach Based on the previous considerations, this approach is a hybrid solution as it accounts for the geometry of the structural detail but not for the weld effects. If the weld configuration corresponds to that of an experimental *S-N* curve, this curve can be used, subject to the following correction:

$$S_G^m N = K_{G0}^m K \qquad (17.3.27)$$

in which S_G = calculated geometric stress range for the actual structure

K_{G0} = geometric stress concentration factor of the specimen

K = constant of the experimental *S-N* curve

If we assume the same nominal stress range S_0 for the specimen and actual detail, the local stress range S_ℓ for the actual detail is

Table 17.6 Coefficient Λ_p

n^a	5	10	15	20	25	30	40	50	100
λ_p	3.5	2.7	2.4	2.3	2.2	2.15	2.05	2.0	1.9

[a] Note: n is the number of samples.

$$S_\ell = K_w K_G S_0$$

As the experimental *S-N* curve is used for the actual detail, this local stress range S_ℓ corresponds to an equivalent nominal stress giving the same local stress for the specimen as for the actual detail:

$$S_\ell = K_{G0} K_w (S_0)_{eq}$$

$$(S_0)_{eq} = \frac{S_\ell}{K_w K_{G0}} = \frac{K_G}{K_{G0}} S_0$$

and we may write

$$(S_0)^m_{eq} N = \left(\frac{K_G}{K_{G0}}\right)^m S_0^m N = K = S_0^m N$$

or

$$N = \left(\frac{K_{G0}}{K_G}\right)^m N_0$$

where N_0 = fatigue life of the specimen
N = fatigue life of the actual detail
K_G = geometric stress concentration factor of the actual detail

Noting that $N_0 = \frac{K}{S_0^m}$ we can say that

$$N = \left(\frac{K_{G0}}{K_G}\right)^m \frac{K}{S_0^m}$$

or $$S_0^m N = \left(\frac{K_{G0}}{K_G}\right)^m K$$

or $$S_G^m N = K_{G0}^m K$$

K_G and K_{G0} have to be calculated according to the same procedure, that is, using the recommendations given in Section 17.2.5 for the determination of hot spot stresses.

Notch Stress Approach By definition, the notch stress includes stress concentrations due to the effects of structural geometry and the presence of welds. Therefore, the design *S-N* curve based on the notch stress approach is such that $K_t = K_w K_G = 1$, that is, it should correspond to the *S-N* curve for unwelded steel components in the as-rolled condition (e.g., HSE "B Curve"). As for the two other approaches, this curve has to be modified, as indicated hereafter.

CORRECTION FOR EFFECT OF RANDOM LOADING (HAIBACH EFFECT)

Experimental *S-N* curves show a threshold level of stress below which the fatigue life is infinite. However, for randomly loaded structures, this fatigue limit has no meaning since the mode of failure of the structure is a combination of crack initiation and crack propagation. Cracks develop while the crack propagation threshold and the fatigue limit decrease. Therefore, the actual fatigue limit cannot be defined. To account for this phenomenon Haibach (1970) proposed to represent the *S-N* curves with a change in slope in the high life range $N = 5 \times 10^6$ or $N = 10^7$cycles:

$$S^m N = K_p \quad \text{for} \quad N \leq 5\times10^6 \text{ or } N \leq 10^7 \qquad (17.3.28)$$

$$S^{m+2} N = K'_p \quad \text{for} \quad N > 5\times10^6 \text{ or } N > 10^7 \qquad (17.3.29)$$

K'_p being obtained to ensure the continuity of the *S-N* curve at $N = 5\times10^6$ or 10^7 cycles. This is reflected in Fig. 17.18.

Table 17.7 gives the mean constant K_{50} and standard deviation of log K_{50} for the eight classes of HSE *S-N* curves.

For randomly loaded structures for which the fatigue limit is obtained for a number of cycles less than 10^8, the IIW (1996) introduces a change in slope of the *S-N* curves at $N = 5 \times 10^6$ cycles with a cutoff at 10^8 cycles. For $5\times10^6 < N \leq 10^8$ cycles the slope m_2 of the *S-N* curve is taken as $m_2 = 2m_1 - 1$. Table 17.8 gives the values of the constants K_p and K'_p in terms of the IIW FAT class.

ALLOWANCE FOR GEOMETRIC AND WELD STRESS CONCENTRATIONS

Experimental *S-N* curves are generally obtained with small specimens and there is a lack of similitude in the fatigue behavior of test specimens and real welded structures, taking into account that the following effects are partly embedded in the experimental curves:

1. *Geometric stress concentrations.* Ship structural details are much more complex than the specimens used to obtain the experimental *S-N* curves. The geometric stress concentration factors of the specimen and actual structure are therefore different.
2. *Local stress concentrations due to the weld geometry.* Conditions of welding are more favorable in laboratories than in shipyards, in particular local residual stresses are smaller, which leads to a

Table 17.7 Basic HSE *S-N* Curves

DEn Class of *S-N* curve	Mean *S-N* curves		Standard Deviation of log *K*	Characteristic *S-N* curves		
	K_{50} $N \times 10^7$	K'_{50} $N > 10^7$		K_p $N \times 10^7$	K'_p $N > 10^7$	Stress Range at N=10^7 cycles
B	1.342×10^{13}	1.633×10^{17}	0.1821	5.802×10^{12}	4.036×10^{16}	83.40
C	8.855×10^{12}	8.165×10^{16}	0.2041	3.459×10^{12}	1.704×10^{16}	70.20
D	3.990×10^{12}	2.162×10^{16}	0.2095	1.520×10^{12}	4.329×10^{16}	53.35
E	3.259×10^{12}	1.543×10^{16}	0.2509	1.026×10^{12}	2.249×10^{15}	46.80
F	1.726×10^{12}	5.350×10^{15}	0.2183	6.316×10^{11}	1.002×10^{15}	39.80
F2	1.237×10^{12}	3.071×10^{15}	0.2279	4.331×10^{11}	5.341×10^{14}	35.10
G	5.666×10^{11}	8.358×10^{14}	0.1793	2.481×10^{11}	2.110×10^{14}	29.15
W	2.171×10^{11}	1.689×10^{14}	0.1846	9.278×10^{10}	4.097×10^{13}	21.00

larger mean fatigue strength and a much smaller scatter.

Consequently, depending on the stress approach, the basic design *S-N* curve modified for effect of random loading may have to be corrected to account for these differences. In particular, for the hot spot stress approach, the basic design *S-N* curve is to be corrected according to equation (17.3.27).

17.3.6 Further Corrections of the Design *S-N* Curve

FACTORS THAT NECESSITATE CORRECTIONS

The basic design *S-N* curve allows for random loading, but it has to be corrected to account for other effects that are not considered in the experimental *S-N* curves. Extensive experiments over the past 20 years revealed the influence of:

1. *Size of specimens and plate thickness* (Gurney (1979) and (1989), Berge (1989), Vosikovsky et al. (1989), Yagi et al. (1991), Niemi (1996)). Increasing the size of the structure or the plate thickness results in higher notch stresses at the weld toe and therefore reduces the fatigue strength. In particular, the defect probability increases with the size of specimens and plate thickness, leading to a reduction of the fatigue strength. Moreover, experiments show that the thickness effect is large when attachment size increases proportionally to the main plate thickness and that fatigue strength of specimens tends to decrease as the weld length per specimen increases.
2. *Residual stresses* (Gurney (1993), Niemi (1996)). Very little is known about residual stresses in ship structures, for example, through thickness distribution, variation through service life, and effect under random loading. Also, modeling and quantifying the effect of residual stresses on the fatigue behavior of welded joints is not an easy task as the joints do not remain constant and can be partly or totally relaxed under random sea loading.
3. *Mean stresses.* High tensile mean stresses tend to reduce the fatigue life while compressive mean stresses tend to improve the fatigue life. This is confirmed by the behavior of existing structures, for example, on existing single hull oil tankers, connections of side shell longitudinals to transverse webs subjected to mean compressive stresses were less damaged than similar connections subjected to mean tensile stresses.
4. *Yield stress.* For the last 40 years, higher tensile steels (HTS) have been more commonly used for the construction of ships. Experience shows that HTS

Table 17.8 Design IIW *S-N* Curves

IIW Class	K_p $N > 5 \times 10^6$	K'_p $N > 5 \times 10^6$	Stress Range at $N = 5 \times 10^6$ cycles
FAT	$2 \times 10^6 (FAT)^3$	$1.085 \times 10^6 (FAT)^5$	0.7368 *FAT*

structures are more sensitive to the risk of fatigue, as already noted in Section 17.1.2.

5. *Environment.* A corrosive environment drastically reduces the fatigue strength as it is observed on existing ships if no measures are taken to protect the structure.

6. *Workmanship.* Since fatigue is the result of two combined actions, structural discontinuities and weld material defects, the fatigue life will be significantly improved by the quality of design and construction.

STRATEGIES TO ALLOW FOR THESE CORRECTIONS

Based on the results of experimental studies, different corrections have been proposed by various authors and regulatory bodies to account for these factors on the fatigue life of actual structures.

1. *Influence of thickness.*

a. For transversely loaded weld joints, Gurney (1989) suggests a modification of the design *S-N* curve according to the following equation:

$$S = S_0\left(\frac{13}{t}\right)^{1/4} \qquad (17.3.30)$$

where S_0 = reference fatigue strength of the design *S-N* curve

t = thickness of the member ($t > 3\,\text{mm}$)

b. The U.K. HSE (2001) modifies the design *S-N* curve according to the following equation for plate thicknesses greater than 16 mm:

$$\log N = \log K_p - m\log S - 0.3\,m\log\frac{t}{16} \qquad (17.3.31)$$

where K_p = constant of the design *S-N* curve

t = thickness, in mm, of the member under consideration, but not taken less than 16 mm

m = slope of the *S-N* curve

c. The IIW (1996) applies a reduction factor $f(t)$ on the fatigue class for plate thicknesses greater than 25 mm:

$$f(t) = \left(\frac{25}{t}\right)^n \qquad (17.3.32)$$

with $0.1 < n < 0.4$ depending on the type of structural detail, $n = 0.2$ for butt welds under any loading and axially loaded T-joints and cruciform joints.

2. *Influence of residual stresses.* A significant contribution to this complex problem has been provided by new tests carried out by Otha et al. (1994) on transverse non-load carrying cruciform joints. These experiments were carried out on small width specimens for $R = 0$ (refer to Fig. 17.25) and stalactitic tests (refer to Fig. 17.26) for which the maximum applied stress is equal to the yield stress.

These experiments show that the fatigue life, measured on specimens for which the residual stress σ_r at the weld toe is equal to the yield stress, is the same for the $R = 0$ test and the stalactitic test. The maximum stress $\sigma_{max} = \sigma_r + S$, which is greater than the yield stress, shakedowns to the yield stress giving the same fatigue strength irrespective of the stress ratio. On the contrary, specimens with residual stresses less than the yield stress exhibit in the low stress range region higher fatigue strength for the $R = 0$ test than for the stalactitic test since, in that case, the maximum stress $\sigma_{max} = \sigma_r + S$ is less than the yield stress. The following conclusions may be drawn from these tests:

a. The experimental *S-N* curves determined for $R = 0$ to 0.1 should be modified to model the influence of residual stresses when they can be close to the yield stress. Unfortunately, there is a lack of experimental data for this modification of the *S-N* curves.

b. For high stress ranges (for which the stalactitic tests may be considered as representative of the behavior of actual structures) the mean stress has no influence on the fatigue life. On the contrary, for lower stress ranges the fatigue strength becomes more dependent on the stress ratio R.

3. *Influence of the mean and residual stresses*

a. Munse (1982) proposes to account for the effect of mean stress according to the following equation:

$$S = S_0\,(1 - 0.25R) \quad \text{for} \quad -1 < R < 1 \qquad (17.3.33)$$

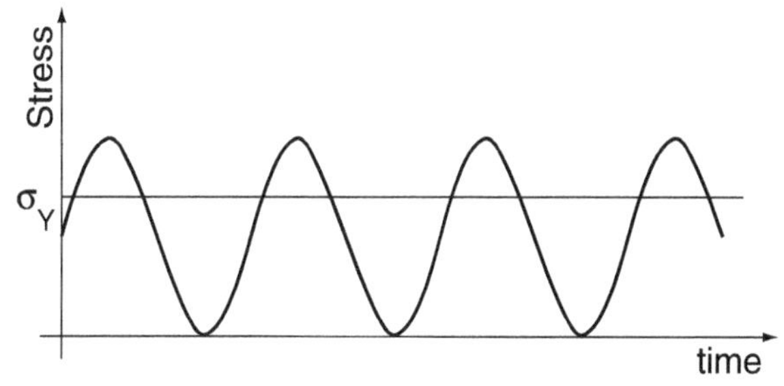

Figure 17.25 R = 0 tests.

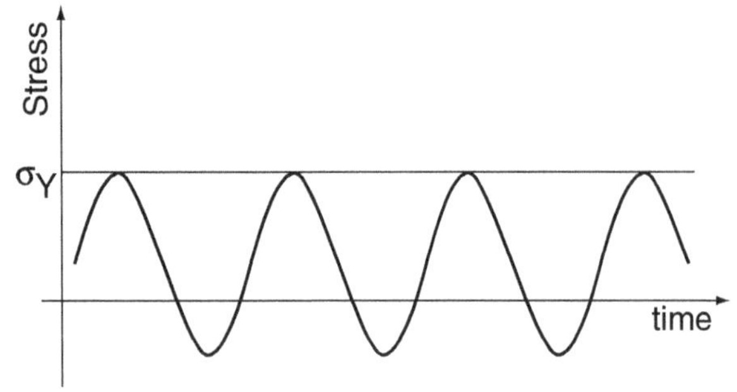

Figure 17.26 Stalactitic tests.

b. The HSE (2001) procedure for fatigue assessment is based on the stress range with no influence of the mean stresses.

c. The IIW (1996) allows for the mean stress in terms of the stress ratio $R = \sigma_{min}/\sigma_{max}$, where σ_{min} and σ_{max} are the applied stresses. An enhancement factor $f(R)$ can be used by multiplying the fatigue class FAT by $f(R)$, depending on the level of residual stresses:

- For low residual stresses ($\sigma_r < 0.2\,\sigma_Y$)

$$f(R) = 1.6 \quad \text{for} \quad R < -1 \qquad (17.3.34)$$

$$f(R) = -0.4R + 1.2 \quad \text{for} \; -1 < R < -0.5 \qquad (17.3.35)$$

$$f(R) = 1 \quad \text{for} \quad 0.5 < R \qquad (17.3.36)$$

- For moderate residual stresses ($0.2\,\sigma_Y < \sigma_r < 0.7\sigma_Y$)

$$f(R) = 1.3 \quad \text{for} \quad R < -1 \qquad (17.3.37)$$

$$f(R) = -0.4R + 0.9 \text{for} -1 < R < -0.25 \qquad (17.3.38)$$

$$f(R) = 1 \quad \text{for} \quad -0.25 < R \qquad (17.3.39)$$

- For high residual stresses ($\sigma_r > 0.7\,\sigma_Y$)

$$f(R) = 1$$

d. Based on the results of Japanese experiments, mean and residual stresses could be accounted for as follows:

- For high residual stresses, experimental *S-N* curves should be translated as follows:

$$\log N = \log K_p - m \log S - \alpha m \text{ with } 0.05 < \alpha < 0.1 \qquad (17.3.40)$$

and the stress range corrected according to equation (17.3.41), where applicable. For stress ratios greater than 0.5, the IIW (1996) suggests a 20% reduction of the fatigue strength at 2×10^6 cycles to account for residual stresses, which corresponds to a coefficient α of about 0.1.

- For low or moderate residual stresses and small stress ranges ($S < \sigma_Y$), the reference fatigue strength of the *S-N* curve could be corrected according to the IIW (1996).
- For large stress ranges, actual stresses are assumed to fluctuate downwards from the yield stress σ_Y to σ_{min} as shown in Fig. 17.26 and the stress range is reduced to account for the less damaging effect of compressive stresses and corrected as proposed by the DEn (1990) according to:

$$S_{cor} = \sigma_Y + 0.6(S - \sigma_Y) \text{ for } \sigma_Y \leq S \leq 2\sigma_Y \qquad (17.3.41)$$

4. *Influence of the yield stress.* Results of fatigue tests carried out for various steels with tensile strength between 400 to 600 MPa show that the crack growth rates for the parent metal, heat affected zone, and weld are within the same scatter band. In particular, crack growth rates are insensitive to the yield stress. Since fatigue of welded joints is essentially a crack growth phenomenon, it may be stated that the fatigue strength of welded structures is not practically improved with increased yield stress. More generally, the analysis of test results leads to the following conclusions:

1. For machined plates the effect of yield stress is large.
2. For as-rolled plates the initiation phase is reduced and the effect of yield strength is small.
3. For welded joints, the fatigue strength is nearly independent of the yield stress.

Therefore, the same design *S-N* curve is generally used for assessment of the fatigue strength of welded ship structures, irrespective of the yield stress, unless special measures, such as post-welded treatment, are taken to improve the weld geometry or welding procedures. This is in direct contrast to static and quasi-static stresses, for which the use of higher tensile steels (HTS) increases the permissible stress in approximately the same proportion. For example, if AH36 steel (yield stress of 355 MPa) is used instead of ordinary steel (yield stress of 235 MPa), Classification Societies allow a 38% increase in permissible (static and quasi-static) stress. Bearing in mind that the fatigue damage is roughly proportional to the stress range cubed, this may result in a possible reduction of the fatigue life from 20 years for a structural detail made of mild steel down to 7.5 years for a similar detail made of HTS if cyclic stresses are a large part of the total stress and no special measures are taken for improvement of the design. This means that particular attention has to be paid to the design of structural details and welding connections (e.g., full penetration welds instead of partial or fillet welds) in structures made of higher tensile steels.

5. *Influence of the environment*. Existing *S-N* curves are generally determined in air. Few data are available on the influence of sea water corrosion. Tests carried out in sea water show that for unprotected specimens there is a detrimental effect on fatigue life, while for cathodically protected specimens the effect of sea water is not very significant. This reduction in the fatigue life, as shown in Fig. 17.27 from Almar-Naess et al. (1985), is more important for unwelded specimens than for welded joints. For plain materials and improved welded joints corrosion pits and grooves act as weld defects and, consequently, may significantly reduce the initiation period. Moreover, these experiments show that the fatigue limit is practically eliminated. These conclusions are also confirmed by the results of crack growth tests carried out in air, with cathodic protection, and under free corrosion. Compared with the growth in air, free corrosion in sea water accelerates the crack growth and, consequently, reduces the fatigue life. In particular, experiments show that the fatigue life reduction under free corrosion depends to a large extent on the type of welded joint, as indicated in Table 17.9 from Almar-Naess et al. (1985).

As pointed out by the TSCF (1997), many factors affect the development of corrosion inside cargo or ballast tanks (e.g., frequency of tank washings, tank washing medium, composition and properties of cargo, time in ballast, humidity in empty tanks, coating breakdown, etc.) and therefore contribute to the reduction of fatigue life measured in air environment, although their effects are generally difficult to quantify.

Based on the HSE (2001), where no corrosion protection systems are provided in compartments prone to corrosion, the design *S-N* curve should be modified by dividing the coefficient K_p by 3 and assuming no change in slope at $N = 10^7$ cycles. For compartments protected from corrosion, for example, by sacrificial anodes, the *S-N* curve for welded joints in air should be modified by dividing the coefficient K_p by 2.5 and assuming a change in slope at $N = 10^7$ cycles.

In addition to the influence of the environment on the parameter K_p, a corrosive environment leads to an increase in the stress range with time. Bea (1992) suggests the following equation for calculation of the cumulative fatigue damage D at any time T of the service life:

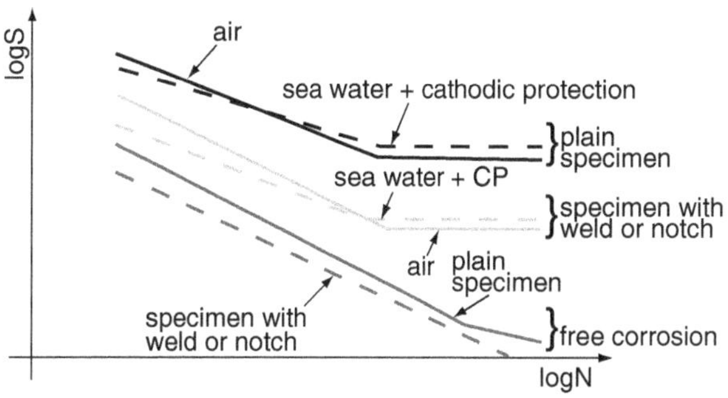

Figure 17.27 Effect of sea water corrosion and cathodic protection on *S-N* curves.

$$D = \frac{t}{r_c\, n_a (m-1)} \left\{ \left(\frac{t}{t - r_c\, n_a} \right)^{m-1} - 1 \right\} D(T) \qquad (17.3.42)$$

where $D(T)$ = cumulative damage ratio at time T calculated with reduced coefficient K_p

n_a = number of years corresponding to time T

r_c = rate of corrosion, in mm/year. As a guidance, Table 17-10 from Bureau Veritas (1998) gives rates of corrosion for various types of ships

t = thickness of the structural element, in mm

m = slope of the *S-N* curve

In conclusion, for joints protected by paint coatings or by anodes, air curves can generally be used provided that it may be ensured that the protection remains effective during the ship's life. Otherwise, the ship's life is divided in two periods for calculation of the damage ratio, considering:

1. Effective corrosion protection during the first period and using a design *S-N* curve for air.
2. No corrosion protection during the second period and using a design *S-N* curve modified as indicated above.

Table 17.9 Fatigue Life Reduction Factors for Various Full Penetration Types of Welds

Type of weld	Reduction factor
Cruciform non-load carrying joints	2.5
Butt welds	3 to 6
T joints	2.5

6. *Influence of workmanship. S-N* curves are assumed to be representative of standard workmanship and welding procedures. Although imperfections cannot be avoided, it is necessary to know what the permissible defects are, as given for example by the IIW (1996), and then to ensure that during the ship construction the defects are kept below the permissible limits or repaired, by implementing a system of inspection and non-destructive measurements,. This corresponds perfectly to the general philosophy of the Quality Control Systems that have been implemented in the past 20 years in most shipyards. Defects may be divided into two categories, fabrication and in-service defects. Table 17.11 gives an overview of the defects that can reduce the fatigue life of welded ship structures and the remedial measures that should be taken.

17.3.7 Design *S-N* Curves for Aluminum Structures

BASIC DESIGN *S-N* CURVE

The basic design *S-N* curve of aluminum welded joints may be selected either among one of the existing sets of *S-N* curves or by prototype testing, subject to appropriate corrections as indicated herein. For randomly loaded structures, design *S-N* curves generally present a change in slope at $N = 5 \times 10^6$ cycles and a cutoff at $N = 10^8$ cycles. Recommendations given in Section 17.3.5 for selection of the design *S-N* curves for steel structures are also applicable to aluminum structures, depending on the stress approach considered:

1. *Nominal stress approach.* Experimental *S-N* curves corresponding to the type of structural detail and welded connection may be used, subject to appropriate corrections as given hereafter.

2. *Hot spot stress approach.* If the actual weld configuration corresponds to that of an experimental *S-N* curve, this curve can be used, subject to the same correction as for steel. In particular, when calculating the geometric stress concentration factor, the greater influence of structural deformations on the level of stresses resulting from the lower Young's modulus has to be carefully taken into account, as pointed out by Violette (1998).

3. *Notch stress approach.* The design *S-N* curve based on the notch stress approach should correspond to the *S-N* curve for plain materials (rolled and extruded products with machined edges). The weld concentration factor K_w may be obtained from Table 17.5 or calculated according to the recommendations given in Section 17.2.5. As for the two other approaches, the design *S-N* curve has to be corrected, as indicated hereafter.

FURTHER CORRECTIONS OF THE DESIGN *S-N* CURVE

1. *Plate thickness.* The IIW (1996) recommends the same corrections as for steel while the ECCS (1992) proposes the following correction:

$$S = S_0(25/t)^{1/4} \text{ for thickness } t \text{ greater than 25 mm} \quad (17.3.43)$$

2. *Mean and residual stresses.* The IIW (1996) and ECCS (1992) recommend the same correction as for steel while, according to Eurocode 9 (1999), no allowance is permitted for the mean stress effect unless experiments carried out for the actual structure show a significant influence of mean stresses on the fatigue strength or improvement techniques are used.

3. *Yield stress.* As for steel, the fatigue of aluminum welded joints is nearly independent of the yield

Table 17.10 Rates of Corrosion For Various Types of Ships

Compartment	Type of cargo	Rate of corrosion mm/year
Ballast tanks	Unprotected	0.40
	Coated	0.20
Cargo tanks		
Deck and bottom	Black products	0.2
	White products	0.35
Elsewhere	Black products	0.10
	White products	0.15
Bulk carriers		0.20
Cargo ships		0.10

Table 17.11 Fabrication and In-Service Imperfections

Types of Defects		**Remedial Measures**
Fabrication defects	*Imperfect shape*	
	Linear misalignment	Account for at the design stage
	Angular misalignment	Idem
	Undercuts	Improve the welding procedures
	Planar discontinuities	
	Lack of fusion	Improve the welding procedures
	Lack of penetration	Idem
	Volumetric discontinuities	
	Slag inclusions	No significant influence
	Lamellar tearing	Select proper materials
	Solidification cracks	Use basic electrodes
	Hydrogen induced cracking	Improve the welding procedures
In-service defects	Corrosion pits	Maintain the corrosion protection
	Stress concentration cracks	Generally repair and improve the design
	Fatigue cracks	Idem

stress. Consequently, the same design *S-N* curve is generally used irrespective of the yield stress, unless special measures are taken to improve the weld geometry or welding procedures.

4. *Corrosion*. Aluminum alloys have a good resistance to corrosion due to a thin layer of alumina. However, this protection may be destroyed in service for various reasons such as erosion and impacts, leading to the development of corrosion pits. Galvanic corrosion resulting from contact with other metals also has to be avoided. It is therefore highly recommended to protect aluminum structures by coating. For unprotected structures, Eurocode 9 (1999) recommends, depending on the alloy, a reduction of the fatigue strength (FAT), for structures subjected to marine environment or immersed in seawater.

5. *Workmanship*. As is the case for steel, the IIW (1996) gives the permissible defects for welded details in aluminum. Moreover, improvement techniques similar to those considered for steel structures may be used to improve the fatigue strength of aluminum welded structures.

17.4 ASSESSMENT OF THE FATIGUE STRENGTH BASED ON THE *S-N* APPROACH

17.4.1 General

The verification of the adequacy of the structure is based on the cumulative damage principle stated by Palmgren and Miner:

If the damage contributed by one cycle of stress range S_i is $1/N_i$, where N_i is the mean fatigue life under a constant amplitude stress range S_i, by superposition the cumulative damage D caused by stress ranges S_1, S_2, . . . S_n applied n_1, n_2, . . . n_k cycles is equal to

$$D = \sum_{i=1}^{k} \frac{n_i}{N_i} \qquad (17.4.1)$$

where n_i = number of cycles of stress range S_i, as obtained from the long term distribution of stresses,

N_i = number of cycles to failure at stress range S_i, as obtained from the *S-N* curve.

From this definition, the structure is considered as failed when the cumulative damage ratio D reaches unity. In assessing the fatigue strength of ship structural details, a ship's service life is generally taken as 20 to 25 years.

Based on the considerations developed above, assessment of the fatigue strength of welded structural members includes the following phases:

1. Determination of the long term distribution of stress ranges.
2. Generation of the design *S-N* curve for the structural detail considered and the stress approach (nominal, hot spot, or notch). The design *S-N* curve should correspond to the relevant survival probability based on the fatigue design strategy and consequences of failure on the ship serviceability.
3. Calculation
   a. of either the cumulative damage ratio D and resulting fatigue life, or
   b. of the probability of failure.

17.4.2 Required Survival Probability

The required survival probability is determined with respect to the risk associated with the failure:

1. For normal welded connections (fail-safe design), a survival probability of 95% (or a failure probability of 5%) is generally acceptable.
2. For special welded connections (safe life design), for example, structural details which cannot be easily surveyed and repaired and whose failure would have serious consequences, a failure probability of 0.5% is more appropriate.

As already mentioned, *S-N* curves are represented by equation (17.3.25):

where $\log K_p = \log K_{50} - \lambda_p\,\sigma_d(K_{50})$ (17.4.2)

$\sigma_d(K_{50})$ = standard deviation of $\log K_{50}$.

λ_p = coefficient depending on the required survival probability which may be taken as:

$\lambda_p = 2$ for $p = 0.05$

$\lambda_p = 3$ for $p = 0.005$

17.4.3 Determination of the Fatigue Damage

GENERAL

According to the Miner's Rule, the fatigue strength is expressed by the cumulative damage ratio D that has to be less than unity for the expected ship's life. However, in practice the maximum allowable damage ratio should be below unity in order to take into account the scatter observed on actual damage ratios at failure:

$$D = \sum_{i=1}^{k} \frac{n_i}{N_i} \leq \frac{1}{\gamma} \qquad (17.4.3)$$

in which γ = safety factor.

For a given fatigue damage D, the expected fatigue life is given by

$$\text{Fatigue life} = \frac{1}{\gamma}\frac{\text{Design life}}{D} \qquad (17.4.4)$$

It is generally sufficient to consider the most severe full load and ballast conditions for assessment of the fatigue strength (refer to Section 17.2.2). Taking into account that fatigue is a cumulative process, the fatigue damage ratio D may be expressed as

$$D = \alpha D_0 + \beta D_0' \leq \frac{1}{\gamma} \qquad (17.4.5)$$

in which D_0 = cumulative damage ratio in full load condition

D_0' = cumulative damage ratio in ballast condition

α = part of the ship's life in full load condition

β = part of the ship's life in ballast

In the equivalent regular wave method (refer to Section 17.2.4) stresses are calculated for each relevant loading condition and load effect. Then, for each loading condition the damage ratio may be taken as the maximum damage ratio for all the load effects or as the mean of the damage ratios calculated for each relevant load effect. In the spectral fatigue analysis, equation (17.4.5) may be used with no other assumptions since, for each loading condition, the load combination is made at the first level, for example, when calculating the stress range transfer functions (refer to Section 17.2.2).

PALMGREN-MINER APPROACH

For calculation of the fatigue damage ratios, each long term histogram is generally converted into a step-curve of at least 40 steps of equal length in log *N*, as shown in Fig. 17.28. Each step *k* is considered independently and the cumulative damage D is given by

$$D = \sum_{i=1}^{n_c} \frac{n_i}{N_i} \qquad (17.4.6)$$

where n_c = number of steps of equal length in $\log N$

n_i = number of cycles of stress range S_i

N_i = number of cycles to failure at constant stress range S_i

CLOSED-FORM FATIGUE LIFE EQUATION

1. Assuming that the long term distribution of stresses may be fitted into a given probability distribution function $p_S(S)$, the number of cycles with a stress range between S_i and $S_i \pm \Delta S_i / 2$ is

$$n_i = N_t\, p_S(S_i)\Delta S_i \qquad (17.4.7)$$

where N_t is the total number of cycles for the expected ship's life and p_S is the probability that the cyclic stress will be in the range $S_i \pm \Delta S_i / 2$

2. For a one-slope *S-N* curve, the cumulative damage ratio D is

$$D=\sum_i \frac{n_i}{N_i}=\sum_i \frac{N_t\, p_S(S_i)\, \Delta S_i}{N(S_i)}$$

$$=\frac{N_t}{K_p}\int_0^{\infty} S^m p_S(S)\, dS=\frac{N_t}{K_p}E\left(S^m\right) \qquad (17.4.8)$$

where $E\left(S^m\right)$ is the expected value, or mean value, of S^m.

As discussed in Sections 17.2.2 and 17.2.3, $p_S(S)$ is represented satisfactorily by the Weibull probability density function given in equation (17.2.13). For a one-slope *S-N* curve, the cumulative damage ratio D_1 is

$$D_1=\frac{N_t}{K_p}\int_0^{\infty}\frac{k}{w}S^m\left(\frac{S}{w}\right)^{k-1} e^{-\left(\frac{S}{w}\right)^k} dS \qquad (17.4.9)$$

where w = characteristic value of the Weibull distribution given by $w=\dfrac{S_R}{\left(\ln N_R\right)^{1/k}}$

S_R = stress range at the probability of $1/N_R$

N_R = number of cycles corresponding to the probability of exceedance of $1/N_R$

k = Weibull shape parameter

If we define $X = (S/w)^{\kappa}$ equation (17.4.9) becomes

$$D_1=\frac{N_t}{K_p}\int_0^{\infty} w^m X^{m/k} e^{-X} dX=\frac{N_t}{K_p}w^m\int_0^{\infty} X^{m/k} e^{-X} dX$$

or

$$D_1=\frac{N_t}{K_p}w^m\,\Gamma\left(\frac{m}{k}+1\right) \qquad (17.4.10)$$

where Γ is the Gamma function given by $\Gamma(a+1)=\int_0^{\infty} t^a e^{-t}\, dt$

Therefore,

$$E\left(S^m\right)=w^m\,\Gamma\left(m/k+1\right)$$

$$=\frac{S_R^m}{\left(\ln N_R\right)^{m/k}}\Gamma\left(m/k+1\right)$$

$$=\left(\alpha S_R\right)^m \qquad (17.4.11)$$

where

$$\alpha=\frac{\left[\Gamma\left(m/k+1\right)\right]^{1/m}}{\left(\ln N_R\right)^{1/k}} \qquad (17.4.12)$$

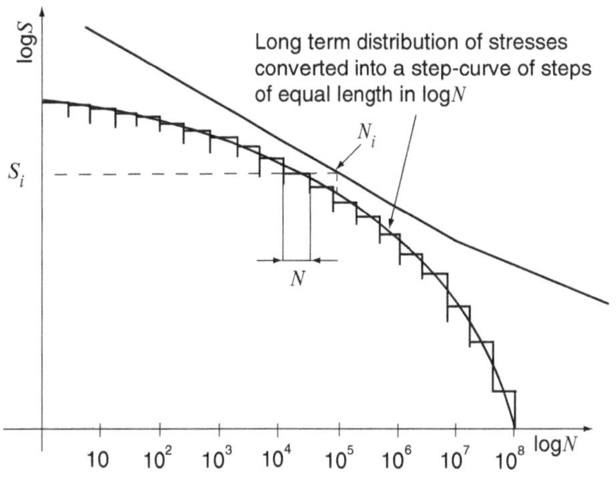

Figure 17.28 Conversion of the long term histogram into a step curve.

3. For a two-slope *S-N* curve, the cumulative damage ratio is

$$D_2=\frac{N_t}{K_p}\int_{S_q}^{\infty}\frac{k}{w}S^m\left(\frac{S}{w}\right)^{k-1} e^{-\left(\frac{S}{w}\right)^k} dS$$

$$+\frac{N_t}{K'_p}\int_0^{S_q}\frac{k}{w}S^m\left(\frac{S}{w}\right)^{k-1} e^{-\left(\frac{S}{w}\right)^k} dS \qquad (17.4.13)$$

or in terms of X:

$$D_2=\frac{N_t}{K_p}\int_{\nu}^{\infty} w^m X^{m/k} e^{-X} dX$$

$$+\frac{N_t}{K'_p}\int_0^{\nu} w^{(m+\Delta m)} X^{(m+\Delta m)/k} e^{-X} dX \qquad (17.4.14)$$

where $\nu=\left(S_q/S_R\right)^k \ln N_R$

S_q = stress range at the intersection of the two segments of the *S-N* curve

Δm = slope change of the upper to lower segment of the *S-N* curve, $\Delta m = 2$

K'_p = constant of the *S-N* curve for the slope $m = 5$

If we again introduce the Γ function, the expression (17.4.14) becomes

$$D_2=\frac{N_t}{K_p}w^m\,\Gamma\left(m/k+1,\nu\right)$$

$$+\frac{N_t}{K'_p}w^{m+\Delta m}\,\gamma\left(\frac{m+\Delta m}{k}+1,\nu\right) \qquad (17.4.15)$$

$$\text{or } D_2 = \frac{N_t}{K_p} w^m \Gamma(m/k+1)$$

$$\times \frac{\Gamma(m/k+1,\nu) + \frac{K_p}{K_p'} w^{\Delta m} \gamma\left(\frac{m+\Delta m}{k}+1,\nu\right)}{\Gamma(m/k+1)} \qquad (17.4.16)$$

where $\Gamma(a+1,\nu)$ = incomplete Γ function given by

$$\Gamma(a+1,\nu) = \int_\nu^\infty t^a e^{-t} dt$$

$\gamma(a+1,\nu)$ = incomplete Γ function, Legendre form, given by $\gamma(a+1,\nu) = \int_0^\nu t^a e^{-t} dt$

Noting that

$$\Gamma(m/k+1,\nu) = \Gamma(m/k+1) - \gamma(m/k+1,\nu)$$

and

$$\frac{K_p}{K_p'} w^{\Delta m} = \left(\frac{S_R}{S_q}\right)^{\Delta m} \frac{1}{(\ln N_R)^{\Delta m/k}} = \nu^{-\Delta m/k}$$

equation (17.4.16) can be written as

$$D_2 = \frac{D_1}{\Gamma(m/k+1)} \times \left\{\Gamma(m/k+1) - \gamma(m/k+1,\nu) + \nu^{-\Delta m/k}\gamma\left(\frac{m+\Delta m}{k}+1,\nu\right)\right\} \qquad (17.4.17)$$

and finally the cumulative damage ratio D_2 is expressed as

$$D_2 = \mu D_1 = \mu \frac{N_t}{K_p} E(S_m) \qquad (17.4.18)$$

with

$$\mu = 1 - \frac{\left\{\gamma(m/k+1,\nu) - \nu^{-\Delta m/k}\gamma\left(\frac{m+\Delta m}{k}+1,\nu\right)\right\}}{\Gamma(m/k+1)} \qquad (17.4.19)$$

The damage ratio D_1 or D_2 may be expressed in terms of a constant amplitude equivalent stress range $\Delta\sigma_{eq}$ giving the same amount of damage as that obtained from equation (17.4.10) or (17.4.18):

—For a one-slope *S-N* curve:

$$D = \frac{N_t}{K_p} E(S^m) = \frac{N_t}{K_p} (\Delta S_{eq})^m$$

and

$$\Delta\sigma_{eq} = \left[E(S^m)\right]^{1/m} = \frac{S_R}{(\ln N_R)^{1/k}} \left[\Gamma\left(\frac{m}{k}+1\right)\right]^{1/m} \qquad (17.4.20)$$

—For a two-slope *S-N* curve:

$$D = \mu \frac{N_t}{K_p} E(S^m) = \frac{N_t}{K_p} (\Delta S_{eq})^m$$

where μ is defined by equation (17.4.19) and

$$\Delta\sigma_{eq} = \left[\mu E(S^m)\right]^{1/m}$$

$$= \mu^{1/m} \frac{S_R}{(\ln N_R)^{1/k}} \left[\Gamma(mk/k+1)\right]^{1/m} \qquad (17.4.21)$$

To achieve a damage ratio $D \le 1/\gamma$ the stress range S_R corresponding to a probability of exceedance of $1/N_R$ should be

$$S_R \le \frac{1}{\gamma^{1/m}} \left(\frac{K_p}{N_t}\right)^{1/m} \frac{(\ln N_R)^{1/k}}{\left[\mu\Gamma(m/k+1)\right]^{1/m}} \qquad (17.4.22)$$

The IACS (1999) proposed the following equation for calculation of the number of cycles for the expected life:

$$N_t = \frac{\alpha_0 T}{4 \log L} \qquad (17.4.23)$$

where T = ship's life, in seconds
α_0 = sailing factor accounting for the time needed for loading/unloading operations, repairs, etc. In general, α_0 may be taken to equal 0.85
L = ship's rule length, in meters.

The damage ratio may also be calculated using the results of the short-term response analysis. In that case, for any sea state of the scatter diagram, the damage ratio is given by

$$D_i = \frac{f_{0i} T}{K_p} E(S_i^m) \qquad (17.4.24)$$

where f_{0i} = frequency of the stress process
$E(S_i^m)$ = expected value of S_i^m

Then, assuming that the stress ranges S_i are Rayleigh distributed, the damage ratio D is obtained by integration of the individual damages D_i for all the sea states of the scatter diagram (refer to Mansour (1994)) and, for a one-slope S-N curve, given by

$$D = \frac{T}{K_p} 2^{m/2} \Gamma\left(\frac{m}{2}+1\right) \sum_{n=1}^{n_k} p_n \sum_{i=1}^{\text{all sea states}} p_i \, f_{0i} \left(\sigma_i\right)^m \quad (17.4.25)$$

where p_n = part of the design life in loading condition n
σ_i = rms value of the stress range in the sea state i
p_i = probability of occurrence of the sea state i
T = ship's life in seconds

17.4.4 Improvement of the Fatigue Strength of Welded Joints

When the calculated fatigue life is significantly less than the required design value, it is necessary to take appropriate measures to improve the fatigue strength. There are several ways to achieve this objective:

1. Modification of the design.
2. Modification of the geometry of the welded joints, for example, by using full or partial penetration welds instead of fillet welds, to reduce the weld shape concentration factor.
3. Improvement of the welding procedures for elimination, as far as practicable, of the weld defects (e.g., weld shape imperfections, undercuts, porosity and inclusions, lack of fusion and penetration).
4. Modification of the ship construction procedure for reduction of the residual stresses (e.g., appropriate sequences for assembly of the pre-fabricated blocks aiming at reducing the locked-in stresses).
5. Use of defect removal methods, for example, by grinding or weld toe remelting.
6. Use of residual stress methods aiming at introducing compressive stresses, for example, by hammer.

With the exception of improvement of welding procedures and workmanship, it is not practical to adopt the last two methods as a normal part of the welding process throughout ship construction. Such methods can only be considered as exceptional measures or applicable to very particular welded joints. Consequently, besides the improvement of welding procedures and strengthening of constructional tolerances, the only practical way to improve fatigue strength consists of improving the design by reducing the nominal and geometric stresses and selecting the appropriate types of welded joints. Depending on the type of structural detail, there are several ways to reduce the geometric stress concentrations, among which the most effective are:

1. Avoidance of abrupt changes in geometry.
2. Adequate structural continuity.
3. Local increase in thickness.
4. Softening of bracket toes.
5. Improvement of the shape of cut-outs.

17.5 FATIGUE RELIABILITY

17.5.1 General

Most of the factors that influence the fatigue life of ship structures have large uncertainties due to the method used to determine:

1. The loads and stresses.
2. The capacity of the structure.

Reliability methods make it possible to account for these uncertainties and to evaluate the probability that a structural detail has failed at the end of the specified fatigue life. In that respect, the Miner rule which states that failure occurs when the cumulative damage ratio D is greater than unity, may be modified as follows:

$$D \geq \Delta \quad (17.5.1)$$

where Δ = random variable representing the damage at failure and accounting for the modeling errors associated with the Miner's rule.
D = cumulative damage ratio, as given by equation (17.4.18)

$$D = \mu \frac{N}{K} E\left(S^m\right) \quad (17.5.2)$$

μ = coefficient as given by equation (17.4.19)
N = total number of cycles
$E(S^m)$ = expected value of S^m

According to the method proposed by Ang and Cornell (1974) for modeling uncertainties, the random variable S can be expressed as follows:

$$S = B_I B_{II} S_0 = BS_0 \tag{17.5.3}$$

where S_0 = value of the random variable S as obtained from a 2D or 3D FEM stress analysis or specified by a design code

$B_I = S_P/S_0$ (B_I is a random variable measuring the statistical uncertainties)

S_P = theoretically predicted value of the variable S

$B_{II} = S/S_P$ (B_{II} is a random variable measuring the approximational uncertainties)

Substituting equation (17.5.3) into (17.5.2) gives

$$D = \mu \frac{N}{K} B^m E\left(S_0^m\right) \tag{17.5.4}$$

and the number of cycles at failure N is obtained by writing $D = \Delta$:

$$N = \frac{K\ \Delta}{\mu E\left(S^m\right)} = \frac{K\ \Delta}{\mu B^m E\left(S_0^m\right)} \tag{17.5.5}$$

where $E(S_0^m)$ is the expected value of S_0^m as given by equation (17.4.11).

17.5.2 Probability of Failure

From equation (17.5.5) the probability of fatigue failure P_f may be defined as

$$P_f = P(N \leq N_t) \tag{17.5.6}$$

where N_t is the required fatigue life.

In (17.5.5) K, Δ, μ, B and, therefore, N are random variables. Based on the analysis of statistical data, especially for K and Δ, Wirshing (1981) and (1984) proposed modeling these random variables by the lognormal distribution which gives

1. Easy use for reliability assessment and providing an exact expression for the probability of failure.
2. Good agreement with service experience of ship structural details.
3. Good correlation with existing data on Δ and K.

If we assume that the random variable μ follows also a lognormal distribution, the safety margin M can be expressed as

$$M = \ln \frac{N}{N_t} = \ln \frac{K\Delta}{\mu B^m} - \ln E\left(S_0^m\right) - \ln N_t \tag{17.5.7}$$

Assuming that the long-term distribution of the stress range is represented by the Weibull distribution, the expected value of S_0^m is given by equation (17.4.11) and the safety margin becomes

$$M = \ln \frac{N}{N_t} = \ln \frac{K\Delta}{\mu B^m} - m \ln\left(\alpha S_R\right) - \ln N_t \tag{17.5.8}$$

with

$$\alpha = \frac{\left[\Gamma\left(m/k+1\right)\right]^{1/m}}{\left(\ln N_R\right)^{1/k}} \tag{17.5.9}$$

The Cornell safety index is given by

$$= \frac{\ln \dfrac{E(K)\ E(\Delta)}{E(\mu)\,E(B)^m} \dfrac{\sqrt{\left(1+V_\mu^2\right)\left(1+V_B^2\right)^m}}{\sqrt{\left(1+V_K^2\right)\ \left(1+V_\Delta^2\right)}} - m\ \ln\ \left(\alpha\ S_R\right) - \ln N_t}{\sqrt{\ln\ \left(1+V_K^2\right)\ \left(1+V_\Delta^2\right)\ \left(1+V_\mu^2\right)\left(1+V_B^2\right)^{m^2}}} \tag{17.5.10}$$

where V_K, V_Δ, V_μ, and V_B are the coefficients of variation of the random variables K, Δ, μ, and B. Note that

$$E(\ln X) = \ln E(X) - \ln\sqrt{1+V_X^2}$$

$$\sigma_{\ln X} = \sqrt{\ln\left(1+V_X^2\right)}$$

The probability of failure is given by

$$P_f = P\left[\frac{M - E(M)}{\sigma(M)} < -\beta\right] = P\left[M_0 < -\beta\right] \tag{17.5.11}$$

where M_0 is the normalized margin. Since the margin M is lognormally distributed, equation (17.5.11) shows that the Cornell safety index β is related to the probability of failure P_f by

$$P_f = \Phi(-\beta) \tag{17.5.12}$$

where Φ is the standard normal cumulative distribution function as shown in Fig. 17.29 (refer also to Table 17.12):

For a given target safety index β_0 the permissible stress range S_{R0} is obtained from equation (17.5.10) by noting $\beta_C = \beta_0$:

$$S_{R0} = \frac{\left(\ln N_R\right)^{1/k}}{\left[\Gamma\left(m/k+1\right)\right]^{1/k}} \left\{\frac{A}{N_t} e^{-\lambda\beta_0}\right\}^{1/m} \tag{17.5.13}$$

in which

$$\lambda = \sqrt{\ln(1+V_K^2)(1+V_\Delta^2)(1+V_\mu^2)(1+V_B^2)^{m^2}} \quad (17.5.14)$$

$$A = \frac{E(K)\,E(\Delta)}{E(\mu)[E(B)]^m}\sqrt{\frac{(1+V_\mu^2)(1+V_B^2)^m}{(1+V_K^2)(1+V_\Delta^2)}} \quad (17.5.15)$$

The reliability of structural elements is frequently expressed in terms of partial safety factors and, in that case, the fatigue design equation is

$$\frac{K^* \Delta^*}{\mu^*(B^*)^m\,E(S_0^m)} - N_t = 0 \quad (17.5.16)$$

where K^*, Δ^*, μ^*, B^* are the coordinates of the most probable failure point (MPFP). Replacing $E(S_0^m)$ by expression (17.4.11) into (17.5.16) gives for the fatigue design equation:

$$K^*\Delta^* - \frac{N_t\Gamma(m/k+1)}{(\ln N_R)^{m/k}}\,\mu^*(B^* S_R)^m = 0$$

or

$$\left(\frac{K_{nom}}{\gamma_K^*}\right)\left(\frac{\Delta_{nom}}{\gamma_\Delta^*}\right) - \frac{N_t\Gamma(m/k+1)}{(\ln N_R)^{m/k}}\gamma_\mu^*\mu_{nom}\left[\gamma_B^* B_{nom} S_R\right]^m = 0$$

(17.5.17)

where γ_K^* = partial safety factor for fatigue capacity of the welded joint or structural detail considered $\frac{1}{\gamma_K^*} = \frac{K^*}{K_{nom}}$

γ_Δ^* = partial safety factor for fatigue damage, $\frac{1}{\gamma_\Delta^*} = \frac{\Delta^*}{\Delta_{nom}}$

γ_μ^* = partial safety factor covering uncertainties on μ' $\gamma_\mu^* = \frac{\mu^*}{\mu_{nom}}$

γ_B^* = partial safety factor covering uncertainties on stresses, $\gamma_B^* = \frac{B^*}{B_{nom}}$

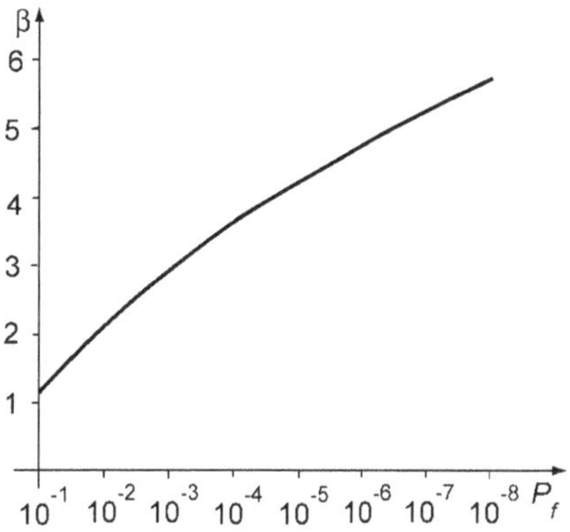

Figure 17.29 Relationship between P_f and β.

An example of calculation of the Hasofer-Lind safety index β_{HL} and partial safety factors is given in Section 17.5.4.

17.5.3 Modeling of Uncertainties

The ISSC (2000) carried out a comparative fatigue strength analysis of a critical structural detail. A Panamax container vessel was selected and a welded pad detail on the top of the longitudinal hatch coaming flat bar was selected for the fatigue analysis. Calculations were carried out according to the fatigue rules and guidelines of eight Classification Societies. Examination of the results shows that the extreme fatigue lives differ by a factor of more than 10 (1.8 to 20.7 years). This example highlights the lack of accuracy of this type of analysis, resulting from the many assumptions, approximations, and judgments that have to be considered and the need for further research.

Uncertainties in any basic variable may be classified in two categories, statistical or random and approximational or modeling uncertainties. Influence on the fatigue strength of random and modeling uncertainties has been thoroughly examined by Wirshing (1981), Guedes Soares (1984), Nikolaidis and Kaplan (1991), and Mansour and Thayamballi (1994). Nikolaidis and Kaplan (1991) concluded that the random uncertainties B_I are small by comparison with the modeling uncertainties B_{II} and, therefore, can be generally neglected in fatigue analyses, which gives

Table 17.12 Relationship Between P_f and β

P_f	10^{-1}	10^{-2}	10^{-3}	10^{-4}	10^{-5}	10^{-6}	10^{-7}	10^{-8}
β	1.28	2.33	3.1	3.7	4.3	4.75	5.2	5.6

$$S = B_I B_{II} S_0 \cong B_{II} S_0 \qquad (17.5.18)$$

Approximational uncertainties that have the greatest influence on the fatigue strength are:

1. Evaluation and combination of loads.
2. Structural response (FEM analyses).
3. Stress concentration factors.
4. Stress approach and selection of the design *S-N* curve.
5. Failure criteria.

Evaluation and combination of loads Any ship structural analysis raises two different problems:

1. Evaluation of the various loads acting on the ship structure.
2. Combination of these loads, taking into account that sea loads are random and their effects not maximum at the same time (phasing is an important parameter). Stochastic or deterministic methods may be used for combination of time-dependent loads.

Assuming that the various approximational uncertainties can be identified, B_{II} may be expressed by the following equation:

$$B_{II} = \prod_{i=1}^{n} B_i \qquad (17.5.19)$$

where the variables B_i represent the approximational uncertainties in loads and FEM model occurring in determination of the random variable *S*, among which the most significant are:

B_1 = environmental description
B_2 = long term wave bending moment
B_3 = local pressures
B_4 = modeling of the structure
B_5 = stress concentration factors (B_{K_G} and B_{K_w})

If the random variables B_i are assumed to be uncorrelated, the bias B_{II} and the coefficient of variation $V_{B_{II}}$ of the random variable B_{II} are given by

$$E(B_{II}) = \prod_i E(B_i) \qquad (17.5.20)$$

$$V_{B_{II}} = \sqrt{\sum_{i=1}^{n} V_{B_i}^2} \qquad (17.5.21)$$

When calculation of the notch stress is based on the nominal stress approach and on the use of geometric and weld shape concentration factors, the local stress range is given by

$$S_\ell = B_{K_G} B_{K_w} B_{II} (K_G K_w S_0) \qquad (17.5.22)$$

and the coefficient of variation of the approximational uncertainties is

$$V_B = \sqrt{V_{K_G}^2 + V_{K_w}^2 + V_{B_{II}}^2} \qquad (17.5.23)$$

where V_{K_G} and V_{K_w} are the coefficients of variation of K_G and K_w.

Another important source of uncertainty comes from the method used for combination of the various wave loads contributing to the fatigue damage, mainly wave-induced loads (hull girder and local loads). This assumes that the following loads can be disregarded:

1. Transient loads due to the small number of load reversals.
2. Vibratory loads. Experience shows that there are only few cases where resonance of the structure within the cargo area leads to structural failures.
3. Impact loads. Classification society rules give requirements to limit the occurrence of impact loads or to minimize their consequences. However, for particular types of ships (e.g., passenger vessels, carferries) the risk of fatigue damage due to impact loads has to be examined. In that case, the damage ratio due to the impact loads may be calculated separately and added to the standard damage ratio, taking into account that fatigue damage is a cumulative process.

In the spectral fatigue analysis, the combination of loads is made at the first level, that is, when calculating the stress range transfer functions (refer to Section 17.2.2). Therefore, the combination of the hull girder and pressure loads is implicitly taken into account in the calculation of the long-term stress ranges. In the deterministic equivalent regular wave method (refer to Section 17.2.4), the safety index may be determined for each relevant loading condition and load effect and the resultant probability of fatigue failure calculated for the mean safety index. In simplified analyses, which are generally based on classification society rules, additional assumptions are to be considered, such as, for example, the relative probability of the relevant basic loading cases.

FEM Structural Analysis Uncertainties in FEM procedures arise from the modeling techniques, boundary conditions, mesh size, etc., quality of the computer program and also from human errors. From comparison and analysis of FEM calculations carried out for other engineering structures, Nikolaidis and Kaplan (1991) concluded that the average bias should be taken as 1.0 and the COV between 0.1 and 0.15.

Stress Concentration Factors (SCF) Wirshing (1981) investigated uncertainties in geometric stress concentration factors for tubular joints of offshore structures and concluded that the average bias in SCF is in the range of 0.8 to 1.2 and the COV between 0.1 and 0.4.

Uncertainties in the weld SCF are much more difficult to appraise since they are influenced by weld imperfections, which depend essentially on the welder's skill. This is highlighted by the different fatigue behavior of sister ships. Application of Quality Control Systems enables the minimization of the human errors, provided that the weld defects remain below the permissible limits. As a first approximation, same biases and COV's could be taken for K_w and K_G.

Stress Approach and Selection of the S-N Curve When starting a fatigue analysis, the first step is to select the type of stress approach and the *S-N* curve associated with the structural detail. It is generally rather difficult to associate a given structural detail to one of the experimental *S-N* curves. It is a matter of judgment and the answer is not unique. To eliminate this uncertainty, it seems advisable to use the notch stress approach with the same *S-N* curve for all types of structural details. In that case, uncertainties come mainly from the calculation of the stress concentration factors and from the procedure used for determination of the experimental *S-N* curve.

Failure Criterion Wirshing (1981), Guedes Soares (1984) and Wirshing and Chen (1987) have investigated the accuracy of the Palmgren-Miner rule and consequences of any inaccuracy on the fatigue reliability. Wirshing and Chen (1987) provides with examples of statistical data on damage at failure and suggests the use of a bias of 1.0 and COV of 0.3 for the modeling error resulting from the inaccuracies of the Miner rule.

17.5.4 Fatigue reliability of the hull girder

GENERAL

Assessment of the hull girder fatigue reliability requires the selection of a critical welded joint subjected primarily to the hull girder bending without any influence of local stresses resulting from the bending of the primary structure and/or secondary stiffeners. As an example, the following calculations are carried out for connections of deck longitudinals to deck transverses of oil tankers, as shown in Fig. 17.30.

FATIGUE RELIABILITY

Assuming that the various random variables are independent and lognormally distributed as proposed by Wirshing (1981) and (1984), the safety margin with respect to fatigue of the hull girder is given by equation (17.5.8) and the Cornell safety index β_C obtained from equation (17.5.10). Calculation of the Hasofer-Lind safety index β_{HL} requires the transformation of the random variables into a set of independent and reduced normal variables. Since the random variables are assumed to be independent and lognormally distributed, the transformation matrix **T** is a diagonal matrix whose elements are equal to $1/\sigma_i$ and the reduced variables are

$$u_i = \frac{\ln X_i - E(\ln X_i)}{\sigma_{\ln X_i}} \qquad (17.5.24)$$

From equation (17.5.8) the safety margin expressed in the reduced space is linear and given by

$$\begin{aligned} g'(u) = {} & \sigma_{\ln K} u_K + \sigma_{\ln \Delta} u_\Delta - \sigma_{\ln \mu} u_\mu - m\, \sigma_{\ln B} u_B \\ & + E(\ln K) + E(\ln \Delta) - E(\ln \mu) \\ & - mE(\ln B) - \ln \frac{(\alpha S_R)^m}{N_t} \end{aligned} \qquad (17.5.25)$$

The coordinates of the MPFP defined as the intersection of the failure surface with the normal drawn by the origin are

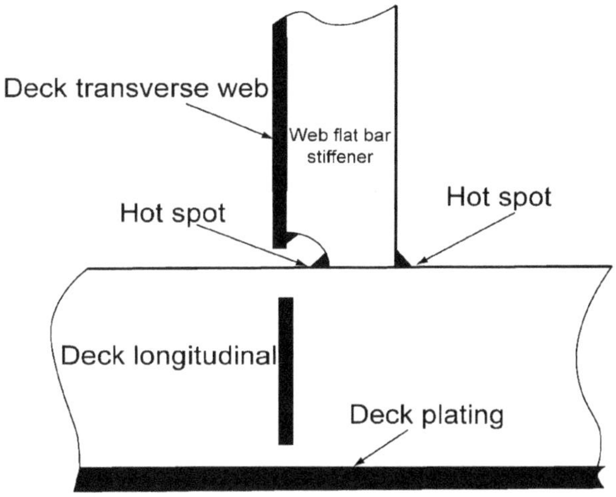

Figure 17.30 Deck longitudinal connection to transverse webs.

$$\frac{u_K}{-\dfrac{\delta g'}{\delta u_K}} = \frac{u_\Delta}{-\dfrac{\delta g'}{\delta u_\Delta}} = \frac{u_\mu}{-\dfrac{\delta g'}{\delta u_\mu}} = \frac{u_B}{-\dfrac{\delta g'}{\delta u_B}}$$

$$\frac{u_K}{-\sigma_{\ln K}} = \frac{u_\Delta}{-\sigma_{\ln\Delta}} = \frac{u_\mu}{\sigma_{\ln\mu}} = \frac{u_B}{m\sigma_{\ln B}} = \lambda$$

where the value of λ is obtained by writing $g'(u) = 0$, which gives

$$\lambda = \frac{E(\ln K)+E(\ln\Delta)-E(\ln\mu)-m\,E(\ln B)-m\ln(\alpha\,S_R)-\ln N_t}{\sigma^2_{\ln K}+\sigma^2_{\ln\Delta}+\sigma^2_{\ln\mu}+m^2\sigma^2_{\ln B}}$$

or

$$\lambda = \frac{\beta_{\mathrm{HL}}}{\sqrt{\ln\left(1+V_K^2\right)\left(1+V_\Delta^2\right)\left(1+V_\mu^2\right)\left(1+V_B^2\right)^{m^2}}}$$

Finally, the Hasofer-Lind safety index is given by

$$\begin{aligned}\beta_{\mathrm{HL}} &= \sqrt{u_K^2+u_\Delta^2+u_\mu^2+u_B^2}\\ &= \frac{E(\ln K)+E(\ln\Delta)-E(\ln\mu)-m\,E(\ln B)-m\ln(\alpha S_R)-\ln N_t}{\sqrt{\sigma^2_{\ln K}+\sigma^2_{\ln\Delta}+\sigma^2_{\ln\mu}+m^2\sigma^2_{\ln B}}}\\ &= \beta_C \end{aligned} \qquad (17.5.26)$$

The partial safety factors are given by $\gamma_i^* = (X_i^*/X_{in})$, in which X_i^* are the coordinates of the MPFP in the original space given by

$$\ln X_i^* = E(\ln X_i)+\sigma_{\ln X_i}u_i^* \qquad (17.5.27)$$

or

$$\begin{aligned}\ln K^* &= \ln\frac{E(K)}{\sqrt{1+V_K^2}} - \lambda\sigma^2_{\ln K}\\ &= \ln\frac{E(K)}{\sqrt{1+V_K^2}} - \ln\left(1+V_K^2\right)^\lambda\\ &= \ln E(K) - \ln\left(1+V_K^2\right)^{\lambda+0.5}\end{aligned}$$

$$\ln\Delta^* = \ln E(\Delta) - \ln\left(1+V_\Delta^2\right)^{\lambda+0.5}$$

$$\ln\mu^* = \ln E(B) + \ln\left(1+V_\mu^2\right)^{\lambda-0.5}$$

$$\ln B^* = \ln E(B) + \ln\left(1+V_B^2\right)^{\lambda m-0.5}$$

$$\frac{1}{\gamma_K^*} = \frac{K^*}{K_{nom}} = \frac{E(K)}{K_{nom}}\left(1+V_K^2\right)^{-(\lambda+0.5)} \qquad (17.5.28)$$

$$\frac{1}{\gamma_\Delta^*} = \frac{\Delta^*}{\Delta_{nom}} = \frac{E(\Delta)}{\Delta_{nom}}\left(1+V_\Delta^2\right)^{-(\lambda+0.5)} \qquad (17.5.29)$$

$$\gamma_\mu^* = \frac{\mu^*}{\mu_{nom}} = \frac{E(\mu)}{\mu_{nom}}\left(1+V_\mu^2\right)^{\lambda-0.5} \qquad (17.5.30)$$

$$\gamma_B^* = \frac{B^*}{B_{nom}} = \frac{E(B)}{B_{nom}}\left(1+V_B^2\right)^{\lambda m-0.5} \qquad (17.5.31)$$

Finally, the design equation expressed in terms of the partial safety factors is

$$\left(\frac{K_{nom}}{\gamma_K^*}\right)\left(\frac{\Delta_{nom}}{\gamma_\Delta^*}\right) - N_t\left(\gamma_\mu^*\mu_{nom}\right)\left(\gamma_B^*B_{nom}\right)^m(\alpha\,S_R)^m \geq 0 \qquad (17.5.32)$$

or

$$S_R \leq \frac{(\ln N_R)^{1/k}}{\left[N_t\,\Gamma(m/k+1)\right]^{1/m}}\frac{1}{\left(\gamma_\mu^*\mu_{nom}\right)^{1/m}} \times \frac{1}{\gamma_B^*B_{nom}}\left(\frac{K_{nom}}{\gamma_K^*}\frac{\Delta_{nom}}{\gamma_\Delta^*}\right)^{1/m} \qquad (17.5.33)$$

NUMERICAL APPLICATION

As we are concerned in this example with the fatigue reliability of the hull girder, we have to select a structural detail subjected to hull girder bending stresses without any influence of local stresses resulting from the bending of the primary structure and/or secondary stiffeners. As an example, calculations will be carried out for connections of deck longitudinals to deck transverses of oil tankers, considering the following characteristics of the random variables.

1. **Main particulars of the ship**

   315 m × 56 m × 29.4 m × 0.835

   High tensile steel at deck and bottom

   $\sigma_Y = 355$ MPa

   Section modulus at strength deck

   $Z_D = 67.7\ \mathrm{m}^3$

   $M_{vw,H}$ (BV rules) 9.476 × 10^6 kN·m

   $M_{vw,S}$ (BV rules) 10.090 × 10^6 kN·m

   $$(S)_{10^{-8}} = \frac{(M_{vw})_S + (M_{vw})_H}{Z_D}$$

   289 MPa

   Weibull shape factor

   $$k = 1.1 - 0.35\,\frac{L-100}{300} = 0.85$$

Stress range at $p = 10^{-5}$

$$10^5 = \exp\left(S_{10^{-5}}/w\right)^{0.85}$$

$$10^8 = \exp\left(S_{10^{-8}}/w\right)^{0.85}$$

$$\frac{\ln 10^5}{\ln 10^8} = \frac{11.513}{18.42}$$

$$= \left(S_{10^{-5}}/S_{10^{-8}}\right)^{0.85}$$

$$S_{10^{-5}} = \left(\frac{11.513}{18.42}\right)^{1/0.85} S_{10^{-8}}$$

$$= 0.575\, S_{10^{-8}}$$

$$= 166.2\,\text{MPa}$$

Local stress range at hot spot

$$S_\ell = K_G\, K_w\, S_{10^{-5}}$$

K_G: geometric SCF taken as K_G = 1.15, based on the results of FEM calculations K_w: weld concentration factor taken as $K_w = 1.55\sqrt{\theta/30} = 1.9$, this gives a notch factor K_f equal to 90% of the value as obtained from Stambaugh et al. (1994)

$$S_\ell = 1.20 \times 1.90 \times 166.2$$

$$= 378.9\,\text{MPa}$$

$$S_R = 335 + 0.6(401.4 - 355)$$

$$= 382.8\,\text{MPa}$$

Stress range for calculation

Taking into account the actual scantlings for this type of structural detail and the calculated value of S_ℓ, following calculations are carried out without any correction for influence of thickness, residual, and compressive stresses:

$S_R = S_\ell = 378.9\,\text{MPa}$ for a probability of exceedance of 10^{-5}

2. **Constant of the *S-N* curve**

nominal constant of the HSE *B* curve

$K_{nom} = E(K)\ 1.342\ 10^{13}$

$\sigma(\log K) = 0.1821$

$$\sigma(\log K) = \sigma\left(\frac{\ln K}{2.3}\right) = \frac{\sqrt{\ln\left(1+V_K^2\right)}}{2.3}$$

$$= 0.1821$$

$\ln\left(1+V_K^2\right) = 0.1754$ which gives $V_K = 0.438$

3. **Damage ratio Δ**

$\Delta_{nom} = E(\Delta) = 1$ and $V_\Delta = 0.3$

4. **Reduction factor μ**

$\mu_{nom} = E(\mu)$ and $V_\mu = 0.15$. $E(\mu)$ is calculated for the mean value of ν according to equation (17.4.19)

$$E\left(S_q\right) = \sqrt{\frac{E(K')}{E(K)}} = 110.3\,\text{MPa}$$

$$E(\nu) = \left(\frac{E\left(S_q\right)}{S_R}\right)^{0.85} \ln N_R = \left(\frac{110.3}{378.9}\right)^{0.85}$$

$$\ln 10^5 = 4.033$$

$$E(\mu) = 1 - \frac{\gamma\left(m/k+1, E(\nu)\right) - \left[E(\nu)\right]^{-\Delta m/k}\gamma\left(\frac{m+\Delta m}{k}+1, E(\nu)\right)}{\Gamma(m/k+1)}$$

$\Gamma(1+3/0.85) = 12.156$
$\gamma(3.53+1, 4.033) = 5.6585$
$\gamma(5.882+1, 4.033) = 72.1668$
$E(\mu) = 0.758$

5. **Uncertainties on stresses**

$B_{nom} = E(B) = E(B_{K_G})\, E(B_{K_w})\, E(B_{II})$ according to equation (17.5.20)

$$E(B) = 0.9 \times 0.9 \times 1 = 0.81$$

$$V_B = \sqrt{V_{K_G}^2 + V_{K_w}^2 + V_{B_{II}}^2}$$

$$V_B = \sqrt{0.1^2 + 0.1^2 + 0.15^2}$$

$$\approx 0.2$$

6. **Cornell safety index**

$E(\mu) = 0.758$

$$\alpha = \frac{\left[\Gamma(1+m/k)\right]^{1/m}}{\left(\ln N_R\right)^{1/m}}$$

$$= \frac{12.156^{1/3}}{11.513^{1/0.85}} = 0.13$$

$$N_t = \frac{0.85T}{4\log L}$$

$$= \frac{0.85 \times 20.365 \times 24 \times 3600}{4 \times 2.498}$$

$$= 5.365 \times 10^7 \text{ cycles}$$

$$\beta_C = \frac{E(M)}{\sigma(M)}$$

$$= \frac{\ln\left(\frac{1.342\,10^{13} \times 0.94}{0.758 \times 0.81^3}\right) - \ln\left(0.13^3 \times 378.9^3 \times 5.365\,10^7\right)}{\sqrt{\ln\left(1.192 \times 1.09 \times 1.0225 \times 1.04^9\right)}}$$

$$= 2.0$$

which corresponds to a probability of failure of 2.3×10^{-2}. This means that on about 2500 connections of this type on a VLCC, about 55 connections have a chance to be damaged in 20 to 25 years.

7. Partial Safety Factors

$$\lambda = \frac{\beta_C}{\sqrt{\ln\left(1+V_K^2\right)\left(1+V_\Delta^2\right)\left(1+V_\mu^2\right)\left(1+V_B^2\right)^{m^2}}}$$

$$= \frac{2}{0.798} = 2.506$$

$$\frac{1}{\gamma_K^*} = \frac{K^*}{K_{nom}} = \frac{E(K)}{K_{nom}}\left(1+V_K^2\right)^{-(\lambda+0.5)}$$

$$= \left(1+V_K^2\right)^{-(\lambda+0.5)} = 0.59$$

$$\frac{1}{\gamma_\Delta^*} = \frac{\Delta^*}{\Delta_{nom}} = \frac{E(\Delta)}{\Delta_{nom}}\left(1+V_\Delta^2\right)^{-(\lambda+0.5)}$$

$$= \left(1+V_\Delta^2\right)^{-(\lambda+0.5)} = 0.77$$

$$\gamma_\mu^* = \frac{\mu^*}{\mu_{nom}} = \frac{E(\mu)}{\mu_{nom}}\left(1+V_\mu^2\right)^{\lambda-0.5}$$

$$= \left(1+V_\mu^2\right)^{\lambda-0.5} = 1.045$$

$$\gamma_B^* = \frac{B^*}{B_{nom}} = \frac{E(B)}{B_{nom}}\left(1+V_B^2\right)^{\lambda m-0.5}$$

$$= \left(1+V_B^2\right)^{\lambda m-0.5} = 1.32$$

In terms of partial safety factors, the design equation for this type of structural detail made of HTS and subjected to the hull girder bending stresses is given by

$$S_R \le \frac{1}{\alpha N_t^{1/m}} \frac{\left(0.6K_{nom}\right)^{1/m}\left(0.8\Delta_{nom}\right)^{1/m}}{\left(1.05\,\mu_{nom}\right)\left(1.3B_{nom}\right)}$$

8. Deterministic cumulative damage ratio

$$D = \frac{N_t}{K_p} \frac{S_R^m}{\left(\ln N_R\right)^{m/k}} \mu \Gamma\left(m/k+1\right)$$

$$S_{10^{-5}} = 0.575 \frac{\left(M_{vw}\right)_H + \left(M_{vw}\right)_S}{Z_D}$$

$$= 166.2\,\text{MPa}$$

Stress range for calculation

$$S_R = S_1 = K_G\,K_w\,S_{10^{-5}} = 378.9\,\text{MPa}$$

$$v = \left(\frac{83.4}{378.9}\right)^{0.85} \ln N_R = 3.18$$

$\gamma\,(3.53 + 1{,}3.15) = 3.5435$
$\gamma\,(5.882 + 1{,}3.15) = 30.3513$
$\mu = 0.873$

$$D = \frac{5.365\times10^7}{5.802\times10^{12}} \frac{378.9^3}{11.513^{3/0.85}} \times 0.873 \times 12.156$$

$$= 0.96$$

9. Influence of the yield stress

Mild steel at deck and bottom

$\sigma_Y = 235$ MPa

$$\left(S\right)_{10^{-8}} = \frac{\left(M_{vw}\right)_S + \left(M_{vw}\right)_H}{Z_D}$$

208.1 MPa

Stress range at $p = 10^{-5}$

$S_{10^{-5}} = 0.575\, S_{10^{-8}} = 119.65$ MPa

Stress range at hot spot

$S_\ell = K_G\,K_w\,S_{10^{-5}} = 272.8$ MPa

Stress range for calculation

$S_R = S_\ell = 272.8$ MPa

$E\,(B) = 0.81$

$$E(v) = \left(\frac{E\left(S_q\right)}{S_R}\right)^{0.85}$$

$$\ln N_R = \left(\frac{110.3}{272.8}\right)^{0.85}$$

$\ln 10^5 = 5.332$

$\gamma\,(3.53 + 1{,}5.332) = 8.3836$
$\gamma\,(5{,}882 + 1{,}5.332) = 116.9165$
$E(\mu) = 0.594$

$$\alpha = \frac{12.156^{1/3}}{11.513^{1/0.85}} = 0.13$$

$$\beta_C = \frac{\ln\left(\frac{1.342\times10^{13}}{0.594\times0.81^3}0.94\right) - \ln\left(0.13^3\times272.8^3\times5.36510^7\right)}{\sqrt{\ln\left(1.192\times1.09\times1.04^9\right)}}$$

$$= 3.5$$

which corresponds to a probability of failure of 2.3×10^{-4}. In that case, on about 2500 connections, less than one connection has a chance to be damaged in 20 to 25 years. This result gives evidence of the better fatigue strength of structures in normal strength steel and the need for improvement of the structural and weld geometry of HTS connections.

Calculation of the deterministic damage ratio gives

Stress range for calculation

$S_R = 272.8$ MPa

$$v = \left(\frac{83.4}{272.8}\right)^{0.85}$$

In $N_R = 4.204$

$$\gamma\,(3.53 + 1{,}4.204) = 6.0415$$

$$\gamma\,(5.882 + 1{,}4.28) = 85.29$$

$$\mu = 0.742$$

$$D = \frac{5.365\times10^{7}}{5.802\times10^{12}}\frac{272.8^{3}}{11.513^{3/0.85}}\,0.742\times12.156$$

$$= 0.3$$

REFERENCES

Almar Naess, A. et al. (1985). Fatigue Handbook, *Tapir*, Trondheim.

American Bureau of Shipping (2002). Rules for Building and Classing Vessels.

Ang, A. H. and Cornell, A. C. (1974). Reliability Bases of Structural Safety and Design," *J. Struct. Div., ASCE*, **100** (ST9), pp. 1755–1769.

Bea, R. et al. (1992). Structural Maintenance for New and Existing Ships," University of California, Department of Naval Architecture & Offshore Engineering, Berkeley, CA.

Beghin, D. and Cambos, Ph. (1997). Fatigue Strength of Knuckle Joints; A Key Parameter in Ship Design," *Proc. ODRA 97*, Poland.

Berge, S. (1989). The Plate Thickness Effect in Fatigue; Predictions and Results (A Review of a Norwegian Work)," *OMAE*, The Netherlands.

Bretschneider, C. L. (1959). "Wave Variability and Wave Spectra for Wind-Generated Gravity Waves," Beach Erosion Board, U. S. Corps of Engineers, Technical Memorandum 118.

Bureau Veritas (1998). *Fatigue Strength of Welded Ship Structures*, NI 393 DSM R01 E, Paris.

Bureau Veritas (2000). *Rules for the Classification of Steel Ships,* Part B—Hull and Stability, Chapter 7 App.3, Paris.

Eurocode 9 (1999). *Design of Aluminium Alloy Structures*, Part 2—Structures susceptible to fatigue, *European Committee for Standardization*, Brussels.

European Convention for Constructional Steelwork (1992). *Recommandations pour le Calcul en Fatigue de Structures en Alliages d'Aluminium*, Institut de la Soudure, Paris.

Fain, R. A. and Booth, E. T. (1979). Results of the First Five Data Years of Extreme Stress Scratch Gauge Data Collected Aboard Sea-land SL-7's. *Ship Structure Committee*, Report SSC-286.

Fricke, W. and Petershagen, H. (1992). Detail Design of Welded Ship Structures Based on Hot-Spot Stresses, *PRAD's 1992*, Newcastle, UK.

Guedes Soares, C. (1984). Probabilistic Models for Load Effects in Ship Structures, *The Norvegian Institute of Technology, Department of Marine Technology*, Trondheim, Norway, Report No. UR-84-38.

Guedes Soares, C. and Moan, T. (1988). Statistical Analysis of Still Water Load Effects in Container Ships, *Trans SNAME*, **96**, pp 129–159.

Gurney, T. R. (1979). Theoretical Analysis of the Influence of Attachment Size on the Fatigue Strength of Transverse Non-Load-Carrying Fillet Welds, *Welding Research International*, **9** (3), pp 49–79.

Gurney, T.R. (1989). The Influence of Thickness on Fatigue of Welded Joints—A review of British Work, *OMAE*, The Hague, Netherlands.

Gurney, T.R. (1993). The Influence of Mean and Residual Stresses on the Fatigue Strength of Welded Joints under Variable Amplitude Loading—Some Exploratory Tests, IIW XIII-1520-93.

Haibach, E. (1970). Modified linear damage accumulation hypothesis considering the decline of fatigue limit at progredient damage (transl.), Laboratorium fur Betriebsfestigkeit, Darmstadt, Germany.

Hobbacher, A. (1993). Stress Intensity Factors of Welded Joints, *Engineering Fracture Mechanics*, **46** (2) pp. 173–182.

Hobbacher, A. (1994). Stress Intensity Factors of Welded Joints, *Engineering Fracture Mechanics*, **49** (2), p. 323.

Hughes, O. F. (1988). Ship Structural Design, A Rationally-Based Computer Aided Design Optimization Approach, *SNAME*.

Hoffman, D. and Lewis, E. V. (1969). Analysis and Interpretation of Full-scale Data on Midship Bending Stresses of Dry Cargo Ships, *Ship Structure Committee*, Report SSC-196.

International Association of Classification Societies, (1994). *Bulk Carriers—Guidelines for Surveys, Assessment and Repair of Hull Structure*, London.

International Association of Classification Societies, (1997). Internal Report on the Development of a Unified Procedure for Fatigue Design of Ship Structures, London.

International Association of Classification Societies, (1999). *Fatigue Assessment of Ship Structures*, Recom. (56-1), London.

International Institute of Welding, (1996). *Recommendations on Fatigue of Welded Components*, IIW Doc. XIII-1539-96/XV-845-96.

International Ship and Offshore Structures Congress, (1985). Report of Committee I-1 on Environmental Conditions.

International Ship and Offshore Structures Congress, (2000). Report of Committee III-2 on Fatigue and Fracture.

Kosteas, D. (1989). Fatigue Behaviour of Welded Aluminium Standards. European Full-scale Tests, E.U. Program COST 506, *European Aluminium Association*, Parts 1 and 3, Munich, Germany.

Lawrence, F. V. (1984). Fatigue Tests Results and Prediction for Cruciform and Lap Welds, *Theoretical and Applied Fracture Mechanics*, **1** (1).

Lewis, E. V. and Zubaly, R. B. (1975). Dynamic Loadings Due to Wave and Ship Motions, *STAR Symposium, SNAME.*

Little, R. S., Lewis, E.V. and Bailey, F. C. (1971). A Statistical Study of Wave-Induced Bending Moments on Large Ocean-going Tankers and Bulk Carriers, *Trans SNAME.*

Maddox, S.J., Lechocki, J.P. and Andrews, R.M. (1986). Fatigue Analysis for the Revision of BS:PD 6493:1980, *The Welding Institute*, Report 3873/1/86, Cambridge, UK.

Mansour A. E. and Thayamballi, A. K. (1994). Probability-Based Ship Design ; Loads and Load Combinations, *Ship Structure Committee*, Report SSC-373.

Miner, M. A. (1945). Cumulative Fatigue in Damage. J. App. Mech., **12**, pp. 159–164.

Munse, W. (1981). Fatigue Criteria for Ship Structural Details, *Extreme Loads Response Symposium, SNAME*, pp 231–247.

Munse, W. (1982). Fatigue Characterization of Fabricated Ship Details for Design, *Ship Structure Committee*, Report SSC-318.

Newman, J. C. and Raju, I. S. (1981). An Empirical Stress Intensity Factor Equation for the Surface Crack, *Engineering Fracture Mechanics*, **15** (1 2), pp. 185–192.

Newman, J. C. and Raju, I. S. (1983). Stress Intensity Factor Equations for Cracks in Three-Dimensional Finite Bodies, *ASTM* STP 791, pp. I-238 – I-265.

Niemi, E. (1992). Recommendations Concerning Stress Determination for Fatigue Analysis of Welded Components, IIW Doc. XIII-1458-92/XV-797-92.

Niemi, E. (1996). Thickness Effect in Welded T-Joints Subjected to Constant and Variable Amplitude Loading, IIW Doc. XIII-1623-96.

Niemi, E. (1996). Residual Stresses in Fatigue Testing of Welded T-joints, Lappeenranta University of Technology, Finland.

Nikolaidis, E. and Kaplan, P. Uncertainties in Stress Analysis on Marine Structures, *Ship Structure Committee*, Report SSC-363.

Ohta, A., Matsuoka, K., Suzuki, N., and Maeda, Y. (1994). Fatigue Strength of Non-Load Carrying Cruciform Welded Joints by a Test Maintaining Maximum Stress at Yield Strength, *Engineering Fracture Mechanics*, **37** (5), pp 987–993.

Palmgren, A. (1924). Die Lebensdauer von Kugellagem, Zeitschrift des Vereins Deutscher Ingenieure, **68**, pp. 339–341.

Paris, P. and Ergodan, F. (1963). A Critical Analysis of Crack Propagation Laws, *Journal Basic Engineering*, pp. 528–534.

Schütz, W. (1981). Procedure for the Prediction of Fatigue Life of Tubular Joints, *International Conference on Steel in Marine Structures*, Paris.

Stambaugh, K., Lawrence, F. and Dimitriakis, S. (1994). Improved Ship Hull Structural Details Relative to Fatigue, *Ship Structure Committee*, Report SSC-379.

Tanker Structure Co-operative Forum, (1997). *Guidance Manual for Tanker Structures*, Witherby and Co, London.

U.K. British Standards Institute (1991). *Structural Use of Aluminium*, Part 1, Code of practice for design, London.

U.K. Department of Energy (DEn) (1990). *Offshore Installations, Guidance on Design, Construction, and Certification*, 4th Edition, London.

U.K. Health and Safety Executive (HSE), (2001). Offshore Technology Report OTO/2001/015.

Urhy, A. (1989). Fatigue Behaviour of Welded Aluminium Standards. European Full-scale Tests, E.U. Program COST 506, *European Aluminium Association*, Part 2, Voreppe, France.

Violette, F.L. (1998). Basic Parameters Governing the Fatigue of Aluminium Ships, *Third International Forum on Aluminium Ships*, Haugesund, Norway.

Vosikovsky, O., Bell, R., Burns, D. J. and Mohaupt, U. H. (1989). Thickness Effect on Fatigue of Welded Joints—Review of the Canadian Program, *OMAE*, The Hague, Netherlands.

Wirshing, P. H. (1981). Probability Based Fatigue Design Criteria for Offshore Structures, *The American Petroleum Institute*, Report PRAC 80-15, Dallas, Texas.

Wirsching, P. H. (1984). Fatigue Reliability for Offshore Structures," *ASCE, Journal of Structural Engineering*, **110** (10), 1984.

Wirshing P. H. and Chen, Y. N. (1987). Considerations of Probability-Based Fatigue Design for Marine Structures," *Proc. Marine Structural Reliability Symposium*, Arlington, VA.

Yagi, J., Machida, S., Tomita, Y., Matoba M. and I. Soya, (1991). Influencing Factors on Thickness Effect on Fatigue Strength in As-welded Joints for Steel Structures, *OMAE*, Norway.

APPENDIX A - U.K. HSE CLASSIFICATION OF WELDED JOINTS IN STEEL (1)

(Excerpt from U.K. Health and Safety Executive Offshore Technology Report OTO/2001/015)

HSE Welded Joint Classification

Joint Classification	Description	Examples
Category 1		
B	1) Parent metal in the as-rolled condition with no flame-cut edges or with flame-cut edges ground or machined.	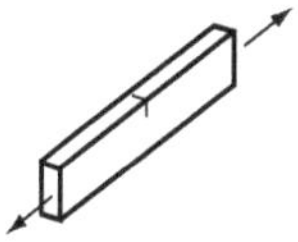
C	2) Parent material in the as-rolled condition with automatic flame-cut edges and ensured to be free from cracks.	
Category 2		
B	1) Full penetration butt welds with the weld cap ground flush with the surface and with the weld proved to be free from defects by NDT.	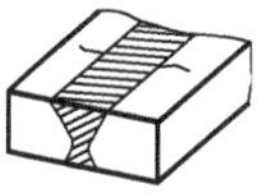
C	2) Butt or fillet welds made by an automatic submerged or open arc process and with no stop-start positions within their length.	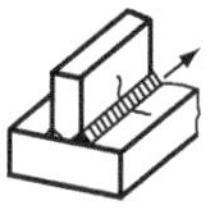
D	3) As (2) but with stop-start positions within the length.	
Category 3	Full penetration butt joints welded from both sides between plates of equal width and thickness or with smooth transition not steeper than 1 in 4.	
D	1) With the weld cap ground flush with the pl ate surface and with the weld proved to be free from significant defects by NDT.	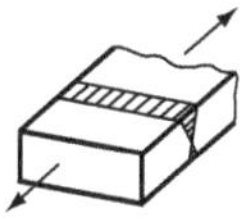
D	2) With the welds made either manually or by an automatic process other than submerged arc provided all runs are made in flat position.	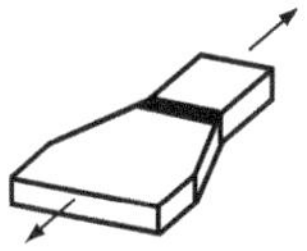
F	3) Parent or weld metal in full penetration butt joints made on a permanent backing strip between plates of equal width and thickness or tapered with a maximum slope of 1/4.	
Category 4	1) Parent material (of the stressed member) or ends of bevel-butt or fillet welded attachments, regardless of the orientation of the weld to the direction of stress, and whether or not the welds are continuous round the attachment:	
F	• a) with attachment length (parallel to the direction of the applied stress) $\ell \leq 150$ mm and with edge distance $d \geq 10$ mm	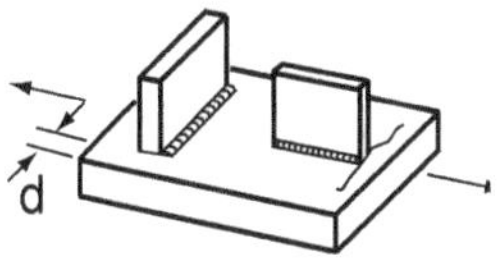

HSE Welded Joint Classification (cont'd)

Joint Classification	**Description**	**Examples**
F2	• b) with attachment length (parallel to the direction of the applied stress) $\ell > 150$ mm and with edge distance $d \geq 10$ mm	
G	2) Parent material (of the stressed member) at toes or ends of butt or fillet welded attachments on or within 10 mm of edges or corners and regardless of the shape of the attachment.	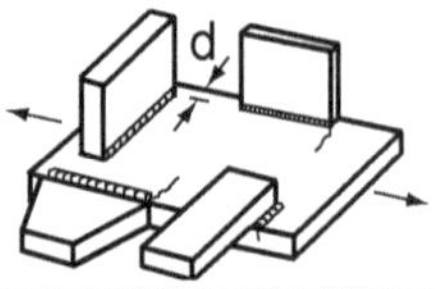
Category 5		
F	1) Parent metal of cruciform joints or T Joints made with full penetration welds and with any undercut at the corners of the member ground out.	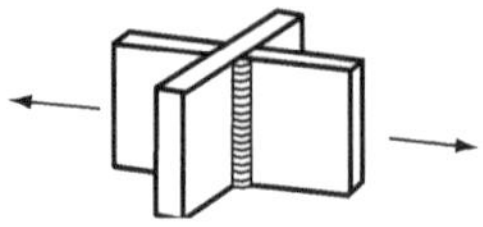
F2	2) As (1) with partial penetration or fillet welds with any undercut at the corners of the member ground out.	
	3) Parent metal of load-carrying fillet welds essentially transverse to the direction of stresses (member X in sketch):	
F2	• edge distance d × 10 mm	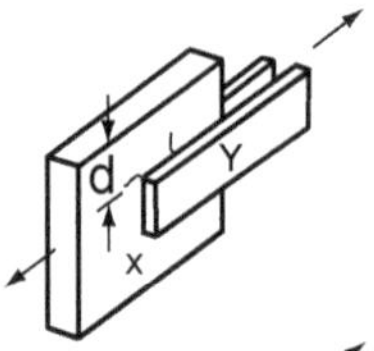
G	• edge dist ance d < 10 mm	
G	4) Parent metal at the ends of load-carrying fillet welds essentially parallel to the direction of stresses, with the weld end on plate edge (member Y in sketch).	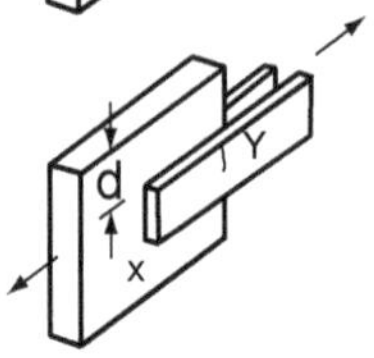
Category 6	1) Parent metal at the toe of a weld connecting a stiffener to a girder flange:	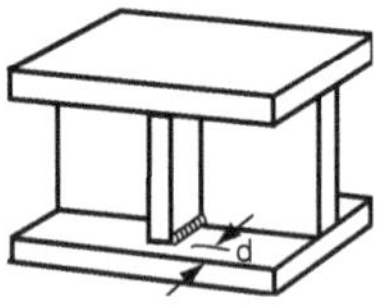
F	• edge distance d × 10 mm	
G	• edge dist ance d < 10 mm	
E	2) Parent metal at the end of a weld connecting a stiffener to a girder web in a region of combined bending and shear.	
E	3) Intermittent fillet welds	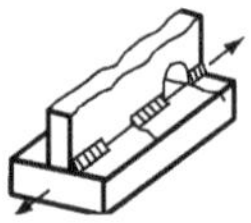
F	4) As (2) but adjacent to cut-outs.	

APPENDIX B - IIW CLASSIFICATION OF WELDED JOINTS IN STEEL

(Excerpt from Recommendations on Fatigue of Welded Components of April 1996)

Structural Detail	Description	FAT
Butt welds transversely loaded		
	Transverse butt weld made in shop in flat position, toe angle < 30°, NDT	100
	Transverse butt weld on permanent backing bar	71
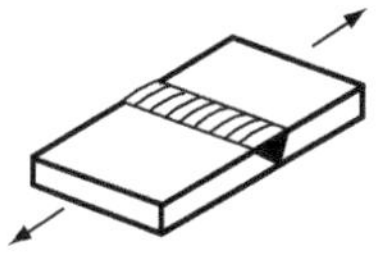	Transverse butt welds welded from one side without backing bar, full penetration: • root controlled by NDT • no NDT	 71 45
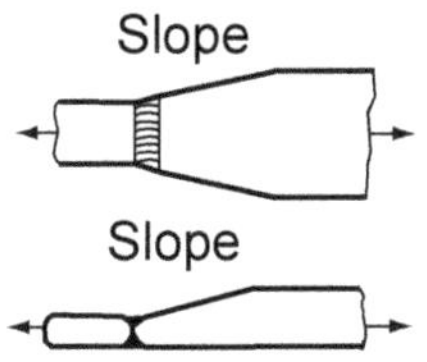	Transverse butt weld ground flush, NDT, with transition in thickness and width: • slope 1:5 • slope 1:3 • slope 1:2	 125 100 80
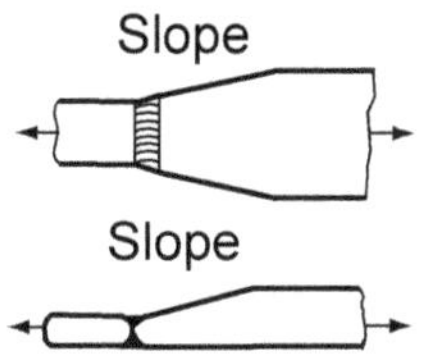	Transverse butt weld made in shop, welded in flat position, weld profile controlled, NDT, with transition in thickness and width : • slope 1:5 • slope 1:3 • slope 1:2	 100 90 80
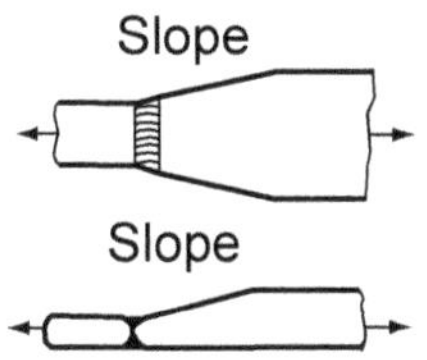	Transverse butt weld, NDT, with transition on thickness and width : • slope 1:5 • slope 1:3 • slope 1:2	 80 71 63
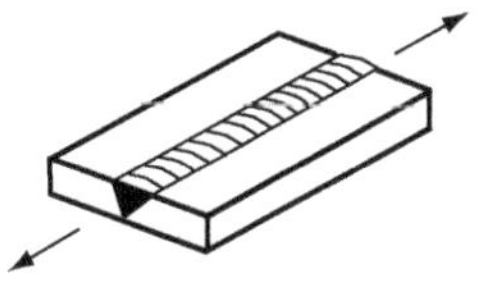	Longitudinal butt weld, both sides ground flush parallel to load direction, 100 % NDT	125
Longitudinal load-carrying welds		
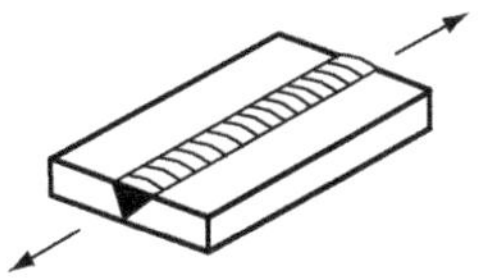	Longitudinal butt weld : • without stop/start positions, NDT • with stop/start positions	 125 90

IIW Welded Joint Classification (cont'd)

Structural Detail	Description	FAT
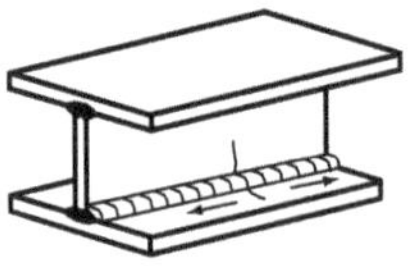	Continuous automatic longitudinal fully penetrated K-butt weld, without stop/start positions (based on stress range in flange), NDT	125
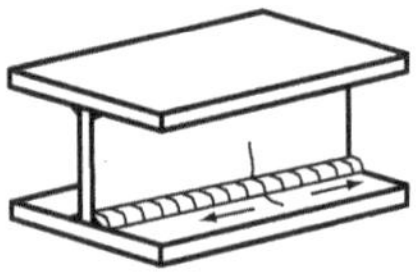	Continuous automatic longitudinal double-sided fillet weld, without stop/start positions (based on stress range in flange)	100
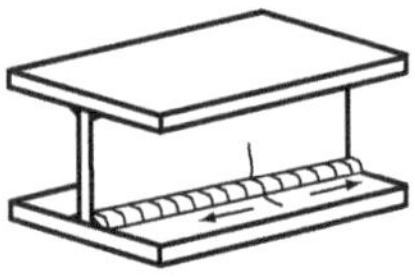	Continuous manual longitudinal fillet or butt weld (based on stress range in flange)	90
Cruciform joints and/or T-joints		
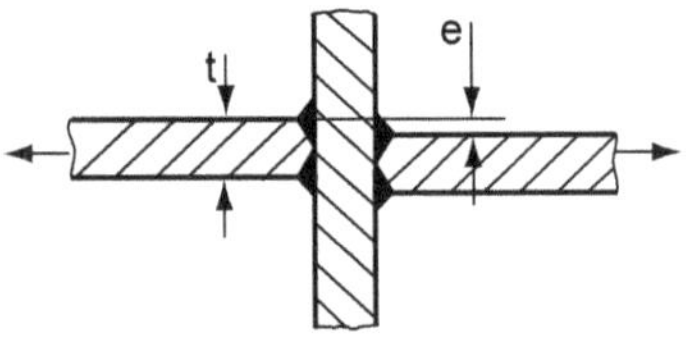	Cruciform joint or T-joint, K-butt welds, full penetration, no lamellar tearing, misalignment $e < 0.15\ t$, weld toes ground, toe crack	80
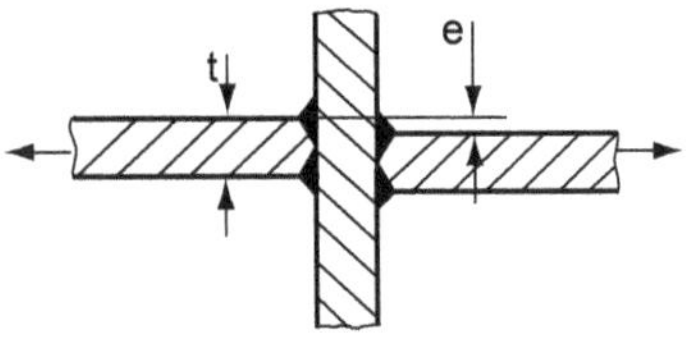	Cruciform joint or T-joint, K-butt welds, full penetration, no lamellar tearing, misalignment $e < 0.15\ t$, toe crack	71
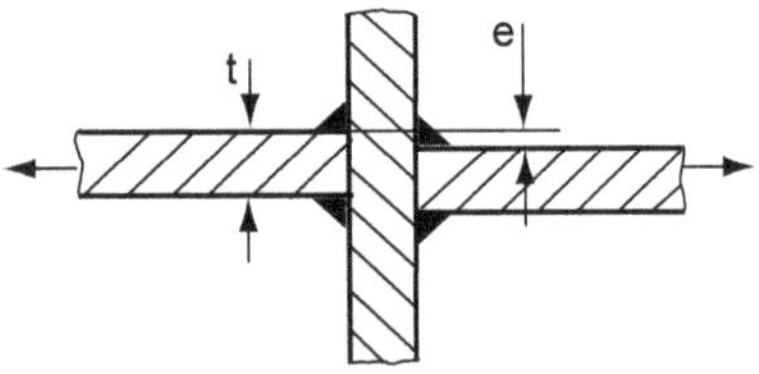	Cruciform joint or T-joint, fillet welds, no lamellar tearing, misalignment $e < 0.15\ t$, toe crack	63
Non-load-carrying attachments		
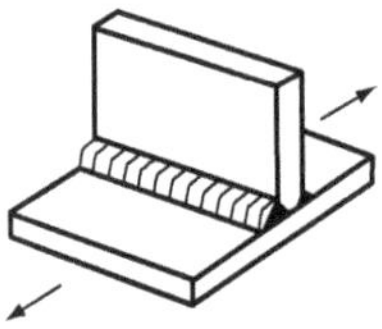	Transverse non-load-carrying attachment not thicker than main plate: • K-butt weld, toe ground • two-sided fillets, toe ground • fillet weld(s), as welded • thicker than main plate	 100 100 80 71

IIW Welded Joint Classification (cont'd)

Structural Detail	Description	FAT
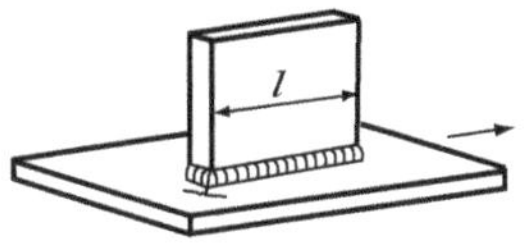	Longitudinal fillet welded gusset at Length *l*:	
	• $l <$ 50 mm	80
	• $l <$ 150 mm	71
	• $l <$ 300 mm	63
	• $l >$ 300 mm	50
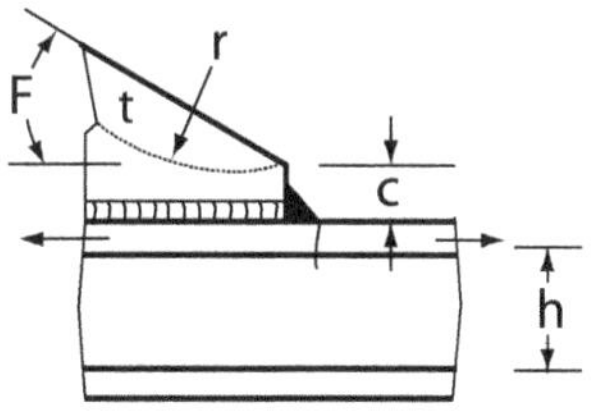	Longitudinal fillet welded gusset with smooth transition (sniped end or radius) welded on beam flange or plate: $c < 2t$, max 25 mm	
	• $r > 0{,}5\ h$	71
	• $r < 0{,}5\ h$ or $\phi < 20^\circ$	63
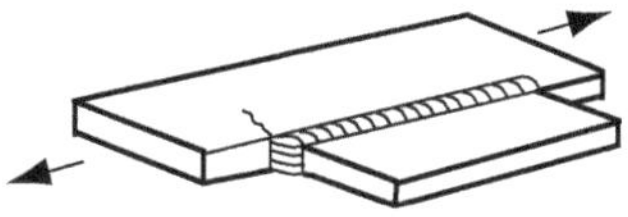	Longitudinal flat side gusset welded on plate edge or beam flange edge, gusset length *l*:	
	• $l <$ 150 mm	50
	• $l <$ 300 mm	45
	• $l >$ 300 mm	40
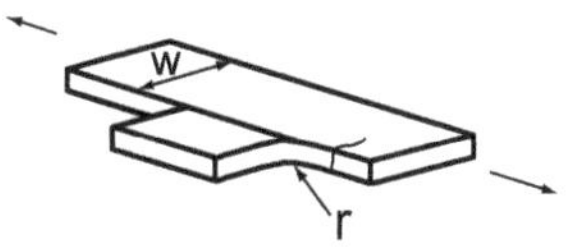	Longitudinal flat side gusset welded on edge of plate or beam flange, radius transition ground:	
	• $r > 150$ or $r/w > 1/3$	90
	• $1/6 < r/w < 1/3$	71
	• $r/w < 1/6$	50
Reinforcements		
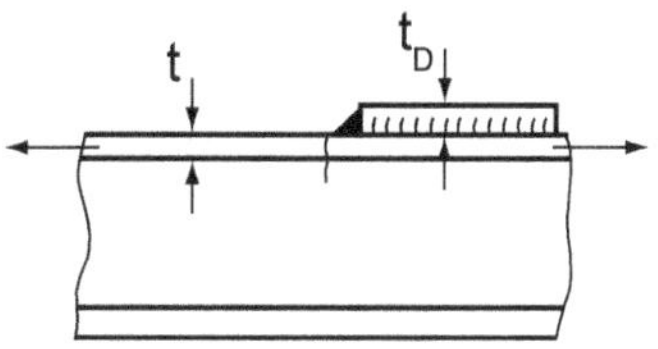	End of long doubling plate on I-beam, welded ends (based on stress range in flange at weld toe):	
	• $t_D \times 0.8\ t$	56
	• $0.8\ t < t_D \times 1.5\ t$	50
	• $1.5\ t < t_D$	45
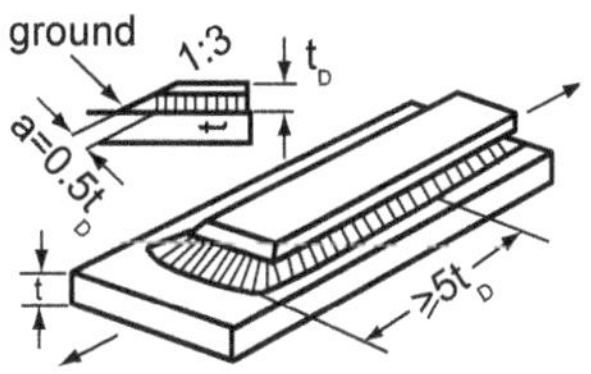	End of long doubling plate on beam, reinforced welded ends ground (based on stress range in flange at weld toe):	
	• $t_D \times 0.8\ t$	71
	• $0.8\ t < t_D \times 1.5\ t$	63
	• $1.5\ t < t_D$	56

APPENDIX C - IIW FATIGUE STRENGTH OF TYPICAL WELDED JOINTS IN ALUMINUM

(Excerpt from Recommendations on Fatigue of Welded Components of April 1996)

Structural Detail	Description	FAT	FAT (Steel)	Ratio Steel /Alu
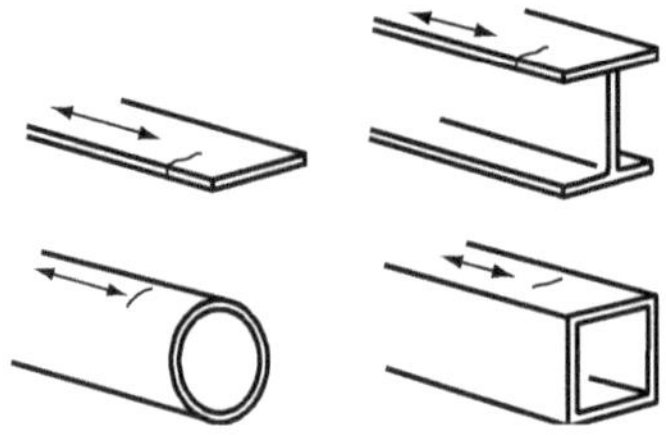	Rolled and extruded products or components with edges machined ($m = 5$).	Series 5000/6000 72 Series 7000 80	160	2.22 2.0
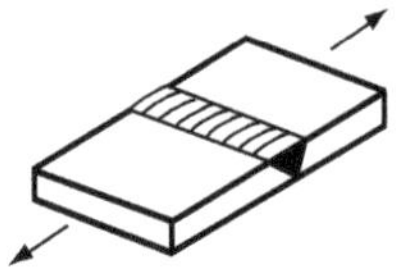	Transverse butt welds welded from one side without backing bar, full penetration • root controlled by NDT • no NDT	 28 18	 71 45	 2.54 2.5
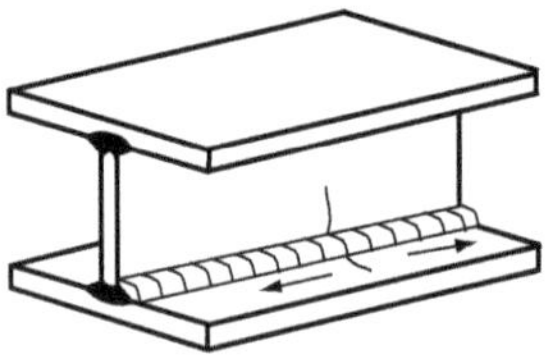	Continuous automatic longitudinal double-sided fillet weld without stop/start positions	40	100	2.5
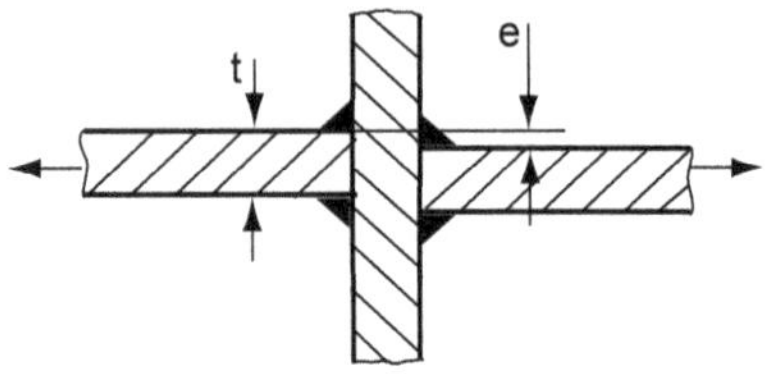	Cruciform joint or T-joint, fillet welds, no lamellar tearing, misalignment $e < 0.15\ t$, toe crack.	22	63	2.86
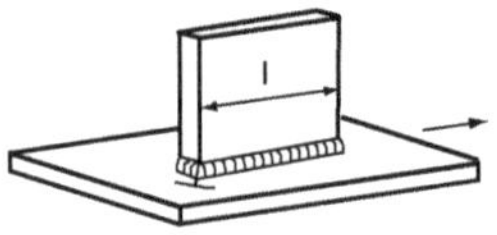	Longitudinal fillet welded gusset of length l: • $l < 50$ mm • $l < 150$ mm • $l < 300$ mm • $l > 300$ mm	 28 25 20 18	 80 71 63 50	 2.86 2.84 3.15 2.78
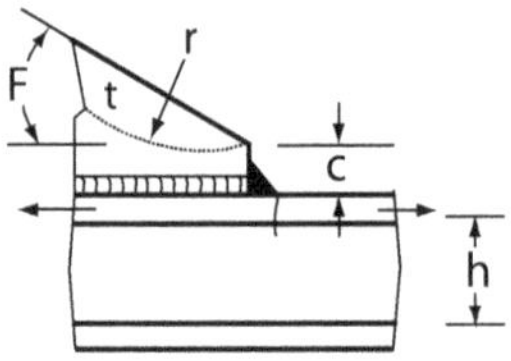	Longitudinal fillet welded gusset with smooth transition (sniped or radiused end) welded on beam flange or plate. $c < 2\ t$, max 25 mm • $r > 0.5\ h$ • $r < 0.5\ h$ or $\varphi < 20°$	 25 20	 71 63	 2.84 3.15

IIW Fatigue Strength Of Typical Welded Joints In Aluminum (cont'd)

Structural Detail	Description	FAT	FAT (Steel)	Ratio Steel /Alu
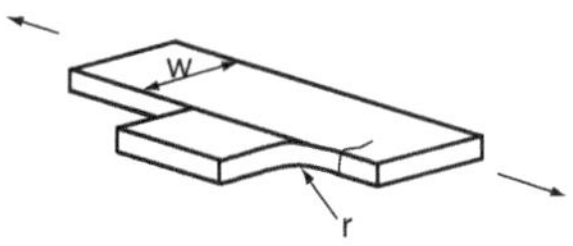	Longitudinal flat side gusset welded on edge of plate or beam flange, radius transition ground.			
	• $r > 150$ or $r/w > 1/3$	36	90	2.50
	• $1/6 < r/w < 1/3$	28	71	2.54
	• $r/w < 1/6$	22	50	2.27

INDEX

CPSIA information can be obtained
at www.ICGtesting.com
Printed in the USA
JSHW010002140722
28055JS00001B/8

9 780939 773824